High-Pressure Science and Technology

Sixth AIRAPT Conference

Volume 1
Physical Properties and Material Synthesis

HIGH-PRESSURE SCIENCE AND TECHNOLOGY
Sixth AIRAPT Conference

Volume 1 Physical Properties and Material Synthesis

Volume 2 Applications and Mechanical Properties

High-Pressure Science and Technology

Sixth AIRAPT Conference

Volume 1
Physical Properties and Material Synthesis

Edited by

K. D. Timmerhaus and M. S. Barber

University of Colorado
Boulder, Colorado

PLENUM PRESS · NEW YORK AND LONDON

Library of Congress Cataloging in Publication Data

International Conference on High Pressure, 6th, University of Colorado, 1977.
High-pressure science and technology.

Includes index.

CONTENTS: v. 1. Physical properties and material synthesis – v. 2. Applications and mechanical properties.

1. High pressure (Technology) – Congresses. I. Timmerhaus, Klaus D. II. Barber, M. S. III. International Association for the Advancement of High Pressure Science and Technology. IV. Title.

TP 156.P75I57 660.2'8429 78-21888
ISBN 0-306-40066-9

Proceedings of the Sixth AIRAPT International High Pressure Conference
held at Boulder, Colorado, July 25–29, 1977

A Division of Plenum Publishing Corporation
227 West 17th Street, New York, N.Y. 10011

Printed in the United States of America

CONTENTS OF VOLUME 1

B. Phase Transitions in Solids

O. Applications in Polymer Chemistry

P. Applications in Biochemistry

Q. Applications in Industry

IV. METROLOGY AND INSTRUMENTATION AT HIGH PRESSURES

R. Pressure Scales and Measurement

S. High Pressure Apparatus and Methods

V. MATERIAL SYNTHESIS UNDER PRESSURE

T. Synthesis and Properties of Diamond and Other Hard Materials

PREFACE

High pressure has become a basic variable in many areas of science and engineering. It extends from disciplines of geophysics and astrophysics through chemistry and physics to those of modern biology, electrical and chemical engineering. This breadth has been recognized for some time, but it was not until the early 1960's that an international group of scientists and engineers established the Association Internationale for Research and Advancement of High Pressure Science and Technology (AIRAPT) for bringing these various aspects of high pressure together at an international conference. The First AIRAPT International High Pressure Conference was held in 1965 in France and has been convened at approximately two to three-year intervals since that time. The past four AIRAPT International High Pressure Conferences have been held in Germany, Scotland, Japan and the U.S.S.R.

Since the first meeting of this kind, our understanding of high pressure behavior of physical systems has increased greatly. Applied research, which is coupled to the expansion of our basic knowledge, has led to the emergence of technologies dependent upon the high pressure properties of materials that have evolved over the past decade or two. In the early 1960's explosive forming, welding, and compaction were rather poorly understood and had limited applications. Today, an expanding variety of products are made possible through this technique. The AIRAPT International High Pressure Conferences have proved to be one of the more effective means of bringing scientists and engineers from many countries together for exchanging new ideas and disseminating new knowledge concerning the properties of materials at high pressures. By contributing to the efficient exchange of new information on high pressure applications and techniques, the AIRAPT International High Pressure Conferences have played an important part in the rapid advances that have occurred in the past two decades, both in our basic understanding of high pressure behavior and in the applied research that has followed.

The Sixth AIRAPT International High Pressure Conference was organized by the University of Colorado and the Association Internationale for Research and Advancement of High Pressure Science and Technology. The conference was held in Boulder, Colorado, July 25 to 29, 1977 and was attended by approximately 500 participants from 24 nations. This technical meeting was sponsored by AIRAPT, The International Union of Pure and Applied Chemistry (IUPAC), the American Institute of Chemical Engineers (AIChE), and the University of Colorado. Gracious support for the conference was provided by the U.S. National Science Foundation, the U.S. Office of Naval Research, the U.S. Department of the Army, the U.S. Air Force Office of Scientific Research, and the U.S. Department of Energy.

The conference was opened with brief introductory and welcoming remarks by University of Colorado Chancellor J. R. Nelson, Dr. K. D. Timmerhaus, chairman of the Conference Planning Committee, and R. N. Keeler, the U.S. member of the AIRAPT Executive Committee. Reminiscences of the late P. W. Bridgman were presented by his daughter, Mrs. Jane Bridgman Koopman, and the first P. W. Bridgman Award Medal was presented by Dr. H. Ll. D. Pugh, president of AIRAPT, to Dr. Harry G. Drickamer of the University of Illinois.

The P. W. Bridgman Award Lecture by Dr. Drickamer pointed up the dual applications of high pressure research in increasing understanding of the behavior and states of matter at high pressure and of the effects of pressure on electronic states and chemical properties of matter. He noted that the basic effect of pressure is to decrease the interatomic distance and thus to increase the overlap between adjacent electronic orbitals. As a consequence, the energy of any one electronic state, in general, changes with respect to that of other states. The information which can be extracted from these energy shifts can conveniently be classified in three categories. (1) It is possible to characterize electronic energy states, electronic excitations, and molecular conformations. (2) It is possible to perform critical tests of theories and hypotheses concerning electronic processes. (3) In a wide variety of materials the relative shift in energy is sufficient to establish a new ground state for the system (or to modify its characteristics greatly by configuration interaction). These new ground states may have new and interesting physical and chemical properties. Studies of pressure-induced electronic transitions have contributed considerably to our understanding of electronic phenomena.

In keeping with the role of AIRAPT, the conference dealt with the full spectrum of high pressure investigations. Because these were so diverse, the Conference Planning Committee was organized along ten main divisions. These divisions and the group of individuals responsible for these divisions are given below.

I. High Pressure Applied to Physical Systems - C. A. Swenson
II. High Pressure Phase Equilibria and Separation Processes - C. A. Eckert

III. High Pressure Kinetics - W. J. le Noble
IV. Metrology and Instrumentation at High Pressures - P. L. M. Heydeman and D. R. Lide
V. Material Synthesis Under Pressure - F. P. Bundy
VI. High Pressure Geophysical/Geological Applications - T. J. Ahrens, A. J. Boettcher and G. C. Kennedy
VII. High Pressure in Energy Resource Recovery - A. S. Kusubov and M. L. Wilkins
VIII. Mechanical Properties at High Pressures - A. L. Ruoff
IX. Industrial Applications of High Pressure - C. B. Boyer
X. Future Directions in High Pressure Applications - R. N. Keeler

Certainly the success of the Sixth AIRAPT International High Pressure Conference was in many ways due to the excellent work of this group of individuals and those who assisted them.

The International Advisory Committee also contributed significantly to the success of the conference by alerting the Conference Planning Committee to new work in high pressures as well as supplying names of younger researchers whose presence at the conference was to be encouraged.

Many other individuals helped in the organization and execution of this conference. Special thanks are due to H. Spetzler and L. M. Christiansen responsible for the local arrangements, Mrs. K. D. Timmerhaus and Mrs. H. Spetzler, cohosts for the spouses activities, and Mrs. E. R. Dillman, Mrs. L. M. Dimond, Mrs. B. Wassan, Mrs. M. L. Woolard and J. D. Smith for their assistance with completion of the program copy.

Finally, the chairman wishes to express his appreciation to the Session Chairmen and to the individuals who reviewed the many manuscripts prior to publication in these Proceedings. These two volumes attempt to present a flavor of the presentations and discussions which highlighted this successful conference.

K. D. Timmerhaus

SPECIAL ACKNOWLEDGMENT

The editors take great pleasure in recognizing the outstanding assistance of many individuals in publishing these Proceedings. Special thanks are due to Mrs. Lenore Dimond, Mrs. Elva Dillman, Mrs. Ruby Fulk, and Mrs. Julie Welker for preparing and compiling the typescript, and to Steve Dyer of Plenum Press for advising and for production.

K. D. Timmerhaus
M. S. Barber

P. W. BRIDGMAN AWARD

Dr. Harry G. Drickamer, Professor of Chemical Engineering and Physical Chemistry at the University of Illinois, is the recipient of the first P. W. Bridgman Award presented by the Association Internationale for the Advancement of High Pressure Science and Technology (AIRAPT). The award is named for the late Nobel laureate, P. W. Bridgman, and was presented to the recipient on July 25, 1977 at the 6th AIRAPT International High Pressure Conference held at the University of Colorado in Boulder, Colorado.

Presentation of the P. W. Bridgman Medal to Dr. H. G. Drickamer (right) by Mrs. Jane Bridgman Koopman, daughter of Nobel laureate P. W. Bridgman, and Dr. H. Ll. D. Pugh, President of AIRAPT.

The fundamental thesis of Professor Drickamer's research program is that very high pressure is a most powerful, indeed an essential, variable for understanding electronic structure and electronic behavior in condensed systems. The central feature of his research has been his selection of those classes of materials and types of measurements which yield the most important information and most useful generalizations about electronic phenomena. Incidental to this process has been the development of techniques for measuring various features of optical absorption and emission spectra and of Mössbauer resonance spectra as well as electrical resistance to relatively high pressures, in some cases to several hundred kilobars.

An overview of one aspect of Professor Drickamer's extensive research program is given in his P. W. Bridgman Award lecture which is published as the first paper in this volume.

6th AIRAPT CONFERENCE PLANNING COMMITTEE

K. D. Timmerhaus, University of Colorado, Chairman
T. J. Ahrens, California Institute of Technology
A. J. Boettcher, University of California, Los Angeles
C. B. Boyer, Battelle Columbus Laboratory
F. P. Bundy, General Electric Company
C. A. Eckert, University of Illinois
P. L. M. Heydemann, National Bureau of Standards
R. N. Keeler, Lawrence Livermore Laboratory
G. C. Kennedy, University of California, Los Angeles
A. S. Kusubov, Lawrence Livermore Laboratory
W. J. le Noble, State University of New York, Stony Brook
D. R. Lide, National Bureau of Standards
A. L. Ruoff, Cornell University
C. A. Swenson, Iowa State University
M. L. Wilkins, Lawrence Livermore Laboratory

INTERNATIONAL ADVISORY COMMITTEE

O. Anderson, University of California, Los Angeles, U.S.A.
D. Bloch, Laboratoire de Magnétisme, France
H. G. Drickamer, University of Illinois, U.S.A.
E. U. Franck, Technische Hochschule Karlsruhe, W. Germany
D. Francois, Commissariat à l'Energie Atomique, France
S. Hamann, Chemical Research Laboratories, Australia
R. N. Keeler, Lawrence Livermore Laboratory, U.S.A.
H. Larker, ASEA, Sweden
J. Osugi, Kyoto University, Japan
H. Ll. D. Pugh, Rolls Royce Ltd., England
S. Saito, Tokyo Institute of Technology, Japan
J. E. Schirber, Sandia Laboratories, U.S.A.
K.-F. Seifert, Universität Bonn, W. Germany
I. Sorgato, Universita di Padova, Italy
N. J. Trappeniers, Van der Waals Laboratorium, The Netherlands
L. F. Vereschagin, Institute for High Pressure Physics, U.S.S.R. (Deceased February 20, 1977)
B. Vodar, LIMHP, CNRS, France
K. E. Weale, Imperial College of Science and Technology, England
B. Whalley, National Research Council of Canada, Canada

CORRESPONDING MEMBERS OF ADVISORY COMMITTEE

B. Baranowski, Polish Academy of Sciences, Poland
B. Buras, Danish Atomic Energy Commission, Denmark
J. B. Clark, National Physical Research Laboratory, South Africa
K. A. H. Heremans, Katholieke Universiteit Leuven, Belgium
V. P. J. Meisalo, University of Helsinki, Finland
V. S. Nanda, University of Delhi, India

CORRESPONDING MEMBERS OF ADVISORY COMMITTEE (CONTINUED)

J. L. Olsen, Eidgenössische Technische Hochschule Zürich, Switzerland
F. E. Prieto, Instituto de Física, México
V. N. Shanov, Institute of Chemistry and Technology, Bulgaria

ACKNOWLEDGMENTS

The Conference Planning Committee is deeply appreciative for the support given by the following organizations to the 6th AIRAPT International High Pressure Conference.

SPONSORS

Association Internationale for Research and Advancement of High Pressure Science and Technology (AIRAPT)
The International Union of Pure and Applied Chemistry (IUPAC)
The American Institute of Chemical Engineers (AIChE)
The University of Colorado

SUPPORTERS

U.S. Air Force Office of Scientific Research (AFOSR)
U.S. Department of Energy (DOE)
U.S. Department of the Army (DOA)
U.S. Office of Naval Research (ONR)
U.S. National Science Foundation (NSF)

COHOSTS

American Instrument Company, Industrial Products Division
ASEA
Autoclave Engineers, Inc.
Harwood Engineering Company, Inc.
National Forge Company
Pressure Products Industries, Division of the DURIRON Company, Inc.
Speciality Engineering Associates
The University of Colorado

BRIDGMAN AWARD LECTURE

HIGH PRESSURE STUDIES OF ELECTRONIC PHENOMENA

H. G. Drickamer

University of Illinois
Urbana, Illinois USA

PROLOGUE

It is a great pleasure and honor to be the first recipient of the Percy Williams Bridgman Award. Many people have asked me whether I was a student of Bridgman. Formally, of course, I was not. However, in a very real sense we are all students of Bridgman. We all share his enthusiasm for high pressure research. We can all hope to reflect something of his intellectual honesty. Today we still use his techniques, or modifications and extensions of those he developed. His book The Physics of High Pressure [1] and his Collected Experimental Papers [2] are still gold mines of useful data and information. Today, sixteen years after his death, his spirit and his ideas still permeate the field. Seldom, if ever, has one man been so significant for any important area of research. At the same time, I'm sure that Professor Bridgman would be most pleased to see the extent to which high pressure has become an integral and essential technique in modern physics, chemistry, geology, engineering, and biology. Today I shall discuss briefly how we have found high pressure a very effective tool in the study of electronic phenomena in condensed systems.

INTRODUCTION

From our viewpoint, the basic effect of pressure is to decrease interatomic distance and increase overlap between adjacent electronic orbitals. As a consequence, any one type of orbital will shift in energy with respect to other types of orbitals. Much information can be extracted from this phenomenon. It is convenient to consider it in three categories: (1) one can characterize electronic states, electronic excitations or molecular conformations; (2) one can perform critical tests of theories; and (3) one can induce electronic transitions - new ground states with different physical or chemical properties. I propose to give an example or two in each of these categories, selected to show how broadly useful high pressure studies of electronic phenomena can be in many areas of science.

CHARACTERIZATION

The example of characterization we shall use involves the conformation of a protein molecule in solution. This is a problem of biochemical importance, and Professor Gregorio Weber is the moving spirit in the work.

A protein is a polymer of amino acids. Its properties depend on the number, type, and sequence of amino acids and on the conformation of the molecule in solution. Any significant change of conformation presumably affects its biological activity, i.e. it denatures the protein. Denaturation is normally effected by changing temperature, pH, or solvent. In general, these techniques yield only modest information about the details of the process. Many questions are unanswered - i.e., does it occur in a series of steps or a single step? - does it involve the whole protein or segments? With temperature, pH, or solvent effects, the protein appears to be either "native" or "denatured." High pressure luminescence studies have revealed considerable detail heretofore unavailable. We discuss two proteins, chymotrypsinogen and lysozyme [3]. For each we have two luminescent probes. The amino acid tryptophan is luminescent; chymotrypsinogen contains eight tryptophans, while lysozyme has six. Each protein also has regions which can complex with anions in neutral solution. Chymotrypsinogen has about twenty such sites and lysozyme has eight. A useful anion is aniline-naphthalene-sulfate (ANS), since as a free ion it emits very feebly, but when it is complexed it has a relatively high quantum efficiency. Tryptophan emits in the UV at 330-350 nm while ANS emits near 520 nm so that the two probes can be studied separately.

We look first at chymotrypsinogen, irradiating in the tryptophan absorption. In Fig. 1 we exhibit the emission of tryptophan in the protein compared with the free molecule in solution. The major feature is the rapid drop in efficiency of emission in the protein in the pressure range 2 to 6 kbars. At the same time the peak shifts from a location typical of a hydrocarbon environment to that characteristic of an aqueous environment. This, then, is one step in the denaturation process.

If we now excite in the ANS (Fig. 2) we see a sharp rise in emission intensity in the range 5 to 9 kbar. The magnitude of rise increases with increasing ANS/protein ratio up to two ligands per protein, and then levels (Fig. 3). Thus, about 10% of the protein (two of twenty sites) is involved in this step.

We see then that pressure denaturation is a process which involves at least two stages in chymotrypsinogen. One can evaluate equilibrium constants from the pressure-intensity curves (the

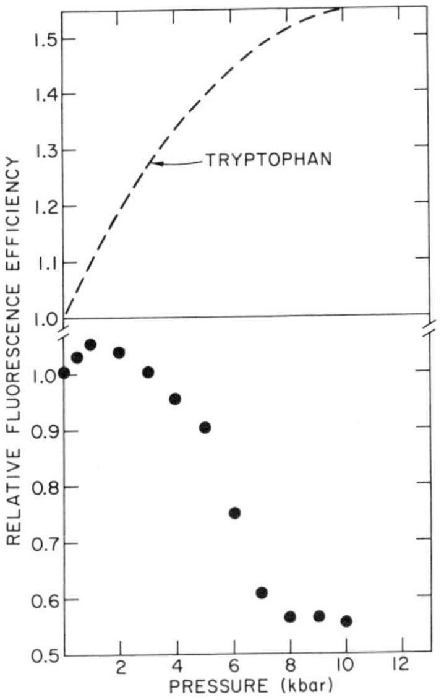

Fig. 1. Change of intensity of tryptophan luminescence with pressure for tryptophan in chymotrypsinogen compared with free tryptophan.

process is reversible) and estimate volume changes of 20 to 60 cc/mol for various steps.

If we turn now to lysozyme, and irradiate in the ANS, we find again a large intensity increase in the pressure range from 5 to 9 kbars, this time involving only one ligand per protein. Again, about 10% (one of eight sites) of the protein is involved.

When lysozyme is irradiated in the tryptophan site (see the upper curve of Fig. 4) there is an initial rise, a definite drop from 2 to 6 kbars, and a second sharp drop from 8 to 12 kbars. Evidently there are two types of tryptophan sites with different resistance to denaturation.

We can extract information about function as well as structure from high pressure studies. Lysozyme attaches to sugar (in a rather strained configuration). In Fig. 4 we see the fluorescence intensity curve for tryptophan in lysozyme attached to a triglucosamide.

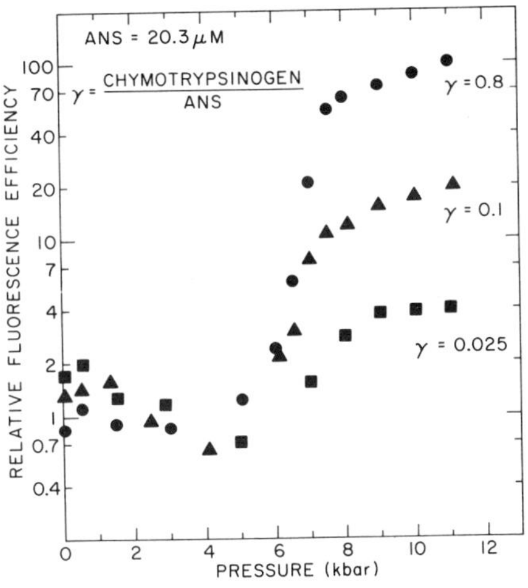

Fig. 2. Change of luminescent-intensity of ANS with pressure in solution with chymotrypsinogen.

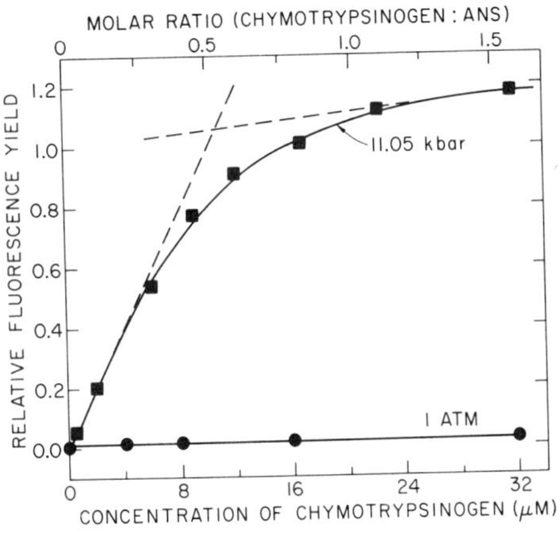

Fig. 3. Relative fluorescence yield for ANS vs. concentration of chymotrypsinogen.

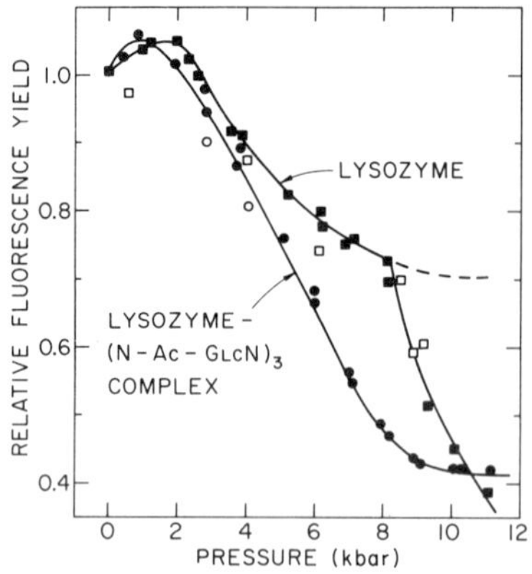

Fig. 4. Pressure dependence of the relative fluorescence yield of lysozyme and lysozyme-(N-AcGℓcN)$_3$ complex, exciting in the tryptophan absorption. The open symbols indicate the degree of reversibility.

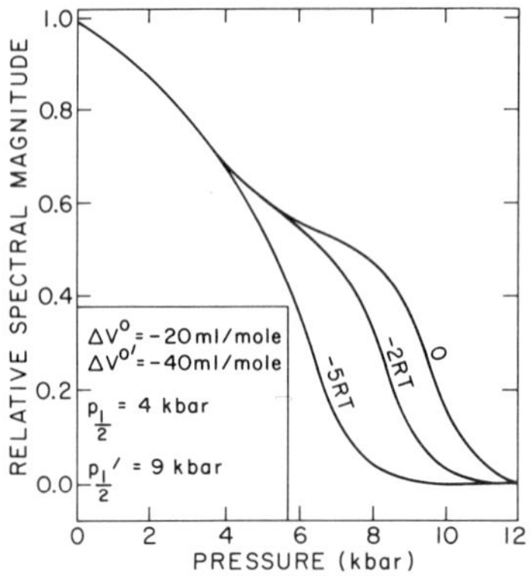

Fig. 5. Calculation of relative spectral efficiency (e.g., fluorescence yield) no pressure, for the conditions shown in the figure.

Now we observe only a single continuous curve. This situation is analyzed in detail by Li et al. [3]. One can develop expressions for the equilibrium constants K_1 and K_2 for the interaction of the protein with the substrate in the regions below and above 8 kbars:

$$K_1 = \exp(P_{\frac{1}{2}}-P)\left(\frac{\Delta V_1^o}{R_B T}\right) \exp\left(\frac{-\delta G_1}{R_B T}\right) \quad (1)$$

$$K_2 = \exp(P_{\frac{1}{2}}'-P)\left(\frac{\Delta V_2^o}{R_B T}\right) \exp\left(\frac{-\delta G_2}{R_B T}\right) \quad (2)$$

where δG_1 and δG_2 are the free energies of binding of the protein to the substrate below and above 8 kbars and the other symbols are defined in the Notation. As can be seen from Fig. 5, using values of $P_{\frac{1}{2}}$ and ΔV^o extracted from data on the protein without substrate, if we assume $\delta G_1 = \delta G_2$ we reproduce the curve obtained for the free protein, but if we assume the partially denatured protein above 8 kbars has a larger negative free energy of binding to the substrate (e.g., $\delta G_2 - \delta G_1 = -5F_BT$ we reproduce the behavior in the presence of the substrate. Evidently the partially unfolded protein binds better to the substrate than the globular native material.

These results illustrate the power of pressure to assist in characterizing a complex biological process.

TESTS OF THEORIES

Phosphor Efficiency

A problem of considerable scientific and engineering interest is the quantum efficiency of luminescence. This is an important factor in the design of fluorescent lamps, cathode ray tubes and lasers, in energy recovery systems, and in organic photochemistry. In the limited picture discussed here we consider only the relative probabilities of photon emission or direct thermal return of the excited electron to the ground state (internal conversion). Intersystem crossing to another, optically forbidden, state such as a triplet state of an organic molecule (intersystem crossing) is also an important means of redistribution of energy. Intersystem crossing has been shown to be pressure sensitive [4,5] but we shall be unable to include a discussion in the time available here.

For localized excitations, the emission vs. internal conversion problem can conveniently be described in terms of a single configuration coordinate diagram (see Fig. 6). Along the vertical axis is plotted energy, while the horizontal axis is some normal mode of vibration of the system. Pressure couples to the volume, so that the configuration coordinate most affected in these experiments is the breathing mode of the system; it is an intermolecular coordinate, i.e., it involves the interaction of an atom or molecule with the environment. This point is discussed in detail elsewhere [6,7].

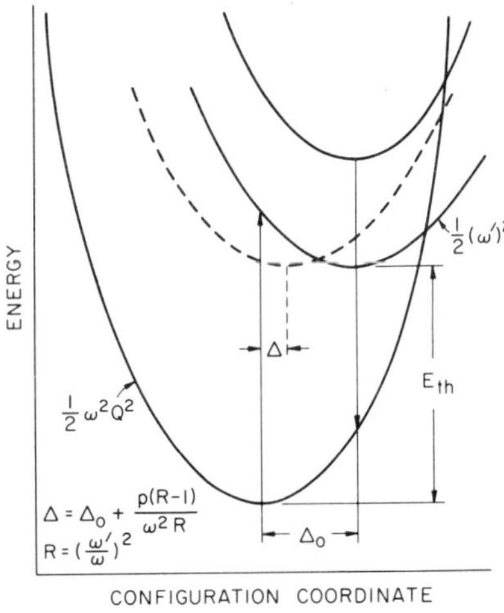

Fig. 6. Schematic configuration coordinate diagram.

Both optical absorption and emission occur vertically on such a diagram (the Franck-Condon principle) because they are rapid compared with atomic or molecular motions. When the electron absorbs a quantum of light and moves to the upper electronic state it rapidly gives off vibrational energy (phonons) and reaches the bottom of the well. In this picture the fluorescence efficiency depends on the relative probability of emitting a photon or crossing thermally to the ground state at the intersection of the two electronic wells.

A quantitative calculation of the emission efficiency requires that one consider the barrier height in the presence of configuration interaction, the tunnelling probability, the probability of crossing to various excited vibrational levels of the ground state, and possibly anharmonicity. Appropriate relationships have been developed by Struck and Fonger [8].

One can, in principle, evaluate the force constants and both vertical and horizontal displacements of the potential wells from the peak shifts and changes of peak shape with pressure for both the absorption and emission peaks. This has been done for a number of systems but requires isolated peaks of simple symmetry [6,7,9]. One can then use the parameters thus obtained in the equations of Fonger and Struck [8] to calculate efficiencies from peak shifts and shape changes. Tyner [10] has done this for some simple inorganic systems.

Here we use the diagram to predict some qualitative changes and compare them with experiment. From Fig. 6 we see that a shift to higher energy would lead to an increased barrier for the thermal process and thus to higher emission intensity, while a shift to lower energy would result in reduced intensity. Similarly, a decrease in relative displacement along the configuration coordinate ($\Delta_o \rightarrow \Delta$) would decrease the probability of radiationless relaxation and increase emission intensity.

For organic molecules in particular, the direction and magnitude of shift will depend on the polarizability and/or dipole moment of the excited state relative to the ground state, and on the polarizability of the medium. If the excited state is more polarizable, it exerts a greater attraction for the environment and a red shift (shift to lower energy) ensues. If it is less polarizable then the ground state one may expect a blue shift. These directions and magnitudes are modified by the polarizability of the environment.

Here we present a study of azulene and an azulene derivative - dicarboethoxychloroazulene (DCECA) in two solvents; polymethylmethacrylate (PMMA), and polystyrene (PS) [11]. Azulene is an isomer of napthalene with one five and one seven membered ring. For most molecules, no matter what the state to which an electron is excited, it returns thermally to the lowest excited state (S_1) before it emits; thus, information is obtained only about S_1, the lowest excited state. Azulene emits from the second excited state (S_2) and some derivatives emit from both S_2 and S_1. It is known that S_2 is significantly more polarizable than the ground state (S_o). For DCECA there is reason to believe that S_1 is less polarizable than S_o. PMMA is a significantly less polarizable solvent than PS.

In Figs. 7 and 8 we exhibit the shift of the emission peak and change in integrated intensity with pressure for the $S_2 \rightarrow S_o$ emission of azulene in PMMA and in PS. The peak shifts to lower energy with increasing pressure as would be expected from the above discussion. It shifts about twice as rapidly in PS as in PMMA, which is consistent with the greater polarizability of the former solvent. By 140 kbars the efficiency in PMMA has dropped to $\sim 2\%$ of its initial value, while in PS at the same pressure the efficiency is less than 1% of the atmospheric value. Thus the shifts and intensity changes correlate with the discussion based on the configuration coordinate diagram. Substituted azulenes give qualitatively similar results for the $S_2 \rightarrow S_o$ emission.

In Fig. 9 we show similar data for the $S_1 \rightarrow S_o$ emission of a substituted azulene in PMMA. For this material the S_1 state is less polarizable than the ground state. As one would predict, the peak shifts to higher energy with increasing pressure. At the same time the intensity increases by a factor of 40 in 130 kbars. This too follows the prediction of our CC diagram. Finally, in Fig. 10 we exhibit data for the $S_1 \rightarrow S_o$ transition in the substituted

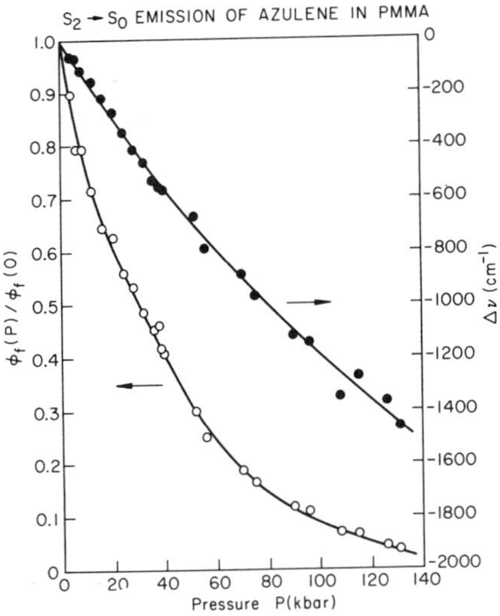

Fig. 7. Pressure dependence of fluorescence peak energy and quantum yield for $S_2 \rightarrow S_o$ transition of azylene in PMMA.

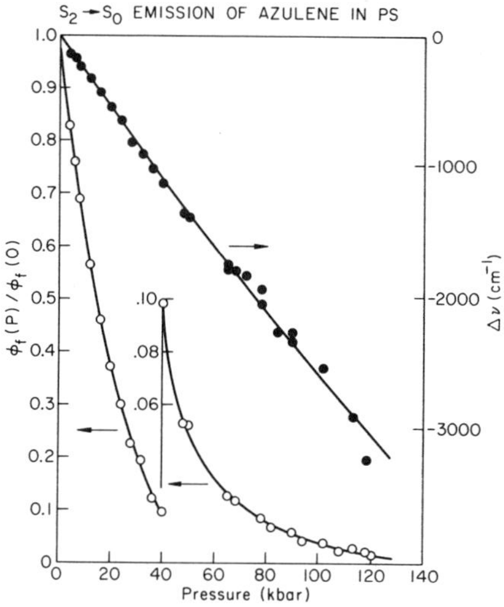

Fig. 8. Pressure dependence of the fluorescence peak energy and quantum yield for $S_2 \rightarrow S_o$ emission for azulene in PS.

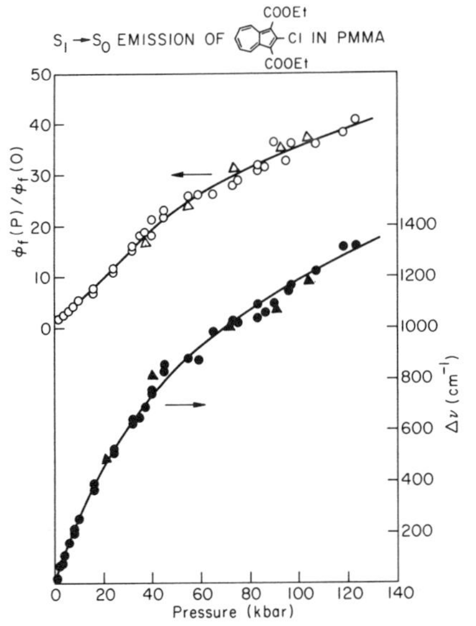

Fig. 9. Pressure dependence of the fluorescence peak energy and quantum yield for $S_1 \rightarrow S_o$ emission for dicarboethoxy-chloroazulene in PMMA.

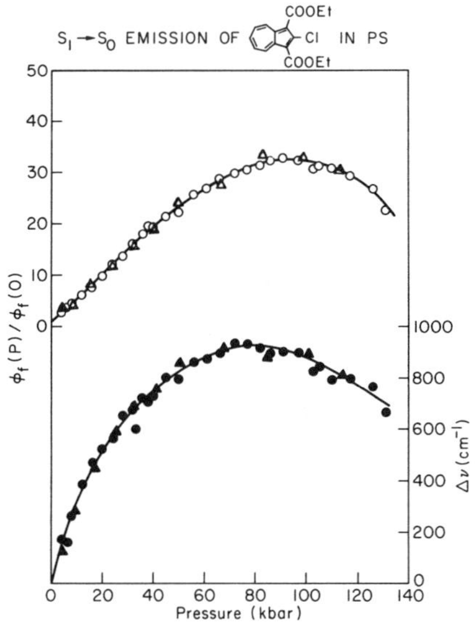

Fig. 10. Pressure dependence of the fluorescence peak energy and quantum yield for $S_1 \rightarrow S_o$ emission for dicarboethoxychloroazulene in PS.

azulene, in PS. Here there are more complicated interactions involving both a vertical and horizontal displacement of the potential wells. In the low pressure region the peak shifts to higher energy; it reaches a maximum near 80 kbars, and then shifts to lower energy. Similarly, the intensity increases at low pressure, maximizes, and then decreases. This is a striking illustration of the relationship between emission energy and luminescent efficiency.

A somewhat different example of the use of pressure to test a theory of phosphor efficiency involves ZnS doped with Cu^+ or Ag^+ and a coactivator such as Cl^- or Al^{+3}. These materials find application in cathode ray tubes, are model compounds for various semiconducting phosphors such as light emitting diodes, and for photoelectric devices used in solar energy conversion. The situation is represented schematically in Fig. 11. The Cu^+ and compensating ion generate levels in the gap between the top of the valence band and the bottom of the conduction band. An electron is excited from the valence to the conduction band. The Cu^+ forms a deep trap for the holes generated in the valence band while the coactivator acts as a shallow electron trap. There is evidence that changes in the energy of the hole trap E_A are relatively unimportant while the coactivator (electron) trap energy E_D is much more significant for phosphor efficiency.

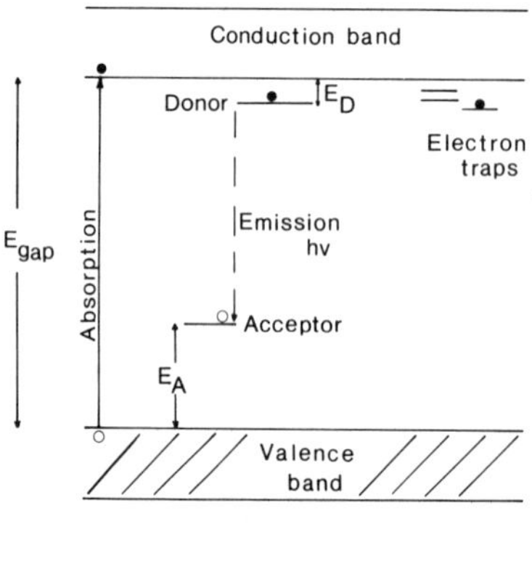

Simple band scheme for ZnS Donor-Acceptor Pair Emission

Fig. 11. Schematic band structure of ZnS with impurity levels.

ZnS involves a complex combination of ionic and covalent binding and there are various theories of the behavior of dopants which are useful in describing one aspect or another of the observed phenomena [12-15]. The theory we apply here is perhaps the most generally useful one. It pictures the luminescence as an electron from the coactivator as donor to the Cu^+ (with trapped hole) as acceptor. The donor wave functions are spread out over a number of lattice parameters with the effective radius of the orbit a strong function of the trap depth E_D. It, in fact, decreases rapidly with increasing E_D. The acceptor wave functions are much more localized. The transfer efficiency depends directly on the overlap of donor and acceptor wave functions, and thus on E_D.

The emission intensity can be expressed in the form

$$I(E_D,r) = \text{const} \times \int \tau\, W(E_D,r)\, G(r)dr \qquad (3)$$

where $W(E_D,r)$, the transition probability depends exponentially on E_D, and $G(r)$ is a distribution function for impurity ions.

In this example [16] we use $A\ell^{+3}$ as the donor and Cu^+ as the acceptor, although a similar analysis would apply to other combinations. The coactivator (donor) depth E_D is well established at one atmosphere. In Fig. 12 we exhibit the shift of the emission peak with pressure, compared with the shift of the absorption edge of the ZnS crystal is given by the dashed curve. To first order the difference represents the change in E_D with pressure. This can be used in equation (3) to calculate the change in luminescent efficiency with pressure. In Fig. 13 we compare this calculation with the measured values. The agreement is remarkably good.

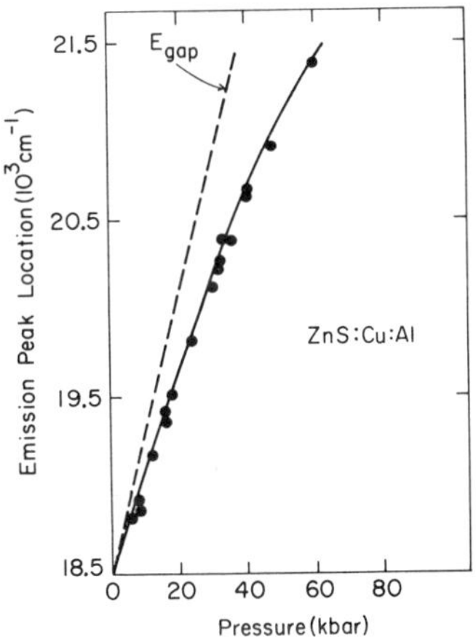

Fig. 12. Shift of emission peak and of absorption edge for ZnS:Cu:Al.

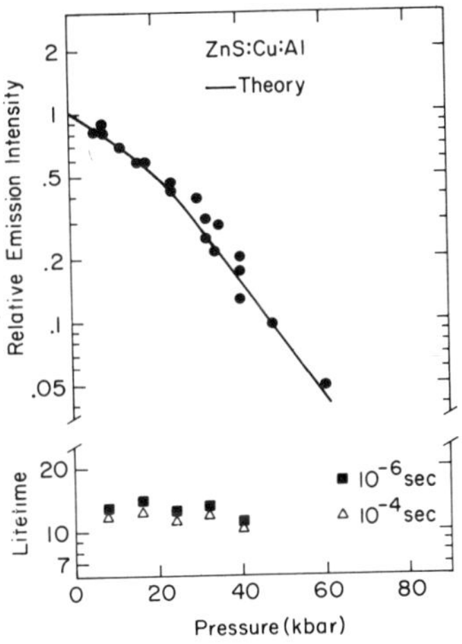

Fig. 13. Comparison of measured and calculated fluorescence efficiencies as a function of pressure for ZnS:CuAl.

In these two studies we see how pressure can be used to test theories of importance in physics, materials science and photochemistry.

Energy Transfer

Energy transfer in phosphors is a topic of widespread interest. Consider a medium (crystal or solution) containing two phosphors which absorb and emit in somewhat different regions of the spectrum. Under some circumstances one can excite one of these phosphors with a definite probability that the excitation can be transferred to the second phosphor by a radiationless process so that both phosphors may emit. This process is important in such diverse areas as fluorescent lighting, the excitation of rare earth lasers, and organic photochemistry including photosynthesis; thus, it is of interest in physics, chemistry, biology, and materials design. If the concentration of phosphors is dilute, the transfer will be by a dipole-dipole process like a van der Waals' interaction. This situation was first discussed by Förster [17] whose analysis was refined by Dexter [18].

The probability per unit time of transfer between donor and acceptor is given by

$$P_{DA} = \frac{\alpha}{n^4}\left(\frac{1}{\tau_D}\right)\frac{1}{R_{DA}^6}\int\frac{f_D(E)\ F_A(E)}{E^4}\,dE \qquad (4)$$

$$= \frac{R_{DD}^6}{R_{DA}^6} \qquad (4a)$$

Here α contains only constants, and the integral is an overlap integral between donor emission and acceptor absorption peaks. R_{DD} is the effective donor-acceptor distance such that there is an equal probability that the donor will fluoresce or transfer its energy to the acceptor. The other quantities are defined in the Notation. The transfer efficiency is given by

$$\eta = \frac{P_{DA}\tau_D}{1+P_{DA}\tau_D} \qquad (5)$$

This can be determined experimentally by measuring the relative areas under the donor and acceptor emission peaks in the mixed crystal. On the other hand, all of the quantities on the right hand side of equation (4) can be determined on the pure medium or on the medium singly doped with donor or acceptor ions or molecules. A comparison between the relative areas under the donor and acceptor peaks and the calculated efficiency is one test of the theory.

A second test depends on measuring the intensity of emission as a function of time. The theory gives this intensity for sensitized luminescence by the relation

$$I_D(t) = I_o\ [\exp-\ t/\tau_D][\exp-\beta R_{DD}^3(t/\tau_D)^{\frac{1}{2}}] \qquad (6)$$

The first bracket gives the lifetime of the donor as a single dopant. The second bracket represents the intensity lost by dipole-dipole transfer to the acceptor. β represents a combination of coefficients independent of pressure. From a comparison of intensity vs. time measurements on the singly and doubly doped medium R_{DD} can be calculated and thus P_{DA} and the efficiency η established.

This theory has been widely used, but most of the tests of the theory have been based on varying R_{DA} by changing the concentration. The use of pressure permits a variation of R_{DA} through the compressibility of the medium, and of the refractive index. The largest potential effect of pressure is, however in changing the overlap integral. We present results for a system [19] KCl:Ag:Tl where there are no important secondary effects and where there is a large change in the overlap integral with pressure. As can be seen schematically in Fig. 14, the Ag^+ ion absorbs at high energy (~ 220 nm) and, at low pressure, the overlap with the Tl^+ absorption, and the consequent Tl^+ emission, is small. By 18 kbars the overlap and the Tl^+ emission have increased considerably. The calculated efficiency of transfer increases from 7 to 28% in 18 kbars.

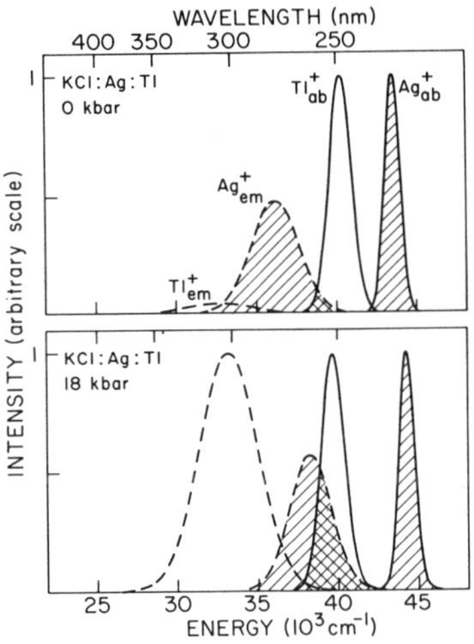

Fig. 14. Schematic representation of spectra for KCl:Ag:Tl at low and high pressure.

In Fig. 15 we compare the calculated efficiency (the solid line) with that extracted from measuring the time-dependent emission intensity of the donor (Ag^+) in KCl:Ag:Tl compared with that in KCl:Ag, using equation (6) to obtain R_{DD} and then calculating η from equation (5). The agreement between theory and experiment is excellent.

As discussed above, it is also possible to test the theory by measuring the relative intensity of donor and acceptor emission compared with values calculated from equation (4). In taking this path it is necessary to make a correction for absorption by the acceptor of photons emitted by the donor and consequent emission from the acceptor (the cascade effect). This correction has been worked out by Dexter [18], and depends on the crystal thickness. (See Bieg et al. [19] for detailed application.) Experiments were performed with crystals 0.75 mm thick and with "thin" crystals 0.20 mm thick to be sure the correction was done adequately. Although there is some scatter due to the number of corrections, the agreement with theory as shown in Fig. 16, is again excellent.

The efficiency can also be varied by changing temperature at constant pressure. The primary effect of lowering the temperature is to decrease the peak width and thus modify the overlap. This effect is most important in regions of small overlap. Figure 17 compares the measured and calculated efficiencies as a function of temperature at two pressures. The agreement is very good.

These results constitute an excellent illustration of the use of pressure to perform a critical test of a theory important in a wide variety of atmospheric pressure applications.

ELECTRONIC TRANSITIONS

Finally we turn to the third class of phenomena, electronic transitions. These have been observed in a variety of materials,

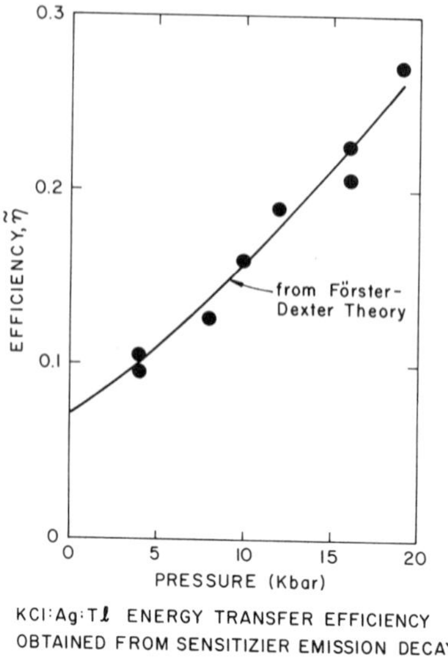

Fig. 15. Comparison of calculated transfer efficiency with that measured from donor lifetimes - KCl:Ag;Tl.

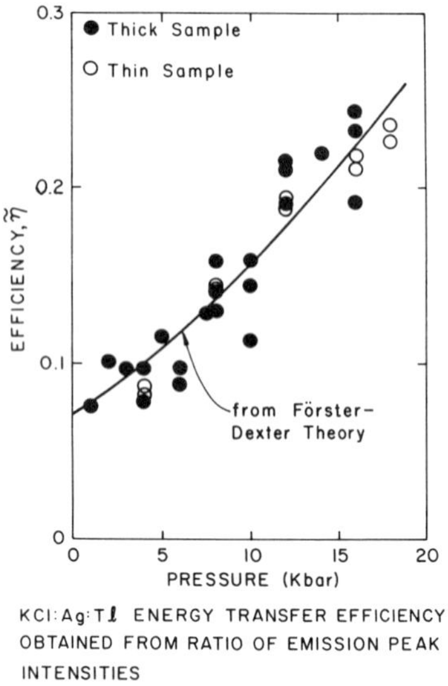

Fig. 16. Comparison of calculated transfer efficiency with that measured from relative emission intensities - KCl:Ag:Tl.

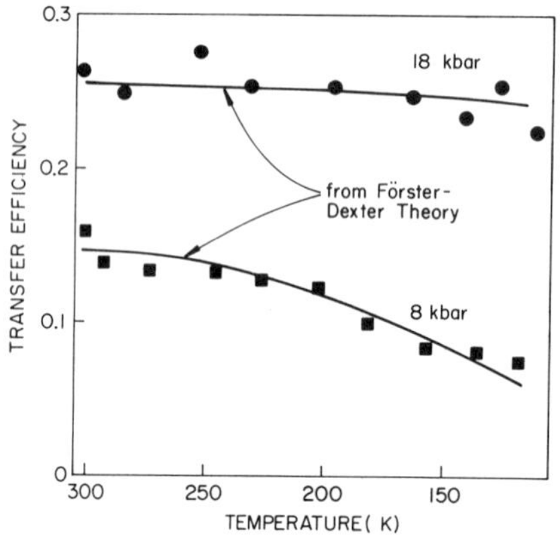

Fig. 17. Comparison of calculated and measured efficiencies as a function of temperature at 8 and 18 kbars - KCl:Ag:Tl.

including alkali, alkaline earth, and rare earth metals. One has also observed insulator-metal transitions in molecular, ionic and covalent materials, and transitions in aromatic hydrocarbons and charge transfer complexes involving induced chemical reactivity of a new type [20].

For our example we discuss changes in spin state and oxidation state of iron with pressure. These results apply to the study of the physics of magnetism in insulators, to organometallic chemistry and to the chemistry of some molecules of biological interest. Some aspects may also be of geophysical importance. The effect of pressure in some ways mirrors and in some ways complements the effect of chemical modification of the ligand.

We concern ourselves with three states of iron: the high spin ferric state with five electrons in five orbitals, a paramagnetic state with a spherically symmetric d shell; the high spin ferrous state, still paramagnetic, but no longer spherically symmetric, and the low spin ferrous state - diamagnetic, with a spherically symmetric d shell.

We use Mossbauer resonance as well as optical absorption to study these processes. The isomer shift, which measures the electron density of the nucleus, and the splitting of the nuclear energy levels by the electric field gradient due to the d shell gives a means of identifying the states. They are characterized by the Doppler velocity necessary to bring them into resonance with Fe^{57} in some fiducial material (e.g. metallic Fe). Fe(II)(HS) has a large positive isomer shift (1.2 to 1.4 mm/sec) and a large quadrupole splitting (2 to 4 mm/sec). Fe(III)(HS) has an isomer shift of 0.3 to 0.5 mm/sec and quadrupole splitting of 0 to 0.8 mm/sec, while Fe(II)L.S. typically has an isomer shift of -0.2 to +0.3 mm/sec and a quadrupole splitting of 0 to 0.5 mm/sec. These characteristic resonances provide "thumb prints" whereby the states can usually be identified.

We turn first to the magnetic state of iron. In a crystal or a molecule the degeneracy of the 3d orbitals is partially removed. In Fig. 18 we show the splitting characteristic of octahedral symmetry. Normally for FE(II) this splitting is 0.5 to 1.0 eV which is not sufficient to overcome the repulsion involved in spin pairing. If the field becomes as large as 1.3 to 1.5 eV spin pairing becomes energetically economical as illustrated in Fig. 18. It is well established that the ligand field increases with pressure [20], usually at a rate somewhat faster than the $\rho^{5/3}$ predicted by simple point charge arguments. Hence a HS→LS conversion with increasing pressure may occur.

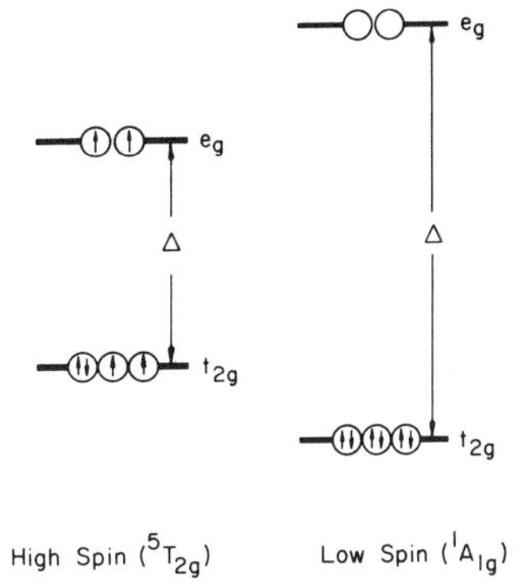

Fig. 18. Ligand field splitting of 3d levels for high and low spin cases - FE(II).

The example we exhibit is Fe(II) as a dilute substitutional impurity in MnS_2, a cubic crystal isomorphous onto FeS_2, iron pyrites. Fe(II) in FeS_2 is low spin at all pressures; however MnS_2 has a somewhat larger lattice parameter so that the local field at the substitutional iron is less than the limit required for the low spin state. The compression supplied by pressure permits the HS→LS transition. Figure 19a shows the Mossbauer spectrum of Fe(II) in MnS_2 at low pressure, with the characteristic large isomer shift and quadrupole splitting of Fe(II)HS. By 65 kbars there is an essentially equal amount of material with characteristics Fe(II)LS parameters, and above 100 kbars (Fig. 19b) the conversion is complete. In a pure magnetically

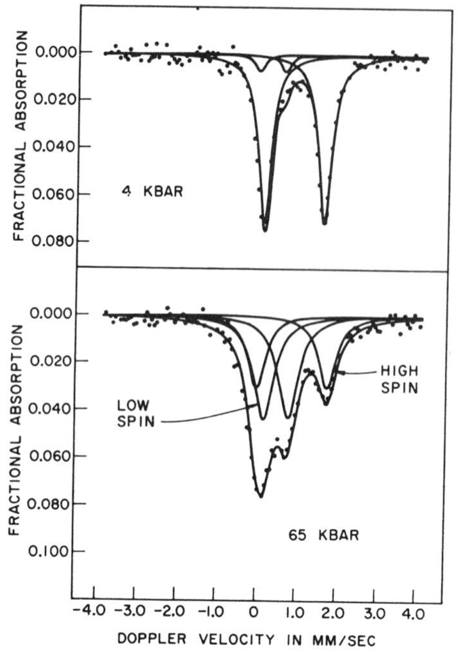

$MnS_2(Fe^{57})$

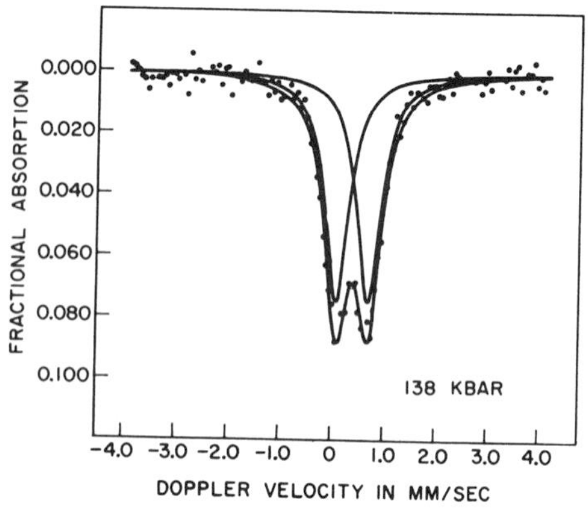

$MnS_2(Fe^{57})$ (Cont'd)

Fig. 19a. Mossbauer spectrum of Fe^{57} in MnS_2 at 4 and 65 kbars.

Fig. 19b. Mossbauer spectrum of Fe^{57} in MnS_2 at 130 kbars.

concentrated crystal the process might well be first order. Here we observe a transition from a paramagnetic to a diamagnetic ground state. These states are important in some catalysts and biometallic compounds. The transition may also occur at depth in the earth.

The final electronic transition discussed here involves the reduction of Fe(III)HS to FE(II)HS by electron transfer from non-bonding ligand orbitals. The process clearly can only occur to any significant extent for ligands with sufficiently low electron affinity. Examples include many organic chelates, the phosphate and sulfate, some hydrates, and the chloride and bromide. It is of interest that the transfer from the iodide ion is so facile that ferric iodide does not normally exist at one atmosphere.

The process is illustrated for ferric acetyl acetonate in Figs. 20a and 20b. By 40 to 45 kbars there is significant reduction, and by 145 kbars the conversion is 85 to 90% complete. As can be seen in Fig. 20b, on release of pressure the process is substantially reversible with some hysteresis. The reasons why the process is not first order are discussed in detail elsewhere [20,21].

We ascribe the reaction to an increase in energy of the non-bonding ligand orbitals relative to the metal 3d orbitals, sufficient to permit the thermal transfer of an electron to the metal

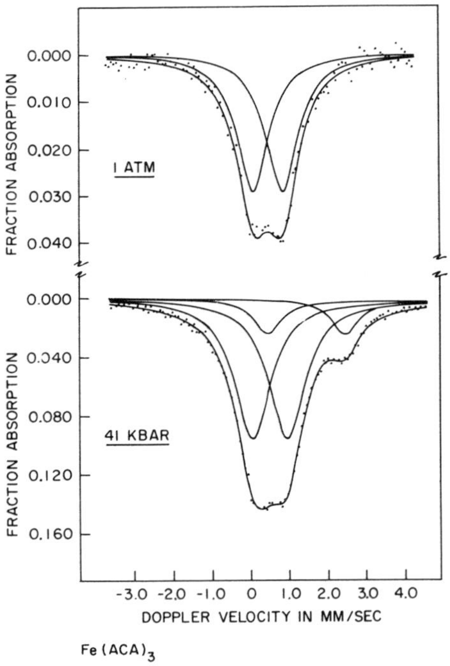

Fig. 20a. Mossbauer spectrum of Fe^{57} in Ferric acetylacetonate-one atmosphere and 41 kbars.

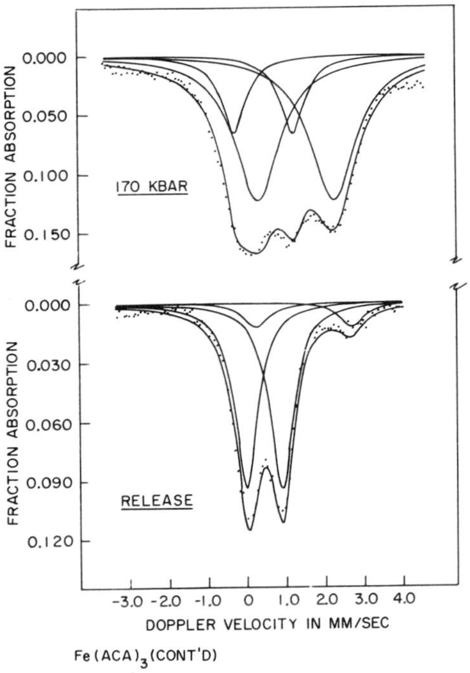

Fig. 20b. Mossbauer spectrum of Fe^{57} in Ferric acetylacetonate-170 kbars and after release of pressure.

orbitals, leaving a hole, probably quite localized, on the ligand.

This behavior is reflected in the optical spectra in several ways. A shift to lower energy of the ligand-to-metal charge transfer peak has been observed for several systems [18]. The area under this charge transfer peak decreases as the amount of ferric ion decreases. This is illustrated in Fig. 21 for several ferric hydroxamates. Hydroxamic acid is an amino acid whose electron affinity can be varied by changing substituent groups. In Fig. 21 the solid line represents the fraction of Fe(III) remaining as a function of pressure as determined from the Mossbauer spectra while the points represent the relative area under the ligand to metal charge transfer peaks. Equally good agreement has been obtained for other iron chelates.

Finally, it is possible to relate the probability of reduction of iron to the location and shape of the charge transfer peak. The analysis, which is discussed in detail elsewhere [20,22] is based on the fact that the optical transition is subject to the Franck-Condon restriction while the thermal process is not. The quantity E_{th}, illustrated in Fig. 6 must approach kT for thermal electron transfer.

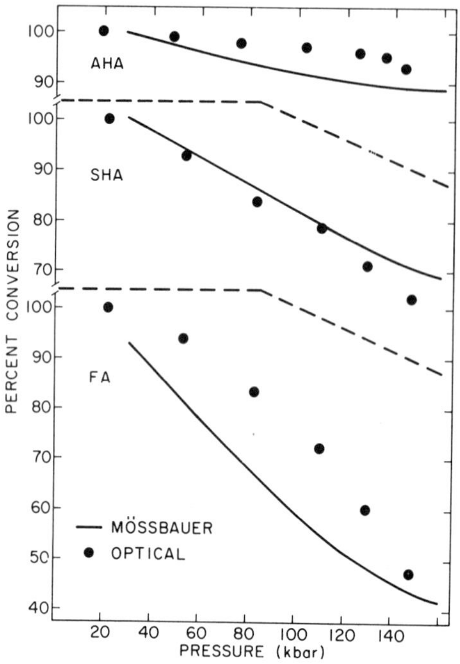

Fig. 21. Comparison of conversion of Fe(III) to Fe(II) with pressure from Mossbauer resonance (———) and from area under the ligand to metal charge transfer peak (●) for ferric hydroxamates.

$$E_{th} = h\nu - 3.6(\delta E_{\frac{1}{2}})^2 R \quad (7)$$

The quantities are defined in the Notation. In the form given, the equation applies at 25°C with all energies in eV. Table I illustrates the application to a series of ferric hydroxamates. The pressures listed in the table are those for which 10% reduction is observed in the Mossbauer spectrum. R = 0.95 to 1.0 for these systems. The calculated values of E_{th} approach zero as one should expect.

The reduction of Cu(II) has also been observed for a number of chelates where the electron affinity of the ligand is low.

These examples are few in number and presented only in outline. However I hope that the picture developed here is sufficiently sharply focused to convey a feeling for the power, in fact the essential nature, of pressure as a tool for studying electronic phenomena.

Table I. Comparison of Optical vs. Thermal Excitation Energy for Ligand to Metal Charge Transfer in Ferric Hydroxamates.*

Compound	Pressure	$h\nu_{max}$	$\delta E_{\frac{1}{2}}$	E_{th} (eV)
AHA	125	2.80	0.900	-0.11
BHA	105	2.70	0.875	-0.06
SHA	70	2.54	0.840	-0.02
FA	37	2.65	0.835	+0.11

*Comparison is made of the pressure for 10% Fe(III)→Fe(II) conversion as determined by Mossbauer resonance.

ACKNOWLEDGMENTS

None of this research could have been performed without the collaboration of my students and colleagues. It is a great pleasure to acknowledge the continuing support of the work by the Energy Research and Development Administration under Contract EY-76-C-02-1198.

NOTATION

$\delta E_{1/2}$ = full width at half maximum of emission or absorbtion peak

E_D = depth of donor level below the conduction band in ZnS

E_{th} = energy difference between bottoms of potential wells (see Fig. 6)

$F_A(E)$ = distribution of intensity with energy-acceptor absorbtion

$f_\Delta(E)$ = distribution of intensity with energy-donor emission

$\delta G_1 \delta G_2$ = free energy of binding of sugar to lysozyme in different pressure regimes

h = Planck's constant

K = equilibrium constant

n = refractive index

P_{DA} = probability of donor-acceptor energy transfer

p = pressure

$p_{1/2}$ = pressure such that K = 1

Q = configuration coordinate displacement (see Fig. 6)

R = $(\omega'/\omega)^2$ = ratio of force constants (see Fig. 6)

R_B = gas constant (Avogadro's number times Boltzmann constant)

S_o, S_1, S_2 = singlet energy states of an electron

t = time

T = temperature

ΔV_1^o = volume change between native protein and protein in state (1)

ΔV_2^o = volume change between protein in states (1) and (2)

α, β = combinations of constants

Δ = displacement of excited state potential well with respect to ground state (see Fig. 6)

Δ_o = Δ at p = 0

ν^o = excitation or emission frequency

ρ = density

τ = emission lifetime

τ_D = donor emission lifetime

ω^2 = force constant for ground electronic state (see Fig. 6)

ω'^2 = force constant for excited electronic state (see Fig. 6)

REFERENCES

1. P. W. Bridgman, The Physics of High Pressure, G. Bell and Sons, London (1949).
2. P. W. Bridgman, Collected Experimental Papers, Harvard University Press, Cambridge (1964).
3. T. Li, J. W. Hook III, H. G. Drickamer and G. Weber, Biochemistry 15, 5571 (1976).
4. D. J. Mitchell, G. B. Schuster and H. G. Drickamer, J. Am. Chem. Soc. 99, 1145 (1977).
5. D. G. Wilson and H. G. Drickamer, J. Chem. Phys. 63, 3649 (1975).
6. B. Y. Okamoto and H. G. Drickamer, J. Chem. Phys. 61, 2870 (1974).

7. W. D. Drotning and H. G. Drickamer, Phys. Rev. B 13, 4568 (1976).
8. C. W. Struck and W. H. Fonger, J. Luminescence 10, 1 (1975).
9. K. W. Bieg, D. Woracek and H. G. Drickamer, J. App. Phys 48, 639 (1977).
10. C. E. Tyner and H. G. Drickamer, J. Chem. Phys. 67, 4103 (1977).
11. D. J. Mitchell, H. G. Drickamer, and G. B. Schuster, J. Am. Chem. Soc. 99, 7489 (1977).
12. M. Schon, Z. Physik 119, 463 (1942).
13. H. A. Klasens, Nature 158, 306 (1946).
14. J. Lambe and C. C. Klick, Phys. Rev. 98, 909 (1955).
15. J. S. Prener and F. E. Williams, J. Chem. Phys. 25, 361 (1956).
16. G. House and H. G. Drickamer, J. Chem. Phys. 67, 3221 (1977).
17. Th. Forster, Ann. Physik 2, 55 (1948).
18. D. L. Dexter, J. Chem. Phys. 21, 836 (1953).
19. K. W. Bieg and H. G. Drickamer, J. Chem. Phys. 66, 1437 (1977).
20. H. G. Drickamer and C. W. Frank, Electronic Transitions and the High Pressure Chemistry and Physics of Solids, Chapman and Hall, London (1973).
21. C. P. Slichter and H. G. Drickamer, J. Chem. Phys. 56, 2142 (1972).
22. H. G. Drickamer, C. P. Slichter, and C. W. Frank, Proc. Nat. Acad. Sci. 69, 933 (1972).

A-1

STATUS OF EQUATION OF STATE OF SOLIDS*

A. L. Ruoff

Cornell University
Ithaca, New York, U.S.A.

and

L. C. Chhabildas

Sandia Laboratories
Albuquerque, New Mexico

INTRODUCTION

Equation of state measurements, namely pressure-volume-temperature measurements enables one not only to evaluate thermodynamic parameters but also to get some insight into the nature of atomic and macroscopic theories. The measurement of volume as a function of pressure can be accomplished as follows: Length change measurement techniques; piston volume displacement techniques; ultrasonic techniques; combined length change and ultrasonic transit time measurement; Brillouin scattering techniques; and shock wave techniques. Each technique has its advantages and serious limitations. A brief discussion of each technique is given elsewhere [1]. The precision and errors associated with these techniques are given in Table I.

Results for the rare gas solids, alkali metals and alkali halides are discussed. Finally, various equations of state which are normally used to represent data are also discussed.

RARE GAS SOLIDS

As the rare gas solids RGS are compressible materials (1.1 GPa for neon to 3.6 GPa for xenon), one can achieve pressures of the magnitude $P \sim B_o$ and volume compressions $V/V_o \sim 0.8$, in a piston cylinder device. Hence, there is considerable interest in

*Invited paper.

Table I. Prescision and Errors Associated With Various Experimental Techniques Used For Equation of State Studies

Method	Max. Pr. (GPa)	Physical Quantity Measured	Precision*	Remarks
Length Change	0.8	L/L_o	3×10^{-8}	% error in B_o, B_o' and B_o'' is 0.05, 3, 50, respectively. B_o'' determined only for very compressible materials.
Piston-volume	4.5	V/V_o	1×10^{-4}	% error in B_o, B_o' is 3 and 20 respectively. B_o'' determined for very compressible materials. Limited by friction and shear strength.
Ultrasonic	1	Transit Time τ	1×10^{-5}	% error in B_o, B_o' and B_o'' is 0.05, 3, 30 respectively. B_o'' determined only for very compressible materials.
Ultrasonic (piston-anvil devices)	3.5 -27.0	Transit Time τ	5×10^{-5}	Small length changes determination and using Cook's analysis limits accuracies. Velocities known to ~ 3%. Hence, $B^S(P)$ is not known to better than 6%.
Brillouin scattering	3.5	Phonon velocity via Doppler shift	1×10^{-3} (P=0) 3×10^{-2} (P>0)	$B^S(P)$ not known to better than (6 to 10%) as velocities are not known to better than 3%. Failure to detect curvature due to large uncertainties.
Shock wave	>100	U_s and U_p	1×10^{-2}	Hugoniot pressures known to (3 to 4%). Isotherms may not be accurate due to uncertainty in thermal conversion and due to neglect of the defect volume.

*This is not to be mistaken for accuracy.

determining the equation of state for rare gas solids with the hope that perhaps they would serve as models for other solids where experiments in that pressure range cannot be done easily. The techniques that have been used are x-ray diffraction [2,3], laser interferometry [4,5], and by applying pressures on crystals (about 300 bars) they have determined the initial bulk modulus of solids. Also, from neutron scattering experiments [6] only the initial bulk modulus is determined. Stewart [7] and more recently, Anderson, Fugate and Swenson [8-11] also have carried out experiments on these solids up to 2 GPa. Within the combined uncertainty of their experiments (5%), the agreement that they obtain is good and is indicated in Table II. Even though the experiments were carried out in piston-cylinder devices, the presence and effect of friction and shear stress for these solids is relatively not too serious as the rare gas solids are low bulk modulus materials.

The interesting results of these experiments as reported by Anderson and Swenson [11] are summarized as follows:

1. The individual isotherms can be represented by P-V relations suggested by the form of the Lennard-Jones potential.
2. A reduced equation of state plot at T = 0 (i.e., V/V_o vs. P/B_o) is superimposed to better than 10^{-3} in V/V_o for argon, krypton and xenon with the neon solid deviating from it essentially demanding a higher bulk modulus.
3. A plot of B vs P appears linear only up to ~ 0.4 GPa. Hence, they list B_o' and B' at 1.7 GPa, thereby indicating a pressure dependence for B'.
4. For $T < \theta_D$, C_V is a constant close to the value 3R. Also, $(\partial P/\partial T)_V$ is found to be independent of temperature and volume for $T > \theta_D$.
5. The results in (4) imply that $\gamma/V = (\partial T/\partial P)_V/C_V$ is a constant for $T > \theta_D$.

We add the following. The value of B_o'' are computed assuming Keane's equation applies. This equation will be described later. It has the realistic feature that B' decreases from B_o' to B_∞'. It is easy to show for this equation that

$$B_o B_o'' = - B_o'[B_o' - B_\infty'] \tag{1}$$

Equation (1) along with the values of B_o, B_o' and B_∞' were used to compute the values of B_o'' in Table II.

For the rare gas solids, we find on the average at 4K that B_oB_o'' is -12.9, B_o' is 7.35, and B_∞' is 5.6. The inverse 5-11 pair potential leads to values of $7.\bar{3}$ and $5.\bar{6}$ for B_o' and B_∞', respectively.

Table II. Bulk Modulus and Its Pressure Derivatives for Rare Gas Solids

Solid	Temp., K	B_o, GPa	B_o'	B'* (P = 1.76 GPa)	B_o'', † $(GPa)^{-1}$	Ref.
Ne	4	1.10	7.8	5.7‡	-12	10
Ne	4	1.11	-	-		3
Ne	5	1.12	-	-		6
Ne	18	0.80	7.0	-		7
Ar	4	2.86	7.2	5.6	- 4.5	11
Ar	77	1.41	8.4	5.6		11
Ar	77	1.26	7.8	-		7
Ar	77	1.27	-	-		2
Kr	4	3.34	7.2	5.7	- 3.9	11
Kr	4	3.45				4
Kr	4	3.44				2
Kr	77	2.17	7.6	6.3		11
Kr	77	2.10				4
Kr	77	2.08				2
Kr	81	1.99	7.5			7
Xe	4	3.63	7.2	5.4	- 3.5	11
Xe	4	3.79				5
Xe	77	2.73	7.7	6.2		11
Xe	77	2.82				5

*Limiting B' values.
†Computed from Keane's equations assuming $B_o' = 7.35$ (average for four elements at 4K) and $B_\infty' = 5.6$ (average for four elements at 4K)
‡This is value given for ~ 0.4 GPa.

ALKALI METALS

A summary of the room temperature results obtained [12-28] by various different techniques are tabulated in Table III. The results obtained by shock wave techniques are not included mainly because the measured Hugoniot points for the alkali metals are above the melting line. There are no known ultrasonic measurements for rubidium and cesium at room temperature, because it is difficult to excite shear waves [23,24] in these crystals. In general the parameters obtained above are fitted to Murnaghan's two or three parameter equations (described later). Swenson's P-V measurements agree remarkably well with ultrasonic values whereas Vaidya et al.'s values seem to be consistently lower than either Swenson's or the ultrasonic measurements. Nevertheless, the

Table III. Bulk Modulus and Its Pressure Derivatives for Alkali Metals at Room Temperature

Solid	B_0, GPa	B_0'	B_0'', $(GPa)^{-1}$	Ref.
Lithium	11.8	3.33		12
Lithium	11.2	3.33		16
Lithium	11.2	3.6		17
Lithium	11.5	3.6		18
Lithium	11.6	-		19
Sodium	6.2	3.5		12
Sodium	6.13	3.7	-0.2	12
Sodium	6.69	3.4		16
Sodium	6.3	3.95		18,19
Sodium	6.15	3.59		25
Sodium	6.40	3.79		15
Sodium	6.46	3.63	-0.69*	13
Sodium	6.31			26
Potassium	3.40	2.99		12
Potassium	3.12	3.65	-0.39	12
Potassium	3.33	3.43		16
Potassium	3.0	3.85		19
Potassium	3.09	3.98		14
Potassium	3.1			27
Rubidium	2.66	3.23		12
Rubidium	2.60	3.37	-0.1	12
Rubidium	3.04	3.29		16
Cesium	1.63	2.84		28

*T = 195K value from Ref. 61 which involved a reanalysis of the original data of Martinson (also involved in Ref. 13 and 15).

agreement is good, considering that the error in B_0' values for PV measurements may be as high as 10%. Swenson finds that a reduced plot (P/B_0 vs. V/V_0) for Li, Na, K overlap (B_0' = 3.85), whereas the data for cesium (B_0' = 2.84) do not. (Whether rubidium does or does not depends on what the correct value for B_0' is. Table III indicates that B_0' is 3.3 ± 0.3.) There has been speculation that this behavior is essentially due to the distortion of the Fermi surface for cesium. Glinski and Templeton [29], by measuring the pressure dependence of the de Hass-van Alphen frequencies in cesium, have calculated the free electron values for B_0. More recently, Beardsley and Schirber [30] dispute the claim of Glinski and Templeton and show that there is no inconsistency between the conventional compressibility measurements and those obtained by them [30]. One possible explanation for this large discrepancy in B_0' value for cesium as compared to other alkali metals is because of the assumption that B' is not a varying function of pressure, and the effect would be more dominant on cesium as its initial bulk modulus is only 1.6 GPa compared to 11.6 GPa for lithium at room temperature. If one makes an assumption that B' varies linearly with pressure up to 2.2 GPa then the values for B_o'' for cesium would be $-0.9\ GPa^{-1}$, taking the B_o' value to be 3.85 as for the other alkali metals.

ALKALI HALIDES

The alkali halides have received considerable attention as is indicated in Tables IV and V. The agreement that one obtains between piston volume experiments and ultrasonic experiments is rather poor for incompressible halides. The B_o values obtained from shock wave experiments in most cases are low when compared to ultrasonic measurements. For the cesium halides, Barsch and Chang [31] have determined B_o'' in addition to B_o and B_o'; this was the first measurement of B_o'' for alkali halides.

If we compute the average value of B_o' and B_oB_o'' for all the data for all the alkali halides shown in Tables IV and V (of both crystals types) we obtain values for B_o' and B_oB_o'' of 5.26 and -10.8, respectively. If only the data of Table IV were used, these values would be 5.21 and -8.1, respectively. For a 1-9 inverse pair potential B_o' and B_oB_o'' have values of $5.\bar{3}$ and $-5.\bar{3}$, respectively.

Sodium Chloride

Chhabildas and Ruoff [32] by using a highly sensitive length change measurement technique, have measured B_o'' isothermally for sodium chloride; this represents the first measurement of B_o'' by length measurement. Their value of B_o' seems to be larger than those obtained by ultrasonic techniques and is listed in Table V. However, most of the ultrasonic measurements were undertaken up to 0.4 GPa only and as they were analyzed on the assumption that B is linear with P, the ultrasonic values of B_o' could perhaps be low by 0.2 which would then tend to raise B_o' up to 5.55 [except for the Spetzler, et al. measurements [50] where non-linear variation has already been introduced]. However, if the ultrasonic techniques do not reproduce B_o to within 2%, as appears to be the case, the error in B_o' must be larger and the ultrasonic B_o' values may not be as accurate as they appear to be.

Also tabulated in Table V are the results obtained by Brillouin scattering techniques [33]. They too happen to be the first measurement of its kind. They fail to detect the curvature (B_o'') that Chhablidas and Ruoff [13] and Spetzler et al. [50] report. Also it is ironic that they fail to reproduce Decker's equation of state ($B_o' = 4.9$), although they use a ruby gauge (which is calibrated against Decker's equation of state of NaCl) to estimate pressures. Instead they obtain a value of 6.0 for B_o'.

Alkali halides have been investigated intensively mainly because the cohesive properties [35] (energy) can be understood (reproduced) in terms of an attractive couloumb interaction and a repulsive interaction either exponential in nature ($Be^{-r/\zeta}$) or by a power law (C/r^n). Smith et al. [36,37] have determined these repulsive parameters for most halides. They also note [38] that LiF, NaCl, NaF and NaI have a common equation of state up to pressures P/B of ~ 0.3 (∿7 GPa). Since Jamieson [39] has used NaCl as an internal standard to estimate pressures, Decker [40] has proposed using the equation of state for NaCl based mainly on the central force model. However, because of the failure of both the Cauchy and Love relation as described in detail by Ruoff and Chhablidas [41], the Born-Mayer model should not be taken as a quantitative model for computing P(V) particularly at pressures as large as B_o. Moreover, it can be shown [1] that the repulsive potential is not unique. One can use for instance a gaussian repulsive interaction, $D\ e^{-r^2/f^2}$, and still reproduce the same cohesive energy and experience the same error magnitude in estimating B_o'. There is, of course, a function representing the repulsive interaction, which if used in formal fashion would give the correct P(V) for the static lattice; however, at present this function is not known. Moreover, even if it were known, its existence does not imply the correctness of a central force approximation. With respect to the equation of state of NaCl, we conclude that no theory at present is approximately satisfactory. The Born-Mayer model can be and will be used to estimate approximate pressures keeping in mind that they are indeed approximate pressures. We note that we have begun the second half century of modern quantum wave mechanics and that it is perhaps time for serious comprehensive work on this problem and we strongly urge that it be undertaken.

Nevertheless, one should not underestimate the role that Decker's equation of state has played. It has been a very important contribution in our understanding of pressure calibration. For instance, the GaP transition which on earlier pressure scales registered 55 GPa [42] is now in the neighborhood of 22 GPa on the NaCl scale. This is a tremendous improvement but perhaps not the final answer.

Table IV. Bulk Modulus and Its Pressure Derivatives for Alkali Halides

Material	Technique	B_o, GPa	B_o'	B_o'', $(GPa)^{-1}$	Ref.
LiF	Length charge	66.7	5.40		43
LiF	Piston volume	62.3	6.82		44
LiF	Ultrasonic	66.5	5.24		37
LiF	Shock	66.8	4.53		34
LiCl	Piston volume	31.9	3.36		44
LiCl	Ultrasonic	29.7	5.63		37
LiCl	Ultrasonic	29.9	5.34		45
LiCl	Shock	29.7	4.71		34
LiBr	Piston volume	24.3	3.50		44
LiBr	Ultrasonic	23.5	5.68		37
LiBr	Ultrasonic	24.1	5.30		45
LiBr	Shock	25.4	4.63		34
LiI	Piston volume	16.8	4.32		44
LiI	Piston volume	13.6	4.94	-0.71	42
LiI	Ultrasonic	17.3	6.15		37
NaF	Piston volume	46.7	5.18		44
NaF	Ultrasonic	46.5	5.25		36
NaF	Shock	47.1	4.70		34
NaCl	(See Table V)				
NaBr	Piston volume	20.3	4.19		44
NaBr	Ultrasonic	19.5	5.44		36
NaBr	Shock	20.4	4.43		46
NaI	Piston volume	15.1	4.15		44
NaI	Ultrasonic	14.9	5.44		37
NaI	Shock	15.0	5.48	-0.6	46
KF	Ultrasonic	30.2	5.38		36
KCl	Ultrasonic	17.4	5.46		36
KBr	Ultrasonic	14.7	5.47		36
KI	Ultrasonic	11.5	5.56		36
RbF	Ultrasonic	26.7	5.69		36
RbCl	Ultrasonic	15.6	5.62		36
RbBr	Ultrasonic	13.2	5.59		36
RbI	Ultrasonic	10.5	5.60		36
CsCl	Piston volume	17.1	5.09		44
CsCl	Piston volume	16.9	5.54	-0.24	44
CsCl	Ultrasonic	16.8	5.98	-0.42	31
CsCl	Shock	17.4	5.50		46
CsBr	Piston volume	14.4	5.32		44
CsBr	Ultrasonic	14.3	5.95	-0.50	31
CsBr	Shock	20.5	4.22		46
CsI	Piston volume	12.5	4.50		44
CsI	Piston volume	11.6	6.10	-0.88	44
CsI	Ultrasonic	11.9	5.93	-0.73	31
CsI	Shock	11.0	5.75	-0.80	46

Table V. Bulk Modulus and Its Pressure Derivatives for NaCl at 29°C

Ref.	Author	Technique	B_o, GPa	B_o'	B_o'', $(GPa)^{-1}$
32	Chhabildas & Ruoff	Length	23.81	5.64	
32	Chhabildas & Ruoff	Length	23.77	5.71	- 1.0
44	Vaidya & Kennedy	Piston volume	23.17	4.92	
47	Haussühl	Ultrasonic	23.73		
48	Ghafelehbashi & Koliwad	Ultrasonic	23.70	5.37	
49	Barsch & Chang	Ultrasonic	23.42	5.39	
50	Spetzler, et al.	Ultrasonic	23.80	5.35	- 0.9
51	Bartels & Schuele	Ultrasonic	23.40	5.35	
33	Whitfield, et al.	Brillouin scattering	23.25	6.00	
52	Fritz, et al.	Shock	23.73*	5.50	- 0.4†

*Value not measured in the shock work; instead Haussühl's value was used.

†Value not given but was calculated from their quadratic U_s vs. U_p relation.

VARIOUS EQUATIONS OF STATE

Various equations of state [53-57] that are often used to fit experimental data are given in Table I by Ruoff et al. [32]. Some of these equations are two parameter (B_o and B_o') whereas others are three parameter (B_o, B_o', and B_o''). ME_1 is the only one for which $B_o'' \equiv 0$; although each of BE_1 and GGKE has two parameters they have an inherent negative value for B_o''. Numerous articles have been prepared on these equations of state (EOS) essentially showing how these are all equivalent for very small compressions ($V/V_o > 0.9$). MacDonald and Powell [58] have provided an excellent article on the precision required to discriminate between these equations of state. For $P/B_o \approx 2 \times 10^{-2}$ the precision in V/V_o required is 1×10^{-6} and hence discrimination at low pressures is nearly impossible.

As ME_1, BE_1, and GGKE estimate pressures that are 5 to 10% larger than presently accepted values [32] for $V/V_o \sim 0.82$, they are therefore not appropriate for accurate extrapolation for $V/V_o < 0.9$. Murnaghan himself [53] proposed this when he realized that he would obtain different parameters B_o and B_o' when he tried to fit Bridgman's Na data (up to 10 GPa) over different pressure ranges and proposed including higher order terms (B_o'').

It is generally established that B_o'' is negative for different groups of solid materials (see Tables II, III, IV and V). However, its magnitude may be in dispute, because it is a very difficult parameter to measure either by length change or ultrasonic technique. Just on physical grounds, one can rule out both the ME_2 and BE_2 equations. The use of these equations with $B_o'' < 0$ leads to a physically and thermodynamically unreasonable condition at very high pressures (although in the present case this is academic because a phase transition occurs before such a condition arises). The bulk modulus increases to a maximum and then decreases and becomes negative and the pressures at which this may happen depends on the magnitude of B_o''. Hence these equations must be used with caution and must be limited in range.

Recently Anderson [59] has proposed using Keane's equation as he found good agreement between extrapolated ultrasonic data and shock wave data. However, he forced the ultrasonic data to agree with the shock wave data to determine B_o''. Chhabildas and Ruoff, on the other hand, used B_o'' to estimate the pressure at the NaCl transition, using Keane's equation. However as the NaCl transition is not known accurately [41] and not established yet, there is no way of commenting on its (KE) validity. However, there is experimental evidence to probably justify its use. Anderson and Swenson [60] note that B' varies with pressure for rare gas solids, helium, hydrogen and deuterium and in each case it appears to approach a constant value above 1 GPa (which is a very high pressure for these solids), thereby agreeing with the concept that B' approaches B_∞' a concept introduced by Keane.

CONCLUSION

The status of EOS presently is not very satisfactory, but is definitely getting better. It is encouraging to note that as experimental precision and accuracies and also computing methods are becoming better, the parameter B_o'' is being measured both ultrasonically and by length change techniques. These two techniques should be combined to increase the accuracy on B_o'' by going first to at least 1.5 GPa. It is quite likely that even by going to 1.5 GPa only, one may be able to grasp a better understanding of material behavior at high pressures.

Some of the problem areas have been mentioned quite critically with the hope that they become the focus of attention and research for the next phase of equation of state study. It is somewhat analogous to an iterative procedure used in computing, where the theorists are looking for good experimental work and vice versa thereby narrowing their differences.

ACKNOWLEDGMENTS

The authors wish to thank Ms. G. Rowe for doing an extensive literature search. One of us, L.C.C., would like to acknowledge B. Butcher for his flexibility and understanding during the course of the work.

A.L.R. wishes to thank the National Aeronautics and Space Administration for their support. The authors also acknowledge the National Science Foundation for their support of the Cornell Materials Science Center, whose central facilities were used by them.

NOTATION

B = isothermal bulk modulus

B_o = isothermal bulk modulus at zero pressure

$B_o' = \lim_{P\to 0} (\partial B/\partial P)_T$

$B_o'' = \lim_{P\to 0} (\partial^2 B/\partial P^2)_T$

B^s = adiabatic bulk modulus

L = length

L_o = initial length at zero pressure

P = pressure

r = interatomic distance

U_s = shock velocity

U_p = particle velocity

V = volume

V_o = initial volume at zero pressure

θ_D = Debye temperature

τ = transit time

REFERENCES

1. A. L. Ruoff and L. C. Chhabildas, Cornell University, MSC Report No. 2853 (unpublished).
2. A. O. Urvas, D. L. Losee and R. O. Simmons, J. Phys. Chem. Solids 28, 2269 (1967).
3. D. N. Batchelder, D. L. Losee and R. O. Simmons, Phys. Rev. 162, 767 (1967).
4. H. J. Coufal, R. Veith, P. Korpiun and E. Luscher, Phys. Stat. Sol. 38 K127 (1970); also J. Appl. Phys. 41, 5082 (1970).
5. P. Korpiun, W. Albrecht, T. Muller and E. Luscher, Phys. Letters 48A, 253 (1974).
6. J. Skalyo, V. J. Minkiewicz, G. Shirane and W. B. Daniels, Phys. Rev. B6, 4766 (1972).
7. J. W. Stewart, J. Phys. Chem. Solids 29, 641 (1968).
8. C. A. Swenson and M. S. Anderson, AIP Conf. Proc. Vol. 3, (1972), p. 105.
9. R. Q. Fugate and C. A. Swenson, J. Low Temp. Phys. 10, 317 (1973).
10. M. S. Anderson, R. Q. Fugate and C. A. Swenson, J. Low Temp. Phys. 10, 347 (1973).
11. M. S. Anderson and C. A. Swenson, J. Phys. Chem. Solids 36, 145 (1975).
12. S. N. Vaidya, I. G. Getting and G. C. Kennedy, J. Phys. Chem. Solids 32, 2545 (1971).
13. P. S. Ho and A. L. Ruoff, J. Phys. Chem. Solids 29, 2101 (1968).
14. P. A. Smith and C. S. Smith, J. Phys. Chem. Solids 26, 279 (1965).
15. R. H. Martinson, Phys. Rev. 178, 902 (1969).
16. P. W. Bridgman, Proc. Am. Acad. Arts. Sci. 76, 71 (1940).
17. C. A. Swenson, J. Phys. Chem. Solids 27, 33 (1960).
18. R. I. Beecroft and C. A. Swenson, J. Phys. Chem. Solids 18, 327 (1961).
19. C. E. Monfort and C. A. Swenson, J. Phys. Chem. Solids 26, 291 (1965).
20. J. P. Day and A. L. Ruoff, Phys. Stat. Sol. A25, 205 (1974).
21. H. C. Nash and C. S. Smith, J. Phys. Chem. Solids 9, 113 (1959).
22. T. Slotwinski and J. Trivisonno, J. Phys. Chem. Solids 30, 1276 (1969).
23. E. J. Gutman and J. Trivisonno, J. Phys. Chem. Solids 29, 805 (1967).
24. F. J. Kollarits and J. Trivisonno, J. Phys. Chem. Solids 29, 2133 (1968).
25. W. B. Daniels, Phys. Rev. 119, 1246 (1960).
26. G. Fritsch, M. Nehmann, P. Korpiun, E. Luscher, Phys. Stat. Sol. (a)19, 555 (1973).
27. G. Fritsch and H. Bube, Phys. Stat. Sol. (a)30, 471 (1975).
28. M. S. Anderson, E. J. Gutman, J. R. Packard and C. A. Swenson, J. Phys. Chem. Solids 30, 1587 (1968).

29. R. Glinski and J. M. Templeton, J. Low Temp. Phys. 1, 223 (1969).
30. G. M. Beardsley and J. E. Schirber, J. Low Temp. Phys. 8, 421 (1972).
31. G. R. Barsh and Z. P. Chang, in NBS Special Publication 326, E. C. Lloyd, ed., Govt. Print. Off., Washington, D.C. (1971), p. 173.
32. L. C. Chhabildas and A. L. Ruoff, J. Appl. Phys. 47, 4182 (1976).
33. C. H. Whitfield, E. M. Brody and W. A. Bassett, Rev. Sci. Instr. 47, 942 (1976).
34. W. J. Carter, High Temp-High Press. 5, 313 (1973).
35. M. P. Tosi, Solid State Physics 16, 1 (1964).
36. R. W. Roberts and C. S. Smith, J. Phys. Chem. Solids 31, 619 (1970), ibid., 31, 2397 (1970).
37. K. O. McLean and C. S. Smith, J. Phys. Chem. Solids 33, 275 (1972); also, 33, 279 (1972).
38. C. S. Smith and K. O. McLean, J. Phys. Chem. Solids 34, 1143 (1973).
39. J. C. Jamieson, Science 139, 1291 (1963).
40. D. L. Decker, J. Appl. Phys. 42, 3239 (1971).
41. A. L. Ruoff and L. C. Chhabildas, J. Appl. Phys. 47, 4867 (1976).
42. A. Onodero, N. Kawai, K. Ishizaki, and I. L. Spain, Solid State Commun. 14, 803 (1974).
43. K. Y. Kim, L. C. Chhabildas and A. L. Ruoff, J. Appl. Phys. 47, 2862 (1976).
44. S. N. Vaidya and G. C. Kennedy, J. Appl. Phys. 32, 961 (1971).
45. L. S. Ching, J. P. Day, and A. L. Ruoff, in J. Appl. Phys. 44, 1017 (1973).
46. M. Van Thiel, "Compendium of Shock Wave Data," Lawrence Livermore Laboratory, Report UCRL 50108.
47. S. Haussühl, Z. Phys. 159, 223 (1960).
48. M. Ghafeleshbashi and K. M. Koliwad, J. Appl. Phys. 41, 4010 (1970).
49. G. R. Barsch and Z. P. Chang, Phys. Status Solidi 19, 139 (1967).
50. H. Spetzler, C. G. Sammis, and R. J. O'Connell, J. Phys. Chem. Solids 33, 1727 (1972).
51. R. A. Bartel and D. E. Schuele, J. Phys. Chem. Solids 26, 537 (1965).
52. J. N. Fritz, S. P. Marsh, W. J. Carter and R. G. McQueen in Accurate Characterization of the High Pressure Environment, NBS Special Publication 326, E. C. Lloyd, ed., U.S. Govt. Print. Off., Washington, D.C. (1971), p. 201.
53. F. D. Murnaghan, Finite Deformation of an Elastic Solid, Dover Publications, New York (1967).
54. F. Birch, Phys. Rev. 71, 809 (1947).
55. F. Birch, J. Geophys. Res. 57, 227 (1952).

56. A. Keane, Aust. J. Phys. 7, 322 (1954).
57. R. Grover, I. C. Getting and G. C. Kennedy, Phys. Rev. B7, 567 (1973).
58. J. R. MacDonald and D. R. Powell, J. Res. NBS 75A, 441 (1971).
59. O. L. Anderson, Phys. Earth Planet Inter. 1, 169 (1968).
60. M. S. Anderson and C. A. Swenson, Phys. Rev. B15, 10, 5184 (1974).
61. K. J. Dunn and A. L. Ruoff, Phys. Rev. B10, 2271 (1974).

A-2

METALLIC HIGH PRESSURE EQUATION-OF-STATE DERIVED FROM EXPERIMENTAL DATA*

R. Grover

Lawrence Livermore Laboratory, University of California
Livermore, California USA

INTRODUCTION

This paper describes a program of making 'global' fits to the large amount of experimental equation of state (EOS) data on monatomic metals that has become available in recent years. The fits are made within the framework of a phenomenological scaling theory for metallic liquids which incorporates recently discovered general theoretical properties of the EOS of liquids. The theory is expected to be applicable to monatomic metals up to high temperatures (~ 10 to 100 x the melting temperature T_m) and at all densities, so long as the metallic bonding does not change character - as, for instance, might be caused by electronic phase transitions or sufficiently large thermal electron excitations. The goal of the present fitting studies is to obtain consistent tabular representations of the EOS of metals over a wide range of densities ($0.5 \leq \rho/\rho_o \leq 2$) and temperatures (up to several eV) which are reliable enough both to be used in applied work (errors < ~5%) and to serve as a guide to future theoretical and experimental studies of metals. Fits to experimental data for examples of three different types of metals - Na, Pb, and Ta - will be discussed.

A lack of a reliable theoretical model for the EOS of metals has frequently led to the need to fit high-temperature, high-pressure experimental data. These fits have been made for the most part with models of limited applicability (e.g., Grüneisen [1]) or to limited ranges of T, P data with special mathematical forms thought to be suitable for extrapolation to more extreme conditions [2]. The wide range of data now available, however, makes such

*Work supported by the U. S. Energy Research Development Administration under Contract No. W-7405-Eng-48.

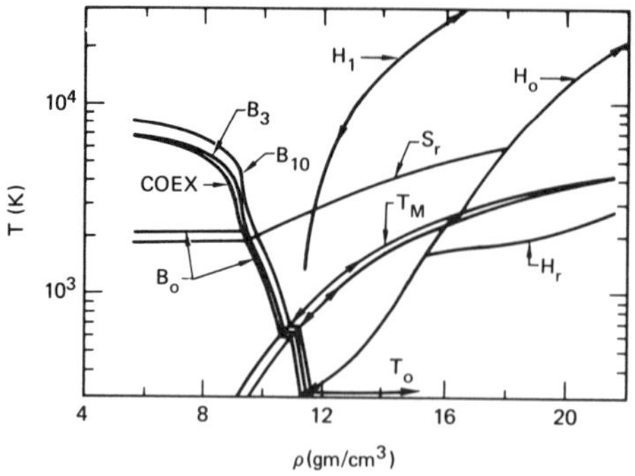

Fig. 1. Experimental T-ρ EOS space for Pb showing thermodynamic paths along which experiments are attempted. Arrows on curves indicate ranges within which data are now available. Legend: T_o - 300 K isotherm; B_3, B_{10} - 0.3 and 1 GPa isobar; H_o - principle Hugoniot; S_r - release isentrope; T_M - melting line; H_1 - porous Hugoniot from ρ_o = 6.79; and H_r - reflected Hugoniot from 35 GPa.

limited fits inappropriate. Figure 1, for example, shows various thermodynamic paths for Pb, extending over a wide range of ρ, T space in all phases along which two or more EOS quantities can now be measured. On some paths having a long excursion from normal conditions, such as the principle Hugoniot compression curve H_o, only a few quantities are currently measurable -- P(ρ) and E(ρ) -- while along some shorter paths, such as T_o, much more data is available. Since the data generally becomes less accurate away from normal conditions, it is desirable to fit experimental data as directly as possible. For this reason the fits to be described were obtained with the use of an interactive display computer code which permits such comparisons of model predictions with experimentally measured quantities.

LIQUID SCALING MODEL

The most serious deficiency in previous equation of state fitting efforts on metals has been the lack of a realistic model for the liquid phase. The EOS scaling model used in this study is a modification of an earlier version [3] based mainly on general EOS properties observed in more recent computer simulations of high-density liquids with a variety of pairwise repulsive interparticle forces. In the computer calculation for these many particle systems it can be recognized [4] that the melting transition and the thermal EOS are determined up to very high temperature (10 to 100 T_M) by some average force constant (or second derivative) derived essentially from the repulsive part of the pair interaction. Such a picture correlates in a simple way the three main computer results [4]: the systematics of the entropy and volume changes at melting, the Lindemann law behavior of the melting temperature, and the universal form for the temperature dependence of the thermal EOS in the liquid phase. Strong effective pair repulsions arise naturally in the pseudo-potential theory of metals from the shielding and polarization properties of the conduction-band, electron gas. It is not surprising that liquid metals have experimental properties in the vicinity of T_M similar to those of the computer model fluids, including a universal temperature dependence for the specific heat [3].

More specifically the model is based on the following assumed properties of liquid metals: (1) The thermal EOS can be written as a sum of independent contributions from atomic motions and electron excitations. (2) The atomic component has a melting transition with a constant entropy of fusion ($\Delta_m S \approx R$) which satisfies the compression form of the Lindemann law

$$\lambda \equiv \frac{d \ln T_M}{d \ln\rho} = 2\gamma_G - 2/3 \qquad (1)$$

γ_G is the usual Grüneisen coefficient $V(dP/dE)_V$ which in the Grüneisen model for a solid is related to the density dependence of the characteristic Grüneisen temperature which determines the high temperature free energy of the solid [5]

$$\gamma_G \equiv - \frac{d\ln\theta_G}{d\ln\rho} \qquad (2)$$

(3) At higher temperatures in the liquid phase the non-ideal atomic part of the thermal EOS is assumed to scale with the melting temperature. The specific form can be derived from the specific heat (C_v) dependence on temperature suggested by Wray [6]

$$C_v = \frac{3}{2} R[1 + \frac{1}{1 + \alpha\tau}] \ , \ \tau \equiv T/T_M \qquad (3)$$

This modification of the original linear specific heat form [3] has been found to fit all computer simulation data adequately. The comparison is shown in Fig. 2 where the nonideal part of the thermal energy is plotted against τ. The line derived from (3) for $\alpha = 0.1$ is a good average representation of all the calculations, including those made for the 'Coulombic' system of positive ions in a neutralizing uniform negative background. (4) The electronic excitations are treated in a low temperature approximation in which the electronic specific heat is linear in the temperature. However, the coefficient of proportionality, g_e, is that appropriate to $T > \theta_G$ and will in general be density dependent. Thus, the electronic gamma, $\gamma_e \equiv + d\ln g_e/d\ln V$ is non-zero (4).

These assumptions restrict the range of application of the model to be less than both $\sim$100 T_M and about one-fifth the electron Fermi temperature of the metal. Most of the experimental data, however, was well within these limits.

It is then possible to represent the metallic EOS in the liquid phase simply as an additive correction to the usual solid EOS of the Grüneisen form. From the specific heat relation (3) simple expressions for the entropy, internal energy, or pressure of a liquid can be derived

$$S_\ell = S_s + \Delta_m S - \frac{3}{2}R\ell n(1 + \alpha\tau) \quad (4)$$

$$E_\ell = E_s + RT_M\left\{\frac{\Delta mS}{R} - \frac{3\alpha}{4} + \frac{3}{2}\left[\frac{\ell n(1 + \alpha\tau)}{1 + \alpha\tau} - 1\right]\right\} \quad (5)$$

$$VP_\ell = VP_s + \lambda(E_\ell - E_s) \quad (6)$$

This form of liquid EOS shares another property with the Grüneisen EOS which considerably simplifies the process of fitting experimental data. Like the solid EOS, the liquid EOS is completely determined by the zero-degree isotherm, $P_o(V)$, and density dependence of γ_G and γ_e. In addition, this simplification permits compact representation of the complete EOS by tables of these quantities [4].

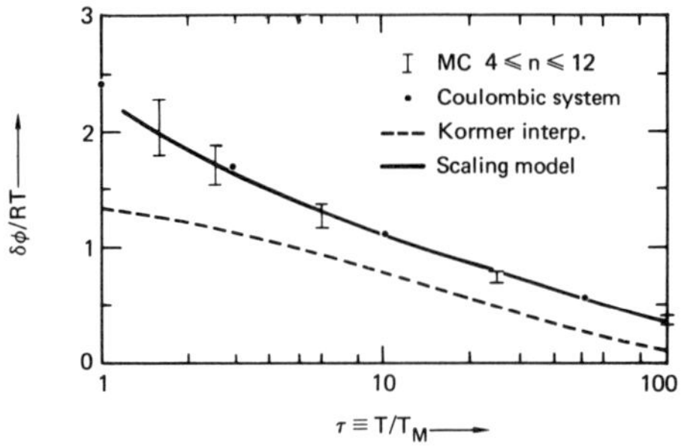

Fig. 2. Comparison of model thermal potential energy with Monte Carlo computer calculations for r^{-n} interaction potentials. The bars indicate the increase of computer results [7] with n in the range $4 \leq n \leq 12$. The circles are computer results for the Coulombic system [8]. The broken line represents the interpolation model of Korner et al. [2].

THREE EXAMPLES OF EOS FITS

In order to test the generality of our fitting model, examples of three different types of metals were chosen -- Na, Ta, and Pb -- for which a substantial amount of data was available. The fitting is done by trial and error adjustments of parameters in functions representing $\gamma_G(\rho)$, $\gamma_e(\rho)$, and the Hugoniot. Separate fits are made for the expanded states where $\rho < \rho_o$ and compressed states where $\rho > \rho_o$ in such a way that all functions are continuously and smoothly joined at ρ_o. The process of finding a fit in this manner, although slower than an automated least squares search, provides some experience for judging the uniqueness of the fits.

A number of fits to experimental data on Pb, Na, and Ta are shown in Figs. 3 through 6. Figure 3 illustrates the wide range of dynamic compression data (up to 2.5-fold compression) that is reproduced by the EOS fits. The comparably wide density range of EOS data now available from isobaric expansion experiments must also be reproduced as shown in Fig. 4. Because of the central role of the Lindemann law in the EOS model, it is necessary to be able to represent well the pressure dependence of the melting temperature when the experimental data covers a significant volume range, as it does in the case of Na shown in Fig. 5. While these fits appear to

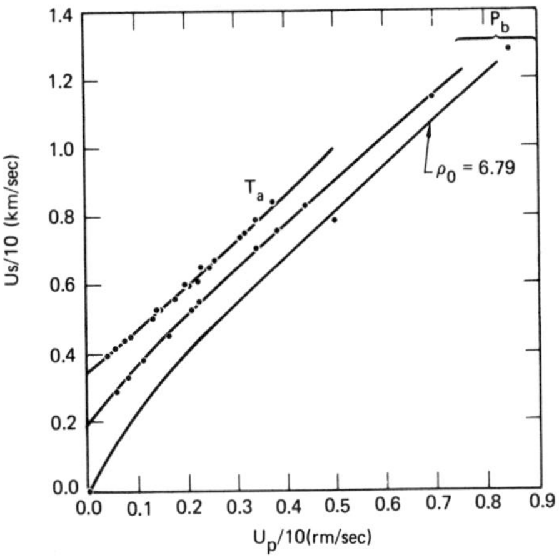

Fig. 3. Comparison of experimental shock velocity and particle velocity data with model calculations for Ta and Pb [9].

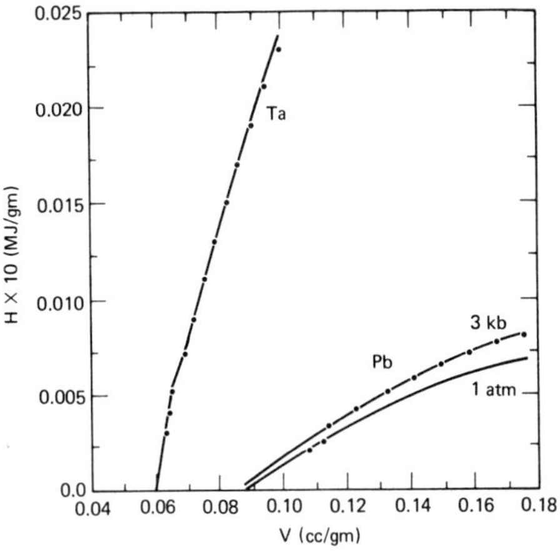

Fig. 4. Comparison of experimental isobaric enthalpy volume data with model calculations for Ta [10] and Pb [11].

be satisfactory on a 'global' scale, they are often not accurate enough to represent the fine structure claimed for the data by the experimentalists. This is particularly true for derivative quantities such as the specific heat and thermal expansion data in the vicinity of melting. However, when direct measurements of EOS derivative quantities such as the sound velocity shown in Fig. 6 are available over a significant range of conditions, it is necessary to obtain good fits.

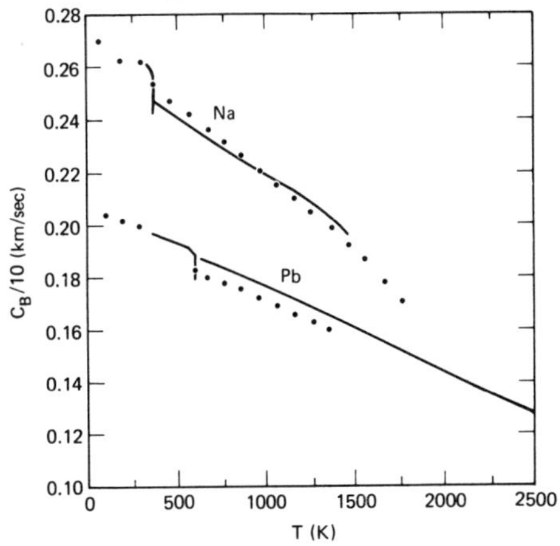

Fig. 5. Comparison of experimental bulk sound velocity temperature data with model calculations for Na [12] and Pb [13].

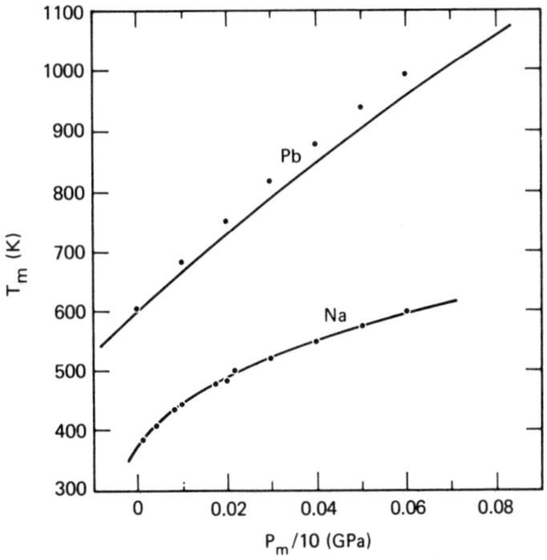

Fig. 6. Comparison of experimental melting temperature-pressure data with model calculations for Pb [14] and Na [14,15].

The basic EOS functions for Pb, Na, and Ta shown in Figs. 7 and 8 from which these fits derive have a number of interesting properties which will be discussed below. In evaluating such

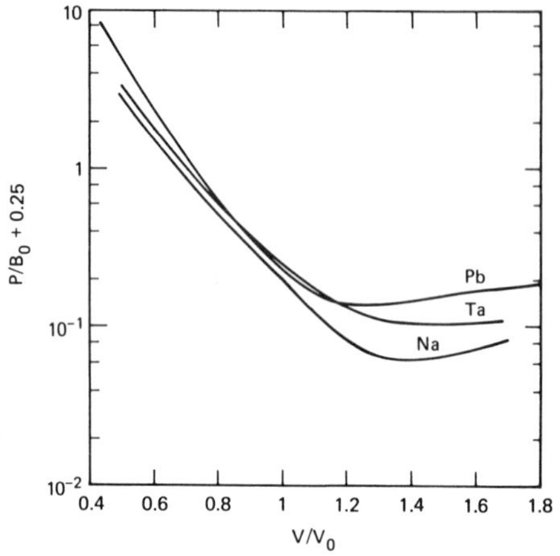

Fig. 7. Zero-degree pressure, scaled by the normal bulk modulus B_o as a function of relative volume from fits to experimental data on Pb, Na, and Ta. The B_o value used for these metals is 44.77, 6.69, and 194.1 GPa, respectively.

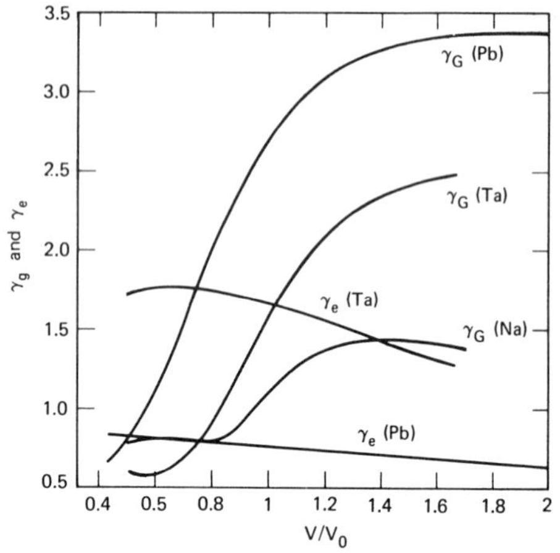

Fig. 8. Relative volume dependence of Grüneisen γ_G and electronic γ_e from fits to experimental data for Pb, Na, and Ta.

fitting results one should first however take into consideration the reliability and uniqueness of the fit in view of the limitations of the experimental data and the fitting procedure. For instance, shock compression data (see Fig. 3) is not adequate to determine $P_o(\rho)$ and $\gamma_G(\rho)$ uniquely at high density. The different types of measurements possible in isobaric expansion experiments make, on the other hand, determination of $\gamma_G(\rho)$, $\gamma_e(\rho)$, and $P_o(\rho)$ more reliable in expansion. Compression values of these functions were then primarily determined by a smooth extrapolation of the expansion values in such a manner that $P_o(\rho)$ appeared well behaved. In the case of Ta and Pb it was found that fits were obtained most simply when g_e was close to its unrenormalized value as estimated theoretically from cryogenic specific heat data [9]. The behavior of $\gamma_e(\rho)$ for both Pb and Ta seemed then to be at least qualitatively fixed by the data fit. Experience has shown, however, that these fits should not be considered absolutely unique. It is probably possible to find other combinations of P_o, γ_G, γ_e, and g_{eo} which will fit the limited available data within the expected uncertainties of the data and model.

COMPARISON WITH METALLIC SOLID STATE THEORY

The functions $P_o(\rho)$, $\gamma_G(\rho)$, and $\gamma_e(\rho)$ obtained from these fits represent properties of the low-temperature solid phase of the metal. Their predicted behavior over a wide range of densities is summarized in Figs. 7 and 8, and will be compared with what is theoretically known about these metals.

The zero-temperature compression curves $P_o(V/V_o)$ in the form shown in Fig. 7 have been scaled by the normal bulk modulus of the metal to permit a clear comparison in one figure. The expected minima of P_o in expansion can be seen to be located over a wide range of relative volumes, from 1.2 to 1.5, reflecting, perhaps, characteristics of the different type of metallic bonding in these metals. Electron band theory calculations of equation-of-state have not been carried out over a wide range of densities for these metals although they appear to be feasible, if lengthy. For free electron-like metals such as Na, pseudo-potential calculations are reliable and are in good agreement with the static compression data [15] and our fits up to 25% compression. The interestingly different behavior of Ta and Pb in expansion, however, is unexplained. The differences in the behavior of the curves at high compression is less pronounced, but also less reliable as mentioned above. Dynamic or static high compression experiments to check these derived curves would be very desirable.

The behavior of $\gamma_G(V/V_o)$ shown in Fig. 8 is similar for all these metals, having an overall S shape caused by an inflection point near normal density and a leveling-off at large expansion. No reliable theoretical methods apparently exist at present for calculating metallic γ_G's [16]. Qualitatively, the shapes of these curves are reasonable, however. At large V volumes a constant γ_G is consistent with the finite size of the ions. Most scaling theories of γ_G such as the Slater-Landau formula [7] predict γ_G will decrease rapidly with compression in agreement with our fitting results. Again the values of γ_G at small volumes are less reliable. For instance, in the case of Pb they can be anywhere below 1.3.

A careful adjustment of the values of g_{eo} gave 1.3, 1, and 2.2 x 10^{-3} J/gm for Pb, Na, and Ta, respectively. These values are close to the cryogenic values when corrected for phonon renormalization [18]. For Pb, the decreasing line for γ_e in Fig. 8 helped fit the porous Hugoniot data in Fig. 3 and the H-V data in Fig. 4 at large V. The small average value of $\gamma_e(\sim 2/3)$ is consistent with the free-electron nature of the electron Fermi surface in Pb. No band theory calculations for either Pb or Ta are available, however, to make a more detailed comparison. Larger values of γ_e were necessary in Ta with its large electronic specific heat in order to fit isobaric expansion data. Such a large value is not surprising in narrow band transition metals and is similar to values that have been calculated for Fe from band theory at normal density [19].

In conclusion, the solid state properties derived from our fits on three distinct types of metals seem at least theoretically consistent with, if not confirmed by, theory. The resulting EOS thus appears to be a reliable guide to material properties suitable for technical applications as well as research.

ACKNOWLEDGMENTS

The author would like to thank J. Shaner and M. Hodgson for use of their data and together with D. Young for many discussions of EOS fitting problems.

NOTATION

B_S = isentropic bulk modulus
C_B = bulk sound speed
C_v = specific heat at constant volume
E = internal energy per gm atom
F = Helmholtz free energy
g_{eo} = electronic specific heat coefficient at T=0
H = enthalpy
P = pressure
P_m = melting pressure
R = gas constant
S = entropy
T = temperature
T_M = melting temperature
U_p = particle velocity
U_s = shock velocity
α = liquid EOS scaling parameter
γ_e, γ_G = electronic and lattice Grüneisen coefficients
λ = melting temperature exponent
ρ = density
τ = scaled temperature T/T_M
θ_G = Grüneisen characteristic temperature

REFERENCES

1. R. G. McQueen and S. P. Marsh, J. Appl. Phys. 31, 1253 (1960).
2. S. B. Kormer, A. I. Funtikov, V. D. Urlin, and A. N. Kolesnikova, J. Exptl. Theoret. Phys. (USSR) 42, 686 (1962); Sov. Phys. JETP 15, 477 (1962).
3. R. Grover, J. Chem. Phys. 55, 3435 (1971).
4. R. Grover, "High Temperature Equation-of-State for Simple Metals," paper presented at 7th Symposium on Thermophysical Properties, NBS, Gaithersburg, Maryland, May, 1977.
5. R. Grover, "Generalized Lindemann Law for Simple Metals," UCRL-76544, unpublished (1975).
6. W. O. Wray, "An Improved GRAY Equation-of-State for Metals," System, Science and Software, La Jolla, California, Rept. #SSS-R-74-2387, unpublished (1974).
7. W. G. Hoover, S. G. Gray, and K. W. Johnson, J. Chem. Phys. 55, 1128 (1971).
8. E. L. Pollock and J. P. Hansen, Phys. Rev. A8, 3110 (1973).
9. M. Van Thiel, ed., Compendium of Shock Wave Data, UCRL-50108 plus suppl. (1967), (available from CFTI, NBS, Springfield, Virginia 22151); also A. C. Mitchell, private communication.
10. J. W. Shaner, G. R. Gathers, and C. Minichino, High Temp. High Press., to be published.

11. J. W. Shaner, G. R. Gathers, and W. M. Hodgson, "Thermophysical Measurements on Liquid Metals Above 4000 K," paper presented at 7th Symposium on Thermophysical Properties, NBS, Gaithersburg, Maryland, May, 1977.
12. M. G. Chasanov, L. Leibowitz, D. F. Fischer, and R. A. Blomquist, J. Appl. Phys. 43, 748 (1972).
13. M. B. Gitis and I. G. Mikhailov, Acoust. J. (USSR) 12, 145 (1966); Sov. Phys.-Acoustics 12, 131 (1966).
14. P. W. Mirwald and G. C. Kennedy, J. Phys. Chem. Solids 37, 795 (1976).
15. I. N. Makerenko, A. M. Nikolaenko, V. A. Ivanov, and S. M. Stishov, J. Exptl. Theoret. Phys. (USSR) 69, 1723 (1975); Sov. Phys.-JETP 42, 875 (1976).
16. R. D. Mountain, "Equation-of-State of Liquid Alkali Metals - 1st Principles Calculation," paper presented at 7th Symposium on Thermophysical Properties, NBS, Gaithersburg, Maryland, May, 1977.
17. J. C. Slater, Introduction to Chemical Physics, McGraw-Hill Book Company, New York (1939), Chapt. XIII and XIV.
18. G. Grimvall, Physica Scripta 14, 63 (1976).
19. M. S. T. Bukowinski, Phys. Earth Planet. Int. 13, 57 (1976).

A-3

GENERALIZED HUGONIOTS AND 0°K CURVES OF ELEMENTS

V. F. Anisichkin

Institute of Hydrodynamics, Siberian Division of the USSR Academy of Sciences
Novosibirsk, U.S.S.R.

INTRODUCTION

Hugoniots are necessary when considering any explosive processes. They constitute the main source of knowledge for an equation of state at high pressures and temperatures. This explains the great interest in their experimental determination. About a thousand Hugoniots are known at present, and among them are approximately 200 Hugoniots for many elements of the periodic system obtained by various researchers. Such a great quantity of data makes it possible to treat them statistically and find definite similarities.

Generalized Hugoniots have been proposed repeatedly, e.g., by Koryavov [1] and Prieto et al. [2]. But in these cases they were considered linear in the U_p and the U_s plane. Generalized Hugoniots of elements which are free of this assumption are considered below.

The 0 K curves of substances can be obtained approximately from Hugoniots if the dependence of the Grüneisen parameter on volume is known. More correctly the 0 K isotherms have been calculated only for hydrogen and deuterium [3,4]. The isotherms of these substances were obtained experimentally at temperatures approaching 4 K [5,6]. This paper will show that a Hugoniot and 0 K isotherm of one substance are adequate to determine, with reasonable accuracy, isotherms at 0 K for many other substances.

TWO GROUPS OF HUGONIOTS FOR ELEMENTS

Experimental Hugoniots are often approximated by portions of straight lines in the U_p and the U_s plane by

$$U_s = C + S\,U_p \qquad (1)$$

The value of S in (1) can be considered as the derivative S = dU_s/dU_p at some intermediate point of a Hugoniot. Generally speaking, this value changes along the Hugoniot curve with an increase in compression of the substance in a shock wave [7].

The distribution of S values shown in Fig. 1 has been constructed on the basis of about 150 experimental shock Hugoniots approximated by portions of straight lines which were basically obtained from van Thiel et al. [7]. The number of Hugoniots having values of derivatives falling within 0.05 of each other is plotted against the vertical axis. As a result of such rounding off of experimental values one obtains a discrete distribution of values for S as n = n(S). It is evident that for most Hugoniots the coefficients of S are close to two specific values, i.e. approximately 1.2 and 1.5. Figure 1 also gives the sum of two normal distributions and shows what the distribution must be if the deviation of S from the two given values can be considered as abnormal. The coincidence is adequate except for the boundaries where the substances undergoing transitions, carbon in particular, are mainly represented. Transitions at shock compression cause changes in the slope of Hugoniots and, consequently, changes in the value of S.

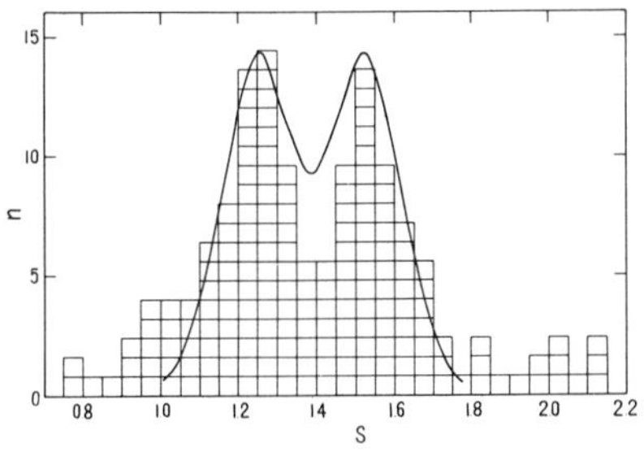

Fig. 1. Histogram of frequencies for experimental values of S and the corresponding Gaussian distribution.

It should be noted that most of the elements in Fig. 1 are represented repeatedly, since several experimental Hugoniots are available for these elements. Therefore, distribution maximums can be narrowed, generally speaking, by increased attention of researchers to certain materials, i.e. to improving the statistical analysis of published results. The possibility of two selected values for S can be eliminated by averaging the data of different authors for the same element; this will be done in constructing the distribution given below.

Deviations of S from the two selected values can be caused by experimental errors when determining Hugoniots which can, in some cases, approach 10% or more of the S value [8]. Besides, the derivative of Hugoniots, as mentioned above, is not constant and changes with the degree of shock compression of the substance.

In order to partially eliminate these possible causes of deviations of S from the two selected values, the distribution shown in Fig. 2 was constructed as follows: regions of evident transitions were not considered; if several Hugoniots for one and the same substance were available, the data of different researchers were averaged and values of S at equal degrees of shock compression

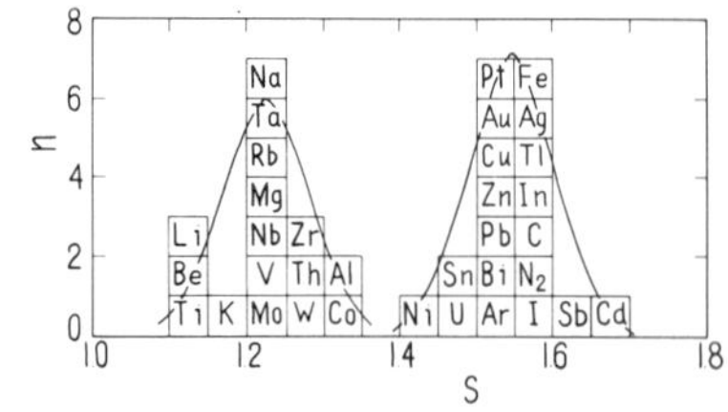

Fig. 2. Histogram of frequencies for averaged values of S at $V/V_o = 0.675$ and the corresponding Gaussian distribution.

of substances $V/V_o = 0.675$ were selected.

The remaining deviations of S from the two selected values can be caused either by experimental errors or by the lack of coincidence of derivatives of Hugoniots for different elements as functions of relative volume; however, it is important to note that the value of such deviations on the average does not exceed possible experimental errors while determining the Hugoniots.

Distributions similar to those shown in Fig. 2 are obtained and at other relative shock compressions of substances. Using the maxima of such distributions, one can construct plots of change in derivatives of Hugoniots for the first and the second groups of elements.

The place which they occupy in the periodic system also favors the existence of the two groups of elements having Hugoniots with equally changing derivatives. Elements constituting the first group are mainly located at the beginning of each period of the system, while the second group includes the remaining elements. This agrees with the assumption that the character of compressibility of substances basically depends on the number of electrons filling the outer shell of the atoms.

GEOMETRIC SIMILARITY OF HUGONIOT CURVES

It follows from the results obtained that within experimental error the derivatives of the Hugoniots can be considered as identical functions of relative volume for each of the two selected groups of elements. For each of these groups the derivatives of the Hugoniots turn out to be equal at the same values of V/V_o, i.e. on straight lines emanating from the origin of coordinates in the U_p, U_s-plane. This means that Hugoniot curves of one group are geometrically similar in the U_p, U_s-plane. In the V/V_o, P-plane Hugoniots of one group can be superposed by extending them along the pressure axis.

Therefore, to obtain one Hugoniot from another, one must know the coefficient of their mutual similarity k_{ij}. This coefficient characterizes a shock-compressible substance relative to another substance and depends first of all on the initial density, molecular weight, and the number of electrons in its atom [9]. The analysis showed that one can develop similarity coefficients for Hugoniots with an accuracy which is sufficient to describe most of the available experimental data, using the following simple analytical dependences:

$$k_{ij} = \left(\frac{\rho_{oi}}{\rho_{oj}} \cdot \frac{M_j}{M_i}\right)^{2/3} \left(\frac{M_j}{M_i} \cdot \frac{Z_i}{Z_j}\right)^2 \quad (2)$$

for the first group of elements and

$$k_{ij} = \left(\frac{\rho_{oi}}{\rho_{oj}} \cdot \frac{M_j}{M_i}\right)^{2/3} \quad (3)$$

for the second group of elements.

A similarity coefficient between two Hugoniots can also be expressed by a ratio of the velocities of shock waves at equal degrees of compression V/V_o:

$$k_{ij} = \frac{U_{si}}{U_{sj}} \quad (4)$$

Figure 3 provides a comparison of the velocity of shock waves in the substances considered in Fig. 2 at $V/V_o = 0.675$ which were obtained experimentally and calculated using (2), (3) and (4). In most cases the coincidence is good and the divergences are partially explained by insufficient accuracy and simplicity of (2) and (3). The more complex interpolating dependencies describe experimental data more accurately.

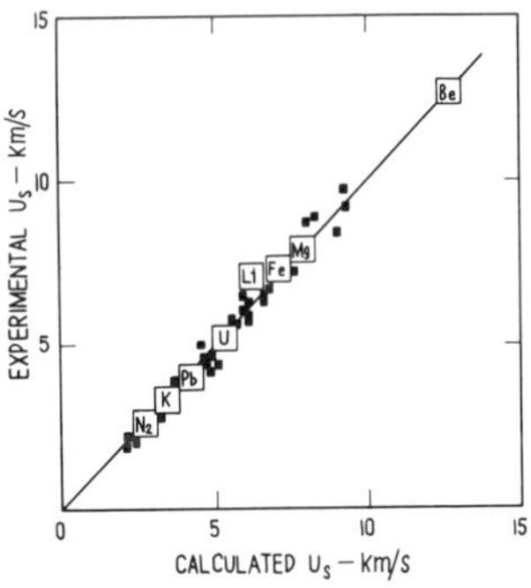

Fig. 3. Comparison of calculated and experimental velocities of shock waves at $V/V_o = 0.675$.

If one uses the analytical representation of Hugoniots proposed previously [10] and using (2) and (3), the generalized Hugoniot of elements for the first group can be approximated in the form

$$U_s = U_p + 3.4\left(\frac{\rho_o}{M}\right)^{1/3}\left(\frac{Z}{M}\right)\sqrt{U_p} + 98\left(\frac{\rho_o}{M}\right)^{2/3}\left(\frac{Z}{M}\right)^2 \quad (5)$$

and the generalized Hugoniot for the second group in the form

$$U_s = U_p + 3.4\left(\frac{\rho_o}{M}\right)^{1/3}\sqrt{U_p} + 9.1\left(\frac{\rho_o}{M}\right)^{2/3} \quad (6)$$

where U_p and U_s are expressed in km/s.

Figures 4 and 5 compare some experimental Hugoniots for both the first and the second group with the Hugoniots calculated

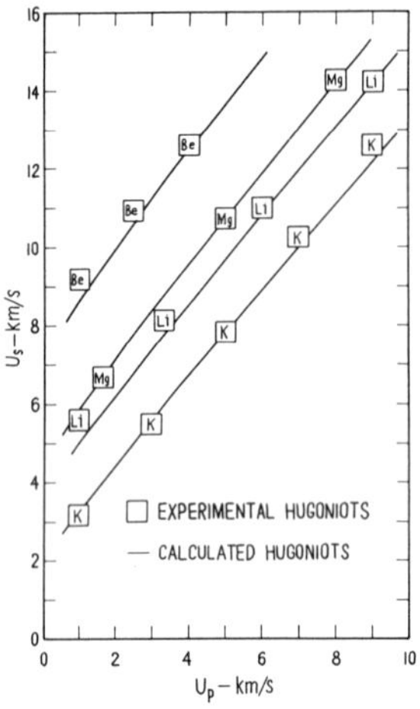

Fig. 4. Comparison of Hugoniots of the first group of elements calculated according to (5) and obtained experimentally.

according to (5) and (6), in particular Hugoniots of lithium and beryllium. It is noted from Fig. 2 that the derivatives of Hugoniots for these latter elements deviates most significantly from the average value. In addition, the Hugoniot of lithium is least accurately described by (2); this is evident from Fig. 3. However, in the cases of lithium and beryllium, divergences on the average do not exceed maximum possible errors when determining the Hugoniots experimentally. Hugoniots of most other elements are described equally well by (5) and (6) as the Hugoniots of magnesium and iron shown in Figs. 4 and 5.

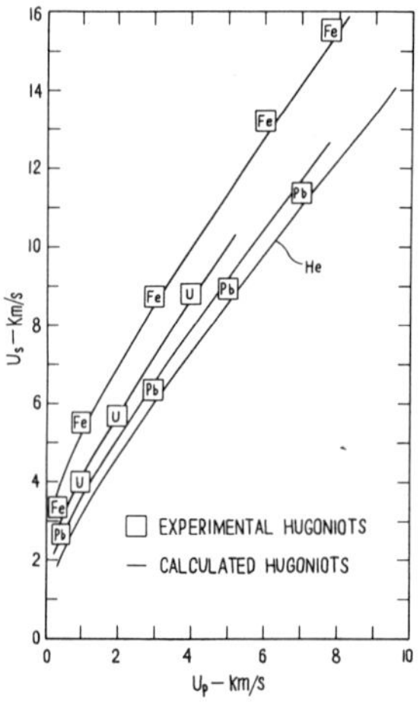

Fig. 5. Comparison of Hugoniots of the second group of elements calculated according to (6) and obtained experimentally.

The apparent consistency makes it possible to calculate Hugoniots of elements which have not yet been obtained experimentally. For example, Fig. 5 shows a Hugoniot path for liquid helium starting from an initial density condition of 0.15 g/cm^3.

HUGONIOTS OF COMPLEX SUBSTANCES

Equation (6) adequately describes the Hugoniots of inert gases such as argon and xenon having atoms with filled electron shells. In this sense the structure of electron shells of multiatom molecules is similar to that of atoms of inert gases. Therefore the coincidence of experimental data shown in Fig. 6 for Hugoniots of liquid nitrogen, hydrogen, and water calculated according to (6) is not surprising. Hence, the Hugoniots of these complex substances can also be attributed to the second group of Hugoniots.

0 K CURVES OF ELEMENTS

Let us for the moment neglect the internal degrees of freedom associated with molecules. In such a situation the virial theorem of classical mechanics allows the pressure of substances to be expressed through the 0 K isotherm, volume, and initial energy by

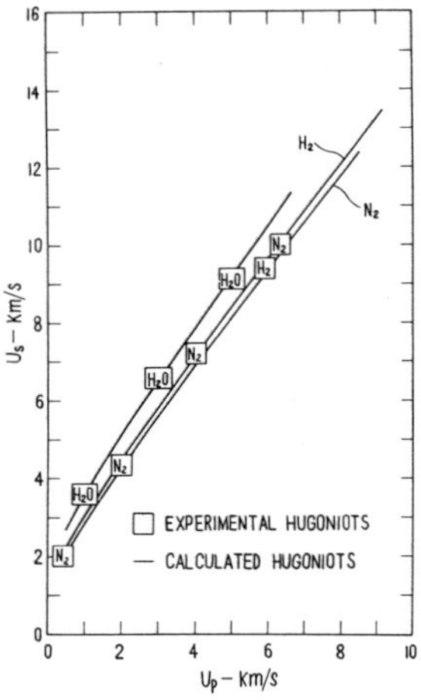

Fig. 6. Comparison of Hugoniots of complex substances calculated according to (6) and obtained experimentally.

$$P = P\left[P_{T=0}\left(\frac{V}{V_o}\right), V, E\right] \qquad (7)$$

The initial energy of such a substance behind the front of a shock wave is equal to

$$E = \frac{P_H + P_o}{2}(V_o - V) + E_o \qquad (8)$$

Let us assume that P_o and E_o are small values and can be neglected. Then from (7) one obtains

$$P_H = P_H\left[P_{T=0}\left(\frac{V}{V_o}\right), V, V_o\right] \qquad (9)$$

According to the dimensional theory, (9) can be written as

$$\frac{P_{T=0}}{P_H} = f\left(\frac{V}{V_o}\right) \qquad (10)$$

From (10) it is evident that Hugoniots of two substances will be mutually proportional in the P, V/V_o-plane if their 0 K isotherms are also proportional. Consequently, 0 K isotherms of elements for each of the groups must also be mutually proportional. Here the coefficient of proportionality between 0 K isotherms can be approximated using (2) and (3).

In order to determine an 0 K isotherm for any element, one must have, besides the Hugoniot of this element, the Hugoniot and 0 K isotherm of at least one element in the same group. In the second group hydrogen can be selected as such an element, the Hugoniot of which is similar to other Hugoniots of the group.

Figure 7 provides, as an example, 0 K isotherms of iron and lead calculated by the above method and compared with the calculated data in the literature [11,12]. For lead the results diverge greatly. A good similarity of experimental Hugoniots for iron and lead makes it possible to conclude that one of the two 0 K isotherms given [11,12] is probably in error. According to our results the 0 K isotherm of lead is incorrect.

For substances of the second group equation (10) relating 0 K isotherms and Hugoniots can be approximated in a simple form by

$$\frac{P_{T=0}}{P_H} = \frac{V}{V_o} \qquad (11)$$

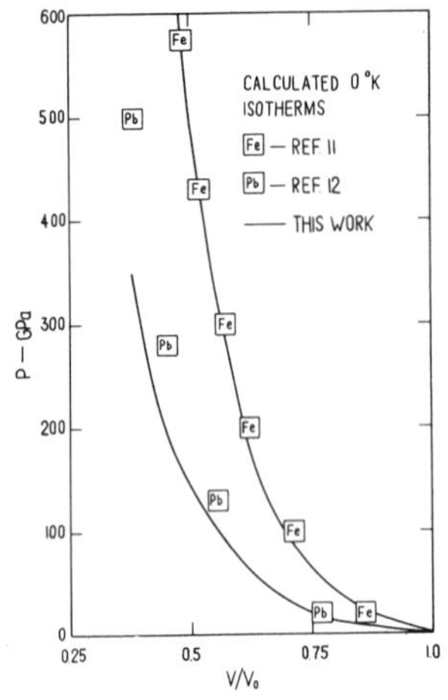

Fig. 7. Comparison of calculated 0 K isotherms of iron and lead and those given in the literature [11,12].

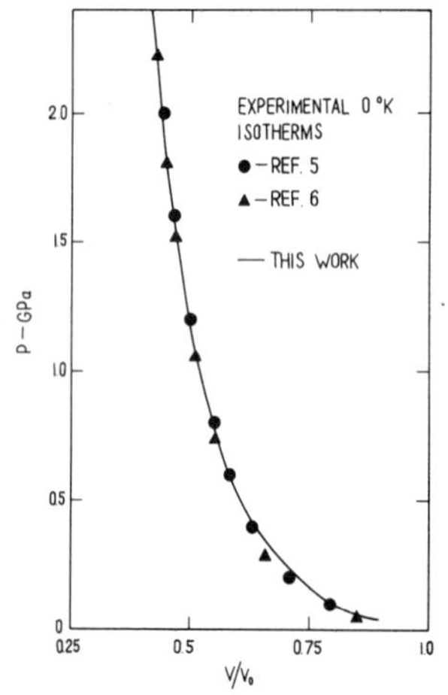

Fig. 8. Comparison of low-temperature isotherms of hydrogen calculated according to (11) and obtained experimentally [5,6].

Figure 8 shows that the 0 K isotherm of hydrogen found from its Hugoniot with the assistance of (11) coincides perfectly with experimental isotherms at 4.2 K [5,6]. Therefore, using the simplified relation in (11), one can determine 0 K isotherms of other substances in the second group with good accuracy.

Unfortunately, no 0 K isotherm of an element in the first group has yet been experimentally obtained in a sufficiently wide range of densities to also check the accuracy of (11) for these substances.

DISCUSSION

The geometric similarity of Hugoniots and 0 K isotherms of substances for the same group makes it possible to determine them using only one experimental point. Thus, one can construct Hugoniots and 0 K isotherms of negligibly compressible elements to high relative densities from data obtained experimentally for easily compressible substances.

The similarity of Hugoniots of substances of one group provides the possibility of considering the similarity of their equations of state. Further, these similarities open up the possibility of simulating shock-wave processes by substances of the same group.

In conclusion it is worth noting that some difficulties in constructing the generalized Hugoniots and 0 K isotherms of substances. Due to the lack of published data not all of the elements in the periodic system were considered above. In addition, the results have a statistical character. Hence deviations from the revealed regularities exceeding an average error of experiments, are, generally speaking, possible.

The preference here is given to equations (5) and (6) due to their simplicity and good agreement with most experimental data. However, to more completely describe all known experimental Hugoniots, it is necessary to use more exact interpolation equations. In particular, while determining the 0 K isotherm of hydrogen according to (11) the Hugoniot of hydrogen passing directly through the experimental point was used.

ACKNOWLEDGMENT

The author is grateful to V. M. Titov for useful discussions.

NOTATION

C, S = constants
E = internal energy
E_o = initial internal energy
$f(V/V_o)$ = relative volume function
k_{ij} = similarity coefficient
i, j = integral indices
n = integer
P = Pressure
$P_{T=0}(V/V_o)$ = 0 K isotherm of substance
$P_{T=0}$ = pressure at 0 K
P_H = Hugoniot pressure
P_o = initial pressure
U_s = shock wave velocity
U_p = particle velocity
V = specific volume of substance
V_o = initial specific volume of substance
Z = number of electrons of substance molecule
ρ_o = initial density of substance
M = molecular weight

REFERENCES

1. V. P. Koryavov, Prikladnaya Mekhanika i Tekhnicheskaya Fizika 5, 123 (1964).
2. F. E. Prieto and C. Renero, J. Phys. Chem. Solids 37, 151 (1976).
3. V. P. Trubitsin, Fizika Tverdogo Tela 7, 3363 (1965).
4. M. Ross, F. H. Ree, and R. N. Keeler, in Proc. 4th Intern. Conference on High Pressure, Kyoto, Japan (1974).
5. J. W. Stewart, J. Phys. Chem. Sol. 1, 146 (1956).
6. M. S. Anderson and C. A. Swenson, Phys. Rev. B 10, 5184 (1974).
7. M. van Thiel, A. S. Kusubov, A. C. Mitchell, and V. W. Davis, UCRL-50108, Lawrence Radiation Laboratory, University of California, Livermore, California (1966).
8. R. D. Dick, J. Chem. Phys. 52, 6021 (1970).
9. N. N. Kalitkin and L. V. Kuz'mina, Trudi III Vsesouznogo seminara po modelyam mekhaniki sploshnoy sredi, Novosibirsk (1976).
10. V. F. Anisichkin, Prikladnaya Mekhanika i Tekhnicheskaya Fizika 1, 83 (1975).
11. V. N. Zharkov, V. P. Trubitsin, and I. A. Tsarevskii, Dokl. Akad. Nauk SSSR 214, 557 (1974).
12. L. V. Altshuler, A. A. Bakanova, and R. F. Trunin, Soviet Physics JETP 42, 91 (1962).
13. M. van Thiel and B. J. Alder, Mol. Phys. 10, 427 (1966).

A-4

A UNIVERSAL EQUATION OF STATE FOR FLUIDS AND SOLIDS AT SHOCK PRESSURES

F. E. Prieto

Instituto Nacional de Energía Nuclear, México
Mexico City, Mexico
and
C. Renero
Universidad Nacional de México
Mexico City, Mexico

INTRODUCTION

The present paper is a summary review of the authors' work on a universal reduced variables formulation of the thermodynamics of very high pressures. It is important to mention that one of the most interesting results of this formalism is an explicit analytic form of the equation of state for shock-compressed matter; this equation seems to be universal in the sense that it is valid for any material and independent of the initial phase. Our earliest papers on this subject were published about seven years ago [1,2] but the main results obtained during the last three years include a law of corresponding states [3], an equation of state for solids [4-6], and an extension of this equation of state to fluids [7,8]. This chronological approach will not be followed here. Instead, only the most important results as well as those unanswered questions and unresolved problems encountered within the formalism will be mentioned. This situation is a consequence of the manner in which the formalism was developed. The main task was first to look for an equation of state of solids; when this equation was obtained, there were already some unresolved problems, but it was necessary to test the universal validity of the equation. Afterwards, efforts were concentrated in extending the validity of the equation to liquids and gases; the number of unresolved problems increased accordingly.

THE FORMALISM

A universal formulation of the thermodynamics of shock compressed matter implies, of course, the use of reduced variables. It was determined [3-5] that a consistent set of reduced variables pertinent to the physics of shock compression can be defined in the following manner:

$$u_s = U_s/A \,, \quad \text{(1a)} \qquad u_p = U_p(A/B) \,, \quad \text{(1b)}$$

$$p = P/\rho_o(A^2/B)\,, \quad \text{(1c)} \qquad x = B\left(1 - \frac{V}{V_o}\right) \,, \quad \text{(1d)}$$

$$e = E/(A^2/B^2)\,, \quad \text{(1e)} \qquad t = \alpha\, BT \quad \text{(1f)}$$

Where A and B are the coefficients of the well known [9] linear relationship between shock and particle velocities

$$U_s = A + B\,U_p \qquad (2)$$

An immediate consequence of the introduction of this set of reduced variables is that the equation for the shock Hugoniot becomes

$$P_h = X\,(1 - X)^{-2} \qquad (3)$$

and is in consequence system-independent or universal. A further development in the formalism [4-6] led to an explicit analytic expression for the pressure - volume - temperature equation of state, given by

$$p(X,t) = F(m,X) + G(m,X) + mC_{vr}t - mC_{Vr}t_o e^{mX} \qquad (4)$$

where

$$2\,F(m,X) = (2-m)(1-X)^{-2} - [2+m(m-4)](1-X)^{-1} - m(3-m)e^{mX} \qquad (5a)$$

$$2G(m,X) = m(4m-m^2-2)e^{-m(1-X)}\,\overline{Ei}(m,X) \qquad (5b)$$

with $m = (BK_rC_{vr})^{-1}$; $k_r = \rho_o kA^2/B$; $c_{vr} = Cv\,B/(\alpha A^2)$; and $\overline{E_i}(m,X) = \overline{E_i}(m) - \overline{E_i}[m(1-X)]$. Here $\overline{E_i}(y)$ is the exponential integral function of supr. negative argument.

SOME CONSEQUENCES AND EXPERIMENTAL EVIDENCE

Since all of the reduction or scaling factors involve one or both of the coefficients of the linear relationship, it is obvious that the formalism can be used and is valid as long as the material under study follows the linear relationship between velocities. The same can be said for the equation of state; its validity for any given material depends on the validity of the linear relationship for that material, and it can be used only if the coefficients for this relationship are known. The universality of both the equation for the shock Hugoniot and the equation of state seem to be related to the conservation of mass, and perhaps to the existence of a limiting volume in shock compression [7,10]. This seems to be a property of any material, but leaves us without clues as to the meaning of coefficients A and B.

The experimental evidence on the validity of the linear relationship for solids is quite large, so that no more comments are needed [9]. There is some evidence [8,9,11-13] for the validity of this relation for liquids and some scarce but conclusive evidence [14-16] for gases. Of course, this implies that for all of these materials the universal equation for the shock Hugoniot must be valid [7]. We have also found some experimental evidence supporting the validity of the equation of state for solids [4,5,17,18], for liquids [7,8,9,11-13] and for gases [7,14-16].

Some of the problems within the formalism that are still unresolved include: (1) A theoretical interpretation and justification of the linear relationship; (2) The physical meaning of the reduction parameters; (3) A justification and interpretation of some regularities found in the values of some of the parameters used in the formalism, such as the constancy of the values of the parameters $m = 4/3$, $C_{vr} = 3/4$ for solids and liquids [5,7], and the constancy on the value of the coefficient $B = 1.67$ for liquids [18]; (4) The importance of the initial temperature appearing in the equation of state and the interest in performing shock experiments using pre-heated or pre-cooled samples [5]; (5) The limit in the validity of the formalism when ionization starts; (6) The extension of the formalism to reactive systems [19]; (7) The temperature dependence of coefficients A, B, ρ_o, α, k, C_v used in the formalism; and (8) If necessary, modification of the formalism to take into account the temperature dependence of some of these coefficients, in particular the specific heat.

CONCLUDING REMARKS

Our approach is still far from perfect, but it provides a good systematization of experimental results and it can be considered as an acceptable prediction tool. For example, if coefficient B is the same for all liquids, then one single shock compression experiment permits the determination of coefficient A for that liquid. Furthermore, it can be said that for solids, rather than more pressure-volume shock compression data, direct measurements of the shock temperature are needed. The same comment can be made relative to liquids although it would be convenient to have shock compression data on more liquids in order to test not only the validity of the formalism, but the constancy of parameter B. For gases the experimental evidence is quite scarce. Accordingly, it would be desirable to have more evidence relative to the validity of the linear relationship and on direct measurement of shock temperatures.

NOTATION

A,B	= constants	ρ_o	= initial density
α	= coefficient of thermal expansion	T	= absolute temperature
C_v	= specific heat at constant volume	T_h	= absolute shock temperature
E	= internal energy	To	= initial temperature
k	= compressibility	U_p	= particle velocity
P	= pressure	U_s	= shock velocity
P_h	= Hugoniot pressure	V	= volume
		X	= scaled relative compression

REFERENCES

1. F. E. Prieto and C. Renero, J. Appl. Phys. 41, 3876 (1970).
2. F. E. Prieto and C. Renero, J. Appl. Phys. 42, 296 (1971).
3. F. E. Prieto, J. Phys. Chem. Solids 35, 279 (1974).
4. F. E. Prieto and C. Renero, Compt. Rend. B 279, 601 (1974).
5. F. E. Prieto and C. Renero, J. Phys. Chem. Solids 37, 151 (1976).
6. F. E. Prieto and C. Renero, in Peaceful Nuclear Explosions, Vol. 4, IAEA Symposium Series, Vienna, Austria (1975).
7. F. E. Prieto, C. Renero, and A. Mondragón, in Peaceful Nuclear Explosions, Vol. 5, IAEA Symposium Series, Vienna, Austria (1976), in press.
8. F. L. Yarger, F. E. Prieto, and C. Renero, Paper H-6-C given at 6th AIRAPT Intern. High Pressure Conference, University of Colorado, Boulder, Colorado, July 25-29, 1977.
9. M. Van Thiel, ed., Compendium of Shock Wave Data, U.S. Atomic Energy Commission (1966).
10. F. E. Prieto and C. Renero, Sci. et Techn. de l'Armement, (France) 49, 325 (1975).
11. R. J. Dick, J. Chem. Phys. 52, 602 (1970).
12. B. C. Craig, Los Alamos Scientific Laboratory Rept. 6 MX-8-Mr-62.4 (1962).
13. W. B. Garn, J. Chem. Phys. 30, 819 (1959).
14. R. H. Christian, and F. L. Yarger, J. Chem. Phys. 23, 2042 (1955).
15. R. H. Christian, R. E. Duff, and F. L. Yarger, J. Chem. Phys. 23, 2045 (1955).
16. W. E. Deal, J. Appl. Phys. 28, 782 (1957).
17. S. B. Kormer, M. V. Sinitsyn, G. A. Kirilov, and V. D. Urlin, Soviet Physics JETP 21, 689 (1965).
18. P. W. Bridgman, Phys. Rev. 60, 351 (1941).
19. F. E. Prieto and C. Renero, Paper presented at 5th Symposium on Combustion Processes, Krakow, Poland (1977).

A-5

ISOTHERMAL COMPRESSION OF V, Nb, AND Ta TO 100 KBAR: CORRELATION WITH ULTRASONIC, SHOCK WAVE, AND OTHER STATIC DATA*

L. C. Ming and M. H. Manghnani

University of Hawaii
Honolulu, Hawaii USA

INTRODUCTION

The transition metals in group VB (V, Nb, and Ta) have bcc structure and easily form bcc alloys with hcp metals in group IVB, and with bcc metals in group VIB. The bcc structure of the VB metals is stable to melting temperatures and is considered to be stable under pressures of the order of several megabars, as no phase transformation has yet been found at these pressures. Their pressure-volume (P-V) relationships have been investigated by three different means: (1) by direct static, isothermal compression measurements up to 30 kbar [1-3] and up to 45 kbar [4,5]; (2) by use of an equation of state utilizing the ultrasonically measured values of bulk modulus (K_o) and its pressure derivative (K_o') [6-9]; and (3) by reduction of the shock wave Hugoniot data to isothermal P-V relationships based on certain thermodynamic considerations [10,11].

In recent years, the diamond-anvil pressure cell has also been successfully employed for investigating the compressibility of several materials under pressures of up to a few hundred kbar [12-14], filling in the pressure gap between ultrasonic and shock wave experiments. Until recently, the P-V data obtained from the diamond-anvil experiments were based on the assumptions that the mixture of sample and an internal standard (e.g., NaCl) are in quasi-hydrostatic environment and that both the sample and the internal standard are subjected to the same pressure. A number of investigators have questioned these assumptions and hence the value of the data thus obtained. In general the discrepancies in the P-V data arise from the anisotropic stresses on the sample [15,16]. The un-

*Hawaii Institute of Geophysics Contribution No. 905.

certainty in the pressure determination can now be mostly overcome by employing a liquid pressure medium (4:1 methanol-ethanol mixture which remains liquid to 104 kbar) and by using the ruby fluorescence pressure calibration technique [17,18]. In this technique the sample, immersed in the liquid, is gasketed between the diamond anvils. In view of these refinements in the use of the diamond-anvil cell and the ruby fluorescence pressure calibration technique [17,18], we have investigated the P-V relations for V, Nb, and Ta under a truly hydrostatic environment to 100 kbar, with an objective to correlate the results with those obtained from the ultrasonic, shock wave, and static experiments mentioned above.

EXPERIMENTAL METHODS

Powdered samples of V, Nb, and Ta were used in the present study and are described in detail elsewhere [8,9,19]. For generating high pressures, the Bassett-type diamond-anvil pressure cell [20] was employed. The cell consists of two gem-quality anvils (1/3 carat, 0.6 mm in diameter) driven by piston and screw assembly. The gasket consisting of T301 hardened steel foil, 0.025 cm thick and having ~ 0.3 mm-diameter hole, was placed between the diamond anvils. The foil was adjusted so that the hole was located centrally between the diamond anvils. The sample, consisting of a mixture of powdered metal (V, Nb, or Ta) + NaCl ± ruby, was then introduced into the hole. In order to maintain the sample in a purely hydrostatic environment, a drop of a mixture of 4:1 methanol-ethanol was added into the hole, and the whole assemblage was quickly loaded to some high pressure to prevent the alcohol from evaporating. A fine-focused, filtered Mo K_α X-ray source and a cassette-type camera with a radius of ~ 50 mm were used for obtaining X-ray diffraction patterns. For calculations of lattice parameter the diffraction lines [110], [200], [211], [220], [310] for Nb and Ta; the lines [110], [200], [211] for V; and the lines [200], [220] for NaCl are used. The lattice parameter and molar volume values thus calculated are accurate to ± 0.1% and to ± 0.3%, respectively. The P-V data were obtained during the decreasing pressure cycles and were corrected on the basis of the results from the run made at atmospheric pressure.

The pressure on the sample (metal + NaCl + alcohol) was determined from the calculated molar volume of NaCl, using the well-established isothermal compression curve of NaCl [21]. The accuracy of the NaCl pressure scale is ± 3% up to 100 kbar [22].

The pressure on the sample (metal + a ruby crystal, 20 to 30 μm in diameter + alcohol) was determined optically by means of the ruby fluorescence technique developed at the National Bureau of Standards. The method is based on the linear pressure dependence of the wavelength of the ruby fluorescence R_1 and R_2 lines. The optical system used is shown in Fig. 1, and is briefly described as follows: a continuous beam of 0.010 w with a wavelength λ = 442 nm is directed from a gas diffusion He-Cd laser to the pressure cell by a dichroic mirror causing the ruby crystal to fluoresce. The fluorescent light, after being transmitted through a dichroic mirror and a red filter, is reflected by a pivoted mirror on to the monochrometer. The wavelengths of the doublet thus recorded can be read to a precision of $\pm$ 0.5 Å. We adapted a $dP/d\lambda$ value of 2.74 $\pm$ 0.016 kbar/Å based on the calibration for the ruby fluorescence (R_1-line) pressure against the NaCl scale up to 195 kbar [17]. Thus, the precision of the ruby pressure determined in the present study is $\pm$ 1.3 kbar and its accuracy, estimated to be the same as that of NaCl pressure scale, is about 3%.

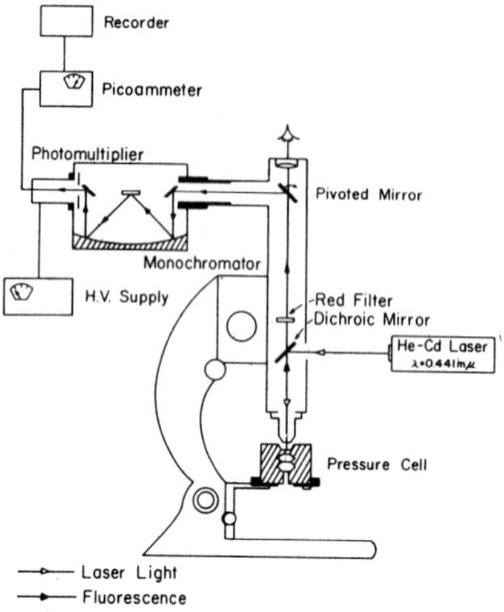

Fig. 1. Pressure calibration system. Schematic diagram of the ruby fluorescence pressure calibration system.

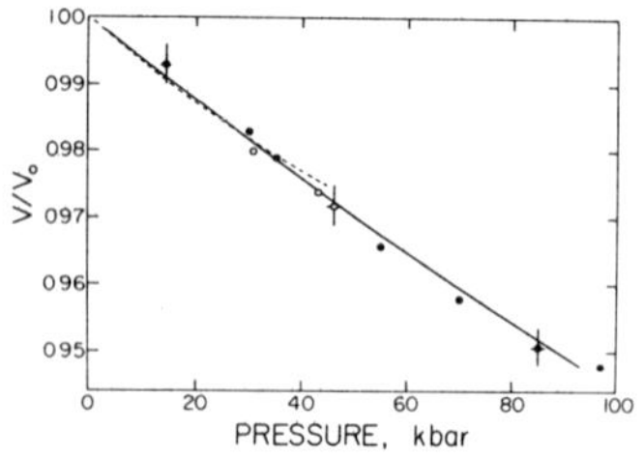

Fig. 2. Isothermal pressure-volume relationship for vanadium (V). Solid curve is the least-squares fit of the data in the present study to the Birch-Murnaghan equation, with K_o = 1.54 Mbar, K'_o = 4.27. Vertical and horizontal bars represent uncertainties in the determination of molar volume and pressure, respectively. The smoothed P-V data of Vaidya and Kennedy [5] are shown by dashed curve.

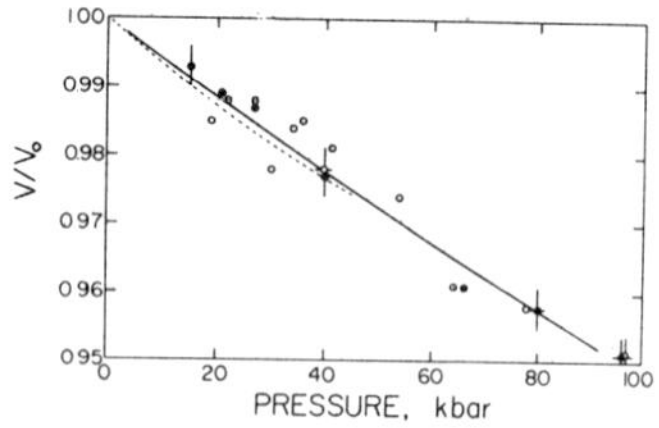

Fig. 3. Isothermal pressure-volume relationship for niobium (Nb). Solid curve is the least-squares fit of the data in the present study to the Birch-Murnaghan equation, with K_o = 1.71 Mbar, K'_o = 4.03. Vertical and horizontal bars represent uncertainties in the determination of molar volume and pressure, respectively. The smoothed P-V data of Vaidya and Kennedy [5] are shown by dashed curve.

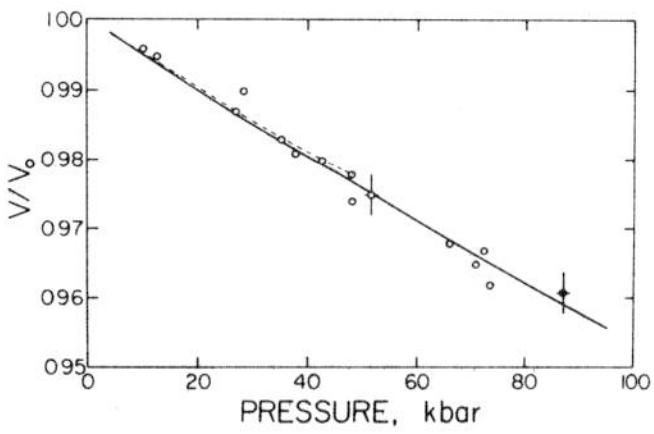

Fig. 4. Isothermal pressure-volume relationship for tantalum (Ta). Solid curve is the least-squares fit of the data in the present study to the Birch-Murnaghan equation, with K_o = 1.95 Mbar, K'_o = 3.80. Vertical and Horizontal bars represent uncertainties in the determination of molar volume and pressure, respectively. The smoothed P-V data of Vaidya and Kennedy [5] are shown by dashed curve.

For the sample of metal + NaCl + ruby + alcohol, i.e., runs 148, 146, and 144 for Nb (Table I), pressure was determined from both the molar volume of NaCl and the shift of ruby doublets. Both pressure values were found to be consistent with one another within experimental errors.

RESULTS AND DISCUSSION

The compression data for V, Nb, and Ta obtained in the present study are listed in Table I and plotted in Figs. 2, 3, and 4, respectively. Isothermal bulk modulus at zero pressure (K_o) can be determined by means of the least-squares fit of the experimental P-V data to the Birch-Murnaghan equation:

$$P = \frac{3}{2} K_o \left[\left(\frac{V}{V_o}\right)^{-7/3} - \left(\frac{V}{V_o}\right)^{-5/3} \right] \left\{ 1 - \xi \left[\left(\frac{V}{V_o}\right)^{-2/3} - 1 \right] \right\} \qquad (1)$$

where V_o and V are the volume at zero pressure and pressure P, respectively, $K'_o = (\partial K/\partial P)_T$ evaluated at P = 0, and $\xi = 3/4 \cdot (4 - K'_o)$. Generally, the high-pressure X-ray diffraction P-V data, including the present data, are not sufficiently accurate to permit a unique set of K_o and K'_o. In order to compare the present X-ray diffraction data with the ultrasonic and shock wave data, we have used the most recently measured ultrasonic values of K'_o in equation (1); i.e., 4.27 for V, 4.03 for Nb, and 3.80 for Ta [9]. The K_o values thus calculated from the resulting one-parameter Birch-Murnaghan equation are 1.54 ± 0.05, 1.71 ± 0.07, and 1.94 ± 0.07 Mbar for V, Nb, and Ta, respectively; the denoted uncertainties consist of the random errors in volume determination and an independent error (± 3.0%) caused by uncertainty in the pressure scale. As can be seen in Table II, the K_o values deduced from the present X-ray diffraction results are in excellent agreement with both the ultrasonic and shock wave data. It is of particular interest to note that our high-pressure X-ray diffraction and ultrasonic data reported here are for the same metal samples. A good agreement between two such kinds of data on the same sample has given us confidence in the results obtained in previous studies [8,9].

For comparison with other static results we also list in Table II the K_o and K'_o values for these metals reported by Vaidya

Table I. Effect of Pressure on the Molar Volume of V, Nb, and Ta*

Metal	Run No.	V/V_o	Pressure, kbar determined by: NaCl	Ruby	Sample Assemblage†
Vanadium (V)	177	0.993		14.5	1
ρ_o = 6.111 g/cm^3	172	0.983		30	1
	137	0.980	31		2
	169	0.979		35	1
	134	0.974	43		2
	132	0.972	46		2
	166	0.966		55	1
	129	0.966	55		2
	164	0.958		70	1
	162	0.951		85	1
	158	0.948		97	1
Niobium (Nb)	R3	0.993		15	1
ρ_o = 8.578 g/cm^3	7	0.985	18		2
	R6	0.989		21	1
	36	0.988	22		2
	R5	0.987		26.5	1
	40	0.978	30		2
	45	0.984	34		2
	49	0.985	35.5		
	97	0.978	40		2
	R1	0.977		40	1
	91	0.981	41		
	12	0.974	54		2
	148	0.961		66	3
		0.961	64		
	146	0.958	78		3
		0.958		80	
	144	0.951		96	3
		0.951	97		
Tantalum (Ta)	72	0.996	10.5		2
ρ_o = 16.683 g/cm^3	82	0.995	14		2
	107	0.987	27.5		2
	84	0.990	28.5		2
	75	0.983	35		2
	77	0.981	38		2
	92	0.980	42.5		2
	80	0.978	47.5		2
	78	0.974	48		2
	83	0.975	51.5		2
	96	0.968	66		2
	103	0.965	71		2
	100	0.962	72.5		2
	105	0.967	73.5		2
	145	0.961		87	1

* ρ_o is the zero-pressure density determined by X-ray diffraction using Debye-Scherrer camera (57.30 mm in diameter).

† Sample assemblage: 1 = Metal + ruby + alcohol, 2 = Metal + NaCl + alcohol, 3 = Metal + NaCl + ruby + alcohol.

Table II. Comparison of K_o Values for V, Nb, and Ta Based on Ultrasonic, Shock Wave, and Other Static Compression Data

Data

Metal	K_o, Mbar	K'_o	Pressure Range, kbar	Equation of State Used*	Reference
V	1.610 (B)	11.90 (B)	0-12	PE	[2]
	1.556 (U)	4.27 (U)	0-5	--	[9]
	1.570 (S)	3.50 (S)	0-1000	LE	[9]
	1.570 (S)	3.80 (S)	0-1000	LE	[11]
	1.319 (V-k)	18.20 (V-K)	0-45	ME	[2]
	1.54 ± 0.05 (x)	4.27 (U)	0-100	B-M	Present study
Nb	1.700 (B)	4.75 (B)	0-30	PE	[1]
	1.690 (U)	4.03 (U)	0-5	--	[8,9]
	1.690 (U)	6.75 (U)	0-0.4	--	[6]
	1.703 (S)	3.60 (S)	0-1800	LE	[8]
	1.442 (V-K)	14.51 (V-K)	0-45	ME	[3]
	1.71 ± 0.07 (x)	4.03 (U)	0-100	B-M	Present study
Ta	2.010 (B)	2.20 (B)	0-25	PE	[3]
	1.942 (U)	3.80 (U)	0-5	--	[8,9]
	1.900 (U)	3.25 (U)	0-3.5	--	[7]
	1.942 (S)	3.60 (S)	0-2300	LE	[8]
	2.056 (V-K)	2.76 (V-K)	0-45	ME	[3]
	1.94 ± 0.07 (x)	3.80 (U)	0-100	B-M	Present study

*PE: Polynomial eq.: $-\frac{\Delta V}{V_o} = aP - bP$, where $a = -\frac{1}{B_o}$, $b = \frac{1}{2B_o^{\,2}}(B'_o+1)$.

LE: Linear velocity eq.: $U_s = C_o + S\,U_p$, where $C_o \approx \sqrt{\frac{K_o}{\rho_o}}$; $S \approx \frac{1}{4}(K'_o+1)$;

and U_s = shock velocity in km/sec; U_p = particle velocity in km/sec.

ME: Murnaghan eq.: $P = \frac{K_o}{K'_o}\left[\left(\frac{V}{V_o}\right)^{-1/K'_o} - 1\right]$

B-M: Birch-Murnaghan eq.: $P = \frac{3}{2}K_o\left[\left(\frac{V}{V_o}\right)^{-7/3} - \left(\frac{V}{V_o}\right)^{-5/3}\right]\cdot$

$$\left\{1 - \frac{3}{4}(4 - K'_o)\left[\left(\frac{V}{V_o}\right)^{-2/3} - 1\right]\right\}$$

Explanation of symbols:

U = ultrasonic data; S = shock-wave data; V-K = values reported by Vaidya and Kennedy; B = values derived from Bridgman's data; and X = this study.

and Kennedy [4,5] and by Bridgman [1-3]. The pressure range of measurement for each metal and the equation of state used to deduce their data are also listed.

For metals, the values of K' are generally in the region of 3 to 7 [23]. The K_o values for V (18.2), Nb (14.5), and Ta (2.75) reported by Vaidya and Kennedy, and values for V (11.8) and Ta (2.2) reported by Bridgman are either too high or too low to be physically true. Assuming K'_o values ranging from 3 to 7, we have deduced from equation (1): the K_o values for V, Nb, and Ta from our present P-V data; from Vaidya and Kennedy's smoothened P-V data; and also from Bridgman's P-V data (except for V). Results are shown in Fig. 5. As seen, there is a fairly good agreement between our present data (solid curves) and Bridgman's data (dash-dot curves), except in the case of V. However, there is a lack of agreement in all cases between our results (solid curves) and those obtained by Vaidya and Kennedy (dashed curves). Referring to Table II, the discrepancies between our values of K_o and K'_o and those of Bridgman for Nb and Ta are due to the differences in the data analysis; and for V are most probably due to the impurity of the sample used by Bridgman (the purity of his sample was 95.02% V) and the small pressure range of his measurements (to about 12 kbar).

The discrepancies between our results and those of Vaidya and Kennedy (Fig. 5, Table II) are significantly large in all cases and cannot be accounted for either by the differences in data analysis or by the reported impurities of the samples used except in the case of V for which Vaidya and Kennedy have not specified the purity. Their reported density of 6.020 g/cm^3 for V is only 1.3% lower than that of our sample [9]. It therefore seems reasonable to say that the discrepancies are caused by some undetermined error factors related to the measurements with the piston-cylinder apparatus.

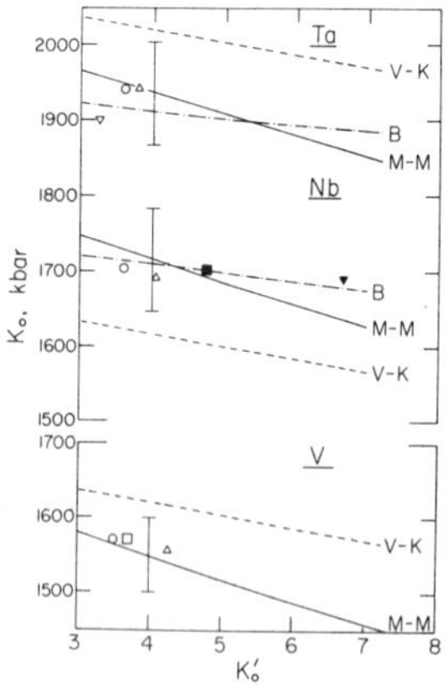

Fig. 5. A comparison of K_o and K'_o values for Ta, Nb, and V reported by various investigators. Symbols: o and ▫ represent shockwave results from McQueen et al. [10] and Al'tschuler [11], respectively; Δ, ▼, and ∇ represent the ultrasonic results of Katahara [9], Graham et al. [6] and Chechille [7], respectively; and ■ represents Bridgman's data [1]. Also shown are the relationships (curves) between K_o and K'_o deduced from the present x-ray diffraction P-V data (solid curves, M-M); from Vaidya and Kennedy's piston-cylinder P-V data (dashed curves, V-K); and from Bridgman's P-V data (dash-dot curves, B) using the Birch-Murnaghan equation with K'_o values ranging from 3 to 7. The vertical bars are the estimated errors in K_o values in the data of the present study.

In summary, most of the discrepancies in the K_o and K'_o values deduced from the P-V data using a two parameter equation of state are either caused by nonunique solutions, or by the impurity in the sample used, or by some unexplained factors related to the measurement with a piston-cylinder apparatus.

Anderson [24] has investigated the relationship between bulk modulus K_o vs. specific volume V_o ($V_o = 2M/p\rho_o$, where M, p, and ρ_o are the molecular weight, number of atoms, and density, respectively) for various elements in groups I, IIA, IIB, IIIA, IVA, and VA. He found that within various groups except for the transition elements the relationship $K_o V_o^{\,x}$ = constants, where x is in the range of 0.8 to 1.15, holds up well. There is no simple K_o-V_o relationship for the group VB transition metals. The K_o values in group VB transition metals from V to Ta increase with increasing atomic weight contrary to the general trend found in the other groups listed above. Since the bulk modulus of a substance is closely related to its internal energy [25], the reversed trend on the values of K_o in the bcc transition metals of group VB is probably due to the additional band structure energy contributed by the electrons in their d-orbitals.

CONCLUSIONS

We have shown that reliable compressibility measurements of metals can now be made under hydrostatic pressure to 100 kbar, using diamond-anvil pressure cell and X-ray diffraction. Reliable equations of state can be established from the P-V data if accurately determined value of K'_o is used (e.g., ultrasonic value of K'_o determined at low pressures). Such reliable equations of state can then be used for extrapolating P-V relationships to higher pressures below the phase transformation pressures. The discrepancies between the present compression data and other static compression data are probably caused by either the impurity of the sample or by some unexplained error factors related to the measurements with the piston-cylinder apparatus.

ACKNOWLEDGMENTS

Thanks are due J. Balogh for expertly designing and maintaining the instrumentation. This research was supported by the U. S. Energy Research and Development Administration Contract No. E(04-3)-235.

REFERENCES

1. P.W. Bridgman, Proc. Am. Acad. Arts Sci. 77, 189 (1949).
2. P.W. Bridgman, Proc. Am. Acad. Arts Sci. 62, 207 (1927).
3. P.W. Bridgman, Proc. Am. Acad. Arts Sci. 70, 285 (1935).
4. S.N. Vaidya and G.C. Kennedy, J. Phys. Chem. Solids 31, 2329 (1970).

5. S. N. Vaidya and G. C. Kennedy, J. Phys. Chem. Solids 33, 1377 (1972).
6. L. J. Graham, H. Nadler, and R. Chang, J. Appl. Phys. 39, 3025 (1968).
7. R. A. Chechille, Office of Naval Research Technical Report No. 10, Contract Nonr-1141 (05), Project NR 017-309, Case Institute of Technology (unpublished) (1967).
8. K. W. Katahara, M. H. Manghnani, and E. S. Fisher, J. Appl. Phys. 47, 434 (1976).
9. K. W. Katahara, Ph.D. Dissertation, University of Hawaii, Honolulu, Hawaii (1977).
10. R. G. McQueen, S. P. Marsh, J. W. Taylor, J. N. Fritz, and W. J. Carter in High-Velocity Impact Phenomena, R. Kinslow, ed., Academic Press, New York (1970).
11. L. Al'tshuler, A. Bakanova, and L. Dudoladov, Sov. Phys.-JET 26(6), 1115 (1968).
12. H. K. Mao, T. Takahashi, and W. A. Bassett, Phys. Earth Planet. Interiors 3, 51 (1970).
13. L. G. Liu, T. Takahashi, and W. A. Bassett, J. Phys. Chem. Solids 31, 1345 (1970).
14. L. G. Liu and W. A. Bassett, J. Appl. Phys. 44, 1475 (1973).
15. D. R. Wilburn, M.S. Thesis, University of Rochester, Rochester, New York (1976).
16. D. R. Wilburn and W. A. Bassett, E.O.S. Trans. Am. Geophys. Union 58, 518 (1977).
17. G. J. Piermarini, S. Block, and J. D. Barnett, J. Appl. Phys. 44, 5377 (1973).
18. G. J. Piermarini, S. Block, J. D. Barnett, and R. A. Forman, J. Appl. Phys. 46, 2774 (1975).
19. L. C. Ming and M. H. Manghnani, J. Appl. Phys. 49(1), 208, 1978.
20. W. A. Bassett, T. Takahashi, and P. W. Stock, Rev. Sci. Instrum. 38, 37 (1967).
21. J. S. Weaver, T. Takahashi, and W. A. Bassett, in Accurate Characterization of the High Pressure Environment, Spec. Publ. 326, E. C. Lloyd, ed., National Bureau of Standards, Gaithersburg, Maryland (1971).
22. H. W. Mao, T. Takahashi, W. A. Bassett, and J. S. Weaver, J. Geophys. Res. 74, 1061 (1969).
23. E. S. Fisher, M. H. Manghnani, and K. W. Katahara, Proceedings of 4th Intern. Conference on High Pressure, Kyoto, Japan (1974), p. 393.
24. O. L. Anderson, in The Nature of the Solid Earth, E. C. Robertson, ed., McGraw-Hill Book Company (1972), p. 575.
25. C. Kittel, Introduction to Solid State Physics, 3rd edition, John Wiley and Sons, New York (1968).

A-6

SHOCK COMPRESSION OF LEAD FLUORIDE AT ROOM TEMPERATURE

D. P. Dandekar and R. M. Lamothe

U.S. Army Materials and Mechanics Research Center
Watertown, Massachusetts USA

INTRODUCTION

Lead difluoride (PbF_2) is known to occur in two distinct crystalline forms; namely, fluorite and $PbCl_2$ structures. The fluorite structure has a cubic symmetry (Fm3m - O^5_h) and $PbCl_2$ has an orthorhomic symmetry (Pbnm - V^{16}). The fluorite and orthorhombic forms of PbF_2 are identified as β- and α-PbF_2, respectively. The prevalent information about the stability of these two forms of PbF_2 may be summarized as follows: (1) α-PbF_2 is the stable form of PbF_2 at ambient conditions; (2) α-PbF_2 transforms to β-PbF_2 at 583 ± 1K; (3) The reverse transformation of β-PbF_2 to α-PbF_2 has not yet been observed; (4) The crystals of PbF_2 grown from its melt are invariably in the form of β-PbF_2; (5) Even though unstable at ambient conditions, β-PbF_2 needs to be subjected to a hydrostatic pressure of around 0.4 ± 0.1 GPa to transform it into α-PbF_2 form. This transformation has also been reported to be rather sluggish; (6) The rate of transformation of β- to α-PbF_2 increases with an increase in temperature; (7) The pressure of transformation P_t of β- to α-PbF_2 decreases with an increase in temperature T. The magnitude of dP_t/dT is approximately -3.6 MPa/K; (8) There is some indication that the β- to α-PbF_2 transformation proceeds readily under shearing stress conditions at room temperature.

The above information is extracted from the works of Byström [1], Jones [2], Schmidt and Vedam [3], Samara [4], Nicol [5], Linsey [6], and references therein. There is no compressibility and thermal expansion data on α-PbF_2.

The present work was initiated as a follow-up on the shock compression work on β-BaF_2 [7]. The unexpected results obtained by Dandekar and Duvall [7] provided the motivation to investigate the shock compression behavior of β-PbF_2, primarily because of its low transformation pressure, i.e., 0.4 ± 0.1 GPa compared to that of β- to α-BaF_2 around 2.5 ± 0.6 GPa, and a speculation that both the transformations may be shear induced [7] as Linsey's [6] work does show that β- to α-PbF_2 transformation is shear induced. The former authors [7] were not aware of the earlier reference [6] at the time they published their work.

The results of the present work on β-PbF_2 show that the transformation of β- to α-PbF_2 is even more stress resistant under shock loading than under static compression. These results also show some features unique to shock compression experiments. However, these results have to be taken as preliminary because only five experiments could be performed due to non-availability of crystals of β-PbF_2 at present.

SHOCK EXPERIMENTS

Shock compression experiments were performed on 2 to 3 mm thick and 14 to 26 mm diameter discs of β-PbF_2. X-ray diffraction pattern of the material indicated it to be well crystallized in the fluorite structure, but we were unable to determine their orientations. The discs were made by Harshaw Chemical Co. The measured densities of the discs were 7.76 Mg/m^3. The disc faces were lapped flat to 5 μm. The mutual parallelity of these faces on a disc was within 3×10^{-4} mm over its entire lateral surface.

The compression behavior of β-PbF_2 has been determined from direct-impact experiments. In these experiments, the discs of β-PbF_2 were impacted upon stationary X-cut quartz gauges. When the stress σ in an experiment was expected to exceed 3.0 GPa, the X-cut quartz gauge used in the experiment was buffered with either a lucalox or a tungsten carbide disc. Details for this type of shock experiments can be found elsewhere [8,9]. X-cut quartz gauges were used in a shorted mode. These gauges were 3.2 mm thick with a diameter of either 12.7 or 25.4 mm. The guard ring width for the smaller and larger diameter gauges were 4.8 and 11 mm, respectively [10,11].

The stress profiles obtained in these experiments were found to attain steady values immediately upon impact for the total recording duration of around 0.5 μs, except in the experiment with a steady state stress of 4.19 GPa (see Fig. 1). In this experiment the stress reached upon impact was 4.62 GPa and within 0.085 μs the stress relaxed to a steady-state value of 4.19 GPa for the remaining 0.415 μs of the recording time. The uncertainties in the reported values of stress and particle velocity u are 3 and 2%, respectively.

RESULTS AND DISCUSSION

The results of five shock compression experiments on β-PbF_2 are summarized in Table I. The observables in these experiments,

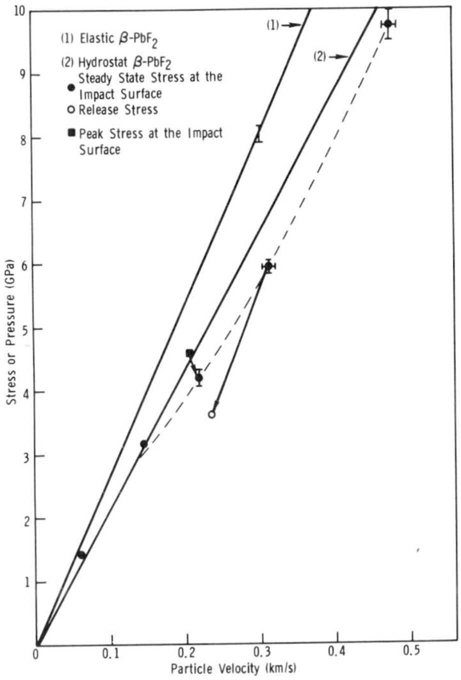

Fig. 1. Stress vs. particle velocity of PbF_2. The starting material is β-PbF_2. The elastic and hydrostatic compression curves of β-PbF_2 are determined from the elastic constant data of Hart. Dashed curve is drawn free hand.

i.e., σ and u are plotted in Fig. 1. The loci corresponding to an elastic and hydrostatic compression of β-PbF_2 are also shown in this figure. These loci are drawn on the basis of the elastic constants values of β-PbF_2 at room temperature as determined by Hart [12].

Figure 1 shows that β-PbF_2 deviates from a linear elastic solid behavior somewhere between 0.5 and 0.75 GPa. With an increase in stress, β-PbF_2 seems to start losing its capability to maintain a shear stress. The loss of shear strength is complete around 3.0 GPa. The evidence for the above behavior of β-PbF_2 is rooted in the fact that the experimentally observed σ,u coordinates of β-PbF_2 in Experiment 1 lie in between the elastic and hydrostatic compression curve of β-PbF_2, and those in Experiment 2 lie on the hydrostatic compression curve of β-PbF_2. The σ,u coordinates above 3.0 GPa tend to lie below the hydrostat. This implies that the phase of PbF_2 above 3.0 GPa does not correspond to that of β-PbF_2. Are these coordinates associated with that of α-PbF_2 or they are associated with some unknown phase similar to those discovered in KCl by Hayes [13] under shock compression? The results of the present experiments tend to indicate that the coordinates corresponding to Experiments 3 and 4 are definitely not associated with that of α-PbF_2. It is difficult to make such a definitive statement about the coordinates of Experiment 5. There are three reasons for the preceding statements about the state of PbF_2 in Experiments 3, 4, and 5. These are: (1) The release wave impedance of β-PbF_2 at 5.96 ± 0.18 GPa is estimated to be 30.2 ± 1.5 $Ggm^{-2}\ s^{-1}$ in Experiment 4. The elastic impedance of β-PbF_2 is 26.8 ± 0.9 $Ggm^{-2}\ s^{-1}$. (2) The x-ray diffraction pattern of the material recovered (by sheer good luck) from Experiment 4 indicated it to be in β-PbF_2 phase. No material was recovered from Experiment 5. (3) If we extrapolate the estimated values of density of PbF_2 at 9.74 and 5.96 GPa to calculate its ambient density, it turns out to be around 8.0 Mg/m^3. The ambient density of α-PbF_2 is 8.52 ± 0.12 Mg/m^3. So we infer that even

Table I. Shock Compressed States of β-PbF_2 At Room Temperature

Experiment	Stress, GPa	Particle Velocity, km/s	Shock Velocity, km/s	Density, Mg/m^3	Remark Code
1	1.44	0.059	-	-	(a)
2	3.18	0.146	-	-	(a)
3	4.19	0.218	1.79	8.57	(b)
4	5.96	0.312	2.13	8.92	(b)
5	9.74	0.474	2.54	9.46	(b)

(a) The values of shock velocity and density for these experiments were not calculated because the value of HEL is not well determined at this time.
(b) The values of shock velocity and density are calculated assuming that P_t = 3.0 GPa. Associated values of density and particle velocity are computed from Hart's data on β-PbF_2, which are 8.11 Mg/m^3 and 0.130 km/s, respectively. It should be noted that the estimates of final densities are insensitive to the value of P_t assumed.

though the experimental points in Experiments 3 and 4 lie below the hydrostatic compression curve of β-PbF_2, these cannot be associated with α-PbF_2 phase. We cannot make a similar statement with regard to the σ,u coordinates of Experiment 5.

The interesting results of the present study may be summarized as follows: (1) β-PbF_2 appears to begin losing its capability to maintain shear stress when shocked above about 0.7 GPa; (2) The above process seems to be completely realized when β-PbF_2 is shocked to about 3.0 GPa; (3) The experimental values of σ,u pairs beyond 3.0 GPa tend to lie below the hydrostatic compression curve of β-PbF_2; (4) These values, however, cannot be associated with α-PbF_2 phase; and (5) The unknown phase of PbF_2 realized at 5.96 GPa gets recrystallized into β-PbF_2 upon release of stress.

We have no explanation as to why β-PbF_2 is so resistant to transform under shock compression. The experiments reported here show that the transformation of β- to α-PbF_2 has not taken place even when stress is twelve times the static transformation pressure of 0.4 ± 0.1 GPa. This is all the more surprising in the view of Samara's work [4] where it is shown that the transition pressure declines from 0.4 to 0.2 GPa as the temperature is increased from 300 to 350 K. At 420 K, the transition pressure of β- to α-PbF_2 becomes around 0.1 GPa. Under shock compression, a material generally gets heated up. In the case of β-PbF_2, the minimum increase in temperature when it is compressed to 3.0 GPa, is estimated to be 30 K. This estimate is based on temperature rise due to the adiabatic compression of β-PbF_2. However, the fact is that the associated rise in temperature does not aid the transformation of β- to α-PbF_2 under shock compression. One could speculate on the reasons for the behavior of β-PbF_2, but these could not be tested for their validity because some of the basic data on α-PbF_2 remains to be collected. Thus it seems that there is a basic need to determine (1) volume compressibility and thermal expansion coefficient of α-PbF_2 and (2) a phase diagram delineating the stability field of β- to α-PbF_2 in pressure-temperature space. These when collected either would provide an impetus to conduct more shock experiments on β-PbF_2 or may even lead to a clearer understanding of the observed behavior of β-PbF_2 under shock compression without further experimentations.

ACKNOWLEDGMENTS

The authors are grateful to J. Fontanella of U.S. Naval Academy for donating five crystals of β-PbF_2 for the shock compression experiments. It is their pleasure to acknowledge the expert assistance of J. Kelley in carrying out the experiments. Efforts of M. Bolt in preparation of this paper are very much appreciated.

NOTATION

P_t = pressure of transformation
T = temperature
u = particle velocity
α-PbF_2 = PbF_2 with $PbCl_2$ structure
β-PbF_2 = PbF_2 with fluorite structure
σ = stress

1. A. Byström, Ark. Kemi. Min. Geol. A 24, 33 (1947).
2. D. A. Jones, Proc. Phys. Soc. B 68, 165 (1955).
3. E. D. D. Schmidt and K. Vedam, J. Phys. Chem. Solids 27, 1563 (1966).
4. G. A. Samara, Phys. Rev. B 13, 4529 (1976).
5. M. Nicol, private communication.
6. C. W. Linsey, Ph.D. Thesis, North Texas University, Denton, Texas (1970).
7. D. P. Dandekar and G. E. Duvall, in Metallurgical Effects at High Strain Rates, R. W. Rohde, B. M. Butcher, J. R. Holland, and C. H. Karnes, eds., Plenum Press, New York (1973), p. 185.
8. W. J. Halpin, O. E. Jones, and R. A. Graham, Symposium On Dynamic Behavior of Materials, ASTM Special Technical Publications No. 336, American Society for Testing and Materials (1963).
9. P. C. Lysene, R. Boade, C. M. Percival, and O. E. Jones, J. Appl. Phys. 40, 3786 (1969).
10. D. B. Hayes and Y. M. Gupta, Rev. Sci. Instr. 45, 1554 (1974).
11. R. A. Graham, J. Appl. Phys. 46, 1901 (1975).
12. S. Hart, J. Phys. D: Appl. Phys. 3, 430 (1970).
13. D. Hayes, J. Appl. Phys. 45, 1208 (1974).

A-7

RESPONSE OF POROUS BERYLLIUM TO STATIC AND DYNAMIC LOADING*

W. M. Isbell

Effects Technology, Inc.
Santa Barbara, California USA

O. R. Walton and F. H. Ree

Lawrence Livermore Laboratory, University of California
Livermore, California USA

INTRODUCTION

The effectiveness of porous materials in attenuating stress pulses and in reducing the thermomechanical stresses arising from rapid energy deposition has been the subject of numerous studies during the past decade. Because of the large number of manufacturing parameters (composition, porosity, pore size, heat treatment, etc.) available to the developers of porous materials, extensive tailoring of properties to meet widely varying requirements is practical, and the materials manufactured and the studies to date now number in the dozens.

The present study involves two porous berylliums of different initial heat treatments and slightly different porosities. A theoretical model was developed and a series of measurements were made to describe the complex equation of state surfaces peculiar to porous materials. The portion of the work to be reported on involves the behavior of the materials under static and dynamic loading. A more complete documentation of the experimental and theoretical results is presented in the literature [1-5].

EQUATION OF STATE SURFACE

Thermodynamic equilibrium properties of nonporous materials may be described by unique relationships between the thermodynamic

*Work supported by the Energy Research and Development Agency under contract W-7405-Eng-48.

variables: pressure, temperature, entropy, energy, volume, etc. This unique relationship between the thermodynamic variables is often represented by a three-dimensional surface showing the allowed thermodynamic equilibrium states of the material in terms of any three of the thermodynamic quantities. Figure 1 shows a schematic representation of the pressure, temperature, and volume relationship for a typical material, including phase changes. Each point on the non-porous equation of state surface represents a unique state-point, i.e., the pressure at a particular volume and temperature is independent of the path (or past history) used to arrive at that volume and temperature.

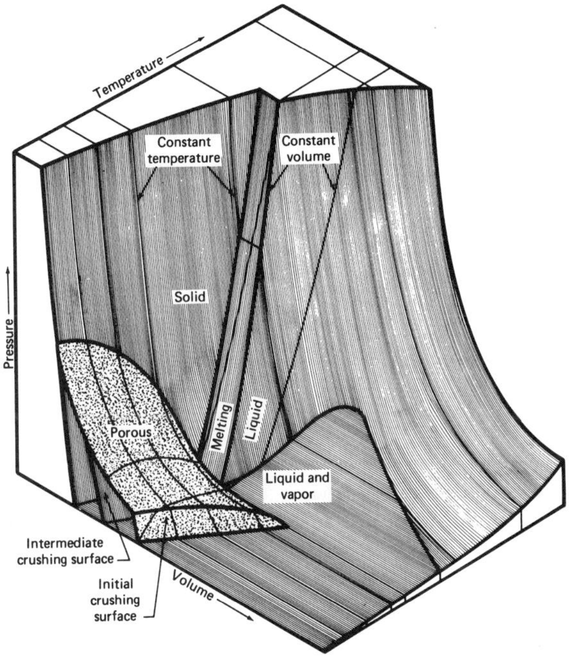

Fig. 1. Schematic representation of the P-V-T relationships for typical solid and porous materials.

No such unique relationship exists for porous materials. The complex nonequilibrium thermodynamic states that can be reached by a porous material are not described by a single equation of state, but rather by a mathematical model that depends on the thermodynamic path by which the material arrived at its current state. The porous "addition" to the surface shown in Fig. 1 represents only an initial crushing surface. Unloading and subsequent reloading of the porous material occurs on another path which penetrates below the initial crushing surface. This is shown as an intermediate crush surface in the figure. Such intermediate crush surfaces are examples of how the pressure in a porous material depends not only on the volume and temperature, but also on the past history of the material.

DYNAMIC BEHAVIOR

The initial compaction surface (and the intermediate unloading-reloading surfaces) shown in Fig. 1 is valid for static or quasi-static loadings and unloadings. If the loading is rapid (as in a shock wave front, for instance) then inertial and viscous resistance to the rapid collapse of the pores will come into play to make the dynamic compaction path lie above the quasi-static path. In both impact tests and energy deposition tests, the initial stresses produce pore-collapse rates that are high enough to lead to temporary overstresses.

Several models have been proposed [6-10] to describe rate-dependent pore collapse. Most of these assume that a dynamic overpressure (dependent on the rate of pore collapse) exists for some characteristic time, τ, (dependent on material properties) with an exponential relaxation to the static crushing surface. Each of the models used can be calibrated, with varying degrees of success, to agree with the rise times and wave velocities from plate impact experiments.

THE POROUS CONSTITUTIVE MODEL

The model we have used to describe the porous beryllium [3] is based on the P-α-τ model of Holt et al. [7] which is, in turn, a rate-dependent modification of Herrmann's model [6]. The new features in the present model are: the inclusion of deviatoric stresses, the use of a porosity-dependent relaxation time for pore closure, an allowance for partial reopening of the pores on unloading, and the use of improved static compaction functions.

Two relations are used in the formulation of the constitutive model. The first relation is an equation which relates the pressure P, volume V, and energy E of a porous material to the equation of state (EOS) of the corresponding fully-compacted solid. The porosity parameter α is defined by the relation

$$P(V,E) = \frac{1}{\alpha} P_s \, (V/\alpha, E_s) \tag{1a}$$

which can be rewritten in an alternate but equivalent form,

$$\alpha = V/V_s = P_s/P \tag{1b}$$

where the subscript s refers to the solid material.

The second relation used in formulating the model is an equation which describes the elastic and plastic portion of paths for α during a time-dependent deformation process. Along the plastic path, we use the rate-dependent pore-closure relation of Holt et al.

$$\alpha = g(P) - \tau \, d\alpha/dt \tag{2}$$

The parameter τ is a time constant describing the rate of plastic flow of material into pores, which reduces α to a final average equilibrium value given by the static compaction function g(P).

To determine g(P), we eliminate V in (1) by using hydrostatic P-V data for both solid and porous beryllium. The resulting data, P and g(P), can be accurately represented by the following expression:

$$g(P) = \alpha_\infty + (\alpha_o - \alpha_\infty)\exp(aP + bP^2 + cP^3) \qquad (3)$$

where α_o is the initial porosity and α_∞ is introduced to account for some residual porosity that appears to persist under extremely high pressures [5,11]. The constants a, b, and c are obtained by fitting the data by a least-squares method.

A modification of the elastic viscoplastic, rate dependent behavior described by Holt's spherical pore model is used to describe the rate dependence of the compaction. A porosity dependent τ of the form

$$\tau = \tau_o \alpha_o (\alpha_o - \alpha_\infty)/[\alpha(\alpha - \alpha_\infty)] \qquad (4)$$

can be calibrated to fit the calculated wave profiles for as-sprayed beryllium.

To obtain an equation for α along an elastic path, the porous and solid equations of state are expressed in incremental form. That is, using Δ to denote the difference in P (and in P_s) at two successive time steps we have

$$\Delta P = K(\Delta\rho/\rho) \qquad (5a)$$

$$\Delta P_s = K_s\,(\Delta\rho_\sigma/\rho_s) \qquad (5b)$$

where K and K_s are the bulk moduli of the porous and solid materials, respectively. After lengthy algebraic manipulation, we obtain an expression for the rate of change of α with P along the elastic path of the form

$$d\alpha/dP = \alpha(K_s/K - \alpha)/(\alpha P - K_s) \qquad (6)$$

where Mackenzie's [12] expressions for K and G are rewritten

$$1/K = (\alpha_\infty/K_s)t + (1/K_o)\,(1 - t) \qquad (7a)$$

$$G = (1/G_s)t + (1/G_o)(1 - t) \qquad (7b)$$

where t is defined as $(\alpha_o - \alpha)/(\alpha_o - \alpha_\infty)$ and the subscript 0 refers to the initial porous state.

Yield Stress

To complete the formulation, the deviatoric stress s must be specified. Then, the axial stress σ_1 (along direction 1) in a 1-D strain deformation is given by

$$\sigma_1 = -P + s_1 \qquad (8)$$

where

$$s_1 = \frac{4}{3} G \ln(V/V_o), \text{ if } |s_1| < \frac{2}{3} Y \quad (9a)$$

$$s_1 = \pm \frac{2}{3} Y \text{ for all other conditions} \quad (9b)$$

and s_1 is positive for loading and negative for unloading.

The shear modulus G in (9a) is calculated from (7b). Y may be calculated from 2/3 Y = σ_1 + P if experimental data are available for P and σ_1. The data can be fitted by a functional form such as

$$Y = \text{Max } [Y_o, Y_1 + Y_2 \varepsilon + Y_3 \varepsilon^2] \quad (10)$$

where $\varepsilon = \ln(V/V_o)$ and small elastic strain is neglected.

DESCRIPTION OF THE POROUS BERYLLIUMS

Two porous beryllium materials were studied, both plasma-sprayed by Union Carbide Corporation from powders supplied by Kawecki Berylco Industries, Inc. The materials were prepared in accordance with Kaman Science Corporation specifications for Models 67 and 68 beryllium.

The plasma spraying process involves the ejection of metallic powder using a jet of inert gas (argon) from a nozzle. The stream of powder passes through an electric arc and is melted to form a stream of molten droplets. These droplets land on a rotating aluminum mandrel (or turntable). The nozzle moves radially in and out across the mandrel face during the spraying process so that the droplets form a spiral pattern in and out across the mandrel, building up a plate, layer upon layer.

Two grades of beryllium powder were used to manufacture the plates, P-1 and P-10. Specimens made from P-10 powder were tested in the "as-sprayed" condition while specimens from P-1 powder were sintered for 2 hrs at 1175°C, producing a less porous and stronger material. Table I summarizes the chemistry of the two powders used and the densities of the resulting plasma-sprayed materials.

Table I. Chemical Analysis of Beryllium Specimens

		P-10 (-325 mesh, not sintered) wt%	P-1 (-325 mesh, sintered) wt%
Chemistry:	BeO	0.66	0.72
	Fe	0.075	0.035
	Al	0.024	0.006
	C	0.029	0.026
	Mg	0.026	0.002
	Si	0.010	0.008
Density, gm/cm^3		1.587	1.647
Porosity		14.2%	11.0%

EXPERIMENTAL DATA

To perform meaningful calculations, a computational model must first be fitted to accurate and reasonably complete experimental data. In the present study, impact and static pressure-volume data provided paths for mechanical loading and unloading and rate-dependency of yielding as well as providing a check on the calculational ability of the model to predict shock wave attenuation. Ultrasonic measurements were used to provide elastic moduli at standard temperature and pressure. Microscopic examination of compressed specimens yielded insight into deformation and mechanisms of the pore collapse process. Effective Grüneisen and expanded volume states were measured using an electron-beam machine to provide nearly constant-volume thermal heating in short deposition times. A large amount of experimental data was generated on the two berylliums during this program. A brief synopsis of the data follows.

Static Compression Data

Throughout the study, quasi-static deformation data [2,13] served to guide the modeling effort. Considerable detail is contained in the data concerning: compression under hydrostatic and one-dimensional conditions, release behavior of the compressed materials, residual porosity at high pressures (> 4GPa), and elastic yield in compression and release. The relatively low cost of this data, when compared to shock wave data, makes the static technique an attractive alternative for characterizing porous materials. The static data, when combined with a modest number of shock wave tests to determine deviatoric and time-dependent behavior, has proven adequate for modeling several materials of interest.

Two types of experiments were performed: loading and unloading under conditions of uniaxial strain and under conditions of hydrostatic pressure. In uniaxial strain loading, an axial stress was applied to a cylindrical sample with the condition that the radial strain remain constant. This was achieved by control of the lateral confining pressure and resulted in a loading path similar, except for time-dependent flow, to that of plane shock-loading. In the hydrostatic testing, an axial stress was applied to a medium surrounding a cylindrical sample. The plastic flow of the surrounding medium (fluid at low pressure and tin at high pressure) insured that the loading was nearly hydrostatic in nature.

Details of the uniaxial strain loading data are shown in Fig. 2. The initial slope of the unsintered curve is seen to be much lower than in the sintered material, showing that the unsintered material is originally more compressible. This may be the result of some or all of the following effects: (1) The sintering process, which produces more spherical and hence stronger pores; (2) the existence of numerous microcracks in the unsintered material; (3) the decreased

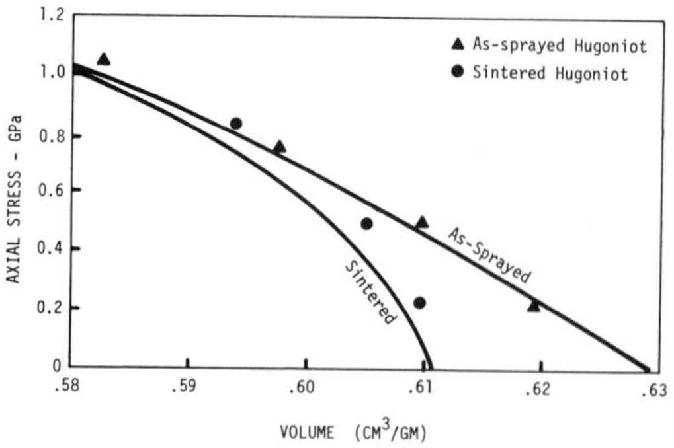

Fig. 2. Comparison of quasi-static uniaxial stress-strain data on two porous beryllium with corresponding Hugoniot data.

porosity and thus greater strength in the sintered beryllium; and (4) the effect of residual stresses in the unsintered material, lowering the applied shear stress necessary to cause yielding.

Also shown in Fig. 2 are Hugoniot points taken over the same stress range as the laboratory uniaxial data. Within experimental error, for the unsintered beryllium the points are coincident with the path that defines the stress-volume curve. In contrast, the shock wave data for the sintered beryllium lie consistently above the static data, most probably indicating increased strain rate behavior for that material.

Shock Wave Tests

Shock wave generation and propagation under conditions of uniaxial strain were measured [1,5] using gas gun-launched impactors and a variety of experimental configurations. Data obtained included: Hugoniots to ~ 3.2 GPa (32 kbars), release adiabats from shocked states, shock wave profiles for attenuated and unattenuated waves, and compressional and release velocities over a range of stresses. A separate series of explosive tests [11] extended the range of Hugoniot measurements to over 33 GPa (330 kbars).

Most data were used directly in developing the material models, although independent check data (primarily attenuated wave profiles) were used to determine the accuracy of the model's predictive capability. This section summarizes the data obtained and experimental techniques used.

Hugoniot data and unattenuated wave profiles were obtained primarily with the three measurement techniques shown in Fig. 3. A 90-mm diameter gas gun was used to launch four porous beryllium specimens into four materials of different impedances mounted on quartz stress gages (Fig. 3a), or reversing the procedure, to launch four materials of different impedances into beryllium specimens mounted on quartz gages (Fig. 3b).

The "direct impact" technique (a) provided Hugoniot data in the form of stress-particle velocity points calculated by knowledge of the Hugoniots of the four impactor materials plus the known characteristics of the quartz stress gages and buffers. This multiple gage technique is particularly efficient in reducing ambiguities among data points in that all specimens are impacted at the same impact velocity and at the same impact tilt. Cross correlation

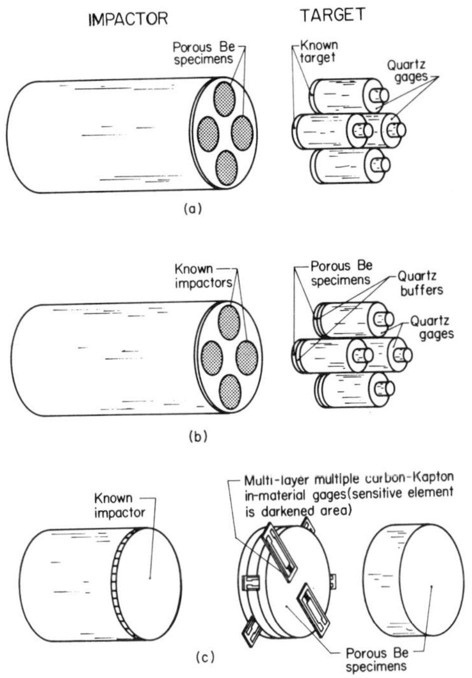

Fig. 3. Schematic of the measurement techniques used to obtain Hugoniot data and the stress-time profiles of transmitted waves. (a) Hugoniot points from four specimens launched into four quartz gages with buffer plates of various shock impedances; (b) Four compressive wave profiles at different material thicknesses; (c) Attenuating wave profiles measured with in-material piezoresistive gages.

between data points is thus facilitated and results in relatively accurate data.

Figure 4 shows Hugoniot data obtained to 4.5 Gpa (45 kbars). The relatively low compaction wave and high initial release wave velocities account for the very rapid attenuation rates customarily found in porous materials.

Although quartz stress gage measurement techniques have advantages in simplicity and reasonably well-known gage characteristics, the (usually) large change in shock impedance at the gage-specimen interface reflects a portion of the shock wave and may considerably complicate analysis. An ideal measurement technique would require in situ measurement of the passage of the shockwave without disturbance of the wave itself. At present, "in-material" piezoresistive stress gages meet this requirement most closely.

For most of the wave profiles used for model development and for checks on predictive capability, carbon-Kapton* in-material stress gages were used in the configuration shown in Fig. 3c. Since the gage is quite thin (0.1 mm), it usually comes into pressure equilibrium with the specimen material within a few hundred nanoseconds by a series of shock reverberations across the gage, even though the shock impedances of the gage and test specimens may be very different. Thus, if a suitable calibration has been performed on the gage, one obtains a direct measure of the stress in the text specimen. It should be noted, however, that the impedance difference does indeed affect the profile shape somewhat, and accurate work requires the ability to calculate profiles using equations of state of the gage and of the material under test.

*DuPont trademark.

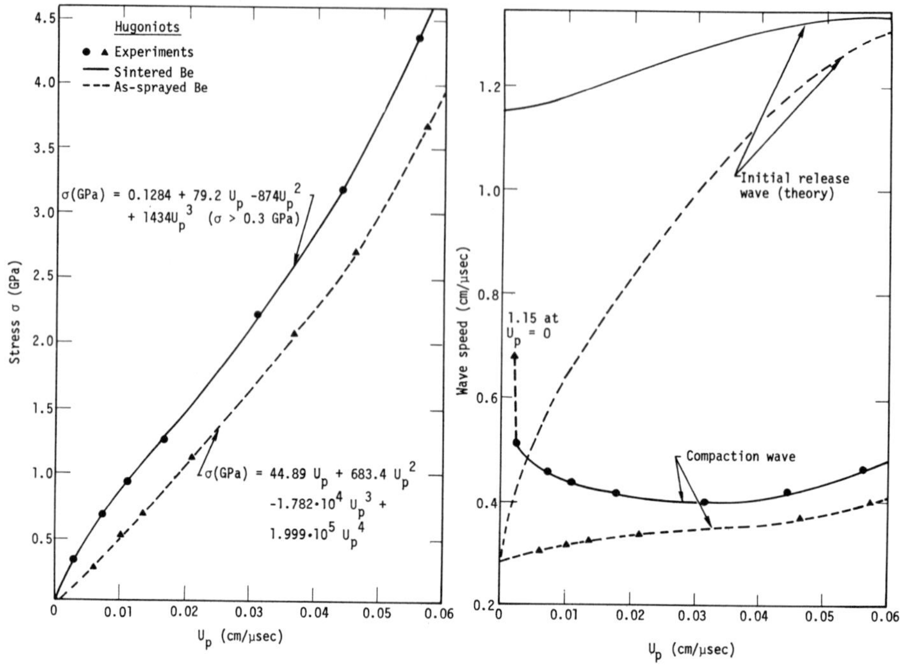

Fig. 4. (a) Hugoniot points and their analytical representations; (b) Compaction wave speeds vs. velocity U_p obtained from Hugoniots and initial release wave velocities calculated using (3) and (7).

Using in-material gages placed between successive layers of the specimen, we studied the evolution of waves as they progressed through the material. Elastic and plastic compressive and release waves were measured and, by using careful timing techniques, wave speeds were obtained.

COMPARISON OF PREDICTIONS WITH DATA

Experimental wave profiles for comparison were obtained at two stress levels, approximately 0.6 and 1.7 GPa. Profiles were obtained by impacting a PMMA plate onto either sintered or as-sprayed porous beryllium having carbon-Kapton piezoresistive gages embedded at as many as six different depths within the material. Thus a single experiment took data on wave profiles at several different levels. The gages were located as deep as 1.9 cm from the impact surface and some measurement times extended up to 4 μs. The resulting 21 wave profiles--both unattenuated and attenuated--represent large variations in the physical parameters and enabled us to extensively test the model.

Measured and computed wave profiles are compared in Fig. 5 for the sintered material and in Fig. 6 for the as-sprayed material. Agreement between the experimental wave profiles and those predicted

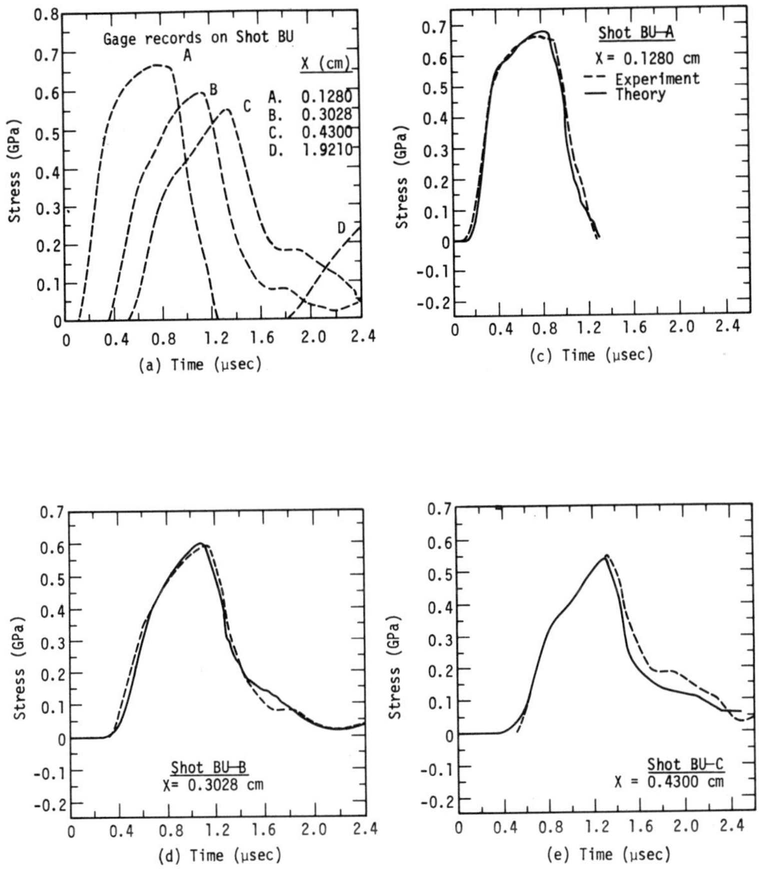

Fig. 5. Full and attenuated wave profiles for a low-stress test in sintered porous beryllium. Theoretical predictions are shown by solid lines. Velocity of the 0.124 cm thick PMMA impactor was 0.0262 cm/μs.

from the present model is good both quantitatively and qualitatively. For sintered porous beryllium (Fig. 5) an elastic precursor about 0.4 GPa high, which corresponds to the "shoulder" in the hydrostat in Fig. 2, precedes the main plastic wave. At the foot of the precursor the velocity is close to the longitudinal sound speed, but it becomes slower at higher stresses. The 0.62 cm/μs value at P = 0.2 GPa chosen for the calculation of the first arrival times agrees reasonably well with the velocity of the precursor at this level.

A faster-rising shock front, the lack of an elastic precursor, and a higher predicted attenuation rate are the chief distinguishing features of compressive profiles in as-sprayed porous beryllium. Other tests at 1.7 GPa (17 kbars) showed similar behavior.

We also note from Figs. 5 and 6 that the arrival times of the shocks (and their precursors in the case of sintered specimens) agree satisfactorily. The relaxation time $\tau = 0.04$ μs matches the observed risetimes of the shocks in the case of the sintered specimens, while the porosity-dependent τ from (4) adequately describes the risetimes of the shocks in the as-sprayed specimens. In the

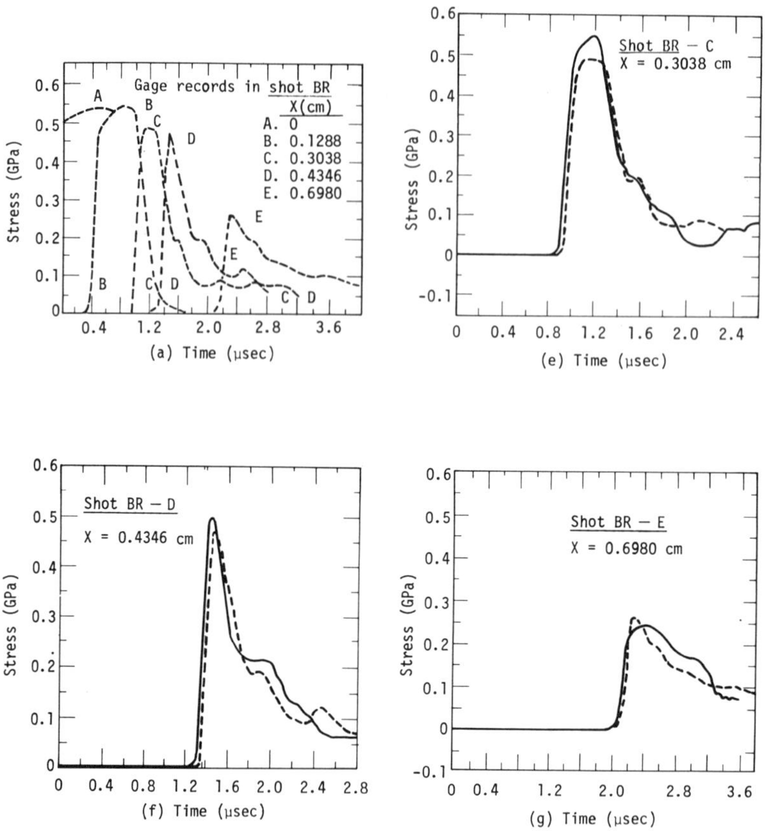

Fig. 6. Wave profiles for as-sprayed porous beryllium at a similar impact velocity. PMMA impactor thickness was 0.1245 cm and impact velocity was 0.0258 cm/μs.

latter case, it is noteworthy that τ changes from 0.060 μs at the foot of the shocks to 0.015 μs at 1.7 GPa. If, instead of (4), a constant value (0.015 μs) had been used for τ, the foot of the shocks would be traveling too fast, resulting in much longer risetimes than those in the observed profiles.

The largest deviation occurs in the calculation of the peak stresses, with the experimental data lying 10% or less below computer predictions. Either experimental uncertainties or approximations in the model or a combination of the two could account for these relatively small differences.

SUMMARY

We have shown that significant differences exist between the porous as-sprayed and the sintered beryllium materials examined in this study. We have demonstrated that the wave propagation properties of both of the materials tested can be described by the porous material model developed. Both the compressive characteristics (first-wave arrival time, risetime of shock, and peak stress) and the release characteristics (arrival time of release wave and general attenuated shape) of shocks have been reproduced satisfactorily for different stresses and pulse durations as well as for different thicknesses of the porous specimens. Slight deviations between the experimental and computed wave profiles lie mostly within the combined errors of the model and the experimental data.

REFERENCES

1. W. M. Isbell, O. R. Walton, and F. H. Ree, Lawrence Livermore Laboratory Rept. UCRL-51682, Part 1 (1977).
2. R. N. Shock, A. E. Abey, and A. G. Duba, Lawrence Livermore Laboratory Rept. UCRL-51682, Part 2 (1974).
3. F. H. Ree, W. M. Isbell, and R. R. Horning, Lawrence Livermore Laboratory Rept. UCRL-51682, Part 4 (1974).
4. J. E. Hanafee and E. O. Snell, Lawrence Livermore Laboratory Rept. UCRL-51682, Part 6 (1974).
5. R. R. Horning and W. M. Isbell, Rev. Sci. Instr. 46 (10), 1374 (1975).
6. W. Herrmann, J. Appl. Physics 40, 2490 (1969).
7. A. C. Holt, A. S. Kusubov, D. A. Young, and W. H. Gust, "Thermo-mechanical Response of Porous Carbon," Lawrence Livermore Laboratory, Rept. UCRL 51330 (1973).
8. M. M. Carroll and A. C. Holt, J. Appl. Phys. 43, 1626 (1972).
9. B. M. Butcher, "Numerical Techniques for One Dimensional Rate Dependent Porous Material Compaction Calculations, SC-RR-710112, Sandia Laboratories Report (April 1971).
10. L. Seaman, R. E. Tokheim, and D. R. Curran, "Computational Representation of Constitutive Relations for Porous Material," Stanford Research Institute, prepared for Defense Nuclear Agency, DNA 3412F (May 1974).
11. W. H. Gust, private communication.
12. J. K. Mackenzie, Proc. Phys. Soc. B63, 2 (1950).
13. R. N. Schock, A. E. Abey, and A. G. Duba, J. Appl. Phys. 47, 53 (1976).

A-8

DETERMINATION OF THE GRÜNEISEN γ FOR BERYLLIUM AT 1.2 TO 1.9 TIMES STANDARD DENSITY*

T. Neal

Los Alamos Scientific Laboratory, University of California
Los Alamos, New Mexico USA

INTRODUCTION

The curve that represents the locus of material states attainable directly by single-shock compression is termed a Hugoniot. Most existing very-high-pressure equation-of-state information is obtained by shock techniques and thus lies on that curve. The shocking process results in an increase in entropy. Thus the isotherm, the isentrope, and the Hugoniot, centered at a common initial-state point, correspond to successively higher energy states for a given material compression. The region in the pressure-volume plane between the principal isentrope and principal Hugoniot will be the subject of this report. That region will be investigated by using flash radiography to directly observe shocks and rarefactions in preshocked materials. The experiments to be described here will be interpreted in terms of the Mie-Grüneisen equation of state. Values of the Grüneisen parameter will be determined by comparing experimentally-attained states with states of corresponding volume on the principal Hugoniot. The results of different experimental techniques and different final compressions can thereby be reduced to a common basis of comparison - the Grüneisen parameter - and some conclusions concerning the high-pressure equation of state of beryllium can be formulated.

FLYING PLATE EXPERIMENT

In the first type of experiment attempted, an explosively-driven stainless-steel flyer was used to impact a beryllium target and induce a shock. The target was backed in one area by copper, in another by uranium-238, in a third by vacuum, and in a fourth by additional beryllium. The first two materials, which have a higher

*Work performed under the auspices of the U.S. Energy Research and Development Administration.

shock impedance than beryllium, were used to reflect the incident shock back into the target. This resulted in beryllium that was shock-compressed in two separate stages. The vacuum and additional beryllium locations were set aside to permit the free-surface and shock velocities at the reflector interface to be measured with shorting pins. These measurements were made to characterize the incident shock in the beryllium. The results and their standard deviations are listed in Table I. All three metals were described by a linear relationship between shock and particle velocity

$$U_{s1} = C_0 + s_0 u_{p1} \quad (1)$$

The coefficients here depend on the particular materials [1]. The free-surface velocity, which is usually very close to twice the particle velocity, can be combined with the shock velocity to yield an experimental point on the principal Hugoniot. For this analysis, however, equation (1), which represents a fit to many such points, will be used in conjunction with the shock-velocity measurement to describe the state achieved. The changes in various thermodynamic quantities across the shock are given by the Rankine-Hugoniot relations

$$P_n = P_{n-1} + u_{pn} U_{sn}/V_{n-1}, \quad V_n = V_{n-1}(1 - u_{pn}/U_{sn}), \text{ and}$$

$$E_n = E_{n-1} + (P_n + P_{n-1})(V_{n-1} - V_n)/2 \quad (2)$$

For the initial shock, n is 1. A flash radiograph was taken with the Phermex machine [2] 0.60 ± 0.03 μs after the incident shock reflected from the copper and uranium-238. The positions of the reflected shocks in the beryllium, and hence their velocities in the laboratory frame of reference, were determined from the radiograph. The reflected-shock velocity is related to that velocity by

$$U_{s2} = D_s + u_{p1} \quad (3)$$

The remainder of the description of the reflected state can be obtained by comparing the pressure with that in the reflector through standard impedance-matching techniques [3]. The graphical solution for this is shown in Fig. 1, where the solid lines are principal Hugoniots and the circles represent the various states achieved in the beryllium. The final state, which is described by (2) with n equal to 2, can be compared to the principal Hugoniot through the Mie-Grüneisen equation,

$$P_n - P = (E_n - E)\,\gamma[V]/V, \quad n > 1 \quad (4)$$

Here, subscripted quantities refer to a state that is not on the reference curve - the principal Hugoniot - and the others refer to one that is. In this formulation, γ is considered to be a function only of volume. To evaluate the Grüneisen parameter it is merely necessary to use (1) and (2) to determine what pressure and energy

Table I. Incident and Reflected States in Be

	Be		Cu		^{238}U	
ρ_0, Mg/m^3	1.851		8.93		18.954	
C_0, km/s	8.005		3.94		2.47	
s_0	1.119		1.489		1.552	
State in Be	Incident		Reflected		Reflected	
U_s, km/s	13.12	$\pm$ 0.08	13.35	$\pm$ 0.75	14.12	$\pm$ 0.66
D_{fs}, km/s	9.06	$\pm$ 0.15	NA		NA	
u_p, km/s	4.57	$\pm$ 0.07	1.94	$\pm$ 0.04	2.56	$\pm$ 0.05
P, GPa	111.0	$\pm$ 2.4	185	$\pm$ 5	214	$\pm$ 6
V, m^3/Mg	0.3520	$\pm$ 0.0018	0.301	$\pm$ 0.004	0.288	$\pm$ 0.004
γ, $[V_1]$	NA		0.99	+ 0.34 - 0.44	0.85	+ 0.28 - 0.35

NA - not applicable

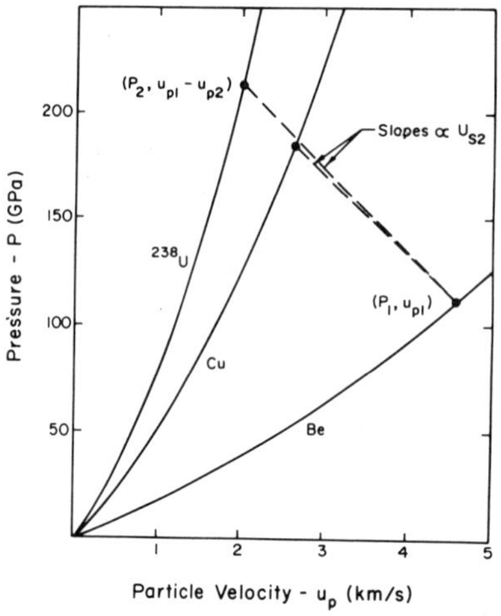

Fig. 1. Graphical impedance-match solution for reflected shocks in Be.

on the principal Hugoniot correspond to a compression to V. The final results are listed in Table I.

BULK SOUND SPEED MEASUREMENT

If a shock impinges obliquely on an interface, an oblique reflected wave will occur. The flow pattern which is discussed in detail elsewhere [4], is depicted in Fig. 2 for the case of a rarefaction. If the reflected wave is a shock, the velocity is given by

$$U_{s2} = U_{s1} (\sin \beta_1/\sin \alpha_1) + u_{p1} \cos (\alpha_1 + \beta_1) \qquad (5)$$

This same equation also describes the velocity of the head of a rarefaction wave. If elastic effects are small and can be neglected, this velocity corresponds to C_1, the bulk sound speed in the preshocked material. In this experiment a shock was induced in the beryllium by detonating the explosive, Composition B-3, that was in contact with it. This shock then obliquely intersected a beryllium-air interface. The incident shock was characterized by a measurement of the shock velocity made with piezoelectric timing pins. The

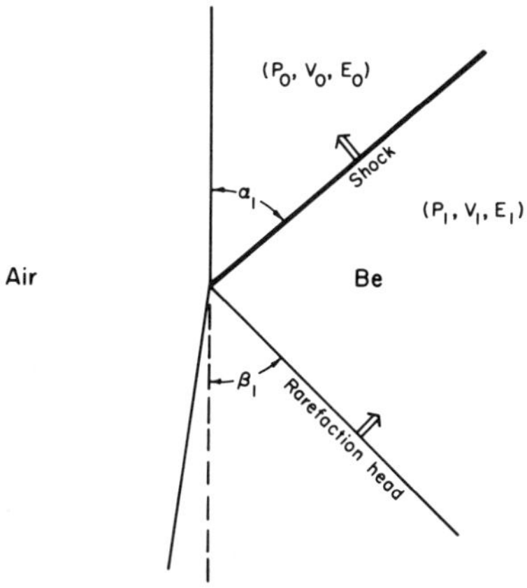

Fig. 2. Schematic of a shock obliquely intersecting a material surface.

collision angle α was established by design of the experiment. A flash radiograph permitted the reflected angle β to be measured. The results of this experiment are listed in Table II. The slope of the isentrope through the shock-compressed state (P_1, V_1, E_1) is related to the bulk sound speed by

$$(\partial P_1/\partial V_1)_S = -(C_1/V_1)^2 \qquad (6)$$

The Mie-Grüneisen equation, equation (4), can be used to relate the slope of the isentrope at this point to the slope of the principal Hugoniot. The result [3] expressed as a function of γ is

$$\gamma[V_1] = 2V_1[(\partial P_1/\partial V_1)_H - (\partial P_1/\partial V_1)_S] \Big/ [P_1 + (\partial P_1/\partial V_1)_H\,(V_0 - V_1)] \qquad (7)$$

Thus the measurement of C_1 also permits γ to be evaluated. That result is presented in Table II.

Table II. Bulk Sound Speed in Be

α, deg	60	β, deg	72.05 ± 0.29
U_{s1}, km/s	9.97 ± 0.04	C_1, km/s	9.78 ± 0.02
P_1, GPa	32.4 ± 0.7	γ [V_1]	1.42 ± 0.08
V_1, m^3/Mg	0.445± 0.002		

SINGLE AND DOUBLE REGULAR REFLECTION EXPERIMENTS

The collision of two shocks to form regular reflection is another method of multistage shock compression that lends itself to radiographic techniques. Regular reflection occurs when two shocks collide with an angle that is less than some critical angle. For solids, this phenomenon is well known and understood [5,6]. Only the case in which two shocks of the same strength collide in the same material will be considered here. The shock configuration is illustrated in the top portion of Fig. 3a. Consider only those shocks for which the angles are labeled. The circled integers indicate the number of times material in a particular region has been shock-compressed, and the small double arrows indicate the direction of motion for the shocks. The shock velocity for the second shock is given by (5), and the particle velocity is given by

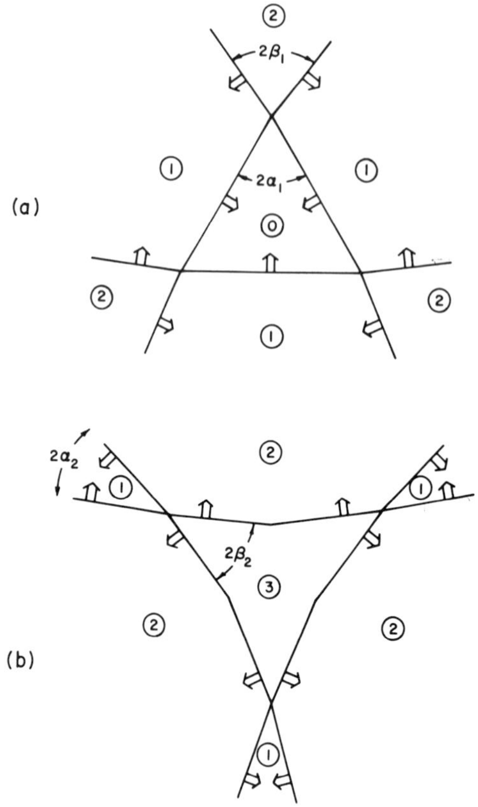

Fig. 3. Creation of double regular reflection.

$$u_{p2} = u_{p1} \cos \alpha_1 / \cos \beta_1 \quad (8)$$

Again, characterization of the initial shock and determination of the collision and reflection angles permit the second-shock state to be completely specified. The two exper ments conducted were identical to that described in the literature [3] Shocks were induced by detonating tw columns of explosive that were in contact with the beryllium. The explosive Composition B-3 was used in the first experiment. In the second, PBX-9404 was used. Measurements were made in the same manner as in the case of the bulk sound speed. The results are listed in Table III.

A three-stage compression may be achieved in a similar fashion using the same phenomenon. This technique was first reported at the 5th AIRAPT International High Pressure Conference and has since been described by Neal [3]. Three plane shocks are caused to converge as illustrated in Fig. 3a. Three separate cases of regular reflection occur as the shocks are moving inward. Eventually the original reflected shocks will themselves collide as shown in Fig. 3b. The angles α_2 and β_2 thus refer to the second occurrence of regular reflection. The system was designed to have threefold rotational symmetry. The two cases of reflection are related by

$$\alpha_2 = 60^\circ - \beta_1 \quad (9)$$

at the time of convergence. Equations (5) and (8) describe the shoc and particle velocities for the second occurrence if all subscripts are simply incremented by one. A flash radiograph corresponding to Fig. 3b is taken as soon after convergence as the angle β_2 can be resolved. The results for these experiments, which were conducted as an extension of the single regular reflection experiments, are also presented in Table III.

CONCLUSIONS

In shock-wave work a simple approximation that works well for metals assumes that the ratio of the Grüneisen parameter to the

Table III. Regular Reflection in Be

Explosive	Composition B-3	PBX-9404
α_1, deg	30	30
U_{s1}, km/s	9.97 ± 0.04	10.31 ± 0.04
P_1, GPa	32.4 ± 0.7	39.3 ± 0.8
V_1, m^3/Mg	0.445 ± 0.002	0.432 ± 0.002
β_1, deg	32.39 ± 0.31	32.06 ± 0.22
U_{s2}, km/s	11.49 ± 0.10	11.91 ± 0.08
u_{p2}, km/s	1.80 ± 0.04	2.11 ± 0.04
P_2, GPa	78.8 ± 2.1	97.3 ± 2.4
V_2, m^3/Mg	0.376 ± 0.002	0.356 ± 0.002
γ $[V_2]$	1.18 ± 0.17	1.26 ± 0.08
β_2, deg	28.46 ± 0.33	28.50 ± 0.17
U_{s3}, km/s	12.82 ± 0.26	13.29 ± 0.19
u_{p3}, km/s	1.81 ± 0.04	2.12 ± 0.04
P_3, GPa	140.7 ± 4.4	176.3 ± 5.0
V_3, m^3/Mg	0.322 ± 0.003	0.299 ± 0.003
γ $[V_3]$	0.96 ± 0.16	1.00 ± 0.07

specific volume remains constant during compression. The results of the various experiments on beryllium are displayed in Fig. 4 in terms of that ratio. The experiments that involved a high-impedance reflector were not as precise as the others. The weighted average value of γ/V is indicated as a dashed line. For all of the higher compression experiments, the determination of γ involved a comparison of an experimentally achieved state with one of the same compression on the principal Hugoniot. Experimental data for the principal Hugoniot, however, are not available for volumes less than 0.35 m^3/Mg. Thus the four highest compression results are really determined from a comparison with the extrapolation of the experimental principal Hugoniot. Fortunately, the principal Hugoniots for many metals remain linear, even at the compressions achieved in these experiments. If future experimental work indicates that this is not the case for beryllium, those results for γ will have to be altered.

The thermodynamic value for γ, as given by Guinan and Steinberg, is also indicated in Fig. 4[7]. None of the shock experiments appear to agree with that result. In fact, extrapolation of the ratio model back to standard density indicates a value of 1.8 for γ.

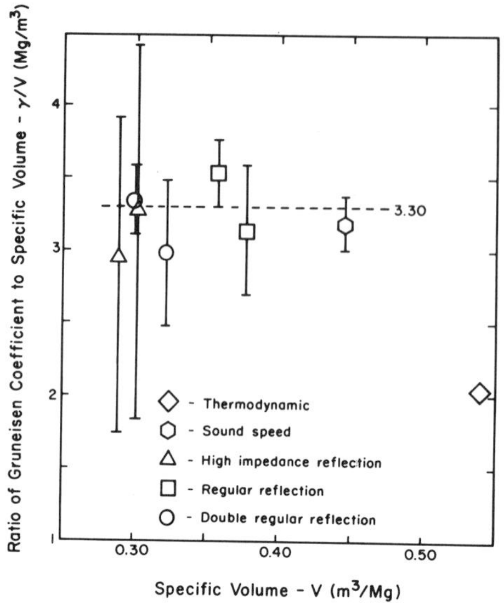

Fig. 4. Results of measurements of the Grüneisen parameter.

Even considering the uncertainties involved, this is quite different from the thermodynamic value of about 1.1. The discrepancy might, in some small way, be due to the difficulties in establishing a thermodynamic value for a material that has exhibited the anisotropic effects seen in beryllium [8]. For the most part, however, it is probably the result of the phase change from hcp to bcc that occurs at 1264°C for atmospheric pressure and at lower temperatures for higher pressures [9]. The phase diagram is pictured in Fig. 5. The solid line is that portion of the phase boundary that has been measured. The dashed lines are the conjectured upper and lower bounds for the extension of that phase line. One is a line of constant slope. The other has been constructed to intersect room temperature at 9.3 GPa, a state for which a discontinuity has been observed in the electrical resistance but not in the x-ray diffraction pattern [10]. In either case the measurements cited here appear to have been taken in the bcc phase; and above 30 GPa the beryllium equation-of-state can be satisfactorily described by the approximation of γ/V having the constant value of 3.30.

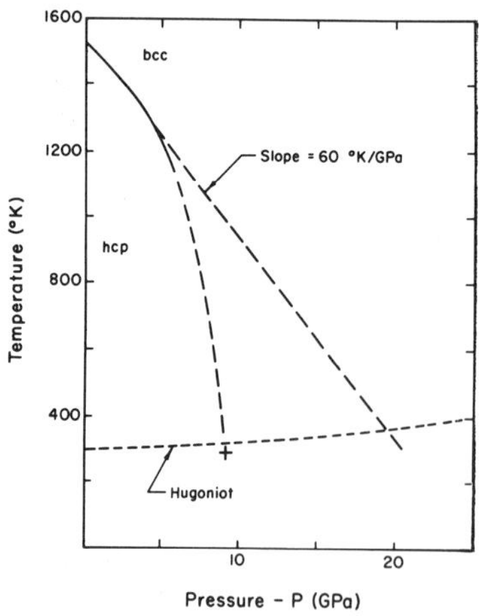

Fig. 5. Phase diagram for Be.

ACKNOWLEDGMENT

The author wishes to thank M. George, who designed and executed the pin measurements for the flyer experiments.

NOTATION

C = bulk sound speed

D_{fs} = free-surface velocity

D_s = shock velocity in laboratory frame of reference

E = specific energy

H = Hugoniot function $[E-E_0+(P+P_0)(V-V_0)/2]$

n = number of times shocked

P = pressure

s_0 = material constant

S = entropy

u_p = particle velocity

U_s = shock velocity

V = specific volume

α = collision angle

β = reflection angle

γ = Grüneisen parameter

ρ = density

REFERENCES

1. T. Neal in Proceedings - Sixth Symposium (International) on Detonation, Cat. No. ACR-221 (U.S. Govt. Printing Office, Washington, D.C., 1976).
2. D. Venable, Phys. Today 17, 19 (1964).
3. T. Neal, Phys. Rev. B 14, 5172 (1976).
4. T. Neal, J. Appl. Phys. 46, 2521 (1975).
5. L. V. Al'tshuler and A. P. Petrunin, Sov. Phys. - Tech. Phys. 6, 516 (1961).
6. T. Neal, J. Phys. Chem. Solids 38, 225 (1977).
7. M. W. Guinan and D. J. Steinberg, J. Phys. Chem. Solids 35, 1501 (1974).
8. L. E. Pope and J. N. Johnson, J. Appl. Phys. 46, 720 (1975).
9. C.W.F.T. Pistorius, Prog. Solid State Chem. 11, 1 (1976).
10. A. R. Marder, Science 142, 664 (1963).

A-9

HIGH-PRESSURE MULTIPLE-SHOCK RESPONSE OF ALUMINUM *

R. J. Lawrence and J. R. Asay

Sandia Laboratories
Albuquerque, New Mexico USA

INTRODUCTION

It is well known that both dynamic yield strength and rate-dependent material response exert direct influence on the development of surface and interface instabilities under conditions of strong shock loading. A detailed understanding of these phenomena is therefore an important aspect of the analysis of dynamic inertial confinement techniques which are being used in such applications as the generation of controlled thermonuclear fusion. In these types of applications the surfaces and interfaces under consideration can be subjected to cyclic loading characterized by shock pressures on the order of 100 GPa or more. It thus becomes important to understand how rate effects and material strength differ from the values observed in the low-pressure regime where they are usually measured, as well as how they are altered by the loading history.

To investigate these problems, both experimental and theoretical techniques have been used in the present study to examine the high-pressure multiple-shock response of a material previously well characterized at low pressures. Aluminum was chosen since both its rate dependence and plastic yielding behavior have been studied extensively in the past. For example, detailed interferometric measurements of complete shock-wave profiles were made in 6061-T6 aluminum alloy by Barker [1]. These data were obtained for shock pressures between 2 and 9 GPa and included the fine structure associated with both loading and unloading. Subsequently Herrmann [2] used these measurements to establish the validity of a strain-rate dependent, multi-element kinematical strain-hardening constitutive relation for the dynamic response of this material. More

*Work supported by the U.S. Energy Research and Development Administration under Contract AT(29-1)-789.

recently, Lipkin and Asay [3] studied both the release and reshock of 6061-T651 aluminum from an initial shock pressure of 2 GPa. Their results indicate that, although rate effects are probably present during shock loading, the fine structure in both the release and reshock waves can be adequately described with a rate-independent, distributed-yield state model.

The experiments performed and analyzed in the present work extend these previous investigations to higher pressures and for the first time include the effects of cyclic loading. The multiple-shock loading histories examined here consisted of an initial shock to states ranging from 8 to 20 GPa, unloading to essentially zero-longitudinal stress, and, finally, recompression from the unloaded state. The reloading shock-wave amplitudes ranged from approximately 3 to 6 GPa.

The next section of this paper describes the experimental configuration and presents the results which were obtained. A method for using these measurements to generate uniaxial-stress stress-strain curves appropriate for the pre-shocked material is also discussed. In the third section, the material constitutive relations used in the numerical calculations are presented. The remaining sections discuss the comparisons between theory and experiment and list the conclusions which were drawn from them. Since the experimental records contain evidence of spall in the aluminum specimens, a brief discussion of the dynamic fracture strength of this material is also presented.

EXPERIMENTAL CONFIGURATION AND RESULTS

To achieve the desired types of loading conditions, copper-backed aluminum disks were mounted on projectiles and impacted against aluminum targets of similar thicknesses. A powder gun [4] was used to accelerate the projectiles to impact velocities ranging from approximately 1 to 2 km/s. The aluminum impactors and targets were prepared from 1100 aluminum alloy plate stock that was first cold-rolled to 30% reduction in thickness and then annealed at 620 K for one-half hour. The specimens were about 75 mm in diameter and were flat on both surfaces to within a few bands of monochromatic light. In addition, the surfaces were parallel to within 0.025 mm. Similar tolerances were maintained for the copper plates. The projectiles were constructed by bonding the copper plates to the aluminum impactors with epoxy. A schematic of the experimental configuration is shown in Fig. 1.

The velocity histories associated with the rear free surfaces of the aluminum targets were measured with a diffuse surface velocity interferometer (VISAR) [5]. The accuracy of this system is about one percent in particle velocity with an overall time resolution of about 10 ns. In addition, the impact velocity was measured with shorting pins with an accuracy better than 0.3%.

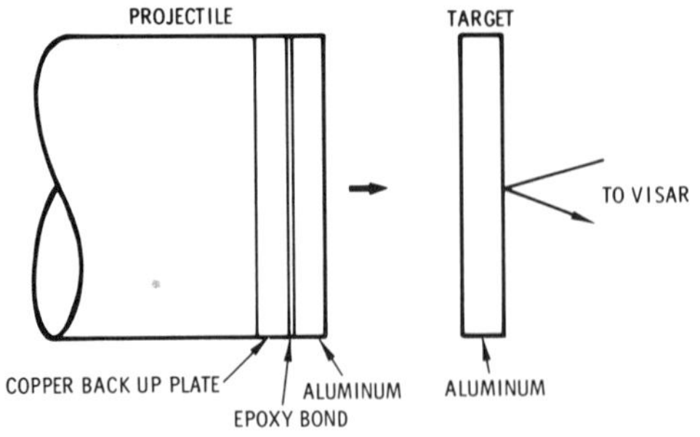

Fig. 1. Experimental configuration.

The parameters associated with the three experiments performed here are listed in Table I. The measured free-surface velocity histories are presented in Fig. 2.

To understand these experimental records, there are a number of wave interactions produced by the experimental configuration in Fig. 1 which should be discussed. These interactions are illustrated in Fig. 3. When the impact-generated shock, state 1, reaches the rear of the target, the free-surface velocity record exhibits an abrupt jump to the value associated with this shock. Because of the wave reflection at the free surface, the target is then unloaded to state 2. The second signal to reach the free surface is the recompression from state 2 to state 4, which originated with the reflection of state 1 from the high-impedance copper interface (producing state 3). This signal consists of an elastic precursor characterized by the current value of the yield strength, followed by a plastic wave. Approximately 1.2 μs later the spall signal, state 6, which originated with the interaction of the unloaded states 5, 7, and 8, arrives at the free surface, resulting in the small decrease in velocity evident at late times on the records.

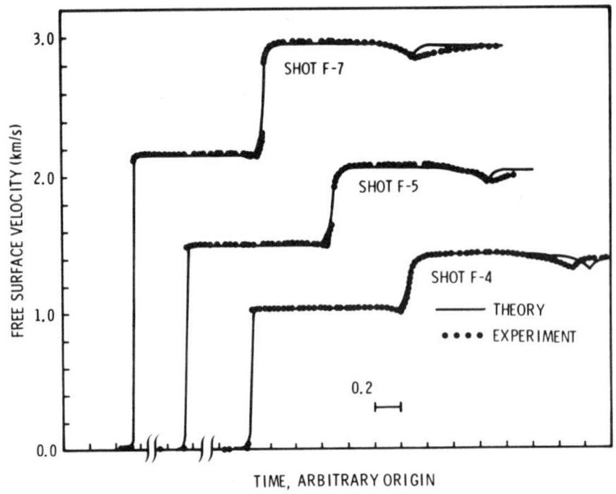

Fig. 2. Free-surface velocity histories for the three experiments conducted in this work. Both experimental measurements and final numerical calculations are shown.

The wave structure observed during recompression can be used to estimate the stress-strain paths achieved during the reloading, and thus provide an estimate of the dynamic yield response appropriate for preshocked aluminum. Although this approach is subject to some uncertainty, the results provide considerable guidance in choosing hardening parameters which can be subsequently used in the constitutive modeling. Basically, for the analysis of these data, it was assumed that the reloading wave (state 4 in Fig. 3) from the unloaded

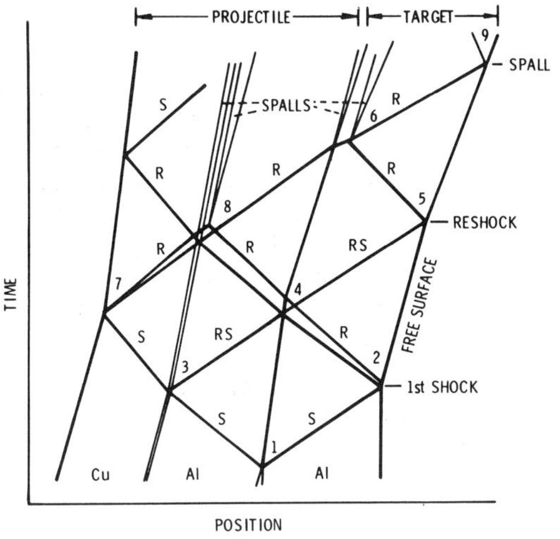

Fig. 3. Position-time diagram showing interactions involved in the experiments performed in this work. S indicates a shock from ambient conditions, R denotes a relief or rarefaction wave, and RS denotes a reshock from some state other than ambient.

state (state 2) was initially a sharp recompression. The velocity doubling approximation [6] was used to convert the measured free-surface velocities to in-situ particle velocities. The self-similar solutions [3] for one-dimensional wave propagation were then used to transform these particle-velocity histories to an estimate of stress and strain accomplished during reloading. Finally, the technique developed by Fowles [7] was used to convert the uniaxial strain results to the stress and strain obtained in a uniaxial-stress experiment. This was accomplished by comparing the longitudinal stress achieved during reloading with an estimated hydrostat, using the constitutive relation presented later.* Using this method of data reduction, an estimate of the dependence of yield stress on strain could be obtained. These results are presented in Fig. 4 for each of the three experiments.

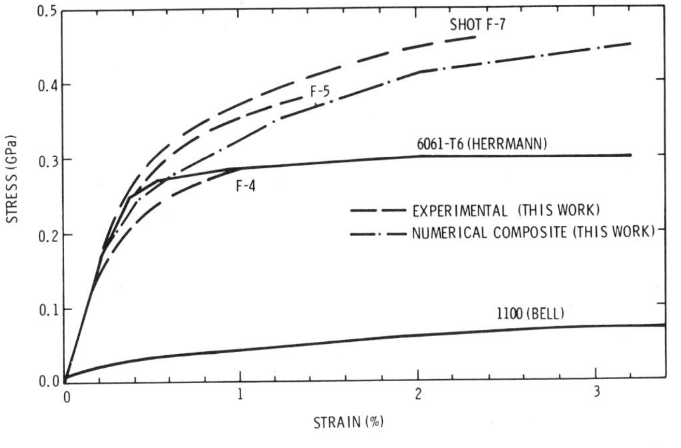

Fig. 4. Uniaxial-stress stress-strain curves for the aluminum studied in this work. The shot numbers refer to the experiments listed in Table I.

*It was assumed that the plastic profiles were steady which resulted in a slight correction in stress due to the fact that loading occurs along the Rayleigh line.

NUMERICAL AND CONSTITUTIVE MODELING

The numerical calculations of the free-surface velocity histories were performed with the one-dimensional, Lagrangian, finite-difference computer program WONDY [8]. This program solves the relations expressing conservation of mass, momentum, and energy in the presence of the appropriate material constitutive equations. The code is capable of treating many independent material layers and allows for a wide variety of boundary and initial conditions, including such complexities as impact, rebound, and spall fracture. Pressure and particle velocity discontinuities associated with shock waves are treated with the method of artificial viscosity.

The material constitutive relations describe the longitudinal stress as a function of strain and internal energy. For the condition of uniaxial strain, the longitudinal stress σ is separated into a spherical component p and a deviator σ',

$$\sigma = p + \sigma' \qquad (1)$$

It is assumed here that all stresses and strains are positive in compression. The pressure response is described by an equation of state of the Mie-Grüneisen form,

$$p - p_H = \Gamma_o \rho_o \left(\mathcal{E} - \mathcal{E}_H\right) \qquad (2)$$

where the subscript 'H' denotes the shock Hugoniot. The Hugoniot is described by

$$p_H = \frac{\rho_o c_o^2 \, \varepsilon}{(1 - s\,\varepsilon)^2} \quad ; \quad \mathcal{E}_H = \frac{p_H \varepsilon}{2\rho_o} \quad ; \quad \varepsilon = 1 - \frac{\rho_o}{\rho} \qquad (3)$$

where the subscript 'o' indicates the reference state. The stress deviator is obtained from a rate-dependent elastic relation of the Maxwellian form

$$\dot{\sigma}' = 2\mu \left(\dot{\varepsilon}' - g\right) \qquad (4)$$

where a dot denotes a time derivative and the shear modulus μ is related to the bulk modulus K in the usual way,

$$\mu = \frac{3}{2}\,\frac{(1 - 2\nu)}{(1 + \nu)}\,K \; ; \; K_o = \rho_o c_o^2 \qquad (5)$$

In one-dimensional strain, the strain-rate deviator $\dot{\varepsilon}'$ reduces to $2/3(\dot{\rho}/\rho)$.

Rate dependence is introduced through the relaxation function g. Here g takes the form

$$g = \frac{1}{\tau}\left(\frac{\sigma' - \sigma'_{eq}}{\sigma^*}\right)^m \qquad (6)$$

Table I. Experimental Parameters

Shot No.	Flyer Dimensions, (Cu/Epoxy/Al) mm	Target Thickness, mm	Impact Velocity, km/s*	Impact Stress, GPa§	Reshock Stress in Target, GPa§
F-4	2.929/0.0046/3.785	3.988	1.039	8.6	3.2
F-5	2.922/0.0016/3.819	3.550	1.497	12.9	4.6
F-7	2.916/0.0027/3.678	3.744	2.151	19.9	6.5

* Measured.

§ From numerical calculations.

Table II. Material Properties

	ρ_o, Mg/m^3	c_o, km/s	s	Γ_o	ν	τ, μs	Yield Strength*
Copper [6]	8.93	3.94	1.489	1.99	0.345	--	Elas. - Perf. plastic Y = 25 MPa
Epoxy [10]	1.194	2.64	1.66	1.0	--	--	-----
Aluminum	2.70	5.25	1.35†	2.0	0.34†	0.1†	Kin. strain hard.†§
			1.40§		0.35§	<< 0.1§	Isotrop. work hard.**

* See Fig. 4 for the various uniaxial-stress stress-strain curves.
† 6061-T6; the additional parameters in Eq. (6) are * = 0.25 GPa, m = 1.6 [2].
§ This work.
**1100 [9].

which Herrmann [2] found to be appropriate for describing the response of 6061-T6 aluminum at relatively low pressures. The equilibrium stress deviator is assumed to result from a multi-element kinematical strain-hardening model in which

$$\sigma'_{eq} = \sum_{j=1}^{n} a_j \sigma'_j \quad (7)$$

where the a_i are a set of normalized weighting factors. Each elemental stress deviator σ'_i is obtained from a rate-independent elastic relation subjected to a von Mises yield condition

$$(\sigma'_j)^2 \leq (\frac{2}{3} Y_j)^2 \quad (8)$$

Utilizing the transformation between uniaxial stress and uniaxial strain [7], the a_i and Y_i can be generated from conventional uniaxial stress measurements [8] such as illustrated in Fig. 4. This model thus represents a superposition of n elastic-perfectly-plastic elements, which has been previously shown to provide a Bauschinger effect on unloading [2]. This response is in contrast to isotropic hardening models which produce no Bauschinger effect.

In addition to the stress-strain curve generated from the present experiments, calculations were also performed using the yielding behavior previously reported for 1100 [9] and 6061-T6 aluminum [2]. The former work applied to low strain-rate response and was interpreted as isotropic work hardening, while the latter work used the multi-element kinematical hardening model described above. The stress-strain curves associated with these two studies are included in Fig. 4.

To provide complete calculations of the experimental records, constitutive models must be provided for both copper and epoxy, as well as aluminum. For these calculations the copper was described according to an elastic-perfectly-plastic model and the epoxy was assumed to behave hydrodynamically. The appropriate parameters for all of these materials are summarized in Table II.

COMPARISON BETWEEN THEORY AND EXPERIMENT

The Hugoniot for aluminum at ambient conditions is well established. Hence there is no difficulty in obtaining numerical agreement with the experiments for the first shock wave. It is only at the arrival of the recompression wave that the differences in the constitutive relations significantly influence the details of the calculated free-surface velocity histories. In fact, it is the fine structure in the recompression waves which provides the basis for distinguishing between the various models discussed earlier. Preliminary analysis indicated that Herrmann's original values for the hydrodynamic parameters [2] would result in calculated arrival times for the plastic portions of the recompression waves which lagged the measured values by small amounts proportional to the final amplitudes. These arrival times depend on the wave velocities at different pressures during three transits of the target and two transits of the aluminum impactor, and thus it was found that a 4% increase in the value of the parameter s in (3) would correct for this discrepancy. This adjustment in s is consistent with the high-pressure value listed by McQueen, et al. [6]. To maintain agreement with the measured elastic-wave velocities in the pre-compressed state, it was also necessary to adjust Poisson's ratio upward by a similar factor.

The initial set of calculations employed the adjusted hydrodynamic parameters with the rate-independent isotropic work-hardening curve reported by Bell [9] for 1100 aluminum. Significant differences were noted between these particular calculations and the experimental measurements. Specifically, the isotropic hardening response did not adequately reproduce the shape of either the elastic or plastic portions of the recompressive wave. It was also found that the incorporation of rate dependence did not materially improve the agreement. Therefore, it was concluded that isotropic hardening did not describe the experimental results, and the kinematical-hardening model was investigated more thoroughly.

To obtain agreement with the experimental results using this latter model it was found that a stress-strain curve more nearly like the 6061-T6 relation shown in Fig. 4 was required. This conclusion is consistent with the experimentally-determined stress-strain curves which are also shown in Fig. 4. The experimental results also suggest that during recompression the material exhibits a slightly softer initial response, but ultimately a greater degree of strain hardening than the 6061 alloy. The resulting composite curve, which was constructed for the preshocked 1100 aluminum and used in the final numerical calculations, is also plotted in the figure.

Calculations incorporating these latter parameters in the constitutive model produced the results presented in Fig. 2. For all three records the differences between the numerical calculations and the experiments are on the order of the experimental uncertainty. Even the small dips in the records just before the arrival of the recompression waves, which are due to the finite thickness of the epoxy bond, are reproduced reasonably well.

One of the major questions posed at the outset of this study involved the importance of rate dependence in a high-pressure multiple-shock situation. Although it has been possible to match the experiments with rate-independent calculations as illustrated in Fig. 2, it would be useful to establish actual limits on the rate dependence of the material under these conditions. Additional calculations using the final aluminum parameters outlined above, but incorporating rate dependence, indicate that agreement as good as the rate-independent results shown in Fig. 2 requires relaxation times in (6) which are considerably shorter than 0.1 μs. Since numerical difficulties impede calculations with shorter relaxation times, it was not possible to reduce further this upper limit on the rate dependence. This lack of rate dependence is not due to temperature-induced melting phenomena, since the calculated temperature rise resulting from the first shock is quite low, ranging from ~20 K for Shot F-4 to less than 100 K for Shot F-7. Lastly, it should be noted that this lack of rate dependence is consistent with the low-pressure results reported by Herrmann [2], which indicate the decreasing importance of rate effects with increasing shock pressure. It is thus concluded that rate effects, as described by (6), must have relaxation times less than 0.1 μs for cyclic loading of 1100 aluminum alloy in the 8 to 20 GPa range.

Finally, additional information can be gained from numerical calculations of the spall-induced pull-back signals exhibited in each of the experimental records. Although it is generally accepted that dynamic fracture does not obey a simple tensile failure criterion [11,12], a single tensile spall strength of 1.5 GPa matches the data very closely for all three experiments.* This value is greater than that often quoted for the softer aluminum alloys [13], but is consistent with the strain hardening postulated here and elsewhere [11] for aluminum.

CONCLUSIONS

In summary, the present work has allowed observations of material response under loading conditions and for pressure regimes which have not previously been studied. For cyclic shock loading of 1100 aluminum alloy to peak stresses of 20 GPa, it was found that the unloaded material exhibits elastic properties which differ considerably from the initial elastic response. In particular, the elastic behavior of this alloy in the shock-deformed state is more nearly represented by that of the harder alloy, 6061-T6. This conclusion is also consistent with the work of Barnes et al. [14] who found that similar yield strengths were necessary to reproduce observed instability growths under dynamic loading in these same two materials. Thus, there is substantial support to the hypothesis that this particular alloy retains little memory of its initial yield properties after the gross plastic deformations that occur in shock-wave experiments.

It was also found in this work that rate-dependent effects in cyclic shock-loaded 1100 aluminum are minimal. This conclusion was derived by comparing numerical calculations based on a Maxwellian description of the rate dependence to the measured recompression profiles. It is estimated that the effective relaxation time in the preshocked state is considerably less than 0.1 μs.

A final comment is appropriate regarding the uniqueness of the model used here. It is well known that different constitutive

*The discrepancy between the calculated and experimental arrival times of the pull-back signal in Shot F-4 is not currently understood

relations can be used to describe experimental data over limited ranges, and therefore it is possible that a strong rate dependence in combination with different yield response could provide similar agreement to the present data. Although such behavior cannot be ruled out from the present data, the approach taken here has been to describe all of the data self-consistently with the simplest possible model. In this respect it is felt that the present description serves to emphasize the phenomenological features associated with the yield response and rate dependence which are most important. In order to determine whether this description applies over a much broader range of pressures and strain-rates, it would be necessary to perform a large number of additional experiments.

ACKNOWLEDGMENT

The authors would like to thank D. Baldwin for his technical assistance in performing the experiments.

NOTATION

a_j, Y_j = kinematical hardening parameters
c_o = bulk sound speed
$\mathcal{E}^o$ = specific internal energy
K = bulk modulus
p = pressure
s = slope of Hugoniot in the shock-velocity particle-velocity plane
Y = yield strength in uniaxial stress
Γ = Grüneisen parameter
$\varepsilon, \varepsilon'$ = strain, strain deviator
μ = shear modulus
ν = Poisson's ratio
ρ = mass density
σ, σ', σ'_{eq} = longitudinal stress, stress deviator, equilibrium stress deviator
σ^*, m = material relaxation constants
τ = relaxation time

REFERENCES

1. L. M. Barker, in Proc. of Symposium on Behavior of Dense Media Under High Dynamic Pressures, Gordon and Breach, New York (1968), p. 483.
2. W. Herrmann, "Development of a High Strain Rate Constitutive Equation for 6061-T6 Aluminum," Sandia Laboratories Rept. SLA-73-0897, Albuquerque, New Mexico (1974).
3. J. Lipkin and J. R. Asay, J. Appl. Phys. 48, 182 (1977).
4. D. E. Munson and R. P. May, AIAA 14, 235 (1976).
5. L. M. Barker and R. E. Hollenbach, J. Appl. Phys. 43, 4669 (1972).
6. R. G. McQueen, S. P. Marsh, J. W. Taylor, J. N. Fritz, and W. J. Carter, in High-Velocity Impact Phenomena, R. Kinslow, ed., Academic Press, New York (1970), p. 293.
7. G. R. Fowles, J. Appl. Phys. 32, 1475 (1961).
8. R. J. Lawrence and D. S. Mason, WONDY IV - A Computer Program for One-Dimensional Wave Propagation With Rezoning, Sandia Laboratories Rept., SC-RR-710284, Albuquerque, New Mexico (1971).

9. J. F. Bell, J. Mech. Phys. Solids 14, 309 (1966).
10. D. E. Munson and R. P. May, J. Appl. Phys. 43, 962 (1972).
11. F. R. Tuler and B. M. Butcher, Int. J. Fracture Mech. 4, 431 (1968).
12. L. Davison, A. L. Stevens, and M. E. Kipp, J. Mech. Phys. Solids 25, 11 (1977).
13. B. M. Butcher, in Proc. of Symposium on Behavior of Dense Media Under High Dynamic Pressures, Gordon and Breach, New York (1968), p. 245.
14. J. F. Barnes, P. J. Blewett, R. G. McQueen, K. A. Meyer, and D. Venable, J. Appl. Phys. 45, 727 (1974).

A-10

VOLUME DEPENDENCE OF THE GRÜNEISEN COEFFICIENT FOR ALUMINUM

J. P. Romain, A. Migault and J. Jacquesson

Université de Poitiers
Poitiers Cédex, France

INTRODUCTION

The volume dependence of the Grüneisen coefficient γ has been the subject of numerous theoretical approaches. Commonly used formulations are those of Slater [1], Dugdale-Mac Donald [2], and Vashchenko-Zubarev [3]. More recently, Migault [4,5] has proposed a generalization of these theories, and this new formulation was shown [6] to be consistent with both shock data and the pressure dependence of the Poisson ratio measured by ultrasonic techniques under static pressure for several elements.

In this paper we first discuss the validity of these theories with respect to available experimental data on the volume dependence of γ for aluminum. We also consider the procedure developed by Pastine and Forbes [7] and by O'Keeffe and Pastine [8]. This procedure is applied to aluminum and compared with the results of other theories and with experimental data. The formulation proposed by Migault is shown to have the best agreement with experimental data. Finally, the possibility of predicting the pressure dependence of the elastic moduli with use of this formulation for γ is examined.

FORMULATION FOR THE VOLUME DEPENDENCE OF THE GRÜNEISEN COEFFICIENT

The Mie-Grüneisen equation of state is generally expressed in the form

$$P - P_K = \frac{\gamma}{V} (E - E_K) \qquad (1)$$

where P is the pressure, V the volume, E the internal energy, subscript K refers to 0 K, and γ is the Grüneisen coefficient. The volume dependence of γ according to the theories of Slater (S), Dugdale-Mac Donald (D-M) and Vashchenko-Zubarev (V-Z) may be represented by the relation

$$\gamma = \frac{3p-4}{6} - \frac{V}{2} \frac{d^2(P_K V^p)/dV^2}{d(P_K V^p)/dV} \tag{2}$$

where p is a parameter, whose value is fixed for all solids and depends on the theory used: p = 0(S), p = 2/3 (D-M) or p = 4/3 (V-Z).

The formulation of Migault (M) is also represented by (2), but under the assumption that p is a parameter characteristic of each material. For a given material, the value of p may be determined by comparison between the experimental Hugoniot curve and a calculated curve $P_H(V)$ derived from the Mie-Grüneisen equation of state and the relation expressing the conservation of energy through the shock

$$P_H = \frac{P_K + \frac{\gamma}{V}(E_0 - E_K)}{1 - \frac{\gamma}{2}(\frac{V_0}{V} - 1)} \tag{3}$$

where V_0 and E_0 represent the volume and energy in the initial state at zero pressure. Assuming that the 0 K isotherm $E_K(V)$ may be described by a Morse potential given by

$$E_K\ (V) = A\ [\exp\ [2a(1 - \frac{V}{V_0})^{1/3}] - 2\exp\ [a(1 - \frac{V}{V_0})^{1/3}]] \tag{4}$$

then the Grüneisen coefficient, equation (2), depends on two constants A, a and on parameter p. Constants A and a may be determined from initial conditions at zero pressure, and setting $V = V_0$, by the relations

$$K_0 = \frac{2Aa^2}{9V_0} \tag{5}$$

$$\gamma_0 = \frac{1}{3} + \frac{a}{2} - \frac{p}{2} \tag{6}$$

Experimental values of the bulk modulus K_0 are known for most materials. For aluminum an average experimental value obtained by different authors [13-15] is 761 kbar. For a given value of parameter p, the constants A and a may be evaluated by means of (5) and (6), the initial value γ_0 of the Grüneisen coefficient being adjusted so that the calculated Hugoniot curve, equation (3), best fits the experimental results. The value of γ_0 obtained by this procedure may then be compared with the thermodynamic Grüneisen coefficient derived from the bulk modulus K_0, the coefficient of thermal expansion α_0 and the specific heat at constant volume C_v using

$$\gamma_0 = \frac{\alpha_0 K_0 V_0}{C_V} \tag{7}$$

For aluminum we have $\gamma_0 = 2.18$ from Gschneidner's data [9]. Conversely, if the $\gamma(V)$ curve is forced to pass through the thermodynamic γ_0 at zero compression state, then parameter p instead of coefficient γ_0 may be evaluated by the same procedure and the value of p so obtained may be compared with the values which characterize the $\gamma(V)$ formulations of Slater, Dugdale, MacDonald and Vashchenko-Zubarev.

In an earlier publication [6] it was shown that p may be calculated also from the pressure derivative of the Poisson ratio and both values of p determined by these independent ways were found in reasonable agreement for several metals. For aluminum, the value of p obtained from shock data is -0.25. This negative parameter corresponds to a decrease in the Poisson ratio with increasing pressure and is confirmed by ultrasonic measurements under static pressure. Previous theories are not consistent with such a change in the Poisson ratio; rather, they predict it to be either constant (S) or an increasing function of pressure (D-M) and (V-Z).

In the relationship developed by Pastine, the Grüneisen coefficient γ_P may be expressed in a series expansion up to a second order in compression by

$$\gamma_P\,(V) = \gamma_0 - \gamma_0'\,(1 - \frac{V}{V_0}) + \frac{\gamma_0''}{2}\,(1 - \frac{V}{V_0})^2 \tag{8}$$

where γ_0 represents the thermodynamic Grüneisen coefficient, equation (7). Relations for γ_0' and γ_0'' were established by Pastine et al. [7,8]. These relations contain parameters whose experimental values for aluminum may be found in the literature: From Gschneidner [9] we have $\alpha_0 = 69.3\ 10^{-6}$, $C_V = 5.48$ cal/g-at/K, and $C_P = 5.82$ cal/g-at/K; slope b of the shock velocity-particle velocity relationship found experimentally to be linear for aluminum [10] is 1.338. A relation for the temperature derivative of b needed in the calculations is established from Pastine [11]. Application of this relation involves the knowledge of the second order elastic constants c_{11}, c_{12} and their temperature derivatives. From the Simmons and Wang data compilation [12], we have:

$c_{11} = 1066$ kbar, $c_{12} = 605$ kbar, $\partial c_{11}/\partial T = -\ 0.364$ kbar/K,

$\partial c_{12}/\partial T = -\ 0.066$ kbar/K. The pressure derivatives of c_{11} and c_{12} are also needed. For these we used average values of data given by Ho et al. [13] and Schmunk et al. [14]: $\partial c_{11}/\partial P = 6.97$, $\partial c_{12}/\partial P = 3.83$. This permitted us to obtain $\gamma_0' = 4.27$ and $\gamma_0'' = 5.22$. The volume dependence of γ could be then determined from (8).

COMPARISON WITH EXPERIMENTAL RESULTS

Determinations of the Grüneisen coefficient at various compressions can be obtained by shock experiments. A review of these techniques was recently presented by Neal [16]. These are: (1) The use of porous samples allowing the determination of Hugoniots for materials in which the initial density has been decreased below normal; (2) the measurement of the bulk sound speed in shock compressed materials; and (3) the determination of second Hugoniots by a two-stage shock compression with regular or Mach reflection of an oblique incident shock on a material of higher shock impedance.

These techniques have been applied to aluminum and some aluminum alloys, whose behavior is not experimentally distinguishable from that of pure aluminum because they contain only small quantities of additional elements. The available data reviewed by Neal [16] are summarized in Fig. 1 and compared with the curves deduced from the different functional forms for γ. The uncertainties accompanying the experimental results are important and show that the volume dependence of γ cannot be obtained experimentally with any degree of precision, particularly at low compressions. However, these measurements can be used, in spite of their poor precision, to perform some verification of the theoretical formulations because the predicted change of γ is strongly dependent on the formulation used.

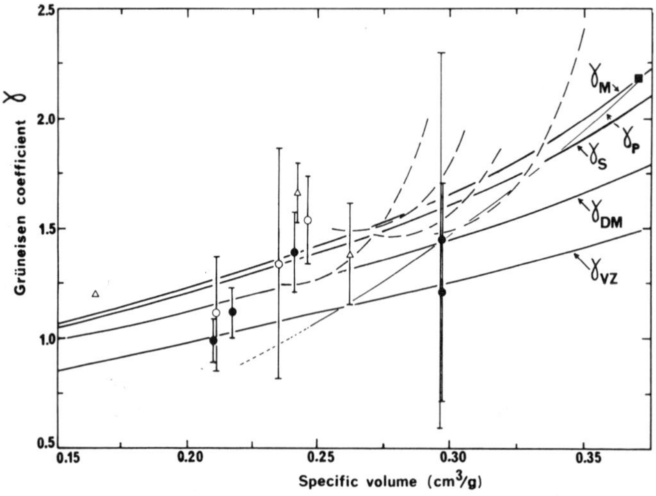

Fig. 1. Experimental [16] and calculated Grüneisen coefficient versus specific volume. Solid lines are calculated from the theories of Slater (γ_S), Dugdale-Mac Donald (γ_{DM}), Vashchenko-Zubarev (γ_{VZ}), Migault (γ_M) and Pastine (γ_P). Dashed lines represent least square fits of measurements made on porous samples of various initial densities. Legend: Point (●), circles (O) and triangles (Δ) are bulk sound speed, regular reflection and Mach reflection data, respectively. Square (■) is the thermodynamic value of γ.

Figure 1 shows that the γ_{DM}, γ_{VZ} and γ_P curves do not yield a good representation for the experimental results, whereas the Slater theory (γ_S) predicts a change for γ which is rather close to the one given by our formulation (γ_M) and which seems to be in better agreement with the experimental data. It may be noted that the calculated γ_P curve does not compare favorably with experiment, although the relations established by Pastine should give an accurate representation of γ. We believe that uncertainties in the evaluation of some parameters needed in the calculations are responsible for this disagreement. In addition to experimental uncertainties, it is possible that coefficient b of the shock velocity-particle velocity relationship does not yield an exact value for the ratio of the first and second derivatives of the Hugoniot curve at the initial state. This point has been discussed elsewhere [6].

Except for the γ_P curve which is forced to pass through the thermodynamic value γ_0 at zero compression state, the initial γ_0 values of the Grüneisen coefficient predicted by the other theories may be compared with the thermodynamic value. Figure 2 presents the variations of γ_0 vs. p and shows that γ_M (p = -0.25) yields the exact thermodynamic value of 2.18, whereas the other theories predict lower values: 2.06 for γ_S (p = 0), 1.76 for γ_{DM} (p = 2/3) and 1.48 for γ_{VZ} (p = 4/3).

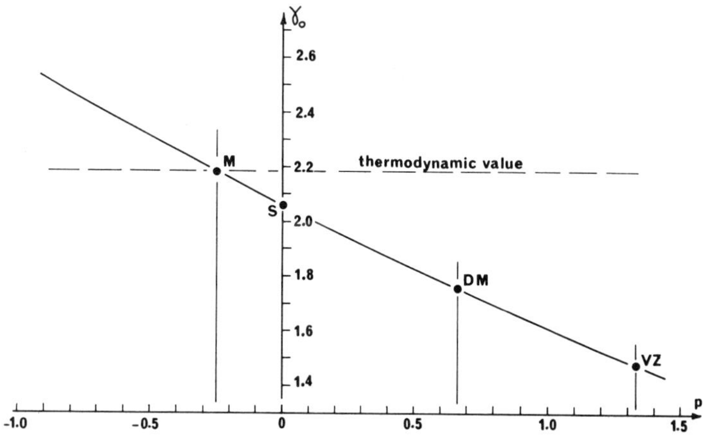

Fig. 2. Initial values of the Grüneisen coefficient γ_0 vs. p.

RELATION WITH THE PRESSURE DEPENDENCE OF THE ELASTIC MODULI

The initial pressure derivatives of the bulk modulus K_0', the shear modulus G_0' and the Poisson ratio σ_0' depend on the parameter p through the following relations:

$$K_0' = 2\,\gamma_0 + \frac{1}{3} + p \tag{9}$$

$$\sigma_0' = \frac{3p(1-\sigma_0^2)(1-2\sigma_0)}{2K_0\ (4-5\sigma_0)} \tag{10}$$

$$G_0' = \frac{G_0}{K_0} K_0' - \frac{2\ \sigma_0'\ (3\ K_0 + G_0)^2}{9\ K_0} \tag{11}$$

Equation (9) is derived from (2) at initial state (zero pressure). Equation (10) was established in a previous paper [6] and (11) is derived from the relation between K_0 G_0 and σ_0, namely

$$\sigma_0 = \frac{1}{2} \frac{3\ K_0 - 2\ G_0}{3\ K_0 + G_0} \tag{12}$$

The calculated values of K_0', G_0' and σ_0' depending on the theory used for the γ trough parameter p, are listed in Table I and compared with various experimental data obtained by ultrasonic measurements under static pressures. Figures 3, 4 and 5 present, respectively, the variations of K_0', G_0' and σ_0' vs. p. Values of the elastic moduli used in these calculations were : bulk modulus K_0 = 761 kbar, shear modulus G_0 = 261 kbar, and Poisson ratio σ_0 = 0.345.

Table I. Calculated and Experimental Pressure Derivatives of the Elastic Moduli for Aluminum

	P	γ_0	K_0'	G_0'	σ_0' 10^{-5} kbar^{-1}
Calculated pressure derivatives	-0.25	2.18	4.44	1.64	- 5.9
	0	2.06	4.45	1.53	0
	2/3	1.76	4.52	1.25	15.8
	4/3	1.48	4.63	0.99	31.5
Experimental pressure derivatives		[13]	5.03	2.06	-18.1
		[14]	5.19	2.01	-12.3
		[13]*	4.41	1.82	-15.6
		[15]	4.75	2.00	-22.4

*Attributed to Thomas.

We observe that the theoretical K_0' values are located in a range delimited by experimental data and are not strongly dependent on p. The reason for this is that the variations of p according to the theory used are almost balanced by opposite variations of γ_0. G_0' and σ_0' are decreasing and increasing functions of p, respectively. Their calculated values are out of the experimental range. However, the results obtained with p = -0.25 are not far from the experimental data and provide better agreement with them in comparison with those predicted from the Slater, Dugdale-Mac Donald and Vashchenko-Zubarev theories. We also note that only p = -0.25 yields a negative σ_0' in accordance with experimental variations of Poisson ratio. Figures 4

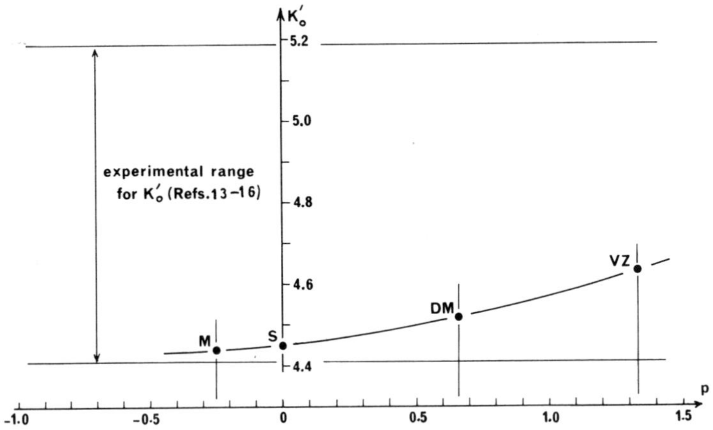

Fig. 3. Pressure derivative of the bulk modulus K_0' vs. p.

and 5 show that experimental values of G_0' and σ_0' would show better agreement with values of p near -0.7. The corresponding γ_0 deduced from the curve of Fig. 2 would then be near 2.4, which is somewhat higher than the thermodynamic value.

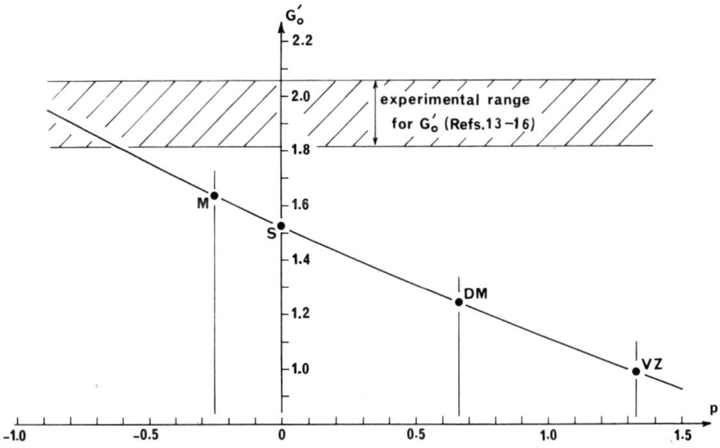

Fig. 4. Pressure derivative of the shear modulus G_0' vs. p.

CONCLUSION

Experimental results on the dynamic determination of the Grüneisen coefficient in aluminum have been compared in this paper to the changes predicted by the formulations of Slater, Dugdale Mac Donald, Vashchenko-Zubarev, Migault and Pastine. Using the Hugoniot curve of aluminum as reference, it was found that Migault's formulation yields the best agreement with experiment. The negative value of parameter p which characterizes Migault's formulation for the Grüneisen coefficient of aluminum also shows the best agreement with the thermodynamic γ.

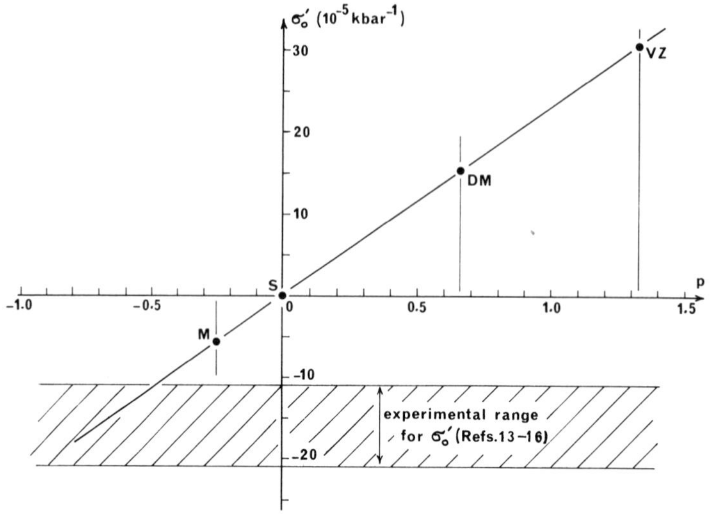

Fig. 5. Pressure derivative of Poisson ratio σ'_0 vs. p.

The consequences on the calculated pressure derivatives of the elastic moduli: bulk modulus, shear modulus and Poisson ratio were examined. Comparison of calculated values with experimental data from ultrasonic measurements under static pressure, also indicate a better agreement when Migault's formulation for γ is used. This generalization of the previous theories of Slater, Dugdale-Mac Donald and Vashchenko-Zubarev may be useful for predicting the pressure dependence of the elastic moduli of various materials.

NOTATION

- b = slope of the shock velocity-particle velocity relationship
- c_{ij} = second order elastic constants
- C_V = specific heat at constant volume
- C_P = specific heat at constant pressure
- E = internal energy
- E_K = internal energy at zero Kelvin temperature
- E_0 = internal energy at zero pressure and volume V_0
- G_0 = shear modulus
- G'_0 = pressure derivative of the shear modulus
- K_0 = bulk modulus
- K'_0 = pressure derivative of the bulk modulus
- P = pressure
- P_H = pressure on the Hugoniot curve
- P_K = pressure at zero Kelvin temperature
- V = specific volume
- V_0 = specific volume at zero pressure
- p = parameter
- α_0 = thermal expansion coefficient
- γ = Grüneisen coefficient
- γ_0 = initial value of the Grüneisen coefficient
- γ_{DM} = Grüneisen coefficient (Dugdale-Mac Donald)
- γ_M = Grüneisen coefficient (Migault)
- γ_P = Grüneisen coefficient (Pastine)
- γ_S = Grüneisen coefficient (Slater)

Notation (continued)

γ_{VZ} = Grüneisen coefficient (Vashchenko-Zubarev)
σ_0 = Poisson ratio
σ_0' = pressure derivative of Poisson ratio

REFERENCES

1. J. C. Slater, Introduction to Chemical Physics, McGraw-Hill Book Company, New York (1939).
2. J. S. Dugdale and D.K.C. MacDonald Phys. Rev. 89, 832 (1953).
3. V. Ya Vashchenko and V. N. Zubarev, Sov. Phys. Solid State 5(3), 653 (1963).
4. A. Migault, Le Journal de Physique 32, 437 (1971).
5. A. Migault, Le Journal de Physique 33, 707 (1972).
6. J. P. Romain, A. Migault and J. Jacquesson, J. Phys. Chem. Solids 37, 1159 (1976).
7. D. J. Pastine and J. W. Forbes, Phys. Rev. Letters 21(23), 1582 (1968).
8. D. J. O'Keeffe and D. J. Pastine, in Metallurgical Effects at High Strain Rates, R. W. Rhode, B. M. Butcher, J. R. Holland, C. H. Kannes, eds., Plenum Press, New York (1973).
9. K. A. Gschneidner, in Solid State Physics, F. Seitz and D. Turnbull, eds., Academic Press, New York (1964).
10. R. G. McQueen, S. P. Marsh, J. W. Taylor, J. N. Fritz and W. J. Carter, in High Impact Velocity Phenomena, R. Kinslow, ed., Academic Press, New York (1970).
11. D. J. Pastine, J. Geophys. Res. 75(35), 7421 (1970).
12. G. Simmons and H. Wang, Single Crystal Elastic Constants and Calculated Agregate Properties, A Handbook, M.I.T. Press, Cambridge, Massachusetts (1971).
13. P. S. Ho and A. L. Ruoff, J. Appl. Phys. 40, 3151 (1969).
14. R. E. Schmunk and C. S. Smith, J. Phys. Chem. Solids 9, 100 (1959).
15. F. F. Voronov and L. F. Vereschagin, Phys. Metall. Metallog. 11, 111 (1961).
16. T. Neal, Phys. Rev. B 14(12), 5172 (1976).

A-11

PRESSURE EFFECT ON COMPRESSIBILITY OF ALUMINUM SINGLE CRYSTALS FROM -196 TO 25°C*

N. J. Trappeniers, P. van't Klooster, and S. N. Biswas

Universiteit van Amsterdam
Amsterdam, The Netherlands

INTRODUCTION

A study of the effect of pressure on the compressibility of metals is important for checking the theories related to the equation of state and lattice anharmonicity as well as those based on interatomic interactions. In the past, several measurements of the effect of pressure on the compressibility of aluminum were carried out as part of a more general program of measurement of the pressure effect on the elastic constants of solids. These measurements were mostly performed by the ultrasonic pulse-echo technique and generally restricted to room temperature. At lower temperatures the only reported measurement of the pressure dependence of the compressibility of Al was that of Ho and Ruoff [1].

Recently, the present authors [2] reported the measurements of the pressure dependence of the elastic shear constants c_{44} and $\frac{1}{2}(c_{11}-c_{12})$ of Al in the temperature range between -196 to 25°C. These measurements were performed by using a torsional-bar-resonance method. The precision of the experimental results was shown to be much higher than that obtained previously by the ultrasonic pulse-echo technique [1]. However, the resonance method was found to be unsuitable for the compressional modes because of the large damping by the pressure-transmitting medium. It was therefore considered worthwhile to measure only the elastic stiffness constant c_{11} of Al, as a function of pressure and temperature, by the ultrasonic pulse-echo technique. Combining the results of the ultrasonic measurements with those obtained previously by the resonance technique yields the values of the elastic constants c_{11}, c_{12}, and hence the compressibility $\beta = 3(c_{11} + 2c_{12})^{-1}$ and the bulk modulus $B = (1/3)(c_{11} + 2c_{12})$.

*239th publication of the Van der Waals Fund.

EXPERIMENTAL PROCEDURE

An aluminum single crystal of right circular cylindrical form, length 2.1 cm, and diameter 1.8 cm, with the cylinder axis in the [100] direction was purchased from Metals Research Limited, England. According to the specifications of the supplier, the crystal was grown from material of 99.99% purity using a Bridgman technique, and the orientation of the cylinder axis was within 1° of the desired [100] direction. The two opposite faces of the specimen were carefully polished with successively finer grades of diamond paste, using a polishing machine, until the faces were parallel to within 0.3 μm. To prevent dislocation effects as described previously by Seely et al. [3], the crystal was subjected to a hydrostatic pressure of 2500 atm and subsequently annealed at 225°C in a vacuum for several hours. Several such compression-annealing cycles were necessary before the specimen could be used for the ultrasonic measurements.

A 10 mc/s X-cut quartz transducer was cemented to one of the parallel faces of the specimen, using Epibond 100 A. The main advantage of using Epibond is that the thermal and elastic properties of the material are fairly well-known and the same bond can be used throughout the temperature range of the present experiment. The resonance frequency of the quartz transducer was determined with an accuracy better than 1 in 10^4, and its change with pressure and temperature was calculated using the data reported by McSkimin and Andreatch [4].

The ultrasonic measurements were carried out by the phase-comparison method first described by Williams and Lamb [5]. The details of the electronic apparatus will be described elsewhere [6]. The transit time τ for the propagation of longitudinal waves in the specimen was determined from the relation

$$2\pi f_n \tau - \tfrac{1}{2}\phi_R = (2n + 2)\,\frac{\pi}{2} \qquad (1)$$

Once pressure and temperature equilibrium was attained, the measurement consisted in recording the null frequencies f_n for successive values of the integer n. These were then fitted to a polynomial by a least-squares analysis, and the value of n_o corresponding to the resonance frequency f_o of the transducer was determined. Finally, the transit time τ was calculated from (1) assuming that, at the resonance frequency f_o of the transducer, $\phi_R = \pi$.

The high-pressure system consists of a beryllium-copper pressure vessel, a mercury gas compressor coupled with an oil press capable of producing a gas pressure of 3000 atm, and a pressure balance to control and measure pressure with an accuracy of 1 in 10^4. The details of the high-pressure system have been described elsewhere [6]. The essential part of the cryogenic system is a metal cryostat designed according to the vacuum type principle. The cryostat consists of the pressure vessel surrounded by a demountable vacuum jacket which is immersed in a bath of liquid nitrogen contained in a dewar vessel. The temperature of the pressure vessel can be maintained constant to within $\pm$ 0.001°C by means of a Thomson bridge circuit. The temperature was measured with an accuracy better than 0.01°C by means of a calibrated Pt resistance thermometer.

RESULTS AND DISCUSSION

The measurement of the transit times was carried out as a function of pressure at several temperatures between -196 to 25°C in steps of 25°C. Helium was used as a pressurizing medium, and the maximum pressure was 2500 atm. The adiabatic elastic constant c_{11} was calculated using the well-known relation $c_{11} = l^2\rho\tau^{-2}$. The length of the specimen at 20°C, as measured by a micrometer, was 2.065 cm. The density of Al used in the calculation was 2.699 g cm^{-3} at 20°C. The change in length and density due to the lowering of temperature was computed from the thermal expansion data of Nix and MacNair [7]. The corrections for the change in length and density due to compression were introduced in the calculation by an iteration procedure, starting with the uncorrected values of the compressibility determined in the present experiment.

In Fig. 1 the values of c_{11}^S and B^S of Al are plotted as a function of temperature at several constant pressures. The reproducibility of the directly determined c_{11}^S data was found to be better than 1 in 10^4, and the average deviation of a single point from the smooth line as calculated from a least-squares analysis did not exceed 0.003%. These figures show that in the present ultrasonic experiment the relative change of the directly-determined elastic constant c_{11}^S can be measured with high accuracy.

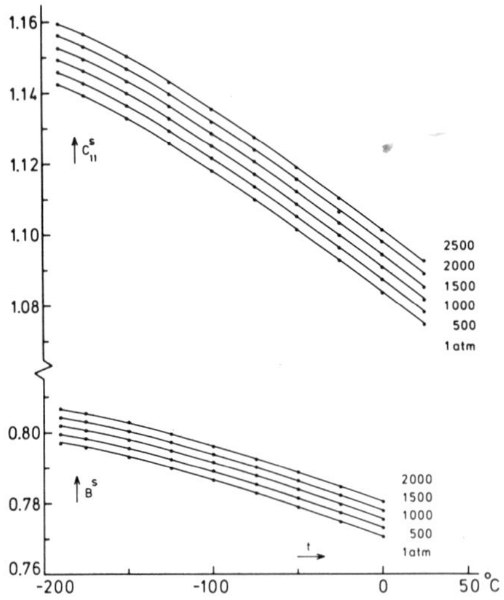

Fig. 1. Adiabatic elastic constant c_{11}^S and adiabatic bulk modulus B^S of Al as a function of temperature at several constant pressures in 10^{12}dyn-cm^{-2}.

To show more clearly the variation of c_{11}^S with pressure, the experimental data along the two isotherms, 0° and -190°C, are normalized to 1.0 at zero pressure and at the appropriate temperature. This is shown in Fig. 2. In both cases and also in the case of other isotherms not shown here, the variation of c_{11}^S with pressure is found to be linear within the accuracy of the present measurement.

The pressure derivatives of c_{11}^S, B^S, and B^T, together with the zero-pressure values of these constants at different temperatures, are given in Table I. The uncertainties in the pressure derivatives as

Table I. Zero-Pressure Values and Pressure Derivatives of c_{11}^S, B^S, and B^T

t, °C	c_{11}^S	B^S	B^T	$(\frac{\partial c_{11}^S}{\partial P})_T$	$(\frac{\partial B^S}{\partial P})_T$	$(\frac{\partial B^T}{\partial P})_T$
	(10^{12} dyn-cm^{-2})					
-190	1.14248	0.79713	0.79294	6.715±0.016	4.771±0.024	4.813
-175	1.13928	0.79589	0.78987	6.794±0.018	4.774±0.023	4.817
-150	1.13292	0.79325	0.78392	6.840±0.010	4.804±0.022	4.849
-125	1.12570	0.79013	0.77733	6.831±0.016	4.788±0.015	4.841
-100	1.11792	0.78669	0.77041	6.827±0.039	4.768±0.054	4.836
- 75	1.10974	0.78300	0.76329	6.860±0.050	4.753±0.061	4.840
- 50	1.10122	0.77907	0.75591	6.918±0.044	4.806±0.033	4.913
- 25	1.09240	0.77494	0.74812	6.967±0.044	4.843±0.044	4.970
0	1.08337	0.77069	0.73965	6.989±0.037	4.820±0.053	4.967
+ 25	1.07428	0.76649	0.73019	7.057±0.033	4.854±0.053	5.006

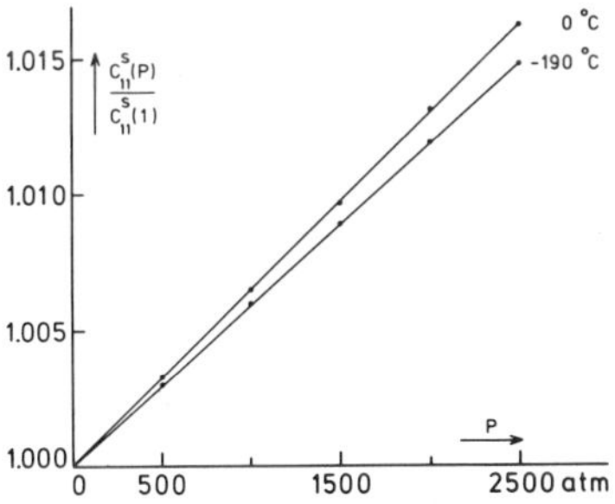

Fig. 2. Variation with pressure of adiabatic elastic constant c_{11}^S for Al at 0 and -190°C.

obtained from a least-squares analysis are also indicated. It should be pointed out here that these uncertainties, which are less than 1% in most cases, do not take into account the systematic errors inherent in the method of measurement. The errors introduced in the pressure derivatives, as a result of the change of phase angle of reflection due to pressure and loss of ultrasonic energy to the pressurizing medium, are estimated to be less than 0.01%.

In Fig. 3 $(\partial B^S/\partial P)_T$ and $(\partial c_{11}^S/\partial P)_T$ are plotted as a function of temperature. As seen, $(\partial c_{11}^S/\partial P)_T$ decreases with lowering temperature, in agreement with the earlier work of Ho and Ruoff [1]. However, in contrast with the earlier work the present experimental results definitely show that $(\partial c_{11}^S/\partial P)_T$ does not become constant at temperatures below $\theta_D/3$. A somewhat similar observation was made previously for the pressure derivatives of c_{44} and $\frac{1}{2}(c_{11} - c_{12})$ of Al, where a nearly monotonic decrease was continued down to liquid nitrogen temperature [2]. It is interesting to note, however, that the $(\partial c_{11}^S/\partial P)_T$ vs. t plot tends to level off at temperatures around $\theta_D/3$, but continues to decrease rapidly again as the temperature is lowered further. The temperature dependence of $(\partial B^S/\partial P)_T$ is found to be very small and, within the accuracy of the present measurement, it remains constant in the available temperature range.

It has been shown previously by the present authors [2] that starting from the relation

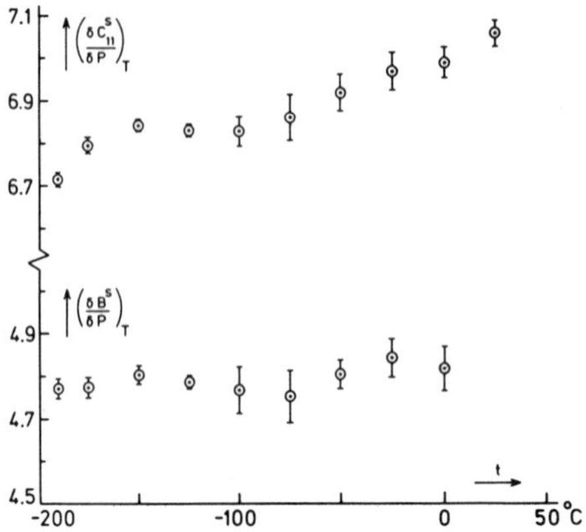

Fig. 3. Pressure derivatives of c_{11}^S and B^S of Al as a function of temperature.

$$C = C_h(1-D\bar{\varepsilon}+dP) \qquad (2)$$

given by Leibfried and Ludwig [8] for the temperature and pressure dependence of the adiabatic elastic constants and using the Debye spectrum for $\bar{\varepsilon}$, one can derive the following relation for the temperature dependence of the pressure derivatives of the elastic constant:

$$\left(\frac{\partial C}{\partial P}\right)_T = -3C_h D k_B \gamma_T \theta_D \beta_T$$

$$[f'(\theta_D/T) + 1/8] + C_h d \qquad (3)$$

with

$$\gamma_T = -(\partial \ln\theta_D/\partial \ln V)_T \quad \text{and} \quad \beta_T = -V^{-1}(\partial V/\partial P)_T$$

This relation predicts that the pressure derivatives of c_{11}^S and B^S of Al should decrease with lowering temperature and tend to reach a constant value at high ($T >> \theta_D$) and low ($T << \theta_D$) temperatures. However, the temperature range of the present experiment does not extend either sufficiently below or above θ_D to draw any definite conclusion.

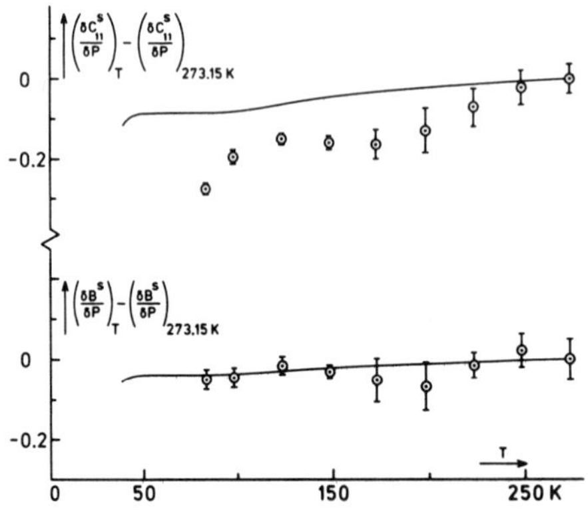

Fig. 4. Comparison of calculated and observed pressure derivatives as a function of temperature. Legend: ——— Calculated, ⊙ observed.

In Fig. 4 the observed changes of the pressure derivatives in lowering the temperature from 0°C are compared with the corresponding quantities calculated from (3). The appropriate values of C_h and D for the two cases are determined by fitting the zero-pressure data into (2) by a least-squares procedure, while those for γ_T and θ_D are obtained from previously reported thermal data [9-10]. As shown, the calculated curve for $(\partial c_{11}^S/\partial P)_T$, although showing the general features of the experimental behavior, indicates a much smaller temperature dependence than found experimentally. The calculated curve for $(\partial B^S/\partial P)_T$ also shows a small temperature dependence which, however, is in agreement with the present work.

NOTATION

B	= bulk modulus	l	= length
C, c	= elastic constant	β_T	= isothermal compressibility
C_h	= elastic constant in the harmonic approximation	γ_T	= Grüneisen parameter
D, d	= constants	$\bar{\varepsilon}$	= mean energy per oscillator
f_n	= null frequency of the nth mode	θ_D	= Debye temperature
$f(\theta_D/T)$	= Débye function	ρ	= density
k_B	= Boltzmann constant	τ	= transit time
		ϕ_R	= phase angle of reflection

REFERENCES

1. P. S. Ho and A. L. Ruoff, J. Appl. Phys. 40, 3151 (1969).
2. N. J. Trappeniers, S. N. Biswas, P. van't Klooster, and C. A. ten Seldam, Physica 85B, 33 (1977).
3. J. L. Seely, G. S. Baker, and P. Gibbs, J. Appl. Phys. 33, 2458 (1962).
4. H. J. McSkimin and P. Andreatch, J. Acoust. Soc. Am. 34, 609 (1962).
5. J. William and J. Lamb, J. Acoust. Soc. Am. 30, 308 (1958).
6. P. van't Klooster, Ph.D. Thesis, University of Amsterdam, to be published.
7. F. C. Nix and D. McNair, Phys. Rev. 60, 597 (1941).
8. G. Leibfried and W. Ludwig, Solid State Phys. 12, 275 (1961).
9. D. B. Fraser and A. C. Hollis Hallett, in Proceedings 7th Intern. Conference Low Temperature Physics, University of Toronto Press, Toronto, Canada (1961), p. 689.
10. W. F. Giauque and P. F. Meads, J. Am. Chem. Soc. 63, 1897 (1941).

B-1

STUDY OF PHASE TRANSITIONS AT HIGH PRESSURE*

A. Jayaraman

Bell Laboratories
Murray Hill, New Jersey, USA

INTRODUCTION

Study of phase transition under pressure has been one of the most active areas of high-pressure research [1-5].† The purpose of this article is to present a critical discussion of the more recent developments on phase transitions.

Pressure-induced structural changes in solids are an ubiquitous phenomenon. Only certain general comments will be made here. Most of the elemental solids, with the exception of high-melting transition elements, undergo pressure-induced phase changes to closer packed structures. Some general trends are evident. Usually, the high-pressure phases of lighter elements simulate the structure of the heavier element in the same column of the periodic table. Many nonmetallic elements transform at high pressure to metallic phases. At high pressures Si, Ge, P, As, Se, and Te become metallic. More recently, S has been shown to undergo a transition to a metallic state near 180 kbar [6]. A very interesting theoretical prediction [7] concerns the behavior of solid hydrogen at very high pressures. In the megabar region, solid hydrogen is expected to turn into a metal. If this prediction is realized, it

*Invited paper.

†1-5 are review articles which contain references to original papers.

will be another significant achievement in high-pressure research.

An interesting phase transformation sequence is the hcp → Sm type → dhcp → fcc, encountered in rare earth elements and some intra-rare earth alloys. These structures represent different stacking arrangements of the atomic layers in the otherwise closely-packed structures. Several explanations [8] have been advanced to rationalize the observed sequence: (1) Correlations with an averaged effective atomic number, (2) influence of 4f electrons, (3) stability due to interactions between pairs of close-packed planes, and (4) transition at a certain critical value of an atomic parameter f which can be related to the pseudopotential. The last approach appears to be very successful and free from most objections.

Compounds crystallizing in simple crystal structures, such as the zinc blende and NaCl-type, show certain definite trends [1]. For instance, a majority of III-V, II-VI compounds which have the zinc blende or wurtzite structure undergo a pressure-induced phase transition to close-packed structures. Phase transition in compounds and pressure-temperature (P-T) diagrams are extensively covered in the review by Pistorius [5].

In recent years, theoretical approaches have been made to understand the experimentally observed structural transitions. Heine and Wearie [9] have reviewed the application of pseudopotential theory to crystal structures including high-pressure phases. The alkaline earth metals Ca, Sr, and Ba have received much attention. These metals have been discussed in terms of pseudopotential theory. There has been some progress in understanding the pressure-induced phase transitions in tetrahedrally coordinated structures. Phillips [10] has used ionicity difference as a criterion for the transformation from tetrahedrally-coordinated to the rock-salt structure. If f_i, the ionicity of the A•B bond, lies within the limits $0 \leq f_i \leq 0.30$, the low-pressure phase is tetrahedral, and the crystal transforms to the white-tin-type, or orthorhombic structure, under pressure. Within the limits $0.3 \leq f_i \leq 0.785$, the low-pressure phase is still tetrahedrally coordinated, but the crystal transforms first to NaCl-type structure under low pressure and then at a higher pressure to the metallic structure. For $f_i < 0.785$, the low-pressure phase is usually NaCl-type which may be expected to transform to a metallic phase. A plot of normalized Gibbs free energy obtained from experimental $P_{tr}\Delta_{tr}$ against $X = (0.785 - f_i)$ yields a straight line plot. It has been suggested that X is the natural parameter which governs the covalent-ionic transition and is more reliable than radius ratio.

ELECTRONIC PHASE TRANSITIONS

The transitions so far discussed were primarily those driven by an instability of the lattice with respect to another atomic rearrangement. There are, however, phase transitions which are caused by an electronic instability in the system and for this reason are described as electronic transitions.

Cerium. The γ-α phase transition [11-14] is probably the most investigated electronic phase transition under pressure. In this, one 4f electron present in Cerium (Ce) is delocalized and as a consequence the valence of Ce changes from trivalent towards the quadrivalent state; in the transition, fcc remains fcc but the volume decreases by about 10%. The phase diagram of Ce in Fig. 1 shows the various phase changes encountered in Ce metal. The γ-α Ce phase boundary terminates at a critical point [14] in the pressure-temperature plane and behaves very much like the vapor-liquid transition, above and below the critical temperature. The occurrence of a broad fusion curve minimum is unique and is consistent with the termination of the γ-α Ce phase boundary at a critical point. A pressure-induced phase transition near 55 kbar [15,16] has been reported for Ce which has been designated α→α'. This has been claimed to be an isostructural transition involving a 4% volume change and an electronic transition in which the Ce undergoes a total conversion to the quadrivalent state; in the γ-α Ce transition, the valence change of Ce is fractional [17]. The structure of Ce above 50 kbar appears to be noncubic [8].

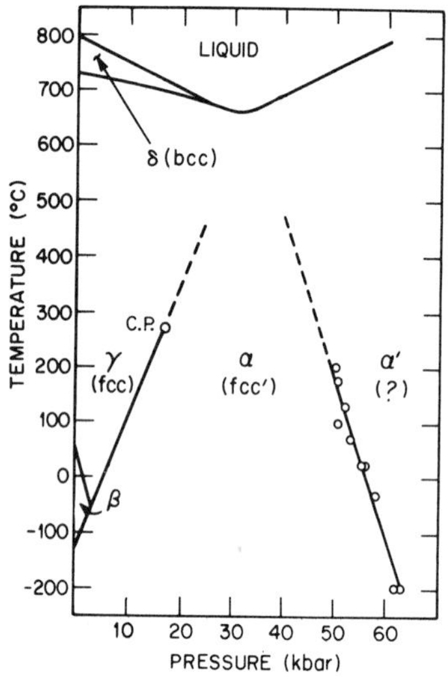

Fig. 1. Phase diagram of cerium metal. The β-phase has the double hexagonal close-packed structure. C.P. is the critical point for the γ → α Ce transition.

Superconductivity has been reported [18] for Ce at pressures above 50 kbar, and the phase in question is presumed to be the α' phase. The absence of a superconducting transition in the α phase has been attributed to a high degree of 4f character present in it, while superconductivity in the α' phase is due to complete loss of 4f character. Thus, Ce is a unique element which can exhibit magnetic ordering, a nonmagnetic state, and superconductivity, depending on the pressure range and its electronic state.

Cesium. Cesium exhibits many polymorphic transitions under pressure, and its phase diagram [19] is shown in Fig. 2. The most interesting transition is the Cs II → III isostructural transition at 42 kbar. The volume decreases by about 8% at this transition and is believed to be due to a change in the electronic state of Cs from 6s→5d. A striking resistance discontinuity has been reported in Cs at about 125 kbar [20]. For the anomalous fusion behavior, a continuous 6s - 5d electronic collapse in liquid Cs has been invoked. Resistance data on the liquid phase, up to 50 kbar pressure, appear to support the above view.

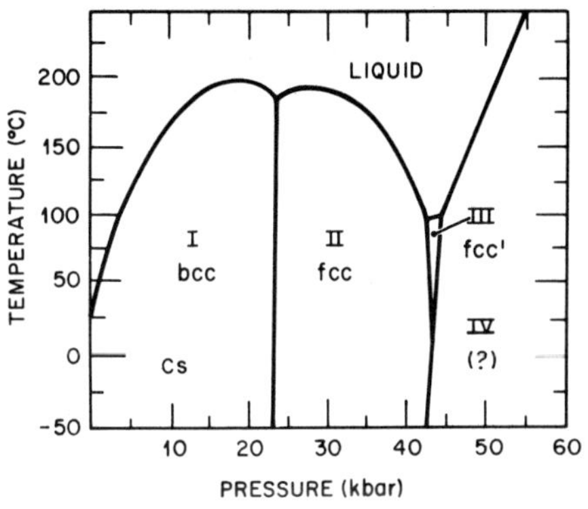

Fig. 2. Phase diagram of cesium.

There are some puzzling questions with regard to the high-pressure behavior of Cs: (1) The resistivity decrease in the Cs III-IV transition is about the same magnitude as the rise in resistivity in the Cs→II-III transition, and the question is what is it that causes such a large decrease. (2) What is the structure of Cs IV? (3) If the Cs II→III transition involves 6s→5d electron collapse, as is believed at this time, what is the nature of the transition observed near 125 kbar? There is the possibility that the 125 kbar transition is the 6s-5d electronic transition. Then we are left with the problem of explaining the Cs II-III transition. The structure of Cs IV and the phase stable above 125 kbar would provide a good starting point for resolving these questions.

Superconductivity around 1.5°K has been reported for Cs near 125 kbar [21], and the data indicate that the T_c decreases with pressure. Superconductivity is attributed to the high pressure polymorph reported by Stager and Drickamer [20]. No other alkali metal has been shown to be superconducting, and that Cs exhibits superconductivity at high pressure certainly indicates something unusual must be happening with pressure, in particular to its electronic structure.

METAL-INSULATOR TRANSITIONS

Phase transitions to the metallic state in semiconductors such as Si, Ge, and many III-V and II-VI compounds [22] are transitions arising essentially out of lattice instability. There are, however, many systems which exhibit metal-insulator transitions in which electronic effects dominate. Pressure experiments are crucial to the understanding of the metal-insulator transitions and to distinguish between various mechanisms.

Among the metal-insulator transitions there is a particularly simple type which is essentially a continuous transition involving band-overlap and nonoverlap with pressure. Mott calles them [23] Wilson band-overlap transition. Such a transition is encountered in the face-centered-cubic divalent metals Yb, Ca, and Sr [24,25]. Iodine is reported to become metallic near 160 kbar, and the transition is continuous [22]. The resistivity of Yb rises strikingly with pressure, and this is attributed to band uncrossing in Yb at about 15 kbar and eventual opening up of an energy gap [26]. The semiconducting fcc phase becomes unstable with respect to the metallic bcc phase at about 40 kbar. In the case of Sr, the resistance rise with pressure is less spectacular and the negative temperature coefficient of resistivity is very much weaker [24]. Calcium behaves similarly but at pressures near 200 kbar [25]. Band structure calculations as a function of pressure by Vasvari, Animalu and Heine [27] have shown that the band overlap decreases in the alkaline earth metals with pressure, and the theory predicts metal-semimetal behavior.

TRANSITION METAL COMPOUNDS AND MOTT TRANSITION

There has been much interest in metal-nonmetal transitions in transition metal oxides [28,23,29]. The band theory of solids fails to predict the electrical properties of oxides such as NiO, CoO, and MnO. A system in which a band is partially filled should exhibit metallic conductivity, and hence these oxides should be metallic, whereas they are very good insulators. Mott [30,31] suggested that when electrons move in narrow bands, Coulomb repulsion effect would dominate their behavior, the system would be highly correlated, and this would lead to a localization of the electron on the transition metal ion in these oxides. Mott made the important suggestion that such a system would exhibit a transition to the metallic state at a certain critical interatomic distance, when the lattice is compressed. Such a transition was envisaged to be a discontinuous first-order transition. This so called Mott-transition is believed to occur in Cr-doped V_2O_3.

Pure V_2O_3 shows metallic behavior at room temperature, but with the substitution of approximately 1% Cr^{+3}, the electrical resistivity of V_2O_3 increases by orders of magnitude. By applying pressure, the system becomes metallic through a first-order phase change in which the long-range order remains the same; corundum type. The P-T diagram for a V_2O_3 sample containing 4% Cr is presented in Fig. 3 [32]. The inset to Fig. 3 shows the phase diagram as a function of composition, where X can be Cr or Ti. Introducing Cr and Ti is equivalent to negative and positive pressure, respectively. The M-I phase boundary terminates at a critical point [33]. There are several facts about the M-I phase transition which support the view that it must be a Mott transition: (1) There is no change in the long-range order accompanying the transition and hence no crystal distortion or doubling of the unit cell to complicate the situation. The applicability of models which involve crystal distortion, such as the Adler-Brooks [28], can be ruled out. (2) The I-phase is the high-temperature phase and hence has larger entropy relative to the M-phase. If the transition had been an "excitonic" metal-insulator transition, the entropy of the phases should have been the opposite of that observed. (3) Further, single-crystal X-ray work has shown that in the M-phase the V-V distance decreases while the V-O distance is practically unaffected. However, a lot of controversy exists regarding the exact nature of the transition. There are other views which stress that electron-lattice interactions cannot be neglected in metal-insulator transitions in vanadium oxides, and the electron-electron Coulomb interaction is not the only important element in this.

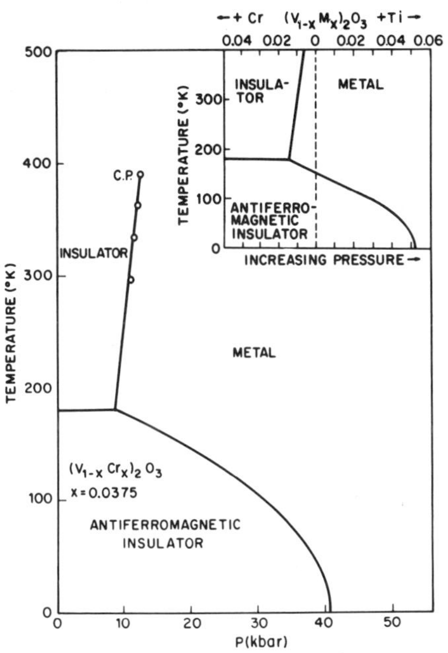

Fig. 3. Phase diagram of a Cr-doped V_2O_3 mixed oxide. C.P. is the critical point for the M-I transition. The inset shows the effect of both pressure and alloying. With addition of Cr, the dashed line moves to the left, and with the addition of Ti, moves to the right.

Other transition metal oxides and sulfides which show semiconductor-metal transition with increasing [38] temperature are VO, VO_2, [36] Ti_2O_3, Ti_4O_7, Fe_3O_4, [37] NiS, TaS_2, and NiS_2 [39]. In many of the above, there occurs a crystal distortion and change of symmetry with the transition. Some of the above have been studied under pressure [37,38,39].

VALENCE CHANGES IN RARE EARTH MONOCHALCOGENIDES

The rare earth monochalcogenides crystallize in the NaCl-type structure and are semiconducting or metallic depending on the valence state of the R.E. ion; semiconducting if divalent and metallic if trivalent. Pressure-induced electronic transitions involving a change in the valence state of the R.E. ion have recently been reported in many divalent R.E. chalcogenides; [40] Sm and Yb monochalcogenides, TmTe and EuO. The electronic transition involves the delocalization of a 4f electron to the 5d-6s conduction band states and results in a change of the valence of the rare earth ion towards trivalent state. The NaCl-type structure remains NaCl in the transition, but there is a large volume decrease ~ 10-12%. Approximately one electron per molecule is delocalized from the 4f state into the 5d conduction state, and hence is a semiconductor-to-metal transition. In Fig. 4 the resistivity vs. pressure behavior for SmTe monochalcogenides are shown. The transition is first-order in SmS and continuous in SmSe and SmTe. The observed (Fig. 5) P-V relationships [40] are consistent with the valence change in SmS (see Fig. 5). In SmS the first-order transition occurs at 6.5 kbar pressure. This phase boundary has been investigated to about 700°C and the behavior suggests that it should terminate at a critical point near 900°C [41]. In optical studies under pressure, a striking change in color from black to metallic gold is observed at the semiconductor to metal transition. Resistivity under pressure has been measured on $SmS_{1-x}Se_x$ compounds [42]. A discontinuous transition has been found up to X = 0.7. The transition pressure increases linearly with the composition. At higher concentrations of Se, only a continuous transition is observed.

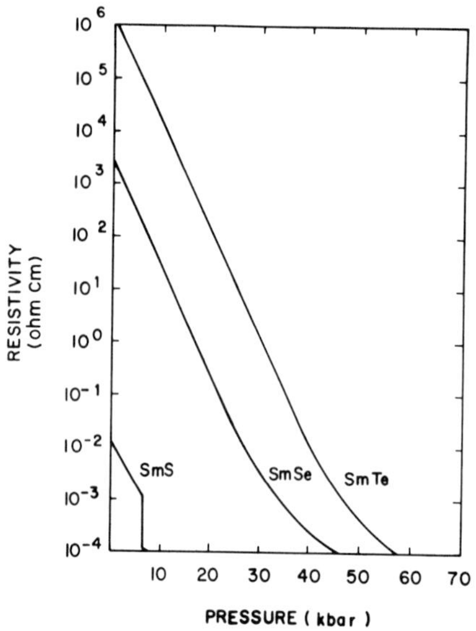

Fig. 4. Resistivity pressure data for Sm monochalogenides. First-order semiconductor to metal transition in SmS at about 6.5 kbar. Continuous transition in SmSe and SmTe.

By chemical substitution of a trivalent R.E. ion for Sm in the SmS lattice, the high-pressure phase of SmS can be stabilized at atmospheric pressure and studied conveniently [40,42].

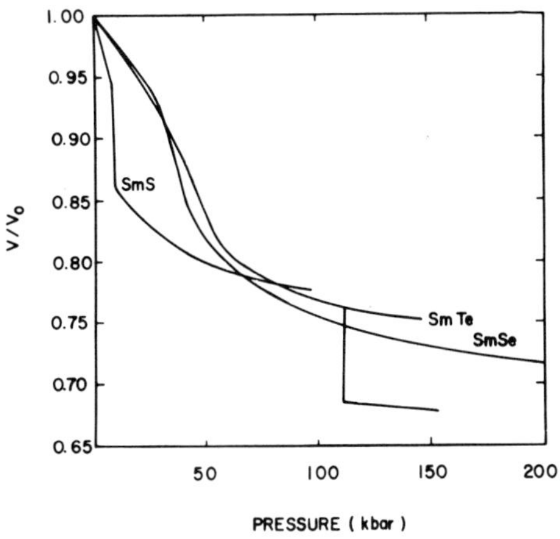

Fig. 5. Pressure-volume relationship for Sm monochalcogenides. The discontinuous change at about 110 kbar in SmTe is due to NaCl to CsCl-type transition. The abnormal compressions for SmSe and SmTe are due to the electronic transition.

Theoretical Aspects

A large amount of experimental work has accumulated on valence change in divalent R.E. monochalcogenides, mainly on SmS and SmS related systems [40,43]. The results have led to a very special aspect of the metallic phase, namely, the fluctuating valence state of the Sm. If the valence change had proceeded all the way to the pure Sm state, the lattice parameter of SmS should have been 5.62 Å, whereas it was found to be 5.70 Å in the metallic phase of pure SmS, and 5.68 Å in the stabilized high-pressure phase. It was this intermediate lattice parameter and the absence of magnetic ordering at low temperature which gave the clue that the electronic transition does not proceed all the way to the +3 state. From the lattice parameter, the valence state of Sm has been deduced as +2.7 in the metallic phase. The interesting question now is how can one describe an intermediate valence state. The description appears to be the following: The 4f levels hybridize with the d states, while retaining a very narrow band width (between 10^{-2} and 10^{-1} ev), so that in the density of states over the smooth s-d background, we have a sharp f-like atomic character. The wave functions near this peak are linear combinations of f-like and d-like wave functions: $\Psi = a\,\Psi_f + b\,\Psi_d$. The coefficients a and b give the proportion of f and d character and vary rapidly near the peak. An essential feature of this model is that the Fermi-level is pinned to the f-peak. A simplified electronic density of state diagram is shown in Fig. 6 for pure SmS in the semiconducting state, for SmS in the metallic phase, and for doped SmS in the nonmetallic state. The Fermi level in the collapsed phase (metallic phase) is located in the f-band [44]. The presence of a high density of states band, namely the 4f band, at the Fermi energy E_F, derives support from several experimental studies [45]. The situation is that in the intermediate valence state, the valence fluctuates between the two ionic states on a time scale of the order of 10^{-12} sec. This situation is often referred to as the interconfiguration fluctuation (ICF).

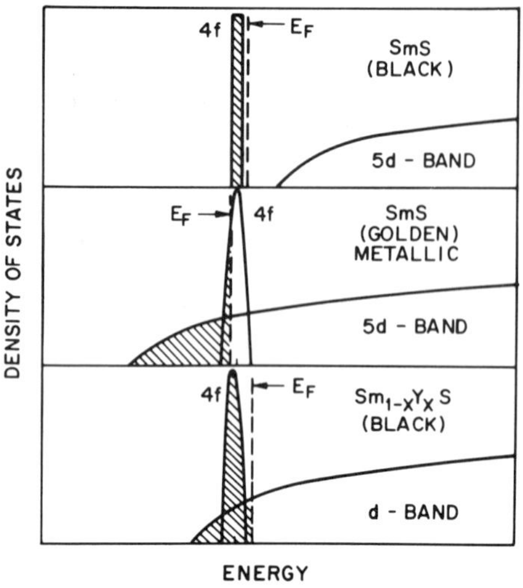

Fig. 6. Electronic structure diagram for semiconducting SmS, metallic phase of SmS, and Y substituted SmS.

There have been many theoretical attempts to explain (1) the continuous nature of the transition, in relation to the various accelerating and decelerating mechanisms acting, and (2) the fractional valence in the golden metallic phase of both SmS and $Sm_{1-x}Ln^{+3}_xS$ compounds. The contribution to the free energy at T = 0 for a system involving $4f^n \rightarrow 4f^{n-1}$ transition should include an electronic energy term E_e, a lattice energy term E_L, and another term to include all other interactions E_{int}. The term E_e includes the promotional energy and correlation energy. Thus,

$$F(v,z) = E_e(v,z) + E_L(v) + E_{int} \qquad (1)$$

where v and z are specific volume and electronic density, respectively. The Falicov-Kimball model [46] originally proposed to explain the metal-insulator transitions in semiconductors having a localized deep donor level, has been applied to the electronic transition in Sm monochalcogenides. This theory uses the short-range Coulomb interaction between the electron-hole pair as the main driving force for the transition. It can be shown from this model that the effective energy gap between the localized level and the conduction band is carrier dependent and would go to zero according to $E_g^{eff.} = E_g - 2Gn$, where E_g is the energy gap between the localized level and the conduction band, G is the screening interaction parameter, and n is the electronic density excited into the conduction band; n can increase continuously or discontinuously depending on the value of G/E_g. This theory has been applied to SmS-SmSe [47] which shows a first-order or continuous transition, depending on the composition. However, the Falicov-Kimball model does not take into account the volume effects, which are very important in Sm monochalcogenides. Other theoretical attempts have included these aspects and have calculated the pressure-volume relationships in Sm monochalcogenides.

Hirst [47] has described the various "accelerating" and "decelerating" tendencies connected with both the E_e term and E_L term. The most important accelerating term, according to Hirst, is the so-called "compression shift mechanism", which originates from the fact that when the lattice is compressed, the d-conduction band edge is lowered with respect to the 4f level, favoring excitation of electrons. Since this in turn decreases the volume, the d-band will be further lowered. The whole mechanism is therefore accelerating and acts independently of the Falicov-Kimball screening interaction, which is also accelerating. The compression shift mechanism of Hirst will drive the system catastrophically through a first-order transition, provided the lattice compressibility is not the limiting factor; the latter can change the transition to a gradual one, if the bulk modulus increases rapidly with the volume change ΔV. The decelerating tendencies arise from: (1) the large density of states difference between the 4f band and the 5d, and (2) the increase in bulk modulus with decreasing volume. The former causes the Fermi energy to rise as electrons are excited from the narrow high density of states 4f band into the 5d conduction band, consequently increasing the promotional energy for further excitation. Anderson and Chui [48] have pointed out that anharmonic terms can make a repulsive contribution to the interaction energy and this can account for the two types of transitions.

In recent work, Penney et al. [49] have taken into account more explicitly the factors governing the electronic energy and the lattice energy which Varma and Heine [50] have already formulated in their work. Parameters of interest such as strain, electronic configuration, and bulk modulus have been calculated as a function of pressure for SmS, and as a function of composition for $Sm_{1-x}Y_xS$ and $SmS_{1-x}As_x$ compounds, using the available experimental data. These results are in good agreement with the measurements. As pointed out earlier, the most interesting behavior is that in the metallic region, very near the critical Y concentration, the bulk modulus becomes very small [49]. This softening has been shown to be due to an electronic term arising from a redistribution of electrons between the two electronic configurations involved and to a lattice term resulting from a renormalization of the lattice which is coupled to the electronic energy by the deformation potential.

The temperature-pressure diagram for SmS and the temperature-composition (T-x) diagram for the system $Sm_{1-x}Gd_xS$ [40] is presented in Fig. 7. In the P-T diagram of SmS, only the phase boundary corresponding to the B-M boundary of Fig. 7 has been determined. The B-M boundary shown in Fig. 7 is believed to terminate at a critical point, as all the isostructural transition boundaries in other systems behave. What is most remarkable is that the metallic phase (M) is unstable with respect to low temperature as well as high temperature. However, the (M) phase does not undergo a first-order transition on the high temperature side, and there is not only a continuous, but an anomalously large thermal expansion. The anomalous expansion strongly suggests that the conduction electron relocalizes on the 4f level with temperature. It is surprising that this change does not lead to a runaway process.

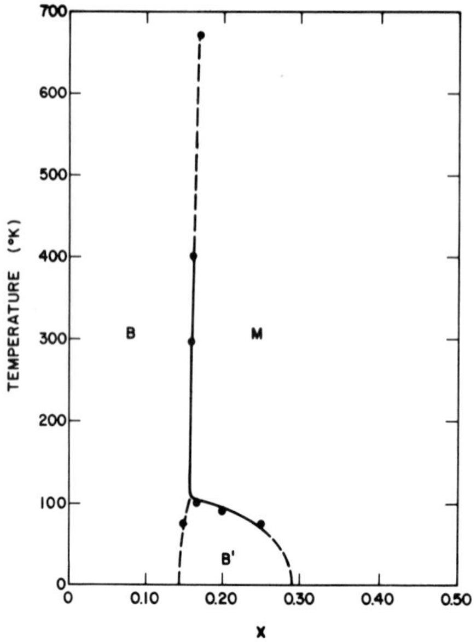

Fig. 7. T-X diagram for $Sm_{1-x}Gd_xS$ system.

The effects of temperature on the metallic phase have been analyzed by Varma [51]. While it is not difficult to explain the first-order transition on the low-temperature side, the mechanism that prevents a first-order transition on the high-temperature side is not completely understood. This has recently been discussed by Jefferson [52].

MAGNETIC TRANSITIONS
Cr-BASED SYSTEMS

In this section, the influence of pressure on the so-called itinerant antiferromagnetism will be discussed. The experiments have contributed substantially to the understanding of this unique magnetic ordering [53,54] encountered in Cr and Cr alloys.

The ordering in Cr systems is due to the spin polarization of the conduction electrons as the system is cooled through the Néel temperature and to the formation of spin density waves (sdw). The itinerant af state has been explained by Lomer and Overhauser [54], using a band model. It is believed that the af state is due to an "excitonic instability" in which the electrons and holes from two parts of the Fermi surface pair, via Coulomb interaction. Energy gaps open up along certain directions in k-space, destroying part of the Fermi surface. This gives rise to the resistivity anomaly associated with the transition, which is a convenient probe to follow the transition in pressure studies. The af state of Cr is sensitively dependent on the band structure, and hence any change in Fermi surface dramatically influences the transition temperature. For this reason, pressure as well as alloying alter the Néel temperature drastically. Theory shows that the situation in Cr is mathematically similar to superconductivity, and in fact the equation for the transition temperature T_N in the so-called Fedders and Martin model [55] is analogous to the BCS-type equation for the superconducting transition temperature. $T_N = T_B \exp(-1/N(0)\bar{V})$, where N(0) is the density of states, $\bar{V}$ is the average coulomb attraction, and $k T_B$ is of the order of band energy. The pressure variation of T_N in Cr fits the equation $T_N = T_N(0) \exp(C \Delta V/V_0)$, where C = -26.5 which has the same form as the equation obtained in the Fedders and Martin Model.

When a transition metal is added to Cr, the T_N increases or decreases, depending on whether the added transition metal increases or decreases the electron-to-atom ratio of 6 in Cr [56]. Neutron studies [56] have shown that when the e/a ratio is raised slightly in excess of 6 by transition metal additions such as Mn, Ru etc., the wave vector of the sdw switches from incommensurate ($q \neq 2\pi/a$) to a commensurate structure ($q = 2\pi/a$). The equivalence of pressure to decreasing e/a ratio suggests that pressure should suppress the commensurate phase and this is in fact borne out by experiments. The magnetic phase diagram of a Cr-Ru alloy is shown in Fig. 8 [57]. Here the commensurate (C), the incommensurate (I), and paramagnetic (P) phases coexist with a triple point in the pressure-temperature plane. The T_N of the P-C transition decreases very rapidly close to the triple point. The sharp break in slope of the phase boundary at higher pressures is due to the P-I transition. More recent experiments in Cr-Fe [58] and Cr-Si [59] alloys under pressure have revealed that Fe and Si donates electrons to the d-band of Cr, stabilizing the commensurate phase. Pressure experiments on Cr alloys have elucidated many aspects of this interesting magnetic ordering: (1) It has been possible to show that the pressure variation of T_N obeys a BCS type of equation, which is a clear vindication of an important prediction of the band model. (2) That the dT_N/dp can be used as a criterion to distinguish the type of ordering in these alloys. (3) They have led to the verification of another important theoretical prediction [60], namely the "depairing" due to ordinary impurity scattering, an effect analogous to a magnetic impurity in a super-conductor. This depairing effect causes a reduction in T_N of the af ordering temperature. (4) In alloys of Cr with Mo, it has been possible to study the vanishing of the antiferromagnetism.

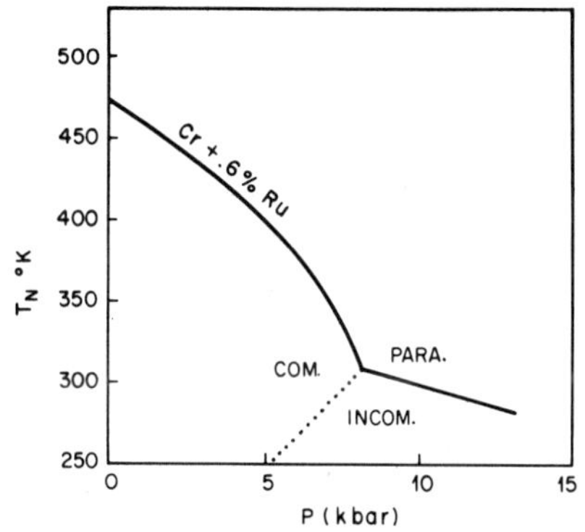

Fig. 8. Magnetic phase diagram of a Cr alloy containing 0.6% Ru. The variation in Neel temperature with pressure is shown. The triple point is between commensurate (COM.), incommensurate (INCOM.), and paramagnetic (PARA.) phases.

MÖSSBAUER STUDIES OF PHASE TRANSITIONS AT HIGH PRESSURE

Using the Mössbauer technique, Drickamer and his colleagues [61] have studied at high pressure a wide variety of inorganic and organic materials containing iron, with Fe [57] as the probe. Among the results, the most interesting effects are perhaps the reduction of ferric iron Fe (III) to ferrous iron (Fe II) and changes in the spin state of iron. These are essentially electronic transitions under pressure. The ferrous ion has nominally six 3d electrons and the ferric iron five 3d electrons. These two states exhibit distinctly different isomer shifts and are easily distinguishable in a Mössbauer spectrum. Also, Mössbauer resonance studies permit a determination of the spin states of iron through a measurement of both the isomer shift and quadrupole splitting. High-spin ferrous ion exhibits a large and positive isomer shift, while the low spin ferrous ion exhibits a markedly smaller shift. The quadrupole splitting which arises from the interaction of an electric field gradient with the nuclear quadrupole moment is large for the high-spin ion, while for the low-spin ion it is much smaller. The relative amounts of the spin states present can be calculated from the areas under the peaks in the Mössbauer spectrum.

Thus, pressure experiments have been useful in a wide variety of situations to discover new mechanisms and in some cases to support and confirm existing ideas.

REFERENCES

1. W. Klement and A. Jayaraman in Prog. Solid State Chem. Vol. 3, H. Reiss, ed., Pergamon Press, Oxford and New York, USA (1966), p. 289.

2. V. V. Evdokimova, Sov. Phys. USP 9, 54 (1966).

3. S. M. Stishov, Sov. Phys. USP 11, 819 (1969).

4. A. Jayaraman and L. H. Cohen, Phase Diagrams Vol. 1, H. Alper, 1, Academic Press, New York, USA (1970), p. 245.

5. C. W. F. T. Pistorius, Prog. Solid State Chem. Vol. 11, J. McCaldin and G. Somarjai, eds., Pergamon Press, Oxford, UK, and New York, USA (1976), p. 120.

6. L. C. Chabildas and A. L. Ruoff, J. Chem. Phys. 66, 983 (1977).

7. N. W. Ashcroft, Phys. Rev. Letters 21, 1748 (1968).

8. A. Jayaraman, references cited in review article, Handbook of Physics and Chemistry of the Rare Earths Chap. 9, North Holland Publishing Co., Amsterdam, Netherlands (1968).

9. V. Heine and D. Weaire, Solid State Physics Vol. 24, H. Ehrenreich, F. Seitz and D. Turnbull, eds., Academic Press, New York, USA (1970), p. 249.

10. J. C. Phillips, Phys. Rev. Letters 27, 1196 (1971).

11. E. G. Ponyatovskii, Sov. Phys.-Doklady 3, 498 (1958).

12. R. I. Beecroft and C. A. Swenson, J. Phys. Chem. Solids 15, 234 (1960).

13. B. L. Davis and L. H. Adams, J. Phys. Chem. Solids 25, 379 (1964).

14. A. Jayaraman, Phys. Rev. 137, A179 (1965).

15. E. Franceschi and G. L. Oleese, Phys. Rev. Letters 22, 1299 (1969).

16. E. King, J. A. Lee, I. R. Harris, and T. F. Smith, Phys. Rev. B1, 1380 (1970).

17. K. A. Gschneidner and R. Smoluchowski, J. Less Comm. Metals 5, 374 (1964).

18. J. Wittig, Phys. Rev. Letters 21, 1250 (1969).

19. A. Jayaraman, R. C. Newton, and J. M. McDonough, Phys. Rev. 159, 527 (1967).

20. R. A. Stager and H. G. Drickamer, Phys. Rev. Letters 12, 19 (1964).

21. J. Wittig, Phys. Rev. Letters 24, 812 (1970).

22. H. G. Drickamer, Solid State Physics Vol. 17, F. Seitz and D. Turnbull, eds., Academic Press, New York, USA (1965), p. 127.

23. N. F. Mott and Z. Zinamon, Rep. Progress in Physics 39, 881 (1970).

24. D. B. McWhan, T. M. Rice, and P. H. Schmidt, Phys. Rev. 177, 1063 (1969).

25. R. A. Stager and H. G. Drickamer, Phys. Rev. 131, 2524 (1963).

26. H. T. Hall, J. D. Barnett, and L. Merrill, Science 139, 111 (1963).

27. B. Vasvari, A. O. E. Animalu, and V. Heine, Phys. Rev. 154, 535 (1967).

28. D. Adler, Solid State Physics, Vol. 21, F. Seitz, D. Turnbull, and H. Ehrenreich, eds., Academic Press, New York, USA (1968), p. 111.

29. I. G. Austin and N. F. Mott, Science 168, 71 (1970).

30. N. F. Mott, Proc. Phys. Soc. (London) 62, 416 (1949).

31. N. F. Mott, Rev. Mod. Phys. 40, 673 (1968).

32. D. B. McWhan and J. P. Remeika, Phys. Rev. B2, 3734 (1970).

33. A. Jayaraman, D. B. McWhan, J. P. Remeika, and P. D. Dernier, Phys. Rev. B2, 3751 (1970).

34. P. D. Dernier, and M. Marezio, Phys. Rev. B2, 3771 (1970).

35. P. D. Dernier, J. Phys. Chem. Solids 31, 2569 (1970).

36. C. N. Berglund and A. Jayaraman, Phys. Rev. 185, 1034 (1969).

37. G. A. Samara, Phys. Rev. Letters 21, 795 (1968).

38. J. A. Wilson, F. J. DiSalvo and S. Mahajan, Adv. Phys. 24, 117 (1975).

39. J. A. Wilson and G. D. Pitt, Phil-Mag. Vol. 23, 1297 (1971).

40. A. Jayaraman, P. D. Dernier, and L. D. Longinotti, Phys. Rev. B. 11, 2783 (1975); also High Temperature: High Pressures, 7, 1 (1975).

41. E. Yu. Tonkov and I. L. Aptekar, Fiz. Tverd. Tela 16, 1507 (1974).

42. E. Bucher and R. G. Maines, Sol. State Comm. 11, 1441 (1972).

43. L. J. Tao and F. Holtzberg, Phys. Rev. B11, 3842 (1975).

44. T. Penney and F. Holtzberg, Phys. Rev. Letters 34, 322 (1975).

45. S. von Molnar, T. Penney, and F. Holtzberg, in Proc. Conf. on Metal--Non-Metal Transitions, J. de Physique 37, C4-241 (1976).

46. L. M. Falicov and J. C. Kimball, Phys. Rev. Letters 22, 997 (1969).

47. L. L. Hirst, J. Phys. Chem. Sol. 35, 1285 (1974).

48. P. W. Anderson and S. T. Chui, Phys. Rev. B9, 3229 (1974).

49. T. Penney and R. L. Melcher, in Proc. Conf. on Metal--Non-Metal Transitions, J. de Physique 37, C4-275 (1976).

50. C. M. Varma and V. Heine, Phys. Rev. 11, 4763 (1975).

51. C. M. Varna, Rev. Mod. Phys. 48, 219 (1976).

52. J. H. Jefferson, J. Phys. (London) C, 9, 269 (1976).

53. T. M. Rice, A. S. Barker, B. I. Halperin, and D. B. McWhan, J. Appl. Phys. 40, 1337 (1969).

54. T. M. Rice, A. Jayaraman, and D. B. McWhan, Supplement to J. de Physique 32, C1-39-45 (1970); D. B. McWhan and T. M. Rice, Phys. Rev. Letters 19, 846 (1967).

55. P. A. Fedders and P. C. Martin, Phys. Rev. 143, 245 (1965).

56. W. C. Koehler, R. M. Moon, A. L. Trego, and A. R. Mackintosh, Phys. Rev. 151, 405 (1966).

57. A. Jayaraman, T. M. Rice, and E. Bucher, J. Appl. Phys. 41, 869 (1970).

58. R. Nityananda, A. S. Reshamwala, and A. Jayaraman, Phys. Rev. Letters 28, 1136 (1972).

59. A. Jayaraman, R. G. Maines, K. V. Rao, and Sigurds Arajs, Phys. Rev. Letters 37, 926 (1976).

60. J. Zittartz, Phys. Rev. 164, 575 (1967).

61. H. G. Drickamer, S. C. Fung, and G. K. Lewis, Advances in High Pressure Research, Vol. 3, R. S. Bradley, ed., Academic Press, New York, USA (1969), p. 38.

B-2

STUDIES OF ELECTRON TRANSITIONS USING SOLID He PRESSURE TECHNIQUES*

J. E. Schirber

Sandia Laboratories
Albuquerque, New Mexico USA

INTRODUCTION

Studies of the pressure dependence of the Fermi surface of metals have proven useful in simplifying the details of the electronic structure of a large number of metals and semimetals in the past few years. Efforts have centered on: (1) the testing of band theoretical models or descriptions of the metal, and (2) studies of electron transitions or changes in the topology of the Fermi surface (FS), the subject of this discussion. Low-temperature pressure studies provide the best method for probing these topology changes because either temperature or alloying tends to smear out effects of these purely electronic transitions. There is a hierarchy of electron transitions of which the simplest situation is the electron or Lifshitz transition [1]. This is a change in FS topology with no change in the crystal structure and no change in the interatomic spacing. The next step in the hierarchy is a change in FS topology accompanied by a change in volume, but no change in structure. This class includes the "mixed valence" and "electron promotion" situations found in an increasing number of narrow band systems which are currently receiving an enormous amount of attention. Included in this class are many of the metal-insulator and metal-semiconductor transitions. Finally, there is the third tier in the hierarchy where the topology change in the Fermi surface is accompanied by (or results from, depending on what is thought to be driving the transition) a change in both structure and volume. In this discussion, emphasis will be placed on the "simplest" situation, the Lifshitz transition, where it is possible in several cases to follow directly the changes in the Fermi surface by use of deHaas-van Alphen or high field magnetoresistance techniques.

*Supported by the U.S. Energy Research and Development Administration under Contract AT(29-1)-789. Invited paper.

In Lifshitz' original treatment of electron transitions [1], he treated four situations: the creation or destruction of an entire sheet of the Fermi surface and the changes in connectivity involved in the creation or destruction of a neck or bridge between larger pieces of Fermi surface. Lifshitz dealt with a free electron density of states (DOS) which caused half integer powers to appear in the free energy, prompting him to call these "transitions of order 2.5." The two in this 2.5 is meant to indicate that that transition is second-order as well as isostructural. Anomalies then are predicted for essentially any properties depending upon the density of states.

Historically, most of the experimental evidence [2] for electron transitions has come from observed nonlinearities in the pressure and concentration (in the case of alloys) dependence of the superconducting transition temperature T_c. This is because T_c depends exponentially upon the density of states in the Bardeen-Cooper-Schrieffer [3] framework and therefore extremely small changes in the DOS are reflected as easily measurable changes in T_c.

A vast amount of work has been published [2] based on such data in the case of Tl. A peak in T_c vs. pressure is observed near 2 kbar. A careful search using deHaas-van Alphen techniques under pressure failed to show any sign of a topology change [4]. Holtham [5] has hypothesized that the small sheet responsible has already been formed at normal volume but that the corresponding peak in T_c occurs at higher pressure. While this is plausible it is perhaps more likely that some feature not simply related to the extremal cross sections measured to date is involved since the excursion in the DOS is only of the order of 0.1% of the total. Similar nonlinearities in T_c vs. pressure have been reported in Re [6], in In and its alloys with Pb and Cd [7,8], in the intermetallic compound [9] Bi_2K, and in VRu alloys [10] near the equiatomic composition. In all these materials, electron transitions have been suggested as the explanation for the anomalies, but there has been no direct confirmation of Fermi surface changes.

There are a number of situations when direct evidence of Fermi surface topology changes have been reported. Itskevich and Voronovskii [11] presented convincing evidence from high field magnetoresistance studies that the arms of the 2nd zone hole sheet of the Fermi surface are formed in Cd by 15 kbar.

The electronic properties of the As structured semimetals have been studied extensively as a function of pressure. Itskevich and Fisher [12] reported a vanishing of the entire Fermi surface of $Bi_{0.95}Sb_{0.05}$ near 10 kbar. This information was obtained from the vanishing with pressure of the frequency of Schubnikov-de Haas oscillations in the magnetoresistance. Brandt and Ponomarev [13] used similar techniques to study pressure-induced electron transitions in a series of Bi-based binary and ternary alloys. An electron transition involving changing the connectivity of the hole sheets of As by pinching off the "α" arms by 1.85 kbar was observed by Schirber and Van Dyke [14]. A volume-dependent band calculation based on the Lin and Falicov pseudopotential for As was found to be consistent with this result. In Sb the cross-sectional area corresponding to the minimum area of the hole sheet is found to initially decrease with pressure, then increase passing through to the normal volume value near 5 kbar [15]. This behavior is probably due to a not-as-yet identified electron transition.

Low-pressure anomalies in the transport properties of Cs are probably related to an electron transition near 7 kbar. Beardsley and Schirber [16] found that the dHvA signal corresponding to the "bellies" of the dimpled spherical Fermi surface disappear near this pressure for field directions where the orbit would be interrupted by the formation of [110] necks joining the spheres. Several band structure calculations indicate that these necks should be formed at modest pressures, but the actual necks were not detected.

Ribault has recently reported [17] that a dHvA frequency disappears in fcc Yb at an extrapolated pressure of 12 kbar. This is very near the pressure at which the electrical properties indicate a change from semimetallic to semiconducting behavior.

In this paper we will discuss briefly what appears to be the best documented case [18] of an electron transition -- $AuGa_2$ -- as an example of the types of anomalies that can be observed with various experimental probes. We will then describe recent studies on InBi and CeAs, as similar analyses have been applied to account for various anomalies in the pressure and temperature dependence of electronic and magnetic properties.

In the following sections the solid He technique [19] is described which has proven invaluable in direct Fermi surface measurements; the $AuGa_2$ situation is reviewed; the results on InBi and CeAs are summarized; the present situation in the study of electron transitions is summarized; and some recommendations for further activity are presented.

EXPERIMENTAL

As mentioned in the introduction, the emphasis in this discussion will be on direct determinations of Fermi surface topologies near electron transitions of the first kind -- those involving no changes in either volume or crystal structure. While much indirect information concerning an electron transition may be obtainable from temperature-dependent properties and from quasihydrostatic pressure studies of superconducting properties, extremely stringent conditions of hydrostaticity are required for direct measurement of the Fermi surface.

Direct measurements of the Fermi surface must be made at low temperatures because very long electron mean-free-paths are required. Above 25 bar, all pressure media are solid at 1 K. While a substantial amount of information has been obtained concerning the initial volume derivatives of cross sections using fluid He as the pressure medium, these techniques are inappropriate if one wishes to effect the transition with pressure. Therefore solid pressure transmitting media are required for these experiments. Itskevich [20] and coworkers, and Brandt [21] and coworkers have been able to use frozen fluid techniques in certain situations to obtain oscillatory magnetoresistance data. These situations usually involve small cross sections with small effective mass ratios. The amplitude factors for the oscillatory phenomena related to external cross sections of the Fermi surface involve products of the effective mass and the so-called Dingle temperature which is proportional to the scattering of the electrons. Thus for very small masses, observable amplitude oscillations may be observed. For more typical values of effective masses, the nonhydrostatic conditions obtained in these frozen water, kerosene, and pentane mixtures usually preclude direct measurements of the Fermi surface. Brandt and coworkers [22] have demonstrated that pulse heating of the sample so that the medium is liquefied or softened can improve the amplitude of the signals in low-temperature high-pressure experiments using frozen iso-n-pentane mixtures, but no additional use of this technique has been reported to date.

On the other hand, it has been amply demonstrated that the solid He technique [19] can give pressures to a maximum of 10 to 15 kbar which are experimentally indistinguishable from truly hydrostatic. This technique involves generation of pressure in fluid He at a temperature just above that at which the He will freeze. At 10 kbar this corresponds to about 60 K. This pressure is then maintained as the sample is cooled slowly from the bottom such that the fluid-solid line moves past the sample at constant pressure. The idea is to maintain a fixed position on the fusion curve until the sample is completely surrounded by solid He. When the fluid solid line reaches the capillary connecting the pressure vessel to the external system, it is considered blocked and subsequent cooling is assumed to be along an isochore. Within the validity of this assumption, the final pressure can be accurately determined from the known pressure-volume-temperature data for He. It is only fair to point out several shortcomings of the solid He technique. There is some danger in handling the high-pressure gas, so volumes must be minimized. This difficulty in handling the gas limits the practical pressure limit of the technique to 10 to 15 kbar. A thermal cycle is required to change the pressure which is fairly time-consuming and which precludes true constant temperature pressure derivatives.

A large amount of data has been accumulated on single crystal metals which indicate that even on the most fragile and anisotropic materials, hydrostatic conditions can be maintained with sufficient care. High-mass, high-frequency oscillations have been observed under solid He pressure conditions in many metals, usually without loss in amplitude. It is therefore possible with this technique to make detailed studies of the cross-sectional areas of the Fermi surface using the standard Fermi surface tools such as deHaas-van Alphen (dHvA) effect, the deHaas-Schubnikov oscillations, and high field magnetoresistance. The field modulation technique for determining dHvA frequencies has in fact proven to be the most powerful technique because it gives cross-sectional areas, electron effective masses, and information on scattering without the necessity of affixing leads to a specimen. The sensitivity with phase-sensitive detection techniques is such that the pickup and modulation can usually be external to the pressure vessel obviating the necessity of introducing leads into the high-pressure environment.

While the emphasis of this discussion is on direct Fermi surface studies and their relation to electron transitions, it is often difficult to obtain the necessary single crystal long mean-free-path samples necessary for such studies. The sensitivity of superconducting properties, particularly T_c, to these changes in the DOS was mentioned in the introduction. Another DOS sensitive probe is the nuclear magnetic resonance frequency shift, or Knight shift, which can be thought of as the product of a hyperfine coupling constant and the susceptibility. We have demonstrated that both pulsed and continuous wave resonance techniques can be employed in conjunction with solid He pressure generation. These techniques have been used to complement direct Fermi surface studies and in situations where the material is neither a superconductor nor can be obtained in single crystal form -- thus these measurements are the best available to probe the system in the region of an electron transition.

PRESENT UNDERSTANDING OF $AuGa_2$

As indicated in the introduction, there is fragmentary information concerning Lifshitz or electron transitions from many types of experiments on many different materials. However, $AuGa_2$ has probably the best understood electron transition from the standpoint of the diversity of experimental evidence and from the detail of the model description available. It seems appropriate then to briefly summarize the situation with regard to this, in some sense, ideal case.

$AuGa_2$ is a free-electron-like, valence 7 fcc metal which originally attracted attention because of anomalies in the magnetic properties which were called by Jaccarino, et al. [23] "the $AuGa_2$ dilemma." Specifically, the Ga nmr showed a large continuous change from positive to negative near 100 K as the temperature decreased. The Au nmr has no such anomaly, nor do any of the nmr signatures in the isomorphous compounds $AuAl_2$ and $AuIn_2$. The dHvA measurements of Fermi surface cross sections in all the three compounds with pressure showed qualitative agreement with Switendick's band calculation [24] as a function of interatomic spacing [25]. This band calculation showed an anomalously flat band having predominantly Ga s character centered at the point Γ in the Brillouin zone and lying completely below E_F. This same band in the related compounds $AuIn_2$ and $AuAl_2$ intersects the Fermi level and gives rise to the second band hole surface. This sheet is easily observable in dHvA work on $AuIn_2$ and $AuAl_2$, but is not seen in $AuGa_2$. Subsequent measurement of the pressure dependence at various temperatures of the Au and Ga Knight shift and spin lattice relaxation time and of the pressure dependence of T_c are all consistent with the following band structure-electron transition picture [18].

The flat Ga s band lies 0.012 eV below the Fermi level at 0 K and normal volumes. It can be pushed through the Fermi level with 7 kbar giving rise to the large abrupt changes in T_c and the Ga Knight shift and spin lattice relaxation. The Au nmr is not affected. With increasing temperature, the band actually drops with respect to E_F but is still sampled because of thermal smearing, thus giving rise to the anomalies in the temperature dependence of the Ga nmr properties and in the elastic and electrical properties. A schematic of this behavior is shown in Fig. 1.

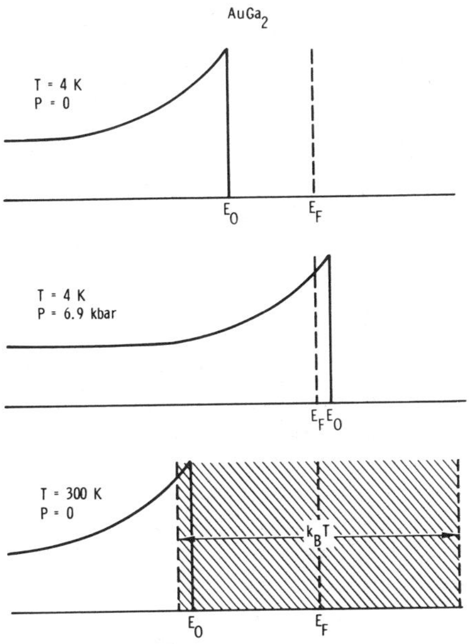

Fig. 1. Schematic representation of relative position of Fermi energy (E_F) and the Ga s-band edge (E_0) in $AuGa_2$ under varying conditions of temperature and pressure. The bottom sketch indicates the energy range sampled in the Knight-shift measurements at 300 K.

This is a nearly ideal situation because the sheet of the Fermi surface involved has a high density of states associated with it. This DOS has a preponderance of Ga s-band character, and Ga has a large moment so nmr is prominent. The large DOS change is reflected in a very large excursion (doubling) of T_c. In addition, $AuGa_2$ is a relatively simple calcium fluoride structured material that can be prepared as good quality single crystals. Finally, this high density of states peak is located very near the Fermi level and can be made to intersect E_F with an easily attainable pressure. This combination of fortuitous circumstances combine to make $AuGa_2$ the showpiece electron transition material.

RECENT RESULTS

InBi

InBi is a semimetal crystallizing in the tetragonal B10 structure instead of the usual cubic structures found in the typical III-V semiconductors. This material was the first compound metal in which direct Fermi surface data were obtained [26]. A number of anomalies in various properties of InBi with pressure and temperature variation have been reported. A change in sign of the In Knight shift and a disappearance of the electric field gradient at the In site were observed at around 230 K and 170 K, respectively [27]. Anomalies in the resistivity were found near 15 kbar by Bridgman [28], and Rapoport et al. [29] reported a jump in the resistance which they attributed to a first-order transition near 19 kbar. Gordon and Deaton [30] saw no volume discontinuity to 30 kbar. Broad anomalies in the pressure dependence of the acoustic mode velocities were observed by Fritz [31] but no phase changes were indicated. Coupled with these conflicting data concerning anomalies, there is a certain amount of controversy about the Fermi surface of InBi because of the likelihood of spurious signals from oriented inclusions of In_2Bi and/or In_5Bi_3. The most likely interpretation [32] is that the largest sheet is an ellipsoid oriented along [001] associated with the γ dHvA oscillation. The β oscillations are associated with a sheet of one-half the volume of the γ sheet, also oriented along [001]. Two much smaller sheets (10^{-2}-10^{-3} the volume of the γ) are observed. The α oscillations correspond to ellipsoids along [001] while the ζ sheet is probably an ellipsoid along [100]. We have measured the pressure derivatives for all of these cross sections using field-modulated dHvA techniques in solid He as shown in Table I. We have also performed relativistic orthogonalized-plane-wave band calculations as a function of interatomic spacing [33]. The first cut band calculation did not agree with the accepted experimental Fermi surface. A modified picture in which bands were shifted small amounts within the calculational uncertainty accounts for the experimental Fermi surface with respect to compensation of holes and electrons and the general orientational features. The model also gives the pressure behavior of the four observed sheets as far as both sign and relative magnitude of the pressure derivatives are concerned. An electron transition which the experimental data indicate will occur near 12 kbar at 4 K as the α sheet disappears is postulated to be the cause of the anomalies in the elastic properties with pressure (see Fig. 2).

The model indicates that the electron pocket near R in the Brillouin zone has substantial In s character and changes in the proper direction to explain at least qualitatively the anomalies in the nmr properties. While some manipulation of the model was necessary to achieve a satisfactory description of the normal volume Fermi surface, this treatment does a good job of explaining a number of heretofore puzzling anomalies in pressure- and temperature-dependent properties of InBi.

Table I. Pressure Derivatives $d\ln F/dP \equiv [F(P) - F(P = 0)]/F(P=0)\Delta P$ for InBi*

Field Direction	Frequency MG	Orbit	dlnF/dP, %/kbar
[001]	0.137	α	-8 + 1†
			-8.3 ± 0.6‡
	3.19	β	0.3 (±0.1)†
			0.3 (±0.1)‡
	7.78	γ	0.03 (±0.02)†
[100]	11.08	γ	0.3 (±0.1)†
	0.63	ζ	9 (±3)†
17 deg. from [110] in (001)	3.8	ε	1 (±0.5)†

*Orbit nomenclature is that of Meyer et al. [32].

†Fluid He phase shift.

‡Solid He to 4 kbar.

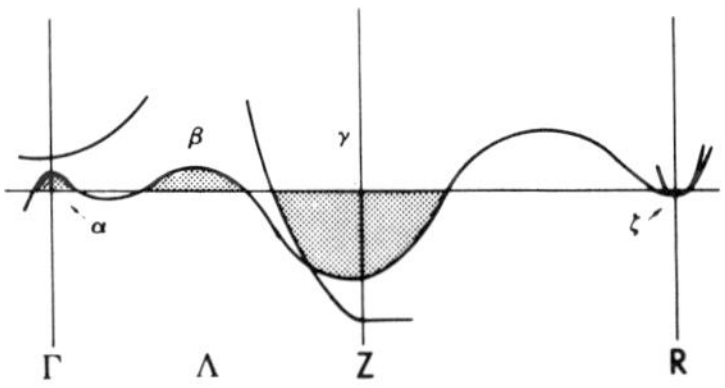

Fig. 2. Schematic energy band model of InBi. Bands not shown are assumed not to cross the Fermi energy. The effect of pressure is to increase the overlap of the bands at Λ and Z but with a net shift of the center of gravity and therefore E_F to higher energy. This decreases the overlap at Γ and increases it at R.

CeP and CeAs

Anomalies in the temperature dependence of the resistivity, thermal expansion [34], and pnictide nmr [35] of CeP and CeAs have been reported. The explanations for these anomalies attempted include promotion of 4f electrons in the Ce, nonlinear hyperfine interactions, and conduction electron effects. In these materials we have no possibility at present of obtaining direct Fermi surface information of any kind. The materials are not superconducting so this probe of the DOS is not useful. Furthermore, there is no detailed band theoretical picture presently available. We therefore have used measurements of the pnictide nuclear magnetic resonance frequency shift as a function of pressure to investigate these materials.

The results for the Knight shifts for the ^{31}P and ^{75}As in CeP and CeAs respectively at normal volume and at 4 kbar are shown in

Fig. 3. The anomalous behavior is the flattening of the Knight shift below 50 K. Based on a large body of data in these types of rare earth compounds [36], the Knight shifts were expected to show Curie-like behavior. The same data are shown in derivative form in Fig. 4.

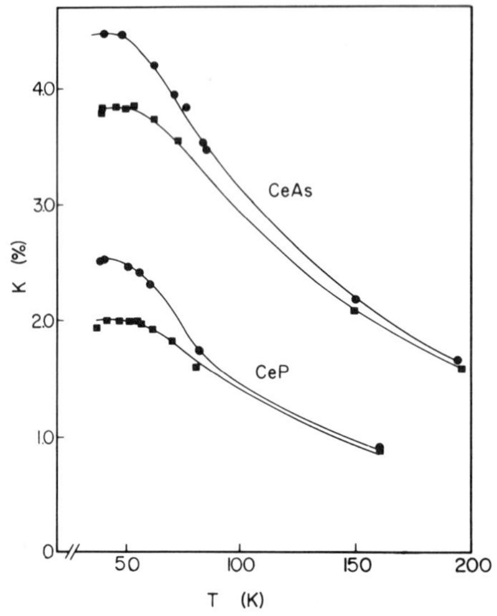

Fig. 3. Temperature dependence of the Knight shift for ^{31}P and ^{75}As contained in CeP and CeAs at 0.3 (■) and 4 kbar (●) pres pressure.

Attempts to interpret these pressure derivatives by varying the crystal field parameters or invoking nonlinear hyperfine coupling give the wrong qualitative behavior with temperature. Instead of showing a decrease in the percentage change per kbar as the temperature increases, they show an increase. A model involving a sharp feature in the DOS which, upon fitting the normal volume and pressure dependence, turns but to be a 30 K wide peak, 85 K below E_F moving towards E_F at a rate of 5 K/kbar. The solid line shown in Fig. 4 is the result of this model calculation and is seen to be in excellent agreement with experiment.

We have obviously leaned heavily on the much better documented cases of the $AuGa_2$ and InBi in evolving such a model. Recently Hirst [37] has discussed fluctuating valence descriptions for Ce which lead to narrow peaks in the DOS very near the Fermi level. Such models may ultimately legitimize our somewhat contrived picture for these materials.

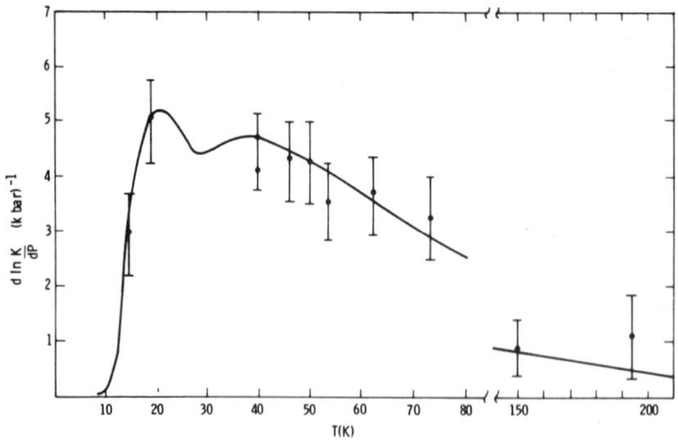

Fig. 4. Pressure derivatives of the ^{75}As Knight shift in CeAs as a function of temperature. The solid curve is derived with the model described in the text.

CONCLUSIONS

This brief review of past activities in the study of electron transitions and discussion of recent and ongoing work are perhaps indicative that progress is being made in the study of electron transitions. However, much needs to be done before a clear picture of the detailed nature of these "simple" transitions will evolve. Other candidate materials that lend themselves to experimental and theoretical investigations as does $AuGa_2$ are sorely needed. A specific case that should be studied in depth is that of Cs as this is potentially the simplest situation we are likely to encounter -- the change from a simple spherical Fermi surface to a multiple-connected array.

A second fundamental question is whether there is any connection between the various levels in the hierarchy of electron transitions. Do the simple electron or Lifshitz transitions serve in any fashion as precursors for more complicated transitions? The Russian investigators [38] have proposed that an electron transition occurs just before the valence change and associated isostructural lattice collapse in Ce. There is evidence, as indicated above, that an electron transition may precede the mixed valence state in CeP [39]. Again, Cs is a good candidate for such considerations.

The intermetallic compounds will probably continue to provide the most opportunities for investigation of electron transitions. Because of the flexibility inherent in the possibility of forming closely related cogeners, these materials are the best hope for providing the vehicles to lead us to the understanding of electron transitions in particular and perhaps phase changes in general.

ACKNOWLEDGMENTS

I thank all my collaborators, especially H. T. Weaver and J. P. Van Dyke, and am grateful to R. L. White for expert technical assistance.

REFERENCES

1. I. M. Lifshitz, Zh. Eksp. Teor. Fiz. 38, 1569 (1960) [Sov. Phys. JETP 11, 1130 (1960)].
2. V. G. Baryakhtar and V. I. Makarov, Zh. Eksp. Teor. Fiz. 49, 1934 (1965) [Sov. Phys. -- JETP 22, 1320 (1966)].
3. J. Bardeen, L. N. Cooper, and J. R. Schrieffer, Phys. Rev. 108, 1175 (1957).
4. J. R. Anderson, J. E. Schirber, and D. R. Stone, Colloque International Du C.N.R.S., Sur Les Proprietes Physiques Des Solides Sous Pression 188, 131 (1970).
5. P. M. Holtham, J. Phys. F: Metal Phys. 3, 1301 (1973).
6. C. W. Chu, T. F. Smith, and W. E. Gardner, Phys. Rev. B1, 214 (1970).
7. T. F. Smith, J. Low Temp. Phys. 11, 581 (1973).

8. R. J. Higgins and H. D. Kaehn, Phys. Rev. 182, 649 (1969).
9. N. E. Alekseevskii, ZhETF Pis. Red. 9, 571 (1969) [JETP Lett. 9, 347 (1969)].
10. C. W. Chu, E. Bucher, A. S. Cooper, and J. P. Maita, Phys. Rev. B4, 320 (1971).
11. E. S. Itskevich and A. N. Voronovskii, ZhETF Pis. Red. 4, 226 (1966) [JETP Lett. 4, 154 (1966)].
12. E. S. Itskevich and L. M. Fisher, ZhETF Pis. Red. 5, 141 (1967) [JETP Lett. 5, 114 (1967)].
13. N. B. Brandt and Y. A. G. Ponomarev, Zh. Eksp. Teor. Fiz. 55, 1215 (1968) [Sov. Phys. -- JETP 28, 635 (1969)].
14. J. E. Schirber and J. P. Van Dyke, Phys. Rev. Lett. 26, 246 (1971).
15. J. E. Schirber and W. J. O'Sullivan, Phys. Chem. Sol., Supplement No. 1 32, 57 (1971).
16. G. M. Beardsley and J. E. Schirber, Bull. Am. Phys. Soc. Series II 17, 693 (1972).
17. M. Ribault, Ann. Phys. 2, 53 (1977).
18. H. T. Weaver, J. E. Schirber, and A. Narath, Phys. Rev. B8, 5443 (1973).
19. J. E. Schirber, Cryogenics 10, 418 (1970).
20. E. S. Itskevich, Cryogenics 4, 365 (1964).
21. N. B. Brandt, I. S. Itskevich, and N. Ya. Minina, Usp. Fiz. Nauk. 104, 459 (1971) [Sov. Phys. -- Uspekki 14, 438 (1972)].
22. N. B. Brandt, S. V. Kuvshinnikov, and Ya. G. Ponomarev, Zh. ETF Pis. Red. 19, 201 (1974).
23. V. Jaccarino, M. Weger, J. H. Wernick, and A. Menth, Phys. Rev. Lett. 21, 1811 (1968).
24. A. C. Switendick and A. Narath, Phys. Rev. Lett. 22, 1423 (1969).
25. J. E. Schirber and A. C. Switendick, Solid State Commun. 8, 1383 (1970).
26. Y. Saito, J. Phys. Soc. (Japan) 17, 716 (1962).
27. D. L. Radhakrishna Setty and B. D. Mungurwadi, Phys. Rev. 183, 387 (1969).
28. P. W. Bridgman, Proc. Am. Acad. Arts Sci. 84, 43 (1955).
29. E. Rapoport, G. D. Pitt, and G. A. Saunders, J. Phys. C 8, L447 (1975).
30. D. E. Gordon and B. C. Deaton, Phys. Rev. B6, 2982 (1972).
31. I. Fritz, Solid State Commun. 20, 299 (1976).
32. R. T. W. Meyer, J. J. A. Hofmans, and A. R. DeVroomen, J. Phys. Chem. Solids 35, 307 (1974).
33. J. E. Schirber and J. P. Van Dyke, Phys. Rev. 15, 890 (1977).
34. T. Tsuchida, M. Kawai, and Y. Nakamura, J. Phys. Soc. (Japan) 28, 528 (1970).
35. S. M. Myers and A. Narath, Solid State Commun. 12, 83 (1973).
36. E. D. Jones, Phys. Lett. 22, 266 (1966); also Phys. Rev. 180, 455 (1969).
37. L. L. Hirst, Phys. Rev. B 15, 1 (1977).
38. L. G. Khvostantsev, L. F. Vereshchagin, and E. G. Shulika, High Temp.-High Press. 5, 657 (1973).
39. A. Jayaraman, W. Lowe, L. D. Longinotti, and E. Bucher, Phys. Rev. Lett. 36, 366 (1976).

B-3

HIGH PRESSURE PHONON DISPERSION OF ZINC CHALCOGENIDES AND THE METALLIC TRANSITION*

B. A. Weinstein†**

Purdue University
West Lafayette, Indiana USA

INTRODUCTION AND EXPERIMENT

The crystal potential responsible for solid cohesion is manifest through the phonon energies. For covalently bonded and partially covalent substances no simple crystal potential has yet been found. Thus, knowledge of the effect of high pressure on the phonon dispersion is important to an understanding in this area. Among other things, it is related to the stability regimes of various structural polytypes. Furthermore, such knowledge can be used to test the microscopic theories presently under development [1,2], which derive the phonon energies from the electronic eigenstates, by requiring that these theories reproduce the phonon dispersion not only at 1 atm, but at an arbitrary pressure. Therefore, a clear and detailed experimental picture of the phonon dispersion under pressure is needed.

Of the various techniques available, Raman scattering is most amenable to high-pressure measurements, especially for the zinc chalcogenides and other tetrahedral semiconductors. This is because the 1 atm Raman spectra and corresponding phonon dispersions of these materials are well understood, due to the extensive research of the last decade [3-7]. In addition, the difficult problems of experimental access presented by high-pressure devices are more readily overcome with an easily directed laser beam than, for example, with a neutron beam.

Several intermediate- (to 10 kbar) and high-pressure (above 100 kbar) Raman investigations of tetrahedral semiconductors have

*Work supported by National Science Foundation Grant No. DMR76-04775 and National Science Foundation/MRL Program No. DMR72-03018A04, and Research Corporation.

†Alfred P. Sloan Research Fellow.
**Current Address, Xerox Webster Research Laboratory, Webster, N.Y.

been reported. These include the work of Mitra [8,9], Cardona [10-12], Weinstein and Piermarini [13,14], and Weinstein [15,16]. The latter experiments, as do the present, employ a diamond-anvil press [17], which appears to be ideally suited for ultrahigh-pressure laser Raman scattering.

This paper discusses the effects of high pressure on the phonon dispersions of ZnS, ZnSe, and ZnTe, as measured by Raman scattering. It is the first detailed experimental study of these effects on a closely related series of materials. As do the other tetrahedral semiconductors, ZnS, ZnSe, and ZnTe undergo pressure-induced transformations to an opaque metallic (or semi-metallic) phase [18,19] at 150, 137, and 95 kbar, respectively [16,17]. Measurements were made up to the threshold of this transition for each material. Accompanying the metallic transition is a structural change to the NaCl polytype for ZnS [20], and also probably for ZnSe and ZnTe. However, in the latter cases insufficient data exist, and the white tin polytype occurring at high pressure for Group IV and III-V semiconductors, or the wurtzite structure are other possibilities [20,21]. In spite of their differences, the diamond and zinc-blende structure semiconductors have qualitatively similar Raman spectra, and their phonon dispersions are influenced similarly by high pressure. Thus, they are well suited for comparative study by these techniques.

The diamond-anvil pressure cell was used in the gasketed configuration with the pressure measured by the Ruby fluorescence technique [17]. At the highest pressures of this study ($\sim$145 kbar), effects due to stress gradients were still minimal. Further details of experimental technique are fully discussed elsewhere [14].

He-Ne (6328 Å) laser excitation was employed in conjunction with a double monochromator and photon counting system. Samples were unoriented single crystal chips $\sim 75\mu$ in diameter, prepared from undoped material. Several different samples were studied at overlapping pressures to provide a consistency check. All measurements were performed at room temperature.

RESULTS

The measured first- and second-order Raman spectra of ZnS, ZnSe, and ZnTe at various pressures are shown in Figs. 1, 2, and 3, respectively. In each case, the 1 atm spectrum is marked by the energies of various Brillouin zone center (ZC) and zone boundary (ZB) phonons at high-symmetry critical points (see Fig. 4) as determined by neutron inelastic scattering [6,7].

Let us first review the 1 atm results [3-5]. The spectra are divided into the sharp, intense, first-order peaks due to ZC optical phonons, labeled TO(Γ) and LO(Γ), and the broad second-order bands due to overtone, combination, and difference modes marked by various distinct critical points. The second-order

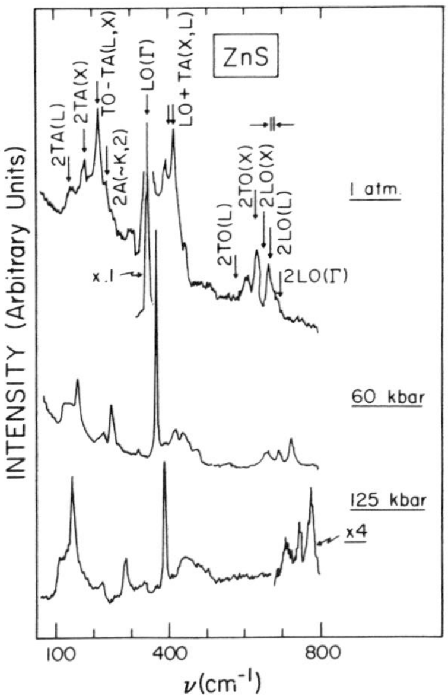

Fig. 1. Measured Raman spectra of ZnS at 1 atm, 60, and 125 kbar. One atm phonon assignments based on inelastic neutron scattering results (see text).

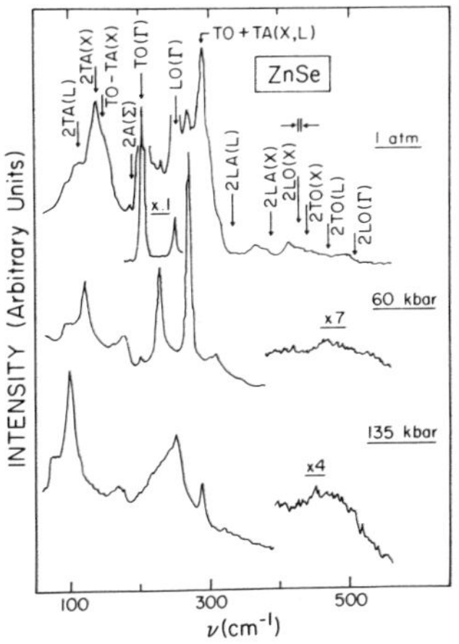

Fig. 2. Measured Raman spectra of AnSe at 1 atm, 60, and 135 kbar. One atm phonon assignments based on inelastic neutron scattering results (see text).

scattering is dominated by ZB phonons, which have the highest density of states. At low frequency, acoustic-mode scattering is prominent; and at high frequency, optic-mode scattering. In the middle, combinations of acoustic and optic phonons contribute strongly.

It has been demonstrated for tetrahedral semiconductors that the two-phonon bands are generally dominated by overtone contributions, and that the spectra have a strong density of states character especially in the acoustic region [12,22]. This fortunate occurrence for these materials makes interpretation of the spectra in terms of the phonon dispersion quite straightforward. However, there are several exceptions to this pattern, such as the TO-TA(L,X) [both L- and X-point phonons may contribute] difference mode at 215 cm^{-1} (1 atm) in ZnS (Fig. 1). Application of high pressure often allows easy identification of these exceptions. There are several key features common to all the spectra in Figs. 1, 2, and 3. Starting at low frequency a shoulder marks the energy of 2TA(L), followed

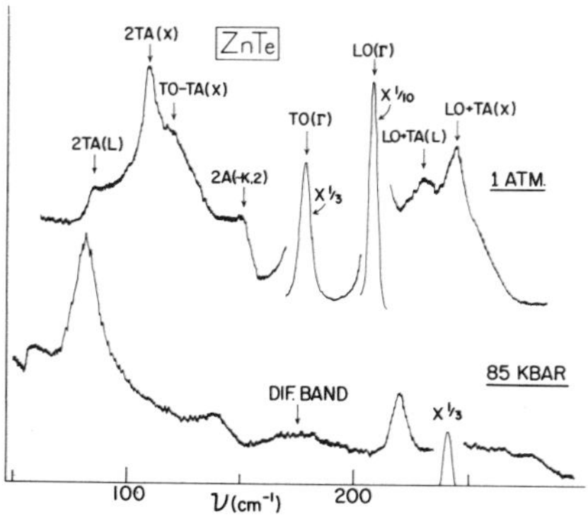

Fig. 3. Measured Raman spectra of ZnTe at 1 atm and 85 kbar. One atm phonon assignments based on inelastic neutron scattering results. The region of optic overtones was too weak to be studied (see text).

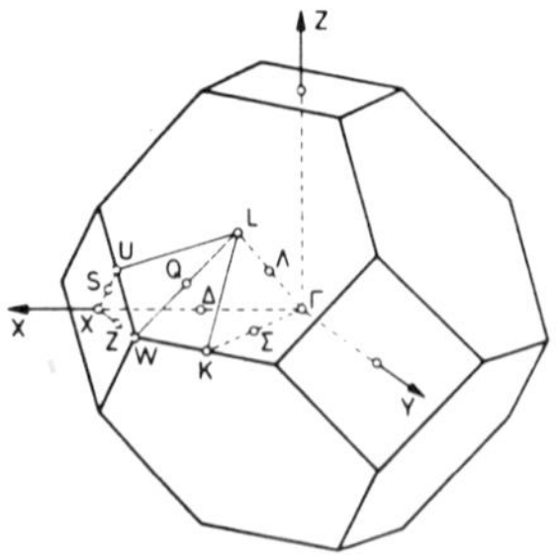

Fig. 4. First Brillouin zone for the fcc structure showing notation for points of high symmetry.

by a strong, sharp peak due to overtones of ZB TA phonons at, and surrounding, the X-point (see Fig. 4). Next, the shoulder labeled 2A(~ K,2) [2A(Σ) in ZnSe] is associated with overtones from the second acoustic branch along the Σ-direction, near the ZB K-point. The first-order TO(Γ) and LO(Γ) peaks follow, and then two second-order peaks probably associated with the combination LO + TA or TO + TA at X and/or L. Sometimes, as for ZnS, the latter are approximately degenerate with LA overtones. Next, two-phonon scattering appears from optic modes. Since the TO and LO bands are usually rather flat, one expects sharp peaks in this region such as in ZnS (Fig. 1), where at 1 atm the peak at 615 cm^{-1} corresponds to TO overtones, than at 670 cm^{-1} to LO overtones, and the peak between these at 640 cm^{-1} to TO + LO combinations. Because of the strong influence of the non-metal ion polarizability on the intensity of second-order optic mode scattering [5], the two-optic-phonon scattering in ZnTe was too weak to study effectively under pressure, and is not shown in Fig. 3.

The effect of high pressure on these spectra is clearly shown in Figs. 1 through 3. For each material, phonons of given origin behave similarly with only the magnitude of the effects varying from one material to another. The energies of ZB (and near ZB) TA phonons from the lowest branch decrease, or "soften", with pressure. The shifts of 2TA(L) and 2TA(X) are typical of the effect on other large $\vec{q}$ phonons from this branch. At the same time, the elastic constants are expected to increase with pressure [23]. The energies of ZB phonons from the second acoustic branch decrease only slightly at high pressure, as evidenced by the small downward shift of the shoulder 2A(~ K,2) [2A(Σ) for ZnSe]. Although not clear from Fig. 2, detailed analysis of the data for ZnSe shows that ZB LA phonons increase in energy with increasing pressure in agreement with previous results on other tetrahedral semiconductors [14,24]. The TO(Γ) and LO(Γ) peaks move to higher energy with pressure; TO(Γ) shifts faster than LO(Γ) so that the ZC LO-TO splitting decreases with pressure. This also agrees with previous findings [8,10,14]. Optic modes from the ZB of both the LO and TO branches also increase in energy as pressure is applied. This is clear for ZnS from the shift of the corresponding second-order features in Fig. 1. Detailed analysis also shows this to be true for ZnSe, and even for ZnTe where some information on ZB TO and LO phonons can be gleaned from the behavior of combination and difference modes. Similar results were previously found for Si and

GaP [14]. For ZnSe the intersection of the LO and TO branches results in considerable spectral distortion as the pressure is changed.

The behavior of combination and difference modes involving ZB TA phonons is particularly interesting due to the negative pressure coefficients of the latter. Combinations of these with optic phonons tend to have small pressure coefficients, whereas difference modes such as TO-TA(X) have large positive pressure coefficients. For each material the TO-TA(X,L) difference mode, occurring (1 atm) just above the prominent 2TA(X) peak, is easily identified by its large pressure shift to higher energy. It is pointed out that the degeneracy of TO-TA(X,L) with other two-phonon modes allows strong anharmonic interactions which can distort the shape of the two-phonon spectrum. Thus for the zinc chalcogenides, only after the TO-TA difference mode has shifted above the region of TA overtones do the spectra resemble the characteristic density of states shape found for Si and Ge [22] (see the high pressure spectra in Figs. 1 through 3). For ZnSe, TO-TA(X) shifts enough to become nearly degenrate with TO(Γ) at high pressure. Thus, the TO(Γ) peak at 135 kbar in Fig. 2 is highly distorted by the anharmonic decay TO(Γ) → TO-TA(X). This will be the subject of a future publication [25]. A similar effect was recently investigated in GaP [15].

The above discussion is summarized in Figure 5, which displays semiquantitative plots (dashed curves) of the phonon dispersions for ZnS, ZnSe, and ZnTe at the threshold pressures of their metallic phase transitions, and compares them to the 1 atm phonon dispersions (circles, squares, and triangles) as measured by neutron inelastic scattering [6,7]. The high-pressure plots were constructed on the basis of the present Raman measurements which accurately establish the positions of key critical points. Since the dispersion curves are analytic and the pressure hydrostatic, it is possible to interpolate reasonably between critical points according to the shape of the 1 atm phonon dispersion. These semiquantitative plots are not expected to be in error by more than 25% for any $\vec{q}$.

The quantitative effects of pressure on the energies of various critical-point phonons are summarized in Table I, which lists the mode Grüneisen parameters, $\gamma_i = -(d\nu_i/\nu_i)/(dV/V)$, measured in this work and by various authors for Si, GaP, ZnTe, ZnS, and ZnSe. The few theoretical calculations of these parameters which exist do not adequately explain the dispersion of γ_i across the Brillouin zone [26-29].

RELATION TO THE PRESSURE-INDUCED PHASE TRANSITION

The "softening" of ZB TA phonons with increasing pressure is a significant property of the tetrahedral structure. By now enough

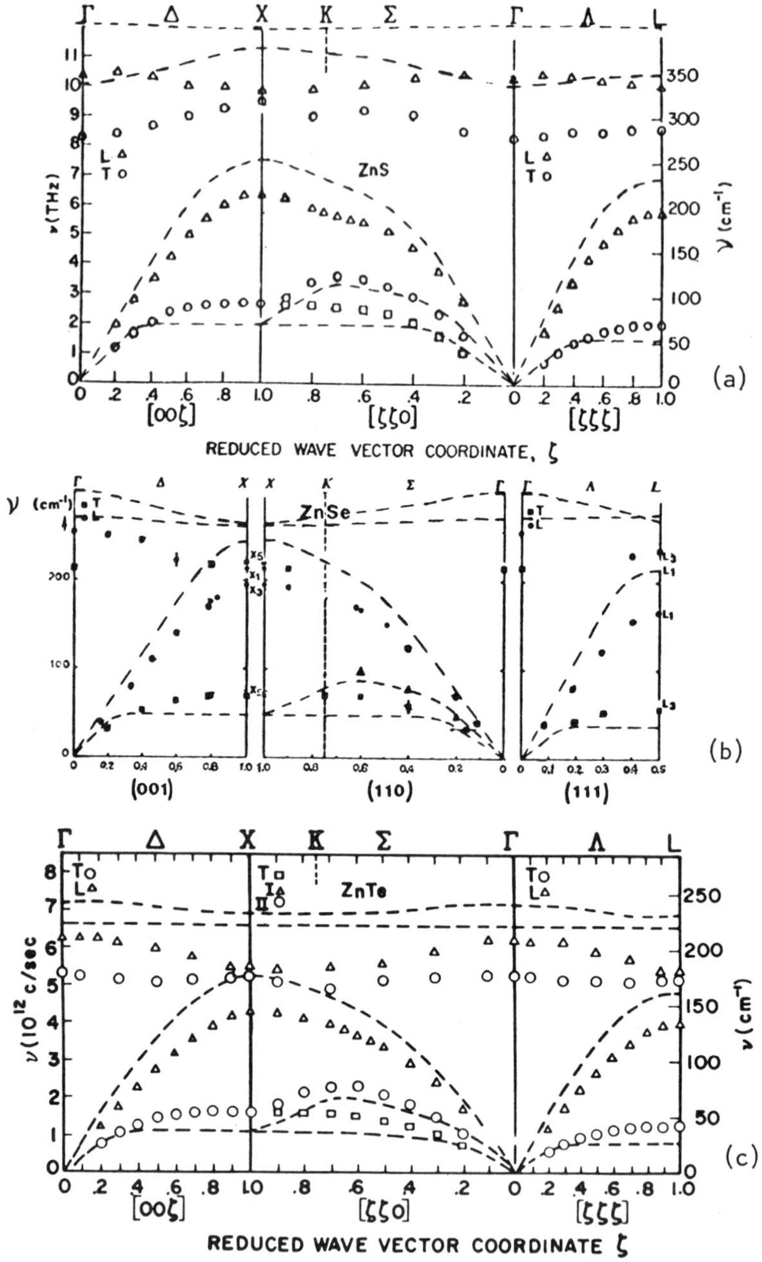

Fig. 5. Semiquantitative plots (dashed curves), based on the present results, of the phonon dispersions for (a) ZnS, (b) ZnSe, and (c) ZnTe at the threshold pressures of their metallic transitions, 150, 137, and 95 kbar, respectively. Corresponding phonon dispersions at 1 atm (squares, circles, and triangles) for these materials from the best available inelastic neutron scattering results (see text).

data has been collected to realize that this characteristic is probably common to all sphalerite and zinc-blende structure semiconductors. It accounts for their anomalous negative thermal expansion at low temperature [30]. Furthermore, it has been shown that the "softening" indicates a weakening of noncentral forces of long-range origin needed to stabilize the tetrahedral lattice against short wavelength shear distortion [31]. Therefore, several authors have conjectured that ZB TA "softening" is a likely driving mechanism for the high-pressure phase transformations in these materials [8,14,32]. The present results provide additional evidence for this.

In Fig. 6 the most reliable mode Grüneisen parameters determined in this and other experiments for the TA(X) phonons of Si[14], GaP[14], ZnS, ZnSe, and ZnTe are plotted as a function of the transition pressures of these materials. The data shows a linear correlation to well within experimental error. Note that this empirical correlation occurs among materials varying widely in ionicity, and in spite of the different polymorphs assumed above the transition. No similar correlation for $\gamma_{TA(L)}$ has been found, nor is there a simple relation between $\gamma_{TA(X)}$ and ionicity. Furthermore, the author would caution against the premature conclusion on the basis of the limited data that all tetrahedral semiconductors will fit into this pattern [33].* Before attempting further conclusions, a detailed calculation of the effect of pressure on the free energy should be carried out. This is now possible on the basis of this and earlier work [14] and will be the subject of a future article [34]. However, for the five materials studied, a clear, albeit empirical, connection has been demonstrated between the effect of pressure on the TA(X) energy and the threshold of the pressure-induced structural and accompanying metallic transition.

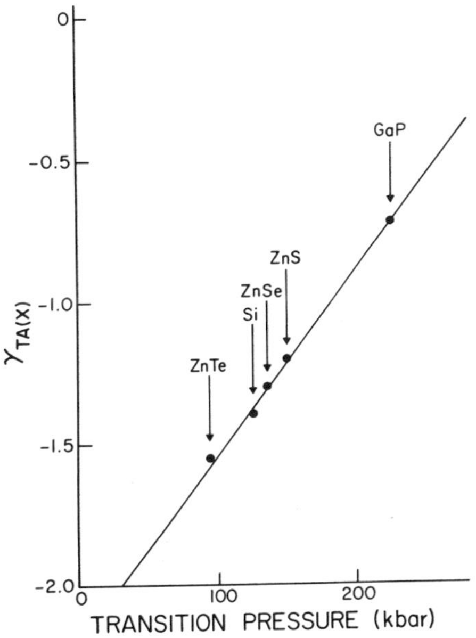

Fig. 6. Observed linear correlation between the mode Grüneisen parameter for the TA(X) phonons in Si, GaP, ZnS, ZnSe, and ZnTe, and the metallic phase transition pressures of these materials.

Table I. Measured Mode Grüneisen Parameters, γ_i

	Si	GaP	ZnTe	ZnS	ZnSe
TO(Γ)	.98 ± .06 [14]	1.09 ± .03 [11]	1.7 ± .1‡		1.4 ± .1‡
	1.02 ± .02 [10]	1.19 ± .04 [12]	1.55± .1 [8]	1.85 ± .2 [8]	1.65± .1 [8]
		1.07 ± .1 [8]			
LO(Γ)	.98 ± .06 [14]	.95 ± .02 [11]	1.2 ± .1‡	.95 ± .1‡	.9 ± .1‡
	1.02 ± .02 [10]	1.16 ± .04 [12]	1.0 ± .1 [8]	.95 ± .1 [8]	1.1 ± .1 [8]
		1.1 ± .1 [8]			
TA(L)	-1.3 ± .3 [14]	-.81 ± .07 [13]	-1.0 ± .2‡	-1.5 ± .2‡	-1.5 ± .2‡
TA(X)	-1.4 ± .3 [14]	-.72 ± .03 [13]	-1.55± .2‡	-1.2 ± .2‡	-1.3 ± .2
	-1.4 ± .2 [11]	-.60 ± .07 [12]	-1.1, -1.6 [9]		- .6, -1.1 [9]
A(~ k,2)	- .3 ± .1 [14]	.15 ± .03 [13]	- .4 ± .1‡	- .3 ± .1‡	- .4 ± .1‡
LA(X)	--------------	--------------	--------------	--------------	1.1 ± .2‡
LA(L)	--------------	--------------	--------------	--------------	1.1 ± .2‡
TO(X)	1.5 ± .1 [14]	1.31 ± .05 [13]	1.8 ± .4‡	1.0 ± .2‡	1.6 ± .3**
TO(L)	1.3 ± .2 [14]	1.40 ± .08 [13]	--------------	1.0 ± .2‡	1.6 ± .3**
		1.50 ± .05 [12]			
LO(X)	--------------	1.3 [12]	1.7 ± .3*	1.1 ± .2	.9 ± .2
LO(L)	--------------	1.3 [12]	1.8 ± .3*	1.1 ± .2	.9 ± .2

† Deduced from shift of TO-TA difference band.
* Deduced from shift of LO+TA combination bands.
** Deduced from shift of TO+TA combination band.
‡ Present study.

It is generally accepted that the decrease in the 1 atm ZB TA phonon energies in the series C, Si, Ge, and Sn reflects the increasing metallicity of these materials, by which long-range forces are weakened due to stronger screening effects [35]. Therefore, one may reasonably expect the pressure induced ZB TA "softening" to reflect a similar metallization in the dielectric properties of these materials.

However, one may take a different point of view, as was done by van Vechten [36]. There the threshold for the pressure-induced metallic transition was predicted with reasonable success by calculating the energy difference between a free-electron gas and a Penn-type semiconductor, according to a scaling theory based on the Phillips ionicity scale [37]. Since the interatomic forces are, after all, a result of the electronic bonding and corresponding eigenstates, one might ask how the broadening of electronic bands and delocalization of states, which are the general effects of compression, change the phonon energies? For ZB TA phonons it is interesting that the mode "softening" which appears to drive the transformation to a state that is invariably metallic is caused by a metallization, viz., increased screening of long-range forces, in the low-pressure phase. It would seem that, even though the transformation is first order, the metallic property of the high-pressure phase is related to changes in the semiconducting phase as the transition is approached. A full understanding of these phenomena will certainly require a detailed microscopic treatment deriving the lattice dynamics of tetrahedral semiconductors from their electronic energies and states. Such models are presently under development [1,2]. The point to be made here is that a proper microscopic theory should both reproduce the 1 atm phonon dispersion, and correctly describe the effect of pressure on it. In particular, the "softening" of ZB TA modes should be well accounted for. In this regard, the semiquantitative plots presented in Fig. 5 can serve as a stringent test of the correctness of such theories, and it is hoped they will stimulate further research in this area.

ACKNOWLEDGMENT

The author wishes to thank A. K. Ramdas for the use of his laser, and for supplying the samples.

*Unpublished results on GaAs indicate that this material would not obey the linear relation in Fig. 6. However, here anharmonic effects may play a larger role.

REFERENCES

1. D. J. Chadi and R. M. Martin, Solid State Commun. 19, 643 (1976)
2. R. Zeyher, Phys. Rev. Lett. 35, 174 (1975).
3. W. G. Nilsen, Phys. Rev. 182, 838 (1969).
4. J. C. Irwin and J. LaCombe, Can. J. Phys. 50, 2596 (1972).
5. R. L. Schmidt, B. D. McCombe, and M. Cardona, Phys. Rev. B11, 746 (1975).
6. N. Vagelatos, D. Wehe, and J. S. King, Chem. Phys. 60, 3613 (197
7. B. Hennion, F. Morissa, G. Pepy, and K. Kunc, Phys. Lett. 36A, 376 (1971).
8. S. S. Mitra, O. Brafman, W. B. Daniels, and R. K. Crawford, Phys. Rev. 186, 942 (1969).
9. O. Brafman and S. S. Mitra, in Proceedings 2nd Intern. Conference on Light Scattering in Solids, M. Balkanski, ed., Flammarion, Paris, France (1971), p. 284.
10. C. J. Buchenauer, F. Cerdeira, and M. Cardona, in Proceedings 2nd Intern. Conference on Light Scattering in Solids, M. Balkanski, ed., Flammarion, Paris, France (1971), p. 280.
11. W. Richter, J. B. Renucci, and M. Cardona, Solid State Commun. 16, 131 (1975).
12. B. A. Weinstein, J. B. Renucci, and M. Cardona, Solid State Commun. 12, 473 (1973).
13. B. A. Weinstein and G. J. Piermarini, Phys. Lett. 48A, 14 (1974).
14. B. A. Weinstein and G. J. Piermarini, Phys. Rev. B12, 1172 (1975)
15. B. A. Weinstein, Solid State Commun. 20, 999 (1976).
16. B. A. Weinstein in Proceedings 13th Intern. Conference on The Physics of Semiconductors, F. G. Fumi, ed., Tipografia Marves, Rome, Italy (1976), p. 326.
17. G. J. Piermarini and S. Block, Rev. Sci. Instr. 46, 33 (1975).
18. G. A. Samara and H. G. Drickamer, J. Phys. Chem. Solids 23, 457 (1962).
19. S. Minomura, G. A. Samara, and H. G. Drickamer, J. of Appl. Phys. 33, 3196 (1962).
20. G. J. Piermarini, private communication.
21. W. Klement and A. Jayaraman, in Progress in Solid State Chemistry, Vol. 3, Pergamon Press, London (1966), p. 289.
22. B. A. Weinstein and M. Cardona, Phys. Rev. B7, 2545 (1973).
23. H. J. McSkimin and P. Andreatch, Jr., J. Appl. Phys. 35, 2161 (1964).
24. R. T. Payne, Phys. Rev. Lett. 13, 53 (1964).
25. B. A. Weinstein, to be published.
26. G. Dolling and R. A. Cowley, Proc. Phys. Soc. (London) 88, 463 (1966).
27. J. F. Vetelino, S. S. Mitra, and K. V. Namjoshi, Phys. Rev. B2, 967 (1970).
28. H. Jex, Phys. Status Solidi (b) 45, 343 (1971).
29. A. Bienenstock, Philos. Mag. 9, 755 (1964).

30. G. K. White, in AIP Conference Proceedings No. 17, Thermal Expansions, R. E. Taylor and G. L. Denman, eds., American Institute of Physics, New York (1974), p. 1.
31. W. Weber, Phys. Rev. Lett. 33, 371 (1974).
32. J. C. Phillips, in Bonds and Bands in Semiconductors, A. M. Alper, J. L. Margrave, and A. S. Nowick, eds., Academic Press, New York (1973), p. 94.
33. P. Y. Yu, private communication.
34. B. A. Weinstein, to be published.
35. G. Nilsson and G. Nelin, Phys. Rev. B6, 3777 (1972).
36. J. A. van Vechten, Phys. Rev. B7, 1479 (1973).
37. J. C. Phillips, Rev. Mod. Phys. 42, 317 (1970).

B-4

THE PRESSURE INDUCED STRAIN TRANSITION IN NiF_2^*

J. D. Jorgensen and T. G. Worlton
Argonne National Laboratory
Argonne, Illinois USA
and
J. C. Jamieson
The University of Chicago
Chicago, Illinois USA

INTRODUCTION

At zero pressure, NiF_2 has the well-known rutile (TiO_2) structure which belongs to the tetragonal $P4_2/mnm$ space group. This structure may be viewed as consisting of sheets of linear F-Ni-F molecules oriented along <110> in the sheet at z = 0 and along <1$\bar{1}$0> in the sheet at z = 1/2. Compounds with this rutile structure have received considerable attention as candidates for pressure-induced phase transitions resulting from a softening of the acoustic mode corresponding to the effective elastic consant $1/2(C_{11} - C_{12})$ [1-3]. However, before the present structural measurements on NiF_2, the second-order strain transition in TeO_2, which has a slightly distorted rutile structure, was the only such transition where the relevant structural parameters, elastic constants, and acoustic phonon modes had been studied in detail at high pressure and compared with Landau's theory [4-8].

Recent high-pressure elastic-constant measurements on NiF_2 by Wu [9] have shown that $1/2(C_{11} - C_{12})$ decreases with increasing pressure to 1.0 GPa, suggesting the possibility of a phase transition at some higher pressure. Subsequent x-ray measurements by Jamieson and Wu [10] confirmed that a tetragonal to orthorhombic transition does occur at elevated pressures. However, they found the orthorhombic strain to be sufficiently small that doublets were resolved only above 4.0 GPa. It was impossible for them to unambiguously determine the space group of the high-pressure phase,

*Work supported by the U.S. Energy Research and Development Administration.

to measure the true transition pressure, or to conclude whether the transition was continuous.

Using time-of-flight neutron diffraction, we have now shown that the transition is continuous and occurs at 1.83 ± 0.1 GPa and that the high-pressure phase is orthorhombic Pnnm. The latter is a subgroup of $P4_2/mnm$ and is consistent with a second-order strain transition involving a soft-transverse acoustic-phonon mode propagating along <110> and polarized along <1$\bar{1}$0> [11,12].

EXPERIMENTAL

The neutron measurements were performed on the time-of-flight powder diffractometer located at beam hole H-8 of Argonne's CP-5 research reactor. The configuration of this instrument is shown in Fig. 1. A polychromatic beam of neutrons from the reactor is chopped into short pulses, scattered from the sample, and detected at a scattering angle of $2\theta = 90^\circ$ in a time-focused array of BF_3 counters. Time focusing allows the use of a comparatively large detector array with no loss of resolution [13]. The fixed 90° scattering angle makes it possible to completely eliminate reflections from the pressure cell using simple collimators in the incident and scattered beams. The diffractometer has a total flight path of 3.7 m. At the chopper speed used for this experiment, the resolution $\Delta d/d$ (FWHM) varies from 1.3% for d-spacings of 0.9 Å to 0.9% at 2.0 Å. Data were collected for about 48 hrs at each pressure.

The pressure cell is a piston-cylinder device consisting of a sintered Al_2O_3 cylinder supported radially in a steel binding ring [14]. The NiF_2 powder was loaded along with deuterated methanol into a 0.69 cm OD x 5.1 cm long teflon tube sealed with an unsupported area type cap. The incident and scattered neutrons pass through 0.32-cm wide x 3.18-cm long slits in the steel binding ring. Although we typically use an internal CsCl pressure calibrant in our high-pressure diffraction runs, the pressure calibration for this experiment is based on a separate set of CsCl measurements using identical sample geometries. This approach was taken because some CsCl reflections happened to fall at spacings of special interest in the NiF_2 pattern. Thus, the overall pressure uncertainties are estimated to be ± 0.1 GPa. Decker's equation of state [15] was used to determine pressures from the measured CsCl lattice parameters. Data were taken at nine different pressures from 0.12 to 3.24 GPa.

Lattice and structural parameters were obtained from the time-of-flight diffraction data using a profile refinement technique [16]. In this method, a calculated profile is compared to the raw data and the difference is minimized by varying the lattice, atomic position, and thermal vibration parameters. If the resolution function of the instrument is well characterized, line positions can be determined far more accurately than might be expected based on their widths, and line broadening, resulting from a small strain, can easily be detected and analyzed.

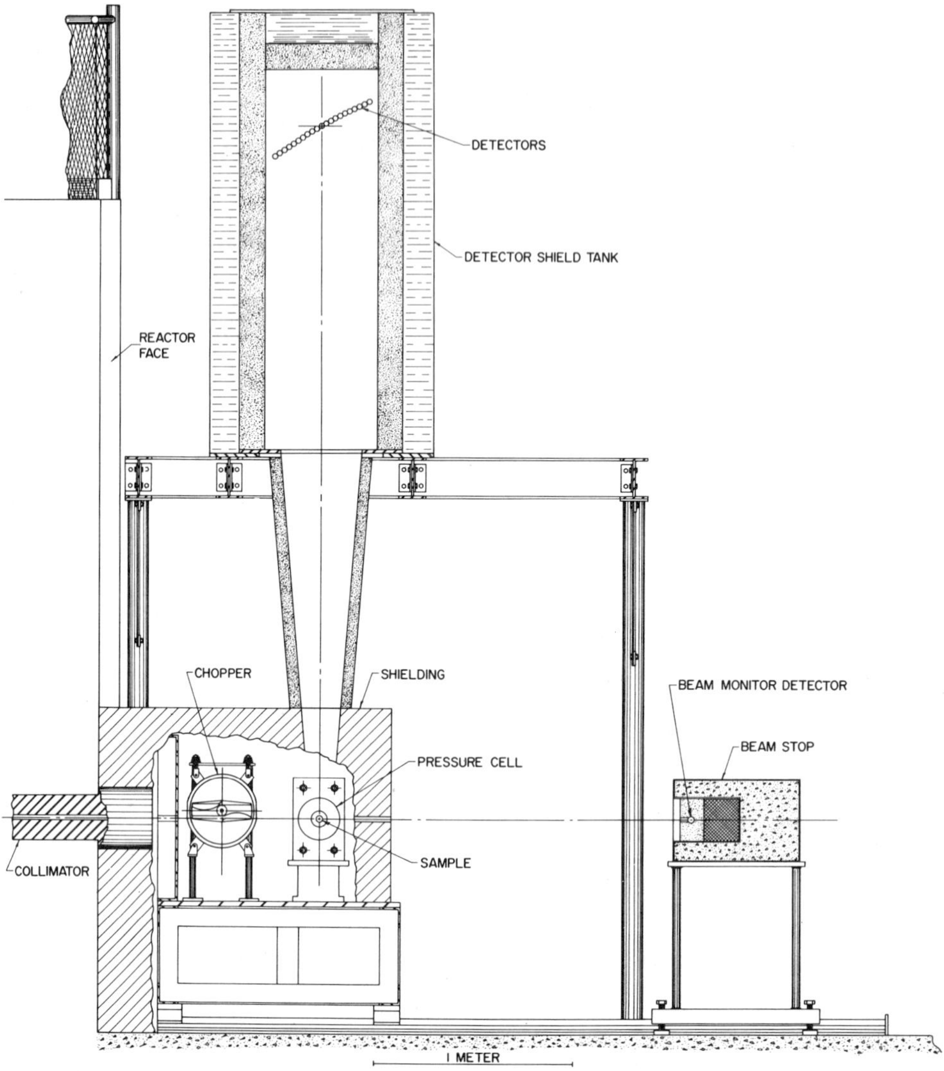

Fig. 1. Schematic drawing of high-pressure time-of-flight powder diffractometer located at the H-8 beam hole of Argonne's CP-5 research reactor.

The small strain associated with the transition in NiF_2 offers an excellent example of the power of the profile refinement technique. Profile refinement fits at three different pressures are shown in Figs. 2, 3, and 4. Figure 2 shows the fit obtained for data at 1.37 GPa using the correct tetragonal $P4_2/mnm$ space group. Figure 3 illustrates the problems encountered if the same tetragonal space group is used to fit data above the transition at 2.87 GPa. Most notably, the (211) reflection is broadened, the (301) (112)

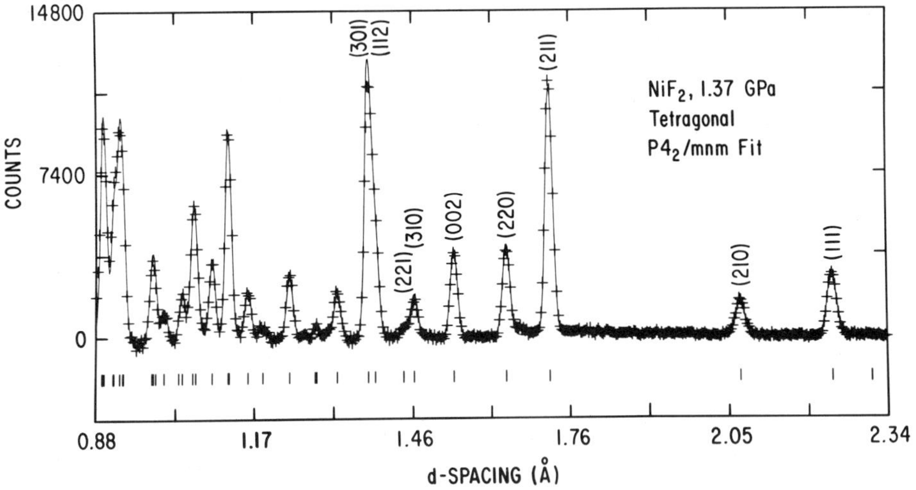

Fig. 2. Refinement profile of NiF_2 diffraction data at 1.37 GPa. Plus marks are the raw data points. The solid line is the calculated profile. Background has been removed before plotting. Tick marks below the profile indicate positions of allowed reflections. A few reflections are indexed in parentheses above the profile.

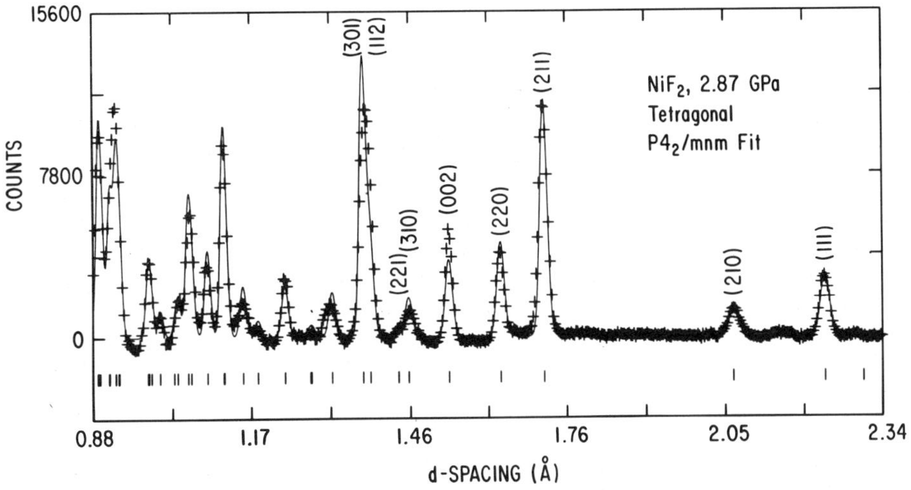

Fig. 3. Refinement profile of NiF_2 diffraction data at 2.87 GPa using the incorrect $P4_2/mnm$ space group. (Format is the same as Fig. 2.) Note especially the poor fitting of the (211), (002), (112), and (301) reflections.

combination appears broadened and displaced toward longer d-spacings, and the intensity of (002) is larger than calculated. (The (210) is also significantly broadened, but is hard to distinguish in this figure.) A perhaps more subtle, but equally important, problem is that several peaks appear displaced slightly either above or below their expected positions based on tetragonal symmetry. Figure 4 shows the much-improved fit obtained when the same 2.87 GPa

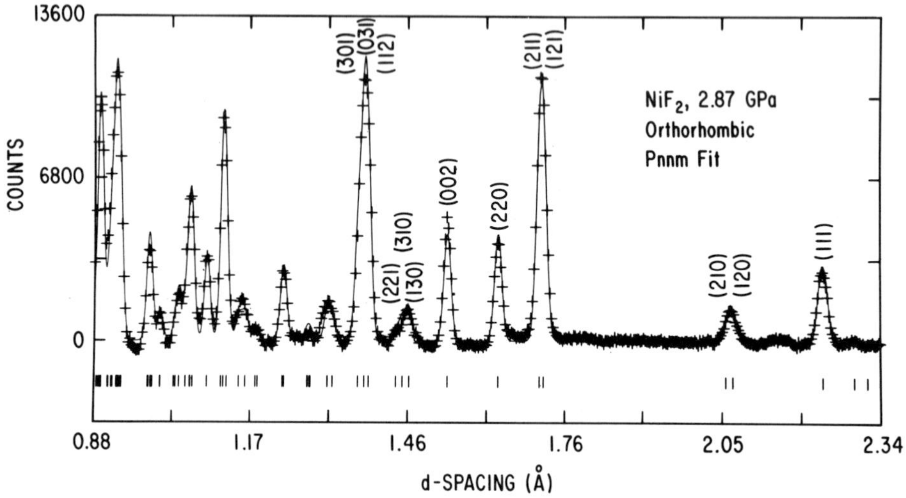

Fig. 4. Refinement profile of NiF_2 diffraction data at 2.87 GPa using the correct Pnnm space group. (Format is the same as Fig. 2.)

data are refined using the orthorhombic Pnnm space group. The (211) (121) doublet matches the observed width, as does the (301)(031)(112) combination. Numerous other improvements are clearly visible in the remainder of the pattern.

The profile refinements in the tetragonal phase were based on 35 allowed reflections with d-spacings between 0.88 Å and 2.34 Å; in the orthorhombic phase 59 reflections were included. For both structures the inclusion of anisotropic thermal parameters gave a small but noticeable improvement in the fits. The refined parameters at the nine different pressures studied are listed in Table I.

RESULTS

From the observed diffraction spectra in the high-pressure phase, it was obvious that the structure is orthorhombic with no change in unit cell size because all lines could be indexed with a slightly distorted tetragonal cell. Since the orthorhombic lattice parameters indicate a continuous transition, the orthorhombic space group must be a subgroup of $P4_2/mnm$. The only orthorhombic subgroups obeying the observed selection rules and having the same unit cell size are Pnnm and Pnn2. When refinements using Pnn2 were

Table I. Structural and Thermal Parameters of NiF_2 vs. Pressure

	Lattice Parameters and Atomic Positions				
P,GPa	a,Å	b,Å	c,Å	x	y
0.12	4.6499(3)		3.0836(3)	0.3040(4)	
0.65	5.6404(3)		3.0801(3)	0.3031(4)	
1.00	4.6348(3)		3.0783(3)	0.3037(4)	
1.37	4.6293(3)		3.0760(3)	0.3031(4)	
1.75	4.6244(3)		3.0735(3)	0.3029(4)	
2.12	4.6063(8)	4.6331(8)	3.0717(3)	0.294(2)	0.312(2)
2.49	4.5967(8)	4.6358(6)	3.0716(3)	0.296(1)	0.310(1)
2.87	4.5884(7)	4.6376(6)	3.0708(4)	0.293(1)	0.311(1)
3.24	4.5808(8)	4.6386(7)	3.0698(4)	0.293(1)	0.312(1)

Thermal Parameters (x 10^3, $Å^2$) Defined Such That the Temperature Factor Takes the Form:

$$\exp[-2\pi^2(\frac{U_{11}h^2}{a^2}+\frac{U_{22}k^2}{b^2}+\frac{U_{33}l^2}{c^2}+\frac{2U_{12}hk}{a\,b})]$$

	Ni				F			
P,GPa	U_{11}	U_{22}	U_{33}	U_{12}	U_{11}	U_{22}	U_{33}	U_{12}
0.12	3(1)		1(1)	-5(1)	11(1)		6(2)	-15(2)
0.65	4(1)		1(1)	-4(1)	12(1)		4(2)	-13(2)
1.00	5(1)		0(1)	-3(1)	12(1)		5(2)	-12(1)
1.37	5(1)		-1(1)	-2(1)	13(1)		4(2)	-12(1)
1.75	6(1)		-1(1)	-1(1)	14(1)		2(2)	-12(2)
2.12	8(3)	4(2)	2(1)	-1(1)	12(4)	17(4)	2(2)	-10(2)
2.49	18(3)	-4(2)	2(1)	-1(1)	24(4)	30(3)	1(2)	-12(2)
2.87	18(3)	-6(2)	1(1)	-1(2)	29(4)	-1(2)	-1(2)	-11(2)
3.24	20(3)	-6(2)	1(1)	-3(2)	30(4)	0(3)	2(2)	-12(2)

Numbers in parenthesis are statistical uncertainties of last significant figures.

attempted, the atomic positions always converged to values consistent with Pnnm symmetry, thus indicating that Pnnm is the correct space group.

The measured lattice parameters and unit cell volume versus pressure are shown in Figs. 5 and 6. The primary feature is the splitting of the tetragonal, a lattice parameter, into the unequal a and b orthorhombic lattice parameters beginning at about 1.83 GPa and increasing smoothly to the highest pressure studied. It is also evident that within the limits of our accuracy (a + b)/2, c, and the unit cell volume are all continuous through the transition, while the pressure derivatives of these quantities all change at the transition.

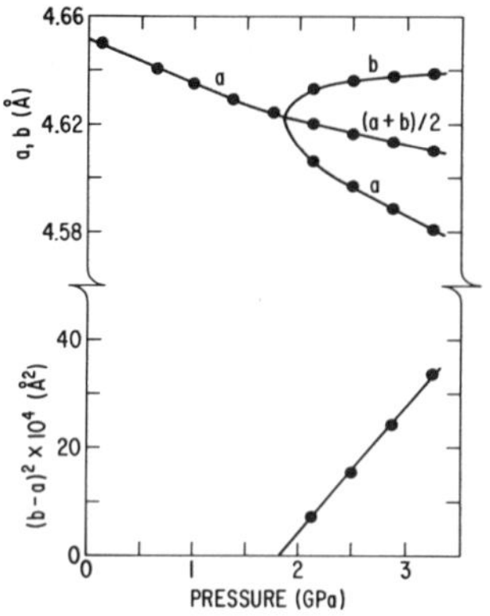

Fig. 5. Lattice parameters a and b of NiF_2 vs. pressure showing the splitting of the tetragonal a into the orthorhombic a and b. The lower curve shows the square of the strain (b - a), vs. pressure. Standard deviations are smaller than the points.

As discussed by Nye [17], an expansion of the elastic energy density in terms of the appropriate order parameters and elastic constants would proceed the same as has been previously done by several authors for TeO_2 [4,6,7,18]. (TeO_2 belongs to space group $P4_12_12$.) The primary order parameter for the NiF_2 transition is the strain (b - a). Figure 5 shows that $(b - a)^2$ vs. pressure is a straight line with a slope of 2.27×10^{-3} Å^2/GPa extrapolating to zero at the transition pressure as predicted by theory. The previous treatments have also shown that when terms of sufficiently high order are retained in the expansion, the observed anomalies in the pressure derivatives of (a + b)/2, c, and V are also predicted.

The atomic motions associated with the transition can be easily understood by referring to Fig. 7. In a given x-y plane the Ni and F atoms lie along the <110> direction in (x,x,o) positions in the low-pressure (rutile) phase. Upon going through the transition, the F atoms move in the directions shown by the arrows to (x,y,o) positions and the b axis becomes longer than a. From the measured atomic position parameters (Table I) the displacement of the F atoms is calculated to be about 0.06 Å. The data seem to indicate that the F atoms move this distance rather abruptly at the transition and then continue to move an additional small amount determined solely by the lattice strain as pressure is increased above the transition. Atomic position data for TeO showed somewhat the same behavior [4]. This abrupt displacement is, however, not surprising when we consider the dimensions and orientation of the thermal ellipsoid for the F atoms. Transforming the anisotropic thermal parameters for F (Table I) to the correct principal axes, we find that the average rms thermal vibration in the tetragonal phase is about 0.16 Å in the <110> direction and roughly 0.07 Å or less in the other two orthogonal directions. Thus, the measured atomic displacement of 0.06 Å at the transition is only about 1/3 as large as the rms thermal displacement in the same direction. It is also significant to note that the observed direction of largest thermal displacement is consistent with the transition mechanism. This is the first case where high-pressure measurements of anisotropic thermal parameters have indicated the displacement pattern involved in a continuous phase transition.

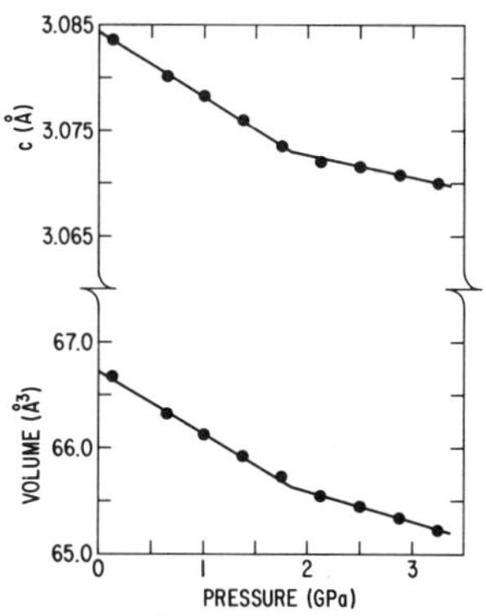

Fig. 6. Lattice parameter c and unit cell volume of NiF_2 vs. pressure. Standard deviations are smaller than the points.

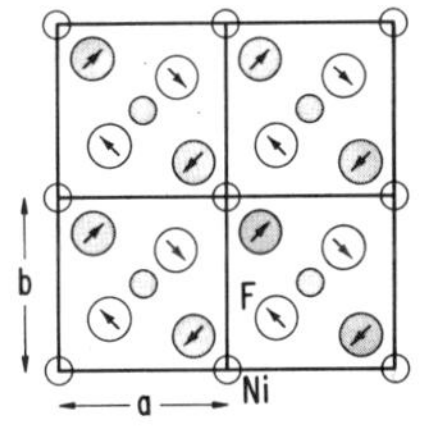

Fig. 7. Projection of the NiF_2 structure onto the x-y plane. Open circles indicate atoms at z = 0; shaded circles indicate z = 1/2. Arrows on the F atoms indicate the atomic motions associated with the transition. In the tetragonal phase a = b and the F atoms lie at (x,x,o); in the orthorhombic phase b > a and the F atoms lie at (x,y,o) with y > x.

DISCUSSION

We have shown that NiF_2 undergoes a continuous pressure-induced tetragonal to orthorhombic phase transition at 1.83 ± 0.1 GPa. The high-pressure space group has been shown to be Pnnm which is consistent with the transition being second order with the strain being the primary order parameter [11,12]. It is relevant to note that Austin [19] has demonstrated that an orthorhombic polymorph of NiF_2 can be produced by quenching from nonhydrostatic high pressures, presumably being stabilized by residual stresses [20]. Our measurements confirm the suggestion by Nagel and O'Keefe [20] that this orthorhombic form of NiF_2 has the $CaCl_2$ structure. However, diffraction spectra taken after our hydrostatic pressure measurements confirmed that our sample had returned completely to the rutile structure.

It is suspected that the transition is driven by a soft transverse acoustic phonon mode propagating along <110> and polarized along <1$\bar{1}$0>; however, this has not yet been confirmed experimentally. The effective elastic constant corresponding to this mode, $1/2(C_{11} - C_{12})$ has been shown to soften upon increasing pressure to 1.0 GPa, but has not been measured up to the transition pressure. The observed atomic motions associated with the transition are identical to those of the B1g (Γ_3^+) optic mode at zero wave vector [21], leading us to speculate that this optic mode may also play a role in the transition.

NOTATION

a,b,c, = lattice parameters
C_{ij} = elastic constants
d = plane spacing
h,k,l = Miller indices
P = pressure
U_{ij} = elements of the thermal displacement tensor
V = unit cell volume
x,y,z = atomic positions
θ = Bragg angle

REFERENCES

1. H. N. Pandey, Phys. stat. Sol. 11, 743 (1965).
2. M. E. Striefler and G. R. Barsch, Phys. stat. Sol. (B) 59, 205 (1973).
3. M. E. Striefler and G. R. Barsch, Phys. stat. Sol. (B) 64, 613 (1974); also 67, 417 (1975).
4. T. G. Worlton and R. A. Beyerlein, Phys. Rev. B12, 1899 (1975).
5. P. S. Peercy, I. J. Fritz, and G. A. Samara, J. Phys. Chem. Solids 36, 1105 (1975).
6. D. B. McWhan, R. J. Birgeneau, W. A. Bonner, H. Taub, and J. D. Axe, J. Phys. C: Solid State Phys. 8, L81 (1975).
7. E. F. Skelton, J. L. Feldman, C. Y. Liu, and I. L. Spain, Phys. Rev. B13, 2605 (1976).
8. L. D. Landau and E. M. Lifshitz, Statistical Physics, Pergamon Press, London (1958), Chap. XIV.
9. A. Y. Wu, Phys. Lett. 60A, 260 (1977).
10. J. C. Jamieson and A. Y. Wu, J. Appl. Phys. 48, 4573 (1977).
11. R. A. Cowley, Phys. Rev. B13, 4877 (1976).
12. N. Boccara, Ann. Phys. 47, 40 (1968).
13. J. M. Carpenter, Nucl. Instr. Meth. 47, 179 (1967).
14. R. M. Brugger, R. B. Bennion, T. G. Worlton, and W. R. Myers, Trans. Amer. Cryst. Assoc. 5, 141 (1969).
15. D. L. Decker, J. Appl. Phys. 42, 3239 (1971).
16. T. G. Worlton, J. D. Jorgensen, R. A. Beyerlein, and D. L. Decker, Nucl. Instr. Meth. 137, 331 (1976).
17. J. F. Nye, Physical Properties of Crystals, 2nd ed., Pergamon Press, Oxford (1969), Chap. VIII.
18. I. J. Fritz and P. S. Peercy, Solid State Commun. 16, 1197 (1975).
19. A. E. Austin, J. Phys. Chem. Solids 30, 1282 (1971).
20. L. Nagel and M. O'Keeffe, Mat. Res. Bull. 6, 1317 (1971).
21. J. G. Traylor, H. G. Smith, R. M. Nicklow, and M. K. Wilkinson, Phys. Rev. B3, 3457 (1971).

B-5

STUDIES OF THE HIGH PRESSURE TRANSITION IN SULFUR

A. L. Ruoff and M. C. Gupta

Cornell University
Ithaca, New York USA

INTRODUCTION

The metallic nature of the elements in Group VI A increases with atomic weight. Oxygen and sulfur are insulators, selenium is a high resistance semiconductor, tellurium is a low-resistance semiconductor or what is often called a semimetal, and polonium is a metal. One of the most interesting high-pressure characteristics of Group VI A elements is that each approaches the metallic state under extreme pressure conditions (except oxygen which has not been studied yet). The results of the studies of high-pressure transition in crystalline orthorhombic sulfur are presented here.

Sulfur [1] is one of the best insulators ($\rho \simeq 10^{16-18}$ ohm-cm) known at room temperature and pressure. Sulfur exists in several allotropic forms. Ordinary sulfur is a yellow solid substance with crystals of orthorhombic symmetry. In orthorhombic sulfur the atoms are arranged in eight numbered puckered rings. Sulfur is a Van der Waals molecular solid with band gap of 2.5 eV.

Static high-pressure resistance measurements on sulfur up to 100,000 atm (old scale) by Bridgman showed no transformation to a metallic state. Under shock compression, David and Hamann [2] measured the resistivity of cast sulfur pellets (of unspecified purity and allotropic state) to 230 kbar and approximately 1500 K. The resistivity varied from about 10^{18} ohm-cm to less than 0.03 ohm-cm, and this change in resistivity was taken to indicate a metal transition. Their results were substantiated by similar measurements reported by Eichelberger and Hauver [3]. Extrapolation of their data indicates that resistance approaches metallic behavior at about 220 kbar.

Suchan et al. [4] extrapolated their static pressure energy gap data for sulfur (of unspecified purity and allotropic state) to the pressure where the energy gap vanishes. It was estimated from the optical data (with appropriate corrections to their old pressure scale) that the energy band gap would extrapolate to zero at a pressure of about 293 kbar.

Vereshchagin et al. [5,6] statically compressed sulfur (of unspecified purity and allotropic state) to a load of 50 kg where a diffuse transition of sulfur to the conducting state was observed; no pressure was given. Notsu in an unpublished result [7] statically compressed sulfur; a sharp transition to metallic state was observed. The transition was observed at two loads differing by a factor of two; no pressure was given. Le Neindre et al. [8] have mentioned in their review paper that the experiment with sulfur was performed in the split sphere and the transition was slightly below 28.5 GPa. Chhabildas and Ruoff [9] statically compressed orthorhombic sulfur at about 175 kbar, the resistance began to drop and proceeded to drop by several orders of magnitude with slight additional loading. The sulfur transition was found to show a considerable time dependence. Dunn and Bundy [10] statically compressed sulfur to 500 kbar where the sulfur was converted to the metallic state with the final electrical resistivity of 0.032 ohm-cm. It appears from their abstract of the paper that a continuous transition was observed rather than a sharp transition as claimed by Notsu [7].

It is apparent from the above discussion of the different works on sulfur that it is uncertain whether sulfur shows a sharp transition or a continuous transition to a metallic state. Secondly, there is disagreement about the transition pressure of sulfur to a metallic state. We believe that the discrepancy about the continuous or sharp transition to metallic state is due, perhaps, to the lack of specification of the initial crystal state and purity in some cases, and to the amount of time the sulfur is left at a given load. So it was decided to study the behavior of the transition in orthorhombic sulfur under static compression to 300 kbar, using a technique in which a diamond indentor with a spherical tip is pressed against a diamond flat. This technique of generating high static pressure will enable us to calculate the transition pressure rather than transition load.

EXPERIMENTAL TECHNIQUES

Pressure Generation

In this experiment high pressures were generated using a diamond indentor with a spherical tip, which was pressed against a diamond flat on which a thin film of material to be studied was deposited as shown in Fig. 1. The stresses developed in the contact region were calculated using Hertz's equation for the elastic

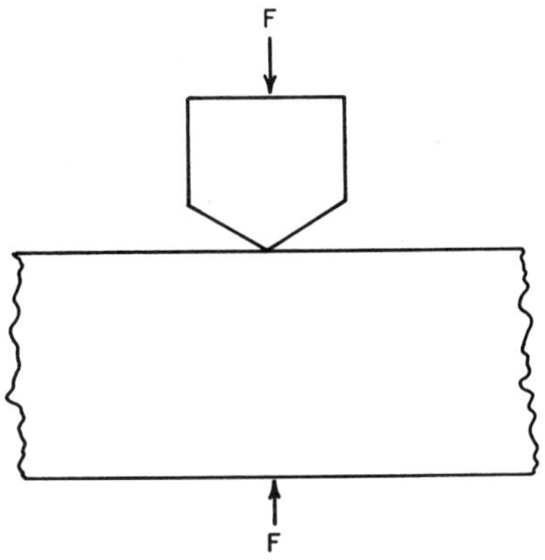

Fig. 1. Indentor with rounded tip pressed against slab [12].

deformation of a spherical surface [11]. If we treat the diamond as an isotropic solid (see Appendix) of Young's modulus E and Poisson's ratio ν and if a load F is applied to an indentor of radius of curvature R, then the maximum pressure at the center will be given by

$$P_o = (\frac{3}{2})^{1/3} \pi^{-1} (\frac{E}{1-\nu^2})^{2/3} R^{-2/3} F^{1/3} \tag{1}$$

and the average pressure over the elastic circle of contact will be

$$\bar{P} = \frac{2}{3} P_o \tag{2}$$

The pressure profile, which is a hemisphere, is given by

$$P = P_o (1 - \frac{r^2}{a^2})^{1/2} \tag{3}$$

A point in the indentor along the z axis far removed from the spherical indentor tip, and a point along the z-axis of the flat anvil far removed from the anvil surface, move toward each other during loading by a distance α, where

$$\alpha = R [\frac{\pi(1 - \nu^2)}{E}]^2 P_o^2 \tag{4}$$

The sulfur film thickness should be kept less than about $\alpha^t/10$. This requirement is necessary in order that the above pressure profile given by (3) and the pressure given by (1) are not altered significantly.

The elastic coefficients (E and G) of sulfur are about 1/100 of the value for diamond. The state of stress near the center of contact region in the sulfur film is essentially hydrostatic as the yield strength of sulfur is much smaller than the pressure at which the transition is observed. A diamond indentor with a radius R of 0.040 cm was used in this study. The accurate profile of the diamond tip was obtained using the Newton's ring technique [13]. The approximate value of the pressure attained was also obtained from the photograph of the imprint of the indentor (which gives the contact region) on the sulfur film as shown in Fig. 2.

The spherical diamond indentor and flat were cleaned by elaborate techniques and then were coated with a very thin film of metal using a sputtering device. The metal film on diamond deposited by the sputtering method was very strong and had a low electrical resistance. The metal coating on the diamond indentor

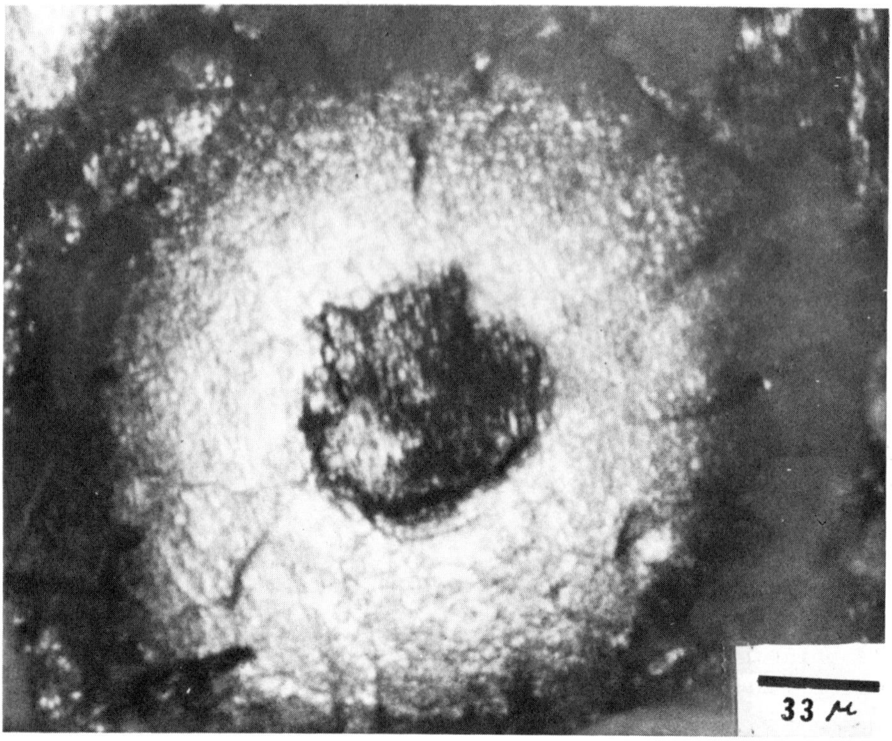

Fig. 2. Photograph of a print of the indentor (which gives the contact region) on a sulfur film (F = 7 kg, R = 0.04 cm). The central dark region is approximately the contact area.

and flat provided the leads for the electrical resistance measurements.

Sample Preparation

A thin film of sulfur, with purity of 99.999 +%, was deposited on the metal-coated diamond flat. The sulfur film was deposited by evaporation under a vacuum of 10^{-6} torr. A crystalline orthorhombic sulfur film was obtained when the substrate was kept at liquid nitrogen temperature during evaporation [14]. Figure 3 shows the surface structure of the film of orthorhombic sulfur and the x-ray diffraction pattern of the crystalline orthorhombic sulfur film. The initial resistances of the deposited crystalline (orthorhombic) sulfur thin films were measured and were greater than $3x10^{12}$ ohms.

RESULTS, DISCUSSION AND CONCLUSIONS

The variation of sulfur resistance with load was determined using two independent techniques: one used an opposed-anvil

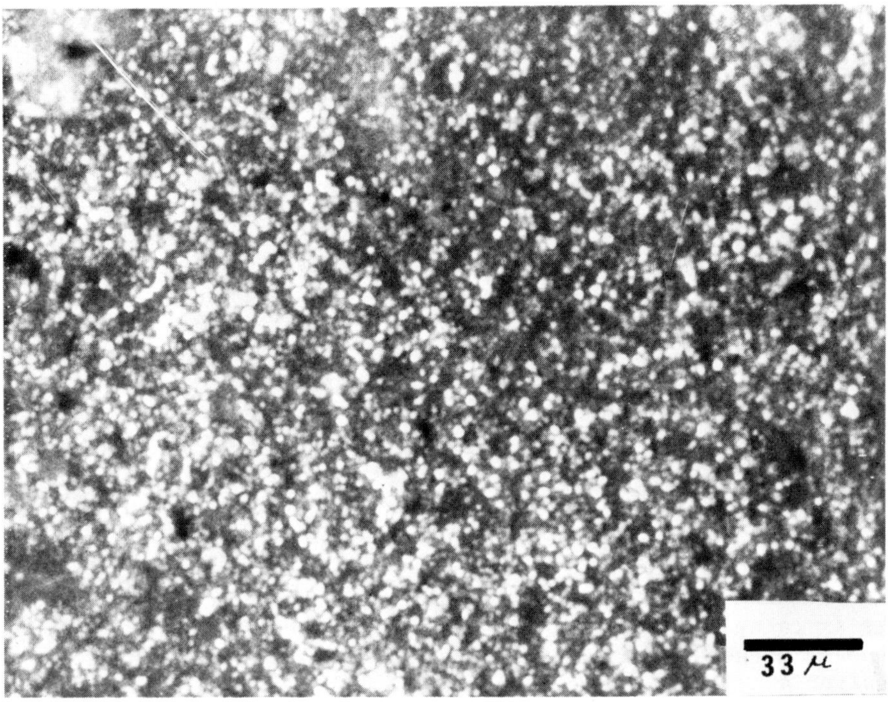

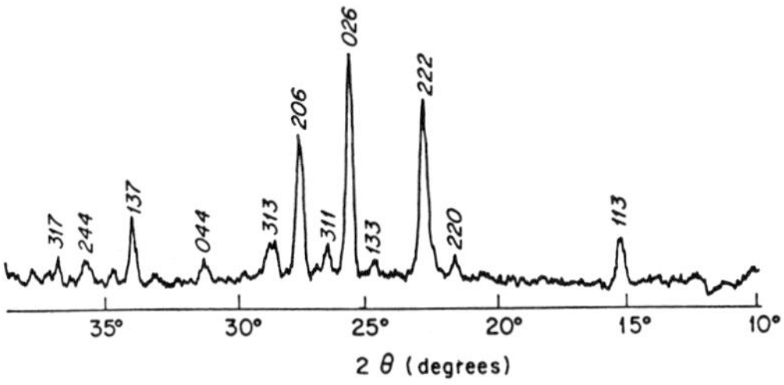

Fig. 3. Surface structure of the orthorhombic sulfur film deposited at liquid nitrogen temperature and its x-ray diffraction pattern.

apparatus with Boron carbide pistons (Fig. 4), and the other used the indentor technique (Fig. 1). Results of the variation of sulfur resistance with load as determined using the opposed-anvil technique are shown in Fig. 5, while results for the indentor technique are shown in Fig. 6. The initial resistance of sulfur can be measured reliably up to 3×10^{12} ohms (the limit of our apparatus for resistance measurements was 10^{14} ohms). Previous resistance measurements have been reported to only about 10^{8} ohms. The first indication of a large resistance drop for sulfur can be seen even when the resistance is 3×10^{12} ohm, which is at a lower pressure than is the case when a less sensitive ohmmeter is used.

When a load is reached such that the sulfur resistance can be measured with our apparatus, then the resistance of sulfur drops with time only. This implies that the sulfur transition is time

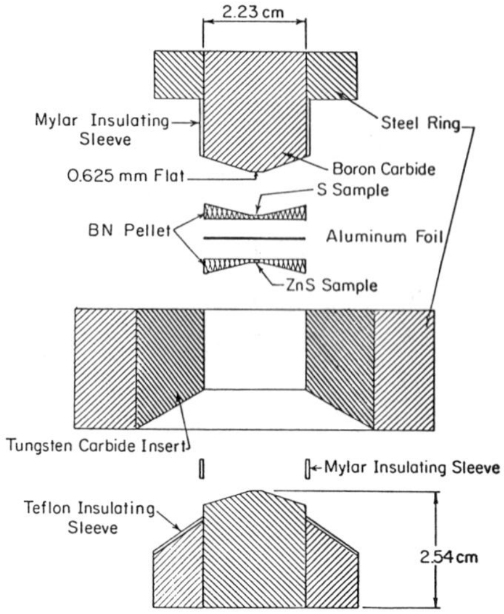

Fig. 4. Supported opposed-anvil device used for simultaneous resistance measurements of two samples. The Boron carbide anvils which are gold coated (by sputtering) are used as individual electrical leads and the center foil is used as a common ground. The pressure versus load relation is nonlinear at higher pressures and particularly when pressure is approaching the saturation value. Failure occurred at 49 tons, probably caused by fracture before the limiting pressure for B_4C of about 275 kbar was reached [9].

dependent. The nature of the transition as shown in Figs. 5 and 6 can be explained as follows. As the sulfur transforms, the volume decreases and hence the load decreases which stops the transformation after a certain period. The load is then increased to the original value and more sulfur is transformed to a new phase. The kinetics of the transition were such that no detectable change in resistance was observed at a given load. Therefore, the load was increased further to accelerate the kinetics of the transition. Finally a load value was reached where all the sulfur apparently had transformed and no change in resistance with time was observed.

The results of our study on sulfur are shown in Table I, along with other reported works on sulfur. The following conclusions can be drawn from this study: (1) The work of Chhabildas and Ruoff [9] has been extended from resistance measurements of 10^8 ohm to 3×10^{12} ohm. Thus this allows us to obtain the resistance vs. pressure curve at lower pressures. This new resistance-measuring apparatus was used with both the supported opposed-anvil technique with Boron carbide postons and with the new diamond indentor technique described here. In both techniques it was found that the resistance drop was observed at a lower pressure. (2) The maximum pressure attainable with supported opposed anvils of Boron carbide (if fracture is prevented) is

$$P_{Max}^{Limiting} = P_H\ (\text{Knoop}) + \Delta P \qquad (5)$$

$$\simeq 265 + 9 \text{ kbar}$$

Table I. Results of High Pressure Studies of Sulfur

Author	Apparatus	Sulfur Purity	Initial Crystal Structure	Transition Pressure	Type of Transition	Comments
Bridgman [16]	static	unspecified	orthorhombic	--	--	No transition observed up to 100,000 atm.
David and Hamann [2]	shock	"	unspecified	230 kbar	--	Temperature at transition ≃ 1500°K, final resistivity ρ=0.03 ohm cm.
Suchan et al. [4]	static	"	"	--	--	Extrapolation of optical data indicates $P_{tr} \simeq 293$ kbar (see Fig. 7).
Vereshchagin et al. [5,6]	"	"	"	--	continuous	Actual pressure unspecified, sluggish.
Notsu [7]	"	"	"	--	sharp	Two transition loads given. Actual pressure unspecified.
Le Neindre et al. [8]	" (split sphere)	"	"	Approx. 28.5 GPa	--	
Chhabildas and Ruoff [9]	static	"	orthorhombic	175 kbar	continuous	Sluggish.
Dunn and Bundy [10]	"	--	--	300 - 500 kbar	"	ρ_{final} = 0.032 ohm-cm at 500 kbar on their extrapolated load vs. pressure curve.
Present study	" (opposed anvil with two samples)	99.999+%	orthorhombic	$P_{ZnS} < P_S^{begins} < P^*_{GaP}$	Perhaps sharp but highly time dependent	Sluggish. Resistance measurements extended to 3×10^{12} ohms.
Present study	static (Indentor)	99.999+%	"	$P_S^{begins} < 200$ kbar		Similar results with two independent techniques used. Resistance drop by factor of nearly 10^{12} observed. (ρ = 0.02 Ω-cm.)

* The GaP transition is at 220 kbar on the ruby scale. Recent more direct determinations by Ruoff [17] indicated the GaP transition is near 180 kbar.

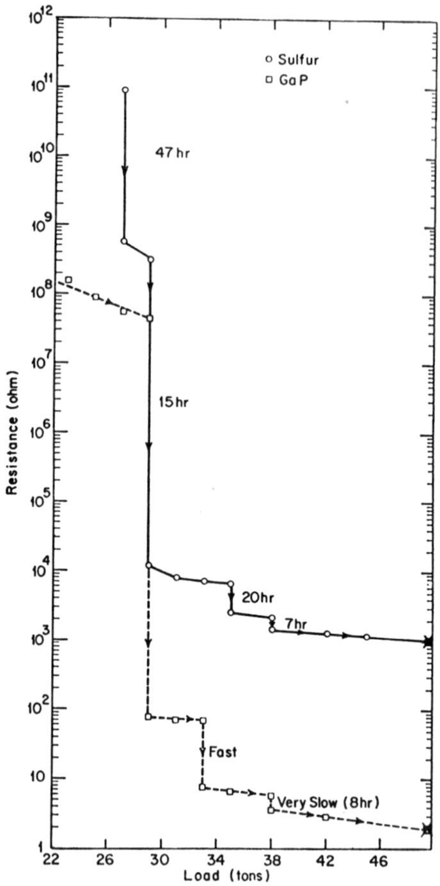

Fig. 5. Resistance vs. force curves obtained simultaneously for GaP and S. The first indication of a sulfur transition is seen when the sulfur resistance is 10^{11} ohm. The sulfur resistance drops with time at a given load until finally no additional detectable change in resistance is observed. The load is further increased and the sulfur resistance drop is noted for several hours at each load value shown in the curve.

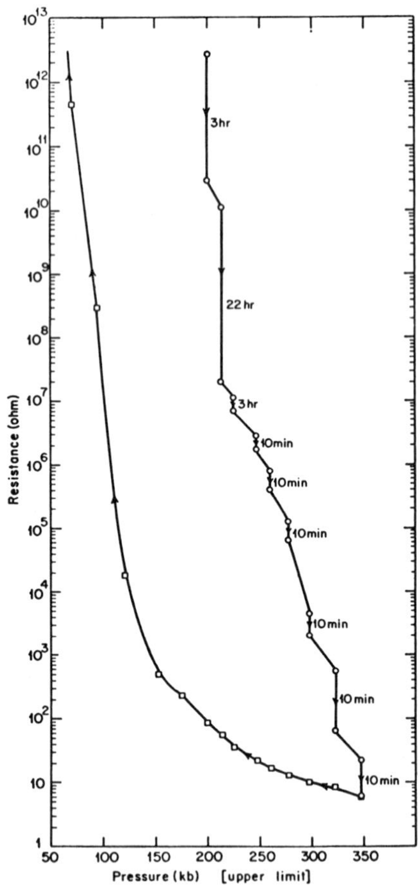

Fig. 6. Resistance vs. pressure curve for sulfur obtained using the indentor technique. The first indication of a sulfur transition is seen when the sulfur resistance is 3×10^{12} ohm. The sulfur resistance drops with time only at the fixed load and finally no additional detectable change in resistance is observed. The load is further increased and the sulfur resistance drop is noted with a time increment (as specified in figure) at each load value shown in the curve. Since the sulfur film thickness is much larger than the $\alpha^t/10$, the pressures calculated from (1) are higher (upper limit) than the actual pressures. (ρ = 0.02 Ω-cm.)

It is likely that fracture occurs sooner in the case of Boron carbide, i.e. it is unlikely that one reaches the limiting pressure of 275 kbar with Boron carbide pistons [15]. (3) With the diamond indentor technique higher pressures can be attained, so the resistance versus pressure relationship can be studied at higher pressures than attained by Chhabildas and Ruoff [9] or by us with Boron carbide pistons. Because the sulfur film thickness (for the data shown in Fig. 6) is larger than required (i.e., larger than $\alpha^t/10$ as discussed earlier), the actual pressure is somewhat less than given by equation (1). (4) The sulfur transition is found to be sluggish. Note that with the indentor technique the measured resistance drop was nearly a factor of 10^{12}. (5) Figure 7 shows the variation

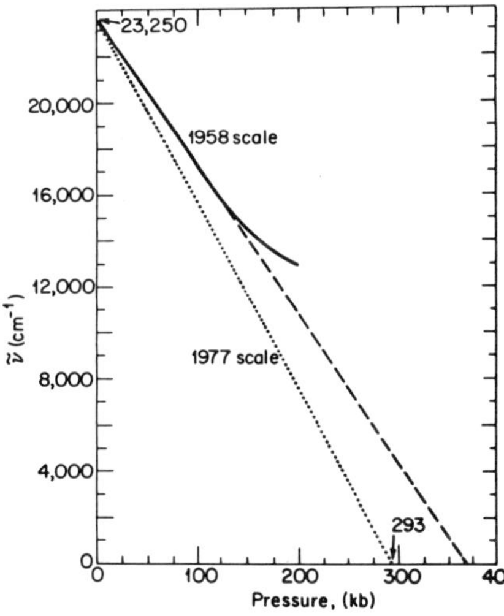

Fig. 7. Absorption edge vs. pressure. Pressure values are based on the AgCl transition at 85.4 kbar. Data points above about 110 kbar deviate from a straight line probably because of serious error in the pressure scale. If the band gap varies linearly with pressure, it will fall to zero at 373 kbar (based on P^{tr}_{AgCl} = 85.4 kbar). However, to account for the present value of the AgCl transition at 67 kbar the slope of the curve was changed as shown by the dotted line. The value of the intercept is 293 kbar. A first-order transition could occur at lower pressures. In the absence of a first-order transition, metallic behavior would be evident when the band gap energy approaches zero.

of band gap energy (absorption edge) with pressure for sulfur that was obtained by Slykhouse and Drickamer [20]. The pressure scale used by Slykhouse and Drickamer [20] was based on the AgCl transition at 85.4 kbar. This pressure scale overestimates the pressure at higher pressures, which probably explains the deviation from linearity above about 110 kbar. The dotted line in Fig. 7 was obtained by moving the AgCl transition to 67 kbar (based on the bismuth transition at 76.7 kbar and on the AgCl calibration of Montgomery [21] where the pressure for this transition was found to be 0.86 P_{Bi}). We note that based on the linear extrapolation (which is in approximation) the energy gap becomes zero at 293 kbar. Of course a first-order transition could occur at any lower pressure. In the absence of the first-order transition, metallic behavior would be evident at about 293 kbar where the band gap energy approaches zero. This expected behavior is consistent with the present results.

ACKNOWLEDGMENTS

This research was performed under a National Aeronautics and Space Administration grant which is gratefully acknowledged. We also benefited from support from the Cornell Materials Science Center via the National Science Foundation. The technical assistance of V. Arnold and R. Terry was helpful.

NOTATION

P_o = maximum pressure
$\bar{P}$ = average pressure
E = Young's modulus
G = shear modulus
K = bulk modulus
ν = Poisson's ratio
R = radius of curvature
F = force
r = radial distance from center

a = radius of contact
α = relative distance of travel under loading, of points in indentor and anvil along z axis
α^t = value of α at transition
ρ = electrical resistivity
P_H = Knoop hardness pressure
$P_{Max}^{Limiting}$ = maximum limiting pressure
ΔP = superposed pressure

REFERENCES

1. B. Meyer, ed. *Elemental Sulfur*, Wiley-Interscience, New York (1965).
2. H. G. David and S. D. Hamann, J. Chem. Phys. **28**, 1006 (1958).
3. R. J. Eichelberger, G. E. Hauver, in *Les Ondes de Detonation*, Centre National de la Recherche Scientifique, Paris (1962), p. 363.
4. H. L. Suchan, S. Wiederhorn, and H. G. Drickamer, J. Chem. Phys. **31**, 355 (1959).
5. L. F. Vereshchagin, E. N. Yakovlev, B. V. Vinogradov, and V. P. Sakun, JETP Lett. **20**, 246 (1974).
6. L. F. Vereshchagin, E. N. Yakovlev, B. V. Vinogradov, and V. P. Sakun, in *Proceedings 4th Intern. Conference on High Pressure*, Kyoto, Japan (1974), p. 860.
7. Y. Notsu, Ph.D. Thesis, Osaka University, Toyonaka, Osaka, Japan (1974).
8. B. Le Neindre, K. Suito, and N. Kawai, High Temp.-High Press. **8**, 1 (1976).
9. L. C. Chhabildas and A. L. Ruoff, J. Chem. Phys **66**, 983 (1977).
10. K. J. Dunn and F. P. Bundy, Amer. Phys. Soc. Bull. **22**, 473 (1977).
11. S. P. Timoshenko and J. N. Goodier, *Theory of Elasticity*, 3rd ed., McGraw Hill Book Company, New York (1970), p. 380.
12. A. L. Ruoff and K. S. Chan, in *Proceedings 6th AIRAPT Intern. High Pressure Conference*, Plenum Press, New York (1978).
13. J. Wanagel and A. L. Ruoff, to be published.
14. W. H. Leighton, Rev. Sci. Instr. **44**, 595 (1973).
15. A. L. Ruoff, in *Proceedings 6th AIRAPT Intern. High Pressure Conference*, Plenum Press, New York (1978).

16. P. W. Bridgman, Proc. Am. Acad. Arts Sci. 76, 4 (1945).
17. A. L. Ruoff, in Proceedings 6th AIRAPT Intern. High Pressure Conference, Plenum Press, New York (1978).
18. H. J. McSkimmin and P. Andreatch, Jr., J. Appl. Phys. 43, 2944 (1972).
19. J. E. Grubernatis and J. A. Krumhansl, J. Appl. Phys. 46, 1875 (1975).
20. T. E. Slykhouse and H. G. Drickamer, J. Phys. Chem. Solids 7, 275 (1958).
21. P. W. Montgomery, ASME paper 64-WA/PT-18 (1964).

APPENDIX

The assumption of diamond as an isotropic solid can be justified as follows: Using the elastic constants as obtained by McSkimmin and Andreatch [18], the value of Young's modulus along the different crystal directions is

$$E_{[111]} = 12.08 \text{ Mbar}$$

$$E_{[100]} = 10.53 \text{ Mbar}$$

$$E_{[110]} = 11.65 \text{ Mbar}$$

$$\bar{E} = \frac{(E_{[111]} + E_{[100]} + E_{[110]})}{3} = 11.42 \text{ Mbar} \qquad \text{(A-1)}$$

The polycrystalline aggregate constants of diamond, computed using the method described by Gubernatis and Krumhansl [19] are found to be E = 11.41 Mbar; G = 5.33 Mbar; $\nu = 0.07$; and K = 4.42 Mbar. We note that E varies with direction in the single crystal by less than 8% from the mean or the polycrystalline.

B-6

BEHAVIOR AND PROPERTIES OF SULFUR AT PRESSURES OF 300 KBAR AND HIGHER

K. J. Dunn and F. P. Bundy

General Electric Company
Schenectady, New York USA

INTRODUCTION

It has been recognized for a long time by most theorists that if covalently bonded elements and compounds are compressed sufficiently their bonding electron systems become unstable and transform over to arrangements with a metallic character resulting from free electrons.

Relatively recent (1974) high pressure experiments on sulfur by Vereshchagin, et al. [1] in Moscow, USSR, and by Notsu [2], at Osaka University, Japan, indicate that this element transforms to a relatively highly conducting state at pressures less than a megabar. Much earlier (1958) shock compression experiments of Hamann [3], and of David and Hamann [4] showed that sulfur and iodine become quite conductive under shock pressures of about 250 kbar and temperatures of approximately 1000 K.

With the advent of an improved Drickamer-type opposed-anvil apparatus [5] in which the centerface part of each cemented tungsten carbide piston was made of polycrystalline diamond, marketed by General Electric under the COMPAX (R) trademark, it has become possible to carry out quantitative electrical measurements on specimens of sulfur up to pressures exceeding 500 kbar and temperatures of a few hundred °C. It has been found that the conductivity of sulfur increases by at least 14 orders of magnitude in going from 150 to about 500 kbar, where it attains a steady semimetallic state. Upon decompression from this state it reverts to an electrically insulating state which x-ray diffraction shows to be crystalline, but this crystal structure has not yet been characterized.

EXPERIMENTS AND RESULTS

The pressure apparatus used in these experiments is illustrated in Fig. 1(a). It is an opposed-piston type with a confined gasket of pyrophyllite stone between the faces. The main bodies of the pistons are of cemented tungsten carbide, and the very highly stressed regions near the tips of the pistons are of very strongly sintered diamond powder.

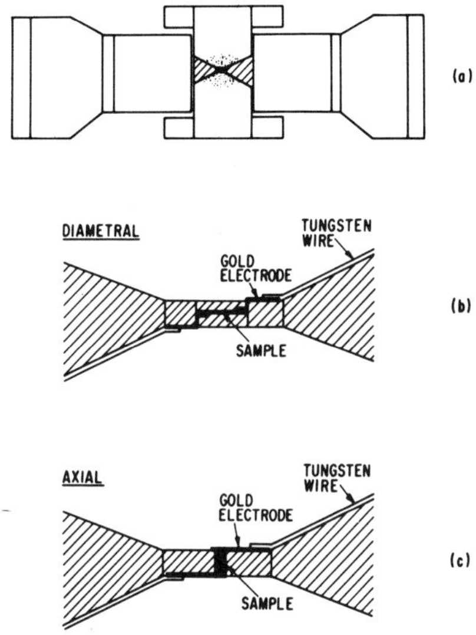

Fig. 1. Drickamer-type apparatus with diamond compact-tipped anvils.

(a) Cross section of diamond-tipped, opposed-anvil ultra-high pressure apparatus.

(b) Section of sample-holder zone showing arrangements for a "diametral specimen."

(c) Section of sample-holder zone showing arrangements for an "axial specimen."

The best available pressure calibration tests indicate that this apparatus can operate routinely up to pressures over 500 kbar without any plastic deformation or degradation.

The specimens were held in the sample holders in the space between the faces of the pistons as shown in Fig. 1(b) and (c). In (b) the tiny flake-shaped specimen of sulfur rests in the equatorial plane between two thin pills of pyrophyllite stone. The ends are connected electrically to opposite pistons by gold foil electrodes and thin tungsten wires. The tungsten wires are connected to the carbide part of the piston because of the poor electrical conductivity of the diamond compact tips.

In the second arrangement (c) the sulfur specimen is on the axis, a geometry which gives a lower resistance and a more uniform pressure in the specimen region.

Figure 2 shows a typical R vs. applied load behavior for an equatorial flake specimen at room temperature. During the first part of the loading the R stays almost constant at about 10^7 ohms because of the background leakage resistance. At about 22 tons (or approximately 300 kbar) the sulfur specimen begins to conduct better than the background insulation, and with increased loading becomes the primary conductor. At about 38 tons loading (almost 500 kbar) the

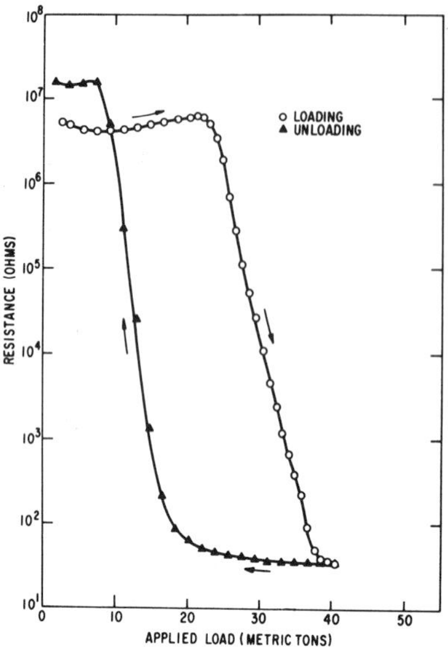

Fig. 2. Sulfur in Di-T. Resistance vs. applied load for a diametral specimen of sulfur at room temp. G_o = 0.163 mm.

R levels out, indicating a fairly saturated semimetallic state with a resistivity of about 0.03 ohm-cm. Upon unloading, the R increases to a value greater than the background resistance. It is known from other experiments using special insulation that the R of the sulfur specimen is at least 10^{14} ohms at about 150 kbar.

It was found that at a given loading a resistance equilibrium of the specimen could be accelerated by increasing the temperature. Figure 3 shows R vs. L behavior when the apparatus was held at about 70°C. In order to determine the activation energy of the conduction mechanism the loading was held constant at certain levels while the temperature was cycled. Figure 4 shows a log R vs. $10^3/T$°K plot of the behavior during three T-cycles at 19.3 tons ram loading. It is seen that the log R vs. $10^3/T$ relationship slope in this case indicates an activation energy of 0.303eV

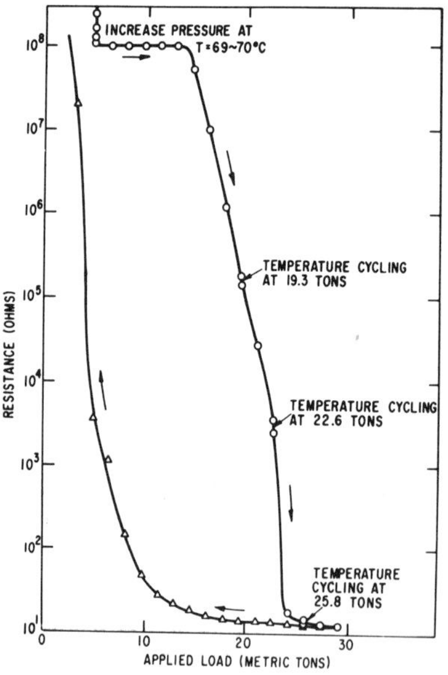

Fig. 3. Sulfur Di-T temperature cycling. Resistance vs. applied load for a diametral specimen at 70°C. G_o = 0.140 mm.

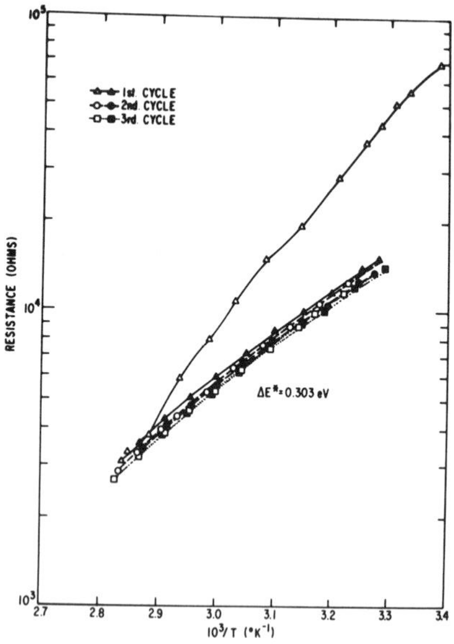

Fig. 4. Log R vs. $10^3/T$(°K) plot for T-cycling of axial specimen of sulfur at about 330 kbar pressure. ΔE* = 0.303 eV.

Figure 5 shows the R vs. T behavior at the pressure at which the R has leveled out to a steady value relative to pressure. It is seen that the slope now indicates a negative activation energy, or rather a positive T coefficient of R characteristic of metals or semimetals. In this case the temperature coefficient of resistance is $7.2 \times 10^{-4}{}^\circ C^{-1}$.

After a number of T-cycling experiments were carried out, the activation energies of conduction were plotted against the cell pressure, with the result shown in Fig. 6. It appears that the activation energy of conduction decreases linearly with the pressure and reaches zero just short of 500 kbar. At this point it becomes semimetallic.

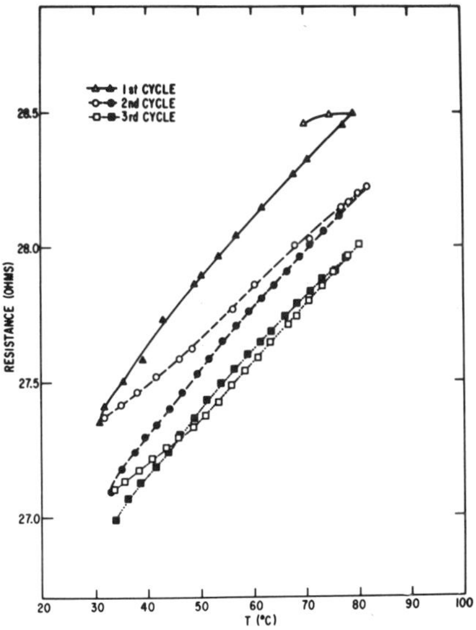

Fig. 5. R vs. T °C plot for T-cycling of diametral specimen of sulfur at a little over 500 kbar pressure.

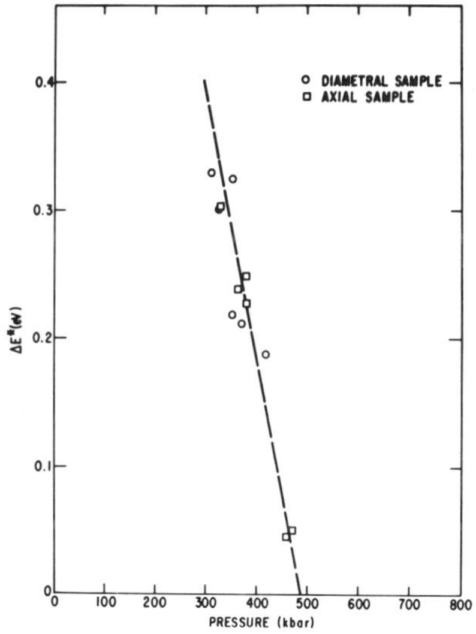

Fig. 6. Activation energy of conduction for sulfur vs. pressure, derived from T-cycling experiments.

Some experiments have been done in which the sulfur specimen was heated while in the semimetallic state by static direct current, or by strong direct current pulses of a few hundred microseconds duration. In the one-shot pulse heating experiments the specimen current and voltage drop were monitored with an oscilloscope, and from the record the R vs. energy input (i.e. temperature) was plotted as shown in Fig. 7. The abrupt drop in R at an energy input which corresponds to about 600°C is reproducible and is most likely the melting point of the semimetallic sulfur.

Experiments are contemplated in which x-ray diffraction patterns of the sulfur under pressure in the semimetallic state will be attempted.

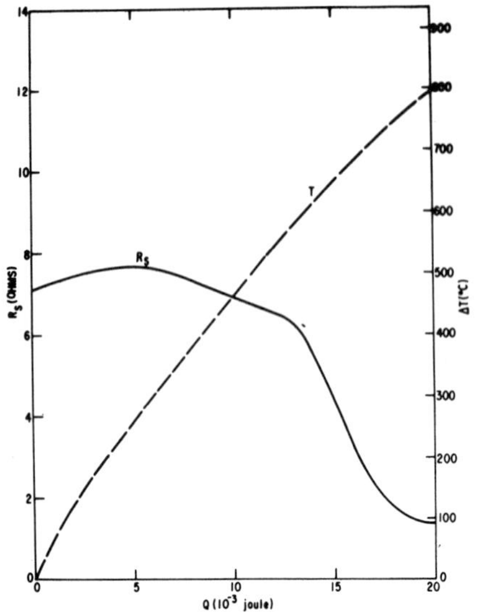

Fig. 7. Resistance vs. heating energy input during a 200 microsecond heating pulse of a diametral specimen of sulfur. The abrupt R-drop occurs at about 600°C.

REFERENCES

1. L. F. Vereshchagin, E. N. Yakovlev, B. V. Vinogradov, and V. P. Sakum, JETP Lett. 20, 246 (1974) [Original ZHETF Pis. Red. 20, 540 (1974)].
2. Y. Notsu, Thesis, Osaka University, Toyonaka, Osaka, Japan (1974).
3. S. D. Hamann, Aust. J. Chem. 11, 391 (1958).
4. H. G. David and S. D. Hamann, J. Chem. Phys. 28, 1006 (1958).
5. F. P. Bundy, Rev. Sci. Instr. 46, 1318 (1975).

B-7

SOME APPLICATIONS OF HIGH PRESSURE STUDIES OF PHASE TRANSITIONS IN SOLIDS*†

G. A. Samara

Sandia Laboratories
Albuquerque, New Mexico USA

INTRODUCTION

High-pressure research has contributed significantly to our understanding of the properties of matter. Perhaps there is no one area to which high-pressure research has contributed more than that of phase transitions in solids. High-pressure studies have improved our understanding of the properties of materials near phase transitions and of the mechanisms for certain types of transitions. Such studies have also led to the discovery of a very large number of new phase transitions which can be induced by pressure alone.

The question might be asked as to why there has been and continues to be this wide interest in the study of phase transitions in solids under pressure. The answer must surely lie in the fact that such studies are important in several areas. These include (1) fundamental understanding of materials properties, (2) synthesis of new phases, (3) geophysics and astrophysics, and (4) devices.

The first of these has been the most widespread area of study, and a wide panorama of interesting and important phenomena have been discovered and explored. These range from subtle electronic transitions [1] to structural transitions [2]. In the synthesis area, the most technologically important developments have been the graphite-to-diamond transformation and the analogous hexagonal-to-cubic transformation in boron nitride. Both transformations are irreversible. Many transformations which are reversible upon releasing the pressure at ambient or at elevated temperature can be retained metastably by quenching to 77 K or some other suitably low temperature and then releasing the pressure at the low temperature.

*Work supported by the U.S. Energy Research and Development Administration under Contract AT(29-1)789.
†Invited paper.

Examples include the recovery of the high-pressure metallic (often superconducting at low temperatures) phases of materials such as Ge and InSb [3].

The importance of high pressure to geo- and astrophysics is well recognized. The earth and the planets are large domains of high pressure with pressures increasing to the megabar range at their centers. Under such conditions many phase transitions can be expected to occur. Laboratory studies of phase transitions in minerals at high pressure have had important implications in connection with a number of different areas. Among these are the nature of (1) the constitution of the earth's core [4,5], (2) the mantle-core discontinuity [5], (3) the conductivity of the mantle [5], and (4) the interior of Jupiter [6].

In the device area the most familiar application involves the use of phase transitions as calibrants of pressure (or stress) gauges. Other applications include electrical switches based on pressure-induced insulator-to-metal transitions and stress wave shaping (or tailoring) in shock-wave research based on the volume change accompanying first-order phase transitions. Among the most unique of all such device applications is a family of shock-wave actuated pulse power sources which employ phase transitions in either ferromagnets or ferroelectrics. These devices have been in existence for many years [7,8], but the materials properties and phase transitions associated with their operation continue to be topics of both fundamental and applied interest.

In this review we shall give a brief account of these pulse power devices and discuss the phase transitions associated with them. Space and time limitations do not allow us to go into many details. We shall therefore only briefly review the basic principles behind the operation of these devices and discuss some of the materials' properties. The relevant materials' responses are those under shock-wave compression, but static high-pressure results have played a complementary role and contributed substantially to the understanding of the phenomena involved. Examples will be given illustrating this point. Some emphasis will be devoted to the displacive structural phase transitions in the $PbZr_{1-x}Ti_xO_3$ system because of the interesting physics involved.

PULSE POWER DEVICES

Two types of shock-wave-actuated pulse-power devices have been developed at Sandia Laboratories, one utilizing ferromagnetic materials [7] and the other utilizing ferroelectric materials [8]. Both devices are simple in concept, small in size (active elements are several cubic centimeters in volume), and are capable of producing peak electrical powers up to hundreds of kilowatts for durations up to several microseconds.

Ferromagnetic Devices

To understand the operation of these devices it is useful to review the main characteristic properties of a ferromagnet. This is done with the aid of Fig. 1. As a function of an applied magnetic field H, the magnetization M of a ferromagnet exhibits the familiar ferromagnetic (FM) hysteresis loop (see Fig. 1a). Upon releasing the field to zero, the material retains a remanent magnetization. This is to be contrasted with the response of a nonferromagnetic, say a paramagnetic (PM), material like copper whose M vs. H response is essentially linear and goes through the origin as shown in Fig. 1a. The magnetization of a ferromagnetic material varies with temperature as shown in Fig. 1b. Magnetization M decreases with T and ultimately vanishes at a critical temperature, the Curie temperature T_c, above which the material becomes paramagnetic.

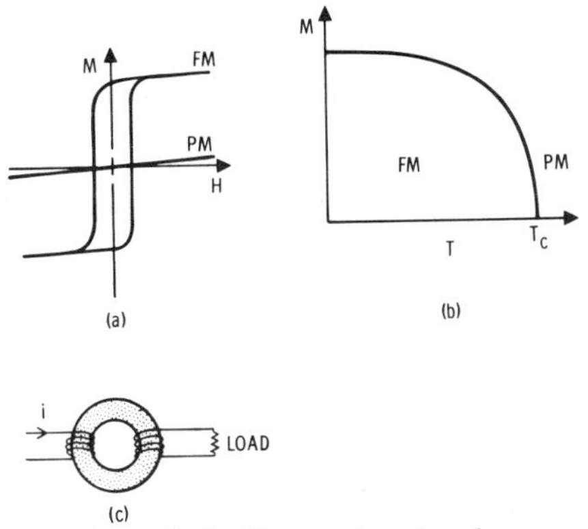

Fig. 1. (a) Characteristic hysteresis loop of a ferromagnet (FM) compared with the linear dependence of magnetization vs. field for a paramagnet (PM). (b) Temperature dependence of the magnetization of a ferromagnet undergoing a second-order phase transition to the paramagnetic state at the Curie temperature T_c. (c) Schematic representation of a ferromagnetic device showing the exciting and pick-up windings.

Figure 1c is a schematic representation of a ferromagnetic device. The device consists of a toroidal ferromagnetic core on which two coils are wound: an exciting coil through which a current, I, is passed to magnetize the core and a pick-up coil across which a load is attached. Any change in the magnetic flux Φ of the core with time t will induce a voltage in the pick-up coil since the induced voltage is given by

$$V(t) = - N \left(\frac{\partial \Phi}{\partial t}\right) \tag{1}$$

where N is the number of turns of the pick-up coil and Φ is given by

$$\Phi = (H + 4\pi M)A \tag{2}$$

where A is the cross-sectional area of the core. The device can be either a voltage generator (open circuit mode) or a current generator, depending on the load. Clearly the larger the rate of change of flux, $\partial\Phi/\partial t$, the larger the electrical response. Thus it is important to change Φ rapidly. In principle this could be done by rapid heating.

More importantly for our purposes, however, is the fact that in some ferromagnetic materials large changes in Φ can be induced in times $\lesssim 10^{-6}$s by the passage of a shock wave through the core. What is responsible for this change in Φ (or essentially in M)?

Studies of the effects of static and shock-wave compression of different classes of ferromagnetic materials have resulted in the identification of three primary mechanisms for shock-induced changes in magnetization. These are (1) second-order phase transition, (2) first-order phase transition, and (3) strain-induced magnetic anisotropy. The first two mechanisms involve pressure-induced phase transitions and are of special interest to us here. The third mechanism does not involve phase transitions and is mentioned here only for the sake of completeness. In this mechanism, depending on the magnetoelastic properties of the material, the uniaxial strain direction in a shock-wave experiment can become an easy axis of magnetization [9]. If this axis is different than that of the applied field H, and if the latter field is relatively small, then the magnetization will rotate and tend to line up with the strain axis. A properly placed pick-up coil will detect this rotation as a change in flux and will produce an electrical output. The magnetoelastic interactions responsible for this effect are present in all ferromagnetic materials, but in some materials such as nickel ferrite and yttrium iron garnet the effect dominates over the other magnetic interactions for shock-wave compression in the elastic range.

The second-order phase transition mechanism is illustrated schematically in Fig. 2. In this mechanism, partial or complete demagnetization results from the lowering of the Curie temperature with pressure (Figs. 2a and 2b). Complete demagnetization is achieved at a pressure (or stress) p_1 which causes T_c to reach a value $T_c < T_1$, where T_1 is the operating temperature. The second-order transition pressure at T_1 is P_1 since the ferromagnetic-to-paramagnetic transition in many ferromagnetic materials is thermodynamically second order. Below this pressure only partial demagnetization is achieved at T_1. Figure 2c depicts the expected pressure dependence of the magnetization at constant temperature.

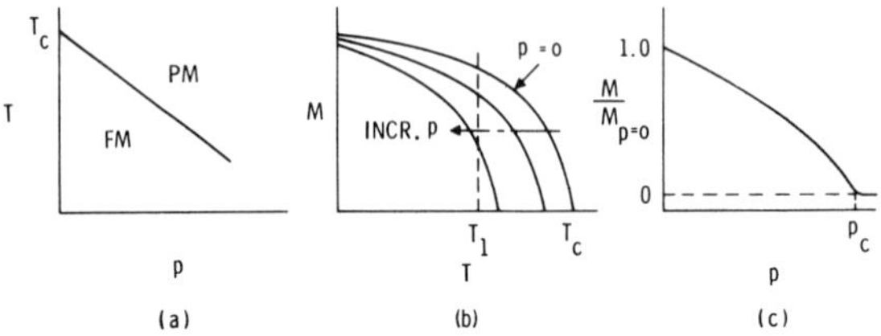

Fig. 2. Demagnetization via a pressure-induced second-order FM-PM phase transition. (a) Decrease of Curie temperature with pressure. (b) Effects of pressure on the magnetization vs. T response. (c) Decrease of the magnetization with pressure at constant temperature; p_c is the transition pressure at this temperature.

Examples of materials which illustrate this mechanism are fcc Ni-Fe alloys in the 30-50 at.% Ni range and MnSb. Some of our results on the Ni-Fe alloys are shown in Figs. 3 and 4 as changes in the saturation magnetization and T_c, respectively, with hydrostatic pressure. The initial pressure derivatives obtained from these data agree well with a variety of other measurements in the literature [10], but the present results generally extend to higher pressures than have been reported earlier. More important, however, is the fact that these results agree fairly quantitatively with the few available shock-wave compression results [10,11], once account is made for the finite shear strength of the materials. (The changes in magnetization with shock compression were measured directly by Wayne [10] and Graham [11]). Complete shock demagnetization via the second-order transition mechanism can also be inferred from the expected large change in compressibility accompanying the ferromagnetic-paramagnetic transition in materials whose Curie temperatures are strongly pressure dependent [12].

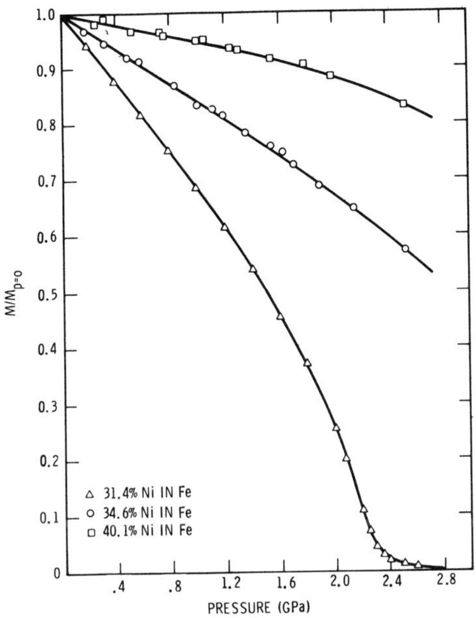

Fig. 3. Fractional change of magnetization with hydrostatic pressure at room temperature for three fcc Ni-Fe alloys. (Percentages are in at.%.) For the 31.4% Ni alloy the FM → PM transition pressure is ≈ 2.4 GPa.

An important conclusion based on the above comparison of static and shock wave results on the fcc Ni-Fe alloys is that the shock-induced change in magnetization in these alloys is primarily a volume effect [10,13]. The comparison also shows that static data (which can be obtained quicker and much less expensively than shock-wave data) can be used to accurately predict the shock wave response of this class of ferromagnetic materials. Alternatively, shock-compression experiments can be used to obtain quantitative information on the magnetic properties of similar alloys to pressures much higher than can be achieved by static techniques. (A review of magnetic properties under shock compression is given by Royce [14].)

In the first-order transition mechanism, demagnetization occurs via an abrupt destruction of the ferromagnetic ordering of the magnetic moments. The high-pressure phase can be paramagnetic, antiferromagnetic, or ferromagnetic, but with lower magnetization than the low-pressure phase. The situation is illustrated schematically in Fig. 5. The Curie temperature and magnetization are usually weakly dependent on pressure, but at the transition pressure, p_o,

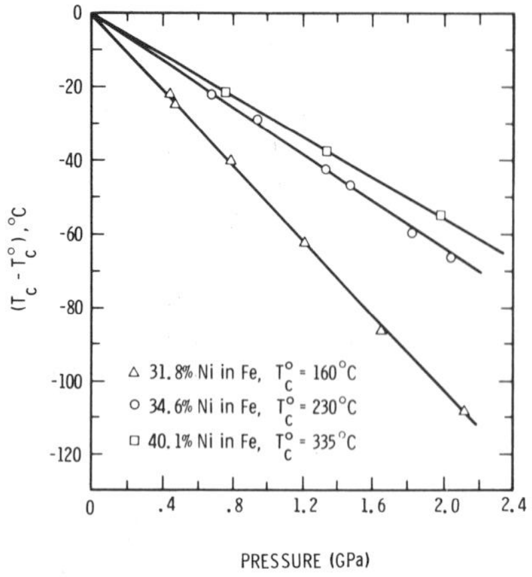

Fig. 4. Decrease of the Curie temperature with hydrostatic pressure for three fcc Ni-Fe alloys. The initial Curie temperatures are designated by T_c^o.

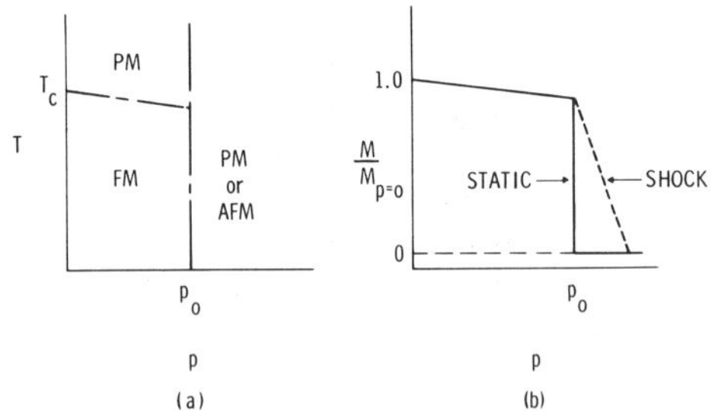

Fig. 5. Demagnetization via a pressure-induced first-order magnetic transition to either a paramagnetic (PM) or antiferromagnetic (AFM) state. (a) Usually the Curie temperature decreases relatively slightly up to the transition pressure p_o. (b) The magnetization also usually decreases only slightly up to p_o.

a rapid decrease in magnetization occurs. Examples of materials which exhibit this kind of behavior include iron, bcc iron alloys, and gadolinium. Iron and its bcc alloys have been studied extensively under static and shock-wave compression. A recent review of the results has been given by Duvall and Graham [13]. As in the case of materials exhibiting second-order transitions, here again there appears to be generally good agreement between the static and shock data after proper accounts are made of the shear-strength effect and the thermodynamic conditions in the mixed phase region in the shock case.

Ferroelectric Devices

These devices are conceptually similar to the ferromagnetic devices. Their operation can be understood with reference to Fig. 6. Ferroelectrics are characterized by the fact that they possess spontaneous polarization (aligned permanent dipole moments) and that the low-frequency polarization, P, vs. electric field, E, exhibits a characteristic hysteresis loop (Fig. 6a) unlike the essentially linear P vs. E response for a normal dielectric or paraelectric (PE) material. The polarization decreases with increasing temperature and ultimately vanishes above some critical temperature T_c via either a second-order (as in Fig. 6b) or a first-order transition. The qualitative similarity of these characteristics to those of ferromagnetic materials (see Fig. 1) is quite evident.

Figure 6c shows a schematic illustration of a ferroelectric device. An electroded ferroelectric plate, originally in the unpolarized virgin state (i.e., random distribution of ferroelectric domains), is polarized by the application of a sufficiently large voltage V across the electrodes. This causes alignment of the

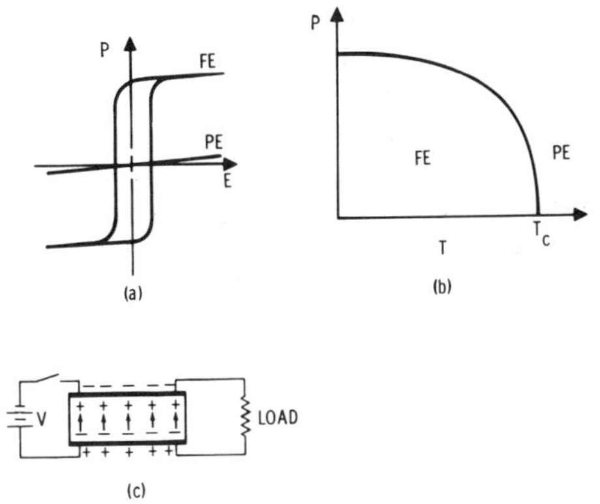

Fig. 6. (a) Characteristic hysteresis loop of a ferroelectric (FE) compared to the linear polarization vs. field for a paraelectric (PE). (b) Polarization decreases with temperature and ultimately vanishes continuously at the Curie temperature T_c in the case of a second-order transition. (c) Schematic of a ferroelectric device showing the bound electrical charges deposited during the polarization process by the applied voltage V. The passage of a shock wave through the FE plate causes depolarization and the release of the bound charge across the load.

electric dipoles with the electric field and results in a net spontaneous polarization. This polarization is compensated by the deposition of an amount of electrical charge (of the opposite polarity to the polarization) on the electrodes as shown. The operation of the device is based on the rapid partial or complete release of this bound charge which then flows through the load. This charge release can be accomplished by reducing or destroying the net polarization. As in the magnetic case, in principle this could be done by rapid heating of the plate, but, alternatively, in some ferroelectrics the polarization can be decreased by the passage of a shock wave through the plate.

The decrease in polarization with shock compression can result from any of several different mechanisms: (1) Pressure- (or stress-) induced second- or first-order phase transition from the ferroelectric phase to either a paraelectric or an antiferroelectric phase. (2) Rotation of dipoles (actually domains of dipoles) due to a stress- (or strain-) induced anisotropy. (3) Change in the magnitude of the effective dipole moment. Mechanism (3) is essentially a piezoelectric response. We note here that all ferroelectrics are piezoelectrics and the passage of a shock wave through the plate will always yield, in addition to other responses, the appropriate piezoelectric response of the plate. We further draw attention to the qualitative analogy between mechanisms (1) and (2) and the response of ferromagnetic materials discussed earlier.

For present purposes we are primarily interested in the first of the above mechanisms; namely, that involving phase transitions. These phase transitions result in complete release of the bound charge for pressures (stresses) above the transition pressures. Examples of ferroelectrics which undergo pressure-induced phase transitions to nonferroelectric phases include KH_2PO_4, SbSI, and the perovskites $BaTiO_3$, $PbTiO_3$, and $PbZr_{1-x}Ti_xO_3$. In most of these cases the bulk of the charge release is associated with the lowering of the Curie temperature T_c with pressure (i.e., a second-order transition mechanism) which is finally interrupted by the onset of a first-order transition. Many such transitions have been investigated under hydrostatic conditions [2,15] and some have also been studied under shock-wave compression [16]. There is generally good qualitative, and in some cases quantitative, agreement between the static and the shock data with respect to the main features of the phenomena involved. Ferroelectrics are materials which exhibit strong electromechanical interactions, and therefore substantial differences in the responses under static and shock conditions can always be expected simply because of differences in the stress states involved. These differences are over and above any kinetic effects that may be present. Nevertheless, static-pressure studies complement shock-wave studies on ferroelectrics and they have provided much of the basis for the interpretation and understanding of the shock studies. The remainder of this review will emphasize hydrostatic pressure results.

Because they can be readily prepared in ceramic form and fabricated in the desired shapes and sizes, the perovskites have been the most widely studied ferroelectrics under shock compression. These materials have also been extensively investigated in both single crystal and ceramic forms under static compression. Mixed crystals of the form $PbZr_{1-x}Ti_xO_3$ constitute a particular interesting group of the perovskite family, because of the variety of phase transitions and phenomena they exhibit. The following section deals with some of the properties of these mixed crystals.

THE $PbTiO_3$-$PbZrO_3$ SYSTEM

The perovskites $PbZrO_3$ and $PbTiO_3$ form a continuous series of solid solutions of the form $PbZr_{1-x}Ti_xO_3$ over the whole composition range. These solid solutions commonly called PZTs, are rich in the variety of ferroelectric and nonferroelectric transitions that can be induced in them by variations in composition, temperature, or pressure (stress). They also exhibit large spontaneous polarizations and piezoelectric coefficients. These factors along with their availability in high-quality ceramic form are responsible for their widespread technological usage in applications based on piezoelectric and ferroelectric properties.

Figure 7 shows the temperature-composition phase diagram for the $PbZr_{1-x}Ti_xO_3$ system [17]. For all compositions the high-temperature phase is paraelectric (PE) and has the ideal cubic perovskite structure. On cooling, this phase transforms either to one of two ferroelectric phases or to one of two antiferroelectric phases, depending on the composition. The rhombohedral ferroelectric region is actually divided into two phases as shown: a high-temperature phase of symmetry R3m and a low-temperature phase of symmetry R3c [18] which has a unit cell double that of the high-

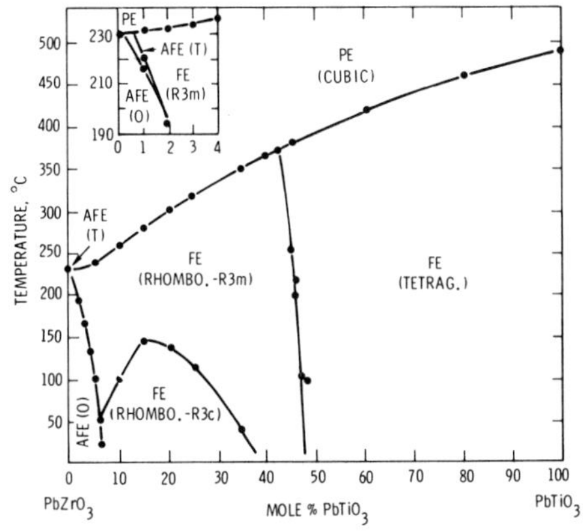

Fig. 7. The temperature-composition phase diagram for the $PbZr_{1-x}Ti_xO_3$ (or PZT) system at 1 bar. (after Jaffe et al. [17])

temperature phase. The tetragonal antiferroelectric phase in Fig. 7 exists over a very limited composition-temperature range at atmospheric pressure and its range of stability is strongly dependent on impurities. For certain PZT compositions this phase is induced by pressure.

Most, if not all, of the phase transitions shown in Fig. 7 are known to be associated with lattice dynamical instabilities or soft phonon modes. By a soft mode we simply mean a normal mode of vibration whose frequency decreases (and thus the lattice literally softens with respect to this mode) and approaches zero as the transition point is approached [2]. From a lattice dynamical point of view, the limit of stability of a crystal lattice is approached as the frequency of any one mode decreases and approaches zero, for then, once the atoms are displaced in the course of the particular vibration, there is no restoring force to bring them back, and they assume new equilibrium positions determined by the symmetry (eigenvector) of this soft mode. Small ionic displacements accompany such transitions and hence they are often called displacive transitions.

In ferroelectrics, identical ions in adjacent unit cells undergo identical displacements, but with negative and positive ions undergoing opposite displacements and thereby producing a macroscopic polarization. The soft mode leading to such a circumstance is thus necessarily a long wavelength (or zone center, $\underset{\sim}{q} = 0$) transverse optic (TO), infrared-active phonon. The static dielectric susceptibility (or dielectric constant ε) is inversely proportional to the square of this soft-mode frequency [2,19], and therefore ε diverges on approaching the ferroelectric transition. In antiferroelectric crystals, on the other hand, similar ions in adjacent cells undergo opposite displacements and there is no net polarization. The characteristic wavelength of a phonon leading to such displacements is thus typically twice the lattice parameter $\underline{a}$, and the wave vector of the soft mode is then $\underset{\sim}{q} = \pi/2a$, i.e., a zone boundary phonon. One consequence of the softening of such a mode is a doubling of the unit cell in the antiferroelectric phase over what it is in the high-temperature phase. This doubling is evidenced by the appearance of superstructure lines in x-ray and neutron diffraction patterns. (The wave vector $\underset{\sim}{q}$ of the soft mode could also be such that the unit cell could enlarge by a factor of

4, 6, 8, etc.). There is no net polarization associated with a soft mode leading to an antiferroelectric transition and there should be no dielectric constant anomaly associated with such a transition if the transition is second-order. For a first-order transition the finite lattice strain at the transition will usually lead to a small discontinuity in ε.

Space limitations do not allow us to discuss the details of the various transitions in Fig. 7 and the effects of pressure on them. As a result, we have selected for discussion three materials whose pressure responses typify the most important features of the transitions in Fig. 7. These materials are $PbTiO_3$, $PbZrO_3$ and, $PbZr_{.65}Ti_{.35}O_3$.

$PbTiO_3$

$PbTiO_3$ is a classic soft-mode ferroelectric. On cooling the crystal from the high temperature cubic phase, the frequency of the soft mode (a T_{1u} (TO) mode which corresponds to the vibration of the oxygen octahedron of the perovskite structure against the Pb and Ti ions) decreases and tends towards zero as the transition temperature is approached. Associated with this decrease in frequency ω_s is an increase in the dielectric constant ε since $\varepsilon \propto \omega_s^{-2}$. This increase obeys a Curie-Weiss law

$$\varepsilon = C/(T - T_o) \qquad (3)$$

where C is a constant ($\approx 4.0 \times 10^5$K at 1 bar) and T_o is the Curie temperature. For a second-order transition, T_o is the temperature at which the soft-mode frequency vanishes and where the transition occurs. For a first-order transition, on the other hand, the transition occurs at a temperature T_c which is higher than T_o. This is the case for $PbTiO_3$ where at 1 bar T_c = 765 K and T_o = 720 K [20]. Figure 8 shows the temperature dependence of ε at a few pressures. The Curie temperature T_c, defined as the temperature where the discontinuity in ε occurs, decreases with pressure and so does T_o which is defined by the extrapolation of the linear ε^{-1} vs. T plot to $\varepsilon^{-1} = 0$. The pressure dependences of T_c and T_o are shown in Fig. 9. An interesting feature of these data is that the difference (T_c-T_o) decreases with pressure. This fact and the associated enhancement of the peak value of ε with pressure (see Fig. 8) imply that the transition is tending towards second order and is expected to become second order at sufficiently high pressure [20]. However, at the highest pressure in Fig. 8 the transition is clearly still first order as evidenced by the thermal hysteresis.

The decrease of T_c (and T_o) with pressure is characteristic of the behavior of many ferroelectric perovskites and implies that these materials can be either partially or completely depolarized (depending on the pressure) under pressure. This pressure effect can be qualitatively understood in terms of the soft mode theory [2] according to which the soft-mode frequency ω_s is determined by

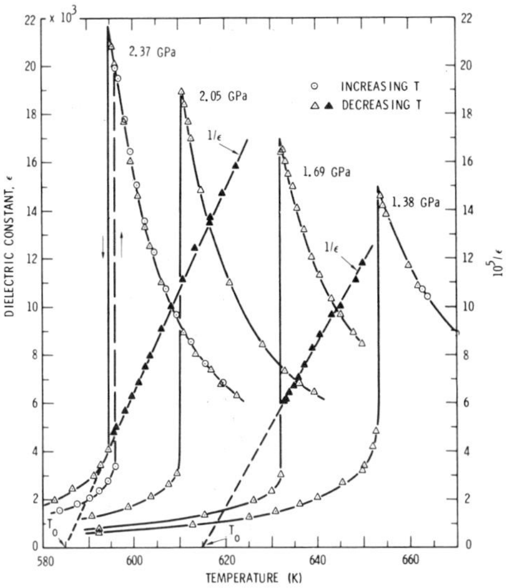

Fig. 8. Temperature dependence of the static dielectric constant of $PbTiO_3$ at different pressures showing the decrease of the transition temperature. Also shown are plots of the reciprocal dielectric constant for the 1.69 and 2.37 GPa isobars. These linear plots define the Curie-Weiss temperature T_o.

a difference between short-range forces and Coulomb forces. Since the short-range forces increase much more rapidly with decreasing interionic distances than do the Coulomb forces, ω_s increases rapidly with pressure. This in turn causes a decrease in T_o (and hence T_c) [2,20].

$PbZrO_3$

$PbZrO_3$ represents a very interesting case. On cooling, this crystal exhibits a first-order transition to an orthorhombic antiferroelectric phase. As mentioned earlier, such a transition should have little if any dielectric constant anomaly associated with it. However, experimental results show a very large anomaly as shown by the 1 bar data in Fig. 10. In fact in the high-temperature phase, ε obeys (3) with $C = 1.60 \times 10^5$ K and $T_o = 475$ K [21]. The transition temperature is $T_a = 505$ K. How does one explain this behavior?

The explanation is that at atmospheric pressure the high-temperature phase of $PbZrO_3$ has two soft and nearly degenerate phonon modes: [21] (1) a zone center TO mode, i.e., a ferroelectric mode which causes the large anomaly and Curie-Weiss behavior of ε, and (2) a zone-boundary mode which causes the antiferroelectric transition. On cooling, the crystal becomes unstable against the zone-boundary mode at T_a just before the instability against the ferroelectric mode sets in at T_o (or slightly below for a first-order transition). While this explanation can be tested by measuring by neutron scattering experiments the phonon dispersion relations of $PbZrO_3$ and their temperature dependence, the measurements have not been made, most likely because of the lack of single crystals of suitable size.

In the absence of such direct measurements, pressure studies gave the first experimental confirmation that there are indeed two soft modes [21]. With increasing pressure the frequency of the ferroelectric mode increases sharply causing a large decrease in T_o, whereas the frequency of the antiferroelectric mode softens further causing an increase in T_a. As a consequence, the large ε anomaly

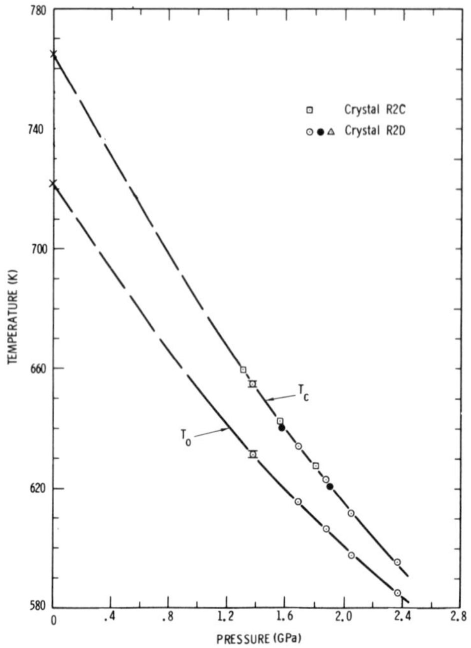

Fig. 9. Pressure dependences of the transition temperature T_c and Curie-Weiss temperature T_o for $PbTiO_3$.

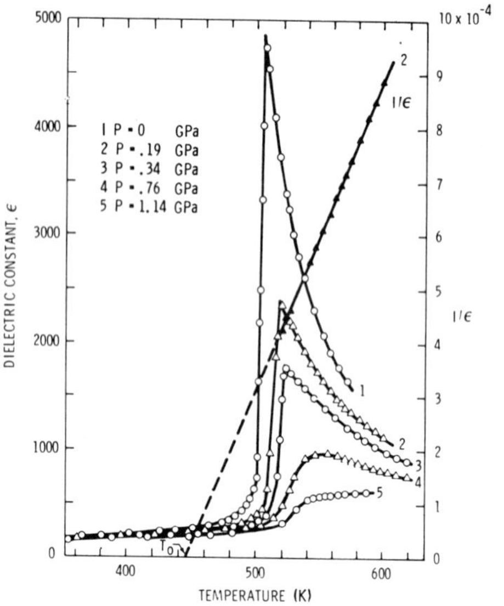

Fig. 10. Temperature dependence of the static dielectric constant of $PbZrO_3$ at different pressures showing the increase of the transition temperature. Also shown for the 1 bar isobar is the linear temperature dependence of ε^{-1}. This response defines the Curie-Weiss temperature T_o.

at the transition is drastically reduced. Some results are shown in Fig. 10 where it is seen that this behavior is in marked contrast to that of $PbTiO_3$ in Fig. 8. The pressure dependences of T_a and T_o are shown in Fig. 11. The increase of T_a with pressure is a characteristic of all displacive antiferroelectric and other transitions involving soft zone boundary (or short wavelength) phonons. We shall discuss the explanation for this effect in the next section.

$PbZr_{.65}Ti_{.35}O_3$

As in the case of $PbTiO_3$, this material, often referred to as PZT 65/35, transforms on cooling into a ferroelectric phase. However, unlike $PbTiO_3$ the ferroelectric phase has rhombohedral instead of tetragonal symmetry (see Fig. 7). This is especially interesting because both ferroelectric phases result from the softening of presumably the same soft mode in the high temperature cubic phase. How can this be? The answer lies in the fact that the soft mode in the cubic phase is doubly degenerate, i.e., it is not a single mode but actually two degenerate modes. The symmetry of the low-temperature phase is determined by which linear combination of the ionic displacements associated with these modes cause the resulting

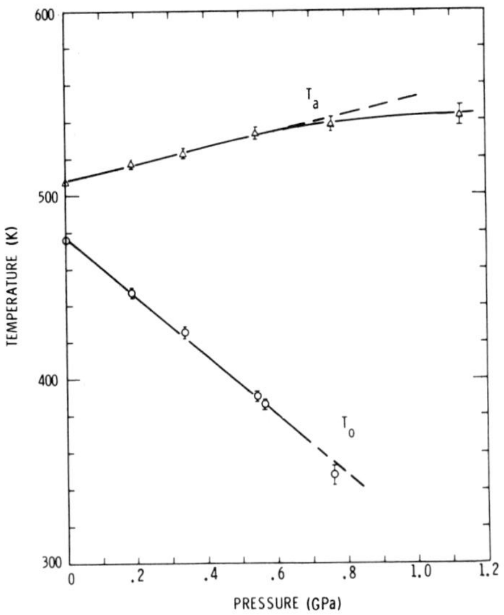

Fig. 11. Pressure dependences of the transition temperature T_a and the Curie-Weiss temperature T_o for $PbZrO_3$.

structure to be most stable. This in turn is determined by lattice anharmonicities [22].

Figure 12 shows the pressure dependence of the cubic-rhombohedral ferroelectric transition temperature T_c of PZT 65/35. Transition temperature T_c decreases as expected in a manner similar to that for $PbTiO_3$. Of special interest in Fig. 12 is the fact that the extrapolated hydrostatic data appear to be in rather good agreement with data deduced from shock wave depoling experiments [16]. This kind of agreement has led to the extensive use of hydrostatic measurements to complement and guide shock-wave studies in PZTs and other perovskites.

As can be seen from Fig. 7, on cooling, the rhombohedral (R3m) ferroelectric phase transforms to a second rhombohedral (R3c) ferroelectric phase at ~ 40°C. This transition is very interesting. The R3c phase has double the unit cell of the R3m phase. This immediately suggests that a soft zone boundary phonon may be responsible for the transition between these two phases. Indeed, a combination of x-ray and neutron diffraction studies has shown that the ionic positions in the R3m and R3c phases are similar, but with an additional frozen-in staggered rotation of the oxygen octahedra about the [111] axes (with respect to the original cubic structure) in the R3c phase [18]. These counter rotations of oxygen octahedra are the motions expected by the condensation of a phonon at the (½,½,½) point of

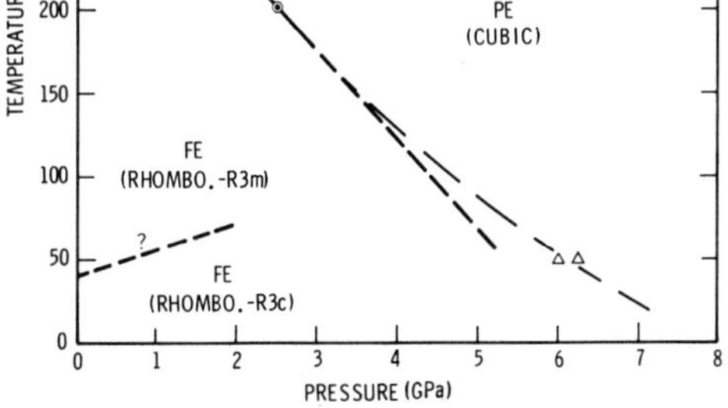

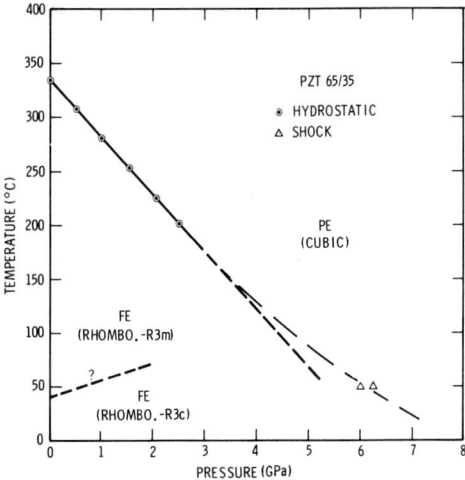

Fig. 12. Pressure dependence of the ferroelectric-paraelectric transition temperature of $PbZr_{.65}Ti_{.35}O_3$ comparing hydrostatic and shock wave data. Also shown is the possible pressure dependence of the transition temperature between the two ferroelectric rhombohedral phases.

the Brillouin zone as is found to be the case in $LaAlO_3$ [23]. The identification of this soft mode in PZT 65/35 does not appear to have been confirmed experimentally, but there can be no doubt that it is correct. (Confirmation requires inelastic neutron scattering measurements which need relatively large crystals which are not available at present.)

The oxygen rotations associated with this soft mode produce no net electric dipole moments, and there appears to be no dielectric constant anomaly accompanying the R3m-R3c transition [17]. There is, however, a weak dielectric loss anomaly. We have not studied the pressure dependence of this anomaly in any detail as yet, but preliminary data suggest that the transition temperature increases at a rate of ~ 2°C/kbar. This is indicated in the Fig. 12 data over the range of the data. A more detailed study is needed to better quantify this result. The indicated increase in transition temperature with pressure is what is expected for a transition associated with a short wavelength phonon as is discussed briefly below.

GENERAL RESULT CONCERNING DISPLACIVE TRANSITIONS

Results on PZTs, other perovskites, and a wide variety of other crystals exhibiting displacive phase transitions driven by soft optic phonons have led us to the following observation [24]: For all cases where the transition involves a long wavelength (zone center) TO soft mode (i.e., a ferroelectric mode) the transition temperature decreases with pressure, whereas for all the transitions involving short wavelength (zone boundary) soft modes the transition temperature increases with pressure. This appears to be a general result to which we know of no exception, and it has important implications concerning the role of competing forces in the lattice dynamics of these transitions [24].

It is generally accepted that the decrease in the frequency of the soft mode as the transition temperature is approached is caused by the cancellation between competing forces. For ferroelectric transitions the physical mechanism is well known. There is a cancellation of the positive short-range forces by the negative Coulomb forces. The frequency of the soft mode can be schematically written as

$$\omega_s^2 \propto [(\text{short-range forces}) - (\text{long-range forces})] \qquad (4)$$

With pressure, the interionic distance r is decreased and this increases the short-range forces ($\propto r^{-n}$, where $n \approx 10$) much more rapidly than the long-range forces ($\propto r^{-3}$) leading to an increase in ω_s and thereby the transition temperature.

In the case of the antiferroelectric and other zone-boundary phonon-driven transitions the origin of the soft mode behavior was not generally known until recently [24]. Here the increase of the transition temperature with pressure can be interpreted as resulting from a softening (i.e., a decrease) of ω_s with pressure. What is the physical origin of this effect? The high-pressure results along with some lattice dynamical calculations which showed that the short-range forces associated with the soft mode are negative (i.e., attractive) suggested the answer [24,25]. For these transitions the softening of ω_s is due to the cancellation of the positive long-range forces by the negative short-range forces. In other words, for the zone-boundary soft-optic modes the roles of the short-range and electrostatic forces are reversed from what they are for the ferroelectric soft mode. Soft mode frequency ω_s can then be written as

$$\omega_s^2 \propto [(\text{long-range forces}) - (\text{short-range forces})] \qquad (5)$$

and ω_s should decrease with pressure. This result provides a ready explanation for the increase of the transition temperature with pressure for the zone-boundary phonon transitions. This can be most easily seen from the observed temperature dependence of ω_s above the transition, namely [24]

$$\omega_s^2 = A(T - T_a) \qquad (6)$$

where A is a constant. A lower ω_s at constant T ($> T_a$) implies a higher transition temperature T_a.

CONCLUDING REMARKS

In this review we have given a brief account of one of the most unique applications of pressure-induced phase transitions in solids, namely that of shock-wave-actuated pulse-power devices employing either ferromagnetic or ferroelectric materials. The objective has been to outline the principles of operation of these devices and to discuss some of the associated materials properties and phase transitions. The discussion, especially of the ferromagnetic materials, was very brief and the reader is referred to the cited literature for details.

Aside from this and other possible applied interests, the study of phase transitions in ferromagnetic and ferroelectric materials at high pressure has been and will continue to be of great fundamental interest. In the case of ferromagnetic materials the exchange forces which govern ferromagnetism are strong functions of interatomic distance and the changes of the magnetization and Curie temperature with pressure reflect changes in these forces. In the case of soft-mode ferroelectrics (and antiferroelectrics), on the other hand, the decrease of the soft-mode frequency leading to the onset of the phase transition is determined by the balance between competing short-range and long-range forces, and this balance is strongly affected by pressure. As discussed in the paper and in some of the references, high-pressure research has already made important contributions to our understanding of these phase transitions and of the associated changes in physical properties. It is certain that such research will always be important in the study of phase transitions.

ACKNOWLEDGMENTS

The author has benefited from discussions with I. J. Fritz, R. A. Graham, and P. C. Lysne at various stages of the work described in this review. B. E. Hammons provided excellent technical assistance in the experimental work.

REFERENCES

1. H. G. Drickamer and C. W. Frank, Electronic Transitions and the High Pressure Chemistry and Physics of Solids, Chapman and Hall, London (1973).
2. G. A. Samara, in Proc. 4th Intern. High Pressure Conference, The Physico-Chemical Soc. of Japan, Kyoto, Japan (1974), p. 247.
3. S. Minomura, B. Okai, Y. Onoda, and S. Tanuma, Phys. Letters 23, 641 (1966).
4. H. Sawamoto, E. Ohtani and M. Kumazawa, in Proc. 4th Intern. High Pressure Conference, The Physico-Chemical Soc. of Japan, Kyoto, Japan (1974), p. 194.
5. A. E. Ringwood, Composition and Petrology of the Earth's Mantle, McGraw-Hill, New York (1975); also The Nature of the Solid Earth, E. C. Robertson, ed., McGraw Hill, New York (1972); G. A. Samara, J. Geophysical Research 72, 671 (1967).
6. W. C. DeMarcus, Astron. J. 63, 2 (1958).
7. F. W. Neilson, Bull. Am. Phys. Soc. 2, 302 (1957).
8. G. W. Anderson and F. W. Neilson, Bull. Am. Phys. Soc. 2, 302 (1957); also R. W. Kulterman, F. W. Neilson, and W. B. Benedick, J. Appl. Phys. 29, 500 (1958).
9. E. B. Royce, J. Appl. Phys. 37, 4066 (1966); also L. C. Bartel, J. Appl. Phys. 40, 661 (1969); R. C. Wayne, G. A. Samara, and R. A. Lefever, J. Appl. Phys. 41, 633 (1970); D. E. Grady, J. Appl. Physics 43, 1942 (1972).
10. R. C. Wayne, J. Appl. Phys. 40, 15 (1969).
11. R. A. Graham, J. Appl. Phys. 39, 437 (1968).
12. R. A. Graham, D. H. Anderson and J. R. Holland, J. Appl. Phys. 38, 223 (1967).
13. G. E. Duvall and R. A. Graham, Rev. Mod. Phys., to be published.
14. E. B. Royce, in Proceedings of the International School of Physics, Enrico Fermi, Course XLVIII, Physics of High Energy Density, P. Caldirola and H. Knoepfel, eds., Academic Press, New York (1971), p. 126.
15. G. A. Samara, in Advances in High Pressure Research, Vol. 8 R. S. Bradley, ed., Academic Press, New York (1969), Chapt. 3.
16. D. G. Doran, J. Appl. Phys. 39, 40 (1967); also P. C. Lysne, J. Appl. Phys. 44, 577 (1973); P. C. Lysne and L. C. Bartel, J. Appl. Phys. 46, 222 (1975); P. C. Lysne and C. M. Percival, Ferroelectrics 10, 129 (1976).

17. B. Jaffe, W. R. Cook, Jr., and H. Jaffe, Piezoelectric Ceramics, Academic Press, New York (1971), Chapt. 7.
18. C. Michel, J-M. Moreau, G. D. Achenbach, R. Gerson, and W. J. James, Sol. State Commun. 7, 865 (1969).
19. W. Cochran, Adv. Phys. 9, 387 (1960).
20. G. A. Samara, Ferroelectrics 2, 277 (1971).
21. G. A. Samara, Phys. Rev. B 1, 3777 (1970).
22. H. Thomas and K. A. Müller, Phys. Rev. Letters 21, 1256 (1968).
23. J. D. Axe, G. Shirane, and K. A. Müller, Phys. Rev. 183, 820 (1969).
24. G. A. Samara, T. Sakudo, and K. Yoshimitsu, Phys. Rev. Letters 35, 1767 (1975); also G. A. Samara, Comments on Solid State Physics 8, 13 (1977).
25. K. Yoshimitsu, Prog. Theor. Phys. 54, 583 (1975).

B-8

STRESS EFFECTS IN FERROELECTRIC CERAMICS

I. J. Fritz and J. D. Keck

Sandia Laboratories
Albuquerque, New Mexico USA

INTRODUCTION

Electromechanical transducers made of ferroelectric ceramic materials such as those based on barium titanate (BT) or on various lead zirconate titanate (PZT) compositions have become widely used for a variety of applications over the past 25 years [1]. Most of these applications utilize the linear piezoelectric effect for the conversion between electrical and mechanical energy, and one of the prime advantages of the BT or PZT materials is that their piezoelectric coupling coefficients are large. Another kind of application of ferroelectric ceramics - one that is perhaps not as generally familiar - is based on non-linear and non-reversible processes that can take place in these materials. One example of this kind of application is the single-shot, shock-activated power supply [2,3]. In this device, electrical energy is stored in a ferroelectric element by the initial poling process, and the passage of a shock wave through the material releases part or all of this stored energy (into an external electrical load) by actually destroying (non-reversibly) the state of initial polarization. There are two important mechanisms by which the destruction of polarization may occur. The first is by domain reorientation processes [4] whereby the directions of the polarization vectors in the individual ferroelectric domains change from one preferred crystallographic axis to another in response to the external stress. This process tends to randomize the domains and dramatically reduce the net polarization of the ceramic. The second possible mechanism for shock-induced depoling is a structural phase transition [3,5] induced by the stress behind the shock front, which transforms the material into a non-ferroelectric state. When such a transformation occurs, the polarizations of the individual crystallites of the ceramic become zero, so that all of the initial poling energy is released.

At present the ability to theoretically predict the response of a ferroelectric ceramic to shock-wave loading is quite limited. The problem is complicated because several mechanisms (piezoelectric response, domain reorientation, phase transitions) may all contribute to the response, and because these mechanisms may cause electrical charge to be released on time scales comparable to the duration of the shock wave, so that rate-dependent effects must be explicitly taken into consideration. Furthermore, the knowledge of the electrical and mechanical equations of state of ferroelectric ceramics at the large electric fields and mechanical strains produced by shock waves is limited. Thus the fundamental parameters that would be needed to predict the response for a specific model of the problem are not known a priori.

The response of ferroelectric ceramics to statically-applied stress is being investigated in order to study processes that might occur under shock-loading conditions. In the following, two typical examples will be discussed. The first involves uniaxial-stress experiments on a modified BT composition. In these experiments domain reorientation effects are observed. The second example is a study of the properties of a PZT composition which exhibits pressure-induced phase transitions from ferroelectric to non-ferroelectric phases. In this latter work, statically-applied hydrostatic pressure is used.

BARIUM TITANATE

The material used in this study was a BT composition modified by the addition of 5 wt. % calcium titanate* (chemical formula $Ba_{0.92}Ca_{0.08}TiO_3$). Upon cooling from high temperature this material transforms at a temperature of 115°C from the cubic perovskite structure to a tetragonal ferroelectric structure with the ferroelectric c-axis aligned along one of the six original cubic axes. The c/a ratio is greater than unity for the ferroelectric phase. The ceramic specimens used in these experiments were macroscopically unpoled, that is, the directions of the ferroelectric axes of the individual crystallites were randomly distributed initially. The application of uniaxial stress to the specimen creates stresses on the individual crystallites and these stresses on the average are along the applied stress direction. A crystallite can partially relieve the local effect of the applied stress by changing the pattern of ferroelectric domains within itself, so that on the average the ferroelectric axes point away from the applied stress direction. Thus, 90° domain switching is to be expected when uniaxial stress is applied to a tetragonal ferroelectric ceramic, and in fact it is well known that a poled ceramic which is compressed along the polar direction becomes depolarized because of this effect [6]. A series of experiments has been performed on unpoled ceramics involving measurements of various physical properties (macroscopic strains, dielectric constant, ultrasonic velocities) as a function of uniaxial stress. The data all exhibit "anomalous" behavior at

*A summary of the properties of modified BT ceramics is available in Chapt. 5 of Piezoelectric Ceramics [1].

low stress levels (0 to 150 MPa) which can be attributed to domain reorientation effects. By using polycrystalline averaging techniques, the overall magnitudes of the observed anomalies can be predicted. A detailed description of the experiments and calculations will appear in a separate publication [7]. In the following, an ultrasonic velocity experiment and the corresponding theoretical calculation will be described.

In the ultrasonic experiment, longitudinal sound waves were propagated along a direction perpendicular to the applied stress direction. The resulting data are shown in Fig. 1. For practical purposes, $f(\sigma)$ is related to $\bar{C}_{11}(\sigma)$ by the equation $f^2(\sigma)/f^2(0) = \bar{C}_{11}(\sigma)/\bar{C}_{11}(0)$. Data are shown for two stress cycles, as indicated in the figure. Upon initial application of stress the elastic constant shows a rapid decrease, and this decrease does not completely recover when the stress is relieved from 150 MPa. On the second stress cycle, $\bar{C}_{11}(\sigma)$ goes through a minimum at ~ 200 MPa, whereafter a normal increase with stress is observed. The overall magnitude of the anomalous decrease of $\bar{C}_{11}$ is ~ 5%. In determining this figure the normal increase in $\bar{C}_{11}$ with stress is accounted for by extrapolating the data at high stress back to the origin. A sample which has been through the stress cycling as indicated in Fig. 1 will, when heated briefly to a temperature just above the Curie point, revert to its initial state, and subsequent stress cycling will reproduce the behavior shown in Fig. 1. This fact is strong evidence that the anomalous decrease in $\bar{C}_{11}(\sigma)$ is due to domain reorientation effects, as heating above the Curie point returns the domain pattern to its original random configuration.

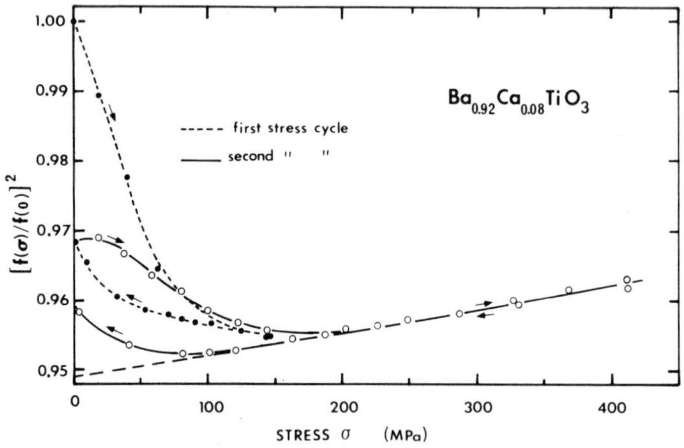

Fig. 1. Ultrasonic repetition rate data vs. applied stress. The effective elastic constant for compressional waves propagating normal to the applied stress direction is proportional to $f^2(\sigma)$. Data are shown for two stress cycles, as indicated.

Since the properties of a ceramic material are ultimately determined by those of the single crystalline material, it is interesting to try to devise suitable averaging techniques to predict the behavior of ceramics. This is a difficult problem because of the complicated distribution of boundary conditions on the individual grains, but it is possible to do approximate calculations [7]. The calculations were based on the averaging procedure of Voigt [8], which successfully predicts the overall magnitudes of the observed anomalies for several of the experiments done in BT. Details of the calculations will be presented elsewhere [7], here the results will simply be given for the elastic constant experiment described above. Before the application of stress, the elastic constant for longitudinal sound propagation is given by the standard Voigt equation, which for tetragonal symmetry reduces to

$$\bar{C}_{11}(0) = \frac{1}{15}\,[6C_{11} + 3C_{33} + 2(C_{12} + 2C_{66}) + 4(C_{13} + 2C_{44})] \quad (1)$$

At large stress levels the following is assumed: (1) All domains whose polar axes are initially within 45° of the applied stress direction undergo a 90° switching, and (2) the switching is non-preferential in the sense that the polar axis switches to either of the two possible new directions with equal probability. With these assumptions it can be shown [7] that the Voigt procedure leads to the following

$$\bar{C}_{11}(\infty) = \frac{1}{1920}\,[627C_{11} + 516C_{33} + 209(C_{12} + 2C_{66}) + 568(C_{13}+2C_{44})] \quad (2)$$

The calculated total anomaly due to domain reorientation turns out to be $[\bar{C}_{11}(\infty) - \bar{C}_{11}(0)]/\bar{C}_{11}(0) = -4\%$, in good agreement with the observed value of -5%.

PZT 95/5

Lead zirconate titanate materials at the zirconia-rich end of the composition range are of interest because they exhibit a variety of structural phases depending on composition, electric field, temperature, and stress [9]. The phase behavior and high electric field properties for $Pb_{.98}Nb_{.02}(Zr_{.95}Ti_{.05})O_3$, PZT 95/5, have been studied. (The niobia doping improves the electrical characteristics of the material.*) This particular composition is of interest because at room temperature it exhibits a pressure-induced phase transition from a ferroelectric to an antiferroelectric phase [9].

To detect phase changes the dielectric constant and dielectric loss of an unpoled sample were measured as a function of temperature and hydrostatic pressure. Typical data are shown in Fig. 2. Here results from pressure scans to 1.2 GPa at three temperatures are shown. At room temperature a drop in dielectric constant at ~0.25 GPa results from the transition from the ferroelectric to antiferroelectric phase. At 125°C this same transition occurs at a slightly higher pressure and a second, broad, transition is

*See Jaffe et al. [1], Chapt. 7, Sec. C.

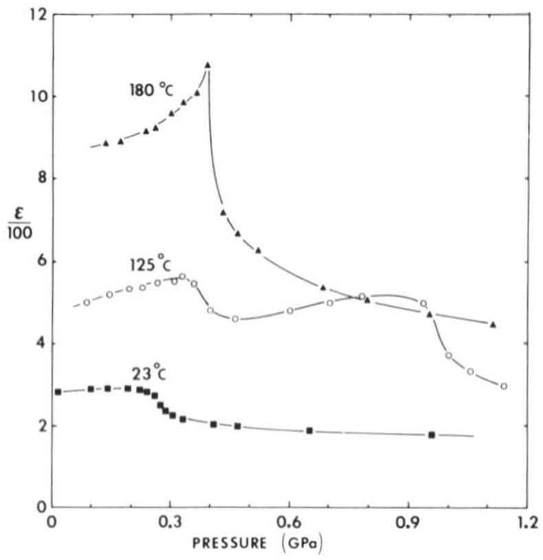

Fig. 2. Typical dielectric constant data showing the effects of pressure-induced phase transitions in PZT 95/5.

observed at ~0.95 GPa. At 180°C a single transition is observed. Scans of this kind and also data taken as a function of temperature at fixed pressure were used to plot the phase boundaries in p-T space.* The results are shown in Fig. 3, whose main features are as follows. At high temperature the material is in the cubic paraelectric (P) phase. At lower temperatures and low pressures the structure is one of two ferroelectric rhombohedral structures, F_{R1} or F_{R2}. These structures are known to be related by a unit cell doubling [11,12]. At high pressures the material is in one of two antiferroelectric phases, one orthorhombic (A_O) and one tetragonal (A_T). It was not possible to map out the phase boundaries in the region indicated by the question mark because the anomalies marking the transitions are rather broad, and because each transition has a fairly wide hysteresis region as indicated by the dashed lines. A previously published phase diagram [9] for a PZT 95/5 composition doped with 1% Nb did not show the A_O phase. That diagram is believed to be in error.

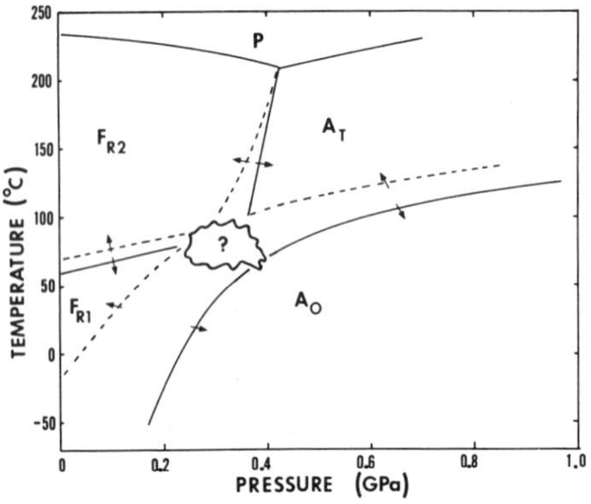

Fig. 3. Pressure-temperature phase diagram for PZT 95/5. Behavior in the region marked by the question mark is uncertain.

In actual device applications there may be large electric fields produced in the sample, and it is therefore important to investigate the properties of the material at high fields as well as at high stress. Figures 4 and 5 show some typical results from this sort of measurement and will illustrate the kind of subtle complexities that are often encountered in PZT materials. Shown in Fig. 4 is a series of high field (~50 kV/cm) D-E hysteresis loops obtained at various pressures near the F_{R2} - A_T phase boundary at a temperature of 110°C. At p = 0.31 GPa an ordinary ferroelectric loop is observed. At p = 0.41 GPa a double loop, characteristic

*Identification of the various phases is done by comparing the characteristic anomalies at the phase boundaries with those previously observed by other workers [9,10].

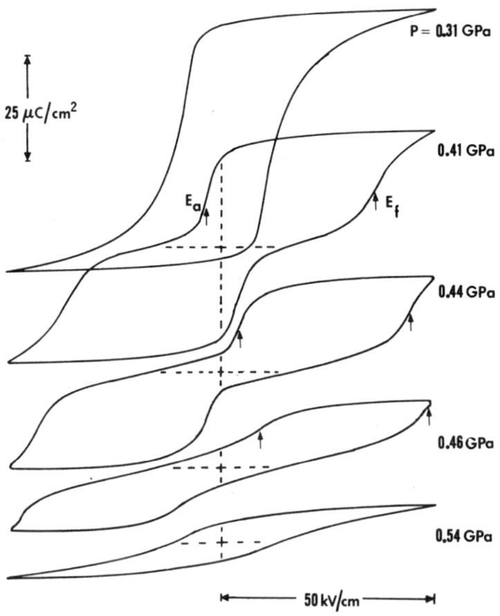

Fig. 4. Typical D-E hysteresis loop data taken at various pressures with a temperature of 110°C for PZT 95/5.

of an antiferroelectric phase, begins to emerge. The field E_f is the field in which the material transforms from A_T to F_{R2}, and the reverse transformation F_{R2} to A_T occurs at E_a which is negative at p = 0.41 GPa. At higher pressures (0.44 and 0.46 GPa curves) E_a is positive. The value of E_f increases rapidly with pressure, and above 0.5 GPa it is larger than the maximum applied field, so that the applied field cannot induce the A_T - F_{R2} transition. It would be expected then that at pressures above 0.5 GPa the material should be in the A_T phase and the D-E curves should be straight lines through the origin (no hysteresis). Nevertheless, a ferroelectric hysteresis loop is observed at 0.54 GPa rather than the expected linear reversible behavior. The reason for this is not known, but a possible explanation is that the A_T phase is actually a mixed phase with a ferroelectric component.

The important parameters obtained from loops such as those of Fig. 4 are plotted in Fig. 5. The top curve shows the remnant polarization P_r (E = 0). Note that P_r is not zero in the A_T phase, but falls to zero in the A_0 phase. In the lower half of Fig. 5 the characteristic fields E_c, E_a, and E_f of the loops are shown. The interesting feature to note here is that the coercive field E_c is the same for the ferroelectric loops observed in the A_T phase as for those observed in the F_{R2} phase. This supports the idea that the A_T phase could be a mixed phase containing regions of the (untransformed) F_{R2} phase.

CONCLUSION

The results presented above illustrate some of the typical behavior observed for ferroelectric ceramics under different conditions of high stress and field. In particular, the observation of domain reorientation effects and of pressure and field enforced phase transitions have been emphasized. These processes represent non-linear responses of the materials to externally applied stresses and are believed to be important factors in determining the response of these materials under shock-wave conditions. A direct comparison of data obtained under statically applied stress with

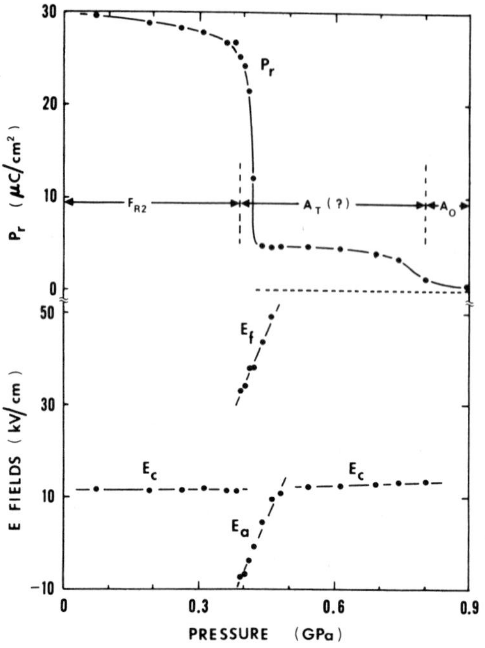

Fig. 5. Electrical parameters obtained from hysteresis loop data at 110°C for PZT 95/5.

data from shock experiments is not possible, at least at present, because the stress states and time scales of the experiments are so different. Nevertheless, there is good evidence from shock wave data that these processes are in fact important. The simple fact that a shock wave can completely depole the PZT 95/5 material [3,5] shows that the ferroelectric to antiferroelectric transition can occur under shock conditions and on a very short time scale. Recently, Lysne [4] has examined in some detail the response of shock-wave-loaded PZT 65/35 ceramic and was able to interpret his results with a model which explicitly included domain reorientation dynamics. This implies that similar physical processes occur under both static and dynamic stress in ferroelectric ceramics. Future work in this area is called for in order to be able to better quantitatively correlate results from the two kinds of experiments.

NOTATION

<u>a</u> = crystallographic <u>a</u>-axis
<u>c</u> = crystallographic <u>c</u>-axis
$\bar{C}_{11}(\sigma)$ = effective elastic constant at σ
C_{ij} = single crystal elastic constants
D = electric displacement
E = electric field
E_a = field for ferroelectric to antiferroelectric transition
E_c = coercive field
E_f = field for antiferroelectric to ferroelectric transition
$f(\sigma)$ = ultrasonic pulse repetition frequency at σ
p = hydrostatic pressure
P_r = remnant polarization
T = temperature
σ = uniaxial stress

REFERENCES

1. B. Jaffe, W. R. Cook, Jr., and H. Jaffe, <u>Piezoelectric Ceramics</u>, Academic Press, New York (1971).
2. F. W. Neilson, Bull. Am. Phys. Soc. <u>2</u>, 302 (1957).

3. P. C. Lysne, J. Appl. Phys. 48, 1020 (1977).
4. P. C. Lysne, J. Appl. Phys. 48, 1024 (1977).
5. P. C. Lysne and C. M. Percival, J. Appl. Phys. 46, 1519 (1975).
6. D. Berlincourt and H.H.A. Krueger, J. Appl. Phys. 30, 1804 (1959).
7. I. J. Fritz, J. Appl. Phys. 49, 788 (1978).
8. W. Voigt, Lehrbuch der Kristallphysik, Teubne, Leipzig (1928), p. 962.
9. D. Berlincourt, H.H.A. Krueger, and B. Jaffe, J. Phys. Chem. Solids 25, 659 (1964).
10. D. A. Berlincourt and H.H.A. Krueger, Annual Progress Report, (unpublished) Sandia Corporation, P.O. 51-9689-A (1963).
11. H. M. Barnett, J. Appl. Phys. 33, 1606 (1962).
12. D. Bäuerle and A. Pinczuk, Solid State Commun. 19, 1169 (1976).

B-9

ELECTRICAL RESPONSE OF SHOCK-WAVE-COMPRESSED FERROELECTRICS*

P. C. Lysne

Sandia Laboratories
Albuquerque, New Mexico USA

INTRODUCTION

Short duration electrical pulses with peak powers approaching a megawatt can be obtained by shock-wave compression of poled ferroelectric ceramics. The polarity of the electrical responses indicates that the physical process involved is not the piezoelectric effect but rather the destruction of the remanent polarization by either a randomization of domains or by a polymorphic phase transformation to a nonferroelectric state. Power supplies based on phase transformations are practical in that their electrical responses are relatively insensitive to the magnitude of the shock as long as the characteristic threshold conditions for the transformation are exceeded.

Before poling, ferroelectric ceramics are isotropic. The poling process introduces a unique axis of symmetry and the orientation of this axis with respect to the shock front strongly influences the temporal characteristics of the electrical response. In "axial mode" devices this axis is normal to the planar shock front. In this case the electric field ahead of the shock tends to depole the ceramic. This depoling causes a strong time dependence of the electrical response as well as electric fields, often exceeding breakdown conditions [1-3]. A somewhat more predictable response is obtained from devices operating in the "normal mode" in which the remanant polarization vector is parallel to the shock front [4,5]. In these devices the magnitude of the electric field is not excessive and it is related to the impedance of the external circuit and to the geometry and material properties of the ferroelectric.

The purpose of this work is (1) to give an analysis of normal mode devices working into general electrical loads and (2) to

*Work supported by U. S. Energy Research and Development Administration under contract AT(29-1)789.

illustrate the normal mode concepts by performing experiments on specimens of the phase transforming ferroelectric ceramic $Pb_{0.98}Nb_{0.02}(Zr_{0.95}Ti_{0.05})_{0.98}O_3$, called herein PZT 95/5. The material parameters of importance are the resistivity, the instantaneous and relaxed dielectric permittivities, and the associated dielectric relaxation time. Dielectric relaxation is introduced in this paper to explain the observed damping of electrical oscillations in experiments utilizing inductive circuit elements.

THEORY

A pictorial representation of a typical normal-mode experiment and a schematic of the equivalent circuit are illustrated in Fig. 1. Shock waves are introduced into the specimen by the detonation of small explosive pellets or, as in the case of this investigation, by projectile-impact techniques. Projectile impact experiments that generate shock pulses that are long with respect to the shock transit time of the specimen are discussed in greater detail elsewhere [4].

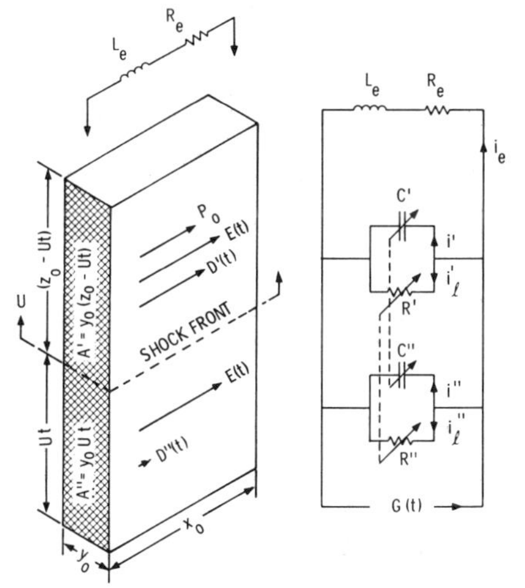

Fig. 1. Pictorial and schematic of the normal mode experiment. The shock is introduced on the lower $x_o y_o$ surface. The cross-hatched surface and its opposite are electroded.

Shock waves traveling in rectangular bars are inherently dispersive due to release waves originating at the sides. However, to a reasonable approximation, the shock front partitions the specimen into two regions containing shocked and unshocked material. In what follows, if an equation is meant to be valid on both sides of the shock front, the parameters that are potentially discontinuous at the front are denoted by asterisks. If one of these parameters is evaluated in the region behind the front, it is denoted by two primes. Likewise, single primes denote discontinuous parameters evaluated in the unshocked region. Other parameters such as dimensions, external circuit elements, and time are not primed.

The general dielectric equation of state that was found to be adequate for this study is

$$D^*(t) = \varepsilon_\infty^* E^*(t) + (\varepsilon^* - \varepsilon_\infty^*)\tau^{*-1} \int_0^t e^{-(t-\hat{t})\tau^{*-1}} E^*(\hat{t})d\hat{t} + P^* \quad (1)$$

Equation (1) is the usual formulation for a relaxing dielectric including the Debye approximation of the exponential relaxation function [6]. The first two terms on the right-hand side represent the field-induced polarization which is common to all materials. Note that if $E^*(t)$ is sinusoidal these terms yield the customary complex dielectric constant. In general, $E^*(t)$ must be determined from the circuit equations and the specimen properties and dimensions.

In this investigation, P^* represents the polarization due to dipole alignment [7]. As long as $E'(t)$ does not exceed the value required to depole the ferroelectric, P^* evaluated ahead of the shock is P_o, the charge per unit area deposited in the poling process. Behind the shock, P^* always vanishes due to the phase transformation to a nonferroelectric state. Thus

$$P' = P_o \text{ (ahead)} \quad ; \quad P'' = 0 \text{ (behind)} \tag{2}$$

The action of the shock is thus to liberate charges which were previously bound by the polarization due to domain alignment. The rate at which this charge is liberated is proportional to the rate at which the area behind the shock increases (see Fig. 1) and is equal to the current source function, $G(t)$. For an ideally supported rectangular specimen in which there is no shock dispersion, G will have the value $-y_o P_o U$ until the shock traverses the specimen, and then G will vanish. In practical experiments the shock is dispersive and G is not discontinuous at transit time, z_o/U. Reasonable agreement with experimental data can be achieved by assuming G is of the empirical form

$$G(t) = -\frac{1}{2} y_o P_o U \left(1 - \tanh\left(\frac{t - z_o/U}{\alpha}\right)\right) \tag{3}$$

Equation (1) means that the capacitors C' and C" may be lossy in the sense that free energy is converted to heat for time-dependent fields. These capacitors may also be leaky in the sense that charge passes through them. Their effective leakage resistance is [8]

$$R^* = \rho^* x_o A^{*-1} \tag{4}$$

where the time-dependent areas are

$$A' = y_o(z_o - Ut) \tag{5a}$$

$$A'' = y_o Ut \tag{5b}$$

Equations (5) neglect the shock-induced strain which is less than $\approx 3\%$ for typical experiments.

Equations (1) through (5) are now used in the usual circuit equations to obtain the electrical functions. In normal mode experiments the polarizations and the electric fields are perpendicular to the electroded surfaces (Fig. 1). Hence the electric field is continuous across the shock and

$$E''(t) = E'(t) \tag{6}$$

(note, in some axial mode experiments, D^* is continuous). Ohm's law relates the product of the electric field and x_o to the currents through the internal leakage resistances,

$$x_o E' = i_\ell^* R_\ell^* \tag{7}$$

and to the current through the external circuit which for the case illustrated in Fig. 1 yields

$$x_o E' = L_e \frac{di_e}{dt} + i_e R_e \tag{8}$$

Obviously (8) can be generalized to any external impedance. The displacement currents flowing into C' and C'' are

$$i^* = \frac{d}{dt} (A^* D^*) \tag{9}$$

Finally, Kirchoff's point rule yields

$$G + i''_\ell + i'' + i'_\ell + i' + i_e = 0 \tag{10}$$

The algebra associated with the solution of the above equations is lengthy but straightforward. The procedure used was to differentiate (1) to obtain

$$\dot{D}^* = \varepsilon_\infty^* \dot{E}^* - (D^* - \varepsilon^* E^*)\tau^{*-1} \tag{11}$$

Equations (6), (9) and (11) were combined to eliminate $\dot{D}^*$ and to obtain di^*/dt in terms of derivatives through $\ddot{E}'$. The remaining equations were invoked to obtain a coupled set of equations for $\ddot{E}'$, di''/dt, and di_e/dt. These equations were integrated numerically using a fourth-order Runge-Kutta scheme. As a check, the currents were integrated to obtain the total charge flow. This integrated current at $t = \infty$ was within $\pm$ 0.7% of the initial charge $y_o z_o P_o$ for all external impedances. The error here may be attributed to uncertainty in the initial conditions which were taken to be

$$\dot{E}'(0) = P_o U (z_o \varepsilon'_\infty)^{-1} \tag{12a}$$

$$E'(0) = 0 \tag{12b}$$

$$i''(0) = y_o P_o U \text{ and} \tag{12c}$$

$$i_e(0) = 0 \tag{12d}$$

These equations were obtained from an approximate solution that suppressed dielectric relaxation, conduction, and the permittivity change across the shock front.

APPLICATION TO PZT 95/5

In previous investigations of PZT 95/5 shocked into nonferroelectric states, the external circuit consisted of pure resistances from 0.1 to 15.0 x 10^3 Ω [4,5]. In this investigation the inductance was 1.06 x 10^{-3} H and the series resistance was 0.1 to 2.1 x 10^3 Ω. The parameters obtained in the manner discussed below and listed in Table I are a fit to all past and present data.

Table I. Material and Geometric Parameters for PZT 95/5

a	0.4 ± 10%	μs
ε'_∞	8 ± 20%	nF/m
ε'	16 ± 20%	nF/m
ε''_∞	7 ± 20%	nF/m
ε''	9 ± 20%	nF/m
P_o	0.29 ± 5%	C/m^2
ρ'	$10^9 \approx \infty$	Ω•m
ρ''	7 x 10^2 ± 50%	Ω•m
τ'	0.7 ± 15%	μs
τ''	0.4 ± 15%	μs
U	3.44 ± 5%	km/s
x_o	9.52	mm
y_o	3.18	mm
z_o	22.2	mm

Previous investigations [4,5,7] had given reasonable values of ρ', ρ'', ε', ε'', P_o, and U (of these only ε' and U were changed in this study). The parameters α and U were evaluated by noting that if the external circuit impedance is negligible ($L_e = 0$, $R_e = 0.1$ Ω), $i_e = -G(t)$. Three experiments at 1.6, 2.2, and 3.2 GPa* were used in this fit; one of these is illustrated in Fig. 2. The remaining parameters (ε'_∞, ε', ε''_∞, τ', and τ'') were obtained by iterative fitting of experiments with $L_e = 1.06 \times 10^{-3}$ H, $R_e \cong 0.1$ Ω, and with $L_e = 0$, $R_e = 15.0 \times 10^3$ Ω. These extreme cases were chosen since they emphasized the effect of dielectric relaxation. The procedure was to estimate numbers for the shocked parameters and to vary the unshocked parameters until good fits were obtained to both the frequency and the relaxation-induced damping of the oscillations for times near zero (the times at which most of

*The stress values were obtained from the Rankine-Hugoniot equations and represent the input shock amplitude in the ceramic.

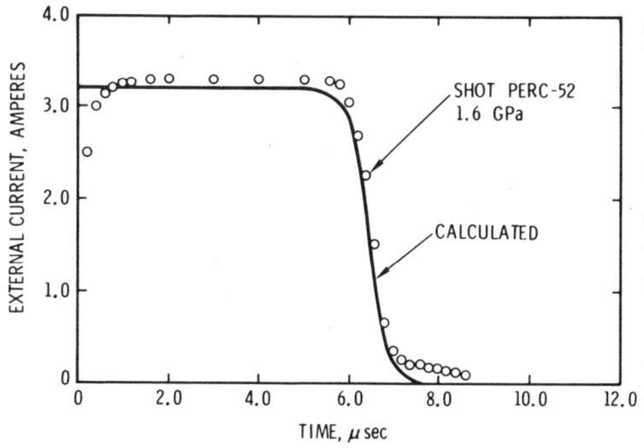

Fig. 2. Calculated and observed external currents for $R = 0.1\Omega$, $L_e = 0$. The calculated curve is a fit to three experiments, one of which is shown as the circles. In the other experiments at larger shock amplitudes, the current approached a discontinuity at $t = 0$ whereas the decrease at $t = z_o/U$ remained about the same as the illustrated data.

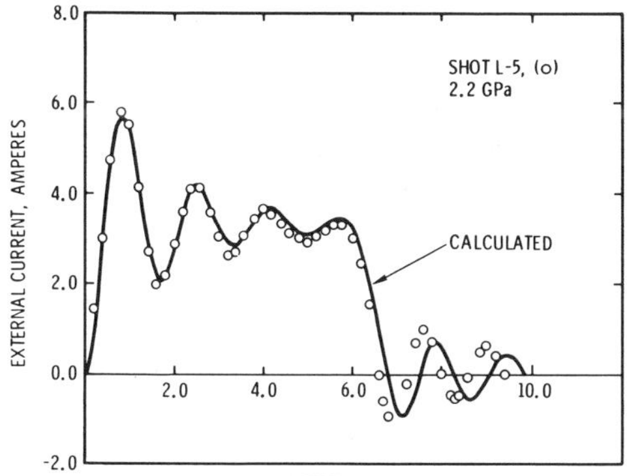

Fig. 3. Calculated and observed external currents for $R = 0.1\ \Omega$, $L_e = 1.06$ mH. The calculated curve is a best fit to all available data. Hence the discrepancy in phase for $t \gtrsim 6.0$ μs.

the material is unshocked). Then a similar procedure was used to refine the shocked parameters by fitting the data for times greater than transit time. Since the numbers for small periods of time are not independent of the shocked parameters, several iterations were needed. These iterations utilized the data from the resistive experiment. Fits to an inductive experiment at 2.2 GPa and a resistive experiment at 1.6 GPa are shown in Figs. 3 and 4, respectively. Fits to other experiments are generally valid to about $\pm$ 10% in amplitude in the stress region 1.6 to 3.2 GPa. Fits to the resonant frequency are somewhat better, namely about $\pm$ 3%. Some problems are apparent for $t > z_o/U \approx 6.5$ μs (see Figs. 2 to 4).

DISCUSSION

The uncertainties associated with the parameters listed in Table I were obtained from the estimated experimental error in i_e (observed) and by varying each parameter individually in the solution for i_e (calculated). It is to be noted that all specimens were of the same size and only one value of inductance was used. Obviously, further experimentation is needed. Nevertheless, the parameters in Table I are felt to be reasonable and they can be used to design useful experiments for their verification, e.g., resonant frequencies $<< 1/\tau^*$ and/or specimen thicknesses $z_o << U\tau^*$.

The relaxing dielectric model developed in this work was chosen to explain the observed damping of current oscillations. Before closing, it is appropriate to discuss other models and to explain why they were excluded.

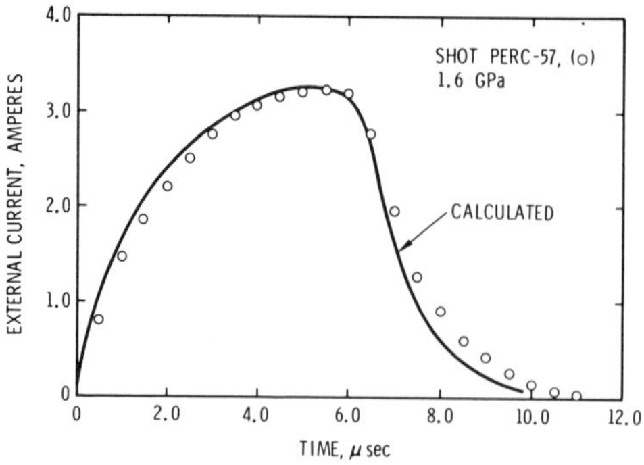

Fig. 4. Calculated and observed external currents for R = 15 KΩ and $L_e = 0$.

By definition, ferroelectrics possess hysteresis in the displacement-electric field plane and hysteretic energy loss will damp oscillatory fields. Calculations of hysteretic damping were made using hysteresis loops obtained in the poling process to represent both the shocked and unshocked material. This procedure maximized the energy loss; however, it was insufficient to explain the observed damping. This was particularly true when specimens were biased into the relatively hysteresis-free first quadrant of the hysteresis loop by using a series resistance in the external circuit.

Another possibility is that electric fields ahead of the shock will induce mechanical disturbances due to the piezoelectric effect. If the material is mechanically lossy, damping will occur. However, the electromechanical coupling is not a large effect and very large mechanical losses would be required to explain the observed damping. Furthermore, damping is observed after transit time when the material is no longer piezoelectric.

In conclusion, hysteretic and mechanical loss are likely present. However, dielectric relaxation apparently dominates the problem. It is thought that this is the first work that demonstrates dielectric relaxation is important in the shock environment.

ACKNOWLEDGMENTS

The author is indebted to S. T. Montgomery for informative discussions and to G. T. Holman and G. R. Lyons for aid in performing the experiments.

NOTATION

A^* = capacitor area
α = constant
C^* = internal capacitance
D^* = electric displacement
E^* = electric field
ε_∞^* = instantaneous permittivity
ε^* = relaxed permittivity
G = source term
i_e = external current
i^* = displacement current
i_ℓ^* = leakage current
L_e = external inductance
P^* = dipole polarization
P_0 = remanent polarization
R_ℓ^* = leakage resistance
ρ^* = resistivity
t = time
$\hat{t}$ = integration variable

τ^* = relaxation time
U = shock velocity
x_o = dimension parallel to polarization
y_o = dimension normal to x_o and z_o
z_o = dimension in direction of shock

Note: Asterisks symbolically denote parameters that may be discontinuous across the shock front. Single and double primes denote these parameters when evaluated in the unshocked and shocked regions, respectively.

REFERENCES

1. W. J. Halpin, J. Appl. Phys. 37, 153 (1966).
2. W. J. Halpin, J. Appl. Phys. 39, 3821 (1968).
3. P. C. Lysne, J. Appl. Phys. 44, 577 (1973).
4. P. C. Lysne and C. M. Percival, J. Appl. Phys. 46, 1519 (1975).
5. P. C. Lysne, J. Appl. Phys. 48, 1202 (1977).
6. H. Frölich, Theory of Dielectrics, Oxford Press, Oxford (1958).
7. P. C. Lysne, J. Appl. Phys. 48, 1024 (1977).
8. P. C. Lysne, submitted to J. Appl. Phys.

B-10

THEORETICAL BASIS FOR UNDERSTANDING THE MIXED PHASE REGION IN SHOCK-COMPRESSED IRON

J. W. Forbes
Naval Surface Weapons Center
Silver Spring, Maryland USA
and
G. E. Duvall
Washington State University
Pullman, Washington USA

INTRODUCTION

Under the influence of shock waves, certain materials undergo polymorphic transformations. The time available in a shock experiment for such a transformation to occur does not exceed a few microseconds, yet the same transformation may require minutes or hours for completion in a static experiment. This large transformation rate is the result of distinct features of nucleation and growth of new phases in shock waves. Predominant among these features are the large number of low energy nucleation sites available on defects created by shock waves. Research on nucleation and growth of the new phase in shock-loaded solids is important for developing quantitative physical models for transformation rates and thermodynamic state paths through the mixed phase region.

Nucleation sites for the iron transformation are sensitive to the amount of applied pressure. The fraction of iron transformed to ε-phase in shock is exponentially related [1] to the difference between Gibbs energies of new and old phases (driving force). A possible explanation for a relationship between the amount transformed and the driving force is that transient embryos of the second phase, which always exist as a result of statistical fluctuations, may be "frozen-in" by sudden application of sufficient pressure to bring the material into the stability field of the second phase. Some of the embryos which existed initially in the homogenous material have suddenly become growing nuclei under stable conditions for the new phase. This type of nucleation would apply particularly in a shock wave and in some martensitic transformations. A relation between the number of frozen-in nucleation sites and the driving force is established in this paper; this fact suggests a basis for understanding both the shock induced α to ε transformation in iron and athermal martensitic transformation. The relation has at best a remote numerical similarity to the exponential experimental relation, but it shows promise of further development.

NUCLEATION MODEL

Consider material in a lattice that is perfect except for thermal fluctuations. In an equilibrium state of the first phase there is a distribution of embryos of the second phase due to normal statistical fluctuations. Figure 1a shows the number of embryos in stable phase 1 as a function of the number of atoms, n, in each embryo. These embryos are created, grow, and shrink through fluctuations. For this case where phase 1 is stable, energy to grow increases monotonically as shown by curve A in Figure 1b where $G_{21} > 0$. If sufficient pressure, P_1, is applied to make the field for phase 2 stable, G_{21} changes sign and curve B of Fig. 1b results. For this case, an embryo for which $n < n_1^*$ requires energy to grow; those with $n_1 > n_1^*$ will continue to grow because energy of the system is diminished by growth.

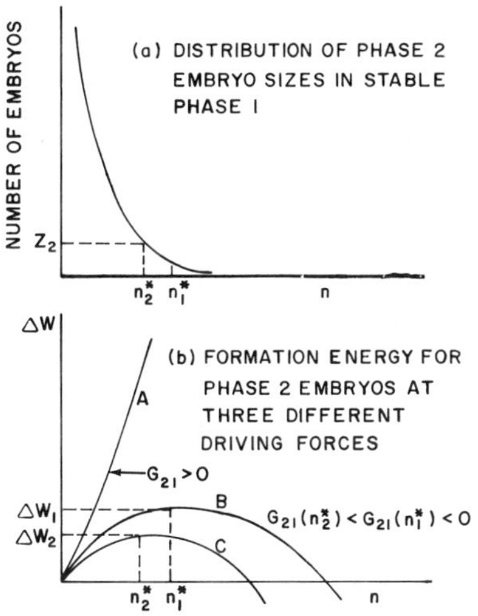

Fig. 1. Spherical embryos in stable phase 1. Curve A is for the stable phase 1 field with $G_2-G_1 = G_{21} > 0$. Curve B is for low pressure sufficient to make the phase 2 field stable with $G_{21}(n_1^*) < 0$. Curve C is for higher pressure, P_2, for which $G_{21}(n_2^*) < G_{21}(n_1^*) < 0$.

Increasing pressure to P_2 produces energy curve C in Fig. 1b. In this case the driving force $G_{21}(P_2)$ is more negative than $G_{21}(P_1)$, so energy required to form a stable nuclei of phase 2 is reduced from ΔW_1 to ΔW_2. For curve C a stable nuclei of phase 2 contains n_2^* atoms: $n_2^* < n_1^*$.

If the distribution of embryo sizes given in Fig. 1a does not change when pressure is increased from zero to P_2, then N* includes all nuclei for which $n > n_2^*$. This condition applies if the time required to apply the driving force is less than the time required to redistribute embryo sizes to conform to the new state.

A relationship between driving force and number of nucleation sites based on these concepts can be derived using nucleation theory. Assume the generally accepted conditions of nucleation; namely, that embryos exist and are internally uniform, having the same structure and properties as the final phase in bulk form. These assumptions about embryos leave shape and size as the only variable parameters. For simplicity, assume that n atoms in an embryo form a sphere. When the spherical embryo is formed, energy of the entire assembly increases by the amount ΔW defined by the expression [2]

$$\Delta W = n\, G_{21} + (36\pi V_2^2)^{1/3}\, \sigma\, n^{2/3} \tag{1}$$

The σ term in (1) makes it necessary for G_{21} to attain a certain negative value before transformation can begin.

Consider now the stability of embryos according to their size. Embryos containing n^* atoms maximize ΔW, where

$$n^* = \frac{-32\pi\sigma^3 V_2^2}{3G_{21}^3} \tag{2}$$

Having reached this size, they continue to grow.

In macroscopic assemblies, fluctuations will lead to local transitory phase transformations. These fluctuations establish a distribution of embryos of different sizes within a stable phase which have the same atomic arrangement as the new phase. Frenkel [3] treated each embryo as a molecule of a particular kind, independent of the others, and randomly present as a dilute solution in phase 1. Size distribution of these embryos was determined by maximizing the Gibbs potential of such a solution, which resulted in the expression

$$N_n = N \exp\left(-\frac{\Delta W(n)}{kT}\right) \tag{3}$$

If the distribution of embryos has insufficient time to change from its initial form when a large stress is rapidly applied, a certain number of embryos are suddenly found to exceed the critical or stable size for the new, thermodynamic state, becoming "trapped" on the side of the energy curve favoring growth. The number of such clusters N^* is given by

$$N^* = N \int_{n^*}^{n^\infty} \exp\left(-\frac{\Delta W^o(n)}{kT_0}\right) dn \tag{4}$$

where $\Delta W^o(n)$ is given by (1) for the initial stable state ahead of the shock front, and n^* is given by (2) for the state behind the shock front. Equation (4) cannot be integrated analytically. However, differentiation of (4) with respect to G_{21} using Leibnitz's rule [4] and eliminating n^* and $\Delta W\ (n^*)$ from (1) and (2) results in the final expression

$$\frac{dN^*}{dG_{21}} = \frac{-32\pi\sigma^3 V_2^2 N}{3G_{21}^3}\left(\frac{3}{G_{21}} - \frac{2}{V_2}\frac{dV_2}{dG_{21}}\right)$$

$$\times \exp\left(\frac{-16\pi\sigma^3 V_2^2}{G_{21}^2}\left(1 - \frac{2G_{21}(P=0,T=295^\circ K)}{3G_{21}}\right)\right) \qquad (5)$$

This equation is significant because it establishes a relation between N^* and G_{21} in the stable field of phase 2.

To calculate values of dN^*/dG_{21} from (5) requires values for σ, V_2, N, G_{21}, and dV_2/dG_{21}. Values for σ found in the literature vary from 20 ergs/cm^2 for a coherent twin interface to 200 ergs/cm^2 for an incoherent interface [2,5]. Examination of (5) reveals that the influence of the exponential function is overriding and if any nucleation is to occur, the argument of the exponential must be less than about 100. This implies that the value for σ must be about 20 ergs/cm^2. Values for G_{21}, V_2, and dV_2/dG_{21} were obtained from a two-phase equation-of-state for iron [1]. The value of N for iron is $8.48 \times 10^{22}/cm^3$.

Inspection of values given in the last column of Table I reveals that as the magnitude of G_{21} increases so does the magnitude of dN^*/dG_{21}. Since dN^*/dG_{21} for G_{21}=A is so small compared to values at a higher driving force, nucleation is initiated for values at a driving force between A and 2A. A smaller value for σ would force the initiation of nucleation nearer A, but the adjustment appears unwarranted considering uncertainties in σ. Values of n^* become unrealistically small for driving forces exceeding 2A which indicates that the model needs improvement.

DISCUSSION

It has been shown [1] that f varies exponentially with G_{21} for the iron mixed phase Hugoniot data giving the expression

$$1-f = \exp\left(\Theta\left(G_{21} - A\right)\right) \qquad (6)$$

where Θ is a fitted constant and A is the value of G_{21} at the onset of transformation. Previous workers [6] have shown that such a relationship describes athermal martensitic transformations; this similarity strengthens the link between martensitic transformations and the shock-induced α to ε iron transformation.

The nucleation model described here does not reproduce (6) but it does provide a relation (equation (5)) between dG_{21} and dN^*. The differential of (6) allows comparison between the two results. In (5), however, the proportionality parameter Θ is not constant but varies over a wide range of values in the mixed phase region. If

Table I. Values of dN^*/dG_{21}

G_{21},‡ ergs/atom	P, Mbar	n^*	$\frac{\Delta W(n^*)}{kT_0}$ †	$\exp\left(-\frac{\Delta W(n^*)}{kT_0}\right)$	$\frac{dN^*}{dG_{21}}$ $\frac{\text{number atom}}{\text{ergs cm}^3}$	$\frac{dN^*}{dG_{21}}$ g/Mbar cm^6
A	0.13	51.1	99.8	4.5×10^{-44}	-8.3×10^{-5}	-7.7×10^{-15}
2A	0.144	6.4	14.4	5.6×10^{-7}	-1.3×10^{32}	-1.2×10^{22}
3A	0.158	1.9	4.84	7.9×10^{-3}	-5.4×10^{35}	-5.0×10^{25}
4A	0.172	0.8	2.28	1.0×10^{-1}	-2.9×10^{36}	-2.6×10^{26}

‡ $A = G_{21}$ (P=0.13 Mbar, T=338K) $= -8.33 \times 10^{-15}$ ergs/atom $= -7.06 \times 10^{8}$ ergs/cm^3

† $\Delta W(n^*) = (6.85 \times 10^{-14}$ ergs/atom$)\, n^* + (4.62 \times 10^{-14}$ ergs/atom$)\, n^{*2/3}$

this transformation is martensitic and the transformation occurs with martensitic plate volume, V_p, of 10^{-8} cm^3, the values of Θ are obtained by multiplying $-V_p$ by entries in the last column of Table I. This gives $\Theta = 7 \times 10^{-23}$ g/Mbar-cm^3 for G_{21}=A and Θ=1.2 x 10^{14} g/Mbar-cm^3 for G_{21}=2A. These values are far from the observed value [1], this is not surprising considering the unreality of the basic assumption that nucleation is occurring in a homogeneous lattice.

Values for Θ calculated from (5) come much closer to the measured value when the homogeneous model is modified by assuming spherical pre-existing embryos of size n_o and that normal statistical fluctuations as in the homogeneous case create a distribution of sizes about the pre-existing sites. The only change to (5) for this case is the replacement of N with N_o. It seems reasonable that N_o be about $10^7/cm^3$ which is the inferred number of twins required to account for all the plastic strain in shocked iron at 130 kbar [7]. If the transformation is martensitic and occurs with V_p of 10^{-8} cm^3, the values of Θ are obtained by multiplying the entries in the last column of Table I by $-V_p N_o/N$. This gives Θ=9.1 x 10^{-38} g/Mbar-cm^3 for G_{21}=A, Θ=0.014 g/Mbar-cm^3 for G_{21}=2A, and Θ=59.0 g/Mbar-cm^3 for G_{21}=3A. These values for Θ approach the measured value of 4.048 kg/Mbar-cm^3 which suggests that the more difficult calculations of nucleation at lattice imperfections should be done.

CONCLUSION

The above calculations show that a relation between dN^* and dG_{21} can be established. Even though the model doesn't reproduce equation (6), the basic concept of freezing-in embryos created by statistical fluctuations appears sound and suggests a basis for understanding of both the shock-induced α to ε and athermal γ to α transformations. Since the weight of observations on athermal martensite suggests that nucleation occurs at lattice imperfections, this calculation should be repeated for the much more difficult problems of embryos formed about such sites. The simple modification of adding preexisting nucleation sites to the homogeneous lattice described above suggests that such calculations will be in the right direction to bring about agreement between measurements and theory. Considerations are also required on the effects on the transformation of (1) strain in the lattice to accommodate the nuclei, (2) use of the hydrostatic Gibbs function in problems dealing with solids, and (3) use of bulk values of surface energy on small surfaces with large curvature. If these modifications can be accomplished, it may well turn out that the too rapid variation of dN^*/dG_{21} with G_{21}, shown in Table I, vanishes and that a reasonable theoretical basis for the observations is produced.

NOTATION

- A = value of G_{21} at onset of transformation
- f = mass fraction of phase 2 in mixed phase region
- G_{21} = difference in Gibbs energy between old and new phase (i.e., G_2-G_1)
- k = Boltzmann's gas constant
- n = number of atoms in an embryo
- n^* = number of atoms in embryo that maximize ΔW
- N = total number of atoms per cm^3
- N_n = number of embryos of size n
- N^* = total number of nucleation sites
- N_o = total number of pre-existing nucleation sites of size n_o
- P = pressure
- P^{TL} = transition pressure under shock loading
- T = temperature
- V = specific volume
- V_p = volume of a martensitic plate in cm^3
- α = body centered cubic lattice
- γ = face centered cubic lattice
- ΔW = increase in system's energy when embryo is formed
- ε = hexagonal close packed lattice
- Θ = proportionality parameter
- σ = surface energy per unit area between two phases

REFERENCES

1. J. W. Forbes, Ph.D. Thesis, Washington State University, Pullman, Washington (1976).
2. J. W. Christian, The Theory of Transformations in Metals and Alloys, Pergamon Press, London (1965).
3. J. Frenkel, Kinetic Theory of Liquids, Dover Publications, New York (1955).
4. W. Kaplan, Advanced Calculus, Addison-Wesley Publishing Co., Reading, Massachusetts (1959).
5. J. Burke, The Kinetics of Phase Transformation in Metals, Pergamon Press, London (1965).
6. C. L. Magee, in Phase Transformations, American Society for Metals (1970), Chapter 3, p. 115.
7. J. N. Johnson and R. W. Rhode, J. Appl. Phys. 42, 4171 (1971).

B-11

A NEW ALLOTROPIC PHASE OF CERIUM ABOVE 122 KBAR

S. Endo and N. Fujioka

Osaka University
Toyonaka, Japan

and

H. Sasaki

Hokkaido University
Sapporo, Japan

INTRODUCTION

It is well known that cerium metal appears in many different allotropic phases. The existence of two fcc phases, γ and α, has been the subject of extensive study particularly relative to the behavior of the 4f electrons in this transition. At pressures higher than 50 kbar, α-Ce transforms to a superconducting α' phase [1]. Conflicting results have been reported on the crystal structure of α'-Ce [2-6]. During the course of an x-ray diffraction study to clarify this situation, we found another allotropic phase around 120 kbar pressure. This paper deals with the crystal structure of the new phase and the resistance change associated with the transition from the α' phase to the new phase.

EXPERIMENTAL

X-ray Diffraction

A high pressure x-ray diffraction apparatus originally designed for liquid helium temperature [7] was used to obtain diffraction patterns at room temperature. This apparatus is based on the Guinier focusing geometry which permits high-resolution patterns with sharp lines in low background. The available range of 2θ is limited within $\pm$ 45°. The Mo-Kα radiation was used, and the lattice parameter was determined within an accuracy of $\pm$ 0.01Å. A powdered

Ce sample filed from an ingot of nominal 99.95% purity was packed into a hole in a boron-epoxy cell and pressed between Bridgman anvils.

Calibration of pressure was made using aluminum as the marker. The pressure-volume relation for aluminum was measured [8] with reference to the equation of state for NaCl [9].

Resistance Measurement

The cerium metal ingot was rolled down to 0.03 mm thickness. A polished foil with 0.60 mm width and 2.00 mm length was placed in a semi-sintered magnesia cell [10] in the shape of an octahedron (Fig. 1). A multi-anvil apparatus modified from the "6-8 type" [11] was used for the high pressure resistance measurement. The eight anvils were forced together by means of a hydraulic press. The pressure in the cell was calibrated by observing the press forces required to produce the γ-α and α-α' transitions in Ce (7 kbar and 51 kbar [4], respectively), the I-II transition in Sn (100 kbar), and the semiconductor-metal transitions in ZnTe (120 kbar), ZnS (150 kbar), GaAs (180 kbar) and GaP (220 kbar) [12]. The pressure was increased up to 220 kbar at a rate of about 0.5 kbar/min. A four-lead measurement was adopted to measure the resistance of a cerium foil.

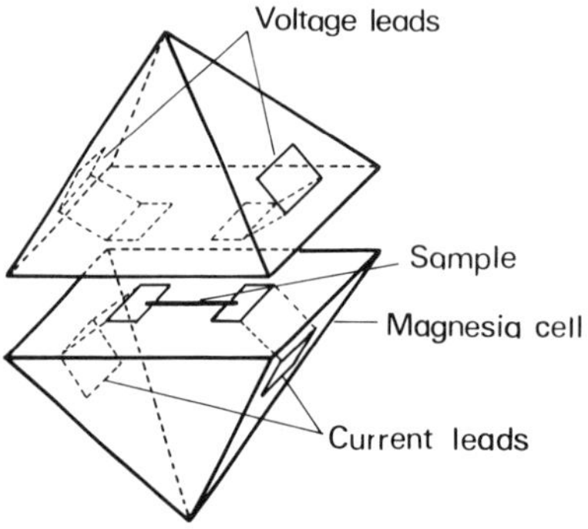

Fig. 1. High pressure cell for resistance measurement.

RESULTS AND DISCUSSION

Figure 2 shows diffraction patterns of Ce, taken as a function of increasing pressure. Gradual structural transitions are observed. After releasing the pressure from 175 kbar to ambient pressure, the sample was again examined with x-ray diffraction. As shown in the bottom pattern of Fig. 2, no measurable change in the diffraction pattern of the γ phase is observed as seen in comparison with the initial state in the top pattern of Fig. 2. Thus it is apparent that the Ce sample was not subjected to oxidation and/or contamination during the high pressure experiment.

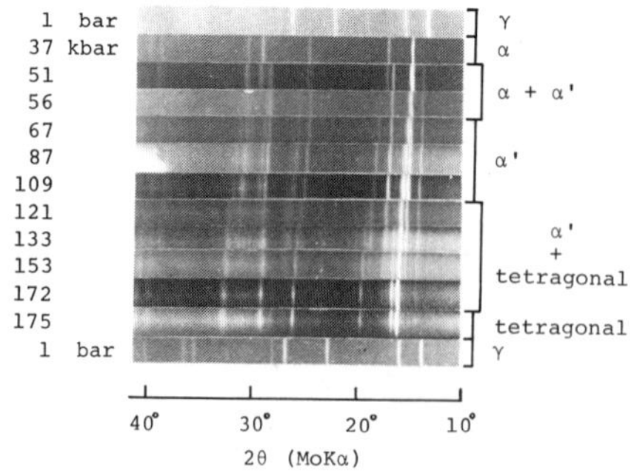

Fig. 2. Powder diffraction patterns for Ce under various pressures at room temperature.

A typical plot of the resistance vs. pressure is shown in Fig. 3. The resistance behavior at the $\gamma - \alpha$ and the $\alpha - \alpha'$ transitions are in good agreement with those obtained by previous investigators [1,4,5].

α'-Ce

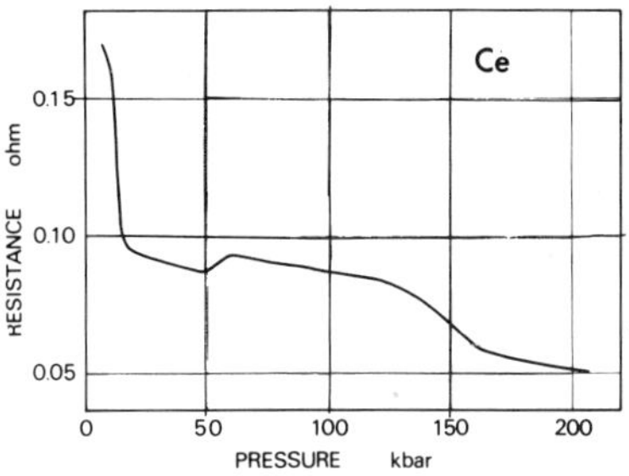

Fig. 3. Resistance vs. pressure for Ce at room temperature.

The diffraction pattern taken at 51 kbar shows that the α' phase appears in addition to the α phase. Table I gives the values of the d spacings and intensities of reflections observed for the α' phase at 67 kbar. A structural analysis was attempted but has not been successful.

In the case of an experiment with a cubic press [13] which generally is believed to generate a more homogeneous pressure than possible with the Bridgman anvils, diffraction patterns different from those with Bridgman anvils were obtained. These appear like the patterns of a distorted fcc structure [14], which seems to be identified as the α'' phase of Ellinger and Zachariasen [3]. Further investigation to study the effect of shear stress is needed to clarify this situation.

New High Pressure Form: Tetragonal-Ce

At pressures higher than 121 kbar, a new phase was detected by x-ray diffraction in coexistence with α'-Ce. At 175 kbar a single-phase pattern was obtained, and the analysis showed that the new phase is body-centered tetragonal with two atoms per unit cell, and has dimensions a=2.92Å and c=4.84Å. The observed d spacings and intensities of the reflections are listed in Table II and compared with the calculated values. The atomic scattering factor of Ce used for the calculation of the intensities was taken from the International Table for X-ray Crystallography [15]. The variation of unit cell dimensions of tetragonal-Ce with pressure is given in Table III. The ratio c/a increases with pressures from 1.63 at 121 kbar to 1.66 at 175 kbar. The average compressibility of the tetragonal phase in this pressure range is $K = (1.1 \pm 0.1) \times 10^{-6}$ cm^2/kg. The variation of atomic volume with pressure for the tetragonal phase and that for the α phase obtained in the present work are shown in Fig. 4. The variation for α' phase obtained by Zachariasen and Ellinger [3] is also plotted for comparison.

Figure 3 shows a slope change in the resistance vs. pressure curve around 120 kbar. The x-ray study clarifies that this decrease in resistance is due to the transition from α' to the tetragonal phase. In order to determine accurately the initial pressure for the transition, dR/dP was plotted against pressure. It turned out to be 122 ± 3 kbar.

Table I. X-Ray Data for Unknown α'-Ce at 67 kbar

Line No.	$d_{obs.}$	$I_{obs.}$
1	2.889	medium
2	2.679	very strong
3	2.628	very strong
4	2.591	weak
5	2.381	strong
6	2.238	medium
7	1.650	medium
8	1.589	weak
9	1.553	trace
10	1.439	medium
11	1.421	weak
12	1.340	weak
13	1.313	weak
14	1.188	trace
15	1.090	very weak
16	1.065	very weak
17	1.017	trace

Table II. X-Ray Data for Tetragonal-Ce at 175 kbar*

hkl	$d_{calc.}$	$d_{obs.}$	$I_{calc.}$	$I_{obs.}$
101	2.500	2.500	100	very strong
002	2.419	2.419	23	medium
110	2.065	2.063	30	medium
112	1.571	1.569	27	medium
200	1.460	1.458	11	very weak
103	1.412	1.411	19	medium
211	1.261 }	1.256	27 }	medium
202	1.250 }		13 }	
004	1.209	1.209	3	very weak
114	1.044 }	1.040	7 }	weak
220	1.033 }		4 }	
213	1.015	1.012	13	weak

*a = 2.92 ± 0.01 Å; c = 4.84 ± 0.01 Å

In 1964, Stager and Drickamer [16] measured the electrical resistance of Ce under pressure. They observed a distinct drop in resistance at about 160 kbar and 296 K, and suggested the existence of a transition. Considering the calibration values of pressure generally accepted in the 1960's, the resistance drop found in the present experiment is assumed to be the same as that

Table III. Unit Cell Dimensions for Tetragonal-Ce

P, kbar	a, Å	c ,Å	c/a	V, $Å^3$
121	2.99	4.86	1.63	21.72
133	2.98	4.87	1.64	21.54
151	2.95	4.86	1.65	21.14
152	2.95	4.86	1.65	21.10
153	2.95	4.84	1.64	21.06
172	2.93	4.84	1.65	20.72
175	2.92	4.84	1.66	20.63

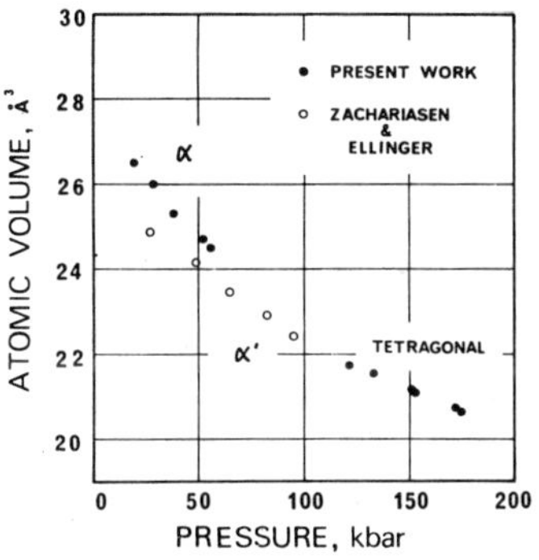

Fig. 4. Atomic volume-pressure relation for α- and tetragonal-Ce obtained in the present work and for α'-Ce obtained by Zachariasen and Ellinger [3].

found by Stager and Drickamer.

The α'-tetragonal transition is very sluggish, as can be seen from Figs. 2 and 3. It ends at pressures near 170 kbar in both experiments carried out with different kinds of pressure apparatus. This sluggishness seems to be an intrinsic property of the transition.

Phase Boundary Between α'- and Tetragonal-Ce

Resistance measurements of Ce are now being carried out at high temperatures and pressures. A tantalum sleeve for an electrical heater is set in the magnesia octahedral cell. Temperature is measured using alumel-chromel thermocouples without correction of the pressure effect on the thermal electromotive force. A preliminary isothermal experiment taken at 200°C indicates that the transition from α' to tetragonal phase begins to occur at 100 kbar. The α'-tetragonal phase boundary tentatively represented by a straight line connecting the two points at room temperature and 200°C has a negative slope of -8.2 °C/kbar. On extrapolation, this boundary line approaches the minimum in the fusion curve. Further experiments are now in progress.

From previous studies of pressure induced transitions in the rare earth metals [17] and the present study of Ce, the structural changes in the following sequence may be expected as the pressure increases:

$$\text{hcp} \rightarrow \text{Sm type} \rightarrow \text{dhcp} \rightarrow \text{fcc} \rightarrow (\text{fcc}') \rightarrow \alpha' \rightarrow \text{tetragonal} \quad (1)$$

where fcc' means a collapsed fcc.

ACKNOWLEDGMENT

The x-ray diffraction study for the present investigation was carried out at Hokkaido University. The authors wish to thank the late T. Mitsui for his sincere encouragement and valuable advice.

REFERENCES

1. J. Wittig, Phys. Rev. Letters 21, 1250 (1968).
2. D. B. McWhan, Phys. Rev. B1, 2826 (1970).
3. F. H. Ellinger and W. H. Zachariasen, Phys. Rev. Letters 32, 773 (1974); also W. H. Zachariasen and F. H. Ellinger, Acta Cryst. A33, 155 (1977).
4. P. H. Schaufelberger and H. Merx, in Proc. 4th Intern. High Pressure Conference, Kyoto, Japan (1974), p. 222; also High Temp. - High Press. 7, 55 (1975).
5. P. H. Schaufelberger, J. Appl. Phys. 47, 2364 (1976).
6. W. H. Zachariasen, J. Appl. Phys. 48, 1391 (1977).
7. S. Endo and T. Mitsui, in Proc. 4th Intern. High Pressure Conference, Kyoto, Japan (1974), p. 824; also Rev. Sci. Instr. 47, 1275 (1976).
8. S. Endo and H. Sasaki, to be published.
9. D. L. Decker, J. Appl. Phys. 42, 3239 (1971).
10. N. Kawai, H. Sakamoto, Y. Notsu and A.Onodera, Proc. Japan Acad. 51, 623 (1975).
11. N. Kawai and S. Endo, Rev. Sci. Instr. 41, 1178 (1970).
12. G. J. Piermarini and S. Block, Rev. Sci. Instr. 46, 973 (1975).
13. S. Akimoto, T. Yagi, Y. Ida, K. Inoue and Y. Sato, High Temp. - High Press. 7, 287 (1975).
14. S. Endo, H. Sasaki and T. Mitsui, J. Phys. Soc. Japan 42, 882 (1977).
15. J. A. Ibers, International Tables for X-ray Christallography, Vol. 3, C. H. Macgillavry and G. D. Rieck, eds., Kynoch Press, Birmingham, England (1962), p. 201.
16. R. A. Stager and H. G. Drickamer, Phys. Rev. 133, A830 (1964).
17. A. Jayaraman and R. C. Sherwood, Phys. Rev. 134, A691 (1964).

B-12

LATTICE PARAMETERS AND VOLUME COMPRESSIBILITY OF YTTERBIUM UP TO 300 KBAR

K. Syassen and W. B. Holzapfel

Max-Planck-Institut Für Festkörperforschung
Stuttgart, W. Germany

INTRODUCTION

The rare earth metal Ytterbium is close to the divalent state under ambient conditions. Pseudopotential calculations by Johansson et al. [1] have predicted an electronic transition towards the trivalent state under a pressure of about 120 kbar. This transition is related to the promotion of an occupied 4f level near to and above the Fermi level, and it is difficult to predict whether this transition is continuous like the γ-α transition of Ce above its tricritical point or discontinuous like the γ-α transition of Ce below its tricritical point.

Information about the electronic transition in Yb can be obtained from the high pressure equation of state (EOS). In the present work we have studied the isothermal compression of Yb up to 300 kbar using high pressure x-ray diffraction techniques. These data are compared with the equation of states of the other rare earth metals.

EXPERIMENTAL TECHNIQUE

The present Yb samples were cut under oil from an ingot which had a purity of 99.8% with the major impurities being H(1400 ppm), O (300 ppm), C (150 ppm), Fe (30 ppm), S (<20 ppm) and less than 10 ppm of any other element.* The samples were mounted in a gasketed diamond anvil device [2] with oil as the pressure transmitting medium. The lattice parameters of Yb were determined from energy dispersive x-ray diffraction patterns. A short discussion of this technique has been published previously [2] and further details will be given elsewhere [3]. Typical spectra are shown in Fig. 1. Least

*The Ytterbium metal was kindly supplied by K. A. Gschneidner, Iowa State University, Ames, Iowa USA.

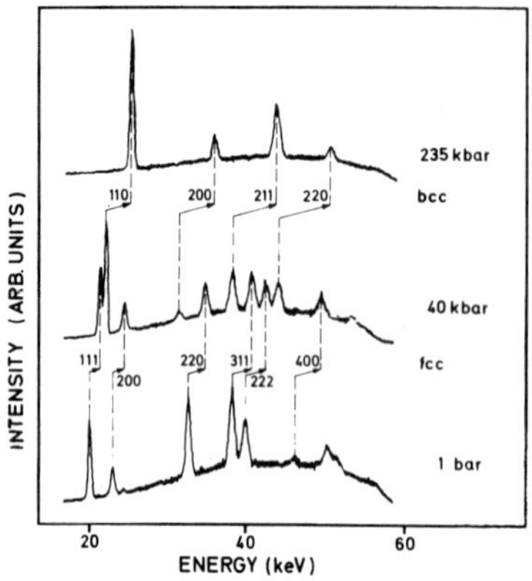

Fig. 1. Energy dispersive x-ray diffraction spectra of polycrystalline Yb under pressure.

squares fits of Gaussian lines to the individual peaks result in sets of energy values E_{hkl}. The latter are related to the lattice spacings d_{hkl} by the Bragg equation $d_{hkl} = C/E_{hkl}$ sin θ) where C is a universal constant and θ is the diffraction angle.

If one follows the procedure which has been proposed recently [4] for the evaluation of x-ray measurements on samples under nonhydrostatic pressures, one finds that the usual plot of $a_{hkl} = (h^2 + h^2 + l^2)^{1/2} d_{hkl}$ with respect to the function $\Gamma_{hkl} = (h^2 k^2 + k^2 l^2 + l^2 h^2)/(h^2 + k^2 + l^2)^2$ shows no systematic variation as long as the pressure does not exceed 40 kbar. The relative mean square deviation $\Delta a/a$ of the a_{hkl}-values from the average value of the lattice parameter a was always less than 1.3 x 10^{-3} for fcc Yb in this pressure range. Nonhydrostatic stresses due to freezing of the oil result, however, in additional systematic variations of the a_{hkl}-values. From plots of these a_{hkl}-values with respect to Γ_{hkl}, one finds that these nonhydrostatic stresses introduce an additional error of $\Delta a/a$ 5 x 10^{-3} in the pressure range above 40 kbar. Additional errors of the same order of magnitude result from small electronic and mechanical drifts during the measurements.

The pressure on the sample was measured by the ruby fluorescence method before and after each x-ray measurement. Whereas the precision of these measurements with respect to the ruby pressure scale [5] was better than ± 2 kbar, the total accuracy is reduced by pressure drifts during the time of the x-ray measurements to about ± 5 kbar.

RESULTS

The present room temperature data for the EOS of Yb are shown in Fig. 2. The solid curve serves only as a guide to the eye and includes the volume discontinuity at the phase transition from the low pressure fcc to the high pressure bcc phase of Yb. Within the accuracy of the present experiments, the EOS of Yb shows no discontinuous change in the high pressure bcc phase up to 300 kbar.

The difference between the equation of state of Yb and the behavior of the other rare earth elements is seen most clearly if one scales the P-V relations of these elements by appropriate powers

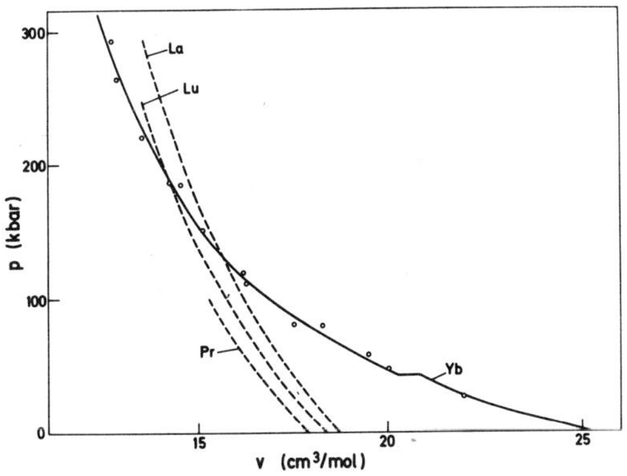

Fig. 2. Equation of state of Yb at room temperature. The behavior of typical trivalent rare earth metals is indicated by the dashed curves which represent appropriately scaled equations of states of other rare earth elements.

of the atomic number Z, according to the universal Thomas-Fermi equation of state [6]. If the scaling is done with respect to the EOS of Yb, one obtains, with the aid of the atomic number Z_{Yb} for Yb, the following scaled variables of the other elements: $v = V(Z/Z_{Yb})$ and $p = P(Z/Z_{Yb})^{-10/3}$.

Within the experimental accuracy, all these scaled EOS for the regular trivalent rare earth metals closely follow the behavior of Lu [7]. The largest deviations from this common scaled equation of state for trivalent rare earth metals are indicated in Fig. 2 by the dashed lines for La [8] on one side and for Pr [9] on the other side.

The comparison of these scaled equations of state of typical trivalent rare earth metals with the EOS of Yb shows clearly in Fig. 2 that the continuous approach of Yb to the behavior of the typical trivalent rare earth metals has been completed largely at 150 kbar. The compressibility of Yb seems to show still an anomalously large value at this pressure and approaches the typical trivalent behavior only at still higher pressures. The small volume discontinuity in the phase transition from the low pressure fcc structure to the high pressure bcc phase seems to account only for a very minor change in the valency of Yb.

ACKNOWLEDGMENT

It is a pleasure to thank E. Böttcher for very skillful help in the data collection and evaluation.

NOTATION

h,k,l	= Miller indices	θ	= diffraction angle
E_{hkl}	= energy of x-ray diffracted on hkl planes	Γ_{hkl}	= Singh-Kennedy function for determination of nonhydrostatic stresses
d_{hkl}	= lattice spacing on hkl planes	a_{hkl}	= lattice parameter value from hkl reflection
C	= 6.199 keV Å	a	= mean value of lattice parameter

Δa = mean squared deviation of a_{hkl}-values from a in the hydrostatic limit plus possible error in a due to nonhydrostatic stresses
Z = atomic number
Z_{Yb} = atomic number of Yb
V = molar volume
v = scaled volume (= V for Yb)
P = pressure
p = scaled pressure (= P for Yb)

REFERENCES

1. B. Johansson and A. Rosengren, Phys. Rev. B 11, 2836 (1975).
2. K. Syassen and W. B. Holzapfel, Europhys. Conf. Abstr. 1A, 75 (1975).
3. K. Syassen and W. B. Holzapfel, to be published.
4. A. K. Singh and G. L. Kennedy, J. Appl. Phys. 45, 4686 (1974).
5. G. J. Piermarini, S. Block, J. D. Barnett, and R. A. Forman, J. Appl. Phys. 46, 2774 (1975).
6. R. P. Feynman, N. Metropolis, and E. Teller, Phys. Rev. 75, 1561 (1949).
7. L. G. Liu, J. Phys. Chem. Solids 36, 31 (1975).
8. K. Syassen and W. B. Holzapfel, Solid State Commun. 16, 533 (1975).
9. P. W. Bridgman, Proc. Am. Acad. Arts Sci. 76, 55 (1948).

B-13

ELECTRICAL CHARACTERIZATION OF PRESSURE-INDUCED PHASE TRANSITIONS*

B. A. Lombos, H. M. Mahdaly, and B. C. Pant

Concordia University
Montreal, Canada

INTRODUCTION

A procedure has been developed for the direct observation of the pressure-induced polymorphic phase transitions. The resistance variation, which is due to the different electronic structures of the two phases of the mercury chalcogenides [1,2], is measured directly on the sample. Small sample volume increases the sensitivity of the technique; therefore, the interaction of its volume variation with the generated pressure is negligible. Thus, the pressure remains an independent variable during the transition. The kinetics and the mechanism of the reaction can be followed directly and independently of the pressure-transmitting medium. Although the measurement of volume variation as a function of pressure seems to closely follow the advancement of the transition reactions, as is detected through the pressure-transmitting medium, the technique is an indirect one. Furthermore, to achieve sufficient accuracy, large sample volumes are required. Consequently, the volume change of the sample generates an auxiliary uncontrollable pressure variation which counteracts the experimentally adjusted pressure. This phenomenon is called retropressure [3,4].

The experimental data obtained by the direct resistance variation technique demonstrate the nucleation dependence of the pressure-induced phase transition in the case of HgSe. A three-dimensional model is developed to convert the directly observable resistance variation, during the polymorphic transition, to volume change. The direct observation of this phenomenon is of help in the explanation and analysis of the kinetics and mechanism related to the pressure-induced phase transitions. It had been suggested as early as 1914 [5] by Fermor and reviewed lately by Birch [6] that the volume change characterizing all polymorphic pressure-induced phase

*Work partly supported by the National Research Council of Canada and by a DGES-FCAC grant of the province of Quebec.

transitions might be related to earthquakes.

EXPERIMENTAL PROCEDURE

A piston-cylinder apparatus capable of generating pressures up to 3000 MPa was employed [2]. The mercury selenide crystals were prepared by a modified Bridgman technique [7]. Samples used for resistance discontinuity measurements had typical dimensions of 2.4 x 2.4 x 5 mm, were cut from the prepared crystals, and had ohmic contacts applied with indium solder. The volumes of these resistance samples corresponded to approximately 5.7×10^{-4} of the working volume (50 cm^3) of the high-pressure chamber; hence, the effect of their variations to the total volume can be neglected. A large sample with a typical volume of 10 cm^3 was used simultaneously in one set of experiments to follow the volume discontinuity.

The pressure was monitored *in situ* by a seasoned manganin coil which was calibrated by the transition pressure of KBr [8]. All data were continuously recorded as dc voltages on a multi-channel strip-chart recorder. The sensitivity of the pressure measurement was better than 100 kPa. The volume variation was followed by monitoring the position, ℓ, of the base plate of the pump, which displaced the pressure-generating mobile piston with a precision of 0.01 μm [9]. The resistance variations of the samples were followed with a sensitivity of 0.003 ohm. All measurements were performed at room temperature.

The experimental results of the simultaneous measurements of volume (solid line) and resistance (dotted line) are shown in Fig. 1. The volume change (as indicated by the base plate displacement ℓ) versus pressure curve clearly depicted the effect of the volume change of the large sample superposed on the experimentally-adjusted pressure variation. Thus, the effect of retropressure could be observed as indicated by A on the upstroke pressure side and by A' on the downstroke pressure side of the volume variation curve given in Fig. 1. The simultaneously recorded resistance variation of a sample with a small cross section (1 mm^2) directly following the reaction, indicated the unper-

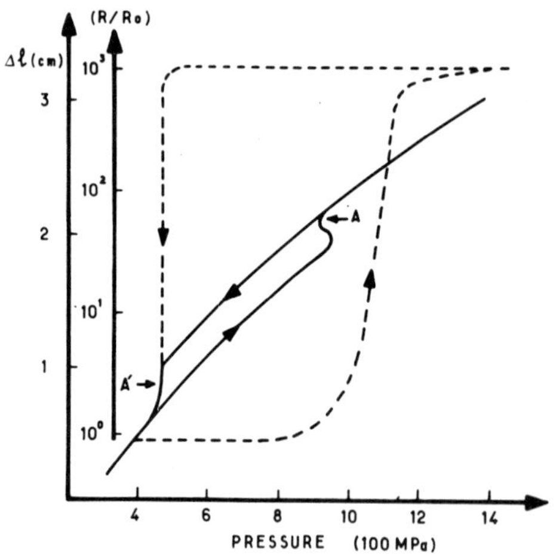

Fig. 1. Experimentally-observed simultaneous variations of the base plate displacement, expressing the volume change due to the polymorphic transition (solid line) and the relative sample resistance R/R_o (dotted line) as a function of the measured pressure in the high-pressure chamber.

turbed continuation of the phase transition in the case of upstroke as well as the downstroke side of the hysteresis curve, given by the dotted line in Fig. 1. The relative change of sample resistance during the cycle was from 1 to 1000.

The effect of the nucleation on the upstroke transition offset pressure was examined using mercury sulfide as impurity. Different concentrations of this material, having the crystal structure of the high-pressure phase, were mixed with the powdered mercury selenide. Sample blocks were prepared from these mixtures by compacting them under 70 MPa at room temperature. Mercury telluride with a transition pressure of more than 1500 MPa [2,10], thus maintaining its low-pressure phase, was used to nucleate the downstroke transition. In Fig. 2 the measured transition pressures (solid line) are given with the calculated one (dotted line) as a function of nucleation concentration. Each experimental curve contains a set of points detected in the same pressure cycle. On the high-pressure side of the central theoretical curve the upstroke transition pressures nucleated by HgS are collected with increasing rates of pressure variations, from 100 kPa to 10 MPa/min. On the low-pressure side of the dotted line the downstroke transition pressures nucleated by HgTe are shown, obtained with increasing rates of pressure change in the same range as above. The dependence of the offset transition pressures of the upstroke as well as of the downstroke sides on the nucleation concentration is clearly noticeable. This forms the basis of the model developed in the following sections.

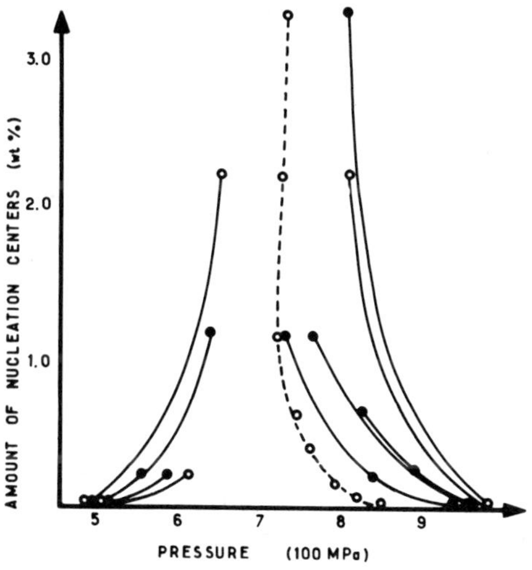

Fig. 2. Upstroke (r.h.s.) and downstroke (l.h.s.) offset transition pressures as the function of heterogeneous nucleation concentrations (solid lines). Computed upstroke offset transition pressures using equation (18) (dotted line). The rate of pressure variation increases approximately by 2500 kPa/min within 100 kPa and 10 MPa/min, considering the curves from the outer to the inner ones.

CONVERSION FUNCTION

The low-pressure phase of mercury selenide is a zero-gap semiconductor [11]; thus, its resistivity is small being of the order of magnitude of 10^{-4} ohm/cm. The high-pressure phase is a large energy gap semiconductor (1) with a resistivity in the range of 10^2 ohm/cm, as determined in this investigation. The total volume under pressure as a function of time and the advancement of the reaction α can be given as

$$V(t) = V_L(t) + (1-\alpha)\ V_1(t) + \alpha V_2(t) \qquad (1)$$

Now taking into consideration the compressibilities, it becomes

$$V(t) = V_L(t)\ (1-K_L P(t)) + V_1(1-\alpha)\ (1-K_1 P(t))V_1(t)\ \alpha \frac{D_1}{D_2}\ (1-K_2 P(t)) \qquad (2)$$

Here the compressibilities are assumed to be pressure independent. Then the degree of advancement of the reaction can be expressed as

$$\alpha = \frac{\Delta V(t) - (V_L K_L + V_1(t)K_1)\ P(t)}{V_1(t)\ ((1-\frac{D_1}{D_2}) + (K_2 \frac{D_1}{D_2} - K_1)\ P(t))} \qquad (3)$$

where

$$\Delta V(t) = V_L(t)+V_1(t) - V(t) \qquad (4)$$

as it also had been discussed by Lacam et al. [3,4].

In this investigation the measured pressure is defined as the sum of the nucleation independent thermodynamical transition pressure (P_{TD}) and the nucleation dependent kinetic pressure (P_K).

$$P(t) = P_{TD} + P_K \qquad (5)$$

Since the analysis of the kinetics and the mechanism of the reaction requires the determination of the degree of advancement of the reaction as a function of time and pressure, a conversion function was developed to transfer the resistance data to volume variations. Consequently, the advancement of reaction can be characterized directly by the resistance change of the solid during the phase transition. The model developed is based on the arguments of Jander [12] and on the results of this work concerning the transition pressure dependence on the concentration of nucleation centers (see Fig. 2).

It is assumed that the nucleation centers are in cross-sectional layers, separated by nonnucleated layers. The resistances representing the high-pressure phase are in parallel within the nucleated layers, and the resulting layer resistances are in series with respect to the end-contact surfaces.

Furthermore, based on this model the following are assumed: (1) The reaction is additive in nature. (2) Instantaneous nucleations are followed by three-dimensional rectangular growth of the nuclei. (3) The nucleation centers are homogeneously distributed in the material. (4) In the metastable state, past the thermodynamical equilibrium transition pressure, the growth of the nuclei is a function of time at any given pressure.

Under these conditions, the sample resistance is given by the resistances in series with a set of surfaces containing the nucleating centers.

$$R_I = \rho_1 \frac{N^{1/3} \ \ell(\alpha)}{ZW - N^{2/3} \ z(\alpha) \ w(\alpha)} \tag{6}$$

On the other hand, the resistance of the growth nuclei transformed in the high-pressure phase is expressed by

$$R_{II} = \rho_2 \frac{N^{1/3} \ \ell(\alpha)}{N^{2/3} \ z(\alpha) \ w(\alpha)} \tag{7}$$

Furthermore, these regions are separated by nonnucleated surfaces and the resistance of these can be given by

$$R_o = \rho_1 \frac{(L-N^{1/3} \ \ell(\alpha))}{ZW} \tag{8}$$

The combination of these three resistances results in the total resistance of the sample as follows:

$$R(\alpha) = R_o + R_I \ R_{II} \ (R_I + R_{II})^{-1} \tag{9}$$

or substituting (6), (7), and (8) into (9) one has

$$R(\alpha) = \frac{\rho_1 \{V_1 + z(\alpha) \ w(\alpha) \ N^{2/3} (\rho_1/\rho_2 - 1) \ [L - N^{1/3} \ell(\alpha)]\}}{ZW \ [ZW - z(\alpha) \ w(\alpha) \ N^{2/3} (1 - \rho_1/\rho_2)]} \tag{10}$$

When the transition is observed by the resistance change of a small sample, the terms dealing with the pressure-transmitting liquid do not enter into the calculation. Therefore equation (1) will be simplified to

$$V(t) = (1-\alpha) \ V_1(t) + \alpha V_2(t) \tag{11}$$

and the advancement of the reaction will be given by the following expression:

$$\alpha = \frac{\Delta V'(t) - V_1(t) \ K_1 \ P(t)}{V_1(t) \ [(1-D_1/D_2) + (K_2 D_1/D_2 - K_1) P(t)]} \tag{12}$$

where

$$\Delta V'(t) = V_1(t) - V(t) \tag{13}$$

Continuing the development of the conversion function one has to note that (10) does not take into account the volume change induced by the phase transition. This can be rectified by replacing L, W, Z, $\ell(\alpha)$, $w(\alpha)$, and $z(\alpha)$ by their pressure- and transition-induced values like

$$L(\alpha,P) = (1-K_1P)^{1/3}\ L \simeq (1-K_1P/3)L \tag{14}$$

Similar expressions can be given for the other dimensions. Substituting these expressions into (10) and keeping only first-order terms in P, it becomes

$$R(\alpha) = \frac{\rho_1[V_1(t)\ (1-K_1P) + z(\alpha)\ w(\alpha)\ A]}{ZW\ [ZW(1-4K_1P/3) - z(\alpha)\ w(\alpha)\ B]} \tag{15}$$

where

$$A = (N\frac{D_1}{D_2})^{2/3}(\frac{\rho_1}{\rho_2} - 1)\ \{L[1-(2K_2+K_1)P(t)/3] - \ell(\alpha)(ND_2/D_1)^{1/3}(1-K_2P)\} \tag{16}$$

and $$B = (ND_2/D_1)^{2/3}\ (1-\rho_1/\rho_2)\ [(1-2(K_1+K_2)P(t)/3] \tag{17}$$

Equation (15) gives the sample resistance as a function of the advancement of transition.

The evolution of the size of the nucleation centers $\ell(\alpha)$, $w(\alpha)$, and $z(\alpha)$ as a function of the advancement of reaction is determined using the experimental curves in Fig. 1. The minimum nucleation concentration is assumed to be of $5.3 \times 10^{18}cm^{-3}$. Then the size of the nuclei as a function of the advancement of reaction is computed to transform the experimental volume variation to the simultaneously-measured resistance variation, based on the retropressure part of Fig. 1. This function, characteristic of HgSe, is used for all computation. It is to be noted, however, that the choice of concentration unit of molecules cm^{-3} does not imply that each impurity molecule would be a nucleus. It can easily be shown that by taking the average physical size of a nucleus to be $8.4 \times 10^5\ \mu m^3$, the 5.3×10^{18} HgS molecules per cc represent about 4000 nuclei per cc.

The offset transition pressure P(t) as a function of nucleation concentration can be deduced from equation (15), neglecting the small difference between the compressibilities of the low- and high-pressure phases. The result is as follows:

$$P(t) = \frac{1}{K_1}\left[1 - \left[\frac{\rho_1V_1 + (\frac{\rho_1^2}{\rho_1} -1)z(\alpha)w(\alpha)\ (N\frac{D_1}{D_2})^{2/3}\{L-\ell(\alpha)(N\frac{D_2}{D_1})^{1/3}\}^3}{R(\alpha)\ ZW\left[ZW + (\frac{\rho_1}{\rho_2} -1)\ z(\alpha)\ w(\alpha)\ (N\frac{D_2}{D_1})^{2/3}\right]}\right]\right] \tag{18}$$

Here a constant value of the nucleus size is used for the computation, taken for a given small value of the advancement of the reaction, corresponding to the onset of the experimentally noticeable resistance variation. The calculated curve for upstroke transition is shown in Fig. 2 (dotted line). It follows reasonably well the experimentally-determined values (solid lines), taking into consideration that they were measured with different rates of applied pressure. These were increased by approximately 2.5 MPa/min within 0.1 MPa/min and 10 MPa/min considering the curves from the outer to the inner ones. The computed pressure using equation (18) at saturation concentration (3.0% at.wt), is 724 MPa. This is close to the one extrapolated experimentally (740 MPa) as discussed in the following section.

KINETICS

It has been a common practice to discuss experimental results of transitions using the transition state theory introduced by Evans and Polanyi [13]. During the pressure-induced phase transition the system is passing over the top of an energy barrier from the initial to the final state. Taking the pressure as a variable, the volume change characterizing the formation of the transition state can be called activation volume, and is given by

$$-\Delta V^{\#} = V_1(t) - V^{\#} \tag{19}$$

Then the well-known expression relating the rate constant of transition to the pressure is as follows:

$$d \log k/dP = \Delta V^{\#}/RT \tag{20}$$

Now, the volume change $\Delta V'(t)$ (see (13)) can be determined from the directly measured change of sample resistance through equation (10). The rate of change of this volume will characterize the transition rate constant, or

$$k = (1/V_1)\ \partial \Delta V'(t)/\partial t \tag{21}$$

The lines in Fig. 3 correspond to the data obtained on samples with increasing nucleation concentrations. The slopes of the straight lines clearly indicate the expected decrease of activation volumes with increasing nucleation concentrations. The data are collected in Table I.

In Fig. 4 the determined dependence of the offset upstroke pressure $P(t)$ as a function of activation volumes $\Delta V^{\#}$ is depicted. Extrapolating the line of this figure to $\Delta V^{\#} = 0$ from equation (5) it is clear that $P(t) = P_{TD} = 740$ MPa. This value fits well within the hysteresis curves published in the literature [2,14].

Thus, following the behavior of the offset transition pressure with the activation volume the value of the thermodynamical transition pressure may be determined. This value would be difficult to obtain by direct measurements since if the zone of indifference as defined by Bridgman [14] contains the thermodynamical equilibrium transition pressure the reaction rate would be undetectably slow.

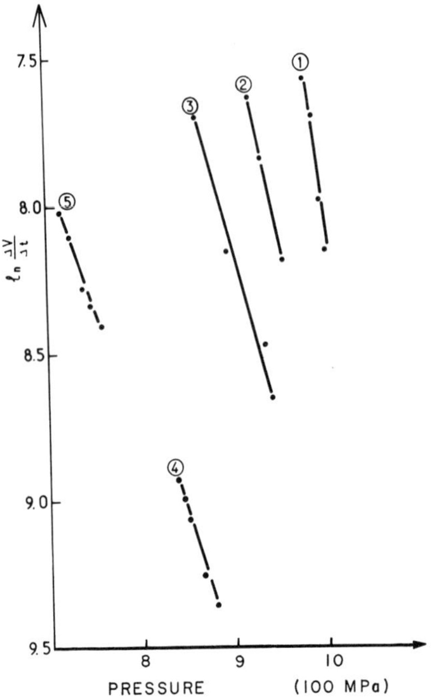

Fig. 3. Experimental values of the logarithm of the rates of volume variations as a function of pressure, determining the activation volumes given in Table I, corresponding to increasing nucleation concentrations. Legend (at.wt.%): (1) 0.03%; (2) 0.05%; (3) 0.1%; (4) 1.0%; and (5) 3.0%.

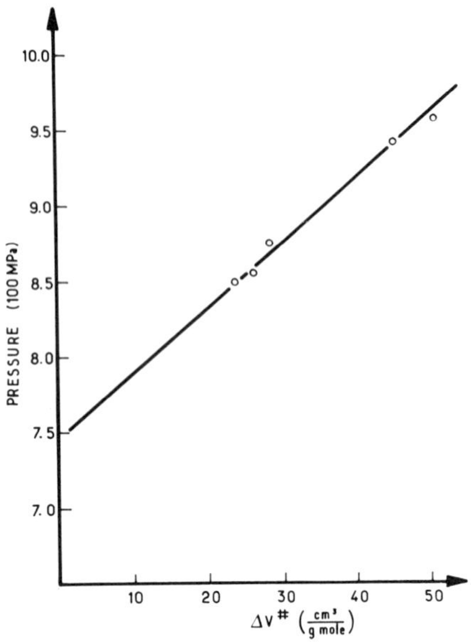

Fig. 4. Calculated activation volumes vs. measured offset transition pressures.

DISCUSSION

Examining the experimental and the calculated values in Fig. 2, one might note that the two sets of curves depicting the width of the hysteresis curve as a function of nucleation concentration demonstrate a saturation effect. That is, increasing the concentration of the nucleation centers does not decrease the kinetic pressure to zero. A range of pressure is indicated around the thermodynamical transition pressure where, although the system is in a thermodynamically metastable state with both phases present, the kinetic pressure for a given nucleation concentration is not reached. Thus, the potential nuclei did not attain the critical size of growth nuclei, and the rate of reaction is undetectably small. This is very similar to the phenomenon observed and originally described by Bridgman [14] who called it the region of indifference.

It is proposed that the width of the region of indifference is related to the sizes of critical nuclei at or above saturation concentration. This suggests that not only the number of nucleation centers influences the kinetic pressure, but a critical size of these potential nuclei must be reached to pass to growth nuclei and initiate the reaction.

Table I. Experimentally-Obtained Kinetics Data with Different Amounts of Nucleation Concentrations

Impurity Concentration, % at. wt	Activation volume $\Delta V^{\#}$, cm^3/g mole	Upstroke offset pressure P(t), 100 MPa	Rate of Transition at offset pressure P(t) x 10^{-4}, cm^3/s
0.03	51.	9.6	5.2
0.05	45.	9.4	4.9
0.1	30.	8.8	4.5
1.0	27.	8.6	2.0
3.0	25.	8.4	2.3

CONCLUSIONS

The direct observation of pressure-induced polymorphic phase transitions by electrical means permitted the study of the mechanism and the kinetics of the reaction.

It has been demonstrated that the increase of the heterogeneous nucleation concentration, up to its saturation level, (1) decreases the activation volume, (2) decreases the offset upstroke transition pressure, (3) increases the downstroke transition pressure, (4) thus decreases the width of the hysteresis curves, and (5) decreases the rate of the reaction. The obtained results were analyzed based on the absolute rate theory and on a three-dimensional nucleation model. It is proposed that the critical size of the nuclei at saturation concentration is related to the region of indifference defined by Bridgman.

Furthermore, extrapolation of the offset transition pressure to zero-activation volume indicates the thermodynamical transition pressure. Finally, if the focal mechanism of earthquakes involves phase transition, as it is proposed in the literature [5,6], then the results obtained in this investigation might be of interest in its control and initiation. It can be shown, based on the work of Richter and Gutenberg [15] that the volume change required to substantiate this assumption is compatible with orders of magnitude characteristic of geophysical changes. At 1000 MPa, for example, to release energy of 10^{17} joules corresponding to an earthquake of magnitude 7 requires about 0.1 km^3 volume decrease. This can easily be envisaged to be supplied by a pressure-induced phase transition. Therefore, transposing some of the conclusions of this work might lead to finding appropriate nucleations, permitting an early, controlled initiation of earthquakes. Furthermore the nucleation-decreased rate of transition could result in a less violent phenomenon.

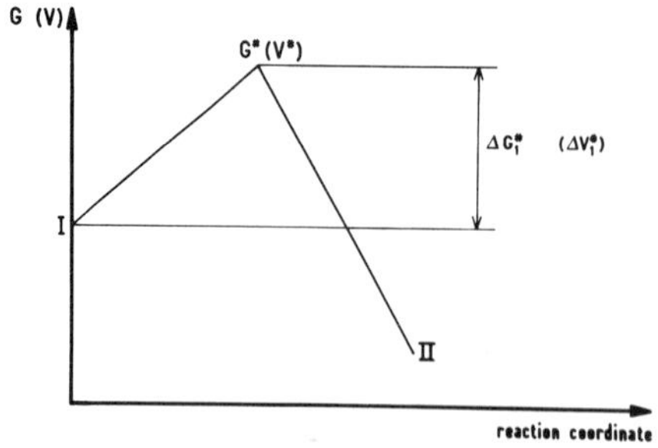

Fig. 5. Variation of free energy, G(volume V) as a function of reaction coordinate.

ACKNOWLEDGMENT

The authors would like to thank A. Lacam and J. Peyronneau of the High Pressure Laboratories of the CNRS of France for performing the volumetric measurements and for informative discussions.

NOTATION

α = advancement of reaction
D_1 = density of low-pressure phase
D_2 = density of high-pressure phase
k = rate constant of pressure-induced phase transition
K_1 = compressibility of low-pressure phase
K_2 = compressibility of high-pressure phase
$\ell(\alpha)$, $z(\alpha)$ and $w(\alpha)$ = length, thickness, and width of nucleation centers
L, Z and W = length, thickness, and width of sample
N = nucleation concentration
P_K = kinetic pressure
P(t)= measured offset transition pressure
P_{TD} = thermodynamical transition pressure
$R(\alpha)$= sample resistance as a function of advancement of reaction
ρ_1 = resistivity of the low-pressure phase
ρ_2 = resistivity of the high-pressure phase
t = time
V(t)= the total volume
V_1 = volume of the low-pressure phase at atmospheric pressure
V_L = volume of the pressure-transmitting liquid at atmospheric pressure
$V_1(t)$ = volume of the low-pressure phase during transition
$V_2(t)$ = volume of the high-pressure phase during transition
$V_L(t)$ = volume of the pressure-transmitting liquid during transition
$V^{\#}$ = volume of the transition state
$\Delta V^{\#}$ = activation volume

REFERENCES

1. J. A. Kafalas, H. C. Gatos, M. C. Lavine, and M. D. Banus, J. Phys. Chem. Solids 23, 1541 (1962).
2. A. Lacam, J. Peyronneau, L. J. Engel, and B. A. Lombos, Chem. Phys. Letters 18, 129 (1973).
3. A. Lacam, J. Peyronneau, and J. L. Kopystynski, J. Physique 34, 1055 (1973).

4. J. L. Kopystynski, J. L. Peyronneau, and A. Lacam, J. Physique 35, 609 (1974).
5. L. L. Fermor, Geol. Mag. 1, 65 (1914).
6. F. Birch, in Solids Under Pressure, W. Paul and D. M. Warschauer, eds., McGraw-Hill Book Company, New York (1963), p. 131.
7. B. A. Lombos, B. Ghicopoulos, S. Bhattacharyya, and C. B. Pant, Can. J. Phys. 54, 48 (1976).
8. F. Birch, in Geol. Soc. Am. Mem., S. P. Clark, Jr., ed. (1966), p. 97.
9. A. Lacam and J. Peyronneau, J. Physique 34, 1047 (1973).
10. A. Lacam, B. A. Lombos, and B. Vodar, Phys. Earth Planet Interiors 3, 511 (1970).
11. B. A. Lombos, E.Y.M. Lee, A. L. Kipling, and R. W. Krawczyniuk, J. Phys. Chem. Solids 36, 1193 (1975).
12. W. Jander, Z. Anorg. Allgem. Chem. 163, 1 (1927).
13. M. G. Evans and M. Polanyi, Trans. Faraday Soc. 31, 875 (1935).
14. P. W. Bridgman, Proc. Am. Acad. Arts Sci. 52, 57 (1916); also 74, 21 (1940) and 76, 55 (1948).
15. B. Gutenberg and C. F. Richter, Seismicity of the Earth and Associated Phenomena, Hafner Publishing Company (1965), p. 10.

C-1

PREPARATION OF METALLIC HYDRIDES BY HIGH PRESSURE GASEOUS HYDROGEN*

B. Baranowski

Institute of Physical Chemistry, Polish Academy of Sciences
Warsaw, Poland

INTRODUCTION

Several metallic components require a high thermodynamic activity of hydrogen for the formation of hydride phases. This can be achieved by several distinctly different methods. The most simple include cathodic procedures, electric discharges in low pressure gaseous hydrogen or implantation techniques. The disadvantage of all these procedures is the non-equilibrium character of the hydrogen supply which often limits the reproducibility of the results and does not provide any valuable thermodynamic information [1]. Gaseous hydrogen under high pressure conditions, on the other hand, offers a unique and reproducible defined thermodynamic activity for hydrogen. With a knowledge of the virial coefficients for the equation of state of gaseous hydrogen [2] the fugacity of this component, corresponding to a given hydrostatic pressure, can be calculated. It is fortunate that the departure from ideality leads to a large increase in the fugacity as a function of the hydrostatic pressure; for example, at 25°C gaseous hydrogen at 12 kbar corresponds to a fugacity of 10^7 bar, about three orders of magnitude higher.

EXPERIMENTAL PROCEDURE

A high pressure device was developed to permit continuous change of gaseous hydrogen pressure from 0.5 to 30 kbar at room temperature [1,3]. It consists of a hydrogen container which is placed in the hydrostatic environment of an organic pressure transmitting medium. Gaseous hydrogen is separated from the organic liquid by a mobile piston. The starting pressure of gaseous hydrogen was about 0.5 kbar. The electrical resistance and the thermopower of the metallic samples investigated were monitored

*Invited paper.

as function of the hydrogen pressure. The same device was also used for measuring the heat conductivity, the diffusion coefficient and the magnetic moment [4]. Recently devices were developed for determining the absorption and desorption isotherms [5] and in situ lattice parameters [6] up to 14 and 9 kbar of gaseous hydrogen, respectively. No liquid transmitting media are applied in these cases.

RESULTS

The formation and decomposition of hydrides were detected by measuring the electrical resistance, the thermopower, the lattice constant or the amount of absorbed or desorbed hydrogen as a function of the stepwise change in hydrostatic pressure while keeping the temperature at a constant value. In the majority of the cases investigated so far, the hydrides were formed in the course of first order phase transitions, exhibiting a more or less discontinuous change in the above mentioned properties. Generally, a distinct hysteresis behavior was observed, giving a higher value for the formation process than for the decomposition process. For the thermodynamic calculations the characteristic values for the decomposition were normally taken into account.

In the Ni-H system the formation of the hydride is well established at 25°C in gaseous hydrogen above 6 kbar. This result was achieved with all of the above mentioned methods with exception of the in situ roentgenographic investigations. A series of Ni-Fe and Ni-Co alloys was also transformed into the hydride state. For the Ni-Fe alloys the miscibility seems to end around 20 at.% iron while in the Ni-Co alloys no indication of such miscibility limitations is even noted for a 40 at.% Co alloy. In both alloy series a linear relationship has been found between the free energy of formation of the hydrides and the atomic percentage of the alloy. An extrapolation to pure iron and cobalt shows values of hydrogen activity not yet available for the achievement of the phase transition [4].

In the Ni-Mn alloy system a systematic shift of the formation pressures to lower values of hydrogen pressure was observed when increasing the concentration of manganese. This indicates clearly the possibility of a hydride formation in pure manganese and this was achieved experimentally in gaseous hydrogen at about 577 K and 8 kbar [7]. An Hcp lattice was found by x-ray analysis.

In pure chromium the hydride and deuteride phases were prepared in gaseous hydrogen or deuterium at about 420 K and 18 kbar. The latter component requires a higher formation pressure than hydrogen. This tendency follows the normal behavior of metal - hydrogen systems [4].

In the Pd-H system the adsorption and desorption isotherms were measured directly by monitoring the volume changes. In no case was the stoichiometric composition of PdH exceeded. From the measurements carried out at 25, 45 and 65°C, the relative partial molar free enthalpies, enthalpies and entropies could be evaluated, exhibiting marked changes when approaching the stoichiometric composition [8].

It seems obvious that new hybride phases can be prepared if the present pressure and temperature ranges experimentally available could be extended. Both changes are now under study in our laboratory and we hope to be able to present new results in the future.

REFERENCES

1. B. Baranowski, Ber. Bunsenges. 76, 714 (1972).
2. A. Michels, W. de Graaff, T. Wassernaar, J. M. Levelt, and P. Louverse, Physica 25, 25 (1959).
3. B. Baranowski and W. Bujnowski, Roczniki Chemii 44, 2271 (1970).
4. B. Baranowski, in Hydrogen in Metals, G. Alefeld and J. Völkl, eds., Springer Verlag, in press.
5. B. Baranowski, M. Tkacz, and W. Bujnowski, High Temp.-High Press. 18, 656 (1976).
6. B. Majchrzak, Roczniki Chemii 51, 1549 (1977).
7. M. Krukowski and B. Baranowski, J. Less Common Met. 49, 385 (1976).
8. M. Tkacz, B. Baranowski, Roczniki Chemii 50, 2159 (1976).

C-2

HIGH PRESSURE STUDIES OF PALLADIUM ALLOY/HYDROGEN SYSTEMS

B. Baranowski

Institute for Physical Chemistry, Academy of Sciences
Warsaw, Poland
and
F. A. Lewis
Queen's University
Belfast, Northern Ireland

INTRODUCTION

At relatively low pressures the only chemical result of an interaction between hydrogen and several of the transition metals is for comparatively small quantities of hydrogen to be adsorbed by the metal. In addition to occupying a limited number of interstitial positions in the metal lattice, this adsorbed hydrogen can also be preferentially located on the internal surfaces of voids, or on less grossly defective regions such as dislocation networks where it can produce mechanically undesirable effects.

In other transition metals, however, such as the lanthanides, actinides, vanadium, niobium, tantalum, titanium, zirconium, hafnium and palladium the reaction between the metals and hydrogen at even relatively low pressures can produce crystallographically distinct but generally non-stoichiometric hydride phases which nevertheless can retain metallic characteristics up to high-hydrogen contents corresponding to ratios of hydrogen-to-metal of the order of 1:1, 2:1 and 3:1 - although the exact equilibrium composition will depend on both temperature and hydrogen pressure.

In addition to a retention of general metallic characteristics, the element palladium also retains a somewhat unique cohesiveness by absorbing hydrogen up to an approximately monohydride composition PdH. This feature has proved to be of experimental advantage in making the palladium/hydrogen system particularly convenient for concurrent studies of physical properties, such as electrical resistance which is a sensitive function of hydrogen content.

The retention of cohesiveness on the adsorption of hydrogen is also a feature shared by alloys of palladium with several other elements. Furthermore, on a gradually changing alloy composition the hydrogen pressure/composition relationships of these alloy systems are generally altered in a very systematic and interesting way. However, with some important exceptions the solubility of hydrogen in the alloys is generally reduced from that in palladium at any given hydrogen reference pressure. Studies at high hydrogen pressures have therefore been invaluable in extending the information available for such palladium alloy/hydrogen systems. This paper will be concerned to a degree with the measurements of the relationship between hydrogen pressure and electrical resistance over a wide range of palladium/platinum alloys at pressures of up to 30,000 atm, and with some uses which have been made of these relationships.

ELECTRICAL RESISTANCE STUDIES

Relationships between Electrical Resistance, Hydrogen Content and Equilibrium Hydrogen Pressure for the Palladium/Hydrogen System

For comparison of the relative magnitudes of changes of electrical resistance, the parameter R/Ro (where R is the measured resistance and Ro is the initial hydrogen-free resistance) has been widely used as a convenient indicator. Figure 1 illustrates the overall form of the accumulated relationships which have been derived at 25°C between R/Ro and the hydrogen content (H/Pd - atomic ration = n, where the composition is expressed as PdH_n) and between the hydrogen content and the steady-state hydrogen pressure under experimental conditions where the hydrogen contents are being both steadily increased and steadily decreased, respectively. The complementary relationships between R/Ro and hydrogen pressure are included in Fig. 2.

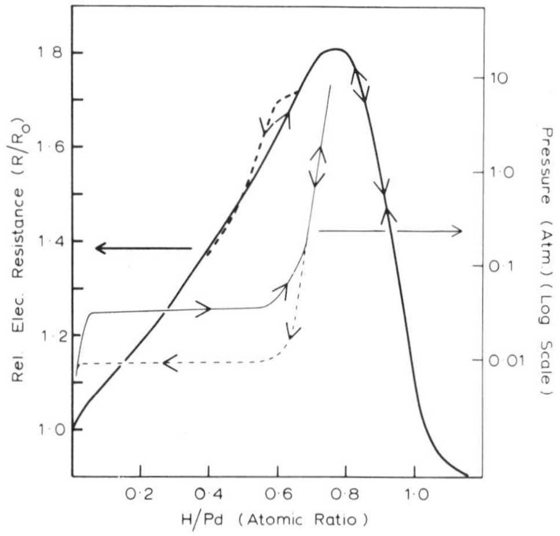

Fig. 1. Changes at 25°C of the relative electrical resistance and the steady-state hydrogen pressure for increasing and decreasing hydrogen contents of palladium.

Comparisons of Figs. 1 and 2 illustrate that the hysteretic differences (associated with $\alpha \rightleftharpoons \beta$-hydride phase transformations) are very clearly reflected by a substantial hysteresis of the pressure-resistance relationships of the PdH_n system in Fig. 2. Both Figs. 1 and 2 indicate that with steadily increasing hydrogen content the electrical resistance increases to a maximum of slightly less than twice its initial value, and then decreases again until at

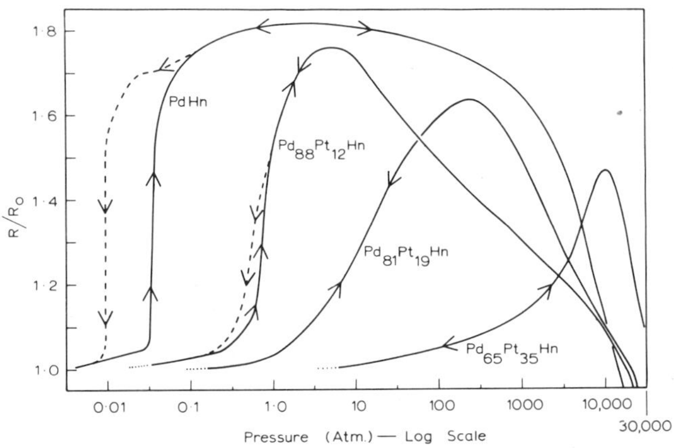

Fig. 2. Relationship at 25°C between relative electrical resistance and both increasing and decreasing steady-state hydrogen pressures for palladium and palladium/platinum alloys.

the highest pressure studied it has in fact returned to slightly below its initial hydrogen-free value.

The flattopped character of the maximum over the intermediate pressure range in the relationship between R/Ro and pressure in Fig. 2, illustrates the importance of the initial high-pressure studies [1-4] in that they were not only necessary to definitely confirm the existence of the maximum which had been suggested by earlier measurements [5], but were also able to demonstrate the steepness of the subsequent decreases of R/Ro which resulted from relatively small further increases in the hydrogen content. Such interesting changes of physical properties at high hydrogen contents that were revealed in these initial high-pressure studies therefore stimulated considerable further work in this general area. Thus, studies [6] of the resistivity changes of specimens with high hydrogen contents over the temperature region of a specific heat anomaly in hydrided palladium near 50 K, led to the observation [7] of superconductivity in these systems for compositions close to $PdH_{1.0}$ which has been the subject of much recent interest [8-12]. Although alternative methods have since been used in order to obtain high values of H/Pd, such as ion implantation techniques [8] and electrolysis under special conditions [10,11], the high-pressure methods still possess advantages in terms of precise content control and sample homogeneity. These features have been taken advantage of in high-pressure studies of changes of hydrogen diffusion coefficients [13] as a function of hydrogen content which is an aspect of the study of metallic hydrides that has recently assumed an increasing significance [14,15].

Relationship between Electrical Resistance and Hydrogen Pressure for Series of Palladium Alloys

With several neighboring elements in the periodic classification palladium can form a series of alloys which approximate to substitutional solid solutions over extensive composition ranges. It has therefore been of interest to carry out systematic studies, as a function of alloy composition, of the adsorption of hydrogen by these alloys at high pressures in order to see whether trends of behavior may be extrapolated to yield information concerning the behavior of the alloying element - which by itself may absorb relatively little hydrogen under conveniently accessible controlled experimental conditions. In these studies changes of electrical resistance (R/Ro) have provided a useful guide to changes of hydrogen content.

The pattern of results at 25°C for the palladium/rhodium series of alloys [4, 16] was similar in one respect to those of many other alloy series in that with increasing rhodium content the changes of R/Ro, indicative of significant hydrogen adsorption, occurred at increasingly higher pressures reflecting a general increase of the endothermicity of hydrogen adsorption. However, the palladium/rhodium/hydrogen system was unusual in that up to very high contents of rhodium there was a continuation of the marked hysteretic differences between the pressure/R/Ro relationships exhibited by pure palladium (see Fig. 2) in cycles of increasing and decreasing pressure. This was in keeping with a persistence of the possibility of regions of $\alpha \rightleftharpoons \beta$-phase hydride transitions and indicated that the critical temperature $T_{c(\alpha,\beta)}$ above which transitions do not occur was not significantly lowered by alloying with rhodium as was found to occur when palladium was alloyed with many other elements.

In cases where $Tc_{(\alpha,\beta)}$ decreases with an increasing content of alloying metal, the hysteretic differences of pressure/R/Ro relationships should gradually decrease; and for $Tc_{(\alpha,\beta)} < 25°C$ the adsorption and desorption relationships should become reversibly coincident. Such a trend of behavior has been observed in studies of the palladium/silver, palladium/gold, and palladium/platinum series of alloys. The changes of R/Ro with pressure changes recorded for a series of palladium/silver and palladium/gold alloys have been found [2,17] to be wholly compatible with extrapolations from studies at lower pressures. The trends of the thermodynamic parameters related to hydrogen adsorption and of the relationships between R/Ro and hydrogen content for these alloy series are however somewhat more complex than those of the palladium/platinum series of alloys where a relatively simple pattern of behavior has been observed. This seems to be a consequence of the fact that palladium and platinum are chemically related elements with closely similar atomic volumes such that their binary alloys represent a relatively ideal series of solid solutions over the entire composition range [18].

The Palladium/Platinum/Hydrogen System

An initial exploratory high-pressure study [19] of the Pd/Pt/H system has recently been extended [20] to determine relationships between hydrogen pressure and R/Ro for fifteen Pd/Pt alloys with compositions varying from about $Pd_{97}Pt_3$ to $Pd_{30}Pt_{70}$ over a range of pressures from 10 to 30,000 atm at 25°C. These high-pressure studies have been complemented by the determination of interrelationships between hydrogen pressure, hydrogen content, and R/Ro at pressures mainly below 10 atm. Full details of these studies are still being prepared for publication [20] but certain salient features of the trends found in the results are illustrated in the plots of relationships between R/Ro and hydrogen pressure which are shown in Fig. 2 for selected alloy compositions.

Comparisons of the relationships in Fig. 2 indicate that increases of platinum content effect a gradual decrease of the hysteretic differences between measurements of R/Ro made during gradual increases and decreases of hydrogen pressure, respectively. This general trend is in keeping with those low pressure/hydrogen content relationships [2,22] which suggested that $Tc_{(\alpha,\beta)}$ should have decreased to below 25°C for alloys containing greater than about 15% Pt. For $Pd_{81}Pt_{19}$ and the alloys with still higher platinum contents, no hysteretic differences could be detected over the whole range of pressures studied. Complementary with the decrease of hysteretic differences, the regions of the most marked increases of R/Ro are shown to occur over increasingly wider ranges of hydrogen pressure. Nevertheless, the same general form of relationship continued to be exhibited by all of the alloys studied. The increases of platinum content, however, resulted in the regions of substantial changes of resistance being systematically shifted towards higher pressure ranges, and this was accompanied by a narrowing of the resistance peak.

Apart from an intrinsic interest in regard to interpretational considerations, [20] the overall form of the family of relationships suggests that they can provide a reliable means of hydrogen pressure gauging over a wide range of pressures. The relationships have indeed already proved experimentally useful for estimating changes of the hydrogen chemical potential from measurements of R/Ro with selected alloys in other experimental environments [23-25]. In regard to these latter practical aspects, the reversible nature of the R/Ro-pressure relationships is additionally valuable in that it is accompanied by a reduction of macrodeformation resulting from cycles of adsorption and desorption of hydrogen. Thus, for example, specimens of the $Pd_{81}Pt_{19}$ composition have been used for numerous experimental runs involving substantial gain and loss of hydrogen [24,25] with virtually no cumulative alteration in the original value of the hydrogen-free resistance, Ro. Moreover, although in common with most other palladium alloy/hydrogen systems, a general decrease of values of internal hydrogen diffusion coefficients is found to occur with increasing platinum content [26]; permeation of hydrogen within suitably surface-activated thin specimens remains quite rapid. Thus, following the changes in experimental conditions, measured values of R/Ro very soon again become accurately representative of the chemical potential of hydrogen throughout the whole specimen and at its surface. These features have been taken advantage of in devices designed for regular periodic measurements of the concentration of hydrogen dissolved in solution [23], and have also been utilized in studies of the changes in the surface hydrogen chemical potential during the course of electrolytic hydrogen discharge processes.

With increasing current densities of electrolysis the component of the total hydrogen overpotential which corresponds to the surface hydrogen chemical potential has been found to attain limiting values. On highly active surfaces these limits tend to be adopted at current densities over which bubbles of gas begin to generally evolve. Values of surface hydrogen chemical potential derived from measurements of electrical resistance during electrolysis have been found to be consistent with estimates derived from measurements of open circuit electrode potentials made after an interruption of electrolysis [24]. Measurements by both of these techniques have indicated that the apparent upper limits of the surface hydrogen potential in acidic electrolytes varies markedly with hydrogen in concentration as indicated diagrammatically in Fig. 3. Moreover, extensions of the resistance measurement technique [25] have provided evidence to suggest that the apparent variations of limits are due to restrictions in the equalization of local cell differences of the surface hydrogen chemical potential, which has general significance in regard to a fuller understanding of the kinetics of electrolytic hydrogen discharge. Measurements of the resistance changes of palladium/platinum electrodes have also provided [25] a means of estimating the high surface hydrogen chemical potentials, equivalent to hydrogen pressures of some thousands of atmospheres, which can develop on catalytically inactive electrode surfaces and which cannot be reliably derived from measurements of the electrode potential.

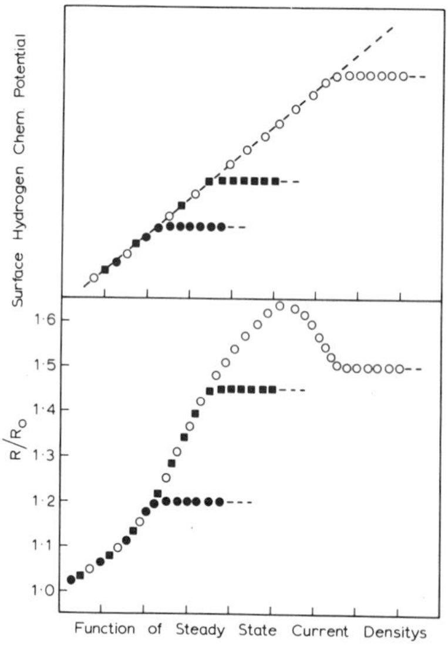

Fig. 3. Diagrammatic illustration of correlations between apparent steady-state values of the electrical resistance of palladium/platinum alloy cathodes and surface hydrogen chemical potential as a function of electrolytic current density for solutions of differing acid strengths.

REFERENCES

1. B. Baranowski and R. Wiśniewski, J. Phys. Chem. Solids 29, 1275 (1968).
2. B. Baranowski and R. Wiśniewski, Phys. Stat. Sol. 35, 593 (1969).
3. B. Baranowski, Platinum Metals Rev. 16, 10 (1972).
4. B. Baranowski, Ber. Bunsenges. Physik. Chem. 76, 714 (1972).
5. J. C. Barton and F. A. Lewis, Z. Physik. Chem. NF 33, 99 (1962).
6. T. Skoskiewicz and B. Baranowski, Phys. Stat. Sol. 30, K33 (1968).
7. T. Skoskiewicz, Phys. Stat. Sol. 6a, 29 (1971).
8. W. Buckel and B. Stritzker, Phys. Lett. 43A, 403 (1973).
9. B. N. Ganguly, Phys. Rev. 14B, 3848 (1976).
10. J.M.E. Harper, Phys. Lett. 47A, 69 (1974).
11. D. S. MacLachlan, R. Mailfert, J. P. Burger, and B. Souffaché, Solid State Commun. 17, 281 (1975).
12. A. W. Szafrański, T. Skoskiewicz, and B. Baranowski, Phys. Stat. Sol. 37a, K163 (1976).
13. M. Kuballa and B. Baranowski, Ber. Bunsenges. Physik. Chem. 78, 335 (1974).
14. M. Nuovo, F. M. Mazzolai and F. A. Lewis, J. Less-Common Metals 49, 37 (1976).
15. G. J. Zimmermann, J. Less-Common Metals 49, 49 (1976).
16. B. Baranowski, S. Majchrzak, and T. B. Flanagan, J. Phys. Chem. 77, 850 (1973).
17. A. W. Szafrański and B. Baranowski, Phys. Stat. Sol. 9a, 435 (1972).
18. J. B. Darby and K. M. Myles, Met. Trans. 3, 653 (1972).
19. B. Baranowski, F. A. Lewis, S. Majchrzak, and R. Wiśniewski, J.C.S. Faraday Trans I 68, 653 (1972).
20. B. Baranowski, F. A. Lewis, W. D. McFall, S. Filipec, and T. C. Witherspoon, to be published.
21. A. W. Carson, T. B. Flanagan, and F. A. Lewis, Trans. Faraday Soc. 56, 363, 1332 (1960).
22. F. A. Lewis, W. D. McFall, and T. C. Witherspoon, Z. Physik. Chem. NF 84, 31 (1973).
23. P. Vignet, R. Gabilly, J. Lutz, and R. Zermizoglou, French Atomic Energy Commission (Saclay) Rept. DPC. IS.SC 1/63-346/PV.DG (1963).
24. J.A.S. Green and F. A. Lewis, Trans. Faraday Soc. 60, 2234 (1964).
25. F. A. Lewis, R. C. Johnston, M. C. Witherspoon, W. F. N. Leitch, A. Obermann, and S. F. Deane, Surface Tech. 4, 89 (1976).
26. E. Wicke, G. Sicking, and B. Huber, private communication.

C-3

IN SITU MEASUREMENTS OF CHANGES OF FERROMAGNETISM CAUSED BY HIGH PRESSURE GASEOUS HYDROGEN

H. J. Bauer

University of Munich
Munich, W. Germany

and

B. Baranowski

Institute of Physical Chemistry, Polish Academy of Sciences,
Warsaw, Poland

INTRODUCTION

Considerable changes in the magnetic moments of metallic components can be caused by the action of high-pressure gaseous hydrogen during the formation and decomposition of hydrides. This is the case especially with nickel and nickel-iron alloys which are ferromagnetic in hydrogen-free states. The possibility of changing the spontaneous magnetization can be used to analyze interesting thermodynamic aspects such as the kinetics of formation and decomposition of hydrides or basically to prove their existence. Accordingly, interest should be given to magnetic measurements carried out under high-pressure conditions. As the available working volumes are of the order of only a few cm^3, a miniaturization of the magnetic devices used under normal conditions is required. Moreover, the inaccessibility of the working area in the high-pressure apparatus requires a measuring device which must be as simple as possible.

From the various known procedures a pull-out method was selected: The magnetized sample is pulled completely out of a stationary induction coil. Thus the signal induced in the induction coil is due to the entire magnetic flux of the magnetized sample. The magnetic field was supplied by a fixed permanent magnet. The first version of this device was used to investigate how the mean magnetization of a nickel sample was reduced during hydrogenation with high-pressure hydrogen [1,2]. The extension of the investigation to nickel-iron alloys [3] made the use of higher hydrogen pressures necessary since the formation pressure increases with increasing Fe-content. The increased pressure required that the working volume of the high-pressure apparatus be reduced and that the volume for the magnetic device be reduced as far as possible. For this reason a new concept was developed using the pull-out method. This concept will be presented here and its operation will be outlined by presenting the results of magnetic measurements made during the formation and decomposition of the hydride phase in NiFe alloys under hydrogen pressures up to more than 26 kbar.

APPARATUS

To accomplish the pull-out method in the new device, a thermal expansive procedure, recently described [1,4], was used which is suitable also for other applications in small inaccessible volumes. The advantage is that the thermal-mechanical procedure works relatively effectively considering the low electrical power used. Moreover, it is accompanied by minimal interference from a magnetic field.

A diagram of the new device is shown in Fig. 1. In contrast to the earlier device [1,2] the sample S is directly combined with a lever l which is jointed next to the pivot p with a thermal expansion wire w. The pivot is located on a tapered point of an elastic ribbon e. The force exerted on the lever is doubled by looping the wire around an insulating roller r. The measuring process is activated by the passage of an electrical current cl through the wire which expands thermally τ_1 and enables the spring sp to expand and move the lever τ_2. The movement of the lever pulls the sample (which lies in an exchangeable tubular holder) out of the stationary measuring coil M. In this arrangement however, the displacement of the sample is only 2 to 3 mm; the sample, in the form of coiled thin foils (5 to 10 μm), and the measuring induction coil (1500 to 3000 windings of Cu wire at 15 to 25 μm ϕ) must be correspondingly small. For this small displacement distance the sample movement is approximately linear. Because of the high sensitivity required in the measuring system the constant magnetic field for the magnetization of the samples is supplied by a fixed permanent magnet PM. For this purpose $SmCo_5$ magnets (with very high magnetic induction values embedded in a hydrogen protective

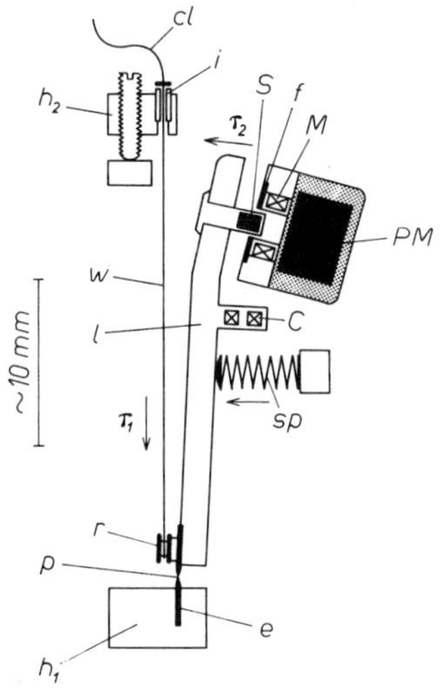

Fig. 1. Diagram of magnetic device.

medium) proved to be successful. At the sample, field strengths of 1500 to 2000 Oe can be obtained in combination with a magnetic flux piece f. Field strength of the magnet was kept constant by careful embedding of the magnet to protect it from the many cycles of high pressure hydrogen. The constancy of the magnetic field is permanently controlled by a second (mobile) induction coil C, connected with the sample holder or the lever 1, respectively. This coil signal serves simultaneously to control the performance of the sample movement. The signals of the measuring and controlling induction coils are simultaneously controlled by an electric integrator. If the sample undergoes changes of magnetic moment the relative intensities of both signals are sensitive to these changes. The height of the cylindrical magnetic device is about 34 mm and its diameter is 17 mm. The high pressure devices for gaseous hydrogen have been previously described [5].

RESULTS

The magnetic device can be monitored by observing <u>in situ</u> the decrease of the spontaneous magnetization of Ni-Fe alloys during the formation of the hydride phase under an increasing high-pressure hydrogen atmosphere. See Fig. 2. In the case of pure Ni the experimental results were obtained using the old device. The results for the alloys, on the other hand, were obtained using the new device. The curves represent relative magnetization of Ni and alloys during hydrogen absorption and desorption cycles as a function of hydrogen pressure. With this information the hydride formation pressure of the Ni-Fe alloys can be determined. The (vertical) curves approximate the corresponding formation pressures of the Ni-Fe hydrides which are known from resistance measurements [6]. The ability of the device to respond very quickly to high pressure changes in the hydrogen atmosphere is shown in Fig. 3 for a Ni alloy with about 16 at.% Fe. Even relatively small pressure changes in hydrogen are indicated by a step change in the magnetization cycles. On

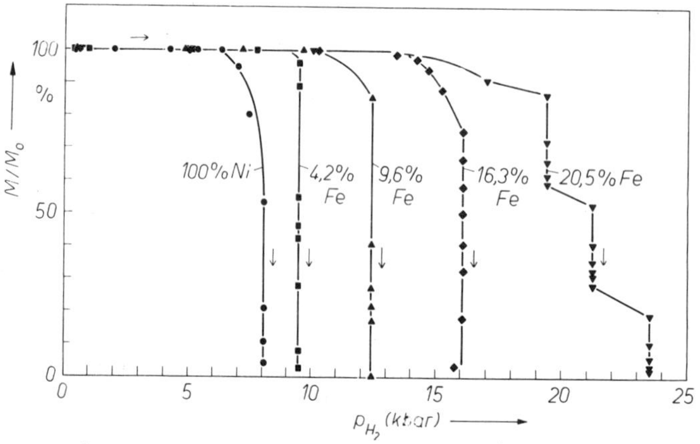

Fig. 2. Relative magnetization M/M_o of Ni and NiFe foils with increasing Fe content (in at.%) as a function of hydrogen pressure at room temperature (M_o=magnetization before the hydrogen absorption).

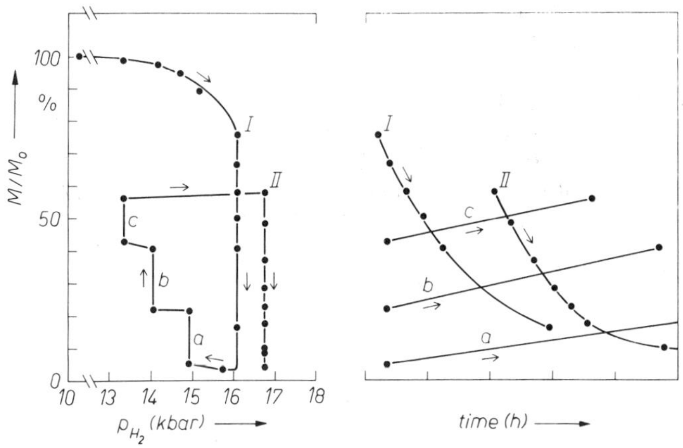

Fig. 3. Relative magnetization M/M_o of a NiFe foil with 16 at.% Fe content (see Fig. 2) as a function of hydrogen pressure and the corresponding time sequence of the magnetization at different constant pressures (I, a, b, c, II).

the right hand side of Fig. 3 one can see the rapidity at which the hydride phase forms under pressures which are higher than the hydride formation pressure of the alloy (14-15 kbar [6]). Also one can see the comparatively slow increase in the magnetization, indicating a slow decomposition of the hydride within or not far below the region of the forming pressure.

The advantage of these direct magnetic measurements, compared to studies of the Hall effects [7] or Curie temperature [8], is obvious. The magnetic flux provides a measure of the nonhydrogenated part of the sample. Its reduction to zero in the course of further hydrogenation can be used as a control for final hydrogen presence in all parts of the sample, i.e. in the α to β phase transformation (see Bauer et al. [2]). Obviously a magnetic device of this type may also be applied to many other problems in the field of high-pressure magnetism where gas can be used as a pressure-transmitting medium.

ACKNOWLEDGMENTS

The participation of A. Ebenböck and K. Niedermaier of the mechanical workshop (Sektion Physik der Universität München) is gratefully acknowledged. The authors also are indebted to C. Schüler from Brown, Boveri for use of the $SmCO_5$ magnet. This work was supported by the Deutsche Forschungsgemeinschaft.

REFERENCES

1. H. J. Bauer and B. Baranowski, in Proc. Europ. High Pressure Research Group (14th Annual Meeting), W.G.S. Scaife, ed., Typografia Hiberniae, Dublin (1976), p. 3.
2. H. J. Bauer and B. Baranowski, Phys. stat. sol. (a) 40, K35 (1977).

3. H. J. Bauer and B. Baranowski, paper presented at Deuxième Congres Intern. L'hydrogène dans les Metaux, Paris, France (1977).
4. H. J. Bauer, J. Phys. E 10, 332 (1977).
5. B. Baranowski and W. Bujnowski, Roczn. Chemii 44, 2271 (1970).
6. B. Baranowski and S. Filipek, Roczn. Chemii 47, 2165 (1973).
7. R. Wiśniewski and A. J. Rostocki, Phys. Rev. B 4, 4330 (1971).
8. E. G. Poniatowskij, W. E. Antonow, and I. T. Belash, Dokl. Akad. Nauk SSR 230, 649 (1976).

C-4

PHASE RELATIONS IN THE In-Bi BINARY SYSTEM TO 40 KBAR

E. Rapoport

Soreq Nuclear Research Centre
Yavne, Israel

and

P. W. Richter and J. B. Clark

National Physical Research Laboratory
Pretoria, South Africa

INTRODUCTION

Three intermetallic compounds, i.e. In_2Bi, InBi and In_5Bi_3 appear in the In-Bi system, while the In-In_2Bi and InBi-Bi eutectic occur at 22 and 52.7 at % Bi respectively. The In-Bi composition temperature diagram shown in Fig. 1 is taken from Hansen [1], with changes between 33-50 at.% Bi [2,3] included.

Some high-pressure phase studies have been reported on the phase transitions in the solid phases of various alloy compositions in the In-Bi system [4-13]. However, no work has been done on the effect of pressure on the various eutectic compositions and the solidus and liquidus temperatures. A phase transition to a so-called X-phase which only reverted back slowly to the low-pressure modification, was observed for alloys with compositions between 55 and 57% Bi [11-12]. The sluggishness of the phase transition enabled a crystallographic study of the X-phase.

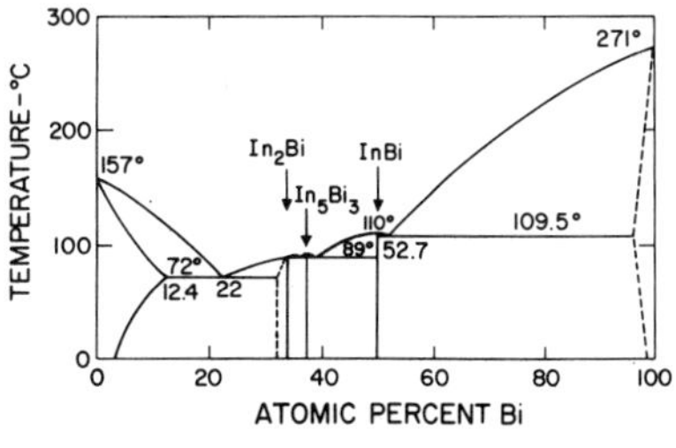

Fig. 1. In-Bi composition-temperature diagram.

This paper describes studies of the effect of pressure on the solidus and liquidus temperatures of six different In-Bi alloys.

RESULTS AND DISCUSSION

Alloys of composition 22, 33, 50, 52.7, 57 and 80 at.% Bi were studied by differential thermal analysis (DTA) and volumetric techniques in a piston-cylinder apparatus to ~ 3.5 GPa [14-16].

The results obtained from the DTA studies on six of the compositions studied are shown in Figs. 2 through 7. The phase relations are summarized in Table I.

The various eutectic compositions in the present system are affected markedly by pressure. In fact, the behavior of most of the phase boundaries observed in the present study can be interpreted in terms of the above phenomenon. The behavior of the In-22% Bi eutectic alloy (Fig. 2) illustrates the point clearly. With increasing pressure the starting composition no longer represents the eutectic composition and therefore two thermal events, which correspond to the temperatures of the eutectic reaction isotherm and the liquidus, are observed. As pressure is gradually increased the eutectic point moves further away from the starting composition and an increasing difference in temperature between the two phase boundaries is observed.

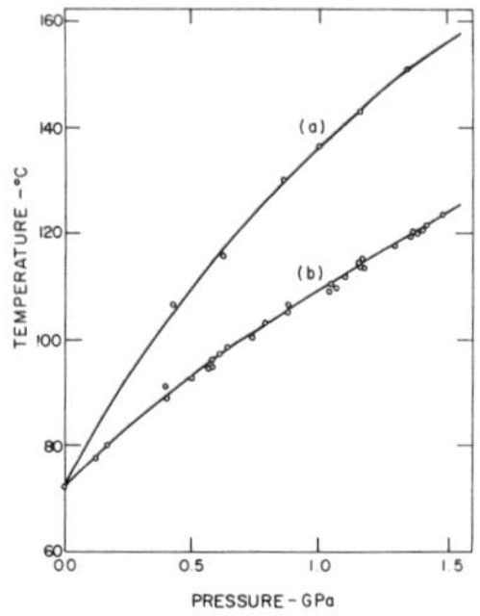

Fig. 2. Phase diagram of the In-22 at.% Bi.

Figure 3 (33 at.% Bi) confirms that this movement of the eutectic composition is towards the Bi-rich side of the x-T diagram. The initial difference of 17° at atmospheric pressure between the eutectic temperature and the liquidus temperature decreases continuously with pressure. This is only possible if, in view of the considerations mentioned for the 22 at.% Bi composition, the eutectic composition moves closer to the starting composition of 33 at.% Bi, i.e. towards a higher Bi content with increasing pressure. At 1.74 GPa, 131°C the starting composition of 33 at.% Bi does indeed become the eutectic composition. This implies a mean change of eutectic composition with pressure of dx/dp = 6.3 at.% GPa^{-1}. If it is assumed that the above value remains constant and that no new phase transitions occur above this pressure range, it may be expected that the In-In_2Bi system will be a limiting eutectic at 2.0 GPa.

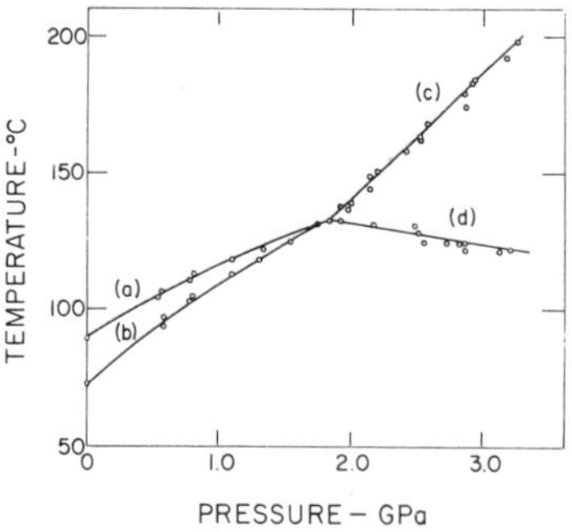

Fig. 3. Phase diagram of the In-33 at.% Bi.

As no clearly established convention exists for the In-Bi system with respect to the designation of solid phases as α, β, etc., the subscript ss

Table I. Phase Relations of the In-Bi System

Boundary		Least-Squares Fit*	Standard Deviation, °C
In-22 at.% Bi	(a)	$t(°C) = 72.0 + 83.5\ P - 18.53\ P^2$	0.8
eutectic	(b)	$t(°C) = 72.0 + 44.8\ P - 7.30\ P^2$	0.9
In-33 at.% Bi	(a)	$t(°C) = 89.0 + 30.1\ P - 3.59\ P^2$	0.8
	(b)	$t(°C) = 72.0 + 42.5\ P - 5.21\ P^2$	1.1
	(c)	$t(°C) = 132.9 + 45.4\ (P-1.83)$	2.2
	(d)	$t(°C) = 132.9 - 8.52\ (P-1.83)$	1.6
In-50 at.% Bi	(a)	$t(°C) = 109.5 + 20.9\ P - 1.92\ P^2$	0.7
(InBi)	(b)	$t(°C) = 142.9 + 42.8\ (P-1.88)$	1.2
	(c)	$t(°C) = 142.9 - 5.58\ (P-1.88) - 2.79\ (P-1.88)^2$	1.3
In-52.7 at.% Bi	(a)	$t(°C) = 109.5 + 17.5\ P$	1.4
eutectic	(b)	$t(°C) = 109.5 + 15.7\ P - 12.8\ P^2$	0.8
	(c)	$t(°C) = 110.9 + 49.9\ (P-1.07) - 15.6\ (P-1.07)^2$	0.7
	(e)	$t(°C) = 140.5 + 44.4\ (P-2.01) - 1.45\ (P-2.01)^2$	1.3
	(f)	$t(°C) = 140.5 - 10.7\ (P-2.00)$	1.1
In-57 at.% Bi	(a)	$t(°C) = 140-38.1\ P$	-
	(b)	$t(°C) = 109.5 + 0.81\ P$	0.2
	(c)	$t(°C) = 114.5 + 20.7\ (P-0.67)$	2.0
	(d)	$t(°C) = 114.5 + 5.21\ (P-0.67) - 14.5\ (P-0.67)^2$	0.4
	(e)	$t(°C) = 113.0 + 49.0\ (P-1.17)$	-
	(f)	$t(°C) = 138.9 + 69.0\ (P-1.68) - 22.9\ (P-1.68)^2$	1.5
	(g)	$t(°C) = 138.9 - 8.43\ (P-1.68)$	0.9
In-80 at.% Bi	(a)	$t(°C) = 222-56.3\ P - 1.67\ P^2$	0.6
	(b)	$t(°C) = 114.7 + 0.30\ P$	0.4
	(c)	$t(°C) = 112.3 - 71.0\ (P-0.98)$	2.0
	(d)	$t(°C) = 112.3 + 63.0\ (P-0.98) - 2.234\ (P-0.98)^2$	0.8

*P in GPa

will be used to indicate solid solution and to facilitate discussion. When this subscript is used with reference to the following phases the following interpretation is intended:

1. In_{ss} - this represents the phase formed as a result of solid solution of Bi in In.
2. In_2Bi_{ss} - this represents the phase formed as a result of solid solution of In in In_2Bi; there appears to be no evidence for any solubility of Bi in In_2Bi. This phase has a narrow composition range.
3. Bi_{ss} - this represents the phase formed as a result of solid solution of In in Bi.

At pressures above 1.74 GPa the 33 at.% Bi composition will be on the In-rich side of the In-In_2Bi eutectic and no longer on the Bi-rich side. Therefore, the solidus and liquidus above this pressure will refer respectively to the following phase reactions on increasing temperature: In_{ss} + In_2Bi_{ss} → In_{ss} + liquid, and In_{ss} + liquid → liquid. Similarly curve (b) of Fig. 2 depicts the boundary of the phase reaction In_{ss} + In_2Bi_{ss} → In_{ss} + liquid and curve (a) that of In_{ss} + liquid → liquid, since the 22 at.% Bi starting composition moves towards the In-rich side of the In-In_2Bi eutectic as soon as pressure is applied. In Fig. 3, curve (b) represents the phase reaction In_{ss} + In_2Bi_{ss} → In_2Bi_{ss} + liquid and curve (a) represents In_2Bi_{ss} + liquid → liquid as the 33 at.% Bi alloy is on the Bi-rich side of the above eutectic at low pressure.

It is to be expected that curves (b) of Fig. 2 and Fig. 3 should be identical as they represent the same phase boundary. An excellent agreement was obtained over the experimental pressure range as may be judged from the least-squares fits in Table I.

An important aspect of the data obtained on the 33 at.% Bi alloy refers to the curves (c) and (d) in Fig. 3. Initially it is tempting to assume that these two curves refer to the liquidus and solidus of In_2Bi. This being the result of the movement of the eutectic beyond the 33 at.% composition. The alternative is that it could be due to a solid-solid transition in In_2Bi at high temperatures and high pressures. However, it is only possible to decide once the nature of the liquidus between In_2Bi and InBi has been clarified.

The congruently melting compound, InBi (Fig. 4), shows a rising melting curve with a triple point at 1.88 GPa, 142.9°C where a solid-solid phase transition breaks off of the melting curve. For convenience we label the new high-pressure - high-temperature phase as Y. Volumetric studies were undertaken in an attempt to clarify the presence of the reported phase transition [17-19]. The present studies showed no evidence of a first order phase transition with a large volume change. The only conclusion that can be drawn from these results is that the changes in resistivity at 1.9 GPa observed by Rapoport et al. [17], are not due to some major structural change. This conclusion is supported by recent additional data [18-19]. A further consideration is the fact that if a phase boundary with a

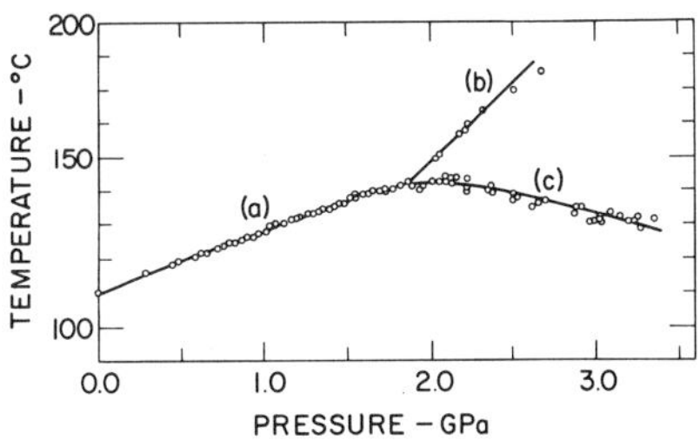

Fig. 4. Phase diagram of the InBi.

large slope was present it would intersect the melting curve and an inflection in the melting curve would result. From the data presented in Fig. 4 it can be seen that there is no clear inflection present.

Studies on alloys between InBi and pure Bi should reflect some of the characteristics of the end-members (InBi and Bi) and the presence of the so-called X-phase [12,13]. This implies a complex set of phase relations, and once again marked changes in the eutectic composition further complicate the phase relations.

The change in eutectic composition is also encountered for the In-52.7 at.% Bi eutectic (Fig. 5) and In-57 at.% Bi (Fig. 6) alloys and by arguments, analogous to that for 22 and 33 at.% Bi alloys, it may be concluded that the 52.7 at.% Bi eutectic initially moves towards higher Bi content with increasing pressure. Indications of an additional solid-solid phase boundary are, however, observed for these compositions (Fig. 5 curve (d), Fig. 6 curve (h) and Fig. 7 curve (e)). It is likely that this is the same phase transition that was observed by Bridgman [11], Gordon et al. [12] and Degtyareva et al. [13] at ~ 2.5 GPa at 25°C where the so-called X-phase is formed. This would imply a mean slope of $dT/dP \simeq -60°C\ GPa^{-1}$ for the phase boundary in alloys with a composition of ~ 52 to 57 at.% Bi. This value depends on the exact value of the transition pressure at room temperature. However, it is in good agreement with the value of $-71°C\ GPa^{-1}$ derived from the least-squares fit of the solid-solid phase transition in the 80 at.% composition (Fig. 7(c)). The disappearance of the DTA signals on decreasing temperature in the 80 at.% case, is in agreement with the sluggishness at lower temperatures as observed by Bridgman [11].

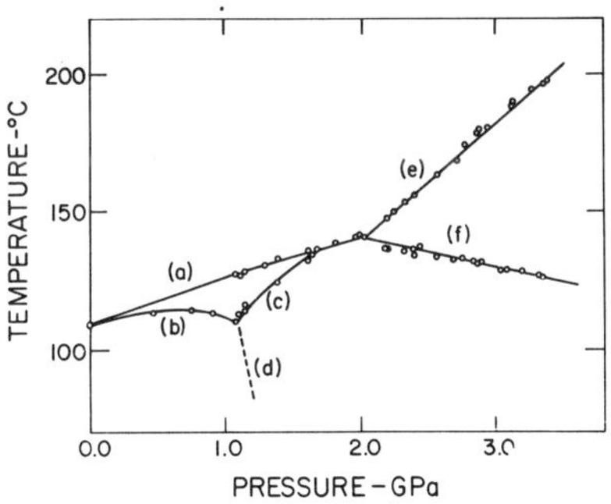

Fig. 5. Phase diagram of the In-52.7 at.% Bi eutectic.

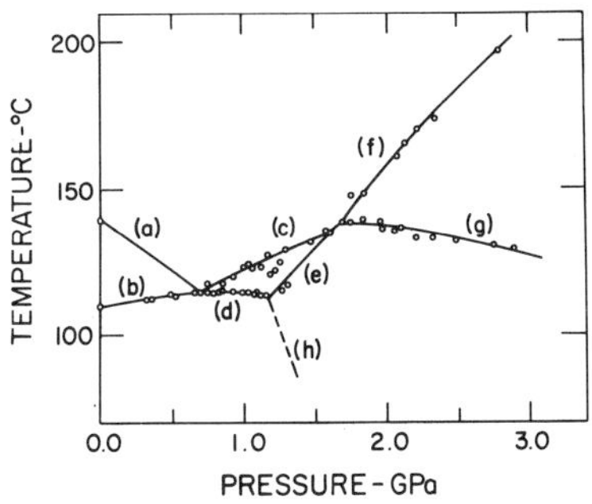

Fig. 6. Phase diagram of the In-57 at.% Bi.

In contrast to the findings of Gordon et al. [12] who observed this transition only for alloys with a Bi content within the range 55 to 57 at.% Bi, indirect indications thereof were clearly observed in the present study

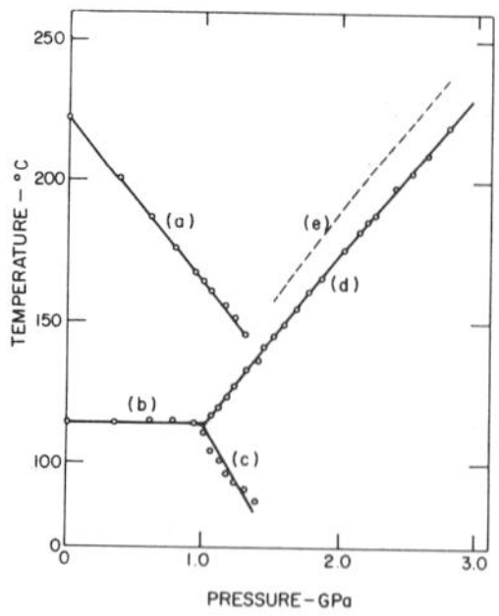

Fig. 7. Phase diagram of the In-80 at.% Bi.

outside this range, i.e. for 52.7 at.% Bi and 80 at.% Bi. It should be noted that certain doubts exist as to the validity of the method used by Gordon et al. [12] to determine the existence range of the X-phase. This region of existence is also supported by the results of Bridgman who observed this transition for some alloys in the 50 to 100 at.% Bi range [11]. No indication of this phase transition was observed in the 50 at.% Bi composition in the present study as the carefully determined melting curve of InBi (Fig. 4(a)) shows no inflection that would indicate the intersection of a solid-solid phase boundary. However, it is possible that the anomalies reported [18,19] for InBi are in some way related to the formation of the X-phase.

Curve (d) of Fig. 5 therefore represents the phase transition m(InBi) + n(Bi_{ss}) → X, while curves (b), (a), (c) and (g) represent the respective phase reactions: (b) InBi + Bi_{ss} → InBi + liquid, (a) InBi + liquid → liquid, (c) X → InBi + liquid, and (g) X → liquid.

It should, however, be noted that curve (g) on Fig. 5 has been attributed to the melting of phase X. This could involve a more complex solidus-liquidus behavior which was not resolved in the present DTA signals.

The liquidus of the 52.7 at.% Bi (Fig. 5 curve (a)) agrees closely with the melting curve of InBi (Fig. 4 curve (a)). The solidus related to the X-phase (Fig. 5 curve (c)) rises so sharply with pressure that for the 52.7 at.% composition it meets the liquidus again at 1.67 GPa, 135.9°C. This seems to imply that the eutectic composition of the above alloy moves from 52.7 at.% at atmospheric pressure to a higher Bi content with increasing pressure, but after the X-phase is formed it moves towards a lower Bi content at pressures above 1.07 GPa until at 1.67 GPa the eutectic point again lies at 52.7 at.% Bi. The corresponding eutectic temperature is 135.9°C.

For pressures below ~ 1.6 GPa the situation for 57 at.% Bi (Fig. 6) is similar to that for 52.7 at.% Bi except that at atmospheric pressure two thermal events are observed. These give rise to two phase boundaries that are the solidus and liquidus, respectively, that are on the Bi-rich side of the 52.7 at.% Bi eutectic. Curves (a) and (b) respectively represent the following phase reactions on increasing temperature: Bi_{ss} + liquid → liquid and InBi + Bi_{ss} → Bi_{ss} + liquid. The difference in temperature between the two curves decreases with pressure as the eutectic composition moves towards higher Bi content, thereby approaching the starting composition. At 0.7 GPa the eutectic composition lies at 57 at.% Bi, where the eutectic temperature is 114.5°C. This implies a mean change in eutectic composition with pressure of $dx/dP \simeq 6.4$ at.% GPa^{-1}, which is the same value that was obtained for the 22 at.% Bi eutectic. This implies that the two eutectic compositions, initially at least, move towards a higher Bi content at approximately the same rate. At pressures above 0.67 GPa the eutectic composition is on the Bi-rich side of the starting composition (57 at.% Bi) and again two boundaries, the solidus and liquidus (Fig. 6, curve (d) and curve (c)), are observed. These respectively represent the following phase reactions on increasing temperature: InBi + Bi_{ss} → InBi + liquid and InBi + liquid → liquid.

The transition to the X-phase is indicated by the sharp change in slope of the solidus curve at 1.17 GPa, 113°C, compared with 1.07 GPa, 110.9°C for 52.7 at.% Bi. The starting composition of 57 at.% again becomes the eutectic composition at 1.6 GPa with a corresponding eutectic temperature of 135.4°C compared with 1.67 GPa and 135.9°C in the case of 52.7 at.% Bi.

In the 80 at.% Bi alloy the eutectic temperature is once again, as for 52.7 and 57 at.% Bi, minimally affected by pressure until the X-phase becomes stable, where it then rises sharply with pressure. Curves (a) and (b) of Fig. 7 respectively represent: Bi_{ss} + liquid → liquid and InBi + Bi_{ss} → Bi_{ss} + liquid. No phase boundary similar to Figs. 4, curve (c), 5, curve (f) and 6, curve (g) was observed. The nature of the thermal event that occurred at pressures above ~1.5 GPa could imply that curves (d) and (e) of Fig. 7 respectively, represent solidus and liquidus phase boundaries.

A comparison of Figs. 5, 6, and 7 with the melting curve of InBi (Fig. 4) and that of pure Bi [4,20] seems to indicate that at higher Bi compositions the behavior is oriented towards that of Bi, whereas lower Bi compositions show far more InBi character.

It should be pointed out though, that Degtyareva [13] quenched at least three high pressure phases to liquid nitrogen temperature from 1.2 to 2.5 GPa in the composition range 65 to 90 at.% Bi. This could imply that more solid phases exist than were observed in the present study and could be stable at pressures higher than ~1.0 GPa, which would further complicate the interpretation of the results at these pressures.

From the present study it is immediately apparent that much work remains to be done before all phase relations within the composition-temperature-pressure volume are understood. It is also apparent that techniques such as high-pressure - high-temperature X-ray diffraction will be invaluable.

REFERENCES

1. M. Hansen, Constitution of Binary Alloys, McGraw Hill Book Company, New York (1958), p. 313.

2. B. C. Giessen, M. Morris and N. J. Grant, Trans. Met. Soc. AIME 239, 883 (1967).
3. R. Boom, P.C.M. Vendel and F. R. de Boer, Acta Metall. 21, 807 (1973).
4. W. Klement, Jr. and A. Jayaraman, Progr. Solid State Chem. 3, 289 (1966).
5. V. E. Faddev, Zh. Fiz. Khim 43, 220 (1969); also Russ. J. Phys. Chem. 45, 1051 (1971).
6. J. C. Haygarth, H. D. Luedemann, I. C. Getting and G. C. Kennedy, J. Phys. Chem. Solids 30, 1417 (1969).
7. N. A. Tikhomirova, E. Y. Tonkoo, and S. M. Stishov, Zh. Eksp. Teor. Fiz. Pis'ma 3, 96 (1966).
8. C. Roux, M. Andreani and M. Rapin, Coll. Intern. CNRS 188, 447 (1970).
9. E. M. Compy, J. Appl. Phys. 41, 2014 (1970).
10. S. Yomo, N. Mori, and T. Mitsui, J. Phys. Soc. Japan 32, 667 (1972).
11. P. W. Bridgman, Proc. Am. Acad. Arts Sci. 84, 43 (1955).
12. D. E. Gordon and B. C. Deaton, Phys. Rev. B6, 2982 (1972).
13. V. F. Degtyareva and E. G. Ponyatovskii, Fiz. Tverd. Tela 17, 2413 (1975).
14. G. C. Kennedy and P. N. Lamori, in Progress in Very High Pressure Research, F. P. Bundy, W. R. Hibbard, and H. M. Strong, eds., John Wiley and Sons, New York (1961), p. 304.
15. P. W. Richter and C.W.F.T. Pistorius, J. Solid State Chem. 3, 197 (1971).
16. C.W.F.T. Pistorius and J. B. Clark, High Temp.-High Press. 1, 561 (1969).
17. E. Rapoport, G. D. Pitt, and G. A. Saunders, J. Phys. C8, L447 (1975).
18. I. Fritz, Solid State Comm. 20, 299 (1976).
19. J. E. Shirber and J. P. van Dyke, Phys. Rev. B15, 890 (1977).
20. C.W.F.T. Pistorius, Progr. Solid State Chem. 11, 1 (1976).

C-5

HIGH PRESSURE PHASE RELATIONS FOR KDP-TYPE COMPOUNDS

E. Rapoport

Soreq Nuclear Research Centre
Yavne, Israel

and

J. B. Clark and P. W. Richter

National Physical Research Laboratory
Pretoria, South Africa

INTRODUCTION

High-pressure phase diagrams have been reported earlier for KH_2PO_4 [1] (KDP), $NH_4H_2PO_4$ [2] (NDP), and KH_2AsO_4 [2] (KDA). The present study reports the phase diagrams of RbH_2PO_4 (RDP), CsH_2PO_4 (CDP), and KD_2PO_4 (DKDP).

RESULTS AND DISCUSSION

The phase diagrams were determined using differential thermal analysis and volumetric techniques in a piston-cylinder apparatus to 45 kbar [3-6]. Samples were contained in Ni and Cu capsules with no contamination or decomposition problems. The phase diagrams of RDP, CDP, and DKDP are shown in Figs. 1 through 3, respectively. The complete phase relations are summarized in Table I.

A clear understanding of the high-pressure crystal chemistry of the KDP-type compounds is hampered by a lack of high-temperature structural data at atmospheric pressure. KDP itself dictates the following pattern: The low temperature ferroelectric orthorhombic phase transforms to the paraelectric tetragonal phase which transforms to a monoclinic high-temperature phase. The monoclinic phase undergoes a further phase transition to an uncharacterized phase which melts at higher temperatures. The present work on RDP shows that the phase diagrams of RDP and KDP are particularily similar to

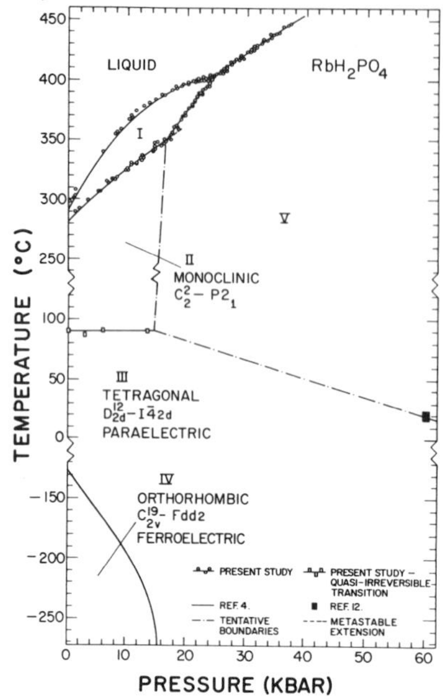

Fig. 1. Phase diagram for RbH_2PO_4 to 60 kbar.

each other. CDP is slightly more complex in view of the monoclinic paraelectric phase and this makes phase correlations difficult.

No entropy change values were available for melting due to decomposition. Differential scanning calorimetry (DSC) studies were performed at atmospheric pressure on CDP and RDP [7]. The RDP III/II transition at 90°C has an entropy change of 12.9 J/mol-K. The observed slope of this boundary is practically 0°C/kbar which implies a zero or very small volume change at the transition. This was confirmed for the CDP III/II transition at 149°C with an entropy of 10.2 J/mol-K. Although no structural data are available for CDP and RDP, some estimates can be obtained from the data on KDP [8-9], which yields a volume change of about 0.6 cm^3/mol for the quasi-irreversible transition. This small volume change in the case of KDP yielded a positive slope on the phase boundary at high pressures, and present work on RDP and CDP indicate very flat phase boundaries.

At higher temperature, CDP undergoes the II/I transition at 230°C with an entropy change of 15.1 J/mol-K. The slope of this boundary appears almost constant at 2.80°C/kbar yielding a volume change of 0.42 cm^3/mol. The volume change of the III/V transition was measured as 1.12 cm^3/mol and is not expected to be any different for the II/V transition leading to a value for $\Delta V_{II/V}$ at the I/II/V triple point. From the additive relations at the triple point $\Delta V_{V/I}$ is found to be 1.54 cm^3/mol. Combining this with the slope of the V/I boundary we can calculate an entropy change of 17.7 J/mol-K from the Clausius-Clapeyron equation. No further calculations are possible as we lack enthalpy data on the melting curve.

The crystal chemistry of the compounds is of interest but the immediate problem is how to fit the KDP II' phase (notation from an earlier paper [1]) with the present RDP II. From Rapoport [1] the complexity of the phase relations concerning KDP II' is apparent. The tentative KDP II/IV phase boundary was not found in high-pressure acoustic measurements [10] confirming the original PVT work [1] and compressibility studies [11]. If this boundary had a lower slope which is thermodynamically possible, it could be the counterpart of the RDP II/III boundary. Similarly, the II/V and III/V boundaries in RDP could not be directly determined. The quasi-irreversible

Table I. Phase Relations

Phase Boundary	Least Squares Fit	Standard Deviation °C
RbH_2PO_4 II/I	$t(°C) = 281 + 5.125\ P - 0.05715\ P^2$	1.7
RbH_2PO_4 V/I	$t(°C) = 350 + 7.143\ (P-16.4) - 0.0919\ (P-16.4)^2$	1.9
RbH_2PO_4 V/liquid	$t(°C) = 402 + 3.909\ (P-24.4) - 0.0270\ (P-24.4)^2$	2.0
RbH_2PO_4 I/liquid	$t(°C) = 292.1 + 9.175\ P - 0.1955\ P^2$	1.9
CsH_2PO_4 II/I	$t(°C) = 230 + 2.80\ P$	3.0
CsH_2PO_4 V/I	$t(°C) = 262 + 8.716\ (P-11) - 0.1107\ (P-11)^2$	2.9
CsH_2PO_4 V/VI	$t(°C) = 384.9 + 2.016\ (P-29.9)$	2.5
CsH_2PO_4 VI/I	$t(°C) = 384.9 + 6.998\ (P-29.9) - 0.0828\ (P-29.9)^2$	2.0
CsH_2PO_4 VI/liquid	$t(°C) = 457 - 0.0411\ (P-42)$	0.7
CsH_2PO_4 I/liquid	$t(°C) = 345.7 + 8.340\ P - 0.1405\ P^2$	4.1
KD_2PO_4 I/liquid	$t(°C) = 250 + 7.334\ P - 0.1081\ P^2$	1.9
KD_2PO_4 V/liquid	$t(°C) = 368.5 + 4.883\ (P-25) - 0.0724\ (P-25)^2$	1.4

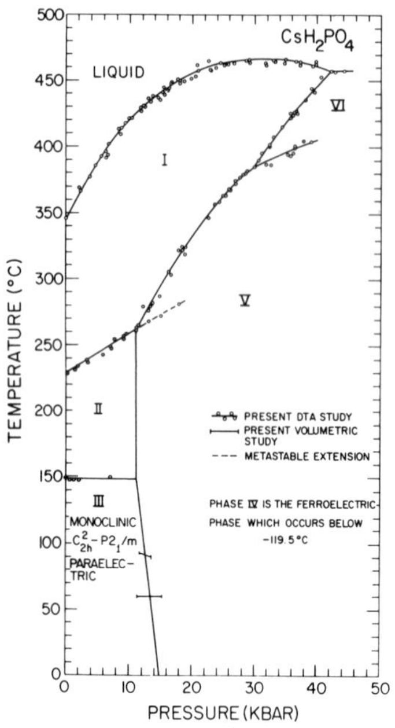

Fig. 2. Phase diagram for CsH_2PO_4 to 45 kbar.

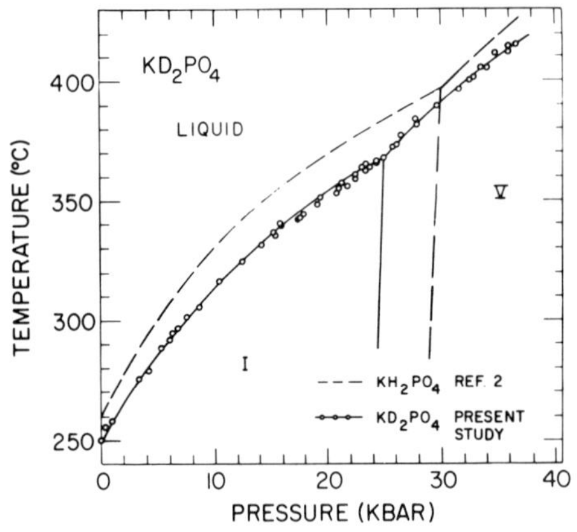

Fig. 3. Melting curve for KD_2PO_4.

nature of these transitions and the crystallographic data available tend to suggest that RDP II and KDP II' are isostructural. However, it is possible that KDP IV is the isostructural phase. Support for this is the fact that at higher pressures (P > 5 kbar) the diagrams are almost identical. The average slopes of the KDP IV/I and RDP II/I boundaries are 3.4 and 4.2°C/kbar, respectively, which adds support for this argument. The isostructural phases are listed in Table II.

The RDP II/III/V triple point could also be expected to have a counterpart involving the KDP IV, II, and V phases. Such a triple point would then produce a KDP II/V boundary which would be at higher pressures than previously thought from the tentative II/IV boundary reported by Rapoport [1]. This would agree well with information presented by Fritz [10] and Blinc et al. [12]. Clearly, additional work is necessary.

CDP has a similar quasi-irreversible transition, but the paraelectric phase is monoclinic and not tetragonal which rules out the possibility of a structural relationship. However, the diagram remains very similar, and CDP I is very possibly related to KDP I and RDP I. This would imply that the CDP II/I boundary is the counterpart of the RDP II/I and the KDP IV/I boundaries. The slope of the CDP II/I boundary fits in well, adding support for this argument. Additional structural data are of vital importance.

The effect of deuteration on the phase relations of KDP is dramatic. The Curie point is 122 K for KDP and 230 K for DKDP. The quasi-irreversible transition is also affected in that the monoclinic phase appears at room temperature after preparation for DKDP. The stability zone thus seems to have been dramatically increased.

Table II. Proposed Isostructural Phases for RDP and KDP

RDP	KDP	Structure
IV	III	Ferroelectric Orthorhombic C_{2v}^{19}-Fdd2
III	II	Paraelectric Tetragonal D_{2d}^{12}-I$\bar{4}$2d
II	II' or IV	Monoclinic C_2^2-P2$_1$
I	I	Unknown

The present work shows that the melting point is substantially depressed by deuteration. The present depression is in agreement with previous work on water and deuterium oxide at high pressures, in particular on the ice VI/VII and VI/VIII boundaries [13,14].

REFERENCES

1. E. Rapoport, J. Chem. Phys. 53, 311 (1970).
2. J. B. Clark, High Temp.-High Press. 1, 553 (1969).
3. G. C. Kennedy and P. N. LaMori, in Progress in Very High Pressure Research, F. P. Bundy, W. R. Hibbard, and H. M. Strong, eds., John Wiley and Sons (1961), p. 304.
4. P. W. Richter and C.W.F.T. Pistorius, J. Sol. St. Chem. 3, 197 (1971).
5. C.W.F.T. Pistorius and J. B. Clark, High Temp.-High Press. 1, 561 (1969).
6. A. B. Wolbarst, J. B. Clark, and P. W. Richter, Rev. Sci. Instr. in press.
7. B. Metcalfe and J. B. Clark, Thermochimica Acta, submitted for publication.
8. A. Thomas, P. Herpin, and J. Doucet, Bull. Soc. fr. Minéral. Cristallogr. 98, 341 (1975).
9. N.S.J. Kennedy, R. J. Nelmes, F. R. Thornley, and K. D. Rouse, Ferroelectrics 14, 591 (1976).
10. I. J. Fritz, Phys. Rev. B13, 705 (1976).
11. P. Morosin and G. A. Samara, Ferroelectrics 3, 49 (1971).
12. R. Blinc, J. R. Ferraro, and C. Postmus, J. Chem. Phys. 51, 732 (1969).
13. J. P. Johari, A. Lavergne, and E. Whalley, J. Chem. Phys. 61, 4292 (1974).
14. C.W.F.T. Pistorius, E. Rapoport, and J. B. Clark, J. Chem. Phys. 48, 5509 (1968).

C-6

PHASE TRANSITION OF InSb

B. Okai and J. Yoshimoto

National Institute for Researches in Inorganic Materials
Sakura-mura, Ibaraki, Japan

INTRODUCTION

InSb is one of the materials which transforms to high-pressure phases at moderate pressures. Partly because of the moderate value of the transition pressure, numerous experimental studies have been made with respect to the phase transition of InSb [1-7].

The theory associated with these transitions has advanced to such a stage as to explain the mechanisms involved at high pressures; the value of the cohesive energies can be obtained with considerable accuracy by using a pseudopotential method which permits the determination of a more stable phase structure as long as transitional elements are not included in the materials under investigation. The covalent-metallic transitions of Si, Ge, and Sn have been investigated in terms of the pseudopotential theory. The calculated values of the transition pressures and volume variations are in good agreement with experimental data [8]. However, when one considers the problem of phase transition in terms of free energy, one cannot explain, for example, the existence of the hysteresis of transition pressures during loading and unloading. A new approach is expected which considers the stability of atoms in a lattice; primarily, phase transition is nothing more than the movement of atoms, with or without electronic transitions.

An experimental study examining the orientation relation between two consecutive phases will provide useful data in this respect for explaining the mechanism of transition. A more detailed account of the behavior at the phase transition may also serve in the establishment of a theory as well.

Two kinds of experiments concerned with the phase transition of InSb are presented in this study; one involves the influence of uniaxial stress on the phase transition and the other is concerned with the orientation relation between the zinc blend structure and the high-pressure metallic forms. Single crystals are used throughout this study.

Some remarks must be made here as to the high-pressure forms of InSb; the P-T phase diagram of InSb is somewhat complicated [7] and the high-pressure metallic form is not uniquely determined [5]. For instance, the x-ray diffraction pattern obtained after rapid compression is that of a β-Sn type structure, whereas a sample after undergoing slow compression has a simple orthorhombic structure with two atoms in a unit cell which can be derived from the β-Sn structure by a simple shear process.

EXPERIMENT PROCEDURE AND RESULTS

Bridgman anvils were used to apply uniaxial stress to the specimens. The center flat of the anvil was 26 mm in diameter. Natural pyrophyllite with a thickness of 2 mm was used for the gasket. A hole 3 mm in diameter was drilled in the center of the gasket to allow for pyrophyllite discs. Calibration curves were obtained beforehand; for the vertical component of the stress, Bi calibrants were placed horizontally and covered with layers of thin (0.02 mm) AgCl film between two discs, 3 mm in diameter and 1.02 mm thick; for the horizontal component, calibrants were inserted vertically between two semicircular discs, 3 mm in diameter and 2.05 mm thick, also covered with AgCl films. Calibration curves showed that the horizontal component was constantly 76% of the vertical component up to 77 kbar.

The carrier concentration of the InSb single-crystal sample used throughout this study was 1.8 to 1.9×10^{16}/cc. Specimens approximately 0.8mm x 0.5mm x 0.15mm in size were sandwiched between two discs in a manner similar to that employed in the measurement of the vertical stress components, but without the layers of thin AgCl films. The sample faces were (100), (110), (111), and (112). Electrical resistance changes were measured. As is often noted, the phase change was quite sluggish, unlike the transition under hydrostatic pressures [9]. Figure 1 shows the time variation of the electrical resistance at the phase transition of the (112) specimen. The change cannot be described with a single relaxation time. It takes more than 24 hrs for specimens to transform completely. Figure 2 shows the resistance change with pressure (the vertical component of the stress). The transition pressure depends on the direction of the uniaxial stress applied to a single InSb crystal with 22, 19.5, 18.5, and 18 kbar for the (100), (112), (111), and (110) specimens, respectively. The hysteresis of the transition pressure is also shown in Fig. 2. The pressure during the unloading process was calibrated by the resistance change of the calibrants; Bi II-I transition was defined as 22.5 kbar from another study.

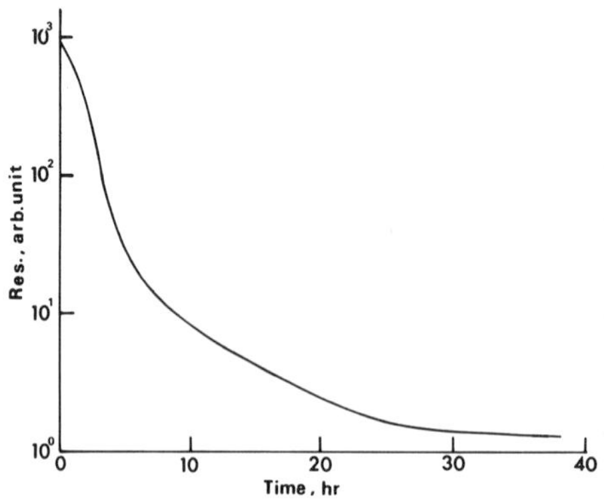

Fig. 1. Sluggishness of phase change under uniaxial stress. Uniaxial stress is normal to (112).

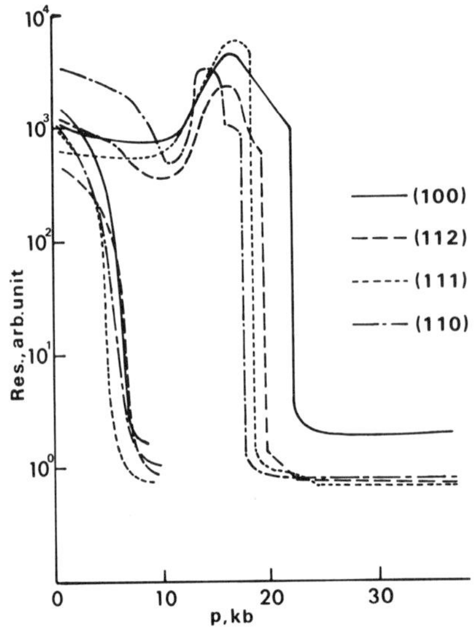

Fig. 2. Resistance change with pressure. Orientations of the surface normal to the uniaxial stress are indicated.

To study the orientation relation between the zinc blende structure and the metallic form, a quenching method was used; after the metallic form was obtained under pressure, the sample was quenched below 210 K before the pressure was reduced to one atm. The metastable metallic forms were retained at one atm at low temperatures. Bridgman anvils were also used and the gasket was identical with the exception of the hole drilled in the center of the gasket which was 4 mm in diameter. The hole was filled with a limited amount of AgCl as shown in Fig. 3 to obtain a possible hydrostatic pressure. Rings of copper phosphide were inserted to prevent extrusion of the AgCl.

The starting samples were single crystals, 3 mm in diameter and 0.6 mm thick. The sample faces were (111), (110), and (100) respectively (hereafter the samples are referred to as I(111) etc.). The size of the quenched samples was about 3.3 mm in diameter and 0.5 mm thick. As several crack lines were observed with an optical microscope, the apparent size itself seemed to have no important meaning.

Although the starting samples were single crystals, the quenched samples were no longer single crystals. No spots were observed in the back Laue patterns. Then with the use of a typical 2θ-θ vacuum-low-temperature diffractometer, diffraction patterns of all the specimens were studied. Copper and aluminum plates served as sample holders. Several scans with respect to one sample were made over the range $20° < 2\theta < 80°$ with Cu Kα radiation. No special care was taken to control the temperature, although in most cases it remained near liquid nitrogen temperature.

Lines of three phases were observed; zinc blende, β-Sn, and orthorhombic. All the lines which were distinct were centered at $2\theta = 30°$. These were the (100), (020), and (011) in the ortho-

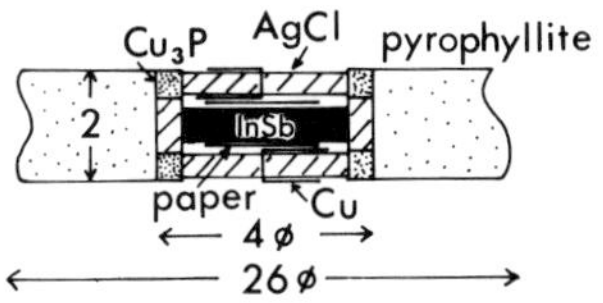

Fig. 3. Sample configuration for quenching.

rhombic phase, whereas in the β-Sn structure the (200) and (011) were distinct. The x-ray diffraction features were not the same. These depended on the samples as shown in Fig. 4; the intensity increased in the order of I(110), I(100), and I(111); besides the lines of zinc blende structure, only those of the β-Sn phase were observed in I(110), whereas both the β-Sn and orthorhombic structures coexisted in I(100) and I(111), although in some cases only lines of β-Sn were recorded in I(100) and those of orthorhombic in I(111). Strong orthorhombic lines were preferentially observed in I(111).

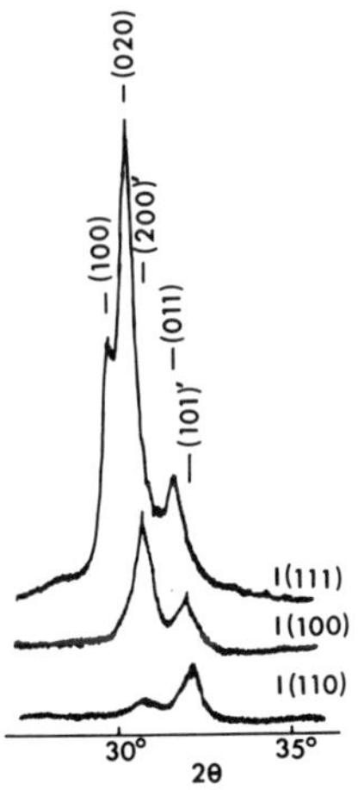

Fig. 4. Diffraction patterns of three specimens. ([hkl] without the prime is orthorhombic and with prime is β-Sn.)

To determine whether the specimen had a preferred orientation, the specimen (θ) was rotated while the detector (2θ) was fixed. Some diffraction patterns obtained in this process are reproduced in Fig. 5. The reflection planes seem to lie around the sample surface.

After the bulk specimens were examined, they were pulverized at liquid nitrogen temperature. The only diffraction patterns noted at that time were those of the β-Sn and zinc blende structures. The orthorhombic lines had disappeared.

In some cases, pressure was released after the metallic transition was observed and samples were returned to 1 atm without temperature treatment. The recovered samples, though no longer single crystals, preserved their initial orientations to a certain extent after undergoing phase transitions back and forth. For example, (111) diffraction lines of the zinc blende structure appeared sharply in I(111), while in I(100) only a trace was observed.

DISCUSSION

The Gibbs free energy under stress is easily calculated. In the present study the strain energy is due to the uniaxial stress p along with the confining pressure of 0.76p. The free energy is of the form ap^2. When the uniaxial stress was applied perpendicularly to the (100), (112), (111), or (110) plane, the coefficient a for each plane was determined to be 8.00, 7.76, 7.65, and 7.74 J/cc, respectively, where the elastic constants are S_{11} = 24.2, S_{12} = -8.55, and S_{44} = 33.1×10^{11} dynes/cm^2 [10]. (The unit of p in

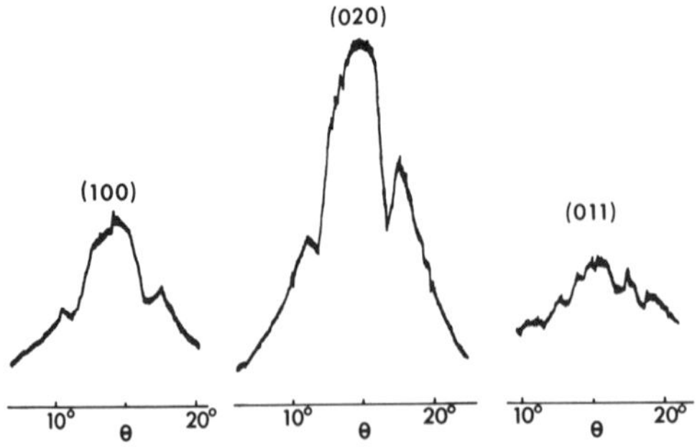

Fig. 5. Diffraction patterns of I(111) when 2θ is fixed and θ is changed.

the above calculation is 10 kbar.) In the case of hydrostatic pressure p, the free energy becomes $10.65p^2$. The relation between the free energy and pressure is shown in Fig. 6 along with the transition points. The curve connecting these points (dashed line) does not represent the free energy of the tetragonal or the orthorhombic phase; to be tetragonal or orthorhombic, its slope must differ by 16% from the slopes of the other curves of the zinc blende structure, corresponding to a volume change of 16% at the transition. Thus the resulting curve does not represent a state of thermodynamic equilibrium. This explains the sluggishness of the phase change as observed in Fig. 1.

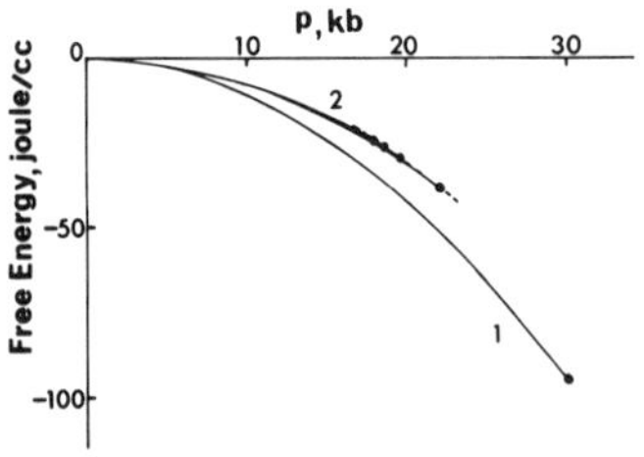

Fig. 6. Free energy of semiconducting InSb: (1) Hydrostatic pressure; (2) uniaxial pressure. Dots are transition points.

The lattice parameters are obtained from the above five diffraction pattern lines; for the orthorhombic phase a=3.02, b=5.89, and c=3.22; for the β-Sn structure a=5.86 and c=3.20. The obvious interpretation would be that the β-Sn structure corresponds quite simply to the zinc blende structure as shown in Fig. 7. At the phase transition, InSb contracts by 50% along the c axis and dilatates by 30% along the a axis. The orthorhombic structure is produced from the β-Sn phase by a simple shear stress as noted earlier. In reality, however, such a simple process does not seem to occur.

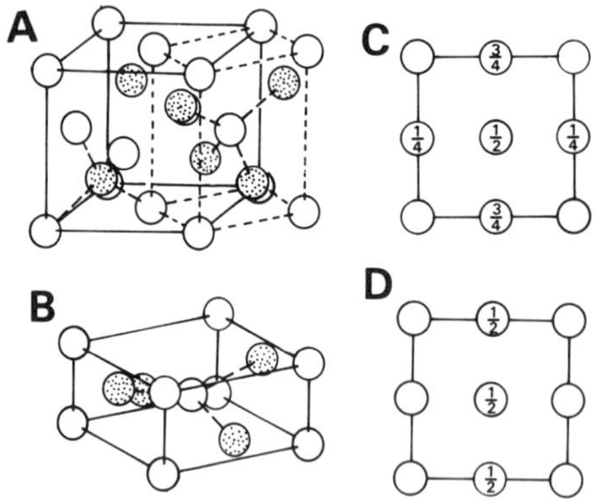

Fig. 7. Relationship of zinc blende (A) and β-Sn (B). Relationship of β-Sn (C) and orthorhombic (D).

The geometrical relationship between the zinc blende structure and the high-pressure phases is shown in Table I. In case a, the (002) line is extinct for the β-Sn structure but not for the orthorhombic phase where the d-value for the (002) line is 1.61 corresponding to 57.2° in 2θ. However, no lines were observed in the vicinity

Table I. Orientation Relationship of Zinc Blende, β-Sn, and Orthorhombic

	in Z.B.	in β-Sn or orth.
a	(100)	(001)
b	(110)	(111)
c	(111)	(011) or (101)

of 57.2°, although the existence of the orthorhombic phase was verified in the I(100) specimen. Thus, the possible occurrence of simple expansion and contraction at the phase transition is very small. In addition to the results of the diffraction patterns, a macroscopic observation also supports this view; a quenched I(100) specimen disc showed only a 10% contraction in thickness and a 10% expansion in diameter.

In case b, the I(110) line corresponds to the (111) line in the β-Sn structure and the (122) line in the orthorhombic phase. The (111) line is extinct in the β-Sn phase, while the orthorhombic structure was not observed in the quenched I(110) specimens. Thus, further discussion is impossible in this case.

In case c, the (011) line in the orthorhombic phase and the (101) line in the β-Sn structure should be carefully observed. These lines, however, were not as strong as the (020) line of the orthorhombic phase. The fact that the orthorhombic (100) and (020) lines appear at all in addition to the orthorhombic (011) line strongly suggests that some rotational movement or nucleation growth must have occurred at the phase transition.

REFERENCES

1. M. D. Banus, R. E. Hanneman, A. N. Mariano, E. P. Warekois, H. C. Gatos, and J. A. Kafalas, Appl. Phys. Letters 2, 35 (1963).
2. R. E. Hanneman, M. D. Banus, and H. C. Gatos, J. Phys. Chem. Solids 25, 293 (1964).
3. A. J. Darnell and W. F. Libby, Phys. Rev. 135, A1453 (1964).
4. J. S. Kasper and H. Brandhorst, J. Chem. Phys. 41, 3768 (1964).
5. D. B. MacWhan and M. Marezio, J. Chem. Phys. 45, 2508 (1966).
6. M. D. Banus and M. C. Lavine, J. Appl. Phys. 38, 2042 (1967).
7. M. D. Banus and M. C. Lavine, J. Appl. Phys. 40, 409 (1969).
8. A. Morita and T. Soma, Solid State Commun. 11, 927 (1972).
9. Y. Kato, Oyo Butsuri (in Japanese) 37, 476 (1968).
10. H. B. Huntington, Solid State Phys. 7, 213 (1958).

C-7

PHASE TRANSITIONS IN INTERMETALLIC COMPOUNDS UNDER PRESSURE: SYNTHESIS OF TUNGSTEN GERMANIDES

S. V. Popova and L. N. Fomitcheva

Institute of High Pressure Physics
Moscow, U.S.S.R.

INTRODUCTION

As is known, tungsten germanides do not exist at ordinary conditions, and the equilibrium phase diagram of W - Ge probably has a simple eutectic form. In the ternary compounds $(Mo,W)_3Ge$, $(Mo,W)_5Ge_3$ and $(Mo,W)_2Ge$ the solubility of tungsten is 10 to 20 at.%; in $W_5(Si,Ge)_3$ and in $W(Si,Ge)_2$ the solubility of germanium is 5 to 15 at.% [1,2]. Thus, tungsten germanides are not stable at atmospheric pressure. One can assume that high pressures may be favorable for the formation of these types of compounds because their synthesis is accompanied by large decreases in specific volume.

METHOD

The high pressure chamber constructed by Vereschagin et al. [3] was used in this study. Bismuth at 25.5, 26.9, and 77 kbar was used in the calibration of the chamber. The experiments were carried out at a constant pressure of 77 kbar and at temperatures from 1500 to 2500°C during the 2 to 5 min. required.

A well-mixed powder consisting of the elements, in different mola ratios, was first pressed and then placed in an ampule of Al_2O_3. The thermocouple was located outside the ampule near its wall. The mixture was heated by a current passing through the sample. The entire system was calibrated by melting pure Mo, Nb and Ge. The phases present were determined from x-ray powder photographs (114 mm diameter camera with CuKα radiation). A mixture with NaCl was used to determine the unit cell sizes of the new phases.

RESULTS

The tungsten germanides W_5Ge_3 and W Ge_2 were prepared at high temperature and pressure. W_5Ge_3 crystallized in two different forms. One had a structure type of W_5Si_3 (space group I42 m, in a tetragonal unit cell with parameters a=9.81 ± 0.02 Å, c = 4.92 ± 0.01 Å, and calculated density of 9.62 g/cm^3) while the other had a structure type of Cr_5B_3 (space group I4/mcm, tetragonal unit cell with parameters a=6.25 ± 0.02 Å, c=11.72 ± 0.03 Å and a calculated density of 9.93 g/cm^3). The psychometric density for the samples which contained mixtures of both phases was 9.8 ± 0.1 g/cm^3. No variation of the cell's parameters with composition was found for these phases.

W Ge_2 has the same structure type as $MoSi_2$. It has a range of homogeneity, with cell parameters varying from a=3.337 ± 0.001 Å, c=8.276 ± 0.003 Å on the tungsten side to 3.320 ± 0.001 Å and 8.192 ± 0.003 Å on the germanium side. X-ray density evaluation provides a value of 12.10 g/cm^3 for the germanium rich alloy while pycnometric density evaluation provides a value of 11.9 ± 0.1 g/cm^3.

The space group for W Ge_2 is I4/mmm with the following atomic positions: W:2(a) and Ge:4(e). The only unknown parameter, z=1/3, was determined from the Patterson function; the disagreement index R was 0.21.

Some of the samples also contained the bcc phase. The powdered tungsten used for the experiments had a cell parameter a=3.166 ± 0.002 Å. After high pressure-high temperature treatment, the bcc phase had a cell parameter ranging from a = 3.162 ± 0.002 Å to 3.147 ± 0.001 Å. This shows that at high temperature and pressure the solid solution W(Ge) was formed.

The thermal stability of tungsten germanides was investigated at atmospheric pressure at 1000°C for 5 hrs. Neither phases of W_5Ge_3 changed under these conditions, but W Ge_2 was decomposed into its elements.

It follows from these experiments that at high pressure and temperature the phase diagram W-Ge becomes fairly complex and similar to the other Mo-Ge diagrams, where Mo is a transition metal from the IV-VI group.

REFERENCES

1. H. Nowotny, F. Benesovsky and C. Brukl, Monatsh. Chem. 92, 365 (1961).
2. P. Stecher, F. Benesovsky and H. Novotny, Monatsh. Chem. 94, 1154 (1963).
3. L. F. Vereschagin, L. G. Chvostansev and A. P. Novikov, United Kingdom Patent No. 1,342,369 (1971).

C-8

PHASE TRANSITIONS IN GROUP III-V AND II-VI SEMICONDUCTORS AT HIGH PRESSURE

S. C. Yu, C. Y. Liu,* and I. L. Spain

University of Maryland
College Park, Maryland USA

and

E. F. Skelton

Naval Research Laboratory
Washington, D.C. USA

INTRODUCTION

Compounds of the group III-V and II-VI elements are technologically important. As a result they have been extensively studied, both theoretically and experimentally. There have been several structural studies of the high pressure phases, pioneered by Jamieson [1,2], but above ~ 100 kbar only a few studies of electrical resistance have been reported. Several "transitions" based on abrupt changes in resistivity have been useful as fixed points.

It is important to note that many transition pressures reported earlier have been based on provisional secondary scales and are subject to revision. For instance, Drickamer [3] revised his "fixed points" above ~ 100 kbar in 1970, so that the zinc-blende-to-rocksalt transition in ZnS was reduced from ~ 240 - 245 kbar to 185 kbar. The most recent value obtained using the NaCl and ruby R-line scale is 150 ± 5 kbar [4].

All pressures recorded in this paper are based on Decker's NaCl scale [5], which is consistent with a linear ruby R-line scale [4]. It has also been shown that the ruby R-line scale is consistent with equation of state data for a number of hard materials to very high pressure [6]. These later studies are in direct conflict with the speculations of Chhabildas and Ruoff [7,8] concerning inaccuracy of the NaCl scale.

*Present address: I.B.M., San José Laboratory, San José, Calif.

Interest in these materials has been intensified because Van Vechten [9] has made model calculations for their high-pressure phases. Using scaling hypotheses in the spirit of the Phillips-Van Vechten quantum dielectric theory in covalent systems [10-12], he estimated the Gibbs free energy difference between the two phases and the pressure at which this difference vanishes (equilibrium transformation pressure, P_t).

For most systems Van Vechten assumed that the high-pressure phases have the β-Sn (A5) structure or its diatomic equivalent. However, he noted that different metallic phases might occur, but free energy differences between metallic structures would be small. He also noted that Jamieson [2] had reported that InP assumes the rocksalt (B1) structure at high pressure. Van Vechten indicates that this structure is only preferred if it is semiconducting, not metallic. This only occurs if the elements have an appreciable difference in electronegativity, so that the compound is ionic in nature.

It is of considerable interest to determine the structures and transition pressures for the high-pressure phases, and to compare them with Van Vechten's predictions. Furthermore, the present measurements can be used to check the scaling formula used by Van Vechten for the volume change at the transition pressure.

$$\frac{\Delta V^{\alpha\beta}(P_t)}{V^{\alpha}(P_t)} = 0.209 - 0.056\, f_i \qquad (1)$$

where f_i is the Phillips ionicity, α the low pressure, and β the high-pressure phase. This predicts a 20.9% volume reduction at the transition pressure for a purely covalent material ($f_i = 0$) and a 15.3% reduction for a purely ionic compound ($f_i = 1$).

Results will be presented for a number of materials and comparisons made where possible with other work. Apart from presenting new results on a number of systems, the first results on a pseudobinary alloy system ($Ga_xIn_{1-x}Sb$) have been obtained across the concentration range, and are presented here.

EXPERIMENTAL

Experiments were conducted in a gasketed diamond-anvil cell developed from a design of Bassett et al [12[(see Fig. 1). Important differences were as follows: (1) In order to reduce the stress at the interface between the diamonds and the rockers, ~ 1/2 carat of diamonds were used. The anvil face was maintained at ~ 0.5-mm diameter. A thin (~ 0.010 mm) shim of Zr separated the diamonds and rockers. (2) In order to facilitate and maintain anvil alignment, the rockers were fabricated with an enlarged diameter of ≈ 16 mm. Adjustments to angular alignment were made with four setscrews mounted in a plate attached to the fixed screw and piston

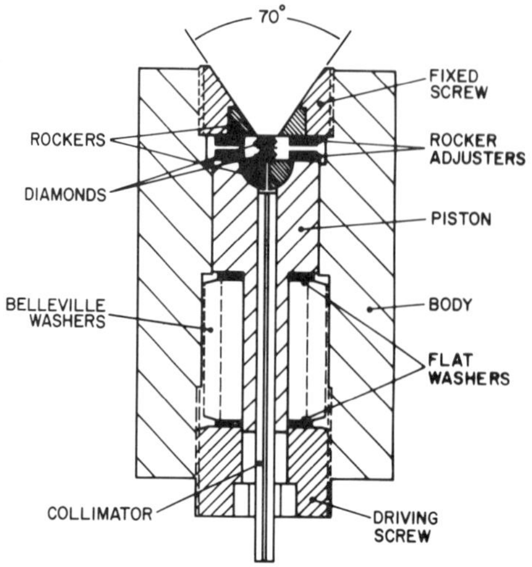

Fig. 1. Schematic of the diamond-anvil cell used in the present work.

(see Figs. 1 and 2). These screws were tightened onto the rocker so that they prevented any motion of the rocker during the course of the experiment. (3) The piston length-to-diameter ratio was kept greater than 1.2 and the piston was lapped into the cylinder with a close, smooth, fit ($\sim 5\mu$in. ($\sim 0.12\mu$)) surface finish, with about 400μin. ($\sim 10\mu$m) clearance between piston and cylinder. (4) The piston was manually driven by the rotation of the screw through a set of Belleville washers. Hardened wear-rings and spring elements were teflon-coated to reduce friction. Rotation of the lower diamond with respect to the upper one was prevented by spring-loading a ball into a hemi-cylindrical groove machined into the side of the piston, parallel to its axis. It was important to pre-tighten the fixed screw onto its seat with sufficient tension so that under application of the highest load from the driving screw no separation occurred between the fixed screw and the seat. Any such separation would inevitably cause misalignment of the diamonds. (5) Tough tool steels were used in the construction of the cell, to minimize the possibility of plastic deformation. The rockers were initially constructed of hemicylinders of sintered boron carbide (B_4C) with a 0.5-mm diameter hole drilled through for optical sighting. More recently the cell has been equipped with rockers fabricated of compacted tungsten carbide/cobalt alloy. A 70° slot allowed the diffracted x-ray beam to exit (see Fig. 2(b)). We believe that the tungsten carbide anvils should be superior to those of B_4C for work at the highest pressures. (6) The incident x-ray beam was collimated by a small hole (~ 0.1 mm) drilled in a lead plug ~ 1 mm long at the mouth of a piece of high-pressure tubing (1/4 in. OD x 1/16 in. ID (~ 6.35 mm x 1.59 mm)). The tubing passed through an axial hole in the piston, nesting in a conically-tapered seat. Initially the rockers were adjusted so that the collimating hole was aligned on their joint axis. However, the collimating hole was always intentionally drilled slightly (≤ 0.05 mm) off the axis of the high-pressure tube. With a gasket drilled with a sample hole of a diameter ~ 0.15 mm, the collimating hole could usually be positioned very close to the center of the sample chamber by rotating it. Several collimators were constructed, each with slightly different geometry so that a collimator could be chosen to match the diamond-gasket arrangement for the cell, thus maximizing the radiation passing through it.

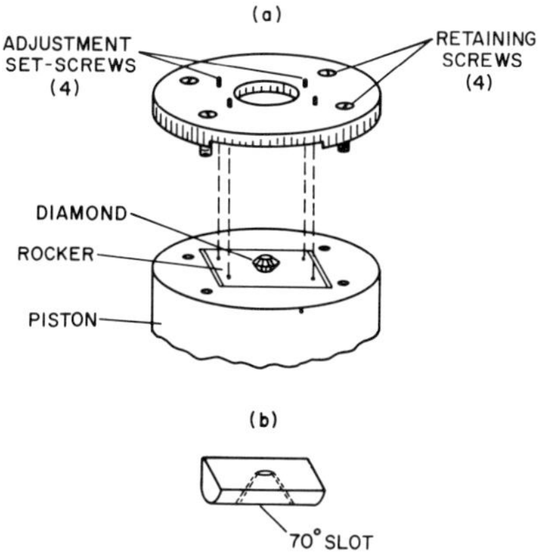

Fig. 2. (a) Sketch of arrangement for adjusting the rockers; (b) Rocker of tungsten carbide.

In an actual experiment the sample to be studied was ground, mixed with NaCl calibrant, and then placed in the sample chamber with a 4:1 methanol:ethanol mixture [14] to ensure hydrostatic conditions up to $\sim$100 kbar. The relative intensity of radiation passing through the cell was monitored with a Geiger counter. As the collimator was rotated, a maximum intensity corresponded to the near centering of the beam in the sample volume. The absence of gasket lines in the diffracted radiation was used as further indication that the beam was not striking the gasket.

With the above collimating arrangement, a wet-film exposure could be generally taken in 10 to 30 hrs - a substantial reduction compared to Bassett's original design (e.g., 100 to 300 hrs). Further details are given elsewhere [15].

Very rapid information was usually obtained using a solid-state, energy-dispersive detector [16-18]. Fine-focused, unfiltered, and polychromatic radiation was directed onto the sample. The diffracted beam was then energy-analyzed with a Si(Li) detector at a fixed angle (typically $\sim$10°). The output from the detector was processed through an analogue-to-digital converter and stored in a multichannel analyzer. The magnetic memory of this analyzer could then be accessed at any time by a remote computer, plotted on an X-Y recorder, or continuously displayed on an oscilloscope.

Stronger x-ray diffraction lines could be detected in a matter of minutes. This enabled the pressure ($\pm$ 3 kbar) as well as the structure of the material being investigated to be monitored in a very short period of time (typically $\sim$10 min). At pressures where more precise information was required, wet-film techniques were used where diffracted angles could be determined to $\pm$ 0.02° in 2θ. This corresponds to about $\pm$ 1 kbar in pressure using the NaCl scale.

The convenience of this energy-dispersive technique cannot be overemphasized. Although pressure can be measured more rapidly and sensitively using the Ruby fluorescence technique, the energy-dispersive x-ray technique allows a very rapid semiquantitative study to be made of a particular material at high pressure. It is sometimes of interest to incorporate a small crystal of ruby for pressure measurement, particularly if the diffraction lines of the sample and calibrant interfere; this procedure has been followed in some instances. In all cases, the NaCl and (linear) Ruby R-line scales have been found to agree, as reported by Piermarini and Block [4].

All experimental results reported here have been obtained at ambient temperature, except for InSb for which a simple heating arrangement was sometimes used. Techniques suitable for x-ray diffraction measurements at low temperature (4 to 300 K) have been reported elsewhere [19]. In addition, facilities have been developed for optical absorption measurements and the measurement of magnetic susceptibility and electrical resistance (electrodeless) [20]. In this paper, the structural measurements will be reported.

The results of the change in unit cell volume with pressure can be interpreted to give bulk moduli (B) for both phases. However, B depends on pressure, and a second order fit to the data is often of limited usefulness since errors in the experimental volumes are relatively large. Where possible, experimental values of elastic constants were used to determine the bulk modulus at P=0, (B_o) and pressure derivatives of these constants to determine $(\partial B/\partial P)^o_{T,P\to 0}=B'_o$. Then, the experimental values of $V^{\alpha}(P)$ were compared to the empirical equation of state values, using V_o, B_o, and B'_o. In all cases agreement was satisfactory.

Since the volume change from the low-pressure, zinc-blende to the high-pressure phase is relatively high (15 to 20%), x-ray diffraction patterns could normally be obtained from two coexisting phases while the piston was being advanced at approximately constant pressure. The width of this two-phase region was usually several times larger than the precision of the pressure measurement, so that the error quoted for P_t is related to the sluggishness of the transition. For technical reasons it was very difficult to observe the reverse transition. Thus, all transitions are reported on the "up-stroke" and questions of metastability of phases after decompression will not be addressed here.

Comparison is made whenever possible with results obtained by earlier workers. Previous estimates of the transition pressures of these materials tended to be too high because of difficulties with pressures assigned to fixed points. Jamieson [1,2] largely used estimates for P_t based on Minomura and Drickamer's work [21]. In 1970, Drickamer [3] revised his transition pressures downwards, so that values quoted in Table I are adjusted to this scale. Piermarini and Block [4] later showed that his 1970 scale was also in error, particularly above 100 kbar. Where comparison is possible, the present values for P_t are lower than Minomura and Drickamer's [21] for III-V compounds and Samara and Drickamer's [22] for II-VI, supporting this contention. However, some differences may also be due to the more nearly hydrostatic pressure conditions realized in the present work.

Table I. Summary of Present Results and Comparisons with other Work on Structure at Ambient Temperature**

Substance	Ionicity f_i	P_t, kbar Calc. [9]	P_t, kbar Exp.	$\Delta V^{\alpha\beta}(P_t)/V^{\alpha}(P_t)$ Calc. [9]	$\Delta V^{\alpha\beta}(P_t)/V^{\alpha}(P_t)$ Exp.	Structure of High-Pressure Phase
InSb	0.321	43	22.5*	0.191	0.185[2] 0.197[43]	See remarks in text and Fig. 3.
			25.5±2†		0.193†	
GaSb	0.261	73	67-84[21] ~60[31] 64±4[32]	0.194	0.169[2]	A5[2]
			62±3†		0.171±.012†	A5†
AlSb	0.426	122	96-105[21]	0.185	0.165[2]	A5[2]
			83±2†		0.20±.01†	B1†
GaAs	0.310	153	185-188[21]	0.192		See remarks in in text
			170±5†		0.15±.02†	
AlP	0.307	269	140 [38]	0.192		See remarks in text
			170±5†		-	
GaP	0.374	216	220[4,40,41]	0.188		See remarks in text
			220±5†		-	
ZnO	0.616	-	80±3†	-	0.166±.012†	B1†
ZnS	0.623	-	185 [22]	-		
			150±5 [4]			B1
			150±5†		~0.15†	B1†

† Signifies present results
* Transition pressure at ambient conditions to orthorhombic phase.
**All pressures measured by Minomura and Drickamer [21] and Samara and Drickamer [22] have been corrected to Drickamer's 1970 scale [3].

RESULTS

A summary of our results is given in Table I and a brief discussion for each compound will be given below.

InSb

InSb is the most widely studied of the III-V materials, and yet the phase diagram is still not completely understood. A study was made to determine more precisely the structures of the high-pressure phases, which can be understood more easily with the phase diagram proposed by Banus and Lavine [23] (see Fig. 3).

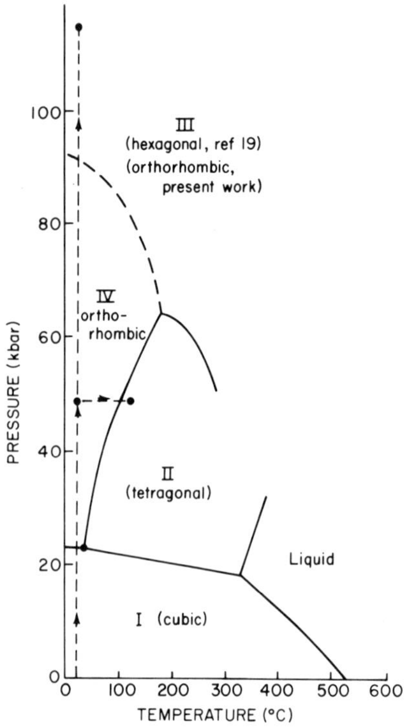

Fig. 3. Phase diagram of InSb proposed by Banus and Lavine [23]. Present experiments are indicated.

At 20°C the transformation to an orthorhombic structure [24,25] was observed at 25.5 ± 2 kbar. From intensity data, the space group was determined to be Pmm2. The positional coordinates for the atoms are sketched in Fig. 4. It is related to the tetragonal structure by a simple displacement of every other (100) plane through a vector b/2 + c/2 in the orthorhombic structure. The above structure differs from that proposed by Kasper and Brandhorst [24] who assigned atoms at (0,0,0) and (0,1/2,1/2) positions in the unit cell, compared to (0,0,0) and (0,1/2,~1/4) in the present work. The unit cell dimensions are a = 2.919Å, b = 5.618Å, and c = 3.066Å at 43.5 kbar. Bond lengths and angles are indicated in Fig. 4.

Above 90 kbar a new orthorhombic phase (III) was observed whose x-ray powder pattern was similar to, but not identical with, that of hexagonal InSb reported by Banus and Lavine [26]. The present pattern could be indexed on the basis of an hexagonal structure, but better agreement was obtained when orthorhombic symmetry was assumed. A sketch of the unit cell is given in Fig. 4. Unit cell dimensions at 115 kbar are a = 5.712Å, b = 5.357Å, and c = 3.063Å. Bond lengths and angles are given in Fig. 4. The density for this second phase was found to be 8.4 gm/cm^3, in close agreement with 8.5 gm/cm^3 found by Banus and Lavine [26] based on their hexagonal interpretation.

Experiments were also conducted on the tetragonal phase (InSb II) after heating InSb IV across the phase line. Smith and Martin [27] reported that this phase is consistent with the space

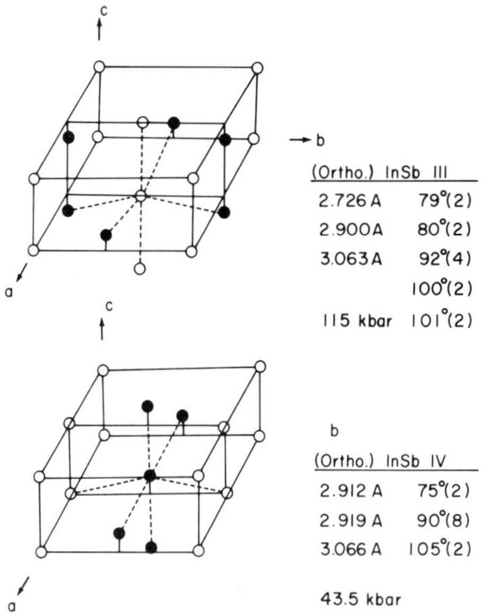

Fig. 4. Positional coordinates for the atoms in the two orthorhombic phases observed in the present work (InSb III and IV). Bond lengths and angles are indicated.

group $I4_1/amd$ to which the β-Sn(A5) structure belongs. However, this space group symmetry would require that both In and Sb atoms be positioned at random. If the tetragonal phase were ordered, the proper space group would be $I\bar{4}m2$.

Previous diffraction patterns obtained by Banus and Lavine [23] contained (311), (410), and (511) reflections which are forbidden for the body-centered cubic structure. Furthermore, McWhan and Marezio [25] found only poor agreement between the observed and calculated d-values for this structure. In both cases it is possible that the sample contained nontransformed, metastable InSb III material.

In the present case only a weak (311) reflection was observed in the pattern, while (410) and (511) reflections were absent. Much better agreement between observed and calculated d-values was obtained, but intensity data could not distinguish between the above two symmetries. However, thermodynamic analysis of the phase diagram indicates that the phase has a relatively high entropy, and is, therefore, probably disordered.

GaSb

GaSb was found to transform to a β-Sn-like, tetragonal structure. The transformation pressure was 62 ± 3 kbar. The uncertainty in this and other measurements of p_t largely arises from the sluggishness of the transition. The results for several runs are summarized in Fig. 5. The dashed line is the prediction for $V^{\alpha}(P)$, based on either the Birch-Murnaghan [28] or Murnaghan [29] equations with experimental elastic constant data [30].* The present value for $\Delta V^{\alpha\beta}(P_t)/V^{\alpha}(P_t)$ is 17.1%, in good agreement with Jamieson's value of 16.9% [2]. Minomura and Drickamer obtained a transition pressure (revised) between 67-84 kbar. The present value for P_t is in good agreement with that of Jayaraman, Klement, and Kennedy [31] and Pitt [32] (see Table I).

*In the absence of data for B_o' a value of $B_o' = 4.5$ is assumed in Fig. 5. If a value of 4.0 or 5.0 were assumed, the pressure at the transition would only differ by ~ 1 kbar.

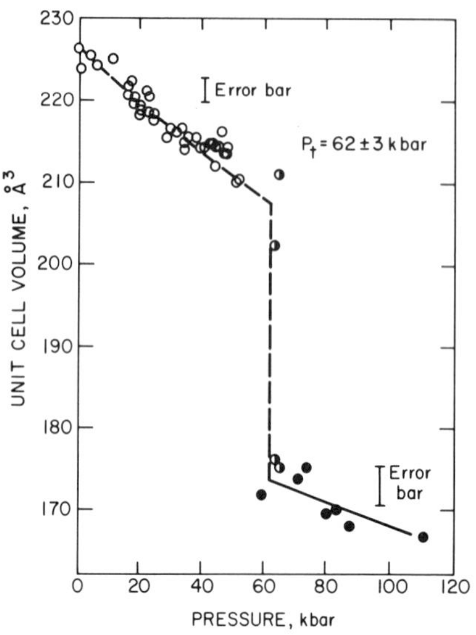

Fig. 5. Plot of the variation of unit cell volume with pressure for GaSb. The dashed line through the points in the low-pressure (α) phase is the prediction of either the Murnaghan or Birch-Murnaghan equation with B_o = 569 kbar, B_o' = 4.5 Open circles are α-phase, filled circles are β-phase, and half-open circles are two-phase.

$(GaSb)_{1-x}$ $(InSb)_x$ Pseudobinary Alloys

The variation of transition pressure for these pseudobinary alloys is given in Fig. 6. The positive bowing of the curve from a linear extrapolation of the values for the end point compounds is in contrast to the results of Adler [33] for alloys with composition close to InSb (x<0.2). The least squares fit to the data gives

$$P_t(x) = 62.0 - 13.4x - 23.6x^2 \qquad (2)$$

The data have been analyzed thermodynamically, using a regular solution model for both phases. If the difference in the enthalpy of mixing for the two phases is given by

$$H_m^\alpha - H_m^\beta = \Delta\Omega x(1-x) \qquad (3)$$

the value of ΔΩ needed to fit the experimental curve from the formula [34]

$$\Delta V_o P_t - 1/2\ \Delta(V_o K_o)P_t^2 + x\Delta g_{oA} + (1-x)\Delta g_{oB} + \Delta\Omega x(1-x) = 0 \qquad (4)$$

is 18 ± 2 kJ/mole. In (4), Δ represents the difference between the α and β phases, K_o is the compressibility at P = 0 ($K_o \equiv B_o^{-1}$), V_o the volume at P=0, and g_o the Gibbs free energy at P = 0. For the α (semiconducting) phase, Ω^α has been measured as 1.99 kJ/mole [35] (nearly ideal). It is surprising that the enthalpy of mixing of the β (metallic) phase should be so high [36,37]. It is perhaps more realistic to attribute the bowing of $P_t(x)$ to kinetic factors. Thus, an "over-pressure" is required to initiate the transformation for the alloys.

This is an unfortunate conclusion, since it implies that thermodynamic factors cannot be used alone to estimate $P_t(x)$ for these alloy systems. Their usefulness for fixed points can be seen immediately, since a transition pressure can be "tailored" in principle by alloying two compounds in certain proportions.

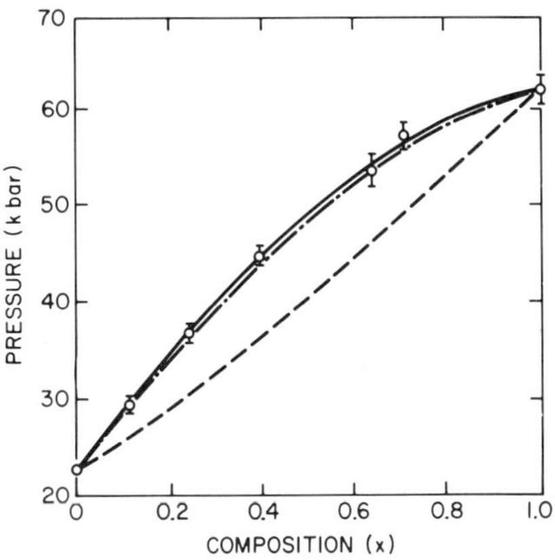

Fig. 6. Variation of the transition pressure of $Ga_{1-x}In_xSb$ alloys with concentration x. Open circles are experimental points, the full line is the least squares fit to the data (2), the dot-dashed line is (4) with $\Delta\Omega$ = 18 kJ/mole, and the dashed line the prediction of (4) with $\Delta\Omega$ = 0 (ideal solution model).

However, the $P_t(x)$ curve must be experimentally determined for each system.

AlSb

Jamieson [2] had earlier found that AlSb transformed to the β-Sn structure at 125 kbar, close to the value of 122 kbar later predicted by Van Vechten's calculations [9]. Surprisingly, the present results showed a transition at 83 ± 2 kbar to the NaCl structure with a volume change of 20.1%.

AlP

Recently, Wanagel, Arnold, and Ruoff [38] showed that this material exhibited a sharp drop in resistance at 140 kbar, a much lower value than that predicted by Van Vechten (269 kbar). The present results confirmed a structural transition at 170 ± 5 kbar). However, the three diffraction lines observed for this phase were not consistent with either the A5 or B1 structure. The three lines could be indexed on the basis of a face-centered cubic structure as (111), (200), and (311) reflections. However, more information is needed to positively identify the structure.

GaP

GaP is of interest since its transition at high pressure is now used as a fixed point at 220 kbar [4]. The transition was first observed in electrical resistance measurements in a split-sphere apparatus [39]. Subsequent studies [4,40,41] showed that the transition pressure of ~500 kbar assigned to the transition was too high by a factor of ~2, since the fixed points on which this estimate was based were also in error [4].

In the present study the transition pressure was observed at 220 ± 5 kbar, in agreement with other quantitative estimates. This is in very good agreement with Van Vechten's estimate (216 kbar). Only four diffracted lines from the polycrystalline sample could be obtained in the high-pressure phase. The data were not consistent with a cubic or A5 structure. However, it is possible that the structure is tetragonal. It was not possible to determine the volume change.

GaAs

Preliminary results for this compound indicate that a transformation to an orthorhombic structure occurs at 170 kbar, with a volume change of ~ 15%. Further experiments are in progress to confirm these results with greater precision. The transition pressure is in reasonable agreement with Van Vechten's estimate (153 kbar) and with Minomura and Drickamer's value [21] when revised using Drickamer's 1970 scale.

ZnO

Bates et al. [42] reported that a high-pressure polymorph of ZnO could be formed with the NaCl structure above about 90 kbar which was metastable at ambient conditions. Only partial conversion could be effected with an ammonium chloride catalyst over long periods of time (~ 36 to 48 hrs) at temperatures above ~ 100°C.

The NaCl form of ZnO was prepared by us at ambient temperature without a catalyst. Conversion was very rapid at 80 ± 3 kbar, with a volume change of 16.6 ± 1.2%. After return to ambient pressure, no trace of this polymorph could be found even if the sample was compressed to 200 kbar to ensure complete conversion.

ZnS

Similar to ZnO, the high-pressure phase has the B1 structure expected of a highly ionic compound. The transition pressure of 150 ± 5 kbar is in good agreement with Piermarini and Block's value [4], while $\Delta V^{\alpha\beta}/V^{\alpha}(P_t)$ was found to be 15%, also characteristic of an ionic material. This latter value is somewhat uncertain since accurate data were not taken in the two-phase region. The value for $V^{\alpha}(P_t)$ was obtained by extrapolating lower pressure data using Murnaghan's equation.

CONCLUSIONS

The high-pressure phases of several III-V and II-VI compounds have been studied. Their structures are often not tetragonal, β-Sn like. The rocksalt phase is preferred for InP [2], InAs [2], AlSb, ZnO, and ZnS [4]. These compounds are more highly ionic in character. However, the ionicity of GaP is greater than that of InAs, and it assumes a metallic structure at high pressure, as do InSb, GaSb, GaAs, and possibly AlP. Of these materials, only InSb and GaSb prefer a tetragonal, β-Sn-like structure at high pressure.

When a metallic structure occurs, the transition pressures are reasonably close to the model values of Van Vechten. His theory needs to be refined to account for the occurrence of the ionic rocksalt structure of several compounds. In addition, as shown in Table I, his scaling values for volume changes at the semiconductor-to-metal transitions do not accurately describe the observed values. In general, if the volume change is smaller than the model value, then also the calculated value for P_t should be increased. However, for InSb, GaSb, and GaAs, where $(\Delta V/V)_{exp} < (\Delta V/V)_{theory}$, this worsens the agreement between $(P_t)_{exp}$ and $(P_t)_{theory}$.

ACKNOWLEDGMENTS

Thanks are due to B. Scesa for his skill in fabricating the diamond anvil cell. This work was supported by a grant from the National Aeronautics and Space Administration, Lewis Research Center.

REFERENCES

1. J. C. Jamieson, Science 139, 762 (1963).
2. J. C. Jamieson, Science 139, 845 (1963).
3. H. G. Drickamer, Rev. Sci. Instr. 41, 1667 (1970).
4. G. J. Piermarini and S. Block, Rev. Sci. Instr. 46, 973 (1976).
5. D. L. Decker, J. Appl. Phys. 42, 3239 (1971).
6. P. M. Bell and H. K. Mao, Trans. Amer. Geophys. Union 58, 518 (1977).
7. L. C. Chhabildas and A. L. Ruoff, J. Appl. Phys. 47, 4182 (1976).
8. L. C. Chhabildas and A. L. Ruoff, J. Appl. Phys. 47, 4867 (1976).
9. J. A. Van Vechten, Phys. Rev. B7, 1479 (1973).
10. J. C. Phillips, Phys. Rev. Letters 20, 550 (1968).
11. J. A. Van Vechten, Phys. Rev. 182, 891 (1969); also 187, 1007 (1969).
12. J. C. Phillips and J. A. Van Vechten, Phys. Rev. B2, 2147 (1970).
13. W. A. Bassett, T. Takahashi, and P. W. Stook, Rev. Sci. Instr. 38, 37 (1967).
14. G. J. Piermarini, S. Block, and J. D. Barnett, J. Appl. Phys. 44, 5377 (1973).
15. E. F. Skelton, C. Y. Liu, and I. L. Spain, High Temp.-High Press., 9, 19 (1977).
16. B. C. Giessen and G. F. Gordon, Science 159, 973 (1958).
17. E. F. Skelton, Rept. NRL Prog. (1972), p. 31.
18. K. Syassen and W. B. Holzapfel, in Europhysics Conf. Abstracts, Electronic Properties of Solids Under High Pressure, Leuven, Belgium (1975), p. 75.
19. E. F. Skelton, I. L. Spain, S. C. Yu, C.Y. Liu and E. R. Carpenter, Jr., Rev. Sci. Instr. 48, 879 (1977).
20. E. F. Skelton, I. L. Spain, and F. J. Rachford, "High Pressure Structural and Lattice Dynamical Investigations at Reduced Temperatures," to be published in Proc. Intern. Conf. on High Pressure and Low Temperature Physics, Cleveland, Ohio (1978).
21. S. Minomura and H. G. Drickamer, J. Phys. Chem. Solids 23, 451 (1962).
22. G. A. Samara and H. G. Drickamer, J. Phys. Chem. Solids 23, 457 (1962).
23. M. D. Banus and M. C. Lavine, J. Appl. Phys. 40, 409 (1969).
24. J. S. Kasper and H. Brandhorst, J. Chem. Phys. 41, 3768 (1964).
25. D. B. McWhan and N. Marezio, J. Chem. Phys. 45, 2508 (1966).

26. M. D. Banus and M. C. Lavine, J. Appl. Phys. 38, 2042 (1967).
27. P. S. Smith and J. E. Martin, Nature 196, 762 (1962).
28. F. Birch, J. Geophys. Res. 57, 227 (1952).
29. F. D. Murnaghan, Amer. J. Math 59, 235 (1973).
30. "Piezo-electric, Piezo-optic and Electro-optic Constants of Crystals," Landolt Bornstein, Vol. 1, Group III, Springer-Verlag, Berlin, Heidelberg, New York (1966).
31. A. Jayaraman, W. Klement, and G. C. Kennedy, Phys. Rev. 130, 540 (1963).
32. G. D. Pitt, High Temp.-High Press. 1, 111 (1969).
33. P. N. Adler, J. Phys. Chem. Solids 30, 1077 (1969).
34. C. Y. Liu, I. L. Spain, and E. F. Skelton, J. Phys. Chem. Solids 39, 113 (1978).
35. L. M. Foster and J. M. Woods, IBM Res. Rept. RC3116 (1970).
36. G. B. Stringfellow, J. Phys. Chem. Solids 33, 665 (1972).
37. G. B. Stringfellow, Mat. Res. Bull. 6, 371 (1971).
38. J. Wanagel, V. Arnold, and A. L. Ruoff, J. Appl. Phys. 47, 2621 (1976).
39. A. Onodera, N. Kawai, K. Ishizaki, and I. L. Spain, Sol. St. Comm. 14, 803 (1974).
40. F. P. Bundy, Rev. Sci. Instr. 46, 1318 (1975).
41. C. E. Homan, D. P. Kendall, T. E. Davidson, and J. Frankel, Solid St. Comm. 17, 831 (1975).
42. C. H. Bates, W. B. White, and R. Roy, Science 137, 993 (1962).
43. R. E. Hanneman, M. D. Banus, and H. C. Gatos, J. Phys. Chem. Solids 25, 293 (1964).

C-9

ON GALLIUM IN THE ALPHA AND DELTA PLUTONIUM LATTICE*

R. B. Fischer

Rockwell International, Atomics International Division
Golden, Colorado USA

INTRODUCTION

The stability of delta-phase plutonium-gallium alloys is a subject of great interest to plutonium metallurgists. A survey of progress on the subject could begin with reference to the paper by Goldberg and Shyne [1] and the use of the references given in that paper. The present effort introduces some general ideas that are offered in an attempt to further an understanding of the nature of the delta-phase plutonium-gallium alloys and their stability. Special attention is given to a hypothesis relating to the role of gallium atom positions in the plutonium lattice.

GALLIUM SITES

One can consider where the gallium atoms might be located in delta-phase alloys. The gallium atom is too large to be an interstitial atom in delta plutonium, so it should substitute for plutonium atoms in the delta unit crystal cell. The limit of this substitution is found to be about 12.5 at. % [2], or the replacement of one-eighth of the plutonium atoms in the structure. (This limit might be considered a gallium-saturated delta phase at 655°C.) Thus, when more than 1 gallium atom per 2 delta unit cells is present, the Pu_3Ga compound† with 1 gallium atom per unit cell can be formed. The

*Work supported by the U.S. Energy Research and Development Administration.

†Cubic $AuCu_3$ type (363-677°C), or tetragonal Sr Pb_3 type (below 363°C) [3].

structural change from the delta cell to the Pu_3Ga cell is interesting to contemplate. One imagined change is shown in Fig. 1.

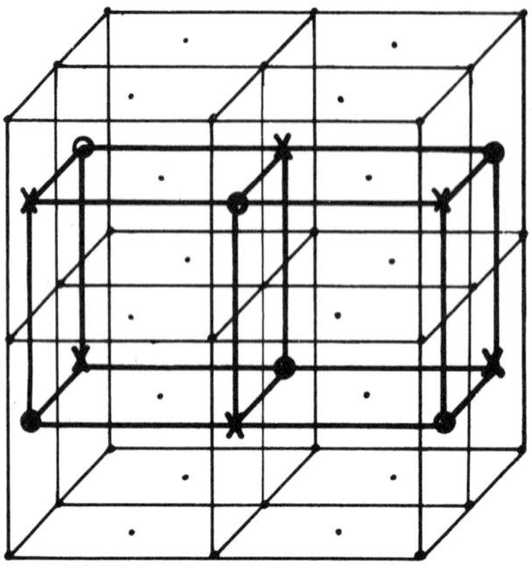

Fig. 1. Hypothetical change from delta to Pu_3 Ga structure on the addition of gallium atoms. Eight delta cells shown, and two Pu_3 Ga cells shown. Lattice-dimensions change slightly on formation of Pu_3 Ga.

In the gallium-saturated delta cell just imagined, the gallium atoms have been located where they are surrounded with plutonium atoms, and the cell symmetry is preserved. Why is this choice made instead of either random or clustered locations? The answer is suggested by considering the volume change of the delta structure as gallium is added.

Data by Ellinger et al. [2] reveal that the addition of gallium to delta plutonium results in a greater decrease in volume than obtained for the calculated additive value of gallium (see Appendix). In effect, this is a negative deviation from Vegard's Law. The negative deviation is an indication that the attraction between gallium and plutonium atoms in the delta-phase structure is greater than the attraction of the similar atoms. Therefore, one would expect a tendency for the gallium atoms to be surrounded by plutonium atoms. This would be a short-range order for these delta-phase alloys. The negative deviation also is taken as evidence of an exothermic reaction as gallium is added to the delta structure; a theoretical point of minor practical importance. For a model, the gallium atoms are imagined to occupy various lattice sites of the face-centered-cubic structure and to tend to have nearest neighbors of plutonium atoms.

EFFECT OF PRESSURE

It is generally recognized that if sufficient isostatic pressure is applied to the plutonium-gallium delta-phase alloys at ambient temperatures, at least some of the structure transforms to alpha. This transformation skips the intermediate phases γ & β and has a reversible tendency when the pressure is released, especially for the alloys with more than 1 at. % of gallium.

The direct transformation from the delta to the alpha phase seems surprising because the face-centered-cubic delta lattice appears so unrelated to the monoclinic alpha lattice. However, as shown by Lomer [4], there is a correspondence of the structures. Spriet [5] diagrammed this shear-movement correspondence, and one can see that a number of atom movements of less than an atomic-diameter distance would also suffice geometrically for the transformation. Whatever the actual mechanism, the skip transformation seems reasonable to imagine. When the delta phase is transformed by pressure to the alpha phase at ambient temperature, 4 delta unit cells could transform to 1 alpha cell. Now the gallium atoms are sited in an alpha cell because diffusion processes are kinetically slow at such temperatures.

While atoms in a face-centered-cubic delta cell have similar geometric environments, those in the monoclinic alpha cells have eight different types of environments. If a gallium atom is substituted for a plutonium atom in an alpha lattice under equilibrium conditions, it is suggested that one of the eight types of sites might be a preferred position.

In the transformation of delta cells to alpha cells, one can readily imagine that a gallium atom sited somewhere in 4 delta cells will not be found necessarily on its preferred site in the alpha cell. In a Pu-1 wt. % Ga alloy, there is a chance of about one in eight that a gallium atom would occupy a delta cell site that could ultimately be a gallium-preferred alpha cell site. Short-range diffusion promoted by heating could result in the movement of the gallium to preferred sites in alpha cells, if the hypothesis is valid. Pressure would be needed to retain the alpha phase in an attempt to use heat to move the gallium to preferred sites.

Of the eight different sites in alpha cells, there may be a range of preference for gallium sites. Also, the alpha cell is made up of sixteen atoms with pairs of atoms having similar environments. This would allow 2 most-preferred sites for gallium atoms per alpha cell. The composition would be 12.5 at. % which, coincidentally, is the saturation limit for delta phase. No significance of this observation is suggested.

SUMMARY

An interesting circumstance dealing with gallium atom positions in delta-and-alpha-phase plutonium-gallium alloys is hypothesized. The hypothesis can be considered as studies of the delta to alpha transformation and its reversion are carried out.

REFERENCES

1. A. Goldberg and J. C. Shyne, J. Nucl. Mat. 60, 137 (1976).
2. F. H. Ellinger, C. C. Land, and V. O. Struebing, J. Nucl. Mat. 12 (2), 226 (1964).

3. O. J. Wick, ed., Plutonium Handbook, Vol. 1 Gordon and Breach, New York (1967), p. 206.
4. W. M. Lomer, Solid State Commun. 1 (96) (1963).
5. B. Spriet, in Plutonium 1965, A. E. Kay and M. B. Waldron, eds., Chapman and Hall, London (1967), p. 88.
6. F. H. Ellinger, C. C. Land, and W. N. Miner, J. Nucl. Mat. 5, (2), 165 (1962).

APPENDIX

In a classic treatment of the study of binary solid solutions, one can plot the volume of 1 gram mole of a solution against the atomic percent composition. Such a plot for Pu-Ga and Pu-Al delta phase alloys is shown in Fig. 2, which is based on Ellinger et al. [1,6] x-ray data. The volume, V, of 1 gram mole of the Pu-1 wt. % Ga alloy, which is 14.8 cm^3, consists of the volume occupied by the Pu atoms plus the volume occupied by the Ga atoms. The apparent volume which would be occupied by 1 gram mole of Pu in this alloy solution, is the partial molar volume of Pu and is termed $\bar{V}_{Pu}$. Similarly, we have $\bar{V}_{Ga}$.

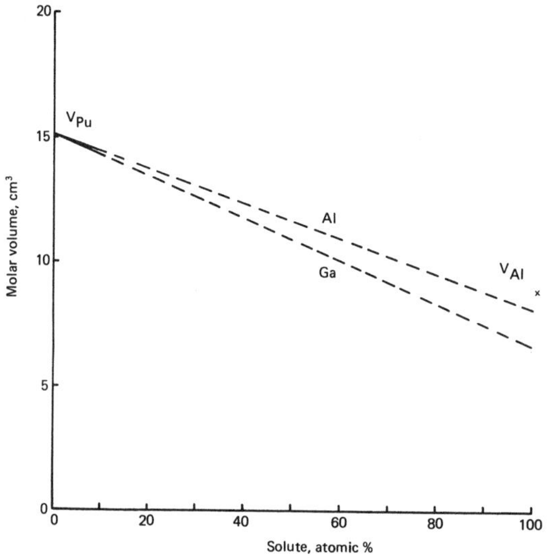

Fig. 2. Molar volumes of Pu-Ga and Pu-Al delta phases. From a computer plot of X-ray data. Data are extrapolated beyond about 10 at. %.

Mathematically,

$$\bar{V}_{Pu} = \left(\frac{\partial V}{\partial n_{Pu}}\right)_{T,P,Ga} \quad \text{and} \quad \bar{V}_{Ga} = \left(\frac{\partial V}{\partial n_{Ga}}\right)_{T,P,Pu} \tag{1}$$

$\bar{V}_{Pu}$ is the rate of increase in the volume of the solution when an infinitesimal amount (dn_{Pu} moles), of Pu is added. When small amounts dn_{Pu} and dn_{Ga} of Pu and Ga are both added, the change in the volume of the solution, ΔV, is

$$\Delta V = \bar{V}_{Pu}\ dn_{Pu} + \bar{V}_{Ga}\ dn_{Ga} \quad (2)$$

Since the values of $\bar{V}_{Pu}$ and $\bar{V}_{Ga}$ depend only on the composition of the solution, and if the constituents are added simultaneously, in the same proportion as they are present in the alloy, then

$$V = \bar{V}_{Pu}\ N_{Pu} + \bar{V}_{Ga}\ N_{Ga} \quad (3)$$

where N_{Pu} and N_{Ga} are the mole fractions of Pu and Ga in the solution.

The values for $\bar{V}_{Pu}$ and $\bar{V}_{Ga}$ can be obtained graphically by drawing tangents to the curves for each composition and finding the intersections at N_{Pu} = 1, and N_{Ga} = 1. In Fig. 2, because of the linearity, we can simply extrapolate the curves. The values found are $\bar{V}_{Pu}$ = 15.05 cm^3, and $\bar{V}_{Ga}$ = 6.66 cm^3. We would like to compare this $\bar{V}_{Ga}$ with the molar volume of pure gallium in the face-centered-cubic structure but pure gallium does not exist in the face-centered-cubic structure. Thus, we turn to a similar review of the situation for the delta-phase Pu-Al alloys.

Referring to Fig. 2, we can obtain a $\bar{V}_{Al}$ of 8.12 cm^3, compared to V_{Al} of 10 cm^3. Aluminum atoms are thus indicated to have a smaller apparent molar volume in the alloy than would be the case for an ideal solid solution where $\bar{V}_{Al}$ would equal V_{Al}. Because of the related nature of aluminum and gallium (valence + 3, amphoteric, corrosion films, periodic table association), it can be argued that it seems reasonable that gallium might also have smaller $\bar{V}$'s in the delta-phase Pu-Ga alloys. Therefore, if plutonium is alloyed with gallium, it is assumed that the volume of the delta-phase alloy is less than the sum of the individual volumes of the metals before alloying, assuming a hypothetical face-centered-cubic volume for the gallium. Assuming this contraction of the molar volume of the mixture, it is suggested that the attractive force between plutonium and gallium atoms in solution may be greater than the attractive force in either pure metal. This is the usual case for binary metal solid solutions. A better expression of this situation is that the total energy of the solution is lower when the atoms are distributed on lattice sites in such a way as to increase the number of dissimilar atoms that form nearest neighbors.

This conclusion, if valid, would mean that in homogenized delta-phase Pu-Ga alloy, one would expect gallium atoms to be surrounded with plutonium atoms rather than adjacent to other gallium atoms. Obviously, this would not be a random distribution (which would have to allow some adjacent gallium atoms). Thus, it can be tentatively concluded that a short-range, even distribution seems reasonable if only in the sense that the gallium atoms might avoid being closest neighbors.

For our best model, we can imagine gallium atoms occupying various lattice sites of the face-centered-cubic structure and having nearest neighbors of plutonium atoms.

D-1

NEUTRON SCATTERING AT 4.5 GPa AND 20 K*

D. B. McWhan

Bell Laboratories
Murray Hill, New Jersey USA

Although the pressure variable is of fundamental physical importance, the technology to use high pressures in elastic neutron scattering experiments above pressures of 1 GPa only became available about 1970, and the first inelastic neutron scattering experiments in this pressure range were done in 1975. Presently, samples of about 1/4 cc can be studied from 1 atm to 4.5 GPa at temperatures from 20 K to room temperature. The relatively small sample volume is the fundamental limitation to the types of inelastic neutron scattering experiments which can be done with currently available neutron sources. It is possible to study the effect of pressure on phonons and crystal field levels up to energy transfers of the order of 10 MeV, and this is sufficient for a number of important problems.

There are basically two types of apparatus, those using helium gas and those using some form of piston-cylinder device to contain a liquid pressure-transmitting medium. It is necessary to balance absorption of the neutrons against the strength of the walls of the pressure cell, and this leads to a further separation according to the maximum pressure attainable depending on the container material and whether or not the cell is externally supported.

The majority of the inelastic neutron scattering studies have been done in helium gas systems in which the pressure cell is made of maraging steel or a high strength Al alloy (7075-T6). The latter has a low absorption cross section but also low strength. The effect of pressure on the helical turn angle of the antiferromagnetic structures of Tb and Ho were determined up to 6 kbar using a maraging steel cell [1]. A series of studies of the lattice dynamics of rare gas solids have been made over the last decade using aluminum cells [2], and the sample volume in some recent designs exceeds 5 cc [3]. Most of these cells have bursting pressures between

*Invited paper.

0.5 and 1.0 GPa. The pressure is almost hydrostatic even at low temperatures because solid helium is quite soft relative to other solids. However, it is possible to cover the same pressure range with only a small loss in hydrostaticity using a piston-cylinder clamp device thereby avoid all the complications of a helium gas system. In recent measurements of the magnetic form factor of SmS in the low and high pressure phases, it was necessary to have a non-magnetic pressure cell which would fit into a superconducting magnet [4]. The beryllium copper cell shown in Fig. 1 was used successfully to a pressure of 0.7 Gpa. The OD to ID ratio was only two and the observed bursting pressure of the autofretaged cell was ≈ 1 GPa.

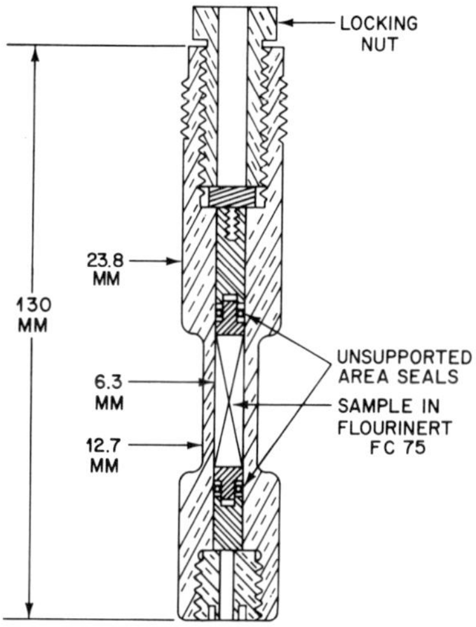

Fig. 1. Beryllium-copper clamp device for inelastic neutron scattering studies up to 0.7 GPa.

The major advance in achieving higher pressures was the use of supported high density Al_2O_3 for the pressure cylinder. In the first experiments a series of slits in the support system provided access to the sample at a few fixed scattering angles [5]. In this geometry, powder patterns were obtained using the time-of-flight method. The most recent example of the use of this method is the phase transition in NiF_2 [6].

In order to do inelastic neutron scattering experiments it is necessary to have access to a scattering plane so as to be able to vary both momentum and energy transfer conveniently. This was achieved at room temperature using a modification of Bridgman's apparatus for measuring compressions up to 4.4 GPa [7]. In our version of this apparatus (which is shown in Fig. 2) an Al_2O_3 cylinder which has a conical exterior surface is compressed radially by forcing it into an external conical ring [8]. The external support and the pressure on the sample were controlled by independent hydraulic rams. Using this apparatus neutron powder patterns were obtained up through the Cs transitions near 4.3 GPa, and the anomalous compressibility of Cs II was studied [9]. Measurements of the soft transverse acoustic phonon mode in TeO_2 were made up to 1.8 GPa using a teflon cell to contain a single crystal and a fluorocarbon [10]. Subsequently it was found that the external support could be clamped, and this eliminated the need for the second hydraulic ram [11].

In order to extend the measurements to low temperatures where a number of interesting phase transitions occur, two independent approaches have been taken. One is the use of a clamp device which

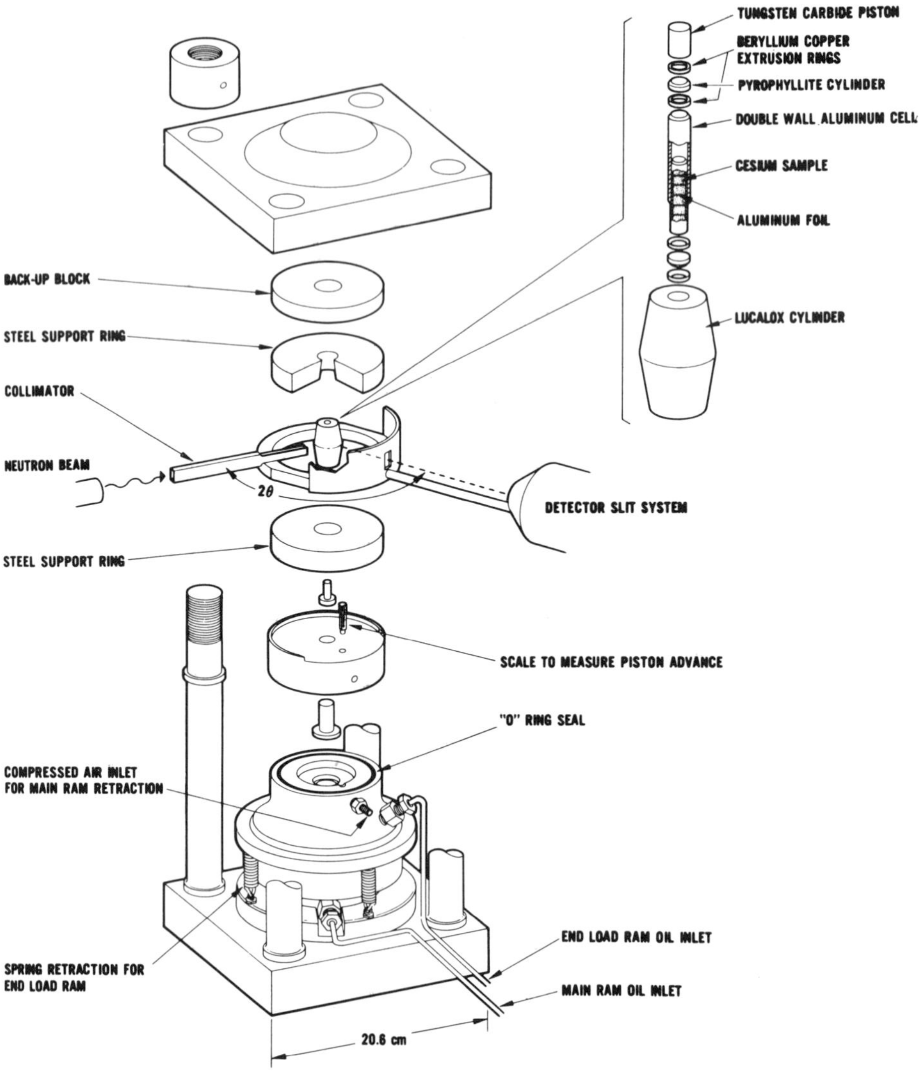

Fig. 2. Apparatus for measurements up to 4.5 GPa at room temperature using externally supported Al_2O_3 cylinders. Inset shows the sample arrangement in experiment on Cs metal [8,9].

is then incorporated into a conventional cryostat [11]. The second is to use a closed-cycle helium refrigerator and to transmit the applied load from the hydraulic ram at room temperature to the pressure cell at low temperature via hollow fiberglass epoxy columns [12]. This apparatus is based on an earlier high-pressure cryostat [13] and is shown in Fig. 3. The vacuum jacket for the cryostat also serves as the frame of the hydraulic press. The

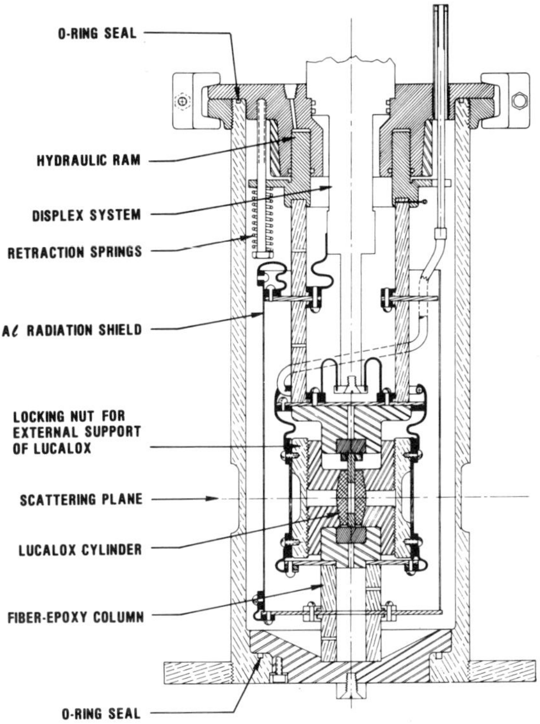

Fig. 3. Low-temperature version of high-pressure neutron scattering apparatus employing a closed cycle helium refrigerator.

closed-cycle helium refrigerator has two cooling stations which are attached through flexible links to outer and inner radiation shields. Temperatures of 20 K have been achieved, but the cool-down time is about 36 hrs. This time can be reduced by at least a factor of two by precooling with liquid nitrogen via the coiled tubing shown in Fig. 3.

To date, Fluorinert FC 75* has been used as the pressure-transmitting medium. However, tests on a number of different crystals indicate that this material becomes extremely viscous above 2 GPa at room temperature leading to an unacceptable increase in the mosaic spread of the crystal being studied. It seems clear that in order to achieve higher pressures a mixture of deuterated methanol-ethanol will have to be used [14]. Both of these fluids turn glassy on cooling and this in principle will produce a nonhydrostatic component to the pressure. However, by cooling at constant load, it was found in studies of the effect of pressure on the crystal field levels in PrSb that the mosaic spread only increased from 0.2 to 0.5° on compressing to 1.6 GPa and then cooling to 28 K[15].

The teflon cell technique also poses problems in studies involving single crystals. As the cell is compressed the walls get thicker, and it is necessary to carefully support the crystal so that it keeps its orientation but also is not subjected to a nonhydrostatic component at the highest pressures. In our latest experiments on the effect of pressure on the phonons in Pb we have put a

*Manufactured by 3M Company, St. Paul, Minnesota.

thin-walled (0.25 mm) hardened beryllium copper liner inside the Lucalox cylinder and used Bridgman unsupported area seals on the tungsten carbide pistons. The liner is necessary because the Lucalox develops cracks at the highest pressures. The diameter of the crystal can then be increased from 4 to 5.3 mm.

The number of major high-flux-reactor facilities which have high-pressure (>1 GPa) apparatus is rapidly expanding. There are alumina cells with fixed windows at Argonne National Laboratory and at Risø and cells with a whole scattering plane accessable at Brookhaven National Laboratory and the Institute Laue Langevin. An apparatus has recently been built at Saclay [16], and a number of other reactor facilities have He gas cells and special cells for studying liquids, and high density gases. Given the large impact that inelastic neutron scattering has already had in the areas of phase transitions and lattice dynamics, this may be one of the more significant areas of high-pressure research during the next decade.

REFERENCES

1. H. Umebayashi, G. Shirane, B. C. Frazer, and W. B. Daniels, Phys. Rev. 165, 688 (1968).
2. J. Eckert, J. D. Axe, and W. B. Daniels, in Proc. Neutron Scattering Conference, Gatlinburg, Tennessee, Conf. 760601-P1 (1976), p. 187.
3. J. Paureau and C. Vettier, Rev. Sci. Instr. 46, 1484 (1975).
4. R. M. Moon, W. C. Koehler, F. Holtzberg, and D. B. McWhan, J. Appl. Phys. 49, 2107 (1978).
5. R. M. Brugger, R. B. Bennion, T. G. Worlton, and W. R. Myers, Trans. Am. Crystallogr. Assoc. 5, 141 (1969).
6. J. D. Jorgenson, T. G. Worlton, and J. C. Jamieson, paper F-1-B presented at 6th AIRAPT Intern. High Pressure Conference, University of Colorado, Boulder, Colorado, July 25-29, 1977.
7. P. W. Bridgman, Proc. Am. Acad. Arts Sci. 72, 45 (1937).
8. D. B. McWhan, D. Bloch and G. Parisot, Rev. Sci. Instr. 45, 643 (1974).
9. D. B. McWhan, G. Parisot and D. Bloch, J. Phys. F 4, L69 (1974).
10. D. B. McWhan, R. J. Birgeneau, W. A. Bonner, H. Taub, and J. D. Axe, J. Phys. C 8, L81 (1975).
11. D. Bloch, J. Paureau, J. Voiron, and G. Parisot, Rev. Sci. Instr. 47, 296 (1976).
12. D. B. McWhan and C. Vettier, to be published.
13. D. N. Lyon, D. B. McWhan, and A. L. Stevens, Rev. Sci. Instr. 38, 1234 (1967).
14. G. J. Piermarini, S. Block, J. D. Barnett, J. Appl. Phys. 44, 5377 (1973).
15. C. Vettier, D. B. McWhan, E. I. Blount, and G. Shirane, Phys. Rev. Letters 39, 1028 (1977).
16. D. Debray, R. Millet, D. Jerome, S. Barisic, L. Giral, and J. M. Fabre, J. de Physique Letters 38, L227 (1977).

D-2

THE SYNTHESIS OF A15-TYPE MATERIALS AT HIGH PRESSURES: COMMENTS AND CRITIQUES

A. W. Webb, T. L. Francavilla, R. A. Meussner

Naval Research Laboratory
Washington, D.C. USA

and

R. M. Waterstrat

National Bureau of Standards
Washington, D.C. USA

INTRODUCTION

Materials with the A15 structure are of interest principally because a number of them are superconductors. Nb_3Ge, which has the highest known superconducting transition temperature T_o, 23.2 K, has this A15 structure [1]. Of 65 known A15 compounds 41 are superconducting. Figure 1 shows the structure of an A15 unit cell. The B atoms, hatched here, form a body-centered cubic network. Woven through this network are three mutually orthogonal chains of A atoms, shown as stippled here. There is evidence that the integrity of these A chains is important to superconductivity [2]. Figure 2 shows the location of the A and B atoms in the Periodic Table. The A atoms (stippled) are from the Ta, V, and Cr groups, with V and Nb forming 30 of these A15 compounds. Technetium has been reported acting as the A element in one compound [3]. The B atoms (hatched) are from the groups of transition metals from Mn through Zn, and the groups of non-transition metals Al, Si and P. Beryllium is included because Mo_3Be has been reported to form with the A15 structure [4]. Nb_3Te has been formed by high pressure, high temperature techniques [5].

These A15 materials have customarily been formed by arc-melting, or in more recent years, by various thin film techniques. Why, then, turn to high pressure-high temperature techniques? One

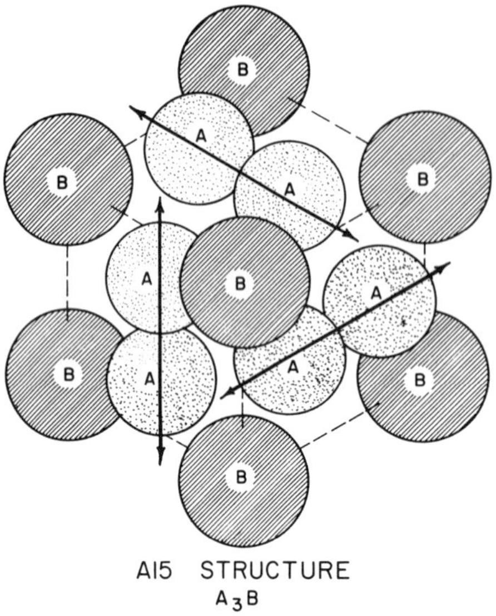

Fig. 1. The A15 structure of an A_3B compound. The hatched B atoms form a bcc array and the stippled A atoms form three mutually orthogonal chains.

reason is containment—the wide disparity in melting points makes it difficult to retain a known quantity of the more volatile constituent. For example, Nb melts at 2468° C while Ge melts at 937° C, Al at 660° C and Ga at only 29.8° C. Of these, Al and Ga both boil before Nb melts! High pressures can contain such systems during high temperature reactions. A second reason is that the desired material may not reach the ideal stoichiometry (such is the case for bulk A15 Nb–Ge) High pressures can modify the free energies to shift phase boundaries to more favorable positions through effects on the volumes and entropies of the constituents. The fact that the constituent elements have different compressibilities causes variation in the ratio of radii as pressure is applied, and thus may introduce new phase fields into the phase diagram. Due to the vast amount of data required, few composition-temperature-pressure systems have been thoroughly studied; none of the systems of interest have had much study.

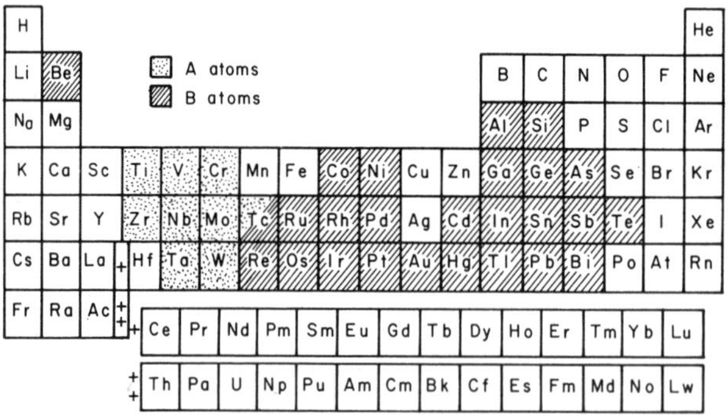

Fig. 2. Periodic Table showing component elements forming A_3B compounds with the A15 structure.

At the initiation of this synthesis program it seemed reasonable to study a known high T_o material to minimize the uncertainties of prediction just noted. A15 Nb–Ge was chosen because of its high T_o, and also because the product obtained in bulk formation is usually Ge deficient—the A15 structure occurs from 17 to 20 at.% Ge at 1800° C [6]. Thus the goal was to increase the Ge content towards the stoichiometric value, as well as to evaluate the procedures used and check for contamination problems, and to begin to delineate the effects of pressure, time, temperature and composition on this A15 system. As the program progressed several systems were added. Figure 3 depicts eight non-transition elements in their normal Periodic Table order. They all form A_3B compounds with Nb as the A element. T_o is seen to increase both in going up the columns and in going to the right. How far does this trend continue? Dew-Hughes, using Gold's relationship of T_o varying inversely with the square root of the mass of the B element, recently predicted a T_o of 30 K for Nb_3Si, if it could be obtained with the A15 structure [7]. The maximum T_o in an A15 material is approached as the compound approaches stoichiometry [8]. Niobium-antimony forms the A15 structure with a T_o of only 2.2 K, but the lattice parameter indicates that it is off stoichiometry [9]. The other extreme members to the upper right don't form in the A15 structure, but rather in the tetragonal Ti_3P-type structure [10]. Thus the systems Nb-Al, Nb-Si and Nb-As were investigated.

Nb_3X

Al *18.8 K* A15	Si *0.29 K* Ti_3P	
Ga *20 K* A15	Ge *23 K* A15	As *0.31 K* Ti_3P
In *9 K* A15	Sn *18 K* A15	Sb *2.2 K* A15

Fig. 3. Nontransition metals in Periodic Table order which form A_3B compounds with A = Nb and showing the observed structure and T_o values.

EXPERIMENTAL

The starting materials used in these studies were from two sources—arc-melted samples and mixtures of the powdered elements. The arc-melted Nb-Ge was about 26 at.% Ge, the eutectic composition. The x-ray diffraction pattern showed both A15 and Nb_5Ge_3 phases to be present, by comparison with the relevant Powder Diffraction File cards 10-296 and 8-354, respectively [11]. The Nb-As and Nb-Si were stoichiometric (A_3B). The Nb_3As was formed by induction melting a stoichiometric mixture of the elemental powders under high pressure argon [12]. For the powder mixes, the elemental powders had an initial purity of at least 99.9%. Powder mixes were used in part of the Nb-Ge study, and exclusively for the Nb-Al work.

The studies were performed using a tetrahedral anvil press which achieved pressures as high as 7 GPa [13]. The sample pressure was estimated by means of a calibration obtained by noting the oil pressures at which the resistive transitions in Ce, Tl, Yb, Ba, and Bi occurred at room temperature [14]. No correction was made for the effect of temperature on the sample pressure. Temperatures to 2600° C were obtained from ac resistance heating of an internal tube furnace. Temperatures were routinely estimated from a calibration of input power against the temperature measured in several runs that had a thermocouple in place of the sample. A check of thermocouple corrections showed the error due to the effect of pressure on the emf to be of the order of +20° C at 1500° C [15]. The uncertainty in temperature is about ± 3% at 2500° C. The anvils were water cooled to minimize heating of the press for those runs which lasted longer than five minutes.

Figure 4 shows the synthesis cell. The container was a tetrahedron machined from pyrophyllite, a natural hydrous aluminum silicate. The parts are shown to scale, with the edge of the tetrahedron being 2.5 cm. The surface of the assembled tetrahedron was painted with an iron oxide/acetone slurry to increase the surface friction and encourage formation of thicker gaskets. The power contact tabs were 0.75 mm thick molybdenum sheet. The graphite end caps and tube formed the internal resistance furnace, which was electrically insulated from the sample volume by the boron nitride caps and tube. The final protection for the sample was provided by a metal tube and caps in which the sample fragments were packed. Thus far the studies have used niobium, molybdenum and tantalum tubes. This system provided a sample volume of 0.02 cc, or about 100 mg of product. This proved sufficient for a series of studies on a given sample, for example, ac susceptibility measurements, Debye-Scherrer x-ray powder diffraction patterns, and metallography. The ac susceptibility measurements were made in a vacuum cryostat which was normally operated over the range of 4.2 to 20 K for determination of the T_o of each sample with the temperatures determined by a calibrated germanium thermometer. For Nb-Si and Nb-As, however, temperatures below 1 K were produced by magnetic cooling techniques employing chromium potassium alum (CPA) as the working substance. Temperatures were determined from the magnetic susceptibility of CPA (calibrated against ^{4}He vapor pressure); thermal contact to the samples was accomplished by embedding the samples in a bundle of copper wires which was attached to the CPA.

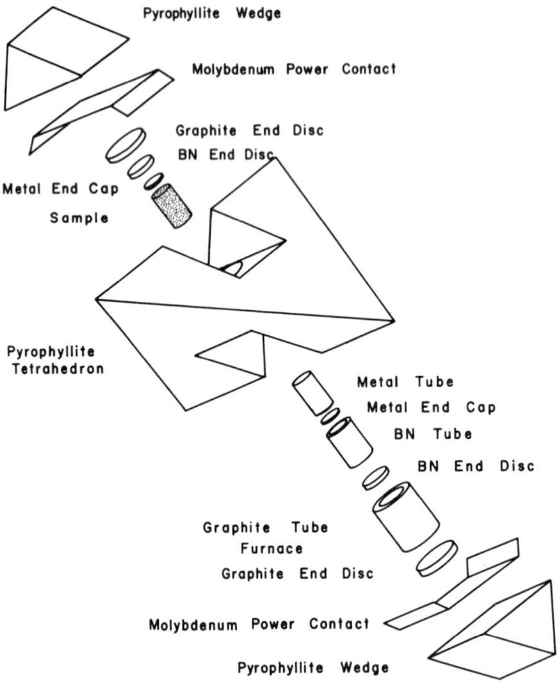

Fig. 4. Assembly diagram of the synthesis cell.

RESULTS

Nb-Ge

The Nb-Ge system has received the greatest effort to date. Figure 5 shows the effect of pressure on T_o of the Nb-Ge arc-melted material for four reaction temperatures: 1200, 1500, 1830 and 2200° C. At all four temperatures T_o increased with increasing pressure to 6 GPa, followed by a sharp decline when pressure was further increased to 7 GPa. This latter effect was so severe as to produce no net change in T_o after treatment at 7 GPa and 1200° C, and to produce an enhancement equivalent to that obtained at 3 GPa after treatment at 7 GPa and 1500° C.

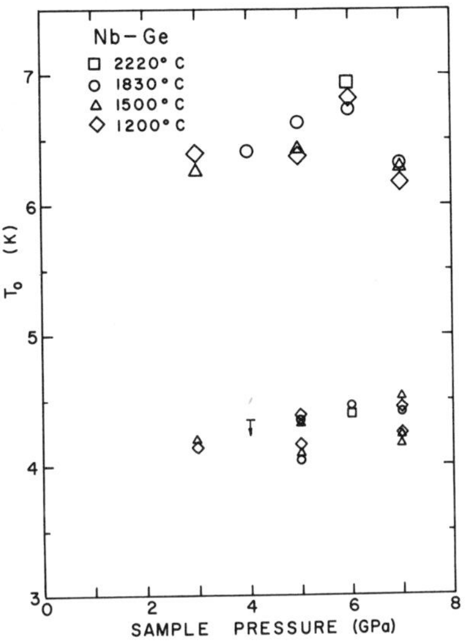

Fig. 5. Effect of pressure on T_o for arc-melted Nb-Ge.

The lower set of points, the small ones, indicate the T_o of the tantalum container after these runs (it wasn't measured for the 4 GPa run). The Ta was monitored as a check on its possible reaction with the sample. Multiple transitions became prevalent for Ta above 4 GPa, indicative of reaction with either the sample or with impurities, such as adsorbed oxygen or the boron nitride. Also, it is noted that there is a general increase in T_o with increasing pressure. Either this indicates a lessening reaction (contrary to the implication of the multiple transformations) or it may be due to increased alloying of Nb with Ta. If the latter is assumed to be the only cause, the amount of increase in T_o of the Ta container would be indicative of an alloy with approximately 20 at.% Nb at 7 GPa. Most likely, however, other reactions also occur, which are also pressure dependent.

Figure 6 summarizes the effect of reaction temperature and container material on T_o at 6 GPa. The T_o for Nb-Ge slowly increased from the as-cast 6.17 K to a maximum of 8.4 K around 1850° C. Niobium containers were found to react with the boron nitride above 1700° C to form NbN, which gave a T_o around 14 K [16]. Metallography indicated that the nitride was dispersed throughout the sample. Molybdenum containers produced the same trend in T_o of Nb-Ge as found with Nb, including the drop-off above 1700° C. However, there was no evidence for either MoN (T_o = 12 K) or Mo_2N (T_o = 5 K) [16]. Tantalum containers were found to react at temperatures above 2300° C to produce a mixed tantalum-niobium nitride with a T_o around 11 to 12 K [16], and complete suppression of the Ta and Nb-Ge transitions. Unlike the results obtained with the niobium and molybdenum containers, however, the T_o of the Nb-Ge did not drop off above

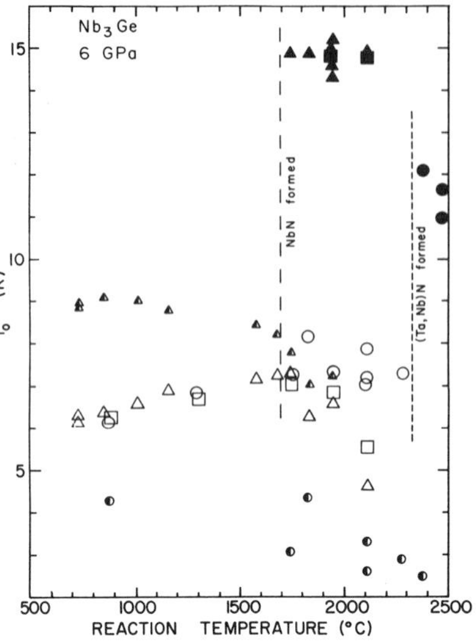

Fig. 6. Effect of reaction temperature and container material (Nb, Mo and Ta) on T_o of arc-melted Nb_3Ge at GPa. NbN was formed above 1700° C, and (Ta,Nb)N above 2300° C.

1700° C with the tantalum containers; rather, it continued to increase.

The niobium and tantalum containers will be considered first. The molybdenum was not monitored because its T_o is only 0.9 K [16], too low for us to routinely observe. The superconducting characteristics of Nb were found to degrade slowly with increasing reaction temperature up to the nitride reaction temperature, above which degradation increased rapidly. The T_o of Ta also degraded slowly with increasing reaction temperature up to its nitride reaction temperature above which it usually reacted totally to form the nitride. Thus, increasing temperature suppressed the T_o of Ta, while increasing pressure (at a fixed temperature) led to enhancement of T_o.

The effect of cooling rates for samples treated at 1830° C and 6 GPa for three minutes was also examined. No effect on T_o was found when the rate was varied from 500°/sec to 2°/sec for samples cooled to 600° C before "quenching." This is contrary to Blaugher's hypothesis which would require an increase of 2 to 3 K for that interval [17]. It also indicates that the shifts in T_o observed in this system were not just due to quenching in the more stoichiometric composition, which is stable at elevated temperatures.

Figure 7 shows the effect of varied reaction time for three temperatures — 900, 1830 and 2200° C, at 6 GPa. Reaction times varied from 18 sec to 5 hrs. For a reaction temperature of 900° C, T_o of the product was still increasing for treatment times of 4 hrs, though at a decreasing rate. At 1830° C, T_o peaked for the product obtained from treatment for 30 min, then dropped off for the 5 hr run. At 2220° C, T_o peaked for the 3 min treatment and was less for a treatment time of 30 min.

Figure 8 gives the results of runs of varying initial composition of the elemental powders. Below 22.5 at.% Ge a double transition was observed. The upper transition (filled symbols) diminished in size as the Ge content increased. This diminishing of the upper transition indicates that the upper transition is due to excess Nb. The lower T_o value (open symbols) would then be due to the Nb-Ge. This interpretation would further indicate that this superconducting

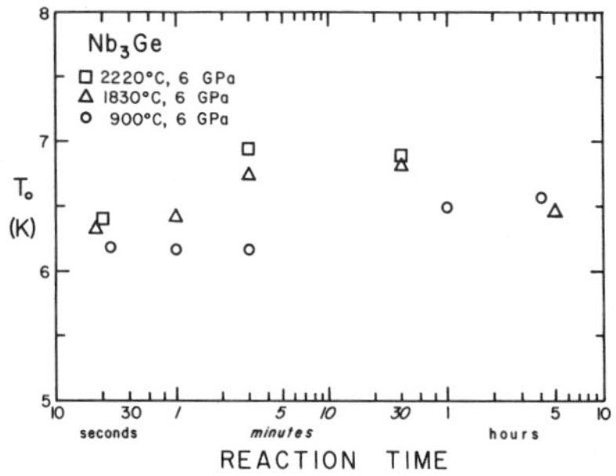

Fig. 7. Effect of reaction time at 6 GPa and 900, 1830 and 2220° C on T_o of the arc-melted Nb_3Ge.

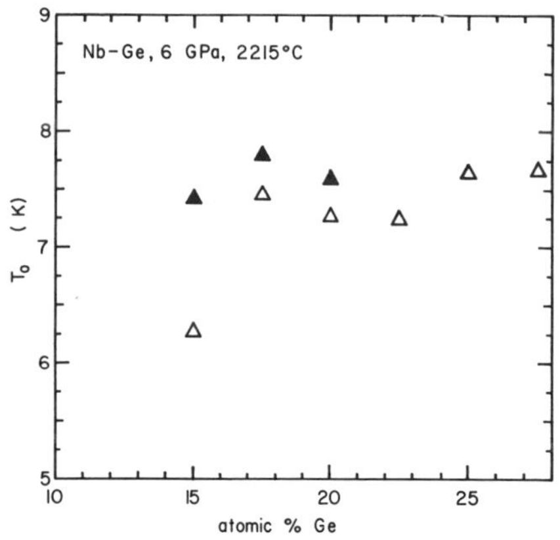

Fig. 8. Effect of composition on T_o of Nb-Ge derived from mixtures of the elemental powders. The filled symbols indicate the transition which vanished with increasing germanium content.

constituent is stable above 20 at.% Ge at 6 GPa, instead of the 17 to 20 at.% observed for the A15 phase at ambient pressure. Regardless of the interpretation used, however, it appears that an excess of Ge yields a higher T_o product, and that a two-phase field exists below 20 at.% Ge.

There have been three other high pressure studies of the Nb-Ge system. The first of these was made by Leger and Hall [18]; it suffers the major defect that no T_o measurements were made, only an x-ray powder study was made. As was the case for part of this study, their starting material was a powder mixture of the pure elements which remained incompletely reacted in all of their experiments. Our problems with nitride contamination call into question their results, since they packed the reactants directly in the boron nitride [19]. Blaugher recently studied this system among others; he also noted problems with boron nitride contamination [17]. Vereschagin et al. recently claimed a 20 K T_o for A15 Nb-Ge reacted with extra Ge at around 2000° C and 7 GPa [18]. To the extent the published details have allowed, efforts to reproduce their results have consistently failed. Blaugher's product exhibited a T_o of nearly 13 K, higher than we obtained. His starting material was pure A15 Nb-Ge [17 at.% Ge) with extra Ge added, whereas our arc-melted material had about 26% of the Nb_5Ge_3 phase present, which seems to have been the only difference between these two studies which can account for the nearly 4 K variation in T_o of the products. This difference may indicate that the A15 phase is metastable at high pressures when it has a Ge composition approaching the stoichiometric value. This would account for it not being obtained when, as in our work, the neighboring phase Nb_5Ge_3 was present to assure equilibrium.

X-ray studies are in progress. So far only the compositional variation series has been completed. This series started with the mixed elemental powders, but there were two runs made with the arc-melted material. The as-cast arc-melted material was clearly two phase. Generally, save for the 25 at.% Ge run, there was a line around 3.4 Å, which may have been due to boron nitride although care was taken to cleanse the samples of this impurity. In most cases there were also other unidentified lines. As to constituents, Nb was present in all runs derived from powder reactants in varying degrees, and many of these runs also showed the characteristic pattern of Nb_5Ge_3. Surprisingly, these latter runs also showed lines characteristic of $NbGe_2$, which could only be present if the system were far from equilibrium when "quenched." Blaugher also noted this result in some runs [17]. Since a number of these lines have dual identity, and since the major line is the major line for all three of the above-mentioned phases, these tentative identifications require further study. The A15 phase was conspicuous by its absence, which raises the question as to what produced the two T_o values. One was most certainly due to impure Nb, but the second, if not A15, presents an enigma.

The product obtained after treatment of the arc-melted material at 7 GPa and 1200° C gave A15 with a lattice parameter a_o of 5.162 Å. The same material reacted with extra Ge at 6 GPa and 2200° C gave an a_o of 5.177 Å. This shrinkage of the A15 cell noted in the first case is encouraging as stoichiometric A15 Nb_3Ge has a lattice parameter of 5.13 Å calculated from the Geller radii [21]. The latter product also showed Nb lines, the only case encountered thus far of the arc-melted reactant decomposing.

Nb-Al

Nb_3Al is reported with a T_o as high as 18.8 K [22]. Like Nb_3Ge it also forms peritecticly, but occurs with the upper limit of its composition stability range at stoichiometry [23]. Figure 9 gives the effect of reaction temperature on T_o at 6 GPa. Multiple transitions were the rule. Evidently no reaction occurred below 1800° C, and thereafter, save for the one point, the samples had a T_o of 10 to 11 K. Note also that the Ta containers were not affected by any of these treatments. The one high point may have been due to formation of NbN, but this is doubtful.

Figure 10 shows the effect of varying the starting composition with treatment at 6 GPa and 1500, 1830 and 2220° C. Save for the one run at 1830° C and 25 at.%, it is seen that the higher reaction temperature, 2200° C, favored higher transition temperatures— definitely above the value expected for Nb. Again, multiple transitions were the rule, indicative of more than one product. The T_o of the Ta containers showed no effect from either the different temperature treatments or from the variation in starting composition.

The x-ray studies always found strong lines for Nb present (or Ta—they are indistinguishable since their lattice parameters differ by only 0.001 Å). Aluminum was present in several runs, with its lattice parameter generally larger after treatment at pressure and

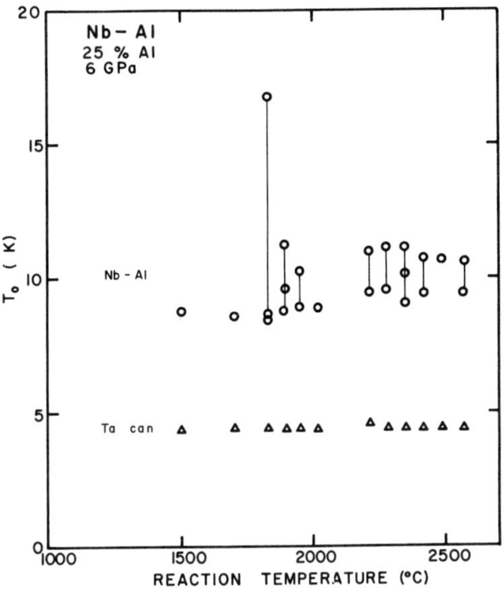

Fig. 9. Effect of reaction temperature on T_c of Nb-Al and of the Ta container at 6 GPa. The tie lines connect the observed temperatures of the product, with the low value generally due to Nb.

temperature. A15 material was present after treatment at 6 GPa and 1830° C with an a_o of 5.187 Å compared to the calculated value of 5.19 Å [21]. $NbAl_3$ was present in two runs, showing a lack of equilibrium similar to that observed in the Nb-Ge work above. Unidentified lines appeared in the patterns of all but one run.

Since only one run gave evidence of A15 material in the x-ray study, the uniformity of T_c values either indicates low A15 content (less than ~ 5%) in all the other products, or enhanced T_c in the Nb solid solution. However, a T_c above 10 K is quite improbable for niobium. Blaugher examined this system starting with A15 Nb-Al, and found T_c was degraded in all cases, and that the product wasn't noticeably affected by container material, or even by contact with boron nitride [17].

Nb-Si

Nb_3Si was included in this study because, although it has the Ti_3P structure as noted above, it was predicted to have a T_c near 30 K if it could be obtained in the A15 phase [7]. Our Ti_3P-type starting material had a T_c of 0.29 K [12]. This material was chemically homogeneous and had the A_3B composition in contrast to the previous studies by Leger and Hall [19] who produced incompletely reacted mixtures of the pure elements. All of our studies in this system were conducted at 6 GPa. For samples treated at 1500 to 1830° C we recovered a body-centered cubic phase with a lattice parameter somewhat larger than pure Nb. There was also a second phase present in the 1830° C products. Only the reactant, The Ti_3P structure, was obtained in the interval 900° to 1300° C. Superconductivity was not observed above 0.01 K in these products. The apparent stabilization of the body-centered cubic structure characteristic of niobium led us to suspect oxygen contamination. This was checked by baking out sample components under vacuum, then assembling samples in an argon dry box. When reacted at 1830° C—well within the body-centered cubic region—neither the bcc nor the second phase was obtained. This supported the contention that these two phases had been the result of contamination by adsorbed O,

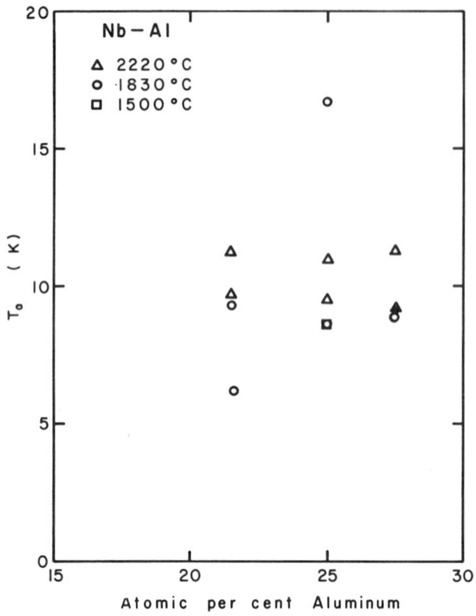

Fig. 10. Effect of composition on T_o of Nb-Al reacted at 6 GPa and 1500, 1830 and 2220° C.

rather than the increased solubility of the Si in the body-centered cubic niobium lattice.

These results would indicate that the Ti_3P phase of Nb-Si is stable to 6 GPa and 1830° C. Recently the results of a shock synthesis of Nb_3Si were published, which found an onset T_o of 18.5 to 19 K. An x-ray powder study of the product was interpreted to give evidence for the presence of a small amount of A15 material with the bulk of the product not identified. Their estimate of the shock pressure was 100 GPa [24].

Recent compressibility results by Skelton et al. on our Nb_3Si starting material and on stoichiometric Nb_3Ge found them roughly equal [25]. Using the Geller A15 radii [21] we calculated the lattice parameter expected for A15 Nb_3Si and found that the A15 structure would be more dense than the Ti_3P structure. If A15 Nb_3Si were to have the same compressibility as A15-type Nb_3Ge (this is reasonable since it has recently been shown that A15-type Nb_3Ir has a similar compressibility [25]), then we would expect the A15 phase to be favored by pressure. However, the results show the Ti_3P, rather than the A15, structure to be stable at elevated temperatures and pressures. This unfavorable situation might be overcome by the addition of an appropriate minor impurity which would favor the A15 phase without degrading T_o too severely, or by the use of higher pressures.

Nb-As

Figure 3 shows Nb_3As as the other non-A15 compound which might be expected to have a reasonable T_o if the A15 structure were to be obtained. We again started with homogeneous, stoichiometric Ti_3P-type material which had a T_o of 0.30 K [12]. At 6 GPa the original structure was obtained everywhere below 1900° C. A second phase was noted in the x-ray pattern of the 2150° C product. The product of this last treatment was not superconducting above 0.03 K, and the structure has not yet been identified.

CONCLUSIONS

The attempted synthesis of an A15 phase in the Nb-Si and Nb-As systems noted that the Ti_3P phase was stable up to pressures of 6

GPa and temperatures of 1800 and 2200° C, respectively. In the Nb-Si system, oxygen contamination was found to favor the bcc Nb solid-solution structure over the initial Ti_3P structure, hence, contamination by oxygen should be avoided in this system.

The Nb-Al and Nb-Ge compounds are similar in that both are peritectic at ambient pressure. With the Nb-Al, however, thus far the investigation has found the pressure effect to be slight to negative on T_o. There was no effect of reaction temperature on T_o of the product from the minimum of 1800° C to nearly 2600° C at pressures of 6 GPa. The Nb-Ge, on the other hand, was affected. The T_o was most improved by reaction at temperatures near 2000° C at pressures of 6 GPa. Longer reaction times were beneficial with the understanding that the optimum reaction time decreases with increasing temperature. T_o of the product was also improved by the addition of extra Ge to the initial charge.

There is a need for further study to improve our understanding of the crystal chemistry of the A15 structure. The fact that pure A15 Nb-Ge mixed with Ge (as used in two other studies) produced only an A15 product with T_o values of 13 to 19 K needs to be corroborated. It is indicative that stoichiometric A15 Nb_3Ge is metastable even at elevated pressures, and the observed effect on T_o may be due only to quench rate in this instance. The Nb-Al system also requires further work. In both of these systems the effect of the starting material on the product has yet to be fully determined. In summary pressure effects appear to depend upon the system under study, and the product obtained to be dependent upon the history and homogeneity of the starting material. The Nb-Si system was much affected by low-level oxygen contamination. Finally, the question of equilibrium at pressure and temperature was found to cast doubts on the observed results, and, related to this, there was no determination that the phases obtained at high pressures were identical with those recovered at ambient conditions.

ACKNOWLEDGMENTS

The authors thank E. Skelton for making their compressibility data available before publication, E. R. Carpenter, Jr. and L. D. Jones for sample parts and upgrading the tetrahedral press for this study, and J. O. Willis and D. U. Gubser for the T_o measurements with the adiabatic demagnetization cryostat.

REFERENCES

1. L. R. Testardi, J. H. Wernick, and W. A. Royer, Solid State Commun. 15, 49 (1974).
2. R. D. Blaugher, R. E. Hein, J. E. Cox, and R. M. Waterstrat, J. Low Temp. Phys. 1, 539 (1969).
3. M. V. Nevitt, in Intermetallic Compounds, John Wiley and Sons, New York (1967).

4. R. M. Paine and J. M. Carrabine, Acta Crystallogr. 13, 680 (1960).
5. J. F. Cannon, D. L. Robertson, H. T. Hall, and A. C. Lawson, J. Phys. Chem. Solids 35, 1181 (1974).
6. R. Flukiger and J.-L. Jorda, paper presented at Workshop on Applications of Phase Diagrams in Metallurgy and Ceramics, National Bureau of Standards, Gaithersburg. Published in NBS Special Publ. 496, 375 (1978).
7. D. Dew-Hughes and V. G. Rivlin, Nature 250, 723 (1974); also L. Gold, Phys. Stat. Sol. 4, 261 (1964).
8. B. N. Das, J. E. Cox, R. W. Huber, and R. A. Meussner, Metallurg. Trans. 8A, 541 (1977).
9. L. Kammerdiner and H. L. Luo, J. Appl. Phys. 45, 4590 (1974).
10. R. M. Waterstrat, K. Yvon, H. D. Flack, and E. Parthe, Acta Crystallogr. B31, 2765 (1975).
11. Powder Diffraction File (Joint Committee on Powder Diffraction Standards, Swarthmore, Pennsylvania (1976).
12. D. U. Gubser, R. A. Hein, R. M. Waterstrat, and A. Junod, Phys. Rev. B 14, 3856 (1976).
13. H. T. Hall, Rev. Sci. Instrum. 29, 267 (1958); also 33, 1278 (1962).
14. P. W. Bridgman, Proc. A. Acad. Arts Sci. 62, 211 (1927); also R. N. Jeffrey, J. D. Barnett, H. B. Vanfleet, and H. T. Hall, J. Appl. Phys. 37, 3172 (1966); D. L. Decker, W. A. Bassett, L. Merrill, H. T. Hall, and J. D. Barnett, J. Phys. Chem. Reference Data 1, 773 (1972).
15. I. C. Getting and G. C. Kennedy, J. Appl. Phys. 41, 4552 (1970).
16. B. W. Roberts, J. Phys. Chem. Reference Data 5, 581 (1976).
17. R. D. Blaugher, IEEE Trans. Mag. MAG-13, 821 (1977).
18. J.-M. Leger and H. T. Hall, J. Less-Common Metals 34, 17 (1974).
19. J.-M. Leger and H. T. Hall, J. Less-Common Metals 32, 181 (1973).
20. L. F. Vereshchagin, E. M. Savitskii, V. V. Evdokimova, V. I. Novokshenov, and V. G. Petrenko, Pis'ma Zh. Eksp. Teor. Fiz. 24, 218 (1976).
21. G. R. Johnson and D. H. Douglass, J. Low Temp. Phys. 14, 565 (1974).
22. R. H. Willens, T. H. Geballe, A. C. Gossard, J. P. Maita, A. Menth, G. W. Hull, Jr. and R. R. Soden, Solid State Commun. 7, 837 (1969).
23. C. E. Lundin and A. S. Yamamoto, Trans. AIME 226, 863 (1966).
24. V. M. Pan, V. P. Alekseevskii, A. G. Popov, Yu. I. Beletskii, L. M. Yupko, and V. V. Yarosh, JETP Lett. 21, 228 (1975).
25. E. F. Skelton, D. U. Gubser, S. C. Yu, I. L. Spain, R. M. Waterstrat, to be published.

D-3

SUPERCONDUCTIVITY AT HIGH PRESSURE*

T. F. Smith

Monash University
Clayton, Victoria, Australia

INTRODUCTION

The first investigation of the effect of pressure upon the superconducting transition temperature, T_c, was made at Leiden by Kamerlingh, Onnes and co-workers [1] some 14 years after their discovery of superconductivity in 1911. In these initial measurements the pressure was applied with He gas which limited the maximum pressure to $\sim$ 200 bars. Nevertheless, the downward displacement in temperature of the transition curves for Sn and In was clearly observed.

Many ingenious techniques and devices have been developed specifically for the study of superconductivity under pressure, and the present day state-of-the-art enables measurements to be made up to $\sim$ 200 kbar and down to $\sim$ 10 mK. Several excellent reviews [2-4] of these techniques have been published and rather than attempting to make a general survey, this paper will concentrate on the details of the specific technique which has been developed at La Jolla and Monash for studies at pressures up to 35 kbar and temperatures down to 0.35 K.

EXPERIMENTAL DETAILS

The 'Clamp' Technique

One of the most significant steps in the study of superconductivity at high pressure came with the introduction of the clamp technique [5] whereby pressure, generated at room temperature with a standard laboratory press, is retained by means of a locking screw thus permitting the relatively compact pressure cell to be removed from the press and transferred to the low-temperature facility. Elegant in concept and simple in practice, this procedure

*Invited paper.

avoids the complications associated with generating the pressure at low temperatures and the attendant problem of achieving hydrostatic conditions. The basic idea is very flexible in application and has been adapted for use with piston and cylinders or opposed anvils.

The particular arrangement which is used at Monash is shown in Fig. 1. As the superconducting state is sensitive to a magnetic field it is important to avoid the use of any strongly magnetic materials in the construction of the pressure cell and clamp body. With the exception of the pistons, Be-Cu alloy full-hardened to Rochwell 40-42C is used throughout. The pistons are cut from 1/4 in.-diameter tungsten carbide rod (Carboloy 883). Care is taken to demagnetize the pistons before each series of measurements to minimize their influence on the superconducting transition.

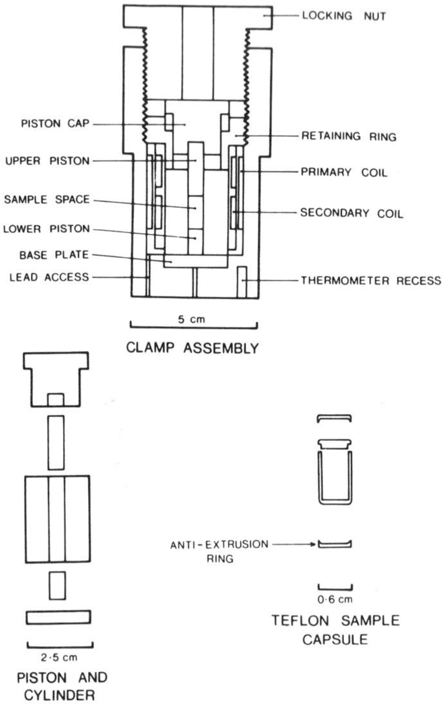

Fig. 1. Details of clamp assembly with piston and cylinder arrangement.

The pressure is generated in a simple piston and cylinder combination. During fabrication, the bore of the cylinder is initially made 0.02 in. undersize. The working volume is then prestressed to ~ 30 kbar with lead as a pressure transmitter. This stretches the bore by 0.010 to 0.015 in. which is then reamed out to the working diameter of 0.250 in. Cylinders prepared in this way may be repeatedly used for runs up to 20 kbar without causing serious deformation of the bore. Pressures up to 25 kbar may be obtained, but only at the expense of a significant reduction in the working life of the cylinder. Pressures up to ~ 35 kbar have been obtained with 3/16 in.-diameter bore tungsten carbide cylinders which are supported by a Be-Cu ring. However, the application of such cylinders has been limited to temperatures above ~ 6 K due to the interference from the magnetic behavior of the tungsten carbide.

Samples are typically 10^{-2}cm^3 in volume and several may be studied simultaneously provided their individual transition curves are sufficiently resolved in temperature. In early investigations of the pressure dependence of T_c, a pressure transmission medium of micron-size Teflon particles was used successfully. However, it was found to be quite unsuitable for measurements on Re [6] a hexagonal material with a superconducting transition which is particularly sensitive to sample condition. This is illustrated in Fig. 2 where the transition curve for the 'as received' material is compared with that for an argon arc cast sample. This figure also shows the relatively large irreversible shift and broadening in the zero-pressure transition curve which occurred after applying a pressure of 18.5 kbar with the solid medium. It was concluded that this degradation of the transition curve was due to anisotropic stresses introduced by departures from hydrostatic conditions. To overcome this problem a self-sealing Teflon capsule, similar to that described by Jayaraman et al. [7], was used to contain the samples in a fluid pressure medium of a 1:1 mixture of n-pentane and isoamyl alcohol. With this arrangement, the zero-pressure transition curve was found to be reproducible to within a millidegree after the application of a pressure of 20 kbar. Subsequent experience has indicated that the departure from pressure homogeneity is of order ± 0.1 kbar [8,9] provided the fluid medium does not freeze during the pressure application.

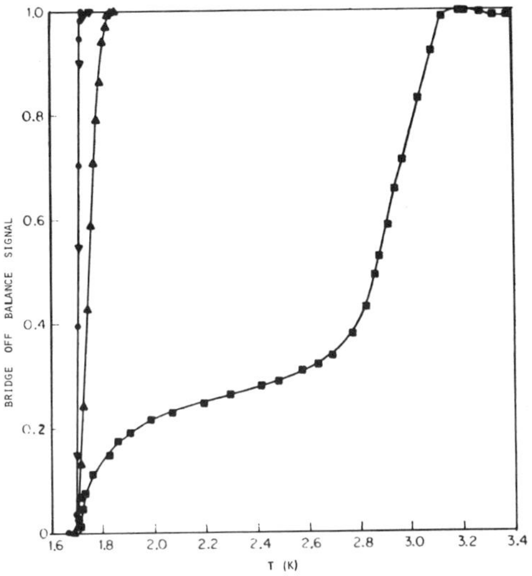

Fig. 2. Effect of sample condition on the superconducting transition curve for polycrystalline Re. ■, as received; ● after arc melting; ▲ after application of 18.5 kbar (Teflon pressure medium); ▼ after application of 17.3 kbar (fluid pressure medium) [6].

The Teflon capsule is positioned in the high-pressure cylinder between two Be-Cu anti-extrusion discs. The pistons are installed and the assembly is lowered into the detection coils, which position it centrally within the clamp body. The retaining ring which prevents the upward movement of the cylinder when the pressure is released is screwed into place. The endcap is placed upon the driving piston and the locking screw is hand tightened. The assembled clamp is supported at the shoulder in a jig and installed in a small laboratory press (10-ton capacity). A removable driving piston passing through the locking screw transmits the press force to the internal piston. When the required load has been applied, the locking screw is tightened down with a wrench and the press is released. The clamp is then removed from the press and installed in the cryostat.

Since differential thermal contraction within the clamp and pressure cell and the freezing of the fluid medium lead to a pressure loss of typically ~ 2 kbar upon cooling, the pressure is determined directly at the low temperature. This is most conveniently done using a superconducting manometer [10], which is included with the samples.

For measurements above 4.2 K a close-fitting slip-on heater is placed around the main body of the clamp, thermal contact being made with a layer of silicon grease. The clamp is then bolted to a support and sealed in a can which is evacuated and then filled with helium exchange gas. The assembly is pre-cooled in liquid nitrogen before being installed in a cryostat consisting of a pair of open-ended glass dewars. After the transfer of liquid helium and the cooling to 4.2 K, the temperature of the clamp is varied by adjusting the heater current and the exchange gas pressure.

The above procedure, which is more complicated and time-consuming than an earlier one [11] in which the clamp was simply suspended over the liquid helium, was adopted when it was found that temperature differences on the order of 0.1 K could not be eliminated between the sample space and the germanium resistance thermometer mounted in the wall of the clamp. Comparisons of zero-pressure transition temperatures measured in the pressure cell with those determined with the sample in direct thermal contact with a germanium thermometer calibrated at the National Measurement Laboratory, Sydney, indicate that with the present arrangement the temperature differential between the thermometer and sample space is not greater than 20 mK above 4.2 K.

For measurements between 4.2 and 1.5 K the clamp is immersed directly into the liquid helium bath whose temperature is varied by pumping. Measurements between 1.5 and 0.35 K are made in a He^3 cryostat with the clamp attached to the base of the He^3 bath.

The ease of magnetically detecting the superconducting transition greatly simplifies such studies at high pressure since it obviates the need for introducing leads into the high-pressure space. The transition to the superconducting state may be readily observed by situating the samples in one of a balanced pair of opposed secondaries of a mutual inductance and using an ac bridge technique to monitor the change in inductance due to the onset of superconductivity. The off-balance signal is amplified by a lock-in amplifier and fed to the Y-axis of a pen recorder. By driving the X-axis with a suitably offset and amplified voltage across the thermometer, transition curves may be traced out continuously.

STUDIES OF T_c AS A FUNCTION OF PRESSURE

As the superconducting transition in a zero magnetic field is the classical example of a second-order transition, the initial studies of superconductivity at high pressure were largely of a thermodynamic nature with dT_c/dP (and the related dH_c/dP where H_c

is the critical magnetic field required to destroy the superconducting state) being treated simply as thermodynamic variables. The development of the theoretical understanding of the superconducting state and the increasing interest in its dependence upon the normal state properties has directed the emphasis more towards regarding pressure (or more significantly, volume) as a physical parameter. Measurements of T_c as a function of pressure have now extended beyond thermodynamics to become a significant aspect of a wide variety of studies related to the superconducting state encompassing basic theoretical models, phase changes, lattice dynamics, structural stability, electronic transitions, magnetic interactions, and Fermi surface topology.

It would not be appropriate or feasible to attempt to review here all the multi-natured aspects of the studies of T_c as a function of pressure; each of them would require an entire review to do them justice. Instead, the present discussion will be restricted to a brief account of some relatively general examples of the applications of T_c measurements as a function of pressure.

Thermodynamic Applications

Traditionally, determinations of the pressure dependence of T_c and H_c have been intimately linked with measurements of the thermal expansion and the volume difference between the superconducting and normal states through the standard thermodynamic relationships [2]

$$\frac{dT_c}{dP} = T_c V \frac{\beta_s - \beta_n}{C_s - C_n} \tag{1}$$

$$V_n - V_s = V_s \frac{H_c}{4\pi}\left(\frac{\partial H_c}{\partial P}\right)_T + \frac{H_c^2}{8\pi}\left(\frac{\partial V_s}{\partial P}\right)_T \tag{2}$$

The validity of these relationships has been firmly established for a variety of superconductors. In the case of non-cubic materials the corresponding expressions to (1) and (2) for length changes along the individual crystallographic axes provided the most convenient means of studying the uniaxial stress components of T_c and H_c [2,12].

A discontinuity also occurs in the elastic constants at T_c and the corresponding relationship to (1) for the discontinuity in the compressibility is

$$\frac{dT_c}{dP} = \frac{k_s - k_n}{\beta_s - \beta_n} \tag{3}$$

While there have been several observations of the change in elastic constants at T_c, there has been no systematic application of elastic measurements to the study of the stress dependence of T_c.

Detailed analyses of the elastic behavior about T_c have been made for Pb, Nb [13], and the A15 compounds V_3Si and V_3Ge [14]. In each case there is serious disagreement between the predicted pressure dependence of T_c and that observed directly [15,16], particularly in regard to the magnitude of the dominant quadratic term d^2T_c/dP^2 which is derived from $d(k_s - k_n)/dT$. This discrepancy between the prediction of the elastic data and the directly-observed pressure dependence, which is illustrated for lead in Fig. 3, may be accounted for to some extent by the relatively large uncertainties which arise when estimating relatively small differences between the temperature derivatives of the elastic constants. Nevertheless, an inconsistency exists which has yet to be resolved.

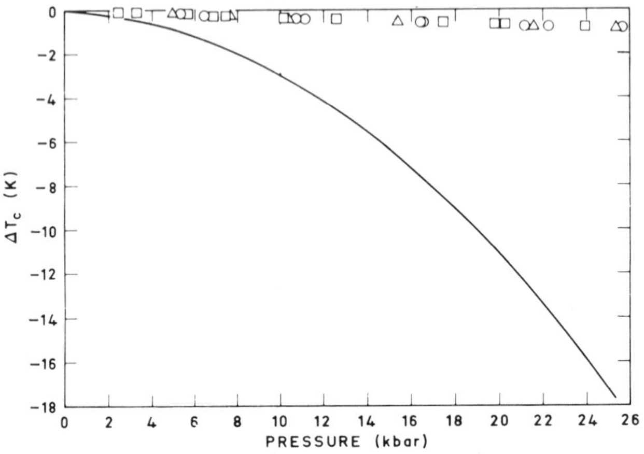

Fig. 3. Change of T_c as a function of pressure for lead. Data points are for several different samples. The solid line represents the variation predicted from elastic measurements [11,13].

Theoretical Applications

Studies of T_c as a function of pressure provide a significant input in the development of the theory of superconductivity. The superconductivity of the non-transition metals is generally regarded as being adequately described by the BCS theory and its extensions to accommodate strong-coupling effects, where pseudo-potential theory is used to derive the required electron-phonon matrix elements [17]. Various approaches have been adopted to calculate the influence of volume change upon T_c, all of which essentially lead to a smooth scaling of the electron-phonon coupling parameter λ with volume such that

$$\frac{\lambda}{\lambda_o} = \left(\frac{V}{V_o}\right)^{\alpha} \simeq 1 - \alpha\,\frac{\Delta V}{V_o} \qquad (4)$$

The corresponding change in T_c which may be calculated from the expressions [18]

$$T_c = \frac{\langle\omega^2\rangle^{\frac{1}{2}}}{1.2}\, e^{-1/g} \qquad (5)$$

$$g = \frac{0.96\lambda - \mu^* (1+0.6\lambda)}{1 + \lambda} \qquad (6)$$

is close to being linear in volume for $T_c/T_{co} \gtrsim 0.5$ in accord with the observed behavior [19]. Figure 4 illustrates the situation for lead where the experimental values for λ/λ_o have been derived from pressure measurements up to 25 kbar [20] and the dotted line represents the calculated variation based upon Ott and Sorbello's [12] theoretical value for α.

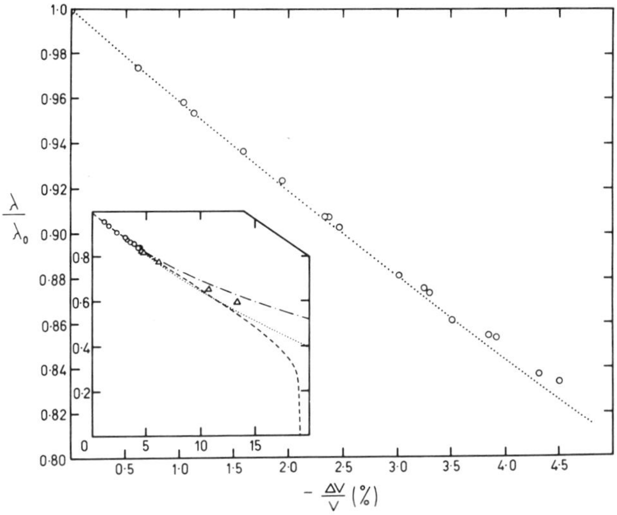

Fig. 4. Variation of λ/λ_o with percentage volume change for lead up to 25 kbars. Dotted line represents calculated variation based upon Ott and Sorbello's [12] theoretical value for α. Insert shows calculated variation up to 200 kbars based upon Ott and Sorbello and extrapolations of the T_c data assuming a linear variation with pressure - - - - - -, and a linear variation with volume -•- •-•- •-•- [20]. Triangles represent data points of Eichler and Wittig [21].

Having established satisfactory agreement between theory and experiment for relatively small volume changes and modest reductions of T_c, it is of interest to determine the extent to which the agreement continues to hold for more substantial changes. The variation of λ/λ_o for lead at volume changes corresponding to pressures up to 200 kbar is shown in the insert of Fig. 4. The three curves are calculated variations based upon possible extrapolations of the low-pressure data. The dotted line is calculated from the Ott-Sorbello theoretical α, the dashed and dot-dash curves represent the variations based on linear fits of T_c to pressure and volume respectively [20]. The experimental values derived from the data of Eichler and Wittig [21] up to 107 kbar fall between the theoretical line and that assuming a linear dependence of T_c on volume.

Calculations based upon (4) predict a progressive weakening of the volume dependence with the departure from linearity becoming evident for $T_c/T_{co} \lesssim 0.4$. Measurements of T_c extending into this region have only been made for Cd [22] and recently Al [23]. The latter measurements which were taken up to a pressure of 60 kbar reach $T_c/T_{co} \simeq 0.07$ ($T_{co} = 1.16$ K) without any appreciable deviation from a linear variation with volume (see Fig. 5). The corresponding volume dependence for λ/λ_o is shown in the insert of Fig. 5. For

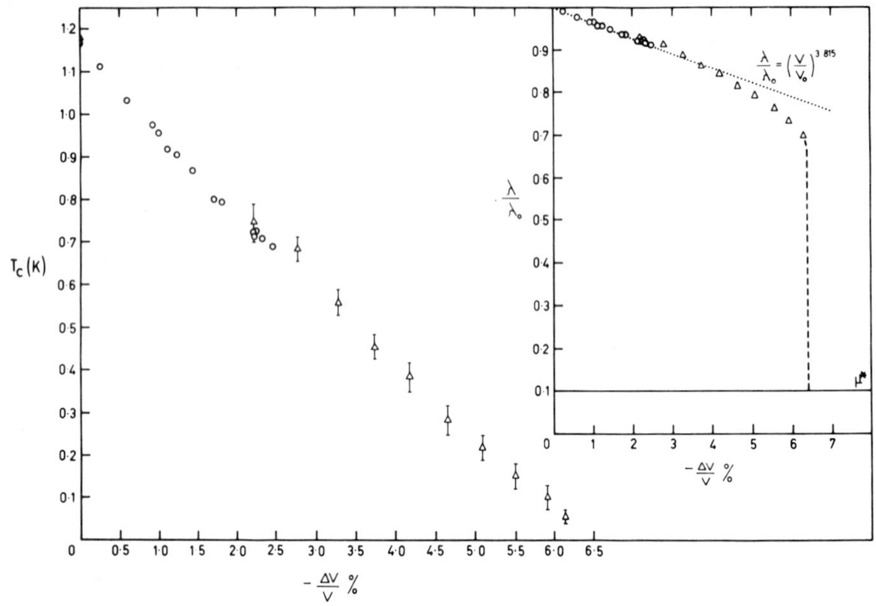

Fig. 5. Variation of T_c as a function of volume change for aluminum [23]. The insert shows the corresponding variation of λ/λ_o. The dotted line represents the calculated variation of λ/λ_o based on Ott and Sorbello's [12] theoretical value for α. The broken line indicates the abrupt drop in λ/λ_o implied by the extrapolation of the T_c data to $T_c = 0$.

small volume changes (< 4%) the measured values agree well with the calculated variation (dotted line) based upon Ott and Sorbello's [12] value for α. At higher volume changes the data points fall progressively lower than the calculated line and extrapolation to $T_c \to 0$ implies a physically implausible cut-off in λ (broken line in insert). However, it must be remembered that the derivation of λ from T_c assumes that the numerical factors in (5) and (6) which arise from the numerical solution to the superconducting gap equation [18] are independent of volume. This assumption may not be justified. Further experimental and theoretical investigations of the consequences of volume change upon superconductivity in the limit that $T_c \to 0$ are still required.

The theoretical description of the transition metal superconductors is inherently more complex [18] and while there has been some degree of success in accounting for the general trends in T_c and dT_c/dP (which can take positive values in contrast to the non-transition metals where it is always negative) there still remains considerable room for improvement in the details of the theory [24].

Fermi Surface Effects

Studies of pressure-induced changes in Fermi surface topology are one example of the application of T_c measurements under pressure to studies other than of the superconducting state itself. By virtue of its exponential dependence upon the density of electron states at the Fermi surface, T_c can serve as a sensitive indicator of changes in this parameter. Discontinuities in the variation of T_c with composition in a number of systems have been attributed to changes in Fermi surface topology [25].

Anomalous variations of T_c with pressure, which have been observed for a number of superconductors [26], have been interpreted in terms of pressure-induced changes in Fermi surface topology. Of these the most dramatic and best established example [9] is that for $AuGa_2$ (see Fig. 6). Note that the data shown in Fig. 6 are a combination of pressure measurements made by the solid helium technique and the clamp method described here. The agreement between the two sets of results is excellent and only in the vicinity of the sharp discontinuity was there any evidence of the inferiority of the pressure conditions in the latter measurements.

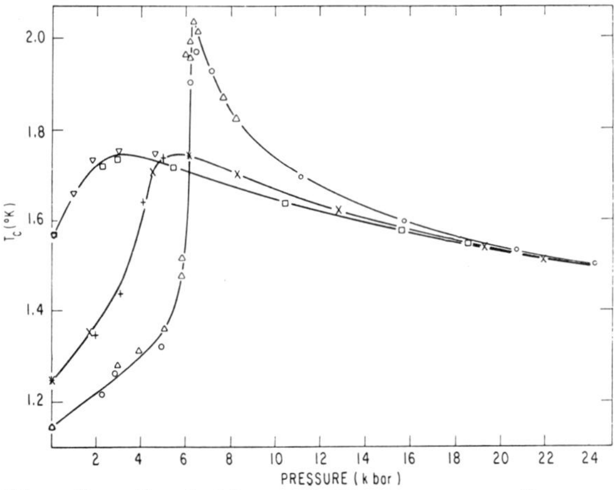

Fig. 6. Variation of T_c pressure for $Au_{1-y}Pd_yGa_2$ alloys. The different symbols distinguish the value of y and the pressure system in which the measurements were made as follows; y = 0, Δ - helium, O - clamp; y = 0.012, + - helium, x - clamp; y = 0.024, ∇ - helium, □ - clamp. The solid lines are smooth curves drawn through the data.

Zero-pressure measurements [27] of the Ga^{71} Knight shift show an anomalous temperature dependence which was attributed to a change in Fermi surface character from predominently p-like to s-like in going from low temperature to room temperature. This interpretation was subsequently substantiated by band structure calculations [28] which revealed a very flat region of the second zone energy band of strong Ga 4s character situated just below the Fermi level. The thermal depopulation of this band, with its large density states, would account for the NMR results. Furthermore, band structure calculations and direct Fermi surface measurements show that the relative separation of the band and the Fermi level decreases with pressure. The almost discontinuous change in T_c at ~ 6 kbar is consistent with the passage of the band through the Fermi level.

Further evidence in support of the above is seen in the changes produced in T_c and its pressure dependence by doping with Pd. The substitution of Pd for Au reduces the electron concentration and therefore is expected to lower the Fermi level towards the high density of states region. This is seen experimentally [29] as increases in T_c and the electronic coefficient of the specific heat. In addition, there is a marked displacement of the peak in T_c to lower pressure indicating a smaller relative energy shift is required for the second zone band to cross the Fermi level. The broadening of the peak is consistent with the smearing out of the strong singularity in the density of states by electron scattering.

The above discussion has only been able to touch on a very small fraction of the areas in which measurements of T_c under pressure have made an impact. With new developments in the art of producing high pressures at low temperature bringing the possibility of Mbar pressures [30], we may expect to continue to see an active interest in high-pressure studies of superconductivity.

ACKNOWLEDGMENT

Financial support of the Australian Research Grants Committee is gratefully acknowledged.

REFERENCES

1. G. T. Sizoo and K. Kamerlingh Onnes, Comm. Leiden 180b, 13 (1925).
2. M. Levy and J. L. Olsen, in Physics of High Pressures, A. F. Van Itterbeek, ed., North Holland Publishing Co., Amsterdam (1965), p. 525.
3. N. B. Brandt and N. I. Ginzburg, Contemp. Phys. 10, 355 (1969).
4. R. I. Boughton, J. L. Olsen, and C. Palmy, in Progr. in Low Temp. Phys., Vol. 6, C. J. Gorter, ed., North Holland, Amsterdam (1970), p. 163.
5. P. F. Chester and G. O. Jones, Phil. Mag. 44, 1281 (1953).
6. C. W. Chu, T. F. Smith, and W. E. Gardner, Phys. Rev. 1, 214 (1970).
7. A. Jayaraman, A. R. Hutson, J. H. McFee, A. S. Coriell, and R. G. Maines, Rev. Sci. Instr. 38, 44 (1967).
8. T. F. Smith, J. Low Temp. Phys. 11, 581 (1973).
9. T. F. Smith, R. N. Shelton, and J. E. Schirber, Phys. Rev. B8, 3479 (1973).
10. T. F. Smith, C. W. Chu, and M. B. Maple, Cryogenics 9, 53 (1969).
11. T. F. Smith, J. Low Temp. Phys. 6, 171 (1972).
12. H. R. Ott and R. S. Sorbello, J. Low Temp. Phys. 14, 73 (1974).
13. J. Trivisonno, S. Vatanayon, M. Wilt, J. Washick, and R. Reifenberger, J. Low Temp. Phys. 12, 153 (1973).

14. L. R. Testardi, Phys. Rev. B3, 95 (1971).
15. L. J. Sham and T. F. Smith, Phys. Rev. B4, 3951 (1971).
16. T. F. Smith, J. Low Temp. Phys. 17, 323 (1974).
17. M. S. Coulthard, J. Phys. F: Metal Phys. 1, 195 (1971); also P. N. Trofimenkoff and J. P. Carbotte, Phys. Rev. B1, 1136 (1970).
18. W. L. McMillan, Phys. Rev. 167, 331 (1968); also J. W. Garland and K. H. Bennemann, in Superconductivity in d- and f-band Metals, D. H. Douglas, ed., American Institute of Physics, New York (1972).
19. T. F. Smith and C. W. Chu, Phys. Rev. 159, 353 (1967); also P. E. Seiden, Phys. Rev. 179, 458 (1969).
20. M. J. Clark and T. F. Smith, to be published.
21. A. Eichler and J. Wittig, Z. Ang. Phys. 25, 319 (1968).
22. N. B. Brandt and N. I. Ginzberg, Zh. Eksper. i Teor. Fiz. 44, 1876 (1963); [Soviet Phys. TETP 17, 1262 (1963)].
23. D. U. Gubser and A. W. Webb, Phys. Rev. Lett. 35, 104 (1975); also M. Levy and J. L. Olsen, Solid State Comm. 2, 137 (1964).
24. A. Birnboim, Phys. Rev. B14, 2857 (1976).
25. E. E. Havinga and M. H. Maaren, Physics Reports 10, 109 (1974).
26. T. F. Smith, J. Low Temp. Phys. 11, 581 (1973).
27. V. Jaccarino, M. Weger, J. H. Wernick, and A. Menth, Phys. Rev. Lett. 21, 1811 (1968).
28. A. C. Switendick and A. Narath, Phys. Rev. Lett. 22, 1423 (1969).
29. J. H. Wernick, A. Menth, T. H. Geballe, G. Hull, and J. P. Maita, J. Phys. C30, 149 (1969).
30. H. K. Mao and P. M. Bell, Science 191, 851 (1976); also S. Block and G. Piermarini, Physics Today 29, 44 (1976).

ANOMALOUS PRESSURE DEPENDENCE OF THE SUPERCONDUCTING TRANSITION TEMPERATURE OF DILUTE ALLOYS OF $LaSn_3$ CONTAINING LIGHT RARE EARTH IMPURITIES*

L. E. DeLong and M. B. Maple

University of California at San Diego
La Jolla, California USA

INTRODUCTION

The pressure dependences of the crystalline electric field (CEF) splittings and certain physical properties ascribed to the scattering of conduction electrons by localized magnetic and non-magnetic states in several rare earth (RE) systems have been found difficult to reconcile with existing theories. Studies of the interesting case of the magnetic behavior of isolated RE ions in a metallic environment are often hampered by a lack of experimental sensitivity when the RE impurity concentration is sufficiently lowered to avoid inter-impurity interaction effects. However, if a suitable superconducting host can be found, information regarding the magnetic state of such impurities in high dilution is easily extracted from measurements of the concentration dependence of the superconducting transition temperature T_c.

EXPERIMENTAL RESULTS

Earlier measurements [1] of the initial depressions of T_c of $LaSn_3$ with RE impurity concentration n revealed an unexpected and dramatic deviation from the behavior of other more extensively studied alloy systems [2]. In an effort to further understand the anomalous behavior of these alloys we have measured the effect of hydrostatic pressure P on the T_c of dilute (<u>La</u>RE)Sn_3 samples (RE = Y, Ce, Pr, Nd, Sm, Eu, Gd, and Lu).

Measurements of the pressure dependence of T_{co} of several pure host samples of $LaSn_3$ yielded results in contradiction with previous

*Supported by the U.S. Energy Research and Energy Administration under Contract E(04-3)-34PA227.

measurements of Huang et al. [3]. It is noteworthy that our samples exhibited a zero pressure T_{c_o} (P=0) of about 6.4 K whereas those of Huang et al. had low values of T_{c_o} (0) of about 6.0 K.

In order to ascertain the strictly nonmagnetic effects of RE additions on the pressure dependence of the host transition temperature, we measured $T_c(P)$ for Y- and Lu- doped samples and found that impurity concentrations $n \lesssim 5$ at.% RE in La had a negligible effect on the high pressure behavior of T_c as shown in Fig. 1, where we have found it useful to define the quantity $\Delta' T_c(P) = \Delta T_c(P) - \Delta T_{c_o}(P)$, where $\Delta T_c(P)$ and $\Delta T_{c_o}(P)$ are the pressure-induced shifts of T_c^o for a dilute alloy and the pure host, respectively. In terms of the Abrikosov-Gor'kov (AG) theory [4] of magnetic impurities in superconductors, we may write

$$\left.\frac{\Delta' T_c(P)}{n}\right|_{\substack{\text{limit}\\ n \to o}} = \frac{-\Pi^2 \Delta\Gamma(P)}{2k_B} (g-1)^2 \; J(J+1) \qquad (1)$$

where g is the Landé factor, J is the total angular momentum of the RE impurity and $\Gamma = N(E_F)\mathcal{J}^2$, where $N(E_F)$ is the bare density of states at the Fermi level for the pure host and $\mathcal{J}$ is the exchange interaction parameter. Note that J = 0 for Y^{3+} and Lu^{3+} impurities which is consistent with the data of Fig. 1 with $n \lesssim 5$ at.% RE in La.

In order to isolate the role of CEF effects in the high-pressure behavior of T_c for (LaRE)Sn_3 alloys, we performed similar measurements on (LaGd)Sn_3 and (LaEu)Sn_3 samples which do not show CEF effects since Gd^{3+} and Eu^{2+} are S-state ions with no orbital contributions

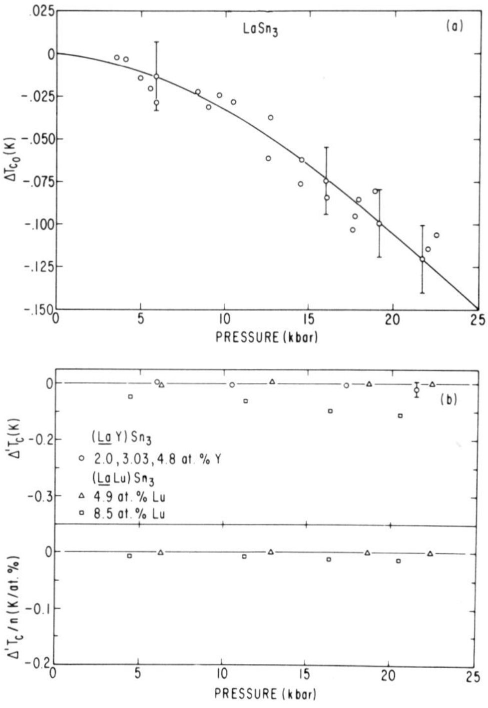

Fig. 1. (a) Change of the pure host transition temperature ΔT_{c_o} of $LaSn_3$ vs. applied pressure P. (b) $\Delta' T_c(P)$ and $\Delta' T_c(P)/n$ vs. applied pressure for several alloys of $LaSn_3$ containing strictly nonmagnetic Y and Lu impurities.

to their total angular momenta. Our results for $\Delta' T_c$ and $\Delta' T_c/n$ vs. P for Gd- and Eu-doped samples are shown in Fig. 2.

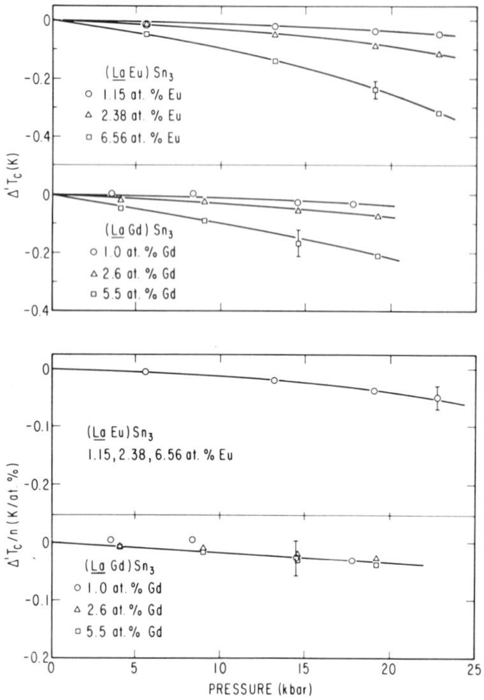

Fig. 2. $\Delta' T_c(P)$ and $\Delta' T_c(P)/n$ vs. applied pressure P for several samples of (LaGd)Sn_3 and (LaEu)Sn_3. Note that three separate Eu-doped samples exhibit identical behavior for $\Delta' T_c(P)/n$ within experimental error.

The AG pairbreaking parameter $1/\tau$ is given by

$$\frac{1}{\tau(P)} = \frac{n\Gamma(P)}{h} (g-1)^2 \; J(J+1) \tag{2}$$

so that for a fixed n, $1/\tau(P) \propto \Gamma(P)$. Representative behavior for the pressure dependence of $1/\tau$ in (LaGd)Sn_3 and (LaEu)Sn_3 is shown in Fig. 3 for a $La_{0.945}Gd_{0.055}Sn_3$ alloy. Thus we conclude that in the absence of CEF splitting of a RE impurity multiplet, the pressure dependence of the initial depression of T_c with increasing n should be independent of impurity potential scattering and should increase in absolute value by about 0.60% $kbar^{-1}$ (note that equations (1) and (2) imply $\Delta[(dT_c/dn)n{=}0] \approx (\Delta' T_c(P)/n)_{n=0} \propto \Delta[1/\tau(P)]$).

Our results for $T_c(P)$ for (LaCe)Sn_3 alloys are shown in Fig. 4. Studies [5] of the concentration dependence of T_c and the heat capacity of (LaCe)Sn_3 alloys at zero pressure have indicated that Ce impurities are nonmagnetic in $LaSn_3$ implying that the large initial depression of T_c with Ce concentration [1,2] is due to the hybridization of localized and conduction electron states. This has also been found to be the case in ThCe alloys [6] where a large, positive pressure dependence of T_c was observed. The data shown in Fig. 4 also show a large enhancement of T_c with increasing pressure.

Kaiser has given a theory [7] for the properties of superconducting dilute alloys containing nonmagnetic impurities, and derived an equation for T_c of the form

$$T_c = T_{c_o} \exp\left[\frac{-An}{1-Dn} \right] \tag{3}$$

where A and D are adjustable parameters. This equation implies

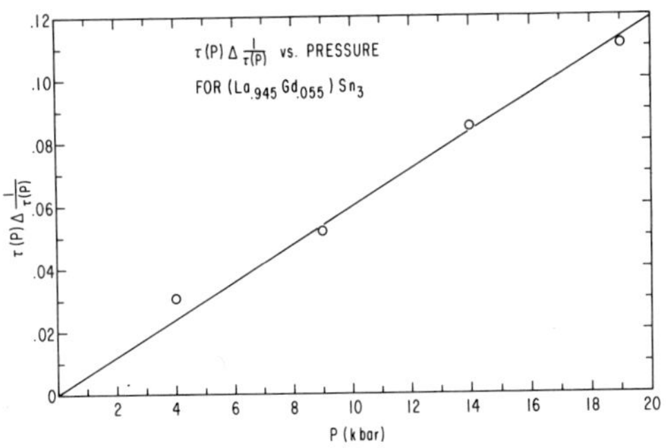

Fig. 3. $\tau(P)\Delta(1/\tau(P))$ vs. applied pressure P for a $La_{0.945}Gd_{0.055}Sn_3$ sample. The data were derived from measurements of $\Delta' T_c(P)/n$ for this alloy shown in Fig. 2.

that $T_{c_o}^{-1}(dT_c/dn)_{n=0} = -A$ so that $A(P)/A(0)$ is the normalized value of the initial depression of T_c/T_{c_o} with n. Data for the change in the depression of T_c and A with pressure for three (LaCe)Sn_3 samples are shown in Fig. 5, where it is noted that there is approximately a 35% decrease in the magnitude of $T_{c_o}^{-1}(dT_c/dn)_{n=0}$ at P = 22 kbar, which is comparable to the same data for ThCe alloys [6]. We therefore conclude that the present data are phenomenologically consistent with a large hybridization of the Ce 4f shell states with the conduction electron states of the $LaSn_3$ host material.

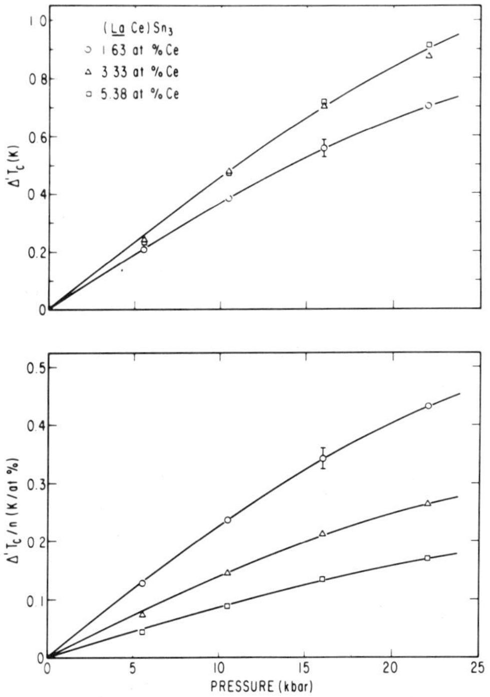

Fig. 4. $\Delta' T_c(P)$ and $\Delta' T_c(P)/n$ vs. applied pressure P for several samples of (LaCe)Sn_3 of varying Ce concentration.

The cases (LaPr)Sn_3 and (LaNd)Sn_3 have been found to exhibit important modifications of their superconducting properties by an energy splitting of their impurity 4f shell configurations by the host CEF [5,8]. The ratios of the energies of the various CEF - split levels and the magnitude of the total splitting are commonly denoted by the Lea, Leask, and Wolf [9] (LLW) parameters X and W, respectively. Fulde and coworkers [10] have developed theories for the superconducting properties of dilute alloys with CEF-split RE impurity multiplets, and have derived an expression for $(dT_c/dn)_{n=0}$ in such alloys

$$\left.\frac{dT_c}{dn}\right|_{n=0} = \frac{-\Pi^2}{2k_B}\,\Gamma(g-1)^2 F(T_{c_o}, X, W) \qquad (4)$$

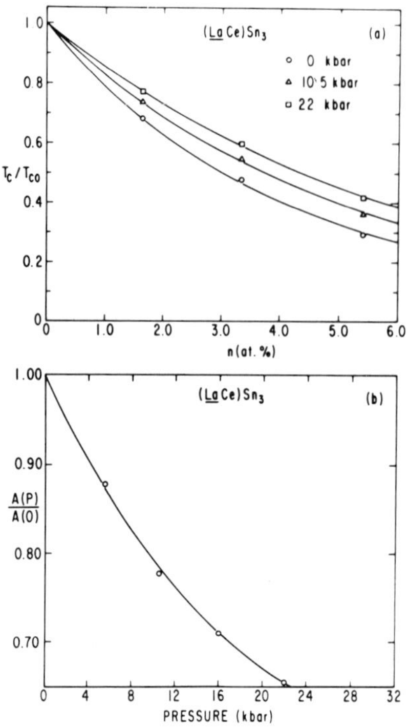

Fig. 5. (a) Data for T_c vs. n for (<u>La</u>Ce)Sn_3 alloys of 1.63, 3.33 and 5.40 at.% Ce at various applied pressures. The solid curves were obtained by a least squares fit of the data to (3). (b) Normalized values of A(P)/A(0) vs. applied pressure P as derived from the least squares fits of (a) above. The plot represents the decrease in absolute value of the initial depression of T_c/T_{c_0} in (<u>La</u>Ce)Sn_3 alloys with pressure.

---

where F is a complicated function of temperature and RE level scheme.

The results of our pressure measurements for (<u>La</u>Pr)Sn_3 alloys are shown in Fig. 6. These results were published previously [11], and here we wish to report on our attempts to analyze these data within the theory of Fulde and co-workers. The zero pressure Pr level scheme parameters of McCallum et al. [8] for (<u>La</u>Pr)Sn_3 alloys were used as a starting point in our calculations. Estimates of the compressibility of $LaSn_3$ [12,13] combined with the point charge model, yielded values for the expected shifts of the LLW parameters X and W with pressure [5]. The predicted shifts in X and W were less than 10% for P < 25 kbar; and the results of computer calculations of $T_c(P)$ employing plus and minus 10% variations in X and W showed that such changes in X and W had a negligible effect on $T_c(P)$. However, the computer calculations were qualitatively consistent with the experimental high-pressure data only if Γ were allowed to have a very large increase with pressure as compared to that implied by the (<u>La</u>Gd)Sn_3 and (<u>La</u>Eu)Sn_3 data of Figs. 2 and 3. It is also noteworthy that the experimental data for $|\Delta' T_c/n|$ for the 2 and 3 at.% Pr samples were approximately half as large as theoretically expected when the theory was forcibly fit to the 1 at.% Pr data by appropriate adjustment of the value of Γ(P) at each pressure studied. An approximate plot of $\tau(P)\Delta\, 1/\tau(P)$ vs. P for the 1 at.% Pr (<u>La</u>Pr)Sn_3 sample is shown in Fig. 7, assuming that the form of (4) is a valid description of the data. (We have used the fact that F(P)≈constant.)

The results of our high-pressure studies of several (<u>La</u>Nd)Sn_3 alloys are shown in Fig. 8. Preliminary measurements [5] of the (zero pressure) Nd concentration dependences of T_c and heat capacity in this system have revealed important manifestations of CEF splitting

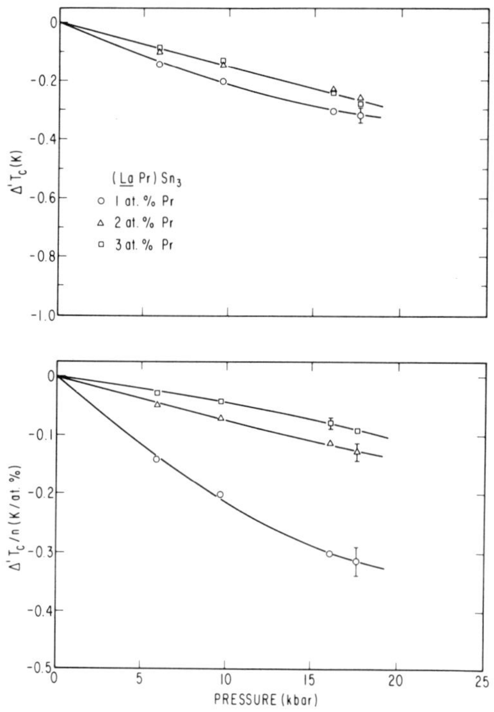

Fig. 6. $\Delta'T_c(P)$ and $\Delta'T_c/n$ vs. applied pressure P for several samples of (LaPr) Sn_3 of varying Pr concentration [11].

of the Nd^{3+} groundstate configuration. A complete analysis of our high-pressure data must therefore await a more detailed understanding of the zero pressure Nd^{3+} level splittings, but the large, concentration - independent value of $|\Delta'T_c/n|$ at each pressure studied is the most important feature of these results.

Our high-pressure data for $T_c(P)$ of (LaSm)Sn_3 alloys are not shown here because of a lack of reproducible behavior in $|\Delta'T_c/n|$ between different samples. The small concentrations ($n < 1$ at.%) of Sm impurities involved, coupled with the relatively high vapor pressure of Sm in the sample melts, make a detailed numerical analysis of our data difficult. The approximate magnitudes of $|\Delta'T_c/n|$ for several (LaSm) Sn_3 alloys were found to fall in a range between those characteristic of Nd and Eu-doped samples, with the larger Sm concentration samples more closely resembling behavior similar to the Eu case. The most important observation which we will make about these data is that these values for $|\Delta'T_c(P)|$ are not extraordinarily large compared to those for the lighter RE impurities, which is surprising in view of the fact that there is significant hybridization of the Sm 4f and conduction electron states in (LaSm)Sn_3 [14,15].

SUMMARY

The results of our measurements of the pressure dependence of T_c for the (LaRE)Sn_3 alloys discussed above can be summarized as follows:

1. The maximum values of $|\Delta'T_c/n|$ observed in (LaRE)Sn_3 alloys show a monotonic decrease, within experimental error, with increasing RE atomic number for light RE impurities.

2. The experimental results for $T_c(P)$ for the case of Pr, and very probably Nd impurities, are consistent with the point charge model

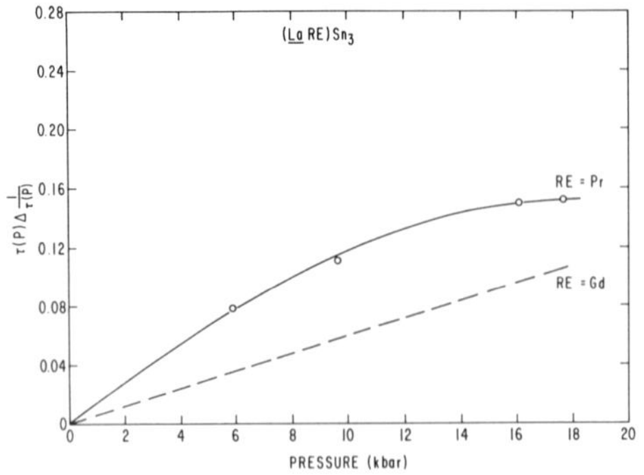

Fig. 7. $\tau(P)\Delta(1/\tau(P))$ vs. applied pressure P for a 1 at.% Pr (LaPr)Sn_3 sample (full line). Data derived from (4) and data of Fig. 6 assuming $F(T_c, X, W)$ approx. constant as a function of pressure. The dashed line represents $\tau(P)\Delta(1/\tau(P))$ vs. P for (LaGd)Sn_3 (from Fig. 3).

coupled with the theory of Fulde and co-workers, providing that there is a very large increase in Γ with applied pressure, contrary to the behavior expected from data for Gd- and Eu-doped samples.

3. The data for $T_c(P)$ for Y, Lu, Gd, and Eu samples indicate that simple alloying effects (pure potential scattering and changes in the electronic density of states) and elastic spin exchange scattering do not alter the high pressure T_c of $LaSn_3$ in any unusual manner, at least in the case of RE substitutions for La.

4. Zero pressure data for (LaSm)Sn_3 [14,15] and (LaCe)Sn_3 [5,16] alloys clearly support the presence of unstable 4f shells due to covalent mixing between localized and conduction electron states in these particular systems. Previously, the large magnitude of $|\Delta' T_c/n|$ observed in (LaPr)Sn_3 alloys has been viewed [11,15] as support for the presence of a Kondo effect in this system in analogy to the large pressure effects observed in many Ce impurity Kondo systems [2]. However, the relatively small values of $|\Delta' T_c/n|$ observed in (LaSm)Sn_3 alloys-in spite of the extremely large value of $(-dT_c/dn)_{n=0}$ in this system [2]-and the substantial values of $|\Delta' T_c/n|$ found in (LaNd)Sn_3 alloys (for which no direct evidence of a Kondo effect has yet been found in the electrical resistivity [17] and heat capacity data [5]) would seem to indicate that large pressure dependences in T_c data are not a reliable indication of unstable 4f shells in appropriate systems.

Several mechanisms, not specifically considered here, may ultimately have to be incorporated into a satisfactory explanation of our data. Although some preliminary work has been done [18], a thorough assessment of the influence of aspherical Coulomb scattering [10] in these alloy systems has not as yet been given. In addition, the potential roles of orbital exchange scattering and the stronger Coulomb correlations present in many-electron 4f shells (contra Ce^{3+} or Yb^{3+}) may be significant in the analysis of our data, particularly in the case of (LaSm)Sn_3 alloys.

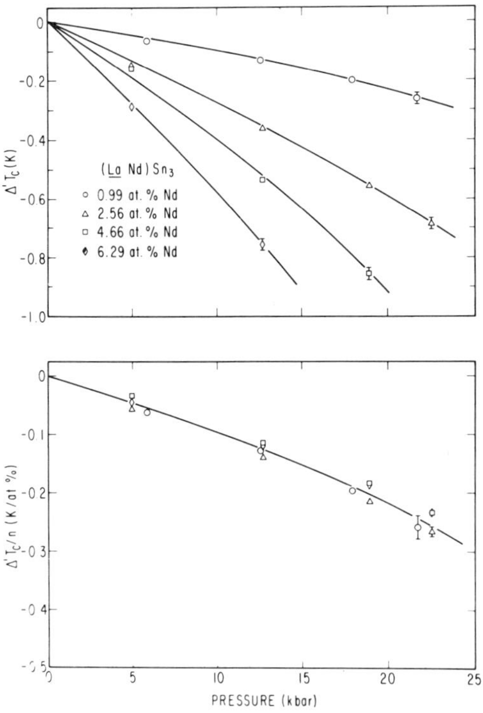

Fig. 8. $\Delta' T_c(P)$ and $\Delta' T_c(P)/n$ vs. applied pressure P for several samples of $(LaNd)Sn_3$ of varying Nd concentration.

ACKNOWLEDGMENTS

The authors wish to thank R. W. McCallum for useful discussions and experimental assistance, R. N. Shelton and D. C. Johnston for experimental assistance in the early stages of this work, L. D. Woolf for help in analyzing some of the data, and J. Keller for valuable discussions and the use of his computer program for calculating $T_c(n)$ in CEF-split impurity systems.

NOTATION

A = Kaiser theory parameter in (3)

AG = Abrikosov and Gor'kov

CEF = crystalline electric field

D = Kaiser theory parameter in (3)

$\Delta T_c(P) = T_c(P) - T_c(0)$, the pressure induced shift of the superconducting transition temperature of a dilute alloy

$\Delta T_{c_o}(P) = T_{c_o}(P) - T_{c_o}(0)$, the pressure-induced shift of the superconducting transition temperature of a pure host metal

$\Delta' T_c(P) = \Delta T_c(P) - \Delta T_{c_o}(P)$

g = Landé g-factor

$\Gamma = N(E_F)\mathcal{J}^2$

J = $|J|$ magnitude of the total angular momentum of a RE ion

$\mathcal{J}$ = exchange interaction parameter

LLW = Lea, Leask and Wolf

n = impurity concentration (usually in at.% RE in La)

$N(E_F)$ = host density of electronic states at the Fermi energy

P = pressure

RE = rare earth

T_c = superconducting transition temperature of a dilute alloy

T_{c_o} = superconducting transition temperature of a pure host metal

$1/\tau$ = AG pairbreaking parameter

W = LLW parameter for overall CEF splitting of a RE multiplet

X = LLW parameter which fixes the ratios of the individual CEF level splittings of a RE multiplet

REFERENCES

1. W. Schmid and E. Umlauf, Commun. Phys. 1, 67 (1976).
2. M. B. Maple, L. E. DeLong, and B. C. Sales in, Handbook on the Physics and Chemistry of Rare Earths, Vol. I, K. A. Gschneidner and L. Eyring, eds., North Holland, Amsterdam (1977), Chapter 11.
3. S. Huang, C. W. Chu, F. Y. Fradin, and L. B. Welsh, Solid State Commun. 16, 409 (1975).
4. A. A. Abrikosov and L. P. Gor'kov, Sov. Phys. JETP 12, 1243 (1961).
5. L. E. DeLong, Ph.D. Thesis, University of California, San Diego, California (1977).
6. J. G. Huber and M. B. Maple, J. Low Temp. Phys. 3, 537 (1970).
7. A. B. Kaiser, J. Phys. C: Solid State Phys. 3, 410 (1970).
8. R. W. McCallum, W. A. Fertig, C. A. Luengo, M. B. Maple, E. Bucher, J. P. Maita, A. R. Sweedler, L. Mattix, P. Fulde, and J. Keller, Phys. Rev. Lett. 34, 1620 (1975).
9. K. R. Lea, M. J. M. Leask, and W. P. Wolf, J. Phys. Chem. Solids 23, 1381 (1962).
10. P. Fulde and I. Peschel, Adv. Phys. 21, 1 (1972).
11. L. E. DeLong, R. W. McCallum, and M. B. Maple, in Proceedings of 14th Intern. Conference on Low Temperature Physics - LT14, M. Krusius and M. Vuorio, eds., Vol. 2, North Holland, Amsterdam (1975), p. 541.
12. H. Damsma and E. E. Havinga, Solid State Commun. 17, 409 (1975).
13. J. Beille, D. Bloch, J. Voiron, and G. Parisot, Physica 86-88B, 231 (1977).
14. L. E. DeLong, R. W. McCallum, W. A. Fertig, M. B. Maple, and J. G. Huber, Solid State Commun. 22, 245 (1977).
15. S. Bakanowski, J. E. Crow and T. Mihalisin, Solid State Commun. 22, 241 (1977).
16. P. Lethuillier, D. Sc. Thesis, University of Grenoble, Grenoble, France (1976).
17. A. I. Abou Aly, S. Bakanowski, N. F. Berk, J. E. Crow, and T. Mihalisin in Proc. 21st Annual Conference on Magnetism and Magnetic Materials, Philadelphia, Pennsylvania (1975), p. 358.
18. J. Keller and P. Holzer, in Proc. 2nd Intern. Conference on Crystalline Electric Field Effects in Metals and Alloys, A. Furrer, ed., Zurich, Switzerland (1976).

D-5

INDICATIONS FOR A PHASE TRANSITION IN CERIUM NEAR 160 KBAR

C. Probst and J. Wittig

Institut für Festkörperforschung
Jülich, W. Germany

It has been confirmed that the room-temperature resistance of Ce shows a faint anomaly at pressures near 160 kbar [1]. The authors have observed a small kink in the dependence of the superconducting transition temperature upon pressure and also a maximum in the residual resistivity at ≃ 160 kbar. The data allow no definite conclusion at present. It is tentatively suggested that a subtle isostructural phase transformation occurs in Ce, as has been postulated for La [2]. Independent of any interpretation, there is a striking similarity of the T_c-P dependences of Ce and Th [3]. This points to the possible existence of a law-of-corresponding states for the electronic band structure of these elements at high pressure [4].

REFERENCES

1. R. A. Stager and H. G. Drickamer, Phys. Rev. 133, A830 (1964).
2. H. Balster and J. Wittig, J. Low Temp. Phys. 21, 377 (1975).
3. W. A. Fertig, A. R. Moodenbaugh and M. B. Maple, Phys. Lett. 38A, 517 (1972).
4. C. Probst and J. Wittig, in Handbook on the Physics and Chemistry of Rare Earths, K. A. Gschneidner, Jr. and L. Eyring, eds., in print.

D-6

STATIC CRYSTAL FIELD FOR ORTHOHYDROGEN IMPURITIES IN PARAHYDROGEN*

J. C. Raich and L. B. Kanney

Colorado State University
Fort Collins, Colorado USA

INTRODUCTION

Ortho-para mixtures of solid hydrogen show a number of unique and interesting properties which have been the subject of numerous investigations during the past decade or so. Among these are measurements of the energy level splitting of the triply degenerate $J = 1$ rotational state of dilute mixtures of orthohydrogen in solid parahydrogen. In the environment of neighboring molecules with rotational quantum number $J = 0$, the triply degenerate $J = 1$ state is split. This splitting is observed in the NMR spectrum and depends on the ortho hydrogen concentration x [1-3].

1. When x is less than about 0.1%, the spectrum from isolated $J = 1$ molecules surrounded by $J = 0$ neighbors dominates. It is found that the triply degenerate $J = 1$ rotational state is split into a two-fold degenerate state with $J_z = \pm 1$ and a non-degenerate state with $J_z = 0$.
2. When x is larger than about 0.1%, the interactions of pairs of $J = 1$ molecules become important in the NMR spectrum, which then consists of three satellites on each side of a central Larmor frequency. The peak-to-peak splittings of the pair spectrum depend on the quadrupolar coupling between pairs of $J = 1$ molecules.

In the present paper, we only concern ourselves with the very low concentration range ($x < 0.1\%$). In that range, the number of $J = 1$ pairs is sufficiently small so that the crystal field representing the distortion of the lattice around the $J = 1$ defect is the dominant feature of the NMR spectrum as a function of pressure.

*Work supported by the National Science Foundation.

The purpose of this paper is to provide a description of the strong anomalous broadening of the central NMR line of solid H_2 as the density is increased at constant temperature. In addition, a fit of the NMR data of Pedroni, et al. [3] to various models of the H_2-H_2 interaction potential should provide an additional criterion for the validity of those model potentials.

CRYSTAL FIELD

The origin of the crystal field in solid hydrogen has been discussed by Van Kranendonk and Sears [4]. This crystalline field arises from the anisotropic intermolecular interactions between isolated J = 1 molecules and their J = 0 neighbors. Here we assume that J = 0 and J = 1 are good quantum numbers so that when the intermolecular potential is expanded in terms of spherical harmonics, $Y_L^M(\theta,\phi)$, it is not necessary to include terms with $L > 2$, since their matrix elements vanish within the J = 0 and J = 1 mainfolds. The intermolecular potential between two H_2 molecules is thus generally written in the form [5]

$$v(R,\hat{\omega}_1,\hat{\omega}_2) = A(R) + (16\pi/5)^{1/2}\, B(R)[Y_2^0(\hat{\omega}_1) + Y_2^0(\hat{\omega}_2)]$$

$$+ 4\pi \sum_M C_M(R)\, Y_2^M(\hat{\omega}_1)\, Y_2^{-M}(\hat{\omega}_2) \qquad (1)$$

where $\hat{\omega}_1 = (\theta_1, \phi_1)$ and $\hat{\omega}_2 = (\theta_2, \phi_2)$ specify the orientations of molecules 1 and 2 with respect to $\vec{R}$, the vector connecting their centers of mass. This truncated potential is considered adequate in the pressure range discussed here; only at pressures in excess of several hundred kilobars does one need to consider the admixture of higher rotational states [6].

For a J = 1 molecule surrounded by J = 0 neighbors, only the first two terms of (1) contribute to the potential energy of the defect. The first term, A(R), is just the isotropic part of the interaction potential between two H_2 molecules. The part of the crystal potential energy which depends on the orientation of the central J = 1 molecule arises from the second part of (1) and has the form [3,4]

$$\mathcal{H}_c = \sum_i (16\pi/5)^{1/2}\, B(R_i)\, Y_2^0(\hat{\omega}_i) \qquad (2)$$

where $\vec{R}_i$ specifies the intermolecular axis from the central J = 1 molecule to the ith J = 0 neighbor, and $\hat{\omega}_i$ is the orientation of the central molecule in the $\vec{R}_i$ frame. The sum in equation (2) runs over all lattice sites of the crystal except the central site.

The crystal structure of solid parahydrogen is generally assumed to be hexagonal close-packed (hcp) with $c/a = \sqrt{(8/3)}$, although there is some evidence that the crystal structure for

$\rho/\rho_o > 1$ becomes complicated [7]. To evaluate the crystal field (2) one transforms from the $\vec{R}_i$ frame to the c frame, a coordinate system with the z-axis along the hexagonal axis of the crystal. Taking account of the three-fold axis of symmetry, the crystal field energy is [4]

$$\mathcal{H}_c = \frac{1}{3} U_c(3J_z{}^2 - 2) \tag{3}$$

where

$$U_c = -\frac{3}{5} \sum_i B(R_i)(3 \cos^2\theta_i - 1) \tag{4}$$

Here $\vec{R}_i = (R_i, \theta_i, \phi_i)$ specifies the lattice vector of site i relative to the central site. If one assumes a uniform deformation of the hcp crystal structure, the crystal field U_c, defined by equation (4), will be the same regardless of the value of J_z for the impurity. In that case, one writes $U_c(J_z = 0) = U_c(J_z = \pm 1) \equiv V_c$.

A value of $|V_c|/k_B \cong 0.02$ K at zero pressure was deduced from the NMR free induction experiments of Constable and Gaines [1] for a hydrogen sample with $x = 2 \times 10^{-4}$. The sign of V_c was determined to be negative, so that the $J_z = \pm 1$ state lies below the $J_z = 0$ state. Pedroni, et al. [3] measured the NMR absorption line over a concentration range $0.015 \leq x \leq 0.030$, a temperature range $1.2\text{K} < T < 4.2$ K, and a density range $\rho_o < \rho < 1.7\,\rho_0$. Here, ρ and $\rho_o = 0.089$ g/cm^3 are the densities at pressure P and zero pressure, respectively. The observed broadening at the higher pressures was found to be an order of magnitude larger than previous determinations at $\rho = \rho_o$ [1]. It was suggested that the anomalously broadened line can be attributed to local orientational alignment of the J = 1 impurities in the crystalline field $\mathcal{H}_c{}^3$. Pedroni, et al. [3] conclude from an analysis of their NMR data that the distribution of crystal field strengths V_c throughout the lattice is approximately Gaussian around a characteristic value $\bar{V}_c$ with a distribution width of about $\bar{V}_c$. At $\rho/\rho_0 = 1.7$, $|V_c|/k_B$ is found to be about 0.6 K, which is about 30 times its value at $\rho/\rho_0 = 1$. The sign of V_c could not be determined.

A finite value of $|V_c|$ indicates that the lattice sum in equation (4) cannot vanish. However, when this lattice sum is evaluated for a perfect hcp lattice with $c/a = \sqrt{(8/3)}$, it is found that the contribution of the first two shells of J = 0 neighbors vanishes. The result is that for an undistorted hcp lattice V_c is negligible. Van Kranendonk and Sears [4] point out that the non-vanishing crystal field strength can, at least in part, be attributed to a distortion from the ideal hcp structure. They argue that a crystal of solid hydrogen is expanded considerably by the zero-point lattice vibrations and since the lattice has uniaxial symmetry, one should expect that this blowing up is not perfectly isotropic. This concept leads to a crystal field strength

$$V_c \cong 45(\Delta c/a) \qquad (5)$$

where Δc is the deviation of c from the value $\sqrt{(8/3)}a$. This model thus requires a uniform long-range lattice distortion corresponding to $\Delta c/a \cong 4 \times 10^{-4}$ at $\rho = \rho_0$. This value is just somewhat smaller than the experimental uncertainty of recent neutron scattering data [8].

In addition to any change of the c/a ratio which may occur as a result of the large zero point lattice vibrations, one should consider the crystal field produced by the local lattice distortion around a J = 1 impurity. If the neighboring J = 0 molecules are allowed to relax to the equilibrium positions determined by the full orientation dependent interaction (1), the lattice sum in equation (4) will no longer vanish. This is indicated qualitatively in Fig. 1, which shows the distortion of a J = 1 defect by displaying the displacements of the neighboring J = 0 molecules from the undistorted lattice positions. In Fig. 1, the displacements are shown for a J = 1 impurity in the $J_z = 0$ rotational state; displacements in the opposite direction would be expected for a $J_z = \pm 1$ impurity. These distortions result from the anisotropic interaction potential (1) in the following way: the pair interaction energy for the J = 1 impurity and a J = 0 neighbor will be slightly different for pairs whose intermolecular axis lies in the hcp basal plane than for those out of the plane. This leads to slightly different equilibrium nearest neighbor separations for J = 0 neighbors of the J = 1 impurity depending on whether these neighbors lie in the hcp basal plane or not. For a local J = 1 defect, the displacements of the neighboring J = 0 molecules depend on the value of J_z for the impurity. Thus, the local crystal field, calculated from equation (4), will be different for $J_z = 0$ and $J_z = \pm 1$, hence $U_c(J_z = 0) \neq U_c(J_z = \pm 1)$.

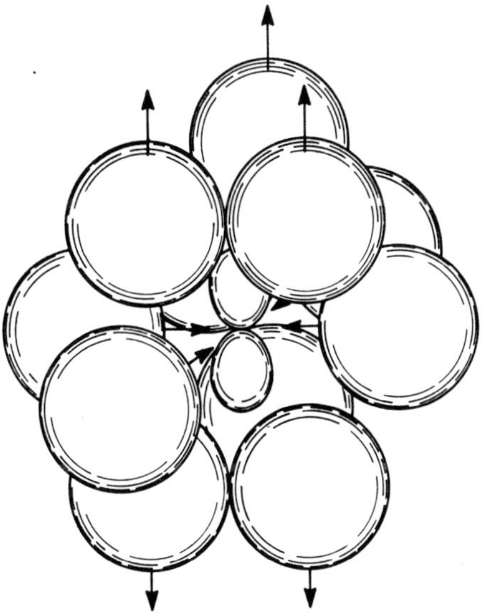

Fig. 1. Local lattice distortion produced by a J = 1, $J_z = 0$ impurity surrounded by J = 0 molecules is shown schematically by arrows indicating the directions of the displacements of the J = 0 molecules from their undistorted hcp lattice positions. The shapes represent the squares of the rotational wave functions. For pictorial clarity, one nearest neighbor molecule has been removed.

LATTICE DISTORTION

The lattice distortion produced by point defects was first discussed by Kanzaki [9] and was subsequently applied to a variety of defect calculations [10-12]. Using the procedures of Kanzaki, one is able to calculate the distortion produced by the substitutional insertion of a J = 1 defect into a J = 0 lattice. The displacement of the J = 0 molecules surrounding the J = 1 impurity is found from the following relation, written in matrix form:

$$\vec{u} = \vec{G}_o \vec{F}^* \tag{6}$$

Here $\vec{u}$ represents the molecular displacements, $\vec{G}_o$ is the static Green's function [13] for the undistorted hcp lattice, and $\vec{F}^*$ represents the Kanzaki forces [9]. These are visualized by imagining the defect being replaced by an arrangement of forces, $\vec{F}^*$, which when applied to the undistorted lattice, produce the same displacements as the defect in the real distorted lattice.

The static lattice Green's function, $\vec{G}_o$, is calculated from the phonon frequencies and polarization vectors for hcp parahydrogen, using only the isotropic interaction potential A(R). For solid hydrogen in the density range $1 \leq \rho/\rho_0 \leq 1.7$ the usual harmonic lattice dynamical treatments [13] are not sufficient because of the large vibration amplitudes of the molecules [14-19]. Improved treatments, including anharmonic corrections, are now available for solid hydrogen [14,15,18,19]. However, for the present work an effective harmonic potential [20], of the Lennard-Jones form, and fit to the observed optical phonon frequencies [8,21,22] was used for A(R) at $\rho = \rho_0$, or a hcp lattice constant of a = 3.756 Å. For $\rho > \rho_0$, the Lennard-Jones potential is unreliable, so the phonon frequencies were obtained by scaling the zero pressure frequencies using the volume dependence of the Grüneisen parameters obtained for the self-consistent lattice dynamical treatment of fcc parahydrogen [18]. This calculation employed two different H_2-H_2 interaction potentials, the Ross potential and the EERD potential [16]. The EERD potential is expected to be more accurate at low pressures, the Ross potential at higher pressures.

The center-of-mass displacements of J = 0 molecules in the defect neighborhood are calculated by considering the response of the crystal, according to equation (6), to a set of Kanzaki forces acting on the molecules. The Kanzaki forces are determined only by the second term of the interaction potential (1) between the J = 1 impurity and the surrounding J = 0 molecules as well as the J_z rotational state of the impurity.

For the anisotropic term, B(R), of equation (1), we have used the anisotropic potentials recommended by Raich, Anderson and England [23]. Of the four model potentials suggested in that paper, we have concentrated on model (a).

The interactions between in-plane and out-of-plane nearest neighbors and the central J = 1 impurity are calculated in the same manner as the crystal field described earlier. The results for the defect interaction, i.e., the second term of equation (1), is $1/5\ B(R_i)\langle 3J_z^2 - 2\rangle_T$, in-plane pairs, $-1/5\ B(R_i)\ [3/4\ (c/a)^2 - 1]\ \langle 3J_z^2 - 2\rangle_T$, out-of-plane pairs. The Kanzaki forces are found to depend on the average $\langle 3J_z^2 - 2\rangle_T$, which is taken as -2 for the $J_z = 0$ state and +1 for the $J_z = \pm 1$ state.

RESULTS AND DISCUSSION

The results of the magnitudes of the displacements u of in-plane and out-of-plane J = 0 neighbors are shown in Fig. 2 for the case of a J = 1, $J_z = 0$ impurity. The directions of the displacements are shown in Fig. 1. The resulting local distortion parameter $\Delta c/a$ is also plotted in Fig. 2. This ratio first increases until ρ/ρ_0 reaches about 1.4, and then levels off to a value of roughly 2×10^{-3}. The value of $\Delta c/a$ at $\rho = \rho_0$, 1.2×10^{-3}, is a factor of four larger than the uniform lattice distortion required by the model of Van Kranendonk and Sears when zero-point lattice vibrations are taken into account.

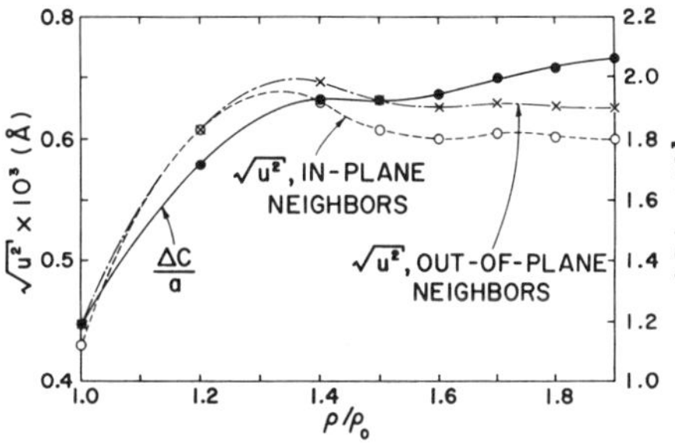

Fig. 2. The magnitudes of the displacements of the nearest neighbor J = 0 molecules for neighbors both in the hcp basal plane (with the J = 1, $J_z = 0$ impurity) and out of that plane as a function of the relative density. The directions of the displacements are given in Fig. 1. The resulting distortion parameter $\Delta c/a$ is also shown.

Displacements in the opposite directions with roughly half the magnitudes are found for the J = 0 neighbors of a J = 1, $J_z = \pm 1$ impurity. Once the molecular displacements are known, the crystal fields, $U_c(J_z = 0)$ and $U_c(J_z = \pm 1)$ are calculated from equation (4). The energy levels are

$$E(J_z = 0) = \langle 0|\mathcal{H}_c|0\rangle = -\frac{2}{3}U_c(J_z = 0) \qquad (7a)$$

$$E(J_z = \pm 1) = \langle \pm 1|\mathcal{H}_c|\pm 1\rangle = \frac{1}{3}U_c(J_z = \pm 1) \qquad (7b)$$

The energy level splitting is

$$\Delta E = E(J_z = \pm 1) - E(J_z = 0) = \frac{1}{3}[2U_c(J_z = 0) + U_c(J_z = \pm 1) \qquad (8)$$

We identify the energy level splitting ΔE with the crystal field strength V_c, determined by Pedroni et al. [3]. In Fig. 3, the

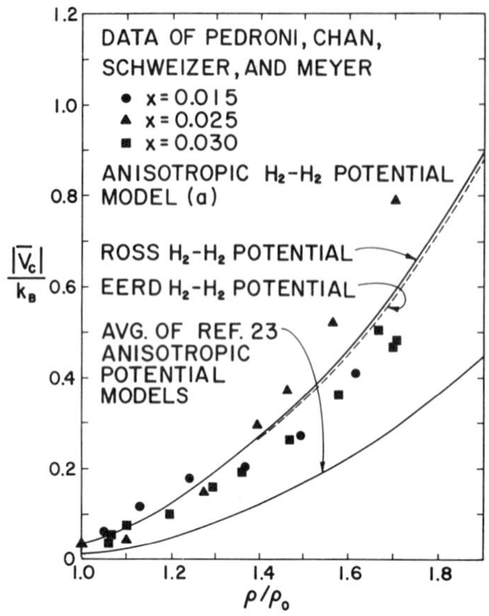

Fig. 3. The absolute value of the calculated energy level splitting ΔE is compared with the average crystalline field $|\bar{V}_c|$ determined from the measurements of Pedroni, et al. [3]. The results for the anisotropic H_2-H_2 model potential (a) of Raich et al. [23] with both the EERD [13] and Ross [14] isotropic potentials are shown.

magnitude of the splitting $|\Delta E|$, determined from (8), is compared to the values of $|\bar{V}_c|$ obtained by Pedroni, et al. [3] in the density range $1 \leq \rho/\rho_0 \leq 1.7$. Results are shown for both the EERD [13] and Ross [14] isotropic H_2-H_2 potentials. The calculation of ΔE is sensitive to the model chosen for the anisotropic part B(R) of the H_2-H_2 interaction (1). It is seen from Fig. 3 that the form (a) of Raich et al. [23] for B(R) is to be preferred over other forms on the basis of a comparison with the data of Pedroni, et al. But regardless of the model, equation (7) then indicates that the $J_z = 0$ state lies below the $J_z = \pm 1$ state in the density range $1 \lessgtr \rho/\rho_0 < 1.7$, in contradiction to the results of Constable and Gaines [1].

From the agreement between the values of $|V_c|$ determined from the measurement of Pedroni, et al. and those calculated using the present local defect model one might be led to the following conclusions:

1. The observed crystal field splitting is adequately explained by the local distortion produced by the J = 1 impurity in the J = 0 environment. No long range deviation of c/a from the ideal hcp value of $\sqrt{(8/3)}$ is necessary to provide a reasonable fit to the data.
2. A contribution to the crystalline field arising from dynamic effects [4], i.e., the molecular zero point motions should be added to the values of ΔE determined in this work. At this time the magnitude of this effect is uncertain, but is expected to decrease in importance for $\rho > \rho_0$.
3. The anisotropic parts of the interaction potential (1) are renormalized somewhat by the zero-point vibrations [5]. This effect has been neglected here. It is most pronounced at $\rho = \rho_0$ and is also expected to be less important at the higher densities.
4. The Ross [17] and EERD [16] isotropic potential forms are found to lead to very similar crystal field values.
5. On the basis of a comparison with the experimental data of Pedroni, et al. and the present model, the anisotropic potential

B(R) seems best represented by potential (a) of Raich et al. [23].

The J = 1 impurity in solid parahydrogen provides an example of a defect produced by a rotational excitation of a molecule in the crystal. Here, this J = 1 excitation has been treated as localized, however, the J = 1 rotational angular momentum can move through the lattice by a slow quantum diffusion which proceeds by a resonant ortho-para conversion (J = 0, J - 1) → (J = 1, J = 0) for a pair of molecules [24,25].

NOTATION

$A(R)$, $B(R)$, $C_M(R)$ = interaction potentials
a, c, Δc = hcp lattice parameters
E, ΔE = energy levels, splittings
F^* = Kanzaki force
G = static Green's function
$\mathcal{H}_c^o$ = crystal field energy
J = orbital angular momentum
J_z = z-component, orbital angular momentum
T = temperature
$\vec{R}$ = lattice vector
ρ, ρ_0 = density
u = molecular displacements
U_c, V_c = crystal field parameters
v^c = total interaction potential
ω = (θ,ϕ) = solid angle

REFERENCES

1. J. H. Constable and J. R. Gaines, Solid State Commun. 9, 155 (1971).
2. N. S. Sullivan and R. V. Pound, Phys. Rev. (1975).
3. P. Pedroni, M. Chan, R. Schweizer and H. Meyer, J. Low Temp. Phys. 19, 537 (1975).
4. J. Van Kranendonk and V. F. Sears, Can. J. Phys. 44, 313 (1966).
5. A. B. Harris, Phys. Rev. B1, 1881 (1970); Phys. Rev. B2, 3495 (1970).
6. W. England, J. C. Raich, and R. D. Etters, J. Low. Temp. Phys. 22, 213 (1976).
7. S. C. Durana and J. P. McTague, Phys. Rev. Lett. 31, 990 (1973).
8. N. Nielsen, Phys. Rev. B7, 1626 (1973).
9. H. Kanzaki, J. Phys. Chem. Solids 2, 24 (1957).
10. R. Bullough and J. R. Hardy, Phil. Mag. 17, 833 (1968).
11. R. Bullough and V. K. Tewary, in Interatomic Potentials and Simulation of Lattice Defects, P. C. Gehlen, J. R. Beeler, Jr., and R. I. Jaffee, editors, Plenum Press, New York (1972).
12. S. Yoshioki, J. Phys. F6, 957 (1976).
13. A. A. Maradudin, E. W. Montroll, G. H. Weiss and I. P. Ipatova, Theory of Lattice Dynamics in the Harmonic Approximation, 2nd edition, Supplement 3, Solid State Physics, Academic Press, New York (1971), p. 353.
14. F. G. Mertens and W. Biem, Z. Physik 250, 273 (1972).
15. M. L. Klein and T. R. Koehler, J. Phys. C. 3, L 102 (1970); Phys. Lett. 33A, 253 (1970).

16. W. England, R. Etters, J. Raich and R. Danielowicz, Phys. Rev. Lett. 32, 758 (1974).
17. M. Ross, J. Chem. Phys. 60, 3634 (1974).
18. A. B. Anderson, J. C. Raich, and R. D. Etters, Phys. Rev. B14, 814 (1976).
19. V. V. Goldman, J. Low Temp. Phys. 24, 297 (1976).
20. W. N. Hardy, I. F. Silvera, K. N. Klump, and O. Schnepp, Phys. Rev. Lett. 21, 291 (1968).
21. I. F. Silvera, W. N. Hardy, and J. P. McTague, Phys. Rev. B5, 1578 (1972).
22. W. Schott, Z. Physik 231, 243 (1970).
23. J. C. Raich, A. B. Anderson, and W. England, J. Chem. Phys. 64, 5088 (1976).
24. R. Oyarzun and J. Van Kranendonk, Phys. Rev. Lett. 26, 646 (1971); Can. J. Phys. 50, 1494 (1972).
25. L. Amstutz, J. Thompson, and H. Meyer, Phys. Rev. Lett. 21, 1175 (1968).

D-7

ACOUSTIC VELOCITY RATIOS IN SOLID ARGON AT 75 K UP TO STATIC PRESSURES OF 150 KBAR

C. G. Homan, J. Frankel, D. P. Kendall, J. A. Barrett & T. E. Davidson

US Army Benet Weapons Laboratory, Watervliet Arsenal
Watervliet, New York USA

INTRODUCTION

In the last decade, a considerable body of accurate high pressure thermodynamic data has been reported for solid argon [1].* Careful measurement of the equation of state (EOS) by piston displacement techniques to 20 kbar, first made by Stewart [2] and recently by Anderson and Swenson [3], agree within 3%. The latter data [3] has an estimated accuracy of ± 0.1% in the T = 0 equilibrium molar volume V_o, and the agreement with the lower pressure isochoric data of Lewis et al. [4] is within their combined experimental uncertainties. This high degree of agreement has led Anderson and Swenson to propose a Lennard-Jones type EOS for the temperature range up to the melting point. Their results suggest that argon obeys a reduced EOS at both T = 0 and near the triple point, which is similar to the EOS of the other rare gas solids, krypton and xenon.

In order to measure EOS information to pressures in excess of 20 kbars, it is necessary to use solid high pressure systems similar to that used recently by Kendall et al. [5] to develop a cryogenic Bi phase diagram from resistometric data [6]. Measurement of the ultrasonic velocity ratio of NaCl to the B_1-B_2 transition at ambient temperatures using a modified Ahrens and Katz ultrasonic interferometric technique [7] has been reported recently by Frankel, Rich and Homan [8]. From it a room temperature EOS isotherm was inferred that was in excellent agreement with the reported x-ray data of Decker et al. [9].

Barker and Dobbs [10] measured the temperature effect on ultrasonic velocities at room pressure in argon using an interferometric technique and reported that the velocity ratio was essentially in-

*This reference contains an extensive review of the current high pressure literature on argon.

variant up to the melting temperature at room pressure.

This paper will present experimental data on the pressure variation of the ultrasonic velocity ratio to about 150 kbars. Using an analysis based on the Ahrens and Katz technique [7], these yield the 75 K isotherm of the argon EOS.

EXPERIMENTAL

Figure 1 shows the details of the sample region of our variably supported Bridgeman anvil device. This device and its associated cryogenic system are described elsewhere [5,6]. The innermost ring gasket, made of boron nitride and coated with parlodian to prevent metallic adsorption, has been subsequently coated with a thin, vapor deposited film of high purity Bi. With this arrangement, simultaneous measurements can be made of the ultrasonic velocity ratio of Ar and the resistometric trace of Bi embedded in the Ar sample. The outer pyrophyllite gaskets are designed to compress the starting liquid Ar sample material from about 40 atm and 95 K to a solid sample entrapped in the sample area at several kbars pressure. The temperature is then lowered to the desired run temperature and measured with thermocouples calibrated by a dip test in liquid nitrogen against an ice point reference junction. The accuracy of the thermocouples are estimated to be ± 1.0 K. The actual experimental data are taken isothermally within a range of ± 2 K of the stated temperature.

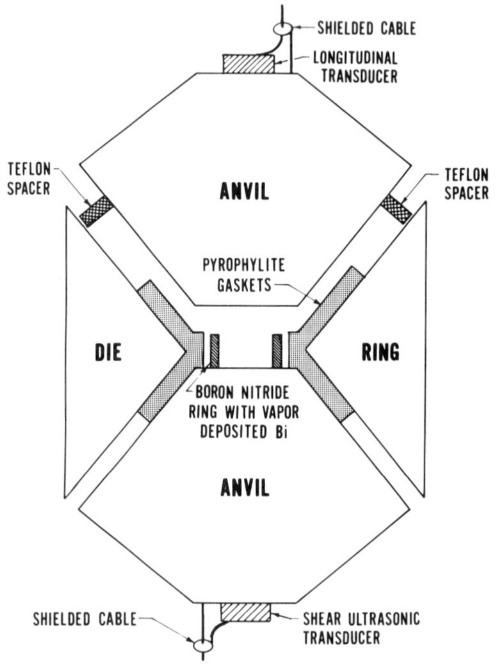

Fig. 1. Schematic of sample area.

When contact is made with the Bi coated boron nitride ring forming the sample, the sample height is 0.122 ± 0.002 cm and the nominal pressure is about 200 bars, the solidification pressure at the sample forming temperature. After several runs on each sample and careful unloading, the boron nitride ring is removed intact. Measurements at room temperature indicate that a permanent deformation of about 10% has occurred in the ring; however, the cylindrical shape has been nearly retained. Typical deformations are a 11% decrease in height and a 10% decrease in diameter even after several runs of nearly 150 kbars at liquid nitrogen temperatures on each sample. These permanent deformations cannot be used to estimate the final high pressure volume, however, since a large recoverable hydrostatic deformation must be occurring as well at these pressures. These deformations are a sample of the pressure environment of the Ar specimen and will be the basis of our assumption of hydrostatic deformation used in the analysis of the data.

Figure 2 represents the smoothed raw data reduced to the fundamental frequencies of each acoustic mode, the ratio of fundamentals (which is equivalent to the velocity ratio), and the pressure calibration determined from the Bi pressure sensor embedded in the solid Ar sample. The pressure is determined from the polymorphic Bi transitions reported earlier [6], whose pressure levels have been redetermined recently [11]. The velocity ratio is determined with an accuracy of about 2%.

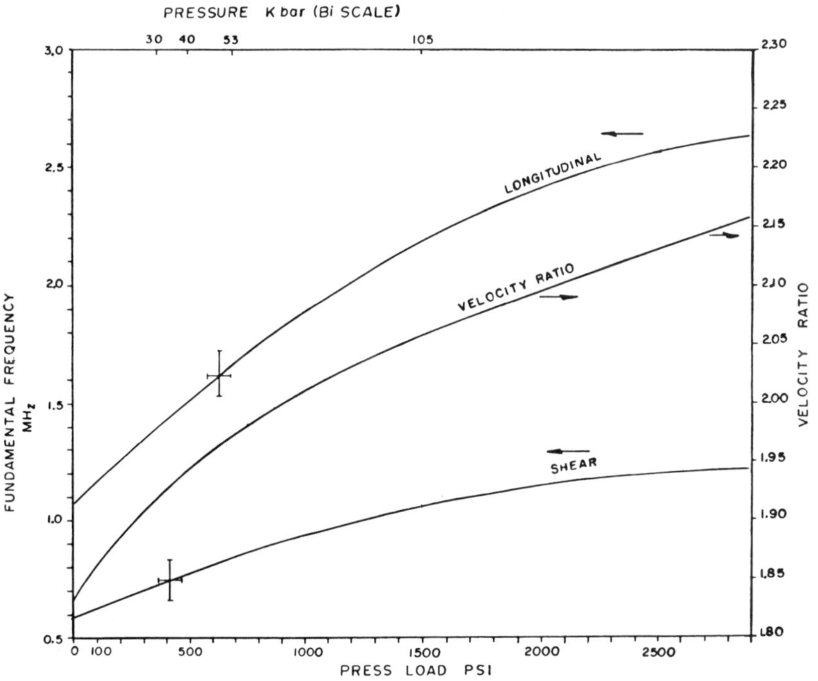

Fig. 2. Smoothed fundamental ultrasonic frequencies and velocity ratio of Ar at 75 ± 2 K. (See text for description of the Bi pressure scale shown.)

THEORY

Assuming hydrostatic deformation, the specimen density and thickness can be determined from the relationship [7,8]

$$\frac{d_o}{d} = 1 + \frac{1}{12 d_o^2 \rho_o} \int_o^P Y dP \qquad (1)$$

The subscript o refers to room pressure values. The function Y is defined by

$$Y = \frac{1 + \Delta}{\Delta f_p^2 - \frac{4}{3} \Delta f_s^2} \qquad (2)$$

where Δf_p and Δf_s are the measured fundamental frequencies and $(1 + \Delta)$ is the conversion between the adiabatic and isothermal bulk moduli and is equivalent to the ratio of specific heats.

The pressure variation of $(1 + \Delta)$ was determined from the relationship

$$\Delta = \frac{TV\, B_T \beta^2}{C_V} \qquad (3)$$

The pressure variation in the molar volume, V, and the isothermal bulk modulus B_T were taken from the data of Anderson and Swenson [3] and the pressure variation in the volume expansibility was estimated from their data. The pressure variation in specific heat C_V was determined by assuming Ar to be a Debye solid. The pressure effect on the Debye temperature can be determined directly from our measured fundamental frequencies using the following relationship [12] in cgs units:

$$\Theta_D^3 = 3.8159\text{E-}7 \frac{\rho_o \ell_o^3}{A} \left(\frac{2}{\Delta f_s^3} + \frac{1}{\Delta f_p^3} \right)^{-1} \qquad (4)$$

assuming hydrostaticity. The values used to determine Δ are tabulated in Table I; however, the net result is that the ratio of specific heats $(1 + \Delta)$ varies from 1.37 at room pressure to nearly unity at 20 kbars.

The resulting smoothed values of $(d/d_o)^3$ vs. P obtained from this analysis are shown in Fig. 3, and under the assumptions of this analysis are equal to the reduced volume V/V_o. Our data begin at about 5 kbar and the dashed line represents our extrapolation to room pressure. The estimated error in reduced volume varies from about 5% at 40 kbar to about 10% at 100 kbar. Also shown are the experimental data points of Anderson and Swenson [3].

Using the value of d calculated from (1), the ultrasonic velocities were calculated and are shown in Fig. 4. Also shown are the RP velocities of Barker and Dobbs [10].

Table I. Summary of Calculation of the Ratio of Specific Heats $(1+\Delta)$ at 75 K

P, kbar	$V^{\dagger}$, cm^3/mol	$B_T{}^{\dagger} \times 10^9$, $dynes/cm^2$	$\beta^* \times 10^{-4}$	$C_V{}^{\#} \times 10^8$, ergs/mole K	$1+\Delta$
0	24.28	14.1	18	2.26	1.37
2	21.67	30	10.3	2.24	1.120
4	20.48	42	6.1	2.22	1.108
6	19.69	60	4.5	2.20	1.080
8	19.08	72	2.5	2.18	1.035
10	18.58	90	2.3	2.15	1.031
12	18.16	100	2.2	2.13	1.031

†Data of Anderson and Swenson [3].

*Calculated from data of Anderson and Swenson [3].

#Calculated from present data (see text).

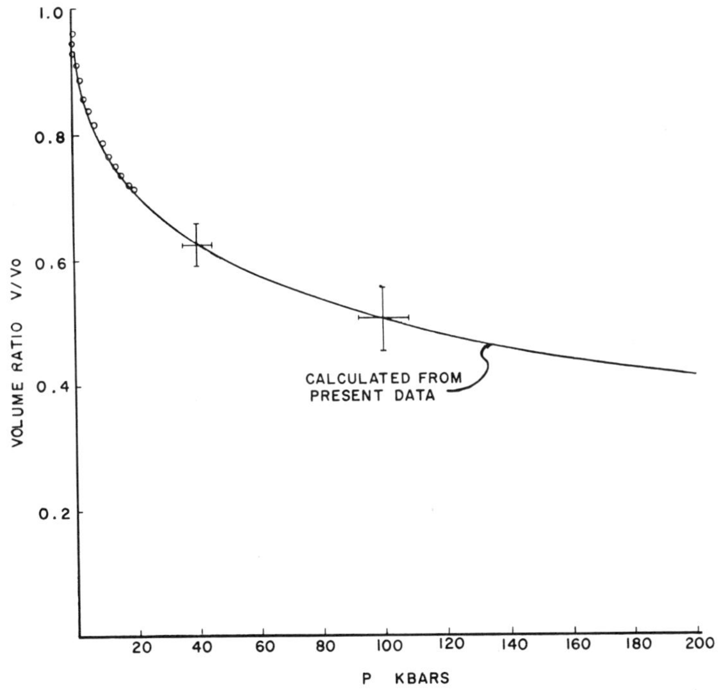

Fig. 3. The EOS of Ar at 75 ± 2 K. The data of Anderson and Swenson [3] to 20 kbars is also shown.

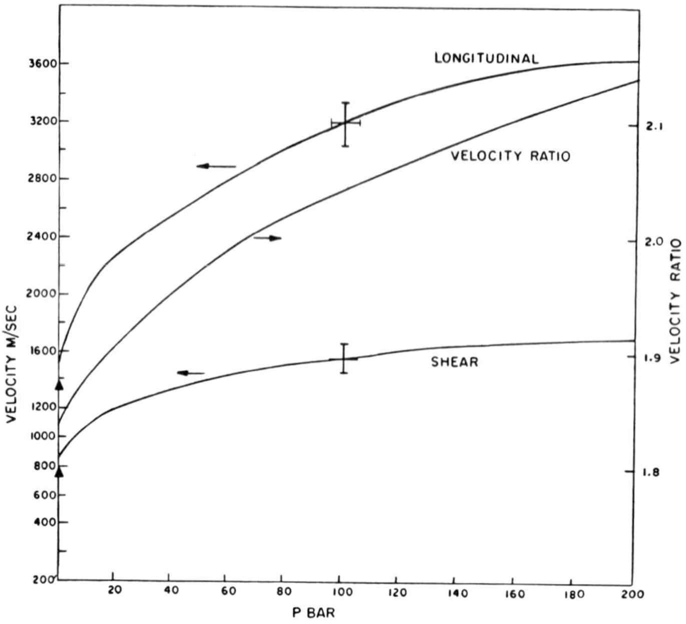

Fig. 4. Longitudinal and shear ultrasonic velocities as a function of pressure determined from the Bi pressure scale [11]. T = 75±2K. The room pressure velocities of Barker and Dobbs [10] and the velocity ratio are also shown.

DISCUSSION

The previously reported Bi pressure scale [6] yielded reasonable agreement with the piston cylinder measurements up to 16 kbar; however, above that pressure the smoothed fit to our data was in poor agreement with the Anderson Swenson data [3].

A small revision was made in the Bi pressure scale, which brought our data fit into excellent agreement with the Anderson Swenson data to the limit of their data. (A full discussion of this pressure reevaluation can be found elsewhere [11], however a downward revision in transition pressures of about 6% for the Bi I-III transition to 12% for the Bi VIII-IX transition was made.)

We are in fair agreement with the 0 K reduced EOS isotherm of Anderson and Swenson above 20 kbar, our data suggesting a slightly more compliant solid.

Comparing our results to the x-ray data of Syassen and Holzapfel [13] on solid Xe at 85 K ($\sim$ Tm/2) who observe a 43% change in volume at 108 kbar, indicates that Ar is more compressible. This result is not surprising, since the modulus of Xe is greater than Ar at lower temperatures [3]. Our data suggests that $B_T \sim 540$ kbar and $dB/dP \sim 5.0$ at 100 kbar in Ar at 75 K which is in reasonable agreement with the x-ray data [13]. We note however, that using the Ar velocity ratio pressure scale, we may state that at v_p/v_s equal to 2.05, the reduced volume is $0.50 \pm .05$.

The room pressure velocity ratio is in excellent agreement with the Barker and Dobbs value [10]. There are no high pressure velocity measurements in Ar near this temperature known to compare the present velocity ratio results to those obtained earlier on NaCl [8]. The surprising result is that the velocity ratios for both solids are remarkably similar, even though the absolute value of the velocities and moduli are quite different. In fact, the velocity ratio pressure scale developed earlier [8] could have been used to evaluate our pressure calibration in Ar with the same precision as the procedure used in this paper. If this observation is accurate, then one must conclude that the pressure effect on the ratio B_s/μ is similar for both materials and is not a strong function of temperature. This observation must be considered highly speculative at this time and additional isotherms for both solids must be obtained and carefully analyzed.

CONCLUSIONS

The simultaneous measurements of the velocity ratio of rare gas solid and the resistometric trace of a pressure calibrant material such as Bi can yield definitive EOS data for these solids. Conversely, other metallic transitions can be studied using the velocity ratio of the rare gas solid as the pressure calibrant material. The latter technique will be invaluable in the study of metastability, since the pressure medium is the pressure sensor and is independent of the frictional hysteresis effects endemic of solid high pressure systems. Our data suggests that solid Ar is a more compressible solid near its melting point than previously believed.

ACKNOWLEDGMENTS

The authors would like to thank M. Anderson and C. A. Swenson for sending their data on Ar and for suggestions on the data analysis. They would also like to thank D. McWhan, who has made very helpful suggestions concerning the Bi pressure scale and I. Spain, who pointed out the Xe reference.

NOTATION

B_T = isothermal bulk modulus
B_s = adiabatic bulk modulus
d = specimen height
P = pressure
v = velocity
V = volume
T = temperature
T_m = melting temperature
Δfp = longitudinal fundamental frequency
Δfs = shear fundamental frequency
ρ = density
μ = rigidity modulus
β = volume expansivity
C_V = specific heat at constant volume
Θ_D = Debye temperature
ℓ = acoustic path length
A^o = atomic weight

REFERENCES

1. R. K. Crawford, W. F. Lewis, and W. B. Daniels, J. Phys. C Solid State 9, 1381 (1976).

2. J. W. Stewart, J. Phys. Chem. Solids 29, 641 (1968).
3. M. S. Anderson and C. A. Swenson, J. Phys. Chem. Solids 36, 145 (1975).
4. W. F. Lewis, D. Benson, R. K. Crawford, and W. B. Daniels, J. Phys. Chem. Solid 35, 383 (1974).
5. D. P. Kendall, P. V. Dembrowski, and T. E. Davidson, Rev. Sci. Instr. 46, 629 (1975).
6. C. G. Homan, J. Phys. Chem. Solids 36, 1249 (1975).
7. T. J. Ahrens and S. Katz, J. Geophys. Res. 67, 2935 (1962).
8. J. Frankel, F. J. Rich, and C. G. Homan, J. Geophys. Res. 81, 6357 (1976).
9. D. L. Decker, W. A. Bassett, L. Merrill, H. T. Hall, and J. D. Barnett, J. Phys. Chem. Ref. Data 1, 773 (1974).
10. J. R. Barker and E. R. Dobbs, Phil. Mag. 46, 1069 (1955).
11. C. G. Homan, J. Frankel, and D. P. Kendall, paper presented at Intern. Conference on High Pressure and Low Temperature Physics, Cleveland, Ohio, July 1977.
12. F. J. Rich, C. G. Homan, and J. Frankel, Bull. Am. Phys. Soc. 21, 244 (1976).
13. K. Syassen and W. Holzapfel, in Electronic Properties of Solids Under High Pressure, Leuven, Belgium (1975), p. 87.

E-1

THE THERMODYNAMICS OF MELTING FOR ALKALI METALS*

I. N. Makarenko, A. M. Nikolaenko, and S. M. Stishov

Institute of Crystallography of Academy of Sciences
Moscow, U.S.S.R.

INTRODUCTION

In spite of the fact that there is no general theory of melting, the status of this phenomenon for simple substances of the argon-type is clear enough. Real and "machine" experiments indicate the prevailing role of the short-range repulsive interactions in the crystallization of rare gases [1,2]. These results indicate the important conclusions concerning the melting thermodynamics of rare-gas systems at very high pressures. In particular, since with increasing density the contribution of the repulsive interaction to the free energy and pressure of the rare-gas system will also increase, we can insist that the following asymptotic relations take place at rare-gas melting:

$$\left.\begin{aligned} \Delta S/R &\to \text{Const} \\ \Delta V/V_s &\to \text{Const} \\ P_m &\to \alpha\, T_m^c \end{aligned}\right\} \qquad P \to \infty \qquad (1)$$

The relations in (1) lead immediately to the derivations of the unlimited increase of the melting temperature on compression and of the impossibility of a critical point on the melting curve.

Unfortunately these results can not be extended to metals, due to the far more complex character of interaction, even in the case of alkali metals. Moreover, from experimental data [3,4] the behavior of the melting curves in the case of alkali metals seems to be different from that of rare gases. In fact, the melting curve of cesium reveals two maxima, and the slope of the melting curve of potassium and rubidium becomes extremely small at high pressures.

*Invited paper.

Thus, the question regarding the melting of metals requires further study. This paper is devoted mainly to the experimental study of the melting of alkali metals at high pressure; but first we shall make some remarks to help clarify the problem and then attempt to obtain preliminary information based on the analysis of simple models and physical speculations.

PRELIMINARY REMARKS

Initially it will be useful to consider the form of the total energy of alkali metals. Later we shall make use of the pseudo-potential approach for this purpose [5].

According to the pseudo-potential concept in the second-order of the perturbation theory for the energy of non-transition metals, excluding the thermal excitations, we have

$$E = E_i + E_e^{(0)} + E_e^{(1)} + E_e^{(2)} \qquad (2)$$

where $E_i = - Z^2e^2/r_a$ is the electrostatic energy of the ions in a background of neutralizing uniform charges (the Madelung energy), $E_e^{(0)}$ is the energy of the interacting electron gas, and $E_e^{(1)}$ is the mean value of the energy due to the non-Coulomb portion of the electron-ion interaction. For a local pseudo-potential and for the condition of electroneutrality of the metal, $E_e^{(1)} = b/V$, where b is a constant determined by the properties of a bare pseudo-potential, and $E_e^{(2)}$ is the correction of the second term to the energy of electrons due to an inhomogeneity of the electron gas in real metals. We can also describe $E_e^{(2)}$ as the energy of the indirect pair interaction of ions.

It is important to emphasize that of all the components of the total energy only the ion energy E_i and the indirect pair interaction energy $E_e^{(2)}$ depend on the ion configuration. With respect to melting as an order-disorder transition [2], one should consider only these parts of the energy as being responsible for the liquid-crystal phase transition in alkali metals.

Since the magnitude of the energy $E_e^{(2)}$ in alkali metals compared with the corresponding magnitude of the ionic energy E_i is rather small, we should expect that the energy of a system of point charges, or the ionic energy, plays an overwhelming role in the crystallization. It should be noted here that a classical system of point charges in a uniform compensating background is of great interest as a plasma model. As is known from Monte Carlo [6-8] calculations such a system exhibits the crystal-liquid transition which occurs when the characteristical plasma parameter $\Gamma = Z^2e^2/r_a kT \approx 150$. We shall need this information later.

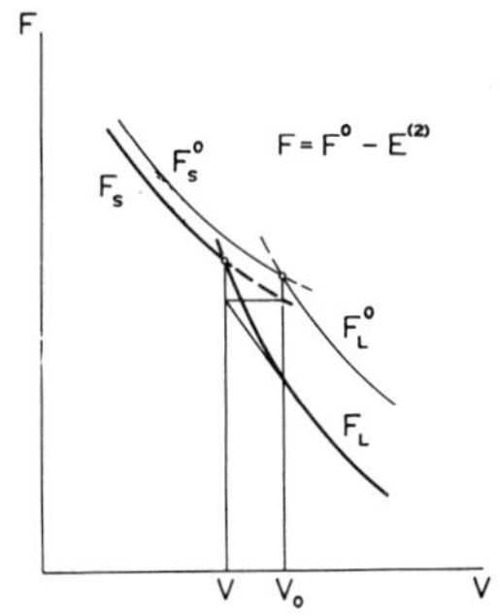

Fig. 1. Effects of the energy of indirect pair interaction of ions on the volume coordinate of melting.

However, we need to remember that the energy $E_e^{(2)}$ can play an important role in the melting of alkali-metals, despite its small contribution to the total energy. The reason for this conclusion is connected with the fact that the thermodynamic properties of phase transitions are determined not by the absolute values of the various quantities, but by the differences in their values in the coexisting phases.

Now we are ready to discuss the basic features of the melting problem in the case of alkali metals. The main contribution to the structure-sensitive energy is given by the ionic energy E_i, i.e. the energy of a system of point charges in a uniform compensating background. But as it follows from the form of the ionic energy $E_i \sim 1/V^{1/3}$, this system possesses negative pressure and compressibility, and consequently the stability of alkali metals is ensured by the pure volume terms of the total energy which are the mean-values of the non-Coulomb portion of the electron-ion interaction $E_e^{(1)} \sim \frac{1}{V}$, and the energy of the interacting electron gas $E_e^{(0)}$. The value of the latter has a small dependence on volume at moderate densities, but at high compression the energy of the interacting electron gas has the form $E_e \sim 1/V^{2/3}$. Thus at unlimited compression the free energy and pressure of alkali metals are determined primarily by $E_e^{(1)}$, whereas the existence of the melting-crystallization phenomenon and the form of the melting curve in V-T coordinates are ensured by the ionic energy E_i and partly by the energy $E_e^{(2)}$. The latter statement will be easy to understand if we recall that pure volume components of energy can not alter the point of volume coordinate at which the intersection of two branches of the free energy occurs.

Thus, due to the situation in the case of alkali metals we cannot write any asymptotic relations based simply on the knowledge of melting in any reference system as we can do in the case of rare gases. Nevertheless, it appears possible using very simple speculations to obtain information about the melting of alkali metals and to compare this with experimental data.

First, we shall consider the problem regarding the form of the melting curve in the V-T plane. If the ionic energy E_i is the only structure-sensitive energy in alkali metals, the melting curves in the V-T plane will have the form $T \sim 1/V^{1/3}$ and the dimensionless parameter $\Gamma = Z^2e^2/r_akT$ will have the same value (~ 150) for all alkali metals. However, the values of the parameter Γ are different for various alkali metals at normal pressure and as a whole they are higher ($\sim$ 180-220) than the value characterizing the melting of the classical one-component plasma.

To explain these results we must take into account the other structure-sensitive component of energy $E_e^{(2)}$. The second-order correction to total energy is always negative [9] and is larger (less negative) for a solid than for a liquid [10]. The latter gives us a clue to explain the observed deviations of Γ for the melting of alkali metals from the corresponding ones for the classical one-component plasma. Figure 1 shows that the energy term $E_e^{(2)}$ changes the volume coordinate of the phase transition in the direction of smaller values and this results in an increase in the value of Γ.

As one can see from Fig. 1 this change in the condition of high pressure can be approximated by

$$V_m^o - V_m \approx \Delta E_e^{(2)}/P_m \tag{3}$$

where V_m^o is the volume coordinate of the phase transition in the pure Coulomb system, V_m is the volume coordinate of phase transition when the second-order term $E_e^{(2)}$ is added, and $\Delta E_e^{(2)}$ is the difference in the values of $E_e^{(2)}$ in the coexisting phases. When $E_e^{(2)} << PV$ we can rewrite (3) in the form

$$\Gamma \approx \Gamma_o + \frac{\Gamma_o \ \Delta E_e^{(2)}}{3 \ P_m V_m} \tag{4}$$

where Γ is the current value of the parameter, characterizing the ratio of the mean Coulomb energy to the thermal one along the melting curve, and Γ_o is the value of the same parameter for the pure Coulomb system.

Since thermodynamic parameters are not independent along the melting curve and taking into account that the right side of equation (4) must be dimensionless, we can transform (4) into the forms

$$\Gamma = \Gamma_o + f \ (V_m/V_c) \tag{5}$$

$$\Gamma = \Gamma_o + \phi \ (T_c/T_m) \tag{6}$$

where the parameters V_c and T_c must be connected with the corresponding ones, characterizing the electron-ion interaction.

To obtain information on the possible trends of Γ along the melting curves we need to make allowances for the forms of $\Delta E_e^{(2)}$ and P.

We shall assume that the term $E_e^{(2)}$ can be roughly approximated with the form α/V, where the constant α has different values for

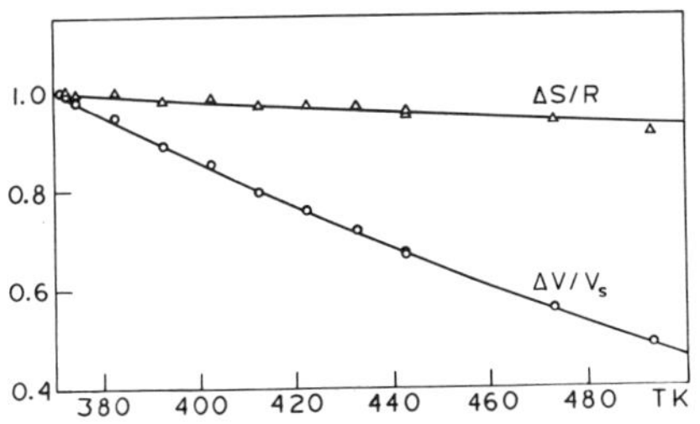

Fig. 2. Relative change of volume $\Delta V/V_s$ and change of melting entropy $\Delta S/R^s$ for sodium as functions of melting temperature. The values of $\Delta V/V_s$ and $\Delta S/R$ are given in reduced units. (The parameters of reducing are the values of corresponding quantities at atmospheric pressure.)

liquid and solid phases. This approximation for $E_e^{(2)}$ means that we have adopted a special form of pseudo-potential and neglected a variation in properties of the electron gas. This doesn't seem to be too unreasonable because of some cancellation effects.

Then approximating the function $P = f(V)$ along the melting curve with the form $P_m \sim V_m^{-n}$ we can obtain the following relation from (4):

$$\Gamma = \Gamma_o + C\, V_m^{n-2} \tag{7}$$

If $n > 2$, which is quite reasonable, we see from equation (7) that the values of Γ will decrease along the melting curve for alkali metals on compression and will tend to a constant Γ_o.

Now we shall deal with the behavior of the relative volume change at melting $\Delta V/V_s$ along the melting curve. For this discussion we shall consider the entropy change during melting as a constant quantity. (Details of this problem have been considered elsewhere [2]. We shall assume that for the high density system, the main contribution to the free energy and pressure is given by $E_e^{(1)}$ in (2); but the V-T coordinates of the melting curve and the energy change at melting are determined by the ionic energy E_i. Then, using the thermodynamic relation $\Delta U + P\Delta V = T\Delta S$ one obtains

$$\frac{\Delta V}{V_s} = \frac{T\Delta S}{V_s P} - \frac{\Delta U}{V_s P} \tag{8}$$

Taking into account that $T \sim 1/V^{1/3}$ and $P \sim 1/V^2$ we see from (8) that

$$\Delta V/V_s \approx C\, V_m^{2/3} \tag{9}$$

Thus, in contrast with the case of rare gases, the relative change of volume occurring during the melting of alkali metals is expected to tend toward zero on unlimited compression. Obviously this result will not be changed if we make use of the more general form for pressure, e.g. $P \sim V^{-n}$ for $n > 4/3$.

It should be emphasized that there is no room for melting maxima in our simple picture of melting for alkali metals and this interesting phenomena should be considered probably to be the result of change of the character of the electron-ion interaction which can occur at high pressures.

EXPERIMENTAL METHOD AND RESULTS

To obtain experimental data, characterizing the thermodynamics of the melting of alkali metals, we have performed measurements of the P-V-T relations along the melting curve for both liquid and solid phases. These measurements were made with the aid of a piston piezometer equipped with a resistive displacement pickup [11,12]. During the taking of measurements, the piezometer was situated inside a thick-walled high-pressure bomb, in which the pressure is created by an outer device. Details of the technique are described in the cited references. The accuracy of the volume measurements was no less than $\pm$ 0.01%. Temperature and pressure were measured with a precision of $\pm$ 0.01°C and $\pm$ 5 bar, respectively, at pressures up to 15 kbar; at pressures above the latter value the limits were $\pm$ 0.01°C and $\pm$ 25 bar.

Using the experimental P-V-T data we calculated changes of volume for the melting process. Then, on the basis of the Clausius-Clapeyron equation it was simple to calculate the discontinuities of enthalpy, internal energy, and entropy. The amount of impurities in the samples under study was not more than 0.01% for Na, K, Rb, and Cs; and not more than 0.05% for Li. The sample of Li was enriched with isotope Li^7 and contained 99.66% Li^7. The piezometers were filled with the samples in a vacuum of $\sim 10^{-6}$ torr.

The results of measurements and thermodynamic calculations for lithium, potassium, and rubidium are given in Table I. The corresponding results for sodium and cesium have been published earlier [13,14]. Figures 2 through 7 illustrate the behavior of various characteristics of melting along the melting curves of alkali metals.

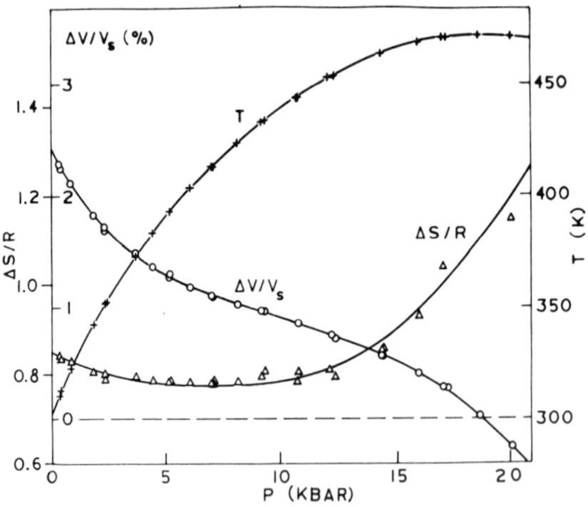

Fig. 3. Melting temperature T, relative change of volume $\Delta V/V_s$, and entropy of melting $\Delta S/R$ for cesium as functions of melting pressure.

DISCUSSION

In the second section we arrived at the conclusion that the expected behavior of the melting curves of alkali metals does not differ very much from the melting curves of rare gases. In fact, if the correction to the total energy, taking into account the inhomogeneity of the electron gas is a monotonic function of volume, the single new feature of the alkali metals melting as compared to the case of rare gases will be a zero asymptotic value for $\Delta V/V_s$.

Table I. Experimental Thermodynamic Data for Melting of Li, K, and Rb

	T°K	P, kbar	V, $\frac{cm^3}{mole}$	V_s, $\frac{cm^3}{mole}$	S/R	U, $\frac{cal}{mole}$
	453.76*	0.001	0.213*	13.353*	0.794	715.6
	454.25	0.149	0.211	13.335	0.791	713.0
	457.62	1.211	0.199	13.212	0.782	705.5
	462.90	3.005	0.182	13.013	0.776	700.7
Li	467.82	4.826	0.167	12.824	0.774	699.9
	472.80	6.836	0.151	12.630	0.766	694.6
	477.95	9.105	0.137	12.427	0.764	695.6
	483.05	11.581	0.124	12.216	0.760	695.2
	485.95	13.122	0.117	12.092	0.762	699.3
	336.96	0.001	1.173	45.982	0.844	565.2
	343.74	0.417	1.095	45.410	0.830	555.6
	359.70	1.488	0.945	44.075	0.811	546.1
	375.50	2.691	0.819	42.758	0.800	543.7
K	394.20	4.334	0.695	41.190	0.794	550.2
	412.78	6.253	0.584	39.660	0.787	558.5
	442.79	10.135	0.438	31.160	0.783	583.0
	452.76	11.693	0.396	36.338	0.782	592.9
	312.12*	0.001	1.450*	56.380*	0.828	513.5
	313.65	0.074	1.435	56.225	0.831	515.3
	332.78	1.077	1.190	54.250	0.815	507.9
	352.67	2.310	1.000	52.210	0.809	511.7
Rb	372.65	3.773	0.840	50.200	0.803	518.6
	392.80	5.518	0.705	48.175	0.799	530.5
	412.70	7.566	0.584	46.203	0.788	540.4
	432.61	10.016	0.482	44.225	0.781	556.0
	452.64	12.995	0.395	42.250	0.778	576.7

Extrapolated values.

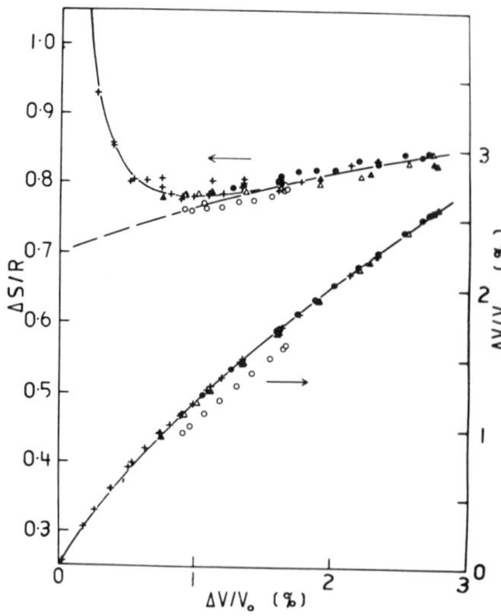

Fig. 4. Relative volume discontinuity $\Delta V/V_s$ and entropy discontinuity $\Delta S/R$ in the melting of alkali metals as functions of reduced volume discontinuity $\Delta V/V_o$. Legend: O - Li; ● - Na; Δ - K; ▲ - Rb; + - Cs. V_o is specific volume of metal at T = 0 K and at atmospheric pressure.

As one can see from Fig. 2, illustrating the typical behavior of $\Delta S/R$ and $\Delta V/V_s$ for Na as functions of temperature, there are no data indicating a limited asymptotic value for $\Delta V/V_s$ but $\Delta S/R$ shows no tendencies different from those noted for the melting of rare gases. Figure 4 is more informative from the point of view of establishing the asymptotic values of $\Delta S/R$ and $\Delta V/V_s$. Here $\Delta S/R$ and $\Delta V/V_s$ are presented as functions of the volume discontinuity ΔV. It follows from Fig. 4 that $\Delta V/V_s$ tends to zero along with ΔV for all alkali metals. Yet we know in the case of cesium that ΔV approaches zero at limited pressure and volume, and thus Fig. 2 cannot distinguish between the two essentially different cases when $\Delta V/V_s$ tends to zero at unlimited compression or when it occurs at some limited pressure.

In this connection it is useful to consider Fig. 5, where $\Delta V/V_s$ is plotted as a function of the molar volume V_s. We see from Fig. 5 that the behavior of the initial portions of the corresponding curves for all alkali metals does not contradict our conclusion that $\Delta V/V_s$ tends to approach zero along with the volume V_s (see equation 9). But the dependence $\Delta V/V_s = f(V_s)$ for Cs changes its form rather sharply at high pressures and $\Delta V/V_s$ for Cs becomes zero at finite volume. This peculiarity of the behavior of Cs gives us some justification to consider the maximum in the melting curve of Cs as an "anomalous" phenomenon.

It is important to know whether this phenomenon is common to all alkali metals or whether it is a specific property of Cs. Searching for an answer we call attention to the behavior of the entropy change along the melting curve of Cs (see Fig. 3). One can see that the melting entropy of Cs increases rather rapidly near the maximum on the melting curve. It is interesting to note also that, as seen in Fig. 5, the behavior of the melting entropy of K and Rb seems to be somewhat different from the corresponding behavior of Li and Na.

Additional information can be extracted from Fig. 6 which demonstrates the interrelationship between the melting entropy $\Delta S/R$ and the relative volume discontinuity $\Delta V/V_s$. The dependence of

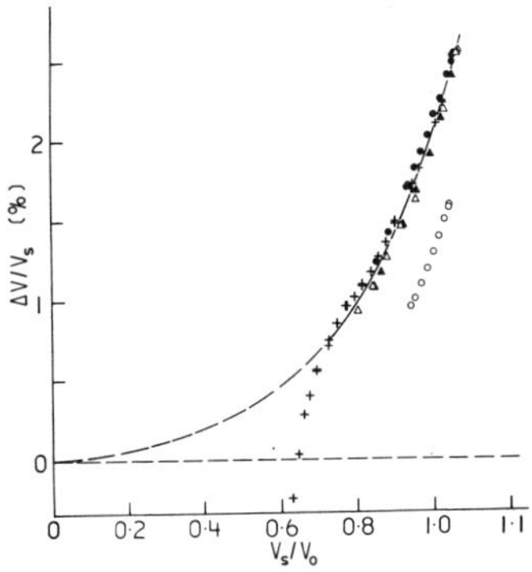

Fig. 5. Relative volume discontinuity $\Delta V/V_s$ as a function of reduced melting volume V_s/V_o. For notations see Fig. 4.

$\Delta S/R = f(\Delta V/V_s)$ is divided into two branches at low values of $\Delta V/V_s$; one branch for Li and possibly for Na, and another one for K, Rb, and Cs. It should be noted here that K and Rb as well as Cs, in contrast with Li and Na, have unfilled d-states at low pressure.

We would like to point out that such a result explains, to some extent, the observed flattening of the melting curves of K and Rb at high pressure [4] and thus the existence of the melting maxima in alkali metals is most probably connected with s-d electronic transitions.

Let us now return to the question of the form of the melting curve of alkali metals in V-T coordinates. Consider Fig. 7, which depicts the temperature and volume dependences of the parameter $\Gamma = Z^2e^2/r_a kT$. The experimental data in this figure confirm our predictions that the parameter Γ decreases and tends to a constant value along the melting curves. We can see that the extrapolation of the corresponding dependences leads to a limited value of Γ which is not over 100. This result is not in agreement with recent estimates for the classical one-component plasma [7,8] and requires a further analysis.

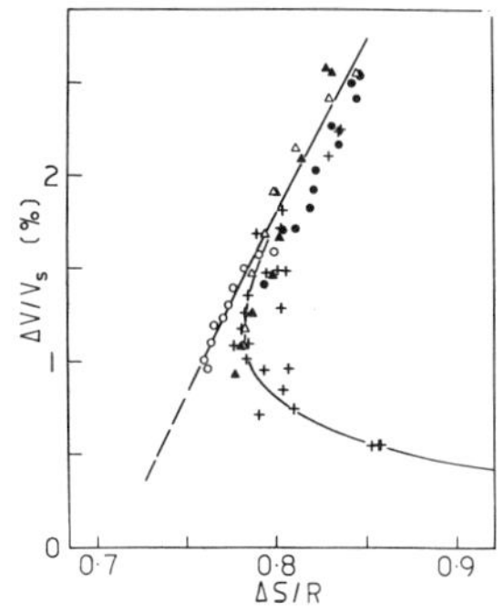

Fig. 6. Relationship between the relative volume discontinuity $\Delta V/V_s$ and entropy discontinuity $\Delta S/R$ in the melting of alkali metals. For notations see Fig. 4.

In conclusion we would like to emphasize that the experimental data on high-pressure melting of alkali metals are in complete accordance with the interpretation of the melting as an order-disorder transition. It is the nature of melting that makes the thermodynamics of the melting of alkali metals and of rare gases so similar. Naturally the basic feature of ideal melting (constancy of the melting entropy and increase of the melting point on compression) can be disturbed by such phenomena as electronic transformations, etc. It can lead to various anomalies in the melting curves but cannot change the nature of melting.

NOTATION

a,b,c = constants
e = electron charge
F = free energy
k = Boltzmann constant
N = Avogadro's number
n = constant

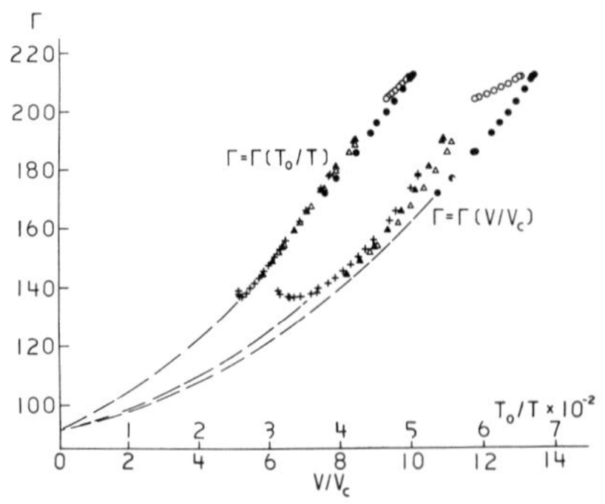

Fig. 7. Parameter Γ as functions of reduced melting temperature T_c/T and reduced melting volume V/V_c for alkali metals. ($T_c = 1/r_c$; $V_c \cong 4/3\ \pi\ r_c^3$, where r_c is the Ashcroft core radius. The numerical values of r_c are taken from Price et al. [15].)

P = pressure
P_m = pressure at melting point
R = gas constant
r_a = radius of a sphere containing the unit charge given by $4\pi r_a^3/3=V/N$
S = entropy
ΔS = change of entropy upon melting
T = temperature
T_m = temperature of melting
U = internal energy
ΔU = change of internal energy upon melting
V = specific or molar volume
V_m = specific volume at melting point
V_s = specific volume of solid phase at melting point
ΔV = change of volume upon melting
Z = valency of ion
α = constant

REFERENCES

1. W. G. Hoover and M. Ross, Contemp. Phys. 12, 339 (1971).
2. S. M. Stishov, Sov. Phys. -Uspekhi 17, 625 (1975).
3. G. C. Kennedy, A. Jayaraman, and R. C. Newton, Phys. Rev. 126, 1363 (1962).
4. H. D. Luedemann and G. C. Kennedy, J. Geophys. Res. 73, 2795 (1968).
5. W. A. Harrison, Pseudopotentials in the Theory of Metals, Benjamin, New York (1966).
6. S. G. Brush, H. L. Sahlin, and E. Teller, J. Chem. Phys. 45, 2102 (1966).
7. E. L. Pollock and J. P. Hansen, Phys. Rev. A8, 3110 (1973).
8. H. E. Dewitt, Phys. Rev. A14, 1290 (1976).
9. L. D. Landau and E. M. Lifshitz, Quantum Mechanics, Addison-Wesley, Reading, Massachusetts (1965).
10. D. Straud and N. W. Ashcroft, Phys. Rev. B5, 371 (1972).
11. I. N. Makarenko, V. A. Ivanov, and S. M. Stishov, Instr. Exp. Tech. (USSR), 3, 862 (1974).
12. I. N. Makarenko, A. M. Nikolaenko, and S. M. Stishov, in Liquid Metals, R. Evans and D. A. Greenwood, eds., Institute of Physics Conference Series number 30 (1977).
13. V. A. Ivanov, I. N. Makarenko, A. M. Nikolaenko, and S. M. Stishov, Phys. Lett. 47A, 75 (1974).
14. I. N. Makarenko, V. A. Ivanov, and S. M. Stishov, JETP Lett. 18, 187 (1973).
15. D. L. Price, K. S. Singwi, and M. P. Tosi, Phys. Rev. B2, 2983 (1970).

E-2

MELTING CURVES OF IRON AT HIGH PRESSURES

E. Boschi, D. Fazio, and F. Mulargia*

Università di Bologna
Bologna, Italy

INTRODUCTION

The melting curve of iron, the main constituent of the Earth's core, plays a fundamental role in geophysics since its knowledge allows us to place a lower limit on the temperature of the outer core, which we know from seismology to be in a fluid state. Moreover, comparisons between melting curves and the adiabatic temperature distributions would make possible speculations about the thermodynamics of the core and the origin of the geomagnetic field.

At present, the problem is to find ways of extrapolating the melting curve up to pressures much higher than those at which experimental data are available.

MELTING THEORY

Thermodynamically rigorous general theory states that a phase transition occurs when the three conditions of equality between the free energies, the temperature, and the pressures of the different aggregation states are satisfied. An exact treatment then should take into account both phases. Lindemann's theory, in its original formulation, is not founded on equilibrium conditions since it considers the vibrations of atoms in the crystal lattice of the solid phase only and says nothing about the liquid nor how the solid changes into liquid. Stishov [1] states that the instability caused by the thermal vibrations of the atoms in the cyrstal lattice coincides with the instability limit of the crystal, but not with the equilibrium point for the solid-liquid transition. However, there is a close relation between the instability point of the crystal and the melting process. For instance, there are theories

*Present address: University of California, Los Angeles, California.

which explain the melting process by the formation of defects or dislocations in the crystal; this is equivalent to comparing the instability of the crystal with the beginning of the melting process. Of course, generally there must be some relation between the melting temperature and the amplitude of the atomic oscillations, i.e., the instability of the crystal, since both quantities may be expressed in terms of the interatomic forces, but surely it is not sufficient to consider only the instability point. Moreover, if we want to have a general melting criterion we must be able to apply it to materials having different interatomic forces; this can be easily done when the interatomic potential explicitly appears in the considered criterion.

In summary, a melting theory must possess the following two fundamental characteristics: (1) melting must be regarded as an equilibrium process, then we must be able to treat both phases the same; (2) the interatomic forces must appear explicitly, since they rule the overall process of the melting. However, at the present state of research these conditions cannot be fulfilled.

A self-consistent approach is given by Ross [2], but his derivation is still based on the assumption of Lindemann's law and is therefore only a one-phase theory. Notwithstanding this limitation, Ross' theory seems to be the most generally available approach since all the other proposed melting criteria [3,4] are special cases of this theory.

CALCULATION OF THE MELTING CURVES

In calculating the melting curve, we will follow the Ross approach [2,5,6,7]. This approach is based on the reformulation of Lindemann's law for a one-molecule cell solid. In this model we can conveniently express the melting relation as a function of some accessory variables. In fact, along the melting curve it should be valid that

$$v_f/v = \text{constant} \tag{1}$$

where v_f is the free volume given by

$$v_f = \int_v \exp\left\{ - [E(\underset{\sim}{r}) - E(0)]/kT \right\} d\underset{\sim}{r} \tag{2}$$

where v is the cell volume, E is the potential energy in the cell, and T is the temperature. As is obvious from (2), the free volume can be easily computed as a function of E, i.e., by definition, from the simple knowledge of the intermolecular potential. For the intermolecular potential we use the Morse, Rydberg, and Lennard-Jones functions, whose constants are elsewhere [8]. Since our theory is applied to iron at the Earth's core conditions, we use the Liu's results [9] in determining the crystal structure diagram.

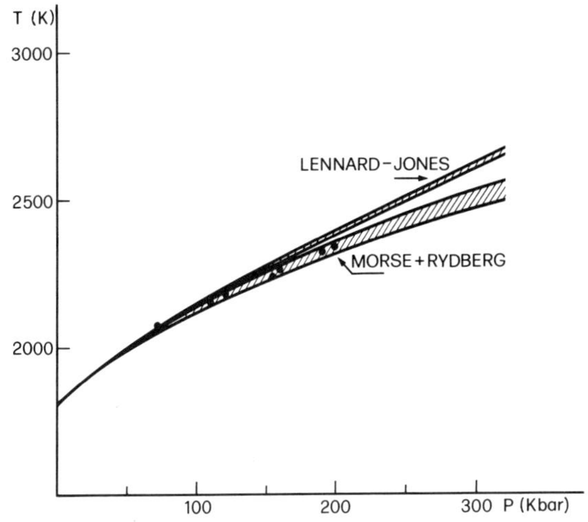

Fig. 1. Melting curves and melting temperature vs. pressure, as a function of the intermolecular potential. The Lennard-Jones, Morse and Rydberg potentials are plotted together with the experimental points at Liu and Bassett.

According to this diagram, iron at core conditions is in the phase ε. We find that the melting curve is sensibly dependent on the choice of the intermolecular potential function. The temperature values at the mantle-core and the inner-core-outer-core boundaries are 3750 ± 100 K, and 5250 ± 200 K for the Morse and Rydberg potentials, and 4450 ± 50 K, 6800 ± 200 K for the Lennard-Jones potentials. Also the average gradient depends strongly on the potential, being .661 ± .066 deg/km for Morse and Rydberg, and 1.036 ± .066 deg/km for the Lennard-Jones functions. These gradients, although different, are both larger than the adiabatic gradient calculated from the thermodynamic Grüneisen gamma function [10,11].

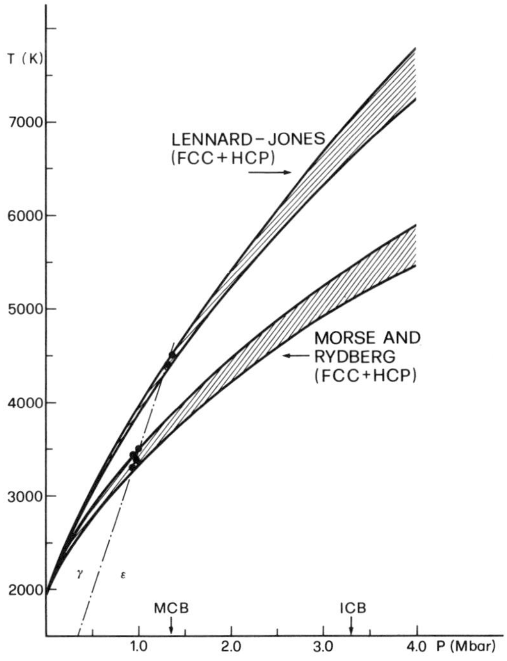

Fig. 2. Idem, for FCC and HCP iron, but with the range expanded up to 4 Mbars. Dots denote the extrapolated triple points by Liu.

ACKNOWLEDGMENTS

This work has been performed with a contribution from Consiglio Nazionale delle Richerche. The authors wish to thank G. Puppi for useful discussions.

NOTATION

C_{VH} = specific heat at constant volume in the quasi-harmonic approximation

E(r)= total potential field

E_H = harmonic internal energy
G = anharmonic coefficient of the free energy
k = Boltzmann constant
P = pressure
r = position of one atom in its cell
T = temperature
U_o = static lattice energy
v = volume of the cell
v_f = free volume
V = molar volume

REFERENCES

1. S. M. Stishov, Sov. Phys. Usp. 11, 816 (1969).
2. M. Ross, Phys. Rev. 184, 233 (1969).
3. E. A. Kraut and G. C. Kennedy, Phys. Rev. Lett. 16, 608 (1966); also Phys. Rev. 161, 668 (1966).
4. F. E. Simon and G. Glatzel, Zeits. Anorg. Alegem. Chem. 178, 309 (1929).
5. E. Boschi, Geophys. J. Roy. Astr. Soc. 38, 327 (1974).
6. E. Boschi, Rivista del Nuovo Cimento 5, 501 (1975).
7. D. Fazio, F. Mulargia, and E. Bòschi, Nuovo Cimento B, 40, 906 (1977).
8. E. Boschi, D. Fazio, and F. Mulargia, paper presented at 6th AIRAPT Intern. High Pressure Conference, Boulder, Colorado, July 25-29, 1977.
9. L. Liu, Geophys. J. Roy. Astr. Soc. 43, 697 (1975).
10. E. Boschi and F. Mulargia, J. Geophys. 43, 465 (1977).
11. D. Fazio, F. Mulargia, and E. Boschi, Geophys. J. R. Astr. Soc. 52, 113 (1978).

E-3

DETERMINATION OF MELTING ENTROPY BY THE MEASUREMENT OF THE VOLUME CHANGE OF MELTING AT HIGH PRESSURE

P. W. Mirwald

Ruhr-Universität-Bochum
Bochum, W. Germany

INTRODUCTION

The determination of phase transitions at high pressures and high temperatures represents an area of major interest in high-pressure research. To date such investigations have been mostly confined to locating the p-T coordinates of the phase boundary studied. Quantitative assessments of changes in the properties of state associated with the transition have been very rare.

For a first-order phase transition a fundamental relation between extensive and intensive parameters is the Clausius-Clapeyron equation $dT/dp = \Delta V/\Delta S$, where $\Delta S = \Delta H/T$. The measurement of either of the extensive parameters along a phase boundary of known slope is sufficient to determine this relation. Presently there are only a few such studies available and these are mainly measurements on melting curves, since this phenomenon has always evoked particular interest. Bridgman [1], Owens [2], and Ivanov et al. [3] have done pioneering work on some low-melting substances by measuring the volume changes caused by melting under pressure. These measurements were limited, however, to temperatures below 400°C and usually to pressures below 15 kbar, or at the most 25 kbar (Ivanov et al. [3]). Successful calorimetric experiments at pressures above 4 kbar or at elevated temperatures have not been reported to date.

Until recently an experimental determination of the volume change of melting using a piston-cylinder apparatus did not appear very promising. Friction phenomena inherent in solid-media presses were considered to be the most serious problems. Recent improvements of the conventional cell assembly to produce a low-friction cell (Mirwald et al. [4]) however, allow this source of error to be essentially eliminated. This substantial improvement seemed to favor the initiation of an exploratory study to 50 kbar in which

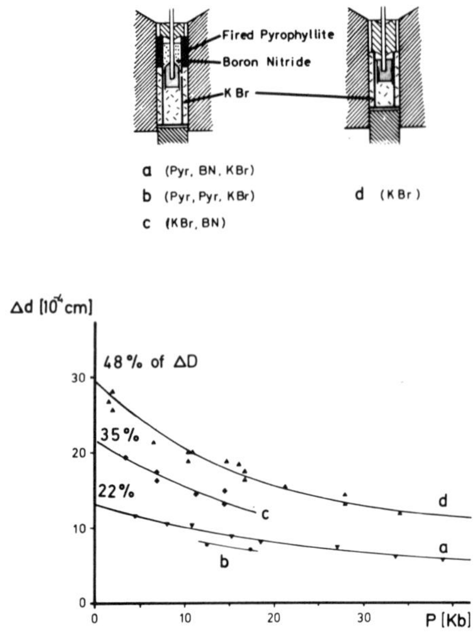

Fig. 1. Various cell assemblies employed in the experimental series (a) to (d) (top) and the corresponding piston displacement Δd (bottom). ΔD is the theoretically expected piston displacement at one atmosphere.

the volume change of melting at high pressure was determined by a piston-displacement technique. Sodium, lead, and silver were chosen as representative samples, since their melting curves cover a wide temperature range, between 100 and 1300°C, which was important with respect to the experimental technique

APPARATUS

The experiments were conducted with a conventional end-loaded piston cylinder high-pressure apparatus in a vessel with a bore diameter of 1.27 cm. The cell assembly was a modified version of the previously developed low-friction cell by Mirwald et al. [4]. The experimental arrangement is schematically shown in Fig. 1. This cell consists of two different cell halves: the upper half of the cell assembly is composed of a steel seal, soft-fired (925°C) pyrophyllite and boron nitride elements; the lower half contains a plastic-behaving pressure-transmitting medium. Various salt media were employed: for the experiments on silver, NaCl and CaF_2 were used; for those on lead and sodium, KBr and AgCl, respectively. In addition, the usual graphite heater was exchanged for a stainless steel furnace tube to accommodate a larger tantalum capsule with a sample of 0.2 cm^3. The temperature of the sample was monitored by two thermocouples in DTA arrangement. The piston displacement was recorded by two dc-transducers mounted on two wings which were attached to the binding ring of the piston. This device resolved piston displacements of the order of 10^{-5} cm. Since the experiments could not be performed at a constant pressure because of experimental limitations, they were all conducted at valved off main ram conditions. This resulted in a pressure change due to thermal expansion as well as to the volume change of melting during the measurement. The initial test runs resulted in the pressure change associated with melting being proportional to the piston displacement. Thus it could be used as a valuable parameter to augment the piston displacement data. The pressure change was monitored by a strain gauge connected to the hydraulic system. Accuracy was estimated to be ± 10 bars. Both parameters, piston displacement and pressure change, were simultaneously recorded on a two-channel strip-chart recorder.

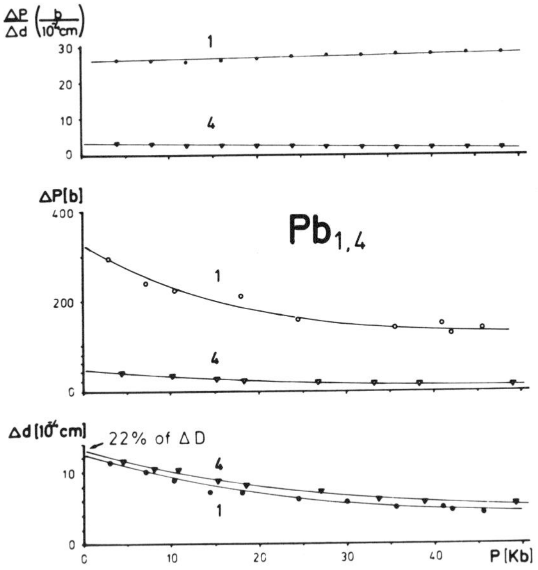

Fig. 2. Piston displacement Δd and pressure change Δp obtained for the volume change of melting of lead. The two data sets, Pb_1 and Pb_4, were performed with different capacities of the hydraulic ram (see text). Plot of Δp/Δd vs. p shows the linear proportionality between the parameters (top).

Before measurements were undertaken the cell assembly was slowly compressed to 10 to 15 kbar at room temperature and left closed off over a period of about 12 hrs. After a thermal stress equilibration of 30 min the heating excursions were usually initiated 40 degrees below the melting curve with a heating rate of 0.2°C/sec. The melting boundary was exceeded by about 15 degrees. After a pause of 10 to 15 min for thermal stress equilibration, the measurement was continued on a reversed temperature path.

While the heating or cooling of the cell causes a continuous change in the recordings of the piston displacement and the pressure change, the sudden volume change due to melting or freezing produces a signal of sigmoidal shape where the step height is the desired information. The precision of the signals was within ± 10%.

METHOD AND RESULTS

The most convenient working temperature for an appraisal of the proposed method seemed to be given with lead as the sample. In Fig. 2, two different sets of piston displacement (Δd) and pressure change (Δp) are displayed. These experiments were conducted with the standard cell assembly consisting of pyrophyllite-boron nitride and KBr elements. In data set Pb_1, the hydraulic system of the ram contained some 10 liters of pressure fluid.

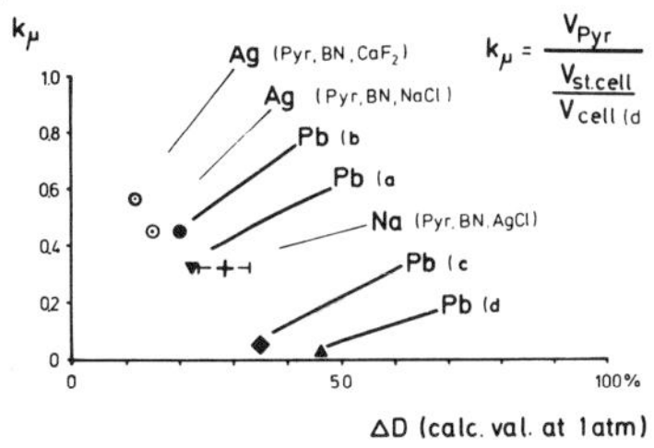

Fig. 3. Plot of the friction coefficient Kμ vs. ΔD. The horizontal bar bracketing the Na-data represents the estimated uncertainty of ± 5% for all data.

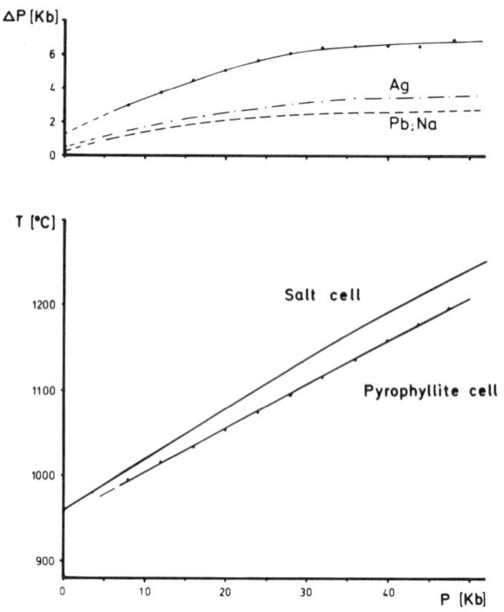

Fig. 4. Comparison of silver melting curves obtained with a low-friction salt cell and a prophyllite cell (bottom). The pressure difference ΔP between the two determinations is considered to represent the friction of the pyrophyllite cell (top). The dashed curves show what corrections must be made to the volume change data of sodium, lead, and silver because of friction losses.

In data set Pb_4 the capacity of the hydraulic system was enlarged by a 50-liter steel bottle. The only significant difference between the two data sets was in the observed Δp-data. The extrapolation to one atmosphere yielded a ratio of $\Delta p^1_o : \Delta p^4_o = 320 : 50$ bars, which reflected in a general way a volume ratio of 60 : 10 liters of the pressure fluid capacity of the hydraulic ram. The Δd-data of the two sets were equal within the precision of the measurements. A plot of Δp/Δd vs. pressure (see Fig. 2) for each data set yielded a straight and approximately horizontal line which clearly demonstrated the linear proportionality between Δp and Δd. However, the extrapolated one-atmosphere values of the two sets of piston displacement data Δd only attained 22% of the theoretical piston displacement ΔD which would have been expected if the volume change of melting of lead at a constant pressure of one atmosphere were fully converted into piston displacement.

This observation initiated a more detailed study in which the cell assembly was subjected to systematic modifications in order to determine the effect of these on the signal magnitude. Four sets of experiments with different cell assemblies, as displayed in Fig. 1, were carried out. The experiments of series (a) were conducted with the standard cell assembly (pyrophyllite-boron nitride and KBr) at normal hydraulic capacity. In a second series (b) the boron nitride plug above the capsule was exchanged for pyrophyllite. This resulted in a small decrease in the signal, thus proving to be unsuitable. The first improvement was achieved in series (c) where the outer pyrophyllite bushing in the upper cell half was replaced with KBr. The extrapolated one-atmosphere value of the piston displacement, Δd_o, increased to 35% of the theoretical piston displacement ΔD. A similar increase was observed for the simultaneously-recorded pressure change Δp. In the fourth series (d) the size of the seal arrangement was enlarged so that the capsule was separated from it by only a very short pyrophyllite plug. In addition, the volume of the pressure-transmitting medium was reduced by 45%. The increase in the extrapolated one-atmosphere value Δd_o to 48% was quite substantial. However, this value was still about 50% below the theoretical piston displacement ΔD expected at one atmosphere.

The existence of simultaneous increases in the correlated pressure-change values Δp_o suggested that in addition to the pressure change which was caused by sample melting and freezing at the valved-off main ram, friction phenomena might play a crucial role in this kind of experiment. It is clear that the main contribution to friction within the cell is related to the amount of pyrophyllite used. By taking a ratio of its volume and that of the entire cell, we have defined a "friction coefficient" $K\mu = V_{pyr}/V_{std}$ cell. Also, to include the measurements with the considerably shortened cell of experiment series (d) a second normalization was necessary. The results are shown in Fig. 3. In addition, the $K\mu$ and Δd_o data derived from the experiments on silver and sodium, using a setup analogous to experiment series (a), are presented. In the experiments on silver in which CaF_2 was employed as the pressure-transmitting medium, the relatively high strength of this halide in such a dynamic experiment became quite noticable. Two facts can be summarized from Fig. 3: (1) there is a clear relation between the Δd_o values and the amount of pyrophyllite present in the cell assembly, thus clearly indicating the presence of friction; (2) even for a theoretically friction-free cell, it is not likely that displacement values at one atmosphere Δd_o will exceed the 50% margin. The resulting difference between the calculated piston displacement and the much smaller observed value must be assumed to be an inherent loss which occurs because of the associated pressure change Δp_o and probably because of the elastic-plastic behavior of the cell and vessel.

The evidence of friction as seen from the Δd_o data in Fig. 3 required finding a way to properly correct this data obtained at high pressure. A crude estimate of the order of magnitude of correction and its dependence on pressure could be obtained by comparing the silver melting curve determined with the low-friction salt cell (Mirwald et al. [4]) with a similar curve measured with a cell consisting entirely of pyrophyllite (see Fig. 4). The pressure difference ΔP is considered to be a measure of the friction of the pyrophyllite cell.

To correct the piston displacement data obtained on silver, the friction established in the above comparison-experiments needed only to be reduced relative to the amount of pyrophyllite employed in the standard cell (50%). For the correction of the Δd data of lead and sodium this reduction factor was increased slightly to 60%, because the pyrophyllite remained in its original soft-fired state in these experiments. In the silver experiment at least a partial hard-firing, >1000°C, had to be accounted for which further increased its strength.

The volume changes of melting for sodium, lead, and silver which were calculated from the determined piston displacement data are

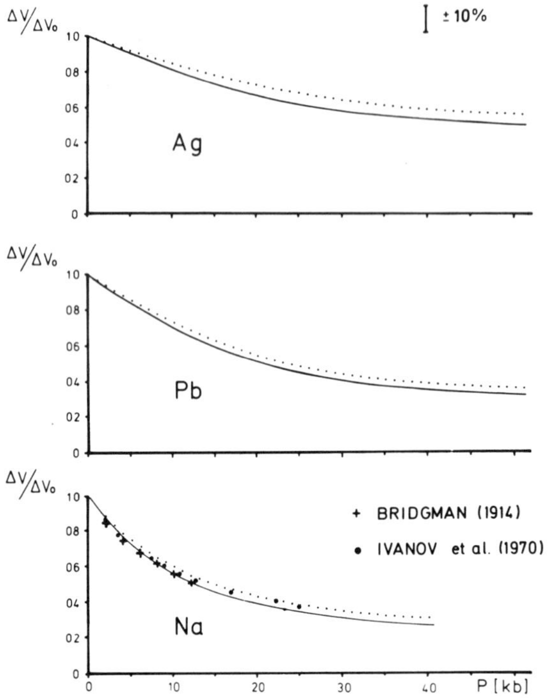

Fig. 5. Relative volume change of melting $\Delta V/\Delta V_o$ for sodium, lead, and silver to 50 kbar pressure. Dotted curves are raw data; solid curves are corrected data.

displayed in Fig. 5. To facilitate the comparison between the three metals, the data were normalized to the corresponding value of volume change ΔV_o at one atmosphere. For each metal the raw data curve (dotted line) and the curve corrected for friction (solid line) are shown. The friction correction applied fortunately does not exceed the experimental uncertainty which is believed to be within ± 10%. For lead and sodium it is almost of no significance. The most pronounced decrease in volume change of melting is observed for sodium which amounts to 60% at 40 kbar pressure. The least affected in the present study is silver which still exhibits a decrease of 45%, however.

The data presented on sodium, whose volume change upon melting has already been studied by Bridgman [1] and Ivanov et al. [3] in hydrostatic high-pressure devices to a pressure of 12 and 25 kbar, respectively, provide good agreement with the previous work. This can be considered as partial confirmation of the reliability of the proposed method.

Based on these volume changes and the melting curve data obtained from recent determinations on sodium and lead [5] and on silver [4], the correlated entropies and heats of fusion were calculated according to the Clausius-Clapeyron equation. From the normalized presentation in Fig. 6 it is evident that the decrease in the volume change of melting with increasing pressure exceeds the decrease in the entropy of fusion in all three cases. It may be inferred from this that even if the volume change of melting approaches zero at very high pressure, the value of the entropy of fusion remains a finite quantity. It then represents solely the difference in the geometrical arrangement of the atoms in the liquid and solid phase at identical specific volume. A comparison of the ΔV-curves with those of the corresponding isobaric compression $\Delta V/V_o$ at room temperature reveals a qualitative similarity in curvature.

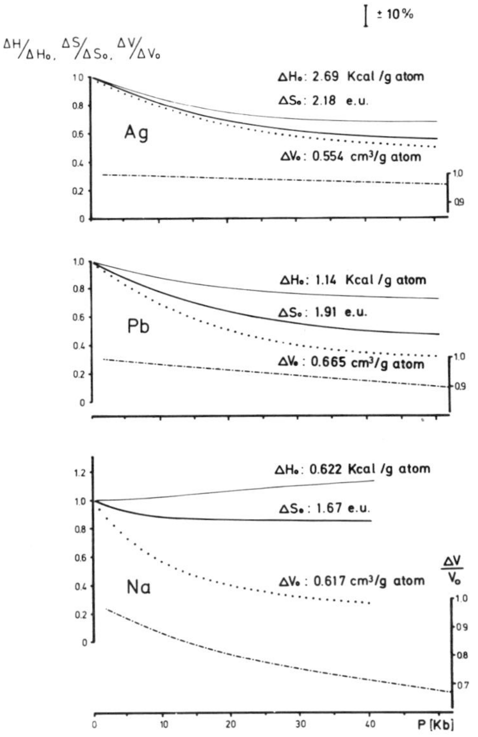

Fig. 6. Relative volume change of melting (dotted line) and the related heat (thin line) and entropy (thick line) of fusion for sodium, lead, and silver. In addition, the relative compression $\Delta V/V$ at room temperature (dot-dash line) is given for each metal.

It is conceivable that the volume change of melting is mainly determined by the higher compressibility of the liquid. From the apparent similarity in curvature it follows that the order of decreasing compressibility from sodium to lead to silver as observed at room temperature is maintained also in the liquid state.

ACKNOWLEDGMENTS

The basic experiments were carried out in the high-pressure laboratory of G. C. Kennedy, Institute of Geophysics, University of California, Los Angeles. The work was completed at the Ruhr-Universität-Bochum. The author wishes to express his sincere gratitude to G. C. Kennedy for the generous support received. I. C. Getting's help in technical discussions is also gratefully acknowledged.

REFERENCES

1. P. W. Bridgman, Phys. Rev. 3, 126 (1914).
2. B. B. Owens, J. Chem. Phys. 42, 2259 (1965).
3. V. A. Ivanov, I. N. Makarenko, and S. M. Stishov, JETP Lett. 12, 7 (1970).
4. P. W. Mirwald, I. C. Getting, and G. C. Kennedy, J. Geophys. Res. 80, 1519 (1975).
5. P. W. Mirwald and G. C. Kennedy, J. Phys. Chem. Solids 37, 795 (1976).

E-4

ENTROPY DETERMINED FROM HEAT OF COMPRESSION AT HIGH PRESSURE FOR BENZENE AND CARBON TETRACHLORIDE

P. Pruzan and L. Ter Minassian
Laboratoire de Chimie Physique
Paris, France
and
A. Soulard
Laboratoire de Minéralogie Cristallographie
Paris, France

INTRODUCTION

The entropy variation ds(T,p) may be determined with accuracy when computed from heat capacity and expansivity measurements from

$$ds = c_p\left(\frac{dT}{T}\right) - \alpha v\ dp \qquad (1)$$

When the variable is only the temperature, the measurement of c_p provides a convenient and sensitive but partial method for the study of the phase transitions. Extension to include the variable pressure requires elaborate techniques and as a consequence the entropy data as a function of pressure are scarce and unreliable. The gap may be filled when the second term in (1), $(\partial s/\partial p)_T$, is measured directly. This term has the meaning of an entropy variation resulting from isothermal liberation of the heat of compression and has been shown to be measurable by calorimetric methods at high pressures [1-3]. Following these methods, successive isotherms are determined as a function of pressure and the result is a full description of the phases and of their mutual transformations. The present paper deals with such piezothermal results obtained with benzene and carbon tetrachloride under their liquid and solid states up to pressures of 4 and 7 kbars.

EXPERIMENTAL

Method

The main features of the piezothermal method were previously described [1-3]. In the present work two distinct calorimetric devices are used leading to different experimental situations. One of the devices consists of a high-pressure tube introduced in a flux calorimeter together with an appropriate thermal shunt. When heat is liberated uniformly in the tube, it has been shown that one measures the quantity of heat liberated from a constant internal volume V_r. The volume V_r can be calibrated and its magnitude has been found to be 0.64 cm^3. The tube is filled with the sample in the molten state and a tight plunger separates the sample from the pressure transmitting medium. When a pressure variation is performed, heat is liberated which is interpreted in terms of the thermodynamic equation of Maxwell

$$\rho(\partial s/\partial p)_T = -\alpha \quad (2)$$

Taking into account the thermal contribution of the steel of the high-pressure tube to the overall heat effect δq, the following equation is obtained:

$$\alpha - \alpha_r = -(1/V_r\ T)(\delta q/dp) \quad (3)$$

The second device is a twin calorimeter Joule system, the calorimeter proper being a high-pressure vessel having an internal volume V_r of 50 cm^3. The sample is contained in a thin envelope and undergoes the hydrostatic constraints exercised by the hydraulic fluid. Due to the very high thermal capacity of the high-pressure vessel (at least two orders of magnitude greater than that of the sample), the quantity of heat measured may be considered as isothermally liberated when an over-pressure is applied. The overall thermal effect δq is the sum of four contributions: (1) the sample: mass m, volume V(p), expansivity $\alpha(p)$; (2) the hydraulic fluid: volume V_r - V(p), expansivity $\alpha_f(p)$; (3) the envelope: volume v_e, expansivity α_e; and (4) the material of the high-pressure vessel.

The expansivity of the sample may therefore be written as

$$\alpha = \alpha^*(1 + \Delta\alpha/\alpha^*) \quad (4)$$

where α^* is a "rough expansivity" defined from the experimental results

$$\alpha^* = -\delta q/V_r\ T\ dp \quad (5)$$

and where $\Delta\alpha$ is a corrective term given by†

$$\alpha = (\alpha^* - \alpha_f)\ (V_r/V(p) - 1) + (\alpha_F - \alpha_e)V_e/V(p) + \alpha_r\ V_r/V(p) \quad (6)$$

Procedure

The usual procedure is followed [2], which consists of an analysis performed by sets of increasing and then decreasing pressure

†The magnitude of the correction $\Delta\alpha/\alpha^*$ in (4) does not exceed a few percent.

steps. The magnitude of one step is around 100 to 200 bars when the region studied is far from a transition, otherwise the step is reduced to a few bars. The measurements are performed at 30°C with the twin calorimeter device and are compared with the results obtained from the other device at the same temperature. The comparison can be done easily because of the almost equivalent accuracy of the instruments which are of the order of 2 to 3%.

Samples

Benzene was supplied by Merck: two samples of different purity (99.5% and 99.96%) were used without any difference in the results. Carbon tetrachloride was of moderate purity (99.5%) supplied by Prolabo. The purity of the samples were tested after the runs by gas chromatography in order to check the absence of impurities.

RESULTS

Tables I and II give typical expansivity α values in sets having two origins. One set is obtained from the device mounted on the flux calorimeter and using equation (3). The other set is computed from the "rough expansivity" α^* given by the twin calorimeter device. The α^* values are corrected using equation (6) whenever the necessary data are available, otherwise they are given uncorrected. The results are plotted in Figs. 1 through 4 also including sets of measurements too lengthy to tabulate. It is important to note that α has the meaning of an expansivity as long as homogeneous phases are concerned, otherwise it has to be interpreted in terms of an entropy of transformation.

In order to compare the results with existing values [4,5] the expansivities computed from volume determinations in the case of benzene at 30°C are given in Fig. 1. As relative volumes at high pressures cannot be measured with an accuracy better than 10^{-3} to 10^{-4}, expansivities cannot be determined from them with an accuracy better than 20 to 30%. Along the isotherms in Figs. 1 to 4, the range of existence of the homogeneous phases may be observed to be limited by peaks. The peaks are of finite width, but their character differs when the transformation is a melting or a solid-solid crystalline transition. The results concerning melting may be compared with those obtained previously in the case of hexamethylbenzene, hexadecane, butane-diol 1-2, cyclohexane,[†] and paradichlorobenzene.[††] With decreasing pressures the melting curves all have a reversed λ form.

On the high-pressure branch reversibility and reproductibility of the measurements are ensured in spite of the very high values attained. Instability and spontaneous melting begin to occur in the vicinity of the sharp maximum. The transformation is complete after

[†] Unpublished.
[††] To be published shortly.

Table I. Thermal Expansivity of Benzene as Function of Pressure*

p,MPa	$10^4\alpha, K^{-1}$	P,MPa	$10^4\alpha K^{-1}$	p,MPa	$10^4\alpha, K^{-1}$
		T = 302 K†		T = 325 K††	
α*eq.(5)		αeq.(3)		αeq.(3)	
I				I	
657.0	2.43	303.6	3.81	378.0	3.75
643.0	2.59	270.7	4.08	349.0	3.93
619.8	2.68	238.0	4.37	321.2	4.16
598.9	2.72	204.4	4.50	294.8	4.46
576.4	2.73	172.4	4.94	266.8	4.74
557.4	2.72	141.2	5.47	238.5	5.52
532.3	2.90	113.0	7.06	214.9	5.84
αeq.(4)					
509.6	2.92	{ 101.1	8.54	{ 199.0	7.09
487.2	2.95	99.4	11.23	187.3	14.65
463.5	3.02	93.2	28.54	181.7	127.88
440.3	3.23	92.2	32.43	178.5	1136.13
418.2	3.25	89.8	46.64	177.5	2744.53
394.9	3.51	86.2	656	L	
		85.2	2839.73	172.1	7.08
				163.7	7.08
αeq.(3)					
378.5	3.28	L			
368.8	3.40	78.9	8.56	155.5	7.23
350.6	3.48	54.2	9.51	144.3	7.16
336.4	3.70	25.7	10.48	115.4	7.68
				86.4	8.42
				57.3	9.47
				28.7	10.60
				10.3	11.73

* I, solid phase; L, liquid phase; { continuous transition region.

† Obtained by two devices (decreasing pressure runs).

†† Obtained by the flux calorimeter.

a few hours duration, and the α values characterize those of the liquid state.

The solid transformation II - I of carbon tetrachloride is known to be of first order [6-10], and its piezo-thermogram may be compared with the II - I transformation in the case of hexamethylbenzene [2]. Both peaks are symmetrical and the experimental points lying on the low-pressure side (determined by decreasing pressures) are as stable as those lying on the high-pressure side. From the experimental point of view we observed a surprisingly large "inertia" of the system corresponding to what Bridgman called the region of indifference [11]. However, the existence of

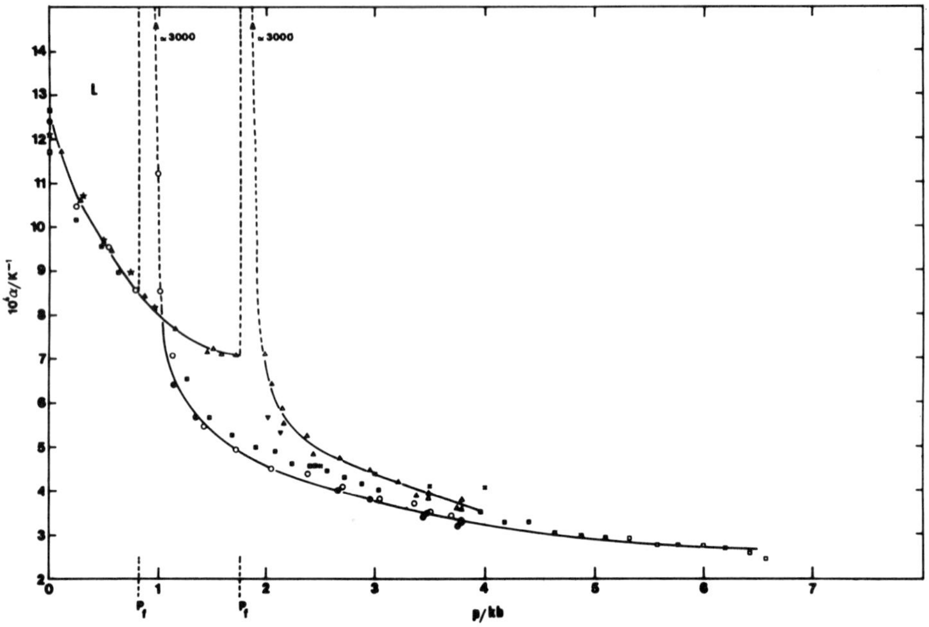

Fig. 1. Expansivity α of benzene. O, ◒, ●, different runs (flux calorimeter) T = 302K; ■ , □ (α*), different runs (adiabatic calorimeter) T = 302.6 K; Δ, ▲, ▼, different runs (flux calorimeter) T = 325K; ⊗, Tables T = 302 K, p = 1 atm; ⊠, Tables T = 325 K, p = 1 atm; * From volume measurements [4]; × from volume measurements [5].

such a region is not a rule and exceptions may be observed, as in the case of the III - I transformation of paradichlorobenzene where the transition occurs spontaneously in a very narrow pressure range.†

When the analysis is performed by increasing the pressure, phase transformation occurs with a systematic but variable delay of pressure. The passage towards the high-pressure phase takes place at a constant pressure with a rapid kinetic rate in the case of a crystallization and at a lower rate in the case of a solid-to-solid transition.

Table III gives the transformation entropies computed by integration of the peaks. The classical procedure used when determining enthalpies from the heat capacity curves has been followed. The areas under the peaks were estimated, taking into account the background determined by smooth extrapolation. This procedure gives the specific entropy jumps $\Delta S/R$ in the case of the twin calorimeter device, and in the case of measurements performed with the flux calorimeter the quantity determined is $\langle\rho\rangle\Delta S/R$, where $\langle\rho\rangle$ is the average density of the phases concerned.

†To be published shortly.

Table II. Thermal Expansivity of Carbon Tetrachloride as Function of Pressure*

p,MPa	$10^4\alpha, K^{-1}$	p,MPa	$10^4\alpha, K^{-1}$	p,MPa	$10^4\alpha, K^{-1}$
III		T = 302 K†		T = 325 K††	
α*eq.(5)		I αeq.(3)		αeq.(3)	
{ 705.7	2.86	316.3	3.88	389.9	3.72
688.7	19.20	284.0	3.66	362.5	3.79
II 669.9	2.81	258.1	4.12	334.6	3.98
640.2	2.90	235.1	4.63	306.5	4.23
613.5	2.71	211.0	4.30	281.5	4.28
591.9	2.86	192.1	5.32	263.0	4.57
589.1	2.98	174.6	5.09	251.1	4.78
551.6	3.06			236.4	4.93
509.7	3.11	{ 159.0	7.23		
458.6	3.25	148.3	52.43	{ 221.0	11.45
418.1	3.49	143.6	542.73	213.5	131.19
389.8	3.77			204.6	32.51
		L 134.7	7.14		
{ 374.7	22.68	113.7	7.48	190.2	5.86
368.5	481.40	83.3	8.11	166.7	6.11
367.1	355	52.0	9.11	136.4	6.64
364.3	8.08	α*eq.(5)		103.3	7.61
		27.1	9.77	69.9	8.49
I 352.9	3.96	6.7	11.20	38.2	9.68
		2.6	11.40	10.9	10.90

*I, II, III, solid phases; L, liquid phase; { continuous transition region.

† Obtained by two devices (decreasing pressure runs).

†† Obtained by the flux calorimeter.

Phase I of carbon tetrachloride presents an interesting feature. Using a piezo-metric method, Trappeniers detected a phase transition he thought to be of first-order [12]. Examination by piezothermal analysis along the 30 and 50°C isotherms allowed us to disclose a jump in the expansivity curve. This jump may be identified with the transition in question (Figs. 2 to 4), however, no thermal phenomenon corresponding to a first-order transition could be detected. It is possible due to the tenuity of the transformation that a singularity in the v-p curve has been confused with a discontinuity.

DISCUSSION

The theories concerning the melting process are unable to describe the phenomenon under all its aspects. This is the case when the thermodynamic properties of a perfect crystal are calculated, with a given structure and interaction potentials, without

Table III. Entropy Jumps of Transition and Fusion of Benzene and Carbon Tetrachloride*

Benzene, Fusion

T,K	p,MPa fusion	$\langle\rho\rangle\Delta S/R$, g cm^{-3}	$\Delta S/R$
302	82	4.17	3.84
325	176	3.73	

Carbon Tetrachloride, Transitions III-II, II-I, Fusion

T,K	p,MPa III-II	$\Delta S/R$ III-II	p,MPa II-I	$\Delta S/R$ II-I	$\langle\rho\rangle S/R$, gcm^{-3} II-I	p,MPa fusion	$\Delta S/R$ fusion	$\langle\rho\rangle S/R$, gcm^{-3} fusion
302	687	0.25	370	2.25	3.79	142	1	1.76
325						210		1.72

*Specific values $\Delta S/R$ are obtained by the adiabatic calorimeter, and $\langle\rho\rangle\Delta S/R$ by flux calorimeter.

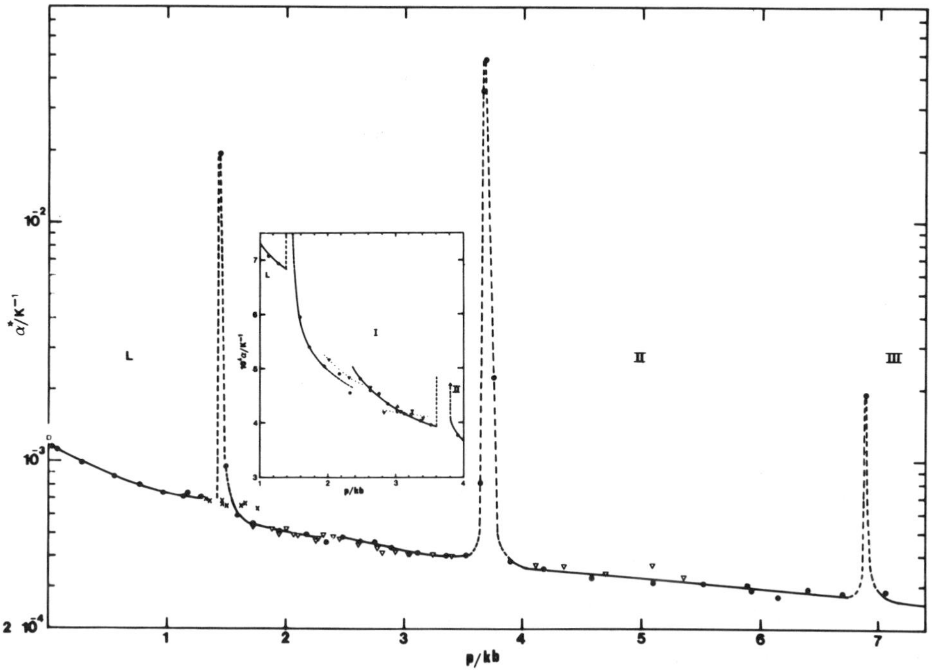

Fig. 2. Expansivity α of carbon tetrachloride (α*, adiabatic calorimeter), T = 301.6 K; ●, different decreasing pressure runs; ∇ increasing pressure runs; **X** surfusion region; □ Tables (p=1 atm). Inset figure: typical runs in phase I, α jump during decreasing pressure run (•) and during increasing pressure runs (∇).

any reference to the defects which may have a major importance in the vicinity of the transformation. It is however useful to pay attention to the inflexion point appearing on the calculated free energy isotherm of a perfect crystal, and which had been interpreted as an instability having some connection with melting [13-15]. A recent computation in the case of krypton [21] allowed a comparison with our experimental results. The variation of the thermodynamic properties near the melting point is similar to the one observed experimentally in our case, although the systems are quite different.

Generally speaking, the theoretical approach to the melting phenomenon may be conceived either under the aspect of a competition between two phases otherwise stable and which leads to a discontinuity, or under the aspect of a phase attaining the limit of its domain of stability and which is replaced by a stable one.

In our case, due to the non isobaric and asymmetric character of the meltings we observed together with other authors [5,11,16-19] in the case of organic solids, we are led to adopt an interpretation founded on the second aspect. The same remarks are valid for the observations at atmospheric pressure: the continuity and increase

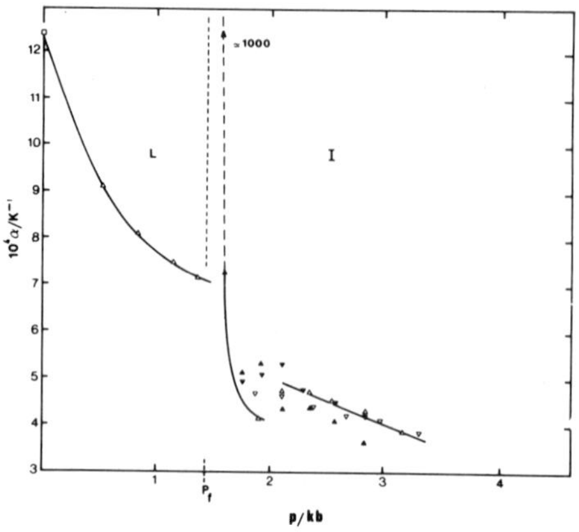

Fig. 3. Expansivity α of carbon tetrachloride, T = 302.16 K (flux calorimeter); ▲, Δ, decreasing pressure runs; ▼, ∇, increasing pressure runs. The scattering of the experimental points is due to the variation of the α jump position with the scanning rate of the analysis.

of the thermodynamic coefficients α, K and c_p in the vicinity of melting, though more or less accentuated by a variety of causes, is a common feature displayed by experimentation [23,24].

The condition for the stability of a phase, is bound to the concavity of the Gibbs function G and as a consequence, to the sign of the Jacobian $D = \partial(T,P)/\partial(v,s)$.

In fact, only the positive values of D are significant, and they physically mean that for any infinitesimal variation of p and T, the response of the system concerning v and s values may be calculated. This is no longer the case for D = 0: variations of dv and ds may be calculated only for definite values of the ratio dp/dT. The system is no longer able to answer any perturbation and may be considered as being on the boundary of its domain of intrinsic stability. The particular value D = 0 determines a curve in the p - T plane in the vicinity of which the following relations hold [20]:

$$dp/dT = (\partial p/\partial T)_s = (\partial p/\partial T)_v \tag{7}$$

$$dp/dT = \alpha/K = c_p/Tv\,\alpha \tag{8}$$

In other words, the slopes of the adiabatics and isochores become identical to that of the boundary. The geometrical solution corresponding to this situation is that of the v and s surfaces getting steeper without discontinuity, to become osculatory to the vertical surface defining the boundary. In the case of melting, the experiment illustrates this scheme (this and other work [5]) in the extreme case where the surface v(T, p) and s(T, p), become vertical at the end of the transformation (see Fig. 5). But the similarity between experiment and theory does not allow yet an interpretation in terms of intrinsic instability, unless one demonstrates that the boundary D = 0 corresponds also to a melting curve. The theoretical calculation on solid krypton [21] gives interesting results in this direction and is worth considering.

Whereas in the solid-liquid direction, the origin of the transformation may be researched in the instability of the solid,

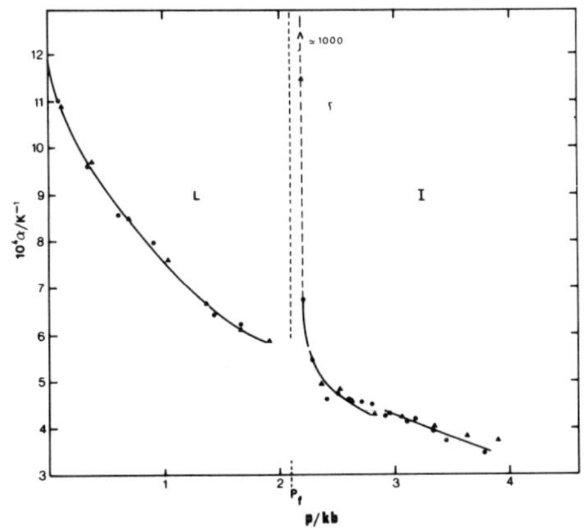

Fig. 4. Expansivity α of carbon tetrachloride, T = 325 K (flux calorimeter). ● decreasing pressure run (scanning rate 200 bars/hr); ▲ decreasing pressure run (scanning rate 100 bars/hr).

elementary considerations show that in the liquid - solid direction a type of relative instability of the liquid, with respect to the solid, has to be considered. The same considerations are also valid in the case of solid to solid transformations with the difference that the substitution of one phase by another is mostly governed by problems of relative stability.

CONCLUSION

The piezothermal method allows a detailed analysis of the phases and of their mutual transformations, much like the Cp determination does as a function of temperature. The accurate determination of entropy variation as a function of pressure brings up a new kind of information in the field of macroscopic thermodynamics, quite suitable for the study of phase transitions.

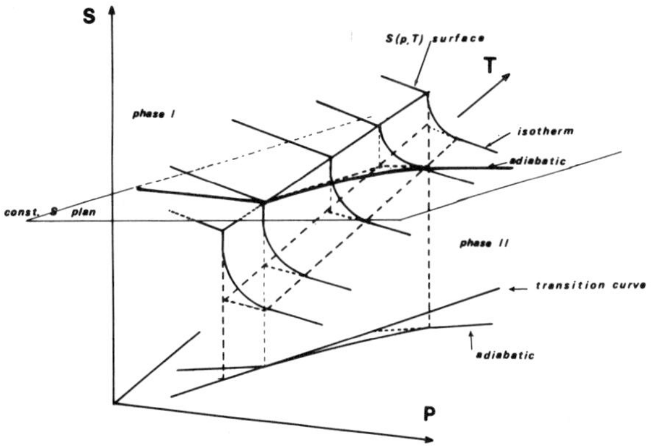

Fig. 5. S (p, T) surface in the vicinity of the melting zone (phase I is the liquid phase, phase II the solid phase). The adiabatics are osculatory to the boundary line where the curvature $(\partial s/\partial p)_T$ is infinite. Permuting S by V the same sketch is obtained in the V, p, T diagram.

NOTATION

c_p	= specific heat capacity at constant pressure
D	= Jacobian = $\begin{vmatrix} (\partial p/\partial v)_s & (\partial p/\partial s)_v \\ (\partial T/\partial v)_s & (\partial T/\partial s)_v \end{vmatrix}$
m	= sample mass
p	= pressure
q	= exchanged heat
R	= gas constant
s	= specific entropy
T	= temperature
v	= specific volume
V_r	= vessel internal volume
V(p)	= sample volume at pressure p

V_e = envelope material volume
α = expansivity
α_r = expansivity of the vessel material
α_f = expansivity of the hydraulic liquid
α_e = expansivity of the envelope material
K = isothermal compressibility
ρ = density

REFERENCES

1. J. C. Petit and L. Ter-Minassian, J. Chem. Therm. 6, 1139 (1974).
2. L. Ter Minassian and P. Pruzan, J. Chem. Therm. 9, 375 (1977).
3. P. Pruzan, L. Ter Minassian, P. Figuiere, and H. Szwarc, Rev. Sci. Instr. 47, 66 (1976).
4. R. E. Gibson and J. F. Kincaid, J. Am. Chem. Soc. 60, 511 (1938).
5. P. Figuiere, A. Fuchs, M. Ghelfenstein and H. Szwarc, J. Phys. Chem. Solids 39, 19 (1978).
6. J. F. G. Hicks, J. G. Hooley and C. C. Stephenson, J. Am. Chem. Soc. 66, 1064 (1944).
7. P. W. Bridgman, Phys. Rev. 3, 153 (1914).
8. C. E. Weir, G. J. Piermarini, and S. Block, J. Chem. Phys. 50, 5, 2089 (1969).
9. R. C. Miller and C. P. Smyth, J. Am. Chem. Soc. 79, 20 (1975).
10. C. P. Smyth, J. Phys. Chem. Solids 18, 1, 40 (1961).
11. P. W. Bridgman, The Physics of High Pressure, Dover Publications Inc., New York (1970).
12. N. Trappeniers, Académie Royale, Belgique, Mémoires in 8è 27, 10 (1952).
13. K. F. Herzfeld and M. Goeppert Mayer, Phys. Rev. 46, 995 (1934).
14. L. Brillouin, Phys. Rev. 54, 916 (1938).
15. M. Born, J. Chem. Phys. 7, 591 (1939).
16. R. R. Nelson, W. Webb and J. A. Dixon, J. Chem. Phys. 33, 6 (1960).
17. M. Ghelfenstein and H. Szwarc, Chem. Phys. Letters 32, 93 (1975).
18. A. Turturro and U. Bianchi, J. Chem. Phys. 62, 5, 1668 (1975).
19. T. Hilczer, Physics Letters, 56A, 4, 330 (1976).
20. A. Soulard, C. R. Acad. Sc. Paris, 282B, 535 (1976).
21. A. Soulard, C. R. Acad. Sc. Paris, 283B, 91 (1976).
22. A. Soulard, C. R. Acad. Sc. Paris, 283B, 331 (1976).
23. E. E. Westrum and J. P. Cullough, Physics and Chemistry of the Organic Solid State I: Thermodynamics of Crystals, Interscience Publishers, New York (1963).
24. A. R. Ubbelohde, Melting and Crystal Structure, Clarendon Press, Oxford (1964).

F-1

TRANSPORT IN DENSE FLUIDS*

D. W. Oxtoby

Université Paris-Nord
Villetaneuse, France

INTRODUCTION

The physics of dense fluids is one of the most fundamental and important areas of application in high-pressure technology. The research in this field can be roughly divided into three areas: (1) Equilibrium properties; (2) Excited state and optical properties; and (3) Transport properties.

Equilibrium studies include determinations of the equation of state and such other properties as sound velocity and heat capacity. Excited state studies include spectroscopy, light scattering, and dielectric response; such investigations should be of increasing importance in the future as more work is performed on high temperature, high pressure thermodynamic states. This paper will, however, be limited to a survey of recent work in the third area: the transport properties of compressed gases and liquids. The other two areas listed above are involved indirectly: the equation of state must be known to relate theories (based usually on temperature T and density n) to experiments (which usually measure transport coefficients as a function of temperature and pressure P); excited states are involved in energy transfer and thermal conductivity of molecular fluids even at room temperature.

The importance of high-pressure studies is evident. Experiments restricted to low pressures can examine only a small range of densities in the liquid phase; high pressures must be used to extend this range, to enable the study of the transition from low density to high density behavior, and to separate the effects of density and temperature. An alternative to direct experimental measurement is computer simulation through the method of molecular dynamics, an approach which has been developed and applied over the last ten years. Extensive comparison of the experimental and simulation results will be made in this paper.

*Invited paper.

The transport properties discussed here are those of pure fluids: self diffusion, shear and bulk viscosity, and thermal conductivity. We will not discuss the important subject of transport in mixtures (which involves, in addition to the transport coefficients of pure fluids, other properties such as mutual diffusion coefficients and thermodiffusion). Transport coefficients are defined [1] as the proportionality constants relating the flux of a property A, Ψ_A, to its gradient dA/dz, where for simplicity we take A to vary only in the Z-direction. The flux of A is defined as the net flow, per unit area, of property A across the plane Z = 0. Specifically, we have

$$\Psi_{nm} = j_z = -D\,\frac{d(nm)}{dz} \tag{1}$$

$$\Psi_{nmv_y} = p_{yz} = -\eta\,\frac{dv_y}{dz} \tag{2}$$

$$\Psi_{nmv_z} = p_{zz} = P - \eta_v\,\frac{dv_z}{dz} \tag{3}$$

$$\Psi_{nC_vT} = q_z = -\lambda\,\frac{dT}{dz} \tag{4}$$

The diffusion constant D relates the current j_z to the gradient of mass density mn; the shear viscosity η and bulk viscosity η_v relate components of the pressure P_{ij} to velocity gradients; finally, λ relates the heat flux q_z to the temperature gradient.

At low densities, transport occurs entirely through the physical motion of molecules across surfaces in the presence of gradients. In this limit, transport coefficients can be rigorously calculated through the Boltzmann equation if the intermolecular potential is known (the bulk viscosity η_v vanishes at low densities). At higher densities, in addition to the "kinetic" contribution to the fluxes there is a "potential" contribution arising from direct transfer of momentum and energy during collisions. This collisional transfer does not contribute to mass flux or to D, but at high densities becomes the dominant contribution to η, η_v, and λ.

Enskog [1] developed a theory for transport in dense gases of hard spheres. His theory is based on the solution of the Boltzmann equation for binary collisions, as in the low density limit, but there are three important differences: (1) Collisional transfer is included; (2) The finite diameter σ of the hard spheres is taken into account; and (3) The collision frequency Y is not that of the dilute gas but is obtained from the hard sphere equation of state

$$\frac{P\tilde{V}}{RT} = 1 + \frac{b_o}{\tilde{V}}\,Y \tag{5}$$

where $\tilde{V}$ is the molar volume and b_o is the second virial coefficient. The resulting transport coefficients can be written as their low density limits multiplied by simple functions of Y and $b_o/\tilde{V}$. For real fluids (which are not hard spheres) Enskog suggested that Y is determined not by the external pressure P but by the "thermal pressure" through

$$T\left(\frac{\partial P}{\partial T}\right)_V = \frac{RT}{\tilde{V}}\left(1 + \frac{b_o}{\tilde{V}}Y\right) \tag{6}$$

This method is referred to as the modified Enskog theory (MET). It allows the calculation of Y from experimental equation of state data, from computer simulation results, or from theoretical treatments.

The Enskog theory is quite successful for moderate density gases. A number of attempts have been made to improve on it (especially for liquid densities); the most important of these is the Rice-Allnatt theory [2] which treats the repulsive part of the intermolecular potential using the Enskog theory and the attractive part using ideas from the theory of the Brownian motion.

MOLECULAR DYNAMICS

In this section we briefly describe the method of computer simulation via molecular dynamics, an approach which has been applied to the calculation of transport coefficients for the past ten years. More detailed descriptions of the method may be found elsewhere [3,4].

The intermolecular potential must be chosen first. Two particular (spherically symmetric) potentials have received the most attention and will be discussed in this paper. The first is the hard sphere potential:

$$V(r) = \begin{bmatrix} \infty, & r < \sigma \\ 0, & r > \sigma \end{bmatrix} \tag{7}$$

σ, the hard sphere radius, introduces a length scale so that density dependence is present, but the absence of an energy parameter eliminates temperature dependence (such dependence may be introduced, however, by allowing the radius σ to vary with temperature). The hard sphere potential is most useful at high temperatures and high densities, where the steeply repulsive part of the real intermolecular potential is dominant. The second important potential is the Lennard-Jones

$$V(r) = 4\varepsilon\left[\left(\frac{\sigma}{r}\right)^{12} - \left(\frac{\sigma}{r}\right)^{6}\right] \tag{8}$$

Here σ introduces a length and ε a temperature scale.

In the computer simulation a certain number N of particles (generally between 100 and 1000) are placed in a box whose size is chosen to give the desired density. The initial velocities of the particles determine the temperature. Rather than putting walls on the box, which would lead to unrealistic physical effects, the infinite system is approximated through periodic boundary conditions in which the finite box is repeated throughout space. Newton's equations of motion are then solved numerically for the force law chosen, and the positions of the N particles are determined as a function of time. The total duration of the computer "experiment" is generally of the order of 10^{-11} sec, though runs as long as 10^{-9} sec have been reported [5].

Two methods of simulation may be used to obtain transport coefficients: equilibrium and non-equilibrium. To explain the first method, an analogy is helpful; the behavior of a mathematical function very close to the origin is determined entirely by the derivative of the function at the origin; similarly, the relaxation of small fluctuations about equilibrium (and therefore the transport coefficients) are determined entirely by equilibrium properties. The mathematical formulation involves the calculation of equilibrium time correlation functions; the resulting expressions for the transport coefficients are given by Alder et al. [6]. The second method that can be used, nonequilibrium simulation, involves imposing a gradient through a fixed external force and evaluating the flux that results. Calculations must generally be carried out for several gradients and extrapolated to zero gradient to obtain the usual (linear) transport coefficients. This method has been used primarily for shear viscosity calculations [7-9], but the thermal conductivity has also been studied in this way [9]. The results of the equilibrium and nonequilibrium simulations should in principle be the same, but differences arise from the finite number of particles N, the finite time duration of the calculation, and the statistical uncertainties of computer sampling experiments.

SELF DIFFUSION COEFFICIENT

Computer simulation results are perhaps most useful (relative to experiment) in the determination of self-diffusion coefficients. There are several reasons for this: first, the different experimental techniques that can be used (tracer diffusion, nuclear magnetic resonance, and neutron diffraction) are often limited in applicability and the uncertainties are large (typically 10 to 15%, as compared to fractions of a percent for viscosity measurements); secondly, modified Enskog theory works particularly poorly for the calculation of D, becoming inaccurate at densities above 0.8 n_c (n_c is the critical density) [10]. Finally, since self diffusion is a one-particle process, the computer calculations are much faster than, for example, viscosity calculations; as a result, D has been determined for a wide variety of intermolecular potentials and thermodynamic states.

The hard sphere (HS) simulations of Alder et al. [6] show two effects which become important as the density is increased. The first, which dominates at moderate densities, is a hydrodynamic effect in which the vortices created by a moving sphere act back on its motion, leading to an increase in the diffusion coefficient.

The second effect, which appears at higher densities, is a back-scattering effect in which a particle bounces around in the cage formed by its neighbors and the diffusion coefficient is reduced sharply from the Enskog prediction. The same effects arise in Lennard-Jones (LJ) potential simulation. Levesque and Verlet [4,5] have compared their LJ results with an HS model where the hard core diameter is a function of both temperature and density and is chosen to fit the computer-calculated structure factor. Even with this rather flexible and complicated fitting scheme, the hard sphere results are only accurate to 10 to 15%. For quantitative results, a more accurate potential than the hard sphere must be used. The following comparisons have been made between simulations and experimental results.

A corresponding states analysis of D for the rare gas liquids near the coexistence curve shows [11] agreement with LJ simulations within experimental error. Calculations with a many parameter Barker-Bobetic potential for argon show [12] agreement with the LJ results and with experiment (we regard the high-pressure extrapolation of Fisher et al. [12] as unjustified). Van Loef [13] has compared the diffusion coefficients for argon obtained from several different techniques with the molecular dynamics data and found agreement within a few percent. On the other hand, the measurements of Carelli et al. [14] on Kr above the critical temperature T_c seem to show deviations from LJ results at densities near critical.

Experimental measurements of D for CH_4 [15] show good agreement with the simulation by Hanley and Watts [16], using a four parameter potential fitted to low density data. These measurements also showed a $T^{0.9}$ temperature dependence close to the linear dependence found in the molecular dynamics calculation of Levesque and Verlet [4]. Powles and co-workers [17] have used NMR to study the behavior of D in liquid N_2; they have also calculated [18] D using an anisotropic potential. The two methods agree to within a few percent.

SHEAR AND BULK VISCOSITIES

The shear viscosity is the transport coefficient which has been studied most precisely and over the widest range of densities and temperatures. The modified Enskog theory is fairly successful for Ar, O_2, and H_2 in two regions [19]: in the gas up to $n = n_c$, and in the "liquid" ($T < T_c$) up to $n = 2n_c$. Recent experiments [20] on F_2 on the saturated vapor pressure curve showed agreement with previous MET predictions over most of the range studied. If the shear viscosity is written in the form

$$\eta(n,T) = \eta(0,T) + \Delta\eta(n,T) \tag{9}$$

then it has been observed for some time that the excess function $\Delta\eta$ is only weakly temperature dependent. Such a dependence has been clearly seen in recent experiments [21], however, which indicate that for moderate and high densities $\Delta\eta$ decreases with increasing temperature. The MET theory does not always predict the correct temperature dependence in $\Delta\eta$ [19].

Alder et al. [6] determined the shear viscosity of a system of hard spheres through computer simulation; they found significantly higher viscosities than those predicted by the Enskog theory (a factor of two near the freezing transition). The quantitative applicability of the HS model seems to be limited to at best to 15 to 20%, however. The LJ simulations of Levesque et al. [5] found a viscosity near the triple point 15% smaller than the HS result. Vermesse and Vidal [22] compared the room temperature viscosity of the rare gases to those of the HS model; even the optimal value of the hard sphere diameter σ predicted viscosities too large by 15 to 25%, and small variations in σ led to even poorer agreement with experiment. Thus it appears that more accurate potentials must be used.

Levesque et al. [5] calculated η for an LJ system corresponding to a state near the triple point of argon, using an equilibrium computer simulation. The result they found was 25% higher than the experimental value (however, see discussion below). A number of non-equilibrium simulation methods have been proposed [23,24]; only Ashurst and Hoover [7-9] have carried out extensive calculations, however, so we limit our discussion to their results. They found significant dependence on the number of particles N near the triple point. By extrapolating (in an approximate fashion) their results and the equilibrium results of Levesque [5], they found [7] that both agreed with experiment to 5%, within the uncertainty of the computer simulation results.

Ashurst and Hoover [7] have calculated the viscosity of an LJ system in a variety of thermodynamic states: along the freezing curve, along the saturated vapor pressure curve, and two isotherms corresponding approximately to room temperature helium and neon. Along the vapor pressure curve they find excellent agreement with experiments on argon. Vermesse and Vidal [22] compared their room temperature measurements of the viscosity of He and Ne with the molecular dynamics results and found very good agreement (within 10%). However, recently Vermesse et al. [25] carried out a similar comparison for Ar and Kr and found very large deviations (the simulation results appeared to be as much as 75% higher than experiment for Kr near the freezing transition). There are several possible explanations for this discrepancy: (1) The viscosities for Ar and Kr were not calculated directly by simulation as they were for He and Ne, but were obtained from an interpolation formula which may not be accurate; (2) The LJ potential may not be adequate for dense Kr (note the discrepancies found in the last section for the diffusion coefficient of krypton); and (3) Three body potentials may contribute significantly at high densities and low temperatures. Vermesse (private communication) has begun some equilibrium simulations to try to resolve some of these questions.

The bulk (or volume) viscosity η_v is the least fully studied of the transport coefficients. It affects the absorption of sound and the linewidth of Brillouin scattered light; however, large effects arising from the thermal conductivity and shear viscosity must be subtracted to obtain the bulk viscosity in both cases, thus increasing the uncertainty of the measurements. At low densities η_v vanishes and at moderate densities the modified Enskog theory is in agreement [10] with the rather limited experimental results for Ar. Clearly, more experimental measurements would be desirable, especially at high densities. The LJ equilibrium molecular dynamics result for η_v of argon near the triple point found by Levesque et al. [5] is $(.95 \pm .1) \times 10^{-3}$ poise; this can be compared to a Brillouin scattering measurement [26] of $(1.2 \pm .1) \times 10^{-3}$ poise and a sound attenuation measurement [10] of 1.6×10^{-3} poise. No non-equilibrium simulations of η_v have been carried out to date.

THERMAL CONDUCTIVITY

The thermal conductivity of dense gases and liquids has been studied over a fairly wide range of densities and temperatures, though measurements are in general neither as extensive nor as precise as shear viscosity measurements. The high density region (near the freezing transition) has not been studied extensively and experimental investigations in this region would be of great interest.

The modified Enskog theory predictions appear [19] to break down at slightly lower densities than in the corresponding shear viscosity case, and temperature dependence in the excess coefficient $\Delta\lambda(\eta,T)$ seems to be more important. It also has the opposite sign from the temperature dependence of $\Delta\eta$:$\Delta\lambda$ increases with increasing temperature [27]. This difference in sign is not predicted by the modified Enskog theory [19].

On the other hand, the hard sphere computer simulation [6] shows only small deviations (less than 7%) from the hard sphere Enskog theory. The equilibrium Lennard-Jones triple point simulation [5] is only 10% larger than the corresponding hard sphere result. Thus λ appears to be less sensitive than the other transport coefficients to the nature of the potential and the dynamics of the dense fluid. These equilibrium simulation results are, however, in serious disagreement with existing experimental data (a factor of two too large for this triple point state [5]). Levesque (private communication) has since found that the error in the formula for λ (noted previously [5]) is more important than previously thought. After correcting it, he finds agreement with experimental results and with the non-equilibrium simulation discussed below.

A non-equilibrium calculation of λ could be carried out by uniformly increasing velocities on one side of the box and decreasing them on the other. Such a direct simulation has not yet been performed. However Ashurst and Hoover in their non-equilibrium simulation of shear flow have shown [7] that a parabolic temperature distribution is generated which allows the determination of the ratio λ/η. Use of the calculated values of η then gives λ. The results are in very good agreement with experiment (in reduced units, Ashurst and Hoover [7] give their calculated value as 6.6 ± 0.4; experiment gives 6.5 ± 0.26; the corrected equilibrium simulation of Levesque et al. [5] gives the same result). Similarly, good agreement (within 10%) is found [9] for the excess thermal conductivity of dense gaseous argon near room temperature, and the positive temperature dependence of $\Delta\lambda$ is predicted correctly.

CRITICAL POINT ANOMALIES

Up to now, we have avoided discussion of the anomalies that occur in transport coefficients close to the gas-liquid critical point. Such anomalies arise because of the increasingly long-range correlations as the critical point is approached. The short-range structure of the fluid does not change appreciably, but the position of a molecule becomes correlated (though weakly) with the positions of other molecules thousands of Angstroms away. The cumulative effect of these correlations leads to the divergence of the compressibility at the critical point as well as divergences in certain transport coefficients. These effects cannot be studied either through Enskog theory (because only short-range, impulsive transfer is considered in it) nor through computer simulation (because of the finite number N of molecules studied). It must therefore be added on as a separate contribution to the background part obtained from these other methods.

The self diffusion coefficient has not been studied for a pure fluid near the critical point. A major anomaly would be surprising, since self-diffusion is governed by motion of molecules over short distances which should be only weakly affected by long-range correlations. Tracer-diffusion studies on binary mixtures near their critical consolute points have shown [28] no anomaly in D.

The shear viscosity has been predicted [29,30] to diverge very weakly near T_c, either as a power law

$$\eta \sim (T - T_c)^{-\phi} \tag{10}$$

or as a logarithmic divergence

$$\eta \sim -\ln (T - T_c) \tag{11}$$

The theoretical prediction of a logarithmic divergence [29] is shown to be in good agreement with experimental data on xenon of Pings et al. [31]. It is clear that the anomaly is weak, giving a 10% increase over the background shear viscosity only when the temperature reaches 0.1 K from the critical. The critical anomaly for N_2 has also been studied [32] and Hanley et al. [33] found a weak critical anomaly for the shear viscosity of methane.

The bulk viscosity appears to be strongly divergent, although the exact form of the divergence is still somewhat controversial.

As a result, sound absorption increases dramatically in the critical region.

The thermal conductivity also shows a strong divergence near T_c [29] of the form

$$\lambda \sim (T - T_c)^{-\zeta} \tag{12}$$

with the exponent ζ near 0.6. Experimental data [34] for CO_2 shows that a 10% increase in λ occurs as far as 20 K from T_c, and closer to T_c the thermal conductivity has been observed to increase by more than an order of magnitude.

CONCLUSION

Most experimental and theoretical studies to date of transport in dense fluids have been limited to the simplest case of atomic fluids. While there is still much work to be done in this area, it is clear that the main emphasis in the future will be on molecular fluids. Many molecules which are close to spherical should exhibit a behavior similar to that of atomic fluids, while more asymmetric molecules can be approximated either as ellipsoids or through atom-atom potentials. The latter method has already been applied successfully to the study of diatomics such as N_2 [18]. An important question for future studies will be the influence of the dipole moment on transport in dense media.

Transport in liquid metals involves both electronic and core effects; it has been reviewed by Nachtrieb [35]. Protopapas et al. [36] have successfully applied a hard-sphere model to diffusion in liquid metals. Computer studies using more realistic potentials should be an important area of future study.

Ionic melts have also been studied by computer simulation. Rahman et al. [37] investigated diffusion in molten LiF and BeF_2; D for F^- was 9×10^{-5} cm^2/sec in the first case and less than 0.5×10^{-5} cm^2/sec in the second. This large difference correlates well with the unusually high viscosity of BeF_2 melts. Lantelme et al. [38] studied diffusion in molten NaCl through molecular dynamics over a wide temperature range. Their calculated anion diffusion coefficients agree with experiment to within 10%, although their cation values are up to 20% too small. The use of better pair potentials should improve the situation still further.

We conclude with a brief summary of some of the advantages and disadvantages of molecular dynamics calculations of transport coefficients. The major limitation is, of course, the size and length of the computer calculations involved. As techniques become more developed, this is becoming less and less of a problem. Secondly, the interaction potential must be known reasonably well to calculate accurate transport coefficients; low density equation of state and/or transport data is usually sufficient to obtain a good pair potential, however.

There are a number of important advantages to the technique. First, parameters may be varied at will in a computer calculation as they cannot be in experiments. As an example, take the question: What is the effect of a dipole moment on the shear viscosity? A series of computer calculations identical except in the dipole moments chosen could answer this question, while experiments on different molecules would of necessity introduce other factors as well. A second useful aspect of computer simulations is that they may be used in cases where experiments are difficult or impossible: for extreme temperatures or pressures, or for corrosive or dangerous materials. A third advantage, especially of non-equilibrium simulations, is that large external gradients and non-linear response can be investigated, allowing the study of transport under the extreme conditions that arise in shock waves and detonations.

ACKNOWLEDGMENTS

I am grateful to J. Vermesse, B. Le Neindre, R. Tufeu and D. Levesque for helpful discussions, and to J. Vermesse and W. Ashurst for providing preprints of their work before publication.

NOTATION

D	= self diffusion coefficient	$V(r)$	= intermolecular potential
n	= number density	Y	= Enskog collision rate
n_c	= number density at critical point	ε	= Lennard-Jones potential well depth
N	= number of particles in computer simulation	η	= shear viscosity
		η_v	= bulk viscosity
P	= pressure	λ	= thermal conductivity
T	= temperature	σ	= pair potential length scale
T_c	= critical temperature	ϕ,ζ	= critical exponents

REFERENCES

1. J. O. Hirschfelder, C. F. Curtiss, and R. B. Bird, Molecular Theory of Gases and Liquids, John Wiley and Sons, New York (1954).
2. S. A. Rice and P. Gray, The Statistical Mechanics of Simple Liquids, John Wiley and Sons, New York (1965).
3. A. Rahman, Phys. Rev. 136, A 405 (1964).
4. D. Levesque and L. Verlet, Phys. Rev. A 2, 2514 (1970).
5. D. Levesque, L. Verlet, and J. Kurkijarvi, Phys. Rev. A 7, 1690 (1973).
6. B. J. Alder, D. M. Gass, and T. E. Wainwright, J. Chem. Phys. 53, 3813 (1970).
7. W. T. Ashurst and W. G. Hoover, Phys. Rev. A 11, 658 (1975).
8. W. T. Ashurst and W. G. Hoover, preprint.
9. W. T. Ashurst and W. G. Hoover, AIChE J. 21, 410 (1975).
10. H.J.M. Hanley and E.G.D. Cohen, Physica 83A, 215 (1975).
11. J. van Loef, Physica 62, 345 (1972).
12. R. Fisher and R. O. Watts, Aust. J. Phys. 25, 529 (1972).

13. J. van Loef, Physics Lett. 35A, 169 (1971).
14. P. Carelli, I. Modena, and F. P. Ricci, Phys. Rev. A7, 298 (1973); P. Carelli, A. de Santis, I. Modena, and F. P. Ricci, Phys. Rev. A13, 1131 (1976).
15. P. H. Oosting and N. J. Trappeniers, Physica 51, 418 (1971).
16. H.J.M. Hanley and R. O. Watts, Mol. Phys. 29, 1907 (1975).
17. K. Krynicki, E. J. Rahkamaa, and J. G. Powles, Mol. Phys. 28, 853 (1974).
18. P.S.Y. Cheung and J. G. Powles, Mol. Phys. 30, 921 (1975).
19. J.H.M. Hanley, R. D. McCarty, and E.G.D. Cohen, Physica 60, 322 (1972).
20. W. M. Haynes, Physica 76, 1 (1974).
21. W. M. Haynes, Physica 67, 440 (1973).
22. J. Vermesse and D. Vidal, Physica 86A, 429 (1977).
23. E. M. Gosling, I. R. McDonald, and K. Singer, Mol. Phys. 26, 1475 (1973).
24. T. Naitoh and S. Ono, Physics Lett. 57A, 448 (1976).
25. J. Vermesse, M. Provansal, and J. Brielles, preprint.
26. B. Y. Baharudin, D. A. Jackson, P. E. Schoen, and J. Rouch, Physics Lett. 51A, 409 (1975).
27. D. E. Diller, J.H.M. Hanley, and H. M. Roder, Cryogenics 10, 286 (1970).
28. J. C. Allegra, A. Stein, and G. F. Allen, J. Chem. Phys. 55, 1716 (1971).
29. D. W. Oxtoby and W. M. Gelbart, J. Chem. Phys. 61, 2957 (1974).
30. D. W. Oxtoby, J. Chem. Phys. 62, 1463 (1975).
31. H. J. Strumpf, A. F. Collings, and C. J. Pings, J. Chem. Phys. 60, 3109 (1974).
32. V. N. Zozulya and Yu. P. Blagoi, JETP 39, 99 (1974).
33. H.J.M. Hanley, R. D. McCarty, and W. M. Haynes, Cryogenics 15, 413 (1975).
34. A. Michels, J. V. Sengers, and P. S. van der Gulik, Physica 28, 1216 (1962).
35. N. H. Nachtrieb, Ber. Bunsenges. Phys. Chem. 80, 678 (1976).
36. P. Protopapas, H. C. Anderson, and N.A.D. Parlee, J. Chem. Phys. 59, 15 (1973).
37. A. Rahman, R. H. Fowler, and A. H. Narten, J. Chem. Phys. 57, 3010 (1972).
38. F.Lantelme, P. Turq, B. Quentrec, and J.W.E. Lewis, Mol. Phys. 28, 1537 (1974); F. Lantelme, P. Turq, and P. Schofield, Mol. Phys. 31, 1085 (1976).

F-2

AN EVALUATION OF THE PNM DENSE GAS TRANSPORT THEORY

R. J. Governale and S. E. Babb, Jr.

The University of Oklahoma
Norman, Oklahoma USA

INTRODUCTION

The current level of theoretical understanding of transport phenomena in dense media in terms of predictive ability is not clearly defined. This paper presents a partial evaluation of one type of such theory, that due to Prigogine and Nicolis and Misguich. The representation of the potential terms is qualitatively correct, but the neglect of the kinetic terms makes a precise comparison impossible. The Enskog hard sphere representation does as well or better than the more sophisticated version.

BACKGROUND

In earlier work [1,2] the Rice-Allnatt theory of dense gas transport phenomena was studied, and found to be incapable of accurately representing observed behavior. Accordingly, the investigation was extended to the PNM theory. The equations for the PNM theory have been given in a number of papers [3-5], no two of which agree on the signs of all terms. Correct equations are given by Governale [6].

This discussion will be limited to the Lennard-Jones potential, both complete and with the central portion replaced by a hard core. The original equations as given are for strictly the hard core case, but a discussion of the extension to more general potentials is given by Scrodt and Davis [7], embedded in a generalization of the entire approach.

For brevity the tables are also limited to Penus-Yevick solutions. Other solutions are very similar [6], and the remarks herein also apply to CHNC and Kirkwood solutions.

*Work supported in part by the National Science Foundation.

Misguich [3] has remarked that from the point of view of numerical accuracy the various integrals involved in the Fourier transforms should be carried as far as possible analytically. Unfortunately the values given by him for the results of this process also contain serious algebraic errors. Again the correct equations are given by Governale [6].

The radial distribution functions used in this work were calculated by Cafky and Babb [2], and are based upon the Kirkwood, Percus Yevick and CHNC theories.

As mentioned above the PNM theory specifically neglects the kinetic contribution to the transport coefficients. Accordingly the first test of the theory should be the comparison with bulk viscosity, where the kinetic contribution vanishes. For experimental data, the work of Madigosky [8] is used. This work was at -38.6°C, and is used without temperature correction for comparison. The comparison, shown in Table I, is well within the admittedly relatively large experimental error. Also shown in this table are calculations from the well-known Enskog theory, and these calculations are discussed below.

Table I. Bulk Viscosity of Argon at 0°C

λ	ρ_N Amagats	PNM	Enskog ϕx10^4poise	Obs.
0.15	139.7	0.2886	0.2220	0.9
0.30	279.4	1.305	0.9614	1.1
0.45	419.1	3.464	2.450	2.5
0.60	558.8	6.995	5.152	5.5
0.75	698.5	12.14	9.826	9.0
0.90	838.2	20.06	17.59	
1.05	977.9	31.93	29.98	
1.20	1118.0	52.80	49.27	

Turning next to shear viscosity, the problem of the neglected kinetic term becomes serious. As a very crude measure of this term one can compare the dilute gas viscosity value to those at the higher pressures, and one finds that this dilute gas value amounts to some 3.9% of the highest viscosity encountered herein, and is thus not negligible when considering the comparison of theory and experiment. The importance of this contribution increases with decreasing pressure.

In an attempt to calculate the kinetic term, the Enskog [9] theory has been used. This theory results in the well-known expresssions:

$$\eta = \eta_o \; [1 + \frac{2}{5} b\rho_N g(\sigma)]^2/g(\sigma) + \frac{48}{25\pi} b^2\rho_N^2 g(\sigma) \quad (1a)$$

$$\phi = \eta_o \; [\frac{16}{5\pi} b^2\rho_N^2 g(\sigma)] \quad (1b)$$

$$\kappa = \kappa_o \; [1 + \frac{3}{5} b\rho_N g(\sigma)]^2/g(\sigma) + \frac{32}{25\pi} b^2\rho_N^2 g(\sigma) \quad (1c)$$

where the first term and half of the second term for η and κ represent the kinetic contributions. Enskog theory is based upon hard sphere dynamics, and only considers binary collisions, and thus the application here is not strictly correct. Attempts to generalize the Enskog theory are not completely satisfactory [10], but as will be seen the basic form of equation (1) actually works quite well.

Table II contains the PNM results plus the estimates from the kinetic terms as well as the full Enskog results. It will be easily seen that the addition of the Enskog kinetic term to the PNM theory results in a considerable improvement, but the errors are still far beyond experimental error.

Table II. Shear Viscosity of Argon at 0°C

λ	ρ_N Amagats	PNM	PNM + Kinetic ηx10^4 poise	Enskog	Obs.
0.15	139.7	0.267	7.533	2.720	2.62
0.30	279.4	1.083	3.496	3.716	3.53
0.45	419.1	2.565	4.990	5.274	5.01
0.60	558.8	4.942	7.376	7.814	7.64
0.75	698.5	8.453	10.90	12.01	12.24
0.90	838.2	13.26	15.75	18.82	20.5
1.05	977.9	19.40	21.97	29.55	32.8
1.20	1118.0	26.70	29.39	46.08	54.3

For the Enskog calculations the g values were evaluated at the position where the Lennard-Jones potential is zero, and then were used in (1). It will be noticed that this procedure results in significantly better agreement with experiment than any other method used herein.

The results for the thermal conductivity are of a similar nature and are shown in Table III. Here there is the added complication of an additional term [5], not present in the original theory, and which must be added in the calculation of the thermal conductivity. This term actually contributes nothing significant to any of the calculations, and is thus not included here. The extra column in Table III, labeled "trunc. potential" shows the effects of truncating the Lennard-Jones potential at the point $r = \sigma$. This modifies the shape of the radial distribution function somewhat, which gives the effects shown on the calculated thermal conductivities. Similar effects are encountered in other transport coefficients.

Table III. Thermal Conductivity of Argon at 0°C

λ	ρ_N Amagats	PNM Full Potential	PNM Trunc. Potential $\kappa \times 10^4$	Enskog cal/cm·sec°C	Obs.
0.15	139.7	0.0367	0.0368	0.5530	0.641
0.30	279.4	0.1399	0.1438	0.7764	0.919
0.45	419.1	0.3197	0.3411	1.117	1.33
0.60	558.8	0.6111	0.6795	1.625	2.01
0.75	698.5	1.044	1.215	2.430	3.21
0.90	838.2	1.630	-	3.705	5.30
1.05	977.9	2.357	-	5.684	8.91
1.20	1118.0	3.1814	-	8.706	14.9

The original PNM theory is based upon such a truncated potential, and, except for the column just noted, the potential used in these calculations is the full potential. Technically this is an inconsistency. Some idea of the seriousness of this inconsistency can be gained by comparing the answers when the differences between g and 0 are neglected for $r < \sigma$, and when this difference is not neglected. In all cases the effects ignoring the values of $g(r)$, $r < \sigma$ is to raise the calculated values, thus improving the agreement. The changes are by no means small, and can rise to as much as 15% at the highest density. This difference is still small in comparison with the divergence with experiment. This indicates that a more careful examination of this point is needed, perhaps along the lines discussed by Schrodt and Davis [7].

In an attempt to consider the effects of the kinetic terms, these terms from the Enskog theory are added the PNM results, with the results also shown in Table II. The addition of these terms does significantly improve the agreement between the calculations and the experimental observations, but not sufficiently.

In terms of data representation, the Enskog theory is closer to the experimental results than any other representations used herein.

In final conclusion the PNM theory appears to - at least qualitatively - represent the potential contributions to the transport phenomena adequately, and perhaps quantitative agreement can be secured when an adequate discussion of the kinetic term contribution is spliced onto the PNM theory.

NOTATION

b = $2\pi/3\ \sigma^3$

$g(r)$ = radial distribution function

η = shear viscosity

η_o = η at zero density

κ = thermal conductivity

κ_o = κ at zero density

λ = reduced density = $\rho_N \sigma^3$

ρ_N = density

σ = hard sphere diameter; characteristic length

ϕ = bulk viscosity

REFERENCES

1. J. W. Cafky, R. M. Winfrey, and S. E. Babb, Jr. in Proceedings IVth Intern. Conference on High Pressures, Kyoto, Japan (1974), p. 530.
2. J. W. Cafky and S. E. Babb, Jr., J. Chem. Phys. 66, 5713 (1977).
3. J. Misguich, J. de Physique 30, 221 (1969).
4. J. A. Palyvos, H. T. Davis, and J. Misguich and G. Nicolis, J. Chem. Phys. 49, 4088 (1968).
5. H. T. Davis in Advances In Chemical Physics, Vol. 24, I. Prigogine and S. A. Rice, eds., John Wiley and Sons, New York, p. 257.
6. R. J. Governale, Ph.D. Thesis, University of Oklahoma, Norman, Oklahoma (1977).
7. I. B. Schrodt and H. T. Davis, J. Chem. Phys. 61, 323 (1974).
8. W. M. Madigosky, J. Chem. Phys. 46, 4441 (1967).
9. S. Chapman and T. G. Cowling, Mathematical Theory of Non-Uniform Gases, Cambridge University Press, Cambridge, England (1970).

F-3

EQUATION OF STATE FOR FLUIDS AT HIGH DENSITIES - HYDROGEN ISOTOPE MEASUREMENTS AND THERMODYNAMIC DERIVATIONS*†

D. H. Liebenberg, R. L. Mills, and J. C. Bronson

University of California, Los Alamos Scientific Laboratory
Los Alamos, New Mexico USA

INTRODUCTION

The high-density properties of hydrogen isotopes H_2, D_2 and T_2 are of current interest in energy programs. Under ERDA sponsorship [1-3] at Los Alamos, we have been developing accurate techniques for studying the hydrogens at high static pressures. Experimental data on these light simple molecules are especially valuable for testing theory.

This report treats the supercritical fluid for which only a very few results have been obtained. For hydrogen there are accurate measurements by Michels and coworkers [4] between 100 and 423 K up to maximum pressures between 0.3 and 2.6 kbar, respectively, a study by Tsiklis et al. [5] between 300 and 423 K at pressures from 0.5 to 6.5 kbar, crude pioneering measurements by Bridgman [6] at 303 and 338 K up to 13 kbar, and, finally, our own recent measurements [7] in the range 75 to 300 K and 2 to 20 kbar. The portion of the P-T plane covered by experiment is shown in Fig. 1.

Even fewer measurements exist for deuterium; these include results of Michels et al. [4], in the same pressure and temperature regime as for hydrogen, and our recent data [8], again over the range 75 to 300 K and 2 to 20 kbar. In addition to the molar volume V, we have measured the sound velocity v_s simultaneously over the P-T surface to provide a set of data that over-defines the thermodynamic system and can be used to improve the accuracy of an equation of state. Our data have an uncertainty estimated to be less than 0.5%.

Real fluids depart from the ideal gas law even at low pressures. While the van der Waals equation describes the qualitative features

*Work performed under the auspices of the U.S. E.R.D.A.
†Invited paper.

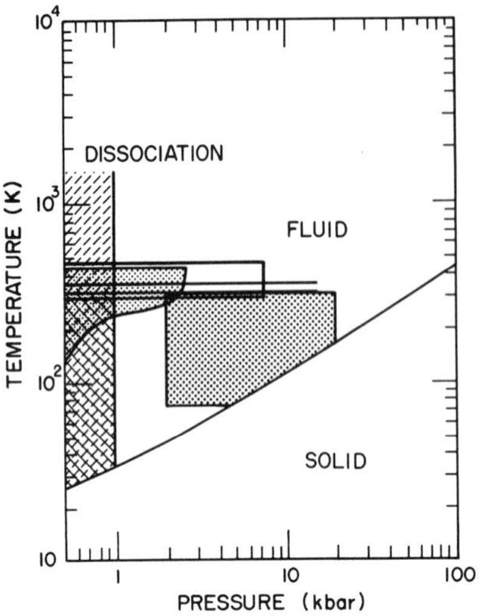

Fig. 1. Log T vs. log P in the higher density n-H_2 fluid region (P $\geq$ 0.5 kbar) and above the freezing line. The areas denoted regions in which measurements have been made and compilations published. The heavy dotted area is the region covered by recent measurements [7]; light dotted area by measurements of Michels et al. [4]; the open region by measurements of Tsiklis [5]; and the two lines the early measurements by Bridgman [6]. Two compilations are shown as cross hatched regions, that by Weber [9] as the solid line cross hatch and that by Vargaftik [19] by the dotted line cross hatch. Dissociation of the molecule occurs at the higher temperatures.

of simple fluids from the subcritical to the supercritical region, it too fails as the pressure is pushed higher. There appears to be no unified equation of state covering the full experimental range of pressures. Theories of dense fluids, having borrowed from gas, liquid, and solid behavior, take various forms.

A virial-type equation of state was expanded to a large number of terms and used by Weber [9] to describe parahydrogen. An empirical equation, devised by Benedict [10] and employing polynomials in P and T, has been found to extrapolate well beyond the range of input data. Haar and Shenker [11] have modified a van der Waals equation of state for use over extended pressures. Theories based on potential energy curves and pairwise interactions have been treated by Barker and Henderson [12] and, recently, by Ashurst [13]. Baker and Swift [14] have applied statistical mechanical equations to high-density fluids. And finally Kerley [15] used a hard-sphere fluid model to obtain results at ultrahigh pressure.

Various of these approaches will be tested by the available P-V-T data for H_2 and D_2. We note briefly transitions to the metallic form, but will avoid reference to higher density atomic, neutronic, and hadronic equations of state.

EXPERIMENTAL

Equilibrium measurements on hydrogen isotopes are difficult at elevated pressures because the gases severely embrittle high-strength steel vessels. At low temperatures these same steels often become extremely brittle. Bridgman recognized the former problem and encapsulated his hydrogen sample in metal surrounded by kerosene in his high pressure cylinder to protect the cylinder walls from direct contact with hydrogen. Corrections for the crushing of the capsule as well as the compressibility of kerosene and bomb material had to be made, so it is not surprising that his reported hydrogen volumes are in error by over 2% at 4 kbar and by 10 and 17% at his highest pressure of 13 kbar and 303 and 338 K, respectively. Vodar and Saurel [16] remark on the latter difficulty "The low temperature study of highly compressed gases is subject to many difficulties of which by no means the least is the fragility of reservoirs." Simple steel cylinders are limited to about 7 kbar, and early methods of measurement are summarized by Levelt-Sengers [17]. Recent improvement in materials and the knowledge of their properties has made it possible to construct a piston-cylinder apparatus that is able to seal and compress hydrogen and deuterium to over 20 kbar. As shown in Fig. 2, a tungsten carbide cylinder to contain the hydrogen is

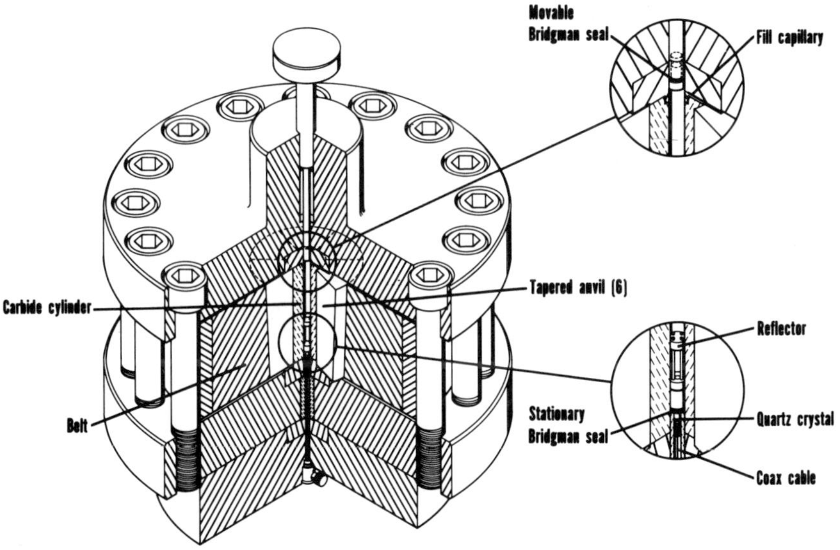

Fig. 2. Cutaway drawing of the piston-cylinder belted anvil apparatus. Major components are identified and the inserts show the upper movable piston seal, retracted so a charge can be admitted through the fill capillary, and the lower seal with the coaxial lead in to the sound transducer on the unsupported area of the Bridgman seal.

supported with tapered segmented anvils pressed into a belt. Axial support of the inner cylinder is provided by bolted flanges. Bridgman unsupported area seals are used on both the movable piston and the stationary plug. A quartz crystal transducer is mounted on the unsupported area of the stationary plug and sound pulses are propagated into the fluid, along a fixed path, and reflected to the transducer through the buffer rod. This design obviates leads into the high pressure chamber. Pulses of 10-MHz or 30-MHz sound are timed with a Tektronix scope to ± 5 ns. Volume changes are determined from piston travel, measured with a dial gauge of 0.0025-mm sensitivity and corrections applied for measured compressibility and dilations of the piston-cylinder components. Lower temperatures are achieved by enclosing the piston cylinder in a cryostat and bathing it in precooled nitrogen gas. Pressures are generated with a Dake press and are determined from Ashcroft free-piston gauge readings of the hydraulic fluid pressure. The cell is charged initially with the piston retracted to expose a capillary opening in the steel cap of the tungsten carbide cylinder. Gas pressure of over 3 kbar is provided from an American Instrument Company mercury-piston compressor. The gas is high purity normal hydrogen or deuterium initially. During the time frame of our measurements no conversion is presumed to occur since there is no catalyst and the self conversion rate is slow.

The apparatus, procedures, and accuracies are completely discussed in earlier references [1,7]. The average errors in the measurements are; ± 0.4% in volume, ± 0.5% in sound velocity, ± 0.5 K in temperature, and ± 0.5% in pressure.

RESULTS

For $n\text{-}H_2$ we measured 1952 complete sets of P, V, T, v_s points along 60 isotherms up to 20 kbar [7]. To these we added the 26 points of Michels et al. [4] determined at ≥ 1 kbar in our temperature range 75 to 307 K. Incomplete data sets of P, V, T, or P, T, v_s were sometimes obtained when accidents and leakage occurred; these data have been used for checks.

The range of data for hydrogen is shown in Fig. 3 by plotting ln V vs. T for five isobars between 3 to 18 kbar. Along the 300 K isotherm the sound velocity measurements, including the calculated lower pressure data points of Michels et al. [4] are shown in Fig. 4. Although the measured volumes are normalized to Michels data at the pressure where the piston seals off the fill capillary there is no further normalization of the absolute sound velocity measurements to the calculated

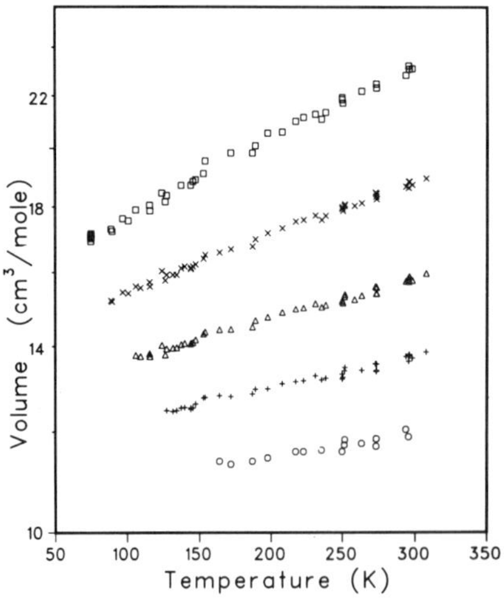

Fig. 3. Plot of log V vs. T for five isobars in fluid $n\text{-}H_2$. Experimental points (interpolated) are shown by square, 3 kbar; cross, 5 kbar; triangle, 8 kbar; plus, 12 kbar; circle, 18 kbar. At the lowest temperatures the points terminate at the freezing line.

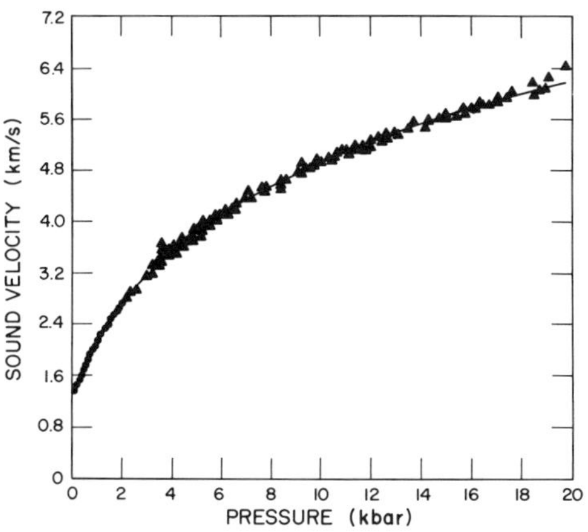

Fig. 4. Sound velocity in n-H_2 vs. P along the 300 K isotherm. The closed circles are from Michels et al. [4]. The closed triangles are measurements from this study [7] at higher pressures.

values of Michels. An interesting feature of the data should be noted: at lower pressure (below the region of validity for our published equation of state) there is an inflection point in the v_s vs. P data, where $(\partial^2 v_s/\partial P^2)_T^s = 0$, which the equation of state to be described does not reproduce. Such a behavior was also noted in the nitrogen data of Lacam [18]. The least-square fit to a power law relation

$$v_s = B_1 + B_2(P + B_3)^n \quad (1)$$

where B_1, B_3 can be set to zero, is shown as the solid line fit to these data in Fig. 4. The temperature dependence of B_2 and n can be determined from the 60 isothermal fits. A polynomial in T is fit to B_2 and n so that the modified equation (1) gives $v_s(P,T)$ over the range of measurements. For hydrogen the smoothed v_s data were used together with graphically smoothed volume data as input to the least-squares fit to the equation of state [7].

The problem of fitting the gas-like equations of state to these data is clearly shown by the temperature dependence of the sound velocity. In Fig. 5 we have plotted the temperature variation of sound velocity at pressures between 0.5 to 2.0 kbar as obtained for deuterium from our extrapolated equation of state and the calculated values of Michels. In this region at the lowest pressure, deuterium shows a gas-like behavior, that is, the sound velocity increases with temperature as required from kinetic gas theory - although the quantitative behavior is not correct. At higher pressures and low temperatures, progressively closer to the melting line, the sound velocity has the liquid-like characteristic of decreasing with increasing temperature. Corresponding results for the lower pressure isobar for hydrogen are also shown for calculated values of Michels and our own equation of state extrapolated to these lower pressures. These curves indicate that isotherms of v_s vs. P will cross each other at lower pressure. This feature of the data is well described by the equation of state fit we have developed for both hydrogen and deuterium. In Fig. 5a further comparison of sound velocity at 0.5 kbar between hydrogen and deuterium is shown where the ratio $v_s(H_2)/v_s(D_2) = \sqrt{M_{D_2}/M_{H_2}}$ is determined in agreement with the prediction of kinetic gas theory.

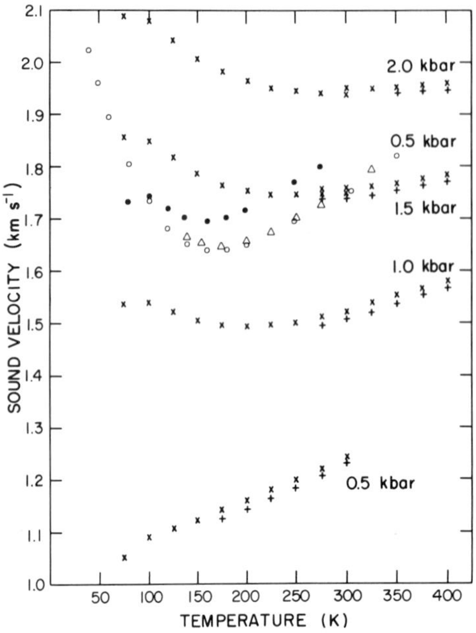

Fig. 5. Sound velocity in n-H_2 and n-D_2 vs. T along several isobars. The crosses refer to n-D_2 calculations from equation (4), the pluses refer to calculations by Michels et al. [4]. The lowest values are at a constant pressure of 0.5 kbar and by steps 1.0, 1.5 and 2.0 kbar. The single isobar at P = 0.5 kbar for n-H_2 is plotted and the filled circles are data from this study [7], the open circles from a compilation by Vargaftik [19] and the triangles data from Michels et al. [4].

For n-D_2, 1340 data sets of P, V, T, v_s were measured over the ranges 75 to 300 K and 2 to 20 kbar [8]. A representative set of volumetric and sound velocity data vs. pressure along one isotherm is shown in Figs. 6a and 6b. The statistical error of these data was sufficiently small that good convergence to the least-squares fitted equation of state occurred with the full data set expanded to 1404 data sets including the lower-pressure data ($\geq$ 1 kbar) of Michels et al. [4].

EQUATION-OF-STATE DEVELOPMENT

For ease of manipulation we have considered an equation of state of the form,

$$V = f\ (P,T) \qquad (2)$$

Many types of closed-form expressions for the equation of state of fluids have been given. Benedict [10] found that an equation of the form

$$V = \sum_{m=1}^{3} \sum_{n=-2}^{2} A_{nm}\, T^{n/2}\, P^{-m/3} \qquad (3)$$

could be simplified, when highly correlated terms were eliminated, to give the following equation which he used for nitrogen:

$$V = (A+BT+CT^{-1/2})P^{-1/3} + (D+ET)P^{-2/3} + (F+GT+HT^{-1/2} + JT^{-1})P^{-1} \qquad (4)$$

We used this equation for the hydrogen and deuterium data, all of

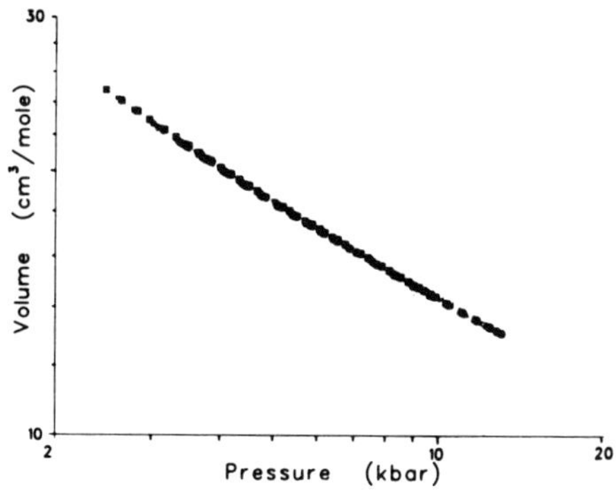

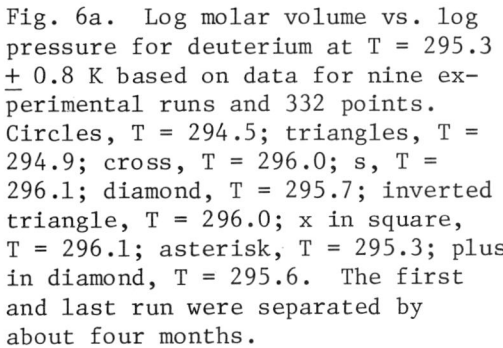

Fig. 6a. Log molar volume vs. log pressure for deuterium at T = 295.3 ± 0.8 K based on data for nine experimental runs and 332 points. Circles, T = 294.5; triangles, T = 294.9; cross, T = 296.0; s, T = 296.1; diamond, T = 295.7; inverted triangle, T = 296.0; x in square, T = 296.1; asterisk, T = 295.3; plus in diamond, T = 295.6. The first and last run were separated by about four months.

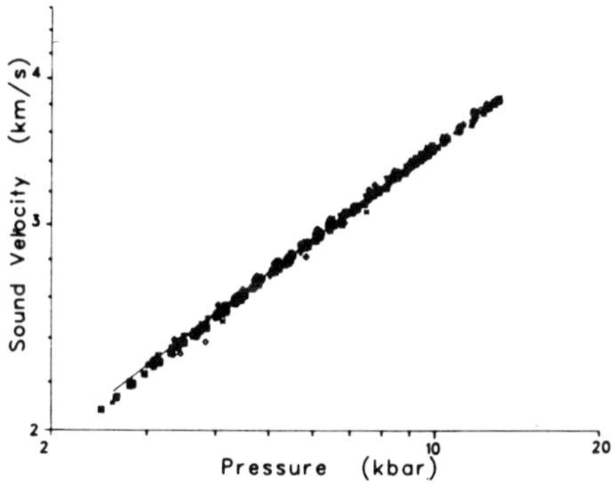

Fig. 6b. Log sound velocity vs. log pressure for deuterium at T = 295.3 ± 0.8 K based on 332 pts and nine isothermal runs; symbols as in Fig. 6a. The smooth curve is a least-squares fit of equation (1) to the data.

which were obtained well into the supercritical fluid region.* The equation has a form that is amenable to use with the double-process least-squares fitting method.

To use P-V-T and v_s data together in a double-process least-squares fitting process, we write the thermodynamic relation

$$\frac{1}{v_s} = \left[\left(\frac{\partial \rho}{\partial P}\right)_s\right]^{1/2} = \frac{M^{1/2}}{V}\left\{-\left(\frac{\partial V}{\partial P}\right)_T - \frac{T[(\frac{\partial V}{\partial T})_P]^2}{C_P}\right\}^{1/2} \equiv g(P,T) \quad (5)$$

and, as discussed [2], minimize the sum S for n sets of points in the following equation:

$$S = \sum_n [V_{exp} - f(P,T)]^2 + \sum_n [\frac{1}{v_s} - g(P,T)]^2 \quad (6)$$

In addition to our own data we need the value of C_p at our normalization pressure over the full temperature range. The lower pressure data of Michels [4] and recently tabulated Russian values [19] (for hydrogen only) were used to obtain C_{P_o} (T) at our lowest pressure of 2 kbar over the 75 to 307 K temperature range. Then the heat capacity

*The critical temperature for H_2 is 33.2 K and the critical pressure is 13.2 bar. For D_2 the corresponding values are 38.2 K and 16.4 bar, respectively.

$$C_p = C_{P_o}(T) - T\int_{P_o}^{P} \left(\frac{\partial^2 V}{\partial T^2}\right)_P dP \qquad (7)$$

can be directly calculated for use in (5) and (6). The expression for n-H_2 in J/g mole-K is

$$C_{P_o}(T) = 1.1230T - 61.468T^{1/2} + 1259.3 - 10512T^{-1/2} + 31638T^{-1} \qquad (8)$$

The resultant fits to (4) for n-H_2 and n-D_2 are listed in Table I.

Table I. Fitted Constants to Equation (4)*

	n-H_2	n-D_2
A	36.716	35.283
B	0.0033003	0.00094703
C	-22.479	3.2843
D	-17.174	-25.090
E	-0.021393	0.0063917
F	-8.9886	13.650
G	0.11001	0.069563
H	69.233	-158.29
J	-31.395	720.00

*As determined by a double-process least-square fit to the n-H_2 smooth data set and the 1340 data point set of n-D_2.

Over the range of temperatures and pressures the average deviation between measured sound velocities for D_2 and values calculated from the equation of state is $\pm$ 0.53% and between measured and calculated volumes is $\pm$ 0.22%. Slightly larger deviations were obtained in the volume data for H_2. Tables of calculated thermodynamic properties are available for hydrogen [20] and similar tables will be published for deuterium. In Fig. 7 the sound velocity surface $v_s(P,T)$ is plotted from the equation of state for hydrogen.

DIRECT COMPARISONS

As we have discussed, there is little data with which our measurements can be directly compared. We observed that Bridgman's H_2 data were in poor agreement. Recently Tsiklis et al. [5] measured hydrogen volumes at 298.2 K and static pressures up to 5 kbar. Comparison shows an average deviation of + 0.22% from our equation of state values. There are no similar measurements for deuterium. Our sound velocity measurements for both hydrogen and deuterium are

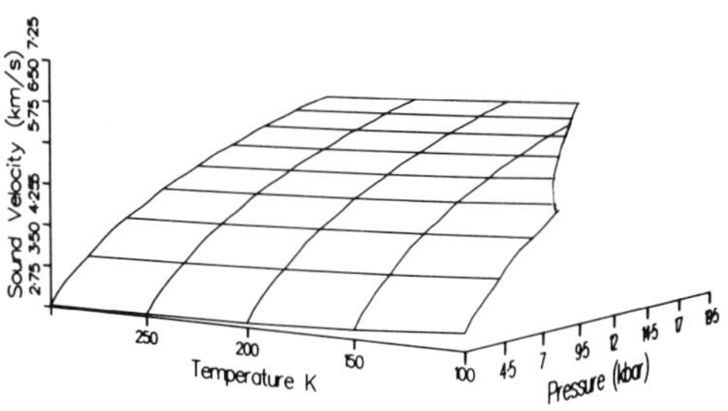

Fig. 7. Sound velocity in n-H_2 vs. T,P. At lower temperatures this surface is truncated by the freezing line.

in a completely new region and can only be compared with Michels' values by extrapolation to low pressure where there is reasonable agreement.

Ashurst [13] recently used potential energy curves for interacting hydrogen molecules to obtain thermodynamic and transport properties. Although the potential was fitted only to the early low pressure data, the model was extended by Ashurst up to the freezing pressure or to 20 kbar from 98.2 to 3273.2 K. His molar volumes deviate from ours at the lower temperatures and pressures by 1 to 5%. For temperatures between 200 to 300 K the deviations are less than 1% at 2 kbar and > 6% at 20 kbar. The calculated sound velocities show a much larger deviation, up to 17%, in this range. We anticipate that, with our new data of 0.5% accuracy these calculations may be refined and provide an improvement of the calculated transport properties as well.

The hard sphere model for deuterium developed by Kerley [15] and modified for the case of hydrogen can be compared with the present measurements. Various data were used to parameterize this theory and our comparison shows good agreement in a range where no previous data existed. In Fig. 8 we show the volume vs. pressure along the 100 and 300 K isotherms. The differences between the measured and theoretical values are at most 2.4% at the freezing point for 100 K and the average deviation is less than 1% throughout the range of the data.

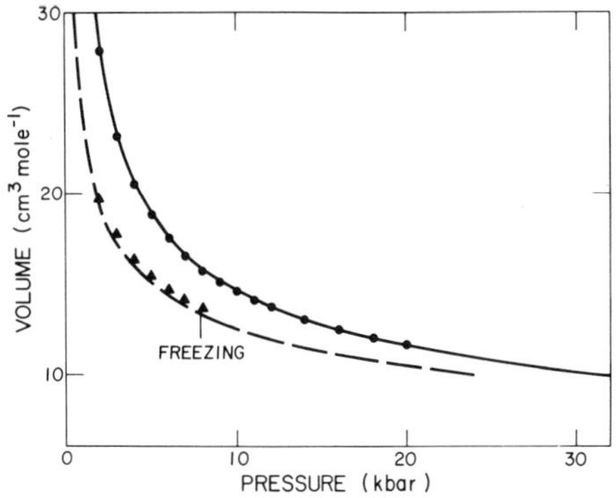

Fig. 8. Molar volume vs. pressure. The full line is the calculation by Kerley [15] at 300 K and results from equation (4) are shown as circles. The dashed line is from Kerley and triangles from this equation of state at T = 100 K.

The sound velocity data are also in satisfactory agreement. A comparison is shown in Fig. 9 along the two isotherms 100 K and 300 K. Average deviation over the temperature and pressure range of the measurements is only 1 to 2%.

EXTRAPOLATED COMPARISONS OF EQUATION OF STATE

The equation of state developed for parahydrogen by Weber [9] was based, at higher densities, upon a 14 parameter surface of a modified Strobridge equation

$$P = RT\rho + (A_1T^2+A_2T+A_3+A_4/T) + (A_5T^2+A_6T+A_7+A_8/T+A_9/T^2)\rho$$
$$+ (A_{10}T^2+A_{11}T+A_{12})\rho^2 + (A_{13}T^2+A_{14}T)\rho^3 \qquad (9)$$

where A_i are constants to be determined. A fit was made to existing data up to 1 kbar in the temperature range of 20 to 300 K. The earlier 17 constant equation of McCarty and Weber [21] extended up to 0.7 kbar. Extrapolation of this equation of state to smaller volumes gives pressures that are too high by over 60% at 20 kbar and 300 K. Sound velocities calculated from this equation disagree with those of Michels by up to 3.5% and only a small fraction of these differences can be ascribed to differences between normal and para-hydrogen.

In comparing results of Baker and Swift [14] with extrapolations of the equation of state for hydrogen at T > 1000 K, we find their molar volumes are larger by 10 to 20% in the pressure range 20 to 100 kbar. On the other hand, a virial-type equation devised by Grigoriev et al. [22] yields molar volumes near freezing that are smaller than those extrapolated from this equation of state by 5, 13, 27 and 31% at pressures of 26, 167, 2440, 4660 kbar, respectively. A measured point at 1 Mbar (estimated temperature 4200 K) during isentropic, dynamic compression gave $V = 3.0\ cm^3/mole$ compared to a value of $4.4\ cm^3/mole$ extrapolated from our equation of state.

We have shown that the isothermal sound velocity and volumetric data are well fit with a power law dependence on pressure,

$$v_s,\ V = B_2P^n \qquad (10)$$

where B_2 and n are constants fit separately for v_s and V. The Lennard-Jones 6, 12 potential

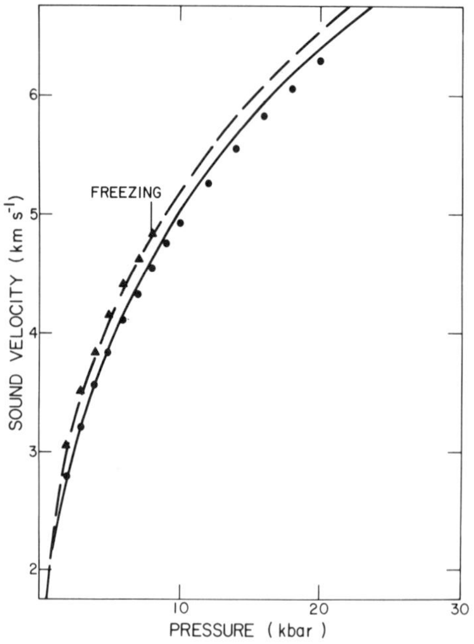

Fig. 9. Sound velocity vs. pressure. The solid line is the calculation by Kerley [15] at 300 K and the results from equation (4) are shown as circles. The dashed line is from Kerley and triangles from this equation of state at T = 100 K.

$$\phi(R) = 4\ \varepsilon[(\frac{R_o}{R})^{12} = (\frac{R_o}{R})^6] \tag{11}$$

in the limit of high density can be simplified in terms of the volume with repulsive forces only,

$$\phi(V) = 4\ \varepsilon(\frac{V_o}{V})^4 \tag{12}$$

and since $P = -(\partial\phi/\partial V)T$ we obtain $P = 16\ \varepsilon(V_o{}^4/V^5)$ and $P = A\rho^5$ where ρ is the density. Using the sound velocity as $v_s^2 = (\partial P/\partial\rho)_s$, $v_s^2 \sim 5\ A\rho^4$ so that $v_s \propto P^{2/5}$ and $V \propto P^{-1/5}$. Since we have also determined experimentally that the product $v_s V \simeq$ const (independent of pressure) as shown in Table II, we compare this potential model and find $v_s V \propto P^{1/5}$. However, by choosing the repulsive term exponent as 9 rather than 12 we obtain

$$v_s \propto P^{3/8} \tag{13a}$$

$$V \propto P^{-1/3} \tag{13b}$$

and $v_s V \propto P^{1/24}$ which is in satisfactory agreement with the measurements.

We have compared this result in Table II and the results for two other heuristic theories, Rao's model for a liquid and the free volume model. Rao's law states that the 'constant' $R_a = v_s^{1/3}\ M/\rho$ where M is the molecular weight. R_a can be interpreted as the molar velocity of sound in analogy with molar refraction. We compare the product $v_s^{1/3}V$ and find large variations in both temperature and pressure.

The free volume model has an interpretation that in the spaces between molecules a gas kinetic velocity of propagation for sound waves is valid and that through the molecules an infinitely fast propagation occurs. The resulting correlation was obtained following Eyring as discussed by Van Dael and van Itterbeek giving $v_s V^{1/3}$ = constant. Values are tabulated in Table II where reasonable temperature independence is indicated.

We have studied the molecular hydrogen isotopes because they are important materials in present energy programs and because they are of theoretical interest. Our new measurements of V and v_s for n-H_2 and n-D_2 in the range $75 < T < 300$ K and $2 < P < 20$ kbar were fitted to a Benedict-type equation of state which reproduced the input data with an average deviation of $\sim 0.5\%$ and gave reasonable extrapolations at both low and high pressures. These new measurements provide the best test to date of models for simple molecules and we find theoretical predictions in general are far outside experimental error. However, the rigid sphere model of Kerley agrees reasonably well with experiment over the full P and T range. Examination of heuristic models yielded a simple correlation allowing extrapolation of V and v_s at high pressures. We are extending our study of the

Table II. Comparison of Rao's Law, a Free Volume Model, and a Parameterization Model

Pressure kbar	300 K	200 K	100 K
Rao rule $v_s^{1/3}$ V = constant			
2	39.37	33.61	28.69
5	29.48	27.08	24.87
10	24.87	23.48	Freezing
15	22.76	21.82	
20	21.43	20.68	
Free Volume model $v_s V^{1/3}$ = constant			
	300 K	200 K	100 K
2	8.45	8.14	8.27
5	10.24	10.17	10.36
10	12.06	12.03	Freezing
15	13.30	13.27	
20	14.25	14.21	
Liebenberg, Mills, Bronson model v_s V = constant			
	300 K	200 K	100 K
2	77.87	67.29	60.42
5	72.42	67.61	64.31
10	72.06	69.11	Freezing
15	72.54	70.18	
20	73.07	70.99	

hydrogens to higher pressures by adapting new technology in an attempt to obtain additional data for further refinement of theory.

ACKNOWLEDGMENTS

We thank L. C. Schmidt (LASL) for the construction and assembly of the piston-cylinder apparatus. We are indebted to M. M. Johnson (LASL) for programming and early computation of the double-process least-square fitting routines for hydrogen. Conversations with many colleagues have been stimulating and we thank R. D. McCarty for information regarding the heat capacity of hydrogen.

REFERENCES

1. D. H. Liebenberg, R. L. Mills, and J. C. Bronson, J. Appl. Phys. 45, 741 (1974).
2. R. L. Mills, D. H. Liebenberg, and J. C. Bronson, J. Chem. Phys. 63, 1198 (1975).
3. R. L. Mills, D. H. Liebenberg, and J. C. Bronson, J. Chem. Phys. 63, 4026 (1975).

4. A. Michels, W. De Graaff, T. Wassenaar, J. M. H. Levelt, and P. Louwerse, Physica 25, 25 (1959); A. Michels, W. De Graaff, and G. J. Wolkers, Appl. Sci. Res. A12, 9 (1963).
5. D. S. Tsiklis, V. Ya. Maslennikova, S. D. Gavrilov, A. N. Egorov, and G. V. Timofeeva, Dok. Akad. Nauk (SSSR) 220, 1384 (1975).
6. P. W. Bridgman, Rec. Trav. Chim Pays-Bas 42, 568 (1923); Proc. Nat. Acad. Sci. (US) 9, 370 (1927); Proc. Am. Acad. Arts Sci. 59, 173 (1924).
7. R. L. Mills, D. H. Liebenberg, J. C. Bronson, and L. C. Schmidt, J. Chem. Phys. 66, 3076 (1977).
8. D. H. Liebenberg, R. L. Mills, and J. C. Bronson, in Proceedings of Seventh Symposium on Thermophysical Properties, Gaithersburg, Maryland, May 10-12, 1977 (to be published).
9. L. A. Weber, US National Aeronautics and Space Administration Rept. NASA SP-3088 (1975).
10. M. Benedict, J. Am. Chem. Soc. 59, 2233 (1937).
11. L. Haar and S. H. Shenker, J. Chem. Phys. 55, 4951 (1971).
12. J. A. Barker and D. Henderson, Rev. Mod. Phys. 48, 587 (1976).
13. W. T. Ashurst, Sandia Laboratories, Albuquerque, New Mexico SAND76-8710, December 1976 and private communication.
14. J. R. Baker and H. F. Swift, J. Appl. Phys. 43, 950 (1972).
15. G. I. Kerley, "A Theoretical Equation of State for Deuterium," Los Alamos Scientific Laboratory Rept. LA-4776 (January 1972) and private communication.
16. B. Vodar and J. Saurel, in High Pressure Physics and Chemistry I, R. S. Bradley, ed., Academic Press, New York (1963), p. 51.
17. J. M. H. Levelt-Sengers, in Physics of High Pressure and the Condensed Phase, A. van Itterbeek, ed., North Holland Publishing Co., Amsterdam (1965), p. 60.
18. A. Lacam, J. Phys. Rad. 14, 351 (1953).
19. N. B. Vargaftik, Tables on the Thermophysical Properties of Liquids and Gases, John Wiley and Sons, Inc., New York (1975), p. 7.
20. D. H. Liebenberg, R. L. Mills, and J. C. Bronson, Los Alamos Scientific Laboratory Rept. LA-6645-MS (April 1977).
21. R. D. McCarty and L. A. Weber, National Bureau of Standards Tech. Note 617 (1972).
22. F. V. Grigoriev, S. B. Kormer, O. L. Mikhailova, A. P. Tolochko, and V. D. Urlin, Zh. Eksp. Teor. Fiz. 69, 743 (1975).

F-4

EQUATION OF STATE OF DENSE FLUIDS: COMPARISON WITH MACHINE CALCULATIONS AND ESTIMATIONS FROM PRESSURE DEPENDENCE OF DIELECTRIC CONSTANT MEASUREMENTS

D. Vidal and M. Lallemand

Centre Universitaire Paris Nord
Villetaneuse, France

INTRODUCTION

There are now very accurate ways to predict thermodynamic data of simple dense fluids at both equilibrium and nonequilibrium and to obtain direct experimental confirmations. In this study of gaseous helium, neon, and argon, experimental data has been compared with predictions of numerical equations of state derived from exact computer calculations for a Lennard-Jones fluid, up to reduced densities $\rho^* = \rho\sigma^3 \approx 1$. In addition, provisional PVT data deduced from dielectric constant measurement of the fluid with respect to pressure are also given.

PVT DATA-EXPERIMENTAL METHOD

An experimental device for absolute density measurement of gases under high pressure has been designed for use up to 12 kbar. Its principle is based on the determination of the mass of gas contained in a piezometer of calibrated volume enclosed in a high pressure vessel (Fig. 1). The calibration of the piezometer was carried out up to 2900 bar by using the Van der Waals Institute's PVT data for argon [1] at 298.15 K. The expansion coefficient of the piezometer found in this range of pressure is then extrapolated up to 12 kbar. Pressure was measured with a manganin gauge previously calibrated by means of a dead weight gauge. For a given pressure the confined gas was recovered and then weighed. Accuracy of the density measurement is estimated to be about 2×10^{-3}.

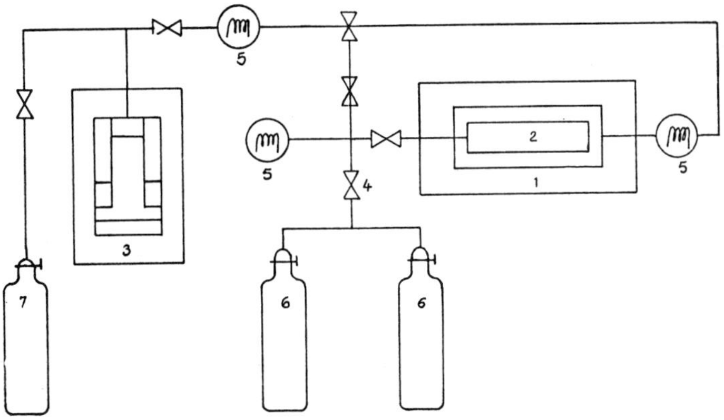

Fig. 1. Experimental high pressure device for density measurements of gases.

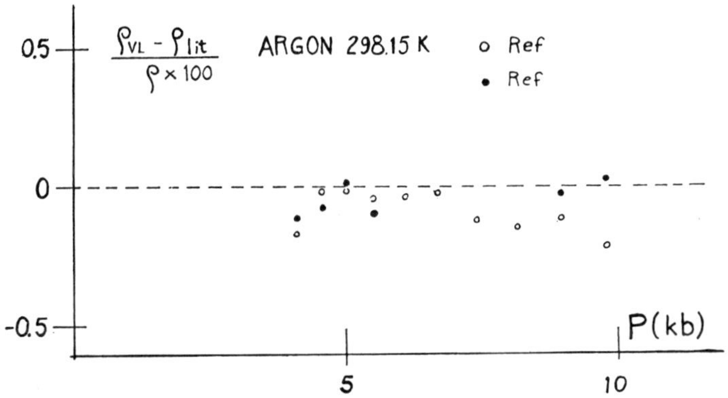

Fig. 2. Deviation of experimental densities for argon with those of Robertson et al. [2] and Stichov et al. [3].

By means of this method we measured molecular density of helium, neon, and argon at 298.15 K up to 10 kbar. Figure 2 compares the values obtained for argon with those of Robertson et al. [2] and Stichov et al. [3]; good agreement up to the high pressure range is noted.

DETERMINATION OF THERMODYNAMIC FLUID DATA BY NUMERICAL METHODS

Several exact computer calculations have been performed, either by the Monte-Carlo method or by the molecular dynamics method for a Lennard-Jones fluid with potential parameters of ε and σ. The emphasis here has been on the numerical equation of state derived by Hansen [4]. These have practical interest for the peculiar case of helium and in connection with the molecular dynamics results obtained by Ashurst [5].

In the calculation of the free energy performed by Hansen, the contribution of the repulsive part of the potential and the first order 1/T term of the attractive part (considered as a weak perturbation) were calculated numerically. Such a division of the thermodynamic potential functions aids in scaling properties. The free energy is dependent only on the product ρ^{*4}/T^{*4} where ρ^* is the reduced density ($\rho^* = N_A\rho\sigma^3$) and T^* is the reduced temperature, $T^* = Tk/\varepsilon$. So, calculation for an isotherm of a property is sufficient to let the experimenter know the value of that property for other reduced densities and reduced temperatures. Accordingly, at sufficient high temperature, the reduced pressure $P^* = P\sigma^3/\varepsilon$, may be expressed as

$$P^* = \rho^* T^* + \rho^{*2}(B_1 T^{*3/4} - C_1 T^{*1/4}) + \rho^{*3}(B_2 T^{*1/2} - 2C_2) + \rho^{*4}(B_3 T^{1/4} - 3C_3 T^{*-1/4}) + \rho^{*5}(B_4 - 4C_4 T^{*-1/2}) + \rho^{*6}(-5C_5 T^{*-3/4}) + \rho^{*11}(B_{10} T^{*-3/2}) \quad (1)$$

with: $B_1=3.629$; $B_2=7.2641$; $B_3=10.4924$; $B_4=11.459$; $B_{10}= 2.17619$; $C_1=5.3692$; $C_2=6.5797$; $C_3=6.1745$; $C_4= -4.2685$; $C_5=1.6841$.

From this equation of state, other thermodynamic quantities can be derived, such as heat capacities, isothermal compressibility coefficient, sound velocity, and so on.

On Fig. 3 the compressibility factor $Z = P/\rho RT$ calculated from Hansen's equation is compared to our experimental determination for helium, neon, and argon at a temperature of 298.15 K.

Choice of the potential parameters is crucial in obtaining agreement with empirical data. For example, if we keep the values given by Hirschfelder et al. [6] for σ and ε, agreement is good only for argon (see Fig. 4). We may observe deviations

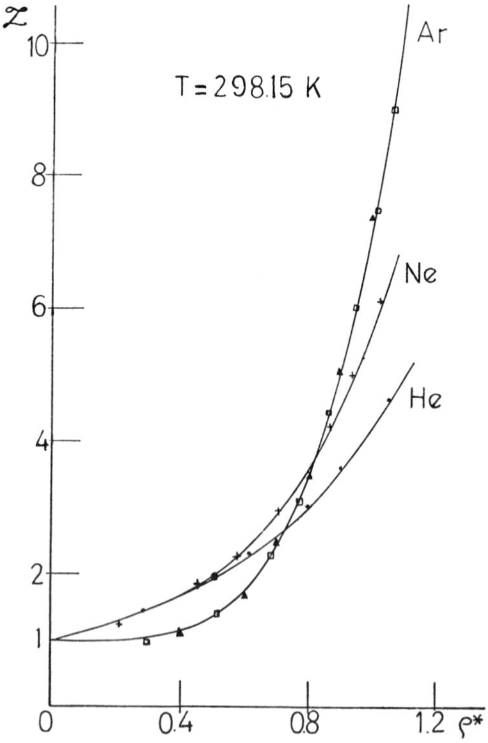

Fig. 3. The compressibility factor of helium, neon, and argon at 298.25 K with respect to reduced density. Square, cross and circle are calculated from Hansen's equation of state, solid lines are experimental results.

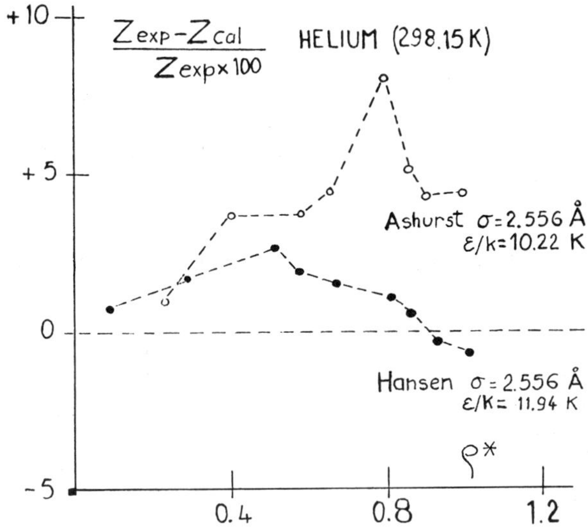

Fig. 4. Deviation of the experimental and calculated compressibility factor for helium at 298.15 K.

of up to 8% in the case of helium. For helium and neon, if we keep the Hirschfelder parameter for σ, a better agreement can be obtained if we take a more pronounced potential depth. Parameters which give a good fit with experimental results up to 10 kbar are then:

	σ, Å	ε/k, K	ε/k, K[6]
He	2.556	11.94	10.22
Ne	2.756	34.90	33.74
Ar	3.405	119.8	119.8

This feature is in agreement with Hansen and Pollock's [7] quantum determination of the ground state properties of compressed helium and neon.

It seems worthwhile to mention the great attention brought to bear on the Lennard-Jones potential parameters. For example, Ashurst [5] derived by the molecular dynamics method results of PVT data for a Lennard-Jones fluid which have reduced parameters very close to those of Hansen. But, with helium he obtained values with conventional σ and ε parameters that showed considerable deviation between the predicted values and experimental ones at 298 K.

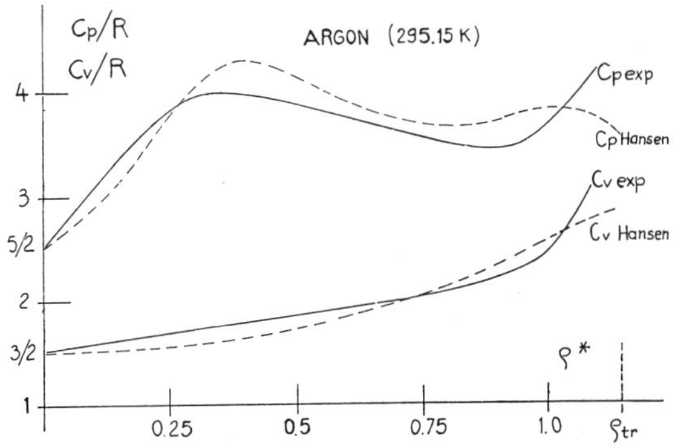

Fig. 5. Evaluation of C_v/R and C_p/R with respect to the density for argon at 298.15 K. Solid line is the numerical result of Hansen, the dashed line represents experimental results obtained from ultrasonic measurements.

Hansen's numerical equation allows prediction of other thermodynamic quantities. Figure 5 plots the evaluation of C_V/R and C_P/R of argon with density. Comparison is made with experimental results from ultrasonic measurements of Liebenberg et al. [8]. Again there is good agreement for this nonuniform variation with numerical results failing only in the vicinity of the fluid-solid transition.

ESTIMATION OF PVT DATA FROM DIELECTRIC CONSTANT MEASUREMENTS

Another quite different way to obtain accurate estimates of PVT data for non-polar dense fluids is from an experimental study of the pressure dependence of the dielectric constant ε. This is generally experimentally more tractable than direct PVT determinations.

For a non polar fluid, molar density and dielectric constant are related by the so called Clausius-Mossotti function

$$C\text{-}M(\rho) \equiv \frac{\varepsilon-1}{\varepsilon+2}\frac{1}{\rho} = \frac{4\pi}{3} W_A \bar{\alpha} \{1+\Delta(\rho^*, T^*)\} \quad (2)$$

We are now in position to determine the value of the right hand member of (2) with an uncertainty of $\pm 0.3\%$ for wide temperature and pressure interval. The term $\Delta(\rho^*, T^*)$, which appears in the statistical mechanics analysis of the C-M function for the dipole-induced-dipole approximation (DID) has been computed exactly by Alder et al. [9] for several reduced densities and reduced temperatures. This was accomplished by using the molecular dynamics method for a Lennard-Jones fluid with constant polarizability value α_o.

A previous study [10,11] reported the experimental dependence of the C-M function for the noble gases at room temperature up to 12 kbar (see Fig. 6). By means of Alder's results it was possible to eliminate the DID contribution and so to deduce the density dependence of the effective polarizability $\bar{\alpha}$. It has been demonstrated under such conditions and within a scatter

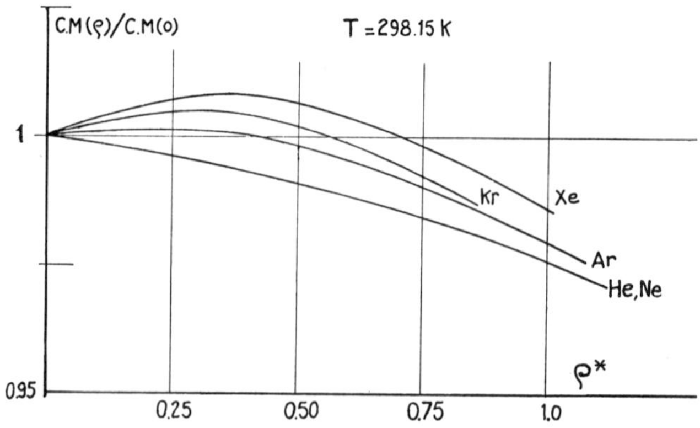

Fig. 6. Evaluation of reduced Clausius-Mossotti function vs. reduced density for the noble gases at 298.15 K.

of 0.6% that $\overline{\alpha}(P^*)$ for noble gases follows an empirical law (up to a reduced density of about unity) of the form:

$$\overline{\alpha} = \alpha_o(1 + B_\alpha \rho^* + C \rho^* + ...) \qquad (3)$$

where α_o is the polarizability of the isolated atoms (see Fig. 7). This expression is almost independent of the temperature and nature of the fluid (a very similar evolution has been found for oxygen and nitrogen) and has the characteristics of the law of corresponding states. In the case of argon, coefficient B_α has a value of -0.004 and C_α has a value of -0.02.

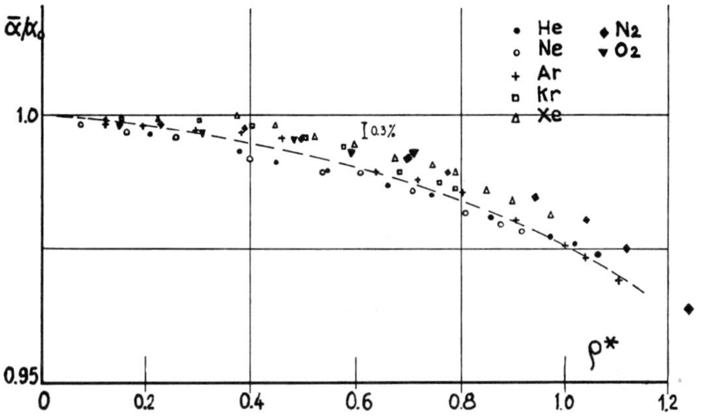

Fig. 7. Variation of the effective polarizability vs. reduced density for the noble gases, nitrogen, and oxygen.

So in a reverse way, the knowledge of the pressure dependence of the dielectric constant ε at fixed temperatures and the combination of (2) and (3) enable us to infer the density of the fluid for the given conditions. The accuracy expected is ± 0.3% up to 10 kbar.

As an application of the proposed method, we give the density of helium at 35°C (σ = 2.62 Å, α_o = 0.2053 Å^3 Table I.

Table I. Density of Helium Gas at 35°C.

P, bar	ε	mole/cm^3
2108	1.0711	0.045_0
4125	1.1035	0.0652_3
6209	1.1260	0.0792_7
8043	1.1414	0.0889_0
10,723	1.1594	0.100_2

REFERENCES

1. A. Michels and H. Wijker, Physica 15, 627 (1949).
2. S. L. Robertson, S. E. Babb, and C. J. Scott, J. Chem. Phys. 50, 2160 (1969).
3. S. M. Stichov, V. I. Fedosinov, and I. N. Makarenko, Institute of Crystallography, Akademia Nauk, U.S.S.R. (1972).
4. J. P. Hansen, Phys. Rev. A 2, 221 (1970).
5. W. T. Ashurst, Sandia Laboratories Energy report, Sandia 76-8710.
6. J. O. Hirschfelder, C. F. Curtiss and R. B. Bird, Molecular Theory of Gases and Liquids, John Wiley and Sons, New York (1954).
7. J. P. Hansen, and E. L. Pollock, Phys. Rev. A 5, 2651 (1972).
8. D. H. Liebenberg, R. L. Mills, and J. C. Bronson, J. Applied Phys. 45, 741 (1974).
9. B. J. Alder, H. L. Strauss, and J. J. Weiss, J. Chem. Phys. 62, 2328 (1975).
10. D. Vidal and M. Lallemand, J. Chem. Phys. 64, 4293 (1976).
11. M. Lallemand and D. Vidal, J. Chem. Phys. 66, 4776 (1977).

F-5

AN EQUATION OF STATE FOR THE STATIC DIELECTRIC CONSTANT OF WATER FROM 0 TO 550°C AND UP TO 5 KBAR

M. Uematsu

Keio University
Yokohama, Japan

and

E. U. Franck

Universität Karlsruhe
Karlsruhe, W. Germany

INTRODUCTION

The static dielectric constant (permittivity) of water in the high pressure region has been investigated experimentally since 1920, and these works have been previously surveyed and summarized by the present authors [1]. The experimental data cover high pressure and high temperature regions to 20 kbar and 600°C, respectively. Based on these data, a new equation of state for the static dielectric constant of water was devised. In this paper, a new method of correlating thermophysical properties as a function of two state parameters is proposed and a new equation of state devised by this method is presented.

METHOD

The empirical equation which is described as a function of two independent parameters, x and y, can be written as

$$Z = \sum_{i=1}^{n} a_i X_i(x) Y_i(y) \qquad (1)$$

The numerical values of the coefficients in (1) are evaluated by the method of least squares to minimize the deviations between experimental data and calculated values. As a result of such a regression analysis, the general standard deviation and the standard deviation of each of the coefficients can be calculated. The standard deviation is a convenient measure of the quality of the fit. The F-value of each of the coefficients, which is defined as

$$F_i = a_i \,/\, \sigma_i \qquad (2)$$

shows the significance of each of the coefficients. As long as F_i exceeds the presupposed value for F which is obtained from corresponding tables, one can be confident that the coefficient a_i is not zero. On the other hand, if F_i is smaller than the presupposed value it may be concluded that a_i is unessential.

If equation (1) has unessential coefficients, the term of (1) whose coefficient has the least F-value is eliminated and again the regression analysis is repeated. Finally the functional form containing only essential terms is obtained. This equation which has such a form is the optimum for the given set of experimental data.

Applying this procedure to the correlation of thermophysical properties we can obtain the empirical equation appropriate to a set of experimental data which is described as a function of two state parameters such as temperature and pressure or temperature and density.

DERIVATION

A survey and critical evaluation of the literature published since 1920, which covers 47 papers, have been made [1]. The experimental data by Akerlof and Oshry (100 to 370°C, for saturated water) [2], Owen et al. (0 to 70°C, up to 1 kbar) [3], Dunn and Stokes (10 to 65°C, up to 2 kbar) [4], Heger (100 to 550°C, up to 5 kbar) [5], Srinivasan and Kay (10 to 40°C, up to 3 kbar) [6], and Lukashov et al. (400 to 600°C, up to 500 bar) [7] were mainly used as the basic data set for fitting in the high-pressure region.

As the first step for the determination of an optimum equation for the static dielectric constant of water based on the available experimental data from 0 to 550°C and up to 5 kbar, the following expression as a function of temperature and density was considered:

$$\begin{aligned}\varepsilon = 1 &+ (a_1T^{-2} + a_2T^{-1} + a_3 + a_4T + a_5T^2 + a_6T^3 + a_7T^4)\,\rho \\ &+ (a_8T^{-2} + a_9T^{-1} + \ldots + a_{14}T^4)\,\rho^2 + (a_{15}T^{-2} + \ldots \\ &a_{21}T^4)\,\rho^3 + (a_{22}T^{-2} + \ldots + a_{28}T^4)\,\rho^4 + (a_{29}T^{-2} \ldots \\ &+ a_{35}T^4)\,\rho^5 + (a_{36}T^{-2} + \ldots + a_{42}T^4)\,\rho^6 \qquad (3)\end{aligned}$$

By the least squares fit for the basic data set which has 664 data points from the available experimental data including those of liquid water at a pressure of 1 bar, the coefficients of equation (3) were evaluated and F-values for each were calculated as listed in Table I. The general standard deviation was calculated as 0.70

Table I. F-values of Coefficients in Equation (3)

	ρ^1	ρ^2	ρ^3	ρ^4	ρ^5	ρ^6
T^{-2}	-0.44	0.42	-0.38	0.20	0.37	-1.35
T^{-1}	0.08	-0.31	-0.11	0.62	-1.35	2.31
T^0	0.46	-0.88	2.24	-0.33	1.09	-2.36
T^1	-1.32	3.04	0.21	-0.31	-0.83	2.17
T^2	2.34	-0.77	-0.78	0.82	0.77	-2.00
T^3	-2.44	-0.33	1.28	-1.06	-0.91	2.02
T^4	1.95	0.51	-1.72	0.84	1.15	-2.16

in ε-units. From the table of the F-distribution, F = 6.63 was found as the F-value for the 99% confidence level [8]. The regression analysis was repeated until each F-value of the coefficients exceeded 6.63. The functional form which has 16 coefficients was obtained as the last step, and each F-value of the coefficients is listed in Table II. This functional form fitted the basic data set with a

Table II. F-values for Final Coefficients in Equation (3)

	ρ^1	ρ^2	ρ^3	ρ^4	ρ^5	ρ^6
T^{-2}		15.24	-19.33			
T^{-1}		-7.01				
T^0		8.49			17.31	
T^1		-8.66			-16.23	
T^2		8.43		-14.37	15.85	-16.34
T^3				13.50	-14.81	14.85
T^4				-12.44	13.48	

general standard deviation of 0.42 in ε-units. By examining the curves of constant ε-value in the temperature-density diagram calculated by this functional form, it was found that some of these curves (i.e., mainly between ε=10 and 20) have some amount of ripple. This kind of ripple is one of the unacceptable characteristics of the empirical equation.

In order to eliminate this ripple, secondary data which were obtained directly by interpolation and extrapolation of the experimental data were added to the basic data set. The number of data points of this extended set amounted to 892. Based on this new data set the regression analysis was carried out several times. Finally, the following expression whose ε-curves were plausible was considered as the optimum equation for temperatures from 0 to 550°C and pressures up to 5 kbar:

$$\varepsilon = 1 + a_1T^{-1}\rho + (a_2T^{-1} + a_3 + a_4T)\,\rho^2 + (a_5T^{-1} + a_6T + a_7T^2)\,\rho^3 + (a_8T^{-2} + a_9T^{-1} + a_{10})\,\rho^4 \tag{4}$$

where
$a_1 = 2.27361 \times 10^3$, $a_1/\sigma_1 = 10.71$
$a_2 = 7.27494 \times 10^4$, $a_2/\sigma_2 = 48.05$
$a_3 = -1.40569 \times 10^2$, $a_3/\sigma_3 = -55.31$
$a_4 = 9.31882 \times 10^{-2}$, $a_4/\sigma_4 = 59.25$
$a_5 = -2.87060 \times 10^4$, $a_5/\sigma_5 = -14.95$
$a_6 = 1.40167 \times 10^{-1}$, $a_6/\sigma_6 = 24.24$
$a_7 = -1.14855 \times 10^{-4}$, $a_7/\sigma_7 = -31.77$
$a_8 = -4.01851 \times 10^6$, $a_8/\sigma_8 = -29.35$
$a_9 = 2.52353 \times 10^4$, $a_9/\sigma_9 = 15.92$
$a_{10} = -3.58644 \times 10^1$, $a_{10}/\sigma_{10} = -15.79$

Equation (4) fits the available experimental data with a standard deviation of 0.33 ε-units.

DISCUSSION

According to each F-value, all of the coefficients in (4) are significant, and the trend of the ε-curves generated using this relation on a temperature-density diagram shown in Fig. 1 appears reasonable. From this figure, equation (4) seems to be valid for a range of temperatures and pressures up to 600°C and 10 kbar.

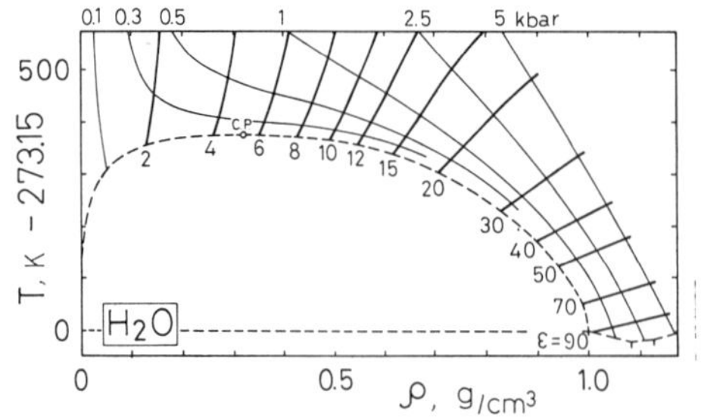

Fig. 1. Plot of constant ε-values for water.

ACKNOWLEDGMENT

The authors wish to thank W. Harder for his many helpful discussions. One of them (M.U.) wants to express his appreciation to the "Alexander von Humboldt-Stiftung" for a "Forschungs-stipendium".

NOTATION

a_i = coefficient
F_i = F-value of the i-th coefficient
T = Temperature, K
ε = static dielectric constant (permittivity)
ρ = density, g/cm^3
σ_i = standard deviation of the i-th coefficient

REFERENCES

1. M. Uematsu, W. Harder, and E. U. Franck, "The Static Dielectric Constant of Water in the Range of Temperatures from 0 to 550°C and Pressures up to 5 kbar," paper presented at meeting of Working Group III of the Intern. Association for the Properties of Steam, Kyoto, Japan, September 1976.
2. G. C. Akerlof and H. I. Oshry, J. Am. Chem. Soc. 72, 2844 (1950).
3. B. B. Owen, R. C. Miller, C. E. Milner, and H. L. Cogan, J. Phys. Chem. 65, 2065 (1961).
4. L. A. Dunn and R. H. Stokes, Trans. Faraday Soc. 65, 2906 (1969).
5. K. Heger, Ph.D. Dissertation, University of Karlsruhe, Karlsruhe, Germany (1969).
6. K. R. Srinivasan and K. L. Kay, J. Chem. Phys. 60, 3645 (1974).
7. Yu. M. Lukashov, B. P. Golubev, and F. B. Ripol-Saragosy, Teploenergetika 22 (6), 79 (1975).
8. P. R. Bevington, Data Reduction and Error Analysis for the Physical Sciences, McGraw-Hill Book Company, New York (1969).

F-6

THERMOPHYSICAL AND DIELECTRIC PROPERTIES OF METHANOL AND RELATED COMPOUNDS TO 350°C AND 5 KBAR*

E. U. Franck, R. Táani, and R. Deul

Universität Karlsruhe
Karlsruhe, W. Germany

The density of liquid and supercritical methanol has been determined experimentally from 25 to 350°C and from 500 to 8000 bar. An externally heated autoclave was used with a Monel metal liner which retarded decomposition at high temperature. Eight isotherms with 194 experimental points were obtained. The mean relative uncertainty of the specific volume is 0.4%. An equation of state in the form of a single extended Tait equation was prepared for the entire temperature and pressure range. It has nine constants determined by a general least squares method. The mean deviation from the experimental points is 0.17%. The new data together with previous low pressure results from other authors were used to calculate functions such as compressibility, thermal expansion, entropy, enthalpy, compressibility factor and fugacity coefficients to 250°C.

With a second apparatus the static dielectric constant of methanol was measured from 30 to 250°C and to pressures up to 3300 bar. Within this region the dielectric constant varied from 36.5 at 30°C and 2000 bar to 6.4 at 250°C and 164 bar. The total uncertainty of the results does not exceed ± 1.7%.

By means of the new density data and the Kirkwood-Fröhlich theory, the Kirkwood correlation factors for methanol were calculated for sub- and supercritical temperatures and high pressures. These correlation factors agree with those of Dannhauser et al. determined previously for saturation conditions. Above 200°C, isotherms of these factors show a maximum at densities of about 0.7 g/cm^3 Dielectric constants and correlation factors of methanol as a function of density have been compared with those determined recently of water, hydrogen chloride and acetonitrile at corresponding states.

*Only this extended abstract is presently available.

F-7

PIEZO-OPTIC BEHAVIOR AND THE EQUATION OF STATE OF LIQUIDS*

K. Vedam and P. Limsuwan

The Pennsylvania State University
University Park, Pennsylvania USA

INTRODUCTION

Consider the piezo-optic behavior of materials in their transparent region of spectrum. In the case of solids such as alkali-halides [1], α-quartz [2], vitreous silica [3], etc., the relationship between the change in refractive index Δn and pressure becomes slightly nonlinear at high pressures (say above 5 kbar), but the same data exhibit perfect linear relationship between Δn and the Lagrangian strain η in the entire range of pressures studied. In the case of liquids such as water and CCl_4, as mentioned in a previous article [4], Δn is grossly nonlinear with pressure, and Δn vs. η is linear only at strains less than 2 or 3%. Motivated by the linearity of Δn vs. η for the solids, one can view the nonlinearity of the Δn vs. η for the liquids as possibly caused by (1) unreliable P-V data used to evaluate η, (2) use of an inappropriate equation of state for the liquids when extrapolating and interpolating literature P-V data to cover the entire 14 kbar pressure range, (3) nonapplicability of Lagrangian strain as a strain measure at the very high strains and (4) perhaps intrinsic nonlinearity of Δn with respect to all strain measures for liquids. Investigation into these possibilities was made with high pressure interferometric measurements on a number of liquids under hydrostatic pressure and the results and conclusions are presented here.

DEFINITION OF VARIOUS STRAINS

Thomsen [5] and Davies [6,7] have recently reviewed the significance and properties of the various representation of

*Work supported by the Office of Naval Research, Physics Program.

strains, in particular the Lagrangian (η) and the Eulerian strain (ε). In brief, for the case of liquids under hydrostatic pressure, we have

$$\eta = \frac{1}{2}\left[\left(\frac{V}{V_o}\right)^{2/3} - 1\right] \tag{1}$$

$$\varepsilon = \frac{1}{2}\left[\left(\frac{V_o}{V}\right)^{2/3} - 1\right] = \eta\left(\frac{V_o}{V}\right)^{2/3} \tag{2}$$

where V_o and V are the initial and final specific volumes. As Truesdell and Noll [8] have shown, the Lagrangian or the "material" strain is rotationally invarient, i.e., invariant under changes of the frame of reference, whereas the Eulerian or "spatial" strain does not in general satisfy this invariancy criterion. Hence, to overcome this limitation of ε, Davies [6] has emphasized that the frame indifferent analogue, E, of the Eulerian strain ε should be used rather than ε. However, for the special case of isotropic bodies under hydrostatic pressure $E \equiv \varepsilon$ and the invariancy condition is trivially satisfied [6]. Thus for the liquids in our hydrostatic pressure experiments, one can conveniently describe a frame-invariant strain at any pressure in either the Lagrangian representation or the Eulerian frame - indifferent analogue representation (which here is identically equal to Eulerian strain). These strains can be evaluated by using an appropriate equation of state to extrapolate and interpolate experimental P-V data to cover our entire 14 kbar accessible pressure range.

EQUATIONS OF STATE

The equations of state, for liquids and solids, that are widely used in the literature [9,10] contain two or three parameters; bulk modulus plus first and sometimes second pressure derivatives of the bulk modulus. Since these constants cannot always be obtained with high accuracy from even the best available data [9], they were evaluated for our studies by least squares fitting each equation of state to literature P-V data for each liquid. The standard error of estimate of each fitting was used as an indication of the ability of each equation of state to describe that particular liquid. In addition, where more than one source of P-V data was available for a liquid, the standard error was used to rank the reliability of each P-V data set.

RESULTS AND DISCUSSION

Table I presents some of the experimental results obtained on a number of liquids by the high pressure interferometric method [4,11]. In every case except n-pentane the maximum pressures listed were determined by the freezing of the liquid at the stated temperature. Figure 1 shows a representative graph of Δn vs. strain, for water evaluated from the P-V data of Adams [12] and the second

Table I. Piezo-Optical Properties of Liquids

Liquid	Temp., °C	λ, nm	n_o	Maximum Pressure, kbar	$\frac{\Delta V}{V_o}$, (%)	Total no. of Fringes Shifted	Δn_{max}, $(n_{max} - n_o)$
water	25	546.1	1.3340	11.43	20.98	369	0.0842
CCl_4	25	546.1	1.4600	1.94	10.85	324	0.0671
n-decane	25	546.1	1.4114	3.23	14.37	402	0.0730
n-octane	25	546.1	1.3968	6.23	20.02	517	0.1116
n-heptane	25	546.1	1.3867	11.50	27.39	788	0.1507
n-hexane	25	546.1	1.3751	11.66	28.62	822	0.1566
n-pentane	30	546.1	1.3535	14.00	33.66	941	0.1809
chlorobenzene	25	589.3	1.5230	6.64	17.86	591	0.1188

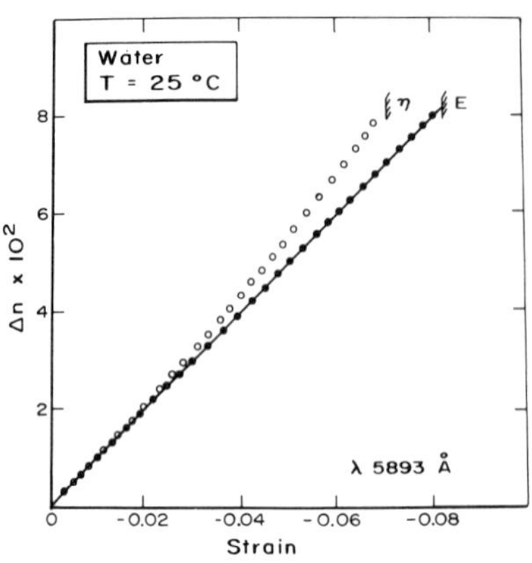

Fig. 1. Variation of refractive index of water with Lagrangian strain η and strain parameter E.

order Murnaghan equation of state. It is seen that Δn increases truly linearly with the strain parameter E for the entire stability field of water in the liquid phase, but that Δn vs. η is linear only below about 3% strain. Similar results were obtained with every liquid studied thus far. It is found that the range of linearity between Δn and strain is much larger when using the strain E instead of η. This is true for each and all equations of state fitted to all P-V data of all the liquids studied. We will later show that three equations of state gave truly linear plots to Δn vs. E over the entire 14 kbar pressure range for all liquids studied).

An important result of this work is that we have experimentally shown for the first time that E is a more useful strain measure than the Lagrangian strain η. Identical conclusions for the usefulness of E over η were theoretically deduced by Davies [6] from both the ultrasonic data on the pressure dependence of the elastic moduli and the Hugoniot shock wave data for MgO.

Table II compares the fit of Δn for water to various degrees of polynomials in E for six of the widely used equations of state discussed in the literature. It is seen that a good linear relation between Δn and E is obtained with the Tait, second-order Murnaghan, second-order Birch, and Keane equations of state as evidenced by the value of the sum of the squares of the residuals as well as the standard error of estimate. The other equations, including those not listed there (such as Bridgman equation, etc.) yield poor linear fit and require higher order polynomials in E.

Similar analysis with the other liquids show that only three equations of state give a good linear relation of Δn with respect to E for all liquids. They are the second-order Murnaghan, second-order Birch, and the Keane equations of state. Anderson [13], and Chhabildas and Ruoff [14] have shown that of all the equations of state reported in the literature, only the Keane's equation yields physically as well as thermodynamically meaningful values on extrapolation to very high pressures. Hence the Keane equation is preferred over the other two that gave linear Δn vs. E plots.

It is one of the more important results of our experiments that a single equation of state, Keane's equation, is found to give an excellent linear fit between Δn and E for all the liquids studied, irrespective of the nature of the liquid, whether polar

Table II. Least-Squares Fitting Results for Water Using Adams P-V Data to Compute E (Δn Measured at 25°C Using λ589.3 nm)

Equation	$\Delta n = A + BE + CE^2$			Sum of Squares of Residuals	Estimated Standard Error of the fit of Δn
	A	B	C		
Tait	-2.50×10^{-4}	-0.990		4.43×10^{-7}	8.74×10^{-5}
	-2.69×10^{-4}	-0.991	-0.016	4.38×10^{-7}	8.77×10^{-5}
1st-order Murnaghan	-1.06×10^{-4}	-0.987		7.54×10^{-7}	11.40×10^{-5}
	-0.07×10^{-4}	-0.980	-0.085	6.26×10^{-7}	10.48×10^{-5}
2nd-order Murnaghan	-1.80×10^{-4}	-0.988		3.47×10^{-7}	7.74×10^{-5}
	-1.45×10^{-4}	-0.986	0.031	3.31×10^{-7}	7.62×10^{-5}
1st-order Birch	-2.90×10^{-4}	-0.990		7.12×10^{-7}	11.08×10^{-5}
	-3.59×10^{-4}	-0.995	-0.058	6.53×10^{-7}	10.70×10^{-5}
2nd-order Birch	-1.82×10^{-4}	-0.989		4.21×10^{-7}	8.52×10^{-5}
	-1.43×10^{-4}	-0.986	0.033	4.01×10^{-7}	8.39×10^{-5}
Keane	-1.85×10^{-4}	-0.989		4.44×10^{-7}	8.75×10^{-5}
	-1.48×10^{-4}	-0.986	0.032	4.26×10^{-7}	8.65×10^{-5}

or nonpolar and whether composed of spherically symmetric, or planar, or long chain molecules and even though the volume strain involved is as high as 33%.

The linear Δn vs. E relationship implies nonlinear relationship between Δn and volume strain and thus failure of the Gladstone-Dale, Drude, Lorentz-Lorenz, Eykman equations since these equations assume constancy of polarizability. The present studies clearly indicate that polarizability is dependent on the volume. This aspect will be dealt with in detail elsewhere.

CONCLUSION

Interferometric measurements on a number of liquids at high pressure (to 14 kbar) show (1) first experimental proof that the Eulerian frame-indifferent analogue strain E is a more useful strain measure than the Lagrangian strain, (2) Keane's equation of state best describes each and all liquids studied, (3) change in refractive index vs. E is linear for all liquids, even though the volume strain involved is as high as 33%.

ACKNOWLEDGMENT

The authors would like to express their sincere thanks to G. R. Mariner for the numerous discussions.

NOTATION

A,B,C = constants
E = frame indifferent analogue of the Eulerian strain ε
n_o = refractive index at stp
Δn = change in refractive index
P = hydrostatic pressure
V = specific volume at pressure P
V_o = initial specific volume
ΔV = change in volume
ε = Eulerian strain
λ = wavelength of light
η = Lagrangian strain

REFERENCES

1. K. Vedam, E. D. D. Schmidt, J. L. Kirk and W. C. Schneider, Mat. Res. Bull. 4, 573 (1969).
2. K. Vedam and T. A. Davis, J. Opt. Soc. Amer. 57, 1140 (1967).
3. K. Vedam, E. D. D. Schmidt and R. Roy, J. Amer. Ceram. Soc. 49, 531 (1966).
4. K. Vedam and Pichet Limsuwan, Phys. Rev. Lett. 35, 1014 (1975).
5. L. Thomsen, J. Phys. Chem. Solids 31, 2003 (1970).
6. G. F. Davies, J. Phys. Chem. Solids 34, 1417 (1973).
7. G. F. Davies, J. Phys. Chem. Solids 34, 841 (1973); also 35, 1513 (1974).
8. C. Truesdell and W. Noll, Handbuch der Physics, Vol. III/3, Springer-Verlag, Berlin (1965), Sec. 19, 26, 29.
9. J. Ross MacDonald, Rev. Mod. Phys. 41, 316 (1969).

10. F. Birch, Phys. Rev. 71, 809 (1947); also A. Keane, Austral. J. Phys. 7, 322 (1954); G. R. Barsch and Z. P. Chang, U.S. NBS Spec. Publ. No. 326 (1971), p. 173.
11. K. Vedam and P. Limsuwan, Rev. Sci. Instr. 48, 245 (1977).
12. L. H. Adams, J. Amer. Chem. Soc. 53, 3769 (1931).
13. O. L. Anderson, Phys. Earth Planet. Interiors 1, 169 (1968).
14. L. C. Chhabildas and A. L. Ruoff, J. Appl. Phys. 47, 4182 (1976).

F-8

WATER HUGONIOT MEASUREMENTS IN THE RANGE 30 TO 220 GPa*

A. C. Mitchell and W. J. Nellis

Lawrence Livermore Laboratory, University of California
Livermore, California USA

INTRODUCTION

Water is one of the most abundant compounds on the surface of the earth as well as in the rest of the solar system [1,2]. Thus, the equation of state of water at high pressures and temperatures is useful for the modeling of planets and their satellites, and for a wide variety of applications on earth [3]. As an example, the interiors of some of the moons of Jupiter are thought to be composed of significant amounts of water as well as ammonia and methane. These satellites are nearly planet-sized and thus these materials are under high pressures by virtue of their own gravitational attraction. For this reason, a program has been undertaken to measure shock Hugoniots of H_2O, NH_3, CH_4, and also D_2, since hydrogen is a major constituent of Jupiter and Saturn [4,5], in order to refine the equation of state data base for planetary modeling.

This paper is a preliminary communication of new data which fills the gap in the water Hugoniot data between the 45 GPa work of Walsh and Rice [6] and the 100 GPa data of Skidmore and Morris [7] and of Bakanova et al. [8]. In addition, data from a double-shock experiment extends the highest pressure of water Hugoniot absolute measurements by a factor of approximately 2. High pressures were achieved by using a two-stage, light-gas gun [9] with a 28 mm diameter launch tube, rather than chemical explosives. A new shock wave detector with subnanosecond time resolution was developed for accurate experiments to be performed in the small geometries dictated by the two-stage gun.

*Work performed under the auspices of the U.S. Energy Research and Development Administration under contract W-7405-Eng-48 and with partial support from the Office of Naval Technology.

EXPERIMENT

The shock wave detectors are electrical self-shorting pins* and are illustrated in Fig. 1. The central conductor is an aluminum wire about 0.75 mm in diameter. The tip is anodized to a depth of about 1 to 2 μm to provide a thin insulating layer. This wire is inserted into a stainless steel tube having an outer diameter of about 1.25 mm. The center conductor is secured in place and electrically insulated from the outer tube by a thin layer of epoxy. A gold film about 1000 Å thick is then deposited over the anodized tip to complete the outer conductor. A 50 Ω resistor is connected in series with the aluminum wire so that when the pin tip is shorted by a shock wave, the cable has a terminal impedance equal to its characteristic impedance. This feature enables several pin signals to be multiplexed, since signal reflections are suppressed.

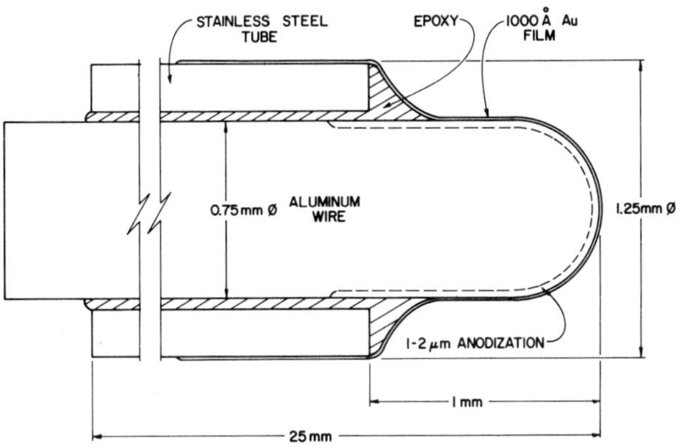

Fig. 1. Self-shorting electrical pin.

Since the shock velocities in typical experiments are of the order ∿ 10 km/s, a shock wave traverses the anodized insulating layer, which is about 1 μm thick, at the tip of the pin in ∿ 0.1 nsec. Thus, whether the pin conducts electrically on the first shock or after reverberations in the end tip, the pin insulating layer shorting time is well below one nsec. Pin signals are displayed on Tektronix 519 oscilloscopes and are recorded on Polaroid film. Total sweeps of 300 nsec are used. Just prior to the experiment, calibration traces having a 5 nsec period are recorded on the same film to be used for the data record. Low-dispersion, 22 mm diameter foam-flex cables are used as delay lines to conveniently space multiplexed signals. The length of the delay lines is measured to ± 0.1 ns. Cross-timing between oscilloscopes is achieved by a fiducial marker recorded simultaneously on each oscilloscope. The total rise time of the system to a step voltage is $\lesssim$ 1 nsec. However, the onset of a sharp pulse can be measured to about 0.3 nsec, as discussed below.

After a shot, the Polaroid film record, illustrated in Fig. 2, is photographically enlarged by a factor of about four. Since the sine wave with the 5 nsec period is also on the film record, non-

*Developed with the assistance of N. Brown.

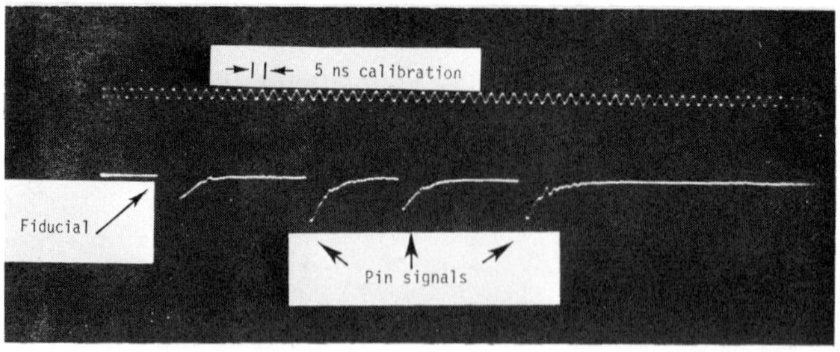

Fig. 2. Typical film record for three, multiplexed, self-shorting pin signals.

linearities of both sweep speed and in the photographic enlargement process are automatically taken into account. The photographic enlargements produce a data record with a resolution of ~ 2 nsec/mm. The results of two sets of 60 arrival time measurements, one each made by the two authors, were compared. The average deviation per arrival time between the two sets of readings was 0.02 nsec; i.e., zero, when the algebraic signs are taken into account. Hence, the reading errors are essentially random when taken over a large sample. The standard deviation between the two sets was 0.3 nsec, which, therefore, is our estimated time resolution. The accuracy of the system is limited by the definition of the signal break on the record, and further photographic enlargement of the film record is not warranted.

Because of the time resolution of the detectors it is now possible to determine very accurately the geometry of the impactor on collision. A series of shots were fired in which 13 pins were arranged in a cross configuration and mounted against a 1 mm thick tantalum target plate. It was found that aluminum and tantalum impactors have a total tilt of the order 10 to 30 nsec and 40 to 80 nsec, respectively, across their 24 mm diameter. In addition the tantalum impactors were found to have a parabolic, concave bow up to 8 nsec deep at velocities of 6.7 km/s across a radius, while bowing of the aluminum impactors is $\lesssim 1$ nsec. It was also found that both the tilt and bowing are reproducible functions of the impactor mass and velocity. Because these distortions from planarity are reproducible and measurable, corrections for the impactor shape are made in treating the data. Since aluminum impactors are essentially free of bowing distortions, corrections to the data are smaller than for tantalum impactors. Thus, the lower pressure experiments using aluminum impactors are inherently more accurate at present. Taking into account all the uncertainties in the system, shock transit times can be measured to $\lesssim 1$ nsec and $\lesssim 2$ nsec for experiments which use aluminum and tantalum impactors, respectively. Since transit times are ~ 200 nsec, uncertainties in shock velocities are $\lesssim 0.5\%$ and $\lesssim 1.0\%$ for experiments with aluminum and tantalum impactors, respectively.

The targets were designed with cylindrical symmetry and the detectors were in regions free from edge effects and unwanted interface interactions. A two-dimensional hydrodynamic calculation using the HEMP code [10] was performed which showed that the edge effect is a shock wave from the cylindrical wall of the water cavity directed radially inward. The calculation showed that this wave makes an angle of about 55° with the direction of the impactor velocity. Thicknesses of the impactors and of the various layers in the target were all checked with the KOVEC hydrodynamic code [10] with planar symmetry. Impactors were either 3.0 mm aluminum or 1.5 or 2.0 mm tantalum. The base plates (front surface of the water cavity) were all 2.0 mm thick aluminum. The water thicknesses were 3 mm for the single-shock experiments and 2.5 mm for double-shock cases. The tantalum anvils for the double-shock experiments were 1.2 mm thick. Pin levels were measured to within 5 μm using a traveling microscope mounted vertically. The velocity of the impactor was measured with a pulsed x-ray technique described elsewhere [11]. The water cavity was thoroughly rinsed before filling the targets with high purity distilled water just before the shot. A copper/constantan thermocouple was attached to each target and the temperature of the water was measured to ∿ 0.1°C in order to determine the initial density. The 160V bias on the detector pins was pulsed a few μsec before impact in order to avoid the possible formation of bubbles in the water sample. That is, a few of the detectors have pin hole leaks in the gold film and anodized layer. Water in these holes can dissociate due to the 160V on the pin and cause gas bubbles to form. Bubbles are to be avoided since they can significantly affect the arrival time measurement on a nsec time scale.

RESULTS

Seven measured Hugoniot points are shown plotted in the shock velocity-particle velocity plane (u_s-u_p) in Fig. 3. These data are linear over the range of measurement. The data of Walsh and Rice [6] and of Skidmore and Morris [7] are also shown for comparison. Our data point at u_p = 4.23 km/s, the lowest particle velocity in this series, agrees very well with Walsh and Rice. The cluster of Walsh and Rice points at u_p = 4.8 km/s is about 1.5% below the straight line through these new data. This difference corresponds closely with the combined experimental uncertainties in the two data sets. The data of Skidmore and Morris has considerably more scatter than these new data. The relatively poor agreement between their data and the extrapolation of our linear u_s-u_p relation is not surprising. The data of Fig. 3 is shown plotted in the pressure-particle velocity plane (P-u_p) in Fig. 4.

Two experiments were performed in which the incident shock wave in the water was reflected off a tantalum anvil across which the shock velocity was measured. The shock Hugoniot of tantalum has been measured in detail [12]. Thus, the pressure and particle velocity in the tantalum can be obtained from the known Hugoniot. Since pressure and particle velocity are continuous across material interfaces, the pressure and particle velocity in the double-shocked water are identical to those in the tantalum.

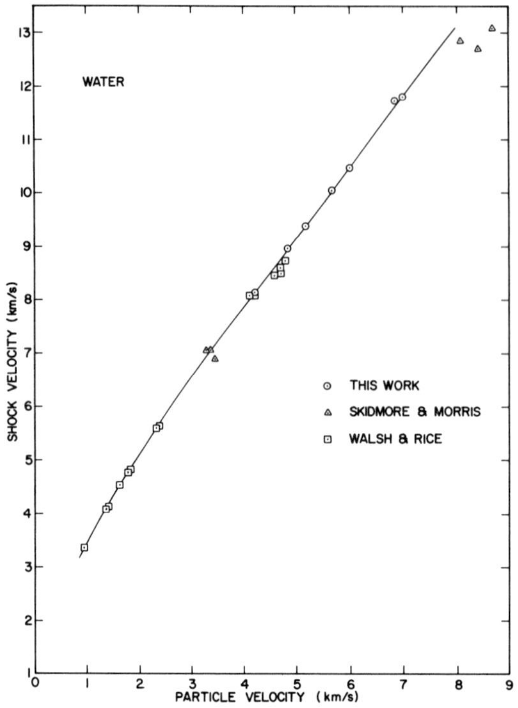

Fig. 3. Shock Hugoniot of water plotted in the shock velocity-particle velocity plane.

The results for all our single and double-shock experiments are shown in Fig. 5 in the pressure volume (P-V) plane. For comparison the cluster of four points by Walsh and Rice in Fig. 3 at u_p = 4.8 km/s are also shown. In addition, the results of Bakanova et al. [8] for a shock in water reflected off a tungsten anvil are also shown. As seen in Fig. 5, there is an offset in the P-V plane between the data of Walsh and Rice and of Bakanova et al. compared to our data. Aside from this offset the states reached in the two reflected shock wave experiments at ~120 GPa are quite close to each other in the P-V plane.

The uncertainties in pressure and volume indicated by the crosses were obtained by assuming that these uncertainties were due to the sum of random experimental uncertainties plus a systematic uncertainty involved in obtaining the particle velocity in the water behind the first shock by the impedance matching method. Calculations by Grover [13] using the GRAY code [14,15] showed that variations in the release isentropes of aluminum using a variety of reasonable models agreed to within $\lesssim$0.3%. Thus, the major uncertainty in the impedance matching is apparently due to the uncertainty in the shock velocity in the water. The calculated uncertainties for the single shocks are 0.7 to 1.4% in pressure and 0.8 to 2.0% in volume. For the reflected shocks the uncertainties are ~3.5% in pressure for both and 3.4 and 7.0% in volume for the lower and upper pressure experiments, respectively.

ACKNOWLEDGMENTS

The authors would like to thank R. N. Keeler, Director of the Office of Naval Technology, for his support and interest in this work. They also wish to acknowledge J. W. Shaner for many interesting discussions on these experiments. Thanks must also go to R. E. Neatherland, who assembled the targets, and to W. D. Barrowman,

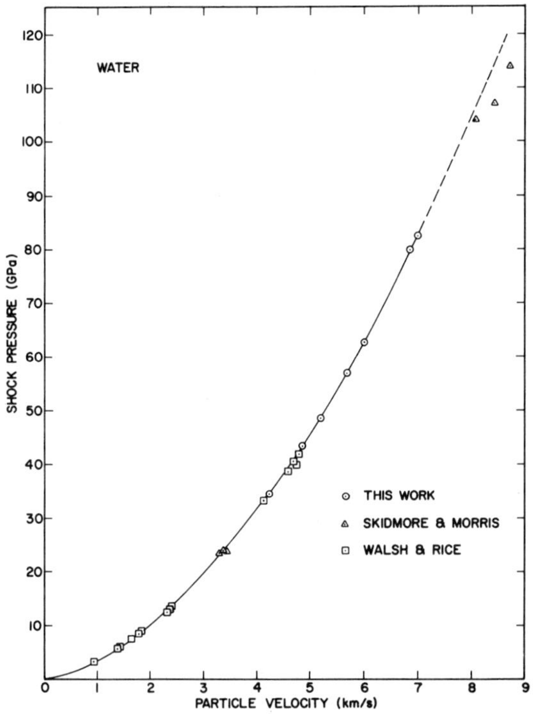

Fig. 4. Shock Hugoniot of water plotted in the pressure-particle velocity plane.

M. A. Benapfl, D. R. Mizer, and J. S. Samuels for their assistance in firing the two-stage gun. They wish to acknowledge N. Brown for providing the precision-made shorting pins.

REFERENCES

1. J. S. Lewis, Sci. Am. 230, 50 (1974).
2. D. P. Cruikshank and D. Morrison, Sci. Am. 234, 108 (1976).
3. F. H. Ree, Lawrence Livermore Laboratory, Rept. UCRL-52190 (1976).
4. W. B. Hubbard and R. Smoluchowski, Space Sci. Rev. 14, 599 (1973).
5. R. Smoluchowski, Am. Sci. 63, 638 (1975).
6. J. M. Walsh and M. H. Rice, J. Chem. Phys. 26, 815 (1957).
7. I. C. Skidmore and E. Morris, in Thermodynamics of Nuclear Materials, Intern. Atomic Energy Agency, Vienna, Austria (1962), p. 173ff.
8. A. A. Bakanova, V. N. Zubarev, Yu. N. Sutulov, and R. F. Trunin, Sov. Phys.-JETP 41, 544 (1976).

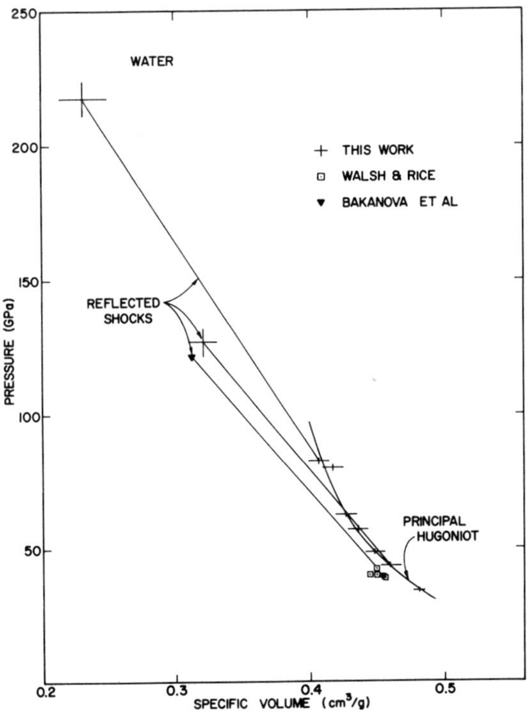

Fig. 5. Hugoniot points for water plotted in the pressure-volume plane.

9. A. H. Jones, W. M. Isbell, and C. J. Maiden, J. Appl. Phys. 37, 3493 (1966); M. Van Thiel, L. B. Hord, W. H. Gust, A. C. Mitchell, M. D'addario, K. Boutwell, E. Wilbarger, and B. Barrett, Phys. Earth Planet. Inter. 9, 57 (1974).
10. M. L. Wilkins, Lawrence Livermore Laboratory, Rept. UCRL-7322, Rev. I (1969).
11. J. R. Long and A. C. Mitchell, Rev. Sci. Instr. 43, 914 (1972).
12. A. C. Mitchell, to be published.
13. R. Grover, private communication.
14. E. B. Royce, Lawrence Livermore Laboratory, Rept. UCRL-51121 (1971).
15. R. Grover, J. Chem. Phys. 55, 3435 (1971).

G-1

THE EFFECT OF PRESSURE ON THE RAMAN SPECTRA IN TRIGONAL Se AND Te

S. Minomura and K. Aoki
University of Tokyo
and
N. Koshizuka and T. Tsushima
Electrotechnical Laboratory
Tokyo, Japan

INTRODUCTION

The crystal structure of trigonal Se and Te consists of infinite helical chains which spiral around the crystalline c axis with three atoms per turn. Each atom within the helical chains is tightly bonded to two neighbors with covalent character. The bonding between individual chains is much weaker. The lattice dynamics of trigonal Se and Te have been studied recently by infrared and Raman spectra [1,2], and neutron scattering [3]. Richter et al. [4] have reported that the first-order Raman frequencies decrease linearly with increasing pressure up to 8 kbar. Trigonal Se and Te under pressure show a covalent-metallic transition accompanied by a discontinuous change in electrical resistivity at about 180 and 40 kbar, respectively [5,6]. The structure of the high-pressure phases has been studied by x-ray diffraction [7-9], but remains unsolved. Jamieson and McWhan [8] have reported another transition to a β-Po structure for Te at 115 kbar.

This paper is concerned with x-ray diffraction and trigonal Raman scattering studies of Se and Te under pressure which provide information on the lattice and electronic instabilities. The high-pressure modification of Te at 45 kbar is described as a puckered layer structure with a monoclinic unit cell. The interference of interchain interactions with the intrachain bonding in trigonal Se and Te is evidenced by the nonlinear softening of the first-order Raman active phonons under pressure.

EXPERIMENTAL

Single crystals of trigonal Se were prepared from aqueous Na_2S solution (1.1 N). Crystals of Te were grown from the melt in a vacuum. The x-ray diffraction patterns for Se and Te under pressure were taken with a flat film cassette using a diamond anvil apparatus or with a diffractometer using a cubic anvil apparatus. The first-order Raman spectra for Se and Te under pressure were recorded on a JASCO R-750 spectrophotometer using a 6471 Å line of Kr ion laser and a diamond-anvil apparatus in the backward-scattering geometry. The pressure of the diamond-anvil cell was determined by the ruby R_1 fluorescence scale [10].

X-RAY DIFFRACTION

The crystal structure of trigonal Se and Te is described as helical chains which are arranged in an hexagonal array. The structural information of Se and Te are given in Table I. The changes in

Table I. Structural Information for Trigonal Se and Te

Parameter*	Trigonal Se	Trigonal Te
a	4.366 Å	4.457 Å
c	4.495	5.929
c/a	1.135	1.330
r	2.237	2.835
θ	103.1°	103.2°
ψ	100.6	100.7
R	3.346	3.495
R/r	1.447	1.232
v	0.984	1.174
u	0.2254	0.2634

*Legend: a,c - lattice parameters of hexagonal cell; r - bond length; θ - bond angle; ψ - dihedral angle; R - interchain atomic distance; v - radius of spiral; and u = v/a - atomic position parameter.

lattice parameters of Se with pressure are shown in Fig. 1. With increasing pressure, the parameter a decreases rapidly while the parameter c increases by a small amount. The increase in the c/a ratio is nonlinear. Trigonal Te under pressure shows a similar behavior. These features are interpreted by the distortions of trigonal structure under pressure which arise predominantly from the rapid decrease in interchain atomic distances and the small increase in bond angles.

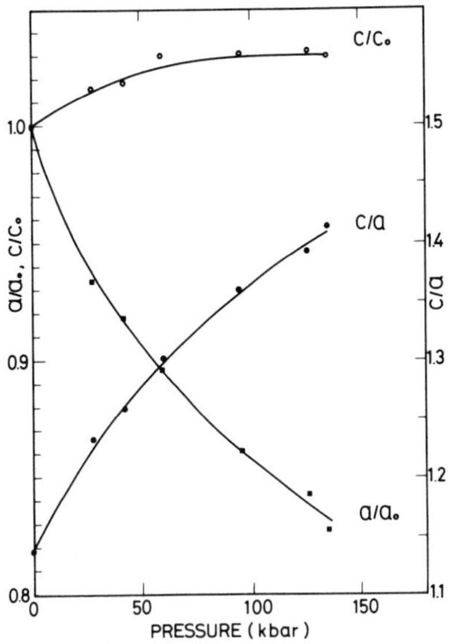

Fig. 1. Changes in lattice parameters for trigonal Se under pressure.

The powder patterns taken for Te at 45 kbar are shown in Fig. 2. These patterns are indexed as monoclinic, space group C_2^2. The lattice parameters are a = 3.104 Å, b = 7.513 Å, c = 4.766 Å, and β = 92.709°. This structure is described as consisting of puckered layers as shown in Fig. 3. Each atom has four neighbors at distances of r_1 = 2.80 Å, $r_{2,3}$ = 3.10 Å, and r_4 = 3.11 Å within the same layer, and four next neighbors at distances of $r_{5,6}$ = 3.47 Å, and $r_{7,8}$ = 3.52 Å within the same double layers. The structural transformation occurs at a c/a ratio of 1.42 both in trigonal Se and Te under pressure. The structure of the high-pressure phase of Se remains unsolved. However, the powder patterns taken for trigonal, α-monoclinic, and amorphous Se under pressure show the transition to the puckered layer structure which is similar to the high-pressure modification of Te.

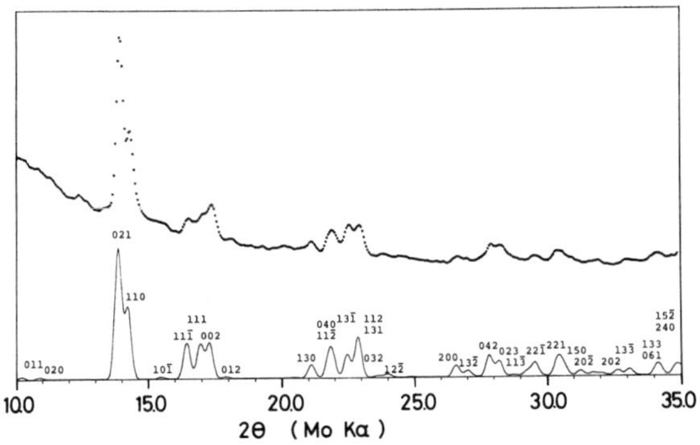

Fig. 2. X-ray diffraction patterns for monoclinic Te at 45 kbar.

RAMAN SCATTERING

Trigonal Se and Te show the first-order Raman spectra of the A_1 mode and the two doubly-degenerated E modes at the zone center. The lattice vibration of A_1 mode is the chain-expansion type. The lattice vibrations of E modes are separated into predominantly angle-bending and bond-stretching types. We denote the low-frequency E mode as E' and the high-frequency E mode as E". The first-order Raman spectra of trigonal Se and Te under pressure are displayed in Fig. 4. The changes in frequencies with pressure for the Raman active phonons of the A_1, E', and E" modes are shown in Figs. 5 and 6. The A_1 mode shows a nonlinear softening with pressure. The E' and E" modes of Se show a frequency shift of about 10 cm^{-1} in the opposite directions up to 25 kbar, but they show a small shift within a few frequencies above 25 kbar. The two E modes of Te under pressure also indicate the small shift.

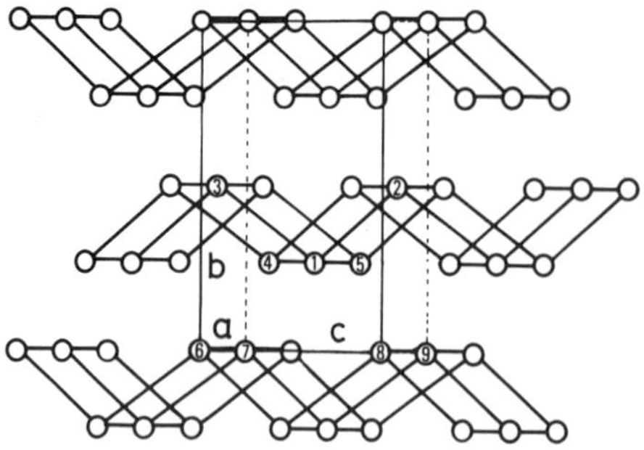

Fig. 3. Puckered layer structure of monoclinic Te.

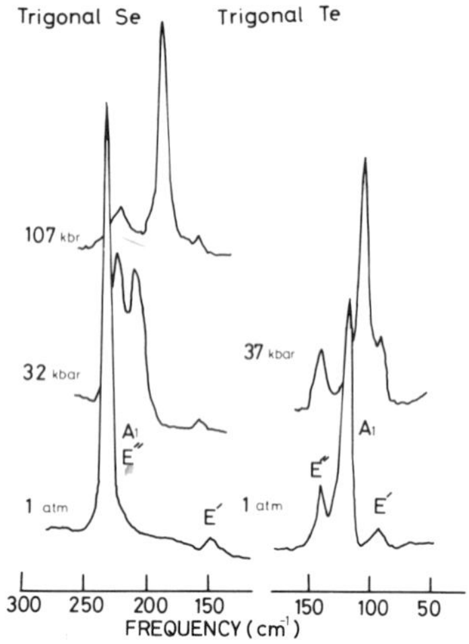

Fig. 4. First-order Raman spectra of trigonal Se and Te under pressure.

The lattice dynamics of trigonal Se and Te have been studied theoretically on the basis of the valence force field models. Nakayama and Odajima [11] represent the lattice energy in terms of five force constants which are fitted to the observed six elastic constants: K_r (bond-stretching constant), K_θ (angular constant), $K_{rr'}$ (cross term of bond stretching with neighboring bond pairs), $K_{r\theta}$ (cross term of bond stretching with bond bending), and K_R (interchain constant). Martin et al. [1] represent the three additional terms of interchain force constants: K_Θ (angular constant between intrachain and interchain bonds), $K_{R\Theta}$ (cross term of interchain bond stretching with bond bending), and K_{rR} (cross term of interchain bond stretching with intrachain bond stretching). The phonon frequency ω of the A_1 mode is given by

$$M\,\omega^2 = \sum_m A_m K_m \qquad (1)$$

where A_m is the geometrical coefficient, K_m is the force constant, and M is the mass. The phonon frequencies ω of the E' and E" modes are given by

$$\begin{vmatrix} \sum_m A_{m,11} K_m - M\,\omega^2 & \sum_m A_{m,12} K_m \\ \sum_m A_{m,12} K_m & \sum_m A_{m,22} K_m - M\,\omega^2 \end{vmatrix} = 0 \qquad (2)$$

For the A_1 mode of trigonal Se and Te under pressure the changes in $M\,\omega^2$ with c/a ratio are shown in Fig. 7. The values of $M\,\omega^2$ decrease nonlinearly with increasing c/a ratio. From equation (1), one can expect that the force constants K_m as well as the geometrical coefficients A_m vary nonlinearly with pressure or c/a ratio. Richter et al. [4] have calculated the initial pressure

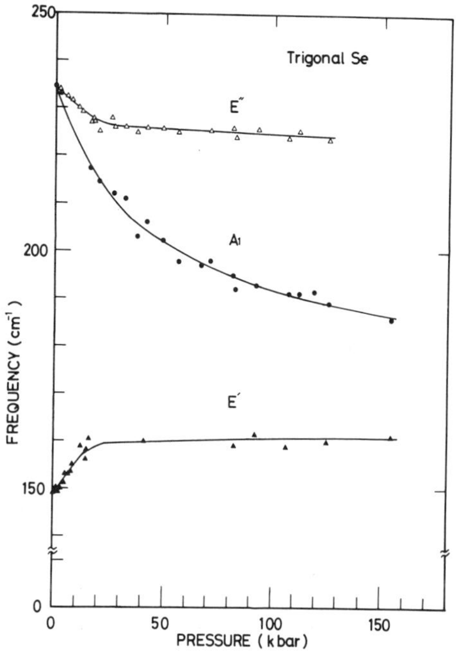

Fig. 5. Frequency shift of Raman active phonons for trigonal Se under pressure.

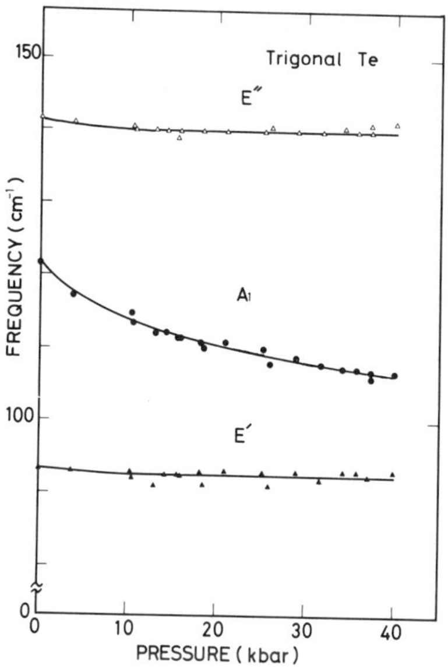

Fig. 6. Frequency shift of Raman active phonons for trigonal Te under pressure.

dependence of infrared and Raman frequencies using five mode force constants which vary linearly with c/a ratio from Se to Te. They have suggested from the calculations that the softening of the Raman active modes is mainly caused by the negative term of K_r, and the smaller frequency shift of the E modes is given by the cancellation among the terms of intrachain and interchain force constants with opposite sign.

The electron states in trigonal Se and Te have been studied by the valence band photoemission spectra [12,13]. Joannopoulos et al. [14] have calculated the band structure, densities of states, and charge densities for trigonal Se and Te using an empirical pseudopotential method. The valence band photoemission spectra consist of three regions arising predominantly from lone pair states, p-like bonding states, and s-like localized states. The charge densities for the p-like bonding states show definitely some covalent-like bonding between neighboring chains. With increasing interchain atomic distance, the charge densities of interchain bonding decrease whereas those of intrachain bonding increase. These studies on the electron states give a reasonable explanation for the lattice and electronic instabilities toward the covalent-metallic transition accompanied by the change in structure from infinite helical chains to puckered layers. The interlayer bonds

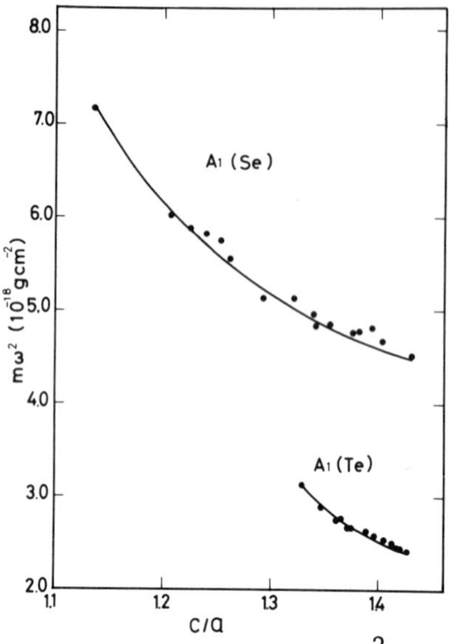

Fig. 7. Changes in $M\omega^2$ for the A_1 mode of trigonal Se and Te under pressure.

of the high-pressure phase are formed by the rearrangement of intrachain and interchain bonds of the low-pressure phase.

NOTATION

A_1 = Raman active mode
A_m = geometrical coefficient
a,b,c = lattice constants
E = Raman active mode
E' = low-frequency E mode
E" = high-frequency E mode
K_m = force constant
M = mass
R = interchain atomic distance
r = intrachain bond length
u = atomic position parameter
v = radius of helix
β = lattice parameter
Θ = angle between intrachain and interchain bond vector
θ = intrachain bond angle
ψ = dihedral angle
ω = Raman frequency

REFERENCES

1. R. M. Martin and G. Lucovsky, Phys. Rev. B 13, 1383 (1976).
2. A. S. Pine and G. Dresselhaus, Phys. Rev. B 4, 356 (1971).
3. W. C. Hamilton, B. Lassier, and M. I. Kay, J. Phys. Chem. Solids 35, 1089 (1974).
4. W. Richter, J. B. Renucci, and M. Cardona, Phys. Status Solidi B 56, 223 (1973).
5. P. W. Bridgman, Proc. Am. Acad. Arts Sci. 81, 165 (1952).
6. S. Minomura, K. Aoki, O. Shimomura, and K. Tanaka, in Electronic Phenomena in Noncrystalline Semiconductors, B. T. Kolomiets, ed., Leningrad, U.S.S.R. (1976), p. 289.
7. S. S. Kabalkina, L. F. Vereshchagin, and B. M. Shulenin, Soviet Physics-JETP 18, 1422 (1964).
8. J. C. Jamieson and D. B. McWhan, J. Chem. Phys. 43, 1149 (1965).
9. D. R. McCann and L. Cartz, J. Chem. Phys. 56, 2552 (1972).
10. G. J. Piermarini, S. Block, J. D. Barnett, and R. A. Forman, J. Appl. Phys. 46, 2774 (1975).
11. T. Nakayama and A. Odajima, J. Phys. Soc. Japan 33, 12 (1972).
12. N. J. Shevchik, M. Cardona, and J. Tejeda, Phys. Rev. B 8, 2833 (1973).
13. M. Schlüter, J. D. Joannopoulos, M. L. Cohen, L. Ley, S. Kowalczyk, R. Pollak, and D. A. Shirley, Solid State Commun. 15, 1007 (1974).
14. J. D. Joannopoulos, M. Schlüter, and M. L. Cohen, Phys. Rev. B 11, 2186 (1975).

G-2

NEW HIGH-PRESSURE PHASE TRANSITIONS IN POTASSIUM CYANIDE

W. Dultz, H. Krause, and J. Ploner

Universität Regensburg
Regensburg, W. Germany

INTRODUCTION

Potassium cyanide, KCN, is an ionic crystal of cubic symmetry at room temperature which is very similar to potassium chloride. In contrast, the ellipsoidal-shaped CN^- molecules rotate more or less freely and independently in their octahedral lattice cells and because of their different orientation from cell to cell the KCN crystal is orientationally disordered. At 168 K the CN^- molecules freeze in with a uniform orientation over the entire crystal, and KCN undergoes an order-disorder phase transition to an ordered orthorhombic phase [1]. There still remains the disorder in the C-N-sequence of the CN^- molecules and it requires another phase transition at 83 K to establish complete order. The CN^- dumbbells order antiparallel in this low temperature phase of space group Pmmn [2].

In addition, an intermediate phase of KCN was found in a temperature region of about 5 K at around 168 K only when the crystal was cycled over the cubic-to-orthorhombic transition [3]. In this phase the uniform orientation of the CN^--molecular axes is a [111]-direction for the original cubic lattice and not a [110]-direction as in the normal orthorhombic low-temperature phases: a [111]-orientation was also proposed for the low temperature phase of RbCN [3]. Since a change from K to Rb in the phase diagram of the alkali chlorides corresponds to increasing pressure, we thought that the intermediate phase in KCN might have a larger stable region at high hydrostatic pressures. In this work we report on Raman experiments under hydrostatic pressure and at low temperatures which indicate the existence of two new high-pressure phases of KCN. The high-pressure optical cell and the cryostat we used are also described.

EXPERIMENTS

We have constructed a steel cell with two coaxial sapphire windows which allow optical measurements of crystals at low temperatures under hydrostatic He gas pressures up to 10 kbar (see Fig. 1). The cylindrical body (1) of the cell is closed by two hollow plugs (2) which carry the sapphire windows (3). A thin gold and a thin aluminum washer are used as seals between the window and the polished plug. The steel hood (4) provides a small force for sealing at low pressures. Teflon rings (5) and Be-Cu rings (6) to prevent extrusion of the Teflon, seal the plugs (2) against the body (1) of the cell. This "unsupported area" seal is held by a disk (7) and a screw (8). The free apperture for the optical pass through the cell is 7 mm which is the diameter of the axial bore of the plugs. The He gas is pressed through the capillary into the chamber. The head (9) is brazed to the capillary and tightly pressed to the cell by the screw (10). In Table I we give a brief description of the different materials used in the cell.

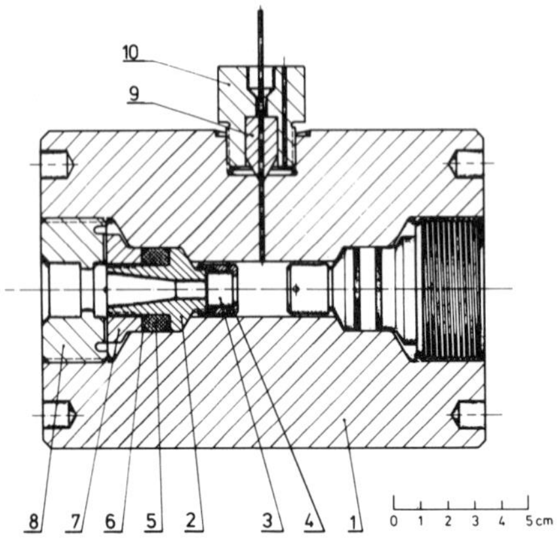

Fig. 1. High-pressure optical cell.

Table I. Materials of Construction for the High-Pressure Optical Cell

Parts	Materials
Steel parts 1, 2, 4, 7, 8, 9, 10	X2NiCoMo 1895 (RHF 32) Röchling-Burbach GmbH Völklingen-Saar
Seal 5	Teflon B 05, Sonderqualität Pampus KG München
Capillary	Almar 362 E. A. Gieben KG München

For measurements at low temperatures, the high-pressure cell is clamped by copper rings (11) to the liquid nitrogen bath (12) of the cryostat (see Fig. 2). To measure the temperature, thermocouples are fixed to the outer surface of one of the sapphire

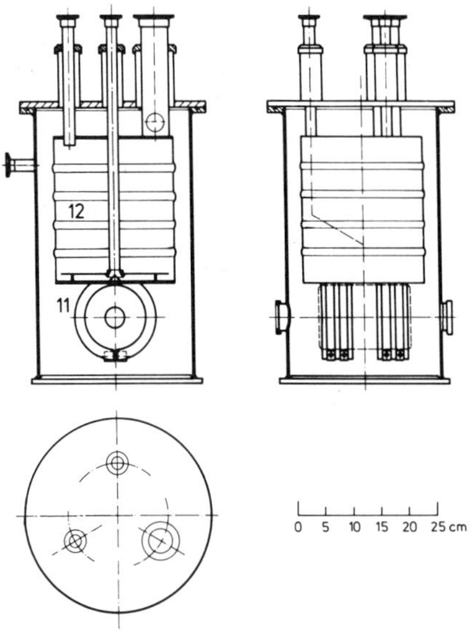

Fig. 2. Cryostat.

windows and to the surface of the cell body. Intermediate temperatures can be obtained by using heaters which are attached to the copper rings (11). By pumping the N_2-bath we reached 70 K at 7 kbar. The accuracy of the temperature and pressure measurement at the sample is ± 1 K and ± 0.1 kbar, respectively. At room temperature the cell was used at pressures up to 9 kbar. The different seals were adequate to establish a vacuum of 10^{-4} Torr in the cryostat during measurements.

We measured the Raman spectrum of KCN crystals grown from aqueous solution at temperatures down to 70 K and at hydrostatic pressures up to 7 kbar. The Raman apparatus consisted of a 1-m double monochromator, a 2-W Argon laser, and photon counting equipment. In the ordered and partly-ordered low-temperature phases the crystals are multidomain and opaque and no polarized Raman spectra can be obtained, but, if necessary, polarized measurements are possible with our cell by mounting a polarizer foil in front of the crystal.

Figure 3 shows the typical Raman spectra we obtain in different regions of the P-T-phase diagram of KCN. At high temperatures the spectrum consists of a broad shoulder of the Rayleigh line at about 110 cm^{-1} and of the band of the CN^--stretching mode at 2078 cm^{-1}. In the low-temperature region the main band at about 120 cm^{-1} represents the librational modes of the CN^--molecules [4]. The feature at 175 cm^{-1} in the two low-temperature phases probably is due to LO-phonons near the X-point of the Brillouin zone of D_{2h}^{25} which appear in the spectrum only because of the disorder in the C-N sequence. The LO(X) phonon becomes Raman active at low temperatures in phase Pmmn when the X-point folds into the Γ-point [2]. Several other bands which appear in the spectrum at low temperatures and low pressures are due to corresponding phonons from other branches [6].

Above 3 kbar the intensity of the 45 cm^{-1} band is higher than in the low-pressure phase Immm. If the crystal under pressure is cooled down below 120 K the Raman spectrum suddenly changes. A new band appears at about 150 cm^{-1}, the 45 cm^{-1} band becomes very intensive and shifts to higher frequencies, and the CN^--stretching mode doubles.

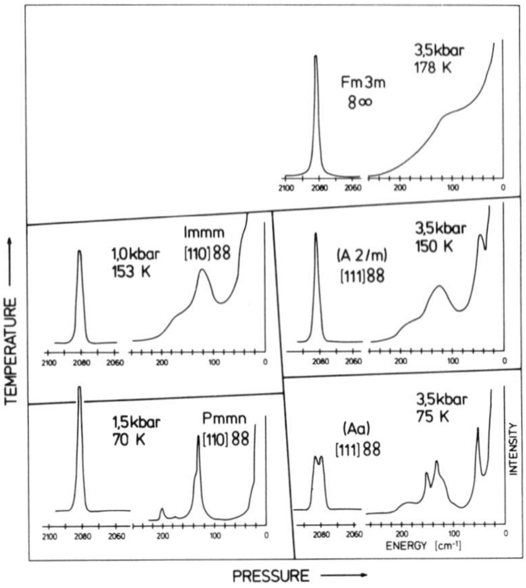

Fig. 3. Raman spectra of KCN in different regions of the phase diagram.

For the low-pressure phase, the space groups and the orientations of the CN^--molecules relative to the axes of the cubic phase are known and shown in Fig. 3. The orientational disorder in the cubic phase and the disorder in the C-N sequence in phase Immm are indicated by two dumbbells, the antiparallel ordering of the C-N sequence at low temperatures and at low pressures is symbolized in the same manner.

Figure 4 shows the actual phase lines of KCN as obtained from our Raman measurements. Arrows at each point indicate the direction in which the phase line was crossed. The phase transition between the phases Immm and Pmmm was shown to be of second order at zero pressure [5], all other phase lines in the diagram were of first order. Large hysteresis occur between the low-pressure phases and the new high-pressure phases and between the two high-pressure phases. For the two new high-pressure phases, the space groups and the orientations of the CN^--molecules are tentatively proposed since no direct analysis of the structures has been performed to date. Arguments for this proposition are given in the discussion.

DISCUSSION

New features in the phase diagram of KCN (see Fig. 1) were not surprising. An intermediate phase of KCN had already been reached by cycling the crystal over the cubic-to-orthorhombic transition at 168 K and zero pressure [3]. The structure of this phase was determined to be Aa with a unit cell twice as large as in the normal orthorhombic phase Immm. In contrast to the orthorhombic low-temperature phases the orientation of the CN^--molecular axes in this intermediate phase is approximately a [111]-direction of the original cubic lattice. We believe that the intermediate phase is a high-pressure phase which is stabilized at zero pressure by internal stresses in the multidomain KCN crystal.

A doubling of the unit cell folds phonons from the zone boundary to the center and we would expect new bands to appear in the Raman spectrum. This is the case in the low-temperature high-

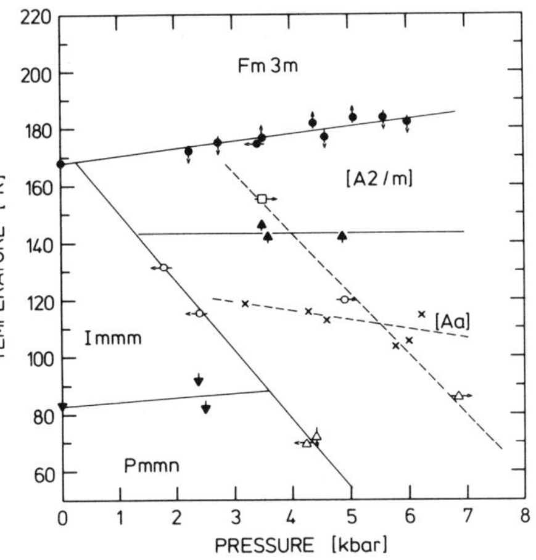

Fig. 4. Phase diagram of KCN as obtained from the Raman measurements. Legend:

- crystal becomes opaque or transparent;
- band at 175 cm^{-1} appears;
- sudden doubling of the CN^-- stretching mode;
- doubling of CN^--stretching mode disappears;
- doubling of CN^--stretching mode appears;
- band at 45 cm^{-1} appears;
- band at 45 cm^{-1} disappears; and
- band at 45 cm^{-1} starts to grow.

pressure phase. In this phase the 45 cm^{-1} line has shifted to higher frequencies and has become very intensive. This supports our assumption that this band is a disorder-allowed one-phonon density maximum which folds into the Γ-point at low temperatures. So we suggest identifying the low-temperature high-pressure phase with the intermediate phase Aa rather than the upper high-pressure phase. The doubling of the CN^--stretching mode then can be explained as a Davydov splitting in the enlarged unit cell; however, a phase mixture is also possible. In the intermediate Aa phase of KCN the doubling of the unit cell is due to alternating shifts of the K^+-ions in the lattice. An obvious suggestion for the structure of the upper high-pressure phase would then be that there are no shifts, thus no doubling of the unit cell occurs.

Since the Raman bands besides the one at 45 cm^{-1} have about the same intensity in the orthorhombic phase Immm and this upper high pressure phase, we suggest a space group with inversion symmetry namely A2/m for this phase. The transition A2/m → Aa folds the Y-point which corresponds to the L-point of the cubic phase Fm3m into the Γ-point, thus the 45 cm^{-1} band seems to be due to acoustic phonons from the L-point of the cubic zone which become Y-point phonons in A2/m. Since the intensity of the 45 cm^{-1} density spectrum is affected by the transition Immm → (A2/m), the corresponding phonons appear to be sensitive to the disorder in sequence for differently oriented CN^--molecules.

The change of the direction of the CN^--molecular axes from [110] to [111] with pressure just below the cubic phase at about 1 kbar raises the question of how the ordering effects in the disordered cubic phase may depend on this near-lying triple point.

Measurements in this region in the ordered phases are difficult because of the domain structure and large hysteresis effects. Nevertheless, we have shown that Raman experiments under high hydrostatic gas pressure are a useful method to determine phase lines and to obtain information about the structure of new phases.

ACKNOWLEDGMENT

The authors are grateful to Prof. Haussühl for providing the KCN crystal. This work is supported by the Deutsche Forschungsgemeinschaft.

REFERENCES

1. H. Suga, T. Matsuo, and S. Seki, Chem. Soc. Jap. Bull. 38, 1115 (1965).
2. W. Dultz, J. Chem. Phys. 65, 2812 (1976).
3. G. S. Parry, Act. Cryst. 15, 601 (1962).
4. W. Dultz, Sol. State Comm. 15, 595 (1974).
5. W. Dultz, H. Krause, and L. W. Winchester, J. Chem. Phys. 67, 2560 (1977).

G-3

RAMAN SPECTROSCOPIC INVESTIGATION OF THE FERROELECTRIC SOFT MODE IN $Pb_5Ge_3O_{11}$ UNDER HIGH HYDROSTATIC PRESSURE*

T. Suski,† W. Müller-Lierheim, W. Dultz, H. Krause, H. H. Otto, and W. Gebhardt

Universität Regensburg
Regensburg, W. Germany

INTRODUCTION

Lead germanate, $Pb_5Ge_3O_{11}$, undergoes a uniaxial, almost second-order, phase transition at 450 K from the ferroelectric phase with the space group P3 to the paraelectric phase with the space group $P\bar{6}$ [1-7]. The spontaneous polarization in the ferroelectric phase is parallel to [001]; there are only two (180°) domains which are enantiomorphous [8-10]. $Pb_5Ge_3O_{11}$ is optically active in [001]-direction, and the handedness of the optical rotation can be reversed by the application of an external electric field [8], which reverses the domain orientation.

Lead germanate has three formula units in the unit cell and, therefore, 171 vibrational degrees of freedom. Considering GeO_4-tetrahedra and Ge_2O_7-double tetrahedra as structural subunits, one can distinguish between the internal modes of these subunits and the external modes of the crystal where the groups GeO_4 and Ge_2O_7 are assumed to be rigid [11]. At the Γ-point (point group C_{3h}), the external modes divide into 27 modes of A-species and 27 doubly degenerate E-modes.

According to the theory of Cochran and Anderson [12,13] the force constant of the atomic motion responsible for a structural phase transformation becomes soft near the transition point. At second-order ferroelectric phase transformations without change of the unit cell, the soft mode is a transverse optic phonon in the center of the Brillouin zone. In the higher symmetry phase, this mode reduces the crystal symmetry to the space group of the lower symmetric phase, and therefore, it belongs to the fully symmetric species of the lower phase.

*Work supported by the Deutsche Forschungsgemeinschaft.
†Present address: High Pressure Research Center, Warsaw, Poland.

Several authors have studied the temperature dependence of the soft mode with Raman spectroscopy [14-16]. Measurements of the static dielectric constant of $Pb_5Ge_3O_{11}$ under hydrostatic pressure show a negative slope of the phase line [17,18]. Measurement of the Raman spectrum of the soft mode was made under hydrostatic pressure at temperatures below the transition point. The pressure dependence of the soft mode frequency can be approximated by a power law; the relationship is similar to the temperature dependence. From this, the Grüneisen function of the soft mode was determined, using the elastic constants from literature values [19,20].

EXPERIMENTS

Single crystals of $Pb_5Ge_3O_{11}$ were grown by slowly cooling the stoichiometric melt in a Pt crucible, as described elsewhere [11]. For the application of hydrostatic He gas pressure, a cylindrical steel cell with axial sapphire windows was mounted inside an oven. As polarized Raman spectra are necessary to distinguish between phonons of different symmetry in $Pb_5Ge_3O_{11}$, a polarizer foil was used inside the high-pressure cell. For details of construction of the cell, see Dultz et al. [21]. The Raman apparatus consists of a 4 W Argon ion laser, a 1 m double monochromator (Jarrel Ash), and photon-counting equipment (Channeltron, Bendix). The laser was operated at 514.5 nm with only 50 mW power to avoid local heating of the crystal by absorption.

A considerable sample dependence of the pressure dependence of the Raman spectrum was found with respect to the ferroelectric soft mode of $Pb_5Ge_3O_{11}$. In the following, data of only one crystal has been used. Figure 1 shows the temperature dependence of the x(zz)y-Raman spectrum (A_{TO}-modes). Note the tremendous softening of the lowest frequency mode as it approaches the phase transition. The intensities of the modes at 39, 53.5, and 118.5 cm^{-1} decrease with increasing temperature, as these modes are Raman inactive above 450K [11]. A quasi three-dimensional plot of the temperature and pressure dependence of the Raman spectrum is shown in Fig. 2. Near the transition point the soft mode is overdamped, and no peak frequency can be obtained from the Raman spectrum.

DISCUSSION

The frequency ω_f of the soft mode is a temperature- and pressure-dependent quantity; thus it can be described as surface over the p-T space.

$$\omega_f = \omega_f\ (p,T) \qquad (1)$$

The contour line $\omega_f = 0$ of (1) is the phase line of the phase transition in the p-T diagram. Generally, the slope of the contour lines can be expressed by

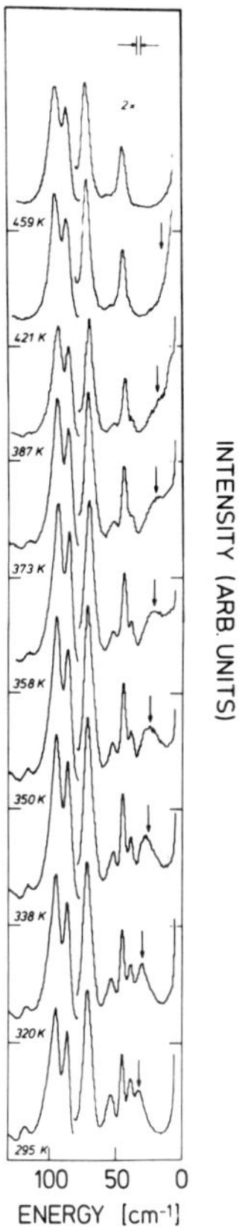

Fig. 1. Temperature dependence of the low-frequency part of the x(zz)y Raman spectrum of $Pb_5Ge_3O_{11}$. The ferroelectric soft mode is marked by arrows.

$$\left.\frac{dT}{dP}\right|_{p,T} = -\left.\frac{\partial\omega_f}{\partial p}\right|_{p,T} \Big/ \left.\frac{\partial\omega_f}{\partial T}\right|_{p,T} \qquad (2)$$

The value of $\partial\omega_f/\partial p$ for three different temperatures at zero pressure was obtained from our measurements of the soft mode frequency (see Fig. 3). The value of $\partial\omega_f/\partial T$ is known over a large temperature range (see Fig. 1 and Hisano et al. [14]). A value of $-(11 \pm 2)$ K/kbar for $\left.\frac{dT}{dp}\right|_{p=0,T}$ was obtained for all three temperatures; thus, this value may also be used for the slope of the phase line at $T = T_c$. This assumption is confirmed by the fact that the resultant value lies between the two values of

$$\left.\frac{dT_c}{dp}\right|_{p=0,T=T_c} = -6.7 \text{ K/kbar and}$$

$$\left.\frac{dT_c}{dp}\right|_{p_c,T_c} = -15.5 \text{ K/kbar}$$

from Gesi et al. [17] and Kirk et al. [18], respectively.

Following Samara [22], a power law is used to describe the pressure dependence of the soft mode frequency. Thus,

$$\omega_f\ (p,T) = A\ (p_c(T) - p)^a \qquad (3)$$

where $p_c(T)$ [kbar] = (450 - T[K])/11 is obtained with the above assumption. A double logarithmic plot of ω_f vs. $(p_c(T)-p)$ is shown in Fig. 4. A straight line is obtained; thus the measurements can be described by a power law (3) with an exponent of 0.40 ± 0.05 and a pressure independent parameter A of 11.1 cm^{-1} $kbar^{-a}$.

As obtained from the generalized Lyddane-Sachs-Teller relation [23], the critical exponent of the pressure dependence of the soft mode should be -1/2 times the critical exponent b of the pressure dependence of the static dielectric constant $\varepsilon_3^{(o)}$ of $Pb_5Ge_3O_{11}$. It is estimated that $b = -(0.7 \pm 0.1)$ for the ferroelectric phase from the measurements of $\varepsilon_3^{(o)}(p,T)$ by Kirk et al. [18]. This gives $a = (0.35 \pm 0.05)$, in good agreement with our result.

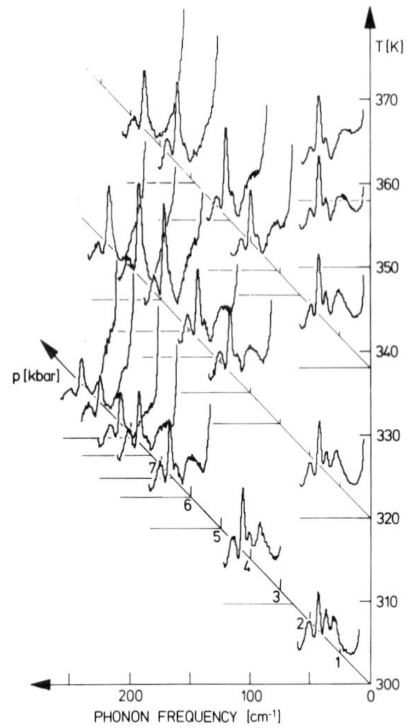

Fig. 2. Temperature and pressure dependence of the low-frequency part of the x(zz)y Raman spectrum of $Pb_5Ge_3O_{11}$.

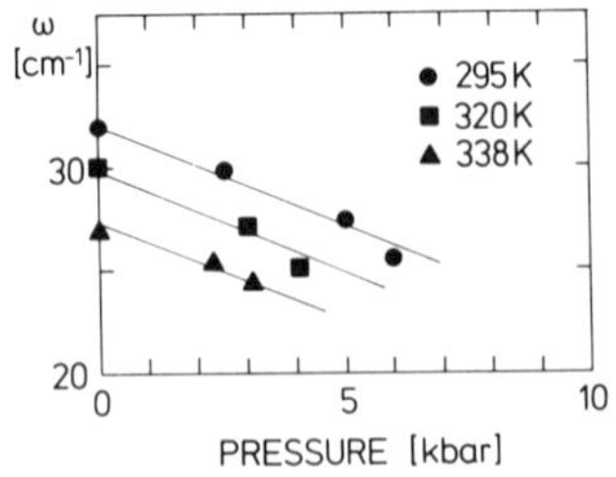

Fig. 3. Plot of the soft mode frequency ω_f vs. pressure for three different temperatures.

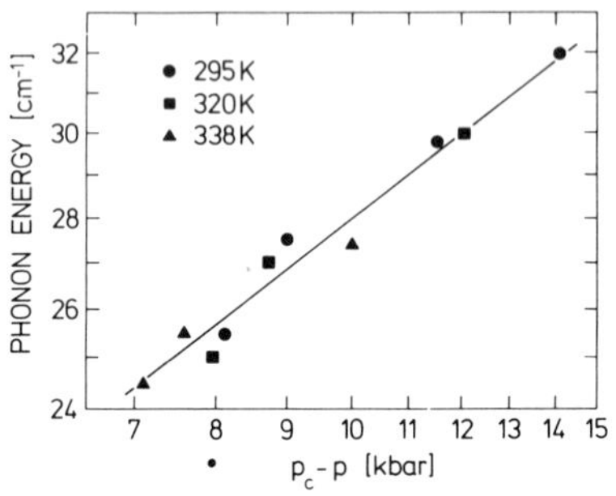

Fig. 4. Double-logarithmic plot of ω_f vs. $(p_c(T) - p)$ with p_c [kbar] = (450-T[K])/11.

Since the mode Grüneisen function γ_i is defined by

$$\gamma_i = - \frac{d \ln\omega_i}{d \ln V} = \frac{1}{\kappa} \frac{d \ln\omega_i}{dp} \tag{4}$$

the soft mode Grüneisen function $\gamma_f(p,T)$ can be obtained from the power law (3).

$$\gamma_f(p,T) = - \frac{160 \text{ kbar}}{(p_c(T) - p)} \tag{5}$$

The compressibility ($\kappa = 2.5 \times 10^{-3}$ kbar^{-1}) was obtained from elastic constants given in the literature [19,20]. As expected for the soft mode Grüneisen function, γ_f is negative below T_c and its absolute value diverges at the transition point.

The pressure dependence of some other A_{TO}-modes of $Pb_5Ge_3O_{11}$ is plotted in Fig. 5. The corresponding mode Grüneisen functions γ_i are listed in Table I. No extraordinary pressure dependence was observed for these modes.

NOTATION

$\varepsilon_3^{(o)}$	= static dielectric constant in direction of the c-axis	p	= hydrostatic pressure
		p_c	= critical pressure

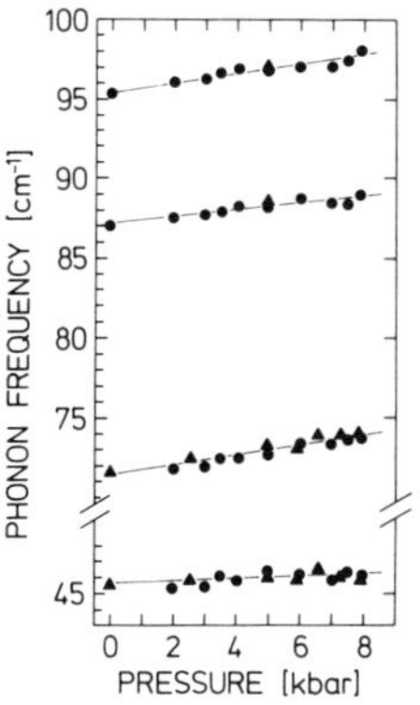

Fig. 5. Pressure dependence of the frequencies of several A_{TO}-phonons at room temperature.

Table I. Mode Grüneisen Functions of Those A_{TO}-Phonons Which Remain Raman Active Above T_c

ω_i, cm^{-1}	γ_i
45.5	0.73
71.8	1.8
87.1	1.0
95.6	1.3

T = temperature
T_c = critical temperature
ω_i = frequency of a mode i
ω_f = frequency of the ferroelectric soft mode
γ_i = mode Grüneisen function of mode i
γ_f = soft mode Grüneisen function
A = constant
a,b = critical exponents
κ = volume compressibility

REFERENCES

1. H. Iwasaki, K. Sugii, T. Yamada, and N. Niizeki, Appl. Phys. Lett. 18, 444 (1971).
2. S. Nanamatsu, H. Sugiyama, K. Doi, and Y. Kondo, J. Phys. Soc. (Japan) 31, 616 (1971).
3. K. Sugii, H. Iwasaki, and S. Miyazawa, Mat. Res. Bull. 6, 503 (1971).
4. T. Yamada, H. Iwasaki, and N. Niizeki, J. Appl. Phys. 43, 771 (1972).
5. H. Iwasaki, S. Miyazawa, H. Koizumi, K. Sugii, and N. Niizeki, J. Appl. Phys. 43, 4907 (1972).
6. Y. Iwata, H. Koizumi, N. Koyano, I. Shibuya, and N. Niizeki, J. Phys. Soc. (Japan) 35, 314 (1973).
7. Y. Iwata, N. Koyano, and I. Shibuya, J. Phys. Soc. (Japan) 35, 1269 (1973).
8. H. Iwasaki and K. Sugii, Appl. Phys. Lett. 19, 92 (1971).
9. H. Iwasaki, K. Sugii, N. Niizeki, and H. Toyoda, Ferroelectrics 3, 157 (1972).
10. R. E. Newnham and L. E. Cross, Endeavor 33, 18 (1974).
11. W. Müller-Lierheim, T. Suski, and H. H. Otto, Phys. stat. sol. (b) 80, 31 (1977).
12. W. Cochran, Advan. Phys. 9, 387 (1960).
13. P. W. Anderson, in Fizika Dielectrikov, G. I. Skanovi, ed., Akad. Nauk SSSR, Moscow, U.S.S.R. (1960).

14. K. Hisano and J. F. Ryan, Sol. State Comm. 11, 119 (1972).
15. G. Burns and B. A. Scott, Phys. Lett. 39 A, 177 (1972).
16. J. F. Ryan and K. Hisano, J. Phys. C 6, 566 (1973).
17. K. Gesi and K. Ozawa, Japan. J. Appl. Phys. 13, 897 (1974).
18. J. L. Kirk, L. E. Cross, and J. F. Dougherty, Ferroelectrics 11, 439 (1976).
19. T. Yamada, H. Iwasaki, and N. Niizeki, J. Appl. Phys. 43, 771 (1972).
20. G. R. Barsch, L. J. Bonczar, and R. E. Newnham, Phys. stat. sol. (a) 29, 241 (1975).
21. W. Dultz, H. Krause, and J. Ploner, in Proceedings 6th AIRAPT Intern. High Pressure Conference, Plenum Press, New York (1978).
22. G. A. Samara, in Advances in High Pressure Research, Vol. 3, R. S. Bradley, ed., Academic Press, London (1969), Chap. 3.
23. W. Cochran and R. A. Cowley, J. Phys. Chem. Solids 23, 447 (1962).

G-4

PRESSURE DEPENDENCE OF THE RAMAN SCATTERING IN CuCl TO 37 KBAR*

R. C. Hanson and M. L. Shand†

Arizona State University
Tempe, Arizona USA

INTRODUCTION

Raman scattering in CuCl has been the subject of many studies which were in part motivated by its anomalous transverse optic (TO) region [1].** CuCl has the zincblende structure; the first-order Raman spectrum should show one TO phonon peak and one longitudinal optic (LO) phonon peak. A peak occurs at about 210 cm^{-1} which can be identified from polarization selection rules as the LO phonon peak. In the region of the TO phonon (140 to 180 cm^{-1}) a two-peak structure occurs with the peaks labeled β and γ. This entire structure shows the overall polarization selection rules for the TO phonon.

A theoretical explanation for this double-peak structure has been proposed by Krauzman et al. [2]. They propose that the TO phonon is strongly coupled by anharmonic interactions to a multi-phonon background. This background is characterized by a P_3 singularity on its high-frequency side (density of states cutoff of square root form). The strong anharmonic coupling leads to two major effects in the TO phonon self-energy: (1) the self-energy is very large ($\sim$ 20 cm^{-1} rather than a more typical 1 to 5 cm^{-1}), and (2) a singularity appears in both the real and imaginary parts of the self-energy. The TO phonon self-energy with these features leads to the double-peak structure in the Raman spectrum.

Previous measurements of the Raman scattering in CuCl under hydrostatic pressure up to 7 kbar at 40K [1] have shown that the

*Partially supported by the National Science Foundation.
†Present address: Allied Chemical Corporation, Morristown, New Jersey.
**This paper compiles the previous lattice dynamical measurements on CuCl with references.

effect of the pressure up to 7 kbar is that the β and γ peak frequencies change only a little, while the relative intensities of the two peaks change dramatically. The intensity of the γ peak relative to that of the β peak increases by 28% per kbar. A fit of the Krauzman formulation to the data up to 7 kbar allowed determination of several theoretical parameters including the quasi-harmonic TO phonon frequency, ω_{TO}, and the cutoff frequency in the multiphonon band, ω_C. These fits indicated that as the pressure increases, ω_{TO} was pushed up towards ω_C. As long as the TO phonon remained within the multiphonon band, the Raman spectrum would still show the observed double structure. An extrapolation of the frequency dependence on pressure showed that at about 18 kbar ω_{TO} would be at ω_C. The Raman spectrum should gradually change over to that expected for a zincblende structure: one peak due to the TO phonon, one peak due to the LO phonon and, perhaps, a less intense broad band due to multiphonon scattering.

In the work presented here we have measured the Raman scattering in CuCl at 77K up to 37 kbar hydrostatic pressure, well beyond the predicted changeover pressure from anomalous to normal TO phonon scattering.

EXPERIMENT

CuCl single crystals were broken into pieces of about 50 μm diameter. One piece of CuCl together with a chip of ruby or silicon for pressure calibration and a pentane-isopentane fluid mixture were placed in the gasket of a diamond-anvil cell. The crystal orientation was unknown and was variable. After setting a pressure at room temperature, the cell was placed in a liquid nitrogen bath.

The optics were arranged in a backscattering configuration so that the incident laser light was incident on the crystal approximately 180° from the direction of the scattered light. A 647nm exciting light was used to avoid strong luminescence in the diamonds and the CuCl. A major problem in this experiment was the finding of CuCl samples that did not luminesce, and then in keeping them from becoming luminescent. Some of this problem was avoided by not applying full laser power to the cell while at room temperature. The laser can be focused on the ruby to determine the pressure from the frequency shift of the R lines [3].

When the silicon chip replaces the ruby, the pressure is determined by the Raman LO phonon shift [4] assumed to be temperature independent. Standard double monochrometer and photon counting electronics analyze the scattered light. The cell is warmed to room temperature before changing the pressure.

RESULTS

Raman scattering in CuCl at 77K and at several pressures is shown in Fig. 1. The high-frequency peak at approximately 210 cm^{-1}

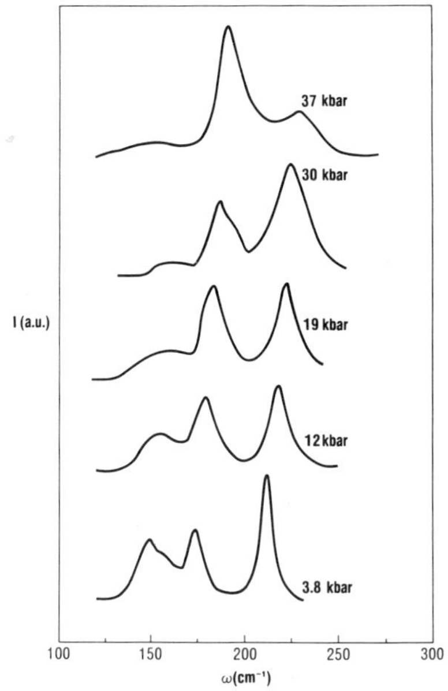

Fig. 1. Experimental Raman spectra from CuCl for various pressures at 77K.

at low pressure corresponds to the LO phonon. Its relative intensity is determined by the crystal orientation which varies from run to run. The TO phonon region at low pressure shows the two-peak structure previously mentioned. As the pressure is raised, the two-peak structure changes to a one-peak structure with only a remnant of the multiphonon background still present in the spectrum. These experimental results verify the predictions based on the low-pressure measurements.

It is evident from the spectra that there is additional broadening at higher pressures. This broadening is attributed to nonhydrostatic pressure in the cell. Trial measurements show that this inhomogeneity becomes much worse above 25 kbar.

ANALYSIS AND DISCUSSION

The theory described elsewhere [1,3] were used to analyze the data. The Raman scattering intensity is given by

$$I(\omega) = \frac{R\,\Gamma\,(\omega)}{\{[\omega-\omega_{TO} - \Delta(\omega)]^2 + \Gamma^2(\omega)\}} , \qquad (1)$$

$\Delta(\omega)$ and $\Gamma(\omega)$ are large and frequency dependent because of the strong anharmonic coupling of the TO phonon to the multiphonon background. $\Delta(\omega)$ and $\Gamma(\omega)$ reproduce our data at 3.8 kbar, as are shown in Fig. 2.

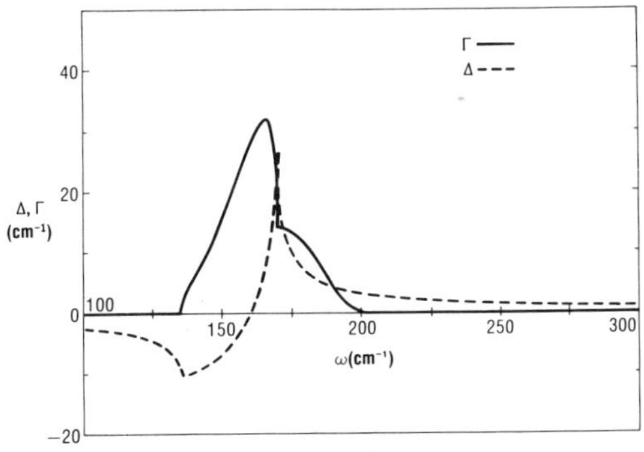

Fig. 2. Real and imaginary part of TO phonon self-energy used to fit the 3.8 kbar data.

The singularity, ω_L, appears in both $\Delta(\omega)$ and $\Gamma(\omega)$ at 170 cm^{-1}. The fact that $\Gamma(\omega)$ is large below the singularity and small above the singularity leads to a broad peak and a sharp peak in the TO phonon region of the Raman spectra. The $\Delta(\omega)$ and $\Gamma(\omega)$ shown, along with ω_{TO} = 163 cm^{-1} yield $I(\omega)$ shown in Fig. 3 along with the experimentally determined $I(\omega)$. The experiment shows also the LO phonon peak. The theoretical curve has contributions only

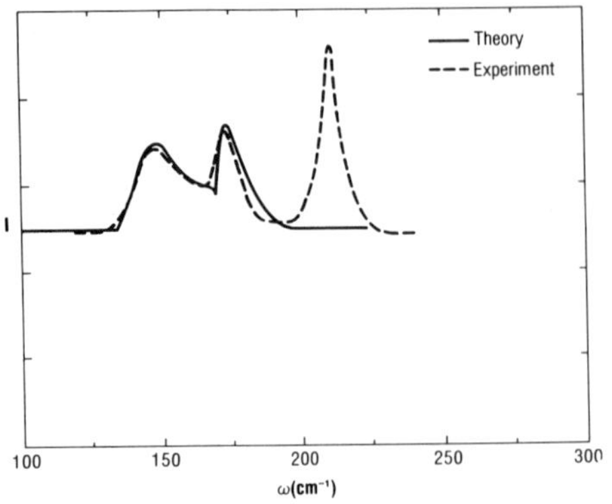

Fig. 3. Experimental and theoretical Raman spectra at 3.8 kbar.

from the TO phonon Raman scattering. The fit of theory to experiment is quite good. This result is comparable to those obtained earlier [1] at similar pressures and 40K.

The fits [1] at several pressures up to 7 kbar showed that ω_C and ω_{TO} both increase with pressure; however, ω_{TO} increases faster than ω_C. When extrapolated to 18 kbar, ω_{TO} should be pushed outside the multiphonon band, and the Raman spectra should show a strong TO phonon peak in the TO region of the spectrum. Referring to Fig. 1, we see qualitatively that at 19 kbar and above the spectrum does show a TO phonon peak, an LO phonon peak, and a broad band which is caused by the multiphonon band. The remnant background multiphonon band may be present due to coupling with the quasi-harmonic TO phonon or to direct phonon Raman scattering.

Several problems limit the usefulness of a quantitative fit of the theory to experiment at high pressures. The major difficulty is the broadening in the spectra. The theory was originally determined [2] for the 0 K case. In order to fit the broadening found experimentally at 40 K it was necessary to add a parameter to the theory which increases the imaginary part of the phonon self-energy. At low pressure and at 77K, this parameter must be further increased to fit theory to experiment. At higher pressures, however, the broadening due to the combination of thermal effects and inhomogenous pressure in the CuCl makes a quantitative comparison of theory and experiment impractical. A qualitative comparison of theory and experiment can be made and is shown in Fig. 4. The experimental curve is taken at 37 kbar. The theoretical curve is produced with $\omega_{C_1} = 170\ cm^{-1}$ and $\omega_{TO} = 188\ cm^{-1}$. The large peak corresponds to the position of the TO phonon. The broad band is due to the multiphonon band, and is present

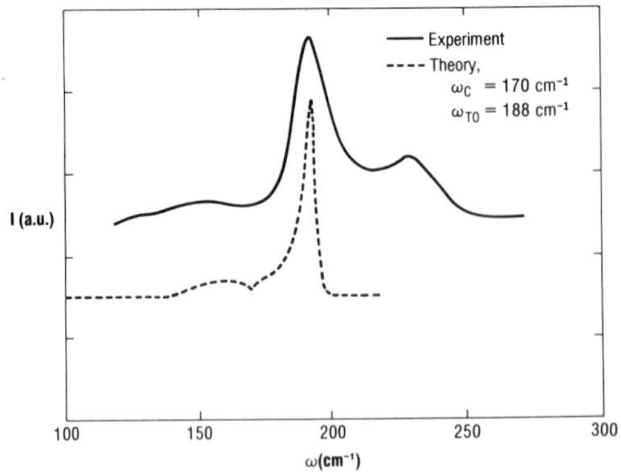

Fig. 4. Experimental and theoretical Raman spectra at 37 kbar.

because of the coupling to the TO phonon at 188 cm^{-1}. Qualitatively, this explanation accounts for the features observed in the Raman scattering in CuCl at 37 kbar.

The pressure dependence of the peak frequencies gives a somewhat less satisfactory fit with the model used here. The model predicts that the pressure dependence of the γ peak should go from a low-slope characteristic of ω_C at pressures below 18 kbar to a higher-slope characteristic of ω_{TO} at pressures above 18 kbar. Expressing the pressure dependence in terms of a Grüneisen parameter, we expect from Shand et al. [1] that the γ peak should be approximately 1.7 at low pressures and 2.4 at high pressures, while the LO peak should be 1.5. In the present work shown in Fig. 5 there is no clear change in the slope of the γ peak, and there is an apparent flattening in the pressure dependence of the LO peak above 20 kbar. However, the region above 20 kbar is just where pressure inhomogeneities begin to set in, so the true pressure dependence may be masked by this experimental difficulty. Further experiments are under way to clear up this difficulty.

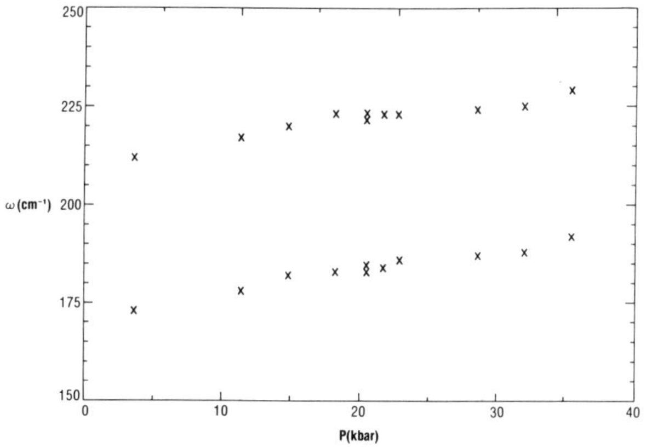

Fig. 5. Experimental Raman peak positions vs. pressure.

In conclusion, it is believed that the experimental data represents the first Raman scattering observation of the quasi-harmonic TO phonon in CuCl.

ACKNOWLEDGMENTS

The authors would like to thank M. Jordan for help in the experiment and M. McKelvey for help in sample preparation.

NOTATION

$I(\omega)$	=	Raman scattering intensity
R	=	one phonon Raman cross section
β	=	peak in TO region at approximately 146 cm^{-1}
γ	=	peak in TO region at approximately 170 cm^{-1}
ω_{LO}	=	frequency of longitudinal optic phonon
ω_{TO}	=	frequency of transverse optic phonon

ω_C =frequency of upper cutoff in multiphonon density of states
$\Delta(\omega)$ = real part of TO phonon self-energy
$\Gamma(\omega)$ = imaginary part of TO phonon self-energy

REFERENCES

1. M. L. Shand, H. D. Hochheimer, M. Krauzman, J. E. Potts, R. C. Hanson and C. T. Walker, Phys. Rev. B14, 4637 (1976).
2. M. Krauzman, R. M. Pick, H. Poulet, G. Hamel, and B. Prevot, Phys. Rev. Lett. 33, 1528 (1974).
3. R. A. Noack and W. B. Holzapfel, paper A-5-B presented at 6th AIRAPT Intern. High Pressure Conference, University of Colorado, Boulder, Colorado, July 25-29, 1977.
4. B. A. Weinstein and G. J. Piermarini, Phys. Rev. B12, 1172 (1975).

G-5

FAR INFRARED SPECTRA OF SOME MOLECULAR SOLIDS AT HIGH PRESSURE AND LIQUID HELIUM TEMPERATURE

J. Obriot, F. Fondère, Ph. Marteau, and H. Vu

Université Paris-Nord
Villetaneuse, France

INTRODUCTION

Far infrared absorption spectra of molecular solids have been intensively studied even at low temperatures for the past fifteen years. Pure phonon spectra due to lattice vibrations as well as localized phonon spectra due to impurity interaction with the lattice lie in that spectral range. Furthermore, by trapping molecules in inert gas matrices, one can expect to investigate their dimer interactions (in fixed configurations).

With a few exceptions, infrared studies have always been performed on thin films condensed from vapor onto a cooled window. In spite of its well-known convenience, this method has some disadvantages: (1) the sample thickness is not known with the accuracy required by intensity measurements; (2) very weak absorption lines cannot be detected; and (3) no variation of the lattice parameter can be produced since, in that case, generally no pressure can be applied to the sample.

Because frequency peak positions appear to be very sensitive to the pressure, a comparison between the experimental shift measurements and theoretical predictions for the density of states as a function of the lattice parameter would be of great interest. For this reason, we have investigated the pressure effect on the far infrared spectrum of relatively thick samples of solid nitrogen. In the first part of this paper, the experimental arrangement is described. The second part presents the results obtained to date.

EXPERIMENTAL

Spectroscopic measurements were carried out on a Coderg interferometer, Model FS 2000, in conjunction with a Varian computer. The experimental setup is schematically shown in Fig. 1. The main

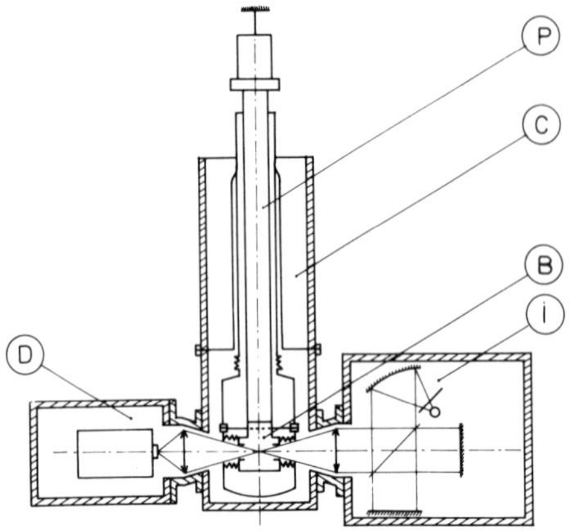

Fig. 1. Experimental setup. Legend: B - high-pressure cell; C - liquid helium cryostat; D - detector chamber; I - Michelson interferometer; and P - Press-arms.

parts include the interferometer, the high-pressure cell in the liquid helium cryostat, and the Golay detector. The cell is immersed in the liquid helium bath and the 4.2 K temperature is effectively reached. The solid sample is an 8-mm diameter cylinder, whose length can be adjusted from 1 to 14 mm. Although the critical problems of far infrared light sources as well as transparent materials are partially compensated for by the use of the multiplex system, special care has been taken to avoid any loss of light.

Dictated by the necessarily large high-pressure cell size, the diameter of the cryostat is also rather large, namely 25 cm. As a result, the use of optical mirrors should result in one of two inadequate solutions: strongly off-axis beams or very large focal lengths inside the vacuum chamber. In the present case, the interferometer beam output is focused through the cell and then to the detector by two zone plate polyethylene lenses. Their minimized thickness provides for maximum transmission of the light beam without considerable distortion. These lenses are also used as a separation between the high cryogenic vacuum and the primary vacuum of the interferometer and the detector chamber. Liquid helium cryostats usually require four windows, two room-temperature windows and two low-temperature windows. One can see that in this arrangement these have been eliminated and replaced by lenses and cell windows, respectively. The latter can be made of diamond or sapphire. Sapphire is normally absorbent in this spectral range at room temperature, but becomes transparent at low temperature up to 330 cm^{-1}.

The molecular solids are obtained as follows: the gas is first introduced in the cell at room temperature and at a pressure of about 1700 bar. The cell is then cooled, and from the high-pressure liquid phase a transparent solid state is obtained. Cooling down to 4.2 K is then achieved. Even at that temperature the pressure on the sample can be easily adjusted between normal pressure and 6 kbar. For that purpose, we use the piston technique with a potassium gasket, previously reported [1]. The acting stress is delivered by an oil press operating at room temperature in the upper part of the cryostat.

The accurate determination of the effective applied pressure is a problem still to be solved, since due to the potassium gasket friction it cannot be simply deduced from the oil pressure value. Nevertheless, a calibrating curve has been prepared starting from "in situ" measurements of the pressure shifts of a fluorescence line of sapphire [2]. For each pressure the corresponding solid density values have been extracted from the data given by Swenson for nitrogen at 4.2 K, between 1 and 10 kbar [3].

RESULTS

The α-N_2 Phase

Solid nitrogen crystallizes at high temperature in the disordered β-phase. A solid phase transition occurs at 35.6 K. The crystal structure of this low temperature α-phase is known to be cubic with four molecules per unit cell, but the assignment of the space group as $P2_13$ or Pa 3 is still controversial. In any case, the deviation from the centrosymmetric structure Pa 3 should be very weak. According to certain measurements, the molecular centers have been reported to be shifted by 0.17 Å parallel to the figure axis, but presently this value is considered as being too strong. With the assumption of the centrosymmetric structure, only three Raman active librational modes and two infrared active translational modes are expected.

Figure 2 shows the pressure effects. The full curve is the spectrum as observed under 0.48 kbar. The absorption peaks lie at 50 and 71 cm^{-1} respectively. By increasing the pressure on the crystal, they are strongly shifted towards higher frequencies. The dotted curve represents the spectrum for 3.25 kbar.

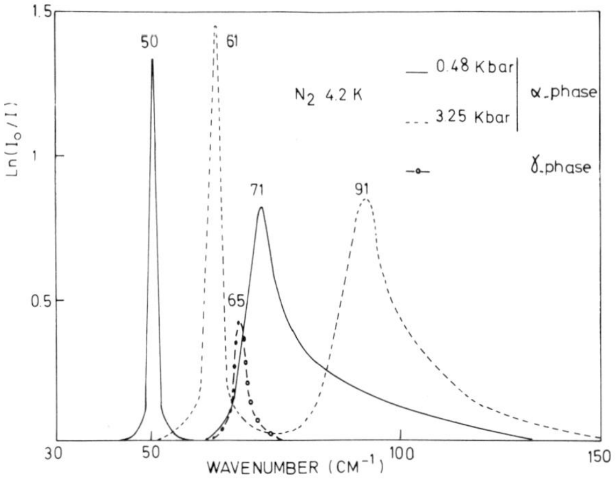

Fig. 2. Pressure effect on the solid N_2 spectrum for path length of 2 mm.

Frequency peak positions as a function of the density (in Amagat units) are shown in Fig. 3. For most of the points, the frequency value has been obtained by averaging over several experiments. Their relative dispersion can be mainly attributed to the lack of accuracy in determining the pressure values. However, linear variations can be reasonably considered in both cases. The extrapolation to 823 Amagat

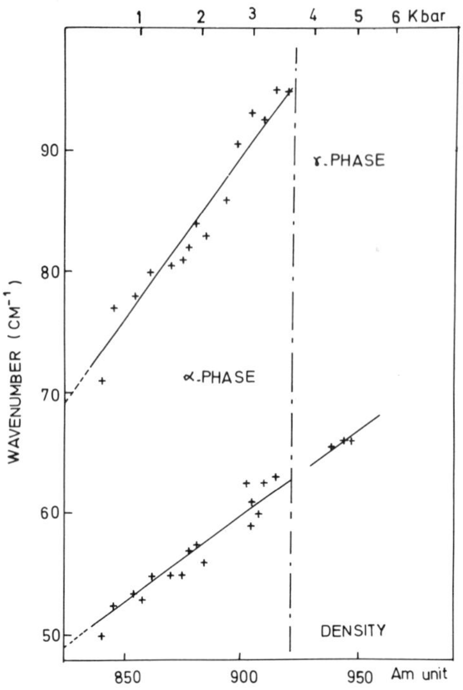

Fig. 3. Frequency shifts vs. density for the two translational modes of $\alpha - N_2$, and for the single translational mode of γ-N_2.

which is the solid density value under normal conditions, leads to the previously reported frequencies, 49 and 69 cm^{-1} respectively [4,5]. From the slope measurements the frequency shifts are found to be 3.7 cm^{-1}/kbar and 7.8 cm^{-1}/kbar.

The γ-N_2 Phase

At 4.2 K and 3.7 kbar a new solid phase transition occurs to give the γ-phase, a tetragonal crystal structure. According to the D_{4h} space group, only one translational mode is expected to be infrared active. Figure 2 shows the γ-phase spectrum recorded at a pressure of 4 kbar. As expected, only one strong line appears. This is the first recorded infrared spectrum of the γ-phase since the solid is not transparent in the near infrared region for which high-pressure measurements have been tentatively made.

Integrated Intensity

The variation of the integrated intensity for the low-frequency line has been investigated. Results are summarized in Fig. 4. The accuracy of our measurements is rather poor, and the uncertainty has not been evaluated. Nevertheless, a strong enhancement of the intensity with increasing pressure can be noticed. (The pure density effect on the absorption coefficient has been taken into account by considering the total intensity per Amagat unit.) The intersection at normal pressure gives the approximate value 10^{-2} cm^{-2}-Am^{-1} which is smaller by a factor of two than that previously reported by St-Louis and Schnepp [6], but agrees very well with RTC (rotational translational coupling) theoretical calculations of Friedman and Kimel [7]. However, the two experimental values could coincide within the experimental error. At the phase transition, a net discontinuity occurs which has not been observed for frequency shifts. However, the break of the curve is not surprising since the symmetry of the two translational modes are not the same for the α and γ phases.

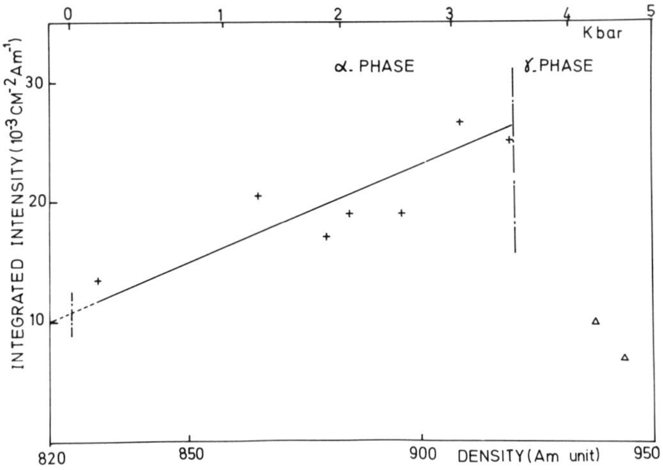

Fig. 4. Variation of the integrated intensity as a function of density of the low-frequency translational mode of α - N_2, and for the single mode of γ - N_2.

REFERENCES

1. M. Jean-Louis and H. Vu, Rev. Phys. App. 7, 89 (1972).
2. D. Fabre, M. M. Thiery, and B. Vodar, C. R. Acad. Sc. Paris B280, 781 (1975).
3. C. A. Swenson, J. Chem. Phys. 23, 1963 (1955).
4. A. Anderson and G. E. Leroi, J. Chem. Phys. 45, 4359 (1966).
5. A. Ron and O. Schnepp, J. Chem. Phys. 46, 3991 (1967).
6. R. V. St. Louis and O. Schnepp, J. Chem. Phys. 50, 5177 (1969).
7. H. Friedmann and S. Kimel, J. Chem. Phys. 44, 3925 (1965); 47, 3589 (1967).

G-6

DOUBLE LIGHT SCATTERING - APPLICATION TO THE DETERMINATION OF THE ISOTHERMAL COMPRESSIBILITY OF PURE FLUIDS NEAR THEIR CRITICAL POINT

Y. Garrabos, R. Tufeu, B. Le Neindre and B. Oksengorn

Université Paris-Nord, Villetaneuse, France

INTRODUCTION

The behavior of the isothermal compressibility of pure fluids in the critical region is of special interest because this parameter is related to several important physical properties of fluids in the vicinity of the gas-liquid transition. The determination of the isothermal compressibility can be made from the analysis of classical measurements of thermodynamics (P V T data) in a range of temperatures not too close to the critical point. Another way is to use optical methods like the interferometric measurements to determine the density profile, the determination of the turbidity (apparent absorption of the incident light beam), or the intensity of the polarized component of the light scattering which is directly related to the isothermal compressibility according to the Einstein-Smoluchowski relation. In the last case it is very difficult to obtain an accurate measurement of the absolute value of the light intensities, with important geometry corrections.

However, the use of experiments on depolarized double light scattering, as proposed by Reith and Swinney [1], can give an accurate method to determine isothermal compressibility from the ratios of light scattering

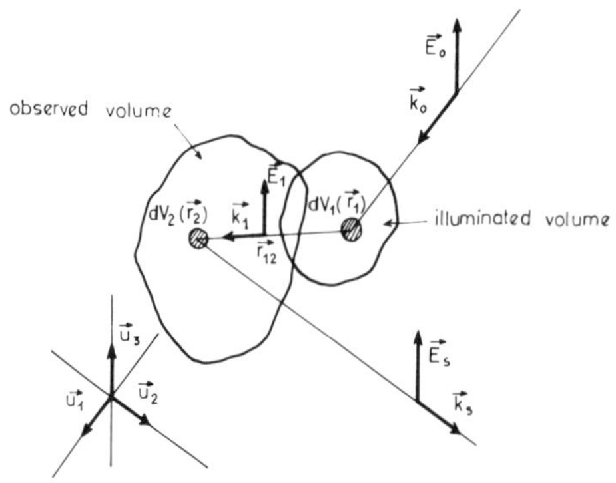

Figure 1. Double light scattering geometry.

intensities which are easier to measure than the absolute values. In this paper we present a brief review of the theories on double light scattering and give some results on the depolarized intensities; from the depolarization ratios we calculate the scattering cross-sections and determine the isothermal compressibilities.

THEORIES

In recent years, many experimental and theoretical investigations have been developed extensively to study the depolarization of Rayleigh light scattered by gases of isotropically polarizable molecules under pressures far from the critical point [2]. The principal physical effect responsible for this phenomenon is the fluctuation in orientation of the effective field seen by a particle, due to the interaction between induced dipoles in a collision of two partners at short distances. However, near the critical point Oxtoby and Gelbart [3] were the first to predict that the main contribution to the depolarization comes from successive single scatterings of undistorted particles separated by macroscopic distances, and not the contribution of the collision-induced effects of two, three, and four interacting molecules. Indeed, these authors have developed a general expression for the depolarized scattering intensity by calculating the effective field seen by a particle at the point $\vec{r}$, as follows:

$$I_R^{VH} = \frac{\alpha^4}{R^2} k_o^4 |E_o|^2 \int_{V_S} d\vec{r}_1 \int_{V_I} d\vec{r}_2 \int_{V_S} d\vec{r}_3 \int_{V_I} d\vec{r}_4 \exp \left[i\vec{k}_o \cdot (\vec{r}_2 - \vec{r}_4) - i\vec{k}' \cdot (\vec{r}_1 - \vec{r}_3) \right] \times T_{k_o}(\vec{r}_1 - \vec{r}_2)_{xz} \; T_{k_o}(\vec{r}_3 - \vec{r}_4)_{xz}$$

$$\langle \Delta\rho(\vec{r}_1)\, \Delta\rho(\vec{r}_2)\, \Delta\rho(\vec{r}_3)\, \Delta\rho(\vec{r}_4) \rangle \qquad (1)$$

where E_o is the amplitude of the incident light wave polarized in the vertical position parallel to the z axis and k_o is its wave number and propagates along the negative x axis. The detector is at a distance R which is large compared to the dimensions of the sample formed of particles with a scalar polarizability α, and the scattering angle is 90° with k' the wave number of the scattered light. The sample volume and the illuminated volume are V_S and V_I respectively. The phase factors and the dipole propagators $T_{k_o}(\vec{r})$ describe the scattered fields; $\Delta\rho(\vec{r})$ is the fluctuation in the instantaneous number density $\rho(\vec{r})$ at $\vec{r}$. By writing the ensemble average in equation (1) in terms of Ursell cluster functions, and by taking into account the Ornstein-Zernike form for the long-range

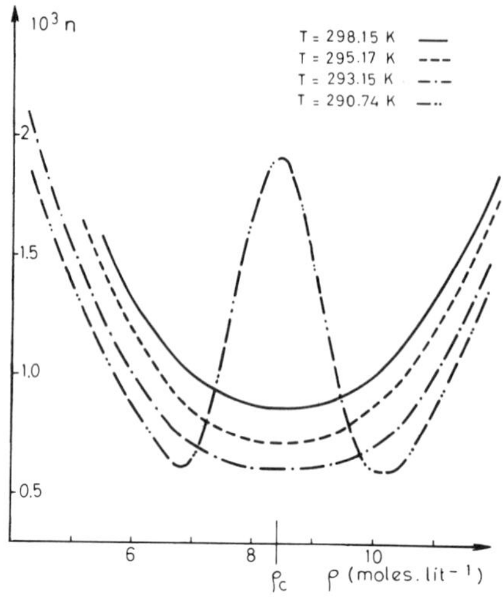

Fig. 2. Depolarization ratio of xenon for several isotherms in the critical density range.

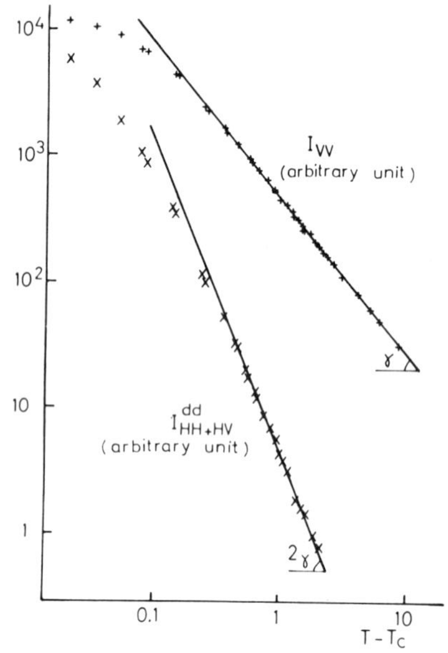

Fig. 3. Xenon variation as a function of T - Tc. Legend: + the polarized scattering intensity and x the depolarized double scattering intensity.

part of the interactions in the critical domain, Oxtoby and Gelbart have shown that the most important contribution to the depolarization comes from the long-range correlations of two particles which in their ideal scattering geometry are proportional to V_I and to a dimension R_S of the sample, with a $(k_o\xi)^4$ dependence, valid in a temperature range for which it is found $k_o\xi \lesssim 0.3$ (ξ is the correlation length in the Ornstein-Zernike approximation). This is the true double scattering, e.g., two successive single scattering events: one in the illuminated volume V_I and the other in the sample volume V_S.

For practical scattering geometries (see Fig. 1), Reith and Swinney [1] have calculated from the polarized single scattering cross-section the depolarized double scattering intensity, and have found for the depolarization ratio the following expression:

$$I^d_{VH}/I_{VV} = \sigma_o \, gh \tag{2}$$

where σ_o is related to the isothermal compressibility χ_T by

$$\sigma_o = \frac{k_o^4}{16\pi^2} \left(\rho\frac{\partial\varepsilon}{\partial\rho}\right)^2_T k_\beta \, T \, \chi_T$$

with ε being the dielectric constant of the medium, T the absolute temperature, k_B the Boltzman constant, and where g is a dimensionless factor which depends on the geometry of the sample and of the observation, and h is the height of the sample seen by the detector. The asymptotic value of g is equal to $\pi/4$, but Bray and Chang [4] have given a general expression for this parameter. However, equation (2) is valid provided h << horizontal dimensions of the sample. A similar relation to (2) has also been obtained by Boots et al. [5].

EXPERIMENTAL RESULTS AND DISCUSSION

The experimental arrangement has been given elsewhere [6]. In a first step, several isotherms have been determined for the depolarization ratio

$$\eta = i/I \simeq \frac{I_{HV} + I_{HH}}{I_{VV}} \tag{4}$$

of the light scattered by xenon in the temperature range $T - T_c \simeq$ 0.5 K and 10 K around the critical density [7]. Figure 2 shows the drastic increase of η when the temperature approaches the critical temperature. This is in qualitative agreement with Oxtoby and Gelbart's theory [3] which predicts a significant increase in the depolarization ratio due to the double scattering [7].

For a quantitative comparison with the theories, we have studied the depolarization of the Rayleigh light scattered along the critical isochore in the one phase range ($0.05 < T - Tc < 10$ K) for gases of isotropic molecules (Xe and SF_6) and gases of anisotropic molecules (CO_2, C_2H_4, and C_2H_6), by using different geometrics of observation: (1) the detector is in the horizontal plane of observation containing the incident laser beam (z = o) and the height h may be varied; (2) h being fixed, one can change the vertical distance of the detector z to the plane of observation hence, one observes only the double light scattering. Figs. 3 and 4 show the temperature

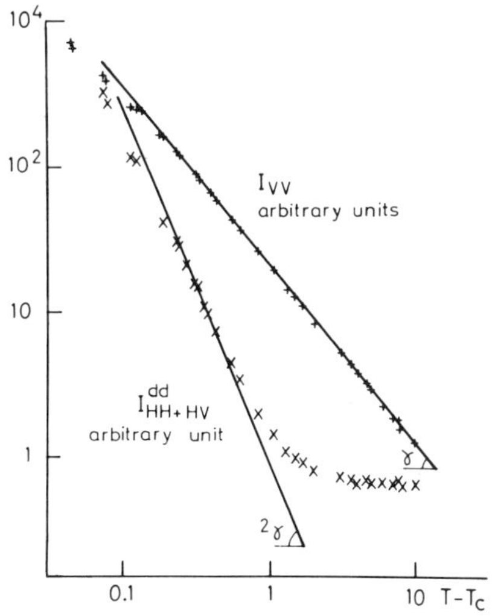

Fig. 4. CO_2 variation as a function of T - Tc. Legend: + the polarized scattering intensity and × the depolarized double scattering intensity.

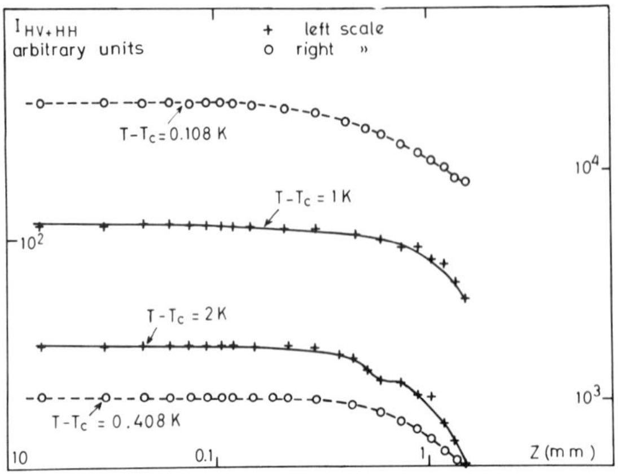

Fig. 5. Space dependence of the depolarized double scattering for C_2H_6.

dependence of the depolarized component for Xe and CO_2, respectively. The dependence as a function of (T - Tc) is found to be in agreement with the predictions given by Oxtoby and Gelbart [3]. Similar results have been obtained for Xe by Reith and Swinney [1], and for CO_2 by Trappeniers et al. [8]. We observed the same behavior of the depolarized intensity for all fluids that we have studied.

For space dependence we have studied the case of C_2H_6. Fig. 5 shows the depolarized intensities as functions of z; one can see that the double scattering has a weak space dependence. A similar result has been obtained by Beysens et al. [9] for a binary mixture near its critical mixing point. Moreover, Fig. 6 shows that inside and outside the illuminated volume the double scattering is the dominant contribution to the depolarized intensity.

According to equation (2), the derivation in terms of h of the experimental depolarization ratio yields to one value of the factor σ_o, since the contributions of the collision-induced and permanent anisotropies do not show a geometrical dependence. This method has been used by Reith and Swinney [1] in their study of xenon. On the other hand, we have determined the respective contributions of the simple scattering (slope - γ) and of the double scattering (slope + γ) (See Fig. 7). Calculating the geometric parameter g from Bray and Chang [4] and for a known value of h, we have found σ_o = 2.47 10^{-4}/mm for xenon at T - T_c = 1 K, while

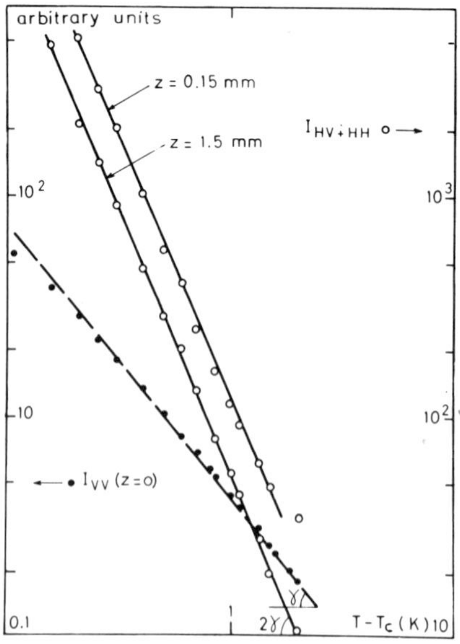

Fig. 6. C_2H_6 variation as a function of T - Tc. Legend: • the polarized scattering intensity and o the depolarized double scattering intensity at different values of z.

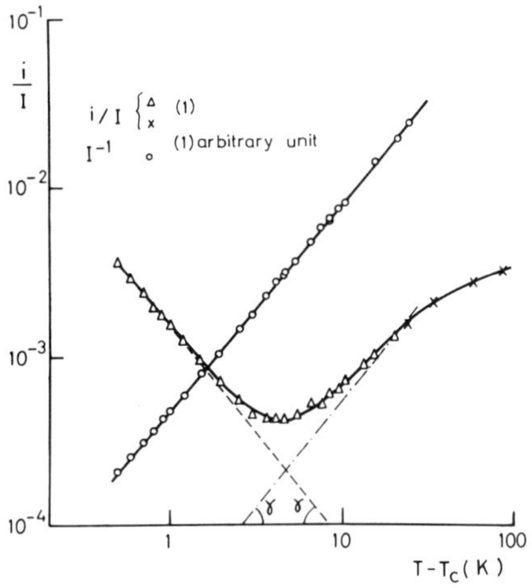

Fig. 7. Xenon variation of the depolarized ratio $\eta = i/I$ as a function of T - Tc.

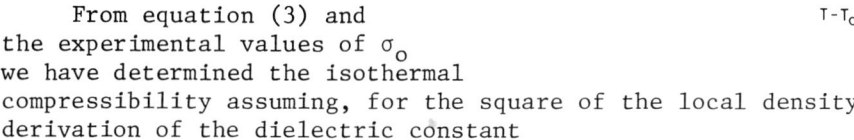

the value obtained by Reith and Swinney was 2.07 10^{-4}/mm. Our result is in good agreement with the measurements of the absolute intensity (σ_o = 2.89 10^{-4}/mm) [10]. In the case of SF_6, we found 1.43 10^{-4}/mm close to the value of 1.30 10^{-4}/mm[11] deduced from measurements of turbidity.

From equation (3) and the experimental values of σ_o we have determined the isothermal compressibility assuming, for the square of the local density derivation of the dielectric constant

$$\left(\rho \frac{\partial \varepsilon}{\partial \rho} \right)^2$$

the value given by the model of a spherical Lorentz local field or the one obtained by Rocard [12]. In Table I we report γ values and those of the parameter Γ_o given by the law of divergence for χ_T:

$$\chi_T = \Gamma_o \left(\frac{T - Tc}{Tc}\right)^{-\gamma} \qquad (5)$$

for all gases studied, obtained by different methods and compared with our values. It must be noted that most of the Γ_o values are in rather good agreement in spite of a large spread in γ values.

CONCLUSION

From the study of the depolarized double scattering in a temperature range not too close to the critical temperature where the higher order multiple scattering may be ignored, we have shown that it is possible to easily determine the isothermal compressibilities of fluids near their critical point with good accuracy. However, for gases of anisotropic molecules the permanent depolarized component is important, and the measurements are less accurate. It is possible to eliminate this contribution by using a Fabry-Perot spectrometer [9] or by making the measurements outside the illuminated volume. Work is in progress in this direction.

Table I

	γ	$10^8 \Gamma_o (m^2\text{-}N^{-1})$	
Xe	1.21 ± 0.03	1.17 ± 0.05	Absolute value from light scattering[10]
	1.232 ± 0.006	1.250 ± 0.066	Absolute value (density profile) [13]
	1.260 ± 0.002	0.959 ± 0.018	Absolute value (density profile) [13]
	1.260 ± 0.06	1.010	Analysis of PVT data of Schneider[14] (NBS equation)
	1.24 ± 0.03		Analysis of PVT data of Schneider[15] (linear model)
	1.19	1.354	Universal scaled equation of state[16]
	1.21	0.856	Depolarization ratio; double scattering[1]
	1.253	0.801	This work (Lorentz local field[6]
	1.253	0.964	This work (Rocard relation)[6]
CO_2	1.219 ± 0.010	0.775 ± 0.046	Absolute value from light scattering[17]
	1.17 ± 0.03	0.970 ± 0.060	Absolute value calibrated on PVT data at T = Tc + 10 K from light scattering[18]
	1.24		Density profile[19]
	1.26 ± 0.02	0.713	Analysis of PVT data of Michels [14] (NBS equation)
	1.21 ± 0.02		Analysis of PVT data of Michels [15] (NBS equation)
	1.19	0.892	Universal scaled equation of state[16]
	1.237	0.593	This work (Lorentz local field)[6]
	1.237	0.686	This work (Rocard relation)[6]
SF_6	1.225 ± 0.020	1.165 ± 0.160	Absolute value from turbidity measurements[11]
	1.223	1.327	Absolute value from light scattering and turbidity[20]
	1.235 ± 0.015	1.26	Relative value calibrated on PVT data at T = Tc + 1.453 K, from light scattering[21]
	1.18 ± 0.03		Density profile[22]
	1.28		Density profile[19]
	1.260	0.950	This work (Lorentz local field)[6]
	1.260	1.076	This work (Rocard relation)[6]

Table I, continued

	γ	$10^8\Gamma_o(m^2\text{-}N^{-1})$	
C_2H_4	1.19	1.53	Universal scaled equation of state[16]
	1.258	0.871	This work (Lorentz local field)
	1.258	1.031	This work (Rocard relation)
C_2H_6	1.245	0.900	This work (Lorentz local field) [6]
	1.245	1.060	This work (Rocard relation)[6]

REFERENCES

1. L. A. Reith and H. L. Swinney, Phys. Rev. A 12, 1094 (1975).
2. W. M. Gelbart, Adv. Chem. Phys. 26, 1 (1974).
3. D. W. Oxtoby and W. M. Gelbart, J. Chem. Phys. 60, 3359 (1974).
4. A. J. Bray and R. F. Chang, Phys. Rev. A 12, 2594 (1975).
5. H. M. J. Boots, D. Bedeaux, and P. Mazur, Physica 84A, 217 (1976).
6. Y. Garrabos, R. Tufeu, and B. Le Neindre, to be published.
7. Y. Garrabos, R. Tufeu, and B. Le Neindre, C.R. Acad. Sci. 282B, 313 (1976).
8. N. J. Trappeniers, A. C. Michels, and R. H. Huijser, Chem. Phys. Lett. 34, 192 (1975).
9. D. Beysens, A. Bourgou, and G. Zalczer, J. de Physique, C1 221 (1976).
10. J. W. Smith, M. Giglio, and G. B. Benedek, Phys. Rev. Lett. 27, 1556 (1971).
11. V. G. Publielli and N. C. Ford, Jr., Phys. Rev. Lett. 25, 143 (1976).
12. Y. Rocard, Annales de Physique 10, 116 (1928).
13. W. T. Estler, R. Hocken, T. Charlton, and L. R. Wilcox, Phys. Rev. A, 12, 2118 (1975).
14. M. Vincentini-Missoni, J. M. H. Levelt-Sengers, and M. S. Green, Phys. Rev. Lett. 22, 389 (1969).
15. W. L. Greer, J. M. H. Levelt-Sengers, and J. V. Sengers, J. Phys. Chem. Ref. Data 5, 1 (1976).
16. J. M. H. Levelt-Sengers and J. V. Sengers, Phys. Rev. A 12, 2622 (1975).
17. J. H. Lunacek and D. S. Cannel, Phys. Rev. Lett. 27, 841 (1971).
18. J. A. White and B. S. Maccabee, Phys. Rev. Lett. 26, 1468 (1971).
19. R. Hocken and M. R. Moldover, Phys. Rev. Lett. 37, 29 (1976).
20. D. S. Cannel, Phys. Rev. A 12, 225 (1975).
21. G. T. Feke, G. A. Hawkins, J. B. Lastovka, and G. B. Benedek, Phys. Rev. Lett. 27, 1780 (1971).
22. D. A. Balzarini, Can. J. Phys. 50, 2194 (1972).

H-1

RAMAN STUDY OF THE CONFORMATIONAL CHARACTERISTICS OF CHAIN MOLECULES UNDER HIGH PRESSURE

P. E. Schoen, R. Priest, J. P. Sheridan, and J. M. Schnur

Naval Research Laboratory
Washington, D.C. USA

INTRODUCTION

As part of a continuing program to study the properties of lubricants and polymers, we have undertaken an investigation of the conformational properties of linear alkanes and polymers as functions of temperature and pressure. We have successfully obtained the Raman spectra of these materials over a range of 1 to 20 kbar from a gasketed diamond-anvil cell while calibrating the pressure with the ruby fluorescence technique [1]. The conformation of heptane (C_7H_{16}) and hexadecane ($C_{16}H_{34}$) have been monitored by the observation of conformationally sensitive Raman bands. Our data indicate that heptane becomes more "kinked" (more gauche bonds) as a function of pressure. This result was surprising to us at first, since at sufficiently high pressure heptane freezes in the unkinked, all-trans conformation. We have developed a theory, based on excluded volume considerations, which is consistent with these observations.

Since the Raman signal from the alkanes was relatively weak and the fluorescence background of diamond [2] and ruby was strong, we have had to employ near infrared laser excitation and pay careful attention to scattering geometry in order to obtain useful spectra. These techniques are described elsewhere [3].

We have benefited in this study from the large amount of work that has been done on the normal mode analysis of the alkanes and polyethylene [4]. From model calculations and Raman and infrared spectroscopic investigations at room pressure the vibrational frequencies of the different normal modes of the alkanes have been identified for the straight chain, all-trans configuration as well as for some of the kinked, gauche conformers. There are three frequency ranges of interest to us: 200 to 550, 1060 to 1140, and 2800 to 3000 cm^{-1}. In the low-frequency region of the Raman spectrum the alkanes have bands known as the longitudinal acoustic modes (LAM) which are caused by accordion-like vibrations of the molecules along their lengths [5]. These vibrations have frequencies which are inversely proportional to the length of the straight-chain segments of the molecule. Kinked-conformer alkanes have LAM-type bands in this frequency region as well and several have been identified by Schaufele [6]. However, only a few of the possible conformations produce distinct Raman bands of measurable intensity, and these only in the case of the low-carbon-number alkanes. (In the longer liquid alkanes the multiplicity of possible gauche conformers produce bands which overlap and are indistinguishable.) Heptane has four distinct LAM-type bands of which the strongest three occur at 309 cm^{-1}, 396 cm^{-1}, and 507 cm^{-1}. Spectra of heptane at room pressure and at 14.7 kbar at 65°C are shown in Fig. 1. Our data consistently show a shift of these bands to higher frequency as a function of pressure. According to Schaufele the band at 309 cm^{-1} is the LAM due to the all-trans TTTT conformation. The band at 507 cm^{-1} results from a molecule with a single gauche bond, TGTT, and the band at 396 cm^{-1} is due to the superposition of bands from a single-kink conformation, GTTT, and a double-kink conformation, TTGG [4,6]. For the longer chain alkanes in the spectral region 1060 to 1140 cm^{-1}, there is a triad of bands referred to as the optical skeletal modes. The 1060/1140 pair is associated with trans bonds and the broad band between them with gauche bonds [7]. The central band indicates the presence of chain kinking without reference to any particular gauche conformation and has been of considerable interest to biophysicists in the study of ordering in model membranes [8]. We attempted to use these bands for an indication of chain kinking under pressure in hexadecane.

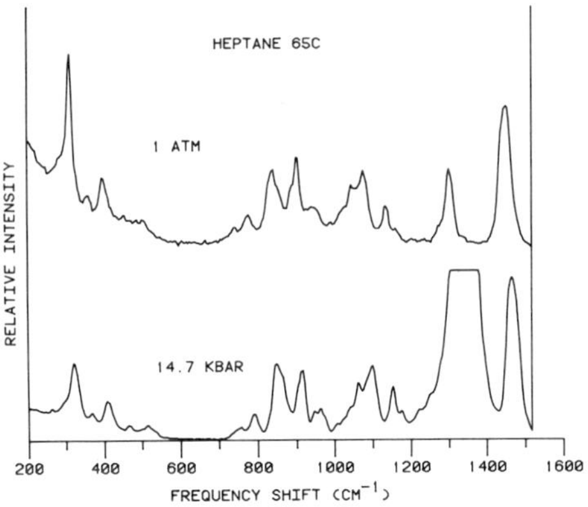

Fig. 1. Spectra of heptane at 65°C at 1 atm and 14.7 kbar. Large band near 1300 cm^{-1} in the lower spectrum is the diamond Raman line.

Finally the region 2800 to 3000 cm^{-1} contains bands caused by the C-H stretching modes which have been shown to be sensitive to crystallinity, local ordering and chain kinking in hexadecane - again as a model for use in biophysical studies [9]. We have observed these bands for evidence of conformational changes under increasing pressure in heptane and hexadecane.

EXPERIMENT

Our samples of heptane and hexadecane were placed in small holes (1/4 mm diameter) drilled in a piece of shim molybdenum 1/4 mm thick. Diamond anvils pressing on either side of the holes sealed in the samples and compressed them to produce high pressure. A small piece of ruby was also placed in the holes so that the pressure sensitive frequency of its fluorescence peak at $\sim$ 14400 cm^{-1} (694 nm) could be monitored to determine the pressure.

We used the 752.5 nm line of a krypton laser for the incident light. This near-IR line was chosen to reduce fluorescence from the diamonds, the ruby, and the sample [3]. The beam was focused into the cell at 15° off normal incidence to avoid specular reflection of the laser light into the spectrometer and to reduce the amount of diamond fluorescence imaged on the spectrometer slits.

Our spectrometer was a Spex 1401 with holographic gratings peaked for performance between 600 and 900 nm. Our detector was an RCA C31034A PM tube which was red-sensitive to about 900 nm.

THEORY AND ANALYSIS OF DATA

To explain the kinking behavior of heptane we have used some concepts which have proved useful in modeling liquid crystal systems. Alben [10] has considered the statistical mechanics of a dense gas of hard rods interacting only through excluded volume effects and Van der Waals attraction. He assumed that the rods have no orientational correlations and are free to rotate end over end as in the isotropic liquid crystal case. Thus if the molecule is made more globular it sweeps out less volume as it rotates. This increases the translational entropy by reducing the excluded volume.

The important parameters of this model are the length-to-breadth ratio (aspect ratio) of the rod and the density. Since the alkane molecules can assume a number of conformations, the hard rod model can be extended by viewing each distinct conformation as a separate species and the liquid as a mixture of these species. Each molecule has an energy depending on the number of gauche bonds it contains and an excluded volume depending on its effective aspect ratio. Minimizing the Gibbs free energy [10] we find that conformations with lower aspect ratios (i.e., more globular) are increasingly favored as the density is increased which is the central result of the experiment. In Fig. 2 these results are presented in a form obtained by eliminating the density as the independent variable in favor of P_o/P_o*, where P_o is the fraction of the molecules which are all trans, and P_o* is the value of P_o at 1 atm. The quantity P_1 is the fraction of molecules in conformations with one gauche bond, and P_{2-4} is the fraction in conformations with 2, 3, or 4 gauche bonds.

The experimental data have been plotted in Fig. 2 under the assumption that the areas of the Raman bands change in proportion to changes in the populations of different conformers. For the data, the abscissa is the ratio of the all-trans band (309 cm^{-1}) at high pressure to its area at 1 atm. The ordinate is the sum of

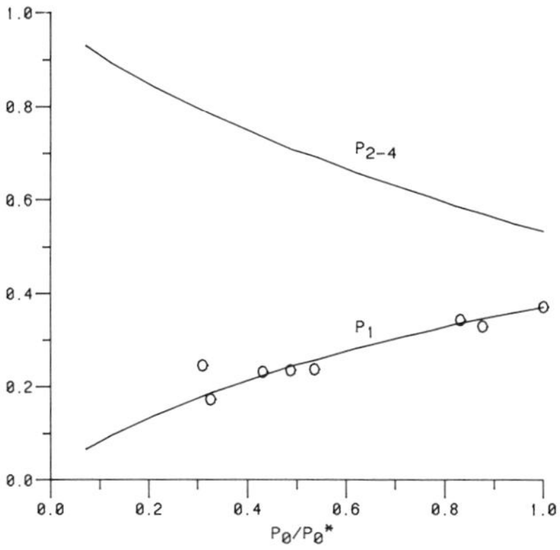

Fig. 2. Solid lines are theoretical curves: upper line is P_{2-4} vs. P_o/P_o^* and lower curve is P_1 vs. P_o/P_o^*. The experimental points represent the sum of the areas of the 357, 396, and 507 cm^{-1} bands (relative to 1450 cm^{-1} band) normalized to agree with theory at 1 atm vs. area of the 309 cm^{-1} band relative to its area at 1 atm.

the areas of the other three bands (356, 396, and 507 cm^{-1}) relative to the area of the band at 1450 cm^{-1}, a band which appears to be relatively insensitive to pressure. The data have been normalized to agree with the theory at room pressure.

Figure 3a shows the area of the 309 cm^{-1} all-trans band, relative to the 1450 band, vs. pressure, and Fig. 3b shows the all-trans band, relative to the sum of the other three low frequency bands, vs. pressure. The theoretical curve is also shown in Fig. 3b and the agreement with experiment is quite satisfactory. In this case the theoretical pressure has been deduced by comparing the theoretically predicted reduction in the all-trans population with the decreases in the LAM intensity shown in Fig. 3a. Both Figs. 2 and 3 indicate that as predicted the population of unkinked and single-kinked conformers drops as the pressure rises.

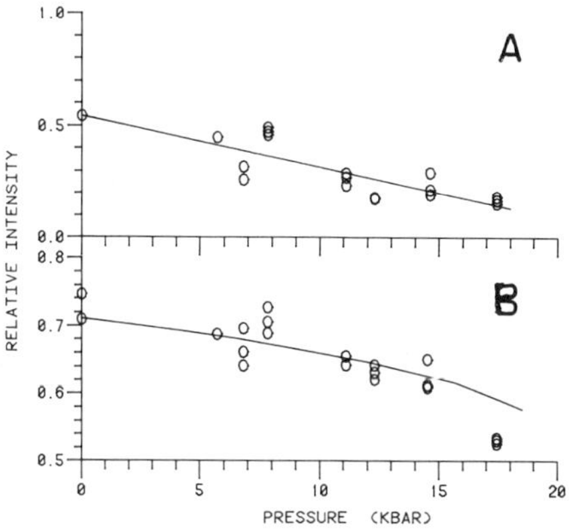

Fig. 3. (a) Area of the 309 cm^{-1} band relative to area of the 1450 cm^{-1} band vs. pressure; solid curve is a least squares straight line fit to data. (b) Area of the 309 cm^{-1} band relative to the sum of the areas of the 375, 396, and 507 cm^{-1} bands vs. pressure; solid line is the theoretical curve.

Unfortunately no bands have been found which correspond to highly kinked conformations. However, the 405 cm^{-1} band which results partially from the two-kink GGTT conformation, according to Schaufele, does experience a considerable increase in intensity relative to the other low-frequency bands and a leveling-off relative to the 1450 cm^{-1} band.

Spectra of the high-frequency C-H stretching bands for heptane at 1 atm and at 14.7 kbar pressure are shown in Fig. 4. The intensity of the band at 2883 cm^{-1} has been correlated with the degree of chain extension and crystallinity in hexadecane [9]. The sharp decrease in the intensity of this band as the pressure rises is consistent with the kinking behavior deduced from the low-frequency bands.

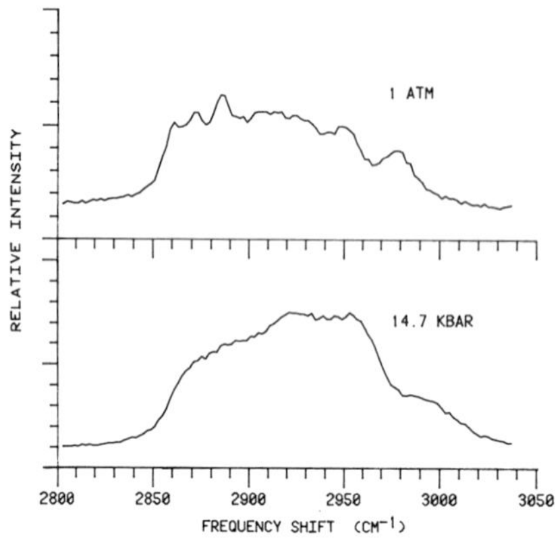

Fig. 4. Spectra of heptane C-H stretching modes at 1 atm and at 14.7 kbar with the temperature in both cases fixed at 65°C.

Similar observations of the LAM and C-H stretching spectra of hexadecane as well as the 1060 to 1140 cm^{-1} optical skeletal region reveal no noteworthy changes in conformational behavior. The spectra show that hexadecane in the liquid phase exists mainly in a number of gauche conformations at room pressure and that this does not change as the pressure is raised. These observations were performed in the ranges 25 to 165°C and 1 atm to 12 kbar pressure.

It is known that for very long chain molecules, i.e. polyethylene, that a pressure of 3 to 5 kbar induces chain straightening as the sample crystallizes from the melt [11]. Thus it may be expected that there would be a crossover value of chain length above which the kinking behavior we have observed in heptane would no longer occur.

We are now studying the longer alkanes and various polyethylenes via light-scattering techniques in an attempt to observe chain straightening, if any, that might occur at higher pressures in the fluid phases of these materials. This study may prove to be quite interesting in view of the recent reports on an ordered 'nematic'-like high-pressure fluid phase of polyethylene [12].

REFERENCES

1. J. D. Barnett, S. Block, and G. J. Piermarini, Rev. Sci. Instr. 44, 1 (1973).
2. D. M. Adams, S. J. Payne, and K. Martin, App. Spectros. 27, 377 (1973).
3. P. E. Schoen, J. M. Schnur, and J. P. Sheridan, J. Appl. Spectros., accepted for publication.
4. R. G. Snyder and J. H. Schachtschneider, Spectrochim. Acta 19, 85 (1963); also M. Tasumi, T. Shimanouchi, and T. Miyazawa, J. Mol. Spect. 9, 269 (1962); R. G. Snyder, J. Chem. Phys. 47, 1316 (1967).

5. W. L. Peticolas, G. W. Hibler, J. L. Lippert, A. Peterlin, and H. Olf, App. Phys. Lett. 18, 87 (1971); also R. F. Schaufele and T. Shimanouchi, J. Chem. Phys, 47, 3605 (1967).
6. R. F. Schaufele, J. Chem. Phys. 49, 4168 (1968).
7. M. Tasumi and T. Shimanouchi, J. Mol. Spectros. 9, 261 (1962).
8. J. P. Sheridan, J. M. Schnur, and P. E. Schoen, in preparation.
9. R. Faiman and K. Larsson, J. Raman Spectros. 4, 387 (1976); also B. P. Gaber and W. L. Peticolas, Biochim. Biophys. Acta 465, 260 (1977).
10. R. Alben, Mol. Cryst. Liq. Cryst. 13, 193 (1971).
11. B. Wunderlich and T. Arakawa, J. Polymer Sci. A2, 3697 (1964); also D. C. Bassett, Polymer 17, 460 (1976).
12. K. Monobe, Y. Fujiwara, and K. Tanaka in Proceedings of 4th Intern. High Pressure Conference, J. Osugi, ed., Kyoto, Japan (1975), p. 63, 865.

H-2

THE CHARACTERIZATION OF POLYMERS UNDER HIGH PRESSURE USING RAMAN SPECTROSCOPY

J. F. Rabolt, S. Block, and G. J. Piermarini

National Bureau of Standards
Washington, D. C. USA

INTRODUCTION

The characterization of polymers under high pressure has become increasingly important because of the expanded number of applications in which polymers are being used. Raman spectroscopy is a unique tool for studying materials in a diamond-anvil cell under pressure because good signal-to-noise spectra can be recorded on minute amounts of material.

Polytetrafluoroethylene (PTFE) is one of the more interesting polymer systems to study because it has a number of first-order solid-solid phase transitions which can be induced by placing it under pressure. For this reason, it has been the subject of many high-pressure studies [1-13]. Figure 1 shows the phase diagram as determined by Hirakawa and Takemura [6] using ultrasonic waves, thermal expansion, differential thermal analysis, and x-ray diffraction. The molecule assumes a helical conformation in phases I, II, and IV while in phase III the PTFE molecule adopts a planar zig-zag conformation similar to polyethylene [8,10,13].

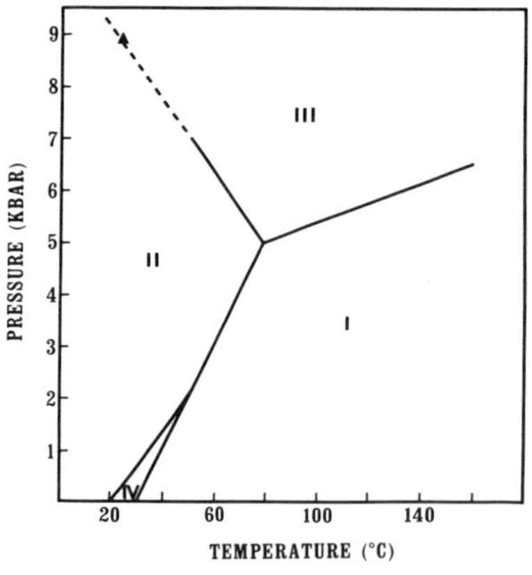

Fig. 1. Phase diagram of polytetrafluoroethylene.

It is the purpose of this study to report a Raman spectroscopic investigation of a perfluoro n-alkane ($C_{20}F_{42}$), a high molecular weight PTFE, and a random copolymer of tetrafluoroethylene and hexafluoropropylene in the high-pressure phase.

A continual change in band intensities has been observed in the phase III spectrum of PTFE with increased pressure. This behavior was markedly different for the oligomer and copolymer, although the II-III transition pressure has been found to remain the same. These intensity changes in PTFE can be interpreted in terms of an increase in transplanar content with increasing pressure.

EXPERIMENTAL

Raman spectra were recorded on samples under pressure using a Waspaloy diamond pressure cell developed at NBS. Using the 5145 nm line of an Ar^+ laser, it was possible to both record the Raman spectra and excite the R_1 ruby fluorescence line whose shift was used to measure the sample pressure [14]. All Raman experiments were performed in the backscattering geometry using an optical system which has been thoroughly described by Weinstein and Piermarini [15].

RESULTS AND DISCUSSION

Our Raman investigation was carried out in the 0 to 52 kbar range. Representative spectra of PTFE at 15 and 28 kbar are shown in Fig. 2. It was observed that the intensity of the 285 cm^{-1} band had increased significantly with respect to the 395 cm^{-1} band. Since the bandwidth at half peak height does not seem to change with pressure, the ratio of the peak intensities was calculated and plotted vs. pressure in Fig. 3. The solid dots represent the variation in peak intensity ratio (PIR) with increasing pressure while the triangles represent the same ratio with decreasing pressure. A considerable hysteresis was observed during the 48 hrs during which the data were taken. This same peak intensity ratio is also plotted for $nC_{20}F_{42}$ and a

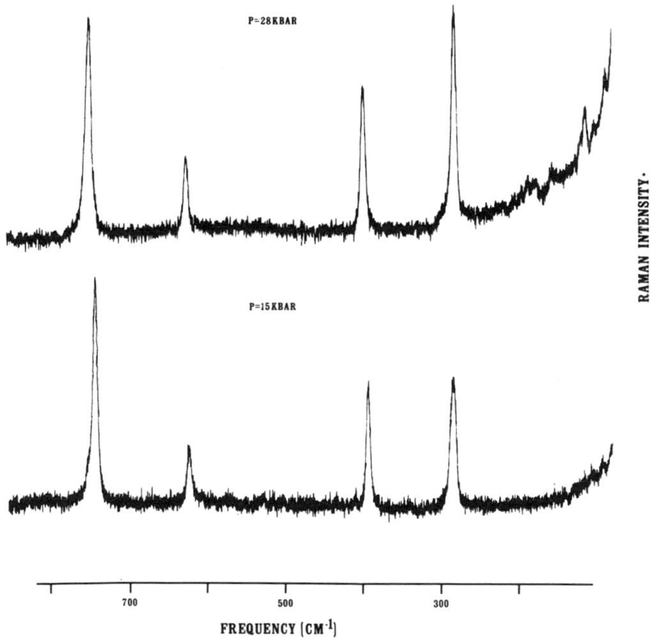

Fig. 2. Raman spectra (100-800 cm^{-1}) of polytetrafluoroethylene at 15 and 28 kbar.

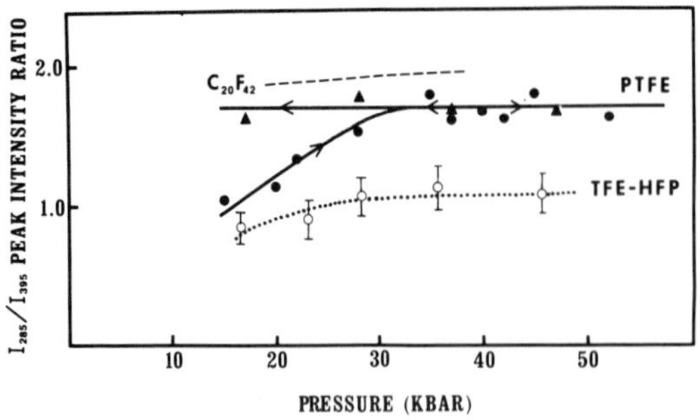

Fig. 3. Plot of I_{280}/I_{395} peak intensity ratio for $nC_{20}F_{42}$, PTFE and the TFE-HFP copolymer as a function of pressure. Legend: ● - increasing pressure; ▲ - decreasing pressure.

random copolymer of tetrafluoroethylene and hexafluoropropylene (TFE-HFP). Although both $C_{20}F_{42}$ and the copolymer are observed to transform to the planar zig-zag form (as evidenced by the appearance of the 625 cm^{-1} band) in the 7 to 9 kbar range, the variation of the PIR is markedly different for all three materials as shown in Fig. 3. Since it is known that the perfluoro n-alkanes have the same hexagonal rod-like packing as is observed in PTFE [16] the dissimilarity in behavior of the PIR with pressure for both $nC_{20}F_{42}$ and the polymer discounts any changes in intermolecular forces as being responsible for the contrast in PIR variation observed in these materials. However, an investigation [17] of the 285 and 395 cm^{-1} bands in the melt of $nC_{20}F_{42}$ indicated that the intensity of the 285 cm^{-1} band could be associated with order in the crystalline regions. This was further supported by a study of an irradiated sample (total dose = 7.03 x 10^8R) of PTFE. X-ray studies [18,19] have shown that radiation doses of this magnitude cause substantial chain scission with a subsequent decrease in crystallinity. It was found that this PIR was significantly larger in the unirradiated higher crystallinity material.

This has, therefore, led us to conclude that in PTFE this variation in PIR with pressure can be attributed to an increase in crystallinity. An explanation of this increase might be understood by considering the nature of the II-III phase transition. This first-order crystal-crystal phase transition involves a conformational change from a helical structure to a planar zig-zag form. All material in the crystalline regions undergoes this transformation at the phase boundary (7 to 8 kbar). With increased pressure it is possible that at some location in the amorphous region random coil conformations may begin to spontaneously convert to planar form, thereby serving as nucleation sites for phase III crystals. A second more probable mechanism involves the propagation of the planar form from the crystalline regions converted at the phase boundary (7 to 8 kbar) into the amorphous zones. Increased pressure then causes a conversion of random coil to planar form with a necessary extension of the planar form from the already established crystalline regions. This model is best described as a pressure driven crystal thickening and it is felt to be the predominant mechanism responsible for the increase in sample crystallinity in PTFE at high pressures.

The less dramatic change in PIR with pressure observed in the TFE-HFP copolymer may be indicative of the higher energy requirements associated with incorporation of perfluoromethyl groups into the transplanar orthorhombic lattice since they are significantly more prevalent on the fold surface [17] than in the crystalline regions.

In light of the proposed mechanism, the relative constancy of the PIR in $nC_{20}F_{42}$ is expected. At the phase boundary all the material is converted to the planar form and a further increase in pressure should have no effect on sample crystallinity.

REFERENCES

1. P. W. Bridgman, Proc. Amer. Acad. Arts Sci. 76, 71 (1948).
2. C. E. Weir, J. Res. Natl. Bur. Standards 46, 207 (1951).
3. C. E. Weir, J. Res. Natl. Bur. Standards 50, 95 (1953).
4. R. I. Beecroft and C. A. Swenson, J. Appl. Phys. 30, 1793 (1959).
5. G. M. Martin and R. K. Eby, J. Res. Natl. Bur. Standards 72A, 467 (1968).
6. S. Hirakawa and T. Takemura, Japan. J. Appl. Physics 8, 635 (1969).
7. C. Nakafuku, S. Taki, and T. Takemura, Polymer 14, 558 (1973).
8. C. Nakafuku and T. Takemura, Japan. J. Appl. Physics 14, 599 (1975).
9. K. Matsushice, R. Enoshita, T. Ide, N. Yamauchi, S. Taki, and T. Takemura, Japan. J. Appl. Physics, to be published.
10. R. G. Brown, J. Chem. Phys. 40, 2900 (1964).
11. C. K. Wu and M. Nicol, Chem. Phys. Lett. 21, 153 (1973).
12. M. Nicol, J. Wiget, and C. K. Wu, in Proceedings of the Intern. Raman Conference, W. Balkanski, R. Laite and S. Porto, eds. (1976), p. 504.
13. D. H. Flack, J. Polym. Sci. 10, 1799 (1972).
14. G. J. Piermarini, S. Block, J. D. Barnett, and R. A. Forman, J. Appl. Phys. 46, 2774 (1975).
15. B. A. Weinstein and G. J. Piermarini, Phys. Rev. B12, 1172 (1975).
16. E. S. Clark, private communication.
17. J. F. Rabolt, S. Block, and G. J. Piermarini, to be published.
18. M. Tutiya, Polymer J. 6, 39 (1974).
19. R. E. Florin, in High Polymers Vol. 25, L. A. Wall, ed., John Wiley and Sons, New York (1972).

H-3

PRESSURE-STRAIN RELATIONSHIPS IN THE CRYSTAL LATTICE OF POLYMERS

T. Ito

Kyoto Institute of Technology

Matsugasaki, Kyoto, Japan

INTRODUCTION

Polymer crystals consist of chain molecules. They are characterized, with regard to the cohesive forces, by the covalent bonds in the fiber-axis direction and the weak intermolecular secondary forces in the transverse directions. All physical properties of polymer crystals are affected by this important feature and are thus highly anisotropic, a fact which will be manifest in a most straightforward way in the study of compression of the crystal under hydrostatic high pressure. In support of this, Müller [1] was the first to find that the hydrocarbon crystals are only 1/10th to 1/40th as compressible in the fiber-axis direction as they are in the directions perpendicular to it.

Since then, however, little experimental evidence has been published on the compressible property of the polymer crystal under hydrostatic pressure. Nevertheless, during these three decades the structures of many polymer crystals have been revealed [2]. It is now well established that the conformation of the individual chain molecule in the polymer crystal differs according to the polymer, from the most simple planar zig-zag to complex helices with various pitches and radii. These should be reflected in the pressure-strain behavior of the crystal. This paper describes experimental results on the pressure-strain relationships in the crystal lattice of many polymers by means of x-ray diffraction.

EXPERIMENTAL

Oriented and highly crystalline polymer specimens obtained by drawing and ample annealing were used. The x-ray diffraction photographs were obtained at 293 ± 1 K with a high-pressure x-ray diffraction vessel of a piston-cylinder type which can be used up to 8 kbar

[3-4]. The sample holder is a small tapered block of beryllium metal with a drilled central hole 1 mm in diameter and 6 mm deep located in a heavy maraging steel cylinder. The cylinder has a horizontal aperture reaching the sample holder and coaxially mounted on a semicylindrical multiple-exposure camera [5] with a radius of 100 mm. A small sample shaped as a rod 0.9 mm in diameter was set in the sample holder and placed under the pressure medium in the cylinder. A portion of the medium was introduced through a pinhole into a high-pressure gauge of the strain-gauge type [6] which had been calibrated with a lever-type controlled-clearance piston gauge [7]. The incident x-ray beam impinged on the sample in a direction perpendicular to the rod-axis. Only the horizontal fraction of the whole diffraction was permitted to emerge and be recorded. The distance between diffraction lines on the film was measured visually with Klug's compensator [8] (use of the photodensitometer was found not to be very effective in our case). The strain of the crystal was obtained from the strain based on the distance of the interplanar spacing of the lattice plane (hkl). It gave the linear strain in the normal direction to that plane and was represented by the symbol ε_{hkl}. The volumetric strain was calculated from these values.

RESULTS AND DISCUSSION

Polyethylene

The diffraction lines of the equatorial planes and the basal (002) plane of polyethylene (PE) under pressure are shown in Fig. 1. It can be seen that the diffraction lines shift toward higher angles

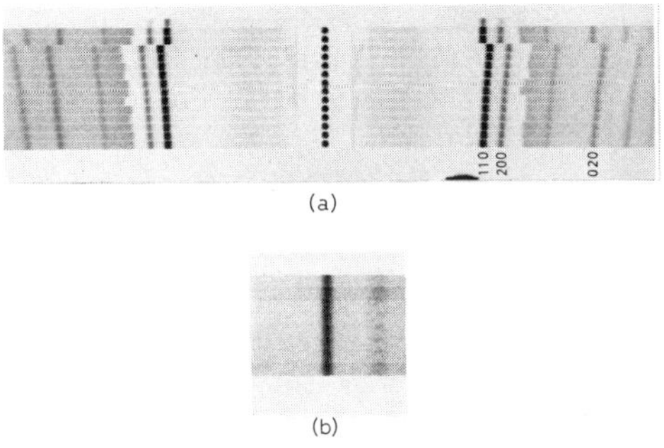

Fig. 1. The equatorial (a) and the basal 002 (b) diffraction lines of drawn and annealed (393 K) high-density polyethylene. Under pressures of (kg/cm^2): from the bottom in (a)—1, 500, 1000, 1500, 2000, 2500, 3000, 4000, 5000, 6000, 7000, 8000, 1, and 167 (residual pressure); from the bottom in (b)—1, 1000, 2000, 3000, 4000, 5000, 1, and 76 (residual pressure). Pressure medium: water.

with increasing pressure, but their widths remain constant throughout the pressure rise. Upon release of the pressure they recovered completely, within limits of the experimental error, to their original positions which had been registered under normal pressure before the pressure was applied. This fact found in polyethylene crystals that the hydrostatic pressure (8 kbar) has little effect on leaving any unrecovered strain was also commonly found in the other polymer crystals studied so far. From the relationships of the interplanar distances deduced from the equatorial lines in Fig. 1, it is further concluded that the crystal of polyethylene strains to conserve its orthorhombic structure under the hydrostatic pressure. Figure 2 gives the strain ε_{hkl} vs. pressure relationships for the (200), (020), (110), (011) and (002) planes. Since polyethylene has an orthorhombic unit cell with the c-axis as the fiber-axis, the strains due to the (hk0) planes give the transverse strains, while those due to the (00l) planes refer to the strain in the fiber-axis direction which represents deformation of the crystal along the chain molecule. In Fig. 2, ε_{hk0}'s are seen to amount to 4.0 to 4.5% under the highest pressure of 8 kbar, with the strain in the a-axis direction ε_{200} being always greater than that in the b-axis direction ε_{020}. In the fiber-axis direction the crystal of polyethylene is essentially incompressible (ε_{002} of 0.07% at 5 kbar) and can well be compared with the incompressible behavior of the diamond (111) plane (see Fig. 5). The strains in the directions of the orthorhombic principal axes and the volumetric strain can be represented by the following conventional polynomial relations as a function of the pressure (bar), which are applicable only within the pressure range below 7850 bar (8000 kg/cm^2).

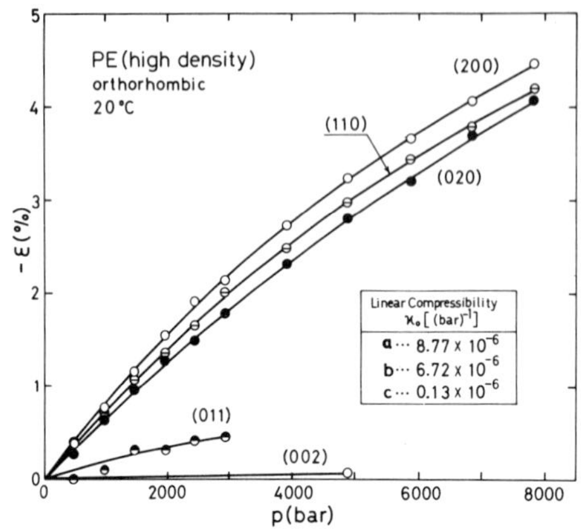

Fig. 2. Pressure-strain curve for the crystal of high-density polyethylene.

$$-\varepsilon_{200} = 8.77 \times 10^{-6}p - 0.54 \times 10^{-9}p^2 + 0.018 \times 10^{-12}p^3 \quad (1)$$

$$-\varepsilon_{020} = 6.72 \times 10^{-6}p - 0.24 \times 10^{-9}p^2 + 0.005 \times 10^{-12}p^3 \quad (2)$$

$$-\varepsilon_{002} = 0.13 \times 10^{-6}p \quad (3)$$

$$-\Delta V/V_o = 15.6 \times 10^{-6}p - 0.84 \times 10^{-9}p^2 + 0.030 \times 10^{-12}p^3 \quad (4)$$

In equation (4), V_o is the volume of the crystal under normal pressure at 293 K and has a value of 0.9994 cm^3/g for polyethylene [9].

The volumetric strain is shown in Fig. 3 where the result recently obtained by Hikosaka, Seto, and Minomura [10] by the use of the x-ray diffraction cell of the diamond-anvil type and Pastine's theoretical prediction [11] are also given. Grüneisen

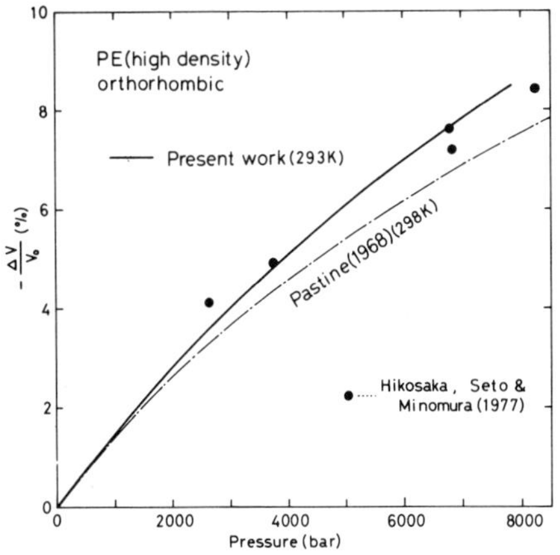

Fig. 3. Experimental and theoretical pressure-volumetric strain curves for the crystal of high-density polyethylene.

parameter calculated from the volumetric strain using equation (5) [11] was found to be 3.0 for polyethylene crystal under normal pressure.

$$\gamma = -\frac{1}{2} - \frac{1}{2}\frac{V(\partial^2 p/\partial V^2)_T}{(\partial p/\partial V)_T} \qquad (5)$$

Polymers Having Helical Skeleton Conformation

It is apparent from the preceding section that the crystal of polyethylene composed of the fully extended carbon-carbon zig-zag chains is essentially incompressible in the fiber-axis direction. How much compressibility is there then in the same direction as the polymer crystals which are composed of chains having helical skeleton conformations such as shown [12] in Fig. 4? This problem is particularly interesting because Young's moduli for these polymer crystals along the fiber-axis, denoted by E_l, have been shown both theoretically [13] and experimentally [14] to be sensitively dependent upon the detail of conformation of the helix. They are reported to vary over a wide range, from 235 x 10^4 bar for polyethylene to 4.0 x 10^4 bar for poly (vinyl tert-butyl ether). Theoretically [15], the hydrostatic strain $\varepsilon_{l,hydro}$ is given by the following equation:

$$\varepsilon_{l,hydro} = -\left(\frac{1-2\nu'}{E_l}\right)p \qquad (6)$$

where ν' denotes Poisson's ratio.

The results produced by the hydrostatic pressure are given in Fig. 5 where the diffraction lines by the planes whose normals are parallel or nearly parallel to the fiber-axis are shown. The polymers selected for our purpose and the planes observed are: $(\bar{1}13)$ of isotactic polypropylene (3_1-helix, E_l=34x10^4 bar), $(10\bar{2})$ of isotactic poly-1-butene (3_1-helix, E_l=24.5x10^4 bar), (007) of isotactic poly(4-methyl-1-pentene) (7_2-helix, E_l=6.6x10^4 bar), (009) of polyoxymethylene (9_5-helix, E_l=53x10^4 bar) and (0013), (0015) of polytetrafluoroethylene (13_6- or 15_7-helix, E_l =153x10^4 bar) [14].

Unexpectedly, as seen from Fig. 5, it was found that the hydrostatic pressures can only induce very small strains along the fiber-axis of the polymer crystals which consist of helical chains. The strains deduced from Fig. 5 are much smaller than the strains

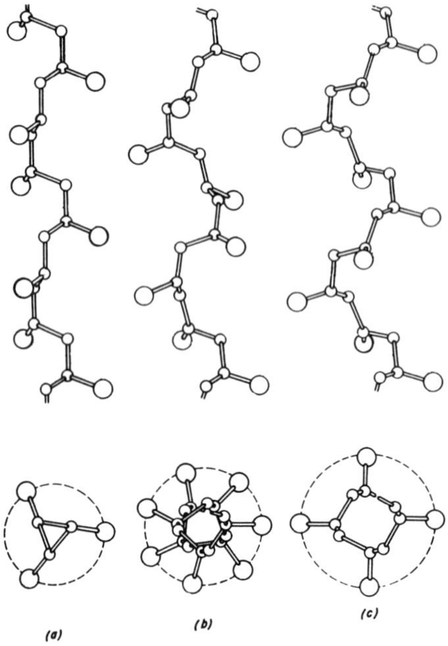

Fig. 4. Helical conformations found in crystals of isotactic hydrocarbon polymers. After Natta and Corradini [12].

calculated from (6) with Poisson's ratio of, say, 0.33. Miyaji also reported [16] the same result for polyoxymethylene (negligibly small ε_{009}'s up to 20 kbar). The only exception to date is poly(4-methyl-1-pentene) which exhibits an ε_{007} of 0.0081 at 2940 bar. This strain amounts to about half of the value calculated from (6) assuming $\nu' = 0.33$, but it is explainable if it is assumed that $\nu' = 0.4$.

For reference purposes, the results for the α-crystal of poly(ethylene oxybenzoate) will be mentioned. In this case the chains in the crystal are not helical but assume a large zig-zag conformation [2] with a much longer arm (3.1 Å) compared with that of polyethylene (0.45 Å) and this is the reason for the low value of E_l (5.9×10^4 bar) [17] of this polymer. Figure 6 gives the strain data where the results by the uniaxial extension [17] are shown for comparison. As seen, the hydrostatic strains in the fiber-axis direction ε_{006}'s are again very small making a typical contrast to the large extensions by the uniaxial stretching.

At present there is no explanation for the fact that polymer crystals consisting of helical chains and/or having low Young's moduli are essentially, or at least unexplainably based on (6), incompressible in the fiber-axis direction.

Effect of Hydrogen Bond

It is well known that the crystal structures of nylons are constructed in the same manner that the hydrogen-bonded sheets of chain molecules are stacked by van der Waals attractions, the conformation of the chains assuming a more or less extended planar zig-zag structure. It was found that the α-crystal of nylon 6 is about four times more compressible in the direction of the stacking (ε_{002}) than in the direction of the hydrogen bonds (ε_{200}), as shown by the following relations (p in bar, $p \leqq 4900$).

$$-\varepsilon_{002} = 10.5 \times 10^{-6} p - 0.68 \times 10^{-9} p^2 \qquad (7)$$

$$-\varepsilon_{200} = 2.6 \times 10^{-6} p - 0.036 \times 10^{-9} p^2 \qquad (8)$$

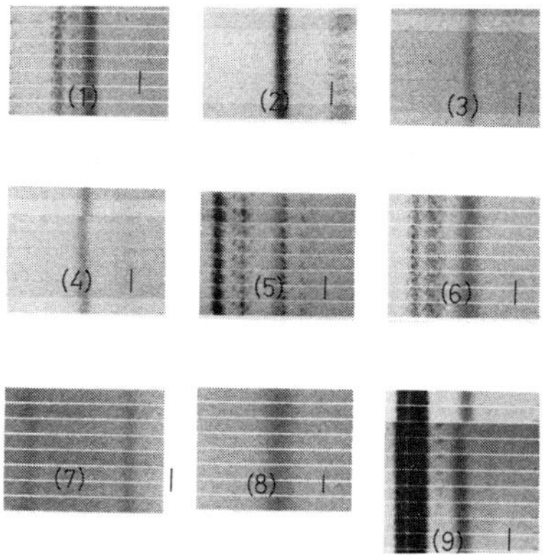

Fig. 5. Pressure-strain curves along the fiber-axis of various polymer crystals. Interval of pressure is 1000 kg/cm^2. The lines in the bottom and the next to the top sections were recorded under normal pressure. Legend: (1) diamond; (2) high-density polyethylene; (3) polytetrafluoroethylene (283 K); (4) polytetrafluoroethylene (297 K); (5) polyoxymethylene; (6) it. polypropylene; (7) it. poly-1-butene; (8) poly(ethylene oxybenzoate)(α); and (9) it. poly(4-methyl-1-pentene).

The same trend in results was obtained with the γ-crystal [18] of nylon 6. In this case the doublet reflection under the normal pressure due to the (001) and (200) planes, whose normals are, respectively, 30 and 90° to the direction of the hydrogen bonds, became well separated by pressure because of the marked anisotropy of their compressibilities.

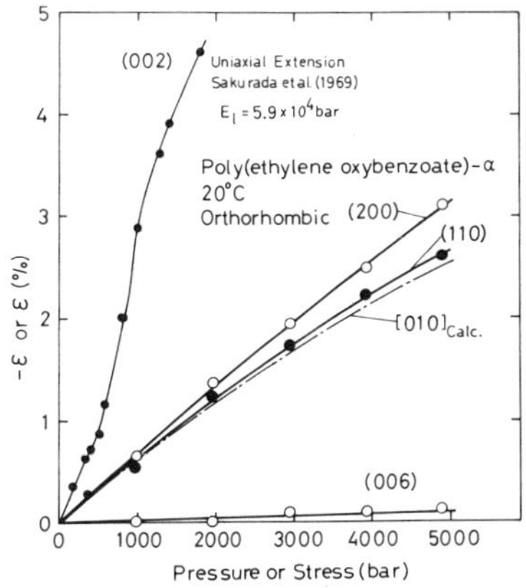

Fig. 6. Pressure-strain and stress-strain curves for the crystal of poly (ethylene oxybenzoate)(α). Pressure medium: water.

Pressure-Induced Phase Transition

Not many examples of pressure-induced phase transition at room temperature are known relative to polymer crystals. Those by polytetrafluoroethylene are well studied. Nakafuku and Takemura recently determined the crystal structure of the high-pressure phase of this polymer by x-ray method [19].

Poly(ethylene oxide)(I) provides another example. This is shown in Fig. 7, where it is seen that the compressibilities of the equatorial (110) and (120) planes change discontinuously at the transition pressure of 4500 bar (at 293 K), both toward much higher compressibilities. These suggest that the transition is of the second order. However, the behavior of the ($\bar{2}$07) plane, whose normal is 7.8° to the fiber-axis, is strange and its spacing begins to increase after the transition. Moreover, the weak triplet [20] of the 007, $\bar{1}$37, and $\bar{3}$27 reflections splits into two reflections at the transition pressure with discontinuous changes of spacings, one giving a smaller and the other a greater distance with reference to the value just below the transition pressure. This inevitably

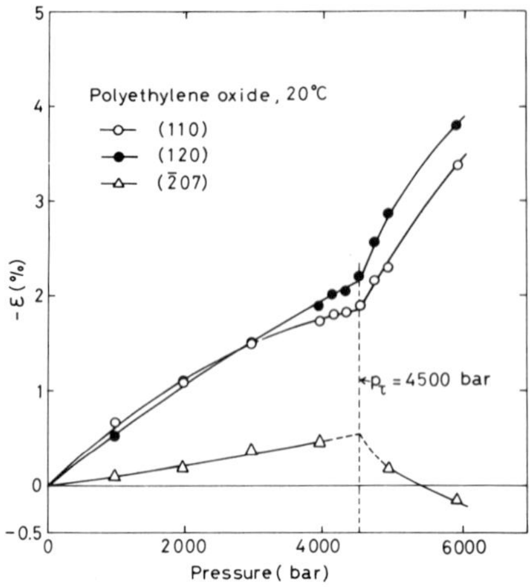

Fig. 7. Pressure-strain curves for the crystal of poly(ethylene oxide)(I), indicating the pressure-induced phase transition at 4500 bar. Pressure medium: n-pentane.

defines this transition to be of the first order. These transitions of poly(ethylene oxide)(I) were shown to be reversible upon release of the pressure.

Compressibility of Polymer Crystals

The coefficient of the first term on the right-hand side of (4) gives the initial compressibility β_o at the normal pressure, as defined by

$$\beta_o = -\frac{1}{V_o}\left(\frac{\partial V}{\partial p}\right)_T \qquad (9)$$

Table I lists data from this study of β_o for various polymer crystals, where the density ρ_o and packing density k [21] of the crystal under normal pressure are also cited. Data for some typical organic molecular crystals were also obtained from our studies. Among the polymer crystals studied to date, the crystal of poly (4-methyl-1-pentene) is the most compressible (β_o =36.7x10^{-6} bar^{-1}) and the carbon fiber derived from polyacrylonitrile fiber having essentially the same structure as graphite is the most incompressible (β_o=3.0x10^{-6} bar^{-1}). (With the carbon fiber as the exception, the crystal of polyoxymethylene may be said to be the most incompressible of the common polymer crystals.)

The compressibility of the polymer crystals decreases with increasing density, as shown in Fig. 8. This is probably reasonable because an increase in density reflects a decrease of vacant space in the crystal when comparing crystals composed of similar atoms. Packing density should give a better parameter in this sense than density. The results of β_o vs. k plots were found to be expressed by (10), with a comment that deviations in the plots are not much improved compared with Fig. 8.

$$\beta_o\ (10^{-6}\ \text{bar}^{-1}) = 7.65(1 - k) + 177\ (1 - k)^2 \qquad (10)$$

Based on β_o vs. k plots, the crystal of polytetrafluoroethylene is extraordinarily compressible.

It should be emphasized that compressibilities of polymer crystals are obtained predominantly by the two-dimensional contractions in directions perpendicular to the fiber-axis. It is interesting to note that in spite of having fewer covalent bonds organic low molecular weight crystals give β_o's that are indistinguishable

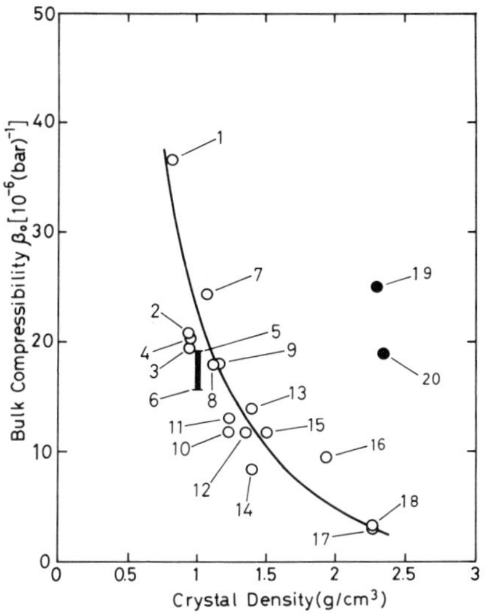

Fig. 8. Plots of the initial bulk compressibility of various polymer crystals against density. Legend: (1) it. poly(4-methyl-1-pentene); (2) it. polypropylene; (3) n-heptacosane; (4) it. poly-1-butene; (5) polyethylene (low density); (6) polyethylene (high density); (7) adamantane; (8) poly(tetramethylene oxide)(I); (9) nylon 6 (γ); (10) poly(ethylene oxide)(I); (11) nylon 6 (α); (12) hexamethylenetetramine; (13) poly(ethylene p-oxybenzoate) (α); (14) pentaerythritol; (15) polyoxymethylene; (16) poly(vinylidene fluoride)(II); (17) carbon fiber; (18) graphite; (19) polytetrafluoroethylene (297 K); and (20) polytetrafluoroethylene (II).

from, or even smaller than, the β_o's of the polymer crystals in both the β_o vs. ρ_o and β_o vs. k_o plots. For this, the large distortions particularly insisted upon for the polymer crystals [22] may give the reason.

REFERENCES

1. A. Müller, Proc. Roy. Soc. A178, 227 (1941).
2. H. Tadokoro, Structure of Polymers (in Japanese), Kagaku Doujin Press, Kyoto, Japan (1976).
3. S. S. Kabalkina and L. F. Vereshchagin, Dokl. Akad. Nauk. SSSR 143, 818 (1962).
4. T. Ito and H. Marui, Polymer J. 2, 768 (1971).
5. T. Ito, Rev. Sci. Instr. 45, 1560 (1974).
6. K. Yasunami, Nippon Kikai-gakkaishi 67, 980 (1964).
7. K. Yasunami, Metrologia 4, 168 (1968).
8. H. P. Klug and L. E. Alexander, X-ray Diffraction Procedures, John Wiley and Sons, New York (1954), p. 322.
9. P. R. Swan, J. Polymer Sci. 56, 403 (1962).
10. M. Hikosaka, T. Seto, and S. Minomura, Polymer Preprints, Japan 26 (2), 421 (1977).
11. D. J. Pastine, J. Chem. Phys. 49, 3012 (1968).
12. G. Natta and P. Corradini, Nuovo Cimento, Suppl. 15, 9 (1960).
13. L. R. G. Treloar, Polymer 1, 95, 279, 290 (1960); also T. Shimanouchi, M. Asahina, and S. Enomoto, J. Polymer Sci. 59, 93 (1962); M. Asahina and S. Enomoto, J. Polymer Sci. 59, 101 (1962); S. Enomoto and M. Asahina, J. Polymer Sci. 59, 113 (1962).

Table I. Compressibility of Polymer Crystals and Some Organic Low Molecular Weight Compounds under Normal Pressure and 293 K

Substance	ρ_{oc}, g/cm^3	k, Packing density	β_o, 10^{-6} bar^{-1}
it. Poly(4-methyl-1-pentene)	0.814, 0.828	0.57	36.7
it. Polypropylene	0.938	0.66	20.7
n-Heptacosane	0.9425	(0.66)	19.3
it. Poly-1-butene (I)	0.951	0.66	20.2
Polyethylene (low density)	0.997	0.7	19.2
Polyethylene (high density)	1.008	0.70	15.6
Adamantane	1.07	0.67	24.3
Poly(tetramethylene oxide)(I)	1.112	0.70	17.8
Nylon 6 (γ)	1.16	0.707	17.9
Poly(ethylene oxide)(I)	1.228	0.72	11.7
Nylon 6 (α)	1.23	0.75	13.1
Hexamethylenetetramine	1.345	0.72	11.7
Poly(ethylene p-oxybenzoate)(α)	1.391	----	13.8
Pentaerythritol	1.396	0.7	8.3
Polyoxymethylene	1.50	0.78	11.7
Poly(vinylidene fluoride)(II)	1.925	----	9.47
Carbon Fiber	2.27	0.887	3.0
Graphite	2.27	0.887	3.2
Polytetrafluoroethylene (297 K)	2.302	0.78	25.0
Polytetrafluoroethylene (II) (283 K)	2.347	0.80	18.9

14. I. Sakurada, T. Ito, and K. Nakamae, J. Polymer Sci. C15, 75 (1966); also I. Sakurada and K. Kaji, J. Polymer Sci. C31, 57 (1970).
15. R. F. S. Hearmon, An Introduction to Applied Anisotropic Elasticity, Oxford University Press, London (1961).
16. H. Miyaji, J. Phys. Soc. Japan 39, 1346 (1975).
17. I. Sakurada, K. Nakamae, K. Kaji, and S. Wadano, Kobunshi Kagaku 26, 561 (1969).
18. H. Arimoto, M. Ishibashi, M. Hirai, and Y. Chatani, J. Polymer Sci., Part A, 3, 317 (1965).
19. C. Nakafuku and T. Takemura, Japan. J. Appl. Phys. 14, 599 (1975).
20. Y. Takahashi and H. Tadokoro, Macromolecules 6, 672 (1973).
21. B. Wunderlich, Macromolecular Physics, Vol. 1, Academic Press, New York (1973).
22. R. Hosemann, Polymer 3, 349 (1962); also Polymer 4, 199 (1963).

H-4

HIGH PRESSURE X-RAY STUDIES OF POLYETHYLENE

B. A. Newman, T. P. Sham, and K. D. Pae

Rutgers University
New Brunswick, New Jersey USA

INTRODUCTION

High-pressure x-ray diffraction and optical studies of polymers using the gasketed diamond-anvil cell have been carried out by several investigators [1-5]. Crystallization and crystal phase transitions can be observed directly using these methods. However, there are two serious problems involved in the utilization of the gasketed diamond-anvil cell for such studies. These are variation of pressure with temperature and the accurate measurement of pressure.

If the pressure remains constant at different temperatures, then a simple method of pressure determination is possible by measuring the melting point at the high pressure of an internal standard. Such a method was used by Jackson, Hsu, and Brusch [5] in optical studies of the crystallization of polyethylene at high pressures. Literature values for the melting points of polyethylene at different pressures were used with the experimental measurement of the melting point to establish the pressure in the diamond cell. Unless the assumption of pressure constancy is involved, this method gives an accurate value for the pressure at the melting temperature only.

In high-pressure crystallization studies it is also usually assumed that the pressure is constant. These studies are assumed to be analogous to those usually carried out at atmospheric pressure, viz., isothermal crystallization at constant pressure and constant supercooling. These conditions will not necessarily be true in the gasketed diamond-anvil cell.

The second and related drawback to the diamond-cell method is the accurate determination of the pressure. The well-known method of incorporating sodium chloride as an internal standard with the sample for x-ray studies is not accurate at pressures less than 10 kbar. Pressure crystallization of polyethylene is often carried out at approximately 5 kbars. At this pressure the error in pressure determination using the sodium chloride technique exceeds 25%. The ruby fluorescence method recently suggested by Piermarini and Block [6] becomes less accurate at elevated temperatures and cannot be used at temperatures greater than 200°C. This is in the temperature range at which pressure crystallization of polyethylene is frequently carried out.

For these reasons an alternative method of measurement of pressure at elevated temperatures has been developed utilizing hexamethylene tetramine (HMT) as an internal standard for x-ray diffraction studies. This method has been used to test the assumption of pressure constancy with temperature in the diamond cell and to investigate the crystallization of polyethylene at high pressures.

EXPERIMENTAL METHOD

The use of sodium chloride as an internal pressure standard is limited, since at low pressures only very small d-spacing changes occur. Molecular crystals with higher compressibilities are more suitable pressure standards at low pressures. Among the many molecular crystals considered, HMT was considered the most appropriate for the following reasons:

1. The highly symmetrical lattice and molecular structure aided and simplified the calculation of the equation of state.
2. There are sufficient experimental data reported for HMT with which the equation of state can be established and evaluated.
3. No polymorphic phase transitions at high pressures and high temperatures have been reported.
4. The melting point (533 K) is higher than that of most polymeric solids.
5. HMT has a cubic lattice and the specific volume can be calculated from the measurement of a single d-spacing.

The pressure-volume relationship at 295 K for HMT has been investigated by Bridgman [7] who determined the specific volume at different pressures up to 40 kbars. This data was supplemented by additional measurements made in the region of interest (0 to 7 kbars). These data are shown in Table I.

These data are shown in graphical form in Fig. 1 combined with the data of Bridgman and were used in the calculation of the equation of state for HMT. The equation of state derived was based on the functional expression proposed by Mie and extended by Grüneisen [8]

$$P = \frac{-1}{V}\,\frac{\partial U}{\partial V} = -\,\frac{1}{V}\,\frac{\partial (U_L + U_T + U_Z)}{\partial V} \qquad (1)$$

Table I. Pressure-Volume Data Obtained at 295 K

Pressure, dynes/cm^2	Volume Change, %
9.806×10^5	0
1.718×10^9	1.54
2.120×10^9	2.05
2.885×10^9	2.64
3.864×10^9	3.46
4.835×10^9	4.25
5.770×10^9	4.93
6.862×10^9	5.67

where U is the total free energy, U_L is the lattice potential energy (assumed to be a function of volume), U_T is the thermal vibrational free energy, and V_Z is the zero-point energy.

The lattice potential energy U_L is the sum of the intermolecular attractive and repulsive energies for all molecules in the lattice, and we assume this depends only on the intermolecular distances between atoms in neighboring molecules. If two atoms in neighboring atoms are separated by a distance r, we assume that the potential energy can be represented by

$$U_r = \frac{-K}{r^6} + B\ \exp(-br) \qquad (2)$$

where K, B, and b are constants which remain to be evaluated. Only atoms in neighboring atoms need be considered since the contributions of other atom pairs are negligible. The crystal structure of HMT has been determined by Cruickshank and Becka [9], and from the structure proposed it can be seen that there are eight CH_2-N nearest neighbor pairs and six CH_2-CH_2 second-nearest neighbor interactions. Different constants for K, B, and b are required for the different interactions.

The thermal vibrational free energy is attributed to the vibrations of atoms and molecules about their equilibrium positions. The following assumptions were made: (1) The contribution of intramolecular vibrations could be neglected. (2) The Debye approximation could be used and the spectral distribution of the intermolecular vibrational frequencies characterized by the Debye temperature θ_D. (The Debye temperature of HMT has been calculated by Haussuhl [10] from the velocities of acoustic waves.) (3) The quasi-harmonic model was used, viz., that the vibrations remain harmonic, but that the Debye temperature is a function of volume. (4) Grüneisen's assumption was used and changes in the characteristic temperature θ_i associated with a frequency ω_i are the same for all θ_i. We write

$$\gamma_i = -\left(\frac{\partial \ln \omega_i}{\partial \ln V}\right)_T = \gamma \qquad (3)$$

where γ is the Grüneisen parameter and is a function of volume and ω_i is the vibrational frequency of the i-th mode. (5) In order to determine the dependence of γ on pressure, Slater's equation was used:

$$\gamma = \frac{-V}{2}\left(\frac{\partial^2 P_p}{\partial V^2}\right) / \left(\frac{\partial P_p}{\partial V}\right) - \frac{2}{3} \qquad (4)$$

where P_p is the potential pressure given by

$$P_p = \frac{-1}{V_o}\left(\frac{dU_L}{\partial V}\right) \qquad (5)$$

With the same assumptions, the zero-point energy U_Z which is $1/2\ \Sigma h\omega_i$ becomes

$$U_Z = 2NZKT\left(\frac{9}{8}\right)\frac{\theta_D}{T} \qquad (6)$$

In order to determine the equation of state for HMT it is necessary to determine seven parameters: K, B, and b for both CH_2-N and CH_2-CH_2 interactions, and the Grüneisen constant γ. Muller [12] determined the attractive and repulsive energies for the CH_2-CH_2 interactions occurring in the linear paraffins, and the constants K, B, and b can be determined from this data. Moreover, the unit cell parameters determined by Becka and Cruickshank at 29.8 K and 298 K can be used, so that the remaining four parameters are not independent. In fact, only two independent parameters remain. These parameters were determined by using the functional form of the equation of state derived and using the least-squares method to best fit this function to the experimental PV data obtained. From this optimization best-fit parameters for the remaining unknown parameters were obtained.

From the equation of state thus derived, both the volume compressibility and thermal expansion coefficient could be calculated directly and were found to be in good agreement with experimental measurements. Furthermore, by measuring the melting points of polyethylene at different pressures and comparing with literature values, the accuracy of this method could be well assessed. It was observed that pressures could be measured to an accuracy of 0.5 kbar at temperatures up to 400°C.

RESULTS

Pressure Calibration

A brass gasket 0.953 cm in diameter and 0.08 cm thick with an axial aperture of 0.08 cm in diameter was used. The gasket was used

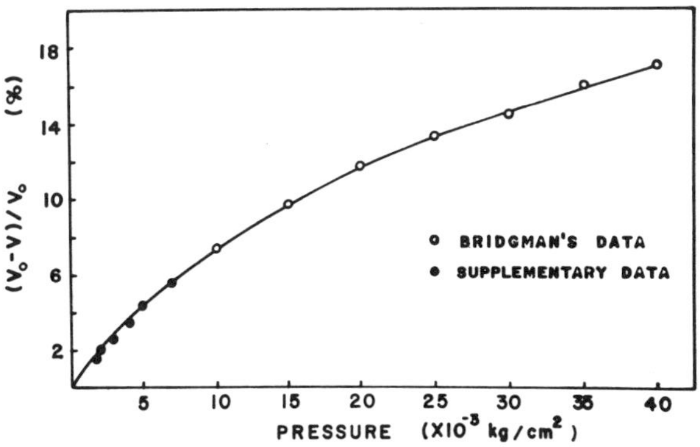

Fig. 1. Pressure-volume relationship of HMT at 295 K.

inside a resistance heater, and an iron-constantan thermocouple together with a current-adjusting silicon rectifier controller was used to measure temperature within ± 0.1°C. Silicone oil was used as a pressure-transmitting fluid. The pressure was measured from changes in the d-spacings of an HMT sample using x-ray diffraction and using the derived equation of state.

A Rigaku-Denki rotating anode 6 kW generator was used with a molybdenum target. Diffracted x-rays were filtered using a zirconium foil to obtain K_{α_1} radiation. The diffraction patterns were recorded on flat film, with a specimen to film distance approximately 8 cm. This distance was determined accurately from the known spacings of HMT at room pressure.

The sample was pressurized to an initial pressure of 4.1 kbars, and the temperature raised in steps. After every temperature increment, the pressure was determined. The results are plotted in Fig. 2. It is clear that the pressure in the diamond cell does not remain constant with temperature, but increases steadily in a nonlinear fashion from 4.1 to 10.6 kbars as the temperature is raised from room temperature to 300°C.

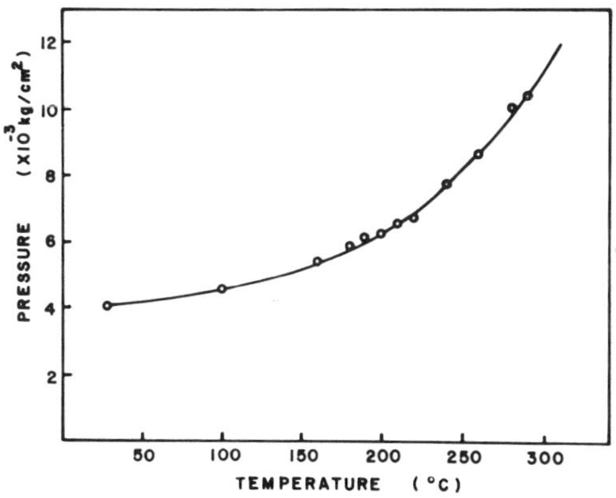

Fig. 2. Variation of pressure with temperature in the diamond-anvil cell.

Crystallization of Polyethylene at High Pressures

The experimental conditions were as described in the above experiment. Linear high density polyethylene (Martex 6009) was used in powder form. A small sample mixed with HMT was molded and annealed at 100°C for 3 hrs at atmospheric pressure to increase the extent of crystallinity.

A sample was pressurized to give an initial pressure of 4.1 kbars at room temperature.

X-ray diffraction revealed that the usual orthorhombic crystal phase for polyethylene was present. The temperature was raised in steps, an x-ray diffraction photograph being taken at each step, until melting took place at 300°C. During this temperature increase the pressure was observed to increase in the fashion indicated by Fig. 2 and, moreover, the orthorhombic crystal structure remained unchanged at all temperatures and pressures until melting occurred. The sample was then isothermally crystallized at 270°C. X-ray diffraction revealed that crystallization into a new form, a hexagonal phase, had taken place. The temperature was then increased and melting was observed to occur at 276°C. Finally, the sample was cooled down slowly to room temperature. At room temperature the sample was observed in the usual orthorhombic phase.

DISCUSSION AND CONCLUSIONS

The results quite clearly show that the pressure in the diamond cell can increase quite significantly with temperature. The initial pressure generation in the gasket is achieved by the reduction in volume of the aperture. As the temperature is increased, the different thermal expansion coefficients of sample, pressure medium, diamonds, and retaining ring combine to give the pressure dependence observed.

The diamond cell can be used to study the crystallization of polymers at high pressures, but it should be noted that changes in pressure are equivalent to changes in supercooling, and the crystallization kinetics and polymer crystal morphology are known to be markedly dependent on supercooling. Therefore, conclusions drawn from crystallization studies carried out with no direct pressure measurement should be treated with caution.

The results show that crystallization of polyethylene carried out at high pressures can give rise to a new crystal form: a hexagonal phase. This result and the relevance of this observation to the question of crystal morphology are discussed in a separate publication. Here, the unusual crystallization conditions imposed by the use of the diamond cell are discussed.

After melting at 300°C, the sample was isothermally crystallized at 270°C. The observed pressure (measured at the completion of crystallization) was 9.6 kbars. Subsequent increase in temperature led to melting at 276°C. Since we have already shown that increasing temperature leads to increasing pressure and an increased melting point, it is clear that the completion of crystallization into the new hexagonal phase occurred only 5°C below the melting point of the new phase. It is apparent that such a small supercooling would be insufficient for solidification to occur in a practicable time period. Therefore we must assume that much greater supercoolings were present at the onset of crystallization. This would be true if the pressure at the beginning of crystallization was high and decreased slowly during the solidification process. This would be a reasonable assumption since during crystallization the specific volume of the polyethylene changes gradually, whereas the total volume available in the diamond cell remains constant at a given temperature.

ACKNOWLEDGMENT

The authors gratefully acknowledge the financial assistance by the Office of Naval Research under Contract N00014-75-C-0540.

REFERENCES

1. H. D. Flack, J. Pol. Sci. (A-2) 10, 1799 (1972).
2. D. Klemper and F. E. Karusz, ACS Pol. Preprints 13(2), 976 (1972).
3. T. P. Sham, B. A. Newman, and K. D. Pae, J. of Mat. Sci. 12, 771 (1977).
4. D. C. Bassett, S. Block, and A. J. Piermarini, J. App. Phys. 45, 4146 (1974).
5. J. F. Jackson, T. S. Hsu, and J. W. Brusch, Poly. Letters 10, 207 (1972).
6. R. A. Forman, A. J. Piermarini, J. D. Barnett, and S. Block, Science 176, 284 (1972).
7. P. W. Bridgman, Proc. Amer. Acad. Arts Sci. 76(3), 55 (1939).
8. E. Grüneisen, Handbuch der physik, Vol. 10, Springer Verlag, Berlin, Germany (1926), p. 1.
9. L. N. Becka and D. W. J. Cruickshank, Proc. Roy. Soc. London A 273, 435 (1963).
10. S. Haussuhl, Acta Cryst. 11, 58 (1958).
11. J. C. Slater, Introduction to Chemical Physics, McGraw Hill Book Company, New York (1939).
12. A. Muller, Proc. Roy. Soc. London A154, 624 (1936).

H-5

EFFECT OF ULTRAHIGH PRESSURE ON SOME QUASI-ONE-DIMENSIONAL CONDUCTORS

A. Onodera

Osaka University
Osaka, Japan

I. Shirotani and Y. Hara

University of Tokyo
Tokyo, Japan

and

H. Anzai

Electrotechnical Laboratory, Mukodai
Tokyo, Japan

INTRODUCTION

The electrical behavior of quasi-one-dimensional conductors at high pressure has been of considerable interest in recent years. Square-planar complexes of transition metals [1-7] and salts of 7,7,8,8-tetracyanoquinodimethane (TCNQ) anion radical [8-17] have been investigated most extensively. Studies have also been carried out on phthalocyanines [18] and polymeric sulfur nitride, $(SN)_x$ [19].

The characteristic feature of the square-planar complexes is that the planar molecules are stacked in columns, with the metal atoms aligned along the columnar axis. The central metal is surrounded by four ligands. The weak bonding between the central metals in the columns can be enhanced considerably by the application of pressure.

Similarly, the TCNQ molecules in the crystals of TCNQ anion radical salts are stacked parallel in columns. The interplanar spacings within the columns are shorter than those calculated from the usual van der Waals distances. The enhancement of overlapping of π electron clouds can be expected under pressure.

In this paper we will present the high-pressure studies on two classes of quasi-one-dimensional conductors: metal glyoximes, and TCNQ anion radical salts with tetrathiofulvalene (TTF) and its derivatives. At ambient pressure the resistivities of metal glyoximes are very high, corresponding to those of insulators. On the other hand, TTF-TCNQ is one of the most conductive materials among organic molecular solids.

The electrical resistivities of polycrystalline samples were measured as a function of pressure, and also as a function of temperature. The effect of substituted components on the electrical behavior under pressure will be discussed.

EXPERIMENTAL

Materials

Figure 1 shows the molecular structures of the platinum glyoximes while Fig. 2 shows the molecular structures of the radical salts studied in this work.

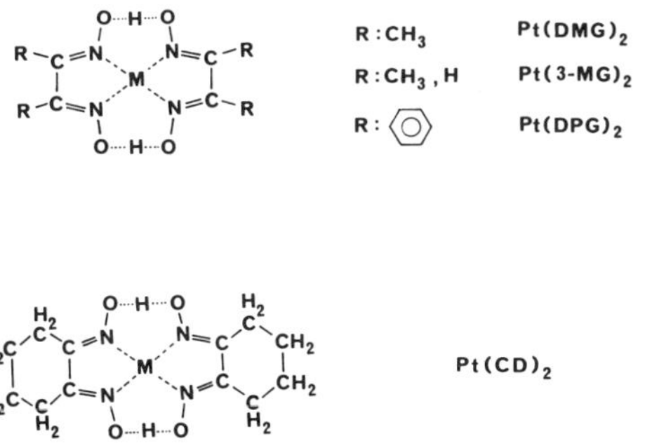

Fig. 1. Molecular structures of Pt glyoximes.

Bis(dimethylglyoximato) Pt(II) ($Pt(DMG)_2$), provided by Ohashi (University of Tokyo), was purified by means of sublimation in a vacuum. Bis(3-methylglyoximato)Pt(II)($Pt(3\text{-}MG)_2$), bis(diphenylglyoximato)Pt(II) ($Pt(DPG)_2$), and bis(1,2-cyclohexanedionedioximato) Pt(II) ($Pt(CD)_2$) were prepared from an aqueous solution of potassium tetrachloroplatinate (II) and an alcoholic solution of 3-methylglyoxime, diphenylglyoxime, and 1,2-cyclohexanedionedioxime, respectively, with the addition of an aqueous solution of sodium acetate.

TTF, dimethyltetrathiofulvalene (DMTTF) and tetramethyltetrathiofulvalene (TMTTF) were synthesized by the coupling of 1,3-dithiolium sulfate, 4-methyl-1,3-dithiolium perchlorate, and 4,5-dimethyl-1,3-dithiolium perchlorate, respectively. The TCNQ was purified by recrystallization in acetonitrile from the commercial grade reagent. The complexes composed of TCNQ and TTF derivatives were prepared by mixing both solutions of TCNQ and TTF derivatives.

The samples were studied in the form of powder pressure-fused into flat platelets.

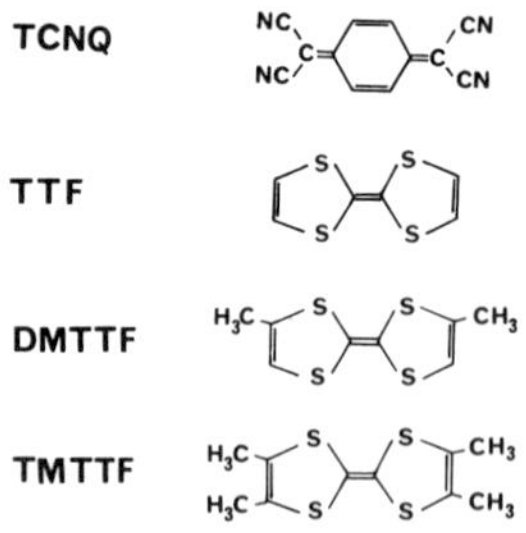

Fig. 2. Molecular structures of TCNQ and TTF derivatives.

High-Pressure Equipment

Pressures were generated in a split-sphere type apparatus [20]. The apparatus was composed of a cube, sphere, and cylinder, split up and assembled in a layered arrangement. A sphere made of hardened steel (H_{Rc} = 64) was equally divided into six segments each with a square face at the front. Three of the six segments were placed in a hemispherical space on the upper part of a cylinder. Similarly, the second assemblage was constructed using the remaining three segments. The two assemblages were set opposed and joined so that a cubic cavity whose < 111 > direction was parallel to the axis of cylinders was formed in the center of the joint.

An assemblage of eight cubic anvils made of cemented tungsten carbide was placed in the cubic cavity. A corner of each cubic anvil was truncated to form a triangular face. When the cubic anvils were put together, an octahedral hollow space was formed in the center of the assemblage. An octahedron made of semi-sintered magnesium oxide [21] was used as a pressure-transmitting medium.

Crushable spacers made of cardboard were sandwiched in the gaps among the eight cubic anvils. When the two cylinders containing a split sphere and cubes were compressed in a uniaxial press, the thickness of the spacers decreased and the high pressure was generated in the octahedral chamber.

The samples to be studied were enclosed in the center of semi-sintered MgO octahedron. Two electrical leads were taken out of the octahedron and were in contact with the top surfaces of two of the eight cubic anvils. Electrical insulation between the anvils was provided by cardboard spacers and also by thin sheets of mica.

RESULTS AND DISCUSSION

Metal Glyoximes

$Pt(DMG)_2$ crystallizes in the orthorhombic structure. Each Pt atom is coordinated to four nitrogen atoms of two dimethyl-glyoxime anions. Planar molecules of the complex are arranged parallel, but staggered, to each other. The distances of adjacent Pt atoms is 3.23 Å, suggesting a weak metal-metal bonding. The resistivity of polycrystalline $Pt(DMG)_2$ at ambient pressure is greater than 10^{15} Ωcm.

In Fig. 3 the solid line shows the change of the electrical resistivity of polycrystalline $Pt(DMG)_2$ as a function of pressure at room temperature. The electrical resistivity decreases significantly up to 40 kbar, above which a minimum appears. This

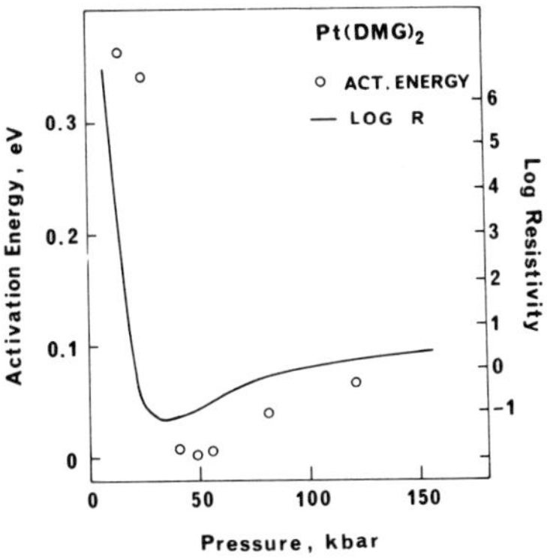

Fig. 3. Electrical resistivity and the thermal activation energy of Pt(DMG)$_2$ as a function of pressure.

behavior is roughly reversible with some amounts of hysteresis, and is reproducible. The resistivity at the minimum is 1 x 10^{-1} Ωcm. Then the resistivity of Pt(DMG)$_2$ exhibits a continuous decrease by a factor of about 10^{-16} with the increase of pressure to about 40 kbar. Next, the electrical resistivity was measured as a function of temperature from 290 to 353 K at constant pressure. The activation energy, E_a, was obtained from

$$\rho = \rho_o \exp(E_a/kT) \qquad (1)$$

where ρ is the electrical resistivity and ρ_o is essentially a temperature-independent constant. The data are illustrated by the open circles. The activation energy also exhibits a minimum value at a pressure slightly above 40 kbar. This suggests that the change of resistivity of Pt(DMG)$_2$ under pressure is ascribed to the change of activation energy for conduction.

Figure 4 shows the effect of pressure on the electrical resistivity of Pt(3-MG)$_2$, Pt(DPG)$_2$, and Pt(CD)$_2$ at room temperature. All the results are reversible and reproducible. The resistivity exhibits a rapid decrease in the three complexes in the low pressure region. A distinct minimum is found for Pt(3-MG)$_2$ and Pt(CD)$_2$ at 127 kbar and 94 kbar, respectively. For Pt(DPG)$_2$ a very broad minimum appears at 193 kbar.

As seen from Figs. 3 and 4, the minimum resistivities of four Pt glyoximes at high pressure are remarkably different. The most conductive among the four under pressure is Pt(DMG)$_2$. The minimum resistivity increases by nearly one order of magnitude from Pt(DMG)$_2$ through Pt(CD)$_2$, Pt(3-MG)$_2$, and Pt(DPG)$_2$, in turn. This behavior seems to arise from the difference in the ability of hyperconjugation of the substituents which are attached to the glyoxime ring. A methyl group behaves as if it were conjugated to the glyoxime ring. A methylene group will also resonate with the glyoxime ring, but not so strong as a methyl group. Pt(DMG)$_2$ is, therefore, most conductive under high pressure. As Pt(3-MG)$_2$ has only two methyl substituents in one molecule, it is less conductive than Pt(DMG)$_2$ or Pt(CD)$_2$. The phenyl group in Pt(DPG)$_2$ does not resonate with the glyoxime ring, and hence this complex is most resistive even at high pressure.

TTF-TCNQ and the Related Compounds

In the TTF-TCNQ crystal, the TTF radical cations and the TCNQ radical anions form homologous columnar stacks with interplanar

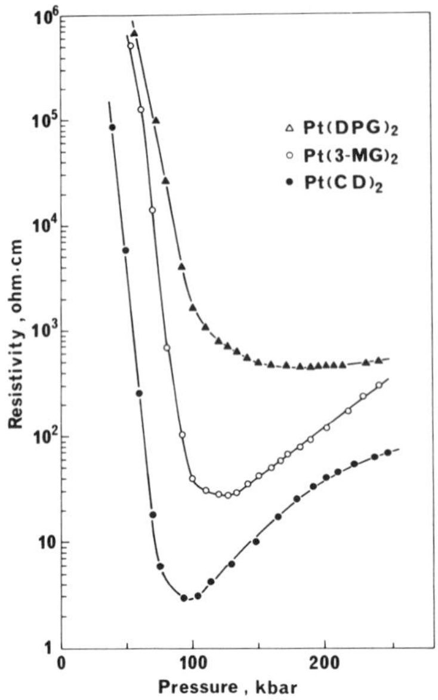

Fig. 4. Electrical resistivity of $Pt(3\text{-}MG)_2$, $Pt(DPG)_2$, and $Pt(CD)_2$ as a function of pressure.

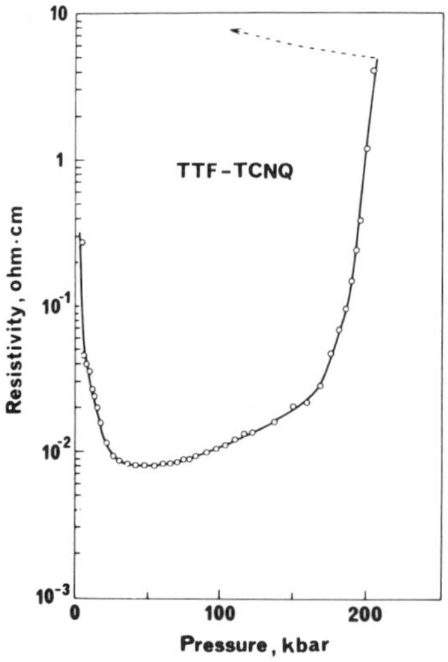

Fig. 5. Electrical resistivity of TTF-TCNQ as a function of pressure.

spacings of 3.47 and 3.17 Å, respectively [22]. Both the TTF cation and TCNQ anion columns are capable of electrical conductivity.

Figure 5 shows the resistivity vs. the pressure curve for TTF-TCNQ at room temperature. The resistivity exhibits a minimum at about 43 kbar: it drifts upwards when the pressure is kept fixed. On decreasing the pressure, the resistivity increases irreversibly as shown by the dotted curve on the figure. This behavior suggests the occurrence of a chemical reaction in the solid state, as observed in some charge transfer complexes [23-27] and TCNQ anion radical salts [7,11]. The recovered specimen after the application of pressure to 100 kbar appeared as a hard lump which was difficult to dissolve using ordinary solvents. Therefore, optical or chemical analyses could not be carried out.

Next, TTF cation radical salts containing halogen anions instead of TCNQ anion radicals were compressed. Figure 6 shows the resistivity vs. pressure curves for $TTF\text{-}Br_{0.8}$ and $TTF_7\text{-}I_5$. The resistivity of each salt exhibits a minimum and drifts upwards during the loading process and changes irreversibly during the unloading process. This behavior again suggests that chemical reaction takes place under pressure. It is likely that the columns of TTF cation radical contribute to the occurrence of chemical reaction under high pressure.

DMTTF-TCNQ and TMTTF-TCNQ were also compressed. The feature of the resistance vs. pressure curves for both DMTTF-TCNQ and TMTTF-TCNQ are very similar to that of TTF-TCNQ. As shown in Table I, the pressure at which minimum resistivity appears is the same as that of TTF-TCNQ within experimental error. In addition, the minimum resistivity of DMTTF-TCNQ and TMTTF-TCNQ was almost the same as that of TTF-TCNQ.

Table I. Minimum Resistivity of TCNQ Salts with TTF Derivatives and Related Compounds

Materials	ρ_{min}, Ωcm	Pressure, kbar
TTF-TCNQ	8.2×10^{-3}	43.0 ± 3.5
DMTTF-TCNQ	7.2×10^{-3}	43.0 ± 3.5
TMTTF-TCNQ	7.0×10^{-3}	43.0 ± 3.5
$TTF-Br_{0.8}$	1.0×10^{-1}	18.0 ± 1.0
TTF_7-I_5	4.4×10^{-2}	41.0 ± 3.0
TTF	6 x 10*	250
DMTTF	9×10^{5}*	250
TMTTF	5×10^{2}*	250

*"Least resistivity" as explained in text.

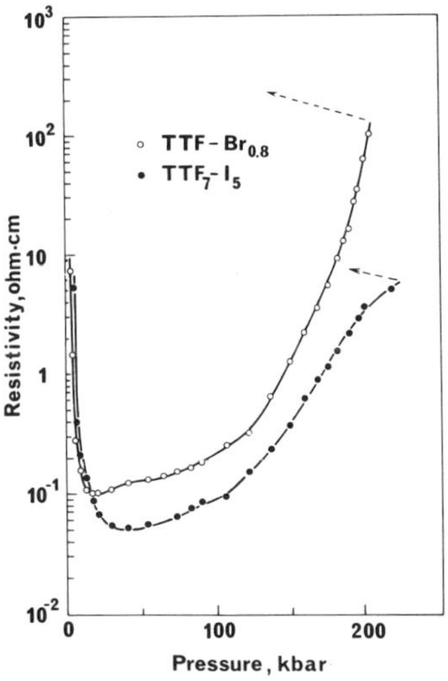

Fig. 6. Electrical resistivity of $TTF-Br_{0.8}$ and TTF_7-I_5 as a function of pressure.

The electrical resistivity of neutral TTF, DMTTF, and TMTTF decreases monotonically with an increase of pressure. Of the three, TTF becomes most conductive, but not so highly conductive as the TCNQ salts of the TTF cation radical. The resistivity of TTF at the highest pressure is about 10^2 Ωcm. Table I also lists the least resistivities of the three neutral TTF derivatives.

CONCLUSION

The electrical resistivity of $Pt(DMG)_2$, $Pt(CD)_2$, $Pt(3\text{-}MG)_2$, and $Pt(DPG)_2$ decreases remarkably with an increase in pressure. The effect of pressure is largest on $Pt(DMG)_2$ whereas it is smallest on $Pt(DPG)_2$. This seems to be due to the difference in the ability of hyperconjugation of the substituents attached to the glyoxime ring.

The electrical resistivities of TTF-TCNQ, DMTTF-TCNQ, and TMTTF-TCNQ at high pressure are very similar. The electrical behavior of these TCNQ anion radical salts under pressure is not affected by the attached methyl groups.

REFERENCES

1. L. V. Interrante and F. P. Bundy, Inorg. Chem. 10, 1169 (1971).
2. R. S. Bradley, D. C. Munro, and D. Singh, J. Inorg. Nucl. Chem. 33, 1981 (1971).
3. L. V. Interrante and F. P. Bundy, Solid State Commun. 11, 1641 (1972).
4. Y. Hara, I. Shirotani, and S. Minomura, Chem. Lett. 579 (1973).
5. W. H.-G. Müller and D. Jérome, J. Physique Lett. 35, L 103 (1974).
6. Y. Hara, I. Shirotani, and A. Onodera, Solid State Commun. 17, 827 (1975).
7. Y. Hara, I. Shirotani, and A. Onodera, Solid State Commun. 19, 171 (1976).
8. R. B. Aust, G. A. Samara, and H. G. Drickamer, J. Chem. Phys. 41, 2003 (1964).
9. I. Shirotani, T. Kajiwara, H. Inokuchi, and S. Akimoto, Bull. Chem. Soc. Japan 42, 366 (1969).
10. C. W. Chu, J. M. E. Harper, T. H. Geballe, and R. L. Greene, Phys. Rev. Lett. 31, 1491 (1973).
11. D. Jérome, W. Müller, and M. Weger, J. Physique Lett. 35, L 77 (1974).
12. I. Shirotani, A. Onodera, and N. Sakai, Bull. Chem. Soc. Japan 48, 167 (1975).
13. J. R. Cooper, D. Jérome, M. Weger, and S. Etemad, J. Physique Lett. 36, L 219 (1975); also Mol. Cryst. Liq. Cryst. 32, 231 (1976).
14. J. R. Cooper, D. Jérome, M. Weger, D. Lefur, K. Bechgaard, A. N. Bloch, and D. O. Cowan, Solid State Commun. 19, 749 (1976).
15. A. A. Adkhamov, R. M. Vlasova, L. D. Rozenshtein, S. K. Kosimi, Kh. S. Karimov, N. Boboev, and A. I. Sherle, Sov. Phys.-Dokldy 21, 90 (1976).
16. J. R. Cooper, D. Jérome, S. Etemad, and E. M. Engler, Solid State Commun. 22, 257 (1977).
17. G. Fujii, I. Shirotani, and H. Nagano, Bull. Chem. Soc. Japan, to be published.
18. A. Onodera, N. Kawai, and T. Kobayashi, Solid State Commun. 17, 775 (1975).
19. W. D. Gill, R. L. Greene, G. B. Street, and W. A. Little, Phys. Rev. Lett. 35, 1732 (1975).
20. N. Kawai, M. Togaya, and A. Onodera, Proc. Japan Acad. 49, 623 (1973).
21. N. Kawai, H. Sakamoto, Y. Notsu, and A. Onodera, Proc. Japan Acad. 51, 623 (1975).
22. T. J. Kistenmacher, T. E. Phillips, and D. O. Cowan, Acta Cryst. B30, 763 (1974).
23. W. H. Bentley and H. G. Drickamer, J. Chem. Phys. 42, 1573 (1965).
24. V. C. Bastron and H. G. Drickamer, J. Solid State Chem. 3, 550 (1971).

25. M. I. Kuhlman and H. G. Drickamer, J. Am. Chem. Soc. 94, 8325 (1972).
26. T. Sakata, A. Onodera, H. Tsubomura, and N. Kawai, J. Am. Chem. Soc. 96, 3365 (1974).
27. A. Onodera, T. Sakata, H. Tsubomura, and N. Kawai, in Proceedings 4th Intern. Conference on High Pressure, Kyoto, Japan (1974), p. 713.

J-1

FLUID PHASE EQUILIBRIA OF MIXTURES AT HIGH PRESSURE*

G. M. Schneider

University of Bochum
Bochum, West Germany

INTRODUCTION

Some recent trends in the investigation of high pressure phase equilibria of fluid mixtures will be treated in the present review particularly with respect to their application in separation processes for organic compounds under high pressure. The pressure dependence and the critical phenomena of liquid-gas, liquid-liquid, and gas-gas equilibria, the methods for their calculation and additionally some other physicochemical properties (η, D) will be briefly reviewed. Finally, the importance of phase studies in fluid mixtures will be discussed for some modern high-pressure methods of separation (e.g., fluid extraction and supercritical fluid chromatography).

PHASE THEORETICAL ASPECTS

Fluid mixtures may exhibit very different and complicated types of phase behavior. Systematic investigations during the last decade, however, have shown that all these phase separation effects can be qualitatively understood from some relatively simple theoretical phase arguments. Schematic phase diagrams of the types that are the most important for their application in separation processes for organic compounds are compiled in Figs. 1a through 1h and Figs. 2a through 2d.

In Figure 1a the well-known three-dimensional pressure-temperature-mole fraction surface is represented for the liquid-gas equilibria of a binary system in a simple case. The dashed lines are the vapor pressure curves of the pure components I and II, respectively; they end at the critical points C.P.I and C.P.II.

*Invited paper.

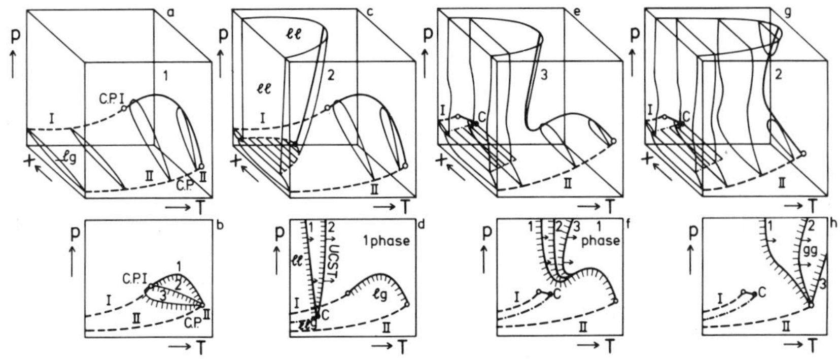

Fig. 1. p-T-x phase diagrams and their p(T) projections of fluid binary mixtures (see text [17]. Legend: ——— p(x) isotherms; ——— critical curve of binary mixtures; - - - vapor pressure curve of pure component; -.-.- three-phase line llg; C critical endpoint).

Some binary pressure-mole fraction isotherms are also schematically represented. For each isotherm the binary critical point is situated at the maximum value of the isothermal p(x) section. The line that connects the critical points of all isotherms is the binary critical curve. Type 1 in Fig. 1b shows the p(T) projection of the schematical three-dimensional phase diagram in Fig. 1a; two other types are additionally given in Fig. 1b. Many examples of these types of phase behavior have been reported in the literature [1]. The paper by Mollerup and Fredenslund [2] predominantly deals with these types of critical behavior.

In the diagrams of Figs. 1c and 1d, additional separation into two liquid phases is taken into account. For the types shown, the critical curve liquid-gas is noninterrupted and extends through the usual pressure maximum. At temperatures far below the critical temperature of pure component I, separation into two liquid phases also takes place; branch ℓℓ of the critical curve corresponds to UCST's that slightly decline (type 1) or rise (type 2) with increasing pressure.

The more the mutual miscibility of the two components I and II decreases the more the branch ℓℓ of the critical curve is displaced to higher temperatures. It can finally penetrate the ranges of temperature and pressure for the critical phenomena liquid-gas and may pass continuously into the critical curve ℓg (see Figs. 1e and 1f). Here the high-pressure branch of the critical curve may have different shapes: it may run to lower temperatures with increasing pressures (type 1 in Fig. 1f), or vertically to higher pressures within the experimental accuracy (type 2 in Fig. 1f), or it may bend back and obtain a positive slope at high pressures (Fig. 1e, type 3 in Fig. 1f).

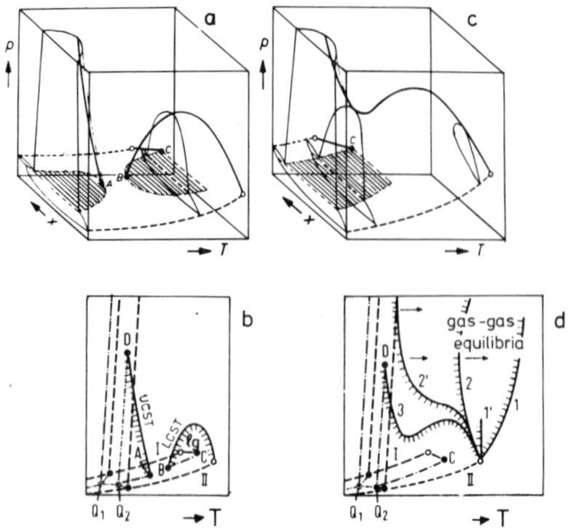

Fig. 2. p-T-x phase diagrams and their p(T) projections of fluid binary mixtures (for symbols see legend of Fig. 1; A, B = critical endpoints).

For still lower mutual miscibility of the components, critical p(T) curves without any pressure maximum or minimum may be obtained (see Figs. 1g and 1h). Here the branch of the critical curve starting from C.P.II may show different shapes: (1) it may start from C.P.II with a negative slope, this slope becoming steeper with increasing pressure; no temperature minimum however being reached (type 1 of Fig. 1h), or (2) it may run through a temperature minimum corresponding to a so-called gas-gas type phase behavior of the second kind (type 2 in Fig. 1h) represented three-dimensionally in Fig. 1g, or (3) it may start from C.P.II with a positive slope and tend directly to higher temperatures and pressures corresponding to a so-called gas-gas type phase behavior of the first kind (type 3 in Fig. 1h).

A modification of the phase behavior such as shown in Figs. 1c and 1d is given in Figs. 2a through 2d. In Figs. 2a and 2b the branch of the critical curve starting at the critical endpoint A has to be attributed to UCST's for liquid-liquid equilibria whereas that starting at B first corresponds to LCST's and then merges continuously into the critical curve liquid-gas. With increasing mutual miscibility (e.g., for smaller differences in size and/or polarity) the branch starting at A is displaced to lower temperatures and may disappear below the crystallization surface whereas on the branch starting at B the LCST's may disappear (for the exhibition of a tricritical point see Creek et al. [3]) resulting finally in a phase diagram of the simple types given in Figs. 1a and 1b. With decreasing mutual miscibility in comparison to Figs. 2a and 2b, types 3, 2', 2, 1', and 1 in Fig. 2d may successively result that are related to similar types in Figs. 1e through 1h. The types given in Figs. 2a through 2d have been shown to be the fundamental types of phase behavior for binary mixtures of hydrocarbons differing considerably in size; e.g., for solutions of polymers in hydrocarbon solvents.

Figures 1 and 2 show that there is a whole pattern of different phase diagrams and critical p(T) curves in fluid binary mixtures with components that may differ gradually in size, shape, and/or polarity and thus demonstrating the existence of continuous transitions between liquid-gas, liquid-liquid, and gas-gas equilibria. A detailed discussion including numerous examples of all types represented has been given elsewhere [4-10]. In the following section some selected examples will be given.

Phase diagrams such as shown in Figs. 1 and 2 have been shown to play an important role in the understanding of fluid extraction processes (especially the types of Figs. 1f, 2b, and 2d). Some examples are treated elsewhere in the presentations by Zosel [11] and Brunner et al. [12].

RESULTS FOR SOME SELECTED FLUID MIXTURES

In Fig. 3 the critical p(T) curves for binary mixtures of squalane (2,6,10,15,19,23-hexamethyltetracosane; component II) with methane, carbon dioxide, ethane, or propane (component I) and for the carbon dioxide + squalene system are plotted. Whereas according to Alwani [13] and Paas [14] the methane + squalane system exhibits a critical curve corresponding to type 2' in Fig. 2d (or to type 1 in Fig. 1h) the systems carbon dioxide + squalane [15] and carbon dioxide + squalene (Alwani [13]) have been shown to belong to type 3 in Fig. 1f with a minimum temperature on the critical curve. For the ethane + squalene system, type 2b was found by Horvath [16]; however, the part of the liquid-liquid immiscibility surface corresponding to UCST's, is displaced to temperatures where solidification or vitrification takes place [14]. The propane + squalane system belongs to the simple type in Fig. 1b [13].

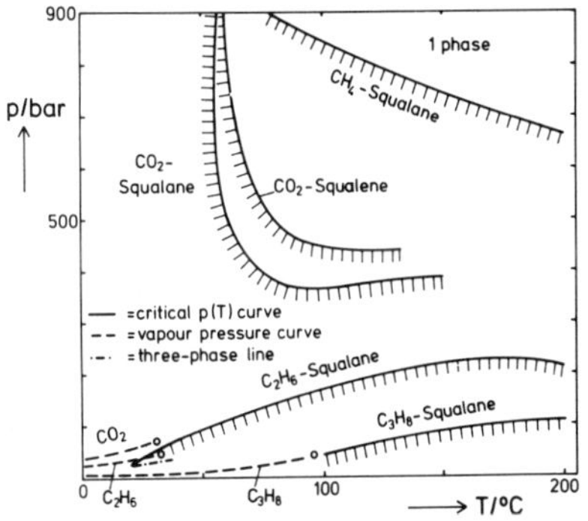

Fig. 3. Critical p(T) curves of binary mixtures of squalane with methane, ethane, propane, or carbon dioxide and of squalene with carbon dioxide [13-16].

Critical curves of some binary systems are shown in Fig. 4 where the less volatile component II is systematically varied (instead of the highly volatile component I in Fig. 3). Whereas the ethane + methanol system [17] belongs to type 2 in Fig. 1f and the ethane + water system [18] to type 2 in Fig. 1h (the latter corresponding to a gas-gas equilibrium system of the second kind) all other ethane binaries in Fig. 4 [17] show intermediate types of phase behavior; e.g., ethane + nitromethane exhibits a phase behavior of type 1 in Fig. 1h.

In Fig. 5a some p(T) curves for constant mole fraction (so-called isopleths) for binary mixtures of water and 1,2,3,4-tetrahydronaphthalene (tetralin) are given [19]. The envelope of the isopleths is the p(T) projection of the critical curve. All isopleths start on the three-phase line liquid-liquid-gas. Thus these

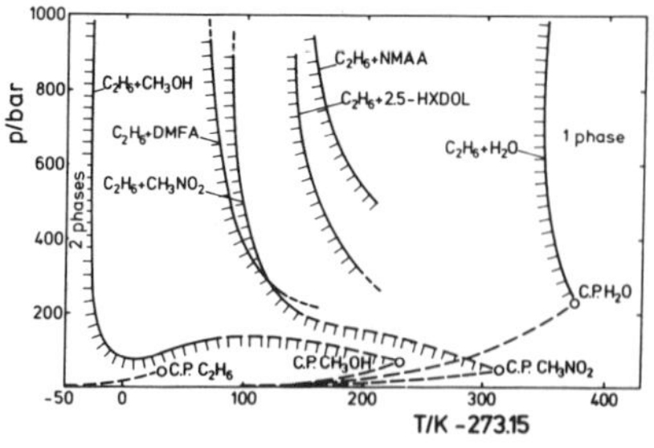

Fig. 4. Critical p(T) curves of binary mixtures of ethane with methanol, NN-dimethylformamide (DMFA), nitromethane, 2,5-hexanediol (2.5-HXDOL), N-methylacetamide (NMAA) [17] or water [18].

isopleths have to be attributed to liquid-liquid equilibria and the critical p(T) curve gives the pressure dependence of UCST's. In Fig. 5b the corresponding isothermal p(x) sections are represented; they exhibit lower critical solution pressures. The critical curve liquid-gas was not determined in these experiments; it is assumed to be noninterrupted and to run through a temperature minimum.

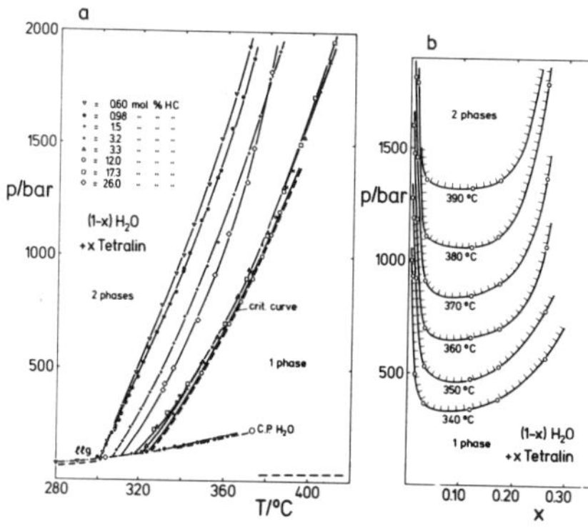

Fig. 5. Phase behavior of the system 1,2,3,4-tetrahydronaphthalene (tetralin) + water. (a) p(T) isopleths; (b) p(x) isotherms.

Similar experiments have been made on many other hydrocarbon + water mixtures in order to study the extractive effect of water on hydrocarbons. In Fig. 6a the critical p(T) curves of several hydrocarbon + water binaries are compiled. For naphthalene + water (curve 5 [20]), 1,2,3,4-tetrahydronaphthalene + water (curve 6 [19]), and decahydronaphthalene(cis) + water (curve 7 [19]) the regions of liquid-liquid and liquid-gas equilibria are completely separated; this type of phase behavior is three-dimensionally visualized in Fig. 6b. For the systems 1,4-difluorobenzene + water (curve 1 [19]), fluorobenzene + water (curve 2 [19]), benzene + water (curve 3 [21]), toluene + water (curve 4 [20]), and cyclohexane + water (curve 9 [22]), however, the regions of liquid-liquid and liquid-gas equilibria intersect resulting in branches of the critical curves that run through temperature minima and thus correspond to gas-gas equilibria of the second kind; this type of phase behavior is represented three-dimensionally in Fig. 6c. For an extensive discussion of the phase diagrams and critical phenomena of hydrocarbon + water mixtures see Jockers [19].

CALCULATION OF FLUID PHASE EQUILIBRIA

The calculation methods start with the fact that conditions for phase equilibria and for critical points are essentially the same for all types of two-phase equilibria in fluid mixtures: liquid-gas, liquid-liquid, and gas-gas. In a binary mixture, two phases (' and ") are in equilibrium if the following equations hold

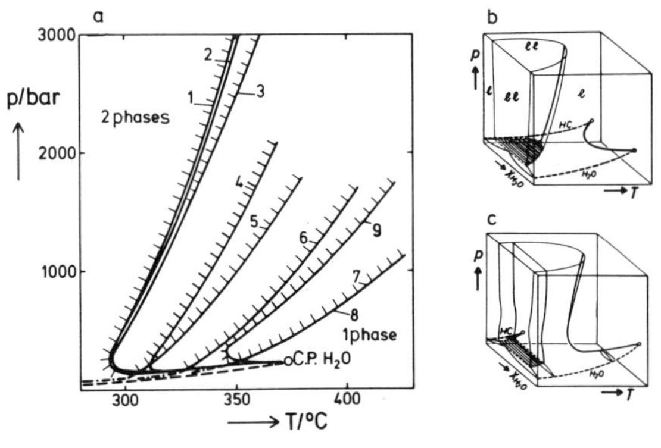

Fig. 6. Phase behavior of hydrocarbon + water systems. (a) Critical p(T) curves. Legend: 1 = 1,4-difluorobenzene + water [19]; 2 = fluorobenzene + water [19]; 3 = benzene + water [20]; 4 = toluene + water [20]; 5 = naphthalene + water [20]; 6 = 1,2,3,4-tetrahydronaphthalene(tetralin) + water [19]; 7 = decahydronaphthalene(cis) (decalin(cis)) + water [19]; 8 = decahydronaphthalene(trans)(decalin(trans)) + water [19]; 9 = cyclohexane + water [22]. (b) p-T-x diagram of the type naphthalene + water. (c) p-T-x diagram of the type benzene + water.

$$p' = p'' \tag{1}$$

$$T' = T'' \tag{2}$$

$$G'_m - x'_i(\partial G_m/\partial x'_i)_{p',T} = G''_m - x''_i(\partial G_m/\partial x_i)_{p'',T} \tag{3}$$

with i = 1 and 2, or the equivalent condition

$$\mu'_i = \mu''_i \tag{3a}$$

with i = 1 and 2. A binary critical point is defined by

$$\left(\frac{\partial^2 G_m}{\partial x_i^2}\right)_{p,T} = 0 \tag{4}$$

$$\left(\frac{\partial^3 G_m}{\partial x_i^3}\right)_{p,T} = 0 \tag{5}$$

From the phase equilibrium conditions (1), (2), and (3) or (3a) the phase diagrams can be deduced, while from the conditions (4) and (5) of a critical point the binary critical curve can be obtained. In most cases, however, the more complicated corresponding conditions for the molar Helmholtz energy A_m have to be used.

The molar Gibbs energy G_m of the molar Helmholtz energy A_m necessary for the calculations are obtained: (1) from an equation of state that is often a modified version of the Redlich-Kwong or the van der Waals equation, sometimes using rather sophisticated combining rules for the parameters of the mixture, or (2) from theories of mixtures with the use of different combining rules for the potential parameters in the mixture, or (3) for liquid-liquid and liquid-gas equilibria using equations describing the deviations from an ideal (or ideally diluted) mixture (e.g., Porter, van Laar, Wilson, NRTL, UNIQUAC, Flory-Huggins etc., or using group contributions, e.g., ASOG and UNIFAC).

In such calculations not only the phase behavior but also the pVT and the caloric properties have to be considered. The first approach has been used by Brunner et al. [26] and by Mollerup et al. [2] with respect to a modified Redlich-Kwong equation. Much of this field has recently been reviewed by Prausnitz [23,24]; for additional citations see Schneider [9,10], Deiters, et al. [25], Twu et al. [27] and Young et al. [28].

Only two examples are presented in Figs. 7a and 7b according to the calculations of Deiters et al. [25] using the Redlich-Kwong equation of state and special combining rules. They show that even systems belonging to type 1 in Fig. 1f or to type 1 in Fig. 1h can be calculated in good agreement with experimental data.

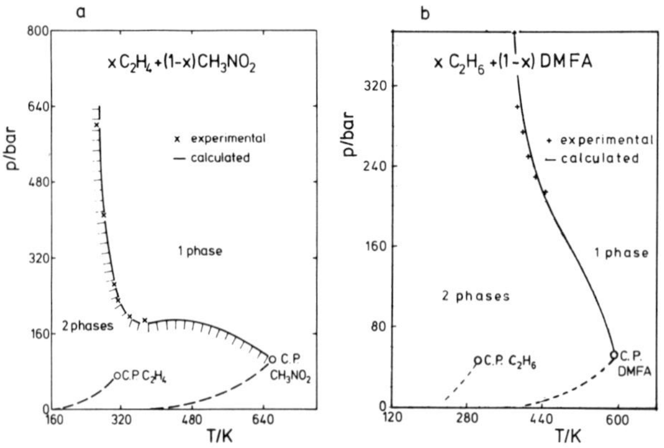

Fig. 7. Comparison of experimental and calculated binary critical curves [25]. (a) Ethylene + nitromethane; (b) ethane + N,N-dimethylformamide.

TRANSPORT PROPERTIES

For separation processes such as fluid extraction or supercritical fluid chromatography the transport properties of fluid mixtures are of primary importance. Since there is a great lack of experimental data in this field to date, only a few remarks will be made in this section.

Phase Rich In Solvent Gas

For very dilute solutions of a substance to be purified, the transport coefficients of the pure solvent gas will give some information about the orders of magnitude involved. In Fig. 8 the density ρ, the viscosity η, and the product $D_{11}.\rho$ of pure carbon dioxide are plotted as a function of pressure up to approximately 500 bar at a temperature of 40°C (about 9 K above the critical temperature) according to data taken from the literature. The $\eta(p)$ isotherm shows a characteristic increase of η in the

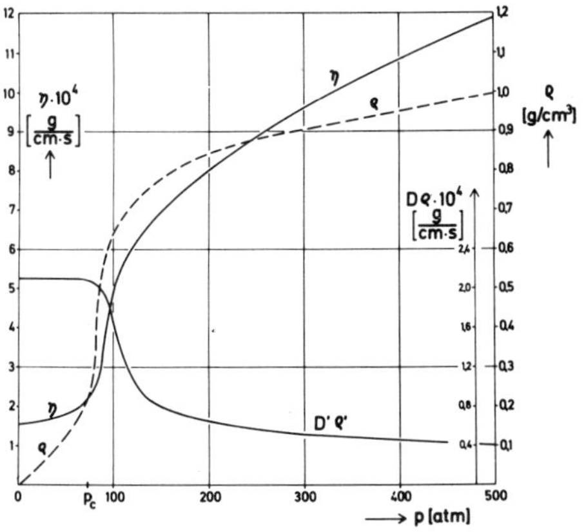

Fig. 8. Density, viscosity, and D•ρ for pure carbon dioxide as a function of pressure at 40°C [36].

neighborhood of the critical pressure p_c. In agreement with simple kinetic theory, η and D_{11}.ρ remain essentially constant up to about 80 bar; with further increase in pressure a rapid increase in η and a decrease in D_{11}.ρ is found. At about 400 bar the following ratios are obtained: $\eta(p)/\eta(1\ \text{bar}) \approx 7$ and $D_{11}(p)/D_{11}(1\ \text{bar}) \approx 5 \times 10^{-4}$.

Some experimental results, however, are also known for D_{12} at high pressures. According to Fig. 9, D_{12} values in the order of 10^{-3} $cm^2\ s^{-1}$ were found for the diffusion of traces of butane in argon between 200 and 1500 bar and at 25°C by Balenovic et al. [29] with a chromatographic technique; these values have to be compared with 1 to 10^{-1} $cm^2\ s^{-1}$ for normal NTP gases and with 10^{-5} to 10^{-6} $cm^2\ s^{-1}$ for normal liquids at ambient pressure. The result is that in the phase rich in solvent gas, relatively low viscosities and high diffusion coefficients can be expected.

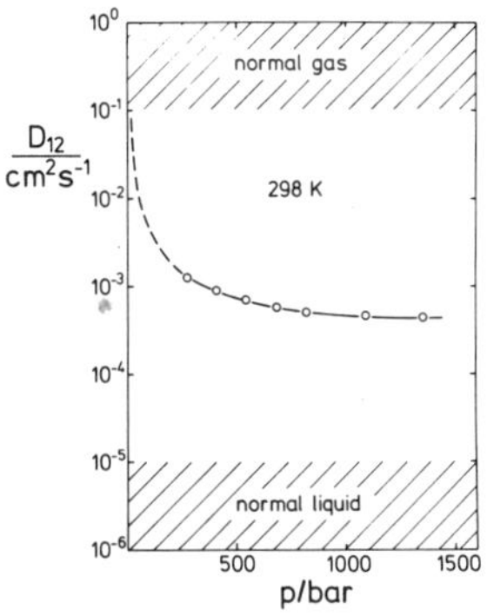

Fig. 9. D_{12} for the diffusion of traces of butane in argon at 25°C (according to Balenovic et al. [28]).

Phase Poor in Solvent Gas

As an example, the viscosity of the model substance squalane saturated with different gases at different pressures is given in Fig. 10. Here the ratio $\eta(p)/\eta(1\ \text{bar})$ is plotted against pressure up to about 1 kbar at 40°C according to measurements of Kuss et al. [30]. It is an interesting result that the increase in viscosity of a liquid with increasing pressure becomes considerably smaller when gases are dissolved in it, and for nitrogen, argon, and carbon dioxide even a decrease in viscosity is found. Since D is approximately proportional to 1/η, relatively high D values are also to be expected. Thus the phases poor in solvent gas may show relatively small viscosities and high diffusion coefficients that are both favorable for a separation process, e.g., in supercritical fluid chromatography.

APPLICATIONS

Among the chemical applications [9,10, 31,32] of the phase behavior of fluid mixtures only fluid extraction and supercritical fluid chromatography will be briefly reviewed here.

Fluid Extraction

In fluid extraction, compressed gases having critical temperatures between about 0 and 200°C (e.g., CO_2, C_2H_6, C_3H_8, fluorocarbons, N_2O, etc.) are used as (mostly supercritical) solvents in extracting highly boiling compounds (preferably organic substances with low volatility and/or low thermal stability) especially in the pharmaceutical and food industry. These problems are discussed in more detail in the presentations of Zosel [11] and Brunner et al. [12]. The importance of fluid extraction is clearly demonstrated by an increasing number of papers and patents in this field [9-12,31,32].

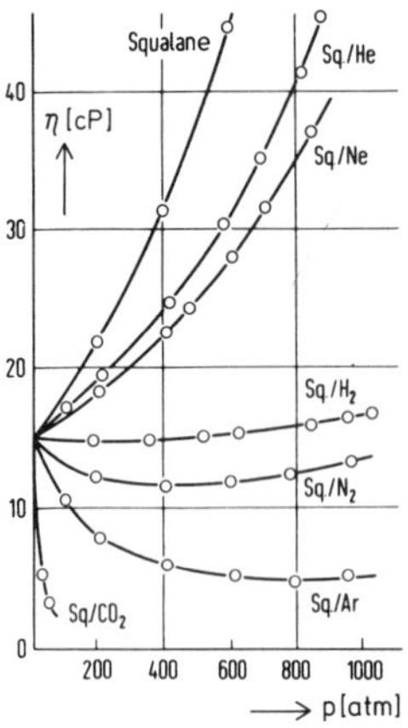

Fig. 10. Viscosity of squalane saturated with different gases as a function of pressure at 40°C (according to Kuss et al. [30]).

Supercritical Fluid Chromatography

Another interesting special application is supercritical fluid chromatography (sfc). In this relatively new technique of chromatography, compressed fluids at temperatures in their critical range are used as mobile phases particularly for the separation of substances with too low a migration rate or too low a thermal stability in order to allow separation by ordinary (eventually temperature-programmed) gas chromatography. The influence of pressure (or density, respectively) is demonstrated in Fig. 11 where chromatograms for the migration of hexadecane using carbon dioxide as a mobile phase at 40°C and at different pressures are given [33]; details are given in the legend of Fig. 11. In the presentations of Gouw et al. [34] and Klesper et al. [35] interesting improvements in experimental technique and new applications, e.g.,

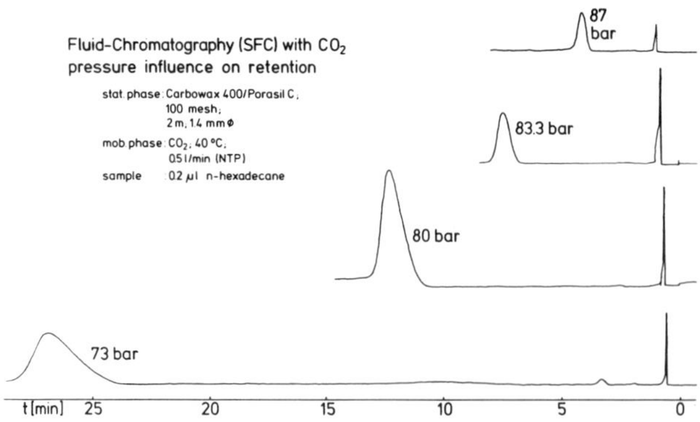

Fig. 11. Supercritical fluid chromatography (sfc). Recorder plots for the migration of hexadecane [33].

for the purification of oligomers or the analysis of exhaust gases, are discussed.

Additionally some physicochemical knowledge can be deduced from sfc experiments by Bartmann et al. [36]. The retention time t^r of a substance under test is related to its capacity ratio k and its partition coefficient K by

$$k = K \frac{V^{stat}}{V^{mob}} = \frac{t^r - t^o}{t^o} \qquad (6)$$

where $K = c^{stat}/c^{mob}$. Thus, k can be calculated from t^r and t^o, and from k the partition coefficient K. In Fig. 12 experimental log k values are plotted against ρ and p of supercritical carbon dioxide as a mobile phase at 40°C for sfc experiments with alkanes having from 10 to 30 carbon atoms [37]; for details see the legend of Fig. 12. The curves show that with increasing density, the retention time t^r and the capacity ratio k decrease and consequently the concentration of the substance under test in the mobile phase increases by factors of ten. Thus Fig. 12 demonstrates particularly well the varying solvent properties of supercritical carbon dioxide and of other fluid phases as a function of ρ and p that are most useful in separation processes.

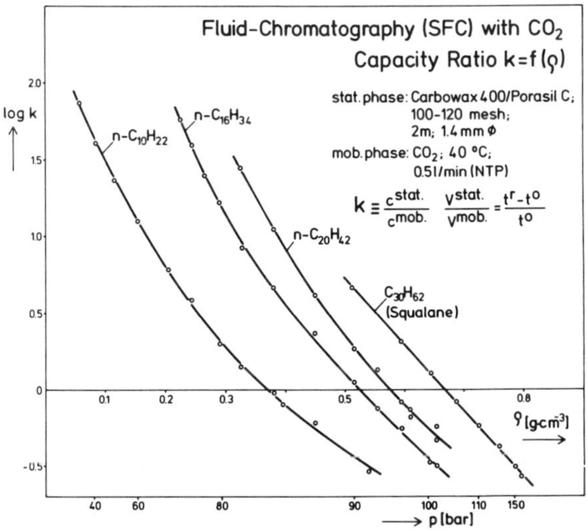

Fig. 12. Supercritical fluid chromatography (sfc); Capacity ratio k as a function of density and pressure [36].

NOTATION

- A_m = molar Helmholtz energy
- c = concentration
- C.P. = critical point of pure component
- D_{11} = coefficient of self diffusion
- D_{12} = binary diffusion coefficient
- G_m = molar Gibbs energy
- k = capacity ratio
- K = partition coefficient
- LCST = lower critical solution temperature
- p = total pressure
- Q = quadruple point
- T = temperature
- t^r = retention time

t^o = retention time of an inert substance
UCST= upper critical solution temperature
V = volume
x = mole fraction
η = viscosity
μ = chemical potential
ρ = density

Subscripts

g = gaseous phase
l = liquid phase
mob = mobile phase
stat = stationary phase
i = component i
I = component I (low boiling)
II = component II (high boiling)

REFERENCES

1. C. P. Hicks and C. L. Young, Chem. Rev. 75, 119 (1975).
2. J. Mollerup and A. Fredenslund, paper J-6-B presented at 6th AIRAPT Intern. High Pressure Conference, University of Colorado, Boulder, Colorado, July 25-29, 1977.
3. J. L. Creek, C. M. Knobler, and R. L. Scott, J. Phys. Chem., in press.
4. J. S. Rowlinson, Liquids and Liquid Mixtures, 2nd edition, Butterworth, London (1969).
5. G. M. Schneider, Ber. Bunsenges. phys. Chem. 70, 497 (1966).
6. G. M. Schneider, Adv. chem. Physics 17, 1 (1970).
7. G. M. Schneider, Fortschr. chem. Forsch. 13, 559 (1970).
8. G. M. Schneider, in Water - A Comprehensive Treatise, vol. 2, F. Franks, ed., Plenum Press, New York (1973), Chapt. 6.
9. G. M. Schneider, Pure appl. Chem. 47, 277 (1976).
10. G. M. Schneider, Spec. Periodical Report: Chemical Thermodynamics, Vol. 2, The Chemical Society, London (1978), Chapt. 4.
11. K. Zosel, paper K-6-A presented at 6th AIRAPT Intern. High Pressure Conference, University of Colorado, Boulder, Colorado, July 25-29, 1977.
12. G. Brunner, S. Peter, B. Tetzlaff, and R. Riha, paper K-6-B presented at 6th AIRAPT Intern. High Pressure Conference, University of Colorado, Boulder, Colorado, July 25-29, 1977.
13. Z. Alwani, University of Bochum, private communication.
14. R. Paas, Ph.D. Thesis, University of Bochum, Bochum, W. Germany (1977).
15. K. G. Liphard and G. M. Schneider, J. chem. Thermodynamics 7, 805 (1975).
16. E. Horvath, M.S. Thesis, Department of Chemistry, University of Bochum, Bochum, West Germany (1965).
17. Z. Alwani and G. M. Schneider, Ber. Bunsenges. phys. Chem. 80, 1310 (1976).
18. A. Danneil, K. Tödheide, and E. U. Franck, Chemie-Ingr-Techn. 39, 816 (1967).
19. R. Jockers, Ph.D. Thesis, Department of Chemistry, University of Bochum, Bochum, West Germany (1976).
20. Z. Alwani and G. M. Schneider, Ber. Bunsenges. phys. Chem. 73, 294 (1969).

21. Z. Alwani and G. M. Schneider, Ber. Bunsenges. phys. Chem. 71, 633 (1967).
22. K. Bröllos, K. Peter, and G. M. Schneider, Ber. Bunsenges. phys. Chem. 74, 682 (1970).
23. R. C. Reid, J. M. Prausnitz, and T. K. Sherwood, The Properties of Gases and Liquids, 3rd edition, McGraw-Hill, New York (1977), Chapt. 8.
24. J. M. Prausnitz, Ber. Bunsenges. phys. Chem., 81, 900 (1977).
25. U. Deiters and G. M. Schneider, Ber. Bunsenges. phys. Chem. 80, 1316 (1976).
26. G. Brunner and H. Hederer, paper J-6-C presented at 6th AIRAPT Intern. High Pressure Conference, University of Colorado, Boulder, Colorado, July 25-29, 1977.
27. C. H. Twu, K. E. Gubbins, and C. G. Gray, J. Chem. Phys. 64, 5186 (1976).
28. C. L. Young et al., J. Chem. Soc. Faraday Trans. II 73, 597, 613, 618 (1977).
29. Z. Balenovic, M. N. Myers, and J. C. Giddings, J. Chem. Phys. 52, 915 (1970).
30. E. Kuss and H. Golly, Ber. Bunsenges. phys. Chem. 76, 131 (1972).
31. V. Pilz, Verfahrenstechnik 9, 280 (1975).
32. W. Sirtl, Chemie-anlagen u. Verfahren 1, 53 (1976).
33. U. van Wasen, University of Bochum, private communication.
34. T. H. Gouw, R. E. Jentoft, and E. J. Gallegos, paper K-6-D presented at 6th AIRAPT Intern. High Pressure Conference, University of Colorado, Boulder, Colorado, July 25-29, 1977.
35. E. Klesper and W. Hartmann, paper K-6-C presented at 6th AIRAPT Intern. High Pressure Conference, University of Colorado, Boulder, Colorado, July 25-29, 1977.
36. D. Bartmann and G. M. Schneider, J. Chromatog. 83, 135 (1973).
37. U. van Wasen and G. M. Schneider, Chromatographia 8, 274 (1975).

J-2

HIGH PRESSURE VAPOR-LIQUID EQUILIBRIA: THE INTERACTION BETWEEN THEORY AND EXPERIMENT

J. Mollerup and Aa. Fredenslund

Instituttet for Kemiteknik
Danmarks tekniske Højskole
Lyngby, Denmark

INTRODUCTION

The development of accurate correlations for the calculation of high pressure vapor-liquid equilibria necessitates an investigation into the type of experimental data needed to test the correlations and the development of proper data reduction methods which the experimental data must be subjected to. First, there will be a discussion of the type of experimental data needed to develop correlations for the calculation of high pressure vapor-liquid equilibria and of how the simplest possible correlation can be developed which also gives an accurate representation of the experimental data. Second, it will be demonstrated that isothermal high pressure vapor-liquid equilibria data can be subjected to a thermodynamic consistency test in accordance with the isothermal Gibbs-Duhem equation. Third, there will be a short review of some of the most recently developed high pressure vapor-liquid equilibria correlations with engineering application.

PURE COMPONENT PROPERTIES AND EQUATIONS OF STATE

Correlations developed to calculate high pressure vapor-liquid equilibria must also be capable of correlating the pure component properties with high accuracy because the accuracy of the correlated mixture properties depends largely upon how well the pure component properties can be calculated. In this connection the saturated surface of the pure components is of great importance and the correlation developed must fulfill the criterion of phase equilibria, that is

$$T^{\ell} = T^{g} \tag{1}$$

$$P^{\ell} = P^{g} \tag{2}$$

$$\phi^{\ell} = \phi^{g} \tag{3}$$

If the correlation is based on a simple two-constant equation of state as the Redlich-Kwong equation or modifications thereof as given by Soave [12] or Hamam et al. [3], the following analysis will show that we are left with three independent variables at the saturated surface.

The Redlich-Kwong equation of state can be written as

$$Z^3 - Z^2 + Z(A-B-B^2) - AB = 0 \tag{4}$$

where Z is the compressibility factor Z defined as the ratio of PV/RT, and A and B are constants defined as

$$B = \frac{bP}{RT} \tag{5}$$

$$A = \frac{ap}{R^2T^{2.5}} \tag{6}$$

From (4) one can calculate the fugacity coefficient of a pure component as

$$\ln \phi = (Z-1) - \ln(Z-B) - (\frac{A}{B}) \ln(1 + B/Z) \tag{7}$$

This leaves us with nine equations because (4) and (7) have to be counted twice: once for the saturated liquid and once for the saturated vapor phase. The variables are T, P, Z and ϕ both for the saturated liquid and the saturated vapor phase in addition to a, b, A and B. This gives 12 variables and nine equations and therefore with equations (1) through (7) we end up with three independent variables. The most reasonable choice of independent variables are the temperature, the vapor pressure and the saturated liquid phase volume.

By specifying these three independent variables we are, by means of (1) through (7), able to calculate the dependent variables including the two constants a and b in the equation of state. While Hamam et al. [3] consider both a and b as variables, Soave [12] and Peng and Robinson [10] assume that b is a known constant. The number of variables is then reduced by one. The most obvious choice of variables is then the saturation temperature and pressure.

With only one adjustable constant, a, we are therefore able to make an accurate correlation of the vapor pressure of a pure substance by means of a two-constant equation of state. If, however, both constants a and b are adjustable we are able to make an accurate correlation of the saturated liquid phase volume as well.

It is common practice to assume that the Helmholtz free energy is an analytical function of T and V at and near the critical point and thus arrive at the condition that

$$(V^g - V^\ell) = 0, \quad \left(\frac{\partial P}{\partial V}\right)_T = 0 \text{ and } \left(\frac{\partial^2 P}{\partial V^2}\right)_T = 0 \tag{8a-c}$$

at the critical point. If these conditions are incorporated in the analyses given above for a two-constant equation of state we end up with five equations, (4), (5), (6), (8b) and (8c), and seven variables, T, P, Z, a, b, A and B, wherefore only two can be chosen as independent variables. We must therefore conclude that a simple two-constant equation of state can not at the same time give a correct representation of the volumetric properties at the critical point and fulfill the conditions in (8b) and (8c). This shows us then that we can only obtain a correct representation of critical properties, P_c, T_c and Z_c and at the same time fulfill the requirements in (8b) and (8c) at the critical point if the equation of state contains at least three constants. If the equation of state contains three constants, which are allowed to vary along the saturation surface, we would be left with four independent variables at the saturation surface below the critical point. Besides specifying the temperature, the pressure and the saturated liquid volume, one could consider the slope of the vapor pressure curve or the saturated vapor volume as the fourth specification. The disadvantage of the equations of state with two constants mentioned above is that they are not able to represent the conditions at the critical point correctly. Only an equation of state with three constants would do this, and at the same time it would be able to correlate the saturated P-V-T surface accurately.

The properties of mixtures can not be calculated from the P-V-T properties of the pure components. It is necessary to introduce in the equation of state parameters which reflect the interaction between unlike molecules. For the equations of state mentioned above it is common practice to calculate a and b from the empirical equations

$$a = \sum_i \sum_j x_i x_j a_{ij} \tag{9a}$$

$$a_{ij} = {}_{ij}\xi a_{ii} a_{jj} \qquad i \neq j \tag{9b}$$

$$b = \sum_i x_i b_i \tag{9c}$$

where a_{ii}, a_{jj} and b_i are the constants of the pure components at the pressure and temperature of the solution and ξ_{ij} represents the deviation from the geometric mean of the energetic interaction. This binary interaction parameter has to be estimated from experimental binary data and can not be calculated from any molecular theory.

Normally it is assumed to be independent of temperature, volume and composition. The best way to estimate this interaction parameter in a vapor -liquid equilibrium correlation is to estimate it as the parameter which gives the best representation of the binary bubble point curve when it is calculated from the specification of the pressure and the liquid phase composition as demonstrated by Mollerup [5].

THE CONSISTENCY TEST

The analysis above demonstrates that the development of correlations for the calculation of vapor -liquid equilibria at high pressures requires knowledge of the saturated P-V-T surface and the P-V-T properties of the pure components above the critical temperature, preferably at high densities, and the binary vapor -liquid equilibria data covering the whole P-T-x-y space.

The only way pure component properties can be tested for "consistency" is by the mutual agreement of independent measurements obtained in different laboratories. This is also one of the best ways to test experimental binary vapor liquid equilibria data although they can be subjected to a thermodynamic consistency test. In this test one calculates the vapor phase composition from isothermal P-x data by solving the following differential equation in the dimensionless excess Gibbs free energy $g = G^E/RT$:

$$P = \sum_{i=1}^{2} (x_i f_i^o/\phi_i) \exp[F_i(g,g')] \tag{10}$$

where x_i is the liquid phase mole fraction, ϕ_i the vapor phase fugacity coefficient and f_i^o the standard state reference fugacity of component i. The choice of standard state which depends upon the system in question is discussed by Christiansen and Fredenslund [1]. The functions $F_i(g,g')$ are given by

$$F_1(g,g') = \ln \gamma_1 = g + x_2 \left(\frac{dg}{dx_1}\right)_\sigma - \frac{x_2 V^E}{RT} \left(\frac{dP}{dx_1}\right)_\sigma \tag{11}$$

$$F_2(g,g') = \ln \gamma_2 = g - x_1 \left(\frac{dg}{dx_1}\right)_\sigma + \frac{x_1 V^E}{RT} \left(\frac{dP}{dx_1}\right)_\sigma \tag{12}$$

where γ_i is the activity coefficient of component i. The subscript σ denotes that the derivatives are taken along the saturation surface. Expression V^E is the liquid phase excess volume with respect to the components in their standard states. Equation (10) can be solved using the method of orthogonal collocation as demonstrated by Christiansen and Fredenslund [1]. The solution contains a set of calculated vapor phase mole fractions which can be compared with the experimental values. The difference between calculated and experimental values serves as a test of the consistency of the experimental vapor-liquid equilibrium data. An example is given in Table I. When the temperature of the system is below the critical temperature of the pure components, the coefficients in the nth order Legendre polynomials, which are used to approximate g, are found by a least squares technique as demonstrated by Grausø et al. [2]. In the above versions the standard state fugacity, the excess volume and the vapor phase fugacity coefficients are calculated by the methods of Prausnitz and Chueh [11]. In this study the volumetric properties are calculated from the equation of state by Hamam et al. and the reference state fugacities and vapor phase fugacity coefficients from the Peng-Robinson equation of state because investigations show that among the two-constant equations of state mentioned above this choice gives the best representation of the thermodynamic properties needed to carry out the consistency test [8]. The details of the consistency test procedure are given by Christiansen and Fredenslund [1] and Grausø et al. [2].

Table I. Thermodynamic Consistency Test for Ethene-Carbon Dioxide System

T, K	P, bar	$\overline{\Delta y}$ *	$\overline{\Delta y}^{\dagger}$
293.15	57-64	0.0039	0.0020
283.15	45-55	0.0030	0.0016
263.15	23-32	0.0021	0.0013
243.15	14-19	0.0026	0.0024
223.15	6-11	0.0032	0.0026

$\overline{\Delta y}$ is the absolute average deviation between calculated and experimental values. (* Prausnitz-Chueh method used to calculate the thermodynamic properties involved. † Thermodynamic properties involved are calculated from the equations of state by Hamam et al. and Peng-Robinson.)

The replacement of the Prausnitz-Chueh method of calculating the standard state fugacity, the excess volume and the vapor phase fugacity coefficient by the method based on the equations of state by Hamam et al. and Peng and Robinson has resulted in closer agreement between the calculated and experimental values as can be seen from Table I. As one would expect the improvement is most pronounced in the high pressure region. The quantity $\overline{\Delta y}$ is the absolute average deviation between the experimental and calculated vapor phase mole fractions. The experimental vapor-liquid equilibrium data are by Mollerup [6]. The calculated estimates of the population variances are 6.7×10^{-6} for the liquid phase mole fractions and 5.1×10^{-6} for the vapor phase mole fractions.

VAPOR-LIQUID EQUILIBRIA CORRELATION

It is not our intention to carry out an evaluation of the various equations of state available for vapor-liquid equilibria correlation. Only four methods will be discussed here, and an example of how they correlate vapor-liquid equilibria at high pressures will be given. No final conclusion can be drawn from the example given; it is merely an illustration. Three of the methods are based on the two-constant equations of state by Hamam et al., Peng-Robinson and Soave.

The equation of state by Soave is a modified Redlich-Kwong equation:

$$P = \frac{RT}{v-b} - \frac{a}{v(v+b)} \tag{13}$$

where $b=b_c$, $a=a_c\ f(T_R,\omega)$ and b_c and a_c are constants determined from the conditions given by (8a-c) at the critical point. The quantity $f(T_R,\omega)$ is a function of reduced temperature and the acentric factor determined from pure component vapor pressures. This function has a value of unity at the critical temperature.

The equation of state by Hamam et al. is also a modified Redlich-Kwong equation

$$P = \frac{RT}{v-b} - \frac{a}{\sqrt{Tv(v+b)}} \tag{14}$$

where $b = b(Tr,\omega)$ and $a = a(T_R,\omega)$ are determined from experimental data from the saturation pressure and temperature and saturated liquid volumes as mentioned earlier. This equation does not meet the requirements of (8b-c) at the critical point.

The equation of state by Peng and Robinson is

$$P = \frac{RT}{v-b} - \frac{a}{v(v+b) + b(v-b)} \tag{15}$$

where $b=b_c$, $a=a_c \cdot \alpha(T_R,\omega)$ and b_c and a_c are constants again determined from the conditions given by (8a-c) at the critical point. The quantity $\alpha(T_R,\omega)$ is a function of the reduced temperature and the acentric factor determined from pure component vapor pressures. At a reduced temperature of unity this function has a value of unity.

The calculation of vapor-liquid equilibria from these equations of state is straightforward and is described in the original papers.

Soave [12] and Peng and Robinson [10] correlate constant a as a generalized function of reduced temperature and the acentric factor and Hamam et al. [3] correlate both a and b as generalized functions of reduced temperature and the acentric factor. All the correlations are for reduced temperatures below unity. Neither Soave nor Peng and Robinson mention how a can be calculated for reduced temperatures above unity. Hamam et al. assume that a and b have the values

Table II. Summary of Results for Vapor-Liquid Equilibrium Correlation of Nitrogen(1)-Methane(2) System

		95 - 120 K, 60 Data Points				130 - 180 K, 84 Data Points			
		JM	SRK	PR	HRK	JM	SRK	PR	HRK
ΔP	AAD	1.4	0.8	1.0	1.3	1.1	1.0	1.2	1.0
	BIAS	1.0	-0.0	0.6	-0.0	1.0	0.8	1.1	0.1
	SSQ	2.9	1.0	1.8	2.8	1.9	2.1	2.6	2.0
ΔK_1	AAD	0.1	0.6	0.2	0.5	1.6	5.7	6.0	3.3
	BIAS	-0.08	-0.6	-0.1	-0.5	-0.6	5.4	5.9	-0.8
	SSQ	0.05	0.5	0.05	0.4	7.6	103.	105.	30.
ΔK_2	AAD	1.0	6.0	2.2	5.5	1.3	5.3	5.4	4.6
	BIAS	0.4	6.0	2.1	5.5	0.8	-4.7	-5.3	-2.3
	SSQ	2.3	53.	8.6	43.	7.3	198.	199.	177.

Legend:

ΔP = relative deviation between experimental and calculated bubble point pressure
$\Delta K_1, \Delta K_2$ = relative deviation between experimental and calculated equilibrium ratio of component 1 or 2
AAD = absolute average deviation per data point x 100
BIAS = average deviation per data point x 100
SSQ = sum of squared deviation per data point x 10^4
JM, SRK, PR and HRK see Table III

obtained at the critical point when the reduced temperature is above unity. This assumption is probably not sound because the change in the P-V-T properties just above the critical point is as pronounced as it is just below the critical point. The constants which are taken to be functions of reduced temperature at the saturation surface must also be correlated as functions of reduced temperature above the critical temperature in such a way that they reduce to the values at the critical point when the reduced temperature is unity.

The fourth model is based on the principle of corresponding states with methane as the reference substance. The theory behind this method will not be presented here but can be found elsewhere [5,7].

Table II shows a summary of results for the correlation of the vapor -liquid equilibria in the nitrogen-methane system. The experimental data are by Parrish and Hiza [9] and Kidnay et al. [4]. Table III shows a comparison between calculated and experimental Henry's law constants. The experimental values are determined by Kidnay et al. [4] using the method developed by Christiansen and Fredenslund [1].

Table III. Henry's Law Constants in Bars for Nitrogen-Methand System*

Temperature	EXP	JM	SRK	PR	HRK
130	46.0	45.5	44.0	42.5	47.6
140	57.7	56.9	54.9	53.3	59.3
150	68.8	67.8	65.0	63.2	69.9
160	79.3	77.6	73.3	71.4	78.7
170	85.8	85.6	78.8	76.9	85.4
180	90.3	90.1	79.2	78.0	89.1

*Legend

EXP = experimental values
JM = calculated from corresponding states theory
SRK = calculated from Soaves equation of state
PR = calculated from Peng-Robinsons equation of state
HRK = calculated from equation of state by Hamam et al.

CONCLUSION

It has been shown that an accurate representation of the volumetric properties at the saturated surface of a pure substance as well as at its critical point requires an equation of state with at least three adjustable constants which are functions of the pure component properties. Further, it has been pointed out that isothermal binary vapor-liquid equilibria data can be tested for consistency in accordance with the isothermal Gibbs-Duhem equation. The reliability of this consistency test is dependent upon how accurate the reference fugacity of the pure components and the vapor phase fugacity coefficient of component i in the mixture can be calculated. Development of new equations of states indicates however that these properties may be calculated with sufficient accuracy. A short account of some of these equations of state is given and an example illustrates the calculation of vapor-liquid equilibria and Henry's law constants by means of these equations and the corresponding states principle.

REFERENCES

1. L. J. Christiansen and Aa. Fredenslund, AIChE J. 21, 49 (1975).
2. L. Grausø, Aa. Fredenslund and J. Mollerup, Fluid Phase Equilibria 1, 13 (1977).
3. S.E.M. Hamam, W. K. Chung, I. M. Elshayal, and B.C.-Y. Lu, Ind. Eng. Chem. Proc. Des. Dev. 16, 51 (1977).
4. A. J. Kidnay, R. C. Miller, W. R. Parrish, and M. J. Hiza, Cryogenics 15(9), 531 (1975).
5. J. Mollerup, in Advances in Cryogenic Engineering, Vol. 20, Plenum Press, New York (1975), p. 172.
6. J. Mollerup, J. Chem. Soc. Faraday Trans. I 71, 2351 (1975).
7. J. Mollerup, Berichte Bunsenges. Phys. Chem. 81(10), 1015 (1977).
8. J. Mollerup, paper to be presented at CHISA'78 Congress.
9. W. R. Parrish and M. J. Hiza, in Advances in Cryogenic Engineering, Vol. 19, Plenum Press, New York (1974), p. 300.
10. D.-Y. Peng and D. B. Robinson, Ind. Eng. Chem. Fund. 15, 59 (1976).
11. J. M. Prausnitz and P. L. Chueh, Computer Calculations for High Pressure Vapor Liquid Equilibria, Prentice Hall, New York (1968).
12. G. Soave, Chem. Eng. Sci. 27, 1197 (1973).

J-3

CALCULATION OF VAPOR-LIQUID PHASE EQUILIBRIA FOR LOW VOLATILE SUBSTANCES IN SYSTEMS WITH COMPRESSED GASES

G. Brunner and H. Hederer

University of Erlangen - Nuernberg
Erlangen, W. Germany

INTRODUCTION

Vapor-liquid phase equilibria of low volatile substances are of interest in separation processes using compressed gases. Furthermore, such phase equilibria are becoming increasingly important for oil recovery and for coal-gasification processes. For rational design purposes it is necessary to calculate the vapor-liquid phase equilibria of compressed gases with high-boiling hydrocarbons of different structure.

It is the aim of this investigation to show that it is possible to calculate such phase equilibria with sufficient accuracy using the equation of state proposed by Redlich and Kwong [1]. This work deals with the generalization of the parameters of the Redlich-Kwong equation and the determination of the interaction constant Θ_{ij} specifically for vapor-liquid equilibrium calculations.

CALCULATION METHOD

The equilibrium between a vapor and a liquid phase is given by

$$RT \ln \frac{1}{K_i} = \int_{P^+}^{P} [\bar{V}_i(y_i) - \bar{V}_i(x_i)]\, dP \qquad (1)$$

According to this equation the K-factor for component i is calculated by integrating the difference of the partial molal volumes of the gaseous and the liquid phase from a reference pressure P^+ up to the system pressure.

We have used the Redlich-Kwong equation with the following modification:

$$P = \frac{RT}{V - b} - \frac{a^+}{V\,(V + b)} \tag{2}$$

where $a^+ = aT^\alpha$. The term for a^+ has been proposed by Hederer, Peter, and Wenzel [2] to take into account the temperature dependence of the parameter a.

There have been various attempts to express the temperature dependence of the parameter a since Wilson [3] first used the RK-equation for calculating phase equilibria [4,5]. These modifications require the availability of the critical data of the substances in question. For most substances of our interest these data cannot be obtained experimentally because of thermal decomposition.

Using the expression $a^+ = aT^\alpha$ it has been possible to calculate one set of parameters for each pure component by fitting the parameters to the vapor pressure, the molal volume of the liquid phase, and to the enthalpy of vaporization. With the aid of this set of parameters, pVT data and enthalpy of vaporization can be calculated with sufficient accuracy over the temperature and pressure range which is generally of interest for phase equilibria calculations.

For calculating the properties of mixtures, the following mixing rules have been employed:

$$a^+ = \sum_{i=1}^{n} \sum_{j=1}^{n} x_i x_j a^+_{ij} \tag{4}$$

$$a^+_{ij} = (1 - \Theta_{ij})\,(a^+_{ii}\; a^+_{jj})^{1/2} \quad \text{for } i \neq j \tag{5}$$

$$a^+_{ii} = a_i T^{\alpha_i} \tag{6}$$

$$b = \sum_{i=1}^{n} x_i b_i \tag{7}$$

CORRELATION OF THE RK-PARAMETERS

In order to be able to determine RK-parameters for low volatile substances from minimum information on the component in question, attempts have been made to establish general correlations for the RK-parameters based on the density or the molal volume of pure components.

Previous attempts to correlate and generalize RK-parameters have been reviewed by Lu et al. [5]. In this review, they present a generalization of the RK-parameters up to about 10 C-atoms, which is based on reduced temperatures and Pitzer's acentric factor.

Therefore it cannot be used for determining RK-parameters for heavy hydrocarbons. Hederer, Peter, and Wenzel [2] reported general correlations for the RK-parameters of the alkanes, the alkenes, and the alkines based on the boiling temperature at normal or reduced pressure. Corresponding correlations could not be established for other groups of substances using boiling temperature as a basis.

For establishing the correlations, the RK-parameters of 214 substances with different structures and different functional groups have been used in this study. Among these 214 substances there are about 50 components with a boiling temperature higher than 300 °C at normal pressure. The RK-parameters have been determined by adjusting vapor pressure data and enthalpy of vaporization to the molar volume of the liquid phase.

Parameters a, b, a^+, α, T_s, and v_L have been submitted to a linear regression analysis resulting in the useful correlations of Table I.

Table I. Overall Correlations of RK-Parameters*

No.	Correlation	Correlation coefficient	No. of components
1	$b = 0.992042 v_{L20} - 0.01361$	0.99862	214
2	$\ln a^+_{20} = 1.735895 \ln b + 7.71744$	0.97192	214
3	$\ln a = 1.863155 \ln a^+_{20} + 0.30181$	0.97110	214

*b = liter/mol and v_{L20} = liter/mole at 20°C and 760 Torr.
$a^+ = (\text{liter})^2 \text{at/mol}^2$; $a = (\text{liter})^2 \text{at/mol}^2\ K^\alpha$

Using the correlations, the following method should be employed to calculate the RK-parameters: (1) Calculate b from the molar volume v_{L20} of the condensed phase using correlation No. 1. (For alcohols multiply b by 1.52.); (2) Calculate a^+_{20} using correlation No. 2 and the value of b from correlation No. 1; (3) Calculate a using correlation No. 3 and the value of a^+ from correlation No. 2; (4) Calculate α from $a^+ = aT^\alpha$. Reference temperature is 293.15 K and

$$\alpha = 0.176035 \ln (a^+/a) \tag{8}$$

The excellent correlation of the parameter b with molar volume v_L confirms the physical meaning of b as a co-volume. More surprising is the fact that for parameters a, a^+, and α correlations can be found which can be used for all substances without using an additional parameter. Yet we found that the alcohols and water had a constant deviation from the main bulk of components in the correlation of a^+_{20} with b. This can be explained by the association of these substances. For v_L and b, values are used which correspond to a non-associated liquid. Adjusting the parameters a and α to vapor pressure data means that these quantities are representative for the associated liquid. To remove this disparity we corrected the co-volume of these substances with their mean deviation from the bulk. The resulting parameters for these substances can now be used for the calculation of phase equilibria.

The reported correlations do not include substances with lower critical temperatures than the reference temperature of 20 °C and it is not recommended that these correlations be used for substances with critical temperatures lower than 50 °C. The RK-parameters of some of these substances are listed in Table II. If the molar volume is not known at 20 °C but at another temperature the correlations of Table III can be used for the conversion. These correlations have been calculated using the same substances as for the parameter-correlations.

Table II. RK-Parameter for Substances with T_c < 50 °C

Component	RK-Parameter		
	a	b	c
Methane	13.02	0.02956	-0.31802
Ethane	53.162	0.043506	-0.37913
Ethylene	45.769	0.038821	-0.39684
Acetylene	88.05	0.03323	-0.51587
Carbon monoxide	10.565	0.027424	-0.38829
Carbon dioxide	89.651	0.027299	-0.55689
Nitrogen	6.7349	0.026526	-0.31575
Hydrogen	0.3	0.01834	-0.01450

Table III. Correlations for Conversion of Molar Volume

Temp., °C	Correlation	Correlation coefficient	No. of substances
0	$v_{L20} = 1.000482\ v_{L0} + 0.00226$	0.99989	187
40	$v_{L20} = 0.999089\ v_{L40} - 0.00238$	0.99993	187
60	$v_{L20} = 0.998126\ v_{L60} - 0.00502$	0.99976	187
80	$v_{L20} = 0.996946\ v_{L80} - 0.00802$	0.99928	187
100	$v_{L20} = 0.994341\ v_{L100} - 0.01065$	0.99907	187

The correlations presented make it possible to determine RK-parameters of any substance from the molar volume only. In addition, there are some other advantages: The calculation of RK-parameters by adjusting to the thermodynamic properties of pure substances often results in very different values for similar substances. With the aid of the overall correlations it is possible to eliminate incorrect parameters due to experimental errors. It is assumed in general, for the procedure of parameter evaluation that the substances in question are really pure substances. This often does not prove to be correct. Sometimes one is more interested in the phase behavior of a mixture, which in many cases consists of narrow boiling components. For such mixtures the RK-parameters can be determined, provided the density and the average molecular weight are known. The RK-parameters determined in this manner can successfully be used to calculate phase equilibria.

The liquid density of a mixture of known concentration has been measured by us using a commercially available apparatus proposed by Kratky et al. which reduces the measurement of density to a measurement of frequency and yields results of high accuracy [6].

NONRANDOM INTERACTION CONSTANT Θ

The constant Θ_{12} characterizes the deviation of the random interaction of unlike molecules. For the calculation of isothermal phase equilibria Θ_{12} was assumed to be independent of pressure. This assumption has been substantiated. For the calculation of isobaric phase equilibria Θ_{12} was found to be temperature dependent (e.g., see Chang and Lu [7]). Using a constant Θ_{12} the temperature range has to be sufficiently small and an average value of Θ_{12} must be employed.

For the determination of the interaction constant Θ_{ij} several methods have been proposed. Chueh and Prausnitz [8] calculated the interaction constants from the critical temperatures. Chang and Lu [7] proposed an equation for calculating the interaction constant which was based on characteristic molecular energies. Because these characteristic molecular energies cannot be calculated theoretically, Chang and Lu used the molar excess volume of an equimolar mixture to evaluate the interaction constant. In most cases however, Θ_{ij} is determined by correlation with binary experimental phase equilibria data. We have applied this procedure where systems of low volatile substances and compressed gases were encountered. For systems consisting of volatile solvents, e.g. benzene and nonvolatile substances, we have applied a method similar to the one used by Chang and Lu.

The interaction constant can be evaluated with the aid of the mixing rule of equation (5) after having calculated a^+ previously using the overall correlations and a measured density of the mixture. Thus,

$$\Theta = 1 - \frac{a^+ - x_1^2\, a_{11}^+ - x_2^2\, a_{22}^+}{2\, x_1 x_2\, (a_{11}^+\, a_{22}^+)^{1/2}} \qquad (9)$$

Because of the reference temperature of 20 °C for the correlations an interaction constant at 20 °C is evaluated by this method.

Conversion to other temperatures is possible with the correlations for a^+ given in Table IV.

Table IV. Correlations for Conversion of a^+

Temperature °C	Correlation	Correlation coefficient	No. of substances
0	$a^+_0 = 1.087104\ a^+_{20} - 1.87901$	0.99999	187
40	$a^+_{40} = 0.924957\ a^+_{20} - 1.54410$	0.99999	187
60	$a^+_{60} = 0.859684\ a^+_{20} + 2.82450$	0.99995	187
80	$a^+_{80} = 0.802430\ a^+_{20} + 3.89451$	0.99990	187
100	$a^+_{100} = 0.751835\ a^+_{20} + 4.79464$	0.99983	187
120	$a^+_{120} = 0.706827\ a^+_{20} + 5.55617$	0.99975	187
T_s (760 Torr)	$a^+_{Ts} = 0.286429\ a^+_{20} + 24.22117$	0.98157	187

APPLICATION OF CORRELATIONS

Ten binary systems at 28 temperatures and in a pressure range up to 1000 bar have been calculated for a check, using the RK-parameters determined with the overall correlations. As examples, Figs. 1 to 3 show phase equilibria of two binary systems and a quasibinary system consisting of a mixture of glycerides (85% by weight triglycerides) and carbon dioxide. The agreement between experimental and calculated values is good.

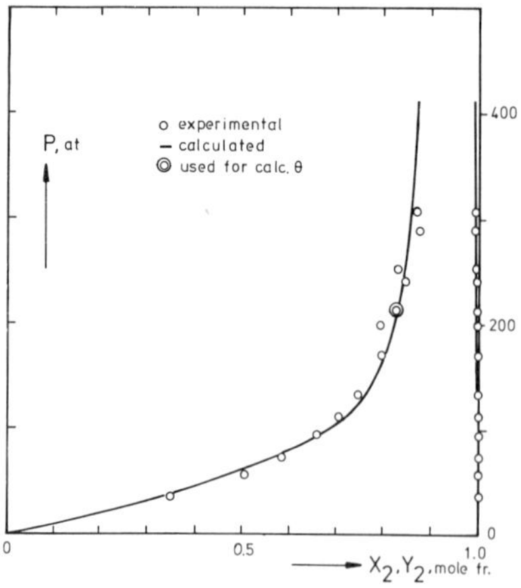

Fig. 1. Isothermal p-x-diagram of the system squalane (1) and carbon dioxide (2) at 343 K, Θ = 0.0741.

For the testing of the evaluation of the nonrandom interaction constant Θ, nine ternary or quasiternary systems at 35 isobaric and isothermal equilibrium conditions have been calculated. Figures 4 and 5 show two examples of the phase equilibria for oleic-stearic acid - dimethylformamide - ethylene and oleic-stearic acid - benzene - carbon dioxide systems which have been calculated using interaction constants determined with the overall correlations.

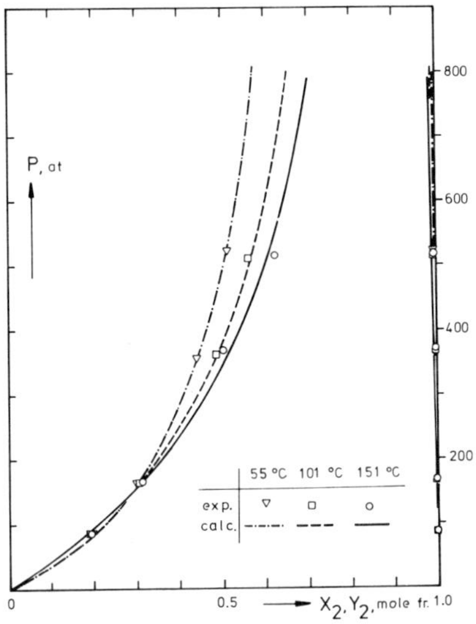

Fig. 2. Isothermal p-x-diagram of the system dibutylphthalate (1) and methane (2).

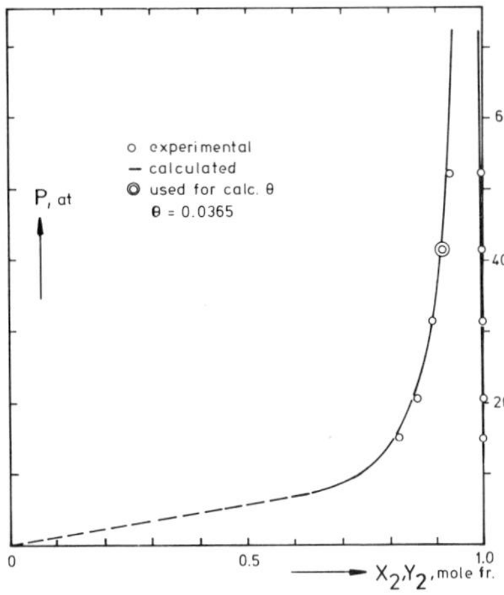

Fig. 3. Isothermal p-x-diagram of a mixture of glycerides (1) and carbon dioxide (2) at 348 K.

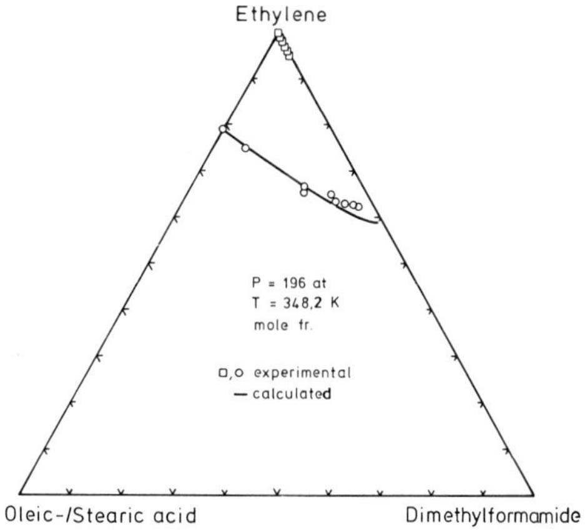

Fig. 4. Phase equilibria of the system oleic-/stearic acid (1), dimethylformamide (2) and ethylene (3).

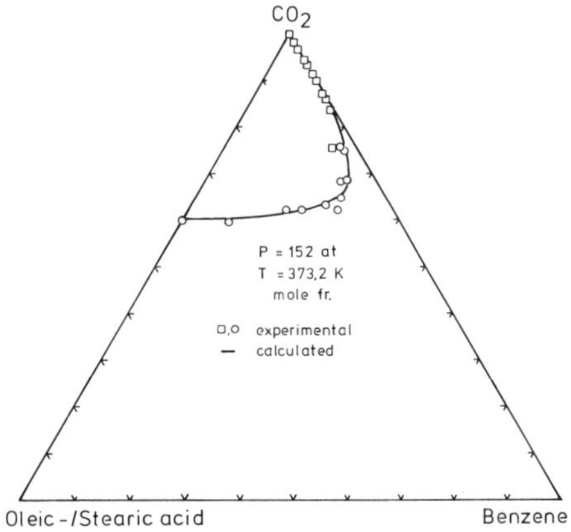

Fig. 5. Phase equilibria of the system oleic-/stearic acid (1), benzene (2) and carbon dioxide (3).

ACKNOWLEDGMENT

The authors would like to thank S. Peter for stimulating discussions and D. Weltle for providing a program for the calculation of the correlations.

NOTATION

a^+_{20} = RK-parameter at 20 °C
a,b,α = constants in the Redlich-Kwong equation of state
K_i = K-factor
P = pressure
p^+ = reference pressure
R = universal gas constant
T = temperature
T_s = boiling temperature
Θ_{ij} = binary interaction constant
v_{L20} = molar volume at 20 °C
$\bar{V}_i$ = partial molal volume
x = liquid phase mole fraction
y = vapor phase mole fraction

REFERENCES

1. O. Redlich and J.N.S. Kwong, Chem. Rev. 44, 233 (1949).
2. H. Hederer, S. Peter, and H. Wenzel, Chem. Eng. Sci. 11, 183 (1976).
3. G. M. Wilson, in Advances in Cryogenic Engineering, Vol. 9, K. D. Timmerhaus, ed., Plenum Press, New York (1964), p. 168; also Vol. 11, K. D. Timmerhaus, ed., Plenum Press, New York (1966), p. 392.
4. G. Soave, Chem. Eng. Sci. 27, 1197 (1972).
5. S.E.M. Hamann, W. K. Chung, I. M. Elshayal, and B.C.-Y. Lu, Ind. Eng. Chem., Proc. Des. Dev. 16, 51 (1977).
6. O. Kratky, H. Leopold, and H. Stabinger, Z. angew. Physik 4, 273 (1969).
7. S.-D. Chang and B.C.-Y. Lu, Can. J. Chem. Eng. 48, 261 (1970).
8. P. L. Chueh and J. M. Prausnitz, AIChE J. 13, 1099 (1967).

J-4

ON THE DETERMINATION OF REACTION VOLUMES FOR COUPLED EQUILIBRIA

D. Thusius, R. Seright and R. Grieger-Block

University of Wisconsin
Madison, Wisconsin USA

INTRODUCTION

The volume change of a chemical reaction is usually determined by measurements of the equilibrium constant at different pressures. This can be a tedious process for coupled equilibria, requiring a large amount of data over a wide pressure range. In this communication we suggest a procedure for determining reaction volumes which has certain advantages over the standard approach. The method consists of determining the equilibrium constants of a system at a single pressure, usually atmospheric pressure, and then measuring the signal change (light absorbance conductivity, etc.) as a function of reactant concentrations for a small perturbation from the initial pressure. This small perturbation condition allows use of linear equations which can be derived by simple inspection of the reaction scheme and solved with linear regression analysis. This method can be used for the quantitative analysis of overall amplitudes observed in pressure-jump relaxation experiments.

THEORETICAL CONSIDERATIONS

It can be shown [1-3] that for small pressure-induced perturbations of equilibrium concentrations, the change in absorbance of a reaction solution is in general given by the following relation*†

---

*The "small perturbation" condition is defined by the relation

$$\delta K_i/K_i = -(\Delta V_i/RT)\ \delta P \leq 0.1$$

where K_i are elementary equilibrium constants.

†ΔA^* is the absorbance change normalized with respect to temperature and the pressure change. The general form of equation (1) is

$$\begin{aligned}\Delta A^* = b_{11}X_{11} &+ b_{12}X_{12} + \cdots\cdots + b_{1R}X_{1R}\\ &+ b_{22}X_{22} + \cdots\cdots + b_{2R}X_{2R}\\ &\qquad\ddots \qquad\qquad\quad \vdots\\ &\qquad\qquad\qquad\quad + b_{RR}X_{RR}\end{aligned} \tag{1}$$

where $\underline{R}$ is the number of reaction steps, the $b_{\alpha\beta}$ are related in a simple way to elementary volume changes and extinction coefficient changes of elementary steps,

$$b_{\alpha\alpha} = \Delta\varepsilon_{\alpha}\Delta V_{\alpha} \tag{2}$$

$$b_{\alpha\beta} = \Delta\varepsilon_{\alpha}\Delta V_{\beta} + \Delta\varepsilon_{\beta}\Delta V_{\alpha} \tag{3}$$

and the X_{ij} are functions of stoichiometric coefficients and equilibrium concentrations at atmospheric pressure.

Expressions for the X_{ij} of (1) are easily derived by the following procedure:

1. Number the reaction steps 1, 2, ... T, and the chemical components 1, 2 ... N.
2. Construct an RxR symmetrical matrix $\underset{\sim}{g}$ whose elements are given by

$$g_{\alpha\beta} = \sum_{i=1}^{N} \frac{\nu_{i\alpha}\nu_{i\beta}}{\bar{C}_i} \tag{4}$$

where $V_{i\alpha}$ is the stoichiometric coefficient of species $\underline{i}$ in reaction α (coefficients are assumed positive for products and negative for reactants), and $\bar{C}_i$ is the molar equilibrium concentration of species $\underline{i}$ at atmospheric pressure.

3. Evaluate the inverse matrix $\underset{\sim}{g}^{-1}$. The elements of $\underset{\sim}{g}^{-1}$ are the X_{ij} of (1).

We may consider (1) a linear regression equation in which the experimental ΔA^* are the dependent variables, the X_{ij} are independent variables and the $b_{\alpha\beta}$ are regression coefficients which can be determined with a least-squares analysis. The reaction volumes are then calculated from the $b_{\alpha\beta}$ and the extinction coefficient changes using (2) and (3). It is assumed that the extinction coefficient

---

independent of the forcing function (temperature, pressure, etc.) and the detection system.

$$\Delta A^* = (RT/\delta P)\,\Delta A$$

differences are independent of pressure.

As an illustration of the method we shall consider the competitive binding of two inhibitors for the active site of an enzyme:

$$E + D \overset{1}{\rightleftharpoons} ED$$

$$E + I \overset{2}{\rightleftharpoons} EI$$

In the general case where no restrictions are placed on extinction coefficient changes or reaction volumes, a pressure-induced shift in equilibrium concentrations will result in an absorbance change described by the following relation:

$$\Delta A^* = b_{11}X_{11} + b_{12}X_{12} + b_{22}X_{22} \tag{5}$$

where the constant terms are

$$b_{11} = \Delta\varepsilon_1 \Delta V_1 \ , \ b_{22} = \Delta\varepsilon_2 \Delta V_2 \tag{6a}$$

$$b_{12} = \Delta\varepsilon_1 \Delta V_2 \ , \ \Delta\varepsilon_2 \Delta V_1 \tag{6b}$$

and the X_{ij} are the elements of $\underset{\sim}{g}^{-1}$, where

$$g_{11} = 1/\bar{E} + 1/\bar{D} + 1/\overline{ED} \tag{7a}$$

$$g_{22} = 1/\bar{E} + 1/\bar{I} + 1/\overline{EI} \tag{7b}$$

$$g_{12} = 1/\bar{E} \tag{7c}$$

In the limiting case where only D and ED absorb significantly at the chosen wavelength, the binding of I to E does not in itself give rise to a signal change. Nevertheless, due to coupling between the two equilibria, the pressure-induced absorbance change is still a function of ΔV_2, as seen by setting $\Delta E_2 = 0$ in (5). The final result is

$$\Delta A^* = \frac{[\bar{E}^{-1} + \bar{I}^{-1} + \overline{EI}^{-1}]^{-1}}{[\bar{E}^{-1} + \bar{I}^{-1} + \overline{EI}^{-1}]^{-1}\,[\bar{E}^{-1} + \bar{D}^{-1} + \overline{ED}^{-1}]^{-1}}\,\Delta\varepsilon_1 \Delta V_1 + \frac{[\bar{E}^{-1} + \bar{D}^{-1} + \overline{ED}^{-1}]^{-1}}{[\bar{E}^{-1} + \bar{I}^{-1} + \overline{EI}^{-1}]^{-1}\,[\bar{E}^{-1} + \bar{D}^{-1} + \overline{ED}^{-1}]^{-1}}\,\Delta\varepsilon_1 \Delta V_2 \tag{8}$$

The reaction volumes for this system can be determined conveniently by carrying out a titration in which ΔA^* is measured following each addition of inhibitor I to an equilibrium mixture of E and D. A simulated titration curve is given in Fig. 1.

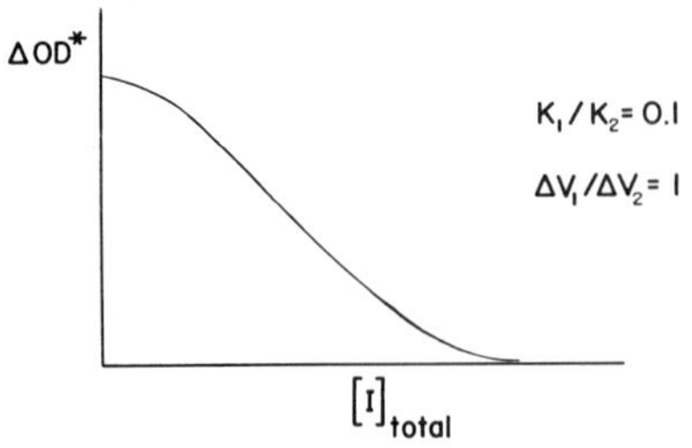

Fig. 1. Simulated titration curve.

Fitting experimental absorbance changes to (8) with a two-parameter least-squares program will yield the products $\Delta\varepsilon_1\Delta V_1$ and $\Delta\varepsilon_1\Delta V_2$. Since $\Delta\varepsilon_1$ can be determined independently, values for both volume changes can be obtained from a single titration.

For more complex coupled equilibria, generation of the inverse matrix by either algebraic or numerical methods will still allow use of linear regression analysis to find all the reaction volumes. Even with several coupled equilibria, it is only necessary to perform one relatively simple titration experiment at high pressure to calculate all the volume changes.

REFERENCES

1. D. Thusius, G. Foucault and F. Guillain, in Dynamic Aspects of Conformation Changes in Biological Macromolecules, C. Sadron, ed., Reidel, Boston (1973), p. 271.
2. D. Thusius, in Chemical and Biological Applications of Relaxation Spectrometry, E. Wyn-Jones, ed., Reidel, Boston (1975), p. 113.
3. D. Thusius, Biophys. Chem., in press.

K-1

THERMODYNAMIC PROPERTIES OF WATER - CARBON DIOXIDE - SODIUM CHLORIDE MIXTURES AT HIGH TEMPERATURES AND PRESSURES

M. Gehrig, H. Lentz, and E. U. Franck

Universität Karlsruhe
Karlsruhe, W. Germany

INTRODUCTION

Natural hydrothermal fluids very often contain carbon dioxide and chloride ions. Thus the investigation of the systems H_2O-CO_2 and H_2O-CO_2-NaCl is fundamental to the understanding of the chemistry of many geothermal systems. The knowledge of the phase behavior and of the activity coefficients of such systems gives more insight into the interaction of rocks with hydrothermal fluids and the deposition of minerals at elevated temperatures and pressures.

In the discussion presented here, the phase boundary curves of H_2O-CO_2-NaCl mixtures up to 3000 bar are described. As one of the boundary conditions for the ternary system, one needs a knowledge of the thermodynamic properties of the binary system H_2O-CO_2. The calculation of activity coefficients in gas mixtures requires data at a lower pressure range. Therefore, pvTx data for the system H_2O-CO_2 up to 500 °C and 600 bar have been redetermined. Greenwood [1,2] has presented pvTx data and activity coefficients up to 500 bar and between 450 and 800 °C.

APPARATUS AND PROCEDURE

The measurements were performed using a so-called synthetic method, with a modified apparatus described in an earlier paper [3]. The high-pressure vessel was a cylinder (OD 50 mm, ID 15 mm) built of a high-alloyed austenitic steel provided with a sapphire window and an O-ring-sealed movable piston. The position of the piston and hence the volume was determined by an inductive readout. The maximum internal volume of the vessel was 52 cm^3. The cell contents were stirred with a magnetic stirrer. The temperature was measured with a calibrated chromel-alumel thermocouple inside the cell, and the pressure was determined by calibrated Heise Bourdon gauges.

A mixture of known composition was heated at a fixed volume and the temperature and pressure values were continually monitored on an X-Y recorder. The phase transition was identified visually and by the change of slope of the pT-isochores. From a series of isochoric pT-curves, one obtains the phase boundary curve at a fixed composition (x=constant) in the pT and vT planes.

RESULTS AND DISCUSSION

Thermodynamic properties of the system H_2O - CO_2

The pvTx data of the homogeneous mixtures have been plotted on isothermal-isobaric diagrams. From these diagrams, excess volumes were derived and plotted as functions of pressure. By integration one obtains the molar excess Gibbs energy

$$g^E = \int_0^p v^E dp \qquad (1)$$

The results for 500 °C and several pressures are presented in Fig. 1.

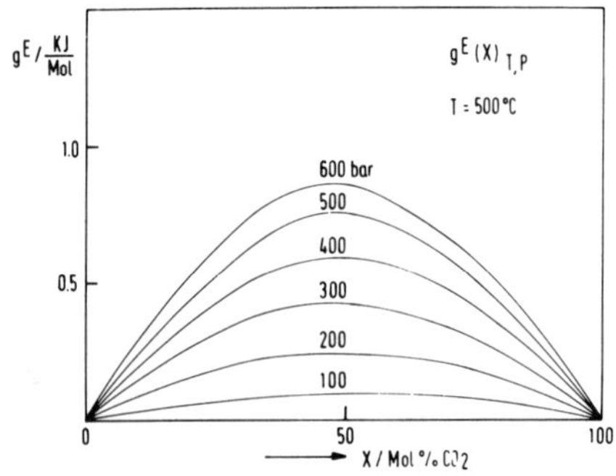

Fig. 1. Excess Gibbs energies of H_2O - CO_2 mixtures at 500 °C as a function of concentration.

These curves can be analytically described by a series expansion of the molar excess Gibbs energy g^E [4,5]. The parameters are chosen by a least squares method. In this range of accuracy, it is sufficient to use only the two parameters A and B

$$g^E = x_1 x_2 \; [A + B \; (x_1 - x_2)] \qquad (2)$$

The activity coefficient γ can be calculated by differentiation

$$RT\ln\gamma_{CO_2} = x_{H_2O}^2 [A+B \; (3-4x_{H_2O})] \qquad (3)$$

$$RT\ln\gamma_{H_2O} = x_{CO_2}^2 [A-B \; (3-4x_{CO_2})] \qquad (4)$$

For a test of consistency, the activity coefficients have been differentiated with the aid of a computer program in order to again obtain excess volumes

$$v^E(T,p) = RT(x_1 d\ln\gamma_1/dp + x_2 d\ln\gamma_2/dp)_T \qquad (5)$$

The agreement between the experimental and the previously calculated excess volumes is satisfactory. With these results, the activity coefficients of Greenwood [1,2] can be compared. The deviation is usually less than 10% and only at very low carbon dioxide concentrations does the deviation of the activity coefficients of CO_2 become larger.

The System H_2O - CO_2 - NaCl

For measurements with salt mixtures up to 500 °C and 3 kbar, a solution of 6% NaCl (by wt.) was used with increasing amounts of CO_2, such that the ratio of water to sodium chloride remained constant. The isoplethic (x = constant) phase boundary curves in the pressure-temperature plane exhibit a temperature maximum and in some cases a pressure minimum, as can be seen for the water-rich part of the system in Fig. 2. The temperature maximum is close to 500 °C, depending on the CO_2 content. At low pressures (larger volumes), the isopleths of mixtures containing low concentrations of CO_2 appear close to the vapor pressure curve of the pure salt solution. At a certain temperature, the influence of the carbon dioxide predominates and the curves proceed to lower temperatures. The shape of these curves resembles phase boundary curves of binary mixtures without salt [6,7], but they are displaced to higher temperatures because of the influence of the salt, as is demonstrated for a mixture containing 4 mol% CO_2 in Fig. 2.

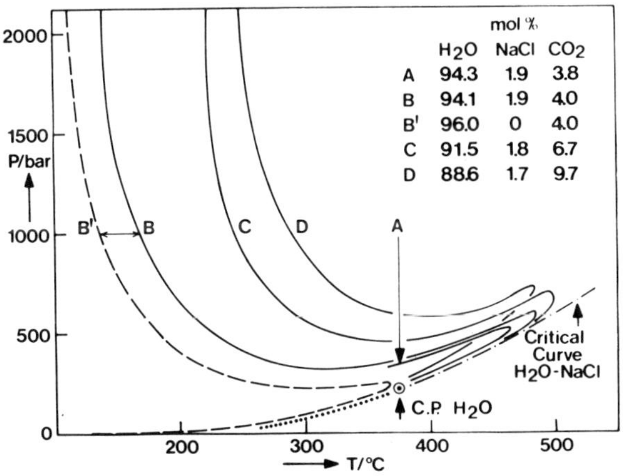

Fig. 2. Isoplethic phase boundary curves of the water-rich side of the ternary system H_2O-CO_2-NaCl in a pressure temperature plane.

The entire phase behavior of these mixtures with a 6% (by wt.) salt solution is described in Fig. 3. At higher pressures (lower volumes), the boundary curve of the heterogeneous region exhibits a temperature maximum at a CO_2 concentration of about 20 to 30 mol%. On both the water-rich side and the carbon dioxide-rich side, the heterogeneous region does not extend to such high temperatures.

The curves describing the phase behavior of the ternary system can qualitatively be compared up to about 15 mol% CO_2 with the results of Kennedy [8]. The agreement is reasonable.

Figure 4 presents the isoplethic (x = constant) phase boundary curves in a volume-temperature plane. For the sake of a better appraisal, only some of the measured curves are shown. The heterogeneous region extends from the boundary curve of pure water

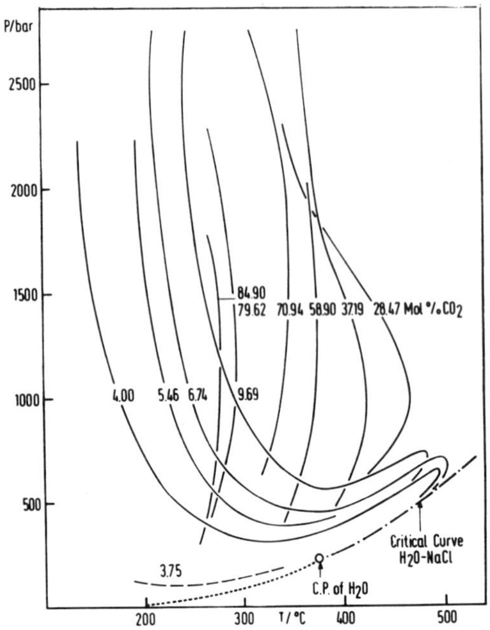

Fig. 3. Isoplethic phase boundary curves in a pressure-temperature plane of a 6% (by wt.) NaCl solution with increasing CO_2 content.

to a temperature maximum and recedes then to lower temperatures for mixtures with a larger CO_2 content. With both representations (the pT and the vT planes) all information of the phase behavior of this mixture with a 6 wt.% salt solution is given.

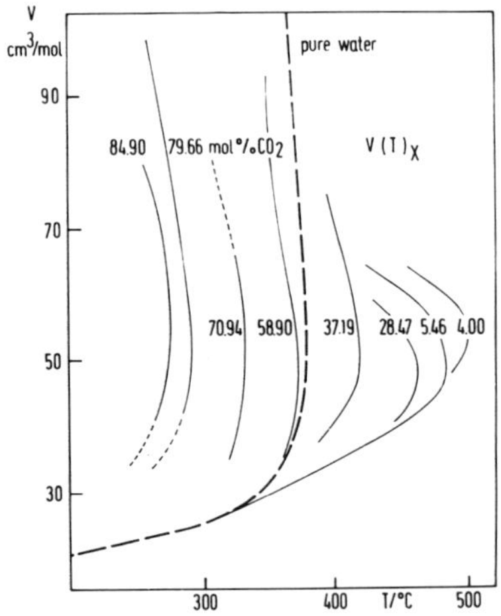

Fig. 4. Isoplethic phase boundary curves in a volume-temperature plane of a 6% (by wt.) NaCl solution with increasing CO_2 content.

REFERENCES

1. H. J. Greenwood, Am. Jour. Sci. 267 A, 191 (1969).
2. H. J. Greenwood, Am. Jour. Sci. 273, 561 (1973).
3. H. Lentz, Rev. Sci. Instrum. 40, 371 (1969).
4. J. M. Prausnitz, Molecular Thermodynamics of Fluid-Phase Equilibria, Prentice Hall, Englewood Cliffs, New Jersey (1968).
5. J. S. Rowlinson, Liquids and Liquid Mixtures, Butterworth and Company, London, England (1959).
6. K. Tödheide and E. U. Franck, Zeitschrift für phys. Chemie neue Folge 37, 387 (1963).
7. S. Takenouchi and G. C. Kennedy, Am. Jour. Sci. 262, 1055 (1964).
8. S. Takenouchi and G. C. Kennedy, Am. Jour. Sci. 263, 445 (1965).

K-2

DETERMINATION OF THE SPECIFIC HEAT CAPACITIES OF AQUEOUS NACL SOLUTIONS AND SEAWATER UP TO 2 KBARS USING THE TEMPERATURE-JUMP TECHNIQUE

K. G. Liphard and A. Jost

Ruhr-Universität Bochum
Bochum, W. Germany

INTRODUCTION

Among thermodynamic properties of aqueous electrolyte solutions, heat capacity is of primary interest. Various measurements of heat capacities as a function of temperature have been made in the past decade, but there are only a few measurements of the pressure dependence [1].

Usually, heat capacities are measured by calorimetric methods. These methods, however, show disadvantages at high pressures: The mass of the autoclave is usually large compared to the mass of the sample. Additionally, temperature gradients and heat losses through pressure connections have to be taken into account.

Using the temperature-jump technique [2], these disadvantages can be avoided. Here the measuring time is below 1 ms so measurements can be made under nearly adiabatic conditions. With this technique, a high-voltage capacitor is discharged through the electrolytically-conducting solution. The electrical energy absorbed causes a temperature rise ΔT given by

$$W_{el} = \frac{1}{2} CU^2 = c \rho V \Delta T \qquad (1)$$

EXPERIMENTAL

During the present measurements, the ΔT (temperature-jump) was measured indirectly by means of the light absorption change of a colored pH indicator (phenol red in very low concentration) dissolved in the buffered solution under test. The temperature dependence of light absorption was calibrated separately in an optical high-pressure cell [3]. The cell was constructed of stainless steel and designed for pressures up to 2.5 kbars. The cell had two sapphire windows for optical absorption measurements and was mounted in a Beckman spectrophotometer. The solutions under test were in a quartz cuvette. The pressure was transmitted by compressed nitrogen. In order to prevent the dissolution of compressed nitrogen in the solution under test, the quartz cuvette was sealed by a flexible tube also filled with the solution. The pressure was transmitted to the solution under test through the wall of the elastic tube. The temperature was measured inside the cell with a steel-sheathed thermocouple. The pressure was measured with Heise gauges.

Heat capacity measurements were carried out in a high-pressure temperature-jump apparatus [4] that had been developed for the investigation of fast reactions in solutions under high pressure. The solution to be measured was filled in a temperature-jump cell with optical sapphire windows for spectroscopic measurements; the cell was placed in a high-pressure autoclave. The autoclave was constructed of stainless steel and designed for pressures up to 4 kbars. It had two optical-sapphire windows for spectroscopic investigations. The pressure was also transmitted by compressed nitrogen and was measured with Heise gauges. To avoid the dissolution of nitrogen in the solution under test, the temperature-difference cell could also be sealed with a flexible tube. The temperature was measured inside the autoclave with a steel-sheathed thermocouple.

The specific heat capacity values can be obtained from equation (1). The values of C and U can be measured directly; the values of ΔT are calculated from the optical absorption change. Density data can usually be obtained from the literature. The values of the effective volume (the volume of the sample in the temperature-difference cell which is only heated) are difficult to measure because of the complicated geometry of the cell. Thus, the effective volume was determined by calibration of the cell with salt solutions of known heat capacity at atmospheric pressure. During these measurements several temperature-jump cells with different volumes were used. As expected, no influence of the cell geometry on the heat capacity measurements could be detected.

Using the temperature-jump technique alternatively, measurements of heat capacities at constant volume, c_V, and at constant pressure, c_p, can be made. When a solution is heated normally, thermal expansion occurs. If the heating time is very short (on the order of 1 μs) the heating takes place under essentially isochoric conditions, i.e, dV = 0. This produces a pressure wave which travels through the solution. If the solution is heated more slowly (for periods longer than about 20 to 30 μs) the pressure is balanced over each period. Therefore, c_V values are measured over short heating periods, c_p data over longer heating periods. These time periods are related to the cell dimensions and can easily be varied by an appropriate choice of high-voltage capacitor.

RESULTS AND DISCUSSION

Measurements have been made for aqueous NaCl solutions [1] in concentration ranges from 0.1 to 2.0 mol/kg at 1 and 2 kbars, and for seawater [5] at 32.6 g/kg salinity up to 2 kbars. The measuring temperature was 293.2 K. The values of the specific heat capacities were obtained from the limiting slopes at zero temperature jump of plots of ΔT vs. U^2, according to equation (1). The accuracy of the specific heat capacity values was estimated to be ± 1%.

Figure 1 shows the pressure dependence of the c_p for aqueous NaCl solutions at different molalities and of pure water [6]. The specific heat capacities decrease with increasing pressure. The influence of pressure on c_p decreases with increasing salt concentrations: The pressure dependence of the specific heat capacity of pure water is about seven times larger than that of a solution of 2.0 mol/kg NaCl in water.

The values of the specific heat capacity of seawater at pressures up to 2 kbars are listed in Table I; the pressure dependence of the heat capacity of the seawater sample is shown in Fig. 2. The specific heat capacities at constant pressure decrease with increasing pressure. The values of c_p and its pressure dependence are similar to the corresponding values of a 0.5 mol/kg solution of NaCl in water.

The unusual high heat capacity of pure water in contrast to other liquids can be attributed to a highly ordered structure. Changes of heat capacities are sometimes

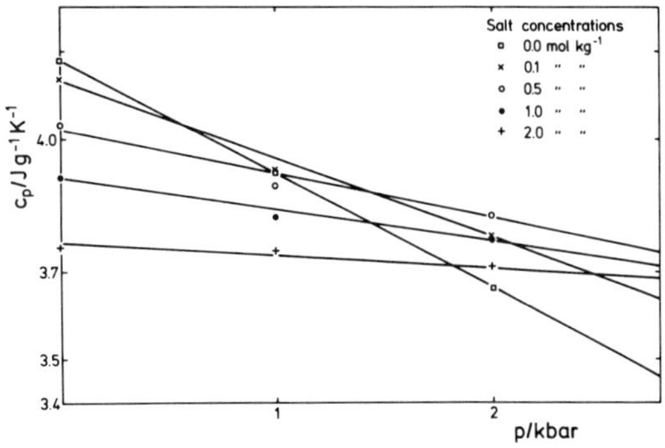

Fig. 1. Specific heat capacity at constant pressure as a function of pressure for aqueous NaCl solutions at different molalities, and for pure water. (T = 293.2 K, m = 0.1 to 2.0 mol/kg.)

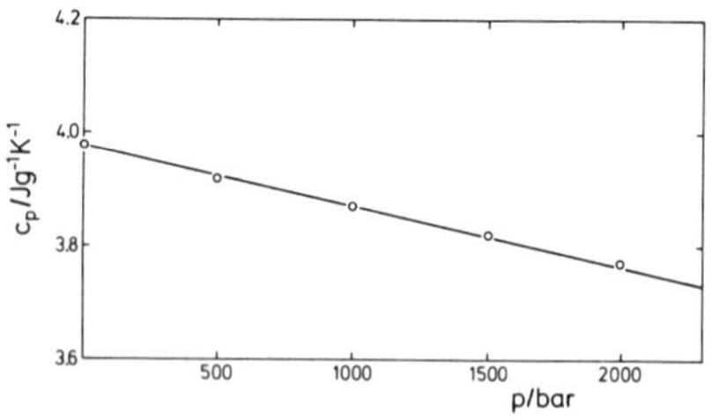

Fig. 2. Specific heat capacity at constant pressure as a function of pressure for a seawater sample at 32.6 g/kg salinity, T = 293.2 K.

explained in terms of "structure making" and "structure breaking". The decrease of the specific heat capacities with increasing pressure could therefore be attributed to the structure-breaking influence of pressure. The decrease of the pressure influence on c_p with increasing salt concentration can possibly be explained by the structure-making effects of NaCl on the structure of water.

Table I. Specific Heat Capacities of a Seawater Sample at 32.6 g/kg Salinity and 293.2 K up to 2 kbars*

p, bar	c_p, J/gK
1	3.98
500	3.92
1000	3.87
1500	3.82
2000	3.77

*Values beyond the actual significant figures have been provided to avoid accumulation of roundoff errors.

In order to obtain more detailed information about the influence of electrolytes on the structure of water at high pressures, measurements with other salts are in preparation.

ACKNOWLEDGMENT

Financial support of the Deutsche Forschungsgemeinschaft and of the Fonds der Chemischen Industrie is gratefully acknowledged. The authors wish to thank the Reedereigemeinschaft Forschungsschiffahrt, Bremen, for a seawater sample.

NOTATION

C = capacity

c = specific heat capacity, J/gK

c_V = specific heat capacity at constant volume

c_P = specific heat capacity at constant pressure

m = molality

p = pressure, bar

T = absolute temperature, K

ΔT = temperature difference

U = loading voltage of the capacitor, V

V = effective volume, cm^3

W_{el} = electrical energy, J

ρ = density, g/cm^3

REFERENCES

1. K. G. Liphard, A. Jost, and G. M. Schneider, J. Phys. Chem. 81, 547 (1977).
2. M. Eigen and L. DeMaeyer, in Techniques of Organic Chemistry, Vol. VIII, A. Weissberger, ed., Wiley Interscience, New York (1963).
3. A. Jost, Ber. Bunsenges. physik. Chem. 79, 850 (1975).
4. A. Jost, Ber. Bunsenges. physik. Chem. 78, 300 (1974).
5. K. G. Liphard, PhD. Thesis, Ruhr-Universität Bochum, Bochum, Germany (1977).
6. A. M. Sirota and P. E. Beljakov, Teploenergetica 6, 67 (1959).

K-3

SOME PHYSICOCHEMICAL PROPERTIES OF WATER AND AQUEOUS SOLUTIONS UNDER HIGH PRESSURE - HYDROPHOBIC HYDRATION

K. Suzuki and Y. Taniguchi

Ritsumeikan University
Kyoto, Japan

and

M. Tsuchiya
The College of Naval Architecture of Nagasaki
Nagasaki, Japan

INTRODUCTION

It is well recognized from the analysis of x-ray diffraction that liquid water at atmospheric pressure retains the four-coordinated open structure of ice-I. On the other hand, the critical pressure for the presence of ice-I is well known to be about 2 kbars in the phase diagram of water. These two facts suggest the possibility that compressed liquid water might retain the structure of dense ice of the corresponding high-pressure phase, and that the physicochemical properties of liquid water are changed dramatically above about 2 kbars.

Indeed, there is much evidence for the latter. The appearance of a minimum in the plot of the viscosity of water vs. pressure, which has been well known for the past century, is one of the observable changes of liquid water under pressure. Analogous behavior has been observed recently by Jonas and his co-workers in the proton NMR chemical shift [1], the spin-lattice relaxation time [2,3], and the self-diffusion coefficient [4] for liquid water or liquid heavy water. For the former we have as yet no experimental evidence. However, the same conclusion as our assumption for the structure of the compressed liquid water has recently been deduced by Konda and Yamamoto [5] from the theoretical consideration on the basis of the mixture model. These structural measurements are now earnestly desired.

A similar inversion is found in the solubility of liquid hydrocarbons such as toluene [6], n-propylbenzene [7], and octanone-4 [8] in water at high pressure; that is, the solubility increases with increasing pressure up to 1 to 2 kbars and then decreases. This fact tells us that the volume change accompanying the dissolution of hydrocarbons in water, ΔV, changes sign from negative to positive at about 1 to 2 kbars.

The presence of a maximum in the solubility-pressure plot may be regarded as a characteristic of hydrophobic hydration together with the presence of a minimum in the solubility-temperature plot at constant pressure [8]. The concept of hydrophobic hydration has been compared to the formation of an "iceberg".* The negative sign of ΔV in the low pressure region means that the partial molar volume of the hydrocarbon molecule is less than the molar volume. A possible explanation is that the volume of the solute molecule itself is eliminated because of the packing of the solute molecule in the open spaces of the iceberg. On the other hand, from the estimated drastic change of the liquid-water structure above about 2 kbars, it might be reasoned that the formation of an iceberg might be impossible or difficult because of the deformation of the water structure by compression. Because of this, it is estimated that the inversion in the partial molar volume of the hydrocarbon molecule in water might occur above 1 to 2 kbars. This estimation is indeed confirmed by the analysis of Kliman's solubility data [8].

This paper presents results of a near-infrared spectral study on compressed-liquid water designed to gain structural information and pressure effects on some physicochemical properties of aqueous solutions of amphiphiles, such as critical micelle concentration (cmc) and low critical solution temperature (LCST). This, in turn, may give information on hydrophobic hydration and the solubility of hydrocarbons in water. (Ionic or polar derivatives of hydrocarbons are called amphiphiles or amphiphilic molecules, according to Tanford [9].)

NEAR-INFRARED STUDIES OF THE COMPRESSED LIQUID WATER

Excellent spectroscopic investigations for compressed liquid water have been performed by Walrafen [10] at 25°C using laser Raman techniques and by Franck and Roth [11] at 30 to 400°C using infrared techniques. These investigations, however, do not cover the interesting region below room temperature.

In the present investigation, the near-infrared absorption band at about 980 nm, which is assigned to the combination band of $2\nu_1 + \nu_3$ [12], was measured at 8, 27, and 75°C to 4.5 kbars, where ν_1 and ν_3 are the fundamental vibrations for the symmetric and antisymmetric stretching modes. A clamp type high-pressure bomb with optical windows and a Shimadzu MPS-50L spectrophotometer were used [13].

*Clathrate-hydrate-like structure formed by the water molecules adjacent to hydrophobic molecules, which was proposed by Frank and Evans [J. Chem. Phys., 13, 507 (1945)].

Figure 1 shows some typical examples of the absorption spectra of liquid water observed between 800 and 1300 nm at each given condition. In this experiment, the influence of pressure and temperature on the absorption band at about 980 nm was measured. It was confirmed by the use of a curve analyzer that the shift of this band was not influenced by the adjacent absorption band at about 1200 nm. Figures 2 and 3 present the influences of temperature and pressure on the absorption-band shift. It is noted in Fig. 2 that the absorption maximum shifts towards the shorter wavelength with increasing temperature at atmospheric pressure, and in Fig. 3 that the pressure effect differs dramatically above and below 27°C. The initial compression at 8°C produces a blue shift; that is, the pressure effect parallels the temperature effect. However, at 75°C or at high pressures above about 2 kbars, even at the lower temperature region of 8 and 27°C, the pressure effect is just opposite to the temperature effect. The spectral observations for a 20 mole % solution of H_2O in D_2O showed the same trend. Such an inversion may support our assumption for the structure of the compressed liquid water.

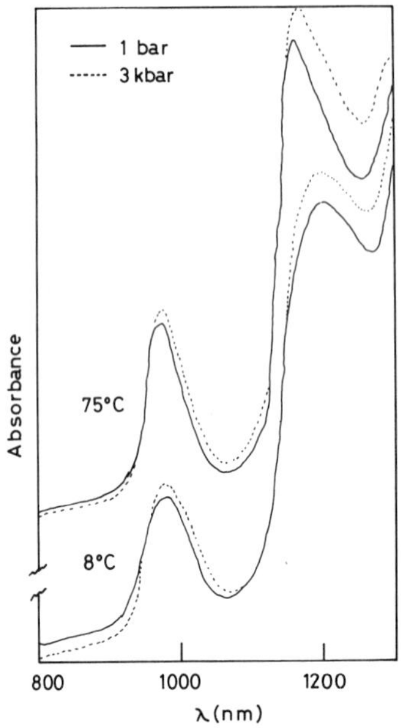

Fig. 1. Absorption spectra of liquid water under different conditions. Solid line shows the spectra at 1 bar while the broken line shows the spectra at 3 kbar at each temperature, respectively.

ANOMALIES IN AQUEOUS SOLUTIONS OF AMPHIPHILES AT HIGH PRESSURE

Influence of Pressure on Critical Micelle Concentration

There are several investigations for the influence of pressure on cmc of surfactants since the first experiment of Hamann [14]. The investigations, however, have been performed only for ionic surfactants of a salt type, using conductivity as the indicator of micellar behavior. It is well known from these studies that a maximum appears at about 1 kbar in the cmc-pressure plot. A plausible explanation, however, has not been given for this curious inversion. Recently, Rodriguez and Offen [15] measured the cmc for sodium dodecyl sulfate, using naphthalene as a probe, and noted that cmc does not invert its behavior above about 1.5 kbars. Therefore, additional studies are required to identify the factors responsible for this discrepancy.

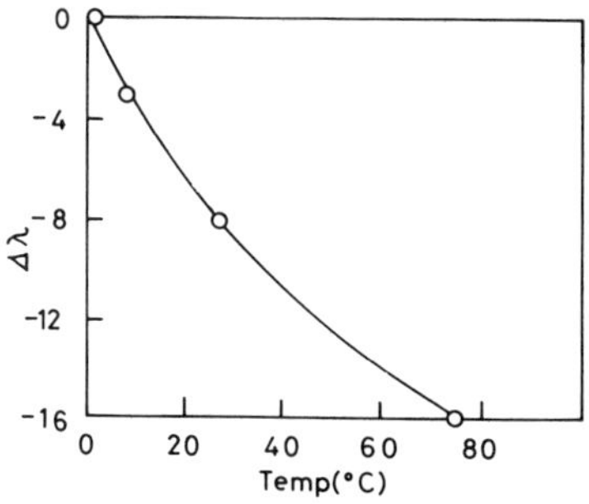

Fig. 2. Influence of temperature on the absorption-band shift at 1 bar. The Δλ is obtained by taking the wavelength of the absorption maximum (about 980 nm) at 2°C as a reference.

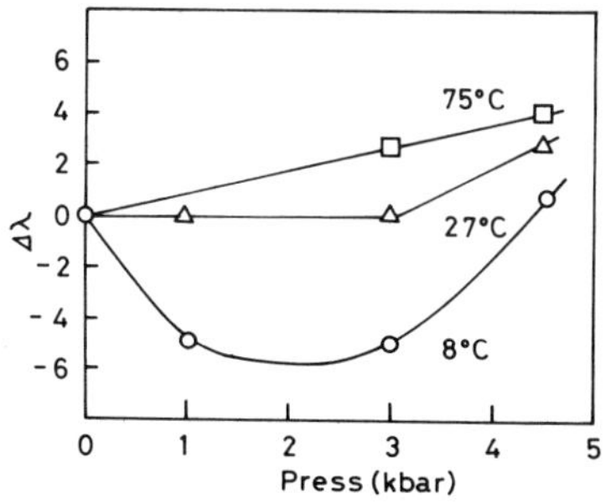

Fig. 3. Influence of pressure on the absorption-band shift at 8°C (-o-), 27°C (-Δ-) and 75°C (-□-). The Δλ is obtained by taking the wavelength of the absorption maximum (about 980 nm) at 1 bar as a reference.

Accordingly, we have measured the cmc for dodecyl sulfuric acid desalted from sodium dodecyl sulfate, using the conductivity method. The cmc-pressure plot is shown in Fig. 4. A maximum appears clearly at about 1 kbar as in sodium dodecyl sulfate. The cmc of the nonionic surfactant pentaoxyethylene dodecyl alcohol, $C_{12}H_{23}O(CH_2CH_2O)_5H$ was measured recently at our laboratory using the I_2-method [16]. In this case, a maximum was also found at 1.5 to 2 kbars. It is well recognized that hydrophobic interactions participate in the micelle formation. Therefore, this inversion must be attributed to the anomaly of the hydrophobic hydration above and below about 2 kbars, as in the case of the solubility of hydrocarbons in water. This assumption is verified by the information presented by Tanaka et al. [17] that the sign for the contribution of the hydrocarbon tail to the partial molal-volume change on micelle formation changes from positive to negative between 1 and 1.5 kbars at 35°C.

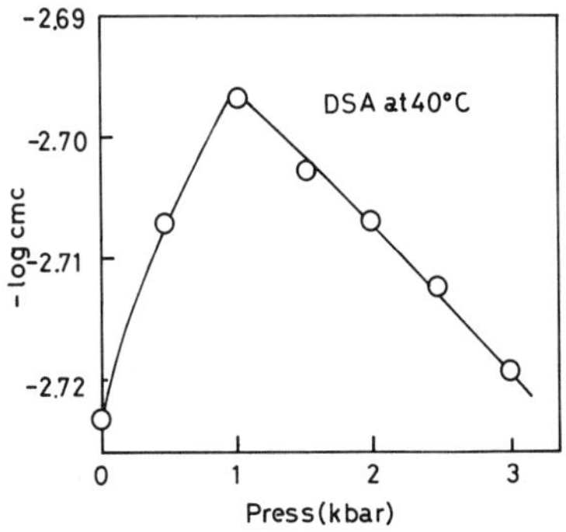

Influence of Pressure on Low Critical Solution Temperature

A low critical solution temperature (LCST) exists in the mutual solubility curve of some organic liquids such as amines, pyridines, piperidines, alcohols, and alcoholic ethers with

Fig. 4. Influence of pressure on the cmc of dodecyl sulfuric acid at 40°C.

water. This signifies that the mutual solubility decreases with increasing temperature. For this unusual dissolution process, Hirshfelder et al. [18] suggested the existence of a hydrogen-bonded complex between the solute and water molecules. Brun et al. [19] confirmed such a complex formation from the self-diffusion measurement of pyridine in aqueous solution. Coop and Everett [20], on the other hand, indicated that the presence of a LCST might be caused by the interaction of the "inert" hydrocarbon group of solute molecules with water. If the LCST is plotted vs. pressure, a maximum appears at a certain pressure in many cases. Schneider [21] had extensively studied such a phase behavior from the standpoint of thermodynamics.

In this paper, we extend the example to polyoxyethylene alcohol, i.e., triethylene glycol n-hexylether (THE), $n\text{-}C_6H_{13}O(CH_2CH_2O)_3H$ and amphiphilic polymer, i.e. partially acetylated polyvinyl alcohol (PVA-VAC), $[CH_2\text{-}CH(OCOCH_3)\text{-}]_x[\text{-}CH_2\text{-}CH(OH)\text{-}]_y$. The experiments were performed in the same way as in the previous paper [22]. The solubility of these compounds in water increases with a decrease in temperature. They also show partial miscibility with water and a LCST, as shown in Fig. 5.

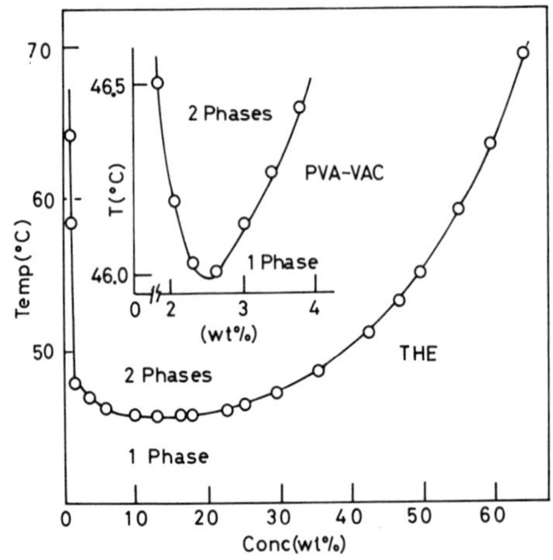

Fig. 5. Phase behavior of triethylene glycol n-hexylether (THE) and partially acetylated polyvinyl alcohol (PVA-VAC) with water at 1 bar. (The PVA-VAC is a sample with a MW of 1.2×10^3 and acetylated 28%.)

The influence of pressure on the LCST of these compounds is shown in Fig. 6, which indicates a maximum at each pressure, respectively. The region above the curve is two phases and the region below the curve is one phase. The solubility of the solute molecules in water initially increases with increasing pressure and then decreases. We assume that the hydrophobic hydration together with the hydrogen-bonded hydration participates in the solubility of amphiphilic molecules in water. The dimer formation of formic acid, in which only hydrogen bonding is assumed to participate, increases steadily with increasing pressure up to 6 kbars [23]. Therefore, the inversion found in the solubility of amphiphilic molecules in water is also understandable from the anomaly of the hydrophobic hydration at high pressure.

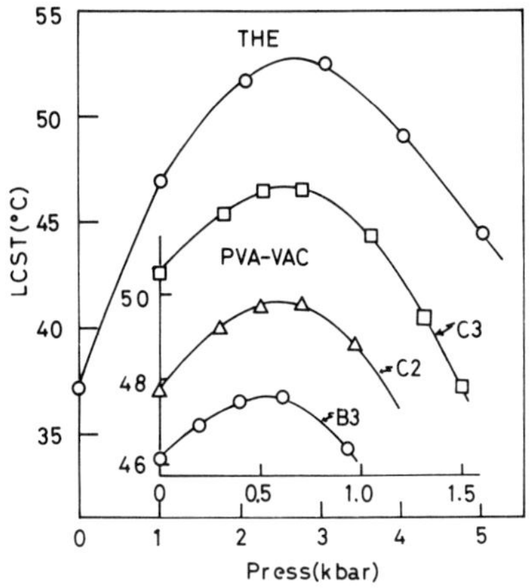

Fig. 6. The influence of pressure on LCST of THE and PVA-VAC with water. The PVA-VAC samples are characterized as follows:

Sample	MW (10^{-3})	Degree of Acetylation,%
B3	1.2	28
C2	1.6	27
C3	1.2	27

CONCLUSION

The combination band of $2\nu_1 + \nu_3$ at about 980 nm for the liquid water changes significantly above and below about 2 kbars (except at relatively high temperature). This observation seems to support our assumption that liquid water compressed above about 2 kbars might retain the dense-ice structure of the corresponding high-pressure phase, differing from the open structure of ice-I.

It is expected from such a structural assumption and from the analysis of the solubility data of hydrocarbons in water that the "iceberg" formation may be impossible or difficult above about 2 kbars. It is assumed from these considerations that the compression below 1 to 2 kbars favors hydrophobic hydration, and further compression above 1 to 2 kbars retards hydrophobic hydration. In the latter case, the term "hydrophobic hydration" might not be adequate for use in the ordinary sense. Thus, the inversion appearing at a certain pressure in the system of amphiphilic molecules and water might be attributed to the anomaly of the hydrophobic hydration at high pressures.

Structural measurements by the scattering of x-rays and neutrons for compressed liquid water are still needed together with some proper structural measurements for aqueous solutions, to completely answer the questions raised in this study.

REFERENCES

1. J. W. Linowski, Nan-I Liu, and J. Jonas, J. Chem. Phys. 65, 3383 (1976).
2. T. DeFries and J. Jonas, J. Phys. Chem. 66, 896 (1977).
3. Y. Lee and J. Jonas, J. Chem. Phys. 57, 4233 (1972).
4. D. J. Wibur, T. DeFries, and J. Jonas, J. Chem. Phys. 65, 1783 (1976).
5. C. Konda and T. Yamamoto, Chem. Phys. Letters, in press.
6. R. S. Bradley, M. Dew, and D. C. Munro, High Temp.-High Press. 5, 169 (1973).
7. A. M. Zipp, Ph.D. Dissertation, Princeton University, Princeton, New Jersey (1973).
8. H. Kliman, Ph.D. Dissertation, Princeton University, Princeton, New Jersey (1969).
9. C. Tanford, The Hydrophobic Effect, John Wiley & Sons, New York (1973).
10. G. E. Walrafen, J. Solution Chemistry 2, 159 (1973).
11. E. U. Franck and K. Roth, Disc. Faraday Soc. 29, 108 (1967).
12. O. D. Bonner and G. B. Woolsey, J. Phys. Chem. 72, 899 (1968).
13. K. Suzuki and M. Tsuchiya, Bull. Chem. Soc. Japan 48, 1701 (1975).
14. S. D. Hamann, J. Phys. Chem. 66, 1359 (1962).
15. S. Rodriguez and H. Offen, J. Phys. Chem. 81, 47 (1977).
16. T. Sugano and K. Suzuki, "Influence of Pressure on the CMC of Nonionic Surfactant," to be published.
17. M. Tanaka, S. Kaneshina, and R. Matuura, Memoirs Faculty of Science (Fukuoka University) 4, 131 (1974).
18. J. Hirschfelder, D. Stevenson, and H. Eyring, J. Chem. Phys. 5, 896 (1937).
19. B. Brun, R. Ganfres, J. Rouviere, and J. Salivinien, C. R. Acad. Sci. Paris 260, 3636 (1965).
20. J. L. Coop and D. H. Everett, Disc. Faraday Soc. 15, 174 (1953).
21. G. M. Schneider, Water, F. Franks, ed., Vol. 2, Plenum Press, New York (1973), p. 381.
22. K. Suzuki and M. Tsuchiya, Bull. Inst. Chem. Res. (Kyoto University) 47, 270 (1969).
23. K. Suzuki, Y. Taniguchi, and T. Watanabe, J. Phys. Chem. 77, 1918 (1973).

K-4

DEUTERIUM/HYDROGEN RATIOS AND WATER CONTENT OF SOME SILICATE MELTS AT HIGH PRESSURE

Y. Hariya

Hokkaido University
Sapporo, Japan

Y. Kuroda

Shinshu University
Matsumoto, Nagano, Japan

S. Matsuo

Tokyo Institute of Technology
Tokyo, Japan

and

T. Suzuoki

Japan Meteorological Agency
Tokyo, Japan

INTRODUCTION

Various approaches have been made by different workers to determine the role of water in the mantle or the lower crust. For example, Kuroda et al. [1] determined the water content of some hydro-silicate minerals under high pressure and high temperature conditions. The stability field of some hydro-silicate minerals was studied by Kushiro [2], Kushiro et al. [3], Lambert and Wyllie [4], Hariya et al. [5,6] and others. The investigations of the role of water on the magma genesis (Allen et al. [7], Kushiro [8], and Burnham and Davis [9]) and the hydrogen and oxygen isotope fractionation of hydrous minerals (Kokubo et al. [10], Sheppard and Epstein [11], and Kuroda et al. [12]) also provide valuable information as to the concentration of water in the upper mantle or the lower crust.

Recently, Suzuoki, and Epstein [13] conducted an experimental determination of the deuterium/hydrogen ratios between water and various hydrous minerals under hydrothermal conditions, and found that δD of hydrogen in the lattice of the hydrous minerals is controlled by the composition of the cations in the octahedral site of minerals, as well as by the equilibrium temperature and δD of the coexisting water. Kuroda et al. [14] reported that the water content and δD of common hornblendes from ultramaficmafic rocks are definitely higher than those from granites in spite of the similarity in δD values. These findings suggest that the appreciable difference in the range of water content between the hornblendes from the two rock types is caused principally by the change in the temperature and pressure conditions under which the hornblendes were formed. Kuroda et al. [1] found that the relationship between water content and vapor pressure on the synthetic amphibole may be regarded as linear.

Since most rocks of deep-seated origin contain only a few percent hydrogen, their D/H ratio is more diagnostic of "juvenile water". Recently, Sheppard, and Epstein [11] investigated the variability of the D/H ratio of hydrous minerals of possible mantle or lower crustal origin. They estimated that the D/H ratio in "juvenile water" is -48 $\pm$ 20 per mil. Kuroda et al. [15] also estimated that the mantle water seemed to have a δD value of -85 $\pm$ 10 per mil on the basis of their study on the olivine nodule of possible mantle origin. It is interesting to note that δD values of the ridge basalts from both the Pacific and the Atlantic oceans show the range from -84 to -75 per mil with water content (H_2O+) varying between 0.2 to 0.45 wt.% [16].

The presence of water in the lower crust and the upper mantle has been suggested by many investigators. It is also widely believed that the water in the lower crust and the upper mantle plays an important role in the origin of magma.

We have investigated the variability of the water content and the D/H ratios of silicate glasses of quartz and albite at high pressures. The quartz and albite with excess water were selected for the following two reasons: (1) several properties of melts of these minerals are very similar to those of natural rock melts in the compositional range between granite and diorite; and (2) a great deal of experimental work has been done on quartz-H_2O and albite-H_2O systems. The objective of this paper is to find a correlation between the water content and the D/H ratios of the silicate melts of albite and quartz.

EXPERIMENTAL TECHNIQUE

High-pressure experiments in the present study were carried out in sealed Pt capsules with a piston-cylinder apparatus. The pressure-transmitting medium was sodium chloride. Reported pressures are believed to be accurate within 1 kbar. Temperatures were measured with $Pt-PtRh_{13}$ thermocouples which were in contact with the Pt capsules containing the samples. No correction was made for the effect of pressure on the emf of the thermocouples. In order to establish that water was not lost during the run, only those samples that showed no weight change during the run were used. All the glass samples were checked with an optical microscope for homogeneity.

Finely crushed samples of pure quartz crystals (for analysis see Table I) collected from Brazil by the Asahi-glass Co. were used for the present study. The starting material of albite was synthesized hydrothermally at 750°C and 1000 bars from homogeneous glass prepared by mixing quartz, Al_2O_3, and sodium disilicate at 1200°C. The water content of the synthetic albite used for the investigation was 0.13 wt.%.

RESULTS

The D/H ratios are reported relative to the standard mean ocean water (SMOW) in per mil units. The δD value of the starting water used for the high-pressure experiment is -75.1 to -78.0 per mil. The δD value and the water content of the quartz and albite glasses at high pressure are shown in Table II. The relationship between the δD value and the water content of all high-pressure runs is shown in Fig. 1 which also includes the δD values of the starting water.

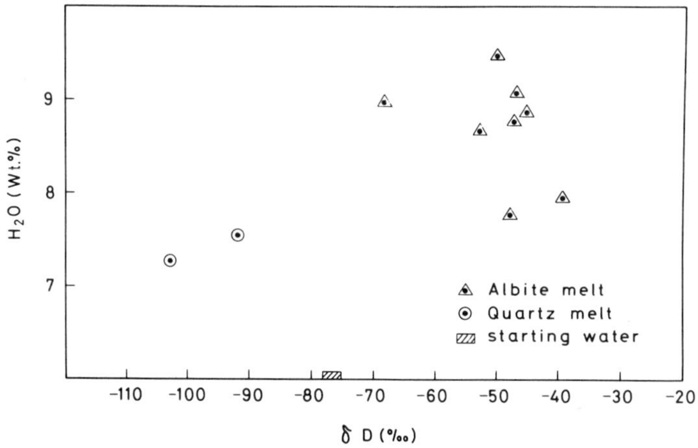

Fig. 1. Plot of δD values for quartz glass and albite glasses at high pressure runs vs. their water contents.

Water content of quartz glass at 1500°C under 15 and 25 kbar was found to be 7.29 and 7.56 wt.%, respectively. The δD values of these quartz glasses are much lighter than the δD value of the starting water. The equilibrated water is enriched in deuterium relative to these quartz melts during the high-pressure experimental runs.

Water content of the albite glasses ranges from 7.4 to 9.5 wt.%. The relationship between the water content of the glasses and pressures or temperatures of experimental runs is not clear. The

Table I. Composition of Quartz Starting Material from Brazil

Constituent	wt. %
SiO_2	99.99
Al_2O_3	0.0009
Fe_2O_3	0.0006
MgO	0.0002
CaO	0.001
Na_2O	0.0010
K_2O	0.0003
Li_2O	0.0010
Total	99.9941

Table II. D/H Ratios and Water Content of Quartz and Albite Glass at High Pressure Runs

Material	Temperature, °C	Pressure, kb	δD (per mil)	H_2O wt.%
quartz	1500	15	-103.1	7.29
quartz	1500	25	-92.1	7.56
albite	1200	15	-48.1	8.8
albite	1200	20	-47.9	7.8
albite	1200	25	-46.8	9.1
albite	1200	30	-45.7	8.9
albite	1000	10	-68.7	9.0
albite	1000	15	-39.2	8.0
albite	1000	20	-50.2	9.5
albite	1000	30	-53.2	8.7

δD value of these glasses varies from -39.2 to -68.7 per mil. As is clearly recognized in Fig. 1 and Table II, the δD value for albite glasses falls in a comparatively narrow range except for the run at 1000°C and 10 kbar. There is no relationship between the δD value and the water content of the albite glasses. The δD values are almost constant for runs at 1200°C and 15, 20, 25, and 30 kbar. It is interesting that the δD value of albite glasses is larger than that of the starting water. Albite liquid is enriched in deuterium relative to these equilibrated waters during high pressure runs. Water contents of quartz glasses are definitely lower than those of the albite glasses made under similar conditions, which suggests that the solubility of water in the silicate melts may be related to their structural changes.

DISCUSSION

In some high-pressure runs on albite, the quenched liquid is a colorless glass containing very small bubbles. This may explain the lack of relationship between the water content and the pressure-temperature conditions of the runs. Boettcher and Wyllie [17] suggested that albite plus vapor melts incongruently to produce jadeite and liquid. In all our experimental runs at 1200°C, we could not find any crystalline phase under the microscope. For the 1000°C runs, the quenched liquid almost always occurred as a clear and colorless glass. Only rarely was it black. The δD value and water content for runs at 1000°C were slightly more variable than those at 1200°C. However, as is clearly recognized from our experimental result, δD values and water contents of albite glass and quartz glass fall in a relatively different range.

Wasserburg [18] suggested in his explanation of the freezing point depression that (1) water may break the

$$-\overset{|}{\underset{|}{\mathrm{Si}}}-\mathrm{O}-\overset{|}{\underset{|}{\mathrm{Si}}}-$$

bridges in the silicate melt to form a chain of the following form:

$$-\overset{|}{\underset{|}{\mathrm{Si}}}-\mathrm{OH\ HO}-\overset{|}{\underset{|}{\mathrm{Si}}}-$$

or (2) it may convert unshared oxygens in the $(SiO_4)^{4-}$ tetrahedra to produce a linkage of the following pattern:

$$-\overset{|}{\underset{|}{\mathrm{Si}}}-\mathrm{O{\cdot}R\ to}\ -\overset{|}{\underset{|}{\mathrm{Si}}}-\mathrm{OH\ HO{\cdot}R}$$

where R represents a cation that is not in a silicon position. Although this model does not satisfactorily explain the solubility of water in the silicate melt, it would appear that in case (1), the addition of not only H_2O but also Na_2O may occur in the silicate systems studied.

Our unpublished data show that the D/H ratios of synthetic amphibole and phlogopite range from -80 to -90 per mil. The δD values of the starting water in the runs of synthetic hydrosilicates at high pressure range between -75.1 and -78 per mil. The δD values of these synthetic hydro-silicate crystals are slightly lighter than those of the starting water. However, the D/H ratios of the synthetic hydro-silicate crystals are different than the values of quartz and albite glass. We thus conclude from our experimental results that the fractionation of the hydrogen isotope between crystal, liquid, and vapor in the upper mantle or lower crust is rather complex.

ACKNOWLEDGMENTS

The authors thank Dr. Gupta of Hokkaido University for his reading of the manuscript. Part of the cost of these experiments was defrayed by a grant from the Takeda Scientific Foundation, which is gratefully acknowledged.

REFERENCES

1. Y. Kuroda, Y. Hariya, T. Suzuoki, and T. Matsuo, Geophys. Res. Letters 2, 529 (1975).
2. I. Kushiro, Carnegie Institute Yearbook 68, 231 (1970).
3. I. Kushiro, Y. Syono, and S. Akimoto, Earth & Planet. Sci. Letters 3, 197 (1967).
4. I. B. Lambert and P. J. Wyllie, Nature 219, 1240 (1968).
5. Y. Hariya and S. Terada, Earth & Planet. Sci. Letters 18, 72 (1973).
6. Y. Hariya, T. Oba, and S. Terada, in Proceedings 4th AIRAPT Intern. Conference on High Pressure (1974), p. 206.
7. J. C. Allen, P. J. Modreski, C. Haygood, and A. L. Boettcher, "The Role of water in the mantle of the earth: The stability of amphibole and mica," paper presented at the 24th IGC (1972).
8. I. Kushiro, Am. J. Sci. 275, 411 (1975).
9. C. W. Burnham and N. F. Davis, Am. J. Sci. 270, 54 (1971).
10. N. Kokubu, T. Mayeda, and H. C. Urey, Geochem. Cosmochim. Acta 21, 247 (1961).
11. S. M. F. Sheppard and S. Epstein, Earth & Planet. Sci. Letters 9, 232 (1970).
12. Y. Kuroda, T. Suzuoki, and T. Matsuo, Contrib. Miner. Petrol. 60, 311 (1977).
13. T. Suzuoki and S. Epstein, Geochim. Cosmochim. Acta 40, 1229 (1976).
14. Y. Kuroda, T. Suzuoki, and T. Matsuo, Geochem. J. 8, 133 (1974).
15. Y. Kuroda, T. Suzuoki, T. Matsuo, and K. Aoki, Contrib. Miner. Petrol. 52, 315 (1975).
16. H. Craig and J. E. Lupton, Earth & Planet. Sci. Letters 31, 369 (1976).
17. A. L. Boettcher and P. J. Wyllie, Am. J. Sci. 257 , 875 (1969).
18. G. J. Wasserburg, J. Geol. 65, 15 (1975).

L-1

SELECTIVE SEPARATION IN THE SUPERCRITICAL GAS PHASE*

K. Zosel

Max-Planck-Institute für Kohlenforschung
Mülheim a.d. Ruhr, W. Germany

INTRODUCTION

A method for the separation of mixtures has been developed in which the mixture is treated with a supercritical gas under such conditions that the resultant phase formed by the gas and the extracted material is also in a supercritical state [1]. The fractionation of an homologous series is found to occur in an order of increasing boiling point and this is overlapped, where different classes of compounds are present, by a separation according to the class of compound. The first is characteristic of a conventional distillation and the second of a conventional extraction and for these reasons the new technique has been termed "destraction".†

PROCESS DESCRIPTION

The advantage of using supercritical gases to separate mixtures is associated with the dramatic increase in the ability of the gas to take-up material which is observed on passing the critical point. The efficiency of the fractionation can be increased by passing the mixture through a column and allowing it to condense onto a "hot finger," whereby a situation similar to refluxing is established. The recovery of the separated material can be carried out either by raising the temperature, by a stepwise release of the pressure (during which additional fractionation is possible), or by the use of a suitable absorption material.

Efficient destraction is obtained in a temperature range of up to 100 °C above the critical temperature of the gas being used and within a pressure range of 50 to 300 atm. Gases which are particularly suitable are those which have neither extremely low nor

†From the Latin destillare and extrahere.
*Invited paper.

extremely high critical temperatures, and satisfactory results have been obtained using such diverse gases as ethane, propylene, carbon dioxide, nitric oxide, and ammonia. The method can be used wherever it is necessary to resort to vacuum distillation. In addition, it has the advantage that it can be applied to mixtures which would normally decompose on distillation either because the components are thermally labile or have a very high molecular weight.

EXAMPLES ILLUSTRATING PROCESS

A system for the separation of a mixture containing equal amounts of C_{16}, C_{18}, and C_{20} olefins using ethane as the carrier gas is shown in Fig. 1. The destraction is carried out at 45 °C (the critical temperature for ethane is about 32°C) and the gas is passed through the olefin mixture at an initial pressure of 60 atm. The resulting ethane-olefin gas mixture is fed through a fractionation column packed with copper rings; part of it condenses onto a hot finger (kept at 85°C using a heat exchanger) and part of it is bled-off and de-pressurized to 30 atm. The olefin fraction condenses out and is collected, while the ethane is repressurized and recirculated. During the course of the destraction, the pressure is slowly raised to 110 atm to keep the rate of separation approximately constant - it is clear that an analogy exists between raising the temperature of a conventional distillation and increasing the pressure of a destraction. A gas chromatograph trace of the first main fraction is shown in Fig. 2.

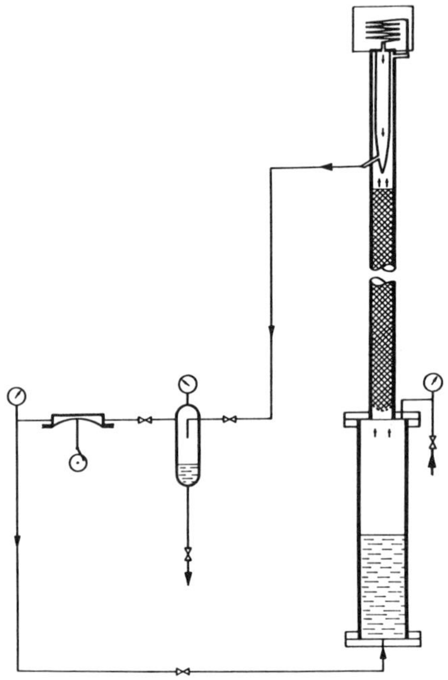

Fig. 1. Destraction apparatus with hot finger.

A similar apparatus has been used to fractionate cod-liver oil. Using ethane as the carrier gas, it has proved possible to separate 20 kg of cod-liver oil into 50 fractions. The analysis of the fractions according to their saponification value and iodine number is shown in Fig. 3. A conventional distillation of this complex mixture of high molecular weight triglycerides is not possible.

Figure 4 shows the flow chart of a pilot plant for the continuous de-asphalting of top residues from crude oil. Propane at 140°C and 130 atm is used as the carrier gas. Seventy-five percent of the feed is destracted by the supercritical gas; the remainder is asphalt and contains most of the vanadium. The highest boiling material, and the

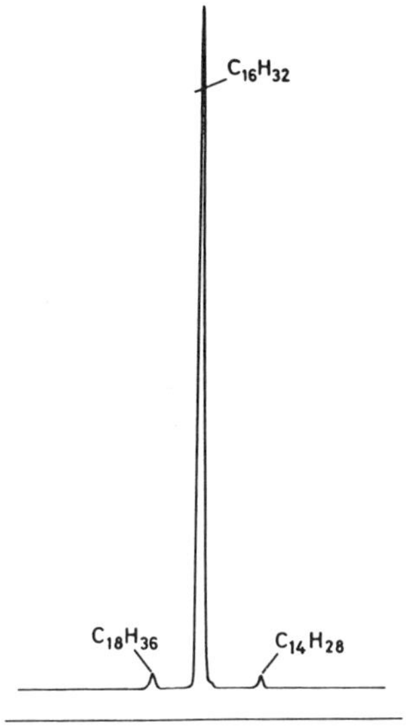

Fig. 2. Fractionation of C_{16}, C_{18}, and C_{20} olefins.

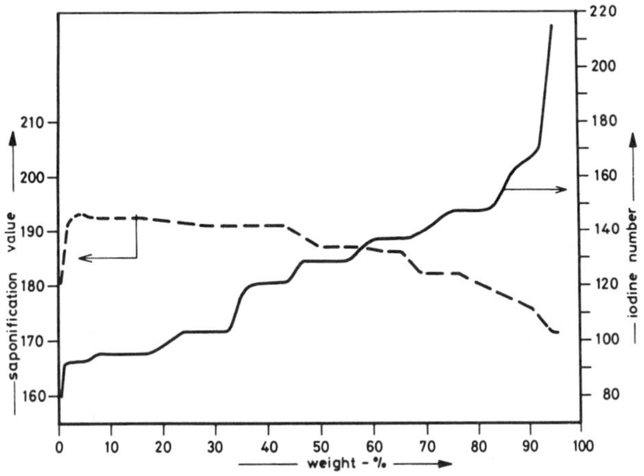

Fig. 3. Fractionation of cod-liver oil.

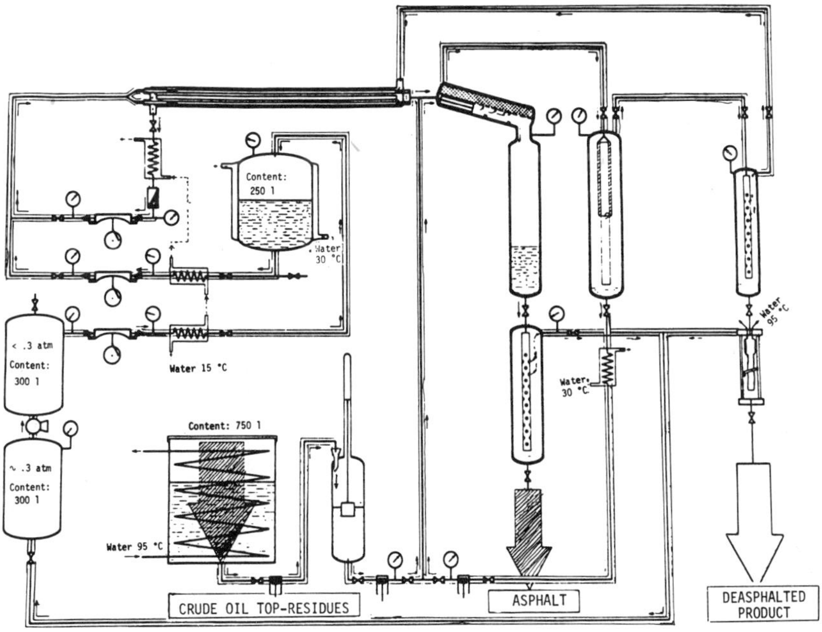

Fig. 4. Pilot plant for the continuous deasphalting of top residues by destraction.

remaining vanadium are condensed out of the gas phase by a partial de-pressurization to 125 atm and recirculated. The quality of the product has been compared in Table I with that obtained by a conventional extraction using liquid propane at 70°C and 70 atm.

Table I. Product Quality Comparison for Two Processes

	Conradson No.	V, ppm
Destraction	1.3	0.15
Extraction	2.0	2.4

In both processes about 75% of the residue is extracted, but the product obtained by destraction has a minimal vanadium content and a lower Conradson number.

The mild conditions under which the destraction can be carried out are of particular interest to the food industry, and it has been demonstrated that fats and oils can be removed from a variety of vegetable and animal matter, whereby a variation in which the extraction is carried out under subcritical conditions and the separation under supercritical conditions can be advantageous. For example, cocoa butter, soybean oil, and corn oil have been extracted from the cocoa bean, soybean and maize, respectively. The essential oils can also be removed from practically all spices and isolated as their concentrates. A further interesting possibility is the removal of alkaloids from plants, and among other applications, it has been used to selectively remove nicotine from tobacco and caffein from coffee. The decaffeination of coffee is a particularly simple process; the moist green coffee beans are treated with supercritical carbon dioxide at 75°C and 180 atm (see Fig. 5). The caffein is then removed from the supercritical CO_2-caffein gas mixture by absorption on activated charcoal pellets which are packed together with the coffee beans in a pressure vessel. Using this arrangement it is possible to reduce the caffein content from an original value which lies between 3 and 0.7% to a value as low as 0.02%. Since the typical coffee flavor is developed by roasting the beans, and only the caffein has been removed, the resulting decaffeinated coffee has a flavor practically identical to that of the untreated coffee.

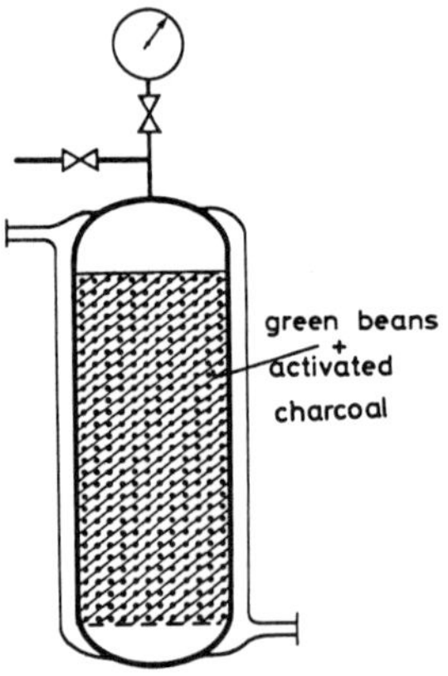

Fig. 5. Decaffeination of green coffee beans.

REFERENCE

1. K. Zosel (Studiengesellschaft Kohle m.b.H), U.S. Patent 3,969,196 (July 13, 1976); Austrian Appl. (April 16, 1963).

L-2

SEPARATION OF LOW-VOLATILE SUBSTANCES USING COMPRESSED GASES

G. Brunner, S. Peter, B. Retzlaff, and R. Riha

Universität Erlangen-Nürnberg
Erlangen, W. Germany

INTRODUCTION

The solubility of components in compressed gases can be used to separate substances with low vapor pressures. Several authors have suggested making use of this effect [1]. The present investigations were concerned with the oleic acid/stearic acid system [2], with the glycerides of oleic acid [3,4] and with mixtures of fatty acids of different degrees of saturation. For example, the separation of monoglyceride from a mixture of glycerides of the oleic acid demonstrated that a separation process using compressed gases is possible without expanding and recompressing the circulating gas. In these investigations we used a semi-commercial apparatus with plate columns.

Subsequent experiments were carried out on a continuously operating laboratory scale apparatus in order to study the separation of mixtures of fatty acids and the suitability of packed columns for extractive distillation at high pressures.

PHASE EQUILIBRIA

The choice of favorable operating conditions depends upon a knowledge of the phase equilibria of the systems under consideration as a function of pressure, temperature and concentration of the different components. Under favorable conditions, the concentration of the non-volatile components in the compressed gas can be more than 10^6 times higher than would be expected from their vapor pressure. Figure 1 shows data obtained on the solubility of oleic acid in different gases at a temperature of 125°C. The

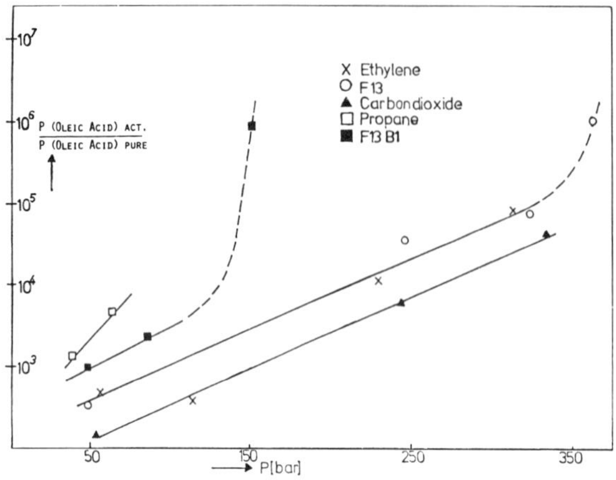

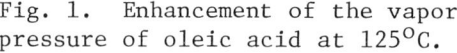

Fig. 1. Enhancement of the vapor pressure of oleic acid at 125°C.

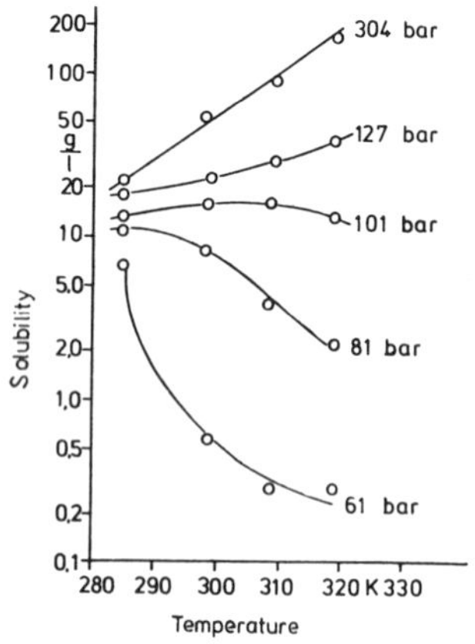

Fig. 2. Solubility of napthalene in compressed ethylene [5].

concentration enhancement of oleic acid in the gaseous phase is plotted using a logarithmic scale on the ordinate and a linear scale for total pressure on the abscissa. The concentration enhancement increases with pressure up to a factor of 10^6 and varies with the nature of the gas. This effect is due mainly to interaction forces between the molecules of oleic acid and the gas.

The influence of temperature on the solubility of a substance in a compressed gas changes with pressure as shown by Fig. 2. This figure shows that the solubility of naphthalene in compressed ethylene is dependent on temperature at various pressures, according to measurements of Tsekhanskaja, Jomtev and Mashkina, [5]. Slightly above the critical temperature of the compressed gas a rise of temperature at moderate pressures causes a decrease in the solubility of the nonvolatile component. At high pressures a rise in temperature causes an increase of the solubility. This can be understood, if the density is considered to be important for the solubility of the nonvolatile component in the gaseous phase.

Figure 3 shows the density of propane as a function of pressure. In general, a rise in temperature at constant pressure decreases the density of the compressed gas, whereas it increases the vapor pressure of the nonvolatile components exponentially. At moderate pressure and at slightly above the critical temperature of the compressed gas, a temperature rise diminishes its density to such an extent that the concentration of the nonvolatile components in the gaseous phase decreases considerably. This effect is used, as will be seen later, to separate the nonvolatile components from the circulating gas. At high pressures the decrease of density caused by the temperature rise is so small that the increase in vapor pressure

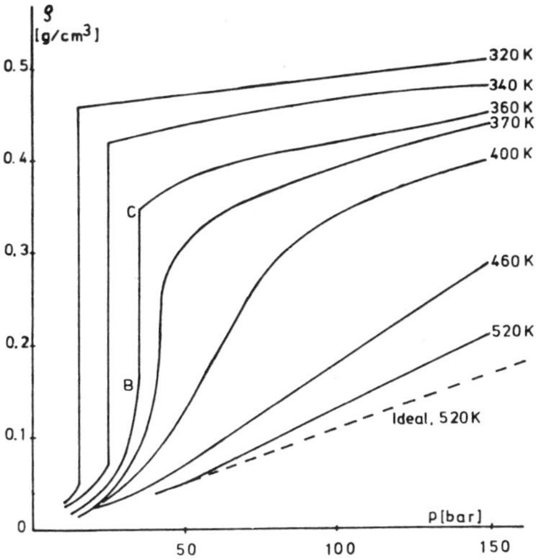

Fig. 3. Density of propane.

leads to a higher concentration of the nonvolatile components in the gaseous phase.

In most cases the relative volatilities of the components to be separated are only slightly influenced by the compressed gas. Therefore, for a mixture of substances with small relative volatilities or separation factors use of an entrainer is recommended. This entrainer has the function of raising the separation factor between the nonvolatile components as well as of separating the circulating gas from the nonvolatile components. One can thereby save the costs of recompression. Furthermore, the entrainer increases the solubility of the nonvolatile components in the compressed gas.

In Fig. 4 the quasi-ternary system CO_2-acetone-glycerides is shown in a ternary diagram at a pressure of 130 bar. At constant temperature, the gaseous phase takes up increasing amounts of glycerides with increasing pressure and with its approach to the critical point of the ternary phase boundary line. In the quasi-ternary system the content of glycerides in the gaseous phase is appreciably higher than in the quasi-binary system CO_2-glycerides. At constant pressure in the temperature region investigated, the glyceride content in the gaseous phase of the ternary system increases with lower temperature.

The binary CO_2-acetone system is supercritical at 130 bar and at both 70 and 100°C. Only the CO_2-glycerides system shows a miscibility gap. In Fig. 4 one can further see that under these conditions it is possible for an enrichment of glycerides to occur in the compressed gas in the region not too far from the critical point. This

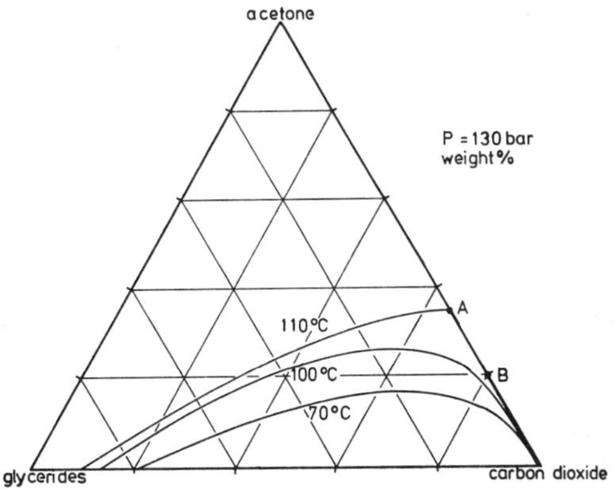

Fig. 4. Phase equilibria of the glycerides - acetone - carbon dioxide system.

condition can be attained with a suitable amount of acetone in the feed. At 110°C the CO_2-acetone binary system is subcritical. The miscibility gap is marked by points A and B. At these conditions the solubility of the low volatile components in the gaseous phase is very low.

Some experimental results of the phase equilibria in the quasi-ternary CO_2-acetone-glycerides system are shown in Table I. These results provide the concentration of glycerides in the gaseous phase as a function of the acetone content. In the last column of the same table the separation factor between monoglyceride and diglyceride is shown. One sees that the separation factor decreases as the amount of entrainer increases, because the critical point at which the separation factors equal unity is approached.

Table I. Phase Equilibria for a Quasi-ternary System at 69.6°C and 132.5 bar*

	Liquid Phase Wt. %			Gaseous Phase Wt. %			
No.	$x_1+x_2+x_3+x_4$	x_5	x_6	$y_1+y_2+y_3+y_4$	y_5	y_6	α_{32}
1	79.1	-	20.9	0.4	-	99.6	-
2	66.3	5.5	28.2	0.3	4.2	95.5	>100
3	57.5	7.7	34.8	2.7	10.0	87.3	30
4	51.1	9.6	39.3	4.3	12.4	83.3	3
5	45.8	11.1	43.1	8.9	14.8	76.3	1.4

*Triglycerides (1), diglycerides (2), monoglycerides (3), glycerol (4), acetone (5), carbon dioxide (6).

APPARATUS AND PROCESS

The arrangement of the apparatus used for the investigation of the process is shown schematically in Fig. 5. It consists of two columns, the first of which is used for the separation of the non-volatile substances and the second is used for the separation of the nonvolatile components from the circulating gas.

The plate columns have an internal diameter of 65 mm and are designed for an operating pressure of 150 bar at 120°C. The main column contains 51 plates, the side column 34 plates. The smaller laboratory packed columns have an internal diameter of 12 mm and a packing of 2 x 0.5 m.

The mixture to be separated is introduced into the middle

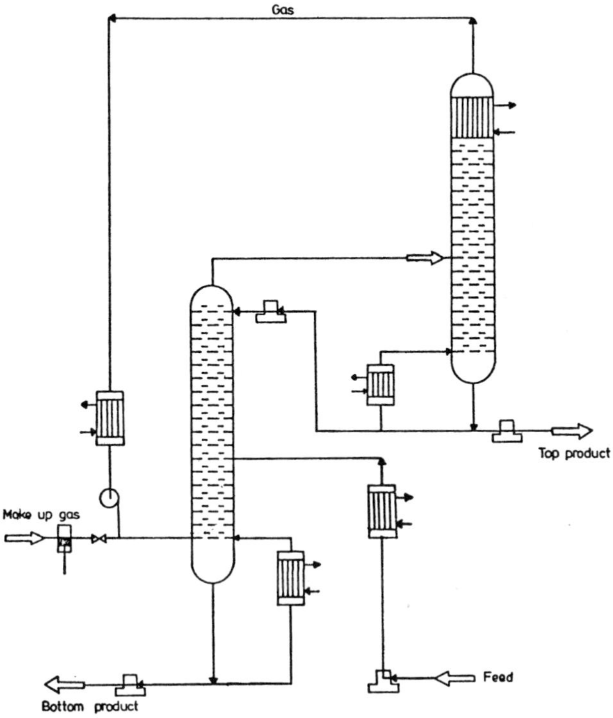

Fig. 5. Flowsheet of the apparatus.

section of the main column together with the entrainer. Compressed gas leaves the rectifying section of the main column at the top and enters the second column in the lower part. After further rectification in this column the circulating gas stream leaves as the overhead stream and is then further heated. Since part of the entrainer is liquified, it is used in the second column as reflux countercurrent to the circulating gas to aid in recovering the nonvolatile components. A reboiler at the bottom of the side column heats the mixture of nonvolatile substances and the entrainer when needed. The product withdrawn from the bottom of the side column constitutes both the top product and the reflux for the main column. The circulating gas, which consists of the compressed gas containing a certain amount of entrainer and which is essentially free of nonvolatile components, is returned after cooling to the bottom of the main column. It is therefore possible to omit the steps of expansion and recompression in the circulating gas. At the bottom of the main column the solution of the nonvolatile substances in the gaseous phase takes place. Here the bottom product is also withdrawn.

The second column has the advantage that its separate internal cycle can be adjusted independent of that in the main column. This aids the enrichment of the nonvolatile components at the bottom of the second column and for purifying the circulating gas.

As is evident from Fig. 4, the binary system of compressed gas and entrainer has to be under supercritical conditions in the main column in order to achieve a high enrichment of nonvolatile substances in the gas phase. For the second column, operating conditions have to be chosen at which the binary system of gas and entrainer mixture is subcritical. Under such conditions, partial condensation of the entrainer takes place and washes out the nonvolatile components from the circulating gas. The entrainer should either have a boiling temperature which is below the decomposition temperature of the nonvolatile components or be removable by extraction.

EXPERIMENTAL RESULTS

The experiments involving the separation of monoglycerides from a mixture of glycerides were carried out at a pressure of 135 bar and a bottom temperature of 80°C in the main column. The temperature at the top of the side column was 110°C. The temperature must not drop below 100°C; otherwise the CO_2-acetone system becomes one phase at this operating pressure. With the above operating conditions, the amount of low volatile components in the circulating phase could be reduced below 0.008 wt. % in the side column (1/500th of the amount in the main column). Practically no diglyceride in the top product was evident when a reflux ratio of 2 was used. This can be seen in Fig. 6 and Fig. 7 which represent the gas chromatograms of the feed and top product, respectively. With columns of these

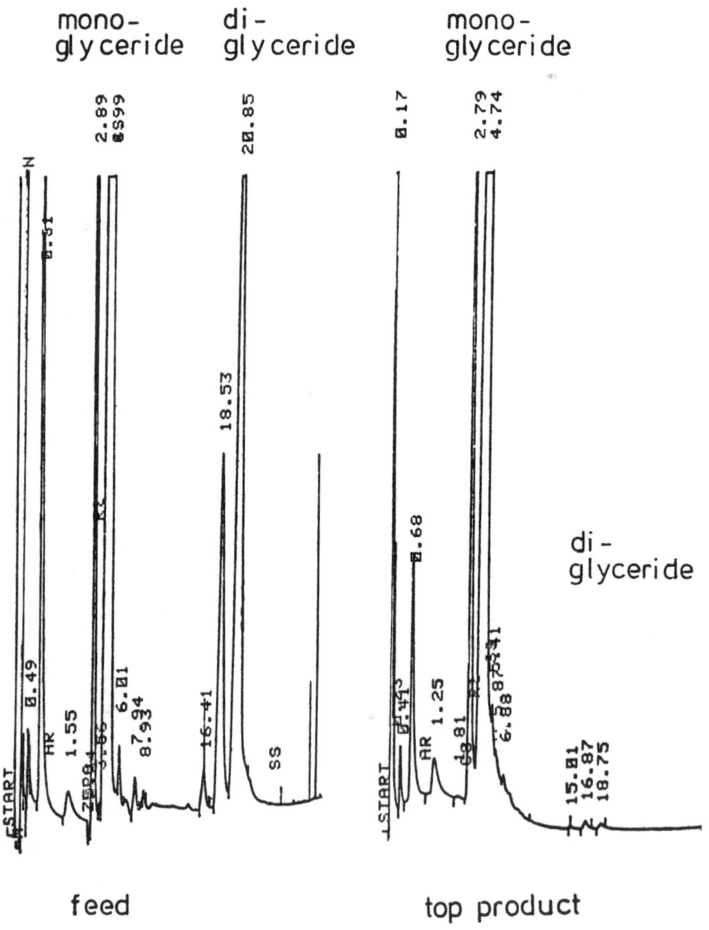

Fig. 6. Gas chromatograms of glycerides.

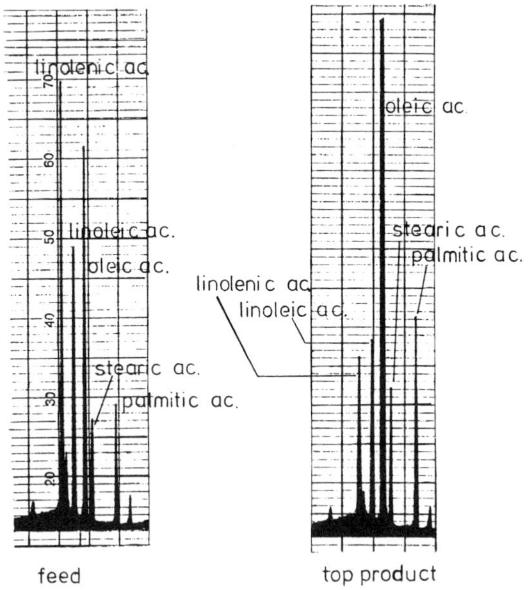

Fig. 7. Gas chromatograms of fatty acids.

dimensions and a reflux ratio of 1.5, a feed of 1.25 kg glycerides/hr could be processed through the column to provide a top product of 0.65 kg glycerides/hr.

Experiments for separating linseed oil fatty acids to fatty acids of different degrees of saturation were carried out in the main column at 140 bar and 50°C with CO_2 as the compressed gas and with methylene chloride as the entrainer. The temperature at the top of the second column was adjusted to 100°C. Gas chromatographic analysis of the products revealed that the top product from the main column and the bottom product from the second column, which are identical with respect to the composition of fatty acids, had been enriched with fatty acids of a higher degree of saturation. Figure 7 presents some of the analytical results. In the top product, the polyunsaturated fatty acids have been reduced from an original 50% by weight to 20% by weight.

DISCUSSION

The density of both phases differs by a factor of about 3 under normal operating conditions. The density difference is therefore appreciably higher than in an extraction. The density ratio of the phases at a given temperature and pressure depends on the concentration of the entrainer. Particular attention must be paid to these parameters when operating a plant. Estimates [4] of the separation costs for different size plants are shown in Table II for a purification process (removal of 5 wt. % impurities as top product) and a separation process (removal of 50 wt. % as top product).

Generally, mixtures of nonvolatile and heat-sensitive substances can be handled with this process at temperatures which are far lower than those in a vacuum distillation. The method described seems especially attractive for processing materials sensitive to oxidation. At the elevated operating pressures there is no danger of oxygen entering the apparatus. Using an entrainer extends the

Table II. Separation Costs in US $/kg

	Plant Size				
	3000 t/a	15,000 t/a		30,000 t/a	
Pressure, bar	Separ.	Separ.	Purif.	Separ.	Purif.
135	0.16	0.064	0.016	0.036	0.012
80	-	0.04	0.012	0.024	0.008

possibility of a separation to mixtures which form azeotropes or contain substances with almost equal vapor pressures. Therefore the new process is not only an alternative to vacuum- and molecular distillation, but also makes it possible to apply the advantages of extractive distillation to mixtures of nonvolatile components, because by adding an entrainer it is possible to influence the relative volatilities. An especially favorable application of this method is the separation of components of low concentration from a mixture of nonvolatile substances. In this case the operating conditions and entrainer can be chosen so that only a small part of the product enters the gaseous phase. An essential part of the plant can then be designed for small quantities.

REFERENCES

1. H. E. Messmore, U.S. patent 2,420,185 (January 4, 1943); also T. P. Zhuze and A. A. Kapelyushnikov, U.S.S.R. patent 113,325 (1955); K. Zosel, German patent 1,493,190 (June 8, 1964); S. Peter, G. Brunner and R. Riha, Chem. Ing. Techn. 46, 623 (1974).
2. S. Peter, G. Brunner and R. Riha, Dechema-Monographien 73, 197 (1974).
3. S. Peter, G. Brunner and R. Riha, Fette, Seifen, Anstrichmittel 78, 45 (1976); also S. Peter, G. Brunner and R. Riha, German patent 2,340,566 (August 10, 1973).
4. S. Peter, G. Brunner and R. Riha, Chem. Ing. Techn. 49, 61 (1977).
5. Yu. Tsekhanskaja, M. B. Jomtev and E. V. Mashkina, Z. fiz. Chim. 38, 2166 (1964).

L-3

FLUID CHROMATOGRAPHY OF OLIGOMERS

W. Hartmann and E. Klesper

Universität Freiburg
Freiburg, W. Germany

INTRODUCTION

Gas chromatography at normal pressures is not used in the separation of oligomers and polymers because large molecules do not possess sufficient vapor pressure. However, gas phases above their critical temperatures and at pressures from 20 to 200 bar, or higher, may possess the ability to dissolve large molecules. This property has long been known [1-2] and extends to many classes of compounds, including inorganic salts. The dissolution ability is particularly pronounced with low-boiling solvents whose critical temperatures are low enough to ensure stability of the substrate, but at the same time high enough to indicate the possibility for sufficiently strong intermolecular forces between the supercritical solvent and the substrate molecules. When the density of the supercritical phase is high enough for sufficient solubility power, this density may still be considerably smaller than the density of the corresponding liquid just below the critical temperature.

Similarly, the mass transport of the substrate in the mobile phase and the viscosity of the mobile phase can be expected to be intermediate between the gaseous and the liquid state. Thus in principle, more favorable conditions are given for chromatographic efficiency with such supercritical fluids as the mobile phase than with the corresponding liquids.

An additional advantage of a supercritical fluid is the possibility of varying by pressure the density and concurrently the solubility power of the mobile phase much more than is possible with liquids or with gases of relatively low pressure. Separations by supercritical fluid chromatography are thus more strongly influenced by pressure than is the case with liquid or gas chromatography.

Previous work in supercritical fluid chromatography utilized low-boiling solvents [3,4] as well as permanent gases [5] as the mobile phase. Pressure levels are generally lower in the first instance, since the solubility power of the supercritical solvents is generally higher than that of permanent gases at the same pressure or with the same number of moles per volume. A wide range of high molecular weight substrates has been subjected to chromatography, e.g., porphyrins [6,7], polynuclear aromatic hydrocarbons [8,9], transition metal cyclopentadienyl complexes [10], oligosiloxanes [11], and oligostyrenes [12,13]. From the numerous separations published so far, it appears that the full potential in resolution has not yet been achieved in many cases, although there are a number of good separations which have been obtained both with supercritical fluid-solid [9] and supercritical fluid-liquid [14] chromatography. Theoretical aspects of supercritical fluid chromatography have also been treated, e.g., the adaptation of the van Deemter equation [15]. Moreover, investigations of phase diagrams which are of immediate interest to fluid chromatography have been carried out [16].

There is little question that the difficulties in designing a full-potential apparatus was one of the major obstacles in the development of supercritical fluid chromatography. A number of designs have been presented to date, however [4,11,17,18]. The capability for pressure programming has been given particular attention recently [4,13] since pressure is a key controlling parameter in supercritical fluid chromatography.

APPARATUS

A chromatographic apparatus has been designed whose characteristic feature is a pressure cascade controlled by feedback loops [13]. This arrangement has the purpose of allowing pressure programming of the chromatographic column without changing the pressure of the supply system or the pressure of the detection part of the apparatus. The apparatus in Fig. 1 possesses two solvent supply systems, labeled 1, both tanks of which can be heated with heating mantles to increase the pre-pressure for avoiding cavitation at the pump inlets. A double-headed membrane pump, MP, 2, allows the feeding of two different liquids simultaneously. The feed rates can be programmed, if desired, 10. Thus, a composition gradient in the mobile phase, and at the same time, a constant or changing total feed rate may be created. The liquid enters the first oven, 3, which can be heated to supercritical temperature and which contains a coil of larger inner diameter to act as a buffering volume for depulsification. It follows a metering valve controlled by an electronic feedback loop, 12, to keep the pressure constant upstream, i.e., in the supply system. This supply system is at the highest pressure level of the pressure cascade. The purpose is to provide feed rates which are not affected by the changing pressure in the column, even with pumps which have a pressure-dependent feed rate.

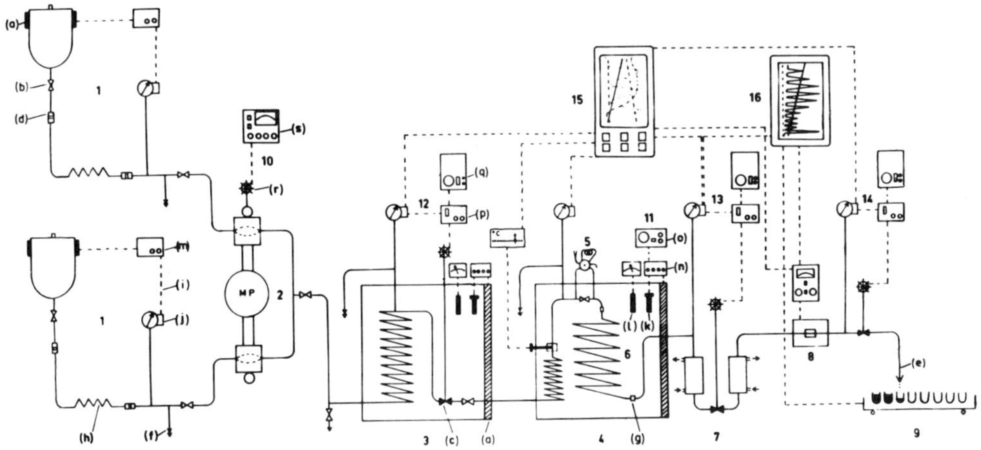

Fig. 1. Apparatus for supercritical fluid chromatography. (See text.) Legend: (a) heating element, (b) shutoff valve, (c) metering valve, (d) sintered metal filter, (e) exit, (f) exit with burst disk, (g) column connection with sintered metal disk, (h) coil of high pressure tubing, (i) electrical connection, (j) manometer device, (k) thermoelement, (l) mechanical mercury manometer, (m) relay for heating current, (n) temperature controller, (o) temperature programmer, (p) controller of feedback loop, (q) programmer of feedback loop, (r) step motor controlling membrane displacement in pump, or driving shaft of a valve, and (s) programmer for feed rate.

The mobile phase enters the second oven, 4, passing first a temperature conditioning coil. This coil is required because the two ovens are not necessarily at the same temperature, particularly if a temperature program, 11, is applied to the second oven. After passing the injection port, 5, the chromatographic column, 6, is entered for sample separation. On leaving this oven, the supercritical fluid is returned to the liquid state by the first of two water-cooled heat exchangers, 7. The metering valve between the heat exchangers is controlled by a feedback loop, 13, which allows control, e.g., programming of the column pressure as the second level of the pressure cascade. The pressure-measuring device of the feedback loop is downstream of the column to prevent slow response.

The second heat exchanger compensates for the decompression and friction effects at the metering valve and eases the thermostatting of the effluent for the adjacent UV-detector, 8. The pressure in the detection part of the apparatus is held constant by a metering valve at the exit of the chromatograph with a corresponding feedback loop, 14. The constant pressure eases detection by not affecting the baseline of the various types of detectors which may be used. The pressure in this third level of the cascade is kept just high enough to retain the liquid state of the effluent.

The effluent is collected in an automatic fraction collector, 9. With mobile phases which boil below room temperature, a higher boiling solvent like acetone enters the stream at the exit (not shown) to maintain the fractions in solution. The operating parameters are monitored by two recorders. A six-channel, point-printing recorder, 15, records: the pressure in the supply system, the temperature of the mobile phase after the conditioning coil, the pressure at the column entrance, the detector trace, and the pressure in the detector part of the apparatus. A two-channel high-speed recorder, 16, records the detector trace and the pressure at the column outlet, i.e., the pressure program. The changing of the vessels for each fraction collected is also recorded.

The maximum supply pressure is 500 bar, the maximum temperature 300°C. Pressure programs of any desired shape can be accommodated within 7.5 min to 36 hr total time.

RESULTS

The chromatogram of a styrene oligomer with an average molecular weight of 2200 is shown in Fig. 2. The mobile phase is 90% n-pentane and 10% methanol at the supercritical temperature of 230°C. The pressure is programmed starting at 20 bar, increasing 6.3 bar/hr, and attaining 130 bar after 17.5 hr (P = trace of pressure). The porous silica employed as the stationary phase caused considerable

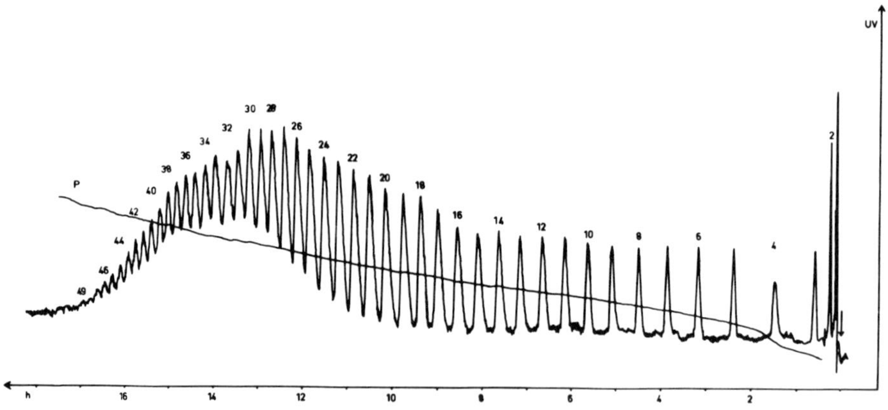

Fig. 2. Chromatogram of styrene oligomer, $\bar{M}_w$ = 2200, shown as UV-absorption vs. elution time. Feed rate was 1.1 ml/min at room temperature in the liquid state. Mobile phase was 90% n-pentane, 10% methanol (v/v). Stationary phase Porasil A (porous silica) 37 to 75 µm particle diameter, pore size < 100 Å. Column dimensions 3 m x 2 mm ID. Pressure program as measured on column exit, starting at 20 bar, increasing by 6.3 bar/hr, ending at 130 bar after 17.5 hr (P = pressure trace). Pressure drop over column ~ 200 bar. Temperature 230°C, sample 20 mg in 100 µl solution of cyclohexane.

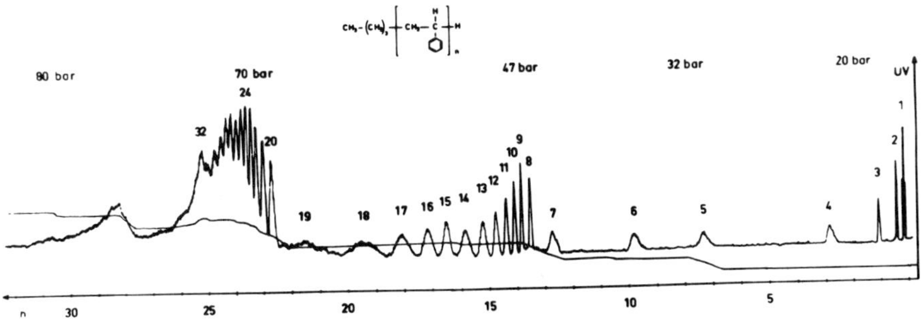

Fig. 3. Chromatogram of styrene oligomer, $\bar{M}_w$ = 2200. Feed rate 2.6 ml/min. Mobile phase 95% n-pentane, 5% methanol (v/v), column dimensions 6 m x 3 mm ID. Pressure program starting at 20 bar, increasing stepwise to 32, 47, and 70 bar. Pressure drop over column 20 bar, temperature 220°C, sample 40 mg/100 µl. Other conditions as for Fig. 2.

pressure drop over the column. A good separation into about 50 peaks appears with the resolution decreasing with the higher peaks, i.e., those which were eluted later. A decreasing resolution would be expected for a homologous series for its higher members. Because of the decreased resolution, there exists an unresolved "background" between peaks 20 and 49. By appropriate change of the parameters for the separation it is, however, possible to resolve this background (see Fig. 7).

The necessity for a pressure program can be seen from Fig. 3. Here the pressure has not been increased continuously but in a few discrete steps. At a given level of pressure, the elution of peaks becomes increasingly slower and finally stops after a certain number of peaks have appeared. Moreover, the peaks broaden quite strongly. It appears that for a given pressure level the substrate of some of the peaks has sufficient solubility in the mobile phase, as has been confirmed by chromatograms which have been run at only one pressure. The higher the level, the more peaks can be eluted. However, the higher the level, the smaller the resolution of the first peaks.

Not only is the starting level of a pressure program of importance, but so is the rate of increase of pressure. In Fig. 4 the rate of increase has been varied, at otherwise comparable conditions, from a high of 6.6 bar/hr in trace III to 2.2 bar/hr in trace I. The decrease greatly enhances the resolution and decreases the amount of substrate which remains unresolved in the background. It has been ascertained that the variations in starting pressure and in flow rate between traces I to III does not alter this result.

The chromatographic behavior in Figs. 3 and 4 indicates that the elution volume of a given peak is primarily governed by the pressure. Moreover, the mode of separation appears to be largely a

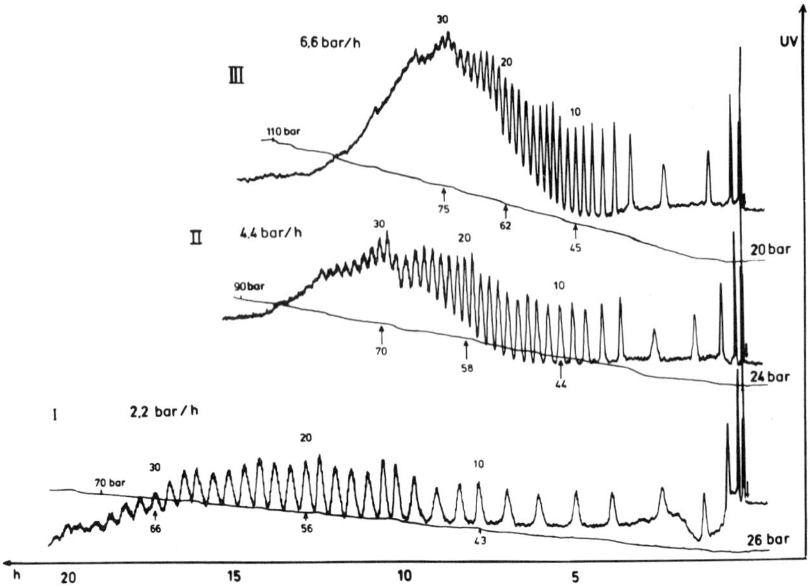

Fig. 4. Chromatogram of styrene oligomer, $\bar{M}_w = 2200$. Column dimensions 6 m x 3 mm ID. Pressure programs starting between 20 and 30 bar, pressure drop 25 bar. T = 220°C, sample 40 mg/100 µl, mobile phase 95% n-pentane, 5% methanol. trace III 6.6 bar/hr, 2.6 ml/min, trace II 4.4 bar/hr, 3.0 ml/min, trace I 2.2 bar/hr, 3.2 ml/min. Other conditions as for Fig. 2.

reprecipitation-redissolution process. The pressure drop over the column has an essential function, inasmuch as it gives rise to the first precipitation of the oligomeric species, as long as the absolute pressure level is low enough. The oligomeric species with the lower degrees of oligomerization are precipitating further down the column, while the higher degrees of oligomerization precipitate further up the column. On programming the pressure upwards, the oligomeric species are redissolving and reprecipitating in separate bands down the column until they are finally eluted. The necessity of a slow pressure program, as seen by Fig. 4, is in accord with this mode of separation, since a reprecipitation-redissolution chromatography is expected to be a particularly time-consuming process if higher resolution is desired.

Larger samples can also be accommodated for preparative separations. Column diameters were increased to 5 mm ID and sample sizes to 100 mg. The chromatogram of Fig. 5, trace I, shows the first 20 peaks for the oligostyrene with an average molecular weight of 2200. Although the resolution is less than with the analytical columns of 2 to 3 mm diameter, it suffices for collecting reasonably pure fractions for the first 15 peaks. For each peak 6 to 20 mg

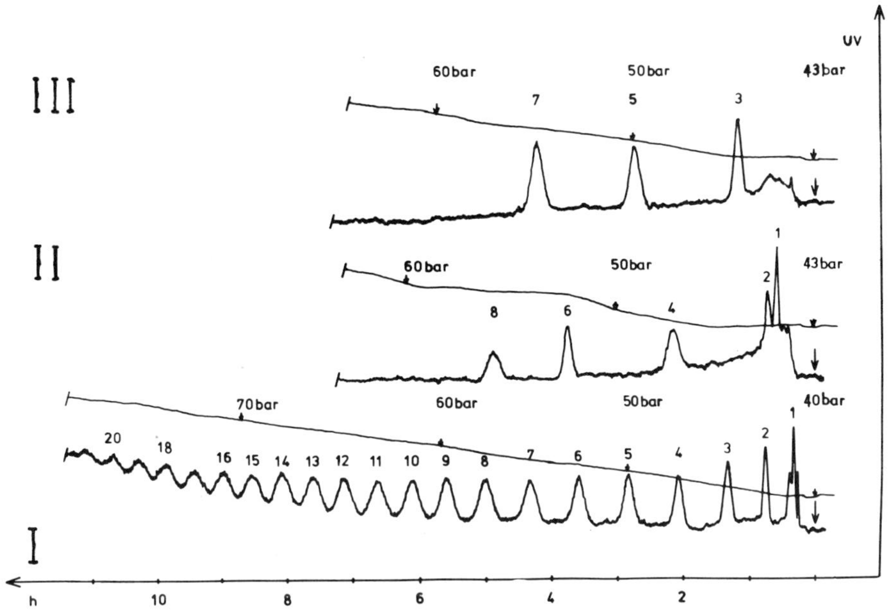

Fig. 5. Preparative chromatogram of styrene oligomer, $\bar{M}_w$ = 2200. Column dimensions 6 m x 5 mm ID. Pressure programs starting at 40 to 43 bar, increasing at 3.3 bar/hr, feed rate 2.4 ml/min, T = 220°C, sample 100 mg/500 µl. Other conditions as in Fig. 2.

were collected by combining the fractions of 16 runs. In order to test for chromatographic purity, 2.0 to 2.5 mg each of peaks 1, 2, 4, 6, and 8 were deliberately recombined and reinjected to yield the chromatogram of trace II. The same procedure was carried out with peaks 3, 5, and 7 for trace III. The empty spaces between the peaks indicate that the fractions do not contain neighboring homologous species, otherwise residual peaks numbered 3, 5, and 7 for trace II or 4 and 6 for trace III should have appeared.

The amount of material collected easily allows further characterization of the fractions, e.g., the determination of molecular weight by mass spectrometry. In Fig. 6 the mass spectrum of peak 8 of Fig. 5 is shown as an example. The highest mass appears at 890 ± 1 which corresponds to the formula given at the top of Fig. 6, representing an oligomeric styrene of a degree of oligomerization of 8 with a n-butyl group attached. The n-butyl group originates from the n-butyl lithium which serves as a starter for the oligomerization. Other fractions give analogous results by mass spectrometry and this indicates that chromatographic separation is indeed by oligomeric species, allowing also the assignment of the degree of oligomerization to other individual peaks.

Having identified the peaks, the relative frequencies of the oligomeric species, i.e., the molecular weight distribution, can

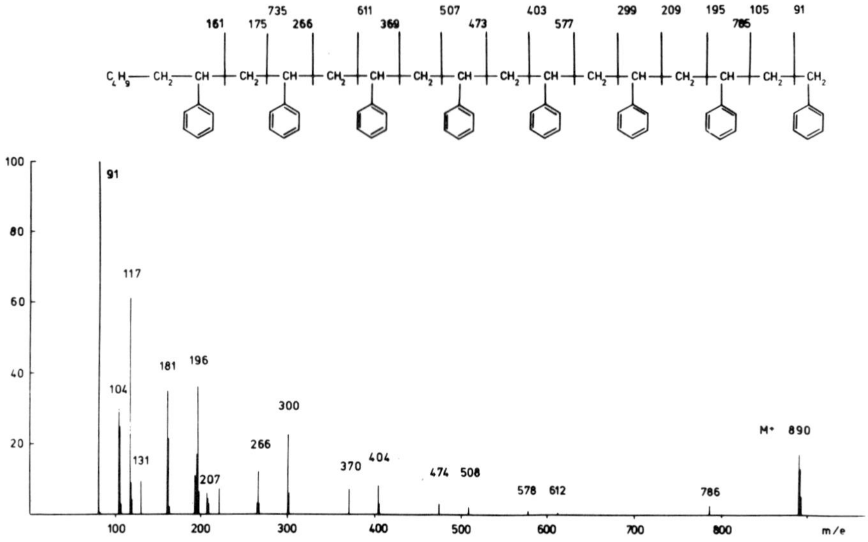

Fig. 6. Mass spectrum of peak 8 of chromatogram of Fig. 5. Relative intensity with height of largest peak set at 100, plotted vs. m/e. Proposed formula for experimentally found highest mass shown on top, together with masses of possible fragments.

be determined. It has been seen in Fig. 2, however, that the resolution of the peaks which are eluted later is less than that of the peaks which are eluted first. To increase the resolution of these higher peaks, a temperature program was applied starting at 210°C and increasing linearly to 260°C within 16.5 hr, as shown in Fig. 7. To decrease the elution time of the first peaks without sacrificing the baseline separation, the starting level of the pressure program was chosen relatively high at 36 bar. This has the effect of compressing the first part of the chromatogram.

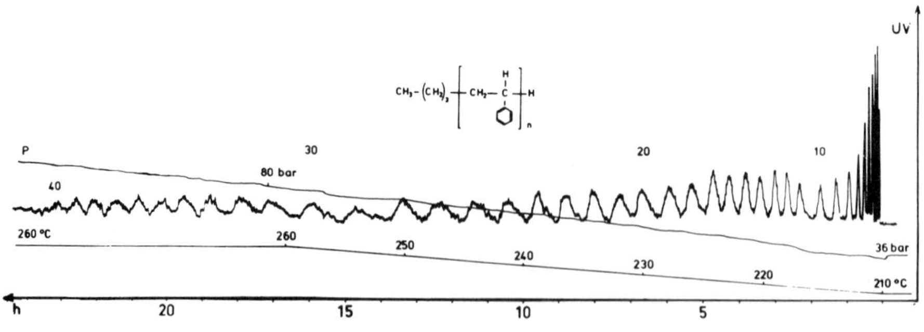

Fig. 7. Chromatogram of styrene oligomer, $\bar{M}_W = 2200$. Simultaneous temperature and pressure programs.

The chromatogram of Fig. 7 was evaluated by cutting and weighting the peaks after drawing a baseline and graphically separating the peaks down to this baseline if necessary. Since the weights divided by the degree of oligomerization n are proportional to the relative frequencies of the oligomeric species they are plotted vs. the degree of oligomerization, n in Fig. 8 in the form of a histogram. Based on the histogram, the mass average molecular weight $\bar{M}_w$, the number average molecular weight $\bar{M}_n$, and the molecular weight heterogeneity $\bar{M}_w/\bar{M}_n$ have been calculated. It is noteworthy that Fig. 7 shows that the molecular weight range includes even the monomeric and the dimeric species in considerable amounts. This leads to small $\bar{M}_n$ and large $\bar{M}_w/\bar{M}_n$ values.

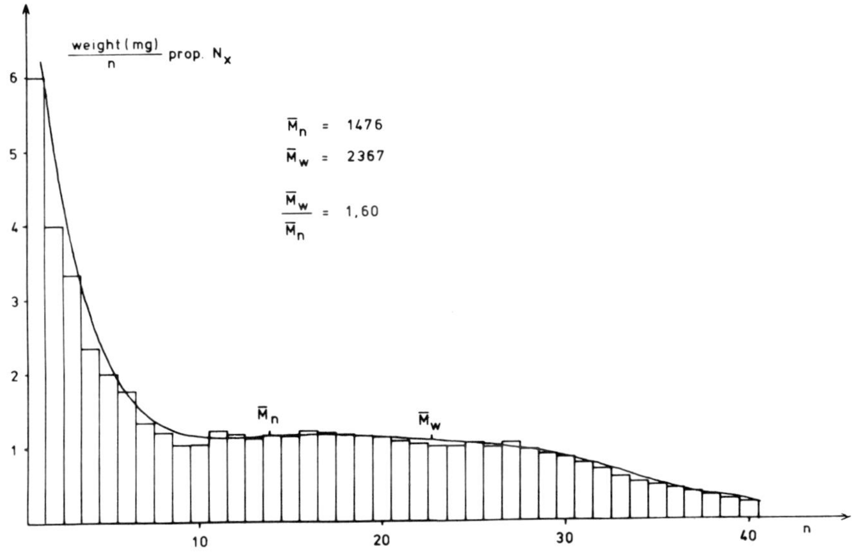

Fig. 8. Molecular weight distribution of styrene oligomer, $\bar{M}_w$ = 2200, n = degree of oligomerization, N_x = relative frequency of oligomeric species at a given n. N_x is measured by weight of cut from peak in mg divided by n which measure is proportional to N_x. Curve fitted to the histogram by a polynomial.

ACKNOWLEDGMENTS

Financial support by the Arbeitsgemeinschaft Industrieller Forschungsvereinigungen and the Deutsche Forschungsgemeinschaft is gratefully acknowledged.

REFERENCES

1. J. B. Hannay and J. Hogarth, Proc. Roy. Soc. London 29, 324 (1879).
2. M. Zentnerszwer, Z. Physik. Chem. 46, 427 (1903).
3. T. H. Gouw and R. E. Jentoft, J. Chromatog. 6, 303 (1972).

4. T. H. Gouw and R. E. Jentoft, Adv. in Chromatog. 13, 1 (1975).
5. M. N. Myers and J. C. Giddings, in Progress in Separation and Purification, Vol. 3, S. Perry, C. J. van Oss, eds., Wiley-Interscience, New York (1970), p. 133.
6. E. Klesper, A. H. Corwin, and D. A. Turner, J. Org. Chem. 27, 700 (1962).
7. N. M. Karayannis and A. H. Corwin, Anal. Biochem. 26, 34 (1968).
8. S. T. Sie and G.W.A. Rijnders, Anal. Chim. Acta 38, 31 (1967).
9. S. T. Sie and G.W.A. Rijnders, Separation Sci. 2, 755 (1967).
10. R. E. Jentoft and T. H. Gouw, Anal. Chem. 44, 681 (1972).
11. J. A. Niemann and L. B. Rogers, Separation Sci. 10, 517 (1975).
12. R. E. Jentoft and T. H. Gouw, J. Polymer Sci., Polym. Lett. 7, 811 (1969).
13. E. Klesper and W. Hartmann, J. Polymer Sci., Polym. Lett. 15, 9 (1977).
14. S. T. Sie and G.W.A. Rijnders, Separation Sci. 2, 729 (1967).
15. D. Bartmann, Ph.D. Dissertation, Ruhr Universität Bochum, Bochum, Germany (1972).
16. G. M. Schneider, Topics in Current Chem. 13, 559 (1970).
17. N. M. Karayannis, A. H. Corwin, E. W. Baker, E. Klesper, and J. A. Walter, Anal. Chem. 40, 1736 (1968).
18. M. N. Myers and J. C. Giddings, Separation Sci. 1, 761 (1966).

L-4

SOME RECENT ADVANCES IN SUPERCRITICAL FLUID CHROMATOGRAPHY

T. H. Gouw, R. E. Jentoft, and E. J. Gallegos

Chevron Research Company
Richmond, California USA

INTRODUCTION

In supercritical fluid chromatography (SFC), the mobile phase is a compound maintained between 1.0 and 1.05 the critical temperature, and at 1 to 10 times the critical pressure. For example, with CO_2 at close to ambient temperatures as the mobile phase this technique can chromatograph compounds which are either too high in molecular weight or thermally not sufficiently stable to be processed by gas chromatography.

In this respect, SFC is more akin to high performance liquid chromatography (HPLC). The rapid growth of the latter has in many respects overshadowed the use of SFC. In most cases, HPLC is more versatile in its ability to solve the variety of problems confronted by the average analytical chemist. This is not valid for all cases. This paper discusses two areas where SFC has been found to be very useful.

ANALYSIS OF POLYNUCLEAR AROMATIC HYDROCARBONS (PNAH)

The carcinogenic properties of several specific PNAH have been known for some time. With the increased emphasis on environmental protection, good analytical methods are necessary to determine these compounds quantitatively in a wide diversity of background materials. The petroleum industry is also interested in the analysis of these substances from a process point of view, since these PNAH can have a deleterious effect upon some catalysts. The isolation and quantification of these compounds in a wide range of matrices is, at best, a fairly formidable proposition. Matters have been made more difficult because over the years the desired detection thresholds have been decreased to the order of a few ppb.

The first major step in each analysis is to obtain a representative sample. This is not always straightforward because the sample can be difficult to collect quantitatively or the mixture is inhomogeneous in nature. In the analysis of PNAH in automobile exhaust, e.g., the capacity of our sampling system and the amount of gas generated necessitate discard of about 92% of the latter. The balance, diverted into the analytical train, had to be representative of the total material at our disposal. For this, we designed an isokinetic exhaust sampling system to fit on the exhaust pipes of the automobile [1]. The equipment that collects all PNAH from the analytical gas stream is also described in this paper.

Taking a sample in waste disposal areas necessitates judicious decisions and some discretion. A large sample should be taken to ensure proportional representation, but what about large size rocks and stones which may be present? In industrial operations there are many areas where waste is produced which is difficult to utilize commercially because of the presence of earth, earth substances, and water. Knowledge of the level of the substances harmful to life as well as the carcinogenic PNAH in this waste will assist in the proper disposal of the material.

Sample Workup and Concentrations

Since in the original material the PNAH of interest are only present at very low levels, several preconcentration steps are necessary to remove the bulk of the extraneous matter. The low levels of these PNAH preclude the direct analysis for these compounds. The design of these steps obviously depends on the composition of the matrix and on the specific PNAH being considered. The emphasis is to remove other classes of compounds.

Figure 1 shows the schematic diagram of the steps involved in the analysis of benzo(a)pyrene [B(a)P] and benzo(a)anthracene [B(a)A] in automobile exhaust. The gases are first cooled, scrubbed in an Oldershaw tower, and then filtered through a Millipore glass fiber filter. After maceration and extraction of the filter, the hydrocarbon fractions of the sample are reduced in volume by distillation. Radioactive B(a)A and B(a)P are added as internal standards to determine the losses in the subsequent procedures. The actual quantity of these tracers is about 10 ng and is below the threshold of detectability in the subsequent analysis by UV spectroscopy. The activity of each of the tracers is about 0.02 μCi.

Several purification steps further, we obtain a concentrate which can be subjected to the final isolation step by SFC. Specific details and the rationale of these separation and concentration steps have been described [1,2].

Figure 2 shows schematically the steps involved in the analysis of a refinery solid waste for the same PNAH. Solid waste is defined here as all material which is waste but does not contribute to air or water pollution. The clay separation involves percolation through

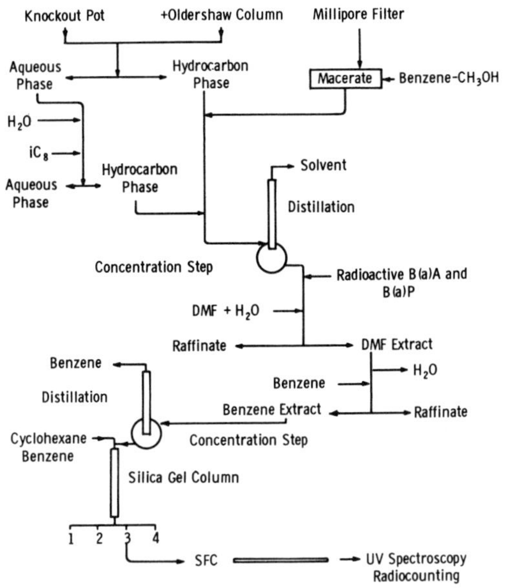

Fig. 1. Schematic diagram of separation and analysis steps.

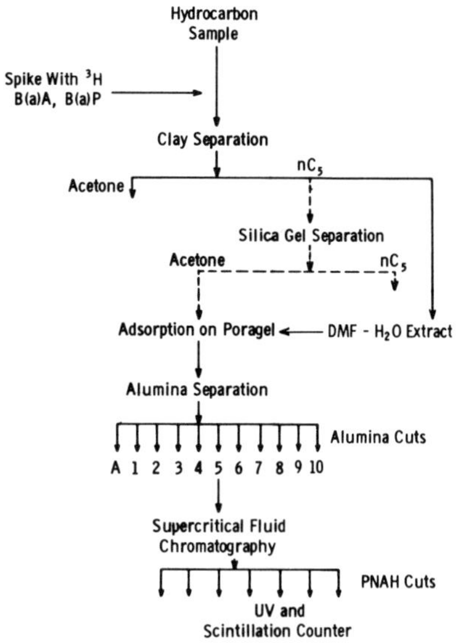

Fig. 2. Schematic diagram for the analysis of PNAH in solid wastes.

a clay column to remove polar compounds and particulate material.

Column chromatography on Poragel is used to separate the PNAH from their alkylated derivatives. This particular separation is not necessary in the analysis of automotive exhaust because the high combustion temperatures effectively reduce the alkylated PNAH in the sample to their parent compounds.

Analysis by SCF

At this point, the levels of the desired PNAH are sufficiently high (0.01 to 0.1%) to allow for their final isolation in measurable amounts and in good purity by SFC. SFC is an excellent technique for this problem because of the high selectivities which can be attained and because trapping and quantitative recovery is fairly simple.

The high selectivity of SFC has been reported in several publications [2-5]. For mixtures of PNAH, data are available on their behavior on columns packed with Du Pont Permaphase-eth [2] or on a VYDAC Reverse Phase* with CO_2 - 1.5% CH_3OH as the mobile phase. The first-named column tends to separate PNAH according to ring type. The parent PNAH and its alkyl substituted derivatives elute together or close to each other with the more highly substituted derivatives tending to elute earlier. Ring systems with a saturated carbon elute earlier than similar molecules without saturated carbon atoms in the rings.

*Separations Group, Hesperia, California.

The VYDAC reverse phase column, on the other hand, separates more or less according to molecular weight. Alkyl substitution increases the retention times. Thus, in most cases, one should expect that the injection of a narrow fraction isolated by chromatography on one of these columns onto a column with the other packing would allow for the isolation of almost any PNAH in a high degree of purity. Sequential or two-dimensional SFC is, therefore, a powerful technique to isolate PNAH compounds from mixtures. It can be further improved by varying the temperature or pressure, or both. Very complex mixtures can be prefractionated by another chromatographic method to reduce the complexity of the initial mixture.

Application to Actual Samples

In a recent publication [3] we described the analysis of automobile exhaust for PNAH by this technique. B(a)P and B(a)P were isolated in one pass through a 5.5-m by 0.21-cm ID column packed with Permaphase-eth. Chromatography was carried out by pressure programming the system from 100 to 200 atm. The fractions corresponding to these peaks are collected in a high pressure fraction collector and subsequently analyzed for purity and for quantitation by UV spectroscopy. Experience has shown that the purity of the recovered B(a)A from this column is comparable to the material purchased as standard. To obtain B(a)P in high purity a single pass is sometimes insufficient. In this case the sample is preseparated on a VYDAC reverse phase column.

By determining the radioactive levels of these peaks, we can determine what aliquot this is from the original solution. This allows for the calculation of the total amount of B(a)A and B(a)P evolved for every engine test.

Figure 3 shows the chromatogram obtained by SFC on a refinery solid waste fraction obtained according to the scheme shown in Fig. 2. The peaks corresponding to B(a)A and B(a)P are indicated. An indication of the purity of the B(a)A fraction isolated by this technique is shown in Fig. 4 where the UV absorption spectra of the isolated sample is compared against that of a standard B(a)A solution.

The curves are identical. Quantification of the level of PNAH in the original sample is carried out by determining the amount of the B(a)A and by determining the radioactivity level of this peak in the isolated fraction.

SFC is shown here as an excellent tool to carry out the final isolation step of specific PNAH in two vastly different materials where they were originally present at ppm or lower levels.

SUPERCRITICAL FLUID CHROMATOGRAPHY - MASS SPECTROMETRY (SFC-MS)

The combination of two different analytical techniques can result in an increase in capabilities far beyond the sphere and scope

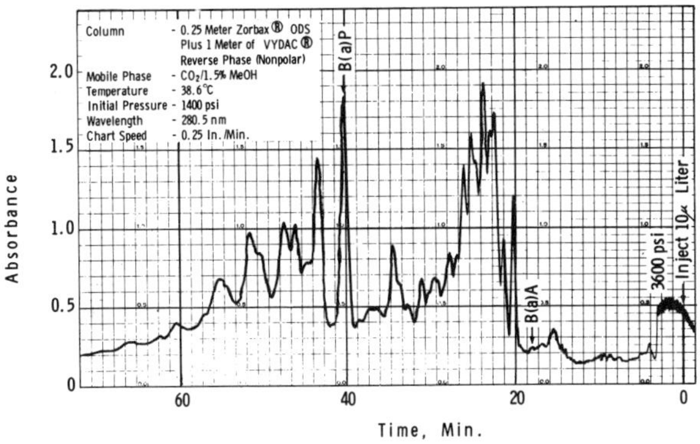

Fig. 3. SFC chromatogram of alumina cut 8.

of each of the component techniques. The most successful "marriage" is that between the gas chromatograph and the mass spectrometer. Considerable effort has, therefore, been expanded on the study of compatible combinations in the expectation of developing a technique of comparable value.

As indicated earlier, high performance liquid chromatography is now an important subject of analytical endeavor. This is understandable if we consider that of the two million known compounds which are currently more or less well characterized only 10% or so can be volatilized without decomposition. This minor fraction falls in the province of gas chromatography. The balance, however, can only be analyzed by liquid or supercritical fluid chromatography.

It is, therefore, not surprising that many investigators have attempted to introduce the effluent from a liquid chromatograph into a mass spectrometer for further elucidation of the separated compounds. Results, so far, have not been too encouraging. The few papers which have reported a modicum of success are based on the principle of the moving wire detector [6,7]. The effluent is contacted with a wire or chain which transports the attached material to an oven to remove the solvent prior to introduction into a chamber which is connected to the ion chamber of the mass spectrometer. These detectors are bulky and necessitate considerable care for good operation.

The most direct approach would be to split a representative side stream into the mass spectrometer. The main problem in this approach is the high molecular weight of the solvent, which will cause problems by drowning out the ion current at the sensitivity necessary to detect the solutes. Horning, et al. [8] solved this by chemically ionizing the eluent in a reaction chamber at atmospheric pressure external to the low pressure region of a quadrupole mass spectrometer. Chemical ionization in the source chamber with the eluent acting as ionizing reagent has been reported by Arpino, et al. [9,10]. The eluate passes through a stream splitter which diverts 10 μl/min. to the spectrometer interface. The pressure in the source region is maintained at 10^{-4} torr.

Interfacing a supercritical fluid chromatograph with a mass spectrometer appears to offer many advantages. SFC is comparable

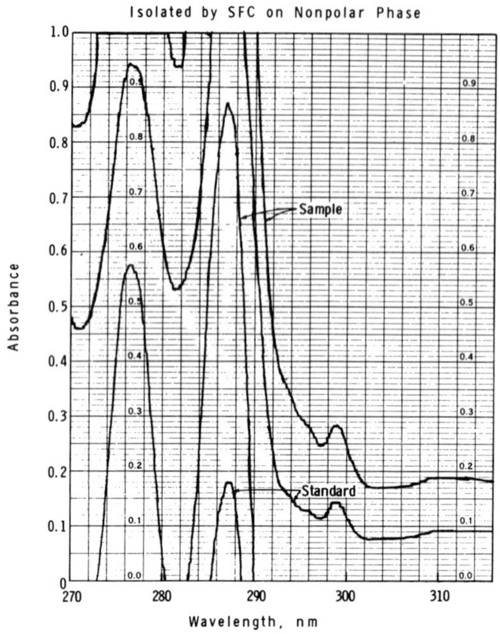

Fig. 4. UV absorption of B(a)A from alumina cut 8 (isolated by SFC on nonpolar phase).

to HPLC in its ability to handle those high molecular weight, thermally labile, and ionic compounds which cannot be processed by gas chromatography. SFC can be carried out with a mobile phase, such as CO_2, which is gaseous at ambient conditions. This combination technique would be quite similar to GC-MS.

Equipment

Figure 5 shows a schematic diagram of the SFC-MS system used in our preliminary investigations. The chromatographic portion consists of a high pressure pump, pressure programmer, chromatograph oven, sample collecting system, and accessory equipment. In the regular configuration detection would have been carried out by a UV detector but for this work, the outlet of the column is connected directly through an

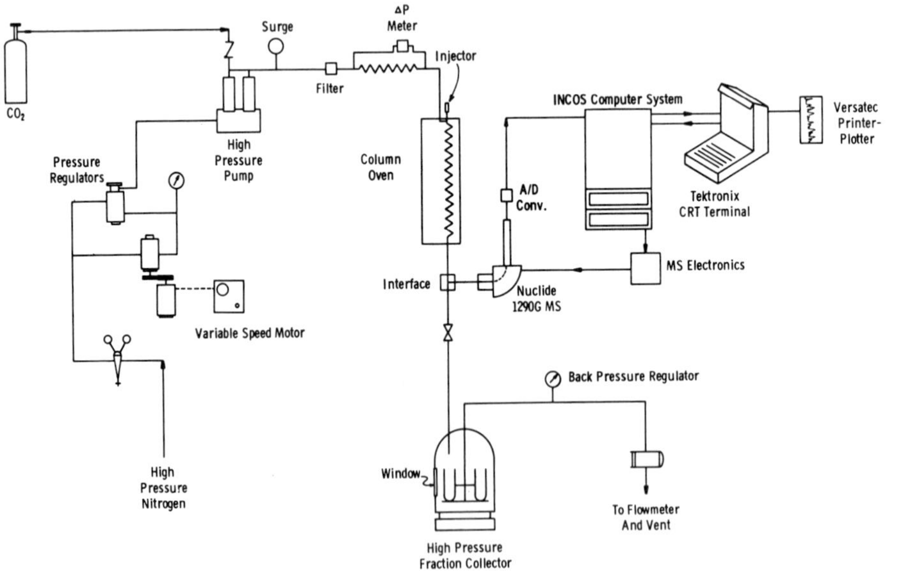

Fig. 5. Assembly for SFC-MS.

interface to the mass spectrometer. The UV detector was taken out during these experiments.

The mass spectrometer is a Nuclide Model 1290G 90° magnetic sector instrument operated under control of an INCOS computer system. To improve the sensitivity of the technique, most of the experiments were carried out in the MID (multiple ion detection mode) which allows for maximum sensitivity and resolution of a selected number of m/e fragments. In the MID operational mode the magnetic field is maintained at a constant value, and the masses are scanned by programming the accelerating voltages. Scan cycle times are usually once per second.

An important accessory in this technique is the interface. In normal operation this interface has to reduce the pressure of 100 to 250 atm at the outlet of the chromatograph by a factor of 10^{10} to obtain the low pressure in the ion chamber. In the first experiment, a fairly simple configuration was tried in which a 0.01 inch ID capillary tube was inserted into one leg of a tee attached to the outlet of the column. The opening in this capillary tube, located inside the tee, is further swaged down to reduce the size of the orifice so as to maintain a pressure of 10^{-5} torr or lower at the ion gauge. This tee was wound with heating tape. At room temperature the peaks showed considerable tailing. Tailing is decreased by heating the interface to a surface temperature of 450°C. The other end of the capillary tube is connected to the sample inlet system of the mass spectrometer.

Operating Conditions

Chromatographic separations were carried out on a 10 ft by 1/8 in. OD stainless steel column packed with Permaphase-eth packing. The mobile phase was either pure CO_2 or CO_2 mixed with 1.5% methanol. The chromatograph oven was controlled by temperatures between 30 and 40°C, depending on the sample. Inlet pressures varied from 75 to 250 atm. The pressure in the sample collecting system was maintained at 60 atm. The flow rate of CO_2 through the column was maintained at about 10 to 20 ml liquid CO_2 per minute.

RESULTS AND DISCUSSION

Figure 6 shows a spectrum map of a synthetic mixture of phenanthrene, fluoranthene, pyrene, benzanthracene, benzo(ghi) perylene, and anthanthrene dissolved in dimethylnaphthalene. Detection by the mass spectrometer is carried out in the MID mode for m/e values of 156, 178, 202, 228, 252, and 276. The spectrum map is an orthogonal projection of the masses scanned as a function of time. The sample size is 5 µl containing about 5 x 10^{-6} g of each of the PNAH. The amount entering the mass spectrometer is estimated at about 10^{-8} g per compound. For this particular mixture, the use of a UV detector would have yielded better chromatograms,

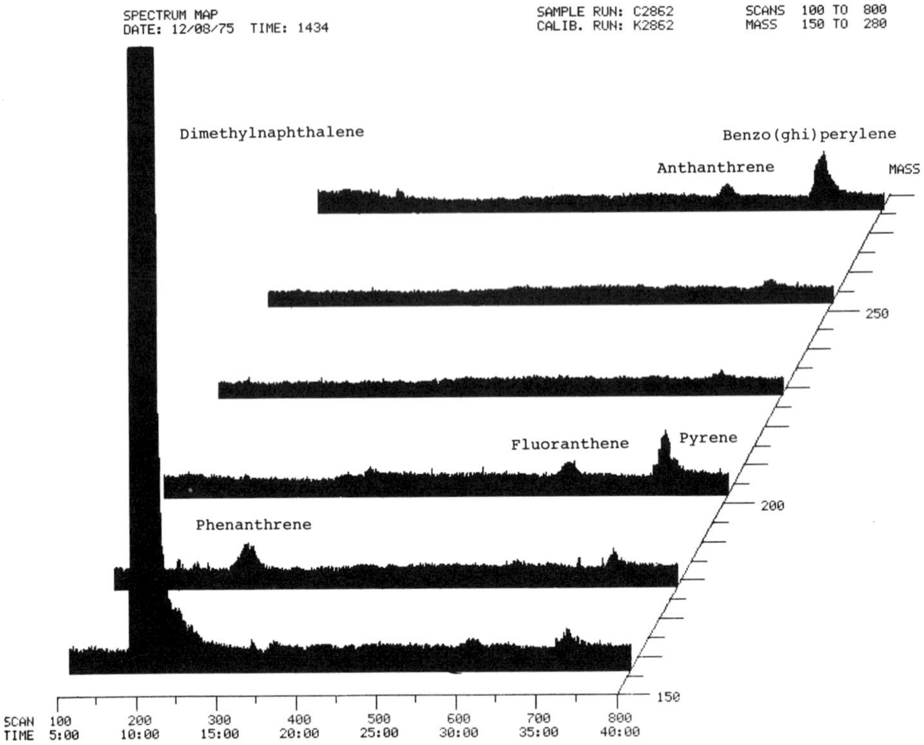

Fig. 6. Spectrum map of synthetic mixture of polynuclear aromatic hydrocarbons.

also because the levels of the solutes in the UV system would be about 500 times higher. The SFC-MS system is very sensitive; it is also able to resolve, very nicely, closely eluting compounds of unequal molecular weights such as pyrene and benzo(ghi)perylene, and anthanthrene and fluoranthene.

We have described earlier the high degree of selectivity which can be obtained by SFC. One phenomenon which could not be clearly demonstrated before is the separation between paraffins and condensed aromatic hydrocarbons. This is because there was no detector capable of monitoring the eluting paraffins. Figure 7 is a composite spectrum map of two consecutive injections of a synthetic mixture of C_9-C_{13} n-paraffins, naphthalene, methylnaphthalene, decalin, and anthracene. The use of a mass spectrometer allows us to detect the paraffins in the effluent. These compounds elute very close to each other before naphthalene comes out. Although not clearly shown in this figure, naphthalene elutes at about twice the retention volume of n-C_9. Anthracene elutes much later. From this spectrum we learn that the cyclic compounds and alkyl benzenes elute close to the paraffins and that the PNAH are retained very strongly on this column.

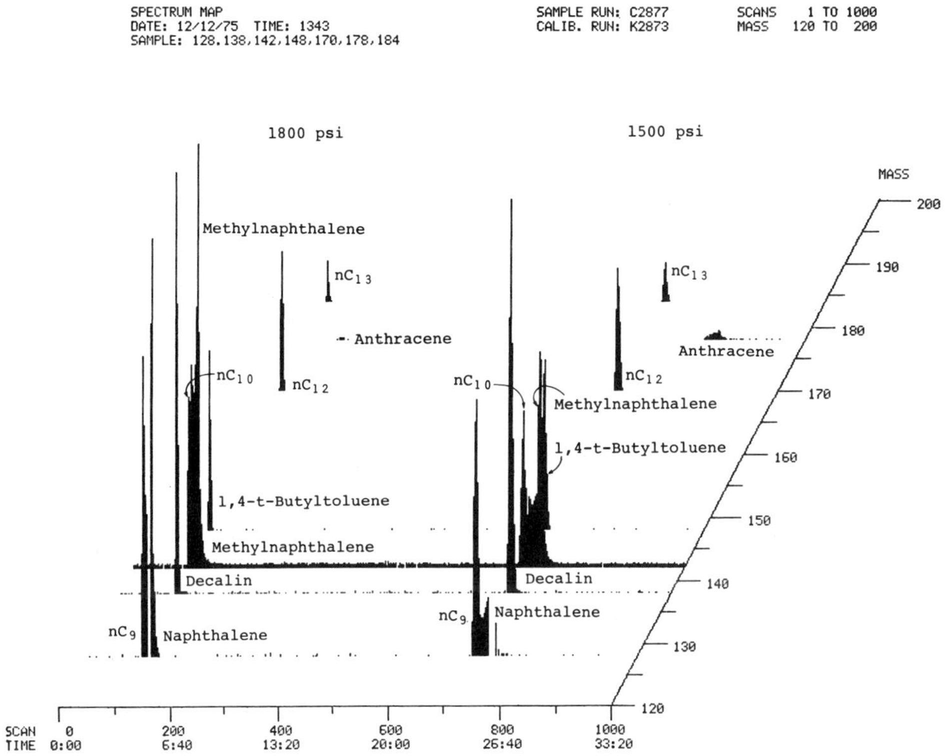

Fig. 7. Spectrum map of two consecutive samples of a synthetic mixture.

Figure 8 shows a spectrum map of a Hiawatha coal oil at two different sample loading levels. The sample size in the second injection (the right portion of the map) is about five times as large as the first injection. One can observe the presence of naphthalene and the alkyl naphthalenes. The detector was not set to scan at m/e of 156, the channel for the C_2 alkyl napthalenes. The signal at m/e equals 178, shown as compound "A" is probably from a condensed ring, three-ring cycloalkane, since anthracene or phenanthrene would have different retention times. Condensed three-ring cycloparaffins have been noted before in coal oil; this signal may be from the parent compound or from a fragment of an alkyl derivative.

CONCLUSION

There are undoubtedly many areas where improvements are possible, such as the design of the interface. One major problem refers to the size of the orifice. This has to be very small to create the required pressure drop. However, if the orifice is too small, plugging becomes a problem, especially with higher molecular

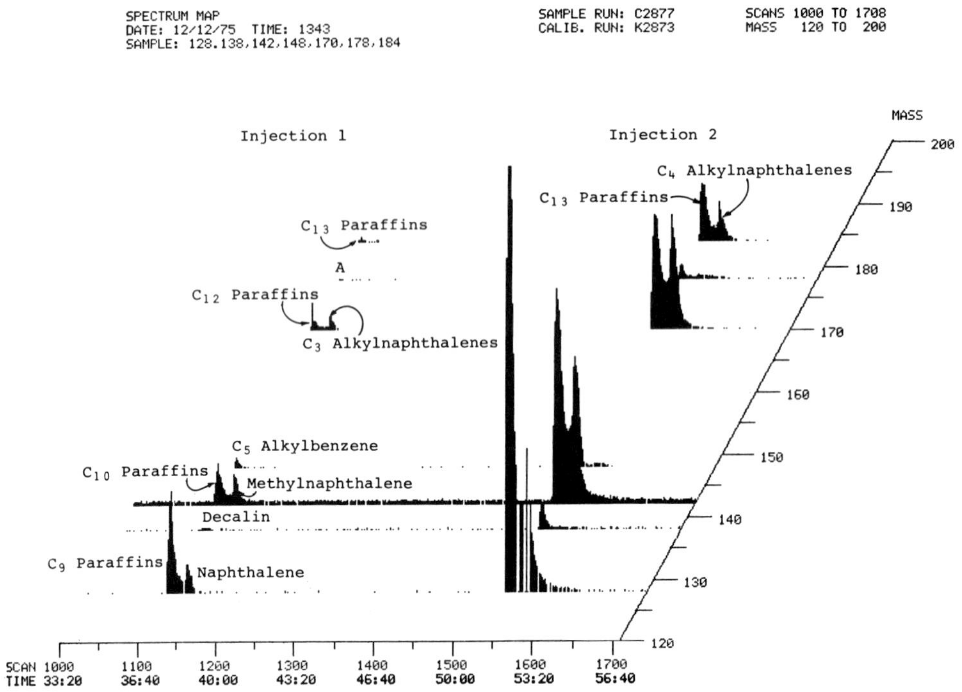

Fig. 8. Spectrum map of Hiawatha oil at two different sample loading levels.

weight compounds which are solid at room temperature. Nevertheless, these series of measurements demonstrate the feasibility of interfacing a supercritical fluid chromatograph with a mass spectrometer.

REFERENCES

1. H. K. Newhall, R. E. Jentoft, and P. R. Ballinger, SAE paper No. 730834 presented at SAE Meeting, Milwaukee, Wisconsin, September 10-13, 1973.
2. R. E. Jentoft and T. H. Gouw, Anal. Chem. 48, 2195 (1972).
3. T. H. Gouw and R. E. Jentoft, Adv. Chromatogr. 13, 6 (1975).
4. S. T. Sie and G. W. A. Rijnders, Anal. Chim. Acta 38, 31 (1967).
5. G. W. A. Rijnders, Chem.-Ing. Techn. 42, 290 (1970).
6. W. H. McFadden, H. L. Schwartz, and S. Evans, J. Chromatogr. 122, 389 (1976).
7. R. P. W. Scott, C. G. Scott, M. Munroe, and J. Hess, Jr., J. Chromatogr. 99, 395 (1974).
8. E. C. Horning, D. I. Carroll, I. Dzidic, K. D. Haegle, M. G. Horning, and R. N. Stillwell, J. Chromatogr. 99, 13 (1974).
9. P. J. Arpino, B. D. Darokins, and F. W. McLafferty, J. Chromatogr. 12, 574 (1974).
10. P. J. Arpino, M. A. Baldwin, and F. W. McLafferty, Biomed. Mass. Spectrum 1, 80 (1974).

L-5

LABORATORY TECHNIQUE FOR STUDYING ADSORPTION FROM SOLUTION AT HIGH PRESSURES AND ITS APPLICATION TO THE METHYLCYCLOPENTANE-ETHANOL-ACTIVATED CARBON SYSTEM

S. Ozawa, K. Kawahara, and Y. Ogino

Tohoku University
Sendai, Japan

INTRODUCTION

Although abundant data on adsorption from solutions have been accumulated [1], little is known about the effects of pressure on it. On the other hand, adsorption characteristics at high pressures seem worth studying, from both a scientific and practical point of view. The data may provide useful information about the structures and properties of liquid-solid interfaces. Moreover, they may provide the basis for the design and operation of advanced separation processes. For these reasons, we have undertaken to clarify these effects.

In the course of this study, some new experimental techniques have been developed. For instance, effective stirring to provide intimate contact of a liquid solution with a solid adsorbent under a pressure of about 5 kbar has been achieved by using a simple electromagnetic *in situ* stirring device. Furthermore, a technique by which a small amount of sample liquid can be withdrawn without disturbing the adsorption equilibrium at high pressure has been developed.

The purpose of this paper is mainly to report the details of the experimental techniques. In order to demonstrate the applicability of the proposed techniques, adsorption data for the methylcyclopentane-ethanol-activated carbon system as well as supporting discussion are presented.

EXPERIMENTAL

Apparatus

The arrangement of the apparatus is shown in Fig. 1. A hand pump A and an intensifier B (10 kbar capacity) serve to generate the pressure, which is measured on a strain gauge C. The pressure is transmitted by a mixture of glycerine and methanol (50-50 vol%) to the high-pressure vessel D. This vessel, containing an adsorption cell as well as an assembly of electromagnets, is capable of withstanding an inner pressure up to 6 kbar; it is maintained at a constant temperature by passing water at constant temperature through jacket E. The top of D is connected to an oil-driven sampling-valve F. The operating oil pressure for F is precisely controlled by a small plunger G. The bottom plug of D is equipped with six electrodes. The electric power for driving the electromagnetic stirrer is supplied through some of these electrodes. Further, the electrodes detect the electromotive force generated in the thermocouple placed in D.

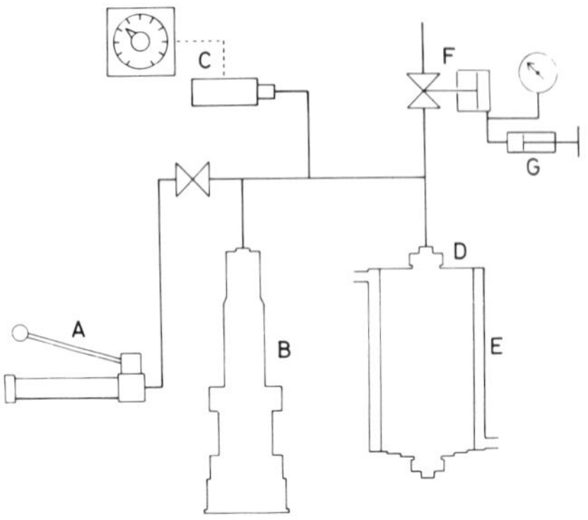

Fig. 1. Apparatus arrangement.

A schematic drawing of the adsorption cell and the built-in stirrer is shown in Fig. 2. The cell body H (stainless steel) consists of two concentric tubes with mercury, I, filling the annular space between the inner and outer tubes. Thus, the cell cap J (stainless steel) can move up and down freely to keep a constant pressure, and a complete seal between the adsorbate liquid and the pressure-transmitting fluid can be maintained during the sampling. The top of the cap is connected to a flexible Teflon tube K (0.5-mm ID) consisting of a part of the sampling line, which runs inside the high-pressure tubing. One end of the tube opens into the adsorption cell and has a cotton filter L at its intake. The other end is connected to the high-pressure mouth of the sampling valve F.

The built-in stirrer is located at the bottom of the adsorption cell. It consists of a permanent magnet bar M and its support N which is covered with a stainless steel mesh O. The rotating motion of L

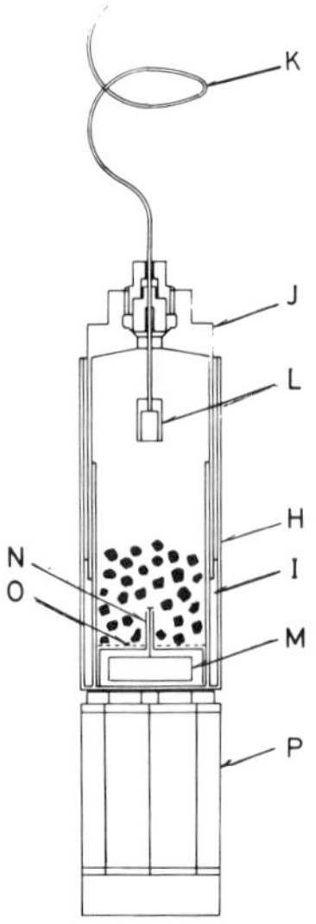

Fig. 2. Schematic drawing of adsorption cell and built-in stirrer.

is controlled by an assembly of electromagnets P located under the adsorption cell.

The electromagnetic assembly is located in the bottom of the high-pressure vessel. The assembly consists of six electromagnets arranged in the manner shown in Fig. 3 where Fig. 3a indicates the electric connection and Fig. 3b indicates the geometrical arrangement, i.e. the top view. Three electrical leads connected to positions i, j, and k of the magnet assembly are routed out of the pressure vessel through three electrodes embedded in the bottom plug of the vessel. Thus, the direction of the electric current to the magnets can be switched successively by operating a motor-driven mechanical switching device located outside the high-pressure system. The schematic representation of the rotating magnetic field generated by this method is also shown in Fig. 3b. Thus, the torque and the speed of the rotation of L can be adjusted separately.

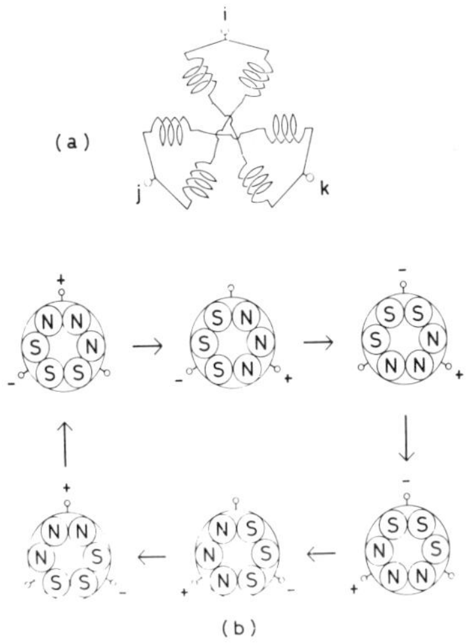

Fig. 3. Electrical (a) and geometrical (b) arrangements associated with the assembly of six electromagnets and the principal of the rotating magnetic field.

Material

Activated carbon was obtained commercially, and granules of 1 to 2 mm were used. The BET surface area measured by nitrogen adsorption was 1350 m^2/g. Special grade methylcyclopentane (MCP) and ethanol were also obtained commercially. Gas chromatographic analyses showed that they contained little impurities. Therefore, no purification was attempted.

Procedure

Granules of a given weight (usually about 5 g) of activated carbon were degassed at 400°C and about 10 Pa for 3 hr, and were then sealed into a thin-walled glass ampule under vacuum. The ampule and a given amount, about 30 g of adsorbate solution of a given composition, were put into the adsorption cell and the ampule broken. The adsorption cell was quickly placed in the pressure vessel and the desired pressure was applied. After an equilibration period of 4 to 5 hr, a small amount of the sample liquid was withdrawn by operating the sampling valve while maintaining the equilibrium pressure. To avoid error caused by evaporation, the sample solution was stored over mercury in a small inverted test tube. The concentration difference between the original solution and the sample solution was measured by a differential refractometer.

Adsorption isotherms at atmospheric pressure were measured separately with a conventional apparatus.

RESULTS AND DISCUSSION

Composite Isotherms

Adsorption isotherms of MCP-ethanol-activated carbon system at 25°C are shown in Fig. 4. A pressure effect on adsorption is evident in this figure. The apparent adsorption which is defined by $n_1^a=(n^o/w)(x_1^o-x_1^e)$ became smaller at higher pressure. Further, it must be pointed out that the adsorption azeotropic points shifted to the left at higher pressures. This fact can be best explained by the aid of the following relation [2]:

$$n_1^a = \frac{n^o}{w}(x_1^o-x_1^e) = \frac{n^s}{w}(x_1^s-x_1^e) \quad (1)$$

where $(n^s/w)(x_1^s-x_1^e)$ is known as the surface excess or the differential adsorption. Since $n_1^a=0$ at the azeotropic point, the surface composition is identical with the equilibrium composition of the bulk solution at this point. Therefore, the shift to the left of the azeotropic point at high pressure means that the surface concentration of ethanol increased at the higher pressure.

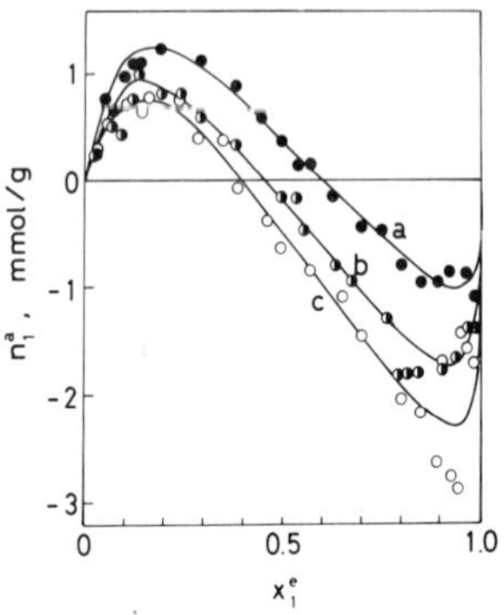

Fig. 4. Adsorption isotherms for MCP-ethanol-activated carbon system at 25°C. Legend: a = 1.0 bar; b = 2.94 kbar, and c = 4.90 kbar.

Application of Sircar and Myers' Adsorption Equation

The experimental adsorption isotherms were analyzed on the basis of adsorption theory proposed by Sircar and Myers [2]. According to this theory, the surface excess is given by

$$n_1^a = m\left(\frac{a_1}{a_1 + Ka_2} - x_1^e\right) \quad (2)$$

This adsorption equation contains two constants, i.e., the equilibrium constant K and the number of moles in the adsorbed phase m. Thus, in order to test the applicability of the theory, values for these two constants were determined as shown in Table I. The constant K could easily be evaluated by applying equation (2) to the adsorption azeotrope in the following manner:

$$K = \frac{(1 - x_1^*)}{x_1^*} \cdot \frac{a_1^*}{a_2^*} \quad (3)$$

On the other hand, no simple method of evaluating m was found. Therefore, it was determined by trial (values for the activity of

Table I. Constants in Sircar-Myers' Adsorption Equation for MCP-Ethanol-Activated Carbon System at 25 °C.

Pressure, kbar	1.0×10^{-3}	2.94	4.90
K	0.83	290	11000
m, mmol/g	4.1	4.7	5.3
$\sigma_1 - \sigma_2$, dyn/cm	-1.4	49	91

components in the equilibrium solution at high pressure were calculated on the basis of literature data on vapor-liquid equilibrium [3], and by the method of Prausnitz et al. [4]). Using the values listed in Table I, theoretical adsorption isotherms were obtained. These are shown in Fig. 4 by solid lines. Agreement between the theoretical and the experimental isotherms is satisfactory.

Individual Isotherms

The individual adsorption isotherms for components 1 and 2 are defined by $n_1^s=(n^s/w)x_1^s$ and $n_2^s=(n^s/w)(1-x_1^s)$, respectively. On the basis of the adsorption theory of Sircar and Myers, the individual adsorption isotherms are given by

$$n_1^s = m \frac{a_1}{a_1 + Ka_2} \tag{4}$$

and

$$n_2^s = m \frac{Ka_2}{a_1 + Ka_2} \tag{5}$$

These two equations are easily obtained by comparing (2) with (1) and setting m equal to n^s/w. With these two equations and the values of K and m listed in Table I, the individual isotherms for MCP and ethanol were evaluated as shown in Fig. 5. It can be seen that the adsorption of MCP remained almost constant while the adsorption of ethanol increased markedly with the increase in the equilibrium pressure.

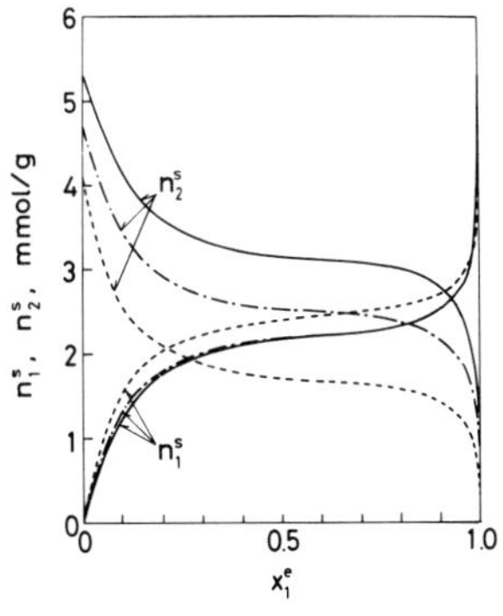

Fig. 5. Individual isotherms. Legend: n_1^s=MCP; n_2^s=ethanol; dotted line=1.0 bar; dashed line=2.94 kbar; and solid line=4.90 kbar.

Constants K and m

According to the Sircar and Myers' theory, m is the number of moles in the adsorbed phase at a given pressure and temperature. As can be seen in Table I, the value of m increases with an increase in equilibrium pressure. This means that the molecular density of the adsorbed phase increased about 23% after compressing the equilibrium fluid from 1.0 bar to 4.90 kbar. The individual isotherms given in Fig. 5 reveal that the increased density of the adsorbed phase is caused mainly by the increase in the adsorption of ethanol. On the other hand, the compression of bulk ethanol is estimated to be about 15%. Therefore, it is considered that a specially dense arrangement of the adsorbed ethanol molecules has resulted from the interaction between the adsorbent surface and the ethanol molecules at high pressures.

The change in the arrangement of the adsorbed molecules seems to be reflected in the change of the adsorption equilibrium constant K. According to the theory of Sircar and Myers, K is given by

$$K = \exp\left(\frac{A(\sigma_1-\sigma_2)}{mRT}\right) \tag{6}$$

Hence, the pressure dependence of the interfacial tension difference $\sigma_1-\sigma_2$ can be evaluated from the pressure dependence of K and m. It can be seen in Table I that the value of $\sigma_1-\sigma_2$ increased greatly with pressure. Since the adsorption of MCP was considered to be affected very little by pressure, the increase in $\sigma_1-\sigma_2$ should mainly be attributable to the decrease in σ_2, i.e. to the interfacial tension between the ethanol and the adsorbent.

It seems interesting to visualize the change in the arrangement of ethanol molecules on the adsorbent surface. The molecules should have a denser two-dimensional packing and they exhibit a smaller interfacial tension at high pressures. Unfortunately, however, the molecular theory of interfacial tension at high pressure is not fully developed. Furthermore, Sircar and Myers derived the adsorption equation by neglecting the change in the PV energy on adsorption. Therefore, considerable errors are assumed to be included in the values of m and K, especially at high pressures. More rigorous adsorption theory has to be established before discussing the molecular arrangement of adsorbed phase, though experimental data of the present work strongly suggest that certain changes in the adsorbed phase occurred at high pressure.

NOTATION

- A = surface area of adsorbent
- a = activity
- K = adsorption equilibrium constant in equation (2)
- m = number of moles in the adsorbed phase in equation (2)
- n = number of moles
- R = gas constant
- T = absolute temperature
- w = mass of adsorbent
- x = mole fraction
- σ = interfacial tension

Subscripts

- 1 = MCP (methylcyclopentane)
- 2 = ethanol

Superscripts

a = apparent adsorption
e = equilibrium solution
o = initial solution
s = adsorbed phase
* = adsorption azeotrope

REFERENCES

1. J. J. Kipling, Adsorption from Solutions of Non-Electrolytes, Academic Press, London (1965).
2. S. Sircar and A. L. Myers, J. Phys. Chem. 74, 2828 (1970).
3. J. E. Sinor and J. H. Weber, J. Chem. Eng. Data 5, 243 (1960).
4. J. M. Prausnitz and P. L. Chueh, Computer Calculations for High-Pressure Vapor-Liquid Equilibria, Prentice-Hall, New York (1968).

L-6

HIGH PRESSURE TERNARY EQUILIBRIUM IN AQUEOUS - HYDROCARBON SYSTEMS

C. A. Irani and D. J. McHugh

Exxon Research and Engineering Company
Linden, New Jersey USA

INTRODUCTION

The possibility of using water as a solvent for aromatics extraction was first demonstrated by Arnold and Coghlan [1] in 1948 when they successfully extracted a nitration grade toluene from reformed naphtha. Experimental and theoretical work performed since then has led to a better understanding of the phase behavior of these systems. It is now possible to predict at least qualitatively and in a few instances quantitatively their binary equilibrium.

Most hydrocarbon-water binaries demonstrate liquid-liquid immiscibility across the entire temperature range. A characteristic feature of such systems is the presence of two distinct critical lines and a three phase liquid-liquid-gas line. The three phase line has been shown to terminate at an upper critical end point [2], which is connected to the critical point of the pure hydrocarbon by a short critical locus. A second locus starting at the critical point of water drops back to a temperature and pressure minimum before changing direction and rising rapidly to very high pressures [3] (see Fig. 1). This critical locus is significant in that it marks the transition from a two-phase to a one-phase system, a condition which arises when the locus is crossed going from left to right.

For water-paraffin binaries the locus starting at the critical point of water has a much shallower temperature minimum before climbing rapidly to higher pressures [4] (see Fig. 1). Essentially, the temperature difference between the aromatic and paraffinic locus dictates the extent of paraffinic solubility when the water-aromatic binary becomes miscible. Thus, by picking extraction conditions in the vicinity of the aromatic locus, the capacity and

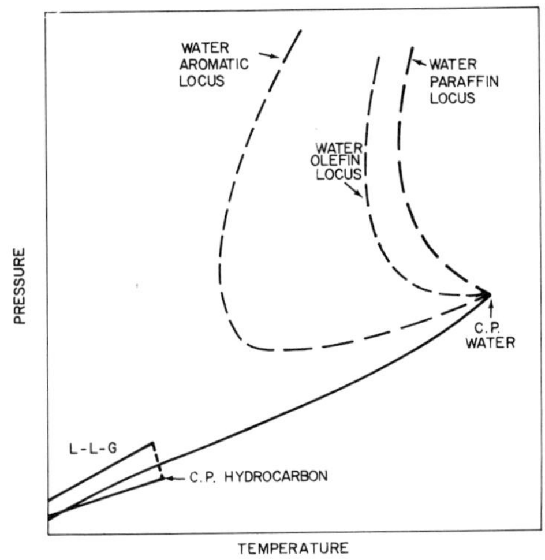

Fig. 1. Pressure-temperature projection of water hydrocarbon binaries.

selectivity of the solvent for aromatics can be optimized. O'Grady [5] studied the water-benzene-n-heptane ternary across a wide temperature range and established the equilibrium phase diagrams for such systems as the water-aromatic critical solution temperature is crossed and that for the water-paraffin system is approached.

The purpose of this study was to extend the results of O'Grady to include both branched paraffins and mono-olefins as the non-aromatic components, and thus evaluate water as a potential solvent for the more complex range of components normally encountered in refinery stocks.

EXPERIMENTAL

The chemicals used were practical quality from Matheson, Coleman and Bell and were not further purified. Distilled water was used as the solvent.

The extractor was a 316 stainless steel autoclave of 600 cc capacity which was completely immersed in a fluidized sand bath heated by resistance heaters. The autoclave was raised and lowered into the sand bath by means of an air driven winch.

The autoclave's contents were mixed by an externally driven Magnedrive unit attached to the main body by means of an extra long neck which permitted it to be positioned outside the sand bath. Stirring speed could be controlled by adjusting the flow of air through the Magnedrive motor.

Calibrated iron-constantan thermocouples were used to measure the temperature of the bath and in the extractor. To measure the temperature in the batch extractor, the autoclave cover was fitted with a thermocouple well which extended into the extraction zone. Temperature could be measured and controlled to $\pm$ 0.2°C. System pressure could be measured to $\pm$ 10 psia using Heise gauges previously calibrated against steam tables.

A new technique was developed for maintaining isobaric conditions during sampling, and is described below. In the past mercury has been successfully used as a liquid piston to preserve isobaric conditions during sampling. Due to safety and convenience considerations mercury was not used; instead, a "quasi-inert" liquid piston was developed for this application. By screwing a cylindrical section into the autoclave cover, as shown in Fig. 2, it was possible to create an annular configuration in the extraction zone, resulting in an inner and outer chamber into either of which fresh hydrocarbon could be introduced by opening the appropriate valves. Entry into the inner chamber was direct, while hydrocarbon entering the outer chamber was heated to system temperature by circulation through a pre-heat coil immersed in the sand bath.

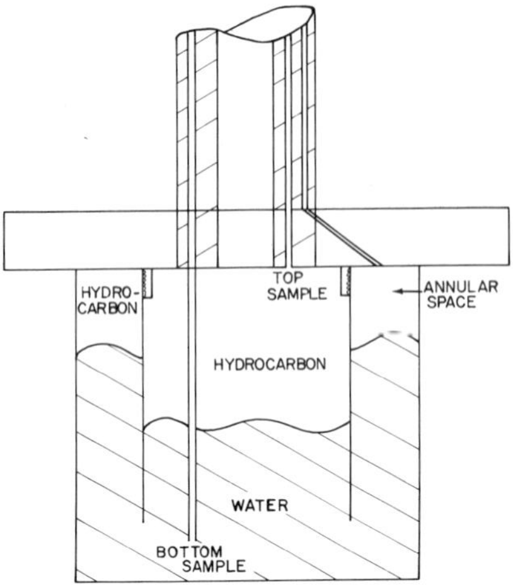

Fig. 2. Batch extractor with annulus.

Helium gas was used to pressurize hydrocarbon feed from the loading bombs into the unit. By introducing a back pressure regulator between the high pressure side of the compressor and the top of the loading bombs, the pressure of the helium gas could be controlled to any desired value ($\pm$ 15 psia). Initially the unit was evacuated and then charged, with the agueous phase first followed by the hydrocarbons. The entire charge was introduced into the inner chamber rather than the annulus. With the stirrer activated, the temperature was raised to the extraction temperature and the system pressure increased to the desired value by introducing fresh hydrocarbon into the inner chamber. After allowing sufficient time for equilibrium to be established, stirring was stopped and the two phases allowed to separate.

Before sampling, the dome pressure on the regulator was first adjusted to the system pressure, and then the valve connecting the loading bombs to the outer chambers of the extractor was opened. The bottom sample line was now purged of 4 cc to eliminate unequilibrated liquid held up in it. The drop in pressure during the purging step was automatically compensated for by the transfer of hydrocarbon from the loading bombs into the top of the outer chamber. Normally the introduction of unequilibrated hydrocarbon feed into the extraction zone would adversely affect the quality of the samples, an affect that is minimized by the use of the outer annular chamber.

The extract phase was sampled immediately following the purge, with the sample removed from the bottom of the vessel. For the unequilibrated feed entering the quiescent upper region of the outer chamber to affect the sample, it would need to diffuse throughout the top hydrocarbon phase, across the hydrocarbon/water interface and then throughout the aqueous phase, a sequence that would take longer than the approximately 2 min required between the start of the purge and the end of the sampling. Next the top sample line was purged and the raffinate sample taken immediately. By taking the raffinate sample from the inner chamber of the annulus, the equilibrated hydrocarbon phase could again be kept isolated from the unequilibrated hydrocarbon make up. Following sampling, mixing was resumed and the system allowed to re-equilibrate, after which a second set of samples were taken using the same procedure.

A sample constituting approximately 10 g. of either phase was collected in frozen tetradecane and stored overnight in an icebox to allow hydrocarbon and water to separate out. The weight and composition of the hydrocarbon phase was determined by gas chromatography using an internal standard procedure. The gas chromatograph was a Hewlett-Packard 5830A equipped with flame ionization detectors.

Reproducibility and accuracy of experimental and analytical techniques (dotted lines) were verified by duplicating O'Grady's data [5] (solid lines) for the water-benzene-n-heptane ternary at 298.9°C and 3600 psia (Fig. 3). Good agreement was found between the two sets of data, both in reproducing the co-existence curve and the slopes of the tie lines between extract and raffinate phases.

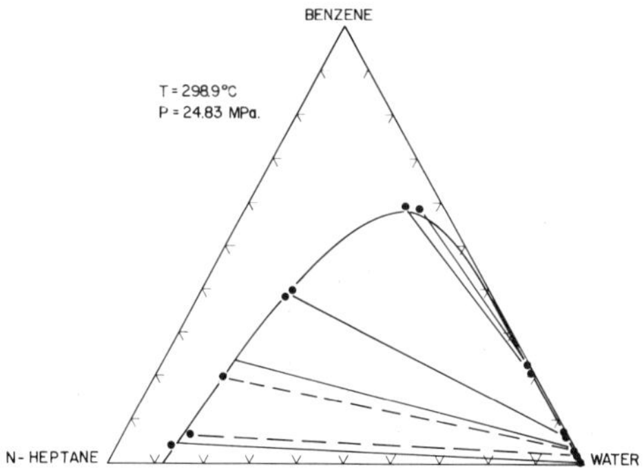

Fig. 3. Water-benzene-n-heptane.

RESULTS AND DISCUSSION

In this work the earlier results with the water-benzene-n-heptane system [5] were extended to include a branched paraffin (3-methylhexane) and two mono-olefins (1-heptene and 1-hexene) as the non-aromatic components. 3-methylhexane was studied because branched paraffins are present in the feed to a conventional BTX (benzene, toluene, xylene) range extraction process. Another feature of feeds going to extraction is the presence of mono- and diolefins and sulfur compounds. The solvents presently used for aromatics extraction from such feeds are sensitive to these components, and consequently the feed needs to be hydrogenated before extraction. However, the chemically inert nature of water should make it more tolerant of such contaminants so that complete hydrofining of the feed may not prove necessary, and only partial hydrofining to remove sulfur and saturate the diolefins could suffice. With this in mind, ternary equilibrium for two systems containing mono-olefins was also determined.

Equilibrium diagrams for the three systems are presented in Figs. 4 to 6, with the equilibrium compositions summarized in Table I. On comparing the 3-methylhexane data with n-heptane at the same temperature and pressure it is clear that branched and straight paraffins show almost identical behavior with regards to the shape of the binodal curves and the tie line slopes. The

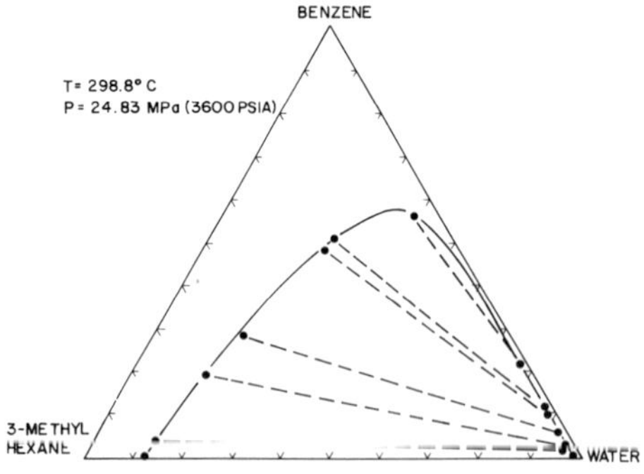

Fig. 4. Water-benzene-3-methylhexane.

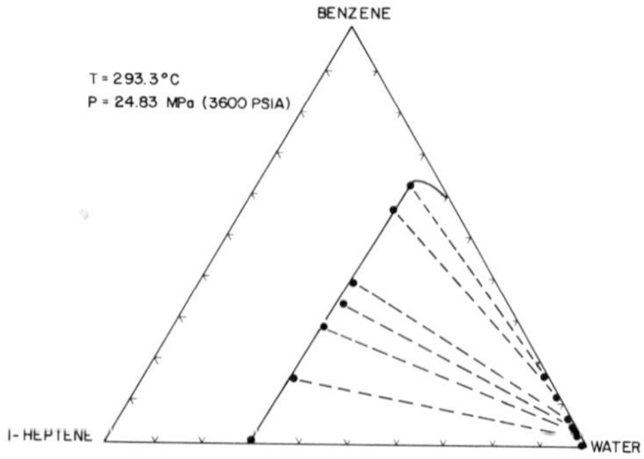

Fig. 5. Water-benzene-1-heptene.

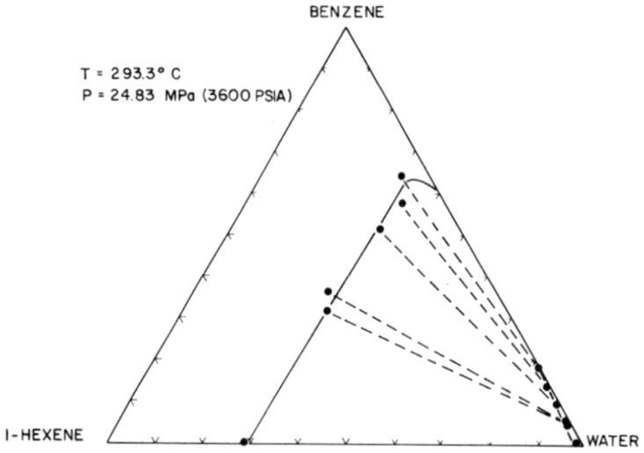

Fig. 6. Water-benzene-1-hexene.

possibility that water binaries with straight and branched paraffins of the same carbon number would show similar solubility behavior was discussed earlier by Connolly [6] and has been demonstrated here to hold for ternary equilibrium.

The higher mutual solubilities of mono-olefin-water binaries relative to the paraffins results in the miscibility locus for mono-olefins being displaced to lower temperatures and hence closer to the water-aromatic binary locus (Fig. 1). This should result in a lowering of the solvents selectivity which is indeed found to be the case (Table I). Additionally, the higher water solubility in the mono-olefins also resulted in a shift to the right of the raffinate section of the equilibrium curve. For feed concentrations containing greater than 90% aromatics, however, the influence of the mono-olefins is minimized and the binodal curve follows the benzene-heptane system.

Decreasing the hydrocarbon molecular weight should result in an additional increase in the mutual solubilities, so that going from C_7 to a C_6 mono-olefin should further decrease the solvent's selectivity. However, on superimposing the ternary equilibrium for the C_7 and C_6 mono-olefins it is found that the two systems demonstrate very similar binodal curves. This discrepancy can be attributed to thermal decomposition of the mono-olefins at the high extraction temperatures, the decomposition products in turn affecting the nature of the ternary equilibrium. Even though no attempt was made to establish the nature of the decomposition products, it was clear that the 1-heptene had undergone more severe degradation (about 5%) then the 1-hexene (about 1%).

Table I. Equilibrium Composition Data

System	Water-Rich Phase, Wt.%			Hydrocarbon-Rich Phase, Wt.%			Distribution Coefficient of Aromatic	Distribution Coefficient of Non-aromatic	Selectivity
	Aromatic	Non-aromatic	Water	Aromatic	Non-aromatic	Water			
	21.65	0.822	77.53	56.35	4.67	38.98	0.3842	0.176	2.182
Water-benzene-	11.26	1.30	87.44	50.23	23.39	26.38	0.224	0.055	4.033
3-methyl hexane	9.81	1.31	88.88	48.13	26.33	24.24	0.204	0.049	4.10
	5.34	1.59	93.07	27.78	53.40	18.82	0.1922	0.0297	6.45
	2.82	1.20	95.98	19.1	65.30	15.60	0.1476	0.0183	8.03
	1.40	2.90	95.7	3.4	83.9	12.70	0.411	0.034	11.89
	--	0.55	99.45	--	87.41	12.59	--	--	--
	16.5	0.69	82.8	62.4	6.21	31.39	0.264	0.111	2.38
Water-benzene-	11.8	1.08	87.12	56.5	12.9	30.60	0.209	0.084	2.49
1-heptene	6.2	1.48	92.32	38.8	29.9	31.30	0.159	0.049	3.23
	4.6	1.17	93.70	33.7	34.0	32.3	0.136	0.0344	3.97
	3.89	1.33	94.78	28.1	41.1	30.8	0.138	0.032	4.27
	2.31	1.74	95.95	15.45	53.06	31.49	0.149	0.032	4.56
	--	1.4	98.6	--	71.14	28.86	--	--	--
	18.21	0.851	80.939	64.08	6.25	29.67	0.281	0.136	2.06
Water-benzene-	14.05	1.09	84.86	57.73	9.37	32.90	0.243	0.116	2.09
1-hexene	9.78	1.2	89.02	51.32	17.01	31.67	0.190	0.070	2.70
	5.30	1.54	93.16	36.06	35.68	28.26	0.147	0.043	3.40
	4.68	1.61	93.71	31.80	38.05	30.15	0.147	0.042	3.48
	2.92	1.61	95.47	22.44	53.81	23.75	0.1301	0.0299	4.35
	--	2.06	97.94	--	71.22	28.78	--	--	--
Water + 15% CO_2-	15	--	85	70	--	30	--	--	--
benzene-n-heptane	10.17	0.658	89.172	59.1	15.8	25.1	0.172	0.042	4.13
	3.4	0.56	96.04	38.37	41.35	20.28	0.089	0.013	6.54

Clearly, the extent of thermal degradation of the feed will be a function of the time during which it is subjected to these high extraction temperatures. In the batch extractor used for this study, the contacting time is in excess of 20 hrs relative to the approximately 30 min required by a high volume, commercial scale process using counter-current extractors. Accordingly, it can be anticipated that the extent of thermal decomposition of the mono-olefins will be considerably lower in a commercial environment, but it may still be of concern.

The results of this study demonstrate that water at high temperatures is an effective solvent for aromatics extraction from partially hydrofined feeds. The branched paraffins have shown similar behavior as straight paraffins with the same carbon number, while the mono-olefins result in a lower aromatics selectivity and some feed decomposition. Partial hydrogenation of the feed to remove sulfur and saturate the diolefins will be required.

Another case which was studied and is presented in Fig. 7 as a pseudo-ternary, was the water + 15% (by weight) carbon dioxide-benzene-n-heptane system. The rationale behind this study was to try and determine the effect on the ternary equilibrium of the introduction of a slightly reactive component like carbon dioxide into the aqueous phase. Figure 7 shows that carbon dioxide has a "salting out" effect expecially on the water-aromatic binary. The mutual solubilities of both water in benzene and of benzene in water are lowered by the introduction of carbon dioxide into the system. By implication, the upper critical solution temperature of the water-benzene binary will be shifted to higher temperature by the addition of carbon dioxide. The influence of carbon dioxide on the water-n-heptane equilibrium appears less pronounced but this binary is also far removed from its critical solution temperature.

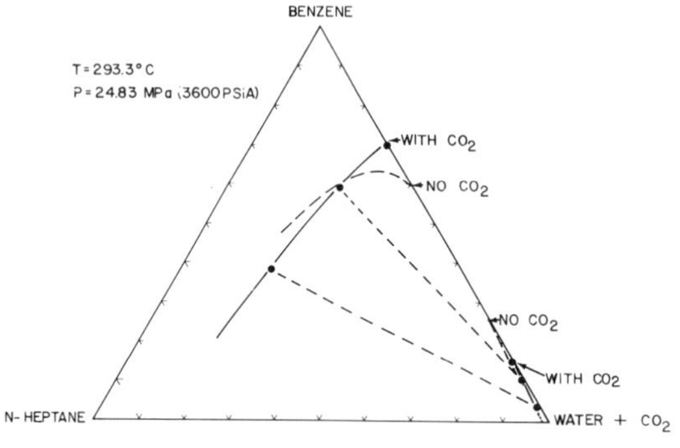

Fig. 7. Water + 15% carbon dioxide-benzene-n-heptane.

LIQUID PHASE ACTIVITY COEFFICIENTS

The Prausnitz-Chueh [7] technique for activity coefficient determinations was applied to the water-benzene binary system. As proposed, the following constant-pressure activity coefficients with experimentally accessible standard-state fugacities for both sub- and supercritical components are defined as

$$\gamma_1^{(P^r)} = \frac{f_1}{x_1\, f_{pure\ 1}^{(P^r)}} \exp \int_P^{P^r} \frac{\bar{V}_1^L}{RT}\, dp \qquad (1)$$

$$\gamma_2^{*(P^r)} = \frac{f_2}{x_2\, H_{2(1)}^{(P^r)}} \exp \int_P^{P^r} \frac{\bar{V}_2^L}{RT}\, dp \qquad (2)$$

The subscript 1 denotes the subcritical component and 2 the supercritical. Use of the asterisk indicates that γ_1 and γ_2 have been normalized by the unsymmetric convention

$$\gamma_1^{P^r} \to 1 \quad \text{as} \quad x_1 \to 1$$

$$\gamma_2^{*P^r} \to 1 \quad \text{as} \quad x_2 \to 0 \qquad (3)$$

At constant temperature the composition dependence of these pressure-independent activity coefficients can be obtained through the use of an unsymmetric molar excess Gibbs energy and a dilated van Laar model, with the necessary computations summarized in the form of two programs, HENRYS and FITTINGS [7]. The parameters required for use with these programs are available for benzene and were calculated for water. Activity coefficients determined by this method are plotted in Fig. 8. Drawbacks for this type of approach are primarily the large amounts of binary data required.

CONCLUSIONS

Ternary equilibrium for the systems water-benzene-3-methylhexane, water-benzene-1-heptene and water-benzene-1-hexene have been measured. Unsaturation in the non-aromatic fraction has been found to have a greater effect on solvent selectivity than molecular size or branching. Thus, 3-methylhexane and n-heptane show identical behavior, while 1-heptene results in lower aromatics selectivity. Addition of a fourth component like carbon dioxide shifts the water-aromatic miscibility boundary to higher temperatures and results in lower hydrocarbon solubility.

NOTATION

$H_{2(1)}^{(Pr)}$ = Henry's constant for solute 2 in solvent 1 at reference pressure P^r and system temperature T

P = pressure

P^r = fixed reference pressure

R = gas constant

T = system temperature

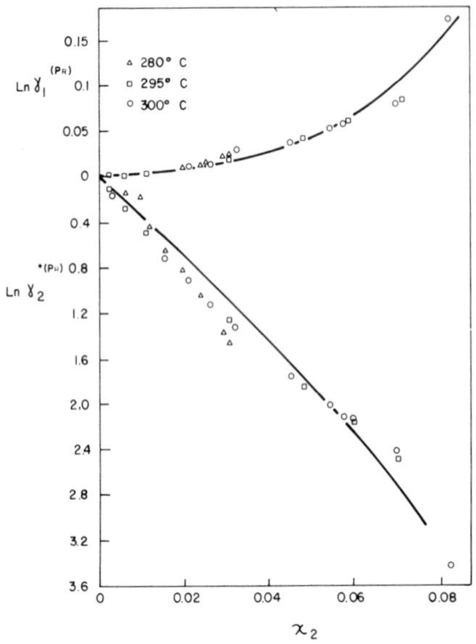

Fig. 8. Liquid phase activity coefficients for water(1)-benzene(2).

f = component fugacity in mixture at system pressure

$\bar{V}^L$ = component partial molar volume in the liquid mixture at system temperature

x = mole fraction

γ = activity coefficient

Subscripts

1 = subcritical component

2 = supercritical component

REFERENCES

1. G. B. Arnold and C. A. Coghlan, Ind. Eng. Chem. 42, 177 (1950).
2. J. S. Rowlinson, Liquids and Liquid Mixtures, Butterworths, London (1969).
3. C. J. Rebert and W. B. Kay, AIChEJ 5, 285 (1959).
4. G. Schneider, Chem. Eng. Prog. Sym. Series 64 (88), 9 (1968).
5. T. M. O'Grady, J. Chem. Eng. Data 12, 9 (1967).
6. J. F. Connolly, J. Chem. Eng. Data 11, 13 (1966).
7. J. M. Prausnitz and P. L. Chueh, Computer Calculations for High Pressure Vapor-Liquid Equilibrium, Prentice-Hall, New Jersey (1968).

M-1

HIGH PRESSURE CHEMICAL KINETICS:

PHENOMENA IN FLUIDS AT HIGH PRESSURES AND TEMPERATURES*

E. U. Franck

Institut für Physikalische Chemie
University of Karlsruhe
Karlsruhe, W. Germany

INTRODUCTION

A great majority of the chemical reactions at high pressures that have been studied or technically used are for obvious reasons reactions in liquids. This means that the applied pressures do not often extend beyond 10 kbar or 1 GPa, which is a moderate pressure range by present standards. Pressures of this magnitude will in most cases only increase the density of the liquid. Deformation of molecules often needs pressures between 10 and 100 kbar, and substantial changes of electronic structure may require pressures above 100 kbar.

Nevertheless, the application of moderate pressures, particularly in combination with elevated temperatures, permits the study of many interesting and possibly useful phenomena. Supercritical dense phases can be produced with continuously variable properties and with high particle mobilities. Binary and ternary systems of very different components can be brought into regions of extensive or complete miscibility. The behavior of the supercritical fluids may often facilitate explanation of properties of the low temperature liquids. Thus, it appears very worthwhile, to use high pressures and temperatures to investigate with sub- and supercritical fluids properties, which are of consequence for the phenomena of chemical kinetics.

A selected number of more recent experimental results of such investigations will be presented here. After some thermodynamic

*Invited paper.

considerations, Raman and infrared spectra will be discussed. The dielectric behavior of certain polar fluids in a wide range of temperatures and densities will be demonstrated. A brief discussion of fluid salts and metals at high temperature will follow, and finally the prospects for further application of the characteristic properties of supercritical fluids will be briefly considered.

PHASE EQUILIBRIA CONSIDERATIONS AT HIGH PRESSURES

The results of most experiments are initially obtained as functions of temperature and pressure. For a correlation and more detailed understanding, however, it is often desirable to discuss such results as functions of temperature and density or of average intermolecular distances. Thus PVT-data are needed, especially for polar fluids, which often provide more interesting environments for chemical reactions. As the most important and also as a typical example, the temperature-density diagram of water is shown in Fig. 1. The full isobars show approximately the range of existing experimental data. The available steam tables reach to about 1000°C and 1000 bar. Static measurements to 10 kbar have been made more recently [1-6]. Above pressures of 25 kbar shock wave measurements are available [7,8]. The region between 10 and 25 kbar must be covered by interpolation. In order to retain the normal water density of 1 g/cm^3 to 500 or 1000°C, pressures of about 9 and 20 kbar, respectively, are necessary. The region most fully investigated extends to about 500°C and 5 kbar, which is also of particular interest for chemistry and for the investigation of "hydrothermal" fluids in geology.

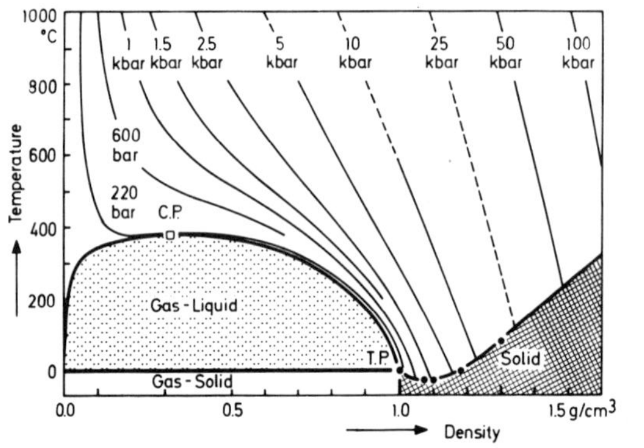

Fig. 1. Temperature-density diagram of water. Legend: Solid lines - measured isobars; Dashed lines - interpolated isobars.

Many minerals have been precipitated from such hydrothermal fluids in nature and analogous mineral svntheses have been accomplished in the laboratory. These fluids are aqueous systems, which often have a high content of salts, silica, carbon dioxide or other components. A high salt concentration affects the total density of the aqueous solutions, especially at high temperature, because of hydration. Theoretical calculations [9-11] and experimental determinations [12] of such effects are at present being made for scientific and technical purposes which extend to about 400°C and 4 kbar. The partial molar volume of sodium chloride for example can assume negative values of considerable magnitude above about 250°C.

Other fluids having small polar molecules have been investigated recently in addition to water. Among these is methanol, which is technically important and which is also interesting to compare with water. PVT-data for methanol have been determined up to 350°C and 8 kbar [13,14]. Figure 2 gives isotherms in a pressure-volume diagram [14]. The data can be very well represented with an extended Tait equation with temperature dependent coefficients. This equation and a combination of data from several authors [13] permitted the calculation of compressibilities, expansion coefficients, entropies, enthalpies and additional functions. Figure 3 gives a comparison of the compressibilities of methanol, water and hexane [14] to 250°C and 8 kbar. It is interesting to observe, that at a "moderate" pressure of 1000 bar, methanol and water behave very different, while the compressibilities of methanol and hexane are rather similar. At 8 kbar, however, the data for methanol and water barely differ by more than the experimental uncertainty of the data

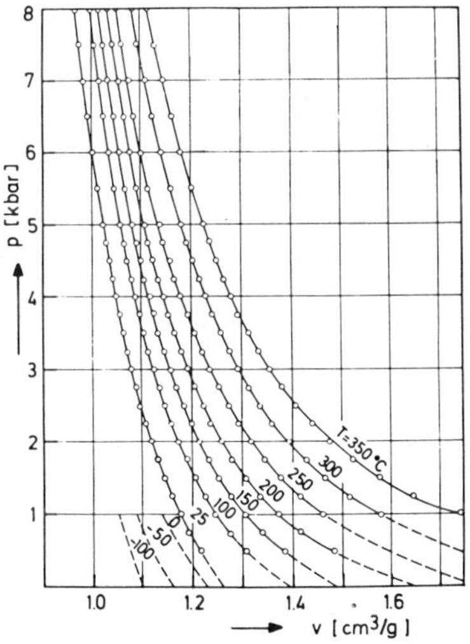

Fig. 2. Pressure-volume diagram of methanol. Legend: Solid lines - new measurements [14]; Dashed lines - previous data from several sources [13].

At sufficiently high temperatures and high pressures, even rather different components can become completely miscible. This can be shown by critical curves. Two-component systems have a critical curve in the temperature-pressure-composition diagram [15]. Some critical curves are divided into two branches. The upper branch begins at the critical point of the pure, higher-boiling component. This upper branch may have a temperature minimum in the pressure-temperature projection before it proceeds to very high pressures and temperatures. Since complete miscibility exists always at the high temperature side of the critical curves, these curves give

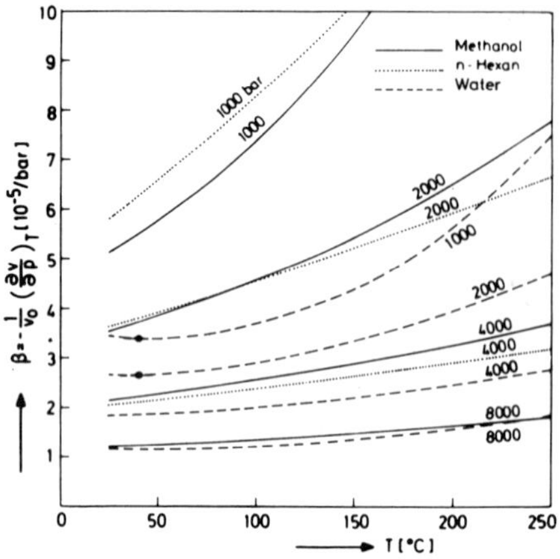

Fig. 3. Compressibility of methanol, n-hexane and water as a function of temperature and of pressures to 8 kbar.

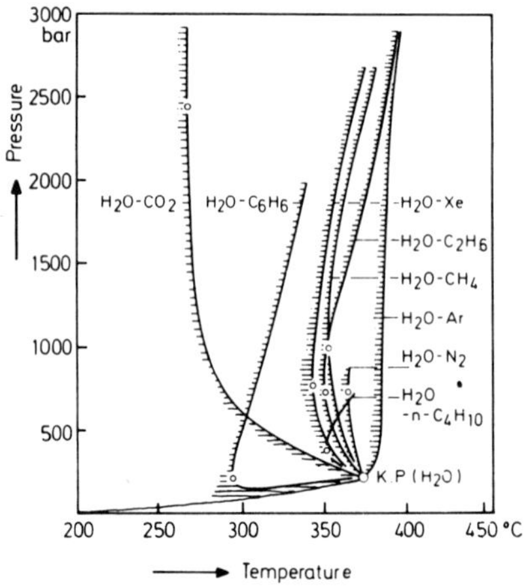

Fig. 4. Critical curves of several binary aqueous systems (upper branches beginning at the critical point of pure water).

a general indication of the range of heterogeneous and homogeneous behavior for practical purposes. Figure 4 shows a number of critical curves, determined by different authors for aqueous systems. Above 400°C all the inert components are completely miscible with water, even at the liquid-like densities obtainable with high pressures. The curves for water-benzene [16] and water-carbon dioxide [17] are remarkable because of the relatively low temperatures to which they extend. Benzene and water become completely miscible at 300°C and pressures of a few hundred bar. The same is true for a few other aromatic compounds in water. Such unusual mixtures can thus be produced at moderate conditions and suggest technical usage. Analogous behavior can be observed in binary systems with ammonia instead of water as the polar component. The peculiar shape of the water-carbon dioxide curve is not surprising, considering the interaction between those two components. Because of the importance of water-dioxide mixtures for technical and geochemical problems, the excess Gibbs energy has been calculated on the basis of several sets of experimental data [18]. Several isobars for 400 and 500°C are shown in Fig. 5. The addition of a soluble salt, for example sodium chloride, as a third component to water and carbon dioxide causes a "salting out" effect and shifts the boundary of the two-phase region and the critical curve towards higher temperatures. This is demonstrated by Fig. 6, which shows several boundary curves for the two-phase region at constant compositions [19]. It can be seen that less than 2 mole % salt can shift the boundary to higher temperatures by as much as a hundred degrees.

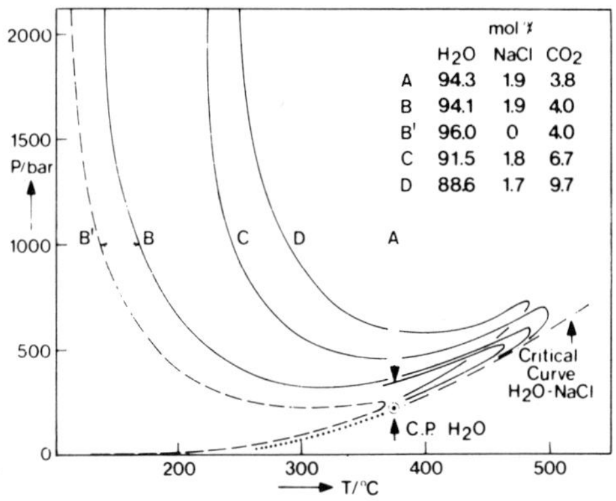

Fig. 5. High temperature shift of the boundary curves for the two-phase region of the H_2O-CO_2 system by the addition of moderate percentages of NaCl [19].

Recently, successful efforts have been made to explain and predict phase behavior and critical curves of binary model systems for a wide range of temperatures and pressures by means of theory. Some of these are based on an extended use of Van der Waals type equations which can essentially be used only for quasispherical inert molecules [20]. Other attempts are based on thermodynamic perturbation theory, which leads to a solution theory that can predict equilibrium surfaces and critical curves for compounds with strong directional intermolecular forces also [21]. These forces may be polar, quadrupolar, or anisotropic shaped. It appears that these procedures, together with appropriately selected experiments, will provide a basis for predictions of phase equilibria for practical purposes in the future.

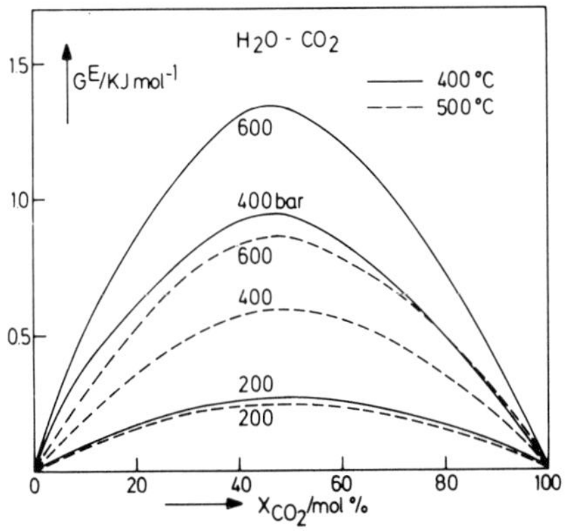

Fig. 6. Excess Gibbs' energy values for supercritical H_2O-CO_2 mixtures at several temperatures and pressures.

RAMAN AND INFRARED SPECTRA

Improved optical cells permit spectrophotometric measurements with fluids which are at high pressures and high temperatures. Band intensities of Raman and infrared spectra are of particular interest. These can provide scientific information about intermolecular interactions within the fluids and also analytic data on species in mixtures. The kinetics of chemical reactions at high pressures and temperatures can also be followed spectroscopically.

One recent example of Raman investigations [22] is given in Fig. 7. The interaction between water and carbon dioxide causes miscibility to begin above 300°C at high pressures, as shown by the critical curve in Fig. 4. An attempt was made to look for signs of this interaction in the Raman spectra. The uppermost spectrum of Fig. 7 shows the pair of bands produced by Fermi resonance between the symmetric valence vibration and the first overtone of the bending vibration of carbon dioxide at high temperature and density. The spectrum below this gives the same bands under similar conditions but for a supercritical aqueous mixture with a small percentage of carbon dioxide only. The relatively small difference between the two spectra shows that the water-carbon dioxide interaction at these conditions is probably only of a nonspecific, "physical" type. The two lower spectra are for room temperature. The high pressure solution of carbon dioxide produces a weak band at a range where there is also one band of the alkali-hydrogen carbonate solution. It is assumed that this weak band indicates a small concentration of carbonic acid, the formation of which should be favored by high pressure. Other aqueous solutions, for example of zinc chloride, have also been investigated in the same range [23].

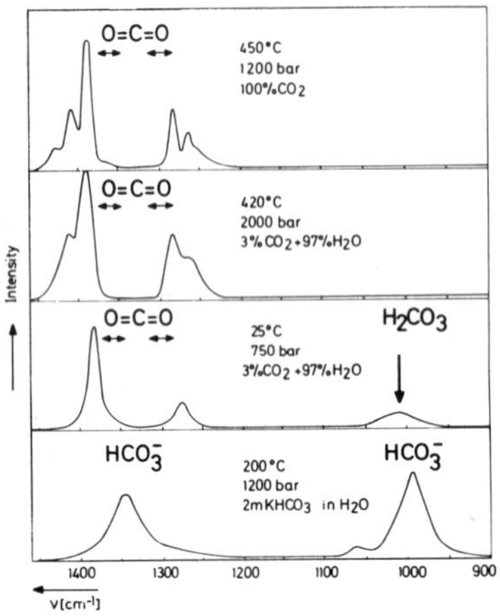

Fig. 7. Raman spectra of high temperature and low temperature dense fluids of pure CO_2 and H_2O-CO_2 mixtures in comparison to the spectrum of an aqueous solution of $KHCO_3$ [22].

Most of the high temperature-high pressure optical cells are equipped with windows of synthetic sapphire. The construction of the cells varies with the specific purpose. Figure 8 shows an example of a cell with two windows for infrared investigations to about 300°C and 3 kbar [24]. The cell contains a stainless steel bellows, filled with a pressurizing fluid, which is used to compress the sample proper without introducing new fluid into the main part of the cell. Analogous designs have been constructed with one window and reflecting mirrors for very thin samples. Cells of the type of Fig. 8 have been used to follow the polymerization of ethylene at short time intervals [25]. Preliminary investigations had shown that the intensity of the bands for CH-vibrations of $CH_{\overline{2}}$ and $-CH_2$ - groups can be used to determine the intensities of monomer and polymer material. This is possible in the range of the fundamentals of the vibration concerned as well as in the overtone range. The bands in the latter range have the advantage of weaker absorption and better separation of monomer and polymer band maxima.

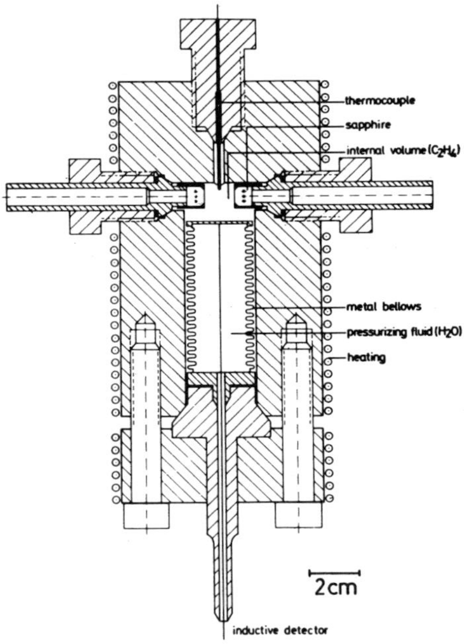

Fig. 8. An optical cell with sapphire windows for temperatures to 300°C and pressures to 3 kbar. The bellows can serve to compress the sample fluid or to keep its density constant.

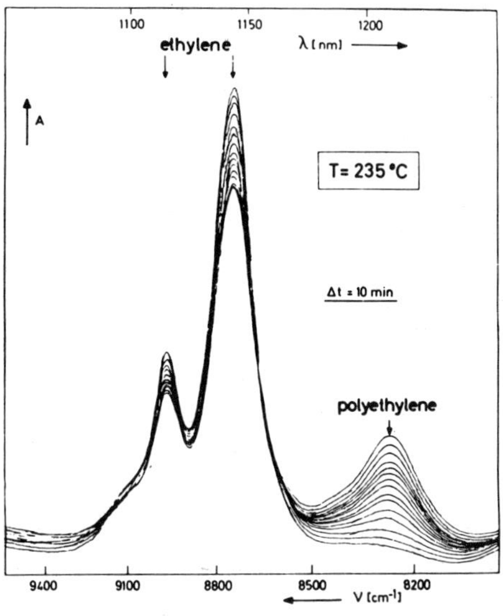

Fig. 9. Absorption spectrum of ethylene and polyethylene in the overtone region of the CH-vibrations.

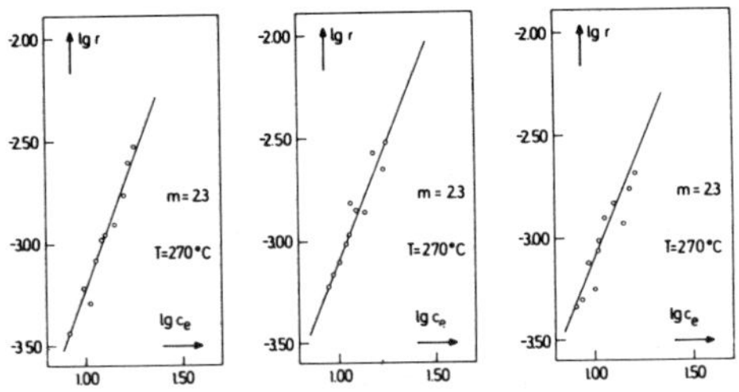

Fig. 10. Logarithm of the polymerization rate, r, as a function of the logarithm of the spectroscopically determined concentrations of monomer and polymer (c_e and c_p). The slope of the curve is m. The temperature is 270°C. The two diagrams with c_e show the reproducibility.

Figure 9 gives an example of the absorption spectra for ethylene and polyethylene in the second overtone region [24]. With suitable procedures, polymer fractions as low as 0.2% can be determined quickly. As an example, plots of reaction rates for ethylene polymerization at 270°C are shown in Fig. 10 [25]. The logarithm of the reaction rate is plotted as a function of the logarithm of the monomer and of the polymer concentrations. Both were determined independently from the same sets of spectra. The agreement, shown by the formally derived "reaction order" m, is good. This type of high pressure fluid state spectroscopy lends itself to further developments and applications in polymerization kinetics and related fields.

DIELECTRIC BEHAVIOR OF SELECTED POLAR FLUIDS

The electric permittivity or static dielectric constant is a particularly useful quantity with which to characterize a polar fluid either as an electrolytic solvent or as a solvent for chemical purposes in general. Thus, it is desirable to obtain dielectric constants of certain characteristic polar fluids for a wide range at sub- and supercritical conditions and to devise procedures to calculate and predict such quantities. At present one depends mainly on measurements. A small group of fluids, however, has been experimentally investigated within a wide range in recent years.

Obviously, the most important polar fluid is water. A critical collection and assessment of all available data published after 1920 together with new experimental values has made it possible to construct the diagram of Fig. 11 [26]. Figure 11 is a temperature-density diagram in which curves of constant values of the dielectric constants are shown. These are calculated by means of an equation of state. This is an empirical equation for the dielectric constant as a function of temperature and density, with the constants based on the critically evaluated experimental data. It is proposed for a region extending to 550°C and 5 kbar and describes the experimental data with a standard deviation of 0.33 dielectric constant units. Figure 11 shows, that the well known very high values are restricted to a narrow region of low temperatures and high densities. At supercritical densities and supercritical temperatures, however, values of the dielectric constant between 10 and 30 can still be expected. These are sufficient to produce ionic dissociation of certain strong dissolved electrolytes. Such are the conditions of many of the "hydrothermal" fluids found at greater depths below the Earth's surface.

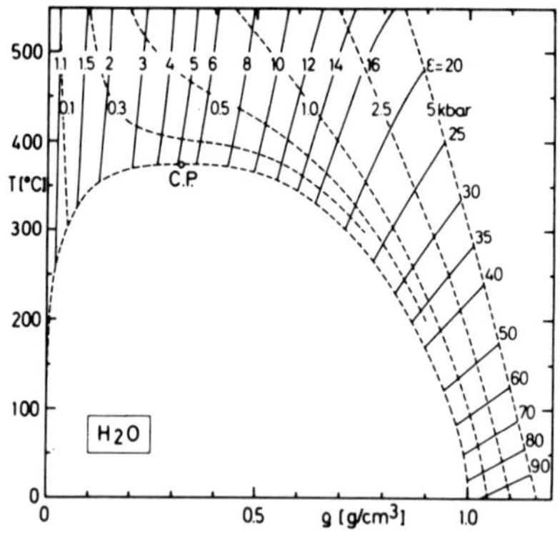

Fig. 11. Temperature-density-diagram of water with curves for constant values of the static dielectric constant ε. The ε-values are calculated with an equation derived from all known experimental data [26].

Experimental determinations usually yield the dielectric constant as a function of temperature and pressure. Increases in temperature and decreases in pressure generally lower the dielectric constant. Figure 12 gives an example of recent isobar measurements with methyl fluoride [27]. This compound has a dipole moment of about 1.8 debye units - similar to the moment of water -

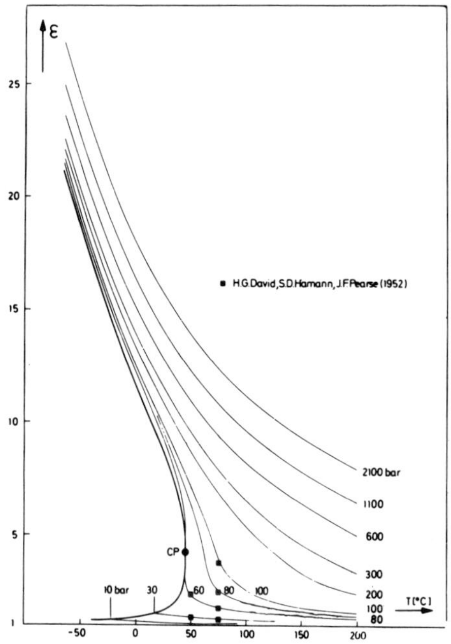

Fig. 12. Experimentally determined isobars for the dielectric constant of methyl fluoride as a function of temperature [27].

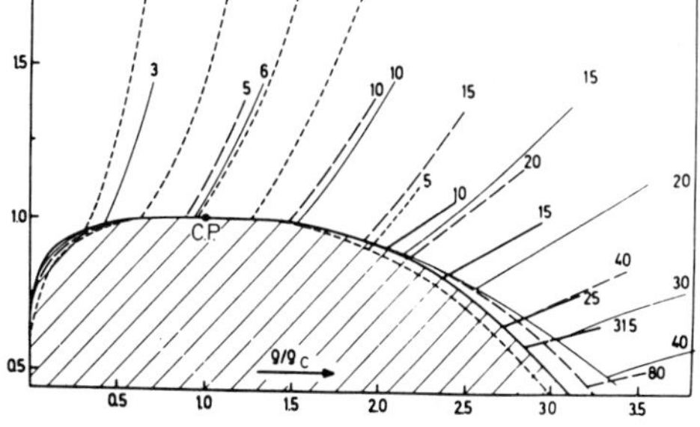

Fig. 13. Reduced temperature-density diagram of HCl, H_2O, CH_3CN and CH_3OH with curves of constant values for the static dielectric constant.

but it does not form hydrogen bonds and is rather typical for similar freons.

In order to compare the dielectric behavior of several substances, a reduced temperature-density diagram such as the one in Fig. 13 can be used. For reduction, the critical temperatures and densities have been applied. Curves for constant values of dielectric constants for HCl, H_2O, CH_3CN and CH_3OH are shown. The first three compounds have dipole moments close to 1, 2 and 4 Debye units, respectively. Methyl cyanide, CH_3CN, is of interest as a solvent in organic electrochemistry; methanol forms hydrogen bonds somewhat like water. While at room temperature the dielectric constant of methyl cyanide is only half as high as that of water, both compounds show about the same values of five to six at critical conditions. Apparently the higher polarity of CH_3CN and the remaining hydrogen bonded structure of H_2O compensate each other. Corresponding state diagrams as in Fig. 13 may serve to roughly estimate unknown dielectric constants of polar fluids at unusual conditions. Improved predictions may be possible with the use of the so called Kirkwood correlation factor. Efforts are also being made to derive radial distribution functions for model polar fluids, which can then serve to calculate Kirkwood factors and eventually dielectric constants at high densities [28,29].

FLUID SALTS AND METALS AT HIGH TEMPERATURES

The combined application of high pressure and temperature makes it possible, to investigate pure ionic fluids, for example fused salts, at various densities. A number of new results have been obtained in this field in recent years [30]. Some salts, for example bismuth trichloride, have been investigated even in the supercritical range with respect to their density and electrical conductivity [31].

A particularly interesting example appears to be ammonium chloride. It has a triple point at 520°C and more recent experiments have shown that it has a normal vapor pressure curve until its critical point is reached at about 882°C. This vapor pressure curve is shown in Fig. 14. It has been possible to determine also the coexisting densities and electrical conductivities of the liquid to about 850°C [32,33]. Small vessels using sapphire and columbium metal had to be used. Examination of the results makes it very probable that ammonium chloride is predominantly or completely ionized near the critical point. This has not yet been found for other salts.

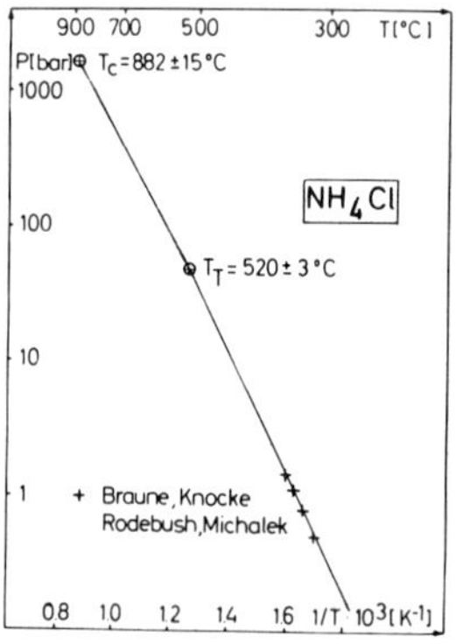

Fig. 14. Experimentally determined vapor pressure for solid and liquid ammonium chloride.

Very high electrical conductivity has also been observed with mercury and the heavier alkali metals in their respective critical regions. The critical temperature of mercury is the lowest of all metals, the critical pressure - about 1500 bar - is, however, rather high. It has nevertheless been possible to measure density, electric conductivity, light absorption and other physical properties in supercritical mercury and to some extent in supercritical alkali metals [34-36]. Figure 15 shows isotherms of the specific electrical conductance of mercury as a function of pressure. It is obvious that the supercritical gas, if compressed to about twice the critical density, has a specific conductance only one or two orders of magnitude lower than that of liquid mercury at room temperature. This means that dense supercritical mercury exhibits metallic, electronic conductance. This conclusion is confirmed by several other observations; for example, the isochoric temperature dependence of conductance, the light absorption, and the thermoelectric power. Similar behavior is observed with cesium or potassium at supercritical temperature beginning at about the critical density. The phenomena can be explained by examining the change of the atomic energy levels when compression forces a close approach of the atoms.

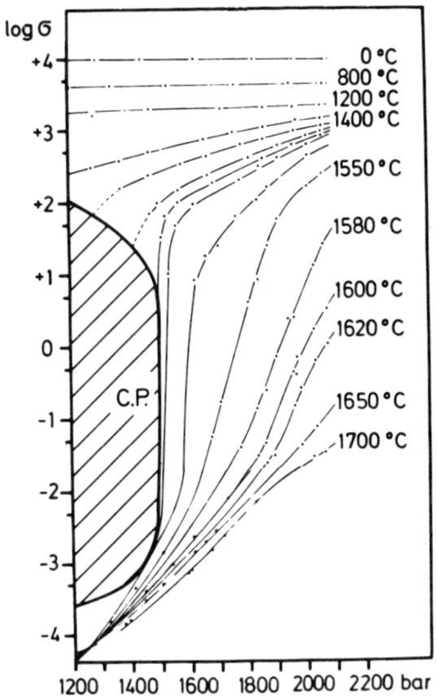

Fig. 15. The specific conductance, σ, of sub- and supercritical mercury.

Other transport properties of high temperature-high pressure fluid metals may be rather normal. The experimentally determined viscosity liquid and gaseous mercury to 1200°C and 1000 bar is shown in Fig. 16. Measurements were made with oscillating small autoclaves of tungsten-rhenium alloy. Because of the limitations of this material, the measurements could not be extended beyond 1200°C. The data to 1500°C had to be calculated. This was possible by means of a modified Enskog equation, using temperature dependent collision diameters for the liquid and gaseous mercury [37]. This half-empirical method proved to be successful not only for mercury, but for other metals as well and may be useful for practical applications [38].

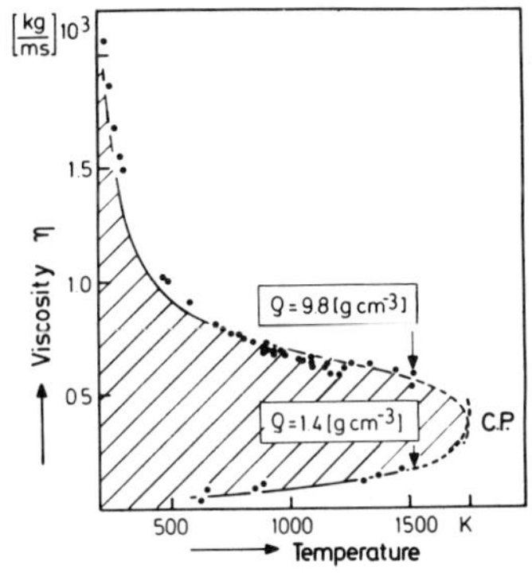

Fig. 16. The viscosity of liquid and gaseous mercury at equilibrium pressures. Legend: Points - experimental data; Curves - calculated, using modified Enskog-equations.

FUTURE CONSIDERATIONS

In conclusion, certain interesting and useful effects can be mentioned again. These can be brought about by the combined application of moderate pressures up to several kbar and elevated temperature: Physical and quite often also chemically relevant properties can be varied widely, continuously and quickly without change of chemical composition. New properties or combinations of properties in certain fluids can be obtained. Because of the relatively high thermal energy, high diffusion rates, ion mobilities and chemical reaction rates may be obtained. Complete extended miscibility of partners can be obtained which are only sparingly soluble at normal conditions. More phenomena could be mentioned. It seems, that continued research with high pressure fluids will be rewarding for pure as well as applied chemistry.

REFERENCES

1. E. Schmidt, "Properties of Water and Steam in SI-Units," Springer, Heidelberg, Germany (1969).
2. K. Tödheide, in Water, F. Franks, ed., Plenum Press, New York (1972).
3. S. Maier and E. U. Franck, Ber. Bunsenges. phys. Chemie, 70, 639 (1966).
4. H. Köster and E. U. Franck, Ber. Bunsenges. phys. Chemie, 73, 716 (1969).
5. R. Hilbert, Ph.D. Thesis, University of Karlsruhe, Karlsruhe, Germany (1977).
6. C. W. Burnham, J. R. Holloway, and N. F. Davis, Am. J. Sci. 267A, 70 (1969).
7. J. M. Walsh and M. H. Rice, J. Chem. Phys. 26, 815 (1957).
8. A. C. Mitchell and W. J. Nellis, in Proceedings 6th AIRAPT Intern. Conference on High Pressures, Plenum Press, New York (1978).
9. H. C. Helgeson and D. H. Kirkham, Am. J. Science 276, 97 (1976); also 1977 in press.
10. R. W. Potter II and D. L. Brown, Geological Survey Bulletin 1421-C, U.S. Government Printing Office, Washington, D.C. (1977).
11. H. L. Barnes, ed. Geochemistry of Hydrothermal Ore Deposits, Holt, Rinehart and Winston, New York (1967).
12. R. Hilbert, Ph.D. Thesis, University of Karlsruhe, Karlsruhe, Germany (1977).
13. V. N. Zubarov, P. B. Pruzakow, and L. W. Zergejewa, in Standard-Values, Moscow, U.S.S.R. (1973).
14. R. Taáni, Ph.D. Thesis, University of Karlsruhe, Karlsruhe, Germany (1976).
15. J. S. Rowlinson, Liquids and Liquid Mixtures, 2nd ed., Butterworths, London (1969).
16. Z. Alwani and G. M. Schneider, Ber. Bunsenges. phys. Chemie, 71, 633 (1967); G. M. Schneider, in Topics in Current Chemistry, Springer, Heidelberg, Germany (1970).

17. K. Tödheide and E. U. Franck, Z. physik. Chem. (Frankfurt) 37, 387 (1968).
18. H. Lentz, private communication.
19. M. Gehrig, Ph.D. Thesis, University of Karlsruhe, Karlsruhe, Germany (1975).
20. R. L. Scott and P. H. van Konynenburg, Discuss. Far. Soc. 49, 87 (1970); R. L. Scott, Ber. Bunsenges. phys. Chem. 76, 296 (1972).
21. C. H. Twu, K. E. Gubbins, and C. G. Gray, J. Chem. Phys. 64, 5186 (1976); K. E. Gubbins and C. H. Twu, Chem. Eng. Sci., in press.
22. R. A. H. Kruse, Ph.D. Thesis, University of Karlsruhe, Karlsruhe, Germany (1975).
23. K. R. Schulz, Ph.D. Thesis, University of Karlsruhe, Karlsruhe, Germany (1974).
24. M. Buback and F. W. Nees, Ber. Bunsenges. phys. Chem. 80, 650 (1976).
25. F. W. Nees and M. Buback, Ber. Bunsenges. phys. Chem. 80, 1017 (1976).
26. M. Uematsu and E. U. Franck, J. Chem. Eng. Data, in press.
27. K. Reuter, Ph.D. Thesis, University of Karlsruhe, Karlsruhe, Germany (1974).
28. V. M. Jansoone and E. U. Franck, Ber. Bunsenges. phys. Chem. 76, 943 (1972); V. M. Jansoone, Acta Physica Austriaca 37, 326 (1973).
29. M. S. Wertheim, J. Chem. Phys. 55, 429 (1971).
30. K. Tödheide, in Molten Salts, U.S. Electrochemical Soc. (1976).
31. G. Treiber and K. Tödheide, Ber. Bunsenges. phys. Chem. 77, 540 (1973).
32. M. Buback and E. U. Franck, Ber. Bunsenges. phys. Chem. 76, 350 (1972).
33. M. Buback and E. U. Franck, Ber. Bunsenges. phys. Chem. 77, 1074 (1973).
34. F. Hensel and E. U. Franck, Ber. Bunsenges. phys. Chem. 70, 1154 (1966); F. Hensel and E. U. Franck, Rev. Mod. Phys. 40, 697 (1968); F. Hensel and E. U. Franck, in Experimental Thermodynamics Vol. II. B. Le Neindre and B. Vodar, editors, Pure and Applied Chemistry, Butterworths, London (1975).

35. F. Hensel, Angewandte Chemie 86, 459 (1974).
36. J. K. Kokoin and A. R. Sechenko, Phys. Metals and Metallography 24, 74 (1967).
37. H. v. Tippelskirch, E. U. Franck, F. Hensel, and J. Kestin, Ber. Bunsenges. phys. Chem. 79, 889 (1975).
38. H. v. Tippelskirch, Ber. Bunsenges. phys. Chem. 80, 727 (1976).

M-2

ACTIVATION VOLUME AS A CLUE TO TRANSITION STATE STRUCTURE*

K. R. Brower

New Mexico Institute of Mining and Technology
Socorro, New Mexico, USA

SOME VOLUMETRIC CHARACTERISTICS OF MOLECULES

The relationship of molar volume to molecular structure and properties is not discussed to any extent in physical chemistry textbooks. One such neglected principle is the additivity of molar volume which was first noted by H. Kopp in 1842. Although it holds best for corresponding states, as for example the boiling points, it is surprisingly reliable for room temperature as shown in Table I. Strong forces of association such as hydrogen bonding cause a slight decrement at temperatures well below the boiling point. Accurate estimates of density can be derived from the additivity principle.

Table I. Additivity of Group Volumes

A	B	Δ	$V_A - V_B$
Ethanol	Methanol	CH_2	17.9
Toluene	Benzene	CH_2	17.6
Ethylbenzene	Toluene	CH_2	16.2
Chloroform	CH_2Cl_2	Cl	16.7
CCl_4	Chloroform	Cl	16.3
Chlorobenzene	Benzene	Cl	13.1
Anisole	Toluene	O	2.9
Methoxyethanol	Propanol	O	4.1
Toluene	Cresol	O(H)	-1.9
Chlorohydrin	Chloroethane	O(H)	-4.4

It might be expected that a liquid mixture of molecules of unequal size would always show a volume decrement because it seems

*Invited paper.

logical that some improvement in packing efficiency would be permitted by mixing. Close-packed spheres occupy only 74% of the available volume, and it would seem that the remaining 26% could be occupied by other molecules if they are small enough. In practice, this phenomenon has not been observed. The volume change of mixing [1] is almost always less than 1 ml/mole, provided there is no chemical reaction and provided the mixture is not aqueous. In an attempt to establish an extreme case we have mixed dibutyl phthalate (266 ml/mole) with methanol (40.4 ml/mole) and observed a volume change of -0.8 ml. It is important to the interpretation of activation volumes that mixing volumes can usually be neglected.

Joel Hildebrand [2] has recently called attention to a simple relation between molar volume and fluidity (the reciprical of viscosity), which was first noted by Batchinski in 1913. At ordinary pressures a plot of fluidity vs. molar volume is linear with nearly the same slope for all liquids. Bridman's measurements of viscosity at high pressures [3] have been used to construct Fig. 1 wherein it can be seen that a single curve accommodates variations in fluidity resulting either from temperature or pressure changes. The intercept of the straight-line portion of the curve at zero fluidity can be interpreted as the intrinsic volume of the liquid, and the fluidity is then directly proportional to the free volume. The apparently negative free volume at very high pressures can be ascribed to the suppression of internal motions which do not contribute to fluidity. The intrinsic volume as defined above should be useful in establishing structure-volume relationships.

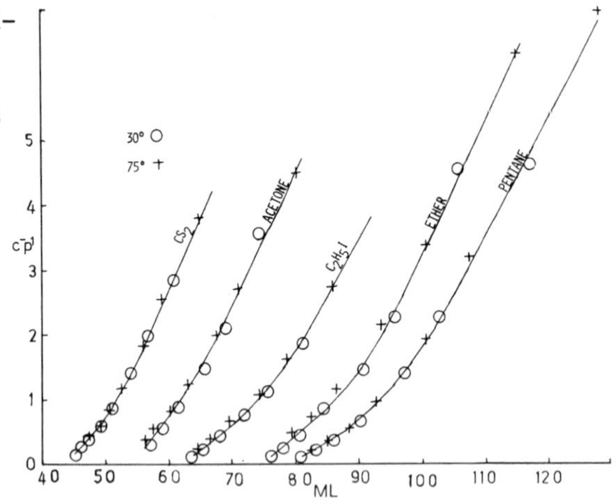

Fig. 1. Fluidity (1/η) vs. molar volume.

Definitions and Generalizations

The effect of pressure on the equilibrium constant K is given by (1) in which ΔV is the difference in volume between reactants and products.

$$d \ln K/ dP = -\Delta V/ RT \quad (1)$$

The effect of pressure on the reaction rate k is given by (2) in which $\Delta V^{\pm}$ is the difference in volume between reactants and the transition state and is called the activation volume.

$$d \ln k/ dP = -\Delta V^{\pm}/RT \quad (2)$$

It has been found empirically [4] that a bond-making step contributes about -10 ml to the activation volume, and conversely a bond-breaking step contributes about +10 ml. Ionization causes electrostriction of solvent which amounts to about -20 ml for water and -40 ml for organic solvents.

Hydroloysis of Epoxides and Acetals

It is very difficult to detect the involvement of solvent molecules in the rate-determining step of any reaction by using standard mechanistic tests. Whalley [5] has shown, however, that activation volumes can reveal the participation of solvent by applying the method to the acid-catalyzed hydrolysis of epoxides. The question is whether the rate is limited by (4), the A_1 mechanism, or (5), the A_2 mechanism, in the sequence below:

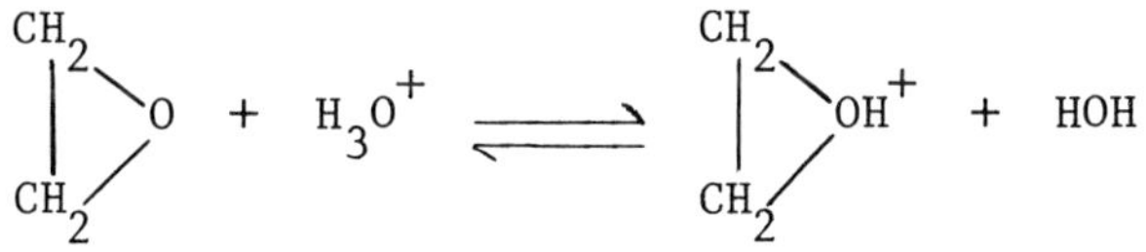
(3)

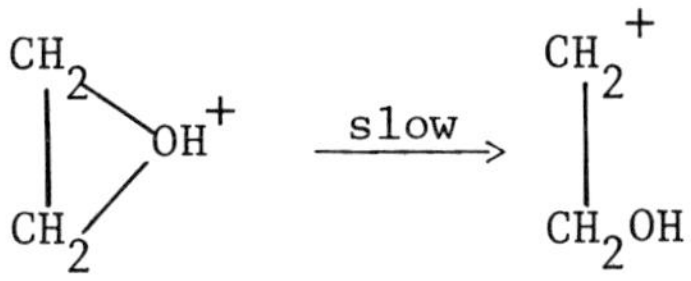
(4)

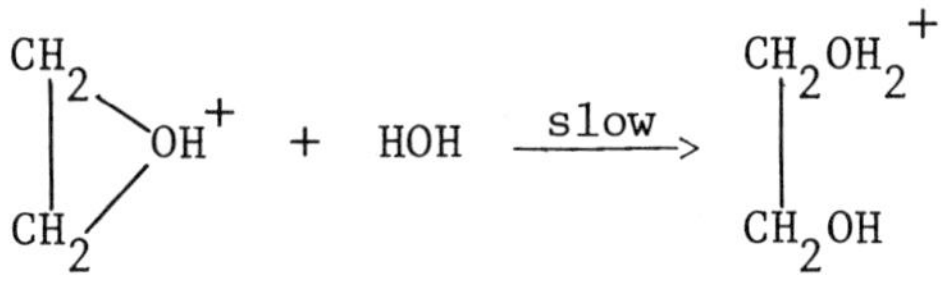
(5)

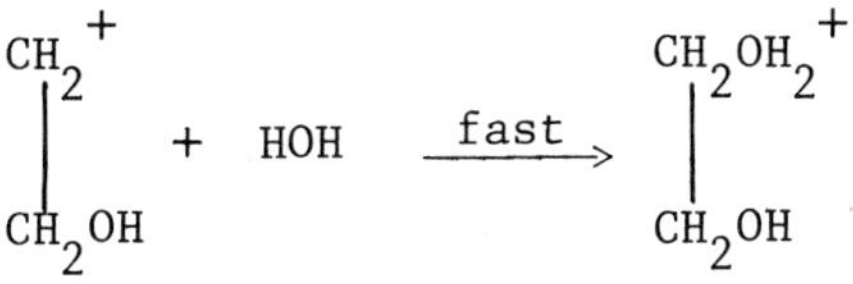
(6)

Since the reaction of a hydronium ion with amines has a small negative volume change, it can be expected that pressure will cause a slight increase in the concentration of protonated epoxide which is common to (2) and (3). The unimolecular decomposition in equation (2) should have a positive activation volume, whereas the bimolecular process of (3) should have an activation volume of perhaps -10 ml/mole. The combined effects should result in a slightly positive or perhaps zero apparent activation volume for the A_1 mechanism of (4) and a distinctly negative value for the A_2 mechanism of (5). The experimental activation volumes range from -9 to -11 ml/mole for various epoxides and clearly favor the A_2 mechanism.

Whalley [6] also studied the acid-catalyzed hydrolysis of acetals for which the A_1 mechanism is more probable on theoretical grounds. The experimental activation volumes ranged from zero to +2 ml/mole exactly as predicted.

The work described above illustrates the use of volume changes for model reactions and activation volumes for reactions of undisputed mechanism in estimating the activation volumes to be expected for alternative mechanisms in an unknown case. When it is not certain that $\Delta V^{\pm}$ will differ in sign for the alternatives, the case will be strengthened if an example of each can be found. This condition was fulfilled by the epoxides and acetals.

Diels-Alder Reaction

The Diels-Alder reaction is represented schematically as

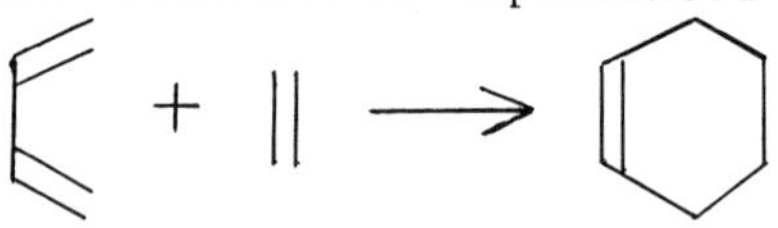

(7)

For many years no one was able to show whether the two new C-C bonds are formed simultaneously or successively. Because the solvent effect is not large it was considered unlikely that the transition state is ionic, but the possibility remained for a biradical intermediate in which only one of the new bonds has formed. Walling [7] was the first to realize that the question could be answered by measurement of $\Delta V^{\pm}$, but the results were clouded by technical problems. Recently Eckert [8] has repeated the measurements with improvements in the state of the art which he originated. The results are shown in Table II. The contraction in volume as the transition state is formed is so large that it forces the conclusion that the two new bonds not only form simultaneously, but that they are essentially complete at the top of the energy barrier.

When maleic anhydride is the dienophile, the transition state is even smaller in partial molar volume than the adduct. Although it is surprising at first, this result can be ascribed to a strong interaction between the pi-electrons of the diene and the carbonyl groups of the dienophile. If the interaction is sterically impossible as in the case of acetylenedicarboxylic ester, the anomaly disappears. This conclusion is supported by other measurements which do not appear in the table.

Table II. The Diels-Alder Reaction

Diene	Dienophile	$\Delta V^{\ddagger}$	ΔV
Cyclohexadiene	Maleic anhydride	-37	-30
Cyclopentadiene	Acetylenedicarboxylic acid dimethyl ester	-30	-34
Isoprene	Maleic anhydride	-39	-36

Detection of Fluoronitrene

Although fluoronitrene (NF) was known as a product of photolysis of fluorides of nitrogen and detected by its spectrum, its participation as a reactive intermediate in a solution reaction was demonstrated by measurement of activation volumes by le Noble [9]. The reaction in question is shown as

$$2\ HNF_2 + 2\ HO^- \longrightarrow N_2F_2 + 2\ F^- + 2\ HOH \tag{8}$$

The mechanistic alternatives for the first step would seem to be either bimolecular displacement of fluoride or proton removal. The apparent $\Delta V^{\pm}$ was measured and found to be +7 ml. This makes it clear that the first step is equilibrium-controlled proton transfer (which has a small volume change as explained earlier) and that the second (rate-controlling) step is decomposition of NF_2^- to NF for which $\Delta V^{\pm}$ should be positive. The behavior of HNF_2 in alkaline solution is thereby shown to be analogous to the much studied hydrolysis of $CHCl_3$. Le Noble [10] had earlier shown that the apparent $\Delta V^{\pm}$ for this reaction is +16 ml. It was already known from hydrogen exchange experiments and analysis of products that its mechanism is as follows:

$$CHCl_3 + HO^- \rightleftharpoons Cl_3C:^- + HOH \tag{9}$$

$$Cl_3C:^- \longrightarrow CCl_2 + Cl^- \tag{10}$$

$$CCl_2 \longrightarrow \text{products} \tag{11}$$

Further information about the hydrogen exchange will be presented in the next section.

ACTIVATION VOLUMES FOR PROTON AND HYDRIDE TRANSFERS

There are many organic reactions in which a hydrogen atom is being transferred from one site to another in the rate-controlling step. The particular one which stimulated the present work is elimination by the E2 mechanism. It was desired to separate the volume changes associated with proton removal and leaving group detachment by measuring $\Delta V^{\pm}$ for H-D exchange on carbon acids. Because the reactions chosen do not involve ionization, and because the rate equations are second-order, it was expected that values near -10 ml would be observed. Chloroform, the first example chosen, had an activation volume of +9 ml. This seems to call for a strong interpretation. In the first place, the contraction associated with bimolecularity is not manifest, and this implies that the transition state is very near to the product side of (9). Furthermore, the only conceivable explanation for an expansion is that the electric charge of $Cl_3C:^-$ is highly delocalized, presumably into the d-orbitals of chlorine. There are additional grounds for believing that the transition state is product-like [11].

In order to find out whether such behavior is general for carbon acids, we performed measurements on the compounds shown in Table III.

Table III. H-D Exchange

Substrate	Base	Solvent	V
$CDCl_3$	HO^-	HOH	+9
CDF_3	HO^-	HOH	+3
$C_6H_5COCD_3$	HO^-	HOH	-1
$C_6H_5COCD_3$	MeO^-	MeOH	-1
$C_6H_5COCD_3$	ArO^-	MeOH	-4
$C_6H_5CD_2CN$	AcO^-	MeOH	-4
$C_6H_5CD_2CO_2Me$	MeO^-	MeOH	-3
CD_3SOCD_3	HO^-	HOH	+2
INDENE-1,1-d_2	MeO^-	MeOH	-4
$C_6H_5C{\equiv}CD$	HO^-	MeOH-HOH	-3

Even fluoroform which ought not to show delocalization in the carbanion has a slightly positive $\Delta V^{\pm}$. Again we are forced to the conclusion that the transition state is product-like. In order to get a good estimate of what value $\Delta V^{\pm}$ would have if the transition state were centrally located and dominated by bond formation, we took advantage of the fact that phenylacetic ester gives two reactions under the same conditions; namely, H-D exchange and ester interchange. The activation volume for the latter is -12 ml as shown in Table IV. This is a normal result for second-order substitution reactions when no ionization occurs. The value of -3 ml for H-D exchange again indicates a product-like transition state.

Table IV. Reference Reactions

	$\Delta V^{\pm}$
$C_6H_5CH_2CO_2ET + MeO^- \longrightarrow C_6H_5CH_2CO_2Me + EtO^-$	-12
$C_6H_5COCH_3 + BH_4^- \longrightarrow C_6H_5(CH_3)CHO^-$	-11
$2\ C_6H_5CHO + HO^- \longrightarrow C_6H_5CH_2OH + C_6H_5CO_2^-$	-27

The importance of electrostrictive changes was gauged by changing the solvent for acetophenone from water to methanol and by changing the base from methoxide to p-chlorophenoxide. The effect of the solvent change is to approximately double any electrostrictive contribution as mentioned earlier. No significant change was observed. The effect of changing from methoxide to aryloxide is to bring about some initial delocalization of charge in the base, probably even more than could be expected for the enolate ion of acetophenone. The change in $\Delta V^{\pm}$ is in the expected direction but rather small. The conclusion is that the near-zero $\Delta V^{\pm}$ for the reaction of acetophenone with strong bases does not result from compensation of a negative bond-forming component and a positive charge-delocalization component.

While this work was in progress we learned of a simple theoretical argument by Swain [12] that hydride transfer should have a centrally located transition state characterized by bond formation and that proton transfer should not. A graphical representation of the relevant molecular orbital for the transition state is shown in Fig. 2.

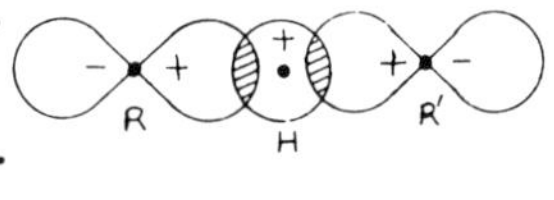

Fig. 2. Molecular orbital for transition state in hydrogen transfer.

In the case of hydride transfer this bonding orbital would be just filled by the electron pair which originally bound the hydrogen. Proton transfer, on the other hand, involves also the electron pair of the base, and the second pair would be forced into an antibonding orbital.

In order to test both implications of this idea we measured $\Delta V^{\pm}$ for borohydride reduction of acetophenone which has a simple second-order rate law [13]. The result was -11 ml (see Table IV) in excellent agreement with the prediction. We are also investigating the Cannizzaro reaction of benzaldehyde which is not quite so simple. The steps are thought to be as follows:

$$PhCHO + HO^- \rightleftharpoons PhCH(OH)O^- \quad (12)$$

$$PhCH(OH)O^- + PhCHO \xrightarrow{slow} PhCO_2H + PhCH_2O^- \quad (13)$$

A volume change of -12 ml should be expected for the formation of $PhCH(OH)O^-$, and the apparent $\Delta V^{\pm}$ of -27 ml as shown in Table IV should have -12 ml subtracted to give approximately -15 ml for the $\Delta V^{\pm}$ of (13).

At the present stage of our investigation, Swain's prediction seems to be valid. Other substrates and conditions are still under investigation.

REFERENCES

1. I. Brown and F. Smith, Aust. J. Chem. 15, 1 (1962).
2. J. H. Hildebrand, Science 174, 490 (1971).
3. P. W. Bridgman, Proc. Amer. Acad. Arts Sci. 61, 57 (1926); also 77, 115 (1949); 66, 185 (1930); 49, 1 (1913).
4. W. J. le Noble, Progr. Phys. Org. Chem. 5, 207 (1967).
5. J. Koskikallio and E. Whalley, Trans. Faraday Soc. 55, 816 (1959).
6. J. Koskikallio and E. Whalley, Trans. Faraday Soc. 55, 809 (1959).
7. C. Walling and H. J. Schugar, J. Am. Chem. Soc. 85, 607 (1963).
8. R. A. Grieger and C. A. Eckert, J. Am. Chem. Soc. 92, 2918 (1970); also 92, 7149 (1970); Trans. Faraday Soc. 66, 2579 (1970).
9. W. J. le Noble and D. Skulnik, Tetrahedron Lett. 51, 5217 (1967).
10. W. J. le Noble, J. Am. Chem. Soc. 87, 2434 (1965).
11. Z. Margolin and F. A. Long, J. Am. Chem. Soc. 95, 2757 (1972).
12. C. G. Swain, R. A. Wiles, and R. F. W. Bader, J. Am. Chem. Soc. 83, 1945 (1961).
13. H. C. Brown, O. H. Wheeler, and K. Ichikawa, Tetrahedron 1, 214 (1957).

M-3

PRESSURE EFFECTS IN INORGANIC SOLUTION CHEMISTRY*

T. W. Swaddle

The University of Calgary
Calgary, Alberta, Canada

INTRODUCTION

A metal cation M in solution can be pictured as being surrounded by a geometrically well-defined first coordination sphere (c.s.) consisting of ligands L and X (which could be solvent molecules, Sol) and a second c.s. of centripetally-oriented Sol and possibly other species Y; beyond the second c.s. lies bulk solvent which may contain free Y. In this paper, we consider the contribution of the study of pressure effects to the understanding of the mechanism of exchange of X, Y, L, and Sol between these three regions.

The centripetal electrostatic forces between the ion M and the anionic or dipolar species in the first and second coordination spheres, considered as acting uniformly over the hypothetical surfaces of these coordination spheres, can readily be shown to be equivalent to a pressure on the order of 1 GPa or more. Thus, L, X, Y, and Sol in these regions are already quite highly compressed, and the application of a further external hydrostatic pressure of (say) 0.4 GPa will produce only a trivial further compression. Consequently, simple exchange of X or Sol from the first c.s. with Y or Sol in the second c.s. will occur with essentially the same volume change at 400 MPa as at atmospheric pressure. On the other hand, transfer of species from these electrostatically-compressed regions to bulk solution will occur with a more positive volume change ΔV at atmospheric pressure than at hydrostatic pressures on the order of 400 MPa, i.e., $(\partial \Delta V/\partial P)_T$ will be non-zero and negative for transfers out of the coordination spheres into bulk solvent (e.g., desolvation), and positive for the reverse process (e.g., increasing solvation). This generalization does not apply to transfers of

*Invited paper.

anionic X or Y, which would be under their own electrostatic self-compression. A different, but equivalent, approach taken by Stranks [1] leads to the same conclusions.

MECHANISTIC CLASSIFICATION

In constructing mechanistic models for overall processes of the type

$$ML_nX + Y \xrightarrow{Sol} ML_nY + X \qquad (1)$$

where any, all, or none of L, X, and Y could be Sol, the question inevitably rises as to whether M-Y bond making is or is not involved in the activation process, i.e., whether the reaction is associatively or dissociatively activated. In organic chemistry, this classic S_N2/S_N1 dichotomy can be redefined operationally in terms of observable kinetic properties (rate laws of first and zeroth order in Y respectively), but for reactions of metal ion complexes in solution the kinetic phenomena lead to a tripartite operational classification, full details of which are given elsewhere [2,3] but which may be interpreted as follows.

<u>D-mechanism</u>. An intermediate ML_n of reduced coordination number forms and survives long enough for the second c.s. to re-equilibrate before ML_nY forms (e.g., reactions of $Co^{III}(CN)_5X$ and of conjugate base species such as $Co^{III}(NH_3)_4(NH_2)X$).

<u>A-mechanism</u>. An intermediate ML_nXY of expanded coordination number forms and is long-lived relative to the relaxation time of the second c.s. (e.g., reactions of square planar complexes of Pt(II), Pd(II), and Au(III)).

<u>I-mechanism</u>. Any intermediate ML_n or ML_nXY (corresponding to dissociative and associative interchange, I_d and I_a, respectively) is short-lived relative to the relaxation time of the second c.s. Thus, the form of the rate law with respect to [Y] simply reflects the formation of the encounter complex (ion pair, etc.) regardless of whether M-Y bond formation occurs before (I_a or S_N2-like) or after (I_d or S_N1-like) the transition state, and hence the I-mechanism becomes operationally distinct, but embraces the equivalents of both the classical S_N1 and S_N2 mechanisms. The substitution reactions of most octahedral cationic complexes proceed via I mechanisms.

Attempts to distinguish clearly between I_a and I_d by traditional methods (effect on rate of nature of Y, steric compression, stereochemistry, etc.) often lead to frustratingly equivocal results [3-5], and give essentially no mechanistic information on the simplest of all exchange reactions involving metal ions, i.e., solvent exchange, for which X = Y = Sol (and even = L) and $\Delta G^\circ = \Delta H^\circ = \Delta V^\circ = 0$.

PRESSURE EFFECTS AS MECHANISTIC CRITERIA

In contrast to the traditional methods, pressure effects can provide unusually clear criteria of mechanism when the reaction is simple (in particular, for solvent exchange), since the essence of mechanistic models is the motion of atoms, or groups of atoms, from one site to another, with changes in volume accordingly. The principles set forth in the Introduction lead to the mechanistic criteria of Table I, for solvent exchange reactions. For I_d processes, one

Table I. Pressure-Dependence Parameters of Solvent Exchange Reactions According to Mechanistic Type

Mechanism	Sign of $\Delta V_{ex}{}^*$	$(\partial \Delta V_{ex}{}^*/\partial P)_T$
D	positive	negative
I_d	positive	insignificant
I_a	negative	insignificant
A	negative	positive

sees only the volume change associated with transfer of a solvent molecule from the first c.s. to the second, and vice-versa for I_a, so that the volume of activation $\Delta V_{ex}{}^*$ will be positive or negative respectively, but not detectably pressure-dependent in either case, since bulk solvent is not involved in an I-process. Conversely, for A or D processes, $\Delta V_{ex}{}^*$ will be pressure-dependent at least to the extent of the molar compressibility of the bulk solvent, since at least one solvent molecule must be transferred between the coordinated spheres and bulk solution.

The $\Delta V_{ex}{}^*$ values listed in Table II are all independent of pressure to within the experimental uncertainty, and indeed estimates of the mean $(\partial \Delta V_{ex}{}^*/\partial P)_T$ values for these reactions over the experimental pressure range fall far short of the mean molar compressibilities of the solvents; for example, the mean $(\partial \Delta V_{ex}{}^*/\partial P)_T$ for the $Co(NH_3)_5DMF^{3+}$/DMF exchange at 328 K is calculated to be $(0.11 \pm 0.17) \times 10^{-2}$ cm^3 mol^{-1} MPa^{-1} over the range 0 to 400 MPa, whereas the molar compressibility of DMF is 2.6×10^{-2} cm^3 mol^{-1} MPa^{-1} at 180 MPa (5.8×10^{-2} at 0.1 MPa). Thus, all of the reactions listed in Table II are I mechanisms and all except those of the cobalt(III) complexes are I_a mechanisms. The assignment of the I_d mechanism to reactions of cobalt(III) complexes (other than anionic complexes and conjugate bases) is in accordance with conventional wisdom [2,3], but Table II shows clearly that the traditional assumption that reactions of cobalt(III) complexes can serve as mechanistic models for those of other M(III) complexes, is quite wrong, even though there is a close parallel between the behavior of Co(III) and M(II) species (the "Eigen-Wilkins" mechanism).

Table II. Volumes of Activation for Solvent Exchange Reactions

Exchange Reaction	$\Delta V_{ex}*$, $cm^3\ mol^{-1}$	
$Co(NH_3)_5OH_2^{3+}/H_2O$	$+1.2 \pm 0.2$	[6]
$Co(NH_3)_5DMF^{3+}/DMF$	$+3.0 \pm 0.2$	
$Co(NH_3)_5DMSO^{3+}/DMSO$	$+10.0 \pm 1.2$	
trans-$Co(en)_2(OH_2)_2^{3+}/H_2O$	$+5.9 \pm 0.2$	
$Rh(NH_3)_5OH_2^{3+}/H_2O$	-4.1 ± 0.4	
$Rh(NH_3)_5DMF^{3+}/DMF$	-2*	
$Ir(NH_3)_5OH_2^{3+}/H_2O$	-3.2 ± 0.1	
$Cr(NH_3)_5OH_2^{3+}/H_2O$	-5.8 ± 0.1	
$Cr(OH_2)_5OH_2^{3+}/H_2O$	-9.3 ± 0.3	
$Cr(DMF)_5DMF^{3+}/DMF$	-6.3 ± 0.2	
$Cr(DMSO)_5DMSO^{3+}/DMSO$	-11.3 ± 1.0	

*Provisional Value

These conclusions hold independently of the nature of the solvent and of the remaining ligands L. It is too early to attempt to account for the absolute magnitudes of the $\Delta V_{ex}*$ values in detail, but we may note that, of the aprotic solvents, the "flat" molecule DMF makes far less stringent spatial demands than the pyramidal DMSO, while the relatively small molecule water, for which hydrogen-bonded structuring is so important, may give rise to surprisingly large volume effects, probably because of disruption of the structure of the second c.s. locally during $M{-}OH_2$ bond forming or breaking [7]. In any event, the mechanistic distinctions made here on the basis of pressure effects are clear cut and are supported by inferences (admittedly often somewhat equivocal) made on other grounds [3]. The anomalous behavior of cobalt(III) among M(III) systems is attributable to steric crowding around the unusually small Co^{3+} ion.

High-pressure studies of solvent exchange reactions of square planar complexes ML_3Sol have not been reported thus far, but Palmer and Kelm [8] report a negative, markedly pressure-dependent $\Delta V*$ value for the common solvent-attack step in substitution reactions of the series of complexes $Pt(dien)X^+$ in water, in accordance with the predictions of the A-mechanism. A D-mechanism is indicated for the isomerization of trans-$Co(en)_2(OH_2)_2^{3+}$ by the strongly positive $\Delta V*$ and negative $(\partial \Delta V*/\partial P)_T$ values [9], which contrast sharply with the relatively small, positive, pressure-independent $\Delta V_{ex}*$ value which establishes an I_d-mechanism for the aqua exchange reaction of this same species (Table II).

For reactions involving changes of charge on going to the transition state, solvational changes can make large contributions to $\Delta V*$, but these will be accompanied by correspondingly large pressure dependences of $\Delta V*$. Indeed, if one assumes (quite reasonably, on the basis of the foregoing) that the compressibility of solvent in the second c.s. is negligible relative to that of bulk solvent, one can represent the pressure dependence of ΔV^* by an adaptation of the Tait equation in which the parameter to be calculated is the number x of solvent molecules added to the second c.s. in the activation process. Thus, for the aquation of Co^{III}-$(NH_3)_5X$, x ranges from 8.0 for X = SO_4^{2-} to 1.9 for X = NO_3^- [10]; evidently, by far the greater part of x represents solvation of the leaving anion rather than the increased net charge on the cationic complex fragment.

Evaluation of the contributions of M-Y bond making and M-X bond breaking to ΔV^* becomes difficult when large solvational contributions are present. Nevertheless, it is found for the aquations of $Co^{III}(NH_3)_5X$ that the volume of reaction $\Delta V°$ is always about 1 cm^3 mol^{-1} more negative than ΔV^*, i.e., that a contraction of the magnitude expected for transfer of one H_2O from the second to the first c.s. occurs on going from the transition state to the final state, and the mechanism is therefore I_d. The analogous quantity for the aquations of $Cr^{III}(NH_3)_5X$, however, is invariably positive but varies with X; this expansion can only mean that a transition state of increased coordination number is breaking up as the reaction goes to completion, i.e., that the mechanism is I_a. Rather similar arguments reveal I_a mechanisms for substitution reactions of $Cr(H_2O)_5I^{2+}$ [11] and $Fe(H_2O)_6^{3+}$ [12,13].

CONCLUSION

Pressure effects on the rates of substitution reactions of metal ion complexes in solution can provide uniquely effective criteria of the mechanism, including information on the associated solvational changes. Indeed, it was on the basis of pressure effects, rather than the extensive but often ambiguous data of the conventional sort then available, that we previously advanced the hypothesis that the I_a mechanism should be regarded as general for simple substitution reactions of octahedral complexes of trivalent metal ions in solution, with the important exception of cobalt(III) [3]. Indeed, if one insists upon strictly operational definitions of reaction mechanisms, then pressure effects can provide an excellent basis for such schemes, as Table I shows. In particular, pressure effects distinguish clearly between the Langford-Gray I_a and I_d mechanisms, which rate-law-based operationism does not, and the original Ingoldian dichotomy is thereby restored to prominence as a basic mechanistic concept.

NOTATION

c.s.	= coordination sphere	$\Delta H°$	= enthalpy of reaction
L	= non-reacting ligand	ΔV	= molar volume change
M	= metal cation	$\Delta V°$	= volume of reaction
Sol	= solvent molecule	ΔV^*	= volume of activation

Notation (continued)

X = replaceable ligand
Y = incoming ligand
ΔG° = free energy of reaction
ΔV_{ex}* = volume of activation for solvent exchange

REFERENCES

1. D. R. Stranks, Pure Appl. Chem. 38, 303 (1974).
2. C. H. Langford and H. B. Gray, Ligand Substitution Processes, W. A. Benjamin, New York (1966).
3. T. W. Swaddle, Coord. Chem. Rev. 14, 217 (1974).
4. S. T. D. Lo, E. M. Oudeman, J. C. Hansen, and T. W. Swaddle, Can. J. Chem. 54, 3685 (1976).
5. T. W. Swaddle, Can. J. Chem. 55, 3166 (1977).
6. H. R. Hunt and H. Taube, J. Am. Chem. Soc. 80, 2642 (1958).
7. D. R. Stranks and T. W. Swaddle, J. Am. Chem. Soc. 93, 2783 (1971).
8. D. A. Palmer and H. Kelm, Inorg. Chim. Acta. 19, 117 (1976).
9. D. R. Stranks and N. Vanderhoek, Inorg. Chem. 15, 2639 (1976).
10. W. E. Jones, L. R. Carey, and T. W. Swaddle, Can. J. Chem. 50, 2739 (1972).
11. M. C. Weekes and T. W. Swaddle, Can. J. Chem. 53, 3697 (1975).
12. A. Jost, Ber. Bunsenges. phys. Chem. 80, 316 (1976).
13. B. B. Hasinoff, Can. J. Chem. 54, 1820 (1976).

M-4

EXPERIMENTAL TECHNIQUE AND DATA REDUCTION FOR HIGH PRESSURE KINETICS*

C. A. Eckert

University of Illinois
Urbana-Champaign, Illinois USA

INTRODUCTION

The experimental technique of high pressure kinetics provides detailed information about the structure and properties of reaction transition states -- data which are often unavailable from any other source. In recent years the use of this experimental tool has become much more prevalent, and data are now available for literally thousands of reactions [1]. Therefore, it is especially important at this juncture to review the experimental possibilities and limitations of high pressure kinetics, with emphasis on the types of uncertainties which are inherent in the experiments as well as in the reduction of data. Moreover, it can be demonstrated that careful planning of both the experimental program as well as the method of data reduction can substantially augment the net information obtained for a given quantum of high pressure kinetics effort.

Transition state theory has been applied to reactions in solution to develop the basic relationships for high pressure kinetics. For example, for a typical bimolecular reaction proceeding through transition state M,

$$A_1 + A_2 \rightleftharpoons M \rightarrow \text{Products} \qquad (1)$$

*Invited paper.

the volume of activation is given rigorously by

$$\left(\frac{\partial \ln k}{\partial P}\right)_T = -\frac{\Delta v^{\neq}}{RT} \tag{2}$$

where $\Delta v^{\neq}$ represents a difference in partial molal volumes

$$\Delta v^{\neq} = \bar{v}_M - \bar{v}_{A_1} - \bar{v}_{A_2} \tag{3}$$

It should be recognized that Eq. (2) is valid only if k is expressed in pressure-independent units [2,3], such as mole fraction or molality. For first-order reactions there is no problem of course, but the use of k in (1/mole-time) for second-order reactions results in a substantial error, which unfortunately has cropped up repeatedly in the literature. In our work, we prefer mole fraction units.

Most often, organic and inorganic chemists have been interested in reaction mechanisms, the overall composition of the transition state. They most often seek the sign and approximate magnitude of $\Delta v^{\neq}$, as these give information about the molecularity of the reaction, the degree of charge formation (or destruction) in the transition state, and perhaps the inclusion of solvent molecules in the transition complex [4,5]. Often the data taken for these purposes might consist of a few determinations of k at one temperature over a pressure range of a few kbars, then fit with some arbitrary analytical function of pressure, most likely a quadratic, to determine the slope at one atmosphere from Eq. (2). If the procedure is reasonably carefully done, the result is usually quite adequate for mechanistic interpretation, but may be inaccurate by 10-20% or more, even though the precision of a replicate procedure would be far smaller.

Now, however, we often wish to view the more detailed structure of the transition state, by separating the differences in large numbers represented by Eq. (3). For this purpose far more accuracy is required in both the experimental determination of k(P) and the data-reduction analysis used to extract $\Delta v^{\neq}$. If such methods are used, then an accuracy of 1-2% becomes feasible, permitting for example the determination of the dipole moments of transition states [6,9], specific solvent interactions with the transition state [10], secondary interactions within the transition state [11,12], the detailed role of a homogeneous catalyst [13], or the compressibility of the transition state [14].

EXPERIMENTAL METHODS

Activation volumes are by their nature a mathematical second derivative: the experimental determinations of concentration as a function of time must be twice differentiated, with the attendant loss of precision. Therefore, it is essential not only to recognize this mathematical limitation in assigning accuracy values to measurements of $\Delta v^{\neq}$, but also to minimize the imprecisions in the experimental technique.

High pressure kinetic measurements in general require the determination of concentration vs. time at a constant pressure and temperature. However, the mere necessity for pressurization can cause considerable uncertainty in the temperature, due to the heat of compression. Moreover, unless some sampling technique or in situ assay method is used, the time requirement for pressurization and depressurization can create considerable uncertainty in the time variable.

These problems have certainly been recognized and various solutions to them proposed and discussed elsewhere [3,15,16]. In work in our laboratories we have used in situ initiation with sampling [17] or continuous conductometric analysis [18] for reactions with half-lives in the range of about 3 min to 30 hrs. For slower reactions, the heat of compression is not important, and any sort of initiation is adequate. Most methods of indirect assay (conductivity, uv or visible spectrophotometry, NMR, dilatometry) may require calibration, but if the heat of compression problem is solved, the limit of accuracy of the determinations approaches that of the concentration measurement. Thus if a reaction is followed over about two half-lives with a concentration measurement of accuracy $\pm$ 1%, it is quite possible to measure moderately large activation volumes (20-40 cc/mole) to an accuracy of better than $\pm$ 1 cc/mole. This is about the limit of accuracy of most conventional analytic techniques (spectrophotometry or vpc, for example) though significantly better precision can be achieved in a few special cases, such as titrations with very precise endpoints, or very precise conductivity measurements in well-defined systems [19-23].

For maximum utilization of $\Delta v^{\neq}$ data, it is, as stated above, frequently desirable to calculate $\bar{v}_M$ from Eq. (3). If an applicable solution theory is available, small variations of $\bar{v}_M$ with respect to solvent medium can be used to deduce the detailed structure of the transition complex as well as its degree of solvation (electrostriction by the solvent medium). Such an application represents in effect yet another differentiation of the experimental data, such that not only are very precise values of $\Delta v^{\neq}$ needed, but measurements of $\bar{v}_{A_1}$ and $\bar{v}_{A_2}$ must be made to at least the same precision. Fortunately, such determinations at ambient pressure are relatively easy to make by injection dilatometry to a precision of $\pm$ 0.1 cc/mole

[24], and with some effort an even better accuracy is obtainable [8].

DATA ANALYSIS

In reducing high pressure kinetic data we seek to evaluate, at 1 atm., $\Delta v^{\neq}$ from Eq. (2). Such a procedure represents a very difficult mathematical procedure: the evaluation of the slope of discrete data points at the boundary of the available data. To date no rigorous functionality of ln k (P) has been discovered, so the result can be highly dependent upon the numerical method chosen.

Various assumptions have been made in the past, including straight line fits over a limited range [13,25], graphical schemes, point-to-point finite difference methods [22,26], and the quadratic series in pressure.

$$\ln k = a + bP + cP^2 \tag{4}$$

Equation (4) is certainly mathematically convenient, and is now currently the most commonly used; it has been suggested as a standard [27].

Unfortunately this form tends to give underestimates of $\Delta v^{\neq}$ [3,4], and the results in any event are highly dependent upon the choice of data points fit. As an example, fairly copious data precise to about 1%, are available for the Diels-Alder reaction of cyclopentadiene with dimethylacetylenedicarboxylate in ethyl acetate at 10°C [12], with six data points up to one kbar and another six up to six kbars. Table I demonstrates the effect of fitting various

Table I. Effect of the Pressure Range of the Data on the Activation Volume Calculated from the Quadratic Expression

No. of pts. used in fit	Upper Pressure, kbar	$\Delta v^{\neq}$ at P=1bar cc/mole	Std. Dev. of fit
6	1.04	-31.3	1.2%
7	1.38	-30.2	1.5%
8	2.07	-28.2	2.2%
9	3.10	-27.0	2.5%
10	4.14	-26.9	2.3%
11	5.17	-25.2	3.8%
12	6.21	-23.3	5.9%

numbers of points, starting of course at atmospheric pressure, on the activation volume calculated from Eq. (4).

From Table I it is clear that not only does $\Delta v^{\neq}$ vary with the range of data fit, but over a large range Eq. (4) simply will not fit the data within the experimental error. A more dramatic demonstration of the same phenomenon is in Fig. 1, where a plot is made for the data for the reaction pyridine-methyl iodide in acetone at 50°C [18] according to a linearization of Eq. (4),

$$\frac{1}{P} \ln \frac{k_P}{k_1} = b + cP \qquad (5)$$

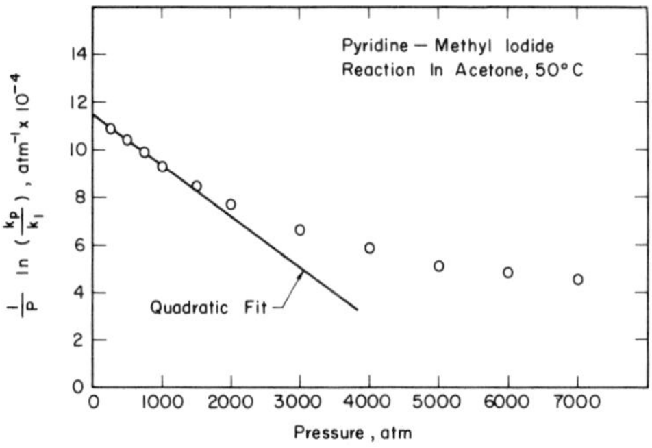

Fig. 1. Limitations of the quadratic fit of high pressure kinetic data in the pyridine-methyl iodide reaction in acetone at 50°C.

Clearly the quadratic fit is valid only over a narrow range, and this range of applicability has been determined to correspond to roughly 10% compression of the solvent [3,28], or effectively 1 or 2 kbars for an organic, but 3-4 kbars for aqueous solutions.

Certainly compression, or density change is a superior, though less convenient, independent variable than pressure. Benson and Berson [29] proposed using the Tait equation of state

$$\frac{\rho_P - \rho_1}{\rho_P} = C \log \left(\frac{B + P}{B + 1}\right) \qquad (6)$$

in conjunction with a number of numerical assumptions to yield a linear plot of a function of rate data vs. $P^{0.523}$, valid only over a limited range (generally about 2-10 kbar, but not below) and requiring an unfortunate degree of extrapolation to yield $\Delta v^{\neq}$. Yet the essence of their concept was eminently valid: the use of density as an independent variable and a valid equation of state for the liquid compression.

Recently new results have been obtained [14] which for the first time permit actual measurement of $\bar{v}_M$ as a function of pressure. For two Menschutkin reactions, for which accurate rate constants were available to high pressure [18], partial molal volumes as a function of pressure were measured for the reactants to a precision of $\pm$ 0.2 cc/mole. These data permit the evaluation from Eqs. (2) and (3) of both $\Delta v^{\neq}$ (P) and $\bar{v}_M(P)$. The resulting partial molal volumes of both reactants and transition states were found to be well represented by the Tait equation. For example, in Fig. 2 there are given the measured compressibilities of both reactants and transition state for the alkylation reactions in acetone by methyl iodide of both pyridine (at 50°C) and tri-n-propylamine (at 30°C). Thus this equation of state can be used, without any limiting assumptions, to yield an analytical expression for $\Delta v^{\neq}(P)$, or in integral form for ln k (P).

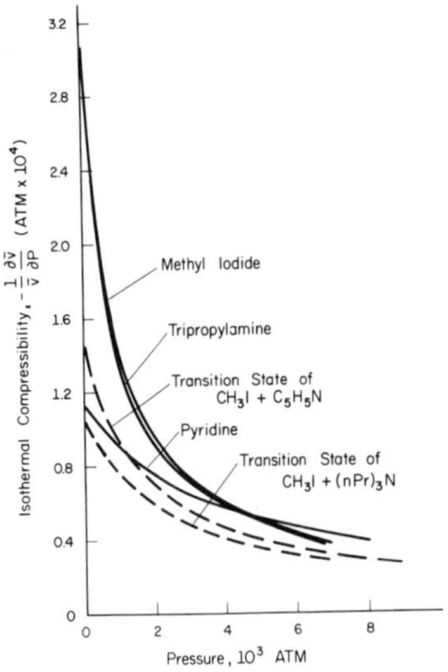

Fig. 2. The isothermal compressibility of the reactants and transition states for two Menschutkin reactions.

It should be stressed that these data are available for only two reactions, and it would be prohibitively tedious to take them in all cases. Rather these two reactions serve as models against which various numerical methods for the evaluation of the volume of activation from high pressure kinetic data may be judged. Using these data, a number of empirical and semi-empirical expressions for $\Delta v^{\neq}(P)$ and ln k (P) were tested [28], and there were two forms which appeared quite promising. Where density data exist (those for the pure solvent were used in every case) an exponential in the density was found to be very effective.

$$\ln (k_P/k_1) = b\,\kappa\,\rho^c \qquad (7)$$

Often density data for the solvent as a function of pressure are unavailable, and in these cases the form recommended was an empirical "Taitlike" equation in pressure.

$$\ln (k_P/k_1) = \ln \left[\frac{Q + P}{Q + 1}\right] \qquad (8)$$

An example of the application of Eqs. (7) and (8) along with the quadratic form Eq. (4) are shown in Fig. 3, where in the lower part of the graph the activation volume at 1 atm. is plotted as a function of the pressure range of data evaluated. The upper part shows the standard deviation of the fits. The dashed line represents the experimental data and shows that using the Tait Eq. (6) for reactants and products, $\Delta v^{\neq}$ at 1 atm. does not depend greatly on the range of data used for evaluation, and Eqs. (7) and (8) are similarly effective over any range. The quadratic fails badly above 1-2 kbar.

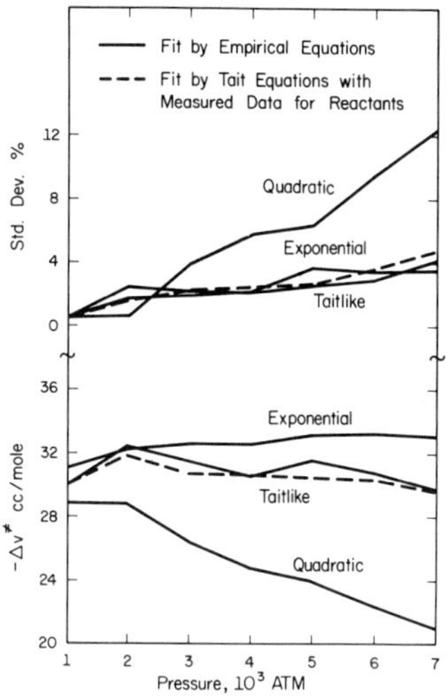

Fig. 3. The activation volume at 1 atm. as a function of the range of data used in its evaluation, compared with measured $\Delta v^{\neq}$ for the tripropylamine-methyl iodide reaction in acetone at 50°C.

All of these equations were tested against several dozen sets of high pressure kinetic data. In general, Eq. (7) was preferred when usable, and if it could not be applied, Eq. (8) gave the best representation. The quadratic is adequate only for a range of compression of less than 10%, and if sufficient precise data are available in this range, any of the forms yield equally good results.

CONCLUSIONS

It is recommended that experimentalists seeking the maximum precision in measurements of the activation volume choose a technique which involves either a very slow reaction or in situ initiation coupled with continual or continuous assay. The analytic procedure used to follow the reaction is critical in that it will probably provide the limiting uncertainty in the data.

Data need not be taken over as wide a pressure range as is accessible; rather a half-dozen or a dozen precise points should be obtained over a pressure range corresponding to 0-10% compression of the solvent. Then any of the expressions quoted above, including the quadratic Eq. (4), will yield a good value of $\Delta v^{\neq}$. If one wishes to view data over a wider range of pressures, density is the preferred independent variable and the exponential expression Eq. (7) should be used. If density data are unavailable, the empirical "Taitlike" form Eq. (8) may be employed. In any event, we as investigators should recognize that even the most precise determinations of the transition state partial molal volume, and especially its change with thermodynamic state, are subject to an accumulation of errors due to successive differentiations. Thus we should strive to be realistic in our assignments of uncertainties to our final values.

ACKNOWLEDGMENT

The financial support of the National Science Foundation and of the Petroleum Research Fund is gratefully acknowledged.

NOTATION

A_1, A_2	= reactant species	Q	= empirical constant
a,b,c	= empirical constants	R	= gas constant
B,C	= constants in the Tait Equation	T	= temperature
k	= rate constant	v	= molar volume
M	= transition state	κ	= compression
P	= pressure	ρ	= density

REFERENCES

1. "Activation Volumes and Reaction Volumes 1967-1976," Special Issue of ΔV News, Oita Univ., Japan (1977).
2. S. D. Hamann, High Pressure Physics and Chemistry, Vol. VII, R. S. Bradley, ed., Academic Press, New York (1963), Ch. 8.
3. C. A. Eckert, Ann. Rev. Phys. Chem. 23, 239 (1972).
4. W. J. leNoble, Progr. in Phys. Org. Chem. 5, 207 (1967).
5. K. E. Weale, Chemical Reactions at High Pressure, Spon. London (1967).
6. H. Heydtmann, A. P. Schmidt, and H. Hartmann, Ber. Bunsenges. phys. Chem. 70, 444 (1966).
7. H. Hartmann, H. D. Brauer, H. Kelm, and G. Rinck, Z. Phys. Chem. (Frankfurt) 61, 53 (1968).
8. J. R. McCabe, R. A. Grieger, and C. A. Eckert, Ind. Eng. Chem. Fund. 9, 156 (1970).
9. J. R. McCabe and C. A. Eckert, Accounts Chem. Res. 7, 251 (1974).
10. Y. A. Ershov, V. B. Miller, M. B. Neiman, and M. B. Gonikberg, Izvest. Akad. Nauk. S.S.S.R., Otdel. Khim. Nauk., 2103 (1960).
11. R. A. Grieger and C. A. Eckert, J. Am. Chem. Soc. 92, 2918 (1970).
12. R. A. Grieger and C. A. Eckert, J. Am. Chem. Soc. 92, 7149 (1970).
13. B. E. Poling and C. A. Eckert, Ind. Eng. Chem. Fund. 11, 451 (1972).
14. J. S. Smith and C. A. Eckert, "The Compressibility of the Chemical Reaction Transition State," to be published.
15. K. R. Brower, J. Am. Chem. Soc. 80, 2105 (1958).
16. D. W. Coillet and S. D. Hamann, Trans. Faraday Soc. 57, 2231 (1961).
17. R. A. Grieger and C. A. Eckert, AIChE J. 16, 766 (1970).
18. C. A. Eckert, S. P. Sawin, and C. K. Hsieh, "An Improved Technique for High Pressure Kinetics," to be published.
19. J. B. Hyne, H. S. Golinkin, and W. S. Laidlaw, J. Am. Chem. Soc. 88, 2104 (1966).
20. D. L. Gay and E. Whalley, J. Phys. Chem. 72, 4145 (1968).
21. B. T. Baliga and E. Whalley, J. Phys. Chem. 73, 654 (1969).
22. D. L. Gay and E. Whalley, Can. J. Chem. 48, 2021 (1970).
23. M. J. Mackinnon, A. B. Lateef, and J. B. Hyne, Can. J. Chem. 48, 2025 (1970).

24. R. A. Grieger, C. Chaudoir, and C. A. Eckert, Ind. Eng. Chem. Fund. 10, 24 (1971).
25. C. T. Burris and K. J. Laidler, Trans. Faraday Soc. 51, 1497 (1955).
26. B. T. Baliga, R. J. Withey, D. Poulton, and E. Whalley, Trans. Faraday Soc. 61, 517 (1965).
27. H. S. Golinkin, W. G. Laidlaw, and J. B. Hyne, Can. J. Chem. 44, 2193 (1966).
28. C. A. Eckert, C. K. Hsieh, J. S. Smith, and S. P. Sawin, "Data Analysis for High Pressure Kinetics," to be published.
29. J. W. Benson and J. A. Berson, J. Amer. Chem. Soc. 84, 152 (1962).

M-5

METHOD FOR MEASURING ACTIVATION VOLUMES OF FAST IRREVERSIBLE REACTIONS: A HIGH-PRESSURE STOPPED-FLOW

K. Heremans, J. Snauwaert, and J. Rijkenberg

Katholieke Universiteit te Leuven
Leuven, Belgium

INTRODUCTION

Flow techniques were introduced more than fifty years ago by Hartridge and Roughton to measure the rapid rate of oxygen binding to hemoglobin. More recently, Chance and Gibson [1] developed the stopped-flow method for the study of enzyme reactions. This method needs smaller amounts of material and therefore has become an important tool for the kinetic study of enzyme reactions and many other fast chemical reactions.

In contrast to the relaxation techniques introduced by Eigen and De Maeyer [2], stopped-flow techniques are not limited to reversible reactions. Both techniques are complementary in the study of enzyme reactions [3]. After the development of a high-pressure relaxation spectrometer [4] we considered the construction of a high-pressure stopped-flow.

We present the description of an apparatus capable of measuring rates of reactions at pressures up to 1200 atm. We also present the first results obtained for a ligand-exchange reaction, a redox reaction, and for the formation of transients in an enzyme reaction.

DESCRIPTION OF APPARATUS

A schematic diagram of the experimental arrangement is given in Fig. 1. Since we will concentrate on the instrumental aspects which are relevant to the high pressure conditions, it is sufficient to state that for illumination and optical detection, modular units of the Messanlagen temperature-jump instrument (Göttingen) were used. Signals were stored in a transient recorder (Biomation 802 or Datalab 905).

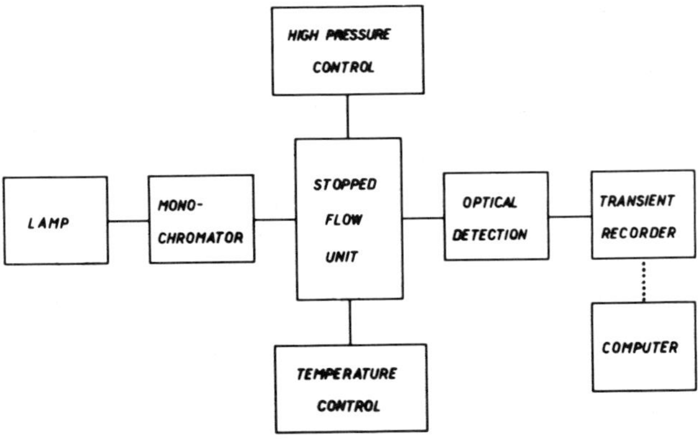

Fig. 1. Experimental arrangement of high-pressure stopped-flow apparatus.

The high-pressure bomb (ID of 40 mm, OD of 120 mm, and total length of 400 mm) is mounted on a support which allows the introduction of the stopped-flow unit in the vertical position and the making of measurements under pressure in the horizontal position. This is of considerable advantage when making runs with solutions of different densities.

The lower part of the bomb contains four feed-throughs for electrical connections insulated with Teflon-sleeves. They are placed in a rectangular position so as to receive the male parts from the stopped-flow unit. Two plexiglas windows allow optical observation.

The stopped-flow unit with syringe-driving mechanism is shown in Fig. 2. The syringe-driving mechanism is a single-step motor A (Kuhnke D39-32 ROR) whose central axis turns over 90° when a voltage of 24V is applied. After a short pulse the axis turns back to its original position by the action of a spring. The screw C rotates in the "on" position of the motor which entails a displacement of the syringes (cf. corkscrew). A freewheel mechanism B disconnects the rotation of the screw from the motor in its "off" position.

Each displacement of the feed syringes D introduces about 0.1 ml into the mixing chamber. The mixing chamber E consists of a tangential 8-jet arrangement. The observation chamber is 1 cm long, closed by plexiglas windows. The fluids coming from the observation chamber go to the waste syringe F. The stopping of the fluids is caused by the internal friction of the syringes and mechanical inertia. This is enhanced by the viscosity of the hydraulic fluid (BP Energol WT). Pressure is transmitted to the reaction mixture through the syringes. Triggering of the transient recorder occurs at the completion of the 90° turn of the stepping motor.

PERFORMANCE

The efficiency was tested by mixing phenolphtalein dissolved in 5 x 10^{-3} M NaOH with 10^{-2} M HCl. No residual absorption due to ionized indicator was seen at the highest sensitivity of the instrument.

Time resolution of the instrument was tested by means of the reduction reaction of dichlorophenol indophenol with sodium

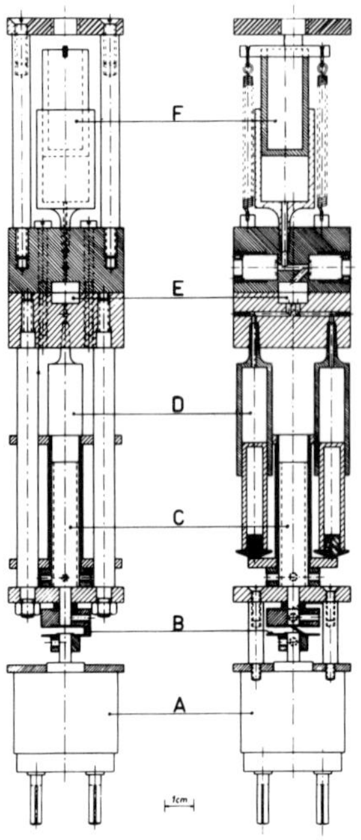

Fig. 2. Stopped-flow unit. Symbols as explained in the text.

ascorbate. Concentration of the dye was 1.25×10^{-5} M after mixing. Observations were made at 600 nm. Concentration of ascorbate was varied as in Fig. 3. It was seen that the pseudo first-order constant remained unchanged up to 90 mM ascorbate corresponding to an observed rate constant of 35 sec^{-1}. The apparatus was thus capable of measuring reactions with a half-life of 20 ms. No appreciable pressure dependence was found.

The dead time of a stopped-flow apparatus is the time needed for the reaction mixture to flow from the mixing chamber to the observation chamber. The dead time measured with the extrapolation procedure [5] gives a value of the order of 20 ms. We found no appreciable pressure dependence on the dead time.

APPLICATIONS

Reaction of Fe(III) Ions With Thiocyanate

Brower [6] was the first to study this reaction with a pressure-jump method performed at different pressures up to 1 kbar. Brower determined a $\Delta V^{\#}$ of 5-6 ml/mole for the overall reaction.

A more detailed study was made by Jost [7] using a temperature-jump relaxation technique up to 2 kbar. He obtained $\Delta V^{\#}$ for the direct and the base-catalyzed pathway. The reaction scheme was as follows:

$$H_2O + Fe^{3+} + NCS^- \underset{}{\overset{k_1}{\rightleftharpoons}} FeNCS^{2+} + H_2O \quad (1)$$

$$K_2 \updownarrow \qquad\qquad \updownarrow$$

$$H^+ + Fe(OH)^{2+} + NCS^- \overset{k_3}{\rightleftharpoons} Fe(OH)NCS^+ + H^+ \quad (2)$$

For such a mechanism the pH dependence of k_{obs} is as follows [8]:

$$k_{obs} = k_1 + k_2/(H^+) \quad (3)$$

where k_1 represents the direct pathway, k_2 the base-catalyzed path and $k_2 = k_3K_2$. Reaction conditions were as follows. Concentrations after mixing: $Fe(ClO_4)_3 = 6.7 \times 10^{-3}$ M and KNCS $= 5 \times 10^{-4}$ M. Ionic strength = 0.4 ($NaClO_4$). H^+ was adjusted with $HClO_4$ between 0.02 and 0.3 M. For such conditions we obtained: k_1 (1 bar) = 150 M^{-1} sec^{-1}; $\Delta V^{\#}$ = -12 ml/mole; k_2 (1 bar) = 27.4 sec^{-1}; $\Delta V^{\#}$ = +8.8 ml. $\Delta V_3^{\#}$ can be obtained assuming values for ΔV_2. This is small

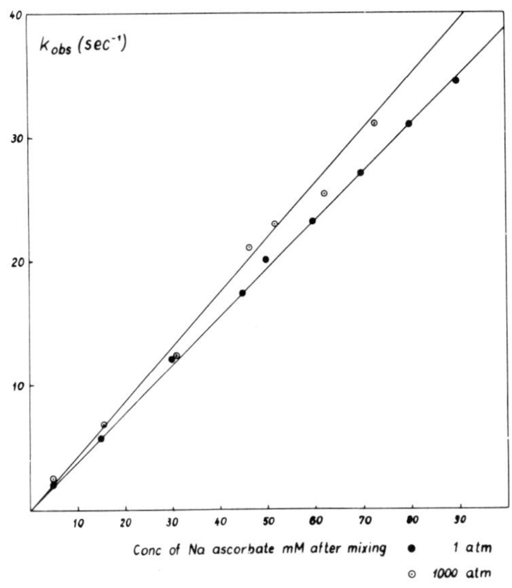

Fig. 3. Time resolution of instrument. Variation of time constant for the reaction of dichlorophenolindophenol (1.25 x 10^{-5} M) with Na ascorbate as a function of Na ascorbate concentration.

(-1 ml [9] to +3 ml [7]). For ΔV_1 we obtain from independent equilibrium measurements a value of 7.5 ml in agreement with Brower and Jost. The latter obtained $\Delta V_1^{\ne} = 0$ ml. Hasinoff [9] obtained $\Delta V_1^{\ne} = -4.5$ ml for the $FeCl^{2+}$ complex formation.

Reduction of Cytochrome C

The structure of both oxidized and reduced Cytochrome C is known. Kinetics of the redox reactions have been studied both by stopped-flow and relaxation techniques.

Ferri-Cyt C + Ferrohexacyanide $\rightleftharpoons$ Ferro-Cyt C + Ferrihexacyanide

In a previous study [10] we obtained activation volumes for this process by chemical relaxation studies under pressure. We obtained $\Delta V^{\ne}_{red} = +13$ ml and $\Delta V^{\ne}_{ox} = -24$ ml. We report now on the reduction of cytochrome with ascorbate: 2 Ferri-Cyt + Ascorbate $\rightleftharpoons$ 2 Ferro-Cyt + Dehydroascorbate + 2 H^+. Cytochrome C (Sigma Type VI) 10 μM was mixed with sodium ascorbate (2-20 mM) (both concentrations after mixing) in Tris-HCl buffer pH 7.5. Ionic strength = 0.04 at 25°C. Observation wavelength was 555 nm (absorption peak of reduced cytochrome).

$\Delta V^{\ne}$ for the reduction is -21 ml. More detailed studies are needed to explain the difference between activation volumes for the reduction by ironcyanides and ascorbate. The observation that ironcyanide form ion pairs with cytochrome C [11] is reflected in the positive activation volume observed with this reductant.

Chymotrypsin Catalyzed Hydrolysis of Esters

The following mechanism has been proposed for the chymotrypsin catalyzed hydrolysis of specific substrates (for a review see [3]):

$$E + S \rightleftharpoons ES \xrightarrow[\searrow P_1]{k_{23}} EP_2 \xrightarrow{k_{34}} E + P_2 \qquad (4)$$

ES stands for the Michaelis complex and EP_2 for the acyl-enzyme intermediate. For esters k_{23} is much faster than k_{34}, so that under steady-state conditions one measures essentially deacylation.

Pressure studies have been performed recently on k_{34} for different substituents by Neumann et al. [12].

We have made a first attempt to obtain activation volumes for the formation of the acyl-enzyme. We have used the proflavin displacement method [13] to observe the formation and the disappearance of the acyl-enzyme intermediate with N-acetyl-L-tryptophanethylester. We had previously shown that the binding of proflavin to chymotrypsin is pressure independent [4].

Experimental conditions were as follows: chymotrypsin, 16 μM; proflavin, 70 μM; and ester, 1 mM. The acetate buffer had a pH of 5.0. The ionic strength was 0.4 (KCl) at 25°C. The observation wavelength was 465 nm. Under these conditions, two first-order processes can be observed which are proportional to k_{23} and k_{34}. We obtain from the pressure dependence: $\Delta V^{\neq}_{23} = -12$ ml and $\Delta V^{\neq}_{34} = -12$ ml. These values are corrected for the effect of pressure on the pK_a of the buffer. It was also assumed that ΔV for the formation of ES is zero. Neumann et al. [14] recently obtained negative activation volumes for the formation of the tetrahedral intermediate in the base-catalyzed hydrolysis of p-nitrophenol esters.

ACKNOWLEDGMENTS

The authors thank L. De Maeyer for advice and encouragement. F. Ceuterick and F. Hennau were of considerable help with data processing and in solving triggering problems.

REFERENCES

1. O. H. Gibson, Methods in Enzymology 16, 187 (1969).
2. M. Eigen and L. De Maeyer, Techniques of Organic Chemistry 8, 895 (1963).
3. H. Gutfreund, Ann. Rev. Biochem. 40, 315 (1971).
4. K. Heremans, J. Snauwaert, H. Vandersypen, and Y. Van Nuland, in Proc. 4th Intern. Conf. High Pressure, Kyoto, Japan (1974), p. 623.
5. H. Gutfreund, Enzymes: Physical Principles, Wiley-Interscience, New York (1972), p. 179.
6. K. R. Brower, J. Am. Chem. Soc. 90, 5401 (1968).
7. A. Jost, Ber. Bunsenges. physik. Chem. 80, 316 (1976).
8. J. F. Below, R. E. Connick, and C. P. Coppel, J. Am. Chem. Soc. 80, 2961 (1958).
9. B. B. Hasinoff, Can. J. Chem. 54, 1820 (1976).
10. H. Vandersypen and K. Heremans, Arch. Int. Physiol. Biochim. 82, 792 (1974).
11. E. Stellwagen and R. D. Cass, J. Biol. Chem. 250, 2095 (1975).
12. R. C. Neuman, D. Owen and G. D. Lockyer, J. Am. Chem. Soc. 98, 2982 (1976).
13. S. A. Bernhard and H. Gutfreund, Proc. Natl. Acad. Sci. (U.S.A.) 53, 1238 (1965).
14. R. C. Neuman, G. D. Lockyer, and J. Marin, J. Am. Chem. Soc. 98, 6975 (1976).

N-1

PRESSURE EFFECT ON THE MECHANISM OF THE THERMAL (2+2) CYCLOADDITION PROCESS*

J. Osugi, M. Sasaki, H. Tsuzuki
Y. Uosaki, and M. Nakahara

Kyoto University
Kyoto, Japan

INTRODUCTION

To explain the mechanisms of chemical reactions, it is very important to have information on the intermediates of these reactions. The thermal (2+2) cycloaddition in one step is forbidden by Woodward and Hoffmann's rule [1]. In other words, the orbital symmetry rule predicts that zwitterion or biradical intermediates may exist in thermal (2+2) cycloadditions. The zwitterionic intermediate or transition state was suggested first by Williams et al. [2] who found that the strong electron acceptor, tetracyanoethylene (TCNE) cycloadded to electron-rich olefins like methyl vinyl ether or p-methoxystyrene under mild conditions. This suggestion was based on the qualitative experimental results that these reactions were markedly accelerated by the polar solvents and the electron donating groups attached, directly or through the benzene ring, to the double bond of the nucleophilic olefins. The mechanism of the thermal cycloadditions by way of the zwitterionic intermediate was reviewed in detail by Gompper [3] and Bartlett [4]. The stepwise mechanism was confirmed by the high but not complete stereospecificity [4], and by the trapping reaction with methanol [5]. The large pressure effect on the cycloaddition of this type also was interpreted as strong evidence for the zwitterion formation [6].

The present authors investigated the system of TCNE and styrene and discovered a new species produced to a small extent at 1 bar and furthered by high pressure [7]. The unstable intermediate was erroneously assumed to be the zwitterion. Recent ^{1}H and ^{13}C NMR studies clarified that this species was the 1,4-adduct of TCNE to styrene, 1,1,2,2-tetracyano-1,2,3,8a-tetrahydronaphthalene; that is the neutral isomer of the zwitterion [8]. Furthermore, it was

*Invited paper.

found that the (2+2) cycloadducts of TCNE to styrene and its derivatives could be obtained at high pressure at room temperature if the polar aprotic solvent, acetnitrile, was used [9]. The structural and kinetic aspects of this 1,4-adduct are discussed here.

EXPERIMENTAL

Styrene, α-methylstyrene, TCNE, dichloromethane, 1,2-dichloroethane, and acetonitrile were purified by normally accepted procedures; p-Chlorostyrene, p-methylstyrene, p-methoxystyrene, β-bromostyrene, cis-stilbene, and trans-stilbene were supplied by Nakarai Chem. Ltd. and used without further purification. Styrene-α-d_1 was obtained from Merck Sharp & Dohme Canada Ltd.

To take the UV, IR, and ^{13}C NMR spectra, Shimadzu UV-200S, JASCO DS-402G, and JEOL FX-100 were employed, respectively. A jacketed piston-cylinder vessel was used for the production of high pressure. To obtain information on the kinetics of the reactions at high pressure, the dilution method (i.e., an abrupt concentration change method) was adopted. The rate was followed by means of spectrophotometry with respect to the 1,4-adduct at 325 nm. Since the reaction between TCNE and α-methylstyrene was completed within several minutes, the in situ mixing technique was applied; details are reported elsewhere [10].

RESULTS AND DISCUSSION

In an earlier paper [7] a new species was reported in the dichloromethane solution of TCNE and styrene in which only the π complex had been recognized in some equilibrium work [11-14] and assumed to be the zwitterion mentioned in the introduction. High-pressure accumulation and low-temperature stabilization allowed this species to be analyzed by the ^{13}C NMR spectroscopy. Figure 2 shows the completely decoupled ^{13}C NMR spectrum of this species at -50°C where the free TCNE has too low a solubility in CD_2Cl_2 to appear. The peaks were assigned on the basis of the off-resonance decoupling spectra of styrene, styrene-α-d_1, and p-chlorostyrene, reacted with TCNE at about 9 kbar. The peak at the highest field, the triplet in off-resonance decoupling, can be assigned to the carbon C-8, indicating that the terminal carbon of the vinyl group in styrene becomes saturated. The four singlet peaks in the range of 108 to 112 ppm is undoubtedly due to the carbons in the nitrile group, and the two singlet peaks at 39.0 and 41.0 ppm are due to the

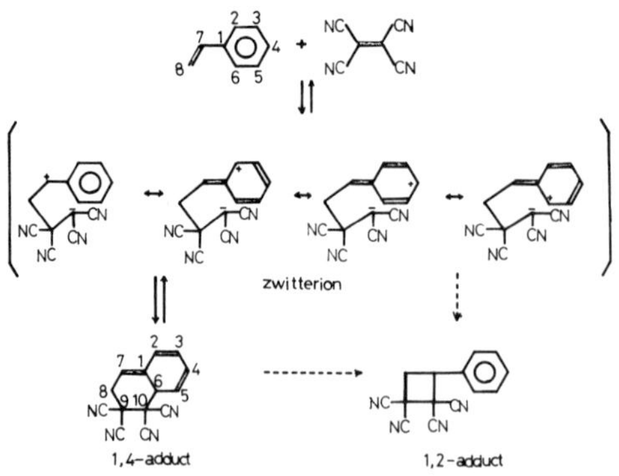

Fig. 1. Reaction scheme.

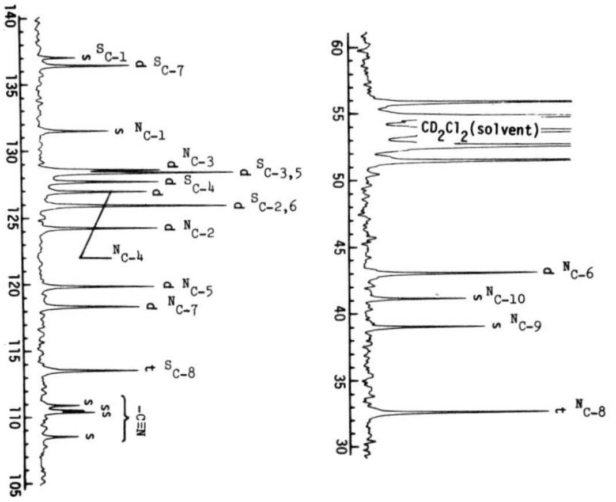

Fig. 2. ^{1}H-Decoupling ^{13}C NMR spectrum of the 1,4-adduct between TCNE and styrene, measured at -50°C. The chemical shifts are expressed in δ (ppm) to TMS. The symbol, e.g., ^{N}C-7 indicates the carbon No. 7 in the 1,4-adduct shown in Fig. 1.

saturated carbons C-9 and C-10, which are unsaturated in TCNE. All these facts suggest an occurrence of an addition reaction between TCNE and styrene. As discussed below, however, the new species is different from the 1,2-adduct, and is identified as the 1,4-adduct. The deuterium incorporation at the position, C-7 in styrene ^{S}C-7, elegantly revealed that the C-7 carbon in the new species (^{N}C-7) is not saturated but is in the sp^2 state, which can arise from the delocalization of the carbonium ion evolved originally at the site C-7 to the benzene ring. In spite of the possibility of ortho and para delocalization, the peak at 43.0 ppm is attributed to the ortho carbon ^{N}C-6 for the following reasons. First, the six carbons formerly belonging to the benzene ring of styrene show quite different chemical shifts; they are not equivalent which is difficult to explain by para bonding. Second, and more decisively, the peak is still doublet in the p-chlorostyrene moiety as shown in Fig. 3a. The above assignment is compatible with the IR spectrum of the new species shown in Fig. 4. (1) When TCNE and styrene are transformed to the 1,4-adduct, the stretching νC≡N at 2220 and 2245 cm^{-1} decreases considerably as a result of the loss of the conjugation of the double bond to C≡N in TCNE. (2) The vibrations, νC-C of TCNE at 955 and 1150 cm^{-1} [15] decrease (and perhaps shift to lower wave numbers because of the reduction

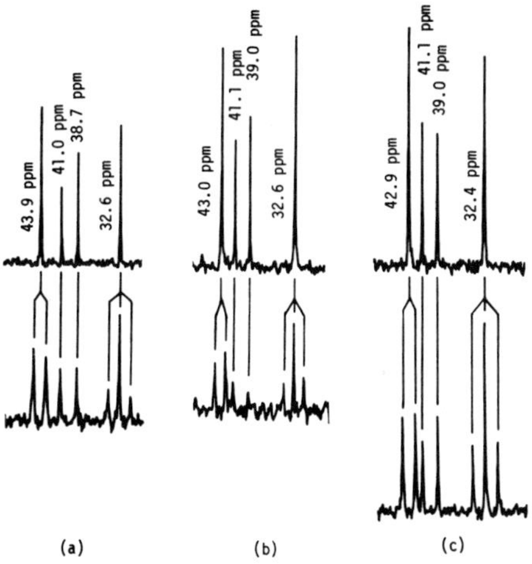

Fig. 3. Off-resonance decoupling spectra of the 1,4-adducts of TCNE to p-chlorostyrene (a), styrene (b), and styrene-α-d$_1$, (c) in CD$_2$Cl$_2$ at -50°C.

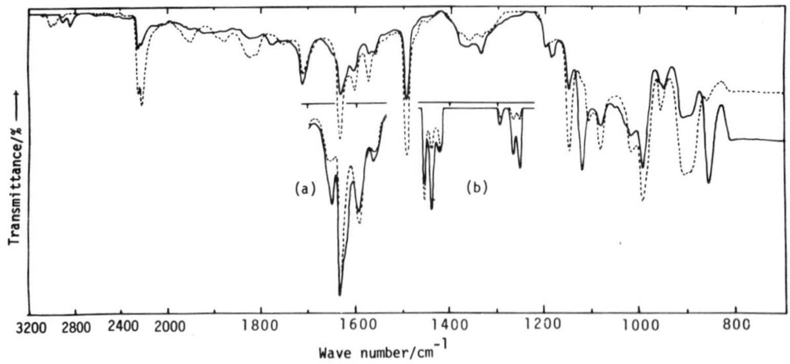

Fig. 4. IR spectrum of the 1,4-adduct of TCNE to styrene in CH_2Cl_2, and to p-chlorostyrene in CH_2Cl_2 (a) and to styrene in CD_2Cl_2 (b).

of its bond order). (3) The absorption νC=C of the vinyl group at 1632 cm^{-1} decreases in the styrene adduct but apparently increases in the p-chlorostyrene adduct (see Fig. 4a) as a result of the two intensified peaks at about 1625 and 1650 cm^{-1}; the absorptions often used for characterizing benzene ring substitution [16] in the range of 1700 to 2000 cm^{-1} also decrease except for those at the lower wave number, indicating the alteration of the ring character. (4) The bending δ CH_2= at 890 ~ 910 cm^{-1} and δ CH= at 990 cm^{-1} decreases, corresponding to the saturation of the vinyl group and the wagging around 1250 cm^{-1} and deformation around 1430 cm^{-1} of the aliphatic $-CH_2-$ are observed in CD_2Cl_2 (see Fig. 4b) where as the liquid cell compensation is not good due to the strong solvent absorption the solvent is changed from CH_2Cl_2 to CD_2Cl_2.

The chromophore of the new species could be attributed to the conjugated triene system rather than to the C≡N groups because the $\pi \rightarrow \pi^*$ transition of the structurally similar triene component in dehydroergosterol has λ_{max} at 320 nm in ether [17] and because the substituent effects on the absorption maximum are really observed as shown in Table I. The substitution at the C-7 position in styrene by methyl or phenyl group causes a red shift, while that at C-4 by methyl group or chlorine atom gives rise to a blue shift.

Once all the species formed in the system of TCNE and styrene have been identified, we can carry out the quantitative equilibrium and kinetic analysis on this system. The system of TCNE and α-methylstyrene in 1,2-dichloroethane is discussed here elaborately assuming the following scheme:

$$D + A \underset{}{\overset{K_1}{\rightleftharpoons}} \pi \underset{k_{-2}}{\overset{k_2}{\rightleftharpoons}} I \qquad (1)$$

where

$$K_1 = \frac{[\pi]}{[D][A]} \qquad (2)$$

$$K_2 = \frac{k_2}{k_{-2}} = \frac{[I]}{[\pi]} \tag{3}$$

The formation and decomposition of I are much slower than the diffusion controlled π-complex formation. When we can observe the charge transfer band immediately after mixing (t=0) the equilibrium constant K_1 can be derived from the Benesi-Hildebrand equation,

$$\frac{[A]_0 d}{B} = \frac{1}{K_1 \varepsilon} \frac{1}{[D]_0} + \frac{1}{\varepsilon}\ ; \qquad [D]_0 >> [A]_0 \tag{4}$$

In case the latter equilibrium (K_2) is accomplished after a long time (t=∞), using Beer's law,

$$B = \varepsilon d[\pi] \tag{5}$$

and the mass balance,

$$[A]_0 = [A] + [\pi] + [I] \tag{6}$$

we obtain

$$\frac{[A]_0 d}{B} = \frac{1}{K_1 \varepsilon} \frac{1}{[D]_0} + \frac{1 + K_2}{\varepsilon}\ ; \qquad [D]_0 >> [A]_0 \tag{7}$$

The values of $[A]_0 \cdot d/B$ at the limit of t=0 and t=∞ are plotted vs. $[D]_0^{-1}$ in Fig. 5. The values of K_1, ε, and K_2 thus obtained are given in Table II. The 1,4-adduct is found to be formed about six times as much as the π complex at 25°C at 1 bar. Tonchéva et al. [13] determined parameters K_1 and ε without knowledge of the existence of the 1,4-adduct. Since the equilibrium (K_2) is almost completed in about 10 min, their measurement perhaps corresponds to the limiting case of t=∞. As a matter of fact, their ε = 200~210 can be compared with the calculated value of $\varepsilon^* = \varepsilon/(1 + K_2) = 290$. For this reason their K_1 and ε are abnormally large and small, respectively.

The rate equation for [I] can be written as

$$\frac{d[I]}{dt} = k_2[\pi] - k_{-2}[I] \tag{8}$$

Eliminating [π] from (8) by using (2) and (6), we have

$$\frac{d[I]}{dt} = a - k_{obs}[I] \tag{9}$$

where

$$a = \frac{k_2 K_1 [D]_0 [A]_0}{1 + K_1 [D]_0} \tag{10}$$

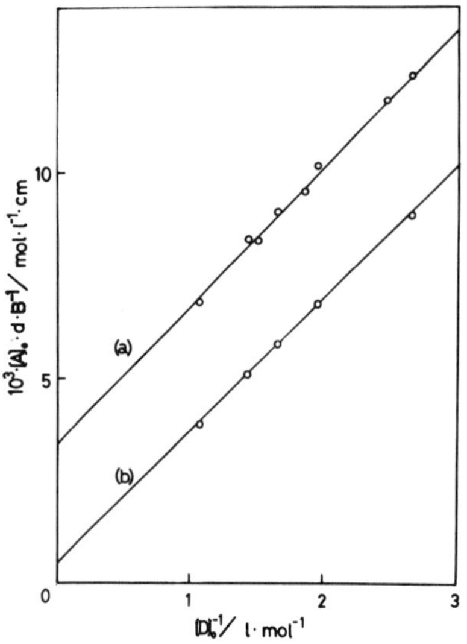

Fig. 5. Plots based on equations (4) and (7). Legend: (a), t=0, (b), t=∞.

and

$$k_{obs} = \frac{k_2 K_1 [D]_0}{1 + K_1 [D]_0} + k_{-2} \qquad (11)$$

Integrating (9) with respect to t, we obtain

$$[I] = \frac{a}{k_{obs}} [1 - \exp(-k_{obs} t)] \qquad (12)$$

Since we can follow [I] by means of the UV absorption around 325 nm, the observed rate constant k_{obs} can be determined experimentally, e.g., by the Guggenheim plot. Equation (11) combined with (3) leads to

$$k_{obs} = k_2 \left(\frac{K_1 [D]_0}{1 + K_1 [D]_0} + \frac{1}{K_2} \right) \qquad (13)$$

In Fig. 6, k_{obs} is plotted against the bracketed factor of the right-hand side of (13). The slope gives k_2 = 6.54 x 10^{-2} sec^{-1}. Putting the values of k_2 and K_2 in equation (3), we obtain k_{-2} = 1.13 x 10^{-2} sec^{-1} which agrees quite well with the directly measured value of 1.12 ~ 1.13 x 10^{-2} sec^{-1} (see Table III).

Table I. Spectroscopic Properties of the π-Complexes and Adducts Between TCNE and Styrene in CH_2Cl_2 at 25°C

	Styrene	p-chloro-styrene	p-methyl-styrene	α-methyl-styrene	1,1-diphenyl-ethylene	p-methoxy-styrene	cis-stil-bene	trans-stilbene	β-bromo styrene
π-complex λ_{max}/nm	480 395	480 360	525 400	492 393	508 ~400	~600 ~400	528 ~390	595 ~390	470 ?
1,4-adduct λ_{max}/nm	320	318	317	325	329	?	-	-	-
1,2-adduct λ_{max}/nm	231	-	232	230	232	242	-	-	-

Table II. Equilibrium Parameters for the TCNE and α-Methylstyrene System in 1,2-Dichloroethane (1 bar)

	T, °C	K_1, l mol^{-1}	ε, l mol^{-1} cm^{-1}	K_2
This study	25	0.155	1970	5.78
Tonchéva et al. [14]	20	2.78-3.18	200-210	--

In principle the pressure effect on each elementary step of equation (1) would be examined according to the spectroscopic measurements at high pressure. According to the reaction scheme discussed in the previous section, the kinetic expression is given as

$$\frac{d\Delta[I]}{dt} = -\left(\frac{k_2K_1[D]_0}{1+K_1[D]_0} + k_{-2}\right)\Delta[I] = -k_{obs}\cdot\Delta[I] \qquad (14)$$

where Δ[I] is the concentration deviation of the adduct from the final equilibrium state and $[D]_0$ the final concentration of α-methylstyrene (donor).

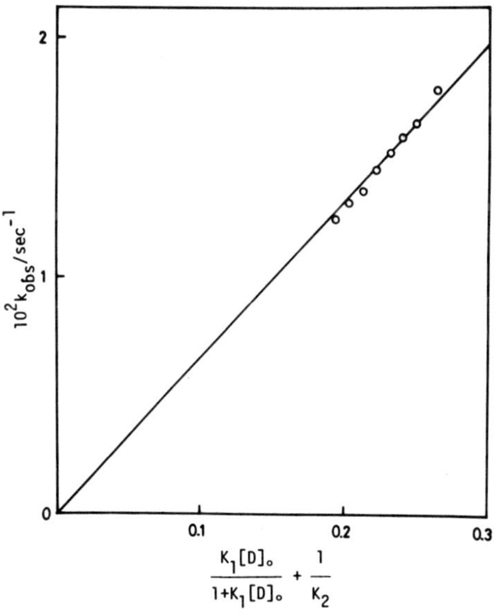

Fig. 6. Plot of k_{obs} to obtain the formation rate constant k_2.

A typical spectrophotometric time-course is shown in Fig. 7a together with the first order plot Fig. 7b. The slope of this plot, which gave k_{obs}, was almost constant within the experimental error even when changing the donor concentration. This fact is justified as follows. The first term in the bracket $\{k_2K_1[D]_0/(1+K_1[D]_0)\}$ will not be substantial compared to the second term k_{-2} because k_2K_1 (at atmospheric pressure) is of the order of 10^{-2} as found in the equilibrium and kinetic studies at 1 atm; thus, the former term is 2×10^{-4} when $[D]_0$ is ~ 2×10^{-2} M as used in the high-pressure kinetic measurement. Then k_{obs} essentially corresponds to k_{-2} itself.

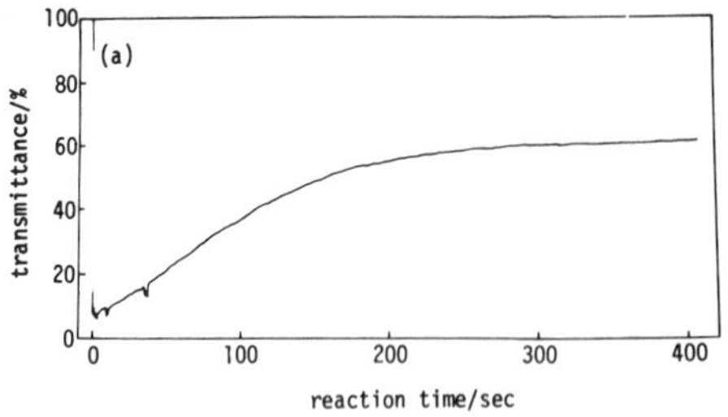

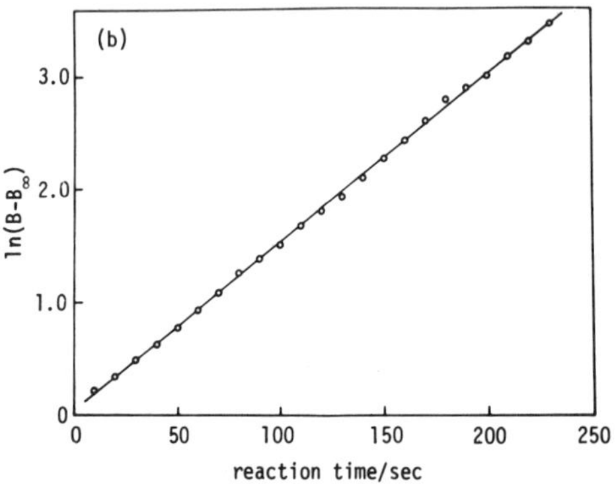

Fig. 7. (a) A typical time-course of color fading by the concentration change method at 325 nm (25°C, 1500kg/cm^2). (b) First-order plot of the change of absorbance B of above trace.

The effect of pressure on the rate constant k_{-2} is given in Table III. Its numerical value at 1 atm coincides well with that determined from another method of calculation using pre-determined values of K_1, K_2, and k_2 as described in the previous section. From the slope of the plot ln k_{-2} vs. pressure (see Fig. 8) the volume of activation at 1 atm was calculated to be -7.4 ± 0.6 cm^3/mol.

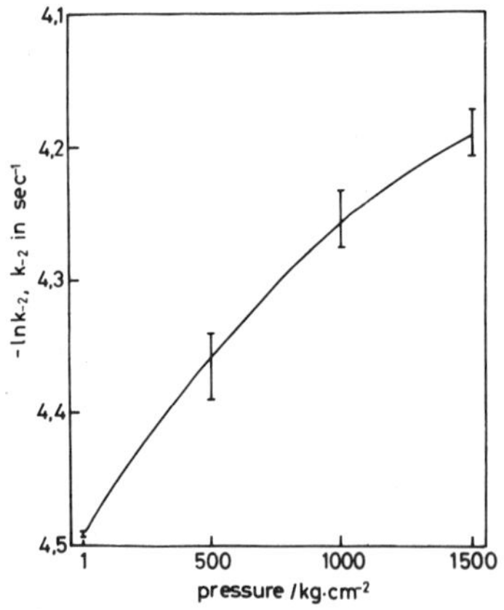

Fig. 8. Effect of pressure on rate constant k_{-2} at 25°C.

Both solvent polarity [2,18] and application of high pressure [6,19] accelerate largely the (2+2) cycloaddition. Even the reverse (2+2) cycloaddition of TCNE-ethyl propenyl ether adduct in methanol accompanies the negative volume of activation [20] which also confirms the existence of the ionic intermediate or transition state. The (2+4) cycloaddition, however, is well known to be a concerted mechanism. The typical one is the Diels-Alder reaction between maleic anhydride and dienes, and its volume of activation is -30 to approximately -40 cm^3/mol [21,22]. In some reactions of maleic anhydride with dienes the transition state is smaller than the adduct due neither to electrostriction nor specific solute-solvent interaction but due to secondary interaction between non-bonding atoms in the transition state [22,23].

Table III. Effect of Pressure on the Decomposition of TCNE-α-Methylstyrene Adduct in 1,2-Dichloroethane at 25°C

Pressure, kg cm^{-2}	$10^2 k_{-2}$, sec^{-1}
1	1.12 ± 0.01
500	1.27 ± 0.03
1000	1.42 ± 0.03
1500	1.51 ± 0.03

The authors reported previously [7] that the formation of the TCNE-styrene adduct, whose character was essentially similar to the present adduct, TCNE-α-methylstyrene 1,4-adduct, was considerably favored by polar solvent and application of high pressure. As mentioned above, the reverse cycloaddition of the 1,4-adduct was also promoted by pressure. These facts may lead to a postulate that the zwitterionic intermediate or transition state will exist even in the 1,4-adduct formation. The volume of activation for k_{-2} is less negative than that of the enol ether adduct -16.7 cm^3/mol [20] probably because in the present system the solvent participation as a reactant is completely ruled out and because the developed charge in the ionic transition state is more widely dispersed over the phenyl group resulting in less electrostriction than the enol ether system.

In conclusion, we have discovered a new species having an absorption maximum at 325 nm, and clarified the structure of this new species as the 1,4-adduct. The new species is changed through the zwitterionic intermediate or transition state with a negative volume of activation.

ACKNOWLEDGMENT

The authors greatly appreciate the assistance of N. Esumi, K. Fujita, and M. Imanari of JEOL Ltd. in measuring the ^{13}C NMR spectra.

NOTATION

K_1, K_2 = equilibrium constants

k_1, k_2, k_{-2} = rate constants

ε = molar absorption coefficient of the π-complexes

d = path length at the measurement of the absorbance

λ_{max} = wavelength at the maximum absorption

$[A]_0$, $[D]_0$ = concentration of the acceptor and donor at t=0

REFERENCES

1. R. B. Woodward and R. Hoffmann, J. Am. Chem. Soc. 87, 395, 2046, 2511, 4388, 4389 (1965); also Angew. Chem. Int. Ed. Engl. 8, 781 (1969).
2. J. K. Williams, D. W. Wiley, and B. C. McKusick, J. Am. Chem. Soc. 84, 2210 (1962).
3. R. Gompper, Angew. Chem. Int. Ed. Engl. 8, 312 (1969).
4. P. D. Bartlett, Quart. Rev. 24, 473 (1970).
5. R. Huisgen, R. Schug, and G. Steiner, Angew. Chem. Int. Ed. Engl. 13, 80 (1974).
6. F. K. Fleischmann and H. Kelm, Tetrahedron Lett. 39, 3773 (1973); also G. Swieton, J. V. Jounne, and H. Kelm, in Proceedings 4th Intern. Conference High Pressure, Kyoto, Japan (1974), p. 652.
7. M. Nakahara, Y. Tsuda, M. Sasaki, and J. Osugi, Chem. Lett. 731 (1976).
8. M. Nakahara, Y. Uosaki, M. Sasaki, and J. Osugi, Chem. Lett. to be published.
9. M. Nakahara, Y. Uosaki, M. Sasaki, and J. Osugi, Chem. Lett. to be published.
10. M. Sasaki, M. Okamoto, H. Tsuzuki, and J. Osugi, Chem. Lett. 1289 (1976).
11. A. R. Cooper, C. W. P. Crowne, and P. G. Farrell, Trans. Faraday Soc. 62, 18 (1966).
12. T. Arimoto and J. Osugi, Rev. Phys. Chem. Jpn. 44, 25 (1974).
13. B. Tonchéva, R. Vélichkova, and I. P. Panayotov, Bull. Soc. Chim. Fr., 1033 (1974).
14. K. Hayashi, P. A. Marchese, S. Munari, and S. Russo, Chim. Ind. (Milan) 56, 187 (1974).
15. D. A. Long and W. O. George, Spectrochim. Acta 19, 1717 (1963).
16. C. W. Young, R. B. DuVall, and N. Wright, Anal. Chem. 23, 709 (1951).
17. L. F. Fieser, M. Fieser, and S. Rajagopalan, J. Org. Chem. 13, 800 (1948).
18. G. Steiner and R. Huisgen, Tetrahedron Lett., 3769 (1973).
19. N. S. Isaacs and E. Rannala, J.C.S. Perkin II, 1555 (1975).
20. W. J. le Noble and R. Mukhtar, J. Am. Chem. Soc. 97, 5938 (1975).
21. C. A. Eckert, Ann. Rev. Phys. Chem. 23, 239 (1972).
22. R. A. Grieger and C. A. Eckert, J. Am. Chem. Soc. 92, 2918, 7149 (1970).
23. K. Seguchi, A. Sera, and K. Maruyama, Tetrahedron Lett., 1585 (1973).

N-2

HIGH PRESSURE STUDIES OF SOLVENT EFFECTS ON A HALIDE EXCHANGE REACTION

P. G. Glugla, J. H. Byon, and C. A. Eckert

University of Illinois
Urbana, Illinois USA

INTRODUCTION

Kinetic solvent effects on chemical reactions are most pronounced when one or more of the reacting species is ionized, and may amount to several orders of magnitude [1]. In spite of the importance of this phenomenon, and after considerable effort to understand it over many years [2-4], no real quantitative or even good qualitative picture has emerged. The advent of high-accuracy high-pressure kinetic measurements provides a new tool with which to attack this problem.

From transition state theory we express solvent effects for a bimolecular reaction

$$A + B \rightleftharpoons M \longrightarrow \text{Products} \qquad (1)$$

in terms of the Bronsted-Bjerrum relationship [5,6]

$$k/k^o = \frac{(\gamma_A/\gamma_A{}^o)\ (\gamma_B/\gamma_B{}^o)}{(\gamma_M/\gamma_M{}^o)} \qquad (2)$$

Such activity coefficients have been correlated or estimated for nonionic reactions to predict kinetic solvent effects [7], but for ionic reactions the ambiguity of evaluating single-ion activities precludes direct application of (2) [8]. However, measurement of the activation volume gives a genuine difference in partial molal volume between transition state and substrate [9,10]

$$\Delta v^{\neq} = \bar{v}_M - \bar{v}_A - \bar{v}_B \qquad (3)$$

The partial molal volume of an ionic species is strongly dependent on its degree of solvation, which is in turn a measure of its activity in solution. Unfortunately, at the present time no exact theory of ionic solutions is available which will link electrostriction volumes to activities. But, there certainly exists a qualitative connection, and the relative changes in partial molal volume along the reaction coordinate will provide good insight into the concomitant kinetic solvent effects.

With this in mind, a data set obtained by a high-pressure kinetic study on a typical ionic reaction would be very valuable in understanding the kinetic solvent effect on ionic reactions. In this work just such a data set was obtained through the use of a new high-volume *in situ* mixing high-pressure reaction vessel and a new injection dilatometer.

EXPERIMENTAL

The reaction chosen for study as a typical ionic reaction is the halide exchange reaction between propyl bromide and iodide ions, namely

$$PrBr + I^- \rightarrow PrI + Br^- \qquad (4)$$

This reaction was chosen for a number of reasons: (1) Its mechanism is well established as S_N2 with only the dissociated ionic species active [11]; (2) The reaction involves only one ion on each side, whereas a reaction involving two ions would complicate the analysis and could confuse the conclusions; (3) The rate of the reaction is experimentally convenient; (4) The reactants are compatible with a wide range of solvents; and (5) The reaction can be followed easily and accurately by titration.

Titrametric analysis, however, requires relatively large samples (several cc). Thus, to obtain the data accuracy required, the *in situ* mixing technique used previously only for microliter sampling [12] was modified by the design and construction of a new reactor large enough (~ 150 cc) to make titration feasible. This reactor, shown in Fig. 1, has the advantages of allowing dissipation of the heat of compresssion before reaction is initiated, along with a highly accurate continual assay. The overall high-pressure equipment is essentially similar to the apparatus described previously [13], except that the high-pressure vessel is enlarged for the high-volume reactor. Pressures are accurate to ± 0.1% and temperatures to ± 0.02°C.

The silver nitrate titration used to monitor the course of the reaction is worthy of further comment. This halide exchange reaction is extremely slow in water. In dilute solutions its half-life is on the order of 48 hrs. As a result, water is a very convenient quench and titration medium. Furthermore, both the iodide ion and the bromide ion can be titrated simultaneously, because of

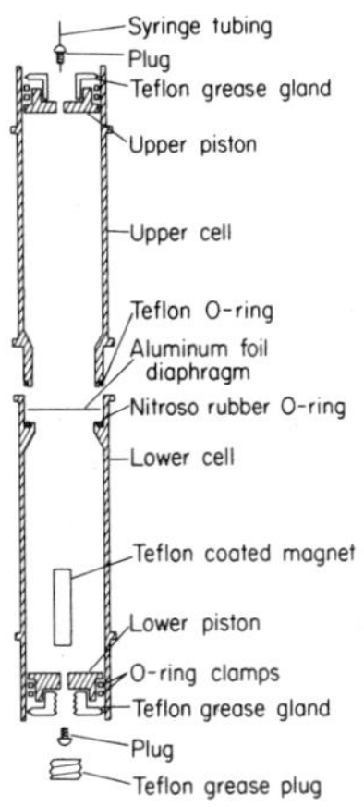

Fig. 1. High volume reaction cell for use with <u>in situ</u> initiation.

their difference in solubility. The titration end point can be determined electrochemically using a carbon and a silver electrode. Because of this simple quench and titration method, reaction rates with half lives as small as 20 min can be measured. The data for a typical kinetic run are shown in Fig. 2. The equilibrium constant for this reaction is on the order of 10; thus, the reverse reaction must be taken into account in analyzing the kinetic data. In Fig. 2 the integrated rate expression, which includes the effect of the reverse reaction, is plotted vs. time.

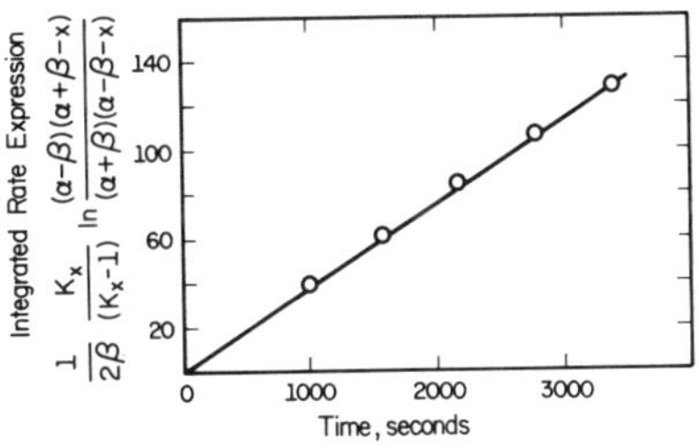

Fig. 2. Experimental data for the halide exchange reaction in acetone at 25° and 187 atm.

In low dielectric constant solvents, salts do not fully dissociate and this fact must be considered analyzing experimental data. This problem was not satisfactorily accounted for in previous studies [14-16]. To account for ion pairing of the reactant salts in the solvents in question, association constants were measured as a function of pressure by a conductance technique that was substantially developed by Fuoss [17]. The high-pressure conductance cell used for this purpose is shown in Fig. 3. The resistance measurements were made using an ac conductance bridge, and experimental precautions were taken in accord with the suggestions of Bollinger [18], Hamann [19], and Wootten [20].

Injection dilatometry was used as a simple, fast method to measure the partial molal volumes of the reactants. The dilatometer is shown in Fig. 4. Ultraprecision syringes were used to determine the volume change on mixing. This quantity is related to the apparent molal volume which in turn is related to the partial molal volume by means of

$$V = n_1 v_1 + n_2 \phi_2 \tag{5}$$

$$\phi_2 = \frac{\Delta V}{n_2} \tag{6}$$

$$\bar{v}_2 = \phi_2 + n_2 \frac{\partial \phi_2}{\partial n_2} \tag{7}$$

The measurement of the partial molal volume of the propyl halide reactant is straightforward and accurate to within $\pm$ 0.2 cc/mole. The measurement of the partial molal volume of the ionic reactant,

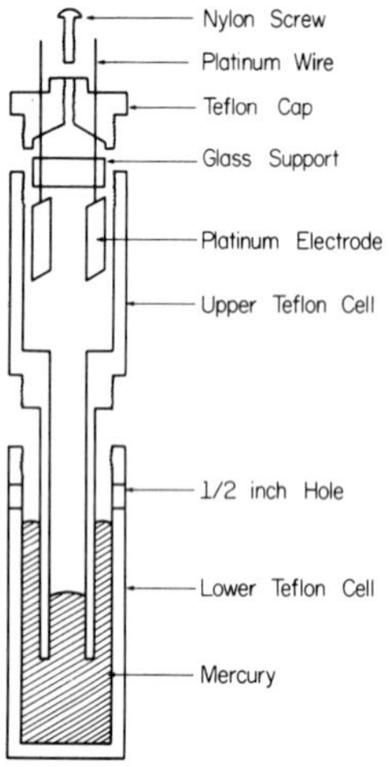

Fig. 3. High pressure conductance cell.

however, is a more difficult problem: First, salts are solids at room temperature and, second, the partial molal volume of the halide ion is desired and not the partial molal volume of the salt. The first problem can be overcome by measuring the volume change when two solutions of the salt are mixed, thus measuring the differences in apparent molal volume between a concentrated solution and a dilute one. The second problem was overcome by using a relationship developed by Hepler [21] which semi-empirically relates an ion's size to its individual partial molal volume. This means that the partial molal volume of a large number of salts must be measured in each solvent to obtain the desired partial molal volume.

Both protic and aprotic solvents were chosen for their variety in physical properties, subject to three practical limitations: One, the solvent viscosity had to be low so that the in situ mixing apparatus could work reliably; secondly, solvents in which propyl iodide decomposed were avoided; and, finally, solvents having dielectric constant of less than 13 were avoided because they would not sufficiently dissociate the salt to allow accurate measurement of the reaction rate. All solvents were prepared for use by well-documented procedures [22], and tested for purity by index of refraction and gas chromatography.

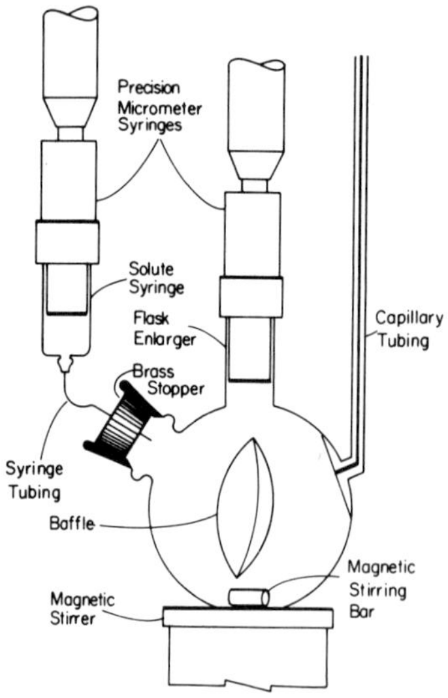

Fig. 4. Large volume injection dilatometer.

RESULTS AND DISCUSSION

Experimental results have been measured in two dipolar aprotic solvents, acetone and acetonitrile; and three protic solvents, methanol, ethanol, and N-methyl formamide (NMF). The data are presented in Table I.

The rate constant is two orders of magnitude greater in the aprotic solvents, and the equilibrium constant as much as two decades lower. The partial molal volume data show clearly that there is little solvent effect due to the propyl bromide, and even the effect on the larger, anionic transition state (which would not affect K, in any event) is not very great. It is the change in the partial molal volume of the ionic reactant I^- which shows the greatest solvent variation. The activation

Table I. Experimental Data for the Reaction of Iodide Ions with n-Propyl Bromide

Solvent	$k \times 10^3$ sec^{-1} at 50°C	K	$\Delta v^{\neq}$, cc/mole	$\bar{v}_{PrBr}$, cc/mole	§$\bar{v}_{I^-}$, cc/mole	$\bar{v}_M$, cc/mole
Acetone	313	0.0129*	-1.4*	92.7*	2*	93*
Acetonitrile	64.9	0.0802*	-6.2*	92.4*	13*	99*
Ethanol	6.58	2.48†	-18†	91.8†	29.3†	103†
Methanol	4.80	2.26†	-14†	92.5†	25.0†	104†
NMF	10.0	1.057†	-17†	90.7†	39.9†	114†

* Data at 25°C.

† Data at 50°C.

§ Partial molal volumes determined in the binary systems are assumed to be equal to those in the ternary.

volumes are much less negative in aprotic solvents, and in a less polar solvent, methylisobutyl ketone, for which partial molal volume data are unavailable ($\Delta v^{\neq}$ = +9 cc/mole).

Another way of viewing these results is shown schematically in Fig. 5, where the partial molal volumes are plotted along the reaction coordinate for all five solvents. The electrostriction of both reactants and products varies by about 30%, even though the solvents do not have vastly different dielectric constants. This effect is almost wholly due to the ionic substrate and not the alkyl halide. Conversely, there is very little solvent effect on the ionic transition state, probably because it is a large charge-dispersed entity.

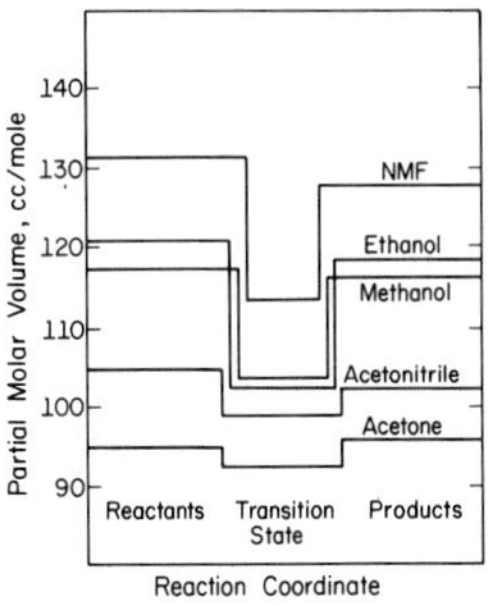

Fig. 5. Volume profile along the reaction coordinate for the reaction $PrI + Br^- \rightleftharpoons PrBr + I^-$.

The protic solvent solvates the reactant anion strongly and easily, without great disruption of structure, by its proton-donating ability, giving a moderate value of $\bar{v}_{I^-}$ and a relatively low activity, due to strong solvation. On the other hand, the aprotic solvent can interact only by dipolar forces, and since it is highly compressible and unstructured, it is strongly electrostricted yet solvates the halide anion much less strongly, yielding a higher halide activity in the aprotic solution. Translated into terms of equation (2) this results in large differences in the γ_A term, while γ_M probably varies only slightly from solvent to solvent.

These conclusions are in accord with those of Parker [8], whose studies of individual species activities for similar reactions led him to assert that the protic solvent's solution of the substrate anion was the major contributor to the kinetic solvent effect. But Haberfield and coworkers [23] measured transfer enthalpies and came to the conclusion that variations in solvation of the transition state were most important. However, both groups were forced to make extrathermodynamic assumptions, which could be questioned.

The conclusion of this discussion then returns to the basic issue of the inadequacy of currently available theories of solution to describe the behavior of ionic species. Rather extensive data have been presented for both the rates and equilibria of a halide-exchange reaction. These have been augmented by the requisite conductivity and partial molar volume studies to describe the behavior of all species in the reaction, yet we are at present limited to a qualitative description of the kinetic solvent effects. Hopefully, future developments in ionic solution theory will permit a more quantitative interpretation of these data.

ACKNOWLEDGMENT

The authors gratefully acknowledge the financial support of the National Science Foundation.

NOTATION

A,B,M = reactants and transition state
K = equilibrium constant
k = rate constant
k^o = rate constant in reference solvent
n_1 = moles of solvent
n_2 = moles of solute
V = solution volume
ΔV = volume change upon injection
v_1 = molar volume of solvent
$\bar{v}_A, \bar{v}_B, \bar{v}_M$ = partial molal volumes
$\Delta v^{\neq}$ = activation volume
$\gamma_A, \gamma_B, \gamma_M$ = activity coefficients of reacting species
$\gamma_A^o, \gamma_B^o, \gamma_M^o$ = activity coefficients in reference solvent
ϕ_2 = solute apparent molal volume

REFERENCES

1. J. Miller and A. J. Parker, J. Amer. Chem. Soc. 83, 117 (1963).
2. A. J. Parker, Advans. Phys. Org. Chem. 5, 173 (1967).
3. A. J. Parker, Chem. Rev. 69, 1 (1969).
4. M. R. J. Dack, Chem. Brit. 6, 347 (1970).
5. J. M. Bronsted, Z. Physik. Chem. 102, 169 (1922).
6. N. Bjerrum, Z. Physik. Chem. 108, 82 (1924).
7. C. A. Eckert, C. K. Hsieh, and J. R. McCabe, A.I.Ch.E. J. 20, 20 (1974).

8. A. J. Parker, Q. Rev. Chem. Soc. 16, 163 (1962).
9. P. Debye, J. Chem. Phys. 1, 13 (1933).
10. K. E. Weale, Chemical Reactions at High Pressure, Spon, London, 1967).
11. S. Winstein, L. G. Savedoff, S. Smith, I. D. R. Stevens, and J. S. Gall, Tetrahedron Letters 9, 24 (1960).
12. R. A. Griegen and C. A. Eckert, A.I.Ch.E. J. 16, 766 (1970).
13. C. A. Eckert, S. P. Sawin, and C. K. Hsieh, "An Improved Technique for High Pressure Kinetics," to be published.
14. V. A. Ershov, V. B. Miller, M. B. Neiman, and M. G. Gonikberg, Izvest. Akad. Nauk S.S.S.R., Otdel Khim. Nauk 2103 (1960).
15. Y. A. Ershov, M. G. Gonikberg, M. B. Neiman, and A. A. Opekunov, Doklady Akad. Nauk. S.S.S.R. 128, 759 (1959).
16. S. D. Hamann, Aust. J. Chem. 28, 693 (1975).
17. R. M. Fuoss and F. Accascina, Electrolytic Conductance, Wiley-Interscience, New York (1959).
18. G. M. Bollinger and G. Jones, J. Amer. Chem. Soc. 53, 411 (1931).
19. S. D. Hamann, Modern Aspects of Electrochemistry 9, 47 (1974).
20. M. J. Wootten, Electrochemistry 3, 20 (1973).
21. L. G. Hepler, J. Phys. Chem. 61, 1426 (1957).
22. D. D. Perrin, W. L. F. Armarigo, and D. R. Perrin, Purification of Laboratory Chemicals, Permagon, London (1966).
23. P. Haberfield, L. Clayman, and J. S. Cooper, J. Amer. Chem. Soc. 91, 787 (1948).

N-3

EFFECT OF PRESSURE ON THE COMPLEX FORMATION OF TRANSITION METAL IONS IN AQUEOUS SOLUTIONS UP TO 2 KBAR

A. Jost

Universität Bielefeld
Bielefeld, W. Germany

INTRODUCTION

The kinetic and thermodynamic behavior of the metal complex formation in aqueous solutions of the divalent metal ions Co^{2+}, Ni^{2+}, Cu^{2+}, and Zn^{2+} with murexide$^-$ * and of Fe(III) with thiocyanate being the new ligands, respectively, were investigated up to 2 kbar at 298.16 K using the temperature-jump relaxation method. The overall reaction scheme in the case of the divalent metal ions is given by

$$M^{2+}_{aq} + L^-_{aq} \underset{k_b}{\overset{k_f}{\rightleftharpoons}} ML^+_{aq} + H_2O \qquad (1)$$

The hydrated metal ion M^{2+}_{aq} reacts with the hydrated ligand L^-_{aq} forming the hydrated complex ML^+_{aq} where one water molecule of the inner coordination sphere of the metal ion is substituted by the new ligand. The overall rate constants of the forward and backward reactions are k_f and k_b, respectively.

As for many metal ion ligand systems the complex formation takes place in two separated steps [2]

$$M^{2+}_{aq} + L^-_{aq} \underset{k_{-o}}{\overset{k_o}{\rightleftharpoons}} (M^{2+}\overline{W}L^-)_{aq} \underset{k_{-i}}{\overset{k_i}{\rightleftharpoons}} ML^+_{aq} + H_2O \qquad (2)$$

In the first, very fast step the hydrated metal ion and ligand form the so-called "outer-sphere" complex where the metal ion and the ligand share one common hydration shell, indicated by $\overline{W}$. k_o and k_{-o}

*Structure of murexide$^-$ [1]: Note that the metal ions bind to the π-electrons of the delocalized system.

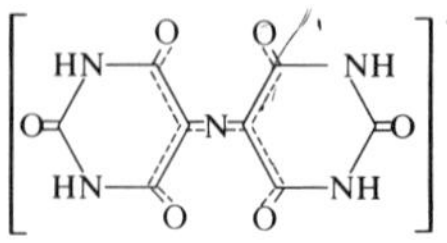

are the rate constants of the recombination and dissociation steps, respectively. In the second rate determining step one water molecule of the inner coordination sphere of the metal ion is replaced by the new ligand forming the so-called "inner-sphere" complex with k_i and k_{-i} being the rate constants of formation and dissociation, respectively.

Analogous to the S_N1 and S_N2 mechanisms of organic chemistry, two different limiting mechanisms for the essential ligand substitution in the inner coordination sphere of a metal ion are shown schematically in Fig. 1. On the left, the outer-sphere complex o-s is shown with an octahedral arrangement of six water molecules around the metal ion and the new ligand. On the right the inner-sphere complex is depicted where one water molecule has been replaced by the new ligand. The two center structures may be regarded as the activated complexes of the inner-sphere substitution steps of S_N1 (top) and S_N2 (bottom) mechanisms, respectively.

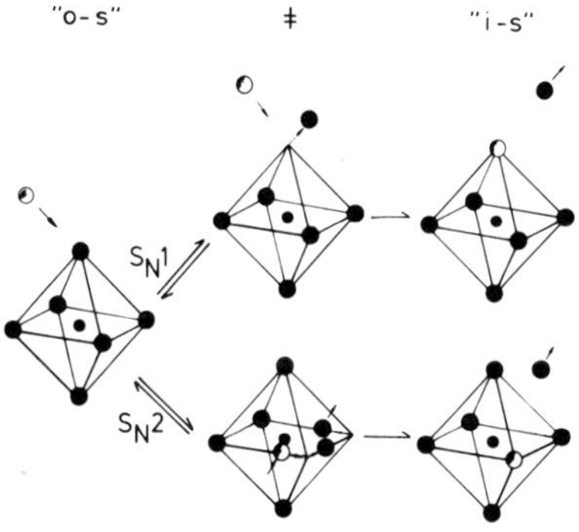

Fig. 1. Schematic showing limiting mechanisms of metal ion complex formation.

In the S_N1 case the water molecule to be exchanged leaves the inner coordination sphere of the metal ion before the new ligand enters. In the S_N2 case the new ligand enters before the water molecule being exchanged leaves the inner coordination sphere. In the S_N1 case the departing water molecule causes a hole in the inner coordination sphere; therefore, the activated complex should occupy more volume than the outer-sphere complex, resulting in a positive activation volume. In the S_N2 case the activated complex requires less volume than the outer-sphere complex, resulting in a negative activation volume. In real systems, probably no pure S_N1 or S_N2 type will be found. In the activated complexes of real systems there will always be a competition between a volume increase due to the stretching of the metal ion water bond and a volume decrease due to the bond formation between the metal ion and the new ligand.

The overall reaction scheme for the Fe(III)/thiocyanate system is given by [3]

$$\begin{array}{ccc} H_2O + Fe^{3+}_{aq} + NCS^-_{aq} & \underset{}{\overset{\mathrm{I}\; k_{12}}{\rightleftharpoons}} & FeNCS^{2+}_{aq} + H_2O \\ \mathrm{II} \updownarrow & & \updownarrow \mathrm{III} \\ H^+ + Fe(OH)^{2+} + NCS^-_{aq} & \underset{\mathrm{IV}}{\overset{k_{34}}{\rightleftharpoons}} & Fe(OH)NCS^+_{aq} + H^+ \end{array} \qquad (3)$$

In this reaction two rate determining steps (I and IV) of the complex formation exist with rate constants k_{12} and k_{34}, respectively, whereas reactions II and III both are rapid equilibrating hydrolytic reactions. The complex formation in reactions I and IV takes place in two separated outer-sphere and inner-sphere complex formation steps analogous to the reaction scheme in (2).

By measuring the pressure dependence of the overall rate constants the overall activation volumes $\Delta V_t^{\neq}$ can be obtained, that is

$$\Delta V_t^{\neq} = -RT \left(\frac{\partial \ln k_t}{\partial p} \right)_T \qquad (4)$$

where T is the absolute temperature and t refers to the total. As can be shown [2], the relation is valid for $k_t = k_i K_o$, with $K_o = k_o/k_{-o}$. Therefore the inner-sphere activation volume is given by

$$\Delta V_i^{\neq} = V^{\neq}_{'i-s'} - V_{'o-s'} = \Delta V_t^{\neq} - \Delta V_o \qquad (5)$$

Values of ΔV_o can be estimated using a theory by Eigen and Fuoss [4-6] permitting the essential inner-sphere activation volumes to be calculated from the experimental values of the pressure dependence of the overall rate constants.

EXPERIMENTAL METHODS

Overall activation volumes and reaction enthalpies were obtained from dynamical measurements using the temperature-jump relaxation method [7] up to 2 kbar at 298.16 K. For all systems the reaction volumes were measured by static methods using a spectrophotometer [6]. Further experimental details are given elsewhere [3,6]. Rate constants and reaction enthalpies were determined from the relaxation times and from the relaxation amplitudes, respectively. The systems Co^{2+}, Ni^{2+}, Cu^{2+}, and Zn^{2+}/murexide$^-$ were investigated at ionic strengths of 0.1 mole kg^{-1} ($NaClO_4$) while the system Fe(III)/thiocyanate was investigated at an ionic strength of 0.2 mole kg^{-1} ($NaClO_4$).

EXPERIMENTAL RESULTS AND DISCUSSION

Figure 2 shows the values of the inner-sphere activation volumes vs. the 3d electrons of the metal ions. Except for reaction IV of the reaction scheme in (3) all inner-sphere activation volumes were found to be positive. Therefore, the mechanisms of the reactions tend to be more of the S_N1 type. For reaction IV the inner-sphere activation volume was found to be $\Delta V^{\neq}_{i34} = -1.2$ cm^3 mole^{-1}. Thus, the mechanism for this reaction tends to be more of the S_N2 type which agrees with results of other authors [8].

Figure 3 shows the pressure dependence of the overall reaction enthalpies for the systems Ni^{2+}/murexide$^-$ and Fe(III)/NCS^-. Both

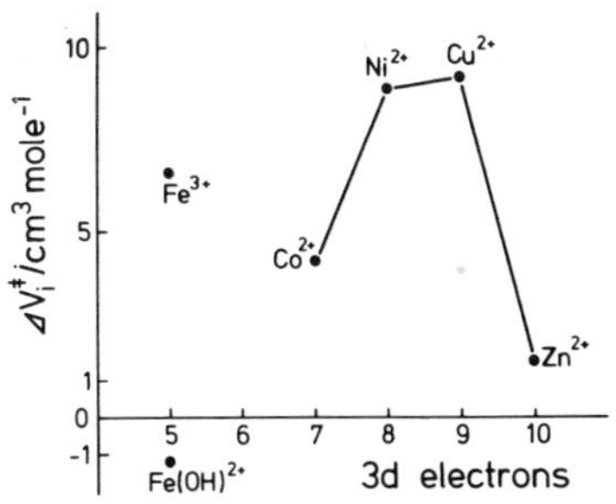

Fig. 2. Plot of $\Delta V_i^{\neq}$ vs. the 3d electrons of Co^{2+}, Ni^{2+}, Cu^{2+}, Zn^{2+}, and Fe(III) at 298.16 K.

systems show a remarkable pressure dependence of the reaction enthalpy which is linear with pressure up to 2 kbar and having similar slopes.

The pressure dependence of the overall reaction enthalpies ΔH_t for several ionic reactions in aqueous solutions is listed in Table I. The first column lists the reactions, the second provides the temperature, the third tabulates the pressure dependence of the reaction enthalpy, and the fourth column shows the temperature dependence of the overall reaction volume ΔV_t, which is correlated with the pressure dependence of the reaction enthalpy by the relation

$$\left(\frac{\partial \Delta H_t}{\partial p}\right)_T = \Delta V_t - T\left(\frac{\partial \Delta V_t}{\partial T}\right)_p \qquad (6)$$

By measuring $(\partial \Delta H_t/\partial p)_T$ and ΔV_t independently, $(\partial \Delta V_t/\partial T)_p$ can be obtained.

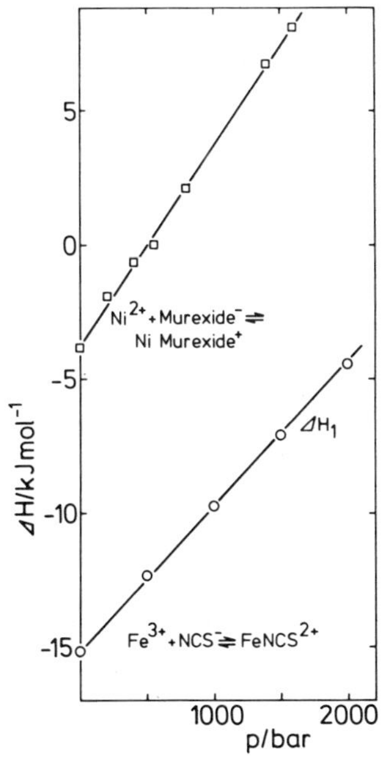

Fig. 3. Pressure dependence of the overall reaction enthalpy ΔH_t for the systems Ni^{2+}/murexide$^-$, and Fe(III)/NCS$^-$ at 298.16 K.

Although the different reactions listed in Table I - as metal ion complex formation, enzyme substrate reactions, ionic reactions like the recombination of water or ammonium hydroxide, and proton transfer reactions - show different pressure dependencies in their reaction enthalpies, they all have similar values of $(\partial \Delta V_t/\partial T)_p$.

All measurements for the reactions listed in Table I were made in the liquid phase with water as the solvent. Thus the temperature dependence of the reaction volumes may result mainly from the solvent. It can be assumed that the temeprature dependence of the reaction volume is a solvent dependent parameter. A similar parameter, the temperature dependence of the activation volume, has been proposed by Laidler [16].

Table I. Pressure Dependence of the Overall Reaction Enthalpy and Temperature Dependence of the Overall Reaction Volume for Different Ionic Reactions

Reaction		T, °C	$(\partial \Delta H_t/\partial p)_T$, J mol^{-1} bar^{-1}	$(\partial \Delta V_t/\partial T)_p$, cm^3 mol^{-1} K^{-1}	Ref.
Ni^{2+} + murexide$^-$		25	7.2	-0.16	6
Fe^{3+} + NCS^-		25	5.4	-0.15	3
proflavine + trypsine		16.5	4.5	-0.09	9
$(Co(NH_3)_6)^{3+}$ + SO_4^{2-}		25	---	-0.12	10
H^+ + OH^-		25	7.1	-0.16	11
NH_4^+ + OH^-		35	6.8	-0.12	12
A^{2-} + H^+	Alizarin$^-$	10	4.7	-0.09	13,15
A^{2-} + H^+	Resorcinol$^-$	10	5.5	-0.12	14,15

ACKNOWLEDGMENTS

Financial support by the Deutsche Forschungsgemeinschaft and by the Fonds der Chemischen Industrie e.V. is gratefully acknowledged.

REFERENCES

1. G. Geier, Ber. Bunsenges. physik. Chem. 69, 617 (1965).
2. M. Eigen and R. G. Wilkins, in Mechanism of Inorganic Reactions, Advan. Chem. Ser. 49, 55 (1965).
3. A. Jost, Ber. Bunsenges. physik. Chem. 80, 316 (1976).
4. M. Eigen, Z. Elektrochem, N.F. 1, 176 (1954).
5. R. M. Fuoss, J. Am. Chem. Soc. 80, 5059 (1958).
6. A. Jost, Ber. Bunsenges. physik. Chem. 79, 850 (1975).
7. A. Jost, Ber. Bunsenges. physik. Chem. 78, 300 (1974).
8. E. P. Cavasino and M. Eigen, Ricerca sci. 4, 3 (1964).
9. H. Vandersypen, Ph.D. Dissertation, University of Leuven (1974).
10. J. Osugi, K. Shimizu, and H. Takizawa, Rev. Phys. Chem. Japan 36, 1 (1966).
11. B. B. Owen and S. R. Brinkley, Chem. Rev. 29, 461 (1941).
12. J. Buchanan and S. D. Hamann, Trans. Faraday Soc. 49, 1425 (1953).
13. K. G. Liphard and A. Jost, Ber. Bunsenges. physik. Chem. 80, 125 (1976).
14. K. G. Liphard and A. Jost, to be published in Ber. Bunsenges. physik. Chem.
15. K. G. Liphard, Ph.D. Dissertation, Ruhr-Universität, Bochum, W. Germany (1977).
16. K. J. Laidler, Disc. Faraday Soc. 22, 88 (1956).

N-4

EFFECTS OF PRESSURE ON THE KINETICS OF STERICALLY HINDERED SQUARE PLANAR PALLADIUM(II) COMPLEXES

D. A. Palmer, M. Cikovic, M. Mares, and H. Kelm

University of Frankfurt
Frankfurt, W. Germany

INTRODUCTION

Generally substitution reactions of square planar complexes of such metal ions as Pt(II), Pd(II), and Au(III) obey a two-term rate law of the form [1]

$$\text{Rate} = k_1\ [\text{complex}] + k_2\ [\text{complex}][Y] \qquad (1)$$

The pseudo first order rate constant k_1 represents the substitution of a ligand X by either a solvent molecule with a subsequent fast replacement of S by Y (reaction scheme (a) or the rate determining dissociation of X from the complex with the formation of a highly reactive, coordinatively unsaturated intermediate (reaction scheme (b) of Fig. 1.) The second order rate constant k_2 reflects a direct substitution of X by Y as depicted in part (a).

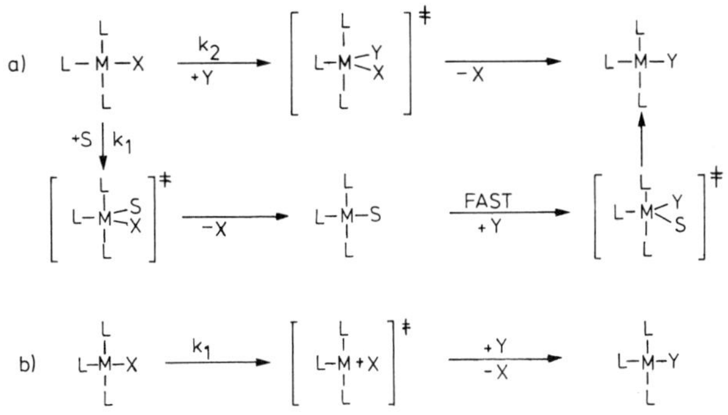

Fig. 1. General scheme of reactions of square planar complexes

In addition to the determination of rate laws, dependencies of rate constants on the departing and entering groups and remaining ligands, temperature and solvent, the variation with pressure provides a valuable parameter for the discussion of the mechanisms involved [2].

$$d \ln k / d P = - \Delta V^{\ddagger}_{exp} / RT \tag{2}$$

In the first approximation, the volume of activation $\Delta V^{\ddagger}_{exp}$ consists of an intrinsic part $\Delta V^{\ddagger}_{intr}$ due to bond making and bond breaking processes, and $\Delta V^{\ddagger}_{solv}$, reflecting the change in solvation during the formation of the transition state. With a reasonable estimation of $\Delta V^{\ddagger}_{solv}$ the sign of the intrinsic contributions can be obtained, thus establishing a criterion for the mechanism of the rate determining step.

SUBSTITUTION REACTIONS OF (DIEN) COMPLEXES

Substitution reactions of bivalent palladium and platinum of the general form

$$M(dien)X^{+} + Y^{-} \longrightarrow M(dien)Y^{+} + X^{-} \tag{3}$$

with dien equal to diethylenetriamine, X and Y representing monovalent anions, were investigated as a function of X and Y. Both types of complexes follow the general rate law and show similar dependencies [3]. An inspection of four typical sets of rate constants, presented in Table I, clearly reveal that the palladium species react much faster. Evidence collected so far from experiments at normal pressure supports the accepted view that both reaction paths are associative in character.

The effect of pressure on a series of reactions of platinum (dien)-complexes (chosen because of their convenient rates) shows

Table I. Rate Constants for the Substitution Reactions of $M(dien)X^{+}$ with the Nucleophiles Y^{-} in Aqueous Solution

M	X	Y	k_1, sec^{-1}	k_2, $l\ mol^{-1}\ sec^{-1}$	μ M	Temp., °C
Pd	Cl^{-}	I^{-}	27	2900	0.05	15
Pd	Br^{-}	I^{-}	19.9	5500	0.05	15
Pt	Cl^{-}	N_3^{-}	0.00011	0.0025	0.20	25
Pt	Br^{-}	N_3^{-}	0.00009	0.0064	0.20	25

[4] that the volume of activation $\Delta V^{\ddagger}_{exp}$ (2) derived from the k_2-path, is negative (see Table II) and is independent of the departing group which strongly supports an associative type mechanism. In view of the fact that charge neutralization during the formation of the transition state should contribute a positive $\Delta V^{\ddagger}_{solv}$ value, a strong negative $\Delta V^{\ddagger}_{intr}$ must prevail which is indicative of a bond making process. However, no definite conclusions can be drawn from the $\Delta V^{\ddagger}_{exp}$ (1) values for the k_1-path, although they are negative and constant within error limits and approach the molar volume of a solvent water molecule, suggesting an associative step.

Table II. Volumes of Activation for the Substitution Processes: $Pt(dien)X^+ + Y^- \longrightarrow Pt(dien)Y^+ + X^-$ (in H_2O) at 25°C)

Substrate $Pt(dien)X^+$	Nucleophile Y^-	$\Delta V^{\ddagger}_{exp}(1)$, $cm^3\ mol^{-1}$	$\Delta V^{\ddagger}_{exp}(2)$, $cm^3\ mol^{-1}$
$Pt(dien)N_3^+$	SCN^-	—	- 12.2 ± 0.1
$Pt(dien)Br^+$	N_3^-	- 15 ± 1	- 8.5 ± 0.2
$Pt(dien)Br^+$	NO_2^-	- 18 ± 1	- 6.4 ± 0.7
$Pt(dien)Cl^+$	N_3^-	- 17 ± 2	- 8.2 ± 1.3
$Pt(dien)I^+$	N_3^-	- 18 ± 2	- 8.2 ± 0.7
$Pt(dien)Br^+$	$OH^-(H_2O)$	- 18 ± 2	—

SUBSTITUTION REACTIONS OF ALKYLATED (DIEN) COMPLEXES

The addition of alkyl substitutents to the (dien) ligand was assumed to effectively block the potential 5th and 6th coordination sites on the metal and thereby forcing a pseudo-octahedral configuration on the metal ion and possibly changing its chemistry [5]. Due to the steric hindrance introduced, predominantly associative pathways are expected to be inhibited whereas dissociative steps should be accelerated.

If one modifies the platinum substrates in the manner mentioned, the relatively slow reactions are in some cases drastically retarded so that experiments become most inefficient. The similar reactivity patterns between the analogous complexes of bivalent platinum and palladium, however, lead to investigation of this effect within the palladium series, the results of which are shown in Table III.

Table III. Effect of (dien) Alkylation on the Rates of Ligand Substitution Reactions of $Pd(R_5dien)X^+$ in Aqueous Solution at 15°C.

Substrate	Nucleophile	k_1, sec^{-1}	k_2, $1\ mol^{-1}\ sec^{-1}$
$Pd(dien)Cl^+$	OH^-	19.1	-
	I^-	27	2900
$Pd(Me_3dien)Br^+$	OH^-	19.9	-
	I^-	29	2900
$Pd(Et_3dien)Cl^+$	OH^-	24	-
	I^-	21	1100
$Pd(Me_5dien)Br^+$*	I^-,OH^-	0.132	<0.1(I^-)
$Pd(Et_4dien)Cl^+$	OH^-	0.0020	0.055
	I^-	0.0019	0.0009
*	$S_2O_3^{2-}$	0.0018	5.6
$Pd(MeEt_4dien)Cl^+$	I^-,OH^-	0.00060	-
*	$S_2O_3^{2-}$	0.00058	0.029
$Pd(Et_5dien)Cl^+$	OH^-	0.00070	-
	$S_2O_3^{2-}$	0.00059	0.019

*Data obtained at 25°C.

Two features emerge from the data: (1) Retardation of the rate of reaction is more effective in the case of ethyl substituents than for methyl and a minimum of four substituents must be in order to influence the rate and (2) the k_1-path is about equally affected, as the k_2-path, which in cases of strong nucleophiles, can be shown to prevail [6].

Using the criterion of steric hindrance, the results obtained support the associative pathway throughout this series of complexes for the formation of a solvento intermediate. Pressure dependencies of the rates can only be studied for the slower reactions with the equipment available. Table IV shows the volumes of activation calculated from the k_1 contribution. The volumes of activation in Table IV are clearly negative, similar to the molar volume of the solvent, and may, in view of their relatively small errors, be interpreted as being modified by the steric effect.

One of the usual indications for a dissociative contribution to a ligand substitution process is the rate dependence on the departing group, a dependence which may also reflect in the corresponding volumes of activation. For the $Pd(Et_4dien)X^+$

Table IV. Effect of (dien) Alkylation on the Volumes of Activation of Ligand Substitution Reactions (in H_2O at 25°C): $Pd(R_5dien)X^+ + Y^- \rightarrow Pd(R_5dien)Y^+ + X^-$

Substrate $Pd(R_5dien)X^+$	Nucleophile Y^-	$\Delta V^{\ddagger}_{exp}$ (1), $cm^3\ mol^{-1}$
$Pd(Et_4dien)Cl^+$	I^-	-14.9 ± 0.2
$Pd(MeEt_4dien)Cl^+$	I^-	-17.5 ± 0.3
$Pd(Et_5dien)Cl^+$	I^-	-18.3 ± 1.0

complex series a variation in X was investigated. The rates, as well as the volumes and free enthalpies of activation, were found to be independent of the nature of the nucleophile used. The results are summarized in Table V.

Table V. Rates and Activation Parameters of the Ligand Substitution Processes: $Pd(Et_4dien)X^+ + Y^- \rightarrow Pd(Et_4dien)Y^+ + X^-$ (in H_2O at 25°C)

Substrate $Pd(Et_4dien)X^+$	10^4 k, s^{-1}	$\Delta V^{\ddagger}_{exp}$ (1), $cm^3\ mol^{-1}$	$\Delta G^{\ddagger}$, $kJ\ mol^{-1}$
$Pd(Et_4dien)Cl^+$	21.6 ± 0.4	-14.9 ± 0.2	86.9 ± 2.0
$Pd(Et_4dien)N_3^+$	3.96 ± 0.04	-13.9 ± 0.5	92.0 ± 2.0
$Pd(Et_4dien)Br^+$	17.4 ± 0.7	-13.3 ± 0.2	86.9 ± 5.8
$Pd(Et_4dien)I^+$	2.65 ± 0.05	-11.5 ± 0.2	91.5 ± 2.5
$Pd(Et_4dien)NCS^+$	1.45 ± 0.03	-10.3 ± 0.2	94.9 ± 3.0
$Pd(Et_4dien)NH_3^{2+}$	0.12 ± 0.01	- 3.0 ± 0.9	101.1 ± 4.6

The relatively constant values of $\Delta G^{\ddagger}$ indicate a similar, if not identical, mechanism for all the reactions studied, while the trend in $\Delta V^{\ddagger}_{exp}$ (1) suggests a certain influence of the departing group.

A linear relationship exists between $\Delta V^{\ddagger}_{exp}$ (1) and $\Delta\bar{V}(1)$, the latter calculated from the measured partial molar volumes of the individual species with a slope about 0.5 indicating the mechanism to be I_a in character. The trend in $\Delta V^{\ddagger}_{exp}$ (1) can be traced to the specific solvation of the potential departing groups in the substrates which progressively affects the contribution of electrostriction more effectively going down the series to the ultimate case of X = NH_3. In that case no change in charge is involved during the reaction and stretching of the $Pd-NH_3$ bond leads only to a positive contribution to $\Delta V^{\ddagger}_{exp}$ (1).

LINKAGE ISOMERIZATION REACTIONS OF ALKYLATED (DIEN) COMPLEXES

Alkylated (dien) complex compounds containing ambidentate ligands like SCN^- undergo linkage isomerization reactions of the type [7]

$$Pd(R_5dien)SCN^+ \rightarrow Pd(R_5dien)NCS^+ \quad (4)$$

Under certain conditions these reactions occur also in the solid state. Therefore intramolecular rearrangements ,as well as intermolecular processes can be discussed. The bulk of the evidence derived from normal pressure exchange studies points towards an intermolecular mechanism similar to that of the substitution reactions, probably involving the same solvento intermediate:

$$Pd(R_5dien)SCN^+ \rightarrow Pd(R_5dien)(H_2O)^{2+} \begin{cases} \xrightarrow{+SCN^-} Pd(R_5dien)NCS^+ \\ \xrightarrow{+Br^-} Pd(R_5dien)Br^+ \end{cases} \quad (5)$$

For two of the substrates the linkage isomerization reactions together with a typical substitution reaction were investigated as a function of pressure. Their rate constants and volumes of activation, $\Delta V^{\ddagger}_{exp}$ (1), are presented in Table VI.

Table VI. Rates and Volumes of Activation of Linkage Isomerization and Substitution Reactions of Alkylated (Dien) Complexes (in H_2O at 30°C)

Reaction	10^4 k, s^{-1}	$\Delta V^{\ddagger}_{exp}$ (1), cm^3 mol^{-1}
$Pd(Et_4dien)SCN^+$	5.48 ± 0.26	-10.1 ± 0.3
$Pd(MeEt_4dien)SCN^+$	4.48 ± 0.15	-10.8 ± 0.3
$Pd(Et_4dien)SCN^+$ + Br^-	5.45 ± 0.25	-10.6 ± 0.4
$Pd(MeEt_4dien)SCN^+$ + Br^-	2.96 ± 0.04	-10.5 ± 0.6

The similar rates of reaction, especially the identical volumes of activation, confirm the existence of a common mechanism. Its character can be described as associative (I_a) in view of the effects discussed in the previous section.

ACKNOWLEDGMENT

Financial support for this research by the Deutsche Forschungsgemeinschaft is greatly appreciated.

REFERENCES

1. C. Langford and H. B. Gray, Ligand Substitution Processes, W. A. Benjamin, Inc., New York (1965).
2. D. R. Stranks, Pure and Appl. Chem. 38, 303 (1974).
3. F. Basolo, H. B. Gray, and R. G. Pearson, J. Am. Chem. Soc. 82, 4200 (1960).
4. D. A. Palmer and H. Kelm, Inorg. Chim. Acta 19, 117 (1976).
5. M. L. Tobe, Inorganic Reaction Mechanism, Nelson, London (1972).
6. D. A. Palmer and H. Kelm, Inorg. Chim. Acta 14, L27 (1975).
7. F. Basolo, W. H. Baddley, and K. J. Weidenbaum, J. Am. Chem. Soc. 88, 1576 (1966).

N-5

REACTIONS OF SMALL COVALENT INORGANIC MOLECULES

A. P. Hagen

The University of Oklahoma
Norman, Oklahoma USA

INTRODUCTION

The use of pressures up to 400 MPa has permitted many organic and inorganic reaction mechanisms to be established with certainty; however, increased pressure is not routinely used in synthetic research even though commercial apparatus is readily available at a modest cost. In our work we have shown that pressures in the region 100 to 400 MPa are important for solving inorganic and organometallic synthetic problems. All of the reactions observed in this study are thermodynamically permitted at 298 K and at 0.1 MPa. The reactions are not observed, however, at 0.1 MPa and at a reasonable temperature. In certain cases when heat alone is used to increase the rate of reaction one of the reactants will undergo thermal decomposition prior to the desired reaction, but when pressure is used to increase the rate of reaction the desired reaction takes place without any side reactions.

EXPERIMENTAL

The samples are normally contained in gold metal ampules. In a typical experiment the nonvolatile reagents are placed into a piece of gold tubing sealed on one end which is then attached to a borosilicate glass vacuum system and evacuated. The volatile reagents are condensed into the tube at 77 K and the tube is then sealed using a H_2-O_2 torch. The ampule is then placed inside of a high-pressure microreactor designed to maintain 400 MPa at 900 K. The ampule is then squeezed by compressing nitrogen gas into the microreactor using a gas pressure booster. At the end of a reaction period the reactor is cooled to 77 K. The pressure is then released and the ampule placed into an opening device attached to the vacuum line. After warming to room temperature condensable materials on the surface of the gold tubing are pumped away and then the tube is opened. The substances which volatilize at room temperature are characterized using standard vacuum line techniques. Nonvolatile material is recovered mechanically from the ampule using an inert atmosphere box when the materials are air sensitive. In addition, the gas pressure booster can be used to intensify directly carbon monoxide.

RESULTS

Sulfur hexafluoride is often regarded as an inert gas even though it has been shown to react at low pressure with $AlCl_3$, SO_3, $LiAlH_4$, and metallic sodium. In our work at pressures up to 400 MPa and at temperatures up to 800 K, SF_6 has been found to react with a variety of materials including water, metal and nonmetal oxides, and organochlorosilanes [1,2]. In each case, sulfur hexafluoride acts as an effective fluorinating agent. For example,

$$SF_6 + 2H_2O \rightarrow SO_2F_2 + 4HF \quad (1)$$

$$SF_6 + (CH_3)_3SiCl \rightarrow SF_5Cl + (CH_3)_3SiF \quad (2)$$

$$SF_6 + 2CS_2 \rightarrow (CF_3)_2S_2 + 3S \quad (3)$$

This study has been expanded to include the reaction chemistry of PF_3 [3-5]. This material readily reacts with sulfides and oxides at elevated pressures, as shown by

$$PF_3 + CO_2 \rightarrow OPF_3 + CO \quad (4)$$

$$PF_3 + SO_2 \rightarrow 2OPF_3 + S \quad (5)$$

$$9PF_3 + MoO_3 \rightarrow 3OPF_3 + Mo(PF_3)_6 \quad (6)$$

$$PF_3 + Se \rightarrow SePF_3 \quad (7)$$

The use of increased pressure to increase the rate of reactions of thermally unstable substances has been demonstrated for the cleavage of the silicon-carbon [6] and silicon-cobalt linkages [7]. These studies are important since increasing the temperature would have destroyed the compounds before the cleavage reaction takes place.

The Si-Co linkages in $CH_3SiCl_2Co(CO)_4$ and $SiCl_3Co(CO)_4$ are not cleaved by HCl or HBr at 360 K (0.5 MPa) which is where thermal decomposition begins. When the pressure is increased to 400 MPa, the bond readily cleaved with very little thermal decomposition of the starting material.

$$SiCl_3Co(CO)_4 + HBr \rightarrow HCo(CO)_4 + SiCl_3Br \quad (8)$$

$$CH_3SiCl_2Co(CO)_4 + HCl \rightarrow HCo(CO)_4 + CH_3SiCl_3 \quad (9)$$

In another study, the silicon-carbon bond of phenylsilanes was cleaved by HCl when the phenylsilane was attached to the $Cr(CO)_3$ grouping and for the uncomplexed ligand. The Si-C linkage readily cleaves at 300 K (0.5 MPa) for $(CH_3)_3SiC_6H_5$ and $(CH_3)_2ClSiC_6H_5$.

$$HCl + (CH_3)_3SiC_6H_5 \rightarrow C_6H_6 + (CH_3)_3SiCl \qquad (10)$$

However, for $(CH_3)_2ClSiC_6H_5$ and $SiCl_3C_6H_5$, much more rigorous conditions are needed. The Si-C bond in $SiCl_3C_6H_5$ cleaves at 425 K (400 MPa), but not at lower conditions. For $(CH_3)_2ClSiC_6H_5$ the cleavage begins at 300 K (400 MPa), but the best yield (50%) was observed at 325 K (400 MPa).

When these compounds are coordinated to the $Cr(CO)_3$ grouping, these cleavages take place under less rigorous conditions, but the high pressure reduces the need for temperatures close to the decomposition point of the complexes.

$$CH_3Cl_2SiC_6H_5Cr(CO)_3 + HCl \rightarrow CH_3SiCl_3 + C_6H_6Cr(CO)_3 \qquad (11)$$

$$SiCl_3C_6H_5Cr(CO)_3 + HCl \rightarrow SiCl_4 + C_6H_6Cr(CO)_3 \qquad (12)$$

The former reaction takes place at 300 K (400 MPa), but 75% cleavage takes place at 325 K (400 MPa). At 325 K, 19% cleavage occurs at 0.5 MPa with a 33% conversion at 100 MPa, and a quantitative conversion at 400 MPa. The latter reaction requires more rigorous conditions, but a 30% conversion occurs at 425 K (400 MPa).

The alkyl-silicon linkage is normally considered as unreactive, even though thermodynamic calculations indicate it should undergo a variety of interesting reactions. In particular, when tetramethylsilane is combined with nonmetal oxides the carbon-silicon linkage is readily cleaved as demonstrated by

$$CO_2 + 4(CH_3)_4Si \rightarrow 2[(CH_3)_3Si]_2O + C + 2C_2H_6 \qquad (13)$$

$$N_2O + 2(CH_3)_4Si \rightarrow [(CH_3)_3Si]_2O + N_2 + C_2H_6 \qquad (14)$$

$$H_2O + 2(CH_3)_4Si \rightarrow [(CH_3)_3Si]_2O + 2CH_4 \qquad (15)$$

These reactions are temperature and pressure dependent. For example, the reaction with carbon dioxide takes place at 560 K (100 MPa) with a 11% yield, but at 475 K (400 MPa) the yield increases to 83%.

The methyl-silicon linkage in methylfluorosilanes, $(CH_3)_xSiF_y$ $(x + y = 4)$, is also cleaved by nitrous oxide at increased pressure. The new siloxane $(CH_3SiF_2)O$ can be synthesized with 81% yield at 190 K (400 MPa) from dimethyldifluorosilane.

$$N_2O + 2(CH_3)_2SiF_2 \rightarrow (CH_3SiF_2)_2O + N_2 + C_2H_6 \qquad (16)$$

Analogous reactions are observed with $(CH_3)_3SiF$ and CH_3SiF_3.

Traditional synthetic methods often are acceptable when starting materials are readily available even though yields are low because of the handling needed for multistep procedure, poor rates, or unfavorable equilibria.

The reaction of carbon monoxide with metal oxides which has been used by other researchers to prepare $Os(CO)_5$, $Tc_2(CO)_{10}$, and $Re_2(CO)_{10}$ has recently been expanded to include the synthesis of $Co_2(CO)_8$, $Fe(CO)_5$, $Mo(CO)_6$, and $W(CO)_6$. While these compounds are commercially available, there is no single step low-loss process for the synthesis of $Fe(CO)_5$ and $Mo(CO)_6$ from isotropically enriched FeO and MoO_3. The reaction with MoO_3 at 500 K (200 MPa) provides an 80% yield of $Mo(CO)_6$

$$9CO + MoO_3 \rightarrow Mo(CO)_6 + 3CO_2 \tag{17}$$

The conversions when cobalt, iron, and tungsten oxides are combined with carbon monoxide are equally high and lower temperatures can be employed. For example, cobalt octacarbonyl is formed at 325 K (300 MPa)

$$11CO + Co_2O_3 \rightarrow Co_2(CO)_8 + 3CO_2 \tag{18}$$

ACKNOWLEDGMENT

This research was supported by National Science Foundation Grant GP-19873.

REFERENCES

1. A. P. Hagen, D. J. Jones, and S. R. Ruttman, J. Inorg. Nucl. Chem. 36, 1217 (1974).
2. A. P. Hagen and B. W. Callaway, Inorg. Chem. 14, 2825 (1975).
3. A. P. Hagen and E. A. Elphingstone, J. Inorg. Nucl. Chem. 35, 3719 (1973).
4. A. P. Hagen and B. W. Callaway, Inorg. Chem. 14, 1622 (1975).
5. A. P. Hagen and E. A. Elphingstone, Inorg. Chem. 12, 478 (1973).
6. A. P. Hagen and H. W. Beck, Inorg. Chem. 15, 1512 (1976).
7. A. P. Hagen, L. McAmis, and M. A. Stewart, J. Organometal. Chem. 66, 127 (1974).

O-1

RADICAL POLYMERIZATION OF N-(4-DIPHENYLAMINO) ACRYLAMIDE UNDER HIGH PRESSURE

Y. Tanaka, K. Noguchi, M. Shiraki, and A. Okada

Research Institute for Polymers and Textiles
Yokohama, Japan

and

M. Ibonai

Kogakuin University
Tokyo, Japan

INTRODUCTION

It is well known [1] that when hindered arylamines are attacked by peroxy radicals they scavenge the radicals to form very stable nitroxy radicals. The stable radicals thus formed further scavenge other radicals. Because of these properties, these compounds are excellent antioxidants for synthetic polymers [2]. Moreover, some unsaturated monomers bearing hindered arylamine groups capable of forming stable radicals have been prepared [3] and their radical polymerizability has been studied [4] with some kinds of radical initiators. The polymerization normally goes only when 2,2'-azobisisobutryonitrile is used, while other initiators such as benzoyl peroxide, cumene hydroperoxide, and tetraethylthiuram disulfide inhibit the polymerization remarkably or completely.

The application of pressures, on the other hand, is well known [5] to provide an additional degree of freedom in controlling the rate and direction of chemical reactions. This paper investigates the initiating ability of these initiators or the polymerizability of an arylamine monomer, N-(4-diphenylamino)acrylamide, under high pressure up to 15 kbar.

EXPERIMENTAL PROCEDURE

Reagents

The monomer N-(4-diphenylamino)acrylamide was prepared from 4-aminodiphenylamine and acryloyl chloride by the method of Braun and Hauge [3] and recrystallized from hexane. The product melted at 152-153°C (lit. [3] mp 155°C). Calculated analysis for $C_{15}H_{14}ON_2$: C, 75.60%; H, 5.92%; N, 11,76%. Experimental analysis: C,75.74%; H, 5.99%; N, 11.54%.

2,2'-Azobisisobutryonitrile (AIBN), benzoyl peroxide (BPO), cumene hydroperoxide (CHP), and tetraethylthiuram disulfide (TETD) used as the initiators were commercially available products and were purified in the standard manner before use. Dioxane, tetrahydrofuran, hexane, and methanol were reagent grade and purified in the usual manner [6] before use.

Polymerization Procedure

Monomer N-(4-Diphenylamino)acrylamide was dissolved in 16 ml of dry tetrahydrofuran followed by the addition of an initiator. The solution was then transfered to a polymerization vessel shown in Fig. 1, degassed and sealed under a stream of nitrogen free of oxygen. The vessel was then placed in a reactor at 60.0 ± 0.5°C for specific intervals under atmospheric or high pressure up to 15 kbar. After the reaction, the mixture was dissolved in a small amount of tetrahydrofuran and poured into anhydrous petroleum ether to remove the unreacted monomer and the initiator. The insoluble material was collected and washed with cooled methanol. A brownish solid was purified by redissolving it twice and reprecipitating it in a large excess of nonsolvent. It was finally dried to a constant weight under vacuum at room temperature.

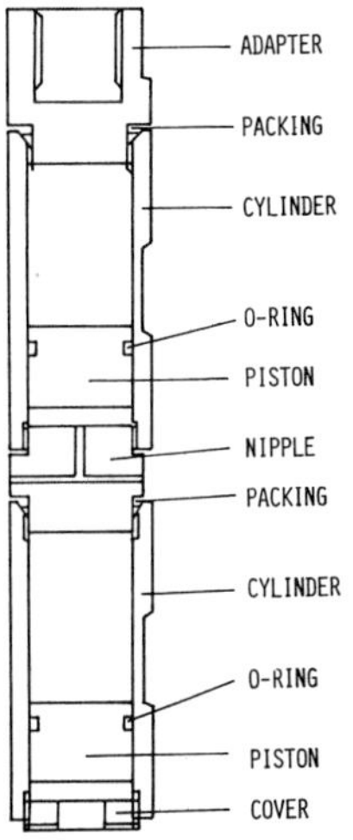

Fig. 1. Internal reaction vessels. Adapter, cylinder, piston, nipple, and cover were made of steel (SUS 27); the packing and O-ring were made of teflon.

The high-pressure apparatus which was used is essentially a device for placing reaction mixtures under hydrostatic pressures up to 15 kbar. It consisted basically of a low-pressure unit, pressure intensifier, reactor vessel, pressure gauge, and auxiliary

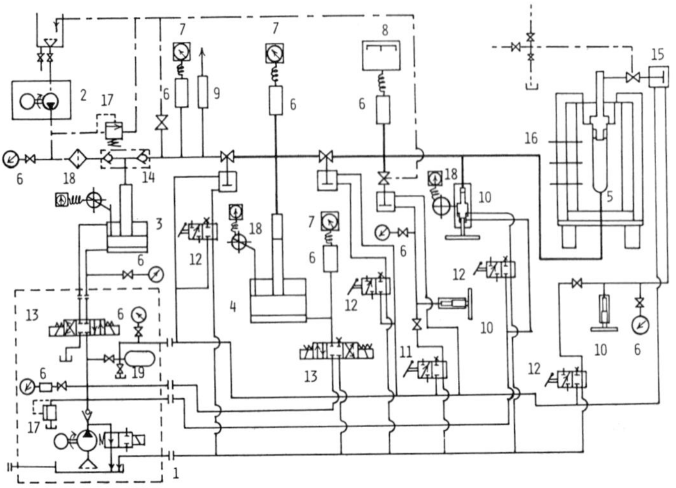

Fig. 2. Schematic diagram of high-pressure equipment. (1) Hydraulic low-pressure unit; (2) supply pump; (3) 1.5 kbar intensifier; (4) 15 kbar intensifier; (5) reaction vessel; (6) pressure gauge; (7) pressure indicator; (8) pressure recorder; (9) rupture cylinder; (10) control valve; (11) release valve; (12) switch valve; (13) directional switch valve; (14) check valve; (15) sampling valve; (16) thermister; (17) relief valve; (18) limit indicator; and (19) accumulator.

equipment. All components were high-pressure fixtures manufactured by Kobe Steel, Ltd. They have been described in detail elsewhere [7]. A schematic diagram is shown in Fig.2.

Test Method

The intrinsic viscosities of the polymers were obtained in tetrahydrofuran with a Ubbelohde type viscometer at 30.00 ± 0.02^{o}C. The points were taken by successive dilutions of the original concentration.

RESULTS AND DISCUSSION

Radical polymerization of N-(4-diphenylamino)acrylamide in tetrahydrofuran under high pressures up to 15 kbar with various initiators such as AIBN, BPO, CHP, and TETD at 60°C for 24 hrs gave blue or brownish polymers. The concentrations of the monomer and an initiator were 2.62 x 10^{-1} and 1.86 x 10^{-3} mole/l, respectively. Results are shown in Table 1. Under pressure higher than 4 kbar, the polymerizations of this monomer occurred normally by using these initiators except TETD which inhibited this polymerization completely up to 13 kbar. The polymerization rates increased uniformly with pressure and reached an equilibrium value at a specific pressure independent of the nature of the initiator except for TETD. The initiators like AIBN and CHP were found to be much more effective and to have larger rates of polymerization than BPO, and the latter was more effective than TETD.

In Fig. 3, the rate of polymerization, R_p in moles/liter/sec

Table I. Effect of Pressure on Polymerization of N-(4-Diphenylamino)acrylamide with Various Initiators*

Pressure, kbar	AIBN		BPO		CHP		TETD	
	Yield, %	[η],† dl/g	Yield, %	[η],† dl/g	Yield, %	[η],† dl/g	Yield %	[η],† dl/g
0	3.78	-	0.00	-	0.00	-	0.00	-
1	9.22	0.160	0.00	-	4.62	-	0.00	-
4	35.6	0.162	20.8	-	39.2	0.163	0.00	-
7	45.5	0.164	41.9	0.163	42.2	-	0.00	-
10	44.2	0.165	42.2	-	43.2	0.165	0.00	-
13	46.2	0.165	43.9	0.162	46.2	-	14.8	0.163
15	47.2	0.166	46.8	0.165	45.8	0.164	29.4	0.165

*0.262 mole/l of monomer was polymerized with 1.86×10^{-3} mole/l of an initiator in tetrahydrofuran at 60°C for 24 hrs. AIBN, 2,2'-azobisisobutryonitrile; BPO, benzoyl peroxide; CHP, cumene hydroperoxide; TETD, tetraethylthiuram disulfide.

†In tetrahydrofuran at 30°C.

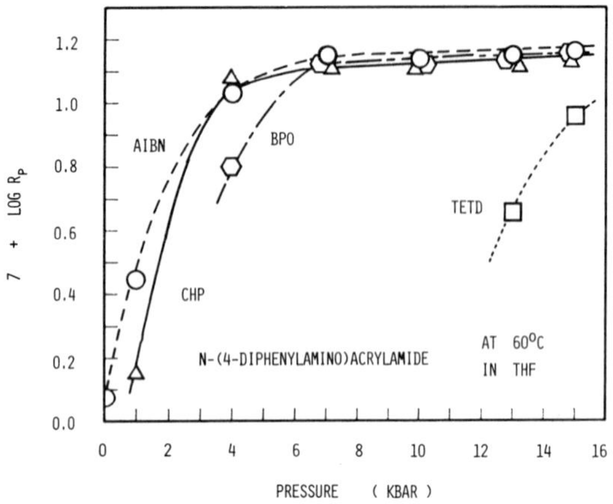

Fig. 3. Pressure dependence of the rate of polymerization, R_p, of N-(4-diphenylamino)acrylamide in tetrahydrofuran at 60°C with various initiators. ○, AIBN; ⬡, BPO; Δ, CHP; □, TETD.

is correlated with pressure. Although the points indicate some curvature in a log R_p vs. pressure plot up to 15 kbar, they give an overall value of the activation volume, ΔV^*. Equation (1) of Van't Hoff [5] or Evans and Polanyi [8] relates the reaction rate constants to pressure, in the form

$$d \ln k/dP = - \Delta V^*/RT \quad (1)$$

Such a curvature in a plot as evident in Fig. 3 might be anticipated since a small transition state could well be less compressible than the larger reactants so that the activation volume becomes a function of pressure. This may indicate a somewhat larger negative value of ΔV^* near atmospheric pressure and the data of R_p, indeed, extrapolate to uncertain but large negative values of ΔV^* at zero pressure (ΔV^* and pressure given in cm^3/mole and kbar): With AIBN, -36, 0; -31, 1; -22, 2; -15, 3; -10, 4; and -6.4, 5. No activation volume has been reported for this monomer but the experimentally determined values can be compared with those [9] for various monomers (monomer and ΔV^* given): styrene, -16 to -18; methyl methacrylate, -17 to -26; allyl acetate, -13; vinyl acetate, -9; acenaphthylene, -6 cm^3/mole.

A simple, chemically-initiated radical polymerization is generally described by the component reactions such as decomposition of the initiator, initiation of the chain growth, propagation of the chain, chain termination, and transfer as shown:

$$I \xrightarrow{k_d} 2\ X^* \quad (2)$$

$$X^* + M \xrightarrow{k_i} P_1^* \quad (3)$$

$$P_m^* + M \xrightarrow{k_p} P_{m+1}^* \quad (4)$$

$$P_m^* + P_n^* \xrightarrow{k_t} P_{m+n}^* \text{ or } P_m^* + P_n^* \quad (5)$$

$$P_m^* + Y \xrightarrow{k_{tr}} P_m + Y^* \quad (6)$$

Equation (2) relates to unimolecular decomposition of the initiator,

I, which produces two radicals, X*, for each molecule of I. M, P_i (i = 1,..m,n), Y and Y* (Y = I, M, P_i) denote monomer, polymer composed of i monomers, any component of the system with which polymer radical, P_i^*, can undergo a chain transfer reaction, and reactive radical of species Y including solvents used.

The rate equations for these reactions, then, are usually written as

$$R_d = 2fk_d[I] \tag{7}$$

$$R_i = k_i[X^*][M] \tag{8}$$

$$R_p = k_p[P_i^*][M] \tag{9}$$

$$R_t = k_t[P_i^*]^n \quad ; \; n = 1 \text{ or } 2 \tag{10}$$

$$R_{tr} = k_{tr}[P_i^*][Y] \tag{11}$$

The rate of chain growth, R_p, is assumed to be independent of the size of the polymer radical, and the kinetic chains terminate in unimolecular (n = 1) or pairs (n = 2), denoted by R_t. The term f is the efficiency of the radicals in initiating chain growth, R_i.

In uncomplicated polymerizations R_i and the rate of chain transfer, R_{tr}, do not affect the overall rate of polymerization. The customary assumption of steady-state conditions, in which the rate of production of radicals is equal to the rate at which they disappear, gives $R_d = R_t$ which then yields an expression for $[P_i^*]$. If the kinetic chains are long, the consumption of monomer occurs almost entirely in chain-growth, so that R_p is equivalent to the overall rate of reaction, -d[M]/dt. Substitution for $[P_i^*]$ in (9) then gives the well-known equation:

$$R_p = -d[M]/dt = k_p(k_d f/k_t)^{1/2}[M]^m[I]^n \tag{12}$$

where m = 1, and n = 1 or 2. If the reaction path remains the same under pressure, the polymerization of this monomer at high pressure should obey (12); the initiating efficiency of an initiator, f, may vary with pressure.

The effect of pressure on the overall rate of polymerization can be formally expressed in terms of an overall volume of activation, ΔV^*, by means of (1). ΔV^* is a composite quantity which is a function of the volumes of activation of the separate rate-determining steps. Equations similar to (1) hold for each of the rate constants and, in conjunction with (12), yield the relation

$$\Delta V^* = \Delta V_p^* + (\Delta V_d^* - \Delta V_t^*)/2 \tag{13}$$

where the subscripts p, d, t, again denote chain propagation, initiator decomposition, and chain termination, respectively.

The volumes of activation for the dissociation of AIBN, BPO, and tert-butyl peroxide are obtained as 3.8 or 9.4 cm^3/mole at 70 or 62.5°C, 4.8 to 9.7 cm^3/mole at 60 to 80°C, and 5.4 to 13.3 cm^3/mole at 120°C in various solvents such as toluene, carbon tetrachloride, cyclohexene, and benzene at 1 atm, respectively [9]. The measurements of k_t at high pressure did not permit an accurate evaluation of ΔV_t^*, but for styrene or methyl methacrylate it appears to be about two or three times as large as ΔV_d^* (monomer and ΔV_t^* in cm^3/mole are given [10]): styrene, 13; vinyl acetate, 16; butyl methacrylate, 18; butyl acrylate, 21; and methyl methacrylate, 25.

From (13), using $\Delta V_d^* = +4$ with AIBN and +9 with BPO, and $\Delta V_t^* = +13$ or +20 cm^3/mole, ΔV_p^* can be calculated to be about -28 or -32 cm^3/mole with AIBN, and -30 or -34 cm^3/mole with BPO, respectively. The mean experimental value in the range 0 to 1 kbar, calculated from accelerating the polymerization at 1 kbar, are -28 to -30 cm^3/mole. The values for various monomers are found [10] to be as (monomer and ΔV_p^* in cm^3/mole are given): styrene, -12 to -18; methyl methacrylate, -19; butyl acrylate, -23; butyl methacrylate, -23; and vinyl acetate, -24. The comparison may be satisfactory in view of the approximations involved, but more precise determinations of the values for the activation volumes ΔV_p^* and ΔV_t^* would be desirable.

The time dependence of the polymerization was investigated with a 2.62 x 10^{-1} mole/l solution of the monomer and the result is shown in Table II. The polymerization was carried out

Table II. Effect of Time on Polymerization of N-(4-Diphenylamino) acrylamide with AIBN and BPO*

Time hrs	AIBN		BPO	
	Yield, %	[η],† dl/g	Yield, %	[η],† dl/g
5	6.93	-	7.20	-
8	22.9	0.161	16.4	0.160
16	35.1	-	33.6	0.161
24	45.5	0.164	41.9	0.163
32	54.7	0.164	53.2	0.164

*0.262 mole/l of monomer polymerized with an initiator (1.86 x 10^{-3} mole/l) in tetrahydrofuran at 60°C under 7 kbar. AIBN, 2,2'-azobisisobutyronitrile; BPO, benzoyl peroxide.
†In tetrahydrofuran at 30°C.

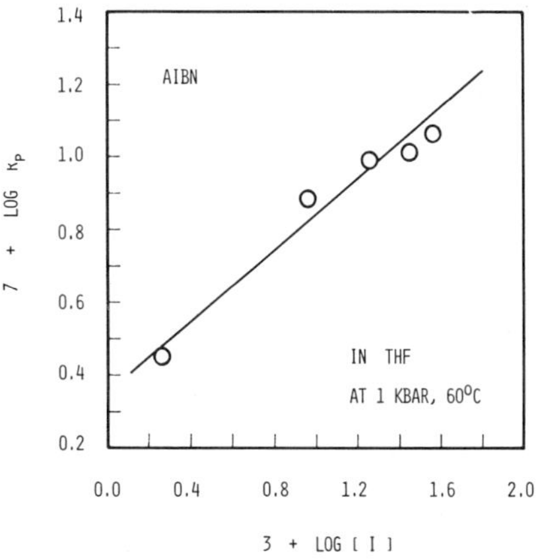

Fig. 4. Effect of AIBN concentration on the rate of polymerization of N-(4-diphenylamino) acrylamide (0.262 mole/l) in tetrahydrofuran at 60°C under 1 kbar.

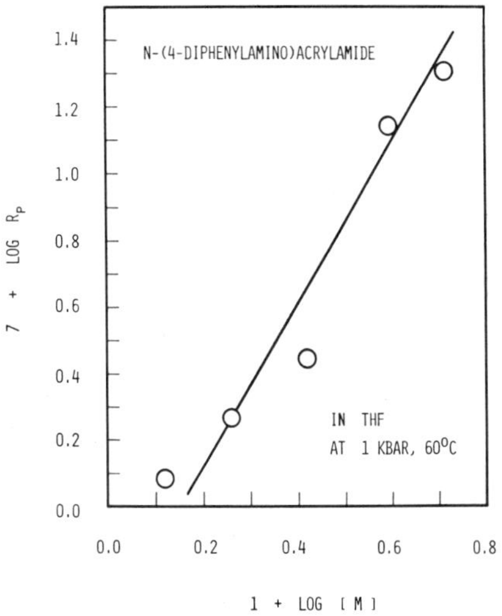

Fig. 5. Effect of N-(4-diphenylamino)acrylamide on the rate of polymerization with AIBN (1.86 x 10^{-3} mole/l) in tetrahydrofuran at 60°C under 1 kbar.

in tetrahydrofuran at 60°C with the use of a solution (1.86 x 10^{-3} mole/l) of AIBN or BPO as the initiator. The rate of polymerization increased uniformly with time and reached a maximum or an equilibrium value at a specific reaction time under atmospheric [4] and a pressure of 7 kbar.

Figure 4 illustrates the effect of initiator concentration on the rate of polymerization. The polymerization of this monomer (2.62 x 10^{-1} mole/l) was carried out with AIBN in tetrahydrofuran at 60°C for 24 hrs under a pressure of 1 kbar. The logarithm of the rate of polymerization is plotted in Fig. 4 against the logarithmic value of the concentration of the initiator. The best straight line was found by the least-squares method and from the slope the order of reaction with respect to the concentration of AIBN is 0.51. The exponent of the initiator concentration remains approximately equal to 0.5 which shows that the kinetic chains are still terminated in pairs under this condition.

The variations of the rate of polymerization at various initial concentrations (0.131 to 0.524 mole/l) of this monomer were investigated, where a 1.86 x 10^{-3} mole/l solution of AIBN was used as the initiator and tetrahydrofuran was used as the solvent. The result obtained under a pressure of 1 kbar at 60°C for 24 hrs is shown in Fig. 5, in which the logarithm of the rate of polymerization is plotted against the logarithmic value of the concentration of the monomer. The fit of the points to a straight line is not very good, but the best straight line was obtained by a least-squares method. From the slope, the order of reaction with respect to the concentration of the monomer is found to be 2.4. To account for this higher exponent of the monomer concentration, a more detailed study of the initiating reaction would be desirable.

The molecular weights of the polymers were not determined, but intrinsic viscosities in tetrahydrofuran were measured at 30°C, and these gave an approximate indication of the pressure effect as shown in Table I. The intrinsic viscosities seem neither to vary with pressure nor to depend on the nature of the initiator. The acceleration of the polymerization can be ascribed to an increase in the rate of chain propagation, and a decrease in the rate of termination of kinetic chains. If there were no other effects, the molecular weight of the polymer would rise with pressure instead of having no variation. The bimolecular transfer reaction between polymer radicals and monomer molecules is probably accelerated nearly as much by pressure as by the chain growth reaction. In this transfer reaction the growing molecular chain is terminated, but a new radical is formed to continue the kinetic chain. Moreover, in the polymerization of this monomer such a chain transfer reaction should be sufficiently large to account for the termination of most of the molecular chains, although the kinetic chains still end by the coupling or disproportionation of pairs of long-chain radicals. No variation in the molecular weight may represent the achievement of an unaltered balance between the rates of chain growth and chain transfer in the range of pressures up to 15 kbar.

In the reaction of oxy and carbon radicals with α-(3,5-ditert-butyl-4-hydroxyphenyl)-N-tert-butylnitrone [11], the former radicals are found to abstract the phenolic hydrogen of the compound to produce the stable radical, whereas the latter add preferentially to the α-carbon of the nitrone to yield a stable nitroxide radical. The results of this polymerization can be explained in the light of the above observation: Under atmospheric pressure, the radicals formed by decompositions of BPO, CHP, and TETD may abstract mainly hydrogen of the amino group of the monomer to inhibit polymerization, while the carbon radicals resulting from AIBN add preferentially to the vinyl double bonds to initiate polymerization, as shown by

$$CH_2{=}CHCONHC_6H_4NHC_6H_5 \begin{cases} \xrightarrow{RO^{\bullet},\,RS^{\bullet}} & CH_2{=}CHCONHC_6H_4\dot{N}C_6H_5 \quad \text{(inhibition)} \\ \xrightarrow{R^{\bullet}} & RCH_2{-}\dot{C}HCONHC_6H_4NHC_6H_5 \quad \text{(polymerization)} \end{cases} \qquad (4)$$

Such selectivity of these radicals, however, would diminish or disappear with increasing pressure. The radicals from BPO and CHO and also the thio radicals resulting from TETD, therefore, tend to add to the vinyl double bonds of this monomer to initiate polymerization under high pressure, although their additivity to the monomer may be very sensitive to an increase in pressure and their initiating efficiencies differ from each other.

ACKNOWLEDGMENT

The authors would like to thank M. Kato and V. M. Zhulin for informative discussions.

NOTATION

k = the specific rate constant

k_d = the rate constant for initiator decomposition

k_p = the rate constant for chain propagation

k_t = the rate constant for chain termination

k_{tr} = the rate constant for chain transfer

$[Z]$ = concentration of reacting species Z (Z = Y or Y*)

P = pressure

R = the gas constant

T = the absolute temperature

t = time

ΔV_d^* = the activation volume for initiator decomposition

ΔV_p^* = the activation volume associated with chain growth

ΔV_t^* = the activation volume associated with chain termination

REFERENCES

1. B. C. Challis and A. R. Buther, in The Chemistry of the Amino Group, S. Patai, ed., Interscience, London (1968), Chapt. 6.
2. J. D. Behun, in Encyclopedia of Polymer Science and Technology, Vol. I, N. M. Bikales, G. O. Schetty, J. Perlman, M. Bickford, J. M. Ricciardi, eds., Interscience, New York (1971), p. 822.
3. D. Braun and S. Hauge, Makromol. Chem. 150, 57 (1971).
4. M. Kato, Y. Takemoto, Y. Nakano, and M. Yamazaki, J. Polym. Sci., Polym. Chem. Ed. 13, 1901 (1975).
5. M. G. Gonikberg, Chemical Equilibria and Reaction Rates at

High Pressures (Moscow 1960), A. Artman, trans., Israel Program for Scientific Translations, Jerusalem (1963).
6. J. A. Riddick and W. B. Bunger, Organic Solvents, Wiley-Interscience, New York (1970).
7. Y. Tanaka, M. Senuma, S. Ebihara, Y. Shimura, A. Okada, and S. Ueda, Bull. Res. Inst. Polym. Text. Japan 96, 21 (1971).
8. M. G. Evans and M. Polanyi, Trans. Faraday Soc. 31, 875 (1935).
9. K. E. Weale, Chemical Reactions at High Pressures E. & F. N. Spon Ltd., London, (1967).
10. M. Yokawa and Y. Ogo, Makromol. Chem. 177, 429 (1976).
11. J. G. Pacifici and H. L. Browning, Jr., J. Am. Chem. Soc. 92, 5231 (1970).

O-2

PRESSURE INDUCED CHANGES IN THE NMR SPECTRA OF A BIOPOLYMER

R. K. Williams, C. A. Fyfe, and D. Bruck

University of Guelph
Guelph, Ontario, Canada

INTRODUCTION

Previous studies of the effect of pressure on the helix and random coil forms of water-soluble polyamino acids by means of specific conductivity, and absorption spectroscopy in the presence of dyestuffs have been reported by K. Suzuki and Y. Taniguchi [1,2]. These authors reported the volume change for helix-to-coil transition to be negative for poly-D-glutamic acid alone and positive for this polymer in the presence of acridine orange. In order to confirm that the helix-to-coil transition does in fact correspond to a negative volume change and that increasing pressure does favor the coil as expected by these authors, the present work has been undertaken. The polymer chosen for study, poly-N^5-(3-hydroxypropyl)-L-glutamine, PHPG, has no ionizable group in its structure and therefore the helix-coil equilibrium for this polymer is wholly dependent on non-electrostatic forces.

EXPERIMENTAL PROCEDURES AND TECHNIQUES

Poly-γ-benzyl-L-glutamate with a degree of polymerization of 685, and molecular weight 150,000 was obtained from Sigma Chemicals. 3-amino-1-propanol obtained from J. T. Baker Chemical Company was dried over barium oxide and redistilled at atmospheric pressure. 1-4-Dioxan, supplied by J. T. Baker Chemical Company, was purified over sodium hydroxide, filtered, dried over sodium, and redistilled at atmospheric pressure. Deutrium oxide, 99.75% D, was used as supplied. Sodium-2,2-dimethyl-2-silapentane-5-sulphonate (D.S.S.) was used as supplied by Stohler Isotopes.

Poly-N^5-(3-hydroxypropyl)-L-glutamine with a degree of polymerization of 685 was synthesized according to the method described by N. Lupu-Lotan et al. [3] from poly-γ-benzyl-L-glutamate and

3-amino-1-propanol in 1-4-dioxan as solvent.

High-pressure glass capillaries with an ID of about 1 mm and an OD about 4 to 4.5 mm were prepared according to the method of H. Yamada [4]. These were connected to a conventional high-pressure pump supplied by the American Instrument Company.

Temperature of the samples was adjusted to within $\pm$ 0.2°C by means of a standard Varian Control unit. Nitrogen gas, precooled by liquid nitrogen and heated by the temperature regulator, was blown over the glass capillary to thermostat the latter. It was found that the NMR spectra observed did not change appreciably: with time after one minute of equilibration, when the temperature was changed by 10°C, or when the pressure was adjusted at any time. Therefore, spectra were observed after five minutes equilibration.

Solutions of the polymer were made by dissolving a 10% (w/v) proportion of the freeze-dried polymer in a 10:1 (v/v) mixture of deuterium oxide and water. For comparison of spectra an internal standard of 1% (w/v) sodium-2,2-dimethyl-2-silapentane-5-sulphonate (D.S.S.) was added to the solution. Spectra are shown with a scale referred to the D.S.S. proton resonance peak.

The numbering convention used by Joubert et al. [5] was used in Fig. 1. The assignments of proton resonance peaks used by these authors was also adopted as is shown in the NMR spectra in Fig. 2. Spectra were recorded on a Varian H.A. 100 spectrometer.

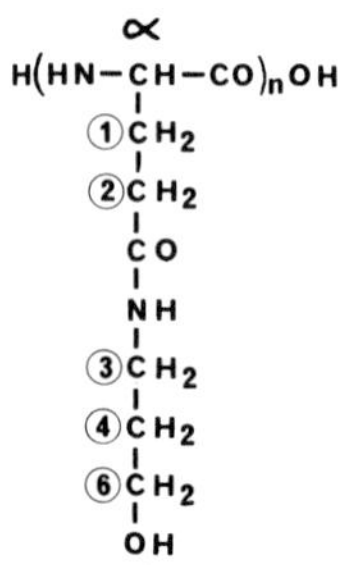

Fig. 1. Structure of poly-N^5-(3-hydroxypropyl)-L-glutamine.

The spectra obtained were limited by the following factors. The capillaries were of small inner diameter, about 1 mm, resulting in a cross section area 25 times smaller than that of a conventional NMR sample tube. It was not possible to spin the sample tube which was attached to the pressure tubing and the pressure pump. These two factors resulted in excessive noise in the spectra obtained and obscured the fine detail of proton resonance splitting. In addition, it was not possible to observe the α (CH) proton resonance peak due to its proximity to the water band which served as the lock signal and which was relatively large.

RESULTS

Figure 2 shows the NMR spectra obtained for the polymer in a mixed deuterium-oxide and water solution at 2°C and at pressures of approximately 1000 and 2000 kg/m^2. For comparison, Fig. 3 shows the spectrum obtained at 10°C and a pressure of 1 atm. By comparison of these spectra it is seen that the effect of an increase in pressure of 2000 kg/cm^2 is roughly equivalent to an increase in temperature from 2 to 10°C, at 1 atm. Figure 4 depicts the spectrum of

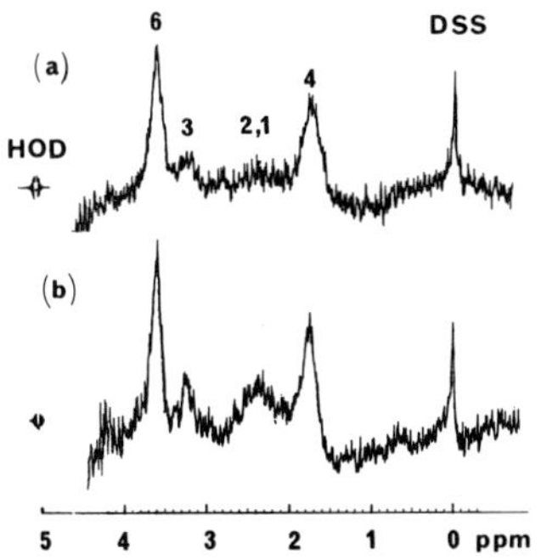

Fig. 2. Nuclear magnetic resonance spectra of the polymer, PHPG, D.P. 685, at 10% (w/v) concentration in D_2O, H_2O (10:1, v/v) solution at 2°C and pressures of (a) 1.03 kg/cm^2 (b) 1968.5 kg/cm^2. [D.S.S.] = 1% (w/v).

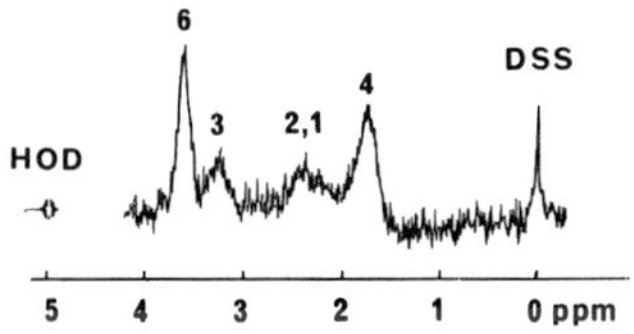

Fig. 3. Nuclear magnetic resonance spectra of the polymer, PHPG, D.P. 685, at 10% (w/v) concentration in D_2O, H_2O (10:1, v/v) solution at 10°C and a pressure of 1.03 kg/cm^2. [D.S.S.] = 1% (w/v).

the polymer obtained at 2° C and 1 atm after releasing the pressure. By comparison with Fig. 2 it can be seen that the original spectrum at 1 atm pressure is recovered, and that the effect of pressure on the polymer is reversible.

In Fig. 2 it is also apparent that the proton resonance peaks corresponding to the side-chain CH_2 residues closest to the α helix (numbers 1, 2, and 3) increase in sharpness and visibility relative to the peaks corresponding to the residues farthest from the helix (numbers 4 and 6) and also to the D.S.S. proton resonance peak.

DISCUSSION

The results obtained herein appear to indicate that the effect of pressure on the polymer is to cause an increased mobility of the CH_2 groups of the side chain which are closest to the α helix. This is similar to the effect of increased temperature on the polymer, and since the latter effect is known to be accompanied by a decrease in percent helix [5], it is reasonable to conclude that an increase in pressure also causes an unfolding of the helix with resulting increase in freedom of movement of the side-chain CH_2 residues. That the proton resonance peaks are not all equally affected indicates that the effect of pressure on the NMR spectra is not simply due to its effect on the rotational diffusion of the polymer molecule as a whole, since in that case all the proton resonance peaks corresponding to the side-chain CH_2 groups would be equally affected. This is clearly not the case.

If the above interpretation is correct, then it follows that the transformation of helix to coil is enhanced by application of pressure, and so, according to Le Chatelier's principle, the volume change corresponding to that change must be negative. Our result therefore confirms the expectation that the unfolding of the helix of a polyamino acid with hydrophobic side chains in aqueous solution should be accompanied by a decrease in volume [6].

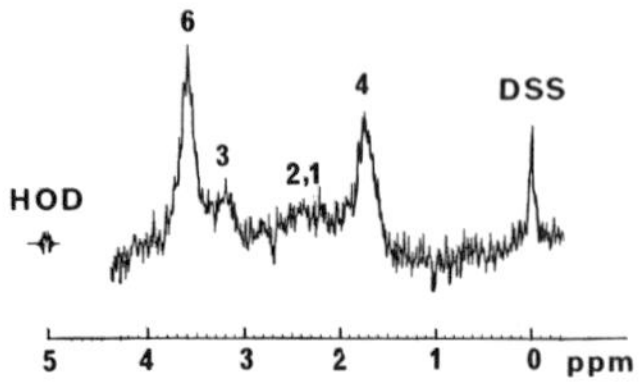

Fig. 4. Nuclear magnetic resonance spectra of the polymer, PHPG, D.P. 685, at 10% (w/v) concentration in D_2O, H_2O (10:1, v/v) solution at 2°C and a pressure of 1.03 kg/cm^2. [D.S.S.] = 1% (w/v). The sample was previously compressed to a pressure of 1968.5 kg/cm^2.

SUMMARY

Measurement of the high resolution NMR spectra of the polyamino acid poly-N^5-(3-hydroxypropyl)-L-glutamine in mixed deuterium oxide and water solvent has been made at pressures from 1.03 to 1968.5 kg/cm^2 at 2°C. The results show an increased mobility of the side chain at higher pressures, and this is interpreted to imply the occurrence of a volume decrease on unfolding of the helix of the polymer to a random coil.

ACKNOWLEDGMENTS

The authors are indebted to L. Van Veen for obtaining the NMR spectra shown herein. This work was supported by funds made available by the Research Corporation, New York, and by the National Research Council of Canada.

REFERENCES

1. K. Suzuki and Y. Taniguchi, Bull. Chem. Soc. Japan 40, 1004 (1967).
2. K. Suzuki and Y. Taniguchi, Biopolymers 6, 215 (1968).
3. N. Lupu-Lotan, A. Yaron, A. Berger, and M. Sela, Biopolymers 3, 625 (1965).
4. H. Yamada, Rev. Sci. Instr. 45, 840 (1974).
5. N. Joubert, N. Lotan and H. A. Scheraga, Biochemistry 9, 2197 (1970).
6. H. L. Kliman, Ph.D. Dissertation, Princeton University, Princeton, New Jersey (1969).

P-1

HIGH-PRESSURE BIOCHEMISTRY: A SURVEY*

K. Heremans

Katholieke Universiteit te Leuven
Leuven, Belgium

INTRODUCTION

Although the use of high pressure in biological systems has a long history, the literature is not voluminous. The older literature is mainly concerned with the irreversible destruction of biological systems, especially proteins. In recent years it has become clear that high pressure as well as low pressure are valuable tools for studying biochemical systems.

Three domains can be discerned: (1) The use of low pressures (max. 400 atm) to perturb chemical equilibria in pressure-jump relaxation techniques. This approach has several advantages over temperature-jump techniques [1]. (2) A second field is a more natural one: deep sea biology. A good review is available [2]. Several examples have been reported of enzymes adapted to the high-pressure and colder temperature environment of the deep sea. (3) The use of high pressure to obtain volume changes and activation volumes for biological reactions. We shall be concerned with this latter domain.

PHASE TRANSITIONS IN PROTEINS, NUCLEIC ACIDS, AND MEMBRANES

Proteins

Since the review of Kauzmann [3] several papers have appeared dealing with pressure effects on the reversible denaturation of proteins. Hawley and Mitchell [4] studied the kinetics of chymotrypsinogen denaturation at pH 2 and room temperature and found evidence for a two-state model. Their results however, according to the authors "do not at all rule out the existence of low levels of intermediate states."

*Invited paper.

Li et al. [5], on the other hand, in their experiments with chymotrypsinogen and lysozyme "clearly disprove the two-state hypothesis." Observing the protein fluorescence they found a first domain of the protein which denatures below 8 kbar. With ANS binding in the case of chymotrypsinogen, and protein fluorescence in the case of lysozyme they found a second independent domain between 8 and 11 kbar. The second domain in lysozyme shifts to lower pressures in the presence of substrate. These studies therefore reveal a plurality of pressure-denatured forms in both proteins. The same authors [7] found only one domain up to 10 kbar for the riboflavin-binding protein of egg white.

Kauzmann [6] recently proposed that the primary effect of pressure on proteins with an open crevice structure such as metmyoglobin, is to shift the equilibrium in favor of the closed crevice. It is easily conceivable that a similar process occurs with enzymes in general. It is known that enzyme denaturation by pressure is retarded by substrate binding.

Hydrophobic Interactions

An interesting aspect of these studies is the magnitude of ΔV for these transitions. They are small compared to the total volumes of the proteins. It is difficult to explain this if one accepts the picture of native proteins being chiefly stabilized by hydrophobic interactions [3]. On the basis of model systems one expects much larger volume changes.

Li et al. [7] have pointed out the possible significance of the methods used to study denaturation. These are ultraviolet, visible and fluorescence spectroscopy and could thus reflect local changes in the protein.

We used another approach while studying volume changes between oxidized and reduced cytochrome C which undergoes changes in conformation as revealed by x-ray studies. Relaxation studies of the redox process under pressure [8] show a 4 ml volume change at pH 7 between oxidized and reduced cytochrome C. Thus again a small change in volume is observed. It would be desirable to validate the conclusion from these studies by other techniques which would consider the protein as a whole and not just one spot on the surface or in the interior.

Another way to explain the discrepancies is to have a closer look at model systems. In this respect it seems justified to make a distinction between aliphatic and aromatic compounds. Several results from the literature support this statement. We must also point out that there is no physical picture to explain the different behavior of these compounds in water. A general conclusion would then be that small volume changes in protein denaturation can simply be explained by compensating effects between aliphatic and aromatic groups [7].

Nucleic Acids

In contrast to proteins, nucleic acids are stable under pressure below the helix-coil transition temperature. If one examines the effect of pressure on the helix-coil transition it is found that the helix is stabilized under pressure. Hawley and Macleod [9] have measured the influence of salt on the pressure dependence of the melting temperature and found that the transition volume increases with the salt concentration. Extrapolation shows that the transition occurs without a volume change at 59°C. All these studies were performed by monitoring the transition in the ultraviolet.

Recently we monitored the helix-coil transition with the intercalating dye proflavin. The melting point then decreased by about 10^{-2} deg/atm. This result can be explained on the basis of the different pressure sensitivity of the dye binding to the helix and to the coil.

There is an important difference which should be kept in mind while studying proteins and nucleic acids. Proteins are globular polyelectrolytes while nucleic acids are rodlike. This entails the condensation of counter ions in nucleic acids [10]. Pressure studies can clarify some important questions in this field. It is sufficient to mention that specific interactions are not considered in polyelectrolyte theories.

Membranes

Synthetic phospholipid vesicles have proved to be very good model systems for the study of biological membranes. With coworkers we have studied the effect of pressure on the transition temperatures, as observed with light scattering. The results are summarized in Table I. For comparison, the effect of pressure on the melting of hydrocarbons is included. The dT/dP values are positive and large (10^{-2} deg/atm) compared to the values obtained for proteins and nucleic acids (see Table II). The addition of drugs, proteins, and salts which in some cases shift the transition temperature considerably, have only small effects on dT/dP. Also dT/dP is pressure independent up to 3500 atm as predicted by the theory [12]. We have also found recently with quasi elastic light scattering that there is no change in the size of the vesicles ongoing through the transition even under pressure (500 atm) [13].

These findings have several interesting applications. The effects of anaesthetics upon phase transitions in lipids is counteracted by hydrostatic pressure [14]. A practical consequence is the separation of subcellular particles in the ultracentrifuge. The experiments of Wattiaux and coworkers [15] on the combined effects of drugs, temperature, and pressure on the intactness of mitochondrial particles in the ultracentrifuge, can be explained by the same principles.

Table I. Effect of Pressure on Phase Transitions in Phospholipid Vesicles

Phospholipid	T_t at 1 atm	dT_t/dP*	ΔT_t
Dilauroyl-lecithine	0.5	17	-
Dimyristoyl-lecithine (DML)	24	20.5	-
Dipalmitoyl-lecithine	41.5	21.8	-
Dilauroyl-phosphatidylethanolamine	31	21.5	-
DML + Na tetraphenylborate	14.9	20.5	-9.1
DML + chlorpromazine	21	21	-3
DML + cholesterol	23	20	-1
DML + cetyltrimethylammonium-bromide	25	20	+1
DML + $UO_2(NO_3)_2$	25.5	20	+1.5
n-C_{18} alkanes [11]	28	25.7	-
DML + 10% phosphatidylserine (PS)	27.1	22	-
DML + polylysine	27	22.9	-
DML + PS + cytochrome C	28.4	21	+1
DML + PS + gramicidine	24.4	21	-3.6

* K/1000 atm.

Table II. Pressure Effects on Transition in Proteins, Nucleic Acids, and Membranes

	dT/dP, deg/1000 atm	
Proteins: ribonuclease	2.2	Brands et al. 1970
chimotrypsinogen	1.7	Hawley 1971
metmyoglobin	6.5	Zipp and Kauzmann [16]
Poly-benzyl-L-glutamate	5.6	Gill and Glogovsky 1965
Helix-coil DNA	0.3-4	Hawley and Macleod [9]
Phospholipid vesicles	21	Desmedt and Heremans 1977

Kinetics of Phase Transitions

The detailed studies by Hawley [4] on the denaturation of chymotrypsinogen which support the two-state model have already been mentioned. Zipp and Kauzmann [16] have also found first-order kinetics under certain conditions for metmyoglobin. From a review by Baldwin [17] on protein folding it is evident that more complicated kinetics can be observed.

With our high-pressure stopped-flow apparatus we have recently started a study on the formation of the Poly A-Poly U double helix starting from the single strands [18]. In the presence of dyes as optical probes, we find high activation volumes. An analysis of the data shows that $\Delta V^{\#}$ equals 4 ml. Assuming a mechanism as proposed by Crothers et al. [19] we find that ΔV for the formation of the first base pair is about +2 ml/mole base pair formed.

No pressure studies have been made on the kinetics of phase transitions in phospholipids. A study made in our laboratory [20] on the formation of dodecylpyridiniumiodide micelles, an analogous process, revealed no pressure effects on the relaxation times. Neither was there any pressure effect on the critical micelle concentration.

ACTIVATION VOLUMES AND ENZYME MECHANISM

The most simple scheme to represent data on enzyme catalysis was proposed by Michael-Menten.

$$E + S \underset{}{\overset{I}{\rightleftharpoons}} ES \overset{II}{\longrightarrow} E + P \qquad (1)$$

(I) refers to the specific binding of the substrate and (II) to the conversion of substrate to products.

The effect of pressure on the kinetics in (I) has been studied by high-pressure relaxation techniques [21]. The study of (II) may be complicated by processes pertinent to the enzyme itself. Denaturation or dissociation into subunits may occur. Only in some cases is it possible to relate the $\Delta V^{\#}$ obtained to the detailed molecular mechanism of catalysis [22]. In a more general approach, Low and Somero [23] have considered contributions from "hydration density" effects of protein groups and "structural" contributions of the protein itself to $\Delta V^{\#}$.

Pressure effects on the allosteric interactions in nitrogenase have been studied. We were able to show that the cooperative behavior vs. ATP resides in the binding of this substrate to the enzyme. This reflects interactions between the subunits. The Hill coefficient is lowered at higher pressure [24]. It is clear that the pressure effect on allosteric behavior has important consequences for deep sea biology.

With the same enzyme we can illustrate another aspect of the possibilities of high-pressure studies. Nitrogenase shows a break in the Arrhenius activation curve for the reduction of substrate. If we look to the pressure dependence of the temperature at which the break occurs one finds $dT/dP = +2.10^{-2}$ deg/atm. This corresponds exactly to the pressure dependence of phase transitions in phospholipids (see Table I). Together with the chemical evidence described in more detail elsewhere [25], we consider this as a strong indication that the break in nitrogenase is a consequence of a phase change in lipids which affect the catalytic activity. If one assumes a change in conformation in one of the proteins, one would expect a much lower value for dT/dP. These findings also explain the observation that the activation volume increases above 250 atm. Pequeux has observed similar phenomena in a different context [26].

CURRENT RESEARCH TOPICS IN MOLECULAR BIOLOGY

Even the most simple schematic model for a living cell, shows many processes all of which could be a possible site of action for pressure. Since the primary site of action for pressure on enzyme synthesis has been found to be the ribosomes [27], pressure studies have concentrated on the subunit interaction in these particles [28,29]. An important but unsettled point is the action of pressure on polysomes, the biological active association of ribosomes with mRNA. In a recent study we found that although polysomes seem equally sensitive to the action of pressure they dissociate much slower than ribosomes, and that mRNA seems to act as a cement between the ribosomes subunits.

Another topic which is very important to consider because of its practical implications are the pressure effects in the ultracentrifuge [30]. From an irritating complexity, pressure effects have become a positive advantage in the study of protein assembly. It should however be mentioned that in some cases volume changes have gone undetected in the ultracentrifuge, while they could be conveniently studied under pressure with light scattering [31].

A final point to be considered is the study of pressure effects on protein assembly. In a very detailed study of the assembly of microtubuli by Engelborghs et al., we showed just how important temperature effects may be in high-pressure studies [32]. While other workers classified microtubuli according to their origin and pressure sensitivity, we showed clearly that temperature has dramatic effects on the pressure dissociation. A more quantitative study revealed that pressure and temperature act primarily on the nucleation process. It is clear that these observations have far-reaching physiological consequences.

ACKNOWLEDGMENTS

It is a pleasure to thank all of the students who contributed to the results presented here. The author also thanks L. De Maeyer and his colleagues for stimulating discussions. The work presented in this paper from our laboratory is supported by the Belgian National Science Foundation (N.F.W.O.).

REFERENCES

1. J. S. Davis and H. Gutfreund, FEBS Letters 72, 199 (1976).
2. A. G. Macdonald, Physiological Aspects of Deep Sea Biology, Cambridge University Press, Oxford, England (1975).
3. W. Kauzmann, in Proc. 4th Intern. Conf. High Pressure, Kyoto, Japan (1974), p. 619.
4. S. A. Hawley and R. M. Mitchell, Biochem. 14, 3257 (1975).
5. T. M. Li, J. W. Hook, H. G. Drickamer, and G. Weber, Biochem. 15, 5571 (1976).

6. G. B. Ogunmola, A. Zipp, F. Chen, and W. Kauzmann, Proc. Natl. Acad. Sci. (USA) 74, 1 (1977).
7. T. M. Li, J. W. Hook, H. G. Drickamer, and G. Weber, Biochem. 15, 3205 (1976).
8. H. A. Vandersypen and K. Heremans, Arch. Int. Physiol. Biochim. 82, 792 (1974).
9. S. A. Hawley and R. M. Macleod, Biopolymers 13, 1417 (1974).
10. G. S. Manning, Ann. Rev. Phys. Chem. 23, 117 (1972).
11. A. Würflinger and G. M. Schneider, Ber. Bunsen-Gesellschaft 77, 121 (1972).
12. J. F. Nagle, J. Chem. Phys. 58, 252 (1973).
13. F. Ceuterick, P. Nieuwenhuysen, H. Desmedt, K. Heremans, and J. Clauwaert, to be published.
14. J. R. Trudell, D. G. Payan, J. H. Chin, and E. N. Cohen, Proc. Natl. Acad. Sci. (USA) 72, 210 (1975).
15. S. Wattiaux-De Coninck, F. Dubois, and R. Wattiaux, Biochem. Biophys. Acta, in press.
16. A. Zipp and W. Kauzmann, Biochem. 12, 4217 (1973).
17. R. L. Baldwin, Ann. Rev. Biochem. 44, 453 (1975).
18. J. Wouters, J. Snauwaert and K. Heremans, to be published.
19. D. M. Crothers, N. Davidson, and N. R. Kallenbach, J. Am. Chem. Soc. 90, 3560 (1968).
20. G. De Rycke, A. Persoons, and L. De Maeyer, to be published.
21. K. Heremans, J. Snauwaert, H. Vandersypen, and Y. Van Nuland, in Proc. 4th Intern. Conf. High Pressure, Kyoto, Japan (1974), p. 623.
22. R. C. Neumann, D. Owen, and G. D. Lokcyer, J. Am. Chem. Soc. 98, 2982 (1976).
23. P. S. Low and G. N. Somero, Proc. Natl. Acad. Sci. (USA) 72, 3014 (1975).
24. J. Peeters, A. Van Rossen, and K. Heremans, Arch. Intl. Physiol. Biochim. 83, 200 (1975).
25. F. Ceuterick, J. Peeters, H. Desmedt, and K. Heremans, to be published.
26. A. Pequeux, J. Exp. Biol. 64, 587 (1976).
27. W. Smith, D. Pope, and J. V. Landau, J. Bacteriol. 124, 582 (1975).
28. E. Schulz, H. D. Lüdemann, and R. Jaenicke, FEBS Letters 64, 40 (1976).
29. P. Nieuwenhuysen, J. Clauwaert, and K. Heremans, Arch. Int. Physiol. Biochim. 83, 983 (1976).
30. W. F. Harrington, Fractions (Beckman) 10 (1975).
31. K. Heremans, in Proc. 4th Intern. Conf. High Pressure, Kyoto, Japan (1974), p. 627.
32. Y. Engelborghs, K. Heremans, L. De Maeyer, and J. Hoebeke, Nature 259, 686 (1976).

P-2

THE EFFECT OF PRESSURE ON THE KINETICS OF SODIUM LAURYL SULFATE MICELLE FORMATION

R. S. Seright and R. A. Grieger-Block

University of Wisconsin
Madison, Wisconsin USA

and

D. Thusius

Laboratorie d'Enzymologie
Physico-chimique et Molèculaire
Orsay, France

INTRODUCTION

In the past decade rapid advances have been made toward the elucidation of the kinetics of micelle formation [1,2]. The micellar processes which occur are relatively fast (half-lives less than one second) and must therefore be studied using special techniques, such as temperature-jump, pressure-jump or ultrasonic absorption methods. In the time range extending from $1/10^9$ seconds to 1 second, two or three relaxations have been observed. Several processes have been postulated to account for the relaxations. For an anionic detergent these include: (1) Association/dissociation of a single detergent molecule to/from a micelle:

$$A_n^z \rightleftarrows A_{n-1}^{z+1} + A^- \qquad (1)$$

(2) Counterion association/dissociation to/from the micelle:

$$A_n^z \rightleftarrows A_n^{z-1} + C^+ \qquad (2)$$

(3) Micellization/dissolution:

$$nA \rightleftarrows A_n \qquad (3)$$

(4) Micelle aggregation:

$$A_n + A_m \rightleftharpoons A_{n+m} \tag{4}$$

(5) Change of micelle shape:

$$A_n \rightleftharpoons A_n^* \tag{5}$$

For pure surfactants the three relaxations can only be detected at concentrations above the critical micelle concentration (cmc). One of the three relaxations has been observed only at high surfactant concentrations. Graber et al. [3] attributed this relaxation (τ_0) to changes of micelle shape (Eq. (5)).

In the microsecond time range a relaxation (τ_1) has been detected by ultrasonic absorption [3,4] and shock tube methods [5]. Yasunaga et al. [4] suggested that this relaxation was due to counterion binding (Eq. (2)). However, Graber et al. [3] postulated that a monomer exchange process was responsible for τ_1 (Eq. (1)).

In the millisecond time range a relaxation (τ_2) has been detected by temperature-jump, pressure-jump and stopped-flow methods [5,6]. This relaxation has been attributed to micelle aggregation (Eq. (4)) by Colen [8] and to a stepwise micellization/dissolution process (Eq. (3)) by Hoffman et al. [1].

The effects of concentration, temperature, ionic strength, type of surfactant head group and counterion, and hydrocarbon chain length on the relaxation times have been investigated [1-7]. However, the effects of pressure have been neglected. We have, therefore, investigated the effects of pressure on the kinetics of sodium lauryl sulfate (also called sodium dodecyl sulfate, SDS or NaLS) micelle formation.

MATERIALS AND METHODS

A temperature-jump apparatus has been constructed which is capable of operating at pressures up to 1200 atmospheres (Fig. 1). The apparatus is equipped to detect light scattered 90° from the incident beam. A Hanovia 901B-11 mercury-xenon lamp was used as a light source. To keep the light intensity as high as possible, no monochromator or filters were used. Scattered light was detected by an EMI 9558Q photomultiplier tube. The current from the photomultiplier was fed through a nulling-filtering circuit to a Biomation 610 transient recorder. The final output could be displayed on a chart recorder, on an oscilloscope, or in digital form on paper tape.

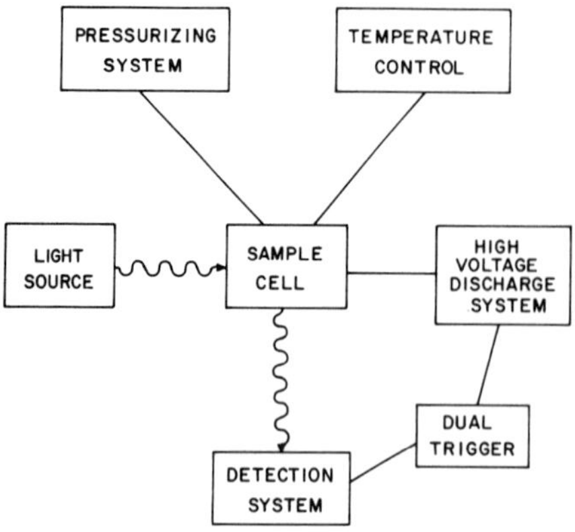

Fig. 1. The temperature jump apparatus.

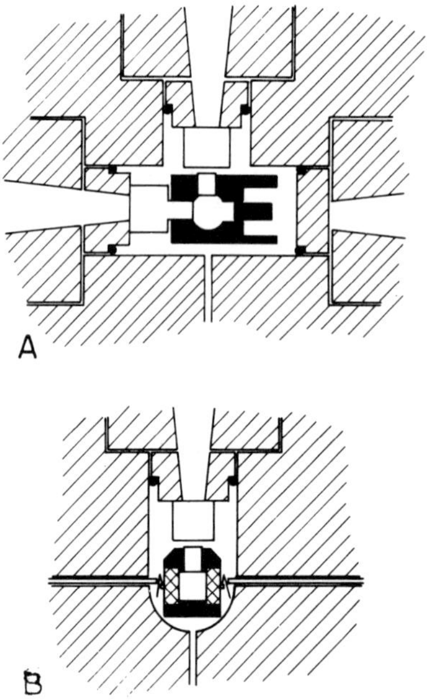

Fig. 2. The sample cell in place within the pressure vessel: a) side view cross-section, b) end view cross-section.

The pressure vessel was constructed from No. 17-4 PH stainless steel and was cylindrical in shape (16.5 cm length, 15.25 cm diameter). A 2.86 cm hole was drilled through the center along the cylindrical axis. A second 2.86 cm hole was drilled perpendicular to the first for observation of light scattered at 90°. Details of the pressurizing system, high-voltage discharge system, and high-pressure electrical leads have been described elsewhere [9].

Sapphire windows were anchored with epoxy to stainless steel seats (Fig. 2). These seats were held in place by plugs which screwed into the pressure vessel. Viton O-rings prevented the pressurizing fluid from flowing past the seats. The sample cell was constructed from black delrin. Stainless steel electrodes (1.27 cm diameter) were situated parallel to the light path. The volume contained by the sample cell was one milliliter. A movable delrin plug compensated for volume changes of the sample solution. The sample cell was surrounded by silicone oil. This liquid acted both as a pressurizing fluid and as an insulator which prevented the high voltage from shorting to the pressure vessel. The electrodes, windows, and movable plug were wrapped with teflon tape and snuggly fitted into the delrin sample cell. To prevent shear stresses from breaking the sapphire windows, the exit window was separated into two pieces. A thin layer of transparent silicone oil transmitted light between the two window parts.

Sodium lauryl sulfate (BDH--Gallard-Schlesinger) was purified by successive extractions with diethyl ether and recrystallizations from water. Evidence of purity was provided by surface tension measurements which showed a monotonic decrease with increasing surfactant concentrations for aqueous solutions. All solutions were prepared from deionized, double-distilled water. All solutions contained 0.2 molal NaCl to raise the conductance. All cell parts which were to be exposed to the sample solutions were thoroughly cleaned and rinsed with filtered water in order to minimize dust. Immediately prior to introduction into the sample cell, the solution was filtered through a 0.22 μm Millipore filter.

Temperature jumps were made from 21 ± .1°C to 25 ± .3°C. The size of the temperature jump was calibrated using a phenol red solution.

RESULTS

The critical micelle concentration for sodium lauryl sulfate in 0.2 molal NaCl is approximately 10^{-3} molal [10]. For surfactant concentrations below 10^{-3} molal, no relaxations could be detected by our apparatus. This is in agreement with the findings of other researchers [1,5].

For concentrations between the cmc and three times the cmc, only one relaxation characterizable by a single exponential was observed. The amplitudes of the relaxation indicated that the intensity of scattered light was decreasing with increasing temperature. This relaxation appears to be the same "τ_2" observed by other investigators [6,11,12]. Figure 3 and Table I demonstrate that pressures up to 680 atm. have no discernable effect on this relaxation time.

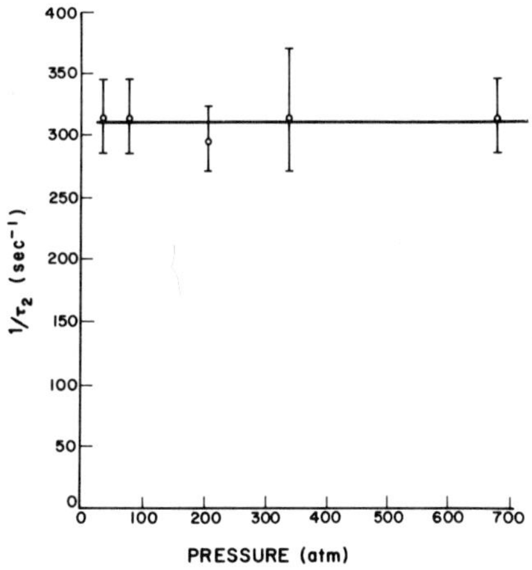

Fig. 3. Plot of reciprocal relaxation time, τ_2^{-1} vs. pressure for 0.003 molal SDS, 0.2 molal NaCl at 25°C.

Hoffmann et al. [1] have developed an expression relating τ_2 to the parameters of a stepwise micellization/dissolution mechanism:

$$\tau_2 = \left[\Sigma\ (1/(k_s^- \bar{A}_s))\right] (\bar{A}_1(1-(\sigma^2/n))) + (\sigma^2 A_T)/n)/(n^2) \qquad (6)$$

The results reported here indicate either that pressure causes each parameter to vary in such a way that the effects cancel, or more probably that pressure has only a small effect on each parameter.

Table I. Relaxation Time τ_2 and τ_3 for SDS at Different Pressures and Concentrations (25°C and 0.2 molal NaCl)*

Pressure (atm) \ Concentration	0.001 molal τ_2 (ms)	0.001 molal τ_3 (ms)	0.002 molal τ_2 (ms)	0.002 molal τ_3 (ms)
34	2.5	N.D.	2.5	N.D.
68	2.6	N.D.	2.4	N.D.
204	2.5	N.D.	2.6	N.D.
340	2.6	N.D.	2.4	N.D.
476	N.D.	N.D.	2.5	N.D.
680	N.D.	N.D.	2.3	N.D.

Pressure (atm) \ Concentration	0.003 molal τ_2 (ms)	0.003 molal τ_3 (ms)	0.01 molal τ_2 (ms)	0.01 molal τ_3 (ms)
34	3.2	N.D.	2.2	38.8
68	3.2	N.D.	2.1	43.6
204	3.4	N.D.	2.0	50.8
340	3.3	N.D.	2.0	68.4
680	3.2	N.D.	1.9	87.3

Pressure (atm) \ Concentration	0.03 molal τ_2 (ms)	0.03 molal τ_3 (ms)	0.05 molal τ_2 (ms)	0.05 molal τ_3 (ms)
34	2.1	19.7	1.6	9.2
204	2.0	30.2	1.6	10.9
340	2.0	39.3	1.5	22.3
680	1.9	60.9	1.6	43.7

*The standard deviations of the relaxations were typically between 10 and 15 percent of the mean value.

N.D. means not detectable.

No relaxation was detected that could have been associated with τ_1. This suggests that the process responsible for τ_1 does not cause significant changes in the intensity of scattered light.

At relatively high surfactant concentrations (0.01 molal and above) a second, slower relaxation was observed. As with τ_2, the amplitudes of the relaxation indicated that the intensity of scattered light decreased with increasing temperature. This can not be the relaxation designated τ_0 since Graber et al. [3] reported that τ_0 was in the nanosecond time range. Thus, we have labeled the new relaxation "τ_3". Table I and Fig. 4 show that τ_3 increases significantly with increasing pressure. At low pressures τ_2 and τ_3 may be so close that it is possible to mistakenly perceive the two relaxations as one. However, the application of pressure effectively resolves the two relaxations. Table I also shows that τ_3 decreases significantly with increasing concentration. At low surfactant concentrations it is possible that τ_3 may occur in a time range too slow to be detected by our apparatus (one second or more).

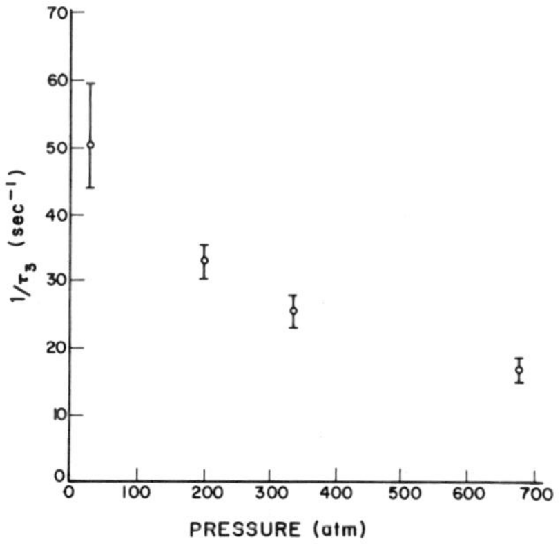

Fig. 4. Plot of reciprocal relaxation time, τ_3^{-1} vs. pressure for 0.03 molal SDS, 0.2 molal NaCl at 25°C.

Sodium lauryl sulfate is known to form spherical micelles at concentrations near the cmc and rod-shaped micelles at high surfactant concentrations [13]. Thus, changes in micelle shape may be responsible for τ_3. Shape changes could occur by a micellar aggregation/dissociation process (Eq. 4). Large, rod-shaped micelles could break up into (or be formed from) two or more small, spherical micelles. This mechanism could account for the concentration dependence of τ_3. Higher surfactant concentrations result in more micelles (of all shapes and sizes), which in turn allow the aggregation process to proceed at higher rates. Assuming that micellar aggregations can be simply described by the process

$$A_n + A_n \underset{k_r}{\overset{k_f}{\rightleftharpoons}} A_{2n} \tag{7}$$

an expression can be developed which relates τ_3 to the rate constants:

$$1/\tau_3 = k_r + 4k_f \bar{A}_n \tag{8}$$

Assuming $\bar{A}_n$ increases linearly with total surfactant concentration, then

$$1/\tau_3 = k_r + K\, k_f\, A_T \qquad (9)$$

Equation (9) is consistent with the observed variation of $1/\tau_3$ with concentration. Using Eq. (9) and the data from Table I, an activation volume of about 70 ml/mol has been calculated for the forward step.

Hoffmann et al. [1] have argued that micellar aggregation is improbable because of the electrostatic repulsion between the micelles. However, under conditions of high ionic strength and high concentrations of micelles, aggregations are more likely to occur. The fact that τ_3 appears to be the slowest of the micellar relaxations may be due to the size and charge of the aggregating species.

Shape changes could also result from a stepwise association (or dissociation) of monomers into (or from) micelles. The concentration and pressure dependence of τ_3 could be rationalized using this mechanism. However, it would be difficult to explain why τ_3 is slower than τ_2. In Hoffmann's model, τ_2 is caused by a stepwise micellization/dissolution process [1]. The concentrations of micelles having aggregation numbers greater than one but less than the mean aggregation number are very low [13]. The rate limiting step in the τ_2 process is believed to be the aggregation of singly dispersed surfactant molecules with these micelles [1]. If a stepwise mechanism is also responsible for shape changes, then the concentrations of all reacting species would be high, and τ_3 would be expected to be faster than τ_2. Thus, the stepwise mechanism for changes of micelle shape does not adequately explain the results.

In summary, a new relaxation is reported for sodium lauryl sulfate micelle formation. Application of pressure effectively resolves the new relaxation from a previously reported relaxation. The new relaxation is tentatively attributed to micellar aggregation.

ACKNOWLEDGMENT

The authors would like to thank Professor P. Mukerjee for advice and aid in the purification of SDS.

NOTATION

A = surfactant molecule
$\bar{A}_n$ = equilibrium for monomer n
$\bar{A}_s$ = equilibrium concentration of species s
A_T = total surfactant concentration
C^+ = counterion
K = constant
k_f = forward rate constant for micelle aggregation
k_r = reverse rate constant for micelle aggregation
k_s^- = rate constant for dissociation of a surfactant molecule from species s
m, n, s = aggregation numbers (monomers per micelle)
z = charge on a micelle
σ^2 = variance of micellar size distribution
τ_0, τ_1, τ_2, τ_3 = relaxation times

REFERENCES

1. H. Hoffmann, R. Zana, E.A.G. Aniansson, et al., J. Phys. Chem. 80, 905 (1976).
2. R. Zana, in Chemical and Biological Application of Relaxation Spectrometry, E. Wyn-Jones, ed., D. Reidel, Boston (1975), p. 139.
3. E. Graber, J. Lang, and R. Zana, Koll. Z. u. Z. Polym. 238, 470 (1970).
4. T. Yasunaga, S. Fujii, and M. Miura, J. Coll. Inter. Sci. 30, 399 (1969).
5. J. Lang, R. Zana, et al., J. Phys. Chem. 79, 276 (1975).
6. T. Yasunaga, K. Takeda, and S. Harada, J. Coll. Inter. Sci. 42, 457 (1973).
7. G. Kresheck, et al., J. Am. Chem. Soc. 88, 246 (1966).
8. A. Colen, J. Phys. Chem. 78, 1676 (1974).
9. A. D.Yu, M. D. Waissbluth, and R. A. Grieger, Rev. Sci. Instrum. 44, 1390 (1973).
10. P. Mukerjee and K. J. Mysels, U.S. Department of Commerce, Washington, D.C. (1971), p. 71.
11. B. C. Bennion, et al., J. Phys. Chem. 73, 3288 (1969).
12. R. Folger, H. Hoffmann, and W. Ulbricht, Ber. Bun. Gesell. phys. Chem. 78, 986 (1974).
13. F. Reiss-Husson and V. Luzzati, J. Phys. Chem. 68, 3504 (1964).

P-3

PRESSURE-JUMP STUDIES OF RIBOSOMAL SUBUNIT ASSOCIATION-DISSOCIATION KINETICS

A. Wishnia, B. Flaig, and F.-L. Lin

State University of New York at Stony Boook
Stony Brook, New York USA

INTRODUCTION

Ribosomes, the complex nucleoprotein particles on which protein synthesis occurs in all organisms, are formed from two specific subaggregates. The particles are referred to by their standard sedimentation coefficients: the Escherichia coli 70S ribosome, molecular weight 2.55 x 10^6 daltons, is composed of a small, 30S, 0.85 x 10^6 dalton subunit, and a large, 50S, 1.70 x 10^6 dalton subunit. The rates and equilibria of the 70S = 30S + 50S reaction depend markedly on the divalent metal ion concentration and less so on the univalent metal ion concentration, as well as on temperature, etc. [1,2]. The reaction is important because a number of the reactions of protein synthesis require partial, if not total, separation of the 30S-50S interfaces.

The dissociation of ribosomes at high pressure was discovered by Infante and Baierlein [3], through an artifact observed in sucrose-gradient centrifugation. Two methods of analysis of the ultracentrifuge data yielded -200 and -500 ml as values of $\Delta V°$ of dissociation of sea urchin ribosomes. Turbidity measurements of the equilibrium [4] and kinetics [5] of dissociation of E. coli ribosomes have been reported. A value of -240 ml for $\Delta V°$, at 22°, 25 mM Mg^{2+}, 60 mM M^+ was computed. The data from pressure down-jumps from 55 to 110 atm to ambient were interpreted as indicating several reactions with different relaxation times. However, the work of this group, as will be seen, is open to serious question.

We built a static-kinetic system capable of up- or down-jumps in any interval between 1 and 1500 atm, following the course of the reaction by monitoring the 90° scattered light (fluorescence and absorbance measurements are also possible). Our first objective was the determination of the equilibrium and activation parameters of the 70S = 30S + 50S reaction with which any proposed detailed model of the molecular interactions involved would have to be compatible. We also intend to study other reactions in the protein synthesis pathway.

METHODS

Preparation of E. coli MRE600 ribosomes was as previously described [1,2]. Experiments were performed in solutions containing 50 mM NH_4Cl, 10 mM tris(hydroxymethyl)aminomethane and the indicated amounts of Mg Cl_2 at pH 7.5 (25°).

The apparatus was adapted from Brower's design [6]. Pressures are monitored with a BLH Electronics Type D-HF 0-30,000 psi pressure transducer, which also supplies the trigger signal. The 90° scattering signal from the 0.5-mm-diameter He-Ne laser beam detected with a red-sensitive photomultiplier tube is ratioed to the signal from the silicon diode beam monitor. Pressure and scattering signals are collected with an IBM 1800 computer. The pressure-jump is initiated by opening a standard high-pressure valve: the pressure change is approximately linear with the entire change taking 10 msec (efforts are being made to reduce the time to 1 to 2 msec). The IBM 1800 computer has a 2 msec minimum data acquisition interval.

Ribosomal particles are sufficiently well-behaved so that the 90° scattering signal, $R_{90°}$, at any time t, is given by the simple equation

$$R_{90°} = K \Sigma n_i M_i^2 \quad (1)$$

where the n_i are the molar concentrations of scatterers (i.e., 30S, 50S, and 70S particles), and the M_i are their molecular weights. Kinetics are determined from the established rate laws [1]. It should be emphasized that, given the molecular weights of the particles, a shift from 100% dissociation (30S + 50S) to 100% association (70S) changes $R_{90°}$ by a factor of 1.8. Our system is sufficiently sensitive and reliable to make absolute comparisons between reference solutions representing the dissociation extremes; further, since the 100% dissociation solution (1 mM Mg^{2+}) shows little dependence on pressure, we obtain the degree of dissociation in any solution at any pressure, unambigously. (A 0.4 mg/ml dissociated ribosome solution gives a noise-to-signal ratio of 1%. The blank scattering-stray light contribution is 10% of the total signal; its reproducibility adds another 1% to the uncertainty in $R_{90°}$. The German apparatus is perhaps 1/10 as sensitive.)

RESULTS

Good "tight" ribosomes at concentrations below 1 mg/ml are completely associated by 5 mM Mg^{2+} at 1 atm ($R_{90°}$ increases by 1.8), and show perhaps 10% aggregation to dimers at 10 mM Mg^{2+} (see Fig. 1 of Wishnia and Boussert [2]).

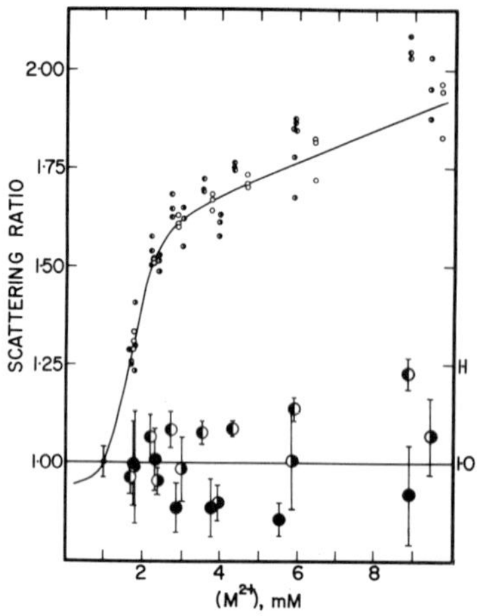

Fig. 1. Effect of alkaline earth cations on the ribosome dissociation equilibrium; abscissa, total divalent cation concentration (1.18 mM Mg^{2+} + increments of Mg^{2+}, ◐ ; Ca^{2+}, ◑ ; Sr^{2+}, ○ ; or Ba^{2+}, ●). Left ordinate, for small circles: corrected 90° scattering relative to solutions at 1.18 mM Mg^{2+}; individual points. Right ordinate: deviation plot, ratio of scattering for other ions to the equivalent point on the curve through the Sr^{2+} values; mean of three determinations with standard deviation bars. Conditions: 50 mM NH_4Cl, 10 mM TRIS, 7 mM mercaptoethanol, pH 7.5, 25°, about 1 mg/ml ribosomes.

An equilibrium pressure run at 3 mg/ml of ribosomes, 22 mM Mg^{2+}, 25° (not illustrated), showed a large, concave decrease in scattering to a steady plateau at 1.8 times the value expected for a 30S-50S dissociation mixture, even up to 1500 atm. What was seen was disaggregation, perhaps of 100S dimers to 70S particles, without further dissociation. Only at 4 mM Mg^{2+} was there no sign of 70S aggregation (see Fig. 2). At 6 mM Mg^{2+} the signal decreased in two waves, the disaggregation at low pressure and the dissociation to 30S and 50S subunits at higher pressure; the aggregation was still visible at 0.4 mg/ml ribosomes (see Fig. 3).

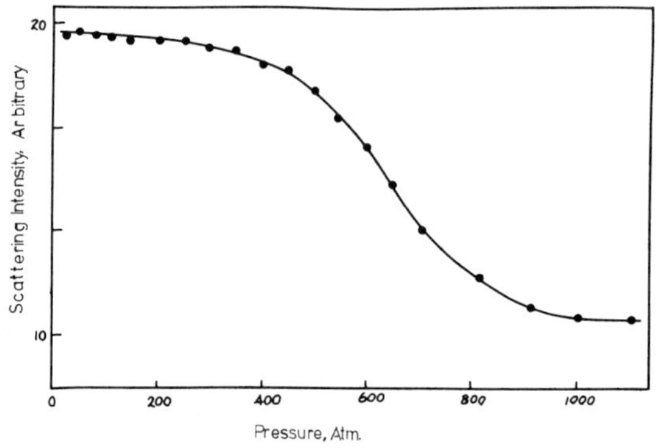

Fig. 2. Effect of pressure on ribosome dissociation. 3 mg/ml ribosomes, 4 mM Mg^{2+}, 60 mM M^+, 20.7°. Lack of inactivation monitored by frequent return to 1 atm. K_D = $4x10^{-10}$ M, $\Delta V°$ = -280 ml.

The results of several runs at different Mg^{2+} concentrations, at 25°, are tabulated below. The parameters are precise enough in terms of the within-run variance, but we are not sure the high-pressure inactivation of ribosomes has been avoided in every case: the difference in $\Delta V°$ between 4 and 5 mM Mg^{2+} which might indicate an undiscovered transition in the ribosomes is not yet to be taken seriously, nor have we enough confidence in the data to assert that $\Delta V°$ is pressure-independent. The time-dependent high-pressure inactivation of ribosomes varies from preparation to

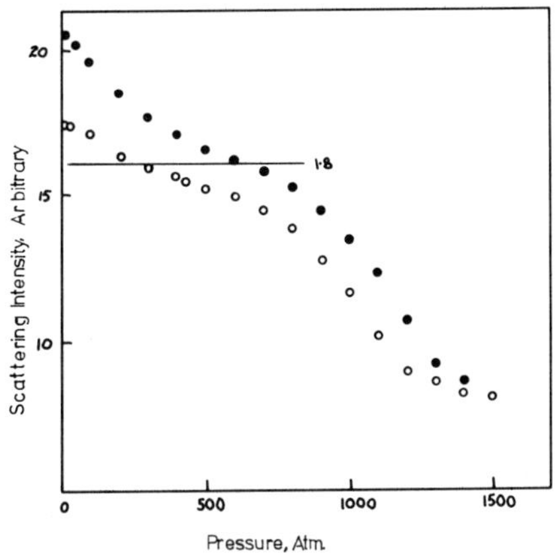

Fig. 3. Effect of ribosome concentration on aggregation. ●, 2.0 mg/ml ribosomes; ○, 0.4 mg/ml. 6 mM Mg^{2+}, 25.0°.

preparation (one showed none after repeated exposures to 1500 atm at 6 mM Mg^{2+}, 25°), and affects the 30S subunit more than the 50S subunit; we are investigating the possibility that a latent ribonuclease is responsible. Note, however, that the 1 atm values of $K°_D$ for 5 mM and 6 mM Mg^{2+} are completely inaccessible from 1 atm scattering data, and that the 4 mM Mg^{2+} value would be subject to gross errors from minor heterogeneity of preparation because of the low degree of dissociation at 1 atm: it is the availability of the entire range of degree of dissociation which permits an accurate two-parameter fit in which one term is $K°_D$ and the other $\Delta V°$. These ribosomes (see Ref. 2) are slightly tighter than in earlier work; the values of $K°_{1,D}$ are, however, a consistent extension of the earlier results at 1.5 to 3.5 mM Mg^{2+} [1].

Table I. $[Mg^{2+}]$-Dependence of Ribosome Dissociation (25°)

$[Mg^{2+}]$, mM	$K°_{1,D}\times10^{11}$, M	$\Delta V°_1$, ml	$K°_{2,D}\times10^{6}$, M	$\Delta V°_2$, ml
3	99 ± 15	-317 ± 6	-	-
4	21 ± 6	-325 ± 11	98 ± 42	-76 ± 93
5	4.0 ± 1.2	-245 ± 7	5.3 ± 0.8	-119 ± 11
6	1.2 ± 0.7	-243 ± 11	2.9 ± 0.2	-122 ± 5

$K°_{1,D} \equiv (30S)(50S)/(70S)$.
$K°_{2,D} \equiv (70S)^2/[(70S)_2]$.

Kinetic studies are also proceeding. In one series at 4 mM Mg^{2+} and a nominal 20.7° (see Fig. 2), 500 atm jumps bracketing the dissociation region (e.g., 123 → 603, 224 → 703, 422 → 903 atm) were recorded. For these fast reactions the adiabatic temperature change is not relaxed; we can take T as 21.2°C throughout. The derived parameters for the forward and reverse reactions at 4 mM Mg^{2+} (see Figs. 2 and 4) are $k°_1 = 5\times10^{-3}$ sec^{-1}, $\Delta V_1^{\neq} = -100$ ml; $k°_2 = 1.3\times10^{7}$ M^{-1} sec^{-1}, $\Delta V_2^{\neq} = +180$ ml; $K°_D = 4\times10^{-10}$ M, $\Delta V° = -280$ ml. These results are consistent with previous stopped-flow and equilibrium values at 1 atm, 25° [1]. Other results have been

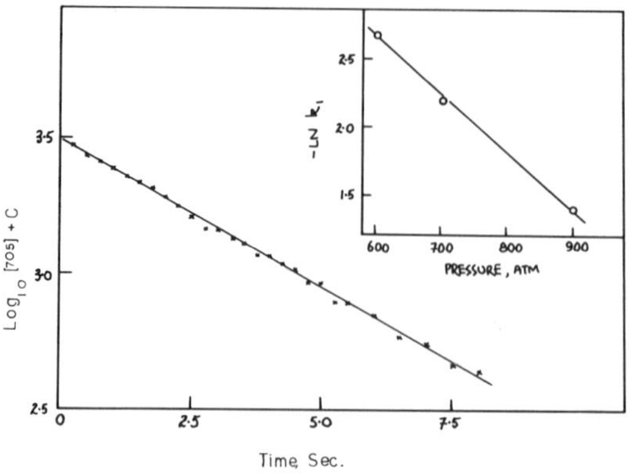

Fig. 4. Kinetics of ribosome dissociation, 4 mM Mg^{2+}, 21.2°. Main curve, X, first-order plot of 70S dissociation; 422 → 903 atm jump, k_1 = 0.248 sec^{-1}. Inset, activation volume plot; $k_1°$ = $5x10^{-3}$ sec^{-1}, $\Delta V_1^{\neq}$ = -100 ml.

analyzed (e.g., at 5.8 mM Mg^{2+}, 1066 atm, 26°, presumed 100S dimers dissociate to 70S ribosomes with $t_{1/2}$ = 20 msec, while dissociation of 70S to 30S and 50S particles had $t_{1/2}$ = 1.7 sec), but it is premature to discuss the mechanistic implications of the rate parameters or their temperature or [Mg^{2+}]-dependence until the work has been repeated at low ribosome concentrations.

DISCUSSION

The result that ΔV° of dissociation of 70S particles is -250 to -320 ml vitiates a model of ribosome association recently proposed [7]. This model assumed that an oligonucleotide interaction involving no more than 3 to 4 nucleotide base-pairs between the 16S RNA of the 30S subunit and the 235 RNA of the 50S subunit was responsible for the association. Such an interaction would produce a ΔV° of at most -12 ml [8,9]; other interactions clearly contribute the lion's share to ΔV° and would presumably contribute substantially to ΔG° and ΔH° at any [Mg^{2+}]. Dissociation of ion-pairs, exposure of hydrophobic groups, are likely sources. Differential binding of Mg^{2+} to 70S particles is not a possible source (cf. [1,2]); a pressure-dependent change in Mg^{2+}-binding to all particles, 30S, 50S, and 70S, is compatible with the observations.

Previous work [4,5] at higher ribosome concentrations (3 to 6 mg/ml) showed substantial aggregation of isolated 30S and 50S particles at high [Mg^{2+}]. The total decrease in scattering between 1 and 1000 atm in 7 mM Mg^{2+} solutions was only 25% of that in 25 mM Mg^{2+}; since tight ribosomes are 100% associated to 70S particles at 1 atm at both concentrations, and since 70S particles should be more sensitive to dissociation at lower [Mg^{2+}] the result can only arise from disaggregation of larger structures than 70S particles. Similarly, the "perturbation" produced by 55 to 110 atm on the 70S = 30S + 50S equilibrium at 12 mM Mg^{2+} is necessarily larger than the equilibrium concentrations of 30S and 50S particles at the final 1 atm pressure. Even without the complication of reassociation to 100S dimers and higher, it is inappropriate to use pseudo-first-order expressions to fit second-order reactions.

REFERENCES

1. A. Wishnia, A. S. Boussert, M. Graffe, P. Dessen, and M. Grunberg-Manago, J. Mol. Biol. 93, 499 (1975).
2. A. Wishnia and A. S. Boussert, J. Mol. Biol. 116, 577 (1977).
3. R. Baierlein and A. A. Infante, Methods in Enzymology 30, 328 (1974).
4. E. Schulz, H. D. Lüdeman, and R. Jaenicke, FEBS Lett. 64, 40 (1976).
5. E. Schulz, R. Jaenicke, and W. Knoche, Biophys. Chem. 11, 253 (1976).
6. K. R. Brower, J. Am. Chem. Soc. 90, 5401 (1968).
7. G. Hui Bon Hoa, M. Graffe, and M. Grunberg-Manago, Biochemistry 16, 1278 (1977).
8. R. E. Chapman and J. M. Sturtevant, Biopolymers 7, 527 (1969).
9. H. Noguchi, S. A. Arya, and J. T. Yang, Biopolymers 10, 2491 (1971).

P-4

IONIC TRANSPORT CHANGES INDUCED BY HIGH HYDROSTATIC PRESSURES IN MAMMALIAN RED BLOOD CELLS

A. Péqueux

University of Liège
Liège, Belgium

INTRODUCTION

Despite the vast amount of experimental data available, the molecular mechanisms concerned with the regulation of ionic passages across biological membranes still remain very obscure. Most of the theories on passive and active transfers involve the existence of chemical reactions, ionization processes, and binding to carriers or enzymes. Being associated with volume changes, these chemical processes must be sensitive to hydrostatic pressure [1]. Hence, the idea of using high pressure as an analytical tool and of applying it to approach the study of the structure of the cell membrane in relation to function.

From this point of view, pressure has already been shown to modify the ionic permeability properties of living tissues implicated in osmoregulatory processes, such as the frog skin and the teleostean gills [2-5].

The purpose of this work is to provide information on the effects of high hydrostatic pressures on ionic transport occurring at the level of mammalian red blood cells, paying special attention to the ouabain-sensitive Na^+ outfluxes and related membrane ATPases.

MATERIAL AND METHODS

Blood was collected from adult pigs and oxen by carotide puncture immediately after killing. It was defibrinated by vigorous shaking with a glass rod. Human citrated blood was supplied by the University Hospital. Red blood cells were separated by centrifugation at 3,000 rpm at 2°C for 15 min.

Experiments With Na^+-Loaded-K^+-Depleted Erythrocytes

A suspension of red blood cells (RBC) was prepared in a solution containing 145 mM NaCl and 10 mM Tris-buffer (pH 7.2). It was stored at 2°C for 3 days in order to reversibly reduce the activity of the pumps and to allow loading of the cells with Na^+ by passive diffusion along the concentration gradient. RBC were washed four times with five volumes of Tris-buffer 155 mM (pH 7.2). After the last centrifugation, 1 volume of RBC was resuspended in 2 volumes of incubation solution (KCl 25 mM, $MgCl_2$ 5 mM, Tris-Cl 125 mM pH 7.2). Ouabain (final concentration 0.2 mM) was added for the determination of the "ouabain-insensitive" fluxes.

The experiments were conducted at 37°C for 60 or 120 min at atmospheric pressure (controls) and under high hydrostatic pressure. Immediately after the incubation, samples were centrifuged at 10,000 rpm for 20 min at 2°C. Ionic content was determined by flame photometry directly in the supernatant and in the hemolysed pellets.

Ionic fluxes were calculated as the difference between ionic contents of the incubation solution before and after the incubation at 37°C. These were expressed as μequiv/ml incubation solution X hour.

Experiments With Depigmented Erythrocytes

After three washings in a medium containing 145 mM NaCl, 10 mM Tris-Cl, erythrocytes were partially hemolysed at room temperature in 10 volumes of an hypotonic solution containing 40 mM NaCl 4 mM $MgCl_2$, and 4 mM Na_2ATP. Isomoticity (310 m,osmoles/l) was restored by adding Tris-Cl 10 min later (pH 7.2). Depigmented erythrocytes were still washed four times in Tris-buffer 155 mM. One volume of separated RBC was finally resuspended in 2 volumes of incubation solution. The experiments were then carried out exactly as described above.

Pressure Vessel

Incubation under pressure was conducted with a high-pressure vessel which is described in detail elsewhere [2 and 3]. Five-ml test tubes containing the samples were placed in a plexiglass cylinder which was then completely filled with silicone oil (Dow Corning Silicone 200, 1 CS) before being fastened to the steel pressure vessel and submitted to pressure. Compression was achieved within 20 to 45 sec using APEX press (Apex Construction Ltd.) manually driven.

ATPase Activity Measurements

Enzymic extracts were prepared according to the method of Post et al. [6]. The reaction medium used to measure enzyme activity contained, in a final volume of 2 ml, 4 mM-ATP, 25 mM Tris-buffer (pH 7.4), 0.25 mM EGTA, 100 mM NaCl, 25 mM-KCL, 5 mM $MgCl_2$, and 0.2 ml of enzyme. Ouabain (final concentration 0.2 mM) was added for estimation of the "ouabain-sensitive ATPase" expressed as the difference between activities with and without the drug.

Incubation was conducted at 37°C for 30 or 60 min at atmospheric pressure (control) and under high hydrostatic pressures. After stopping the reaction with 0.2 ml trichloroacetic acid 50% and centrifugation at 10,000 rpm for 20 min, inorganic phosphate was determined in the supernatant by the method of Fiske and Subbarow [7]. Protein content of the extract was determined by the method of Lowry et al. [8]. Results were expressed as micromoles of inorganic phosphate per mg of proteins per hour (μmoles Pi/mg prot x hr).

RESULTS AND DISCUSSION

Preliminary flux measurements performed at atmospheric pressure on pig erythrocytes in a Na^+-free medium containing 25 mM KCl give a Na^+ outflux of about 1.5 to 1.8 m.equiv/l RBC x hr. About 80% of this flux (i.e. about 1.2 to 1.4 m.equiv/l RBC x hr) can be inhibited by ouabain 2.10^{-4}M. Similar flux values have been obtained by measuring either the appearance of Na^+ in the incubation solution (supernatant), or the output of Na^+ from the RBC pellets. Those data are in good agreement with estimations made by Sorenson et al. [9] by means of radioisotopes. While it was extremely difficult to measure K^+ fluxes by our method (due to the high K^+ content of the RBC or to the high concentration of the incubation medium), it was possible to record also an "ouabain-sensitive" K^+ influx of about 0.5 mequiv/l RBC x hr. This corroborates the idea that the "ouabain-sensitive" Na^+ flux corresponds to an exchange of internal Na^+ against external K^+ resulting in net fluxes of both ions against an electrochemical gradient (Na^+/K^+ pump).

The question of the "ouabain-insensitive" fluxes is not so clear. The simplest interpretation likens them with a simple leak [10].

In the analysis of the passive mechanisms for Na^+ and K^+ transports, it has thus been considered that all active components are eliminated when incubation is done in K^+ free solutions containing ouabain.

The effects of pressure steps up to 1,000 kg/cm^2 on the Na^+ outfluxes of pig intact erythrocytes are summarized in Fig. 1. It clearly appears that Na^+ total output diminishes when pressure is raised. The flux inhibition is higher the higher the pressure, reaching 80% at 1,000 kg/cm^2. A very similar picture is obtained by considering the "ouabain-sensitive" Na^+ flux which has been related to the Na^+ pump. Application of pressure steps higher than 500 kg/cm^2 also seems to inhibit the Na^+ flux more significantly. This could tentatively be related to the fact that pressures of 500 kg/cm^2 and more decrease the percentage of "ouabain-sensitive" Na^+ outflux related to the total Na^+ outflux (see Fig. 2).

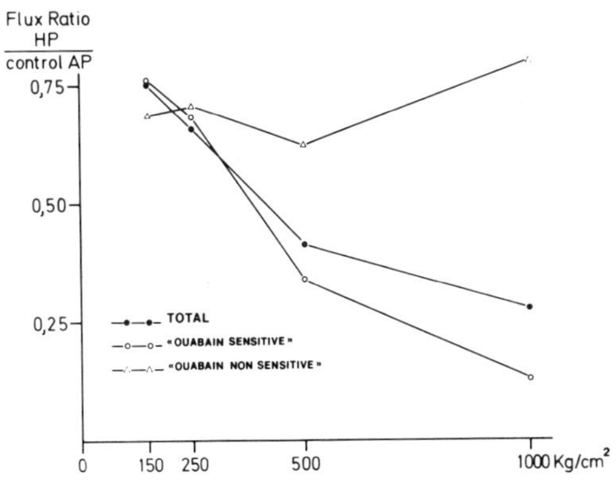

Fig. 1. Effects of hydrostatic pressure on the Na^+ outfluxes of pig erythrocytes (i.e., total, ouabain-sensitive and non sensitive). Ordinate - ratio between the fluxes measured at elevated pressure (HP) and at atmospheric pressure (control AP). Abscissa - hydrostatic pressure.

The K^+ "ouabain-sensitive" influx has also been observed to be concomitantly inhibited by pressure. The inhibition reached about 20% at 250 kg/cm^2 but appeared to be dependent on the K^+ content of the incubation medium. Those findings well support the idea of the linkage between both the "ouabain-sensitive" Na^+ and K^+ fluxes implicated in the Na^+/K^+ exchange pump.

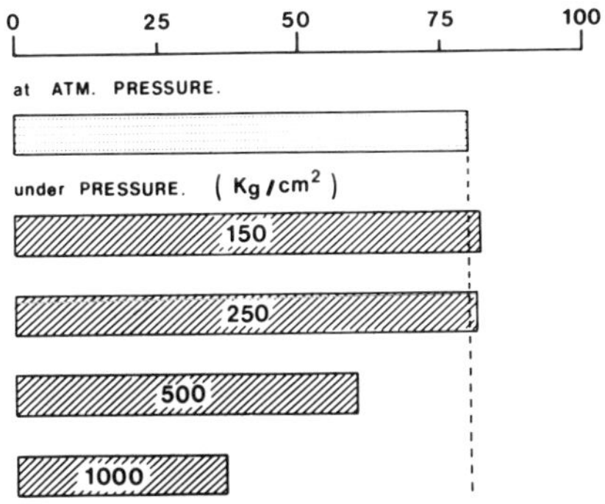

Fig. 2. Effects of hydrostatic pressure on the percentage of "ouabain-sensitive" Na^+ outflux. Total Na^+ outflux is always considered as being 100% for each pressure condition.

The inhibition of the active Na^+ transport suggested by our results has been related to the effects of pressure on the activity of the (Na^+ + K^+) ATPase extracted from the RBC membranes. In a previous paper, it had been established that the (Na^+ + K^+) ATPase extracted from tissues implicated in iono- and osmo-regulation is little affected by low pressures but strongly inhibited at 500 kg/cm^2 and more (up to 80% inhibition at 1,000 kg/cm^2) [2,11]. Similar results have been obtained with the RBC. While the activity of the RBC (Na^+ + K^+) ATPase is very low (about 0.02 µmoles Pi/mg proteins x hr), an inhibition of 60% occurs at 500 kg/cm^2. On the contrary, there is no significant effect at 250 kg/cm^2. Hydrostatic pressures higher than 500 kg/cm^2 would thus induce inhibition of the active processes by acting directly on the enzyme linked to the pumps. The picture of the enzyme sensitivity to pressure is indeed very close to the pressure effect on the percentage of "ouabain-sensitive" Na^+ outflux shown in Fig. 2. The situation is not so clear when pressure remains lower than 500 kg/cm^2 as the membrane ATPase does not seem to be affected. This could suggest that the enzyme is not the only mechanism involved in the Na^+ pump, another step of the process being more sensitive to low pressure. These effects have been observed to remain quite reversible. The Na^+ permeability characteristics of RBC assayed at atmospheric pressure after compression up to 1,000 kg/cm^2 was indeed not significantly different from that recorded without applying any pressure step.

Pressure appears to have little effect on the "ouabain-insensitive" Na^+ outflux. A constant flux decrease of about 30% is observed till 500 kg/cm^2; that decrease seems to be reduced at 1,000 kg/cm (only 20%) (Fig. 1). When incubation is performed in K^+ - Mg^{++} free solution containing ouabain, the "ouabain-insensitive" Na^+ outflux which has then been considered as a passive Na^+ leak has been observed to be more inhibited by pressure (70% inhibition at 250 kg/cm^2). These results indicate that pressure effects on the passive mechanisms seem to be determined by the ionic environment and hence the gradients across the cell membrane. This could perhaps be related to an effect of pressure on the transmembrane potential difference similar to which has been observed on the frog skin [2,4]. Experiments with sulfate salines are now being pursued in the laboratory to study the pressure effects when the potential difference has previously been changed.

Experiments performed on depigmented erythrocytes (ghosts) at atmospheric pressure and under pressure give quite similar results as for intact RBC. They thus indicate that the inhibition of the ionic active permeability well results from a direct effect of pressure on membrane-bound mechanisms. As ATP has artificially been supplied when preparing the ghosts, an effect of pressure on the metabolism producing energy can be discarded. It may thus now be tentatively concluded that, beside an inhibition of the Na^+/K^+ active transports, hydrostatic pressure would induce a decrease in passive permeability of Na^+ and K^+.

Experiments actually in progress in this laboratory indicate that the permeability of human and ox erythrocytes is similarly affected by pressure while quantitative differences occur (see Table I). It can be noted that the percentage of "ouabain-sensitive" Na^+ outflux of ox erythrocytes is particularly low (23%) and relatively sensitive to a pressure step of 250 kg/cm^2. Those experiments however need further development.

The present results could tentatively have been explained by considering that pressure acts either directly or indirectly upon the kinetic constants of the passive transport systems or of the enzymes linked to the active processes. In the first view, pressure would act on the volume changes which accompany the different steps of the catalytic process. On the other hand, pressure would induce conformational changes in the transport-bounded proteins leading to activity changes. According to the last view, changes in protein-protein or protein-lipid interactions could occur. More particularly, the requirement of the (Na^+ + K^+) ATPase for phospholipids [12] and

Table I. Effects of 250 Kg/cm^2 Pressure on the Na^+ Outfluxes of Pig, Human and Ox Erythrocytes*

Material	Incubation Conditions: Ouabain	Incubation Conditions: Applied pressure, kg/cm^2	Na^+ Outflux, μequiv/ml inc. sol x hr	Ouabain-Sensitive Na^+ Outflux: (1) - (2), μequiv/ml inc. sol x hr	Ouabain-Sensitive Na^+ Outflux: Percent of total Na^+ outflux, %	Percent Inhibition by Pressure: Total Na^+ outflux, %	Percent Inhibition by Pressure: Ouabain-sensitive Na^+ outflux, %
Pig	-	-	0.788 (1)	0.669	84.9	-34.8	-36.5
	+	-	0.119 (2)				
	-	250	0.514	0.425	82.7		
	+	250	0.089				
Human	-	-	0.757	0.315	41.6	-13.1	-31.4
	+	-	0.442				
	-	250	0.658	0.216	32.8		
	+	250	0.442				
Ox	-	-	0.380	0.089	23.4	-16.6	-53.9
	+	-	0.291				
	-	250	0.317	0.041	12.9		
	+	250	0.276				

*Incubation conditions at 37°C: KCl 25 mM
$MgCl_2$ 5 mM
Tris-Cl 125 mM pH : 7.2

the fact that the temperature of the break in the Arrhenius plot of that enzyme is pressure dependent [13,14] support the idea that permeability changes could be linked to variations in the lipidic architecture of the membrane. As a conclusion, it would be tempting to explain our results in terms of a phase transition in the lipidic components of the membrane affecting the conformation of the enzyme proteins associated with the active processes and of the proteins specifically involved in the passive ionic transports. Such an effect could occur solely or together with a pressure action on the state of ionization of the fixed charges in the membrane which are considered as playing an important role in the selective regulation of the ionic permeability of biological membranes.

ACKNOWLEDGMENTS

The author is greatly indebted to A. Distèche and R. Gilles for their constructive discussions and criticism throughout this work. He also wants to thank A. Cambron for practical cooperation.

REFERENCES

1. A. M. Zimmermann, High Pressure Effects on Cellular Processes, Academic Press, New York (1970), p. 324.
2. A. Péqueux, J. exp. Biol. 64, 587 (1976).
3. A. Péqueux, Comp. Biochem. Physiol. 55A, 103 (1976).
4. A. Brouha, A. Péqueux, E. Schoffeniels, and A. Distèche, Biochim. biophys. Acta 219, 455 (1970).
5. A. Péqueux and R. Gilles, Experientia 33, 46 (1977).
6. R. L. Post, C. R. Merrit, C. R. Kinsolving, and C. D. Albright, J. biol. Chem. 235, 1796 (1960).
7. C. H. Fiske and Y. Subbarow, J. biol. Chem. 66, 375 (1925).
8. O. H. Lowry, N. Y. Rosebrough, A. L. Farr, and R. J. Randall, J. biol. Chem. 193, 265 (1951).
9. A. L. Sorenson, L. B. Kinschner, and J. Barker, J. gen. Physiol. 45, 1031 (1962).
10. R. L. Post, C. D. Albright, and K. Dayani, J. gen. Physiol. 50, 1201 (1967).
11. A. Péqueux and R. Gilles, Comp. Biochem. Physiol., in press.
12. R. Tanaka and A. Teruya, Biochim. biophys. Acta 323, 584 (1973).
13. J. L. Dahl and L. E. Hokin, Annu. Rev. Biochem. 43, 327 (1974).
14. F. Cauterick, J. Peeters, K. Heremans, H. De Smedt, and H. Albrechts, Arch. Internat. Physiol. Biochim. 84, 587 (1976).

Q-1

CATALYST DEVELOPMENT FOR HIGH PRESSURE OXO PROCESS

A. Matsuda, S. Shin, J. Nakayama, and K. Bando

National Chemical Laboratory for Industry
Tokyo, Japan

INTRODUCTION

Catalyst recovery and recycling in homogeneously catalyzed oxo processes are of the utmost importance from an industrial point of view, and there have been many reports concerning the separation of the cobalt catalyst from the reaction product. For example, Moffat has attempted to fix $HCo(CO)_4$ as solid poly-2-vinylpyridine complex [1,2].

Previously, we have reported that a cobalt carbonyl-pyridine complex for homogeneous hydroesterification spontaneously separates from the reaction product [3]. The separated catalyst phase consisted of an ionic complex, $H_2Co_3(CO)_9(Py)_5$, where Py denotes pyridine, which is insoluble in hydrocarbons and ether, and soluble in alcohol, ketone, aldehyde, and pyridine. Although this complex can also be used as a catalyst for hydroformylation, it does not separate from the hydroformylation product since it is soluble in aldehyde.

The present paper deals with a new oxo catalyst, which is derived from $Co_2(CO)_8$ and ethylene glycol di-3-(2-pyridyl)-propionate (hereafter denoted as compound (I)), and which is soluble under the reaction condition, but which spontaneously separates from the reaction product on cooling to room temperature. An even more interesting feature of our newly developed catalyst may be that it is active over a low temperature range, 50 to 70°C.

EXPERIMENTAL

Reactions in a low pressure range (10 to 50 kg/cm^2) were carried out in a 400 ml stainless steel autoclave equipped with a vertical agitator and windows for observation during the reaction.

Reactions in a high-pressure range (50 to 240 kg/cm^2) were carried out in another 300 ml autoclave of the same type, but without windows.

Compound (I) (bp 220°C, 1 mmHg) was prepared by trans-esterification of ethyl (or methyl) 3-(2-pyridyl) propionate and ethylene glycol at 200°C in the presence of sodium methylate (0.02%). Ethyl (or methyl) 3-(2-pyridyl) propionate was prepared by the hydroesterification of 2-vinylpyridine with carbon monoxide and ethanol (or methanol) at 150 to 170°C, 70-100 kg/cm^2 in the presence of $Co_2(CO)_8$ (ca. 1%). The ethyl 3-(2-pyridyl) propionate (bp 122°C, 10 mmHg) and methyl 3-(2-pyridyl) propionate (bp 118°C, 10 mmHg) obtained in this manner were identified by IR, and ^{13}C-NMR spectra.

RESULTS

Reactions Under Low Pressure Conditions

The hydroformylation of 1-decene (0.2 mol) was carried out in diethyl ether (30 g) in the presence of $Co_2(CO)_8$ (8 mmol) and compound (I) (16 mmol). The autoclave was charged with carbon monoxide, heated to reaction temperature, and the pressure adjusted to attain the desired carbon monoxide partial pressure; then hydrogen (or nitrogen) was introduced to attain the desired hydrogen partial pressure and total pressure. Synthesis gas (H_2:CO=1) was supplied during the reaction to keep the total pressure constant.

When the autoclave was cooled after the reaction, a dark-red viscous catalyst phase which consisted of a complex of cobalt carbonyl with compound (I) appeared at the bottom of the autoclave. The remaining gas was discharged and the upper liquid phase consisting of solvent and reaction product was drained off; then 1-decene (0.2 mol), diethyl ether (30 g), and the gas were charged again and the reaction was repeated. Thus, experiments were performed eight times by recycling the same catalyst complex under various conditions. The results of this work are listed in Table I.

The reaction rate increases as the P_{H_2}/P_{CO} ratio is increased; the reaction at a low temperature of 50°C is possible at a P_{H_2}/P_{CO} ratio of 9. The selectivity to linear aldehyde (A_1) increases as the total pressure is increased (compare Runs No. 1 and 7); however, the increase in total pressure due to indifferent gas pressure does not affect the selectivity (compare Runs No. 7 and 8). Catalyst recovery also increases as the P_{H_2}/P_{CO} ratio is increased, and reaches 98% when this ratio is increased to 9.

Reactions at Higher Pressure Conditions

The hydroformylation of 1-decene (0.2 mol) in diethyl ether (30 g) was carried out in the presence of $Co_2(CO)_8$ (4 mmol) and compound (I) (10.67 mmol) in a procedure similar to that for reactions in the low pressure range. Experiments were performed six times by recycling the same catalyst complex under various condi-

Table I. Hydroformylation Under Low Pressure Conditions

Run No.		1	2	3	4	5	6	7	8
P_{H_2}, kg/cm^2		25	10	40	20	45	45	10	10
P_{CO}, kg/cm^2		25	40	10	10	5	5	10	10
P , kg/cm^2		50	50	50	30	50	50	20	50
Temp., °C		70	70	70	70	60	50	70	70
Time, hr		4	20	2	2	3	15	5	10
Conv., %		51.2	73.4	70.0	68.4	64.5	59.1	70.0	52.3
Selectivity %	A_1	73.0	68.8	69.8	63.4	66.0	73.1	54.1	54.1
	A_2	18.8	21.1	20.1	23.4	22.4	18.5	28.6	28.5
	A_3	4.3	5.1	4.7	6.3	5.5	4.4	7.7	8.0
	A_4	3.9	5.0	5.4	6.9	6.1	4.0	9.6	9.4
$k \times 10^2$ min^{-1}		0.30	0.11	1.00	0.96	0.58	0.10	0.40	0.12
Dissolved Co, mgatom		1.92	2.24	0.94	1.01	0.19	0.22	1.54	0.87
Co recovery %		88.0	84.1	92.0	90.7	98.1	97.7	83.8	89.0

Legend: Run No. is the progressive order of catalyst recycling; P is the total pressure equal to $P_{H_2} + P_{CO}$ (except in Run No. 8, where nitrogen pressure (P_{N_2} = 30) is employed); Conv. is the conversion of decene; A_1 = undecanal; A_2 = 2-methyl decanal; A_3 = 2-ethyl nonanal; A_4 = 2-propyl octanal and/or 2-butyl heptanal; k is the first-order rate constant, calculated from conversion x(%) and reaction time t(min), using the relation $k=1/t \ln(100/100-x)$; Dissolved Co is the cobalt which remained dissolved in the reaction product; Co recovery is the percent of cobalt recovered after each run.

tions, the reaction time in each run being fixed at 4 hrs. The results of these experiments are listed in Table II.

Cobalt recovery in the higher-pressure range is generally higher than for the lower-pressure range (compare Tables I and II). It is evident from the results of both Tables I and II that a hydrogen partial pressure of more than 45 kg/cm^2 is preferable in order to achieve good separation of the catalyst after the reaction.

Pressure Effect on Reaction Rate and Selectivity

The selectivity to linear aldehyde (represented by O), and the rate constant k (represented by Δ) is plotted vs. hydrogen partial

Table II. Hydroformylation Under Higher Pressure Conditions

Run No.		1	2	3	4	5	6
P_{H_2}, kg/cm^2		47	94	141	127	47	75
P_{CO}, kg/cm^2		33	66	99	33	10	5
P , kg/cm^2		80	160	240	160	57	80
Temp., °C		70	70	70	70	70	50
Conv., %		24.2	16.8	12.3	45.6	86.0	20.1
Selectivity %	A_1	76.4	79.2	79.3	77.6	68.0	75.0
	A_2	16.5	14.5	14.5	15.0	20.3	17.2
	A_3	3.7	3.5	3.6	3.9	5.9	4.0
	A_4	3.4	2.8	2.6	3.5	5.8	3.8
$k \times 10^2 min^{-1}$		0.12	0.077	0.055	0.25	0.82	0.093
Dissolved Co, mgatom		0.092	0.28	0.16	0.13	0.23	0.22
Co recovery %		98.8	96.5	97.9	98.3	96.9	96.9

Abbreviations: See footnote of Table I.

pressure (see Fig. 1), carbon monoxide partial pressure (see Fig. 2), and total pressure (see Fig. 3). The rate constant increases with an increase in P_{H_2}, increasing more significantly in a low P_{H_2} range of 10 to 20 kg/cm^2. The selectivity to linear aldehyde increases with increase in P_{H_2} in a lower pressure range, but it levels off in the higher pressure range. The rate constant decreases, but the selectivity to linear aldehyde increases with an increase in P_{CO}. The rate constant decreases with increases in total pressure, probably because its decrease due to an increase in P_{CO} is larger than its increase due to an increase in P_{H_2}. The selectivity to linear aldehyde, on the other hand, increases with increasing total pressure in the lower pressure range as a co-effect of P_{H_2} and P_{CO}, but it levels off in the higher pressure range.

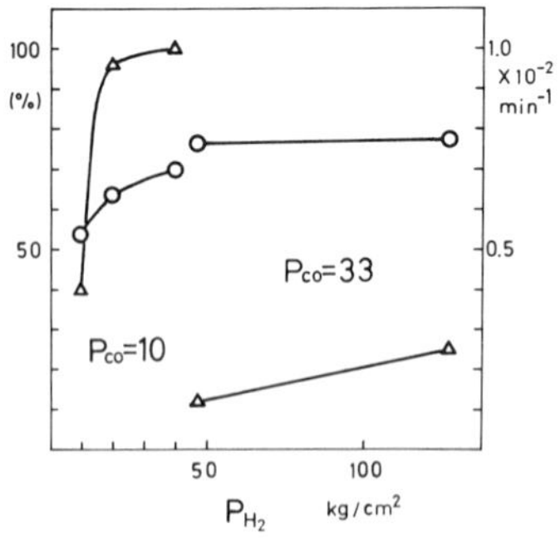

Fig. 1. Effect of hydrogen partial pressure.

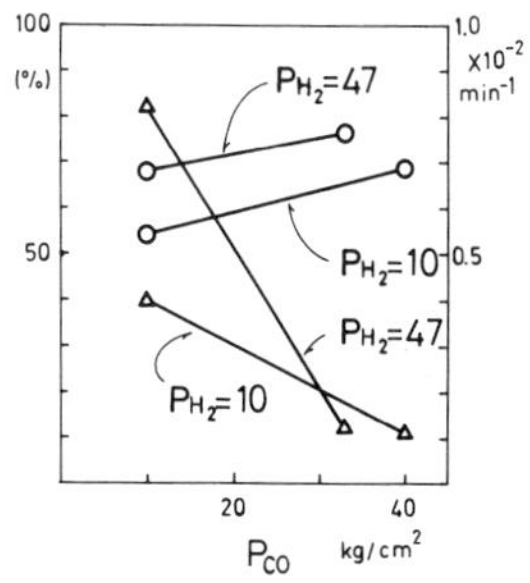

Fig. 2. Effect of carbon monoxide partial pressure.

Purification of Hydrogen by Means of Oxo Process

The hydroformylation of 1-decene (0.1 mol) was conducted in diethyl ether (30 g) in the presence of $Co_2(CO)_8$ (4 mmol) and compound (I) (10.67 mmol). The autoclave was pressurized with hydrogen containing 4 to 6% carbon monoxide to 100 kg/cm^2 (the amount of gas corresponds to about 1 mol), heated to 60°C and agitated at this temperature for 5 hrs; then it was cooled and the remaining gas phase and liquid phase were analyzed. The reaction was repeated with the same catalyst. Results are shown in Table III.

The carbon monoxide initially contained in the hydrogen was almost completely (98%) consumed by the hydroformylation, and the hydrogen containing about 0.1% carbon monoxide remained in the gas phase. Selectivity to linear aldehyde was 62%, and Co recovery was 98 to 99%.

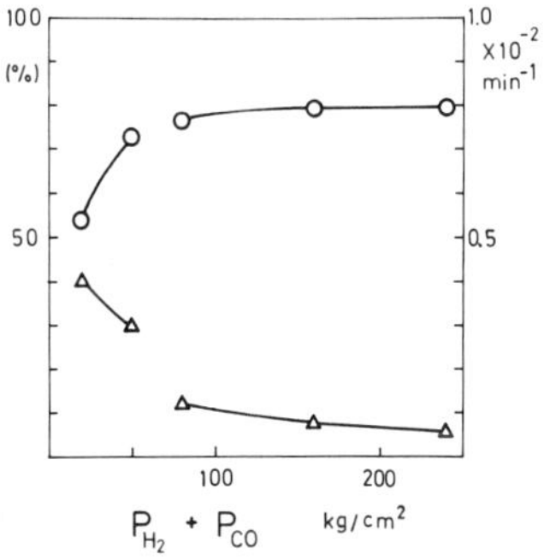

Fig. 3. Effect of total pressure.

DISCUSSION

Since the complex, $H_2Co_3(CO)_9(Py)_5$, is selectively produced by the reaction of $Co_2(CO)_8$ and pyridine in a molar ratio of 1:4 at 100°C, 50 kg/cm^2 (H_2:CO=1) [4], a complex of an analogous structure may also be formed from $Co_2(CO)_8$ and compound (I) under similar conditions. The catalyst structure and the mechanism of catalyst separation may be understood by a model shown in Fig. 5 where compound (I) is represented by N ~ N.

The higher the unit complex concentration is during the reaction, the higher the probability becomes that the unit complex will be linked by N ~ N bridges with neighboring complexes after being cooled to room temperature, thereby becoming insoluble. This agrees with experimental results (see Fig. 4). When the temperature is raised, on the other hand, the linking between complexes by N ~ N bridges may be broken; then the complex dissolves into the solution and catalyzes the hydroformylation. This is in good agreement with what was visually observed through the windows during our experiments.

Since the ionic complex, $H_2Co_3(CO)_9(Py)_5$, has been found to be in equilibrium with such non-ionic species as $Co_2(CO)_8$ at an elevated temperature [3], the following equilibrium may exist during the reaction, where n may be 4 (or 3) and I represents compound (I).

$$H_2Co_3(CO)_9(I)_n + 3CO \rightleftharpoons H_2 + 3/2\ Co_2(CO)_8 + n\ I$$

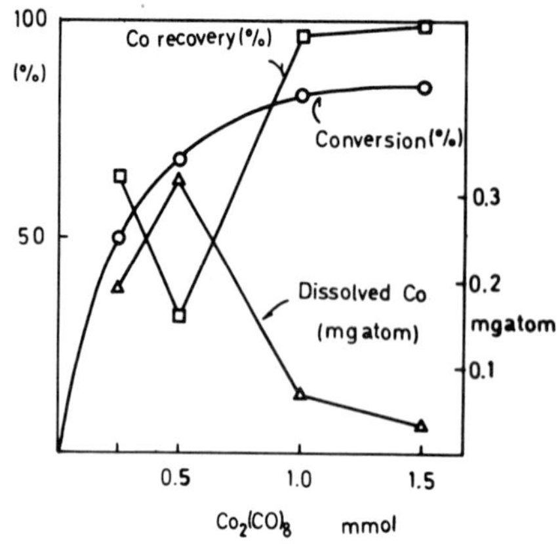

Fig. 4. Effect of catalyst concentration I/Co = 1.33.

As the complex, $H_2Co_3(CO)_9(Py)_5$, reacts with an olefin even at room temperature [4], the complex, $H_2Co_3(CO)_9(I)_n$, may be an active catalyst for hydroformylation. Then, according to the above equilibrium the concentration of active catalyst, $H_2Co_3(CO)_9(I)_n$, will increase as the hydrogen partial pressure is increased, but will decrease as the carbon monoxide partial pressure is increased. The accelerating effect of hydrogen and the retarding effect of carbon monoxide on the reaction rate are thus explained. The effect of hydrogen on the separation of the catalyst may also be understood, since such soluble species as $Co_2(CO)_8$ will increase as the hydrogen partial pressure is decreased, according to the above equilibrium. The reason why the selectivity to linear aldehyde increases with increases in respective partial pressures of hydrogen and carbon monoxide, however, remains unexplained.

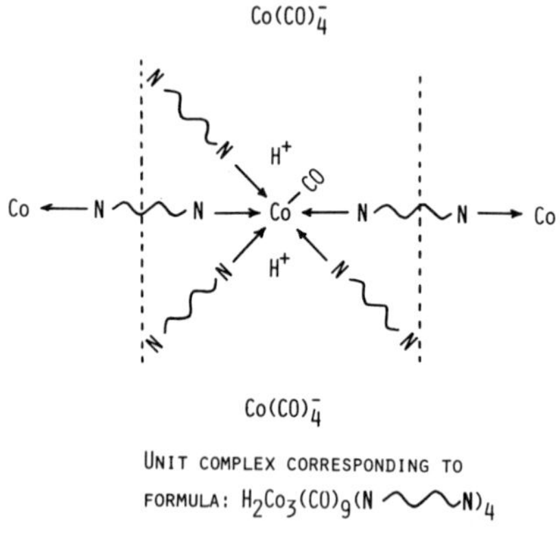

Fig. 5. Possible catalyst structure.

REFERENCES

1. A. J. Moffat, J. Catalysis 18, 193 (1970).
2. A. J. Moffat, J. Catalysis 19, 322 (1970).
3. A. Matsuda, K. Bando, S. Shin, and Y. Horiguchi, in Proceedings 4th AIRAPT Intern. High Pressure Conference, Kyoto, Japan (1974), p. 725.
4. S. Shin, A. Matsuda, J. Nakayama, and K. Bando, Chem. Lett. No. 2, 115 (February 1977).

Table III. Purification of Hydrogen by Means of Oxo Process

Run No.	CO in H_2, %		Conv.,	Selectivity, %				Co re-
	Initial	Final	%	A_1	A_2	A_3	A_4	covery, %
1	3.8	0.06	37.5	62.2	23.0	6.6	8.2	99.3
2	6.2	0.11	60.9	61.9	21.9	7.0	9.2	97.8

Abbreviations: See footnote of Table I.

Q-2

HIGH-PRESSURE SYNTHESIS OF POLYACOHOLS BY CATALYTIC HYDROGENATION OF CARBON MONOXIDE*

R. Fonseca, G. Jenner, A. Kiennemann, and A. Deluzarche

Université Louis Pasteur de Strasbourg
Strasbourg, France

INTRODUCTION

Future efforts to explore the various energy resources should consider the use of coal and heavy fractions of petroleum oil to produce large amounts of carbon monoxide and then reactivating the Fischer-Tropsch process involving CO and H_2 which was developed in Germany during World War II. The CO-H_2 chemistry may lead to various products: Unbranched and branched hydrocarbons, primary monoalcohols, ethylene, α-olefins, polyalcohols, esters, and polymethylenes. Among these reactions, the synthesis of polyhydroxylated compounds is a noticeable example of the homogeneous catalysis, but it is difficult and requires drastic conditions [1,2]. In recent years Union Carbide [2] has actively investigated the synthesis, but despite considerable research, the results described by the patents are difficult to understand and are not very clear regarding the conversion and selectivity. We have investigated the same reaction in order to define the various conditions enabling the direct synthesis of polyalcohols from carbon monoxide and hydrogen.

TECHNOLOGY

The synthesis involving CO and H_2 can be achieved in special autoclaves with operating pressure up to 4 kbar and temperatures up to 350°C. It was therefore necessary to find a high strength steel able to work satisfactorily under high stresses and yet compensate for the deleterious effects of three parameters: (1) High temperatures; (2) Intercrystalline corrosion due to hydrogen diffusion through the steel walls. To counteract this powerful corrosion, it

*Invited paper.

is necessary to select steels containing metallic elements giving easily stabilized carbides (namely tantalum, niobium, vanadium, and titanium); (3) Corrosion due to carbon monoxide which shows its highest activity under pressure between 170 and 350°C. Beyond this region, no alloys seem to be available which are fully inactive against CO attack. Austenitic stainless steels appear to be the best because an attack proceeds selectively by the superficial reaction of CO with the nickel contained in these alloys leading to an increase in the concentration of chromium which is more resistant to carbon monoxide.

We have found [3] that the best material that combines good behavior under high stresses and an excellent inertness against corrosion by hydrogen, carbon monoxide, and organic compounds formed in the reaction is stainless steel Z5 NCTD 26-15 (French AFNOR standard). However, because of the rather poor mechanical properties of this steel when compared to the high resistance steels, the calculation of the stresses for pressures as high as 4 kbar forced us to develop a multilayered vessel consisting of an external jacket of (35 NCD 16) steel and a liner of (Z5 NCTD 26-15) steel. Tightness was achieved without any seal by direct contact cone over cone.

Permeability of titanium steel to hydrogen can reach sufficiently high values at high temperatures to cause a desorption of the walls of the vessel. This desorption occurs when the impregnation coefficient χ for each reaction vessel tends towards a maximum value, χ_m, depending on the thickness of the walls:

$$\chi = \sum_{i=1}^{n} \mu_n \theta_n \qquad (1)$$

where θ is the contact time and μ is the permeability at (P,T) of the steel hydrogen. Desorption is carried out under high vacuum and prolongated heating (200°C for several hours).

RESULTS

Various experimental parameters may influence the course of the synthesis of polyalcohols: The nature of the catalyst, ligand, and solvent; temperature; pressure ratio of CO/H_2; contact time; and additives. Table I sums up, in a very simplified form, the elements used by Union Carbide in the production of ethylene glycol.

From the production elements which will be defined below, the reaction generally provides three main products: ethylene glycol, methanol, and glycerol. Secondary products consisting of mono- and diacetates of ethylene glycol, ethanol, methyl formiate, methyl acetate, water, methane, and CO_2 are present only in small amounts. Table II shows some outlines of the synthesis showing the effect of catalyst, ligand, and solvent respectively.

Table I. Elements Used by Union Carbide in the Production of Ethylene Glycol*

Catalysts	Rhodium catalysts alone or associated with organic salts of Li, Na, Ca, Ba, Sr, Zn
Ligands	Hydroxylated, alkylated or acetylated pyridine or piperidine compounds
Solvents	Tetraglyme, lactones, T.H.F.
Pressure	500 to 3200 bar
Temperature	200 to 260°C

*The amount of ethylene glycol varies according to the experimental conditions.

Considering the catalyst effect, it may be noted that ethylene glycol and/or glycerol are formed in appreciable amounts only with the use of rhodium catalysts, including rhodium foam and rhodium clusters. A cobalt cluster associated with a rhodium catalyst is also effective. Even ruthenium clusters which are found to be inactive, in the synthesis of polyalcohols produce ethylene glycol, though in lesser amounts. Iron catalysts are poor catalysts for this kind of reaction.

The ligand effect is essential to obtain good selectivity in polyalcohols and, as shown in Table II, considerably modifies the course of the reaction. Efficient ligands are pyridine and its hydroxylated derivatives, particularly 3-hydroxy-pyridine giving more than 90% selectivity in ethylene glycol. This high stereospecificity remains obscure.

Solvents containing C-O bonds are favorable for the synthesis of polyalcohols, while aliphatic hydrocarbons are not. Tetrahydrofuran affords the maximum conversion, but limits the reaction to ethylene glycol. The most important yield in glycerol is obtained when tetraglyme is the solvent. Table III shows the influence of temperature, pressure, and CO/H_2 ratio, respectively.

With respect to the temperature effect, it seems that there exists an optimal temperature (230°C) for the synthesis of ethylene glycol. At 266°C the selectivity of polyalcohols decreases, although conversion is higher than at 230°C. At 300°C the conversion decreases markedly, suggesting that the active species of the catalyst are destroyed. At 180°C, the reaction is too slow.

The pressure effect is important. A minimal pressure of 900 bar is required for the formation of ethylene glycol under our conditions. Although Union Carbide reported lower pressures for the synthesis, the reaction times become prohibitive. The higher the pressure, the higher the total yield and the yield of polyalcohols. When the pressure is raised, the yield of glycerol becomes important.

Table II. Influence of Catalyst, Ligand, and Solvent*

Catalyst	Ligand	Solvent	Conversion, %	Selectivity, % EG	GL	ME	Yield of polyols, %
$Rh(CO)_2AcAc$	2-pyridinol	TGM	35.2	65.9	3.7	20.2	24.5
$Rh_4(CO)_{12}$	2-pyridinol	TGM	36.9	56.0	0	29.8	20.6
$Rh_6(CO)_{16}$	2-pyridinol	TGM	35.5	67.8	6.7	17.5	26.4
Rh(foam)	2-pyridinol	THF	24.4	60.2	10.5	10.3	17.2
$Rh_2CO_2(CO)_{12}$	2-pyridinol	TGM	27.4	68.3	0	18.5	18.7
$Ru_3(CO)_{12}$	2-pyridinol	TGM	12.9	22.9	0	16.1	3.0
$Fe(CO)_5$	2-pyridinol	TGM	8.0	4.2	0	4.2	0.4
$Rh(CO)_2AcAc$	none	TGM	19.8	5.2	0	33.1	1.0
	pyridine	TGM	31.4	51.2	0	38.7	16.1
	2-pyridinol	TGM	37.8	67.9	3.5	19.2	27.0
	3-pyridinol	TGM	34.4	91.2	0	5.1	31.4
	4-pyridinol	TGM	30.1	82.4	0	14.0	24.8
	pyrocatechol	TGM	23.9	14.7	0	30.7	3.5
	triphenylphosphine	TGM	8.7	0	0	traces	0
$Rh(CO)_{16}$	2-pyridinol	TGM	35.5	67.8	6.7	17.5	26.4
	2-pyridinol	THF	40.0	64.6	0	23.1	25.9
	2-pyridinol	ME	32.6	56.9	3.1	35.2	19.6
	2-pyridinol	decane	26.3	0	0	28.3	0

EG = ethylene glycol; GL = glycerol; ME = methanol; $Rh(CO)_2AcAc$ = rhodium carbonyl acetyl acetonate; TGM = tetraglyme; THF = tetrahydrofuran.

*Conditions: Temperature (230°C); pressure (1650 to 1750 bar), ratio CO/H_2 (1/2), contact time (5 hrs)

Table III. Influence of Temperature, Pressure, and CO/H_2 Ratio*

Temperature, °C	Pressure, bar	Ratio, (CO/H_2)	Conversion, %	Selectivity, % EG	GL	ME	Yield in polyols
180	1720	1/2	5.6	50.2	0	5.6	2.8
230	"	"	37.8	67.9	3.5	19.2	27.0
266	"	"	45.3	48.5	2.0	48.4	22.0
300	"	"	27.5	6.7	4.9	11.6	3.2
230	945	1/2	7.3	32.1	0	38.6	2.3
"	1660	"	37.8	67.9	3.5	19.2	27.0
"	2270	"	35.9	65.8	12.5	13.2	27.3
"	3645	"	42.5	56.5	16.2	15.8	30.8†
230	1700	1/8.4	41.6	54.2	0	26.9	22.6
"	"	1/2	37.8	67.9	3.5	19.2	27.0
"	"	1/1.5	29.0	80.6	12.5	1.1	27.1
"	"	5/1	11.6	62.0	10.6	7.7	8.4

*Conditions: Catalyst (rhodium carbonyl acetyl acetonate), ligand (2-pyridinol), solvent (tetraglyme), contact time (5 hrs).

†+ 6.3% of heavier products.

EG = ethylene glycol; GL = glycerol; ME = methanol.

Partial pressures of CO and H_2 can also influence both the conversion and the selectivity. Elevated partial pressures of hydrogen lead to high conversions, but the selectivity in polyalcohols is low with the reaction giving mostly methanol. Optimal selectivity and yield are obtained with the stoichiometric CO/H_2 ratio within the limits of 1/1.5 and 1/2.0.

On the basis of our results, it is difficult to draw sufficient conclusions about a precise reactional scheme. We first anticipated that there would be an association between the water produced in the reaction and a primary hydroxylated species. Such an association would be favored by an increase in pressure; however, the addition of water to the starting medium decreases the selectivity of ethylene glycol. Such an association is thus improbable.

With certain cluster catalysts ($Co_2(CO)_8$, $Rh_4(CO)_{12}$, $Rh_6(CO)_{16}$, and $Ru_3(CO)_{12}$ we have shown in the homologation of methanol [4] the existence of two active reaction sites which could interfere. It is possible that in the synthesis of polyalcohols with the same catalysts as cited above, such sites would also exist, but the preceding hypothesis we made [4] (propagation through bridged CO) has so far not been demonstrated. It has been shown [2] that a complex species $Rh_{12}(CO)_{30}^{--}$ is present in the medium if experimental conditions of pressure and temperature allow it. This species would be formed through the equilibria

$$2\ Rh_6(CO)_{15}H^- \underset{}{\overset{H_2}{\rightleftharpoons}} Rh_{12}(CO)_{30}^{--} \overset{CO}{\rightleftharpoons} Rh_{12}(CO)_{34}^{--} \qquad (2)$$

and should be responsible for the synthesis of polyalcohols.

REFERENCES

1. U.S. Patent 2,451,333 (du Pont de Nemours).
2. Belgian Patents 793,086 (Union Carbide);
   Ger. Offen. 2,426,495 (Union Carbide);
   Ger. Offen. 2,426,411 (Union Carbide); etc.
3. G. Jenner and A. Deluzarche, Chem. Ing. Techn. 49, 420 (1977).
4. A. Deluzarche, G. Jenner, A. Kiennemann, and F. Abou Samra, Erdöl und Kohle, in press.

R-1

A SYSTEM OF PRESSURE CALIBRATION FOR THE RANGE 0.05-1.0 MBAR BASED ON SHOCK WAVE EQUATIONS OF STATE FOR Cu, Mo, Pd, AND Ag

H. K. Mao and P. M. Bell

Carnegie Institution of Washington
Washington, D. C. USA

and

J. Shaner and D. Steinberg

Lawrence Livermore Laboratory, University of California
Livermore, California USA

INTRODUCTION

With modern experimental techniques it is possible to conduct a wide range of scientific experiments at static pressures near 1 Mbar and at higher pressures [1,2]. Recent primary calibration [3] based on the pressure-volume functions of Cu, Mo, Pd, and Ag was done for measurement of pressure to 1 Mbar, but the only secondary scale currently available for use at these pressures is the ruby fluorescence pressure gauge. Although the ruby scale is now well calibrated, there is a great need for intercomparison of internal pressure standards that can be adopted more generally in high-pressure research where the ruby fluorescence method cannot be used. Furthermore, the new primary calibration standard can be used to correlate previous data of all types.

This paper is a review of the problems involving various calibration methods that have been used in high-pressure research. There are obvious advantages to interpolating a pressure scale between known experimental points compared with extrapolating to a range in which the uncertainties are unknown. A particular function, whether theoretical or empirical, may apply well for a given material but not for others. There is great danger in extrapolating the pressure-volume relations from low-pressure values, but interpolation is always superior regardless of how well a given empirical curve coincides with low-pressure data.

The concepts of internal and external calibration should be clarified because intercomparison of data may suffer from differences in these procedures. External calibration techniques, such as measurements of electrical resistance correlated with external load, employ extensive properties that cannot be correlated with the internal pressure on the sample in any simple way. Internal calibration methods, however, utilize intensive properties of materials at high pressure that depend only on the stress conditions in the sample. The present choice of using the four metals as internal standards was made to avoid calibration procedures whose accuracy cannot be evaluated.

The four metals Cu, Mo, Pd, and Ag were selected as primary calibrants because reliable shock-wave data were available and their physical properties are representative of many materials. These metals undergo no known phase changes in the experimental range, and they are cubic, thereby reducing problems of orientation and anisotropy in shock-wave experiments. These metals have a broad range of compressibility and strength, but their bulk moduli are all greater than 1 Mbar.

PRIMARY CALIBRATION

The primary pressure-volume curves on which the present system is based are given in Fig. 1. The isotherms are taken from the study of Carter et al. [4]. It was estimated that the isotherms are systematically too high in pressure by a maximum of 2% for Cu and Ag and 4% for Mo because of the strengths of the metals. The addition of uncertainties due to error of the Hugoniot data and to the thermal corrections resulted in the following estimations of the uncertainties of the isotherms: Mo, +2%, -6%; Cu, +2%, -4%; Pd, +3%, -4%; Ag, +3%, -5%. As indicated below, static pressure experiments, in which the specific volumes of three metals were measured simultaneously, gave the maximum error in pressure between any two metals of 5%.

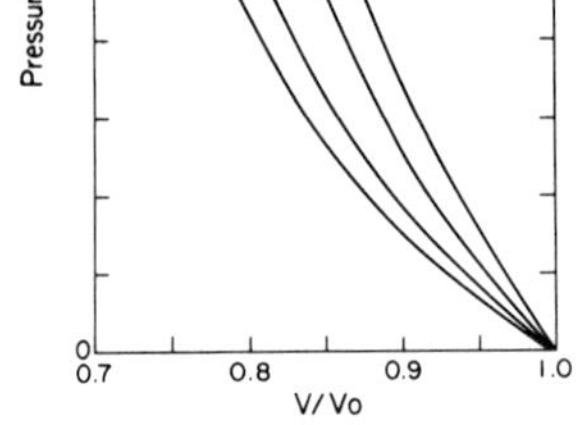

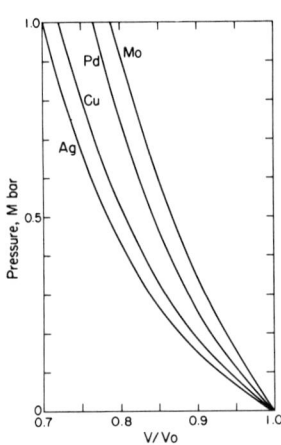

Fig. 1. Pressure-volume curves for Cu, Mo, Pd, and Ag after Carter et al. [4].

RUBY-FLUORESCENCE SCALE

Figure 2 is a plot of the wavelength shift of the ruby R_1 fluorescence line with pressure. Data for the four metals are also plotted because the R_1 wavelength was measured in the same experiments in which sets of three of the metals were x-rayed for specific volume measurements. The near-linearity of the ruby line shift and, indeed, the

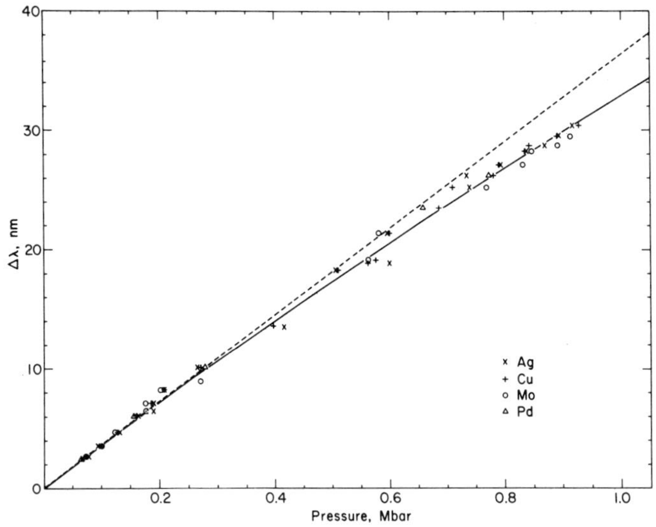

Fig. 2. Primary calibration of the ruby pressure gauge. Experimental points were determined in the diamond-window, high-pressure cell (after Mao et al. [3]). Dashed line is a linear extrapolation of the low-pressure data of Piermarini et al. [9]; curved line is a least-squares fit to the points according to P (Mbar) $= 3.81[(1 + \Delta\lambda/\lambda_0)^5 - 1]$.

theoretical basis for the shift are unexplained, but the use of the shift as an indicator or pressure gauge depends only on its reproducibility.

The errors of the scale are discussed in detail by Mao et al. [3]. The pressures obtained by this method are low compared with the mean pressure because of nonhydrostatic effects and effects of the arrangements for the x-ray diffraction technique used in the study. The maximum random error of the calibration was estimated to be ±6% of the pressure from 0.5 to 1.0 Mbar. The data can be represented by various functions with equally suitable fits to the experimental points. An empirical Murnaghan-type equation that assumes

$$\lambda \frac{\partial P}{\partial \lambda} = a + bP \qquad (1)$$

(where λ is the wavelength of the R_1 line, P is pressure, and a and b are constants) is the best fit to the data and is given by

$$P(\text{Mbar}) = \frac{a}{b}\left[\left(\frac{\lambda_0 + \Delta\lambda}{\lambda_0}\right)^b - 1\right] \qquad (2)$$

where a = 19.04, b = 5, λ_0 is the wavelength of the R_1 line at 1 bar, $\Delta\lambda$ is the change of wavelength from 1 bar to that at pressure P of the R_1 line. It should be emphasized that the quality of such an equation in predicting the pressure depends on how well the parameters fit the data. This equation can be used to 1 Mbar within the stated accuracy. Obvious dangers of extrapolation to higher pressures mentioned above exist, and so the estimated uncertainty at

1.5 Mbar is 15% of the pressure.

MgO AND ε-Fe SCALES

Mao and Bell [2] measured the specific volumes of MgO and ε-Fe to 1 Mbar in experiments in which MgO and ε-Fe could be x-rayed simultaneously and in which ruby crystals were included with the charge. The pressure-volume curves can now be used as secondary pressure calibrants, based on the calibrated ruby gauge.

The data for MgO are given in Fig. 3. The best curve fit to the data is given by a first-order Murnaghan equation of state as follows:

$$V/V_0 = [1 + 2.81P(\text{Mbar})]^{-0.222} \quad (3)$$

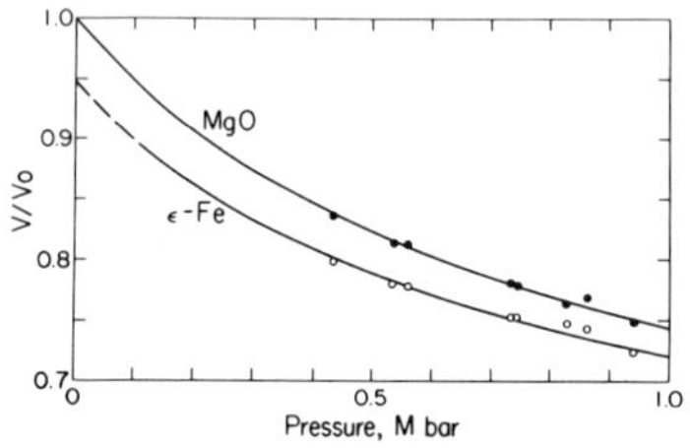

Fig. 3. Pressure-volume curves for MgO and ε-Fe, after Mao and Bell [2].

where V_0 is the specific volume of MgO at 1 bar and V is the specific volume at pressure P. It is significant that this equation gives a value for the zero pressure derivatives of the bulk modulus of MgO that agrees with values determined independently by ultrasonic techniques [5]. This agreement indicates that extrapolation of this function (based on the $(B_t)_0$ of Anderson and Andreatch [5] for MgO would have been justified within experimental uncertainties.

The ε-phase of metallic iron is unstable at 1 bar, and thus curve-fitting to zero-pressure ultrasonic data was not possible, but the present data fall on a curve extrapolated from previous results obtained at lower pressure. Plotted in Fig. 3 are the present data and the data of Mao et al. [6] fitted to the following first-order Murnaghan equation of state:

$$V(\text{cm}^3/\text{mol}) = 6.7155[1 + 3.08\ P(\text{Mbar})]^{-0.196} \quad (4)$$

It is clear that extrapolation based on this form of equation would have been correct to 1 Mbar.

SCALES BASED ON NaCl, METALS, ALLOYS, AND OXIDES

The use of sodium chloride as a secondary standard is limited at this time to pressures below approximately 0.3 Mbar, where a phase transition occurs [7]. The present data on MgO and ruby are the basis of its calibration. In other studies, Weaver et al. [8] made simultaneous volume measurements on MgO plus NaCl and Piermarini et al. [9] and Yagi [10] carried out experiments in which the shift of the ruby R_1 fluorescence line was correlated with the volume compression of NaCl.

The resulting secondary calibration scale for NaCl (based only on the present MgO and ruby data) is plotted in Fig. 4. If the volume equation of state for NaCl that results from the present calibration is compared with the semi-empirical equation of state of Decker [11], the agreement is remarkable.

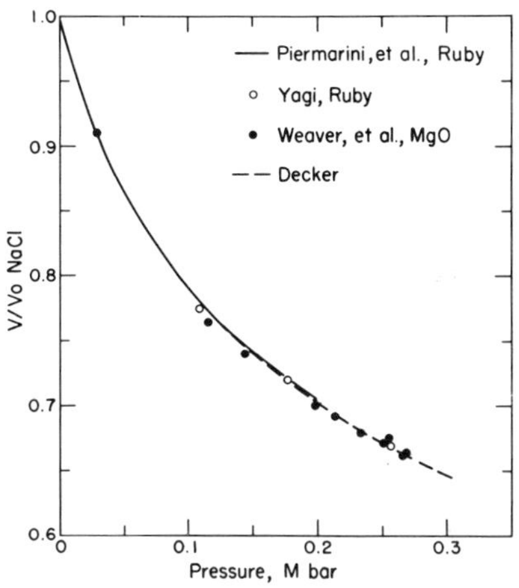

Fig. 4. The NaCl pressure scale. Open circles, after Yagi [10], based on the present ruby pressure scale; solid circles, after Weaver et al. [8], based on the present MgO pressure scale; solid curve, after Piermarini et al. [9], based on the present ruby pressure scale; dashed curve, Decker's [11] equation of state.

Confusion has arisen from Decker's equation because he made reference to a central force model in his formulation [12]. The central-force approximations, however, are known to have inherent difficulties [13], and unless a fundamental theory is proposed, one would not expect that such a model could be used to predict the volume-pressure relationship under an infinite range of conditions. Ruoff and Chhabildas [12] also observed that Decker's model for NaCl could not be extrapolated infinitely, but the fact that this semi-empirical model does not satisfy 0 K conditions is an inadequate basis for criticism, because such objections do not indicate where the model departs from the absolute pressure-volume function of the material. It is important only that the P-V relation predicted by Decker's equation of state be accurate from 1 bar to 0.3 Mbar, which is the useful range for NaCl.

The equation of state of Chhabildas and Ruoff [14] does not agree with the present calibration, but this is not surprising because their equation is based on data obtained at 0.0075 Mbar, and thus a 40-fold extrapolation is required to 0.3 Mbar. Chhabildas and Ruoff's extrapolated curve is approximately 10% too low in pressure at 0.3 Mbar. The uncertainties attached to this extrapolation by Chhabildas and Ruoff could be as high as ±30% of the pressure at 0.3 Mbar.

The NaCl scale has been used as a tertiary reference for numerous materials. For example, the data for materials listed in Table I (after Bassett and Takahashi [15]) can be considered additional secondary pressure calibrants with reference to the present scale. All the above materials will serve for internal calibration. Figure 5 is a summary of the system showing the generations of calibration, and showing the standards that were studied simultaneously.

Table I. Metals, Alloys and Oxides Whose Pressure-Volume Functions Can Be Used as Secondary Pressure Scales with Reference to the Sodium Chloride Scale (after Bassett and Takahashi [15]

Fe	Magnetite, Fe_3O_4
Pb	γ-Fe_2SiO_4
Re	γ-$(Fe_{0.9}Mg_{0.1})SiO_4$
Fe-5.2Ni	γ-$(Fe_{0.8}Mg_{0.2})_2SiO_4$
Fe-10.3Ni	γ-CO_2SiO_4
Fe-7.2Si	γ-Ni_2SiO_4
Fe-25Si	Pyrope, $Mg_3Al_2Si_3O_{12}$
Wüstite, $Fe_{0.924}O$	Pyrope-almandite, $(Mg_{0.6}Fe_{0.3})_3Al_2Si_3O_{12}$
Hematite, Fe_2O_3	Almandite-pyrope, $(Mg_{0.2}Fe_{0.8})_3Al_2Si_3O_{12}$
Ilmenite, $(Mg,Fe)TiO_3$	Almandite, $Fe_3Al_2Si_3O_{12}$

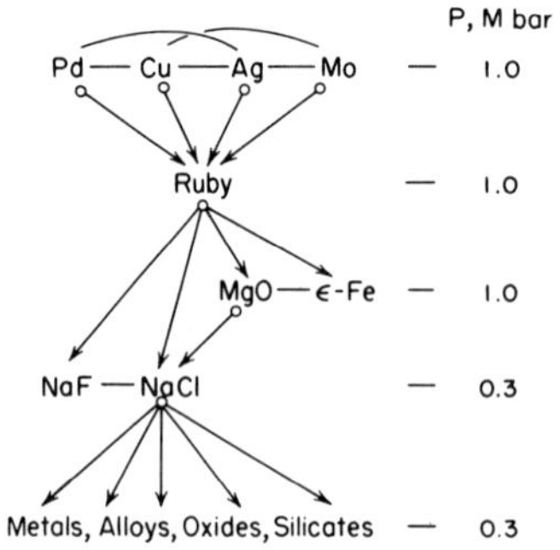

Fig. 5. Schematic diagram of the present system of calibration in the diamond-window, high-pressure cell. Lines connecting two materials indicate that they were studied in simultaneous experiments. Arrows point to material calibrated; circled materials are calibrants. Primary calibration of the four metals Cu, Mo, Pd, Ag is the shock-wave data of Carter et al. [4].

FIXED-POINT CALIBRATION

Fixed points include phase changes, electrical resistance changes, and other abrupt phenomena that are used for pressure calibration. The changes observed at fixed points are physical properties of the materials and are related to thermodynamic functions, but the observations include a range of pressures because of effects outlined below.

The most common source of error in comparing fixed points is instrumental hysteresis that occurs independently from other errors. Even in comparisons of the same reaction in essentially the same type of apparatus, significant differences in pressure are observed [16], but corrections can be applied to reduce the errors in routine experiments.

Reaction hysteresis is a function of kinetic effects and also of the ability of the observer to detect reaction. Under hydrostatic conditions, the thermodynamic equilibrium value for a transition is within the hysteresis bracket of the first appearance of the high-pressure assemblage and the first appearance of the low-pressure assemblage. This type of observation is valid for a particular experimental design if the fixed point is calibrated and reproducible.

Nonhydrostatic stress causes significant departure of the accepted values of fixed points from those determined under hydrostatic conditions. The Gibbs free energy under such conditions is not a state function and is not defined generally. Only local equilibrium conditions along a specified direction and path can be treated by rigorous analysis because a phase transition depends on the orientation of the stress field. Factors such as the shape and distribution of the new phase formed with respect to the old phase and the crystallographic orientation of both phases with respect to the stress field must be considered [17-21]. The equilibrium pressure under hydrostatic conditions may not be within the experimental bracket obtained under nonhydrostatic conditions, and a sharp, nonsluggish transition can produce a two-phase assemblage over a range of pressure. The precision of fixed-point calibrations may be high, but because of nonhydrostatic effects the accuracy is usually unknown; therefore, application of fixed-point data above 0.1 Mbar must be made with caution.

Four fixed-point calibration transitions can be correlated with the present primary and secondary scales at pressures above 0.1 Mbar at 25°C, subject to the errors and uncertainties described above.

α-ε PHASE TRANSITION IN Fe

The data of Mao et al. [6] indicate that the α-ε transition in Fe is between 0.11 and 0.13 Mbar (no data are available between these pressures) calibrated against the NaCl and ε-Fe scales. These observations are based on the first appearance of the ε phase, and it is noted that some of the α phase persisted to 0.16 Mbar as pressure was increased. On release of pressure some of the ε phase persisted down to 0.07 Mbar. The coexistence of the α and ε phases in this range of pressure is caused by nonhydrostatic stress, not by pressure gradients or by reaction kinetics (sluggishness). The reaction is fast, nearly instantaneous. Clendenen and Drickamer [22] noted that the α phase persisted to 0.3 Mbar in their experiments.

FCC-HCP PHASE TRANSITION IN Pb

Mao et al. [23] observed the fcc-hcp transition in Pb at 0.13 Mbar, based on the first appearance of the hcp phase calibrated against the fcc Pb, α and ε-Fe scales. The fcc-phase persisted up to 0.16 Mbar on increasing pressure, and the hcp phase, down to 0.1 Mbar on decreasing pressure, again because of nonhydrostatic stress. Mao et al. [23] observed that the fcc phase persisted to higher pressures in their experiments than did the α-Fe phase. The reaction is rapid.

B1-B2 PHASE TRANSITION IN NaCl

The B1-B2 phase transition in NaCl has been widely studied, and for the reasons stated above, one would expect a range of values from different experimenters. Experiments in the diamond-window, high-pressure cell by Bassett et al. [7] and by Piermarini et al. [24], based on observations of the first appearance of the B2 phase calibrated against the NaCl scale place the transition pressure at 0.30 Mbar. The B1 phase persists to pressures above 0.33 Mbar; and the B2 phase, below 0.25 Mbar. Hugoniot data for the (100) and (111) directions indicate transition pressures of 0.26 and 0.23 Mbar, respectively, at the high temperatures of the shock-wave experiments [25].

The use of the B1-B2 transition in NaCl is subject to the same difficulties as those described above for all solid-solid transitions. Ruoff and Chhabildas [12] proposed an "upper bound" of estimated pressure for the NaCl transition based on the temperature corrections of the shock-wave value for the (111) direction, and on the systematic variation of entropy vs. volume for other materials taken from Bassett et al. [7]. It should be noted, however, that in shock-wave experiments the stresses are uniaxial. The equilibrium pressure may be different from the values measured for the (100) and (111) directions, and thus the values cannot be considered bounds.

The systematic variation of entropy-volume as originally employed by Bassett et al. [7] is a purely empirical formulation whose thermodynamic basis was not explored by Ruoff and Chhabildas [12]. The properties of sodium chloride were extrapolated to a point outside the range of the entropy volume function, by a factor of two. The uncertainties of this extrapolation are too great ($\pm$ 1 cal/deg/mol, corresponding to $\pm$ 40 kbar) and thus, combined with the uncertainties in the (111) shock pressure stated above, Ruoff and Chhabildas' estimate of the transition at 0.257 Mbar carries two significant figures more than it should. Within the uncertainties, therefore, the estimate does not differ from the calibrated value of 0.30 Mbar.

ANTIFERROMAGNETIC-PARAMAGNETIC TRANSITION IN Fe_2O_3

The only fixed-point calibration for pressures above 0.5 Mbar that can be correlated with the present system is the magnetic transition in Fe_2O_3 observed by Mao et al. [3]. This transition was observed from ^{57}Fe Mössbauer resonance. The first appearance of the transition is at 0.60 Mbar with the ruby R_1 gauge. Some of the low-pressure phase of Fe_2O_3 persisted to 0.67 Mbar in these experiments.

CONCLUSION

The present calibration system has its primary base on the shockwave equations of state of Cu, Mo, Pd, and Ag. The pressure shift of the ruby R_1 fluorescence line was calibrated in the same experiments and is used as a secondary calibration for other continuous internal standards, and for the less accurate, fixed-point calibrations. If the accuracy of the primary shock-wave data is improved, appropriate adjustments can be made in the secondary scales.

REFERENCES

1. H. K. Mao and P. M. Bell, Science 191, 851 (1976).
2. H. K. Mao and P. M. Bell, Carnegie Inst. Washington Year Book, 76, 519 (1977).
3. H. K. Mao, P. M. Bell, J. Shaner, and D. Steinberg, J. Appl. Phys. 49, 3276 (1978).
4. W. J. Carter, S. P. Marsh, J. N. Fritz, and R. G. McQueen, NBS Spec. Publ. 326, 147 (1971).
5. O. L. Anderson and P. Andreatch, Jr., J. Am. Ceram. Soc. 49, 404 (1966).
6. H. K. Mao, W. A. Bassett, and T. Takahashi, J. Appl. Phys. 38, 272 (1967).
7. W. A. Bassett, T. Takahashi, and Y. K. Campbell, Trans. Am. Cryst. Assoc. 5, 93 (1969).
8. J. S. Weaver, T. Takahashi, and W. A. Bassett, NBS Spec. Publ. 326, 189 (1971).
9. G. J. Piermarini, S. Block, J. D. Barnett, and R. A. Forman, J. Appl. Phys. 46, 2774 (1975).
10. T. Yagi, Carnegie Inst. Washington Year Book, 76, 528 (1977).
11. D. L. Decker, J. Appl. Phys. 42, 3239 (1971).
12. A. L. Ruoff and L. C. Chhabildas, J. Appl. Phys. 47, 4867 (1976).
13. L. Thomsen, J. Phys. Chem. Solids 31, 2003 (1970).
14. L. C. Chhabildas and A. L. Ruoff, J. Appl. Phys. 47, 4182 (1976).
15. W. A. Bassett and T. Takahashi, Adv. High Pressure Res. 4, 165 (1974).
16. W. Johannes, P. M. Bell, H. K. Mao, A. L. Boettcher, D. W. Chipman, J. F. Hays, R. C. Newton, and F. Seifert, Contrib. Mineral. Petrol. 32, 24 (1971).
17. J. W. Gibbs, in Collected Works of J. Willard Gibbs, Yale University Press, New Haven (1906).
18. W. B. Kamb, J. Geol. 67, 153 (1959).
19. W. B. Kamb, J. Geophys. Res. 66, 259 (1961).
20. A. McLellan, J. Geophys. Res. 71, 4341 (1966).
21. Y. Ida, J. Geophys. Res. 74, 3208 (1969).
22. R. L. Clendenen and H. G. Drickamer, J. Phys. Chem. Solids 25, 865 (1964).
23. H. K. Mao, T. Takahashi, and W. A. Bassett, Carnegie Inst. Washington Year Book 68, 251 (1970).
24. G. J. Piermarini and S. Block, Rev. Sci. Instrum. 46, 973 (1975).
25. J. N. Fritz, S. P. Marsh, W. J. Carter, and R. G. McQueen, NBSSpec. Publ. 326, 201 (1971).

R-2

CALIBRATION OF THE RUBY-PRESSURE-SCALE AT LOW TEMPERATURES

R. A. Noack and W. B. Holzapfel

Max-Planck-Institut für Festkörperforschung

Stuttgart, W. Germany

INTRODUCTION

Development and accurate calibration of the ruby manometer [1-4] led to rapid progress in diamond anvil cell high-pressure techniques [5]. Applications of the ruby manometer at low temperatures [6], however, suffered from the fact that the pressure dependence of the ruby R_1 luminescence line had been calibrated accurately only at room temperature [4], and the same physical reasons which are responsible for the temperature dependence of the R lines at normal pressure could also lead to a variation in the pressure dependence of these lines with temperature.

A first attempt to calibrate the ruby manometer at low temperatures by the use of fixed points and cooling of a diamond anvil cell at constant load actually indicated [7] a small decrease in the pressure dependence of $d\nu_{R1,2}/dp$ from a value of $-0.757(4)cm^{-1}/kbar$ at 293K [4] to a value of $-0.73(2)cm^{-1}/kbar$ at 166.8 K. Since this low temperature calibration was based, however, on the assumption of a constant pressure shift upon cooling at constant applied force, a more direct calibration was desirable.

EXPERIMENTAL TECHNIQUE

Figure 1 illustrates the pressure vessel which has been used in the present study with helium gas pressures up to 12 kbar at temperatures between 4.2 and 360 K. The vessel is connected to a commercial two-stage gas compressor [8] by 3M-0.025 capillaries and is clamped to the bottom of the liquid helium tank of the cryostat by copper flanges. The body of the pressure vessel is supported by a binding ring which is forced over the slightly conical wall of the pressure vessel by the nut which is shown in Fig. 1

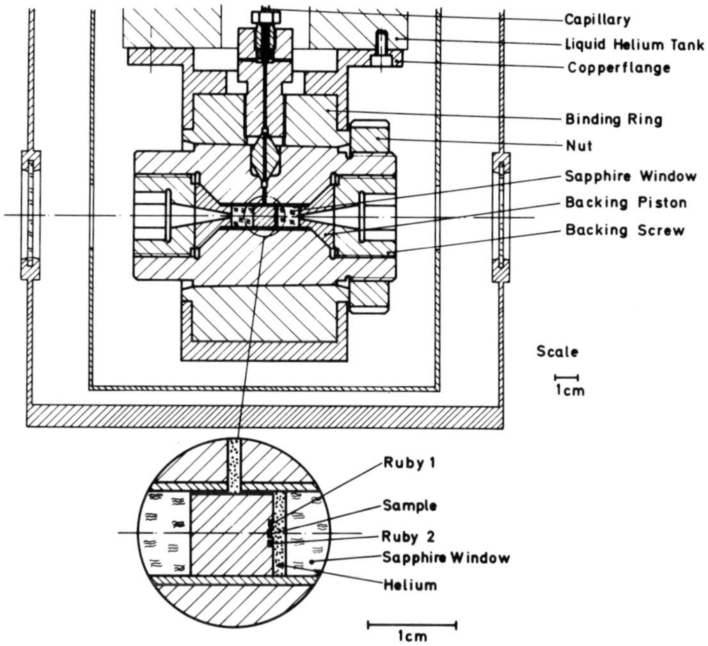

Fig. 1. Helium gas pressure vessel for optical studies at low temperatures.

as a ring around the right-hand side of the pressure vessel. The two sapphire windows of the pressure vessel are glued to their backing piston with a thin layer of epoxy resin. The pistons are sealed with packings of brass, copper, and Viton rings. The conical central openings in the steel plugs as well as in their backing screws determine the optical aperture of 1/6 for the sample chamber. The insert in Fig. 1 gives an enlarged view of the sample chamber. The samples are usually inserted into small cavities on one side of the sample holder and a spring (not shown in Fig. 1) pushes the sample holder against the window to hold the samples in their cavities free of any uniaxial stress. Two ruby samples were used in the present experiments. Ruby 1 contained 0.5 wt.% Cr_2O_3 in the Al_2O_3 [4], and Ruby 2 less than 0.05 wt.% Cr_2O_3 [9].

The luminescence of the ruby samples was excited with the green (19436 cm^{-1}) light of the focused beam of an argon ion laser. The spectra were recorded with a 1/2 m double monochromator and with standard photoncounting techniques. The reproducibility of the wavenumber scale was checked in each run by simultaneous measurements of sharp atomic emission lines from gas lamps using either the 14431.1 or 14219.9 cm^{-1} lines of neon or the 14356.7 cm^{-1} line of argon. This technique allowed measurement of shifts of the ruby luminescence lines at low temperature with an accuracy of typically $\pm$ 0.2 cm^{-1}. Typical spectra are shown in Fig. 2.

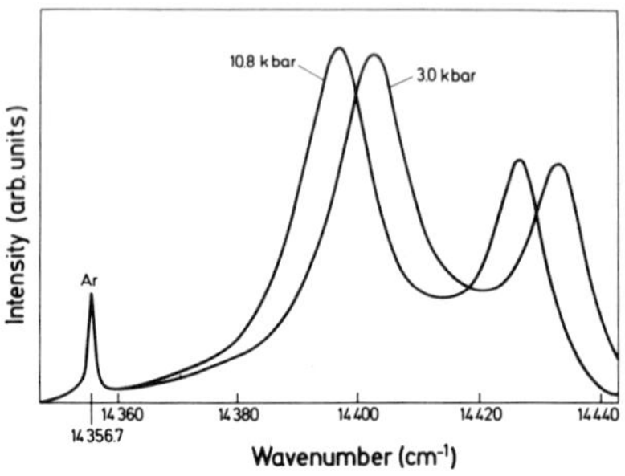

Fig. 2. Typical ruby luminescence spectra at 283 K.

Since the ruby luminescence lines depend not only on pressure but also on temperature [10], the temperature of the high-pressure vessel was carefully controlled by thermocouples. During the pressure runs at elevated temperatures, the temperature of the vessel changed at the most by $\pm$ 0.2°K. On the other hand, the power of the exciting laser light was always kept constant at 10 mW and measurements at higher power levels showed that even 20 mW did not produce any noticeable shift due to possible over heating.

The helium gas pressure P was measured with a commercial manganin resistance gauge [8] which had been calibrated against a calibrated Heise-Bourdon manometer with an accuracy of $\pm$ 3 bar up to 4 kbar. Within this pressure range, the manganin pressure reading P_M shows systematic nonlinear deviations from the gas pressure P of up to $\pm$ 8 bar. This nonlinear variation was determined more accurately by a calibration in which the freezing of mercury was recorded by the resistivity change of mercury at constant temperature as a function of P_M. The literature values [11] of 7.61(1) kbar and 12.13(1) kbar were used for the melting pressures at 0.25(5)°C and 22.85(5)°C, respectively. A quadratic interpolation of the form $P = P_M(1-\delta \cdot P_R) + \delta \cdot P_M^2$ resulted in a value $\delta = 4 \pm 1$ $Mbar^{-1}$ which compares reasonably with the literature values [11] of $3 \pm 1/Mbar^{-1}$. The linear coefficient $(1-\delta \cdot P_R)$ adjusts the scales to coincide at the reference pressure P_R. The value P_R = 4 kbar was chosen in the present calibration. The maximum error in the quadratic coefficient δ amounts to $\pm$ 70 bar at 10 kbar. The total error in the present pressure determinations is therefore smaller than $\pm$ 1% between 1 and 10 kbar and typically $\pm$ 10 bar at lower pressures.

In the ruby luminescence measurements at 4.2 K, the samples were embedded in solid helium. Therefore, the pressure on the sample was different from the gas pressure measured by the manganin gauge at room temperature. From the construction of the pressure vessel, it is plausible that the freezing of the helium occurred first in the capillary and, therefore, resulted in a freezing at constant volume in the sample cavities. If this assumption is made, one can derive the pressure at the sample from the measured pressure before freezing by the use of the tabulated equation of state of helium [12]. In any case, this procedure will give at least lower and upper limits for the pressure at the sample in the solid helium.

RESULTS

The effect of pressure on the R_1 and R_2 ruby luminescence lines is shown in Fig. 3. Since both lines shift at the same rate within the present experimental accuracy, the shift of both lines is represented by the same experimental points referring either to the left-hand scale for the R_1 line or to the right-hand scale for the R_2 line. The crosses represent the results for Ruby 1 and the heavy points for Ruby 2. The results for Ruby 2 at 4.2 K are represented by open circles since these data involve the assumption that the helium pressure medium freezes in the sample cavities at constant volume and not at constant pressure.

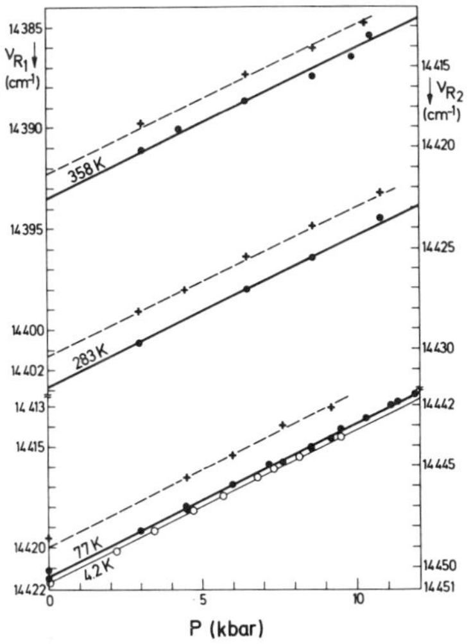

Fig. 3. Effect of pressure and temperature on the ruby R_1 and R_2 luminescence lines. Legend: + Ruby with 0.5 wt.% Cr_2O_3; •,o Ruby with less than 0.05 wt.% Cr_2O_3.

The constant slope $d\nu_{R1}/dp=-0.76(2)cm^{-1}/kbar$ of all the broken lines for Ruby 1 as well as of the continuous lines for Ruby 2 fits all the present data within the experimental accuracy. The estimated uncertainty of 3% includes the possible errors from both the pressure and wavenumber determinations.

Two facts should be noticed: (1) The present value $d\nu_{R1}/dp=-0.76(2)cm^{-1}/kbar$ is in close agreement with the literature value $d\nu_{R1}/dp=-0.757(4)cm^{-1}/kbar$ [4] which had been determined by comparison with the lattice parameter changes of NaCl in the pressure range from 0 to 200 kbar at room temperature. Since the equation of state of NaCl includes an uncertainty of about 2% in this pressure range [4,11], both the previous [4] and the present calibration of the ruby R^1 line shift have about the same absolute accuracy of $\pm$ 3%. (2) In contrast to the previous low-temperature calibration of the ruby R_1 line shift [7], which indicated an increase of $d\nu_{R1}/dp$ between 300 and 163 K from -0.76 to -0.73(2)cm^{-1}/kbar, the present measurements do not confirm this temperature dependence but give a temperature independent value of -0.76(2)cm^{-1}/kbar between 0 and 300 K within the given 3% accuracy.

The possible effect of higher temperatures on the ruby R_1 line pressure dependence can be estimated from the following considerations: The theoretical expression for the temperature shift of the ruby R_1 luminescence line is given [10] by

$$\Delta\nu_{R1}(T) = \alpha(T/T_D)^4 \int_0^{T_D/T} (x^3/(e^x-1))dx \qquad (1)$$

where higher order terms have been neglected, T_{D_1} = 760 K may be used as an effective Debye temperature and α=-400 cm^{-1} represents the electron phonon coupling parameter. From this expression, one obtains for the temperature dependence of the pressure derivative

$$d\nu_{R1}/dp\,\big|_T = d\nu_{R1}/dp\big|_{OK} + (\gamma_\alpha/B)\Delta\nu_{R1}(T) - (\gamma_D/B)\alpha(T/T_D)^4 \int_0^{T_D/T} (x^4e^x/(e^x-1)^2)dx, \qquad (2)$$

where γ_α = -d lnα/d lnV represents the unknown volume dependence of the coupling parameter α, γ_D = -d lnT_D/d lnV is the Debye-Grüneisen parameter and B is the bulk modulus of ruby. For low temperatures $T \ll T_D$, the last term in this expression can be replaced by $-(\gamma_D/B)4\Delta\nu_{R1}(T)$. With reasonable estimates for γ_D, one finds that this term increases from 0 to 0.05(2) cm^{-1}/kbar between 0 and 300 K. The fact that the experimental data (Fig. 3) show no temperature dependent slope within the present accuracy of $\pm$ 0.02 cm^{-1}/kbar indicates therefore, that the last two terms in equation (2) must compensate each other at least partly. This means that $\gamma_\alpha \simeq 4\gamma_D$. From this point of view, ruby luminescence measurements under pressure with an order of magnitude higher precision and preferably an even wider temperature range would be very useful for an accurate determination of γ_α and for a test of the theoretical equation (2).

ACKNOWLEDGMENT

The authors would like to thank E. U. Franck for the permission to calibrate one Heise-Bourdon gauge on the free-piston gauge of his laboratory. The authors benefited from the skillful technical assistance by W. Böhringer.

NOTATION

ν_{R1} = wavenumber of the ruby R_1 luminescence line

p = pressure

p_M = pressure reading of the manganin cell

p_R = reference pressure for manganin cell calibration

δ = parameter for the non-linear (quadratic) variation of the maganin pressure reading

T = temperature

$\Delta\nu_{R1}(T) = \nu_{R1}(T) - \nu_{R1}(0)$ = temperature dependence of ν_{R1}

α = electron-phonon coupling parameter for ruby R_1-luminescence

T_D = effective Debye temperature for ruby R_1-luminescence

V = specific volume of ruby

γ_α = negative logarithmic volume derivative of α

γ_D = Debye - Grüneisen parameter

B = bulk modulus of ruby

REFERENCES

1. R. A. Forman, G. J. Piermarini, J. D. Barnett, and S. Block, Science 176, 284 (1972).
2. J. D. Barnett, S. Block, and G. J. Piermarini, Rev. Sci. Instr. 44, 1 (1973).
3. G. J. Piermarini, S. Block, and J. D. Barnett, J. Appl. Phys. 44, 5377 (1973).
4. G. J. Piermarini, S. Block, J. D. Barnett, and R. A. Forman, J. Appl. Phys. 46, 2774 (1975).
5. S. Block and G. J. Piermarini, Physics Today 29, 44 (1976).
6. R. S. Hawke, K. Syassen, and W. B. Holzapfel, Rev. Sci. Instr. 45, 1598 (1974).
7. D. M. Adams, R. Appleby, and S. K. Sharma, J. Phys. E. 9, 1140 (1976).
8. Harwood Engineering, private communication.
9. H. d'Amour, D. Schiferl, W. Denner, Heinz Schulz, and W. B. Holzapfel, to be published in J. Appl. Phys.
10. D. E. McCumber and M. D. Sturge, J. Appl. Phys. 34, 1682 (1963).
11. D. L. Decker, W. A. Bassett, L. Merrill, H. T. Hall, and J. D. Barnett, J. Phys. Chem. Ref. Data 1, 773 (1972).
12. I. L. Spain and S. Segall, Cryogenics 11(1), 26 (1971).

R-3

LINEAR RUBY SCALE AND ONE MEGABAR?

A. L. Ruoff

Cornell University
Ithaca, New York USA

INTRODUCTION

This paper is about the ruby R_1 fluorescence, its nonlinear pressure dependence at high pressures, and matters related to this, such as claims of one megabar in opposed anvil diamond devices. Experiments will be described which give upper bounds on the pressures available using supported uniaxial opposed cemented tungsten carbide anvils and which give upper bounds on the pressure attainable using opposed diamond anvil devices. Direct static determination of the transition pressures of GaP by two methods will be described; both lead to a pressure of 17 to 18 GPa, substantially less than the 22 GPa found on the linear ruby scale. Then an example is given in which the use of shock-based 'marker materials' such as silver, when used as the basis of pressure measurement in x-ray diffraction experiments, leads to bulk moduli of cubic carbides (at pressure) which are in extreme disagreement with the expected values. Moreover, use of the marker method has led to indications of pressures of 55 GPa in uniaxial supported opposed anvil devices of 3% cobalt cemented tungsten carbide; however, the attainable pressures in such devices do not exceed about 20 GPa.

More recently the ruby shift has been calibrated by use of x-ray techniques using several cubic metals as 'markers' at higher pressures; again the equation of state of these markers comes from shock data. Again, the pressures indicated by the marker method (100 GPa) are very much higher than those calculated for the yielding of the diamond anvils (50 GPa), larger plastic deformation of diamonds (70 GPa), or even the calculated limiting pressure (involving extreme plastic deformation) for opposed flat diamond anvil devices (85 GPa), none of which were observed.

The implications of all of the present results are: (1) The linear ruby scale increasingly overestimates the pressure as the pressure rises above 10 GPa. (2) Since the linear ruby scale calibration is a shocked-based scale, making use of marker materials via x-ray diffraction, this discrepancy must be introduced either via the x-ray measurements or by way of the conversion of Hugoniot data to isothermal P-V curves, or by both procedures.

At pressures up to 2.3 GPa the ruby R_1 shift has been calibrated against known melting points which in turn were obtained in free piston devices [1]. For purposes of interpolation in this range the linear representation or any number of nonlinear representations as described later are valid. At higher pressures the ruby scale calibration [2] is based on the sodium chloride pressure scale which is basically based on shock results; it should be noted that the shock data of NaCl [3] is not absolute but in turn is based on impedance-matching techniques with 2024 aluminum. Therefore the ruby scale calibration of Bloch and Piermarini [4] is based definitely although indirectly on the shock equation of state of 2024 aluminum. The more recent calibration of Bell, Mao, Shaner, and Steinberg [5] also makes use of the marker method but utilizes directly shock data of cubic metals.

While Hugoniot data can be obtained with reasonable accuracy, there can be major difficulties associated with the conversion of this data to isothermal P-V curves. The problems are twofold: (1) It is usual to assume that the Grüneisen parameter varies only with volume in some specified relation. (2) It is usual to assume that the material in the shocked state is, except for a difference in temperature and pressure, the same material as the initial material; no allowance is made for the defects created during shocking which may on their own account change the volume by several percent (and consequently the pressure enormously). While assumption (1) has been much discussed, assumption (2) has been virtually ignored even though its effects may be much larger than those of (1). Perhaps major understanding in this area will be brought about by the next generation of shock work; there is a relevant paper on this issue at this conference [6].

It is useful to review the results of direct static determinations. There is the quasi-free piston result for the bismuth III-V transition at 7.75 GPa [7] somewhat above an earlier determination (not direct) of 7.38 GPa [8], since changed to 7.6 GPa [9]. The transition in bismuth at 7.75 GPa is the highest direct static determination made (based on P = F/A), and only two pressure cycles with one piston were made. The determination by Jeffery et al. [8] was based on lattice parameter measurements and an assumed equation of state based on shock data. The much studied transition of iron [10] is not a direct static determination. It will be shown in this paper that, at the present time, pressure calibration must be based on direct absolute static pressure determinations and not on shock data via the marker method.

SUPPORTED, OPPOSED, CEMENTED TUNGSTEN CARBIDE ANVIL APPARATUS

Device

A typical supported opposed anvil device utilizing 3% cobalt cemented tungsten carbide pistons is shown in Fig. 1 [11]. Different methods of using this device are described elsewhere [11,12].

Yield Strength

When the above device is used, the pressure distribution across the circular tip face is nearly uniform [13]. The maximum shear stress in the piston can readily be attained in terms of the pressure [14]. For 3% cobalt cemented tungsten carbide, Poisson's ratio is 0.185 [15]. Hence the yield pressure P_o^* is related to the compressive yield stress σ_o according to

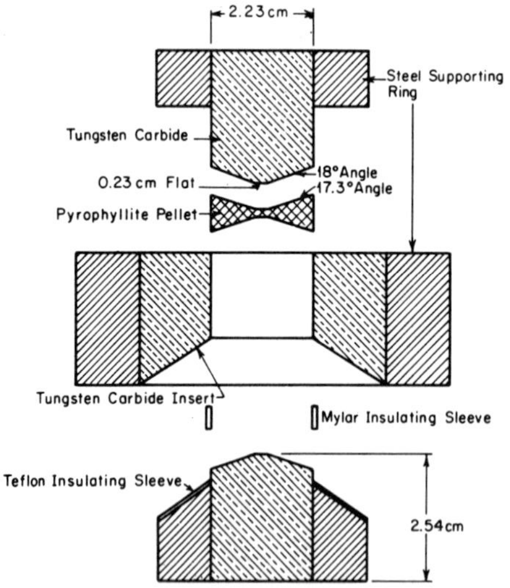

Fig. 1. Supported opposed anvil device.

$$P_o^* = 1.39\ \sigma_o \qquad (1)$$

If a superposed pressure ΔP is everywhere present then

$$P_o^* = 1.39\ \sigma_o + \Delta P \qquad (2)$$

The compressive yield stress can therefore be determined from

$$\sigma_o = 0.72\ (P_o^* - \Delta P) \qquad (3)$$

It should be noted that even if the pressure distribution is drastically different, the coefficient is altered little. Thus even for a hemispherical pressure distribution on a circular area of radius a, $P = P_o\ (1 - r^2/a^2)^{1/2}$, the coefficient is 1.48 instead of

1.39. Hence little error is introduced by minor variations in the pressure distribution because it is essentially only the pressure along the axis of the pressure distribution which determines yielding. To measure the yield stress, we proceed as follows. The surface profiles are measured in several known directions, using a submicron profilometer capable of resolving 10 Å displacements. Next a sample with a known transition pressure is used, e.g., bismuth. As soon as the transition is reached, the load is removed. Then surface profiles are measured in the same directions as before. A result at the bismuth transition is shown in Fig. 2. Note that initially the flat tips are not quite flat but are slightly rounded. Note that the maximum residual deflection, given by the difference in the curves, is in this case about 7000 Å. Larger pressures cause larger residual deflections while at much lower pressures the residual deflection becomes immeasurable. The permanent deflection of 500 Å at the middle of the 0.2286-cm-diameter tips is taken here as the definition of the yield pressure.

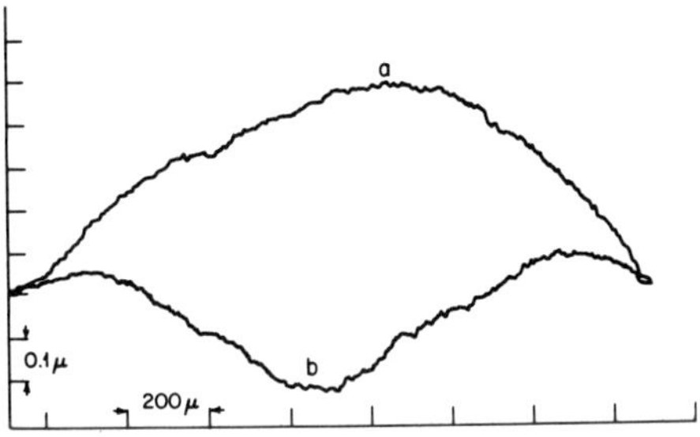

Fig 1. Profiles of 3% cobalt cemented tungsten carbide piston tip (a) before loading, (b) after loading to bismuth III-V transition.

For purposes of calibration we used for the observance of the bismuth point 7.75 GPa [7], although the indirect value of 7.6 also exists [9]. Also, the barium transition at 5.5 GPa [16] was used. No perceptible deformation (less than 100 Å) was seen at the barium transition. Consequently, since the pressure P vs. applied force F varied approximately as $P \propto F^{1/2}$, we used intermediate forces between the fixed points and used the above relation to determine the yield pressure. The resultant yield stress determined from equation (3) is 5.2 GPa. This method as presently used is over 100 times as sensitive in measuring residual deflections as one used earlier [17].

The present value of the yield stress agrees well with values found by other techniques. For example, one well-known technique is to simply divide the Knoop hardness pressure by 3.0 [18]. The Knoop hardness for 3% cemented tungsten carbides is 1650 - 1700 kg/mm^2 [19]. Using the average value gives σ_o = 5.47 = 5.5 GPa. Both methods have also been used with 6% cemented tungsten carbide, resulting in values about 0.15 GPa less in each case. Shock results of the Hugoniot elastic limit are also available for the 6% material, from which the dynamic compressive yield stress can be obtained. The value is 5.4 GPa [20]. It should be noted that prior to yielding in a shock experiment, the behavior is purely elastic, i.e., reversible thermodynamics properly describes the behavior of the system up to the catastrophic event of dynamic yielding. Hence there is no fundamental problem with using shock data to obtain accurate dynamic yield results. It should also be observed that the loading rate in a shock experiment is extremely rapid compared to the usual static experiment so that the 'dynamic' yield stress in general might be expected to be an upper bound for the 'static' yield stress. It is also interesting to note that Boyd [21] found that a compression sample of a 6% cobalt cemented tungsten carbide, in the presence of a superposed radial stress of 2.0 GPa, bulged when the overall stress state reached

$$\begin{pmatrix} 7.y & 0 & 0 \\ 0 & 2.0 & 0 \\ 0 & 0 & 2.0 \end{pmatrix}$$

where y was 0 to 5, so that the compressive yield stress was 5.0 to 5.5 GPa. Finally, Doi et al. [22] directly measured the compressive yield strength for 0.2% offset for fine grained 6% cobalt cemented carbide and obtained 5.2 GPa. The results are summarized for the 6% material in Table I. The reader may also wish to refer to the

Table I. Compressive Yield Strengths of Fine-Grained 6% Cobalt Cemented Tungsten Carbide

Method	σ_o, GPa	Reference
Profilometer	5.1	This study
Knoop	5.3	19
Dynamic	5.4	20
Supported piston	5.0-5.5	21
Direct compression (0.2% Offset)	5.2	22

paper of Brew and Crossland [23] who measured the ultimate compressive stress for 5 and 9% cobalt cemented carbide and who found extensive permanent deformation with large superposed hydrostatic pressure. The present results are used in the next subsection and they also serve to provide the basis for the technique used in the next section.

Maximum pressures attainable in supported uniaxial opposed anvil devices.

Two approaches are used. In the first we recognize that the maximum pressure is essentially the Knoop hardness pressure [18] plus a superimposed pressure (i.e., if the Knoop test were carried out in a hydrostatic environment),

$$P^{Limiting}_{Max} = P_H(\text{Knoop}) + \Delta P \text{ (Superimposed)} \tag{4}$$

The superposed pressure due to the support material surrounding the tip can be taken as the average pressure in the cylinder (see Fig. 1).

The second method is to use Hill's equation [24] for the maximum pressure needed to cause unlimited expansion of an initially infinitesimal spherical cavity. Here

$$P^{Limiting}_{Max} = \frac{2}{3}\sigma_o \{1 + \ln [\frac{E}{3(1 - \nu)\sigma_o}] \} + \Delta P \tag{5}$$

These methods will now be used for a steel with a compressive yield stress of 2.1 GPa and Knoop hardness of 6.3 GPa (at 500 g load), and a 3% cobalt tungsten carbide with a yield stress of 5.2 GPa and a Knoop hardness of 17 GPa. (See Table II.)

Table II. Maximum Attainable Pressure in Supported Uniaxial Opposed Anvil Devices

Piston Material	P_{max}(GPa) Eq. (4)	P_{max}(GPa) Eq. (5)
Steel (σ_o = 2.1 GPa)	6.3 + ΔP	6.7 + ΔP
3% Co Tungsten Carbide (σ_o = 5.2 GPa)	17.0 + ΔP	17.1 + ΔP*

*For elastic constants, see Day et al. [15].

Ruoff and Wanagel [25] using such steel pistons have, with ΔP = 1.2 GPa, reached the bismuth transition (77.5 GPa). However, in about half of such experiments, the bismuth transition was not reached, indicating that the pistons were extremely close to their limiting pressure when the transitions were obtained. The experimental results are therefore in reasonable agreement with the theoretical predictions.

With 3% cobalt tungsten carbide pistons, the GaP transition is usually (but not always) obtained in the author's laboratory. The transition is obtained in our laboratory with ΔP = 0.6 GPa. The transition in these cases is not nearly as sharp as when it is carried out with the much stronger boron carbide pistons [12] or with diamond pistons. This fact and the fact that it is not always obtained suggests that we are approaching very near to the limiting pressure. Hence the transition pressure for GaP must be near 17.7 GPa (see Table II).

These results also require us to conclude that the upper bound on pressures attainable in single stage uniaxial supported opposed anvil device with 3% cobalt cemented tungsten carbide pistons is about 20 GPa, because the highest average pressure ΔP used is about 2.4 GPa [26].

SECOND METHOD FOR DETERMINATION OF TRANSITION PRESSURE OF GaP

Boron carbide has a yield strength approximately twice that of tungsten carbide so much higher pressures can be attained if support is furnished to prevent fracture. The dynamic compressive yield stress of the material provided by the Norton Co. is 11.8 ± 0.8 GPa [27]. We have used pistons of similar material in an opposed anvil device as shown in Fig. 1. For this material (ν = 0.188) the yield pressure due to a uniform pressure distribution over a circular area is as given by equation (2). We expect the 'dynamic' yield stress to be an upper bound for the 'static' yield stress, although in the case of the tungsten carbides the two values seem remarkably close. We therefore expect yielding to commence when

$$P_o^* = 1.39\ (11.8) + \Delta P \qquad (6)$$

With GaP we find the transition occurs prior to yielding. Only when the external load is increased slightly beyond the appearance of the GaP transition (a load increase of 2 to 3% more) do we subsequently observe yielding (a permanent maximum residual deflection of the surface of 500 Å). At this time ΔP = 0.6 GPa. In this way we obtain 16.7 GPa for the transition. It is perhaps useful to consider further the value of σ_o. The number given by Gust and Royce [27] was obtained from the Hugoniot elastic limit using linear elasticity. Because of the high stresses involved, nonlinear elasticity should be used [14]. We would expect the results to be similar to those found for germanium and silicon, namely about a 4% decrease in σ_o to 11.4 GPa [14]. However, let us ignore this reduction. If we allow for two standard deviations of error (1.6 GPa), (92% confidence limit) the dynamic yield stress still does not exceed σ_o = 13.4 GPa and with ΔP = 0.6 GPa, the phase transition does not exceed 18.9 GPa.

The errors introduced by using a coefficient in (6), which is incorrect because the pressure distribution is not uniform, is of the order of 2%. With the new exceedingly sensitive method of obtaining surface profiles and hence the onset of yielding, we are confident that we now observe yielding at a pressure within 3% of its onset. If both of these potential absolute errors are added to the most probable 16.7 GPa value we obtain a value of 17.5 GPa at the GaP transition.

It should be noted that pressure transitions are affected by shear stress and these deviatoric stresses can vary from one experiment to another. Hence one cannot say, unless working in a truly hydrostatic environment, that the transformation pressure of this material is this pressure.

DISCUSSION OF X-RAY DETERMINATION OF EQUATION OF STATE OF CUBIC CARBIDES USING SHOCK BASED MARKERS AND CEMENTED TUNGSTEN CARBIDE PISTONS

Minomura and Drickamer [28], using the opposed anvil device with 3% cobalt cemented tungsten carbide pistons, failed in 22 experiments to obtain the GaP transition. We have noted previously that we usually obtain this transition under similar circumstances although it is clearly near to or at the limit of the device. It is likely that there has been a slight improvement in the carbides and it is possible that this accounts for the difference. Variations in pyrophyllite could also have caused this. Their failure to obtain the GaP transition, taken by itself, suggests they were unable to reach the 18 GPa required (as indicated by the present work) or 22 GPa (as indicated by Piermarini and Block [4]). However, using the same technique of generating pressures and using silver and other marker materials whose equation of state was obtained from shock experiments, Champion and Drickamer [29] measured the lattice parameter of the silver. From this they deduced the pressure. At the same time, they measured the lattice parameter of a cubic carbide. In all, three carbides were studied to the highest pressure.

In view of the discussions of the previous section it is interesting that the maximum pressure indicated by the markers was 35 GPa. The accuracy of the lattice parameter measurements is impressive inasmuch as the initial values of the bulk modulus K_o (calculated directly from the data provided in their table) agree extremely well with values obtained by ultrasonic techniques [30-32] shown in Table III. It is interesting to point out that for K_o itself the mixture rule closely holds, e.g., K_o(NbC) = 3.00 GPa $\doteq$ 0.5 K_o(Nb) + 0.5 K_o(C) = 3.05 GPa.

Table III. Initial Values of Bulk Moduli by Different Methods

Material	Lattice Parameter K_o, GPa	Ultrasonic K_o, GPa
NbC	307	300 (Ref. 30)
TiC	247	242 (Ref. 31)
ZrC	226	209 (Ref. 32)

From Table 1 of Champion and Drickamer [29] the bulk modulus at their highest pressure also can be obtained (in exactly the same manner as K_o was obtained). The results are shown in Table IV. We know that

Table IV. Computed Values of K (GPa) at Maximum Pressure and of K_o'

Material	$K_{max\ pressure}$	Measured K/K_o	K_o' *	Expected † K/K_o
NbC	972	3.24	19.2	1.27
TiC	664	2.74	12.1	1.33
ZrC	408	1.95	5.7	1.38

*Here K_o' is computed from (11) with P = 35 GPa. If P is at most 20 GPa (as the evidence of this paper suggests) then the values K_o' would be multiplied by 1.75.

†Expected on basis of $K_o' = 4$, $P_{Max} = 20$ GPa.

$$K = K_o + K_o'P + \sum_2^\infty \frac{K_o^{[n]}P^n}{n!} \tag{7}$$

can at low pressures ($P < K_o/20$) be represented by

$$K = K_o + K_o'\ P \tag{8}$$

to a reasonable approximation. This of course leads to Murnaghan's equation of state

$$P = (K_o/K_o')\ [(V/V_o)^{-K_o'} - 1] \tag{9}$$

As the pressure increases

$$K = K_o + b(P)\ P \tag{10}$$

where $b(P) < K_o'$ [33].

Let us use (8) as the basis for calculating the K_o' implied by the second column of Table IV, i.e., we have

$$K_o' = K_o\ [\frac{K(P)}{K_o} - 1]/P \tag{11}$$

The results are shown in Table IV.

In the last few decades a great deal of information has come forth on the value of K_o'. It has a value of 3 to 6 for a wide range of materials. The highly covalent solids germanium, silicon, and diamond have K_o' = 4.54 [34], 4.24 [35], and 4.03 [36], respectively. The metals Nb, Ti, Zr have K_o' = 4.06 [37], 4.35 [38], and 4.08 [39], respectively. The only published value for a carbide (cobalt cemented tungsten carbide) is near four [15]. There is reason to believe that K_o' for the highly covalent cubic carbides will be four.

However, the values of K_o', obtained by use of the 'marker method', are so drastically different from the expected values, that we conclude that a major discrepancy exists. Figure 3 helps to explain this discrepancy. It shows for NbC the Champion and Drickamer P-V curve extending to $V/V_o = 0.9383$ and to P = 35 GPa (on the 'marker' scale). If one uses $K_o = 300$ GPa (see Table III), $K_o' = 17.4$, and $V/V_o = 0.9383$ in Murnaghan's equation one obtains P = 35 GPa. Figure 3 also shows the curve obtained using Murnaghan's equation with $K_o = 300$ GPa, $K_o' = 4$; for $V/V_o = 0.9383$, this yields a pressure P = 21.8 GPa and a value of K = 387 GPa at the maximum pressure (instead of K = 972 GPa, see Table IV). The pressure of 21.8 GPa is also only slightly above the maximum pressures that were predicted for this device in an earlier section (see Table II). Note also that if $K_o' = 4$ and P = 20 GPa, then $K/K_o = 1.27$ at the maximum pressure instead of the highly unlikely value shown in Table IV.

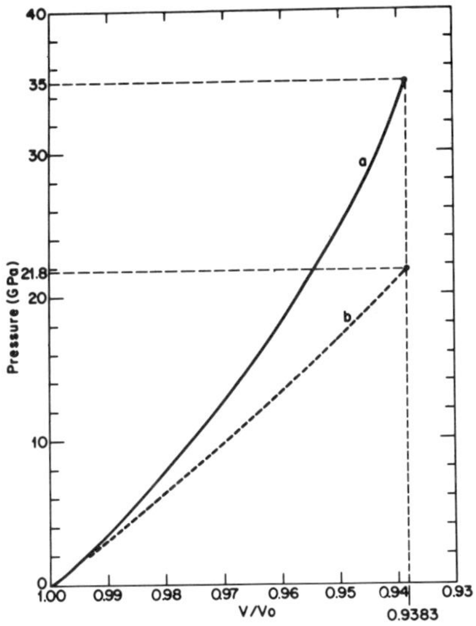

Fig. 3. P vs. V/V_o for NbC. (a) Experimental curves of Champion and Drickamer (b) Curve with same initial slope but with $K_o' = 4$.

It is quite clear that the marker method leads to serious overestimates of the pressure. Another example of this divergence is shown in Fig. 6 of the paper of Perez-Albuerne, Forsgren and Drickamer [26], where palladium serves as the marker. Here the highest marker pressure is about 57 GPa (in a system incapable of reaching the 18 GPa transition). The authors [26] give a table of P vs. V/V_o for the palladium marker (derived from shock work); hence the pressure corresponding to a given V/V_o can be obtained from the more recent shock work. There is little change.

It should be noted that the state of stress in these x-ray measurements is undoubtedly extremely nonhydrostatic inasmuch as the sample was mixed with up to 80% amorphous boron, which is stiff and supports high shear stresses. Part and perhaps much of the error of the marker method appears to be a result of the nonhydrostatic pressure.

The effect of nonhydrostaticity is shown in a recent paper by Wilburn and Bassett [40]. They noted that when the bulk moduli of cubic solids were measured by x-ray techniques in hydrostatic medium, the corresponding K_o values were significantly less than the corresponding values of K_o based on measurements in a solid pressure-transmitting medium between opposed diamond anvils. Inasmuch as

$$K_o = \lim_{V \to V_o} \left(-V \frac{\partial P}{\partial V}\right) \tag{12}$$

we can conclude that the apparent pressure in the nonhydrostatic case appears to be significantly higher than it actually is.

RUBY SHIFT VERSUS PRESSURE

Ruby is basically Al_2O_3 (with dopant) for which the bulk modulus is 254 GPa. We can expect the bulk modulus shift, $\Delta K = K - K_o$, to vary with pressure (for pressures not too large) according to

$$\Delta K = (K_o' + K_o''P/2)\, P \tag{13}$$

or

$$\Delta K = K_o' \left(1 + \frac{K_o K_o''}{2K_o'} \frac{P}{K_o}\right) P \tag{14}$$

The variation will be linear if

$$\frac{K_o K_o''}{2K_o'} \left(\frac{P}{K_o}\right) << 1. \tag{15}$$

For most solids $|K_o K_o''/2K_o'| \approx 1$ [33]. Thus, if equation (14) were fitted over some pressure ranges 0 to P by a linear expression, the deviation of the slope would be about $P/2K_o$; for 2.3 GPa this is 0.5%.

The bulk modulus shift with pressure is determined primarily by the interaction of the outer electrons of the ions, as is the ruby shift. We therefore should not find it surprising if nonlinearities of similar magnitude were to appear in the ruby shift versus pressure. (This discussion is given because of the absence of any reasonable theory for the ruby shift.)

The ruby shift has been calibrated to 23 kbars (2.3 GPa) against melting points which in turn had been calibrated against free piston gauge pressure with the result that on this range P (GPa) = $(0.2746 \pm 0.0016)\ \Delta\lambda$ (Å). Thus we cannot exclude at the 57% confidence limit nonlinearities in the slope of less than 0.0016/0.2746 or 0.6%, nor at the 92% confidence limit deviations in slope up to 1.2%. This means that when we measure a pressure of 2.3 GPa, there is a 57% probability that it does not fall outside the range

$2.2866 \leq P \leq 2.3134$. For purposes of interpolation this behavior can be represented by the following functions with equal justification:

$$P_L = 0.2746\ \Delta\lambda \tag{17}$$

$$P_Q = 0.2776\ \Delta\lambda - 4.56\text{x}10^{-4}\ \Delta\lambda^2 \tag{18}$$

$$P_C = 0.2763\ \Delta\lambda - 9.94\text{x}10^{-7}\ \Delta\lambda^3 \tag{19}$$

$$P_t = 51.67\ \tan^{-1}\ [0.005314\ \Delta\lambda] \tag{20}$$

All of these functions can be used for small extrapolations beyond 2.3 GPa, but large extrapolations are not justified for any of them unless we have more scientific information. Let us consider, however, such an extrapolation, as shown in Table V.

Table V. Values of Pressures Given by Various Expressions

Transition	$\Delta\lambda$*	P_L	P_Q	P_C	P_t
†	8.376	2.3	2.287	2.313	2.298
Bi	28.22	7.75	7.47	7.77	7.69
ZnS	54.63	15.0	13.8	14.9	14.6
GaP	80.12	22.0	19.3	21.6	20.8
NaCl	106.0	29.1	24.3	28.1	26.5
**	209.4	57.5	38.1	48.7	43.4
††	364.2	100.0	40.6	52.6	56.5

*$\Delta\lambda$'s are computed from $\Delta\lambda = P_L/0.2746$. This is not to imply that $\Delta\lambda$ can be obtained experimentally with this sensitivity.

†The values of P's given by various expressions for the case where $P_L = 2.3$ GPa.

**Shift reported by Piermarini and Block [4].

††Shift reported by Mao and Bell [42].

Note how even tiny nonlinearities which cause only minor deviations at the ZnS transition (up to 8%) or at the GaP transition (up to 12%) become exceedingly important at higher ruby shifts. Ruoff and Chhabildas have recently given arguments which suggest 258 kbars as an upper bound for the sodium chloride transition [41]. Also shown are other reported shifts [4,42].

What Evidence Is there for the Nonlinear Term With the Negative Sign?

First, using the temperature dependence of the ruby shift at atmospheric pressure, and assuming that the quadratic coupling was via the acoustic phonons, Feher and Sturge [43] estimated that the quadratic coefficient in the expression

$$P = A\,\Delta\lambda + B\,\Delta\lambda^2 \qquad (21)$$

should be approximately $B = -9.14 \times 10^{-4}$ GPa $/\text{Å}^2$. This is more than twice the value given in equation (18) and described as P_Q in Table V.

Second, Feher and Sturge have measured the R_1 and R_2 shift with uniaxial stress parallel to the c-axis to a stress of 8.6 kbars; the quadratic coefficient in the equation for uniaxial compression is $-(3.6 \pm 1.9) \times 10^{-4}$ GPa/Å^2 for R_1 and $(-4.9 \pm 1.8) \times 10^{-4}$ GPa/Å^2 for R_2; they also measured the R_2 shift with a uniaxial stress perpendicular to the c-axis; the quadratic coefficient is $(0.9 \pm 1.0) \times 10^{-4}$ GPa/Å^2 [44]. The deviations are for one standard deviation, sigma. They conclude on the basis of this data that they do not have conclusive evidence (at the two sigma level or 92% confidence level for the negative term). However, the best fit to the data does involve a negative coefficient of about the magnitude used in Eq. (18).

A research project in which the uniaxial stress experiment was performed to much higher stresses (using, e.g., ruby filaments), with more data points, with higher accuracy because of the improvements in technology, might be capable of directly determining the nonlinear term.

Third, the direct determination of the GaP transition (17 to 18 GPa) described in the present paper requires a major deviation of the ruby scale from linearity inasmuch as the linear ruby scale places this transition at 22 GPa.

Fourth, the linear ruby scale is based on the use of the 'marker method' as described earlier and it appears that this method when used in a highly nonhydrostatic environment overestimates the pressure. If this method indicates a pressure of 100 GPa, the actual pressure may be much less. The pressure environment used, by Piermarini et al. [4] and Mao et al. [42] is highly nonhydrostatic.

PRESSURES ATTAINABLE IN UNIAXIAL OPPOSED DIAMOND ANVIL DEVICES

The yielding and strength of diamonds is discussed in detail in another paper at this conference [14] and so will only be summarized here.

Analysis of opposed flat anvil diamond devices indicates that if fracture is prevented, yielding of the diamonds will begin at pressures of about 50 GPa; that if fracture is prevented macroscopic plastic deformation will be clearly evident at about 70 GPa; and that again if fracture is prevented the limiting pressure which can be reached equals the true or corrected Knoop hardness pressure (about 85 GPa) in which case extremely extensive plastic deformation will have occurred. None of these three events have been observed (and the latter two would be obvious), so that it is necessary to conclude that pressures of 70 GPa have not been reached. In fact, until plastic deformation of diamond anvils (in opposed flat anvil diamond devices) is clearly shown, claims of pressures as high as 50 GPa should leave doubt [14].

SOME FORCE CONSIDERATIONS

Piermarini and Block [4] correctly point out the potential errors possible when load vs. pressure extrapolations are attempted in supported anvil devices; the difficulty is because the major fraction of the load is carried by the support, and this fraction changes. The purpose of this section is to point out the unfortunate neglect of direct load measurements in spring-loaded and leveraged opposed anvil diamond cell devices which are essentially unsupported. It would be relatively easy to measure the load. The levers can have attached strain gauges (after the two friction pins) and these can be directly calibrated against force at the anvils. The direct measurement of load would add considerably to our understanding of gasket materials. This will now be discussed.

The opposed diamond anvil configuration of Piermarini and Block is shown in Fig. 4. The load applied by the springs to the lever arm is 3/4 (1360) = 1020 kg. The mechanical advantage is two, but there are two friction pivots. The coefficients of friction of lubricated steel on steel [45] is 0.10; so the effective load is 1020 x 2 x 0.90 x 0.90 = 1625 kg. The tip area is given as 0.5 mm^2 but inasmuch as the gasket squashes and does slightly support the sides of the anvil, we assume that contact area is 1.05 times as large. Consequently the average pressure is 31.5 GPa.

If the gasket is made of diamond whose bulk modulus K_o = 442 GPa, the pressure distribution is [46]

$$P = P_o(1 - r^2/a^2)^{-1/2} \tag{22}$$

If the gasket is made of NaCl (K_o = 23.8 GPa), the pressure distribution at high loads is approximately [47]

$$P = P_o(1 - r^2/a^2)^1 \tag{23}$$

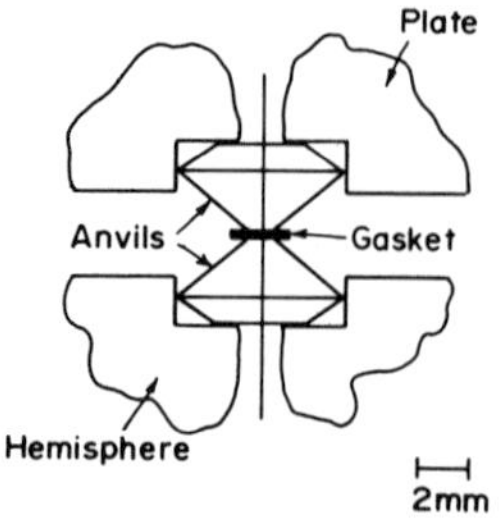

Fig. 4. Enlarged cross section of the opposed diamond-anvil configuration (after Piermarini and Block, [4]).

If the gasket is stainless steel (K_o = 160 GPa) then

$$P = P_o(1 - r^2/a^2)^n \qquad (24)$$

where the expected value of n (note the relative K_o values) lies between the above two cases and we expect $0 < n < 1/4$. Then, inasmuch as

$$P_o = (n + 1)P_{Pav} \qquad (25)$$

we expect in the Piermarini and Block experiment $31.5 < P_o < 39.3$ GPa.

This deserves two comments. First, these expected pressures are considerably less than the predicted metallic transition in BP at 40 GPa [48]. While, in general, the transformation pressures predicted by VanVechten are somewhat high, perhaps the reason that Block and Piermarini do not observe the BP transition [49] is that their pressure is less than the 40 GPa needed and not the 57.5 GPa they observe based on the linear ruby scale. Second, measurement of the load will help us to understand eventually the behavior of gasket materials. The calculation of this section is definitely not a proof that P_o is less than 40 GPa. A detailed finite element plasticity analysis would be needed to determine the actual pressure profile and P_o. In the absence of that we make the most reasonable estimates we can.

CONCLUSIONS

1. The maximum pressure attainable in a uniaxial opposed anvil device can be obtained in two ways: From the true Knoop hardness pressure plus superposed pressure, or the Hill limiting pressure plus superposed pressure: This is the limit of full local plasticity and corresponds to extreme plastic deformation.

2. The maximum attainable pressure in uniaxial supported opposed anvil devices with 3% cobalt cemented tungsten carbide pistons (commercially available for the past decade) is about 20 GPa. (The piston tips yield above the barium transition but below the bismuth transition and show obvious macroscopic deformation at 12 GPa.)

3. The transformation was observed by two different methods to occur at 17 to 18 GPa.

4. The use of marker-substances such as Ag and Pd in conjunction with x-ray lattice parameter measurement, with the equation of state of the 'markers' based on shock experiments has lead to an extremely serious overestimate of the pressures in highly non-hydrostatic situations.

5. Since the ruby scale indicates the GaP transition at 22 GPa, and since it is based completely on the markers method, we conclude that the ruby scale is highly nonlinear at high pressures. The linear ruby scale increasingly overestimates the pressures, as the pressure is increased in the same manner as the marker method on which it is based.

6. The maximum pressure attainable in uniaxial opposed flat anvil diamond devices prior to yielding is about 50 GPa. Such anvils (if not fractured first) would show obvious macroscopic deformation (dishing) by 70 GPa, and would if not fractured first reach a limiting pressure equal to the true Knoop hardness (85 GPa), at which time extreme plastic deformation would be evident; major side support would have to be applied to diamonds to reach either of the last two conditions, and possibly the first as well.

7. The absence of observance of yielding of diamonds in uniaxial opposed flat anvil diamond devices suggests that pressures of 50 GPa, and most certainly 70 GPa, have not been reached in such devices.

8. Yielding in side-supported and gasketed uniaxial opposed diamond anvil devices is expected at 45 GPa and resultant residual deflections then should be observable, using a surface submicron profilometer.

ACKNOWLEDGMENTS

The support of this work by the National Aeronautics and Space Administration is gratefully acknowledged. The use of the facilities of the Cornell Materials Science Center, supported by the National Science Foundation, is appreciated. The author also wishes to thank D. A. Nelson for the least squares analysis of the data of E. Feher and M. Sturge as well as the latter for kindly providing their original data. Finally, the extremely capable technical assistance of V. Arnold is acknowledged.

NOTATION

A,B	= coefficients in quadratic ruby gauge
a	= radius of circular area over which pressure acts
E	= Young's modulus
F	= force
K	= bulk modulus
K_o	= bulk modulus at zero pressure
K_o'	= $\lim_{P\to 0} dK/dP$
K_o''	= $\lim_{P\to 0} d^2K/dP^2$
n	= exponent in equation (24)
P	= pressure
P_C	= pressure assuming a cubic ruby gauge
P_H	= Knoop hardness pressure
P_L	= pressure assuming a linear ruby gauge
$P_{Max}^{Limiting}$	= limiting pressure attainable with uniaxial opposed anvil devices (after extreme plastic flow)
P_o	= pressure at center of pressure distribution having circular symmetry
P_o^*	= value of P_o at yielding
P_Q	= pressure assuming a quadratic ruby gauge
P_t	= pressure assuming an inverse tangent ruby gauge
r	= radius from axis
V	= volume
V_o	= initial volume
$\Delta\lambda$	= ruby R_1 shift
ΔP	= superimposed pressure
ν	= Poisson's ratio
σ_o	= compressive yield stress

REFERENCES

1. R. A. Foreman, G. J. Piermarini, J. D. Barnett, and S. Block, Science 176, 284 (1972).
2. G. J. Piermarini, S. Block, J. D. Barnett, and R. A. Foreman, J. Appl. Phys. 46, 2774 (1975).
3. J. N. Fritz, S. P. Marsh, W. J. Carter, and R. G. McQueen, Accurate Characterization of the High-Pressure Environment, E. C. Lloyd, ed., National Bureau of Standards Special Publication 326, Washington, D.C. (1971), p. 201.
4. G. J. Piermarini and S. Block, Rev. Sci. Instr. 46, 973 (1975).
5. P. M. Bell, H. K. Mao, J. Shaner, and D. Steinberg, Proceedings of 6th AIRAPT Intern. High Pressure Conference, Plenum Press, New York (1978).
6. S. A. Raikes and T. J. Ahrens, Proceedings of 6th AIRAPT Intern. High Pressure Conference, Plenum Press, New York (1978).

7. J. C. Haygarth, H. D. Luedemann, I. C. Getting, and G. C. Kennedy, Accurate Characterization, E. C. Lloyd, ed., National Bureau of Standards Special Publication 326, Washington, D.C. (1971), p. 35.
8. R. N. Jeffery, J. D. Barnett, M. R. Vanfleet, and H. T. Hall, J. Appl. Phys. 38, 3172 (1966).
9. D. L. Decker, W. A. Bassett, L. Merrill, H. T. Hall, and J. D. Barnett, J. Phys. Chem. Data 1, 773 (1972).
10. J. F. Cannon, J. Phys. Chem. Ref. Data 3, 781 (1974).
11. H. G. Drickamer and A. S. Balchan, in Modern Very High Pressure Techniques, R. H. Wentorf, Jr., ed., Butterworth, London (1962).
12. J. Wanagel, V. Arnold, and A. L. Ruoff, J. Appl. Phys. 47, 2821 (1976).
13. K. F. Forsgren and H. G. Drickamer, Rev. Sci. Instrum. 36, 1709 (1965).
14. A. L. Ruoff, Proceedings of 6th AIRAPT Intern. High Pressure Conference, Plenum Press, New York (1978).
15. J. P. Day and A. L. Ruoff, J. Appl. Phys. 44, 2447 (1973).
16. J. C. Haygarth, I. C. Getting, and G. C. Kennedy, J. Appl. Phys. 38, 4557 (1967).
17. A. L. Ruoff and J. Wanagel, J. Appl. Phys. 46, 4647 (1975).
18. D. Tabor, The Hardness of Metals, Oxford University Press, Oxford, England (1951).
19. P. Schwarzkopf and R. Kiefer, Cemented Carbides, The Macmillan Company, New York (1960).
20. W. H. Gust, Bull. Amer. Phys. Soc. 19, 5, 660 (1974); also private communication.
21. R. F. Boyd in Modern Very High Pressure Techniques, R. H. Wentorf, Jr., ed., Butterworth, London (1962), p. 157.
22. H. Doi, Y. Fujiwara and K. Miyake, Trans. AIME 245, 1457 (1969).
23. P. S. Brew and B. Crossland, Proceedings of 6th AIRAPT Intern. High Pressure Conference, Plenum Press, New York (1978).
24. R. Hill, The Mathematical Theory of Plasticity, Clarendon Press, London (1950), Chapt. V.
25. A. L. Ruoff and J. Wanagel, Rev. Sci. Instr. 46, 1294 (1975).
26. E. A. Perez-Albuerne, K. F. Forsgren, and H. G. Drickamer, Rev. Sci. Instr. 35, 29 (1964), see Figure 4.
27. W. H. Gust and E. B. Royce, J. Appl. Phys. 42, 276 (1971).
28. S. Minomura and H. G. Drickamer, J. Phys. Chem. Solids 23, 451 (1962).
29. A. R. Champion and H. G. Drickamer, J. Phys. Chem. Solids 26, 1973 (1965).
30. H. L. Brown, P. E. Armstrong, and C. P. Kempter, J. Chem. Phys. 45, 547 (1966).
31. J. J. Gilman and B. W. Roberts, J. Appl. Phys. 32, 1405 (1961).
32. H. L. Brown and C. P. Kempter, Phys. Stat. Solidi 18, K21 (1966).
33. A. L. Ruoff and L. C. Chhabildas, Proceedings of 6th AIRAPT Intern. High Pressure Conference, Plenum Press, New York (1978).
34. H. J. McSkimin and P. Andreatch, Jr., J. Appl. Phys. 34, 651 (1963).

35. H. J. McSkimin and P. Andreatch, Jr., J. Appl. Phys. 35, 2161 (1964).
36. H. J. McSkimin and P. Andreatch, Jr., J. Appl. Phys. 43, 2944 (1972).
37. K. W. Katahara, M. H. Manghnani, and E. S. Fisher, J. Appl. Phys. 47, 434 (1976).
38. E. S. Fisher and M. H. Manghnani, J. Phys. and Chem. Solids 32, 657 (1971).
39. E. S. Fisher, M. H. Manghnani, and T. J. Sokolowski, J. Appl. Phys. 41, 2991 (1970).
40. D. R. Wilburn and W. A. Bassett, High Temperatures-High Pressures 8, 343 (1976).
41. A. L. Ruoff and L. C. Chhabildas, J. Appl. Phys. 47, 4867 (1976).
42. H. K. Mao and P. M. Bell, Science 191, 851 (1976).
43. E. Feher and M. Sturge, Phys. Rev. 172, 244 (1968).
44. D. A. Nelson, private communication.
45. R. E. Bolz and G. L. Tuve, eds., Handbook of Tables for Applied Engineering Science, the Chemical Rubber Co., Cleveland (1970).
46. S. P. Timoshenko and J. N. Goodier, Theory of Elasticity, McGraw-Hill Book Co., New York (1970), p. 408.
47. G. Piermarini, S. Block, and J. D. Barnett, J. Appl. Phys. 44, 537 (1973), see Fig. 2.
48. J. A. VanVechten, Phys. Rev. B 7, 1479 (1973).
49. S. Block and G. Piermarini, Physics Today (September 1976), p. 51.

R-4

MATERIALS AND TECHNIQUES FOR PRESSURE CALIBRATION BY RESISTANCE-JUMP TRANSITIONS UP TO 500 KBAR

K. J. Dunn and F. P. Bundy

General Electric Corporate Research and Development
Schenectady, New York USA

INTRODUCTION

Equipment capable of generating the highest pressures usually consist of pistons of truncated-cone or truncated-pyramid geometry which are forced toward each other with their pressure faces against the pressure cell and the spaces between their tapered flanks filled, or partially filled, with a gasket material. Of the force applied to the base of a piston, the fractions which apply to the face against the pressure cell and to the flanks against the gasket are not only unknown, but also vary with the loading. Hence there is no accurate way of calculating the cell pressure from the force applied to the piston base in relation with the area of the pressure face. In such equipment the cell pressures may be calibrated in terms of the force applied to the pistons by monitoring selected calibration materials in the cell which have known physical behaviors with pressure, such as lattice compression, shifts of optical bands or lines, or resistivity jumps associated with first-order phase changes. Such pressure calibration is not absolute, and its accuracy depends upon the more basic, or absolute, calibration of the phenomena being monitored. A discussion of the various more or less absolute methods, and the history of pressure calibration of very high pressure equipment will not be presented here. Rather, the purpose of this article is to present information on, and techniques for using, materials wich have first-order resistance jumps in the 200-to-500-kbar range. In particular, these are the alpha-epsilon transitions in the iron-cobalt and the iron-vanadium alloys which were first observed in shock compression experiments and reported by Loree, et al. in 1966 [1].

EXPERIMENTAL WORK

The apparatus used has been described in detail elsewhere [2] and is shown in partial section in Fig. 1. The pressure generated at the center of the space between the piston faces depends upon the initial gap G_o which is determined by the gasket thickness, and by the force L applied to the pistons.

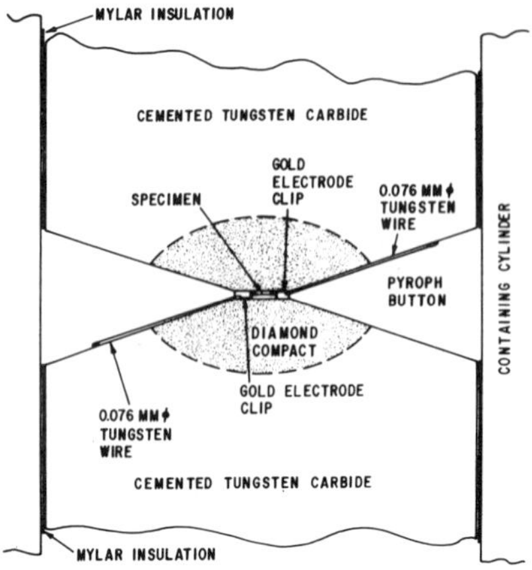

Fig. 1. Cross section of diamond-tipped opposed-anvil apparatus used in the calibration experiments.

Figure 2 shows the resistance vs. force loading of Fe6V, Fe12V, Fe16V, and Fe20V specimens for about the same initial gap conditions in the apparatus. It is seen that the resistance rises associated with the α-ε transition are reasonably sharp, and that the usual hysteresis is evident in the unloading parts of the curves.

Figure 3 shows a series of experiments performed with Fe12V using different initial gaps. The greater the initial gap thickness, the greater the force that must be applied to the pistons to generate a given cell pressure between the faces. This is because, for the fixed taper angle of the thickness of the gasket, any change of the gap at the center changes the radial pressure distribution in the gasket. In the procedure of studying pressure calibration of the apparatus it was instructive to carry out a series of experiments for each calibration metal for a range of gaps G_o. In this way a family of lines could be generated on a chart showing gap vs. loading pressure required to start the transition for each different calibrant, as shown in Fig. 4. Developing such organized sets of data is important because even with the most careful and uniform techniques of preparing and loading the cells there is some scatter of the results from the ideal functional relationship, as is evident in the chart. However, an adequate number of tests and data points establish quite well "ideal" lines for each calibrant in a given apparatus.

REDUCTION OF EXPERIMENTAL DATA

If the transition pressure of each calibrant is known quite accurately in absolute terms, those pressures may be assigned to the lines in Fig. 4, and from that chart a cell pressure calibration chart of P vs. L and G_o may be constructed as shown in Fig. 5. However, if the values are not known with satisfactory accuracy, one can arbitrarily set up some kind of a formal analysis procedure to develop closer probable values of the higher pressures. One

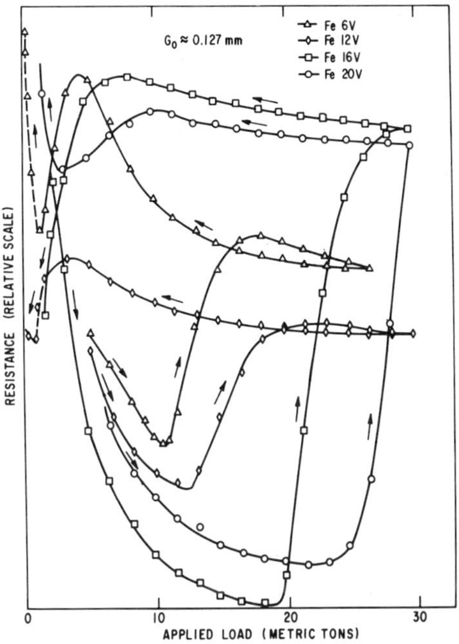

Fig. 2. Resistance vs. force loading in the apparatus for specimens of Fe6V, Fe12V, Fe16V, and Fe20V, all at about the same G_o in the same apparatus.

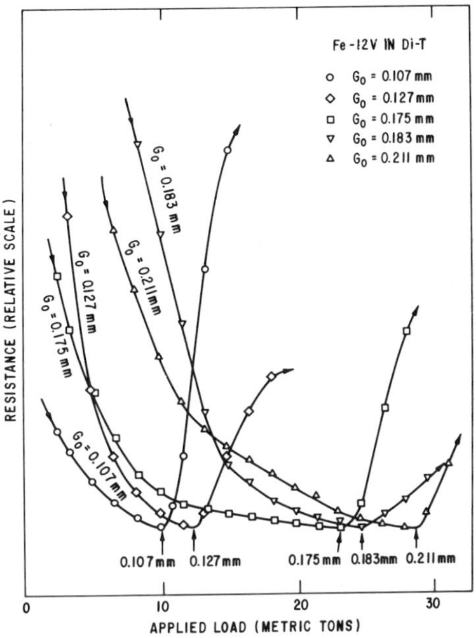

Fig. 3. Resistance vs. loading for Fe12V in the apparatus for a range of initial gaps G_o.

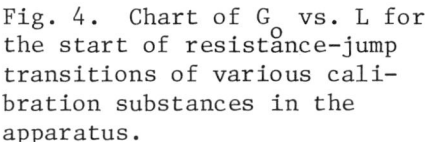

Fig. 4. Chart of G_o vs. L for the start of resistance-jump transitions of various calibration substances in the apparatus.

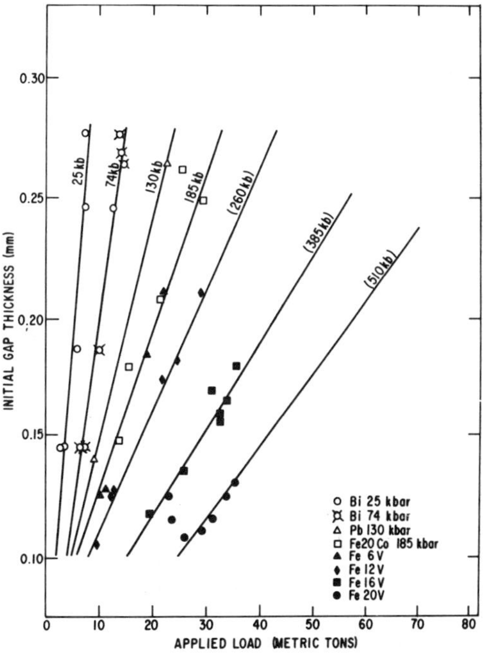

such procedure is as follows:

First construct a chart of P vs. L for various G_o's from the lines of Fig. 4 which correspond to the lower, fairly well-established transition pressures, for example up to about 250 kbar, as shown by the solid lines of Fig. 5. Second, extend each G_o line linearly (which is known to be fictitious). Third, on each

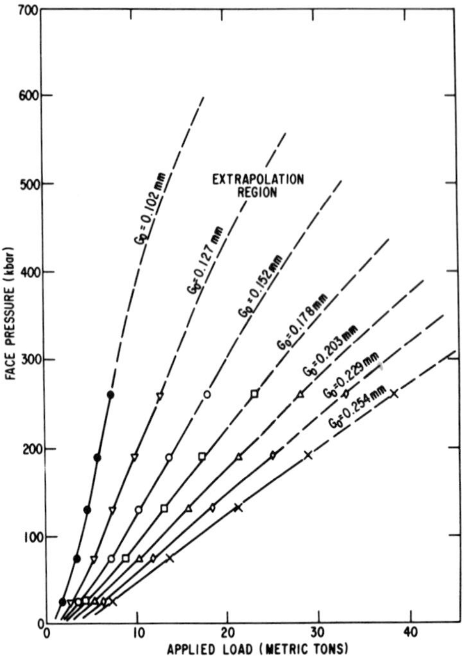

Fig. 5. P vs. I, G_o calibration chart based on well-recognized pressures below about 200 to 250 kbar with extrapolations to higher pressures.

extended G_o line place a point (with error bars) corresponding to the L at which the transition of a given calibrating substance occurs. This yields a P vs. L chart, like that shown in Fig. 6 in which the indicated pressures for a given transition are obviously far too high at the small G_o values, but which level out to fairly definite asymptotic values for the larger G_o's. It is likely that this asymptotic value is nearly correct, and if so, the other values should be corrected down to the same pressure level at the observed loading, as shown in the construction of Fig. 6. The error bands are indicated in Fig. 6, and at the extreme right the asymptotic pressure transition values for the four FeV alloys are shown and are compared with the values taken from the shock compression data of Loree, et al. [1]. It is seen that there is quite good agreement for the Fe6V (at about 190 kbar) and the Fe20V (at about 510 kbar), but not so good for Fe12V (250 vs. 310 kbar) or Fe16V (385 vs. 430 kbar).

These corrected points for a given G_o then become the basis of the most probable P(L) curve for that G_o. Figure 7 shows the final adjusted P(L) chart from the test data for the apparatus. The solid parts of the lines correspond to pressures actually reached in the tests while the dashed parts indicate extrapolations beyond that.

DISCUSSION

It now appears that there are available resistance-jump pressure calibrants which cover the pressure range to slightly over 500 kbar. These can be used to calibrate any apparatus amenable to electric resistance monitoring of the specimen. The pressure values have been cross-checked against the "NaCl scale" up to nearly 300 kbar [3], and against shock compression values up to over 500 kbar. The values of the pressure numbers assigned to the various resistance-jump transitions can be debated in terms of the correctness of the model (such as NaCl) or the shock pressure derivations and temperature corrections.

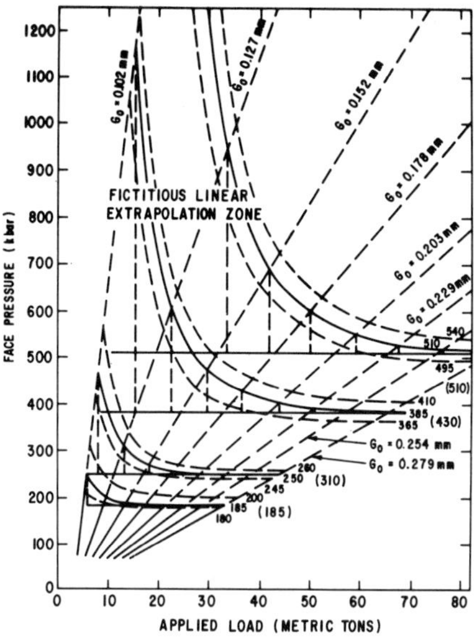

Fig. 6. Chart similar to Fig. 5, but ficticiously extrapolated in a linear manner.

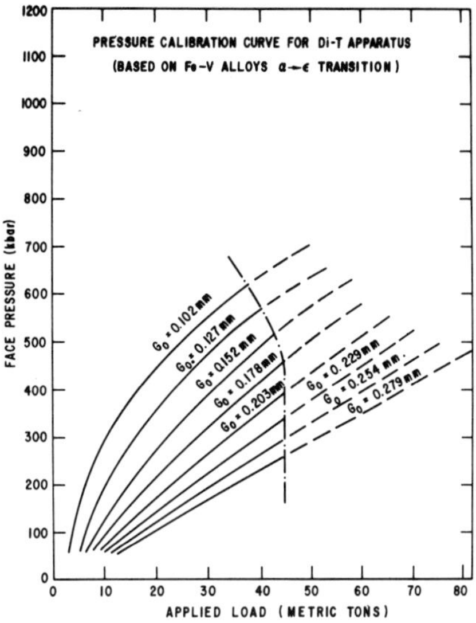

Fig. 7. Corrected, or adjusted, P(L) curves for various values of G_o.

One feature of the methodology we have presented is that it can also be applied to the establishing of more reliable values for some other pressure-dependent phenomena. For example, Fig. 8 shows a G_o vs. L plot for the apparatus with the lines corresponding to the various resistance-jump calibration transitions as a background. Also plotted are eight points corresponding to sulfur going into the semimetallic state [4], eight points at which the conduction of a sulfur specimen begins to exceed the leakage conduction of the cell and apparatus insulation ($\sim 10^8$ ohms), and three points corresponding to the transformation of GaP from diamond-cubic to the metallic form. The scatter of these points can be averaged out in a rational way on this diagram to yield pressure values of about 470, 300, and 220 kbar, respectively.

Table I provides a proposed up-to-date list of resistance-jump transitions which may be used for calibration of apparatus.

This appears to be the limit for the FeV alloy series, because the alloy solubility limit is just above 20%V. For higher pressure calibration points the metallic transitions in the III-V compounds suggested by Van Vechten [5] (such as BP, 420 kbar; SiC, 660 kbar; diamond C, 1800 kbar; etc.) may have to be explored when apparatus of 600 to 1800 kbar capability becomes possible.

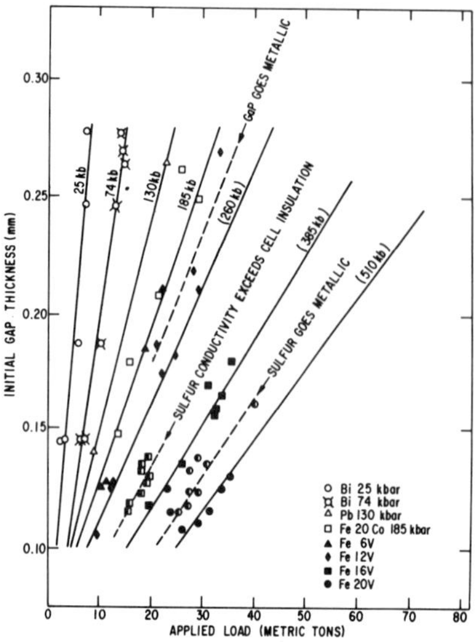

Fig. 8. G_o vs. L plot showing location of characteristic points for sulfur in comparison to the transition points for various calibrating materials.

Table I. Proposed Transitions Useful for Apparatus Calibration

Bi(1-2)	25 kbar
Ba(1-2)	53 kbar
Bi(5-6, Homan notation)	74 kbar
Fe(α-ε)	112 kbar
Ba(2-3)	120 kbar
Pb(1-2)	130 kbar
Fe20Co and Fe6V(α-ε)	~190 kbar
Gap (metallic transition)	~225 kbar
Fe12V(α-ε)	~260 kbar
Fe16V(α-ε)	~385 kbar
Fe20V(α-ε)	~510 kbar

ACKNOWLEDGMENT

The authors gratefully acknowledge the strong support of M. Aven and A. W. Urquhart for this exploratory research project.

REFERENCES

1. T. R. Loree, C. M. Fowler, E. G. Zukas, and F. S. Minshall, J. Appl. Phys. 37, 1918 (1966).
2. F. P. Bundy, Rev. Sci. Instr. 46, 1318 (1975).
3. D. Papantonis and W. A. Bassett, J. Appl. Phys. 48, 3374 (1977).
4. K. J. Dunn and F. P. Bundy, this conference, Session B-1-D.
5. J. A. Van Vechten, Phys. Rev. B7, 1479 (1973).

R-5

TRANSFORMATION PRESSURE OF ZnS BY A NEW PRIMARY PRESSURE TECHNIQUE

A. L. Ruoff and K. S. Chan

Cornell University
Ithaca, New York USA

INTRODUCTION

The transformation of zinc sulfide to the conducting state under high pressure has been observed previously by several different techniques [1-3]; however, the transformation pressures were obtained from hypothetical or extrapolated pressure scales. The present experiment was designed to measure the transformation pressure by a primary pressure method using an indentor-anvil system made of single-crystal diamond.

Samara and Drickamer [1] used the supported opposed-anvils developed by Balchan and Drickamer [4] to obtain a ZnS transition pressure of 24.0 GPa (based on the 1961 fixed-point scale [4]) by observing a sharp resistance drop. Calibration of this opposed-anvil system was based on an extrapolation of Bridgman's data [4] (obtained at a maximum pressure of 3 GPa for lead, platinum, and indium) up to 50.0 GPa. This is a gross extrapolation of data. This scale was later revised in 1970 by Drickamer, with the ZnS point being revised to 18.5 GPa.

Le Neindre et al. [3] used a split sphere to obtain a ZnS transition pressure of 15.0 GPa by observing a sharp resistance drop. Their apparatus [5] was calibrated against the 1970 fixed-point scale proposed by Drickamer [6]. In this new scale pressure was calibrated mainly against the shock pressure-volume data of Rice et al. [7] for Al and Ag interpolated to room temperature. The Al and Ag data were also checked against Decker's calculated p-v values for NaCl [8]. The accuracy of the shock-wave data and Decker's calculated data are discussed elsewhere [9,10].

Piermarini and Block [2] used the diamond-anvil cell to obtain a ZnS transition pressure of 15.0 GPa by observing the change from the transparent phase to a black, opaque phase under visible light.

Pressure was calibrated [2] against Decker's calculated p-v value for NaCl up to 29.1 GPa.

METROLOGY

For a spherical-tip indentor pressed against a flat anvil of the same material, the pressure can be measured in two ways: (1) From the known R and a given F, the pressure can be calculated from elasticity theory [11-13]. The pressure profile is a hemisphere

$$P(r) = P_0[1-(r/a)^2]^{1/2} \qquad (1)$$

where

$$P_0 = (3/2)^{1/3}\pi^{-1}\ [E/(1-\nu^2)]^{2/3}\ R^{-2/3}F^{1/3} \qquad (2)$$

This applies to the isotropic situation. Young's modulus for diamond varies from the mean value by at most 8% [14], so the assumption of isotropy is reasonably well satisfied for single crystals. We use $E = 1141$ GPa and $\nu = 0.07$ [14]. Since

$$P_0 = [E/\pi(1-\nu^2)]\ a/R \qquad (3a)$$

$$P_0\ (\text{GPa}) = 365\ a/R \qquad (3b)$$

(2) From a known applied force and by directly measuring the contact area using interference techniques (Newton rings), P_{av} is directly measured. This agrees closely with the average pressure computed directly from the elasticity theory, thus again verifying the contact theory. From contact theory

$$P_0 = 1.5\ P_{av} \qquad (4)$$

Equation (4) does not depend on the actual sphere radius. For the indentor-anvil system, the distance of approach of two points far from the contact is

$$\alpha = R[\pi(1-\nu^2)/E]^2\ P_0^2 \qquad (5)$$

When a thin film of thickness t is sputtered or evaporated onto the surface, the above equations will still closely apply if $t < \alpha^t/10$.

In order to study the thin film in the high-pressure region, a new non-shorting measurement technique was developed. A non-shorting electrode-grid is shown in Fig. 1. (This can be placed directly on the anvil using optical lithography.) The thin-film sample, an electrical insulator, is then deposited on it. When the indentor is pressed against the film, the contact pressure is established over an area of radius a, shown by the dashed circle in Fig. 1. If the pressure near the center is sufficiently high, the

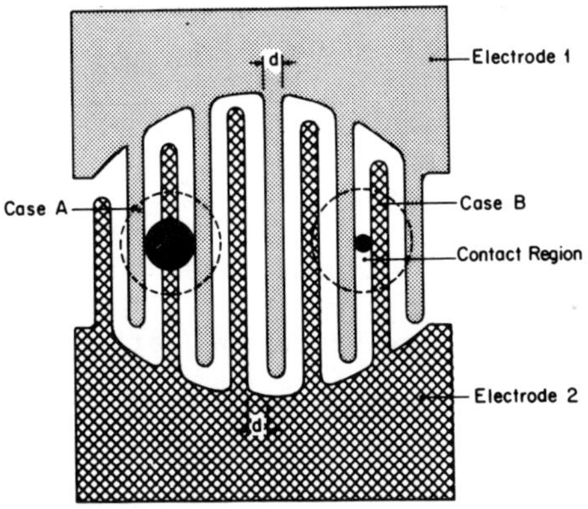

Fig. 1. Schematic of non-shorting electrode-grid. Actual grids contain 60 or more fingers. Indentor contact perimeters are shown by dashed circles. The dark center spots show samples which have undergone an insulator to conductor transition. In each the conducting phase is just large enough to close the circuit.

film converts to an electrically conductive phase. Under this condition there can be two extreme cases in the positioning of the indentor tip relative to the fingers of the grid. This relative positioning (together with the accuracy of the values of R and F) determines the accuracy by which the transformation pressure can be characterized. The small solid circles in Fig. 1 represent the minimum area of transformed ZnS film at which the resistance drop becomes detectable. If the width and the spacing of the fingers are both d, then the radii of these small circles will be 1.5 d (case A) and 0.5 d (case B). The percentage error due only to this relative positioning therefore satisfies

$$e_L \leq e \leq e_U \qquad (6)$$

where

$$e_L = \frac{P_0 - P(0.5\ d)}{P_0}(100) \qquad (7)$$

and

$$e_U = \frac{P_0 - P(1.5\ d)}{P_0}(100) \qquad (8)$$

The appropriate values for P(r) can be computed from equation (1); with R = 2640 μm, d = 12.5 μm, and P_0 = 13.8 GPa, e_U = 1.8% and e_L = 0.2%.

EXPERIMENTAL PROCEDURES

Two procedures differing only in the order the ZnS film and the electrode-grid were deposited onto the diamond (100) surface will be described.

Electrode-Grid on Diamond

A photoresist grid pattern was laid down on a single-crystal diamond (an electrical insulator) anvil having parallel (100) faces by microelectronics technology. The diamond was cleaned in an ultrasonic cleaner using trichloroethylene (TCE), acetone, methanol, and deionized water as solvents in sequence. After bake-drying, photoresist was coated on it. The coating was then bake-dried and a mask of the grid pattern was placed on it. It was exposed and subsequently developed to get the photoresist grid pattern. Nickel was then sputtered at low power (to avoid excessive hardening of the photoresist at high temperature) onto the grid pattern and the photoresist dissolved away to get a non-short electrode-grid of 250 Å thick. A magnified portion of such a grid is shown in Fig. 2. Nickel was chosen because it adheres strongly to diamond, even when sputtered at low power. Zinc sulfide was then vacuum deposited on the diamond surface bearing the Ni electrodes.

Fig. 2. Magnified portion of actual electrode-grid, d = 6 μm.

Electrode-Grid on ZnS

A thin film of ZnS (about 3000 Å thick) was evaporated in a vacuum of 10^{-5} to 10^{-6} Torr onto a diamond anvil having parallel (100) faces. A photoresist grid pattern was then laid down onto the ZnS film as described above. Silver was then vacuum evaporated onto the grid pattern and the photoresist which acted as a mask in evaporating the silver was dissolved away in acetone to obtain, finally, the silver electrode-grid.

This development completely eliminated the possibility of shorting which exists when both the indentor and the anvil are electrical conductors [15]. The alignment problem is also eliminated since there will always be a pair of fingers close by to act as electrodes (see Fig. 1).

The crystallinity of the ZnS film evaporated on micro-slides was examined by an x-ray Debye-Scherrer camera and by a transmission electron microscope (TEM). One film of about 2.5 μm thickness was scraped off the micro-slide and examined by x-ray, and another of 750 Å thickness was examined by a TEM operating at 200 kv. Both films were found to be polycrystalline and of sphalerite structure.

The ZnS film was then subjected to high pressure by pressing a single-crystal diamond (an electrical insulator) spherical-tip indentor against it. An x-y stage was used to move the indentor laterally so that the tip falls within the finger region of the electrode-grid. The resistance of the film was monitored by a Keithley multimeter. Several loading cycles were made on the same spot. Applied force was measured by a load cell. If the yield strength of ZnS is not a significant fraction of the applied stress in the ZnS film at a given position inside the contact area, the state of stress there is essentially one of hydrostatic pressure. We expect the yield stress to be small compared to the transition pressure.

DISCUSSION

An example of an experiment is shown in Fig. 3. At the present time this work is in a preliminary stage so we will not quote a specific figure for the ZnS transition point (although we hope to do so in the near future). The error due to positioning can be made a small fraction of 1% as can the error due to force measurement. We believe that eventually the maximum error in measurement of the tip profile (R) can be reduced to less than 5%. Finally, because we are dealing with a thin film, it is necessary that we take extreme care to characterize the stoichiometry and attain a stoichiometric structure. Work on this is now underway. The method is restricted to materials whose initial resistivity is significantly higher than that of the electrode material. It is also restricted to materials for which good thin films can be prepared and characterized.

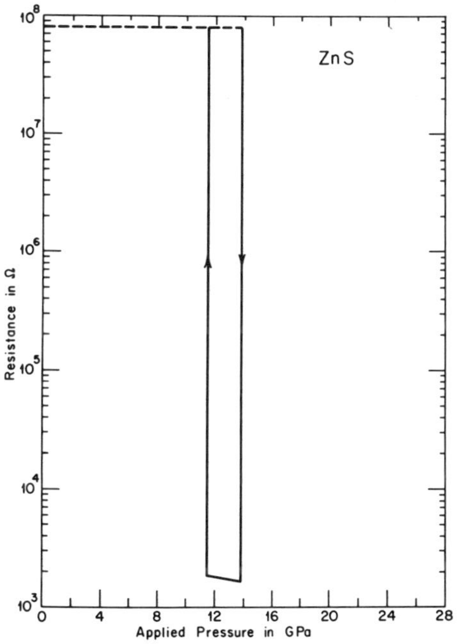

Fig. 3. Example of transition and reverse transition in ZnS using a diamond indentor with R = 2640 μm.

ACKNOWLEDGMENTS

The authors wish to thank the National Aeronautics and Space Administration for support of this work and also the National Science Foundation for their support through the Cornell Materials Science Center. They also wish to thank J. Frey and J. M. Ballantyne for making available their microelectronics laboratory, and R. Terry for his technical assistance.

NOTATION

a = contact radius

d = width (or spacing) of the fingers of the electrode-grid

e = percentage error in insulator-to-conductor transition pressure due to positioning of indentor

e_L = lower limit of e
e_U = upper limit of e
E = Young's modulus of diamond
F = applied force
P_{av} = average pressure
P_0 = maximum pressure
R = indentor tip radius
r = horizontal distance from z-axis
t = total thickness of the electrodes and the ZnS film
α = distance along z-axis by which two points, one in the anvil and the other in the indentor, far removed from the contact plane, approach each other
α^t = value of α at transition
ν = Poisson's ratio of diamond

REFERENCES

1. G. A. Samara and H. G. Drickamer, J. Phys. Chem. Solids 23, 457 (1962).
2. G.J. Piermarini and S. Block, Rev. Sci. Instr. 46, 973 (1975).
3. Bernard Le Neindre, Kaichi Suito, and Naoto Kawai, High Temp-High Press. 8, 1 (1976).
4. A. S. Balchan and H. G. Drickamer, Rev. Sci. Instr. 32, 308 (1961).
5. N. Kawai and S. Endo, Rev. Sci. Instr. 41, 1178 (1970).
6. H. G. Drickamer, Rev. Sci. Instr. 41, 1667 (1970).
7. R. H. Rice, R. G. McQueen, and J. M. Walsh, Solid State Phys. 6, 1 (1958).
8. D. L. Decker, J. Appl. Phys. 36, 157 (1965).
9. A. L. Ruoff and L. C. Chhabbildas, J. Appl. Phys. 47, 4867 (1976).
10. A. L. Ruoff, paper A-5-C presented at 6th AIRAPT Intern. High Pressure Conf., University of Colorado, Boulder, Colorado, July 25-29, 1977.
11. S. Timoshenko and J. N. Goodier, Theory of Elasticity, McGraw-Hill Book Company, New York (1970), p. 409.
12. A. L. Ruoff and K. S. Chan, J. Appl. Phys. 47, 5077 (1976).
13. K. S. Chan, Thesis, Cornell University (1977).
14. A. L. Ruoff, Cornell Materials Science Center Report No. 2855, Cornell University, Ithaca, New York, June 1977.
15. L. F. Vereschchagin, E. N. Yakovlev, B. V. Vinogradov, V. P. Sakun, and G. N. Stepanov, High Temp.-High Press. 6, 505 (1974).

R-6

DETERMINATION OF THE TRANSITION PRESSURES OF SOME ALKALI HALIDES UNDER HYDROSTATIC PRESSURE WITH X-RAY DIFFRACTION METHOD

M. Nomura, Y. Yamamoto, N. Nakagiri
Y. Shirai and H. Fujiwara

Hiroshima University
Hiroshima, Japan

INTRODUCTION

Although many reports have been presented concerning the polymorphic transition pressures of alkali halides, there have been few reliable studies with respect to the hysteresis phenomena. Recently Lacam and Peyronneau [1] observed the definite hysteresis loops of several rubidium halides controlling such conditions as purity of the specimen, heat treatment, and pressurization rates. For the potassium halides, on the other hand, which have higher transition pressures [2] than those of rubidium halides, no reliable hysteresis loop was reported. Mizouchi [3] measured the transition pressures of potassium chloride as a test of his new pressure apparatus both in compression and decompression runs, but his result is unlikely to be regarded as a hysteresis loop. In order to explain the hysteresis phenomena, therefore, the lattice constants of the potassium halides were measured using the x-ray diffraction technique at pressures including the phase transition. The specimens studied were KCl, KI, KBr, and KF.

EXPERIMENTAL

The specimens of potassium halides used were of commercial grade purity. Each specimen was pulverized in an agate mortar and was made to pass through the 325-mesh screen. Except for KCl, care was taken to prevent the specimen from absorbing moisture. For KF the pulverization was carried out in a liquid of 1:1 pentane-isopentane, but the sieving was not.

The x-ray was MoKα radiation. The measurement of the intensity of the diffracted beam was made in step scanning at every 0.05° in 2θ. The diffraction lines except KBr were those from the (200) and

(220) planes in the low-pressure phase and the (110) plane in the high-pressure phase.

The pressure cell used is shown in Fig. 1, similar to a unit described elsewhere [4]. The cell was made of amorphous boron solidified with epoxy resin and was an 8 mm cube at the center of which a cylindrical hole was drilled. The hole was plugged with copper or teflon end plugs. The middle space thus formed was used as a sample chamber. In order to apply hydrostatic pressure, two kinds of pressure transmitting liquids were used according to the pressure range; isopropanol was used for KCl, KBr, and KI and 1:1 pentane-isopentane for KF. The sample pressure was determined by the use of the pressure marker mixed with the sample to be studied. The pressure cell was pressurized with a cubic-anvil press.

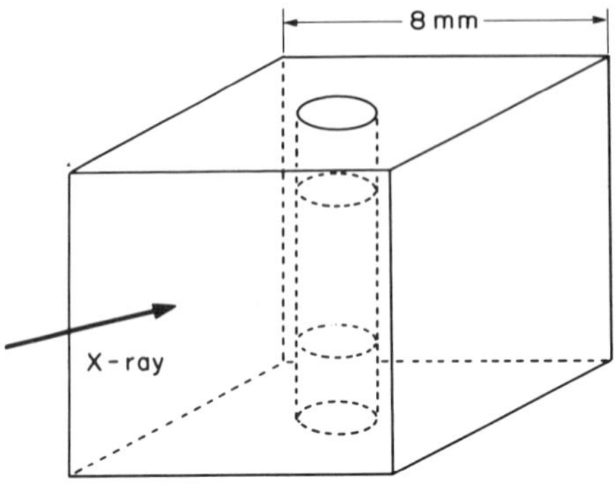

Fig. 1. Pressure cell for the x-ray diffraction observation under hydrostatic pressure. The cylindrical hole in the center of the cube is the sample chamber and was plugged with copper or teflon end plugs.

For the purpose of determining the diffraction angle precisely, the diffraction profiles obtained were fitted to the Gaussian function superposed on the quadratic background with the regression method of least square.

RESULTS AND DISCUSSION

Potassium Chloride KCl

For measurement of the pressure-volume (P-V) relation, NaCl as a pressure marker was mixed with KCl in an equivolume ratio. The diffraction lines observed were those from the (200) and (220) planes in the low-pressure phase and (110) in the high-pressure phase of KCl, and those from the (200) and (220) planes of NaCl.

In the compression run, a definite range in the press ram load was detected where the low-pressure phase of KCl was observed to coexist with the high-pressure phase. Over the above mentioned range, the lattice constants of the low and high-pressure phase of KCl, and the lattice constants of NaCl were kept unchanged. This behavior could reasonably be accepted by the interpretation that the pressure in the sample chamber was kept constant due to the discontinuous decrease in volume of KCl at phase transition, compensating for the effect of the increment of the press ram load.

The constant-pressure region was also detected in the decompression run, but not as marked as in the compression run. This is due to the fact that fine adjustment of the pressure in the sample chamber was rather difficult in the decompression run.

The hysteresis loop in the P-V relation is shown in Fig. 2. As for the pressure dependence of the molar volume both in the low and high pressure phases in the compression run, the present results are in good agreement with other results obtained both with the liquid [5] and the solid pressure medium [6]. The transition pressures P_+ in the compression run and P_- in the decompression run were 2.53 and 1.66 GPa, respectively. The value of P_+ is considerably larger than published results obtained under hydrostatic pressure such as 2.19 GPa by Mizouchi [3] and 2.12 GPa by Yagi [5]. The transition pressure P_-, however, was in good agreement with the 1.69 GPa of Mizouchi [3]. As a result, the hysteresis width ($P_+ - P_-$) was fairly large.

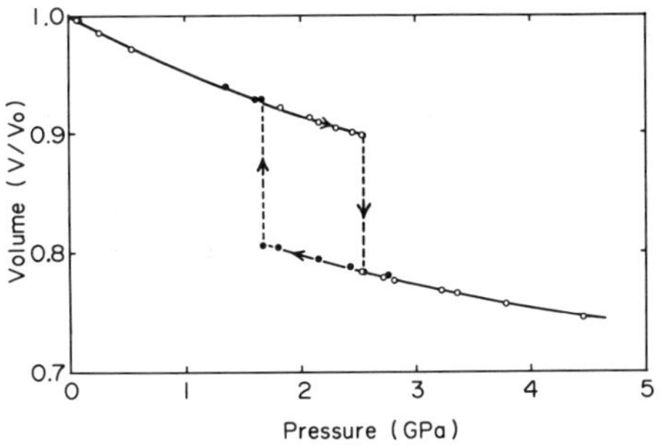

Fig. 2. Hysteresis loop for KCl observed under hydrostatic pressure. Open circles and solid circles indicate the compression and decompression runs, respectively.

Potassium Iodide KI

Potassium iodide has a higher x-ray absorption coefficient than KCl, so that the pressure marker, NaCl, was mixed with KI in the volumetric ratio of 2:1. The diffraction lines observed were the same as those for KCl.

Although the constant range of the sample pressure was observed both in the compression and decompression runs, the range was not as wide as in the case of KCl. This is partly due to the small quantity of KI in the sample chamber.

When the press ram load was fixed in the constant-pressure region in the compression run, namely the region where the low-pressure phase coexistence with the high-pressure phase was observed, it was found that the high-pressure phase increased very gradually for about 12 hrs. The rate of the transition in the decompression run was much smaller than that in the compression run. Such a sluggish transition seems to be characteristic of KI.

The hysteresis loop obtained experimentally is given in Fig. 3. For the P-V relation good agreement was also obtained between the present result and the results obtained by various methods [2,6]. In the present work P_+ was 2.05 GPa and P_- was 1.70 GPa. The center pressure of the hysteresis, $\bar{P} = 1/2(P_+ + P_-)$, was in good agreement with results obtained using the solid pressure medium [6].

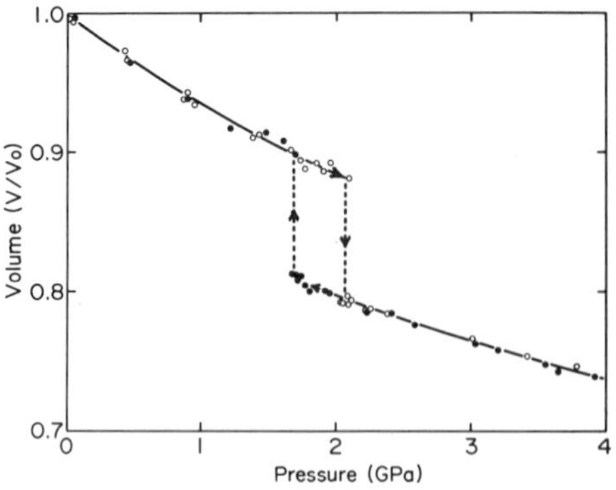

Fig. 3. Hysteresis loop for KI observed under hydrostatic pressure. Symbols used are the same as in Fig. 2.

Potassium Bromide KBr

For the MoKα radiation presently employed, bromine has a high absorption coefficient. It was practically impossible to detect the constant-pressure range monitoring the diffraction line of KBr, because the experimental method in the present work requires a large amount of the alkali halide. With respect to the transition pressures, it should be possible to determine them without observing the diffraction lines of the specimen by using the following procedure.

KBr and NaCl as a pressure marker were packed in the sample chamber in a different manner than in the case of KCl or KI. The NaCl powder was sandwiched with KBr powder and the x-ray beam was made to pass only through the NaCl. Therefore, the diffraction lines observed were those from the (220) and (200) planes of NaCl. At the phase transition the pressure in the sample chamber remained constant due to the discontinuous volume change of KBr by varying the press ram load by a small increment, leading to the constant pressure as described earlier.

The pressure in the sample chamber thus measured is plotted in Fig. 4 against the press ram load both for the compression and decompression runs. In the compression run a noticeable stagnancy of pressure associated with the phase transition of KBr was observed, whereas the stagnancy was not so clearly observed in the decompression run for the reason already mentioned. The transition pressures P_+ and P_- thus obtained were 2.12 GPa and 1.52 GPa, respectively.

Potassium Fluoride KF

There is scatter in the published data on the transition pressure of KF in the compression run. The transition pressures P_+'s reported are 1.73 GPa [6], 1.46 GPa [7], and 3.8-4.0 GPa [5]. The scatter may result from some kind of chemical reaction pointed out by Yagi [5].

As KF may be chemically active, the pulverization and the packing of the powder into the sample chamber were carried out in a mixture of 1:1 pentane-isopentane. As a pressure transmitting medium 1:1 pentane-isopentane was used.

Using the mixture with NaCl, some extra lines other than NaCl or KF were observed occasionally. This result suggests that NaCl is inadequate as the pressure marker in the case of KF. Then, instead of NaCl, Al powder was used as the pressure marker, and no

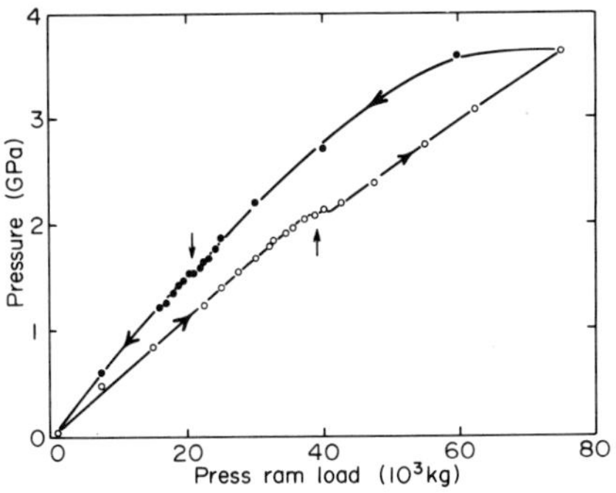

Fig. 4. Pressure in the sample chamber for the case of KBr against the press ram load. Symbols are the same as in Fig. 2. Stagnancies in the sample pressure due to the phase transition of KBr are indicated by arrows.

undesired line was detected. Although the pressure dependence of the lattice constant of Al was available [8], the calibration experiment was made independently using NaCl. In the preliminary experiment, the constant-pressure range similar to the cases of the other potassium halides was observed in both the compression and decompression runs.

Influence of Hydrostatic Pressure on the Phase Transition

When the solid pressure medium is employed, the transition pressure accompanies an inevitable broadening or ambiguity as was observed by Inoue and Asada [9] for KCl. The reason for this behavior is likely to be in the uniaxial stress component and the distribution of the uniform pressure component. Sodium chloride mixed as a pressure marker, however, tends to release a uniaxial stress component. With respect to the uniform-pressure distribution, the difference in pressure [10] actually applied to the potassium chloride and sodium chloride due to the difference in their compressibilities should be small, because the compressibility of potassium chloride is quite close to that of sodium chloride. A fairly reliable result may, therefore, be expected to be obtained, even when the pressure transmitting liquid is not used.

According to this consideration, the mixture of KCl and NaCl was packed in the sample chamber without any liquid, and the phase transition was observed in the compression run. As might be expected, the result was similar to that of Inoue and Asada [9]; that is, there was a distribution in the transition pressure, but the compression rate with pressure was very close to that in the hydrostatic pressure.

Figure 5 gives the pressure dependence of the volume of the specimen estimated from the integrated intensity of diffraction lines of the low- and high-pressure phases in the compression run. At the beginning of the phase transition, a slight but sharp drop in the volume was detected followed by a tail probably due to the inherent distribution of the pressure. Furthermore, it is interesting to note that the pressure where the sharp drop in volume takes place was in good agreement with that obtained under the hydrostatic pressure. At the initial stage of the phase transition

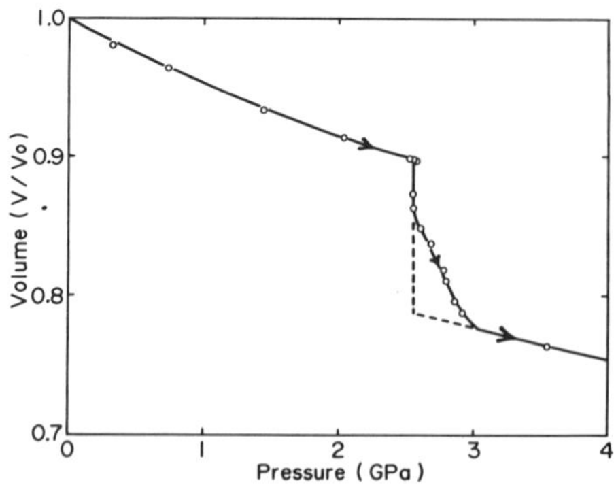

Fig. 5. P-V relation for KCl observed in the compression run without any liquid.

where the amount of the high-pressure phase is small, the release of the pressure due to the volume change of KCl takes place in the very small region of the sample chamber. As the high-pressure phase increases or the transition proceeds, the sample portion where the pressure is released increases. In order to attain a uniformity in pressure a long distance propagation of the pressure is needed, which is interrupted because of the absence of the pressure transmitting liquid. As a result, the tail, as is shown in Fig. 5, should appear at the phase transition.

The transition pressures of the potassium halides presently obtained are listed in Table I. It has been found that as a rule the P_+'s are high, P_-'s are low and the hysteresis widths ($P_+ - P_-$) are large compared with other results.

Table I. Transition Pressures of KCl, KI and KBr in GPa Measured Under Hydrostatic Pressure (GPa)

References		P_+	P_-	$\bar{P}$	(P_+-P_-)
KCl	Present work	2.53	1.66	2.10	0.87
	Mizouchi [3]	2.19	1.69	1.94	0.50
	Yagi [5]	2.12			
KI	Present work	2.05	1.70	1.88	0.35
KBr	Present work	2.12	1.52	1.82	0.60

REFERENCES

1. A. Lacam and J. Peyronneau, C. R. Acad. Sc. Paris B273, 997 (1971); also J. Phys. (Paris) 34, 1047 (1973); Rev. Phys. Appl. 10, 295 (1975).
2. P. W. Bridgman, Proc. Amer. Acad. Arts Sci. 76, 1 (1945).
3. N. Mizouchi, Rev. Sci. Instr. 44, 28 (1973).
4. Y. Yamamoto, M. Nomura, and H. Fujiwara, Japan. J. Appl. Phys. 16, 397 (1977).
5. T. Yagi, Tech. Rep. ISSP, Ser. A, No. 784 (1976).
6. S. N. Vaidya and G. C. Kennedy, J. Phys. Chem. Solids 32, 951 (1971).

7. C. W. F. T. Pistorius and H. C. Snyman, Z. Phys. Chem. Neus Forge 43, 1 (1964).
8. M. Senoo, H. Mii and I. Fujishiro, J. Phys. Soc. (Japan) 41, 1562 (1976).
9. K. Inoue and T. Asada, Japan J. Appl. Phys. 12, 1786 (1973).
10. Y. Sato, S. Akimoto, and K. Inoue, High Temp.-High Press. 5, 289 (1973).

INTENSIFIER DEAD-WEIGHT PISTON GAUGE WITH CONTROLLED-CLEARANCE FOR PRESSURES TO 1500 MN/m^2

R. Wiśniewski

Warsaw Technical University
Warsaw, Poland

INTRODUCTION

In all primary standards the measurement of any physical quantity is referred directly to the standard of length, mass and time. All primary standards should be so simple in principle that systematic measurement errors can be either evaluated or eliminated. There are two commonly used primary standards of pressure; the mercury column and the dead-weight piston gauge.

Zhokhovsky [1] has indicated that Parrot and Lenz [2] were the first designers of a dead-weight piston gauge. Amagat [3] was the first to use the so-called packingless piston - cylinder system in a wide range of laboratory investigations, particularly in pressure measurements. Wagner [4] has defined the effective area of a piston-cylinder system as the arithmetic mean of the areas for the cylinder and piston. Michels [5] somewhat later analyzed the effective area and showed its dependence upon pressure.

There are three different types of piston-cylinder systems: (1) the simple piston [2]; (2) the re-entrant cylinder, developed by Bridgman [6]; and (3) the controlled-clearance piston-cylinder system, developed by Johnson and Newhall [7]. The effective area for this last system may be determined mainly from geometric measurements of the piston dimension under normal conditions coupled with a knowledge of the Young's modulus and Poisson coefficient and their dependence on pressure.

There are three independent ways to minimize the standard weights: (1) the differential piston system, developed originally by Michels [5]; (2) the pressure intensifier system used for high pressures by Zhokhovsky [8] and for low pressures by Amagat [3]; and (3) the single-or double-lever system used for high pressures by Basset [9] and for low pressures by Wiebe [10].

EFFECTIVE AREA

If we want to obtain a general formula for the effective area of a dead-weight piston gauge we must consider all of the forces acting on the rotating or oscillating piston. There are some approximate methods of calculating the effective area. All of them eventually lead to the following useful relation:

$$A_{eff} = A_{o,eff}\,[1 + \lambda^* f(p)] \approx A_{o,eff}\,(1 + \lambda p) \qquad (1)$$

where $A_{o,eff} = Q/p = \pi r^2 (1 + 2H/r)$ is the effective area under zero pressure (low pressure value), λ^*, λ are coefficients whose magnitudes are different for differently constructed dead weight piston gauges and are inversely proportional to the elastic modulus, and $f(p)$ is a complicated function of pressure and difficult to evaluate. Dadson [11] has pointed out that it is possible to determine values experimentally of λ and function $f(p)$ by applying (1) to two geometrically identical dead-weight piston gauges constructed from different materials.

CONSTRUCTION OF GAUGE

The device consists of several parts. The main part is a column with a simple primary dead-weight piston-gauge and a pressure intensifier. The primary piston and cylinder constructed of steel has a nominal effective area $A^{1,nom}_{o,eff} = 50.0 \times 10^{-6} m^2$. The maximum pressure generated by this gauge is 15 MN/m^2 and the maximum standard mass needed for this is ≈ 75 kg. The rotating motion of the piston permits hand operation of loading and unloading of weights. The pressure measuring intensifier consists of a two piston-cylinder system: a large unit, constructed of steel, has a nominal effective area $A^{2,nom}_{o,eff} = 0.50 \times 10^{-3}\ m^2$ and a small unit, with a piston machined from carbolloy, has a nominal effective area $A^{3,nom}_{o,eff} = 5.00 \times 10^{-6}\ m^2$. The magnitude of the multiplication coefficient for the pressure intensifier is thus equal to 100. Both pistons were connected by an elastic, specially constructed coupling and were rotated with an experimentally determined optimal angle velocity either clockwise or counterclockwise by means of a pulley using a special electrical motor arrangement. Four additional parts include the intensifier for generating the jacketing pressure, the intensifier of the pressure being measured (both applied by hand operated pumps), the initial-pressure pumps, and the pump generating the primary, but accurately known, pressure. Figure 2 shows a cross section of the column. One can see most of the important details of construction.

METROLOGICAL PROPERTIES

The effective area of the dead-weight gauge presented here has the following form:

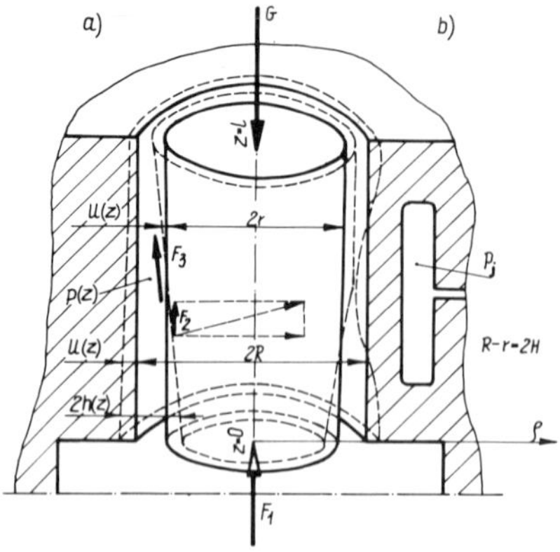

Fig. 1. Diagrammatic sketch of piston-cylinder systems (clearance greatly exaggerated). a - for simple piston, b - for controlled-clearance.

$$A_{eff} = \frac{(A^1_{o,eff})(A^3_{o,eff})}{A^2_{o,eff}} \frac{[1 + \lambda^*_1 f_1 (P_1)] [1 + \lambda^*_3 f_3 (P_3)]}{1 + \lambda^*_2 f_2 (P_2)} \qquad (2)$$

If we assume that the first piston-cylinder system and the primary side of the intensifier are constructed of the same materials and are geometrically similar (the pressure medium remains the same), we obtain: $1 + \lambda^*_1 f_1 (P_1) = 1 + \lambda^*_2 f_2 (P_2)$; $P_1 = P_2$ and because $p_3 = p$

$$A_{eff} = A_{o,eff} [1 + \lambda^*_3 f_3 (P)] \qquad (3)$$

This means that the change in the effective area of the device with pressure is the same as the change in the effective area of the third piston-cylinder system. For our situation, when the leakage past the controlled clearance piston is nearly equal to zero, we may consider the procedure developed by Newhall et al. [12] for deterinining the effective area of the third system. At constant temperature we have

$$A^3_{eff} = A^3_0 (1 + \frac{3\mu - 1}{E_p} p_o) [1 + \frac{2k^2}{(k^2 - 1) E_j} \Delta p_j] \qquad (4)$$

The numerical data for the parameters used and the percentage of errors are shown on Tables I and II. For this data and for $\Delta p_j = 20$ MN/m^2 we obtain

$$A_{eff} = 10^{-9} (499.86 \pm 0.44) [1 - 5 \times 10^{-7} (\frac{MN}{m^2})^{-1} p] \ m^2 \qquad (5)$$

One can see that in this situation the correction due to the change in the effective area of the high pressure piston-cylinder system with the controlled clearance can be neglected up to pressures of about 1000 MN/m^2.

Table I. Experimental Data

$2\ r_1 = (7.9700 \pm 0.0005)$ mm	$2\ H_1 = (3 \pm 1)$ μm
$2\ r_2 = (25.210 \pm 0.0010)$ mm	$2\ H_2 = (6 \pm 2)$ μm
$2\ r_3 = (2.5230 \pm 0.0002)$ mm	$2\ H_3$ -
$E_{carb} = (5.5 \pm 0.2)\ 10^5\ \frac{MN}{m^2}$	$E_{steel} = (2.05 \pm 0.05)\ 10^5\ \frac{MN}{m^2}$
$\mu_{carb} = 0.22 \pm 0.01$	$\mu_{steel} = 0.30 \pm 0.02$ $k = 8 \pm 0.5$

Warsaw acceleration due to gravity: 9.81227 ± 0.00020 m/s^2

Normal acceleration due to gravity: 9.80665 ± 0.00015 m/s^2

Table II. Relative Uncertainties

Errors	%
Load of D W T	$\pm$ 0.003
Effective area $A^1_{o,eff}$	$\pm$ 0.035
Effective area $A^2_{o,eff}$	$\pm$ 0.020
Effective area $A^3_{o,eff}$	$\pm$ 0.020
Temp. uncertainty	$\pm$ 0.010
Overall uncertainty	$\pm$ 0.088

Finally, we obtain a working pressure relationship given by

$$p = Q/A_{eff} + \Delta p$$

and thus

$$P = \left\{\left\{\left\{2.0004\ [1 + 5 \times 10^{-7}\ (\frac{MN}{m^2})^{-1}\ p]\ Q \pm 0.1\%\right\} + \right.\right.$$

$$\left.\left. + 0.80\right\}\right\}\ \frac{MN}{m^2} \qquad (6)$$

REMARKS

The controlled-clearance dead-weight gauge described here operates very well with a mixture of glycerin and ethylene glycol (minimum 15 volume % of glycol). Over the entire pressure range it was possible to reduce the leakage to a very small value by use of the jacketing pressure without producing friction or loss

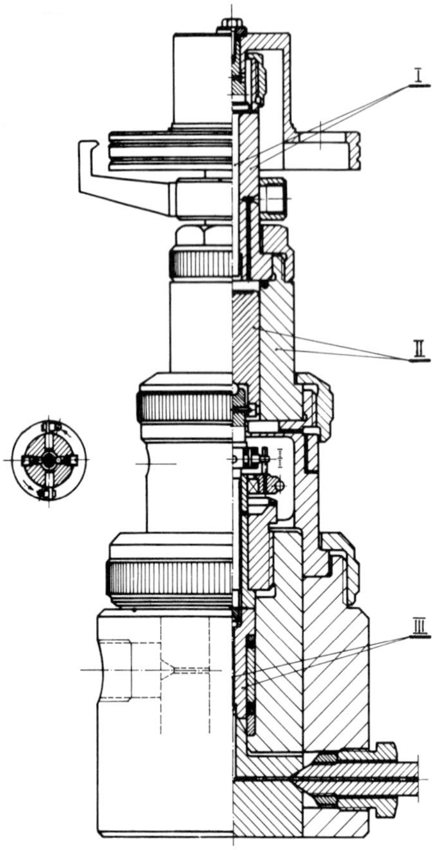

Fig. 2. Cross-section of the simple dead-weight piston gauge (standard of pressure, I - piston-cylinder system) and of the pressure intensifier (II and III - piston-cylinder systems).

of sensitivity. The sensitivity in our experiments was of the order of 10^{-4}. For example, it was about 0.025% for a pressure of 1300 MN/m^2 with a pressure jacket providing practically zero leakage.

The total weight of our gauge is about 2.67 kN and in its commercial form it may be reduced to approximately 2.2 kN. The maximum dimensions of the unit are 0.6 m x 0.72 m x 0.5 m.

The aim of our work has been to build a portable and acceptable version of a dead-weight piston gauge, suitable for investigations in a scientific and technical laboratory.

ACKNOWLEDGEMENTS

This work was carried out with the support of the Physical Chemistry Institute of the Polish Academy of Science, represented by J. Koszewski. The author would like to thank the Rector of the Warsaw Technical University S. Pasynkiewicz for his continuous support.

NOTATION

$A_o^{1,2,3} = \Pi r^2$ = area of the non-deformed pistons

E_j = Young's modulus (jacket cylinder)

E_p = Young's modulus (piston)

H2 = clearance of not deformed piston-cylinder system

k = wall ratio (jacket cylinder)

p = standard pressure

Fig. 3. Photograph of 1500 MN/m^2 piston gauge.

Δp = pressure generated by gauge at $Q = 0$

Δp_j = difference between the intercept and the actual jacket pressure during measurements

Q = effective total floating weight

r = piston radius

μ = Poisson's ratio

λ^*, λ = coefficients of dead weight gauge

REFERENCES

1. M. K. Zhoshovsky, "Theory and Calculation of Apparatus with Nonsealed Piston" (in Russian) I.K.S.M. i I. P., Moscow, U.S.S.R. (1966).
2. Parrot and Lenz, Met. St. Pet. Acad. 2, 595 (1833).
3. E. M. Amagat, Ann. de Chemic et de Physique 29, 68, 505 (1893).
4. E. Wagner, Ann. Phys. ser 4 15, 906 (1904).
5. A. Michels, Annalen der Physik 72, 285 (1923); also 73, 577 (1924).
6. P. W. Bridgman, Proc. Am. Acd. Arts and Sci. 47, 321 (1912).
7. D. P. Johnson and D. H. Newhall, Trans. ASME 75, 303 (1953).
8. M. K. Zhokhovsky, Yu. S. Konyaev and V. G. Levheva, Pribory i Tehn. Eksperim. 3, 118 (1959).
9. I. Basset, Chim. et Ind. 53 (5), 303 (1945).
10. H. Wiebe, Z. Instrumentenkumte 30, 205 (1910).
11. R. S. Dadson, Nature 176, 188 (1955).
12. D. H. Newhall, L. H. Abbot and R. A. Dunn, "A Redetermination of the Freezing Pressure of Mercury using Improved Apparatus and Technique," in High Pressure Measurement, Butterworths, London (1963).

R-8

A SOLID-DIELECTRIC CAPACITIVE PRESSURE TRANSDUCER

J. H. Colwell

National Bureau of Standards
Washington, D. C. USA

INTRODUCTION

We are attempting to develop a direct-reading pressure transducer that will be as precise and stable as the piston gauge so that it may be used as a pressure transfer standard with these gauges. The development has centered on a capacitive transducer in which the change in capacitance of a solid-dielectric capacitor is measured. The capacitors consist of discs of dielectric material with metallic electrodes and guard rings deposited directly on their surfaces, forming, in effect, parallel-plate capacitors. When subjected to hydrostatic pressure, the capacitance changes as a result of the combined effects of the change in size of the capacitor and the change in the dielectric constant of the material. Andeen, Fontanella, and Schuele [1] were the first to demonstrate the feasibility of such a device. They used single-crystal CaF_2 as the dielectric medium with which a pressure change of 100 MPa produces a capacitance change of only 0.37%. Although this change is small, by using the three-terminal capacitor method, modern ac bridge techniques, and accurate ratio transformers, capacitance measurements can be made with accuracies in excess of $1:10^7$ and with resolution approaching $1:10^9$. It should, therefore, be possible to measure pressures with an uncertainty of less than 1 kPa, which is about the uncertainty of piston gauges intended for the 100 MPa range. The drawback with using CaF_2 capacitors is their large temperature dependence. To realize the pressure resolution of 1 kPa with this material requires that it be thermostated to within 0.1 mK which is clearly impractical. To overcome this pronounced temperature dependence, we conducted an extensive search for materials having a more favorable ratio of temperature-to-pressure dependence. Several materials were found which were an order-of-magnitude better than CaF_2, but this was not considered a sufficient improvement.

We have now turned to the use of a combination of two capacitors which effectively circumvents the temperature dependence problem.

CAPACITANCE DATA

During the course of this study we have made capacitance measurements as a function of temperature and pressure on samples of more than one hundred different materials. Most of the samples were single crystals, some were glasses, and a few were ceramics. Nearly all were discs, 25 mm in diameter and 1 mm thick. The electrodes were evaporated aluminum coatings. The low-voltage electrode was 17.5 mm in diameter and was separated from the guard ring by a narrow gap of 10-20 μm, formed by a masking ring during the evaporation. The high-voltage electrode on the opposite side of the disc covered nearly the entire surface. The adherence of the coatings was usually very good, but was a problem with some materials (see below). The samples were mounted in the pressure vessel in sample holders which insured a completely guarded circuit for the capacitance measurements. The capacitances of the various samples ranged between 5 and 80 pF. Data were taken at four temperatures between 5 and 50°C and at pressures from 0-140 MPa. The data could, in general, be fitted to the expression

$$C = C^\circ\ (1 + AP + BP^2) \tag{1}$$

where C° and A are linear functions of temperature and B is temperature independent.

In Table I are data on a few materials which are exemplary and will serve our discussion below.

Table I. Temperature and Pressure Characteristics of Capacitors

Material	$\frac{1}{C^\circ}\frac{dC^\circ}{dT} \times 10^6/K$	$A\times10^{12}/Pa$	$\frac{dA}{dT} \times 10^{12}/(PaK)$	$B\times10^{20}/Pa^2$
CaF_2	263	-37.8	-0.025	0.27
$CaCO_3$ ⊥	331.5	12.0	0.0149	0.42
$CaCO_3$ ∥	335	71.0	0.0169	1.25
As_2S_3	76.7	110.6	0.030	-1.4
$Bi_{12}GeO_{20}$	73.7	-102.8	0.037	2.0

PRESSURE RESOLUTION AND TEMPERATURE EFFECTS WITH CAPACITOR PAIR COMBINATIONS

The pronounced temperature dependence of the individual capacitors has been overcome by using a device where two different capacitors, placed in opposite arms of an ac ratio arm bridge, are both within the pressure vessel. The resulting bridge balance gives the ratio of the two capacitances

$$R = C_1/C_2 \tag{2}$$

The observed fractional change of the ratio measurement as a function of temperature or pressure is

$$\frac{1}{R}\frac{dR}{dT} = \frac{1}{C_1}\frac{dC_1}{dT} - \frac{1}{C_2}\frac{dC_2}{dT} \tag{3}$$

and

$$\frac{1}{R}\frac{dR}{dP} = \frac{1}{C_1}\frac{dC_1}{dP} - \frac{1}{C_2}\frac{dC_2}{dP} \tag{4}$$

The appropriate capacitor combinations for use in this type of pressure transducer, therefore, should have nearly identical temperature dependences but widely differing pressure dependences. It should be pointed out (see equation (3)) that it is the fractional temperature dependences of the capacitances that must be equated; this is a property of the material and does not depend on the size of the capacitance.

Initially, no appropriate combinations were known, and we began our search by looking at anisotropic materials with the idea that we might be able to tailor a crystal with a particular temperature dependence. The capacitance along different axes in anisotropic materials are usually quite different, and intermediate values can be obtained with samples cut at the various angles between the principle axes. This approach did not directly produce any useful combinations, but it did lead to the combination of parallel and perpendicular cuts of calcite ($CaCO_3$) which could be used as a pair in a pressure transducer. Subsequently, a second useful combination was found which consists of $Bi_{12}GeO_{20}$ and As_2S_3.

The data on the capacitor pair combinations can be analyzed in the same fashion as those of the individual capacitors given in (1) by writing

$$R = R^\circ \ (1 + A_r P + B_r P^2) \tag{5}$$

This can be evaluated directly or from the data on the individual capacitors since

$$R = \frac{C_1{}^\circ}{C_2{}^\circ}\left(1 + (A_1 - A_2)\ P + \left[(B_1 - B_2) - A_2(A_1 - A_2)\right] P^2 + \ldots\right) \tag{6}$$

Similarly, the pressure and temperature derivatives of R leads to expressions involving the differences of the corresponding terms for the individual capacitors such as

$$\frac{1}{R}\frac{dR}{dT} = \frac{1}{R^\circ}\frac{dR^\circ}{dT} + \frac{dA_r}{dT}P + \ldots \tag{7}$$

$$= \frac{1}{C_1^\circ}\frac{dC_1^\circ}{dT} - \frac{1}{C_2^\circ}\frac{dC_2^\circ}{dT} + \left(\frac{dA_1}{dT} - \frac{dA_2}{dT}\right)P + \ldots$$

and

$$\frac{1}{R}\frac{dR}{dP} = A_r + (2B_r - A_r^2)\,P + \ldots \tag{8}$$

$$= (A_1-A_2) - [2(B_1-B_2) - (A_1^2-A_2^2)]\,P + \ldots$$

Data for the calcite pair and the $Bi_{12}GeO_{20}$-As_2S_3 combination are given in Table II. The data obtained from the capacitor combinations agree with that derived from the individual measurements using equations (6) through (8).

Table II. Temperature and Pressure Characteristics of Pair Combinations

Material	$\frac{1}{R^\circ}\frac{dR^\circ}{dT}\times10^6/K$	$A_r\times10^{12}/Pa$	$\frac{dA_r}{dT}\times10^{12}/(PaK)$	$Br\times10^{20}/Pa^2$
$CaCO_3\perp$-$CaCO_3\parallel$	3.6	59.0	0.0020	0.76
$Bi_{12}GeO_{20}$-As_2S_3	3.0	213.4	-0.007	-1.2

Temperature influences the effectiveness of the capacitor pair combinations in three ways. The first is the requirement of thermal equilibrium between the two capacitors. This requirement is determined by the ratio of the pressure sensitivity of the pair to the temperature sensitivity of the individual capacitors. The second effect may be termed the short-term stability. It is the requirement that the temperature of the capacitors at pressure must be within certain limits of their temperature when the zero pressure ratio, R°, is measured. This limit is determined by the temperature and pressure dependences of the combination. The third effect depends on long-term temperature accuracy in that the temperature dependence of A_r requires that the transducer be used within certain limits of the temperature at which it was calibrated. It is the ratio of the pressure sensitivity to the term $(dA_r/dT)P$ that determines this temperature limit. Values of these three temperature limits for the two capacitor combinations are given in Table III.

These thermal restrictions are, in general, rather easily met. The thermal equilibrium value for the calcite pair is small, but causes no difficulty as the two capacitors are in close proximity and symmetrically located at the center of the pressure vessel (see below). The thermal compensation of both capacitor combinations is equally good at high pressure as at zero pressure. This is a

Table III. Thermal Stability Required for Accurate Pressure Resolution

Materials	Thermal Equilibrium Between Capacitors for 1 kPa Resolution	Short-Term Temperature Stability for 1 kPa Resolution	Long-Term Temperature Accuracy for Pressure Accuracy of 1 ppm
$CaCO_3\perp$-$CaCO_3\parallel$	0.17 mK	16 mK	42 mK
As_2S_3-$Bi_{12}GeO_{20}$	2.9 mK	80 mK	44 mK

consequence of the small value of the last term in (7) for both combinations. This is a particularly fortunate occurrence in the As_2S_3-$Bi_{12}GeO_{20}$ combination since of all the materials studied $Bi_{12}GeO_{20}$ is the only example where A and dA/dT are of opposite sign (see Table I).

ADIABATIC HEATING AND TRANSDUCER THERMOSTATING

Precise thermostating of large pressure vessels is usually a problem because of the large heating effects which occur when the pressure in the vessel is changed. The large heat capacity of the vessels then results in excessively long times for the equilibrium temperature to be restored.

The heating which occurs in material under compressive or tensile stress is given by

$$Q = - T \bar{V} \bar{\beta} \Delta P \qquad (9)$$

When a pressure vessel is pressurized, the contents of the vessel warm, but the vessel itself, being under tensile stress, cools. We have taken advantage of this by equating these heating and cooling effects so that no net heating occurs. The cooling that occurs in the pressure vessel itself, when internally pressurized, is equal in magnitude to the heating that would occur upon externally pressurizing a piece of steel of the same composition as that of the vessel and having the same volume as the internal volume of the vessel. In other words, if we could completely fill the inside of the vessel with a steel plug and then pressurize the system using a negligibly small amount of oil, there would be no temperature change when thermal equilibrium within the vessel is reestablished. To accomplish this, it was necessary to equate the sum of the $\bar{V}\bar{\beta}$ terms of the contents of the vessel with the product of the internal volume of the vessel and the β for the steel of the vessel. Since the β of the paraffin oil we are using as a pressurizing fluid is about 15 times that of the steel vessel, it was necessary to fill up nearly 95% of the internal volume with Invar, which has a very low thermal expansion, in order to get a balance. Figure 1 shows a partially disassembled transducer. The inner assembly, fabricated from six Invar plates, completely fills the pressure vessel except for the two wells for the capacitors and space for the electrical contacts and feedthroughs.

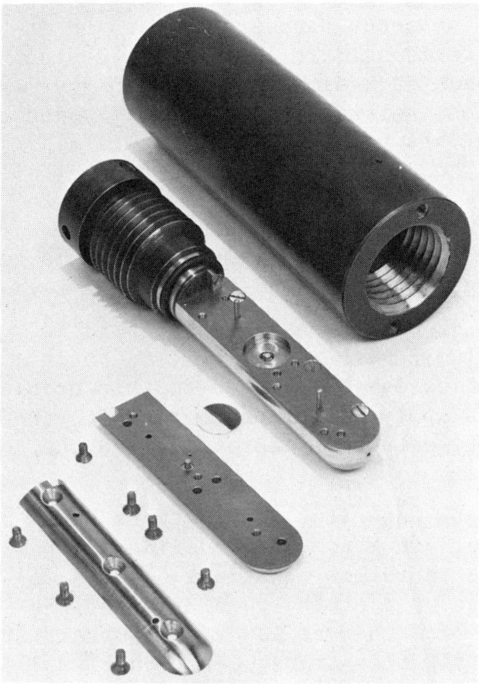

Fig. 1. Partially disassembled pressure transducer with 140 MPa pressure vessel.

Two of these devices have been built, one for a 140 MPa vessel and one for a 700 MPa vessel. In both cases they allow space for about 1 cm^3 too much oil so that there is a temperature change of about 4 mK when the pressure is changed by the full amount. This small temperature change is well within the permitted temperature limits given in Table III so it is not necessary to wait for the transducer to equilibrate with the thermostat after a pressure change. Thermal equilibrium within the vessel itself is reestablished in 6 to 7 min after a full range pressure change. This is about the same length of time it takes thermal effects in a piston gauge to dissipate. A simple air thermostat has been all that is required in the laboratory to hold the transducer within a few millikelvin of its operating temperature.

CURRENT STATUS OF TRANSDUCER

Although we have overcome the thermal dependence of this transducer, we have still not solved all of its problems. With both capacitor combinations we frequently experience precipitous capacitance changes when the pressure is cycled. These changes are equivalent to pressure changes of the order of 1 kPa. The source of the instabilities is not known, but is thought to arise from a slight shifting of the capacitors with the flow of oil in and out of the sample holders. There is some fringing field outside the dielectric material at the gap between the low-voltage electrode and the guard ring which could be effected by a slight change in the environment outside of the gap. This is especially true considering that the total capacitance change is only of the order of $1{:}10^7$.

A second, and possibly more fundamental difficulty, is a small relaxation which is observed with calcite and $Bi_{12}GeO_{20}$. This relaxation is observed when the capacitors are held at pressure for long periods. In both cases, the apparent pressure appears to increase 1 to 2 kPa when the pressure is held at 100 MPa for several hours. The capacitance relaxes back at a similar rate when the pressure is returned to zero. These relaxations are possibly due to impurities in the dielectric materials. Calcite is a natural material and inherently impure. The $Bi_{12}GeO_{20}$ samples studied have all contained small areas of "core material". This is thought to be a region of non-stoichiometric material and occurs along the axis of the boule as it is pulled from the melt. It is possible to cut samples from the edges of large boules, avoiding the core region.

Another problem with both $Bi_{12}GeO_{20}$ and As_2S_3 capacitors is that we have been unable to get aluminum to adhere well to their surfaces. The observed instabilities may be related to this problem. We are having commercial producers apply different coatings to these materials to determine if this will effect an improvement in the measurement.

The calcite and $Bi_{12}GeO_{20}$-As_2S_3 capacitor combinations have both been measured to a pressure of 700 MPa. The individual capacitors and the combinations continue to follow equations (1) and (5), respectively. There was no indication that the resolution of the device was depreciated in any way from that at lower pressures. It is in the higher pressure ranges that this transducer probably will be most useful. The accuracy of the piston gauge falls off at higher pressures, so that the demands on a transfer standard will be less stringent.

R. S. Kaeser of this laboratory has developed a self-balancing, digital-readout capacitance bridge for use with this transducer. The bridge far exceeds the precision demands of the transducer.

NOTATION

A = linear pressure coefficient of capacitance
B = quadratic pressure coefficient of capacitance
$\bar{\beta}$ = average thermal expansion coefficient over range of pressure
C = capacitance
C° = capacitance at zero pressure
ΔP = pressure change
P = pressure
Q = heating due to pressure change
R = ratio of two capacitances
R° = ratio of two capacitances at zero pressure
T = absolute temperature
$\bar{V}$ = average volume over range of pressure
$||$ = electric field parallel to crystal axis
$\perp$ = electric field perpendicular to crystal axis

Subscripts
1,2 = designation of individual capacitors
r = ratio of two capacitors

REFERENCE

1. C. Andeen, J. Fontanella, and D. Schuele, Rev. Sci. Instr. 42, 495 (1971).

R-9

ZnO VARISTORS AS PRESSURE MEMORY AND PRESSURE SENSOR

J. Wong and F. P. Bundy

General Electric Corporate Research and Development
Schenectady, New York USA

INTRODUCTION

Metal oxide varistors formed by sintering a mixture of ZnO with small additions of Bi_2O_3, transition, and post-transition metal oxides at high temperature constitute a novel class of electronic ceramics, which exhibit a highly non-ohmic behavior in current-voltage characteristics [1]. The functional dependence of current I on applied voltage V in the non-ohmic regime may be expressed empirically by $I = KV^{\alpha}$ where α, the nonlinear exponent, can take values from 5 to 50 or higher, and K is a constant. An idealized varistor V-I curve plotted on a log-log scale is shown in Fig. 1. This property, very similar to that of a back-to-back Zener diode, finds wide application for this class of materials as circuit protectors against voltage transients and power overloads [2,3].

The ceramic microstructure [4-10] and the conduction mechanisms [11-15] in ZnO varistors are found to be strongly correlated and have been the subjects of intensive investigation in recent years. The former, resulting from a liquid-phase, reactive sintering consists of a matrix of conductive ZnO grains (resistivity ~ 1 Ω-cm) [16] isolated from each other by a continuous network of intergranular, insulating thin-film barriers rich in Bi_2O_3 [7,8]. The varistor microstructure shown in Fig. 2, is a scanning electron micrograph of a thin-section (~ 100 μ thick) of the sintered material in which the ZnO grains have been etched out preferentially with perchloric acid, thus revealing a three-dimensional network skeleton of the threadlike intergranular material. The cavities, of course, correspond to sites occupied by ZnO grains that were etched away by the $HClO_4$ treatment. At low applied field, the conduction process follows a thermally-activated Schottky-type law, giving rise to the so-called leakage current I_L shown in Fig. 1. At high applied

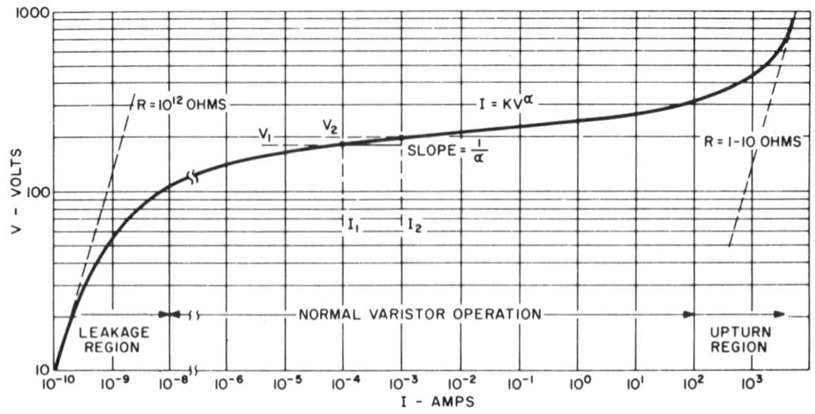

Fig. 1. Idealized voltage-current curve plotted on log-log scale showing the non-ohmic behavior of ZnO varistors over a wide range of current.

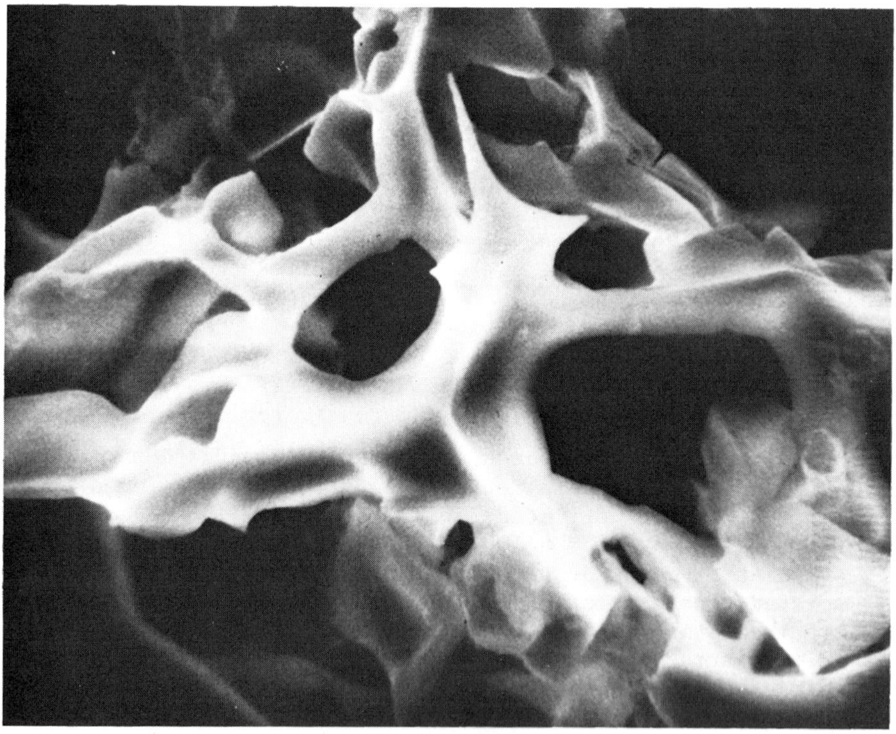

Fig. 2. Microstructure of a typical ZnO varistor. Scanning electron micrograph of the intergranular network material in ZnO varistor after the ZnO matrix has been etched out preferentially with $HClO_4$ acid.

field, the onset of varistor action is attributed to a Fowler-Nordheim tunneling process in the intergranular barrier between conductive ZnO grains [11,15].

Recently it has been observed that the leakage current I_L in ZnO varistors exhibits a strong dependence on pressure [17]. Upon release to atmospheric condition (1 bar), only a partial recovery to the original V-I characteristics takes place. The percent of recovery varies monotonically with applied pressure, indicative of a pressure-memory effect. This phenomenon together with the use of these novel materials as pressure sensors in the range of 0.5 to 10 kbar are reported in this paper.

EXPERIMENTAL

Varistor samples were prepared by sintering powder mixtures of ZnO (~ 97 mol %) with about 1/2 mol % each of the oxides of bismuth, cobalt, manganese, antimony, and tin. Details of powder preparation have been given elsewhere [1,4,7]. The pressed pellets were sintered in air for 1 hr at 1325°C. Two varistor systems were studied: a medium-voltage GE-MOV® varistor* and a low-voltage GE-MOV® varistor* The former has a varistor voltage of 400 V at 1 mA, while the latter exhibits a varistor voltage of 40 V at 1 mA [10] (see Fig. 4). The sintered specimens were approximately 14 mm in diameter by 1 to 3 mm thick and are coated with a silver electrode on each face.

Two methods were used to generate pressure within a varistor sample: (1) Uniaxially, by simply squeezing a sintered disc between two opposing Carboloy® metal* anvils in a calibrated hydraulic press (see Fig. 3a); (2) Hydrostatically, by encasing the sintered disc in a near-hydrostatic cell (see Fig. 3b) and using a belt-type apparatus. Pyrophillite was used as a pressure-transmission medium within the cell. Electrical connections were made from each varistor face via a thin copper foil to the Carboloy metal caps and finally to the opposing Carboloy metal anvils which were electrically insulated from the rest of the hydraulic press. Current-voltage curves were measured using a point-by-point procedure with a Keithley picoammeter and a Kepco 1.5 kV power supply. Measurements were also taken after recycling to atmospheric pressure before subjecting the sample to a higher load.

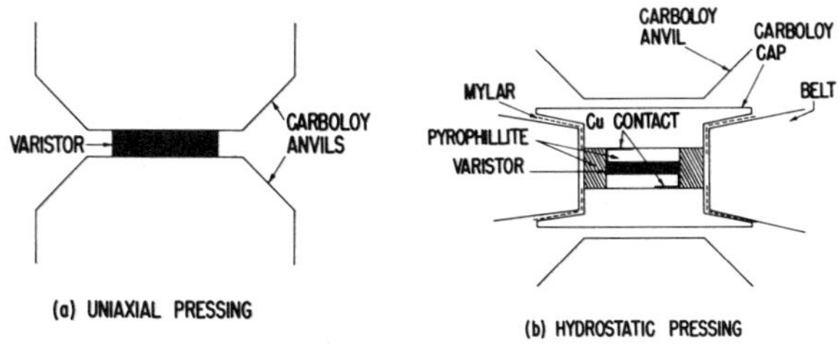

Fig. 3. Schematics of generating pressure within a varistor specimen: (a) Uniaxial pressing, and (b) Hydrostatic pressing.

* ® Registered trademark.

RESULTS AND DISCUSSION

In Fig. 4 the room temperature dc voltage-current characteristics of the above two varistor systems are plotted as a function of the applied load up to 10 tons for the case of uniaxially pressing. A strong dependence on pressure is evident in both systems. With an increase in pressure, the approximately ohmic prebreakdown region at low current moves to higher currents over a few orders of magnitude, accompanied by a decrease in the nonlinear exponent α. In other words, the leakage current I_L, defined as that current at half the initial varistor voltage at 1 mA, varies monotomically with the applied load. Thus, for the case of the medium-voltage varistor shown in Fig. 4a, the leakage current is defined at ~ 200 V and its variation with applied load is plotted in Fig. 5. The solid line labeled ON denotes the I_L values obtained under pressure, while the solid line labeled OFF corresponds to those at the same voltage measured at atmospheric pressure after the specimen had been subjected to the corresponding load indicated in Fig. 5. It is noted that reversibility is obtained upon recycling from ~ 2 tons (nominal pressure 1.1 kbar) or lower, above which only partial recovery to the initial V-I characteristics is obtained, as shown in Fig. 6 for the case of the medium-voltage varistor. Similar behavior of leakage current with applied load was observed for the case of the low-voltage varistor shown in Fig. 4b. In both systems the leakage current varies exponentially with load P, according to $I_L \sim P^7$. Beyond 12 tons both varistors lose their ceramic integrity and crumble due to the large shear stresses set up at the free edges of specimens in the uniaxial compression configuration. Finally, similar qualitative results were obtained for both varistor systems under hydrostatic conditions.

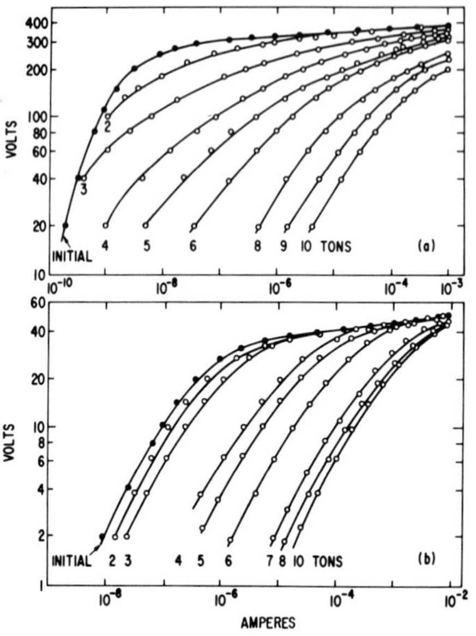

Fig. 4. Pressure variations of room temperature V-I characteristics of ZnO varistors: (a) Medium-voltage and (b) Low-voltage varistors. The nominal pressure at a load of 10 tons as calculated by load/area is 5.5 kbar.

Measurements carried out at 100°C on a medium-voltage varistor specimen showed that the I_L values increased by a factor of two, relative to room temperature, over the full pressure range as shown by the dotted line in Fig. 5. The temperature coefficient of I_L at

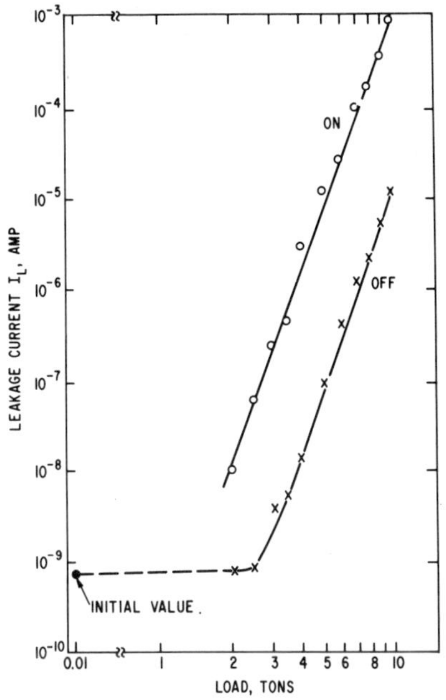

Fig. 5. Leakage current I_L measured at 200 V in medium-voltage varistor vs. load at 25 and 100°C. ON and OFF denote data obtained under pressure and upon recycling to 1 bar after being stressed to a given load, respectively. See actual V-I plots in Fig. 6.

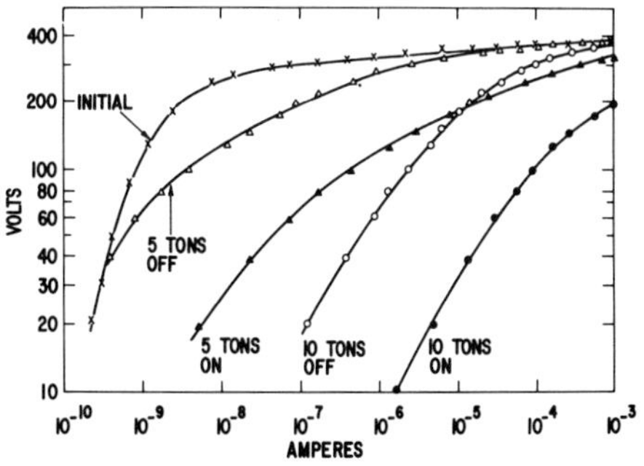

Fig. 6. Room temperature dc V-I characteristics of medium-voltage varistor showing partial recovery to original characteristics, recycling to 1 bar after being stressed to a given load.

constant pressure $(1/I_L \times \partial I_L/\partial T)_p$ for this class of varistor is about 0.012 per °C at 25°C and around 0.008 per °C at 100°C. When using these varistors as pressure-monitoring devices the temperature should be kept quite constant, or it should be observed and corrections made to the I_L in accordance with the temperature coefficient. From Fig. 5 it can be seen that a 75° change in temperature produces about the same I_L change as does a pressure change of 1/4 to 1/2 kbar at the lower and higher parts of the pressure range, respectively.

The observed partial recovery to the initial V-I characteristics shown in Fig. 6 is indicative of the fact that the varistor "remembers" the maximum load (or pressure) it has undergone. At room temperature the partially recovered V-I characteristic is stable with time (no detectable change after a month) so that the varistor can be used as an electronic pressure memory. An obvious application of this phenomenon is to use ZnO varistors for mapping out pressure gradients in a pressurized environment (whether static or transient) by simply measuring the I_L values of the device before and after a high-pressure event and using a calibration curve of the type shown in Fig. 5. The pressure memory may be erased by thermal annealing. Complete recovery to the original, unstressed V-I characteristics can be brought about by annealing at 450°C for 10 min in air [17].

ZnO varistors can also be used as a pressure sensor. The procedure consists of prestressing a pristine varistor to some high pressure, about 10 kbar or less, and releasing the pressure to 1 bar to obtain a partial recovery of the V-I characteristics. This is best exemplified in the case of a low-voltage varistor. In Fig. 7, curve (a) is the V-I curve of the original, unstressed characteristic up to 1 mA. Under a pressure of about 8.6 kbar, the leakage current (at 10 V) increases by over 3 orders of magnitude as shown by curve (b) in Fig. 7. Upon release of pressure to 1 bar, partial recovery to curve (c) is observed. For pressure sensing application, the stressed varistor is again subject to pressure in the range 0.5 to 8.6 kbar and the leakage current at constant voltage (at 10 V) is now measured as a function of pressure in this range. In Fig. 8, the current through the device at 10 V is plotted vs. pressure. Excellent reproducibility is obtained upon pressure cycling up and down from 8.6 kbar. The pressure coefficient defined by $(\Delta I/I_o)\ 1/P$ is calculated to be ~ 1 $kbar^{-1}$ and is ~ 10^3 larger than that for manganin wire in the same pressure range [18]. For varistor pressure sensor, I_o denotes the leakage current at a given voltage of the device at 1 bar after it has been prestressed, and ΔI is the current range used for pressure sensing as shown in Fig. 7. For a medium-voltage varistor the pressure coefficient is slightly higher, ~ 3 $kbar^{-1}$.

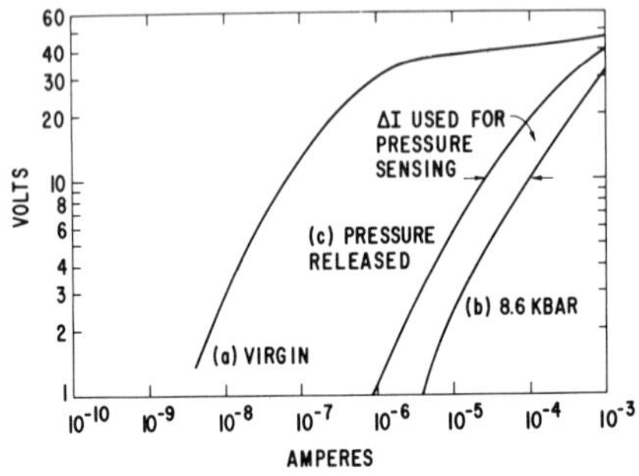

Fig. 7. Room temperature dc V-I characteristics of low-voltage varistor prestressed hydrostatically to 8.6 kbar for pressure-sensing application (see Fig. 8).

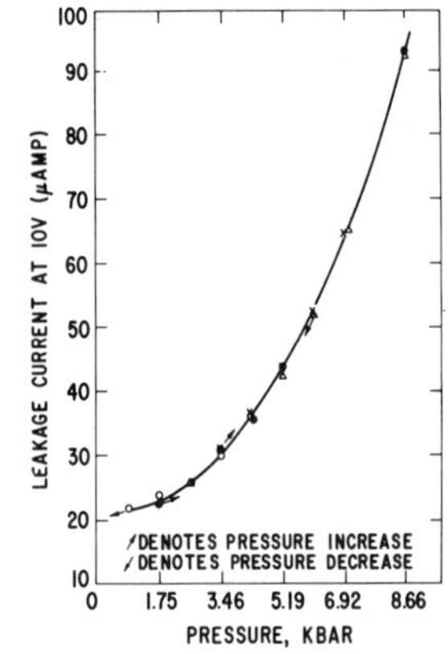

Fig. 8. Leakage current through a low-voltage varistor initially prestressed to 8.6 kbar (see Fig. 7), vs. subsequent pressure cycling.

CONCLUDING REMARKS

Pressure is not only an important extensive variable in physicochemical and thermodynamic investigations of matter, but is also an essential process parameter in certain specialized technologies utilizing a high-pressure route in manufacturing the final product. In recent years extensive efforts have been devoted to the measurement and standardization of high-pressure environments [19]. In the range > 10 to 300 kbar, a series of reference points - the so-called "fixed points" indicated by phase changes and/or resistance jumps in selected materials, such as those listed in Table I, are commonly used for pressure calibration as well as sensing.

Table I. Reference Points

Material and Transition	Fixed Point Pressure, kbar	Ref.
Hg freezing pt. at 22°C	11.8	20
Bi I - II at 25°C	25.5	20
Tl I - II at 25°C	36.7	20
Cs II - III IV	42	21
Ba I - II at 25°C	55	20
Bi III - V at 25°C	74	20
Sn I - II	94	21
Fe α-ε	110	21
Ba II liq (?)	120	21
Pb I - II	130	21
Rb II - liq (?)	146	21
$Fe_{15}Co$ α-ε	150	21
$Fe_{20}Co$ α-ε	190	21
$Fe_{40}Co$ α-ε	290	21

In the range below 10 kbar (0.5 to 10 kbar), however, such fixed point calibrants are rare: Ce at 7 kbar [22] and Hg freezing point at 0°C, 7.6 kbar [19]. Manganin (84 Cu-12 Mn-4 Ni) gauges are often used in the range 2 to 14 kbar and higher, but have a pressure coefficient of electrical resistance $K = \Delta R/R_o \; 1/P$, where R_o is the resistance at atmospheric pressure, of the order 10^{-3} per kbar and is not constant [18].

It is clear from the results shown in this investigation that ZnO varistors provide a more sensitive pressure sensor in the range up to 10 kbar by virtue of their high pressure coefficient. Effects of pressure on the ceramic microstructure and their correlation with the observed electrical properties and annealing kinetics will be reported in a subsequent publication.

REFERENCES

1. M. Matsuoka, Japan J. Appl. Phys. 10, 736 (1971).
2. J. D. Harnden, F. D. Martzloff, W. G. Morris, and F. D. Golden, Electronics 45, 91 (1972).
3. E. C. Sakshaug, J. S. Kresge, and S. A. Miske, IEEE Trans. Power Apparatus Syst. PAS-96, 647 (1977).
4. W. G. Morris, J. Am. Ceram. Soc. 56, 360 (1973).
5. J. Wong, J. Am. Ceram. Soc. 57, 357 (1974).
6. J. Wong and W. G. Morris, Am. Ceram. Soc. Bull. 53, 816 (1974).
7. J. Wong, J. Appl. Phys. 46, 1653 (1975).
8. J. Wong, P. Rao, and E. F. Koch, J. Appl. Phys. 46, 1827 (1975).

9. W. G. Morris and J. W. Cahn in Grain Boundaries in Engineering Materials, J. L. Walter, et al., eds., Claitors, Baton Rouge, Louisiana (1975), p. 223.
10. W. G. Morris, J. Vac. Sci. Technol. 13, 926 (1976).
11. L. M. Levinson and H. R. Philipp, Appl. Phys. Letter 24, 75 (1974); also J. Appl. Phys. 46, 1332 (1975).
12. J. D. Levine, Critical Reviews in Solid State Science 5, 597 (1975).
13. J. Bernasconi, H. P. Klein, B. Knecht, and S. Strassler, J. Electronic Mater. 5, 473 (1976).
14. J. Wong, J. Appl. Phys. 47, 4971 (1976).
15. H. R. Philipp and L. M. Levinson, J. Appl. Phys. 48, 1621 (1977).
16. H. R. Philipp and L. M. Levinson, J. Appl. Phys. 47, 3177 (1976).
17. J. Wong and F. P. Bundy, Appl. Phys. Lett. 29, 49 (1976).
18. Y. A. Atanov and E. M. Ivanova in Accurate Characterization of the High Pressure Environment, E. C. Lloyd, ed., NBS Special Publication 326 (1971), p. 49.
19. Accurate Characterization of the High Pressure Environment, E. C. Lloyd, ed., NBS Special Publication 326 (1971).
20. E. P. Lloyd and F. R. Boyd, in Accurate Characterization of the High Pressure Environment, E. C. Lloyd, ed., NBS Special Publication 326 (1971), p. 1.
21. F. P. Bundy in Accurate Characterization of the High Pressure Environment, E. C. Lloyd, ed., NBS Special Publication 326 (1971), p. 263.
22. P. W. Bridgman, Proc. Am. Acad. Arts Sci. 74, 425 (1942).

R-10

METROLOGICAL PROPERTIES OF HIGH PRESSURE MANGANIN TRANSDUCERS

E. Czaputowicz

Warsaw Technical University
Warsaw, Poland

INTRODUCTION

On the basis of data collected over many years by this laboratory and by other investigators, the experimental data included calibration of different kinds of manganin sensors on dead weight pressure balances up to 1000 MPa. The metrological properties of manganin high-pressure transducers have been examined.

DISCUSSION

The relative change of resistance vs. temperature in terms of pressure p can be presented to a first approximation by parabolic equations. Symbols used are as follows:

$$(R_{pt} - R_{pm})/R_o = -0.5\ E\ (t_{pm} - t)^2 \qquad (1)$$

$$(R_{ot} - R_{om})/R_o = -0.5\ E\ (t_{om} - t)^2 \qquad (2)$$

where $E = (0.6 - 1.1) \times 10^{-6}K^{-2}$ and $t_{om} = (0 - 40)\ °C$.

The pressure coefficient of temperature for maximum resistance is independent of pressures up to 1000 MPa. That is,

$$a_m = (t_{pm} - t_{om})/(T_{om}p) = const \qquad (3)$$

where $a_m = (1.9 - 2.4) \times 10^{-5}MPa^{-1}$.

The pressure coefficient of maximum resistance increases linearly with increasing pressure as given by

$$k_m = (R_{pm} - R_{om})/(R_{om}p) = k_{om} - k_{1m}p \qquad (4)$$

where for manganin obtained in the United States

$$k_{om} = 2.337 \times 10^{-5} MPa^{-1} \quad (5)$$

$$k_{1m} = 2.14 \times 10^{-10} MPa^{-2} \quad (6)$$

The pressure coefficient of resistance is given by

$$k_{pt} = (R_{pt} - R_{ot})/(R_o p) \quad (7)$$

and may be evaluated using the quantities introduced in equations (1) through (4).

$$k_{pt} = k_o - k_1 p \quad (8)$$

where $k_o = k_{om} + Ea_m T_{om} (t - t_{om})$; $k_1 = k_{1m} + 0.5\, Ea_m^2 T_{om}^2$. Our data and that found in the literature provides ranges of values of $(2.0 \div 2.4) \times 10^{-5} MPa^{-1}$ and $(0.9 \div 3.5) \times 10^{-10} MPa^{-2}$ for k_o and k_1, respectively. The calculated values

$$\Delta k_o/(k_o \ \Delta t) = (Ea_m T_{om})/k_o = (1.3 \div 4.1) \times 10^{-4} K^{-1} \quad (9)$$

are in good agreement with our data and other experimental data.

According to the proposition of Atanov and Ivanova the measured pressure may be found in the first approximation from

$$p = \Delta R/(R_o k_{pt}) \approx \Delta R/\{R_o k_o \ [1 - (k_1/k_o^2)(\Delta R/R_o)]\} \quad (10)$$

Because commonly used laboratory methods have many disadvantages, a new version of manganin gauges was developed. The principle of operation is based on an unbalanced bridge circuit supplied with direct current. There are similar manganin transducers in the four arms so that the measuring couple is placed in the high pressure area and the compensating one on the outside. A digital voltmeter calibrated in the same pressure units as the pressure indicator was used. For I = 20 mA, R_o = 500 ohm, and k_o = $2.0 \times 10^{-5} MPa^{-1}$, the pressure sensitivity of the gauge is

$$c = U/p = 0.5 \ IR_o k_o = 10^{-4} V/MPa \quad (11)$$

The main differences in pressure readings (Δp) are due to temperature differences (Δt) between the measuring couple and the compensating ones. From equations (1), (3) and (11)

$$\Delta p = 0.25 \ Ic^{-1} \ ER_o \ (2 \ t_{om} + 2a_m T_{om} p - 2t_o - \Delta t) \ \Delta t \quad (12)$$

For t_o working conditions of (20 ± 2) °C, the best transducer is one using manganin with t_{om} = 20°C. If one does not have this type of manganin it is possible to reduce Δp considerably by connecting two coils of manganin wire with t'_{om} and t''_{om} in series for which $t''_{om} - t_o \approx t_o - t'_{om}$, for example German manganin with $t'_{om} = 15.4°C$ and U.S. with $t''_{om} = 32.7°C$. If one has only one type of manganin for which $t_{om} = 0°C$ or $t_{om} = 40°C$ the thermocouples may be introduced into the electrical system to compensate for the additional voltage caused by the changes in resistance due to the result of temperature differences. The values of reduced errors may be calculated from

$$\Delta p_r = \Delta p - nc^{-1}S\,\Delta t \qquad (13)$$

The temperature differences Δt following from the changes of pressure vs. time τ are given by

$$\Delta t = \Delta t_o \exp(-\tau/\theta) \qquad (14)$$

The initial value of Δt_o depends on the value of the changing pressure, the rate of pressure change, and the compressibility of the liquid used. The time constant θ is typical for a high-pressure chamber. If Δp from (12) and Δt from (14) are introduced into (13), we can find values of reduced errors Δp_r as a function of time τ after a rapid change of pressure.

CONCLUSION

The metrological properties of high-pressure gauges with manganin transducers up to 1000 MPa were analyzed. The high-pressure gauges developed herein may successfully replace the spring-type gauges currently used and eliminate the commonly used laboratory methods of high-pressure measurements by means of separate manganin coils.

NOTATION

a_m = pressure coefficient of temperature for maximum resistance
c = pressure sensitivity of gauge
E = temperature coefficient of resistance
I = electric current supplied to high pressure gauge
k_m = pressure coefficient of maximum resistance
k_{pt} = pressure coefficient of resistance
n = number of thermocouples
p = pressure
R_o = resistance under ambient conditions
R_{om} = maximum resistance under atmospheric pressure
R_{ot} = resistance under atmospheric pressure and temperature t
R_{pm} = maximum resistance under pressure p
R_{pt} = resistance under pressure p and temperature t
S = thermoelectric power
t = temperature
t_o = ambient temperature
t_{om} = temperature for maximum resistance under atmospheric pressure
t_{pm} = temperature for maximum resistance under pressure p
T_{om} = t_{om} + 273
τ = time
θ = time constant of high pressure chamber
U = voltage

R-11

FRICTION MEASUREMENTS IN PISTON-CYLINDER APPARATUS USING QUARTZ $\leftrightarrows$ COESITE TRANSITION

J. Akella

Lawrence Livermore Laboratory, University of California
Livermore, California USA

INTRODUCTION

The configuration for a high-pressure apparatus is generally a cylinder that is closed at one end and has a movable piston in the other. A specimen to be pressurized is placed inside the cylinder and is compressed by a piston, which is driven by the ram of a hydraulic press. A number of high-pressure apparatuses are in use for static high-pressure and high-temperature studies to 6.5 GPa and 2073 K. Of these, the end-loaded piston-cylinder apparatus has the simplest geometric configuration for directly measuring the pressure inside the cylinder. The pressure in the cylindrical WC pressure chamber is the product of the pressure on the ram driving the piston into the chamber and the ratio of the ram cross-section area to that of the piston. This pressure inside the cylinder is uncorrected for friction and is the nominal pressure. The basic apparatus is described by Boyd and England [1].

The effectiveness of solid-media high-pressure piston-cylinder apparatuses for studying phase equilibria has been limited by the uncertainty in determining the pressure on the sample. This uncertainty is mostly due to friction in the system. The relationship between load on a piston and the pressure on the sample is a complex function of friction between the cylinder wall and the piston, the mechanical properties of the pressure medium, and the differential loading of the high-pressure cell. In the cell, high-strength cell elements and large differential compaction of elements can cause one region to bear a disproportionate fraction of the total load applied.

Friction can be estimated from observations of some pressure-dependent variable within the cell over a range of applied pressure, both while the piston is moving in (compression) and moving out (decompression). Williams and Kennedy [2] have suggested that the "hysteresis loops" obtained from plotting the piston displacement for both compression and decompression cycles as a function of nominal pressure could be used in determining the actual pressure on the sample. The mid-values of such hysteresis loops at any nominal pressure will give the true pressure on the sample.

A second approach is to make a series of compression and decompression runs, traversing the pressure of an equilibrium transition or reaction. The true pressure of the transition will be midway between the compression- and decompression-run equilibrium points, assuming the friction acts symmetrically in both runs. Mirwald et al. [3] studied the melting of Ag, Au, and Pb by a differential thermal analysis technique using low-friction salt cells and concluded that the double value of friction is negligibly small in a salt cell.

In the present investigation, the value of friction was determined by monitoring the piston displacement as a function of nominal pressure on compression and decompression cycles at 1273 K. This value was compared with the friction value obtained by reversing the quartz ⇌ coesite transition at 1273 and 1073 K in a talc-glass-alsimag cell described by Akella and Kennedy [4] (see Fig. 1a), and a low-friction salt cell developed by Mirwald et al. [3] (see Fig. 1b).

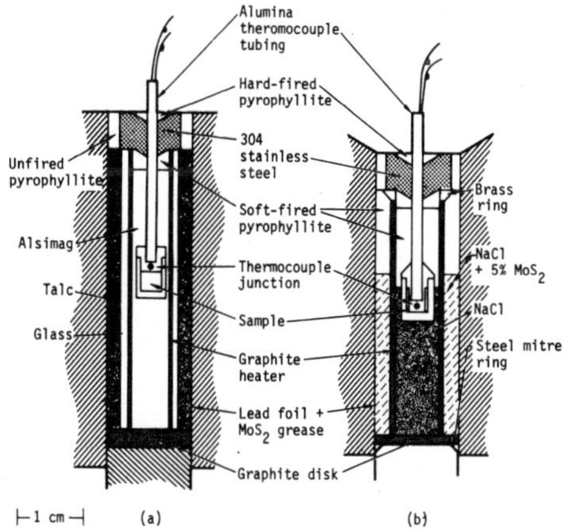

Fig. 1. Comparison of the talc-glass-alsimag cell and salt cell.

EXPERIMENTAL METHOD

All experiments were conducted in pressure vessels fitted with new WC cores. The cores used in this study had 6% Co binder and the pistons 3% Co binder. Pistons with 3% Co binder were used to reduce friction due to dilation. The talc-glass-alsimag cell has been described previously by Akella and Kennedy [4], and a detailed description of the low-friction salt cell was given by Mirwald et al. [3]. In the present study, Mirwald et al.'s salt cell (Fig. 1b) was used with minor modifications.

For the quartz ⇌ coesite transition work, platinum capsules with tight lids or graphite capsules with 25.4-μm-thick platinum-foil liners were used as sample containers. Dummy capsules made out of platinum were used in the piston-displacement runs. Starting materials for the quartz ⇌ coesite phase-equilibrium study were 90% quartz + 10% coesite and 90% coesite + 10% quartz. To enhance the rate of reaction, we moistened the starting materials with water before loading them into capsules.

Temperatures were measured with Pt and Pt 10% Rh thermocouples. No correction for the effect of pressure on thermal emf was applied. All of the quartz ⇌ coesite experiments lasted 8 hrs.

FRICTION MEASUREMENT BY PISTON DISPLACEMENT

Piston displacement measured at 1273 K in a talc-glass-alsimag cell is plotted as a function of nominal pressure in Fig. 2. Piston displacement was measured for three consecutive compression and decompression cycles. The pressure was maintained at a set value for at least 10 min before the displacement reading was taken with a dial gauge: then the pressure was lowered or increased to a new set value. It is evident from Fig. 2 that the cell itself is compressed to a large extent during the first compression cycle. The displacement readings for the successive compression and decompression cycles do not change significantly.

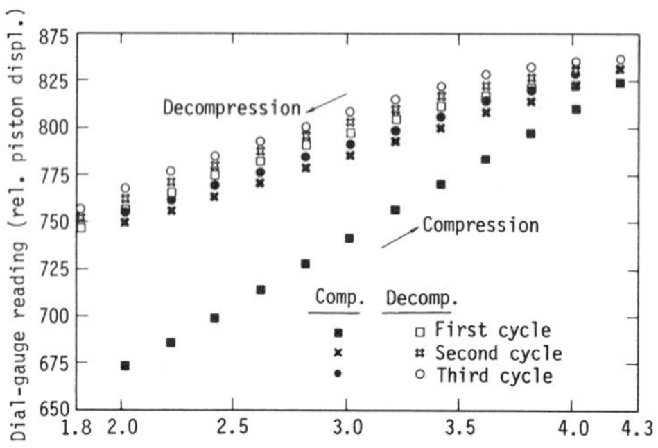

Fig. 2. Nominal pressure vs. relative piston displacement for talc-glass-alsimag cell: sensitivity of the dial gauge was 0.001 in.

Displacement readings versus nominal pressure for the third cycle are plotted in Fig. 3. The double value of friction, which is defined by the width of the hysteresis loop, is 0.4 GPa at 3.1 GPa nominal pressure. Assuming that all the friction effects are symmetrical and fully reversed, the middle of the hysteresis loop can be taken as the true pressure. To obtain the actual pressure on the sample, half of the double value of friction should be subtracted from the nominal pressure on the compression cycle or added on the decompression cycle. Williams and Kennedy [2] and Mirwald et al. [3] pointed out that the hysteresis loop obtained by this piston-displacement technique can indicate the magnitude of friction in some cases. The accuracy of the piston-displacement technique to determine friction will be discussed later.

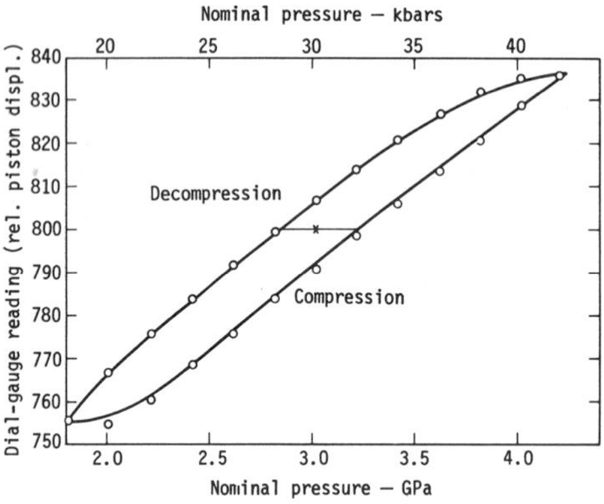

Fig. 3. Typical hysteresis loop in the third cycle of piston-displacement experiment at 1273 K: dial-gauge sensitivity was 0.001 in.

QUARTZ ⇌ COESITE TRANSITION

The quartz ⇌ coesite transition has many necessary qualitites to recommend it as a potential calibration reaction [5]. Quartz can be easily obtained in a very pure form. Chemical hysteresis in the quartz ⇌ coesite reaction seems to be negligible in quenching experiments lasting 1 hour or more. The slope of the quartz-coesite curve is 90 to 100°C/kbar so that minor errors in temperature measurement will have negligible effect on the results.

Coes [6] first synthesized coesite and MacDonald [7] reported on the first detailed study of the quartz ⇌ coesite stability field. Boyd and England [1] and Boyd et al. [5] carefully studied the quartz ⇌ coesite transition and suggested this transition as a potential calibrant for the piston-cylinder high-pressure apparatus. Kitahara and Kennedy [8], Boettcher and Wyllie [9], and Böhler and Arndt [10] also investigated the quartz ⇌ coesite reaction at high pressures and temperatures. All of the above studies used different kinds of high-pressure cells and different techniques; thus, their results cannot be compared for validity.

The objectives for studying the quartz ⇌ coesite transition are: (1) To accurately locate the quartz-coesite transition pressure at 1073 and 1273 K by using the two most widely used high-pressure cells (talc-glass-alsimag cell and low-friction salt cell). This will be done on both the compression and the decompression cycles. (2) To compare the double value of friction thus obtained with that from the piston-displacement technique.

Results of the quenching experiments on the quartz ⇌ coesite transition are presented in Figs. 4 and 5. With the talc-glass-alsimag cell at 1273 K, the quartz ⇌ coesite transition curve is located at 3.4 and 3.15 GPa on the compression and decompression cycles, respectively. However, with the salt cell, the transition pressures are lower (see Table I). On the compression cycle at 1073 K, the measured transition curve is located at 3.18 GPa with the talc-glass-alsimag cell, 2.98 GPa with the salt cell.

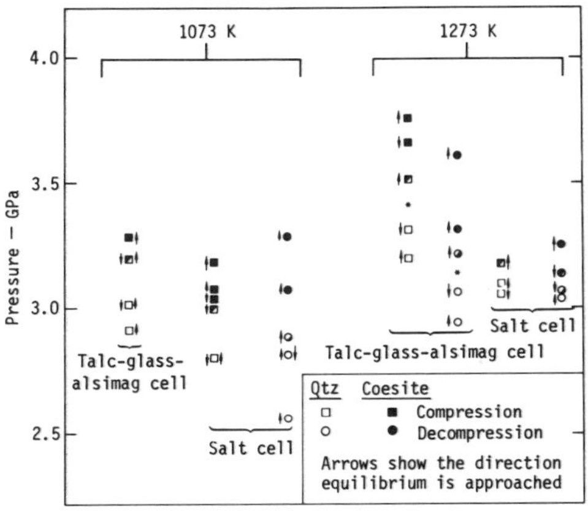

Fig. 4. Quartz ⇌ coesite transition data from compression and decompression runs in the two cells at 1073 and 1273 K.

The actual quartz ⇌ coesite transition pressure at 1273 K with talc-glass-alsimag cell is 3.28 GPa. In a similar study using a talc-boron nitride-pyrophyllite cell, Boyd et al. [5] obtained 3.75 ± 0.02 GPa

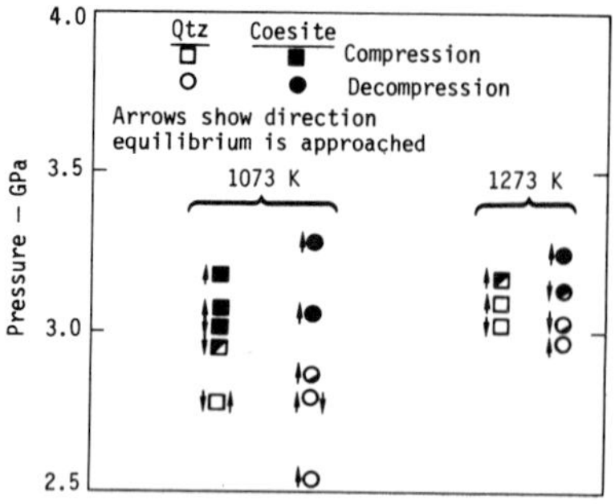

Fig. 5. Quartz ⇌ coesite transition data obtained with the salt cell at 1073 and 1273 K.

as the transition pressure at 1673 K. These two values are plotted in Fig. 6. The equation for the quartz ⇌ coesite transition curve calculated from both data points is P = 20.07 + 0.012 T, where P is in kbars and T is in °C. Similar data, obtained at 1073 and 1273 K using the salt cell, are also plotted in Fig. 6. As mentioned before, the transition pressures measured with the salt cell are lower than from the talc-glass-alsimag cell, so the transition curve based on the salt-cell data is displaced slightly to lower pressures. The equation obtained for this curve is P = 21.2 + 0.010T. The equation for the transition pressure obtained from the salt cell in the present study is remarkably close to that reported by Kitahara and Kennedy [8].

Table I. Quartz ⇌ Coesite Transition Pressures at 1073 and 1273 K

Temperature, K	Transition Pressure, GPa			
	Compression		Decompression	
	T-G-A* cell	Salt cell	T-G-A* cell	Salt cell
1073	3.18	2.98	n.d.†	2.86
1273	3.4	3.12	3.15	3.09

*Talc-glass-alsimag cell.

†Not determined.

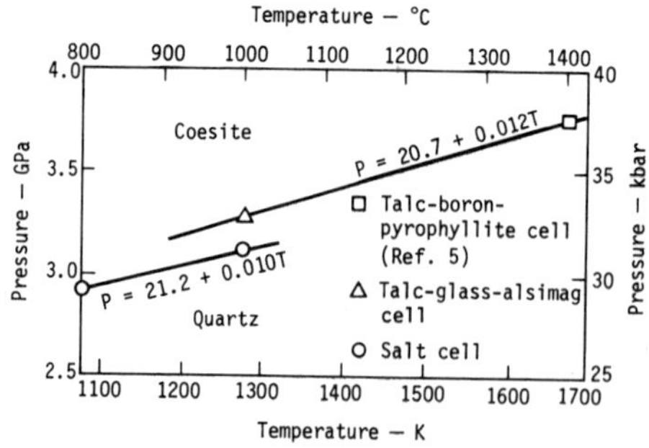

Fig. 6. Data on the quartz ⇌ coesite transition from the present investigation and that of Boyd et al. [5]: P and T in the equations for the curves are in kbar and °C, respectively.

DISCUSSION

The double values of friction obtained from the quenching runs at 1273 K are: with the talc-glass-alsimag cell 0.25 GPa, and with the salt cell 0.03 GPa. The piston-displacement technique gives a double value of friction of about 0.4 GPa, which is higher than the value obtained from the quenching runs at 1273 K. The discrepancy between the results from piston-displacement studies and those from phase-transition studies could be due to the hysteresis in the pressure medium which may have failed to attain equilibrium before the displacement reading was taken in the piston-displacement experiments. Alternatively, the discrepancy may be due to pressure gradients in the pressure medium that did not reverse symmetrically when the direction of piston motion changed. A third source of this discrepancy could be that even if the friction and gradient effects are symmetrical and fully reversed, differential loading could contribute a substantial absolute error in pressure determination. In this case the true pressure values may lie well outside the hysteresis loop. The present results suggest that if piston-displacement hysteresis loops were used in evaluating the actual pressure on the sample, pressures could be overestimated for decompression runs and underestimated for compression runs.

Mirwald et al. [3] concluded from their study that the salt cell exhibits extremely low friction over a range of 1 to 6 GPa and 573 to 1673 K. In the present study, the friction value obtained at 1273 K is 0.03 GPa. At 1073 K it is about 0.12 GPa. Both of these values are lower than the friction values reported for the talc-glass-alsimag or talc-boron nitride-pyrophyllite high-pressure cells. The relatively high friction value obtained with the salt cell at 1073 K is in conflict with the observation of Mirwald et al. It is possible that salt may not really be a weak pressure medium at very low temperatures as it was thought to be.

ACKNOWLEDGMENTS

The experimental work reported here was started at the NASA-Johnson Space Center, during the author's tenure there as a National Research Council and National Academy of Sciences Senior Resident Research Associate and Visiting Scientist at the Lunar and Planetary Institute of the Universities Space Research Association, under Contract No. NSR-09-051-001 with the National Aeronautics and Space Administration. Thanks are due to O. Mullins for his technical help.

REFERENCES

1. F. R. Boyd and J. L. England, J. Geophys. Res. 65, 749 (1950).
2. D. W. Williams and G. C. Kennedy, J. Geophys. Res. 74, 4359 (1969).
3. P. W. Mirwald, I. C. Getting, and G. C. Kennedy, J. Geophys. Res. 80, 1519 (1975).
4. J. Akella and G. C. Kennedy, Amer. J. Sci. 270, 155 (1971).
5. F. R. Boyd, P. M. Bell, J. L. England, and M. C. Gilbert, in Annual Rept. Geophysics Laboratory, Carnegie Institute, Washington, D.C., Yearbook, Vol. 65 (1966), p. 410.
6. L. Coes, Science 118, 131 (1953).
7. G. J. F. MacDonald, Amer. Min. 41, 744 (1956).
8. S. Kitahara and G. C. Kennedy, J. Geophys. Res. 69, 5395 (1964).
9. A. L. Boettcher and P. J. Wyllie, Contr. Mineral, and Petrol. 17, 224 (1968).
10. R. Böhler, and J. Arndt, Contr. Mineral, and Petrol. 48, 149 (1974).

S-1

APPARATUS FOR DEFORMATION TESTS AT INTERMEDIATE STRAIN RATES - 10^{-2} TO 10^{2} s^{-1} - AND AT HIGH PRESSURE*

A. E. Abey and L. L. Dibley

Lawrence Livermore Laboratory, University of California
Livermore, California USA

INTRODUCTION

Few stress-strain data are available at intermediate strain rates (up to 10^2 s^{-1}) and at confining pressures above 100 kPa. Even though at least three apparatuses work in this regime of strain rate and pressure [1-3], our laboratory's interest in obtaining equation-of-state data in this regime dictated that we develop a new apparatus.

Design criteria dictated by our one-dimensional stress experiments were: strain rates of about 10^{-2} to 10^2 s^{-1} for a 2.54-cm-long sample, confining pressures of up to 1 GPa, and maximum strength of a 2.54-cm-diam sample of 2 GPa at a confining pressure of 1 GPa. Also, the strain-time history for both loading and unloading has to be programmable within reasonable limits.

The strains and stresses have to be measured internally so that frictional and compliance corrections can be eliminated. To complement standard strain-gage techniques for strain measurement, a non-contact method had to be developed for measuring the radial and longitudinal strains. The apparatus had to be designed so that it could be modified to do pressure-volume and quasi-one-dimensional strain tests as well.

*Work performed under auspices of U.S. Energy Research and Development Contract W-7405-Eng-48.

Because many materials, particularly some geological materials, are anisotropic and inhomogeneous, the apparatus has to be capable of using fairly large samples, about 10 cm long and 10 cm in diameter. The criteria of confining pressure and maximum material strength will decrease for larger samples.

APPARATUS

The apparatus in Fig. 1 was designed to meet the criteria given above. The sample is deformed when the cam, held in the shaft, rotates and depresses the cam-follower. The strain-time history of the sample is determined by the shape of the cam and the height at which the pressure vessel is placed below the shaft. The total time of deformation is determined by the rotational speed of the shaft holding the cam. The maximum strain on a 2.54-cm-long sample is about 17%. The strain can be varied by changing the lift of the cam. Large changes in strain rate are obtained by varying the rotational velocity of the shaft.

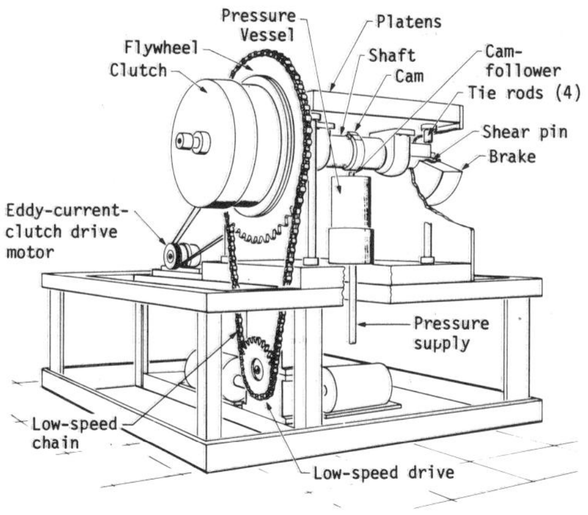

Fig. 1. Intermediate-strain-rate apparatus. Overall height is about 2 m.

At the start of an experiment, the shaft is pinned with the cam almost a full revolution away from striking the cam-follower. The flywheel (700 kg, 1-m diam) is then brought up to the desired speed. (The maximum angular speed is approximately 145 s^{-1}.) With the flywheel at the desired speed, the air clutch (maximum torque = 32 kN·m) is engaged. This takes about 100 ms. While the clutch is engaging, the torque in the shaft (15-cm diam) is increasing. When the torque reaches a certain level (determined by the pin size), the pin holding the shaft shears and the shaft and cam begin to rotate. At the maximum angular velocity of the flywheel, the shaft reaches the speed of the flywheel in about 1/8 revolution. The moment of inertia of the system is such that energizing the shaft and deforming the sample decrease the angular velocity of the flywheel by about 5%.

The flywheel can be driven by either of two systems: For low speeds, the flywheel is connected to a motor-gearbox combination by a chain similar to a very large bicycle chain. For high speeds, the chain is removed and the flywheel is brought up to speed by a belt-drive combination of electric motor and magnetic clutch.

After the cam depresses the cam-follower, a strap holding the cam to the shaft is sheared. The cam then exits at the back of the apparatus. Thus the cam-follower is struck only once and the sample is deformed a single time.

The strain rates generated are accurate to about 20% of the measured value; e.g., $50 \pm 10\ s^{-1}$. The confining pressure is brought into the high-pressure vessel through its bottom. The volume of the vessel was chosen large enough that the pressure rise from the cam-follower will be less than 5% under all conditions when argon is used as the pressure medium.

The sample holder is shown in Fig. 2. The sample is placed immediately below a load cell. A number of experiments with load cells above and below the sample showed that the stress-time history is the same both above and below the sample. The load cell is touching the cam-follower so that the cam-follower depresses the load cell and shortens the sample.

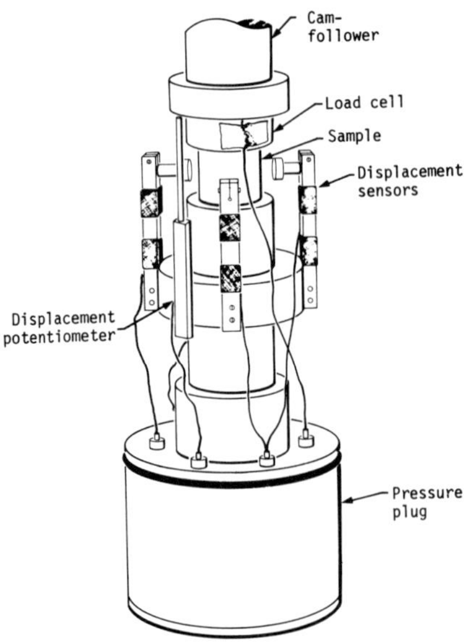

Fig. 2. Sample holder for apparatus. Holder fits inside pressure vessel.

Samples are held concentric with the load cell and bottom support by various means during deformation, depending on the type of sample. Metallic samples have a small hole (~1 mm deep and ~1 mm in diameter) in the center of both faces. Pins on the axis of the load cell and bottom support fit into these holes and hold the sample. Dry, porous rock samples are cemented to the load cell and bottom support with epoxy resin. Wet samples with hardened-steel end plugs are held concentric by rings that hold the end plugs to the load cell and bottom support.

Two basic strain-measuring systems have been used. (Because the number of electrical feed-throughs into the high-pressure vessel is limited, only one strain-measuring method can be used at a time.) Conventional strain gages are used for small strains, and a non-contact system is used for large strains.

For a material (such as Indiana limestone) that will undergo only small strains, conventional strain gages are used to determine the longitudinal and radial strains.

For a material like 6061-T6 aluminum that will undergo large strains, a unique noncontact strain-measuring system is used. The longitudinal displacement of the lower end of the load cell, and therefore the upper end of the sample, is measured with a high-speed rectilinear potentionmeter. The radial displacements are determined at two different heights on the sample by displacement sensors at the middle of the sample and one-fourth of the length above the bottom of the sample. These displacement sensors (part of a balanced bridge circuit) are coils in which an electric current oscillates at 1 MHz and sets up eddy currents either in the sample, if metallic, or in a metallic jacket on the sample. These eddy currents are detected by the coils and produce an imbalance in the bridge which is directly proportional to the radial displacement of the sample.

The data are recorded on transient recorders that have a variable sampling rate with a maximum of 10 MHz. The sampling rate is chosen so that the experiment takes approximately 1000 data. This gives adequate resolution of the data taken during sample deformation. After the raw data have been collected by the transient recorder, they are recorded on magnetic tape for later reduction and plotting.

SAMPLE PREPARATION

Three basic approaches were used for sample preparation. For metallic samples such as aluminum, the samples were machined to the desired dimensions and the centering holes drilled. The flat surfaces were coated with a layer of moly-disulfide type grease to minimize friction.

Samples of porous rock with small grain size such as Indiana limestone are machined to size and then coated with several layers of epoxy. After each layer of epoxy has dried, the sample is lightly sanded. (Microscopic examination of the samples indicates that epoxy intrusion into the sample is minimal.) Conventional strain gages are then applied on the surfaces of the rock specimen. Before the test is run, samples are epoxied to the load cell and to the lower support. This gives a known boundary condition to the test; namely, no motion of the sample surface relative to the load cell or lower support.

Saturated or wet samples such as coal are jacketed in thin copper, and both ends are capped with hardened steel plugs. The jacket is longer than the sample and so extends partly over the tapered sides of the end plugs. Tapered rings seal the ends of the jacket to the sides of the plugs. Conventional radial strain gages are then placed on the outside of the metal jacket. Strain in samples of this type could also be measured by eddy-current displacement sensors.

EXPERIMENTAL RESULTS: ONE-DIMENSIONAL STRESS

Indiana Limestone

The axial stress-time history for an Indiana limestone sample is shown in Fig. 3. The strain rate for this experiment was approximately 10^2 s^{-1}, at which the sample failed in the brittle mode after approximately 0.3 ms. Even after the sample failed, the stress did not drop instantly to zero, but decreased rapidly as the sample fragments were pushed away. The axial stress vs. longitudinal strain is shown in Fig. 4. The stress-strain relationship is, on the average, linear up until sample failure but does contain some unusual slope changes which correlate with the axial stress in Fig. 3.

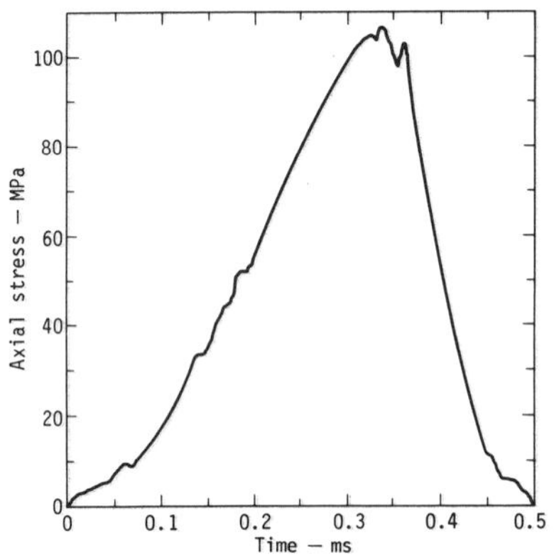

Fig. 5. Axial stress vs. time for an Indiana-limestone sample; strain rate $\dot{\varepsilon} \approx 10^2$ s^{-1}.

The strains plotted in Fig. 4 were measured with strain gages attached to the lateral surface of the sample. Because the deformation was highly inhomogeneous, the deformation experienced by the strain gages was only about 10% of the total sample shortening. The sample failed in a highly localized section along cones which remained attached to the load cell and lower sample support. The shape of the sample fragments recovered after the experiment confirmed the localized nature of the failure. Indiana limestone showed this mode of deformation and

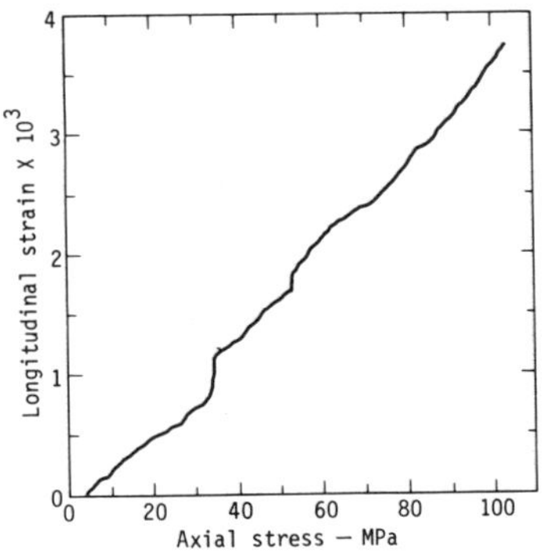

Fig. 4. Longitudinal strain vs. axial stress for an Indiana-limestone sample; $\dot{\varepsilon} \approx 10^2\ s^{-1}$.

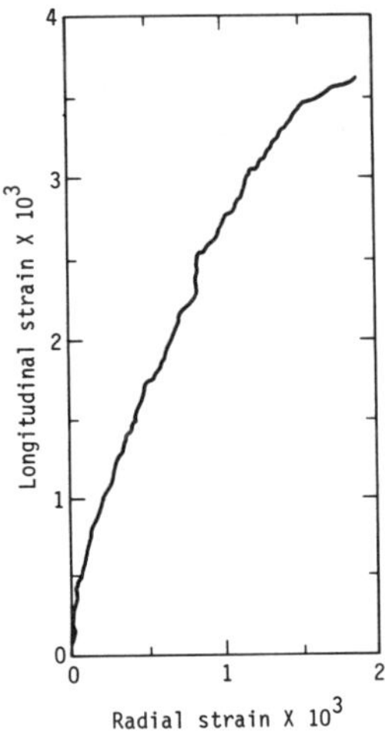

Fig. 5. Longitudinal strain vs. radial strain for an Indiana-limestone sample; $\dot{\varepsilon} \approx 10^2\ s^{-1}$.

failure in all of our experiments. In Fig. 5 the longitudinal strain is plotted vs. the radial strain. The strain ratio (Poisson's ratio) increases with longitudinal strain, approaching 0.5 at failure. The axial stress at failure vs. ln of the strain rate is shown in Fig. 6 for Indiana limestone at atmospheric pressure and at a confining pressure of 12 MPa. Linear least-squares fits give ratios of failure-strength change to the change in ln of strain-rate. These ratios are 5.1 MPa at atmospheric pressure and 3.26 MPa at a confining pressure of 12 MPa. The failure data for Indiana limestone show little scatter.

6061-T6 Aluminum

Our rather limited experience indicates that conventional strain gages are unreliable (errors of up to 10%) when used at high-strain rates for materials undergoing fairly large strains. Therefore, we used the eddy-current displacement sensors in Fig. 1 to measure the radial strain on aluminum. These sensors have proven to be accurate (strain error $\pm$ 0.0005) and suitable for measuring radial strains at strain rates up to 130 s^{-1}. However, these sensors are very fragile, and while they have generally not failed during the actual tests, they have during preparations. More rugged versions are now being built.

We determined longitudinal strain by using the rectilinear potentiometer. This was not entirely satisfactory as the potentiometer has a tendency to hang up, making the elastic part of the stress-strain relationship difficult to measure. The potentiometer will be replaced by an eddy-current displacement sensor for future work.

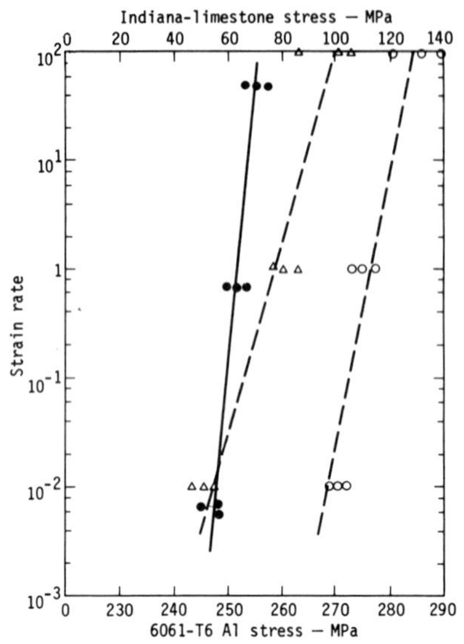

Fig. 6. Axial strain vs. ln $\dot{\varepsilon}$ for: failure stress of Indiana limestone, confining pressure ≈ atmospheric (Δ); failure stress of Indiana limestone, confining pressure = 12 MPa (o); yield stress of 6061-T6 aluminum, confining pressure ≈ atmospheric (●).

As an example of the data that can be obtained, the stress-vs.-time curve for 6061-T6 aluminum at a strain rate of $7 \times 10^{-3}\ s^{-1}$ is shown in Fig. 7. The elastic and plastic regions are clearly visible. The radial strains vs. time, as determined from the two sets of displacement indicators, are shown in Fig. 8. The strains at the two different heights are very similar. The use of Molycoat grease held the barreling of the sample to a minimum — less than 3%. This was the case for aluminum samples at all strain rates. The yield strength of the 6061-T6 aluminum is given in Fig. 6 as a function of strain rate. The change in yield strength with strain rate is quite small, less than 5% over about four orders of magnitude in strain rate.

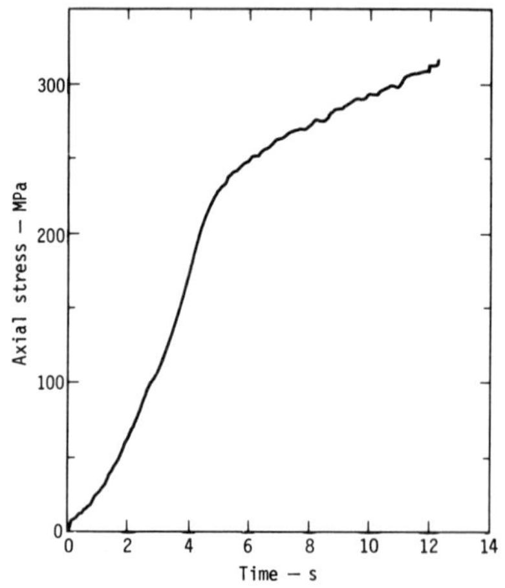

Fig. 7. Axial strain vs. time for 6061-T6 aluminum; $\dot{\varepsilon} \approx 7 \times 10^{-3}\ s^{-1}$.

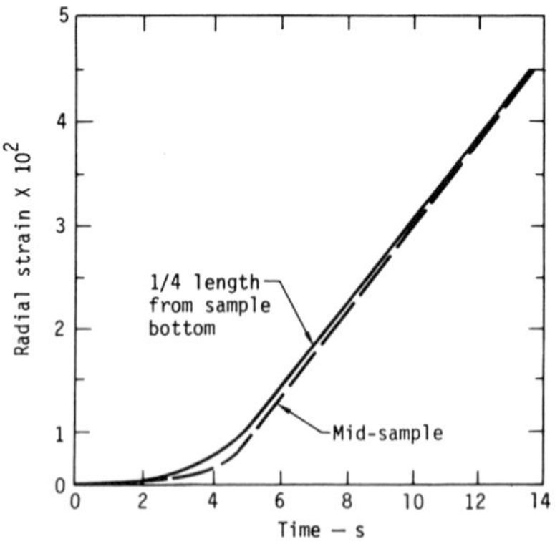

Fig. 8. Radial strain vs. time for 6061-T6 aluminum; $\dot{\varepsilon} \approx 7 \times 10^{-3}\ s^{-1}$.

CONCLUSIONS

Not all the problems of this apparatus have been solved. However, the apparatus has been developed to the point where useful stress-strain data taken internally can be obtained at strain rates up to 10^2 s^{-1}.

ACKNOWLEDGMENTS

The authors would like to thank K. T. Keller for his help in carrying out the experiment. They would also like to thank H. C. Weed, A. G. Duba, and H. C. Heard for reading and suggesting improvements in the manuscript.

REFERENCES

1. S. J. Green, J. D. Leasia, R. D. Perkins, and A. H. Jones, J. Geophys. Res. 77, 3711 (1972).

2. J. M. Logan and J. Handin, Proc. Twelfth Symp. Rock Mechanics, AIME, New York, USA (1971), p. 167.

3. S. Serdengecti and G. D. Booyer, Fourth Symp. Rock Mechanics, Pennsylvania State Univ., University Park, Penn., USA (1961), p. 83.

S-2

SAFETY CONSIDERATIONS IN HIGH PRESSURE WORK*

G. Saville

Imperial College
London, England

INTRODUCTION

High-pressure work is hazardous whatever its scale, but this does not mean that the operator need, or even should be allowed to, take risks. Individuals can of course avoid risks only when they know how to assess the adequacy of their equipment and the magnitude of the hazard in the work which they are doing. My concern here is to try to help the user in making this assessment, but it must be realized that it is impossible to make blanket generalizations which apply equally to the inexperienced technician and the highly-skilled professional designer of high-pressure equipment.

I divide safety considerations into essentially three aspects, although the divisions are not watertight. First there is the necessity for good design - is the vessel strong enough for the purpose? etc.; then there is the question of safe operating procedures; and finally the desirability and design of protective barricades. My ideal purpose would be to adequately cover each of these aspects to be useful to the high-pressure user. This is not possible in the time available and in particular I shall say nothing about safe operating procedures, but will restrict myself to considering the philosophy behind the 'High Pressure Safety Code' produced by the United Kingdom High Pressure Technology Association and to pointing out some of the more important statements. For a

*Invited paper.

fuller discussion on the whole subject I must refer you the complete Code.

ASSESSMENT OF PRESSURE VESSELS

Pressure vessel design is an exceedingly complex subject as evidenced by the size of design codes such as the ASME Pressure Vessel Code and it must be realized that even this is geared primarily to the lower end of the pressure scale. Whereas it is essential for the professional designer to be fully conversant with a code such as this, one cannot say the same for a technician using a small cylindrical pressure vessel in a research laboratory. Such an individual needs to be able to assess in a simple manner, the suitability of the vessel on hand. It is our view that an assessment of this sort is possible in many straightforward situations provided that the criteria used are sufficiently conservative. If assessment is not straightforward then it is necessary to direct the user towards the professional designer and to dissuade the user from trying to do it himself.

To begin with, in view of the difficulty in assessing a vessel made from one of the very high strength steels where the most likely mode of failure is brittle fracture, we decided that it would be unwise to allow the inexperienced user to make his own assessment. On the other hand, vessels made from ductile, medium strength steels, provided they are not misused, are unlikely to fail in a brittle manner and we believe that it is permissible for the user to make his own assessment of this class of vessel. For the purposes of the Code, a ductile steel is defined as one which simultaneously has an ultimate tensile strength of less than 10 kbar (140 000 psi), a tensile yield strength of not more than 0.85 of the UTS, a reduction in area of more than 30%, and a Charpy V-notch impact value of more than 27 J (20 ft-lb), although there are in addition certain size restrictions specified.

Provided that the pressure vessel will not experience more than 10^3 pressure cycles, we base the criteria for safe operation on the yield and bursting pressures both of which are couched in terms of tensile rather than shear properties since values of these are more readily obtained by the user. We use the maximum shear stress criterion for the yield pressure and the empirical mean diameter formula for the ultimate bursting pressure, both chosen because they provide a conservative result. Thus, we recommend the calculation of the yield and ultimate bursting pressures for a thick-walled monobloc vessel from

$$P_y = \frac{\sigma_y}{2}\left(\frac{K^2 - 1}{K^2}\right) \qquad (1)$$

and

$$P_b = 2\sigma_u\left(\frac{K - 1}{K + 1}\right) \qquad (2)$$

where σ_y and σ_u are the yield and ultimate stresses for the material in tension and $K = OD/ID$ of the vessel.

The way in which these calculated pressures are used in determining the maximum allowable working pressure depends on what is known about the vessel. Thus, for example if the results of mechanical tests (UTS, proof stress, Charpy) are not available for the actual billet out of which the vessel was made, then for vessels whose outside diameter does not exceed 15 cm the maximum allowable working pressure is specified as the lesser of $0.67P_y$ and $0.25P_b$. On the other hand, if the billet has been thoroughly tested during manufacture and found to be satisfactory it is permissible to increase the maximum allowable working pressure to $0.42P_b$. The net effect of these restrictions is to place an upper limit of 3 kbar on vessels made from steel of unknown origin and 5 kbar on ones made from a tested, ductile steel.

Of course, not all pressure vessels are of simple monobloc construction, and unfortunately from the point of view of pressure vessel assessment all deviations from the basic design make the task of the assessor more difficult. Since it is our aim to enable the nonexpert to do this task, it follows that only a few, well-studied variations can be included in the Code. Perhaps the most important of these is the possibility of fatigue failure, particularly in a vessel with a cross-bore. Here, we advise the user that if the vessel is to have an 'infinite' life the dominating range of the pressure cycle, ΔP, should satisfy

$$\Delta P \leq \left(\frac{K^2 - 1}{K^2}\right)\frac{\tau_f}{J} \qquad (3)$$

where for ductile materials the allowed range of shear stress τ_f may be approximated as $\sigma_u/4$. The fatigue strength reduction factor J depends on the geometrical design of the vessel and by way of example for a simple cross-bore in a cylindrical vessel we give the value 2.7.

BARRICADES

The barricading necessary in a given situation depends on the magnitude of the hazard and also on how sure one can be that the vessel will not fail. Thus a vessel made from a ductile steel that was thoroughly tested during manufacture and which is operated well below its maximum allowable working pressure is most unlikely to fail and complete protection against brittle fracture is probably unnecessary. At the other extreme, we have thought it necessary to make complete protection mandatory for vessels made of nonductile steels and also for vessels made from ductile but untested material.

Barricade design is very badly covered in the scientific and engineering literature and much that is found there is often difficult to reconcile with experience. We have therefore reassessed the information available and expressed it in a more easily usable form.

When a pressure vessel fails it produces both fragments traveling at high velocity and a shock wave, the latter being particularly important when the pressure vessel contents are gaseous. However, the way in which the stored energy is divided up between them depends critically on the mode of failure. Thus, for example, if a large vessel fails catastrophically due to brittle fracture, of the order of 80% of the energy goes into the shock wave and only 20% into the kinetic energy of the fragments. On the other hand, if the failure is the ejection of an electrical lead-through or probe, the vessel itself remaining intact, most of the stored energy in the vessel will be released relatively slowly; the shock wave will be of low energy, but the probe will be ejected with very high velocity. In deciding what barricading to use it is first necessary to define the likely failure modes, and clearly the safest possible solution and the most expensive is to protect against both catastrophic failure and the high-velocity ejection of probes.

Having evaluated the shock wave energy and the fragment velocities (details for doing this are given in the Code) one should consider separately what protection is needed for each item. By way of example I quote the empirical formulae to be used for computing the thickness of mild steel required to stop a given fragment (provided its velocity is below 1000 ms^{-1}):

$$t = 6 \times 10^{-5} m^{0.33} V \qquad (4)$$

$$t = (0.5 \times 10^{-4} m/A) \log_{10}(1+5 \times 10^{-5} V^2) \qquad (5)$$

$$mV^2 = 3 \times 10^9 d^3 (t/d)^{1.41} \qquad (6)$$

where t is the thickness of steel plate required (m), m is the mass of the fragment (kg), A is the fragment presented area (m^2), V is the fragment velocity (ms^{-1}), and D is the diameter of the missile (m). Equation (4) is for compact blunt steel fragments of mass not exceeding 1 kg, (5) is for large blunt fragments of mass exceeding 1 kg, and (6) is for rod-shaped missiles.

Damage caused by a shock wave from a rupturing pressure vessel can be estimated by assuming that it is the same as that caused by the detonation of such amount of TNT as will release the same energy. This procedure is practicable in view of the extensive literature on the effects of explosives on structures. We use the equivalence: 1 kg TNT $\equiv 4.5 \times 10^6$ J. To prevent eardrum damage, protection against a shock wave is required whenever the distance between the operator and the pressure vessel is less than $E^{1/3}$, where E is the shock wave energy in joules.

The design of a closed cubicle may, for convenience, be done on the basis of the 'equivalent static pressure'. This is the pressure which when maintained indefinitely in the cubicle will produce the same deflection in the walls as the shock wave in question. It is given approximately by

$$P = 7.6[(E/10^6)/V]^{0.72} \text{ bar} \tag{7}$$

where E is the shock wave energy in joules, and V is the volume of cubicle in m^3. Thus, what was initially a dynamic problem has been transformed into one of designing a cubicle to withstand a static pressure. Hence normal pressure vessel design techniques can be used whether the cubicle is made of mild steel or reinforced concrete.

If the cubicle is vented, as for example will often be the case when the contents of the pressure vessel are toxic or flammable, although it is in principle possible to reduce the strength of the cubicle below that of the closed type, the difficulty in determining this reduction is such that it has led us to suggest that vented cubicles should be built to the same specifications as closed ones. The main disadvantage of a vented cubicle is that it allows the shock wave to 'escape'. At a distance of R meters from the cubicle the incident peak blast pressure is given approximately by the equation

$$P = 8.45 \times 10^{-5}(E^{1/3}/R)^{2.30} - 0.16R/E^{1/3} + 0.060 \text{ bar} \tag{8}$$

where E is the shock wave energy in joules. The equation is valid for $0.004 < R/E^{1/3} < 0.2$. If eardrum damage is to be avoided, this pressure must not exceed 0.07 bar and if no damage to windows is required, then it must not exceed 0.03 bar. Values for other structures are given in the Code.

It is the opinion of many members of the Association that all too often pressure vessels are operated without any real quantitative assessment of the hazard. We hope that our endeavors in this last section of the Code will help to remedy this.

S-3

LUBRICATION AND SAFETY IN COMPOUND PRESSURE VESSELS

I. C. Getting*

University of California at Los Angeles
Los Angeles, California USA

INTRODUCTION

Various types of high-pressure apparatus utilize compound cylinders to contain pressures in the range up to about ten GPa (100 kbar). Compound construction vessels usually consist of a high strength innermost die preloaded in compression by surrounding concentric support rings. Figure 1 shows a cross-section of such a vessel typical of a piston-cylinder apparatus for use to 7 GPa (70 kbar). The compressive preload on the innermost ring is achieved by inserting the die and its adapter shim into the tapered ID of the binding ring assembly with a substantial interference fit. The preload pressure on the OD of the die ranges from about 1.5-2.0

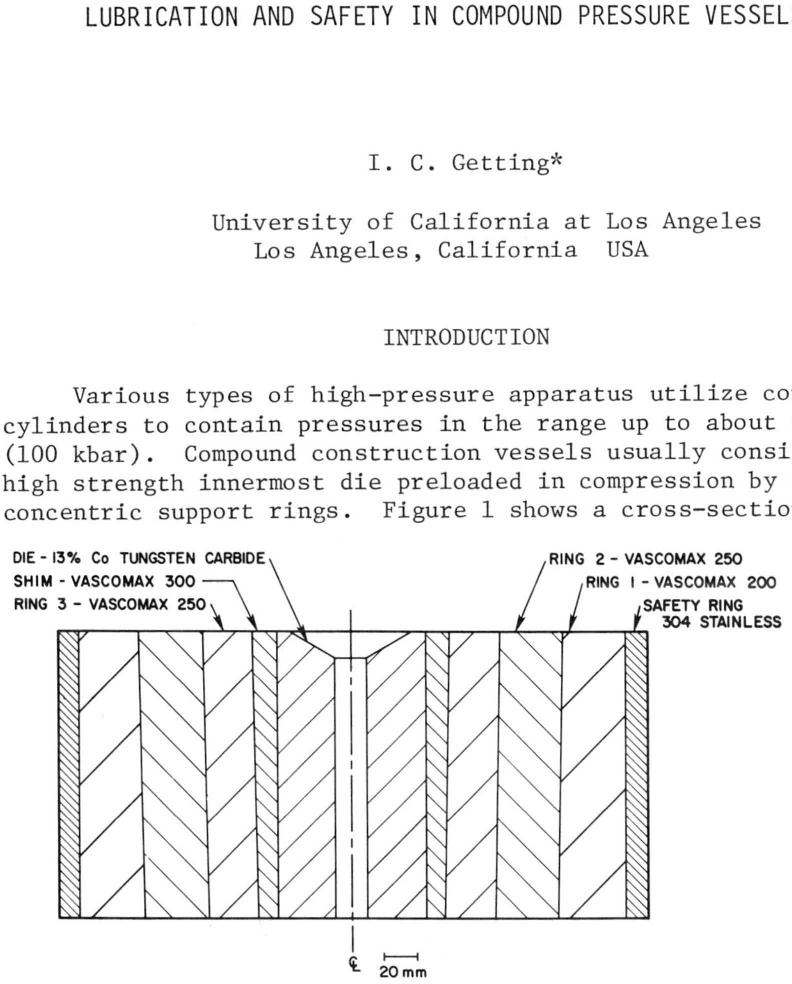

Fig. 1. Cross-section view of a compound pressure vessel.

*Present address: CIRES, University of Colorado, Boulder, Colorado.

GPa (15-20 kbar) and produces bore closure of about 1%. The various binding rings are also assembled on tapered surfaces with interference fits designed to produce a stress field capable of containing the die preload. An extensive bibliography and an excellent review of the pertinent stress analysis and design criteria for these vessels is given by Davidson and Kendall [1].

Figure 2 shows the stress field for the vessel in Fig. 1. The three support rings have undergone substantial plastic strain in their inner regions. In such a vessel the tungsten carbide die suffers reverse yielding with each pressure cycle. Its useful life can range from several tens of cycles to several thousands depending on internal pressure, temperature, and chemical environment. The binding rings, however, are expected to be much more long-lived. While in an extreme design such as Fig. 2 the binding rings may suffer reverse yielding when the die is removed, they should withstand many tens of die replacement cycles and many years of use. This paper is principally concerned with the lubrication problems in assembling and using such vessels, some problems of assuring their longevity, and safety techniques to tolerate their failure.

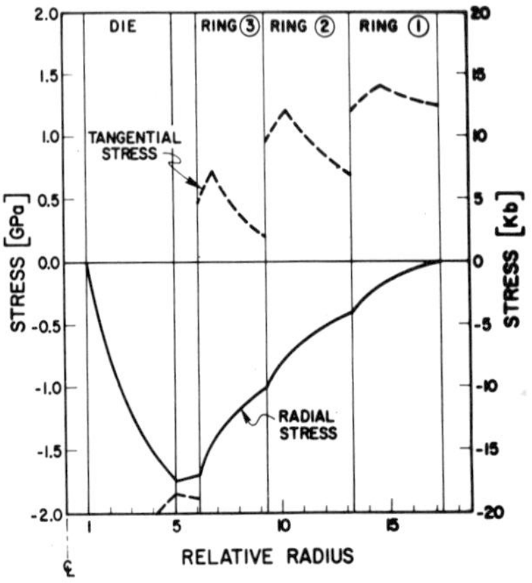

Fig. 2. Stress field upon assembly in a compound pressure vessel. The binding rings consist of three rings with an interference fit on the inside of each. Application of pressure to the inside of the die alters the die stress field tremendously, but has relatively little effect on the binding rings.

ASSEMBLY LUBRICATION

Measurement of Coefficient of Friction

Compound pressure vessels are usually assembled by pressing each successive ring into an assembly of those outside it. When a given ring is set into place, it seats in the taper with substantial protrusion. It is then pushed in causing one tapered surface to slide over another. This process requires critical lubrication. Not only must the lubrication be effective in the sense of preventing galling, scratching, and stick-slip motion, but the coefficient of friction must lie within an acceptable range. To prevent the rings from spontaneously extruding on their tapers, the coefficient of friction must be greater than the tangent of the taper angle (measured from the center line to one side). The minimum practical coefficient of friction is about 0.02. Values above about 0.04 at 2 GPa interface pressure cause the assembly forces to become intolerably large.

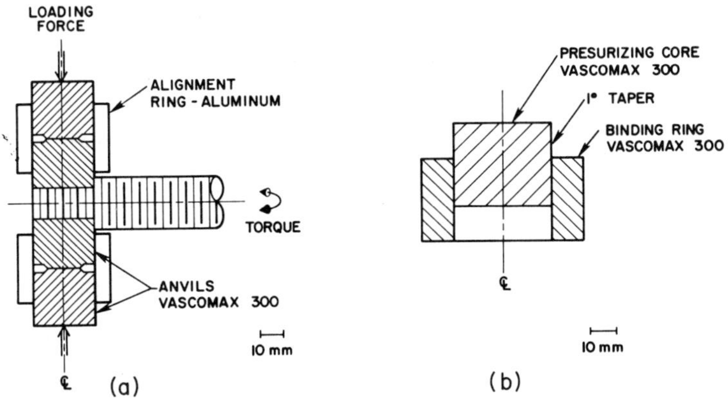

Fig. 3. Cross-section view of friction-testing devices: (a) double-opposed anvil device for measuring coefficient of friction to 2.0 GPa (20 kbar), and (b) model binding ring assembly, shown in position for core insertion.

The coefficient of friction between two hard-steel surfaces was measured for various lubricants over the pressure range from 0.2 - 2.0 GPa (2-20 kbar). The steel used was 18% Ni maraging steel, grade 300 (Vascomax 300), with a yield strength of 1.9 GPa. The coefficient of friction at various pressures was calculated from the torque required to rotate a pair of opposed anvils, shown in Fig. 3a.

Before each run the anvils were ground flat and their corners given a radius of 130 μm (0.005 in.). After the lubricant was applied, the anvils were aligned and rotated about one turn at extremely low load to assure distribution of the lubricant. Data was then taken at 0.1 GPa (1 kbar) steps, on compression and decompression.

As is common with friction determinations, the data showed considerable scatter. In some cases this was due to outright failure of the lubricant, with resultant galling. Figure 4 shows detailed data for several cases. The solid symbols are data from the opposed-anvil device. The lowest friction and the greatest reproducibility were given by the use of 25 μm (0.001 in.) thick lead foil. Data for two runs with a well-burnished film of technical grade MoS_2 powder are shown as solid triangles. Boyd and Robertson [2] and Hyde [3] have also measured the coefficient of friction for MoS_2 in opposed-anvil apparatus. Their results show the same trend but with values 1-1/2 to 2 times higher.

A single run for Molykote G paste (a commercial mixture of MoS_2, mineral oil, and stearates) is shown. The scatter for an individual run was much greater for this material, with the scatter between runs being about three times that for an individual run.

Figure 5 gives a summary of the data from the rotating-anvil apparatus for various lubricants. Lead foil and the MoS_2 preparations all showed a smooth decrease in coefficient of friction with

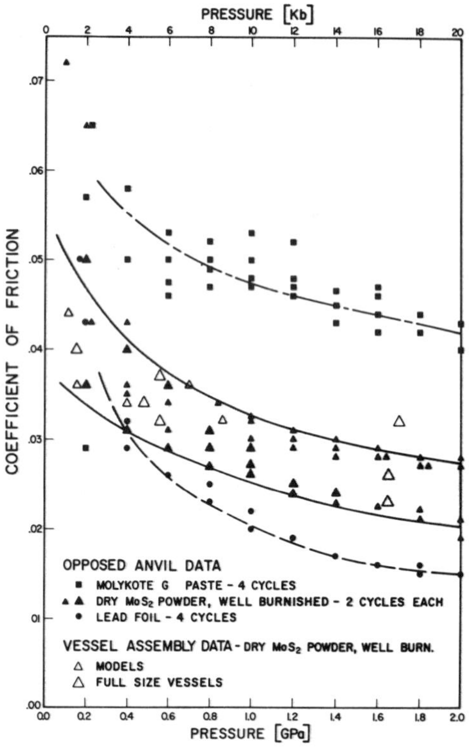

Fig. 4. Coefficient of friction between hard-steel surfaces as a function of pressure for three different lubricants. The shaded band shows the range of expected values for a well-burnished film of technical grade MoS_2 powder.

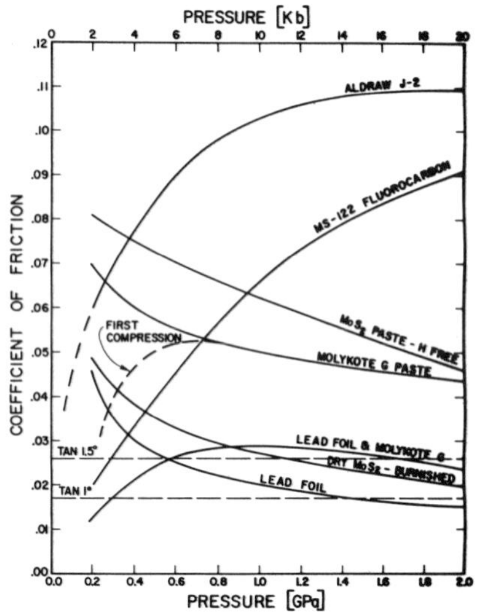

Fig. 5. Summary of the coefficient of friction as a function of pressure, as measured in the opposed-anvil device. Reproducibility at moderate pressure ranged from ± 0.003 for lead foil to ± 0.01 for Molykote G and Aldraw J-2.

pressure. The addition of Molykote G paste to lead foil resulted in lower friction than the lead alone below 0.6 GPa on the first compression, probably due to the presence of mineral oil, but raised the friction at higher pressures. Two commercial preparations showed dramatic increases in the coefficient of friction with pressure: Aldraw J-2, a liquid drawing die lubricant, and MS-122 Fluorocarbon, a spray preparation of tetrafluorethylene polmer solids.

Also indicated in Fig. 5 are the tangents of candidate taper angles: 1° and 1-1/2°. For a well-burnished film of technical grade MoS_2 powder (4-7 μm grain size) a taper angle of 1° on a side satisfies the post assembly stability condition. Above 1.2 GPa interface pressure, 1-1/2° does not.

Model Vessel Assembly

Successful vessel assembly depends not only on selecting and designing to an appropriate coefficient of friction, but also on achieving reliable lubricant performance under actual assembly conditions. As a preliminary study of this problem, Vascomax 300 model vessels were constructed with an outside diameter of about 70 mm (2.7 in.) as illustrated in Fig. 3b. The solid-steel core was inserted into the binding ring with an interface which generated 0.7-0.9 GPa (7-9 kbar) pressure in the binding ring. This ranges substantially above the yield pressure of the binding ring.

Approximately 60 assemblies of models were made, the vast majority with MoS_2 preparations, in search of a technique which would prevent scratching and/or galling of the sliding surfaces, and provide long-term reliability. Most of the preparations failed to lubricate adequately. Some of the preparations which did permit reasonable vessel assembly acted as stress corrosion agents and resulted in the catastrophic failure of the binding ring. A summary of the results follows; μ is the coefficient of friction at 0.8 GPa.

Aldraw J-2. μ about 0.09, assembly very smooth, no galling, occasional light longitudinal scratches. Spontaneous and very slow extrusion of the core occurred at pressures of a few hundred MPa (a few kbar) as the Aldraw J-2 remains a viscous liquid at these lower pressures and exhibits essentially no static friction.

Molykote G Paste. μ about 0.06, assembly generally smooth but with occasional stick-slip, light to medium galling and longitudinal scratching.

Molykote Gn Paste. μ about 0.05, assembly generally smooth, no galling, severe longitudinal scratching.

Mineral Oil Film with Technical Grade MoS_2 Powder Dusting. Assembly unsuccessful, severe stick-slip halfway in, fine galling, light longitudinal scratching.

MoS_2 Technical Grade Powder Painted On in a Perchloroethylene Slurry. μ about 0.04, assembly smooth, moderate galling and longitudinal scratching, thick (20 μm) unevenly-distributed plates of MoS_2 formed.

MoS_2 Technical Grade Powder Painted On in an Isopropyl Alcohol Slurry. μ about 0.03, assembly smooth, no significant galling or scratching, and MoS_2 plates developed. Binding ring failed within 2-1/2 days. Numerous longitudinal cracks formed on the inside surface.

MoS_2 Technical Grade Powder Painted On in a Water Slurry. Assembly smooth, no significant galling or scratching, MoS_2 plates developed. Binding ring failed within 2-1/2 days. Numerous longitudinal cracks formed on the inside surface.

Molykote 321R Spray Bonded MoS_2 Lubricant, Air Dried 4 Hours. μ about 0.03, assembly smooth; thick, continuous coating formed; metal surfaces in excellent condition; no galling or scratching.

Binding ring failed in 10 weeks. Numerous longitudinal cracks formed on side surface.

Technical Grade MoS_2 Powder, Well Burnished. μ about 0.03, assembly smooth, no galling or scratching. Very thin uniform film of MoS_2 formed, nearly invisible. No model failures over a one year test period at internal pressures from 0.2-0.9 GPa (2-9 kbar). Assembly rates of up to 15 mm/s (35 in./min), 700 times normal, were explored. No problems or failures were encountered. This was the only lubrication technique which worked well. Numerous efforts with MoS_2 powder did give problems, however. It appears to be the establishment of a thin uniform film before assembly which is critical. This is accomplished by extensive, hard, slow rubbing of the powder onto the clean surface with a clean felt cloth. No residue powder is left on the surface as this tends to produce scratches.

The lubricating performance of MoS_2 depends on a large number of subtle factors such as exposure to humidity, degree of oxidation, particle size, surface cleanliness, etc. Such factors are discussed in review articles by Winer [4] and Farr [5]. Extensive bibliographies are also given.

A variety of other MoS_2 preparations were explored briefly and all found to be unsatisfactory, usually because they resulted in galling, scratching, stick-slip movement, or high friction. They were Molykote, Spray-kote, Molykote G rapid spray, Molykote M-8800, Dow Corning Corp.; Molylube spray, Bell-Ray Co., Inc.; Moly-Dee spray, Arthur C. Withrow Co.; and various homemade mixtures of different grain size MoS_2 powders with hydrocarbon and fluorocarbon oils.

Full Scale Vessel Assembly

It was a long history of lubrication failures and difficulties with pressure vessels which led to this entire study. In the foregoing sections, results were described which would indicate that a well-burnished film of technical grade MoS_2 powder would give an acceptable coefficient of friction, see Fig. 4, and perform well in the geometry and stress field required. In practice, the following procedure has been used and found to work extremely well in vessels with outside diameters ranging up to 700 mm (30 in.). All the vessels assembled by these instructions have had binding rings constructed of 18% Ni maraging steels (Vascomax), grades 200, 250, and 300.

1. For smaller vessels, tapered surfaces are machined after heat treatment and polished with 320 grit SiC paper in a lathe. Otherwise the tapered surfaces are ground to a 0.3 μm (16 microinch) finish. All corners are given a radius of about 0.5-1.5 mm (0.020-0.060 in.), depending on vessel size.

2. The sides of the conical surfaces must be straight to within 10 μm (.0004 in.). This can be tested with a "knife" straightedge. Nonstraight sides have resulted in galling problems. After each ring is inserted into the assembly of outside rings, its inside surface has sometimes required refiguring to achieve the required straightness. The ends tend to close a little more than the center. Refiguring can be done by proper grinding, or satisfactorily with a fine flap wheel sander and a drill motor.

3. All surfaces are cleaned first with acetone and then with perchloroethylene. The presence of hydrocarbons inhibits the formation of a good MoS_2 film.

4. Dry, fresh, technical grade MoS_2 powder, e.g., Molykote Z powder, is applied with a clean felt cloth. After a brief rubbing with a felt cloth, some power burnishing technique is generally used. Small parts can be burnished in a lathe. Larger parts were burnished with a MoS_2 dusted felt cloth glued to a buffing wheel and turned by a drill motor. The powder is applied and well burnished three times. The importance of this step cannot be overemphasized. The film which is established is extremely thin, probably less than 1 μm (4×10^{-5} in.). It has a slight luster and gray tone, but is nearly invisible. No way of checking the film before assembly has been found.

5. The mating cylinders are carefully seated, set into a press, and pushed together at a displacement rate of 20 μm/s (.050 in./min). A very well established MoS_2 film results in smooth assembly with no squealing. If the sliding parts emit a high squeal as they slide, high frequency stick-slip, the film is probably satisfactory. If the stick-slip develops into separately discernable steps, however, assembly should be terminated and the MoS_2 application repeated.

It is recommended that a record of insertion force vs. displacement be plotted as the process goes on. Final insertion force can be calculated from the interface area, final pressure, and the appropriate coefficient of friction. Any force significantly exceeding a straight line terminating at that calculated point indicates likely problems. Figure 6 shows a family of such curves for the assembly of the vessel shown in Figs. 1 and 2. The coefficient of friction for well-burnished technical grade MoS_2 powder is shown in Fig. 4. Values from the assembly and a number of different vessels are shown (open triangles) and agree well with the data from the opposed anvil device.

VESSEL BORE LUBRICATION

The data in Fig. 4 suggests lead foil as an effective bore lubricant. To evaluate bore friction, plots of piston displacement versus nominal pressure were made for an end-loaded piston-cylinder apparatus cycled to 5 GPa (50 kbar). The interior of the vessel was

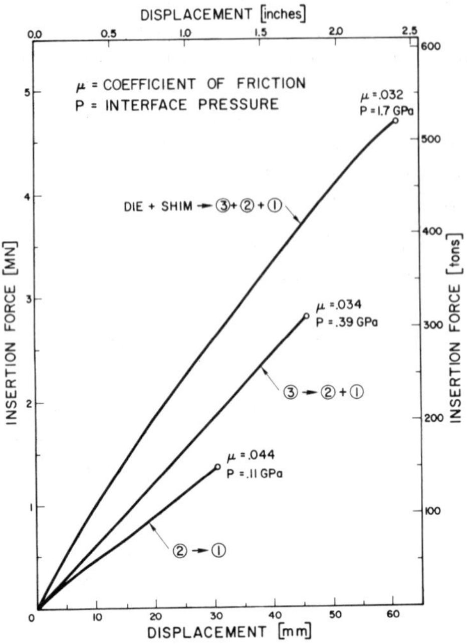

Fig. 6. Assembly curves for the vessel shown in Fig. 1, insertion force vs. displacement. Each ring is pushed into an assembly of those outside it. Circled numbers refer to ring numbers in Fig. 1.

12.7 mm in diameter x 50.8 mm long (0.500 in. diameter x 2.00 in. long). It was filled with a pressed pellet of ground KBr which has a first-order phase transition at about 1.75 GPa. The large piston displacement associated with the transition is readily observed, see Fig. 7. The relative performance of bore lubricants is displayed by the relative shapes of the displacement curves.

Six bore lubrications were examined: (1) 25 μm (.001 in.) thick lead foil wrapped around the KBr pellet; (2) lead foil with technical grade MoS_2 powder burnished onto the carbide bore of the vessel and rubbed on the foil; (3) lead foil with Molykote G paste applied to the foil and the bore; (4) technical grade MoS_2 powder alone burnished on the bore; (5) Molykote G paste alone applied to the bore and the pellet; and (6) no lubricant. A new carbide core was used for the study. The run with no lubricant was performed first. The bore was carefully cleaned with perchoroethylene and 500 grit aluminum oxide paper between runs. The piston displacement rate was 20 μm/s (.050 in./min). About one-half hour was required for each pressure cycle.

The results fall into two completely distinct groups: those with lead foil and those without. A representative cycle for each is shown in Fig. 7. The data plotted is for first cycles. The narrower of the two loops is for lead foil alone. It is characteristic of all the curves with lead. On compression the transition started at a nominal piston pressure of 2.12 GPa and ended at 2.55 GPa. The transition, at this relatively rapid rate of piston displacement and long sample aspect ratio, was spread out over a 0.43 GPa increase in nominal pressure. On decompression, it was spread out over 0.29 GPa at a considerably lower average pressure in the cell.

For the other case plotted in Fig. 7, technical grade MoS_2 powder, the transition onsets occurred at similar nominal pressures

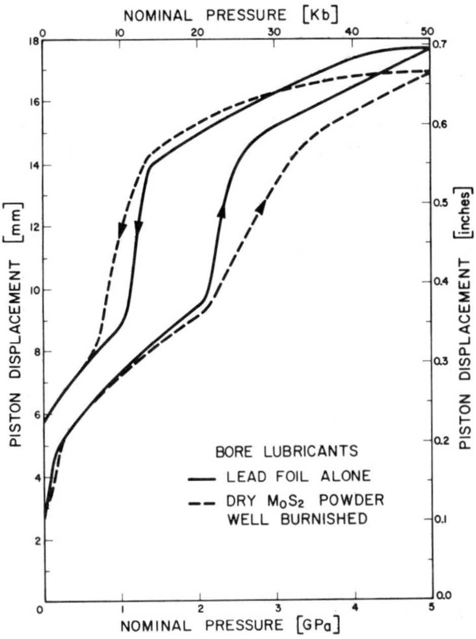

Fig. 7. Piston displacement vs. nominal pressure for a piston-cylinder apparatus filled with KBr. The vertical offset centered at 1.75 GPa is due to the change in volume associated with a phase transition in KBr. All cases with lead foil were very similar to the solid curve; all cases without, to the dashed curve.

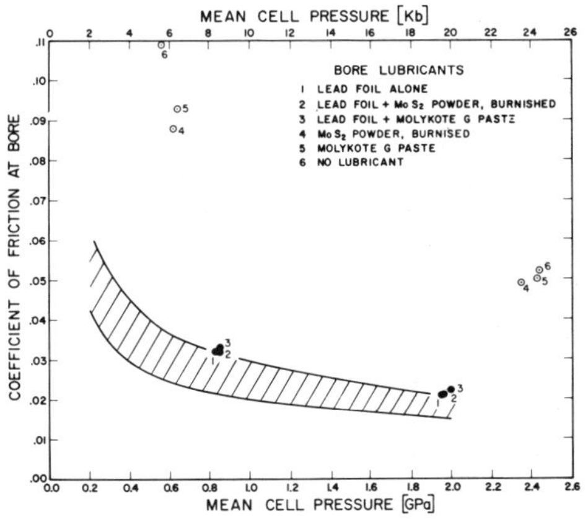

Fig. 8. Coefficient of friction at the bore of a piston cylinder apparatus vs. mean pressure in the cell. These values were calculated from the piston displacement data as in Fig. 7. Solid circles represent cases with lead foil as a lubricant. The two curves delineate the range of values observed for lead foil in the opposed anvil device.

indicating that the piston friction was nearly the same in both cases. However, on compression, the transition was spread out over a 1.23 GPa increase in nominal pressure and in decompression, over 0.59 GPa. These figures are three and two times, respectively, the corresponding figures with lead foil.

An approximate calculation of the coefficient of friction against the bore of the vessel can be made by assuming that the entire pressure drop down the KBr pellet is due to bore friction. The results are shown in Fig. 8. The effectiveness of lead foil and the ineffectiveness of the MoS_2 preparations are shown clearly. Even in the absence of lead foil the MoS_2 lubricants had very little effect with KBr, which is relatively weak itself.

Below about 10 kbar the addition of Molykote G paste made a very slight improvement on the first compression, but increased the friction at higher pressures. Thus for bore lubrication, 25 μm (.001 in.) thick lead foil emerges as an effective and critical lubricant.

PRESSURE VESSEL SAFETY

The large amount of stored elastic energy in compound pressure vessels makes them potentially quite dangerous. A number of steps should be taken to reduce this hazard.

The outermost binding rings experience the highest tensile stresses (see Fig. 2) and are therefore the most prone to brittle failure. Relative to the inner rings, the outermost ring should be made from a lower strength material with higher fracture toughness. In larger 18% Ni maraging steel vessels, grade 200 is recommended for the outer ring: for carbon steel, Rockwell-C 40 maximum. Inner rings should be made from 250 grade maraging steel or Rockwell-C 50 maximum carbon steel.

The vessel should be surrounded by an energy-absorbing safety ring capable of absorbing all of the radially-released energy should the binding rings fail.

Several materials have been explored for this purpose. Martensitic low carbon steel (Grade B pipe) and aluminum have both proven unsatisfactory. Their ductibility is known to be highly strain rate dependent even at room temperature. Small models and vessles up to 300 mm (12 in.) in diameter have broken such "safety rings" in failure. Austenitic stainless steel grade 304, however, has worked very well. Six vessels from 70-350 mm (2.7-14 in.) in diameter were failed into 304 safety rings without any failure of the safety rings. The safety ring wall thickness was 2-4% of its diameter: 4% is recommended.

The surfaces of highly stressed steels must be protected from stress corrosion agents such as hydrogen and water. Stress corrosion at these stress levels can proceed so rapidly as to generate tens of subcritical cracks and to cause vessel failure in a matter of days. The fractures often lack the branching morphology typical of lower rate stress corrosion and appear in analysis to be caused by simple overstraining.

Plating of binding rings is not recommended due to hydrogen embrittlement. A thick coating of silicone grease, carefully maintained, has proven successful in protecting models submerged in water and vessels subjected to cooling water spills. New lubricants should be tested for stress corrosion effects before use in vessels.

Mechanical care of binding ring surfaces is also important. Longitudinal scratches on the surface of rings, likely to be made during assembly, are particularly undesirable. End surfaces of binding rings should not be ground after assembly. Grinding or turning of inside surfaces of assembled binding rings has not caused any problems; corner radii should be maintained, however. All surfaces must be protected from chuck jaw indentation in such operations.

Finally, the tungsten carbide die eventually suffers multiple fractures. In this condition especially, fragments of it may explode out the ends of the vessel. All persons working with such vessels should avoid the potential trajectory of such fragments and be adequately protected by safety shields at all times.

SUMMARY

A solution has been found to the long persistent problems of assembling highly stressed compound pressure vessels. Reliable lubrication has been achieved by the use of an extremely well burnished film of technical grade MoS_2 powder. Factors critical to its use such as surface finish, powder freshness, powder dryness, burnishing techniques, assembly test criteria, and appropriate taper angles have been discussed. For the lubrication of pressure vessel bores, lead foil has been demonstrated to be superior to MoS_2 preparations.

An adequate safety ring design to absorb radial energy on vessel failure has been developed and tested. Grade 304 stainless steel with a wall thickness 4% of its diameter has proven successful. Various mechanical and chemical considerations important to vessel longevity and personnel safety have been delineated.

ACKNOWLEDGMENTS

This work was performed in the laboratory of G. C. Kennedy at UCLA. His support and encouragement are gratefully acknowledged. Assistance in analyzing and understanding the stress corrosion failures was rendered by E. G. Kendall of the Aerospace Corporation, El Segundo, California. Generous support in the form of steel for the pressure vessel models and general technical assistance was provided by M. J. Granger of the Vasco Pacific Division, Vandium Pacific Steel Co.

REFERENCES

1. T. E. Davidson and D. P. Kendall, in Mechanical Behavior of Materials Under Pressure, Vol. 55, H. Ll. D. Pugh, ed., Elsevier Publishing Co., Ltd., Amsterdam, Netherlands (1970), p. 118.
2. J. Boyd and B. P. Robertson, ASME Trans. 67, 51 (1945).
3. G. R. Hyde, M. S. Thesis, Brigham Young University, Provo, Utah (1957).
4. W. O. Winer, Wear 10, 422 (1967).
5. J. P. G. Farr, Wear 35, 1 (1975).

S-4

MODIFICATION OF BELT-LIKE HIGH-PRESSURE APPARATUS

O. Fukunaga, S. Yamaoka, T. Endoh, M. Akaishi and H. Kanda

National Institute for Researches in Inorganic Materials
Ibaraki, Japan

INTRODUCTION

A belt apparatus [1] is a useful and convenient high-pressure generator in the pressure range of 1 to 6 GPa. Essential features of a belt apparatus are the tapered disks and anvils and compressible gaskets which are inserted between the tapered walls. The gasketing function, however, is still complicated and puzzling. Gasket properties determine the amount of stroke and support pressure on the die-anvil members. A larger stroke and support pressure are desirable.

We present herein a new design of the compressible gasket and of the die-anvil geometry in order to increase the stroke limit of anvils for maintaining sufficient supporting pressure inside the gasket. This paper places major emphasis on the use of a compressible gasket of an unusually large size to meet the above requirements.

Composite pyrophyllite-laminated paper sheet gasketing has been used successfully. Experimental results showed a remarkable increase in the maximum generating pressure. Preliminary results on diamond synthesis indicate that the newly designed apparatus is useful as a routine apparatus with a large sample volume.

DESIGN OF DIE AND ANVILS

Analysis of the geometrical proportions of dies and anvils have not previously been analyzed for belt-like apparatuses. To do this, geometrical parameters for our die-anvil members are defined as shown in Fig. 1. The parameters given in the figure can apply to annular dies and tapered anvils of a variety of designs. The parameters of previously designed belt-like apparatuses and those of the present design (FB-13 and FB-25) are listed together in Table I.

Table I. Geometrical Parameters of Belt-Like Apparatuses

	θ_c	ϕ_c	d_c	l_c	m	R_c
Hall [1]	83°	11°	10.16	24.23	0	
Sclar [2]	86°	4°	12.7	21.66	0	4.06
Young [3]	90°	35°	12.7	25.4	6.35	0
Bundy [4]	83°	38°	5.02	23.15	1.77	16.83
Lorent [5]	40°	35°	10.0	60	3.0	
FB-13	72°	0	13.0	12.22	5	6.0
FB-25	72°	7°	25.0	35.12	0	16.0

	l_c/d_c	d_a	l_a	ϕ_a	l_a/d_a
Hall [1]	2.39	8.89	6.35	30°	0.714
Sclar [2]	1.70	7.62	10.6	30°	1.39
Young [3]	0.5*	12.7	13	35°	1.02
Bundy [4]	0.353*	6.5	23	27.5°	3.54
Lorent [5]	0.3*	7.0	33.28	35°	4.75
FB-13	0.94	10.0	9.2	16.7°	0.92
FB-25	1.40	20.0	19.0	20°	0.95

*m_c/l_c value was sited. All units are in mm.

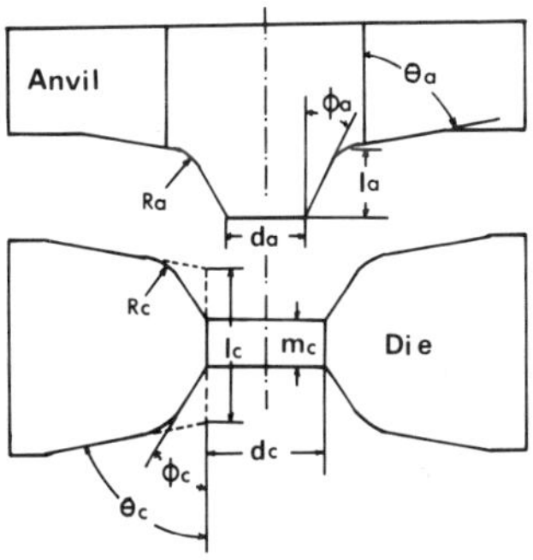

Fig. 1. Geometrical parameters for the die and anvil of the belt-like apparatus.

The major difference between the original belt apparatus and the present design is in the l_c/d_c values. When we choose larger values of l_c/d_c, the length of the sample is relatively larger than the stroke limit of the anvils. Maximum pressure of the apparatus is determined by the stroke limit rather than the failure of the die-anvil members. To increase the maximum pressure attainable in the apparatus the present l_c/d_c parameters were chosen to be less than 1.4. The compressible gasket in the previous design was inserted between the steep tapered sides of the die-anvil members only (e.g., at the wall with an angle ϕ_a, and ϕ_c in Fig. 1). This type of design is based on the need to produce a larger gasket stroke limit.

As described later, we adopted gaskets of larger diameter and these extended to the flat wall portion of the di-anvil. In this case smaller l_c/d_c ratios are preferable. The design procedure is as follows: First, determine d_c, l_c/d_c, and θ_c for the die. Second, determine the diameter of the anvil flat d_a and taper angle θ_a. When we choose a small l_c/d_c value, the angle ϕ_c is selected as zero. The radii R_c and R_a are designed to obtain a smooth variation in the thickness of the gaskets during loading. Two typical die-anvil designs characterized by the relatively small value of l_c/d_c are shown in Fig. 2. We have designated these designs as "flat belt".

DESIGN OF COMPRESSIBLE GASKET

It has been suggested that gaskets of very large diameter are preferable for belt-like apparatuses [6]. A simple model of the pressure distribution in the gasket shows that gaskets of large diameter and thickness can generate sufficient support pressure to the die-anvil member. If 2a is the outer diameter of the gasket then less than 0.2 of $d_c/2a$ is the standard size of our gasket. As an easy method of making such a large size gasket involved using machined pyrophyllite blocks for the inner part and laminated paper sheets for the outer parts of the gaskets as shown in Fig. 2.

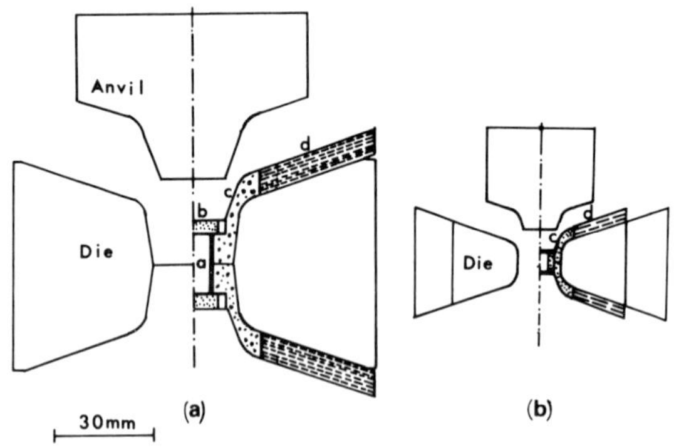

Fig. 2. Schematic diagram of the present apparatus. (a) FB-25, (b) FB-13 apparatus. Legend: a - sample space; b - disk; c - inner pyrophyllite gasket; and d - outer paper gasket.

Compression tests for outer gasket materials were performed using a flat opposed anvil, 26 mm in diameter. The maximum residual thickness h_{rc} (mm) at an average load pressure of 1.44 GPa was measured. The results are shown in Table II. Among the various laminated paper sheets, Hi-Pigeon has almost the same behavior of compression as the pyrohyllite and therefore was selected as the material for the outer gasket.

PRESSURE GENERATION AT ROOM TEMPERATURE

FB-13 Apparatus

The cell assembly for the pressure calibration is essentially as illustrated in Fig. 2b. The results of the pressure calibration using Bi and Ba as pressure sensors are shown in Fig. 3. The thickness of the inner pyrophyllite gasket was constant (4 mm) throughout the experiments, whereas the thickness of the outer laminated paper gasket was modified from 2.4 to 4.4 mm. Efficiency of pressure generation E varied from 59 to 26.6%. (Efficiency is defined as

Table II. Maximum Residual Thickness of Laminated Paper Disk at 1.44 GPa

No.	Type of Paper	$h_{rc}/2a$
1	Hi-Pigeon (Honshu PCC)	0.065
2	PB#1 (Abegawa PCC)	0.057
3	Tai-Coat #400 (Taiko PCC)	0.052
4	#600	0.051
5	S-Bowl (Senju PCC)	0.050
6	Union	0.050
7	Blanch (Chuetsu PCC)	0.050
8	San-Coat (Sanko PCC)	0.049
9	Coat-Bowl	0.047
10	Manila-Bowl	0.045
11	Craft-Bowl (Fuji PCC)	0.044
12	LG#38	0.042
13	File Paper	0.042
14	PB#2 (Abegawa PCC)	0.035
15	Davis (Abegawa PCC)	0.034
16	Gloss-Coat A (Shibakawa)	0.029
17	S-Bowl H (Senju PCC)	0.029
18	Fiber-Red (Hokuetsu PCC)	0.026
	Raw African Lava A	0.05-0.06
	Fired Lava A	0.09

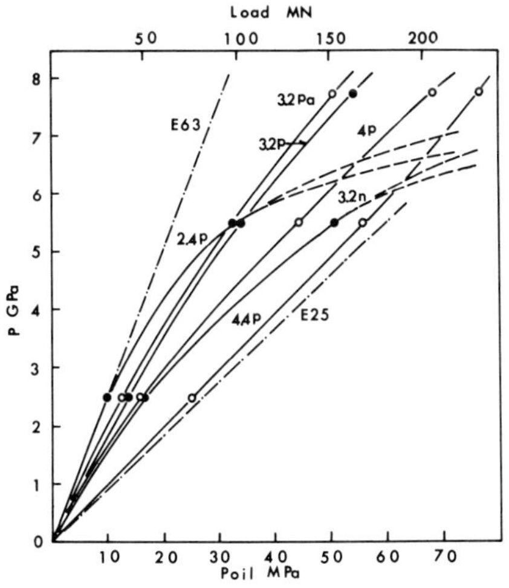

Fig. 3. Pressure calibration in the FB-13 apparatus. Numeral denotes thickness of paper gasket or percentage of efficiency E. Legend: p_a = Al_2O_3 disk and n = NaCl cell.

$E = \pi d_a^2 P/4L$, where P is the pressure inside the chamber and L is the load required to generate pressure P.)

When thin outer gaskets were used the efficiency decreased rapidly with increasing pressure and we could not attain the pressure of the high Bi-fixed point (7.7 GPa). This result is similar to that experienced when using the small-diameter gasket previously adopted for many belt-like apparatuses. Present use of large-diameter laminated-paper gaskets show a clear improvement in attaining maximum pressure. When we chose the proper gasket thickness, a smooth load vs. pressure curve of moderate efficiency was obtained up to the high Bi pressure fixed point. Maximum pressures of more than 10 GPa can be generated in this apparatus.

FB-25 Apparatus

Another apparatus which we successfully used was a vessel with a diameter d_c of 25 mm (see Fig. 2a). It had the same design principle but a larger l_c/d_c value than that of the FB-13 apparatus. Results of the pressure calibration at room temperature are shown in Fig. 4. The standard thickness ratio of the internal to external gasket was about 0.7.

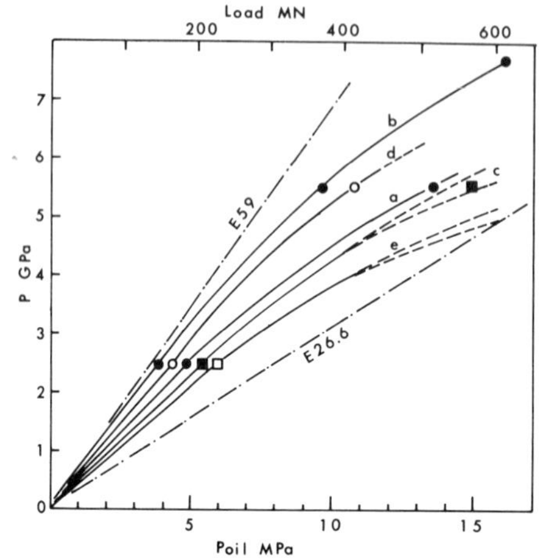

Fig. 4. Pressure calibration in the FB-25 apparatus. Circles and squares represent the horizontal and vertical setting of the sensor, respectively. Thickness of the paper gasket is 6.4 mm (a, b, and c) or 4.8 mm (d and e). Pyrophyllite disk (a, d, and e) and Al_2O_3 disk (b and c) are used.

The use of sintered Al_2O_3 disks (curve b in Fig. 4) instead of pyrophyllite (curve a) at the top of the tungsten carbide anvil was effective in increasing the maximum pressure. The orientation of the pressure sensor changed the load vs. pressure curve markedly. Setting the wire parallel to the loading axis (curves c and e) showed less efficiency for pressure generation.

PRESSURE AND TEMPERATURE CORRECTION

Accurate measurements of pressure and temperature at elevated temperatures are presently not possible in our laboratory. We checked the deviation of temperature and pressure using the intersection of three phase boundary lines. For this purpose the diamond/graphite equilibrium and the Ni-C or Co-C eutectic lines were chosen.

The diamond/graphite equilibrium has been studied extensively [7-9]. (Recent data by Kennedy, et al. [9] are especially useful for an approximate correction of experimental curves.) The pressure coefficients of the eutectic lines of Ni-C and Co-C have also been studied and are fairly well established [10,11]. The minimum point of diamond formation in the system Ni-C is very near the intersection between the diamond/graphite equilibrium and the Ni-C eutectic lines (see d in Fig. 5). In the system Co-C it is about 50 MPa and 20 °C below the intersection [11] (see f in Fig. 5). The diamond/graphite equilibrium line given by P(GPa) = 1.94 + T/400 (°C) [9] (see line c) and the pressure coefficient of the eutectic lines (for Ni-C (see line Ni in Fig. 5), dP/dT = 0.141 °C/GPa [10], and for Co-C (see line Co), dP/dT = 0.048 °C/GPa [11]) were used to estimate the minimum condition of diamond formation. These are 5.42 GPa and 1394 °C for Ni-C, and 5.22 GPa and 1314 °C for Co-C.

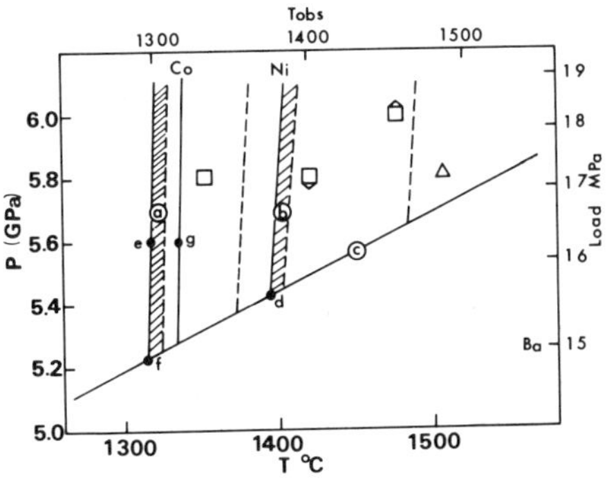

Fig. 5. Diamond formation region in the system Co-C. Legend: (1) Crystal habit of diamond: square = cubic, square plus triangle = cubo-octahedral, and triangle = octahedral diamond. (2) Observed minimum condition of diamond formation: a = Ni-C and b = Co-C. (3) Fixed points for temperature and pressure correction: c-g (see text).

Experiments for diamond synthesis were performed using the FB-25 apparatus at various pressure-temperature conditions near the boundary line. The cell assembly used in the experiments is shown in Fig. 2. A disk of metal catalyst solvent of about 3 mm in thickness was placed between two spectroscopic grade graphite disks of the same size. Temperature measurement was at the top surface of the catalyst metal using a Pt/Pt13Rh thermocouple, 0.3 mm in diameter.

The results of Co-C system are shown in Fig. 5. Hatched curves a and b show the minimum conditions of diamond formation for the Co-C and Ni-C systems, respectively. The results suggest about a 15 to 20 °C lower shift of emf at about 5 to 5.5 GPa and 1300 to 1400°C. This degree of correction in the thermocouple emf is about one third of the correction reported by Hanneman and Strong [12], but agrees basically with the data of Getting and Kennedy [13]. Correction of the emf will decrease with an increase of temperature for the pressure seal point. It may be assumed that in the present apparatus the pressure seal point is at a higher temperature than in the original belt apparatus due to the effect of the larger gasket.

The pressure deviation from the room temperature calibration was about -4% based on curve c in Fig. 4 at 1300°C. Strong and Bundy [14] have pointed out that the pressure in pyrophyllite cells decreases quite rapidly at high temperatures and eventually falls below the pressure calibrated at room temperature. Present results for the pyrophyllite cells show a similar tendency.

REMARKS ON THE LIFE OF APPARATUS

About 200 runs at high temperature-pressure conditions were made before replacing the tungsten carbide material of the die. Small hair cracks parallel to the loading axis had gradually developed after 100 runs. The top portion of the anvil in the present design is somewhat steeper than that of conventional designs. We did experience one cap and cone type failure of the anvil. It seems, however, that the failure of the anvil was greatly affected by thermal shock during quenching procedures.

An apparatus with a smaller l_c/d_c ratio has a higher pressure limit for hair crack generation; but die failure eventually occurs with the tungsten carbide part being entirely pulverized. Before being destroyed, vertical cracks occurred at the outer margin of the die wall. This suggests a shear stress effect between the die wall and the laminated gasket. Lubrication by MoS_2 powder on the surface helps to prevent crack propagation.

REFERENCES

1. H. T. Hall, Rev. Sci. Instr. 31, 125 (1960).
2. C. B. Sclar, L. C. Carrison, and C. M. Schwartz, in High Pressure Measurements, Butterworth, London (1963), p. 286.
3. A. P. Young, P. B. Robbins, and C. M. Schwartz, in High Pressure Measurements, Butterworth, London (1963), p. 262.
4. F. P. Bundy and R. H. Wentorf, Jr., J. Chem. Phys. 38, 1144 (1963).
5. R. E. Lorent, Rev. Sci. Instr. 44, 1691 (1973).
6. O. Fukunaga, in Proceedings 4th AIRAPT Intern. High Pressure Conference, Kyoto, Japan (1974), p. 798.
7. R. Berman and F. Simon, Z. Electrochem. 59, 333 (1955).
8. F. P. Bundy, H. P. Bovenkerk, H. M. Strong, and R. H. Wentorf, Jr., J. Chem. Phys. 35, 383 (1961).
9. C. S. Kennedy and G. C. Kennedy, J. Geophys. Res. 81, 2467 (1976).
10. H. M. Strong and R. E. Hanneman, J. Chem. Phys. 46, 3668 (1967).
11. H. M. Strong and R. E. Tuft, Technical Information Series Report No. 74CRD118, G. E. Corporate Res. and Development, Schenectady, New York (1974).
12. R. E. Hanneman and H. M. Strong, J. Appl. Phys. 36, 523 (1965).
13. I. C. Getting and G. C. Kennedy, J. Appl. Phys. 41, 4552 (1970).
14. H. M. Strong and F. P. Bundy, "Accurate Characterization of the High-Pressure Environment," NBS Special Publication 326 (1971), p. 283.

S-5

CIRCULATION PUMP FOR HIGH-PRESSURE GAS SYSTEMS

L. P. Baudoin

Sandia Laboratories
Albuquerque, New Mexico USA

INTRODUCTION

Constant fugacity studies in high-pressure systems most often require a device to move a gas or liquid in a recirculation mode within the high-pressure system itself. The differential head pressure required is often small, 7 to 14 kPa (1 to 2 psi), yet even a small displacement pump rated for system pressure can be cost prohibitive.

METHOD

Presented here is a relatively inexpensive device that functions with a double-acting piston pump action and constant velocity flow, except for the fraction of a second lost when the piston reverses direction. Figure 1 shows the assembly of the circulation pump.

The piston is "free-floating" in the cylinder, but held within the fields of four permanent magnets grouped around the outside of the pump cylinder with like poles together. The magnet group is attached to the housing of a commercial ball-reversing mechanical actuator, a device that converts rotary motion into reciprocating constant-velocity linear motion. The rotary motion is supplied by a variable-speed-electric-gear motor, allowing for a wide selection of flow rates.

The pump cylinder is a straight length of nonmagnetic stainless steel tubing rated for system pressure; length optional, but within the limits of the linear actuator travel. A standard high-pressure cone connection couples each end of the cylinder to a tee block. Installed in the other two openings of each tee are ball check valves used to control flow direction.

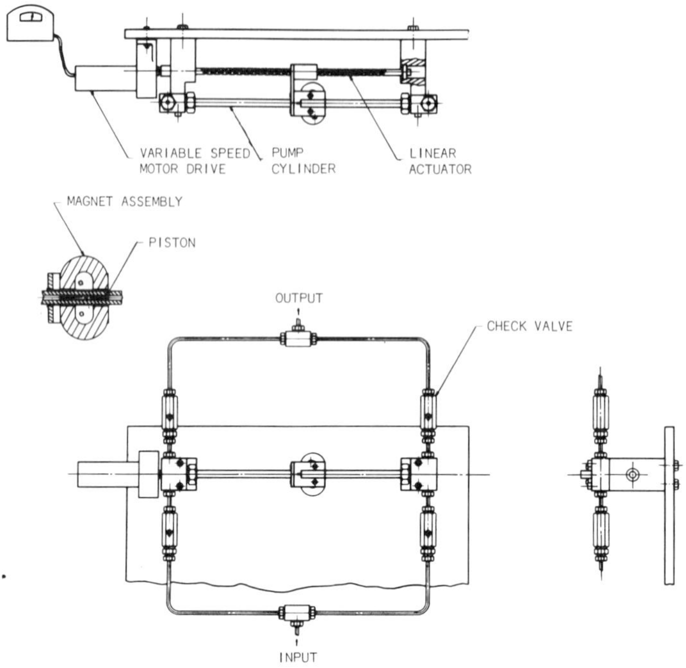

Fig. 1.

EQUIPMENT

The system pressure for which this design evolved is 415 MPa (60,000 psi). The pump cylinder utilized is 14.28 mm OD x 4.76 mm ID (9/16 in. x 3/16 in.) 316 stainless steel tubing (60,000 psi rating). The inside diameter is not altered and the piston OD is sized to "free travel" with a diametral clearance of 0.050 mm to 0.076 mm (0.002 to 0.003 in.).

The most desirable design of the piston has been found to be an assembly of a mild steel core within a thin wall tube of 316 stainless steel, the ends being plugged and welded. The length of the mild steel core was optimized at 75% of the outside dimension of the horseshoe magnet poles.

The check valves are commercial high-pressure surge valves, operated in the inverted position. The size selected for the interconnecting tubing is 6.35 mm OD x 2.1 mm ID (0.250 x 0.083 in.) and rated for 415 MPa (60,000 psi).

RESULTS

The force required to push the piston out of the magnetic field is 272 g (0.6 lb), resulting in an actual developed pressure of 158.5 kPa (23 psi). The flow at zero head is 0.023 m^3/hr (6 gph) when the frequency of stroke is 45 (90 motions) per minute over a travel length of 406 mm (16 in.).

S-6

VISCOMETER FOR USE AT HIGH TEMPERATURES AND PRESSURES

S. I. Smedley, I. Torrie, and G. W. Bird

Victoria University of Wellington
Wellington, New Zealand

INTRODUCTION

The study of liquids at high temperatures and pressures necessitates the use of gas as the pressurizing medium. A sealed collapsible container, or a rigid container with flexible bellows is often used to isolate the fluid from the pressurizing gas. These conditions make it very difficult to measure viscosity. Even more difficult is the study of fluids containing dissolved volatiles such as, in our case, water rich magmas. These fluids must be sealed in a container to keep the water in and the pressurizing gas out. Our viscometer allows one to measure the viscosity of a fluid within a sealed container, and is based on the work of Robinson and Smedley [1]. The device is a development of the torsionally oscillating quartz crystal technique used previously at low temperatures and high pressures [2,3].

DESCRIPTION OF VISCOMETER

The instrument described in this paper is illustrated in Fig. 1. It consists of two α-quartz piezoelectric crystals joined end to end, and this assembly is joined to a glass rod. Each piezoelectric crystal has four electrodes plated longitudinally and opposite pairs of electrodes are electrically connected. An especially developed crystal driver applies an ac voltage, V_d, to one pair of electrodes on the drive crystal, forcing it into torsional oscillation. The other crystal, the gauge crystal, is fixed to the drive crystal so it must also oscillate in the torsional mode. The voltage, V_g, induced by this vibration is fed back to the crystal driver making the device a closed loop oscillator. At the condition of minimum drive voltage the frequency of oscillation will be the fundamental frequency of the

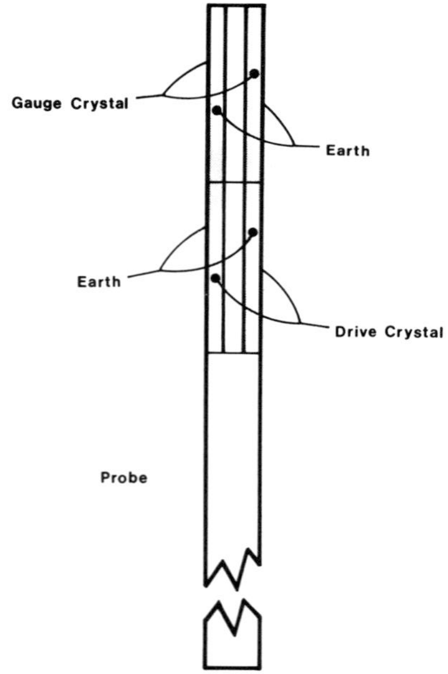

Fig. 1. Dynamic elements of viscometer, which consists of two piezoelectric crystals and a probe. The electrodes are painted on to the four quadrants of the crystals with a conducting paint, and then drive and earth leads are attached at the nodes. The lead positions are distorted on the diagram for clarity of illustration.

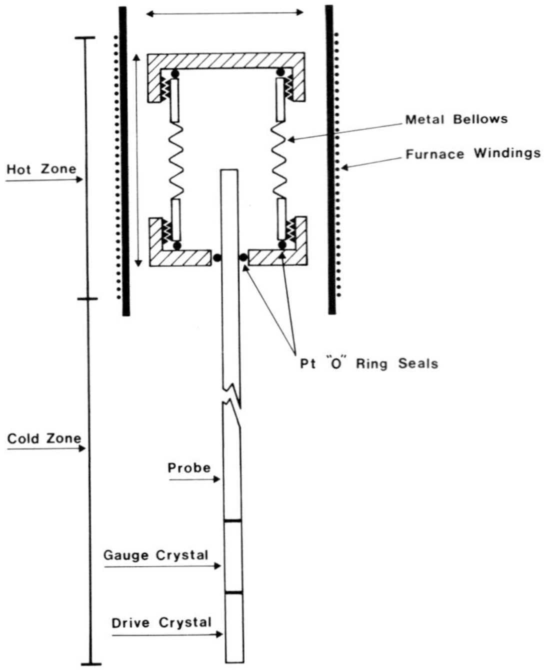

NOT TO SCALE

Fig. 2. Proposed sample container design for high pressure use. The system differs from the high temperature version only in having bellows instead of a solid container. Ceramic discs used to insulate drive and gauge crystals and rods used to hold the apparatus in the furnace are not shown.

crystals, in this case 40kHz. The rod attached to the crystals is tuned to a resonance length, so that standing torsional waves are generated within it. It is this rod that can be used as a probe and immersed in the fluid to a depth of $\ell\lambda/4$.

When the probe is immersed $\ell\lambda/4$ into a viscous fluid the fluid dampens the vibration and increases the resonance period, τ, of the oscillating assembly. To determine the viscosity of a fluid one measures V_d, V_g, and τ in vacuum, and then repeats the measurements when the probe is immersed $\ell\lambda/4$ in the fluid. For non-elasticoviscous fluids the viscosity is given by

$$\rho\mu = C(Q^{-1}(\mu))^2 \tag{1}$$

or

$$\rho\mu = 4C\left[\frac{\tau_{fluid} - \tau_{vacuum}}{\tau}\right] = 4C\left(\frac{\Delta\tau}{\tau}\right)^2 \tag{2}$$

$$C = \frac{\pi}{4\tau}\left[\frac{ma}{a^3(4\ell\lambda + \frac{\pi a}{2})}\right]^2 \tag{3}$$

$Q^{-1}(\mu)$ is the mechanical damping due to the fluid and can be calculated via

$$Q^{-1}(\mu) = K\left[\frac{V_{d_{fluid}} - V_{d_{vacuum}}}{V_g}\right] = K\left(\frac{\Delta V_d}{V_g}\right) \tag{4}$$

where K is a constant that can be calculated from the electromechanical properties of the system. The constant K can also be calculated by measuring the half width resonance response and (V_d/V_g), and then substituting them into

$$Q^{-1}(\mu) = [n^2 - 1]^{-1/2}(\tau_2 - \tau_1)/\tau_r \tag{5}$$

For elasticoviscous fluids, i.e. viscous fluids that exhibit elasticity, the complex viscosity is written as

$$\mu^* = \mu_1 - i\mu_2 \tag{6}$$

where

$$\rho\mu_1 = 2C\left(\frac{\Delta\tau}{\tau}\right)(Q^{-1}(\mu)) \tag{7}$$

$$\rho\mu_2 = \frac{C}{2}\left[(Q^{-1}(\mu))^2 - 4\left(\frac{\Delta\tau}{\tau}\right)^2\right] \tag{8}$$

The crystal driver is designed to maintain V_g at a preselected value and to increase V_d in response to mechanical damping. Thus the amplitude of vibration, maximum shear velocity, and therefore shear rate are proportional to V_g. For non-shear rate-dependent fluids the ratio V_d/V_g will be constant with varying shear rate, for shear thinning fluids it will decrease with increasing shear rate; and conversely for shear thickening fluids.

For use at high temperatures and pressures the probe must be sealed into a closed container. This can be achieved quite readily by gripping the probe at a node, located $\lambda/4$ from the free end (see Fig. 2). A platinum O-ring is squeezed tightly around the rod at the node, and this ring is then pressed tightly into the bottom of the container using a specially designed jig.

Figure 2 depicts the probe sealed into a plate which is then fixed to a collapsible metal bellows. Once the bellows have been filled with the test sample (solid or liquid) the container can be sealed.

For high temperature operation the piezoelectric crystals must be separated from the hot zone. This is accomplished by passing the probe through a shield of refractory discs so that the gauge and drive crystals are maintained at room temperature. We are currently working at temperatures in the range of 350 to 400°C on $ZnCl_2$ melts and intend to extend the temperature range to 1300°C using B_2O_3 before any high-pressure tests are attempted.

Throughout our development work we have scaled the apparatus so that the viscometer, viscometer holder, furnace, and furnace insulation will fit inside a 5-cm bore internally-heated pressure vessel. We are building such a vessel with a 5-cm bore and 21-cm OD which should be suitable for 5 kbar pressure.

To study the viscosity of hydrous silicate melts (magmas) we intend to use a platinum container as shown in Fig. 2. The container will be filled with sufficient powdered rock sample and water to completely fill it when the sample is molten. A small volume of air will be trapped in the sample but this will dissolve in the silicate melt. By coupling the bellows to a linear potentiometer we will be able to measure $(\partial V/\partial P)_T$ and/or $(\partial V/\partial T)_P$ for our samples and from the results calculate PVT data for the melts. When this data is combined with the viscosity data, it will substantially increase our knowledge of silicate chemistry.

ACKNOWLEDGMENT

The authors would like to thank W. H. Robinson of the Physics and Engineering Laboratory of the Department of Scientific and Industrial Research, New Zealand, for his help and guidance.

NOTATION

a = radius of rod

ℓ = an integer

m = total mass of the resonating assemblage

$n = \dfrac{V_{d_1}}{V_{d_r}} = \dfrac{V_{d_2}}{V_{d_r}}$

Q^{-1} = mechanical damping due to the fluid

V_d = drive voltage

V_g = gauge voltage

V_{d_1} = drive voltage at period 1

V_{d_2} = drive voltage at period 2

V_{d_r} = drive voltage at resonance

λ = wavelength of vibration

μ = viscosity of fluid

μ^* = complex viscosity of an elasticoviscous fluid

μ_1 = viscosity at resonant frequency

μ_2 = a function of elastic shear modulus of the fluid

ρ = density of the fluid

τ = period of vibration

τ_1 = period of drive voltage V_{d_1}

τ_2 = period of drive voltage V_{d_2}

REFERENCES

1. W. H. Robinson, S. I. Smedley, J. Appl. Phys., to be published.
2. A. F. Collings, E. McLaughlin, Trans. Farad. Soc. 62, 340 (1971).
3. R. E. Rein, T. T. Chorny, C. M. Sliepcevich, and W. J. Ewbank, NASA Report CR-120786 (1971).

S-7

VISCOSITY MEASUREMENTS IN THE DIAMOND-ANVIL PRESSURE CELL

G. J. Piermarini, R. A. Forman, and S. Block

National Bureau of Standards
Washington, D. C. USA

INTRODUCTION

Comparatively few experiments have been reported concerning the effects of pressure on the viscosity of liquids at pressures above 30 kbar. Bridgman measured the viscosities of several organic liquids to about 30 kbar utilizing a swinging vane apparatus [1]. More recently, Barnett and Bosco described a capillary-type viscometer for measuring viscosities between 10^3 and 10^{10} poise up to 60 kbar [2]. We have developed a method of measuring the pressure dependence of the viscosity of liquids in the diamond-anvil cell utilizing a simple falling-sphere technique in conjunction with the ruby method of pressure measurement. In this report we describe the details of the method and include a preliminary discussion on the accuracy of the method. Viscosity data on a 4:1 methanol:ethanol mixture to 70 kbar are reported and discussed in relation to its hydrostatic properties and glass transition pressure.

EXPERIMENTAL METHOD

A diamond cell similar to one described earlier [3] is used with a modification designed to produce essentially linear displacement of the pressure plate in order to minimize wear and distortion in the piston and cylinder. This modified design utilizes a yoke-type pressure plate that positions the pressure plate bearing and the lever fulcrum in a line parallel to the long dimension of the pressure cell body. A description of the details of this design will be published elsewhere.

A solid nickel alloy sphere approximately 0.035 mm in diameter is confined in the gasket hole containing the liquid along with a small piece of ruby, approximately the same size as the Ni sphere, which acts as the pressure sensor. The assembled pressure cell is mounted on a modified optical goniometer, the translational and rotational motions of which are used to position the cell for viewing the pressurized sample through the diamond window with a microscope. The rotational feature permits rapid inversion of the cell while maintaining optical alignment. The solid nickel sphere can be observed through the microscope while falling freely through the liquid in a gravitational field. The microscope contains a one-dimensional linear scale graticule calibrated against a micrometer slide (0.01 mm graduations) so that the position of the sphere can be determined accurately as it falls through the liquid. The image of the sample can also be displayed on a TV monitor as well as recorded on video tape as described in earlier reports [3,4]. The rate-of-fall of the sphere through the liquid and thus its terminal velocity can be determined either with an electronic digital timer or by counting time elapsed frames on the video tape as has been found necessary when measuring very high velocities (very low viscosities).

Absolute viscosity is calculated using a modified Stokes' equation

$$n = \frac{2r^2 g\ (\rho-\rho_o)}{9v}\ \gamma \qquad (1)$$

where n is the absolute viscosity, r the radius of the sphere, g the acceleration due to gravity, v the terminal velocity of the sphere in the liquid, ρ the density of the sphere, ρ_o the density of the liquid, and γ a geometric factor.

To accurately calculate viscosities at any given pressure the pressure dependence of r, ρ, and ρ_o must be taken into account as well as the wall effect which is expressed in the geometric factor, γ.

RESULTS

As a check on the accuracy of this method, the viscosity of a Dow-Corning 200 fluid of known viscosity was measured both in the pressure cell at 1 atm and room temperature (20 to 22°C) and by a conventional falling-ball technique outside the pressure cell under the same conditions of pressure and temperature. The conventional method gave a value of $(1.1 \pm 0.1) \times 10^3$ poise. The pressure cell method gave an apparent viscosity of $(4.2 \pm 0.2) \times 10^3$ poise which is considerably larger than the true value. As Stokes' law is accurate only for a sphere in a medium of infinitely large extent, a correction for the boundary conditions for the system used is necessary. In our system the sphere experiences an appreciable drag due to the close proximity of the walls, i.e., the two opposed diamond faces and the circular gasket wall. The wall effect decreases the velocity of the sphere and thus accounts for the larger apparent viscosity. In the experiments with the Dow-Corning 200 fluid, Ni spheres approximately 0.05 mm in diameter were used and were situated between two essentially parallel walls (diamond faces) approximately 0.18 mm apart. Thus, the Ni sphere is about one diameter away from each wall. Furthermore, because of the circular shape of the gasket wall and its close proximity to the Ni sphere, the velocity of the sphere is affected in a complex manner.

Although the influence of wall effects on viscosity has been studied for a variety of boundary conditions, such as cylindrical and infinite parallel walls, none has been reported which accurately describes the geometry of the gasketed anvil arrangement. For the purpose of this report we apply an approximate wall correction which includes extremes in boundary conditions to show the maximum and minimum possible corrections to the viscosity. It follows from the work of Hill and Power [5], that extreme values for the correction factor can be determined. We consider a concentric spheres model and assume that the solid Ni sphere is confined within the boundary of a larger rigid sphere containing the liquid. Although the gasketed anvil arrangement is not well approximated by a sphere, two spherical boundary conditions provide the two extreme limits. One limit defines an enclosing sphere whose diameter is equal to the distance between the opposed anvil faces and represents a maximum wall effect, and the second limit is a larger enclosing sphere whose diameter is equal to the diagonal distance between two opposite points on the gasket wall where it comes into contact with the diamond face. This larger sphere represents the maximum boundary of the liquid giving a minimum wall effect and thus a minimum correction to the viscosity. Other known corrections such as those due to infinite parallel walls or a cylinder are within these limits.

For the Dow-Corning 200 fluid, corrected viscosities for the concentric sphere and infinite parallel walls models are tabulated in Table I. The correction factor due to the smaller sphere is 0.24,

Table I. Observed Viscosities for Dow-Corning 200 Fluid $(1.1 \pm 0.1) \times 10^3$ Poise

Apparent Value Diamond Cell, poise	Corrected Values Sphere Maximum	Sphere Minimum	Infinite Wall
$(4.2 \pm 0.2) \times 10^3$	2.8×10^3	1.0×10^3	2.4×10^3

reducing the viscosity to slightly less than that measured by the conventional falling-ball method outside the pressure cell. The correction factor due to the larger sphere is 0.67. The correct factor is likely some value between the two extremes and from the geometry of the system one can conclude that it is much closer to the minimum value. The parallel wall correction does not appear to be suitable in this particular case. We estimate that absolute viscosities with a 10% uncertainty in principle can be measured by this technique provided that accurate corrections due to a wall effect are applied. A detailed analysis of the boundary conditions and their effect on modifying the velocity of the falling sphere in a diamond cell is presently being completed and a detailed discussion of the treatment including quantitative corrections which permit a meaningful evaluation of the accuracy of this method will be published elsewhere.

Viscosity measurements were made on a 4:1 mixture (by volume) of methanol:ethanol to 70 kbar to demonstrate the utility of the method and at the same time to provide data on the hydrostaticity of this pressure-transmitting fluid. The results for this specific mixture corrected for a parallel wall effect are shown in Fig. 1. The infinite parallel wall correction was used here because it represented more correctly the geometry of the system as used at these high pressures where the gasket separation was reduced appreciably. The parallel wall correction reduces the apparent viscosity by a factor of 0.46. Although the precision of the pressure measurement is 0.5 kbar, the accuracy is estimated to be 1.5 kbar. There are several important general features regarding these results which should be mentioned. First, a typical exponential behavior of the pressure dependence of the viscosity is displayed by the data, in agreement with the general observations of Bridgman [1] and Barnett and Bosco [2] for other liquids. Second, in the region of 15 kbar an inflection point is observed which is similar to previously reported results for 1:1 pentane:isopentane, methanol, and other liquids. Third, viscosities ranging from 10^{-1} poise at 10 kbar to 10^5 poise at 70 kbar are observed. Fourth, at 70 kbar the terminal velocity of the solid Ni sphere through the liquid medium is about 10^{-8} cm sec^{-1}. The 4:1 methanol:ethanol mixture is obviously within the hydrostatic regime over the pressure range of concern here, i.e., 70 kbar.

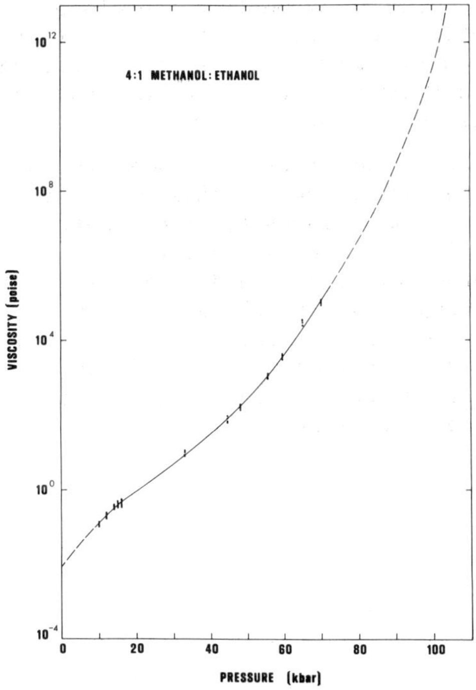

Fig. 1. Viscosity as a function of pressure for 4:1 methanol:ethanol (20-22°C) corrected for an infinite parallel wall effect. The dashed lines represent extrapolations to the glass transition (10^{13} poise) and to the viscosity of pure methanol at 1 atm (6×10^{-3} poise).

CONCLUSIONS

The pressure dependence of the viscosity of a liquid can be measured in a diamond-anvil-type pressure cell by utilizing a falling-sphere technique and the ruby method of pressure measurement. The accuracy of the method has not been fully determined because of the present uncertainty of the wall effect. Maximum and minimum corrections representing extremes in the wall effect in the gasketed anvil system have been applied and bring the apparent viscosity of a known fluid more in agreement with the expected value. As expected, the maximum correction overcompensates slightly and the minimum correction undercompensates greatly. The exact correction lies somewhere between the extremes and is probably quite close to the maximum correction. A theoretical analysis of the wall effect in the diamond-anvil cell is presently under study and the treatment along with quantitative correction factors to permit an evaluation of the accuracy of the method will be published elsewhere.

The pressure dependence of the viscosity of a 4:1 mixture (by volume) of methanol:ethanol has been determined by this method to 70 kbar. The exponential behavior of the viscosity of 4:1 methanol: ethanol as a function of pressure is as expected and agrees with previous studies reported on numerous organic liquids which show similar exponential behavior. Extrapolation of the data to 10^{13} poise (the generally accepted value of the viscosity of a liquid at its glass transition) gives a glass transition pressure of about 105 kbar in agreement with our earlier reported value as determined by broadening of the R_1 ruby fluorescence line [6].

REFERENCES

1. P. W. Bridgman, Collected Experimental Papers, Vol. VI Harvard University Press, Cambridge, Massachusetts (1964), p. 3903.
2. J. D. Barnett and C. D. Bosco, J. Appl. Phys. 40, 3144 (1969).
3. G. J. Piermarini and S. Block, Rev. Sci. Instr. 46, 973 (1975).
4. S. Block and G. J. Piermarini, Physics Today 29 (9), 44 (1976).
5. R. Hill and G. Power, Quart. J. Mech. Appl. Math. 9, 313 (1956).
6. G. J. Piermarini, S. Block, and J. D. Barnett, J. Appl. Phys. 44, 5377 (1973).

S-8

EXPERIMENTAL ASPECTS OF HIGH PRESSURE SPECTROSCOPY OF MOLECULAR INTERACTIONS*

H. Vu

Université Paris-Nord
Villetaneuse, France

INTRODUCTION

The physics of molecular interactions is the study of forces which exist between molecules. It is obvious that these interaction forces determine the structure of matter and that their knowledge is an essential prerequisite for the understanding of the physical properties of matter and for the discovery of new materials useful in applied physics. This double aspect, both fundamental and applied, is probably the origin of the tremendous development of this field throughout the world during the last decade. It is so widespread that it is hard to define at present the exact limits of this field since it also covers thermodynamics, statistical physics, quantum mechanics, material physics, and the multiple specialities of spectroscopy: neutron, XR, RMN, IR and R. We will restrict ourselves to the domain of high-pressure IR and R spectroscopy, even though we cannot go into details at this time. Only a brief review of the principal technical, experimental, and theoretical advances will be given and we refer to the original papers for further details.

TECHNICAL ASPECTS

High-pressure technology in spectroscopy concerns essentially the vessel containing the substance studied. Up to 2 kbar, the pressure vessel is usually relatively simple. It consists essentially of a metallic cylinder constructed of stainless steel, beryllium copper, or more-or-less hardened steel, depending on the pressure and temperature ranges and the chemical properties of the molecules studied. It is provided with two windows for an absorption cell, or with three or four windows for a scattering cell. The material chosen for the windows must match both the pressure and temperature

*Invited paper.

resistance and the transparency in the spectral range studied: pyrex, fused or crystalline quartz, CaF_2, MgO_2, Si, Ge, NaCl, KBr, sapphire, diamond, or other sintered materials. Figure 1 [1,2] shows a diagram of such a conventional cell (for up to 2 kbar) and its thermostat-cryostat that allows a temperature range from 500 to 77 K. The cell is linked by high-pressure tubing to a gas compressor which is at room temperature. We have built a series of such cells with absorption path lengths ranging from a few microns (for strong absorption of polar solids) up to 10 m (for weak induced absorption) [1-3] and a diameter of about 6 to 15 mm for the window through which the light beam passes (the unsupported area of the window). The window, usually made of oriented sapphire, has the following dimensions: diameter 15 to 25 mm and thickness 6 to 15 mm.

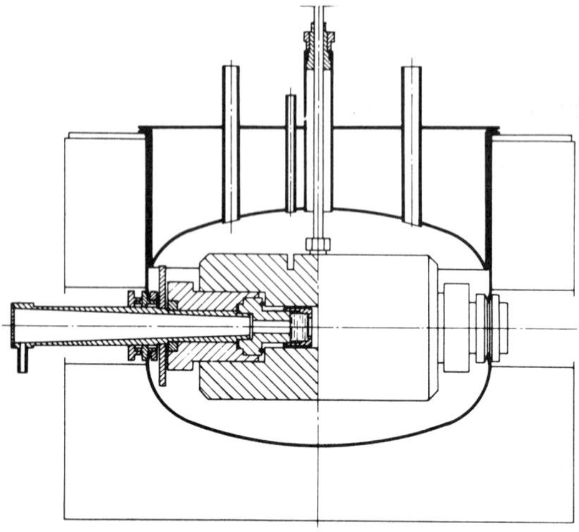

Fig. 1. Conventional pressure vessel for 2 kbar and 500 to 77 K [2].

From 2 to 10 kbar the cell must be a little more sophisticated. usually requiring a double wall with a hardened steel outer wall. The diameter of the light beam must be reduced (4 to 8 mm for the diameter of the unsupported area); otherwise the thickness of the window has to be very large and the dimensions of the cell increasingly large. We have built a cell for 10 kbar with the inner wall constructed with BeCu with conical sapphire windows having a diameter of 8 mm for the unsupported area.

Temperatures in the range of 500 to 4.2 K can be obtained easily by plunging the cell into a thermal or cryogenic bath contained in a thermostat-cryostat. Usually the large mass of the pressure vessel compared to the sample in the working chamber provides sufficient thermal inertia to stabilize the temperature of the sample for hours. For extreme temperature ranges more sophisticated devices must be designed. In our laboratory a pressure vessel with an internal furnace and an outer water-cooling jacket has worked successfully around 1700°C and 3 kbar [4]. At the low end of the temperature scale the range down to 1.5 K can be obtained with pumped liquid He^4 in a conventional cryostat. Even lower limits are technically possible but need expensive equipment: a dilution He^3-He^4 refrigerator for temperatures near 0.01 K or an adiabatic demagnetization device. It is obvious that the overall dimensions and the total mass of the cell are crucial factors in the low temperature potential of the apparatus.

For pressures above 10 kbar there are several different devices each presenting advantages and disadvantages; the choice is usually dictated by the physical problem being studied. The first one is a piston-cylinder device in which the body of the cell is a cylinder provided with 2, 3, or 4 windows. The sample contained in the cylinder is compressed directly by a gasketed piston activated by a hydraulic press which is at room temperature. Several versions of this device have been built: (1) Figure 2 gives the plan of an apparatus designed in our laboratory for IR and R studies of solid H_2 [5] and other molecular solids up to 15 kbar and temperatures down to 2 K. The cylinder is constructed of BeCu and also with maraging steel in another version. Sapphire windows are used with the Poulter device. The BeCu body has worked safely up to 15 kbar while the maraging steel body has worked up to 20 kbar but has cracked at 15 kbar after 2 years of use. The pressure limit was imposed by the low resilience of the materials at 4.2 K and also by the relatively large diameter (4 mm) of the unsupported area of the window.

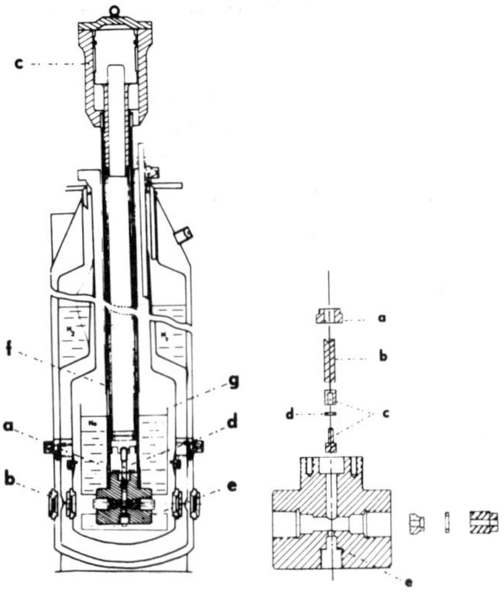

Fig. 2. Optical cell for 10 to 15 kbar at 77 to 2 K using the piston cylinder device (for IR, far IR and R spectroscopy [5].

(2) Figure 3 shows the schematic of the well-known Drickamer cell [6] with its NaCl windows cleverly designed for very high pressure: a sequence of 4 windows with increasing diameters: the inner one having a diameter of 0.7 mm and the outer one a diameter of 6.5 mm. This cell has worked up to 60 kbar and even up to 200 kbar and 700 K (with an internal furnace) with a tapered piston which acts as a multiplier in a double-stage device. This apparatus has been reproduced with some variations by Sherman [7] for low temperatures down to 77 K and by Klyuev [8] for IR studies with a larger hole for the light beam. The disadvantage of this device is that it requires sophisticated technology and an expert staff for manipulations. The piston, which is not completely supported, suffers from very frequent fracture, especially at low temperature;

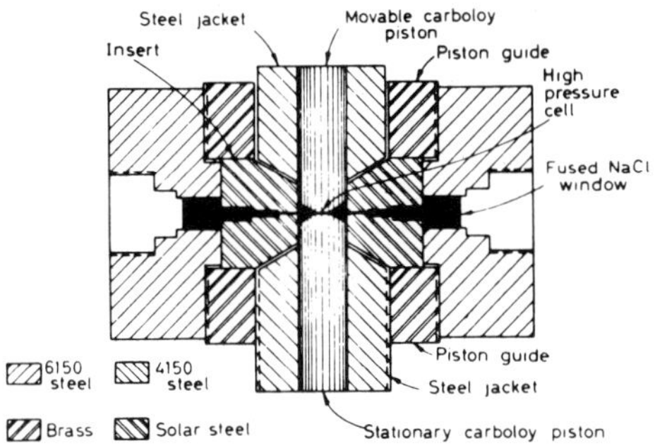

Fig. 3. Drickamer optical cell for 60 to 200 kbar [6].

but this inconvenience is general for any piston-cylinder device. (3) Another version has been proposed by Wong and Whalley [9] in which a diamond is used as a transparent cell body while two opposite pistons compress the sample. This cell has been reported to work up to 8.2 kbar. The disadvantage of this system is a probable low limit for the pressure and also the risk of fracture of the diamond when the piston fails. This very compact version is easily heated or cooled and seems to be useful for scattering work, especially for experiments at moderate pressures allowing the use of a separated compressor instead of an internal piston.

A second device, based on the Bridgman double-anvil system, was proposed by Weir et al. [10] in 1959. Two opposite anvils made with diamond or sapphire or other hard crystal serve both as the pistons transmitting the compression strength and as windows through which the light path is focused. Solid samples can be directly compressed between the anvils. For liquids or solutions a gasket made of an appropriate material is necessary and in this case the sample is a very thin circular disc, 0.5 mm in diameter and a few microns thick. Pressures of about 100 kbar are routinely reached and in some improved versions pressures as high as 500 kbar have been reported [11]. The compression strength is performed either by means of a calibrated screw as in the original version, or by means of a hydraulic ram. The diamond alignment is crucial in very high pressure performance: this can be improved by mounting one diamond in a hemisphere while the other diamond is mounted on a plate which is translationally positioned for axial alignment by a screw adjustment. A microscope is useful for manipulations, while a beam condenser is necessary to focus the light beam through the small optical area without too much intensity loss. This device has several advantages and also some disadvantages compared to the piston-cylinder device: (1) The apparatus is very simple and can be handled by nontechnical personnel. Since it is very compact and small, it can be easily heated or cooled and allows very high pressures without necessitating high strength or sophisticated equipment: 300 kbar and 1000°C by Chien Min Sung [12]; 100 kbar and 0.03 K by Webb et al. [13]. (2) The small dimensions of the working chamber, however, may be a handicap in some experiments, especially since the thickness is difficult to control and to measure; for instance, it seems to be inappropriate for quantitative measurements of absorption by low-absorbance substances. Furthermore a pressure gradient exists across the diameter of the sample [14] although hydrostatic pressures up to about 104 kbar should be achievable in gasketed systems.

Another device based on the hexahedral anvil system has been adapted by Bradley et al. [15] for far IR spectroscopy. The light beam travels along the axis of two opposite anvils through conical holes with 2.5 mm as the smallest diameter. The working chamber is in a cylinder of aluminum terminated by two quartz windows. This cell has been used in the study of far IR spectra of chlorobenzene and of carbon disulfide up to 35 kbar at 500 K. It seems that this apparatus has not been exploited further. The device probably suffers from requiring complicated high-pressure equipment and expertise for its use. It seems to be appropriate for studies of thick samples (0.5 to 1.5 mm thick) as is the piston-cylinder device, although somewhat less practical.

The fourth device is the ballistic compressor [16]. It consists of a long tube containing the gas to be studied at low pressure (< 1 bar), provided at one end with optical windows for spectroscopic observations and at the other end with a locked gasketed piston. The piston is backed by a chamber filled with helium at a pressure of about 10 bar. When the locking mechanism is released, the helium pressure accelerates the free piston along the tube. At the end of the piston stroke the pressure in the reduced working chamber can attain, during several milliseconds, values as high as 10 kbar with temperatures in the range of 10,000 K. So far this device has been used for studies of emission spectroscopy but not in molecular spectroscopy, probably because of the transient nature of the pressure and temperature, preventing equilibrium conditions. Nevertheless the apparatus has proved to be useful for specialized studies of transient phenomena where both high pressure and high temperature are needed. An extreme limit of this device is the shock wave cell [17] which utilizes high explosive to obtain, during a few microseconds, pressures in the megabar range and temperatures higher than 20,000 K.

EXPERIMENTAL ASPECTS

In the following we will try to summarize the principal experimental and theoretical advances in the field of high pressure molecular spectroscopy without attempting to go into details. We will restrict ourselves to very simple molecules, namely monoatomic and diatomic ones for which detailed quantitative experimental results have been obtained and given rise to theoretical investigations.

Rare Gas Atoms

Rare gas atoms have a spherical symmetry and constitute the simplest interaction system. They have no dipole moment nor quadrupole moment; thus the interactions are reduced to the simplest form. They are essentially isotropic and well represented by the 6-12 Lennard Jones potential with overlap forces in the repulsive part and dispersion interactions in the attractive part.

Even under pressure the pure gas has no IR absorption, but in binary mixtures Fig. 4 [18] shows that there is a broad induced absorption spectrum located in the far IR, with intensity obeying the following law:

$$\alpha \equiv \int_0^\infty A(\nu)\, d\nu = \alpha_1 \rho_a \rho_b + \alpha_{2a} \rho_a^3 + \alpha_{2b} \rho_b^3 + \alpha'_{2a} \rho_a^2 \rho_b + \alpha'_{2b} \rho_a \rho_b^2 + \dots \quad (1)$$

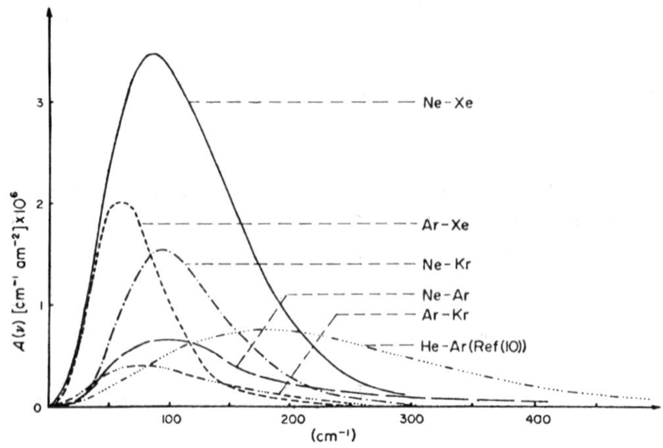

Fig. 4. Pure translational band induced in binary mixtures of rare gases [18] in ordinate: $A(\nu)$ the absorption coefficient.

where $A(\nu)$ is the absorption coefficient and ρ is the amagat density. This formula clearly shows that the first term is induced by binary collisions of unlike atoms, while the four following terms come from triple collisions and so on. These induced bands were in part observed for the first time by Kiss et al. in 1959 [19]. The complete band has been obtained by Bosomworth and Gush [20]. Later Marteau et al. [18], in our laboratory, made a systematic study of the five rare gases Ne, Kr, Xe, Ar, and He for densities up to 600 amagat.

According to the theory of Poll and Van Kranendonk [21], this pressure effect is due to a dipole moment induced primarily by overlap forces between two unlike colliding atoms (for the binary term). This dipole moment depends upon the intermolecular distance and therefore is modulated by the relative translational motion of the collision pair. It is called the pure translational band. Other authors, Sears [22], Levine and Birnbaum [23], and Marteau and Schuller [24] have contributed to the derivation of an analytic expression for the absorption profile.

An interesting property of the translational band is the existence of an intercollisional interference effect which is a correlation effect between successive collisions introducing a diminution of the absorption intensity preferentially at the low frequency tail of the band. This effect, predicted by Poll et al. [21] has been observed by Marteau et al. [25]. It arises from the fact that at increasing densities a colliding atom tends to go back and forth along the same direction, so that the induced dipole moment tends to be aligned along this direction while changing sign alternately. This oscillation causes a cancellation effect which diminishes the net amplitude of the induced moment. In other words, this intercollisional correlation effect introduces, for long periods, a negative part into the auto-correlation function of the dipole and this negative part gives rise to a depression at low frequencies of the Fourier transform of the correlation function, that is the absorption profile.

Obriot et al. [26] have shown that these translational bands become the phonon spectra of the solid rare gases when the mixtures are solidified at low temperatures (77°K). The absorption profile is then quite similar to the phonon spectrum calculated by Mannheim and Cohen [27] for the bcc and fcc systems at 0 K (see other studies [28]). One can conclude that the induced spectra in rare gas solid mixtures are able to give useful information about the phonon spectrum which so far is given only by neutron spectroscopy.

In R scattering, the equivalent of the induced translational band can be observed in pure gases in the depolarized part of the inelastic diffusion spectrum. It extends beyond 50 cm^{-1} on each side of the Rayleigh line and results from a second-order effect on α (the polarizability), the first-order effect being responsible for the polarized part. These R spectra have been studied by Mc Tague and Birnbaum [29], Thibeau et al. [30] at this laboratory, Lallemand [31], Gelbart [32], and more recently by Fromhold [33]. A detailed review of this effect has been given by Gelbart [32] who also gives a general statistical and quantum mechanical treatment of the phenomenon. In the solid phase of pure rare gases the R spectrum is forbidden because of the cubic structure with only one atom per unit cell, as predicted by the theoretical studies of Werthamer et al. [34]. Only the Brillouin and the second order R spectra are allowed. The latter have been observed by Fleury et al. for Ar, Kr, and Xe [35] and the observed spectra are in good agreement with those calculated by Werthamer et al. [34]. Among the rare gases, solid He plays a particular role; it can exist under two crystalline phases: the bcc phase and the hcp phase with two atoms per unit cell. Therefore, the hcp phase displays a sharp first-order Raman spectrum alongside the broad second-order band. These spectra have been studied under atmospheric pressure by Slusher et al. [36] for both He^3 and He^4 and the experimental results confirm the theory of Werthamer et al. [37].

Although they could be very useful, high-pressure experiments on solid rare gases have never been attempted; the very weak intensity of the absorption spectra or of the scattering spectra requires very thick samples and therefore the technical problem becomes very difficult at high pressures.

Homopolar Diatomic Molecules

The interaction potential is relatively well known. For the long range attractive part, the isotropic contribution is relatively well represented by the Lennard Jones R^{-6} term while the anisotropic contribution is essentially due to quadrupolar interaction. The short range repulsive part is not as well known: the isotropic contribution is usually assumed to be represented by the Lennard-Jones R^{-12} term while the anisotropic contribution which is not known at all is usually disregarded in theoretical considerations.

The IR pressure-induced spectra show spectacular effects, giving interesting information about the interactions. In the non-perturbed state H_2 has no IR absorption. But under pressure, for both the case of pure H_2 and H_2 in solutions in a foreign gas, an induced dipole moment gives rise to the absorption spectra in the pure translational region (far IR), the pure rotational region (mean IR), and the vibration-rotational region (near IR). This dipole moment is, of course, dependent on the intermolecular distance and therefore the absorption profile shows prominent translational features. This effect was first discovered by Welsh and coworkers [38] and then extensively studied in their laboratory and also in our laboratory [39]. A detailed theory of the effect has been carried out by Van Kranendonk [40]. The principal characteristics of this effect can be summarized as follows:

The variation of the absorption intensity as a function of the density follows a formula similar to the one previously given for rare gases where one can recognize the contributions of binary collisions and of multiple collisions.

If the translational transitions are not quantized, the rotational and the vibration-rotational transitions obey a selection rule of the Raman type $\Delta J = 0, \pm 2$. Figure 5 [39] shows the vibration-rotation band of H_2 perturbed by Ar on which one can recognize the Q branch ($\Delta J = 0$) and the S(0) and S(1) branches corresponding respectively to the $(v = o, J = o) \rightarrow (v = 1, J = 2)$ transition and the $(v = 0, J = 1) \rightarrow (v = 1, J = 3)$ transition.

The separation of the interaction potential into two contributions arising respectively from overlap forces and quadrupolar interactions is well reflected in the induced dipole and reproduced on the absorption profile, which shows well separated overlap and quadrupolar structures. In Fig. 5, particularly in the profile of the Q transition, one can observe that each transition shows a threefold structure: the central part at the frequency of the vibration-rotation transition is due to quadrupolar induction, while the two others, much broader and expanding on each side of the central line, are due to overlap induction. The shape of the quadrupolar line is similar to that of the pure rotational lines observed in the mean IR region. In fact the rotational band is due essentially to quadrupolar interaction, the anisotropic part of the overlap interaction being negligible, and therefore their width is not very sensitive to the pressure. The shape of the two overlap structures on the contrary is very dependent on the pressure and resembles the pure translational band which has also been interpreted as due essentially to overlap induction. These two structures can be considered as being satellite bands of the Q transitions arising from coupling with translational motion.

The intercollisional interference effect exists here also and is responsible for the depression observed in the low frequency tail of the pure translational band, and for the splitting, in the

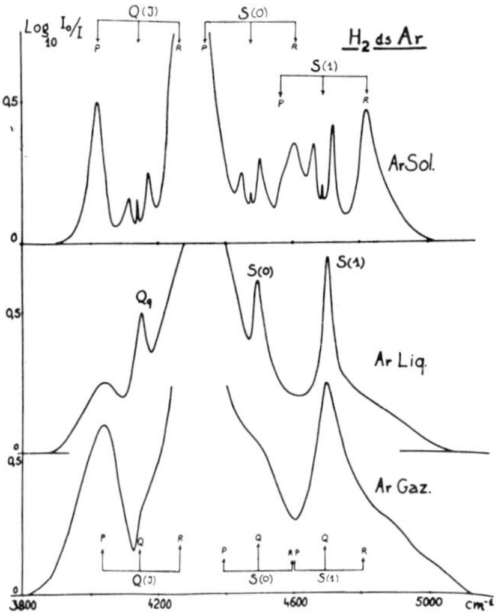

Fig. 5. Pressure induced vibration-rotational band of H_2 in solution in Ar. Lower curve: gaseous Ar (100 bar); middle curve: liquid Ar (100 bar); upper curve: solid Ar (1.7 kbar, 77 K) [39].

vibration-rotation band, of the overlap band into two structures.

In the solid phase of argon (upper curve of Fig. 5), the satellite overlap structures become phonon structure with localized modes of H_2 molecules in the argon lattice. On the other hand, the quadrupolar line collapses suffering from a "cancellation effect" resulting from the symmetry of the argon environment around the H_2 molecule.

Another characteristic of this induction effect is the existence of simultaneous transitions in which two neighboring H_2 molecules each simultaneously undergoing a transition while sharing one single photon, the frequency of which is the sum of the frequencies of the two individual transitions. In other words, the quadrupole moments induce a dipole which is common to the two H_2 molecules and is modulated by the motion of both molecules and therefore at a frequency equal to the sum of the two frequencies. Figure 6 shows the simultaneous band observed at this laboratory for a solid mixture $H_2 + D_2$ [41]. This effect is quite general and multiple combinations between translational, rotational, and vibrational motion have been observed [1,2,38,41]. We shall see later that it exists also in mixtures of polar molecules. Van Kranendonk [40] has shown that these double transitions do not suffer from the "cancellation effect" which is effective only for single transitions and therefore in the overtone spectral region the simultaneous band by far dominates the actual overtone band.

Raman scattering of H_2 and D_2 molecules has also been studied extensively up to 2000 bar by the Toronto group [42]. These spectra are not forbidden under normal conditions and the pressure effects are less spectacular than in IR absorption. They consist chiefly in a broadening and frequency shift of the rotational lines and of the Q branch. Fiutak and Van Kranendonk [43] and later Gray and Van Kranendonk [44] have explained the broadening effect by a theory based on the Anderson theory while the frequency shift has been studied theoretically by May and Poll [45].

The solid phase of H_2 and D_2 molecules has been investigated in detail under various experimental conditions: in IR absorption

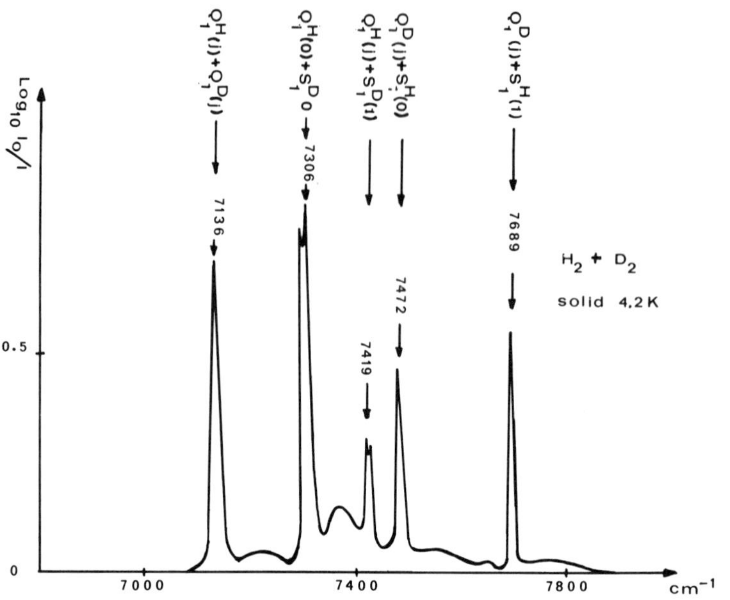

Fig. 6. Simultaneous transition in a solid mixture (50%) of $H_2 + D_2$ at 4.2 K [4].

[46] and in R scattering [47], at low temperatures from 11 to 1 K, with polycrystalline and monocrystalline samples, and at different ortho-para concentrations. In the solid phase, the translational band becomes the phonon spectrum, as we have seen previously, while the quadrupolar features (Q_q branch and pure rotational bands) split into several components under the effect of the crystalline field giving interesting information about the reorientational motion and the vibrons and the librons traveling in the crystal. At our laboratory, Jean-Louis et al. [43] have studied the pressure effect up to 15 kbar at 4.2 K (see Fig. 7) for both normal and pure para-H_2 and for H_2 in solution in argon. Their results show that at 15 kbar the overlap interaction is predominant and that the quantum character of the solids diminishes with increasing densities. Libron and phonon spectra have also been studied both in IR absorption [49] and in R scattering [37,50] for hcp and fcc phases but mainly at normal pressures. They show that this spectral region is quite interesting for the study of phase structure and lattice dynamics. They have shown also that the space group of the fcc phase is still questionable for solid H_2.

Polar Diatomic Molecules

Four molecules, HCl, HBr, HF, and CO, have been extensively studied in our laboratory by Vu and co-workers, but chiefly in near IR absorption under various experimental conditions: under pressures up to 1.8 kbar and recently up to 8 kbar, at different temperatures ranging from 500 to 4.2 K, either in pure substances or in dilute solutions in inert simple gases (N_2, H_2, and Ar), and in the gaseous, liquid, and solid states [1,2,3,51,52,53, and 54]. Other groups [55] have particularly studied liquid solutions in complex or organic solvents, both in IR absorption and in R scattering.

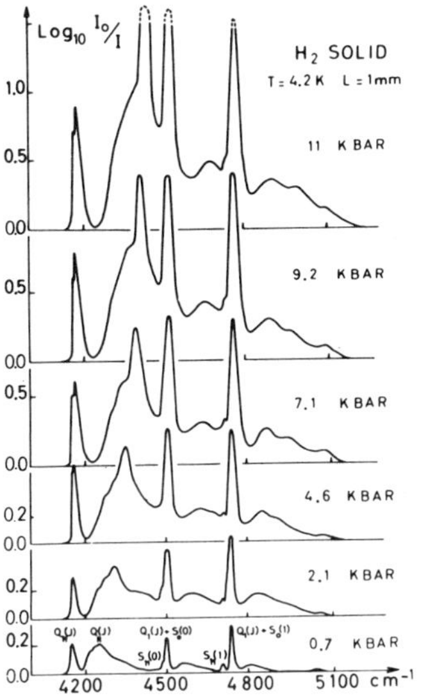

Fig. 7. Effect of pressure on the fundamental virbration-rotation band of pure solid H_2 [48].

Several theoretical semiclassical approaches exist to interpret the results. The most rigorous is the theory given by the Besancon group (Robert, Galatry, Bonamy, and Girardet) [56] in which the vibration and the rotational motion are treated quantum-mechanically. Gordon [57] has proposed another theory in which the rotational motion is treated classically based on the correlation function formalism. This approach also gives satisfactory results and interesting information about the molecular dynamics. The theory proposed by Bratos et al. [58] uses the stochastic process approach, also with classical approximation for rotational motion. It can be considered as an intermediate between the two former ones. Less sophisticated, it has the advantages of a simple theory easily applicable to IR and R spectra for comparison with experimental results. It has proven to be satisfactory for hydracid molecules in solution in liquid complex solvent. We summarize as follows the principal results obtained:

These polar molecules have allowed spectra in the non-perturbed state. At low perturbing densities the pressure effect consists only in the broadening and frequency shift of the rotational and vibrational structures [59]. In the case of spherical perturbers, these effects can be accounted for within the frame of the impact Anderson theory by the dispersion terms (in R^{-7}) in the isotropic part of the interaction potential [60]. Giraud et al. [61] have shown that the non-rigidity of the molecule also plays an important role through the anharmonicity of the molecular vibrations. For diatomic perturbers the situation is more complex and the broadening effect is related through the vibration-rotation coupling to the anisotropic term of the long range part of the potential [61].

At higher perturbing densities drastic changes can occur in the absorption profile depending upon the nature of the perturber and the interaction strength. Figure 8 shows the actions on the vibration-rotational fundamental of HCl for several perturbers. Essentially, a change in the selection rules introduces new transitions of the Raman type $\Delta J = 0 \pm 2$, ... For Ar as the perturber for example, one can observe in the perturbed profile a weak Q branch between the P and R branches. For N_2 as the perturber, the Q branch becomes predominant while the P and R branches collapse. In the liquid and solid phases of a N_2 perturber the P and R branches almost completely disappear. This pressure effect is very sensitive

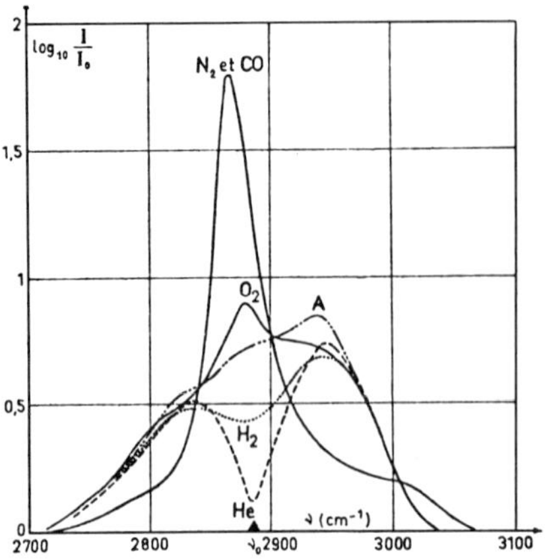

Fig. 8. Fundamental vibration-rotation band of HCl perturbed by He, H_2, O_2, N_2, or CO (400 amagat, 200 K) [2].

to the temperature: at a constant density of N_2 the Q branch can disappear again, giving rise again to the P and R branches when the temperature is increased up to 500 K [2,51].

This effect is now fully explained from a theoretical viewpoint. According to the theory [56,57], the induction of the Q branch is mainly due to that part of the anisotropic (angular) long range interaction potential which depends upon cos θ, θ being the angle between the internuclear axis of the HCl molecule and the intermolecular axis joining the centers of mass of the collision pair HCl-perturber. For N_2 perturber such a term exists already in the well-known electrostatic angular potential arising mainly from the dipole-quadrupole interactions, but for Ar perturber such a term is unknown so far and has been only recently introduced [62]. It arises from the fact that the center of the interactions does not coincide with the center of mass and its discovery can be considered as a useful contribution from molecular high-pressure spectroscopy. It gives rise not only to the induction of the Q transitions but also to an "intensity shift", shifting the intensity from one rotational line to another and producing the collapse, at high densities, of the P and R branches. Figure 9 gives a comparison between the "bar spectrum" calculated by

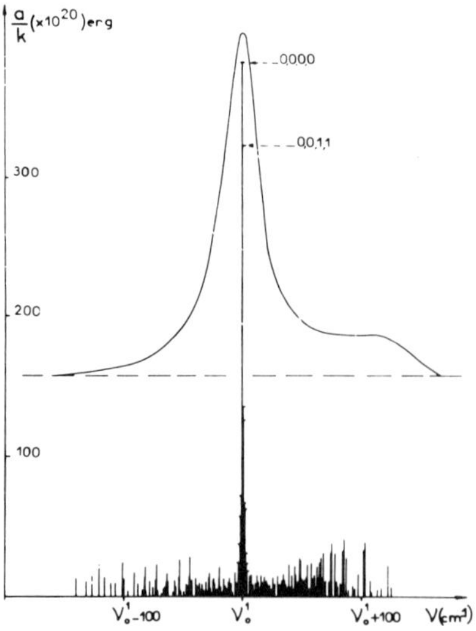

Fig. 9. Comparison between experimental profile [2] and calculated "bar spectrum" [56] for HCl in gaseous N_2 (see Fig. 8).

Girardet et al. [56] and the experimental profile observed for the HCl + N_2 mixture. It is worth mentioning that the agreement is fairly good.

One can also observe simultaneous transitions in a mixture of two polar molecules or a "polar + nonpolar" mixture in either the gaseous, liquid, or solid phase [51,53]. The induction mechanism is similar to the one mentioned previously for non-polar mixtures except for the fact that the effect arises now from dipole-dipole or dipole-quadrupole interactions.

The solid state, cubic and orthorhombic phases of pure hydracids have also been investigated under pressure in our laboratory up to 5 kbar at 77 and 4.2 K, mainly in IR absorption, in the fundamental band and in the first and second overtone band [3,53,54]. Raman and far IR spectra have also been studied by other groups [63] but without high pressure and only in the fundamental band. These results show that the anharmonicity constant changes considerably in the solid phase and that the spectrum profile splits into several components reflecting the lattice structure and the effects of the crystalline field. The combination of R and IR results appears to be necessary in helping to determine the crystalline structure of hydracids, although this problem is still open to question and needs further studies to be resolved. Simultaneous transitions also play an important role in the overtone band, at least for the HCl molecule [53].

Polymer spectra have also been studied [64]. In our laboratory [65] these spectra are obtained by rapidly cooling under pressure (1.7 kbar) a dilute solution of HCl molecules in Ar, from 100 K, a temperature where the solution is homogenous, down to 77 K. At this temperature, diffusion of HCl solute molecules occurs giving rise to the formation of dimers, trimers, etc. The absorption spectra then undergoes evolutions as a function of time from the monomer spectra and the dimer spectra up to the polymer spectra. By controlling the concentration of the original solution, one can stop the polymerization process upon dimer formation, trimer formation, other formation, or finally at the stage of large aggregates or crystallites of pure HCl embedded in an argon lattice (see Fig. 10).

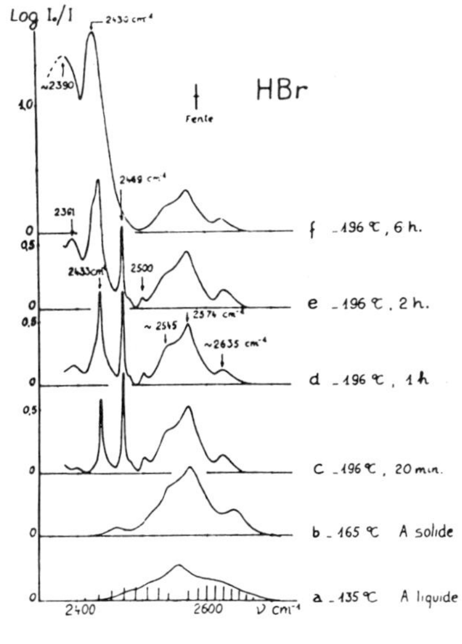

Fig. 10. Polymerization spectra of HBr in solid Ar [65]. Legend: (a) Homogeneous solution of HBr in liquid Ar at 140 K; (b) Homogeneous solution of HBr in solid Ar at 110 K; and (c), (d), (e), (f): Evolution of the profile as a function of time up to 6 hrs (1.7 kbar and 77 K).

The theory of this effect has been formulated by Girardet et al. [66] and given a qualitative agreement with the observed spectra. For more quantitative comparison, further experimental results especially in the far IR spectral region are needed. Such a study up to 6 kbar at 4.2 K is presently in progress in our laboratory.

Another interesting result is the discovery of Van der Waals molecules in a gaseous system at low temperature which earlier had been predicted theoretically by Stogryn and Hirschfelder [67]. The spectroscopic evidence has been given for the first time by Vu in mixtures of HCl and foreign gases [2,51] (see Fig. 11). Welsh and co-workers [68] have obtained well resolved rotational spectra of H_2-H_2, H_2-Ar ... molecules in the near IR. Fromhold [33] has studied the contribution of Ar dimers in the R translational induced spectra which is nevertheless not important (10% at 300 K) and not apparent. More recently, Prengel and Gornall [69] have succeeded in observing this contribution (unresolved dimer rotational band) for CH_4 at 115 K (43%).

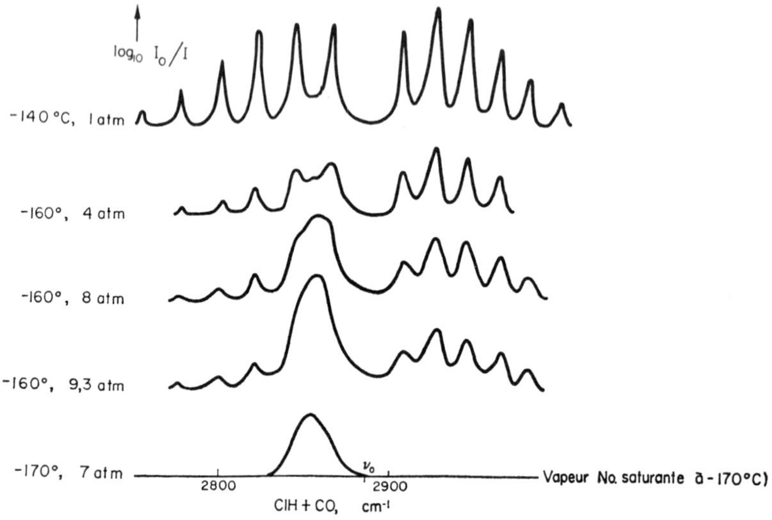

Fig. 11. Near IR spectra of Van der Waals molecules HCl-CO obtained at relatively low (7 atm) pressure and low temperature (100 K) [51].

CONCLUSION

We have given only a very brief review of the principal results obtained in the field of molecular interaction spectroscopy. Numerous omissions have been necessary because of space limitations. Nevertheless, we hope to have demonstrated the useful contribution of high-pressure spectroscopy to the study of molecular interactions. We have seen that although high-pressure technology has made considerable progress and is exploring the range of 50 to 500 kbar, quantitative spectroscopic investigations have not yet gone beyond the range of 10 to 15 kbar. We hope that in the near future technological improvements will help to explore the next high-pressure range where unknown overlap interactions predominate, and where new phase transitions and significant changes in spectroscopic properties can be reasonably expected.

REFERENCES

1. R. Coulon, J. Rech. CNRS 9, 305 (1958).
2. H. Vu, J. Rech. CNRS 11, 313 (1960).
3. P. Khatibi and H. Vu, J. Chim. Phys. 69, 654 (1972).
4. R. Granier, Ann. Phys. 4, 383 (1969).
5. M. Jean-Louis and H. Vu, J. Phys. Appl. 7, 89 (1972).
M. M. Thiery, D. Fabre, M. Jean-Louis, and H. Vu, J. Chem. Phys. 59, 4559 (1973).
6. A. Fitch, T. E. Slykhouse, and H. G. Drickamer, JOSA 47, 1015 (1957).
7. W. F. Sherman, J.S.I. 43, 462 (1966).
8. Yu A. Klyuev, Pribory Tekh Eksper. SSSR 9, 164 (1964); Translation 9, 174 (1964).
9. P.T.T. Wong and E. Whalley, Rev. Sci. Instr. 45, 904 (1974).
10. C. E. Weir, E. R. Lippincott, A. Van Valkenburg, and E. N. Bunting, J. Res. NBS A63, 55 (1959).
11. G. J. Piermarini and S. Block, Rev. Sci. Instr. 46, 973 (1975).
12. C. M. Sung, Rev. Sci. Instr. 47, 1343 (1976).
13. A. W. Webb, D. U. Gubser, and L. C. Towle, Rev. Sci. Instr. 47, 59 (1976).
14. G. I. Piermarini, S. Block, and J. D. Barnett, J. Appl. Phys. 44, 5377 (1973).
15. C. C. Bradley, H. A. Gebbie, A. C. Gilby, V. V. Kechin, and J. H. King, Nature 211, 839 (1966).
16. G. T. Lalos, Rev. Sci. Instr. 33, 214 (1962).
17. S. D. Hamann, in Advance in High Pressure Research, Vol. 1, R. S. Bradley, ed., Academic Press, New York (1966), p. 85; E. F. Green and J. P. Toennies, Chemical Reactions in Shock Waves, Arnold, London (1964).
18. P. Marteau, H. Vu, and B. Vodar, JQSRT 10, 283 (1970).
J. Quazza, P. Marteau, H. Vu, and B. Vodar, JQSRT 16, 491 (1976).
19. Z. J. Kiss and H. L. Welsh, Phys. Rev. Lett. 2, 166 (1959).
20. D. R. Bosomworth and H. P. Gush, Can. J. Phys. 42, 751 (1965).
21. J. D. Poll and J. Van Kranendonk, Can. J. Phys. 39, 189 (1961).
22. V. T. Sears, Can. J. Phys. 46, 1163 (1968).
23. H. B. Levine and G. Birnbaum, Phys. Rev. 154, 86 (1967).
24. P. Marteau and F. Schuller, J. de Phys. 33, 645 (1972).
25. P. Marteau, H. Vu, and B. Vodar, C. R. Acad. Sc. Paris 266B, 1068 (1968).
26. J. Obriot, P. Marteau, H. Vu, and B. Vodar, Spectrochim. Acta 26A, 2051 (1970).
27. Ph. D. Mannheim and S. S. Cohen, Phys. Rev. B4, 3748 (1971).
28. G. O. Jones and J. M. Woodfine, Proc. Phys. Soc. London 86, 101 (1965).
J. Grindlay and R. Howard, in Proc. of Intern. Conf. on Lattice Dynamics, R. F. Wallis, ed., Pergamon, New York (1965), p. 129.
29. J. P. McTague and G. Birnbaum, Phys. Rev. A3, 1376 (1971).

30. M. Thibeau, B. Oksengorn, and B. Vodar, J. Phys. Rad. 29, 287 (1968); also 30, 47 (1969); Molec. Phys. 15, 579 (1968); JQSRT 10, 839 (1970).
31. P. Lallemand, Phys. Rev. Lett. 25, 1079 (1970); also J. Phys. Rad. 32, 119 (1971).
32. W. M. Gelbart, in Adv. Chem. Phys., Vol. 26, I. Prigogine and S. A. Rice, eds., John Wiley and Sons (1974), p. 1.
33. L. Fromhold, J. Chem. Phys. 61, 2996 (1974).
34. N. R. Werthamer, R. L. Gray, and T. R. Koehler, Phys. Rev. B2, 4199 (1970).
35. P. A. Fleury, J. M. Worlock, and H. L. Carter, Phys. Rev. Lett. 30, 591 (1973).
36. R. E. Slusher and C. M. Surko, Phys. Rev. Lett. 27, 1699 (1971); also Phys. Rev. B13, 1086 (1976).
C. M. Surko and R. E. Slusher, Phys. Rev. B13, 1095 (1976).
37. N. R. Werthamer, Phys. Rev. 185, 348 (1969).
N. R. Werthamer, R. L. Gray, and T. R. Koehler, Phys. Rev. B4, 1324 (1971).
38. M. F. Crawford, H. L. Welsh, and J. L. Locke, Phys. Rev. 76, 580 (1949).
D. A. Chisholm and H. L. Welsh, Can. J. Phys. 32, 291 (1954).
W.F.J. Hare and H. L. Welsh, Can. J. Phys. 35, 88 (1957).
H. L. Welsh, M.T.P. International Review of Science, Physical Chemistry, Spectroscopy, Vol. 3, Butterworth, London (1972), p. 33.
39. H. Vu, M. R. Atwood, and E. Staude, C. R. Acad. Sc. Paris 257, 1771 (1963); also H. Vu, J. Phys. 25, 741 (1964).
40. J. Van Kranendonk, Physica 23, 825 (1957); also 24, 347 (1958); 25, 337 (1959); 25, 1080 (1959); Can. J. Phys. 38, 240 (1960).
41. M. Jean-Louis, M. Bahreini, and H. Vu, C. R. Acad. Sc. Paris 268B, 479 (1969); also J. P. Colpa and J.A.A. Ketelaar, Molec. Phys. 1, 14, 343 (1958); R.P.H. Rettschnick, Ph.D. Dissertation, University of Amsterdam, Amsterdam, Netherlands (1962).
42. A. D. May, V. Degen, J. C. Stryland, and H. L. Welsh, Can. J. Phys. 39, 1769 (1961); also A. D. May, G. Varghese, J. C. Stryland, and H. L. Welsh, Can. J. Phys. 42, 1058 (1964).
43. J. Fiutak and J. Van Kranendonk, Can. J. Phys. 40, 1085 (1962); 41, 21 (1963).
44. C. G. Gray and J. Van Kranendonk, Can. J. Phys. 44, 2411 (1966).
45. A. D. May and J. D. Poll, Can. J. Phys. 43, 1836 (1965).
46. H. P. Gush, W.F.J. Hare, E. J. Allin, and H. L. Welsh, Can. J. Phys. 38, 176 (1960).
47. S. S. Bhatnagar, E. J. Allin, and H. L. Welsh, Can. J. Phys. 40, 9 (1962).
E. J. Allin, A. H. McKague, V. Soots, and H. L. Welsh, Can. J. Phys. 26, 615 (1965).
48. M. Jean-Louis, M. M. Thiery, H. Vu, and B. Vodar, J. Chem. Phys. 55, 4657 (1971).
49. W. N. Hardy, I. F. Silvera, K. N. Klump, and O. Schnepp, Phys. Rev. Lett. 32, 291 (1968).

K. N. Klump, O. Schnepp, and L. H. Nosanow, Phys. Rev. B1, 2496 (1970).

50. I. F. Silvera, W. N. Hardy, and J. P. McTague, Phys. Rev. B4, 2724 (1971); also Phys. Rev. B5, 1578 (1972); Disc. Far. Soc. 48, 54 (1969).
W. N. Hardy, I. F. Silvera, and J. P. McTague, Phys. Rev. Lett. 26, 127 (1971).
51. H. Vu, M. R. Atwood, and B. Vodar, J. Chem. Phys. 38, 2671 (1963). H. Vu and B. Vodar, JQSRT 3, 397 (1963).
H. Vu and B. Vodar, Zeitsch f. Elektroc. 64, 756 (1960).
52. M. R. Atwood, H. Vu, and B. Vodar, Spectres Chimica Acta 23A, 553 (1967).
M. R. Atwood, H. Vu, and B. Vodar, J. Phys. 33, 495 (1972).
53. P. Khatibi and H. Vu, J. Chim. Phys, 662 (1972); also 674 (1972).
54. S. Avrillier, S. S. Mitra, and H. Vu, J. Chem. Phys. 64, 2202 (1976).
55. J. Lascombe, P. V. Huong, and M. L. Josien, Bull. Soc. Chim. 1175 (1959); also M. Perrot, P. V. Huong, and J. Lascombe, J. Chim. Phys. 68, 619 (1971); M. O. Bulanin and N. Orlova, Opt. and Spectr. 15, 208 (1963); G. M. Barrow and P. Datta, J. Phys. Chem. 72, 2259 (1968); S. Abramowitz and R. P. Bauman, J. Chem. Phys. 39, 2757 (1963).
56. D. Robert and L. Galatry, J. Chem. Phys. 55, 2347 (1971).
L. Bonamy, D. Robert, and L. Galatry, J. Molec. Struct. 1, 91 (1967); also 1, 139 (1967); G. Girardet, D. Robert, and L. Galatry, J. Chem. Phys. 55, 5304 (1971).
57. W. B. Neilson and R. G. Gordon, J. Chem. Phys. 58, 4131 (1973); also 58, 4149 (1973).
58. S. Bratos, J. Rios, and Y. Guissani, J. Chem. Phys. 52, 439 (1970); also S. Bratos and E. Marechal, Phys. Rev. A4, 1078 (1971).
59. R. A. Toth, R. H. Hunt, and E. K. Plyler, J. Mol. Spectr. 35, 110 (1970); also M. Giraud, and J. Pourcin, C. R. Acad. Sc. Paris 266, 1593 (1968); W. S. Benedict, R. Herman, G. E. Moore, and S. Silverman, Can. J. Phys. 34, 850 (1956); J. P. Bouanich, M. Larvor, and C. Haeusler, C. R. Acad. Sc. 270, 1220 (1970); D. H. Rank, D. P. Eastman, B. S. Rao, and T. A. Wiggins, J. Mol. Spect. 10, 34 (1963); A. Levy, E. Piollet Mariel, and C. Boulet, JQSRT 13, 673 (1973); 13, 897 (1973).
60. P. W. Anderson, Phys. Rev. 76, 647 (1949); also A. Ben Reuven S. Kimel, M. A. Hirschfeld, and J. H. Jaffe, J. Chem. Phys. 35, 955 (1961); F. Schuller and B. Oksengorn, Mol. Phys. 5, 573 (1962); R. M. Herman, Phys. Rev. 132, 262 (1963); R. H. Tipping and R. M. Herman, JQSRT 10, 881 (1970).
61. D. Robert, M. Giraud, and L. Galatry, J. Chem. Phys. 51, 2192 (1969); also M. Giraud, D. Robert, and L. Galatry, J. Chem. Phys. 53, 352 (1970).

62. R. M. Herman, J. Chem. Phys. 44, 1346 (1966); also J. F. Le Men E. Sci. Dissertation, Rennes (1963).
63. M. Ito, M. Suzuki, and T. Yokohama, J. Chem. Phys. 50, 2949 (1969); also R. Savoie and A. Anderson, J. Chem. Phys. 44, 548 (1966); R. E. Carlson and H. B. Freydrich, J. Chem. Phys. 54, 2794 (1971); H. B. Freydrich and R. E. Carlson, J. Chem. Phys. 53, 4441 (1970); T. S. Sun and A. Anderson, Spectrosc. Lett. 4, 377 (1971); R. Savoie and M. Pezelot, J. Chem. Phys. 50, 2781 (1969).
64. H. E. Hallam, J. Mol. Spectrosc. 22, 329 (1968); also Ann. Rept. Chem. Soc. (London) 6, 117 (1970).
65. M. R. Atwood, M. Jean-Louis, and H. Vu, J. Phys. 28, 31 (1967).
66. C. Girardet and D. Robert, J. Chem. Phys. 58, 4110 (1973); also 59, 5020 (1973).
67. D. E. Strogryn and J. O. Hirschfelder, J. Chem. Phys. 31, 1531 (1959).
68. A. K. Kudian, H. L. Welsh, and A. Watanabe, J. Chem. Phys. 43, 3397 (1965); also A.R.W. McKellar and H. L. Welsh, Can. J. Phys. 52, 1082 (1974).
69. A. T. Prengel and W. S. Gornall, Phys. Rev. A13, 253 (1976).

S-9

APPLICATION OF HOLOGRAPHIC INTERFEROMETRY TO HIGH PRESSURE EXPERIMENTS

H. A. Spetzler, I. C. Getting, and R. J. Martin III

University of Colorado
Boulder, Colorado USA

INTRODUCTION

All high-pressure experimentation involves non-hydrostatic stresses. Even when the sample itself is in a hydrostatic environment, the sample vessel, the loading frame etc., are subject to non-hydrostatic stresses. When designing equipment or performing an experiment, it is often important to know the strain field resulting from the applied stresses.

Optical holographic interferometry can be used to measure the deformation of surfaces. The surface deformation which accumulates between two exposures of a hologram is represented as a topographic map with a contour interval of approximately 300 nm. Any part of a test apparatus that is visible, or that can be made visible through the use of mirrors, can, in principle, be examined (quantitatively) for small surface deformations by this technique. Inhomogeneities or concentrations in deformations due to stress concentrations because of faulty design, material flaws, heat concentrations, etc. are readily identified. In this paper we give some examples of the use of holography in high-pressure experimentation.

OPTICAL HOLOGRAPHIC INTERFEROMETRY

Viewing a hologram is optically equivalent to looking through a window (the hologram) at an object (the three-dimensional image) that is illuminated by a laser. The three-dimensionality of the image is due to the different views presented by the object when one is looking through different parts of the window. More than one image can be stored on one hologram, i.e., several scenes can be viewed through the same window (hologram).

When two images of the same object are recorded on one hologram, and the object was strained between the two recordings (exposures), the two images interfere with one another. The interference pattern (fringe pattern) which results can be interpreted in terms of the strain field on the object. Heflinger et al. [1] have devised an optical system which makes this interpretation very simple and useful. A slightly modified optical arrangement [2] has been used in our laboratory and is shown in Fig. 1.

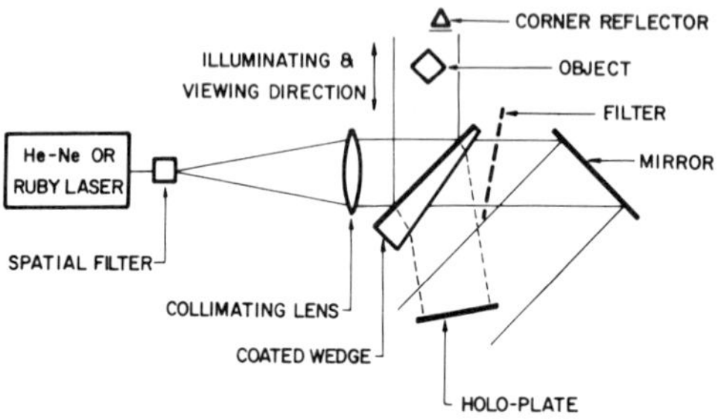

Fig. 1. Optical arrangement which allows for coincident viewing and illuminating directions. The corner reflector aids in achieving coincidence.

Because of the coincident illuminating and viewing directions and the parallelism of the illuminating beam, rigid body translations of the object between exposures do not contribute to the fringe pattern. The fringe pattern represents a topographic map of the surface elevation change in the direction of viewing. The contour interval is equal to one half of the wavelength of the illuminating laser light. The wavelength must be that appropriate to the optical medium in which the change in optical path length occurs (e.g., the high-pressure fluid, air, vacuum, etc.). A rotation of the object between exposures results in straight fringes which are parallel to the axis of rotation. The fringe patterns for some simple deformations and the use of mirrors to simultaneously obtain several views was explored by Meyer et al. [3]. In the following sections we present some examples of the use of this powerful technique.

DESIGN OF EQUIPMENT AND EXPERIMENT

The use of optical holography in high-pressure experimentation is limited only by the experimentalist's imagination. It is finding increasing use not only in static applications but also in dynamic experiments [4,5]. Exposure times to record an image vary from the nanosecond range to many seconds. Here we will give several examples of how we used optical holography in the design and checkout of experiments.

One of the functions of our laboratory is to investigate material properties while a sample is under hydrostatic pressure. We have designed optical windows [6] which allow us to use optical holography to measure strain fields under confining pressure. Figure 2 shows the design of two windows as they are installed in the high-pressure closure. To check their deformation as high pressure was applied, we painted the entire flat surface of the window which was exposed to the high-pressure fluid and painted a cross on the front (opposing) flat surface. At 6.9 MPa (69 bar) intervals we made sets of double exposures and observed the deformation of both surfaces simultaneously. A typical set of fringes is shown on the photograph in Fig. 2. We used finite element analysis in conjunction with the holographic analysis in the final design [6]. At present we have a pressure vessel with an optical window (50-mm-diameter clear view) capable of withstanding more than 300 MPa (3 kbars).

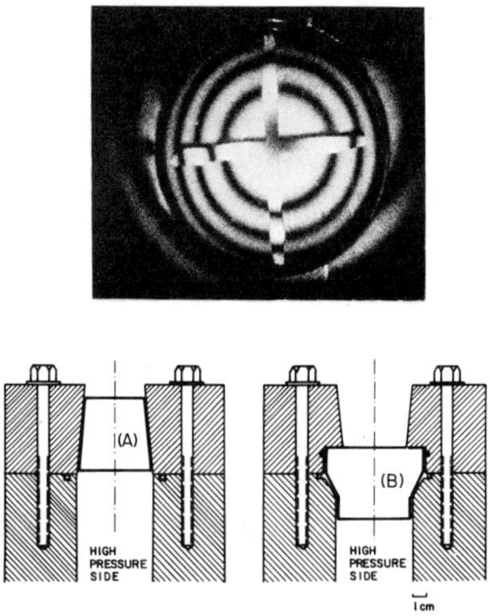

Fig. 2. Two optical window designs are shown. The photo insert is a photograph of a doubly-exposed hologram of the window (design A). The circular fringe patterns resulted from a 6.9 MPa (69 bar) pressure change between the two exposures.

The wavelength of the laser light in the pressure medium must be known if we are to measure strains by optical holography. We placed a parallel-plate interferometer within the pressure vessel and counted the number of fringes as the index of refraction of the fluid between the parallel plates changed in response to the hydrostatic pressure. The index of refraction as a function of pressure for our pressure fluid (Dow Corning Silicone Fluid, 20 centistokes viscosity) is plotted in Fig. 3. For comparison we show the curve obtained by the Lorentz-Lorenz law from the initial index of refraction and pressure dependence of the density.

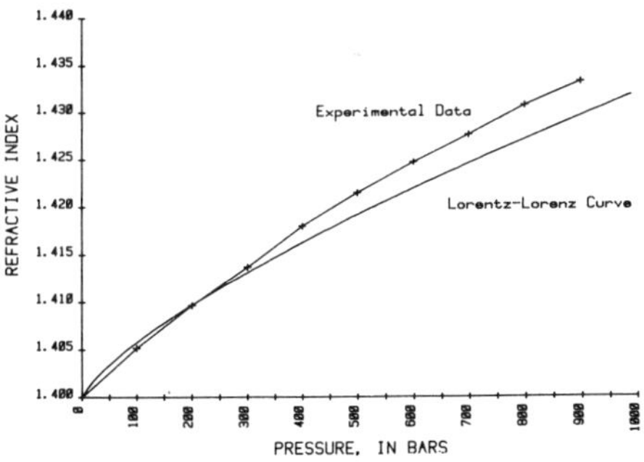

Fig. 3. Refractive index vs. pressure for the pressure fluid (Dow Corning 200 fluid). The manufacturer's compression data (ρ = density) was used to calculate the index n according to the Lorentz-Lorenz law $K\rho = (n^2 - 1)/(n^2 + 2)$. Maximum error in holographic strain measurements introduced by the use of the Lorentz-Lorenz extrapolation is less than 0.5%.

When holography is used to measure a strain field on a sample, it is often convenient to also measure linear average strains by this technique. As an example, let us assume we are observing the uniaxial compression of a sample. The viewing and illuminating directions are coincident as shown in Fig. 1 and perpendicular to the axis of compression. The doubly-exposed holograms give us only the components of strain in the viewing direction and therefore contain no direct information about the amount of strain in the direction of compression. The addition of a leaf spring between the top and bottom piston can be used to convert the strain in the direction of compression to displacement in the viewing direction. Figure 4 illustrates this idea. The buckling of a thin-hinged beam (the leaf spring) may be approximated by a sine curve such that

$$y = a \sin \frac{\pi x}{s} \tag{1}$$

where y is the amplitude of deflection in the viewing direction, s is the length of the sample, and x is the coordinate of the compression axis. The change in the maximum of y (i.e., a) due to a change in the sample length s depends upon the length L of the spring. The sensitivity of displacement conversion for a buckling thin beam and a triangular configuration are shown in Fig. 4 as a function of leaf spring-length to sample-length ratio. In practice it is best to count the total number of fringes that are visible over the entire length of the leaf spring. The change in sample length Δs may then be obtained from

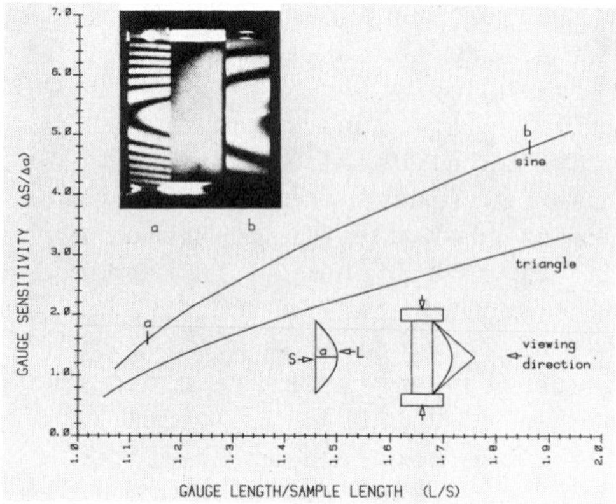

Fig. 4. Gauge sensitivity vs. gauge length to sample length change gauges. Photo insert shows the response of two sinusoidal gauges to a nominal sample shortening of 3 μm. Sample length is 44 mm and gauge-length to sample-length ratios for a and b are 1.13 and 1.86, respectively. Non-parallelism of fringes is the result of a tilt of less than 0.2 μm across the sample.

$$\Delta s = \frac{m\lambda}{4} \left(\frac{\Delta s}{\Delta a}\right) \tag{2}$$

where $(\Delta s/\Delta a)$ is the conversion factor obtained from the graph in Fig. 4, and λ is the optical wavelength in the medium in which the spring is located, i.e., the medium within which the change of optical path length occurs between the two exposures of the holoplate. The above technique is not only applicable in atmosphere and high-pressure environments but is especially advantageous when ultrahigh vacuum and/or temperatures are involved. Since no special materials such as epoxy are needed for mounting the gauges, no contamination of a vacuum should occur nor should high temperatures pose problems.

Another important use of holographic interferometry is illustrated in Fig. 5. A series of photos shows the rotation of (mainly) the top piston and the bending of the sample during the early stages of a supposed uniaxial compression test. It is clear from this series of pictures that the sample had to undergo considerable bending before complete contact with the top piston was achieved.

Fig. 5. Photos of doubly-exposed holograms show behavior of piston and sample during initial stages of a supposedly uniaxial compression test. The sample is Westerly Granite. It has top and bottom steel end pieces attached with epoxy. From top left to bottom right: (1) the sample bends in response to the top piston contacting the top end piece. The top end piece rotates considerably more than the bottom end piece. The rotation is about an axis which is parallel to the fringes; (2) the top piston is slightly distorting and rotating in response to the resistance offered by the sample; (3) the top piston and the top sample end piece are now in intimate contact and rotate about the same axis. (Note the higher fringe density on the top end piece than on the bottom end piece.) The sample is bending to accommodate the difference in the rotation between the top and bottom end pieces; (4), (5), and (6) during later phases of the loading, the nonuniform compression is maintained and bending remains a significant part of the sample deformation.

EXAMPLE OF AN APPLICATION

In rock mechanics we are very interested in any phenomenon that precedes failure of a stressed rock sample. The importance of being able to recognize any precursor is self-evident when one considers the possibility of predicting earthquakes and rockfalls in mines.

We now view brittle failure in rocks to be due to the coalescence and interaction of microcracks. The precursory phenomena should then be related to the formation and distribution of these microcracks. The close observation of the surface deformation (strain) that accompanies the formation of microcracks can give valuable information about their spatial distribution. We have used optical holography to study the deformation of a jacketed (with copper foil) rock sample (Westerly Granite) while it was stressed to failure

[7,8]. It had a square cross-section 17.5 mm on a side by 50 mm long. The sample was subjected to 50 MPa (0.5 kbar) confining pressure with a superimposed axial stress. The latter was increased slowly until failure occurred. An optical window (described earlier) in the pressure vessel enabled us to produce topographic maps (by optical holography) of strain intervals as the sample slowly deformed. The total surface deformation can be obtained by adding the strains measured from consecutive topographic maps.

By observing strain accumulation on the surface which was to contain the diagonal fault trace, we discovered a premonitory bulge. This bulge first became evident at approximately 60% of the final axial stress at failure. As the stress was increased above this value, the bulge became better defined. A strain interval map is shown in Fig. 6 when the cumulative axial strain had reached about 96% of its final value. Figure 7 shows the progressive development of the bulge in relation to the ultimate failure plane.

Fig. 6. Photo of a doubly-exposed hologram shows the growth of the intense deformation zone in a sample of Westerly Granite under axial differential stress. The hologram was made shortly before failure, when the axial strain had reached 96% of its failure value.

The bulge is associated with an intense deformation zone within which the microcracks are concentrated and finally coalesce, leading to brittle failure. In related experiments we have observed the time dependence of the development of the intense deformation zone and associated bulge in polymethylmethacrylate [9] and pyrophyllite [10].

ACKNOWLEDGMENTS

Much of the recent work on optical inferometry has been done by the students currently working in the laboratory: W. W. Bayne, N. S. Brodsky, C. N. Gerlitz, W. K. Heinzman, and P. L. Swanson. Financial support was provided by NASA Grant NAG 9036 and the U.S. Geological Survey Contract 14-08-0001-15256.

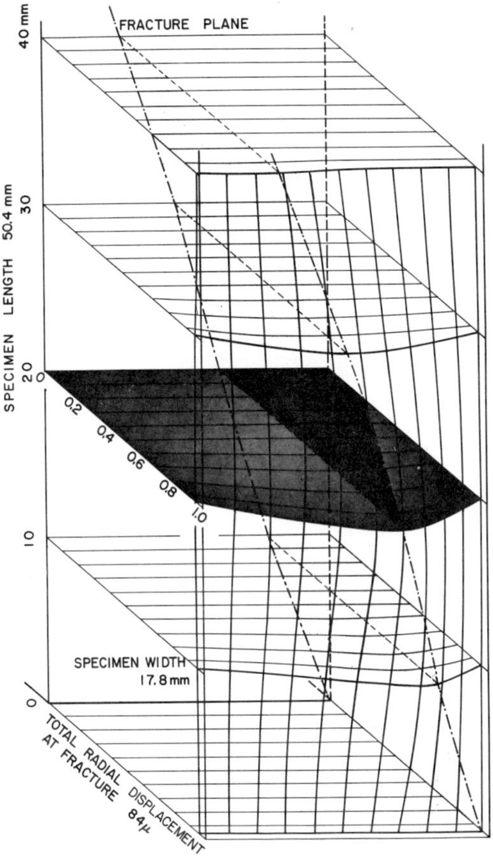

Fig. 7. Composite of the information gathered from a number of consecutive holograms as shown in Fig. 6. The progressive development of the dilatant bulge in relation to the ultimate failure plane is shown. Confining pressure = 0.5 kbar; axial differential stress at fracture = 5.7 kbars.

REFERENCES

1. L. O. Heflinger, H. Spetzler, and R. W. Wuerker, Rev. Sci. Instr. 44, 629 (1973).
2. H. Spetzler, C. H. Scholz, and J. Chi-Ping Lu, Pure and Applied Geophysics 112/3, 573 (1974).
3. M. D. Meyer and H. A. Spetzler, Experimental Mechanics 16, 434 (1976).
4. D. C. Holloway and W. L. Fourney, in Proc. 18th U.S. Symposium on Rock Mechanics, Fun-Den Wang, ed., Colorado School of Mines, 3C4-1, Golden, Colorado (1977).
5. D. E. Munson, private communication.
6. N. Soga, D. Holcomb, and H. Spetzler, Rev. Sci. Instr. 47, 1453 (1976).
7. H. Spetzler, N. Soga, H. Mizutani, and R. J. Martin III, in High Pressure Research, Academic Press, New York (1977), p. 625.
8. N. Soga, H. Mizutani, H. Spetzler, and R. J. Martin, submitted to J. Geophys. Res.
9. H. Spetzler, G. A. Sobolev, N. Soga, and B. G. Salov, National Academy of Sciences, U.S.S.R., Dushanbe, U.S.S.R. (in Russian) (1976).
10. G. Sobolev, H. Spetzler, and B. Salov, J. Geophys. Res. 83, 1775 (1978).

S-10

FOCUSING HIGH PRESSURE X-RAY DIFFRACTION CAMERA

G. Hägg, I. Engström and B. Törmä

University of Uppsala
Uppsala, Sweden

INTRODUCTION

Several techniques have been developed to obtain x-ray powder diffraction data at high pressures. Reviews of different techniques currently utilized have been given by Bradley [1] and McWhan [2]. Amorphous boron is often used as the gasket material between Bridgman anvils of sintered tungsten carbide [3]. The x-ray system can be of either Debye-Scherrer or Guinier geometry [4-7], preferably with MoKα-radiation. Constructions have been described where the whole press frame and the anvil assembly can be oscillated [4], but in general the diffraction equipment is designed for fixed samples.

This paper describes a modified Guinier type high-pressure x-ray powder diffraction camera for work up to somewhat above 10 GPa (100 kbar). The load is uniaxially applied on a pressure gasket of amorphous boron. The main difference between this camera and earlier designs is that the sample and the high-pressure system remain stationary, while the x-ray system can be oscillated around the sample. The camera is also equipped with load-regulating systems, permitting a more exact load application. Some experiences with the apparatus are presented and some test results are discussed and compared with literature data.

CAMERA DESIGN

High Pressure System

Figures 1 and 3 show the high-pressure system. It consists of a hand pump for applying the load to a standard hydraulic ram equipped with an ordinary load cell and Bridgman anvils. It also contains four manually-operated valves for coarse stabilization and two Peltier elements for fine stabilization of the load.

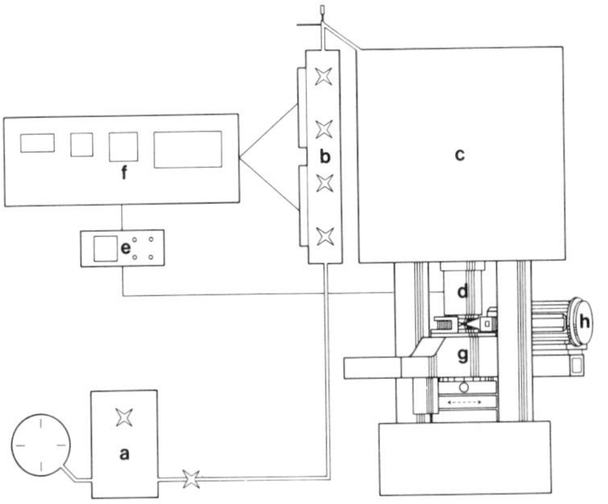

Fig. 1. Schematic diagram of high-pressure x-ray powder diffraction equipment. Legend: (a) Hand pump; (b) coarse and fine stabilizing systems; (c) thermally insulated hydraulic ram; (d) load cell (pressure transducer); (e) load indicator; (f) control system for fine stabilizing; (g) oscillating stage, and (h) x-ray system.

The load cell connects the upper anvil to the hydraulic ram. The output signal of the cell is used to indicate the load and at the same time control the fine stabilizing system for constant load. Pressure variations caused by room temperature fluctuations are reduced by the thermal insulation of the ram. The hydraulic system is also equipped with a manually-operated evacuating valve to remove gas from the system, which further reduces the thermal sensitivity of the high-pressure system.

X-ray System

Initial experiments with quartz monochromators and MoKα-radiation gave less satisfactory results since strictly monochromatic $MoK\alpha_1$-radiation is not attainable with quartz monochromators if high intensity is desired [8]. In the present design much better results are obtained with a germanium monochromator of the Johann type. The x-ray source is a fine-focus tube in line-focus orientation. This choice was made after comparative tests of the diffraction line equality as obtained from Philips fine-focus, normal-focus, and high-power tubes with each tube examined in line-focus as well as point-focus orientation. Of the different tubes which were also tested for their focal intensity distributions by direct pinhole recording of their foci, the fine-focus tube chosen was most suited for the cylindrical sample.

Figures 2 and 3 show the different parts of the x-ray system. The x-ray tube with the monochromator and the film holder are both mounted on an oscillating stage. The oscillation range can be varied up to 90°, or the film can be exposed with the x-ray system fixed in any position within this range. Between the monochromator and the sample there is a slit used to reduce the diffuse background scattering caused by the boron gasket. The oscillating x-ray system is aligned with the pressure axis by means of two screws that actuate a mechanical stage mounted on roller bearings and positioned below the oscillating stage (see Fig. 1).

The sample container is an amorphous boron disk of 3-mm diameter with a 0.25-mm hole for the fine-powdered sample. The sample diameter does not change appreciably under pressure. The height of the

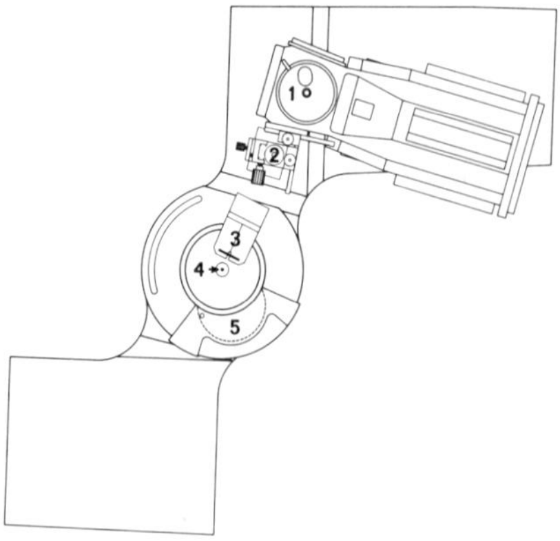

Fig. 2. Projection of the x-ray system. Legend: (1) X-ray tube; (2) monochromator; (3) sample slit; (4) sample in its boron gasket; and (5) film holder. Parts 1, 2, 3, and 5 are mounted on the oscillating stage.

sample cylinder is 1.0 mm before load application and about 0.3 mm after pressure release from above 10 GPa. Good reproducibility has been obtained by using gaskets of very fine-ground boron without any binder and pressed in a die fitted with a wire to form the hole for the sample.

The sample is centered on the lower anvil with a micromanipulator while being observed with a low-power microscope. The sample position is then checked by taking radiographs with the x-ray system fixed in several positions within the oscillation interval.

Diffraction patterns are usually exposed with NaCl as an internal pressure calibration standard. The actual sample pressure is then deduced from the lattice parameter variation of the NaCl with pressure according to Decker et al. [9]. The ratio of standard to sample is adjusted to give about the same diffraction line intensity for both. Heavier elements may need further dilution with boron to give an optimum x-ray thickness for a specific pressure range. Well-exposed films are obtained with about 2 hr exposures at normal dilution and ambient pressure. About ten times longer exposures are necessary for films of the same optical density at 10 GPa due to changes in the geometry of the sample and the boron gasket with pressure.

EXPERIENCE

Figure 4 shows that the two most important factors for good diffraction line quality in the high-pressure camera are the oscillation of the whole x-ray system around the sample and the introduction of the sample slit in the x-ray beam close to the sample. By oscillation, spotty diffraction lines can be avoided (see Fig. 4a and 4b), while the slit appreciably improves the quality of the diffraction pattern (see Fig. 4b and 4c).

As an example of the line quality it may be pointed out that the diffraction line at scale unit 27 in Fig. 4 actually consists of two very close doublets from the four reflections (311), (140), (212), and (231) of a mineral sample of $BaSO_4$ (containing a small amount of $SrSO_4$ in solid solution). The corresponding d-values are 2.119, 2.115, 2.105, and 2.100 Å, respectively. The doublets are not resolved on the film exposed without the sample slit in the

Fig. 3. Close up of the x-ray optics. The sample in its boron gasket is centered on the lower anvil with the upper anvil and the load cell above. The sample slit, the monochromator, and the x-ray tube are in the background. The film holder is in the right-hand foreground. The pin in its groove on the left-hand side actuates the oscillation switch.

x-ray beam (see Fig. 4b), while they are partially resolved on the film exposed with the slit inserted in the beam (see Fig. 4c). The background of the latter film is lower, especially in the low angle region, although the original film was exposed twice as long. The sample slit clearly increases the resolution and the contrast of the diffraction pattern by reducing scattering from the boron gasket. The films of $BaSO_4$ were exposed on fine-grained, single-coated x-ray film (CEA Test-X L) to show the differences more clearly. For a routine work standard, single-coated x-ray film (CEA Test-X H) is used.

The approximate pressure during an exposure is obtained from the applied load on the boron gasket and the experimentally-found relation between the applied load and the sample pressure. A more reliable pressure determination is made when using an internal pressure calibration standard by calculating the actual sample pressure from its observed lattice parameters. Both methods have been checked in the camera against the phase transitions of KCl, Te, and Bi at about 2, 4, and 8 GPa respectively, showing good consistency.

If the sample position remains fixed while the pressure is changed, the initial camera calibration curve, obtained before the load application, can be used at any pressure [5]. If this is not the case, however, a systematic change in the camera calibration curve occurs.

To investigate the effect of sample displacement on the calculated lattice parameters, some experiments with different sample positions were made. The parameters are usually affected to the same extent for both the sample and the standard. For accurate work one should be able to observe any significant sample displacement, and all the more so when the internal standard is used for both pressure and camera curve calibration.

The sample slit is adjusted to provide an x-ray beam slightly wider at the sample position than the sample itself. The position of the sample in the x-ray beam can then be observed directly by taking several radiographs of the sample after the pressure has been changed.

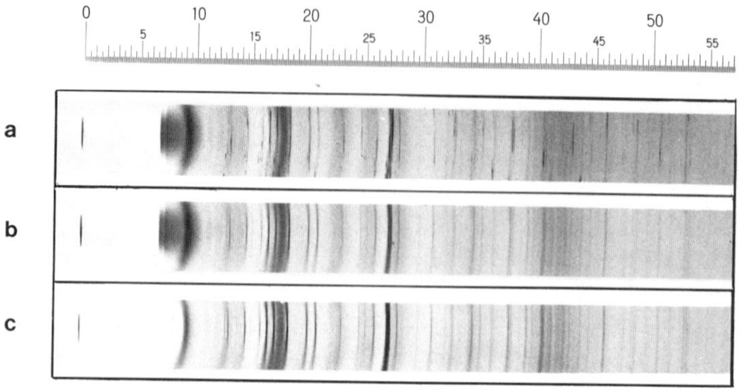

Fig. 4. Examples of films of $BaSo_4/B_4$(1/1 by wt.) under different conditions at normal pressure (0.1 MPa). (a) Nonoscillating camera without the sample slit, (b) oscillating camera without the sample slit, and (c) oscillating camera with the sample slit inserted in the x-ray beam. (The original film of Fig. 4c was exposed twice as long as the other two of Fig. 4).

Even small sample displacements can be observed by taking diffraction patterns of the sample with the oscillating stage fixed in several positions within the oscillation range. If there is no sample displacement, the calculated lattice parameters from the various exposures should be equal within the limits of error.

Figure 5 shows the diffraction patterns of NaCl above 10 GPa and after pressure release. The shifts of the diffraction lines between the two pressures are seen in the double-exposed film (see Fig. 5b). As an example, the line shift is 1.8 mm for the (200)-reflections (scale units 20.3 and 22.1) and 4.6 mm for the (422)-reflections (scale units 50.4 and 55.0). Given in 2θ the shifts are 1.30° and 3.28°, respectively. The diffraction line shifts in the figure indicate a pressure of 11.4 GPa according to Decker's pressure calibration curve for NaCl [9]. A simultaneous sample displacement of about 0.1 mm (~1/2 sample diameter) would cause

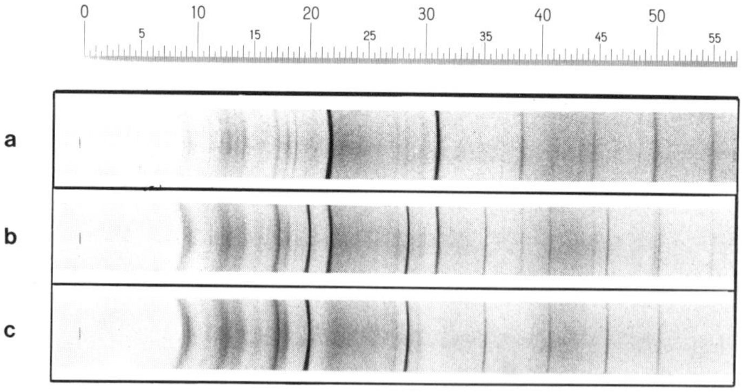

Fig. 5. Examples of films of NaCl/B (1/1 by wt.) at different pressures. (a) 11.4 GPa, (b) 11.4 and <0.01 GPa (double-exposure), and (c) <0.01 GPa.

a line shift of less than 0.04 mm for the (200)-reflection and less than 0.10 mm for the (422)-reflection, contributing about 2% of the total line shift. Such a sample displacement will cause an error in the pressure determination of about 4%. When a pressure change in the sample and a sample displacement occur at the same time, a corrected calibration curve must be used. The location of the sample position and the control of any significant sample displacement with pressure are therefore essential. With the present experimental arrangement it is possible to reduce the uncertainties in the determined lattice parameters to less than 0.1% [5].

Figures 6 and 7 show some results from an investigation of tellurium at high pressures. Similar investigations of tellurium have been made earlier [10-11]. We considered it worthwhile to check the reproducibility of our results from different runs and also to compare these results with those from earlier works. As Fig. 7 shows, the agreement is quite good between our results from runs at different sample dilutions and those deduced from Jamieson and McWhan's work [10].

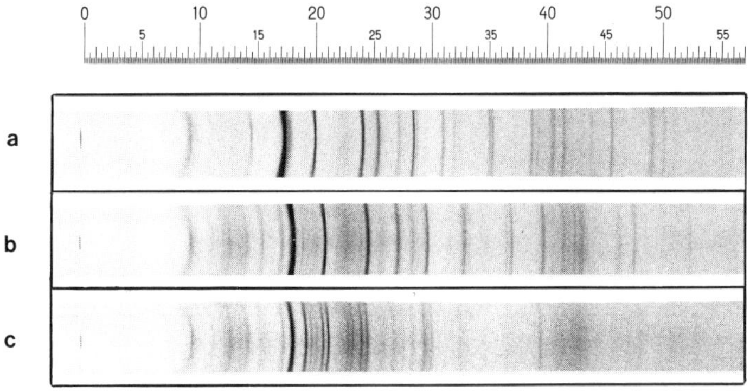

Fig. 6. Examples of films of Te/NaCl (1/1 by wt.) at different pressures. (a) 0.1 MPa - one Te phase, (b) 4 GPa - beginning phase transformation, and (c) 5 GPa - two Te phases.

In the experiment with the highest tellurium content (Te/NaCl/B in the ratio 1/1/0 by weight) there were some additional, weak diffraction lines below 2 GPa, possibly indicating a phase transformation at low pressures as Kabalkina, Vereshchagin, and Shulenin have reported [11]. Further experiments with tellurium are in progress to investigate this effect.

ACKNOWLEDGMENTS

This work has been financially supported by the Swedish Natural Science Research Council, which is gratefully acknowledged. The authors are greatly indebted to I. Olovsson and S. Rundqvist for the facilities placed at their disposal and also to K. A. Wilhelmi, Stockholm, for valuable help and advice. They also want to thank B. Ericson, B. Nilsson, K. E. Oskarsson and F. Zackrisson for skillful technical assistance and finally J. H. Loehlin for valuable linguistic help.

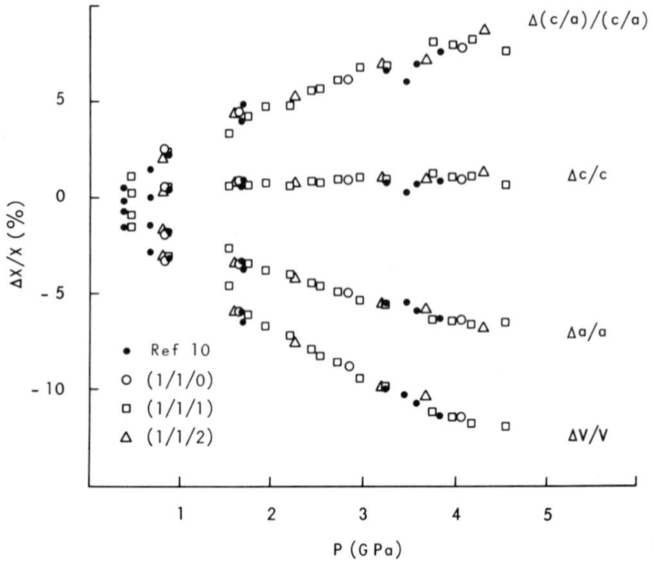

Fig. 7. Relative lattice changes with pressure for Te. The numbers within parentheses indicate the parts by weight of Te, NaCl, and B, respectively. For comparison, the results deduced from Jamieson et al. [10] are also plotted in the same diagram (filled circles).

REFERENCES

1. C. C. Bradley, High Pressure Methods in Solid State Research, Butterworths, London (1969).
2. D. B. McWhan, Trans. Am. Cryst. Assoc. 5, 39 (1969).
3. J. C. Jamieson and A. W. Lawson, J. Appl. Phys. 33, 776 (1962).
4. D. B. McWhan and W. L. Bond, Rev. Sci. Instr. 35, 626 (1964).
5. D. B. McWhan, ASME Publication 64-WA/PT-22, New York (1964).
6. T. N. Kolobyanina, S. S. Kabalkina, L. F. Vereshchagin, M. F. Kachan, and V. G. Losev, High Temp.-High Press. 4, 207 (1972).
7. J. Singh, Rev. Sci. Instr. 45, 1475 (1974).
8. R. E. Scott, Rev. Sci. Instr. 35, 118 (1964).
9. D. L. Decker, W. A. Bassett, L. Merrill, H. T. Hall, and J. D. Barnett, J. Phys. Chem. Ref. Data 1, 773 (1972).
10. J. C. Jamieson and D. B. McWhan, J. Chem. Phys. 43, 1149 (1965).
11. S. S. Kabalkina, L. F. Vereshchagin, and B. M. Shulenin, JETP, 45, 2073 (1962); also Soviet Phys. JETP 18, 1422 (1964).

S-11

HIGH PRESSURE CONTACTLESS VISCOUS SEALS

N. Ahmed

Western Electric Company
Princeton, New Jersey USA

INTRODUCTION

The development of high-pressure equipment is often hindered by the non-availability of seals that can withstand the extreme environment to which they are subjected. Besides containing the high pressure, these seals are often called upon to withstand extremes of high temperature and severe wear.

In this investigation, the development and construction of a high pressure contactless viscous seal is described. Such a seal depends for its operation on the resistance to flow offered by a fluid at high pressure. It is well known that the viscosity and shear strength of fluids increase remarkably with pressure [1,2]. For instance, the shear strength of AC-polyethylene at atmospheric conditions is approximately $2.49 \times 10^5 N/m^2$ (360 psi). At a pressure of $6.9 \times 10^5 N/m^2$ (100,000 psi) and room temperature, this shear strength increases to $5.5 \times 10^7 N/m^2$ (8000 psi). While much is known about the effect of pressure on the shear strength of materials, there does not seem to have been sufficient emphasis on the exploitation of this increased shear strength for high-pressure sealing.

PRINCIPLE OF A VISCOUS SEAL

To illustrate the principle of a viscous seal, an example from an operational item of high-pressure equipment is offered. Consider an endless-chamber hydrostatic extruder such as that developed by Fuchs [3]. In this extruder which is built to operate up to $2.07 \times 10^9 N/m^2$ (300,000 psi) four sets of segments each moving on a doughnut-shaped track are brought together to form a bore. As the segments move forward they clamp the rod to be extruded and push it towards the extrusion die.

The surface of the rod may be coated with a high shear strength substance such as wax just before the rod enters the endless chamber. This wax in addition to acting as a lubricant in the extrusion die serves to seal the space between the segments.

To examine the viscous seal in detail consider the contact length (land) between two adjacent segments (see Fig. 1). Let L denote the length of sealing lands and let δ be the gap between them. If τ is the shear strength of wax then for an incremental length ΔL the pressure drop is given by force balance

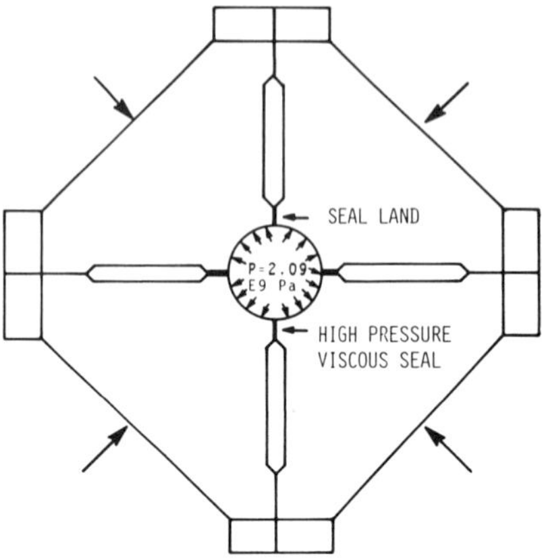

Fig. 1. Viscous seal across lands in MOD III extruder.

$$\Delta P = 2\tau\Delta L/\delta \quad (1)$$

The shear strength τ varies with temperature and pressure. Let this dependence be expressed by

$$\tau = \tau_o e^{\beta P-\gamma\Delta T} \quad (2)$$

where β and γ are constants; τ_o is the shear strength at RTP. The pressure drop then becomes

$$\Delta P = 2\tau_o \Delta L/\delta \; e^{\beta P-\gamma\Delta T} \quad (3)$$

Equation (3) has to be solved incrementally. It should be noted that as the gap δ between the two segments decreases the pressure drop ΔP in an incremental length ΔL increases. For segments that come into physical contact, the gap $\delta = 0$ and $\Delta p \to \infty$. It is therefore impossible to squeeze out the wax from between the segments. In other words, the wax acts to seal the space between the two surfaces. This is the principle of a viscous seal.

In the endless-chamber extrusion machine [3], the bore pressure is 2.07×10^7E9 N/m^2 (300,000 psi). The length of land is 3.9 mm (0.156 in.). The dependence of shear strength of a wax such as AC-polyethylene 656 on pressure is known [1] to be

$$\tau = 2.4 \times 10^6 \; e^{(2.31 \times 10^{-11} \; P\Delta T)} \quad (4)$$

where P is the pressure in N/m^2 and ΔT is the temperature in °C. This relation is valid up to 4.8×10^8 N/m^2 (70,000 psi). Above that value τ equals 5.5×10^7 N/m^2. Using this information, the isothermal pressure profile on the sealing land for a gap of 0.05 mm (0.002 in.) is shown in Fig. 2. If the gap is smaller, the pressure profile drops off more rapidly from the bore. If it is higher, the pressure drops off more slowly. If the gap is over 0.15 mm (0.006 in.) then leakage of the wax occurs and sealing will

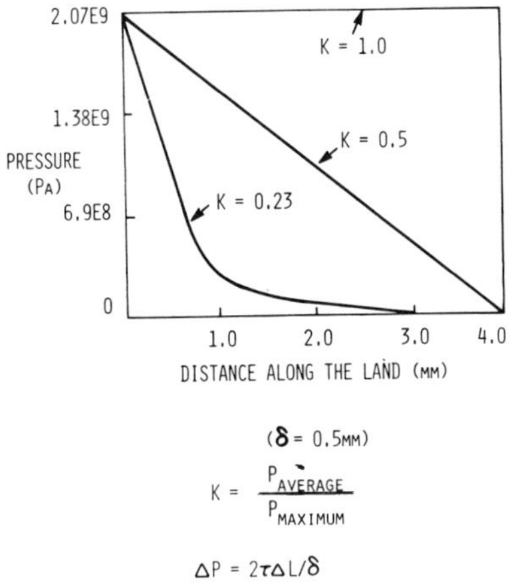

Fig. 2. Pressure profile along a MOD III segment land. $K = P_{ave}/P_{max}$.

require the use of a wax with a higher shear strength. It may be appreciated that for all gaps smaller than 0.15 mm (0.006 in.) the wax serves as an effective high-pressure seal because of its intrinsic shear resistance.

CONSTRUCTION OF A VISCOUS SEAL

From the above discussion it is seen that the important variables in a viscous seal are (1) length of sealing land, (2) gap between the mating surfaces, and (3) shear characteristics of the sealing fluid. Described below is an apparatus constructed to study the effect of these variables on the performance of a viscous seal.

Referring to Fig. 3, a circular cavity C of diameter d is machined into a block B of diameter D and fastened to the ram A of a press. The cavity C is supplied with high-pressure fluid at pressure P from an intensifier. Surface S_1 of block B is mated with surface S_2 of block D. Block D is attached to a load cell LC. The gap between surfaces S_1 and S_2 is measured by a low-voltage distance transducer (LVDT) hooked onto a calibrated digital display. Figure 4 shows a photograph of the equipment used.

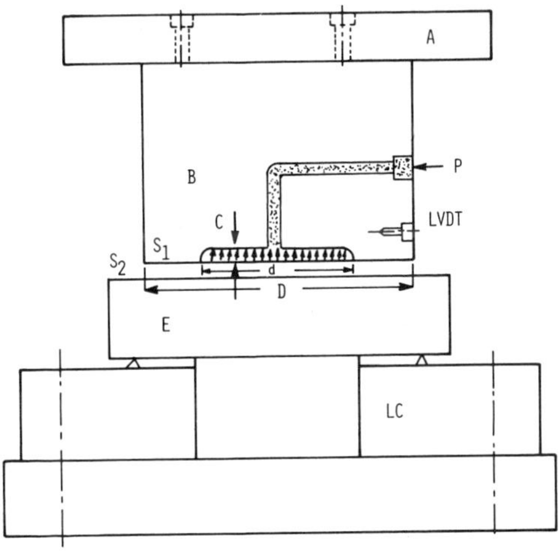

Fig. 3. Diagram of apparatus.

Referring to Fig. 5, the pressure equilibrium across an incremental distance δr of the mating surfaces is defined by

$$\Delta P/\delta r = [P_1 \delta + 2r_1 \tau (P,T)]/(r_1 + \Delta r) \qquad (6)$$

where $\tau(P,T)$ is the shear strength of the fluid at pressure P and temperature T. If data on the pressure and temperature dependence of a certain fluid are available then (6) can be used directly to design a high-pressure viscous seal.

The importance of a high shear-strength fluid for sealing may be appreciated from the nature of the pressure profile across the

Fig. 4. Photograph of apparatus used.

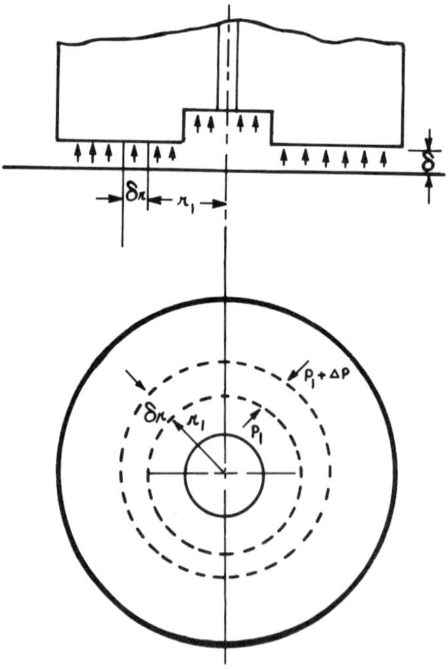

Fig. 5. Pressure equilibrium for a viscous seal.

sealing land. The higher the shear strength of the fluid the faster the pressure drops off from the bore. The nature of the pressure profile across the sealing land may be described by a factor K which represents the ratio of the average pressure across the lands to the maximum pressure in the bore.

It is of primary importance in viscous seal design to keep the K-factor as close to zero as possible. A low K-factor enables the maintenance of a high bore pressure with a minimum countervailing force. As will be pointed out later, this fact is of enormous advantage in the construction of hydrostatic bearings that are required to carry large loads.

EXPERIMENTS

In an experiment the two mating surfaces are clamped together with a predetermined force F. The pressure P of the fluid is increased gradually until leakage occurs between the mating surfaces. The separation of the two surfaces at incipient leakage is measured by the reading on the LVDT. The clamping force F and the line pressure P are also recorded.

Experiments were conducted with various fluids at room temperature. Several sizes of cavity C and several different lengths of sealing land were used and the sealability of fluids for these land lengths and cavity sizes were determined.

RESULTS

The nature of the pressure drop across the sealing surfaces

may be examined from the ratio of average fluid pressure across the land to the pressure P in the cavity (the K-factor). The faster the pressure drops off from the bore the smaller this ratio will be and the more effective the seal will be.

From force equilibrium

$$K = \frac{F-\Pi d^2/4}{P\Pi(D^2-d^2)/4} \quad (7)$$

With this definition the combined effects of the variables that influence viscous seal design may be expressed in terms of the factor K and gap δ. The measured K-factors and gaps for some of the fluids tested are given in Tables I and II. The important results from these tests may be summarized as follows: (1) The larger the size of the cavity, the larger is the K-factor; (2) The greater the length of the land, the larger is the K-factor; (3) The gap δ is smaller for smaller cavity areas; (4) Smaller lengths of sealing lands require smaller gaps to seal; (5) Certain inorganically gelled, synthetic fluorocarbon damping fluids demonstrate extremely high capability to seal and may therefore be ideal as viscous sealing fluids; (6) Teflon impregnated polymers are good lubricants. A mixture of 50% of such polymers with 50% of the highly viscous damping greases may provide good sealability and minimum wear in high-pressure viscous seals. Such mixtures were, in fact, tried in the pressure pads of the Mod III extruder up to $3.45 \times 10^8 N/m^2$ (50,000 psi) with satisfactory results.

APPLICATIONS

A viscous seal may be used in high-pressure applications where the mating surfaces are to be sealed with a small gap between them. An example is the sealing of space between two adjacent segments of an endless chamber continuous extruder built to operate up to $2.07 \times 10^9 N/m^2$ (300,000 psi).

An area of application that can use high-pressure viscous seals is the construction of hydrostatic bearings that are required to carry large loads. It may be seen from a study of Tables I and II that a 5.08-cm (2-in.) diameter cavity with a 1.27 cm (0.5 in.) land containing a pressure of $3.45 \times 10^8 N/m^2$ (50,000 psi) using a fluid that gives a K-factor of 0.1 serves to carry a load of 87 tons. The apparatus shown in Fig. 4 serves as a hydrostatic bearing to carry up to 9.1×10^4 kg (100 tons). The ability to carry such large loads with a hydrostatic bearing using a viscous seal is of great advantage in applications where leakage between the bearing surfaces cannot be tolerated. A previous design [4] for hydrostatic bearings uses a continuous flow between the bearing surfaces to sustain the load between the bearing surfaces. If leakage between the bearing surfaces is allowed, then a recirculating pump needs to be furnished to replenish the fluid with consequent increase in equipment cost. Furthermore, such recirculating fluid bearing systems cannot be used to support large pressures because of the pressure

Table I. Measured K-Factors*

Fluid	K-Factor	Gap, mm
Mobilux grease	1.0	0.075
Teflon impregnated polymer	0.45	0.070
Nye gel 779†	0.21	1.76
Nye gel 773B†	0.10	0.05
50% Teflon impregnated polymer + 50% Nye gel 773B	0.10	0.05

* Diameter of fluid cavity = 5.08 cms;
Length of land = 1.27 cms;
Pressure to be sealed = 3.45×10^8 N/m^2 (50,000 psi).

† An inorganically gelled synthetic fluorocarbon damping fluid.

Table II. Measured K-Factors*

Fluid	K-Factor	Gap, mm
Mobilux grease	0.5	0.05
Teflon impregnated polymer	0.42	0.06
Nye gel 779	0.18	0.15
Nye gel 773B	0.06	0.05
50% Teflon impregnated polymer + 50% Nye gel 773B	0.1	0.05

* Diameter of fluid cavity = 1.27 cms;
Length of land = 1.15 cms;
Pressure to be sealed = 3.45×10^8 N/m^2 (50,000 psi).

limitations of recirculating pumps currently on the market. In practice such pressures are of the order of $2.07 \times 10^7 N/m^2$ (3000 psi). Certain other designs [5,6] use a contact seal between the bearing surfaces to contain the fluid pressure. Such seals are subject to wear when relative movement between bearing surfaces takes place. A hydrostatic bearing with a viscous seal overcomes both of these problems and provides an effective low cost method of carrying large loads.

ACKNOWLEDGMENT

The author is indebted to E. Kovacs for his assistance with the experimental work.

REFERENCES

1. N. Ahmed, "Rheological Characteristics of Certain Waxes at Pressures up to 200,000 psi," to be published in J. Lub Engg.
2. L. C. Towle, J. Appl. Phys. 44, 1611 (1973).
3. F. J. Fuchs, Jr., U. S. Patent 3,985,011.
4. O. A. Carnahan, U. S. Patent 2,347,663.
5. J. G. C. De Gast, U. S. Patent 3,442,560.
6. J. L. Lebach, U. S. Patent 3,322,473.

S-12

SOME MECHANICAL PROPERTIES OF 6-8 TYPE ANVILS

J. Yoshimoto and B. Okai

National Institute for Researches in Inorganic Materials
Sakura-mura, Ibaraki, Japan

INTRODUCTION

A 6-8 type anvil apparatus is a double-stage pressure-generating system studied first by Kawai and Endo [1], the first stage being composed of 6 anvils and the second of 8 anvils. One corner of each of the 8 cubic anvils is truncated so that when the anvils are assembled an octahedral specimen chamber is produced.

The 6-8 type anvil apparatus is somewhat complicated in terms of pressure generation compared with, for example, a Drickamer cell, or a piston-cylinder or the Bridgman anvil in which the choice of gasket configurations is quite limited. With 6 to 8 type anvils many choices of sample configuration exist including the size of an octahedral gasket, the number of spacers, their location etc. The characteristic of the 6-8 type anvil apparatus is its pressure generating ability. Very high pressures can be generated by its appropriate use.

The mechanical properties of the apparatus depend partly on the mechanical properties of the gasket which were dealt with in our previous study [2] in terms of pyrophyllite between Bridgman anvils. The results, however, cannot be applied directly to the present study; the gasket of the Bridgman anvils is a singly-connected disc, while those of the 6-8 type anvil apparatus are separate pieces resulting in deformation features which are not quite the same as in the Bridgman anvils.

EXPERIMENTAL

The 6 anvils of the first stage, whose flats were squares of 20 mm per side, were driven simultaneously in a 2,500 ton press. The inner anvils were cubes of 10.5 mm per side, one corner of which was truncated, leaving an equilateral triangle face. With these eight triangles an octahedral specimen chamber was formed. An octahedron of natural pyrophyllite was placed in this chamber. Twelve pieces of gasket in trapezoidal form were placed in contact with the octahedron between the eight cubic anvils. Natural or fired pyrophyllite was used. The size of various parts varied, including truncated corners, and pyrophyllite octahedrons and trapezoids. Parallel to the trapezoid-spacer gasket a paper gasket of 1 mm width was placed over the diagonal of the anvil face to prevent the initial flow of pyrophyllite gaskets and to obtain stability of the assembled anvils. The efficiency of pressure generation depends on the size, configuration, and nature of the anvils and gaskets.

The pressure scale used in this study was Bi I-II, 25.5 kbar; Ba I-II, 55 kbar; Bi III-V, 77 kbar; Ba II-III, 120 kbar; Pb I-II, 130 kbar; ZnS I-M, 150 kbar; GaAs I-M, 190 kbar; and GaP I-M, 220 kbar [3,4].

From Fig. 1 it is observed that for anvils having a truncated corner of 3 mm per side, a trapezoid gasket of 1.5 mm width and 1.5 mm thickness is more appropriate than other configurations. Natural pyrophyllite was used for the gasket, the pressure being limited to 130 kbar.

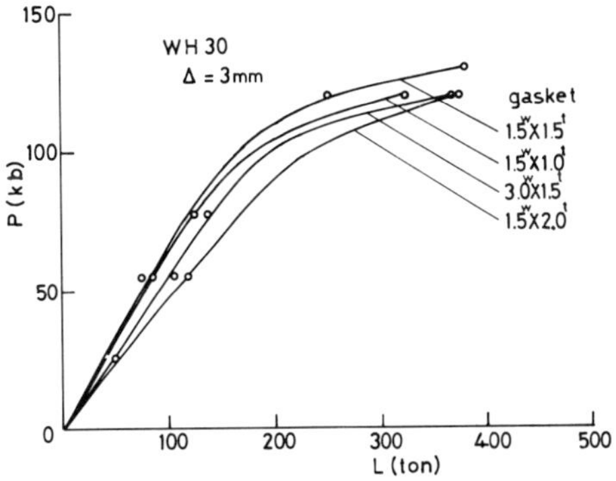

Fig. 1. Relation between pressure and applied load for various types of gaskets. Anvils are WH30 and truncated corner is 3 mm per side.

In Fig. 2 the relation between the pressure and the applied load is obtained for various types of anvils and anvil configurations. Fired pyrophyllite (600°C for 3 hr) was used to obtain improvement in terms of strokes. The side of a truncated corner of the inner anvils was reduced to 2 mm. The octahedron gasket was 5.4 mm per side. Pressure producing efficiency depends upon the mechanical properties of the anvil; it has increased in the order of WH30, WH10, and WH05. Their mechanical properties along with chemical constituents are shown in Table I. The elastic constants were determined by measuring the acoustic wave velocities by a pulse-echo-overlap method [5]. The increase in Lamé's constants (λ,μ) and Rockwell hardness correspond to the increased efficiency of pressure generation.

The upper pressure limit was increased by tapering the inner anvils. Three surfaces of the inner anvil in contact with the

Table I. Property Data for Four 6-8 Type Anvils*

	V_L km/s	V_T km/s	ρ g/c.c.	λ+2μ kb	μ kb	B kb	H.R.A.	CH.C.
WH05	6.86	4.09	14.8	6969	2477	3666	93.1	WC-TaC-Co
WH10	6.86	4.14	14.9	7011	2554	3605	92.3	WC-TaC-Co
WH20	6.82	4.17	14.75	6861	2565	3441	90.6	WC-Co
WH30	6.82	4.17	14.7	6837	2556	3429	90.2	WC-Co

* V_L = longitudinal wave velocity; V_T = transverse wave velocity; ρ = density; λ and μ - Lamé's constants; B = bulk modulus; H.R.A. = Rockwell hardness; and CH.C. = chemical constituents for various anvils.

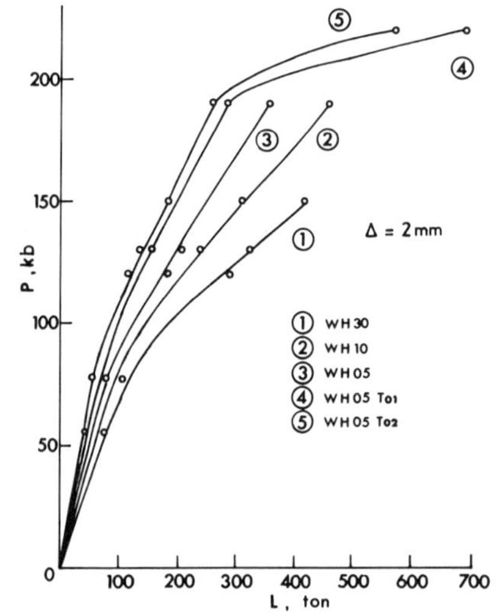

Fig. 2. Relation between pressure and applied load for various types of anvils. T_{OI} and T_{02} are tapered anvils of 1/100 and 2/100, respectively. Truncated corner is 2 mm per side.

truncated corner were tapered 1/100, 2/100 or 4/100. WH05 anvils were used. The efficiency increased accordingly.

The advance of an inner anvil was measured with the use of a displacement meter set at the outer anvil. A correction was made by subtracting a small amount due to the elastic strain of the anvils from the apparent displacement. The final stroke was calibrated by the residual width of a piece of wire fuse inserted between the inner anvils. The relation between the remaining stroke and the load is shown in Fig. 3. A difference is observed between tapered and untapered anvils.

The residual size of gaskets was measured after unloading, as was the volume of the octahedron and the thickness of the "wings" which had originally been trapezoid gaskets. Figure 4 shows the relationship between the residual volume V of the octahedrons and the remaining stroke. A remarkable characteristic is that the final volume is independent of the pressure attained in the pressure range from 77 to 190 kbar, but depends upon the types of anvils. The thickness of the wings t was measured as a function of distance from

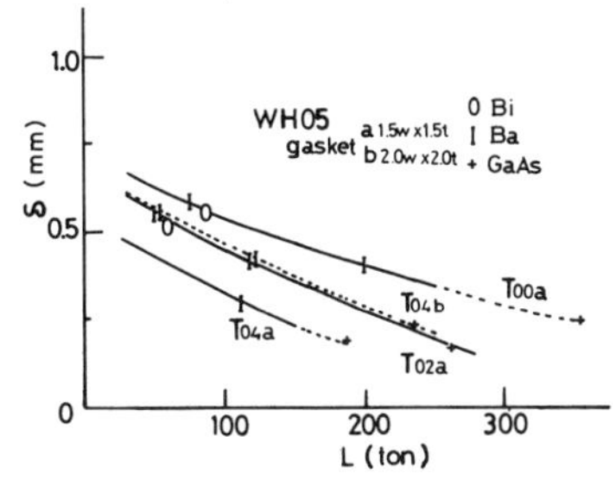

Fig. 3. Relation between stroke δ and applied load. WH05 anvils are used. T_{0n} is a tapered anvil of n/100. Bi, Ba, and GaAs refer to the fixed points of pressure of 77, 120, and 190 kbar, respectively.

the edge of the truncated corner which is shown in Fig. 5. A difference is also observed between the tapered and untapered anvils.

DISCUSSION

The distribution of pressure within the wing may be estimated from Fig. 5; the difference Δt between the thickness of a wing and that of a wire fuse can be related to the lateral pressure $p_{\perp}$.

The value Δt is due to the compression of pyrophyllite Δt_G and the contraction of the compressing anvils 2ΔL

$$\Delta t = \Delta t_G + 2\Delta L \qquad (1)$$

The first term Δt_G is given by

$$\Delta t_G = \frac{t}{B} p_{\perp} \qquad (2)$$

where B is the bulk modulus of pyrophyllite and known to be 400 kbar [6]. As for L, the following relations apply:

$$\Delta L = (1-2\sigma)\,\Delta L'$$

$$= (1-2\sigma)\,\frac{L}{\lambda+2\mu}\,p_{\perp} \qquad (3)$$

where σ is the Poisson ratio, L is the side length of the cubic anvil, and ΔL' is the contraction of the anvil when the pressure is applied on one surface instead of three surfaces meeting at right angles. From Table I, σ is taken as 1/4. Combining (1), (2), and (3), p can be calculated directly from Δt. The results are shown in Fig. 6. The effect of tapering is apparent; it reduces the lateral support. As noted from Fig. 3, the stroke is nearly the same for all of the cases. The problem of how to increase the stroke, however, is still unresolved.

The present method of estimating lateral pressures is justified by other data; the values of loads are easily calculated and they do not deviate greatly from the measured values as long as they are relatively small. This relationship is shown in Fig. 7.

A similar estimation can be made with regards to the octahedral part (see Fig. 4). Final pressures estimated from the residual volume are in good agreement with measurement for WH05 anvils, but not for WH30 anvils. The discrepancy in the case of the WH30 anvils

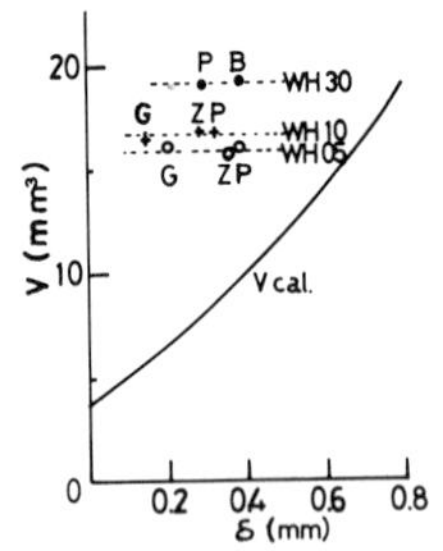

Fig. 4. Relation between residual volume and stroke δ. Calculated volume as a function of δ is also indicated V_{cal}. WH30, WH10, and WH05 anvils are used. G, Z, P, and B are applied pressures of 190, 150, 130, and 77 kbar, respectively.

may be caused by a large strain on the anvil at the truncated corner due to large lateral support. However, why a large lateral support appears in WH30 anvils is still unknown.

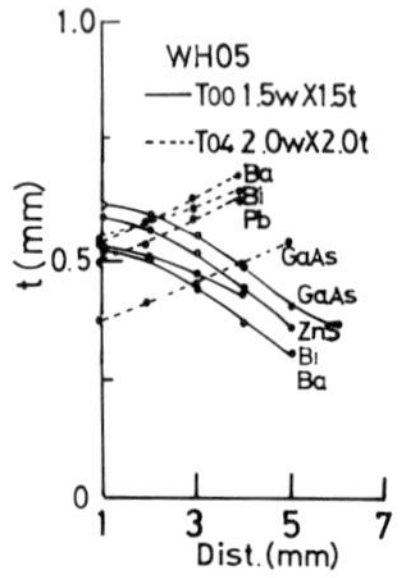

Fig. 5. Thickness of a wing as a function of distance from the edge of a truncated corner. Anvils are 4/100 tapered (T_{04}) and untapered (T_{00}) WH05. Applied pressures indicated are: Bi, 77 kbar; Ba, 120 kbar; Pb, 130 kbar; ZnS, 150 kbar; and GaAs, 190 kbar.

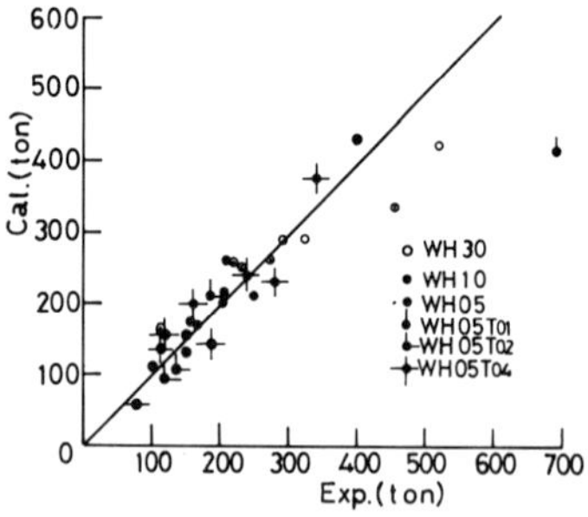

Fig. 7. Calculated load vs. experimental value. Types of anvils and tapering are indicated in the figure.

ACKNOWLEDGMENT

The authors would like to thank N. Kawai, A. Onodera, and K. Suito of the Osaka University for informative advice and encouragement.

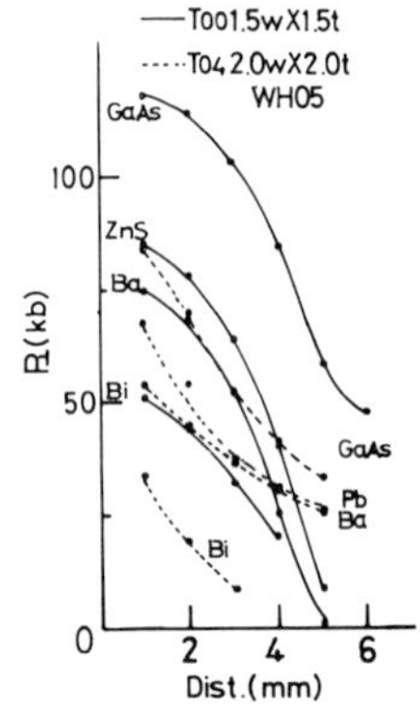

Fig. 6. Estimated lateral pressure in a wing as a function of distance from the edge of a truncated corner. See Fig. 5 for pressures.

REFERENCE

1. N. Kawai and S. Endo, Rev. Sci. Instr. 41, 1178 (1970).
2. B. Okai and J. Yoshimoto, High Temp.-High Press. 5, 675 (1973).
3. T. Yagi and S. Akimoto, J. Appl. Phys. 47, 3350 (1976).
4. G. J. Piermarini and S. Block, Rev. Sci. Instr. 46, 973 (1975).
5. E. P. Papadakis, J. Appl. Phys. 35, 1474 (1964).
6. S. N. Vaidya and G. C. Kennedy, J. Phys. Chem. Solids, 31, 2329 (1970).

T-1

HIGH PRESSURE, HIGH TEMPERATURE TRANSFORMATIONS FROM LABORATORY TO INDUSTRY

P. D. St. Pierre

General Electric Company
Worthington, Ohio USA

INTRODUCTION

The invention of apparatus and processes for synthesizing diamond, cubic boron nitride, and sintered abrasives is an outstanding technical triumph. But the plaudits for such an achievement are short-lived unless economically viable products that meet the real needs of society are produced commercially. The glamour of the original discovery has, however, obscured somewhat the significance of a whole series of sequential developments that were necessary for commercial success and service to the market. The intent here is to redress the balance, and examine the requirements for a successful transformation.

THE TRANSFORMATION

The motivation for the early, and unsuccessful, attempts to produce diamonds was the synthesis of gems. The motivation in the present case, however, was the desire to secure a vital, strategic, industrial material. Up to the middle of this century, virtually all industrial diamonds came from Africa. The continued troubled international political situation made the supply of diamonds, vital for a modern technological society, tenuous. Thus, a successful laboratory synthesis in 1953 gave rise to hopes that an industrial product could be developed.

The first diamonds to be made in quantity (Fig. 1) were vastly different from anything that had previously been used for industrial purposes. It was quickly discovered, however, that this product substantially out-performed natural diamond in grinding tungsten carbide. Thus, it was no longer a matter of convincing users that the product was truly a diamond; the results spoke for themselves, and suggested the potential for tailoring the product for specific applications.

Fig. 1. Early diamond suitable for grinding tungsten carbide tools.

Fig. 2. A diamond grade suitable for sawing concrete and rock.

Key process parameters of time, temperature, and pressure, as well as raw materials were identified; then began the slow, costly, challenging studies of scale-up and product differentiation. This was most effectively done in the "men, machines, materials and cost" atmosphere of an operating department rather than in the fundamentally oriented research laboratory. This resulted in the toughest synthesized diamond currently available having a quality, consistency, and property level that was unattainable in the early days (Fig.2).

Today, 20 years later, tons of this product have been made, sold, and satisfactorily used. In order to qualify for the industry today, it is necessary to have the capability to ship withn 24 hrs., orders drawn on nearly 400 product offerings that differ in physical and chemical properties. These include the various sizes and shapes of Man-Made (T) diamond, BORAZON (R)* cubic boron nitride abrasive, as well as tool, die, and drill blanks made from consolidated diamond and cubic boron nitride particles.

DIFFERENTIATION

Typical properties which serve to differentiate products are size, shape, toughness, thermal properties, crystallite size, and chemical composition. Many processes are needed to produce the wide variety of superabrasive products available today. This contrasts sharply with the natural mineral, where we are limited to product modification by dressing processes on the run-of-mine material. With manufactured products, we can pinpoint the exact properties desired before synthesis is even begun and make adjustments throughout all the phases of the process. For this reason, the various size particles in the same product line may actually be made by different processes because of the specific property requirements predesignated by the user. We are concerned, therefore, not merely with making the species, but also producing output concurrently with the correct properties and distribution of sizes and shapes. Failure to achieve

*(T) (R) Trademarks of General Electric Company

the correct distribution places a heavy economic burden on the activity in the form of excess inventory and non-beneficial use of capacity.

Several processes have been devised and proposed for making industrial diamond. For example: It is possible to convert carbon to diamond by the use of high-pressure shock waves. Unfortunately, this technique provides only a very fine size product. The crystal size is measured in angstroms, and the potential for product differentiation and wide ranging applications is, therefore, very limited.

There have been reports of the epitaxial growth of diamond from gaseous phases; the so-called "metastable formation of diamond." The establishment of whether real diamond has, in fact, been made by the process may still be questioned. But, given the truth of the proposition, the fact remains that growth is unacceptably slow and limited. It is hard to conceive of the development of lines of differentiated products from such a process.

The catalytic processes and their associated equipment on the other hand, have served well - not only for the purposes cited, but they have also led to the discovery and development of new chemical and mineralogical species (Fig. 3). In addition, they have permitted the manufacture of complex shapes and large aggregates, such as compacts. Concerning the latter, for instance, the introduction of the COMPAX (R) tool products illustrates the point (see Fig. 4). Both diamond and cubic boron nitride compacts are available to serve a variety of applications and industries that range from metal turning and wire drawing to rock drilling in oil and gas wells.

Fig. 3. BORAZON cubic boron nitride suitable for grinding hard steel.

MEETING THE CHALLENGE

We do not lessen the importance of the original scientific and engineering achievement by claiming that the developments and discoveries that followed the first synthesis were equally significant in the transformation from laboratory to industry. Heavy capital investment and the ongoing commitment to timely delivery of the quantity and quality of an ever-enlarging product line were also essential factors to success. New products, for instance, have been introduced at an average rate of one every 18 months, and this has been the result of a substantial investment in development programs.

New products require educational programs for original equipment manufacturers and final product users in order to assure successful

Fig. 4. A selection of Man-Made diamond and BORAZON CBN compacts for metal turning, wire drawing, and rock drilling.

and cost effective applications of the products. Conferences and publications are obvious and useful vehicles for such eduation, but practical demonstration in the shop is the most effective means. Thus, application laboratories and trained personnel are needed to show users and equipment manufacturers the products in action, and give them the opportunity to have their own systems tested.

The maintenance of quality and the assurance thereof is vital in the production of sophisticated, costly products. All must be tested, often destructively, in order to ensure that appropriate quality standards are met and that continuity and consistency are maintained.

The costs of inventory are a factor not always fully appreciated or understood. Man-Made diamond is a complex and costly material. Neither the end user nor the equipment manufacturer wants to tie up working capital by carrying large inventories. They want only enough product to meet their immediate needs. Thus, shipment within 24 hrs on all standard products is a necessity in order to serve the demands of the market. This, in turn, puts a burden on technology to optimize production and throughput. It further creates a heavy dependence on the reliability of the production cycle.

TODAY'S REALITIES

Today's high pressure production technology bears a relationship to the earlier laboratory techniques similar to that of modern steel making to the thermodynamic studies one might make on iron oxide in a laboratory furnace. We can no longer say diamond is diamond, is diamond, anymore than steel is steel, is steel. Single word names are inadequate to convey the differentiation of modern products that may have similar generic derivations. The key to success lies in assuring reliable, fast delivery of consistent, differentiated products. This means the commitment of millions of dollars to plant, equipment, inventory, receivables, and customer services.

THE FUTURE

The industrial diamond industry today ranks in complexity with steel, non-ferrous metals, and ceramics - and all this has been achieved in only 20 years. Truly, "the transformation" from laboratory to industry is the most significant one that has taken place in the whole field of high-pressure-high-temperature technology, and it stands unique in its economic contribution to our technological society. We may expect progress in the future to be along the lines of continuing product differentiation and the qualification of synthesized products into areas of application not currently served.

T-2

THE STRENGTH OF POLYCRYSTALLINE DIAMOND COMPACTS

P. D. Gigl

General Electric Company
Worthington, Ohio USA

INTRODUCTION

Polycrystalline diamond compacts are important tools for new material-working techniques such as turning difficult-to-cut materials, drilling, and wire drawing. Consequently, the properties of such materials are needed to formulate their proper use. This paper deals with the measurement of the mechanical properties of synthetic diamond compacts and other hard materials, both synthetic and natural. The transverse rupture strength (TRS), ultimate compressive strength (UCS), and Young's modulus of elasticity (E) in tension and compression have been measured. Density and resistivity were also obtained.

In order to take advantage of the unique properties of diamond such as its hardness and strength, but to minimize the undesirable aspects such as brittleness, size, and availability, diamond-to-diamond bonding within the mass must be attained in order to form an extremely tough compact. The polycrystalline compact of particular interest in this paper consists of diamond powder sintered in the presence of a diamond catalyst at the high pressure-high temperature (HP-HT) conditions at which the catalyst promotes the bonding of the diamond. As will be shown, this type of sintered compact does have good diamond-to-diamond bonding which is the main contributor to its strength.

This type of compact was compared to such hard materials as Co-cemented tungsten carbide, natural carbonado, natural single crystal, and another synthetic compact. Cemented tungsten carbide is a composite structure of tungsten carbide with Co as the cementing agent. A 6% Co grade was tested because it is one of the strongest grades. The carbonado is a naturally-occurring polycrystalline diamond mass which has voids and included minerals in varying concentrations. The single crystal was gem quality diamond which was shaped into the desired configuration. The other synthetic compact was a commercially available material which is sintered at high pressures and high temperatures with noncatalytic sintering aids of Si and/or SiC.

EXPERIMENTAL

Two types of samples were used for the compression and tension tests. The tension test consisted of a three-point beam-loading technique [1] which required a flat rectangular sample, and the compression test consisted of a uniaxial compression-loading technique [1] which required a cylindrical sample. The catalytically-sintered samples were initially fabricated with the diamond layer bonded to cemented tungsten carbide as shown in Fig. 1. The carbide was removed so that the tests could be performed on the polycrystalline masses only. The other sample types consisted of the material in question without any substrates.

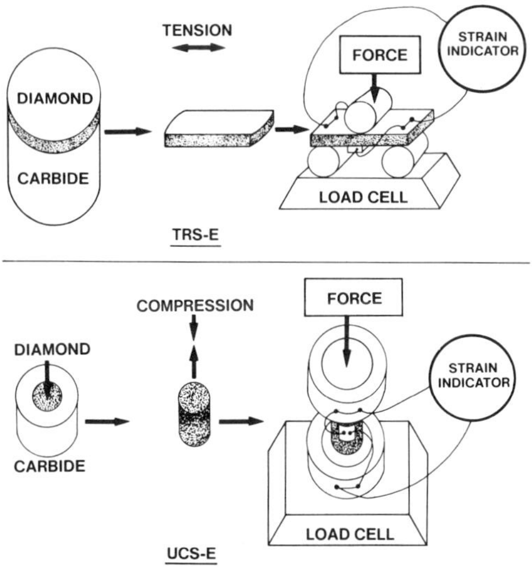

Fig. 1. Catalytically-sintered compact sample configurations before and after carbide removal and schematics of TRS-E and UCS-E tests.

The three-point beam technique was used to obtain the modulus of elasticity in tension and TRS. The sample rests on two lower hardened-steel rollers and is stressed by one upper roller that is centered and parallel to the two lower rollers (Fig. 1). The fixture and sample were centered in an Instrom TT-D Universal Testing Machine with a 454 kg load cell. The beam had a span of 5.21 mm and rollers were 3.23 mm in diameter. The carbide backing was removed from the sintered diamond compacts by grinding and all surfaces were lapped with 15 μm diamond. The nominal dimensions were 0.5 mm thick by 5 mm wide. A strain gage (0.38 mm x 0.51 mm) was centered and attached on the tension side of the samples with methyl-2-cyanoacrylate cement (Fig. 1). The stress-strain curve was obtained by loading the sample to the specific stress, recording the strain on a Baldwin SR-4 strain indicator, and then repeating this sequence at higher stress levels until the sample failed. The stress-strain curve was plotted and the slope was taken as the E value. The point of failure was taken as the TRS value. The reported values are the average results of two or three tests.

The uniaxial compression technique was used to obtain E in compression and UCS. The cylindrical sample is placed between two anvils which were COMPAX® die blanks covered with a 0.025-mm steel foil which was used to reduce friction at the ends (Fig. 1). A 9,072 kg load cell was used. In order to minimize unequal stress across the samples, the fixture was aligned in the testing machine as used in the TRS method. The samples were prepared by OD grinding to remove the tungsten carbide and end lapping with <30 μm diamond powder to assure parallelism. The nominal sample dimensions were 2.97 mm in height by 3.30 mm in diameter. A similar strain gage was attached as in the TRS method. It was located on the circumference at mid-height and oriented to measure the axial compression (see Fig. 1). The stress-strain curve was obtained as in the TRS method and the slope of the linear portion of the curve was taken as E. The point of failure was taken as the UCS value. The reported values are the average of two tests.

The density was obtained by the sink-float method using thallium malonate-formate heavy liquids for the diamond samples. The liquids were standardized by a pycnometric method. The cemented carbide density was obtained by volume and weight measurements.

Resistivity measurements were obtained on the catalytically-sintered-diamond cylindrical samples. The test configuration consisted of the samples being pressed between two copper electrodes. The pressure was increased until a constant resistance was measured by a L and N Kelvin bridge ohm meter. Resistivity was calculated based on the length and area of the sample. The contact resistance of the system which was measured by substituting a copper sample for the diamond sample was subtracted from the resistance of the compacts.

RESULTS

Catalytically-Sintered Type

Typical stress-strain curves as obtained in tension are given in Fig. 2. The curves are linear until failure. The proportional limit was not exceeded before failure and, therefore, plasticity was not considered to occur. A compilation of the average stress-strain curves is given in Fig. 3. The catalytically-sintered compacts had a modulus of elasticity of 89×10^{10} Pa which was very high and approaching the value expected for single-crystal diamond.

The low resistivity of 35 μΩ·cm (see Table I) indicated that at least a portion of the metallic phase was continuous. By taking advantage of the chemistry difference of the catalyst and diamond

(R) Registered trademark of General Electric Company.

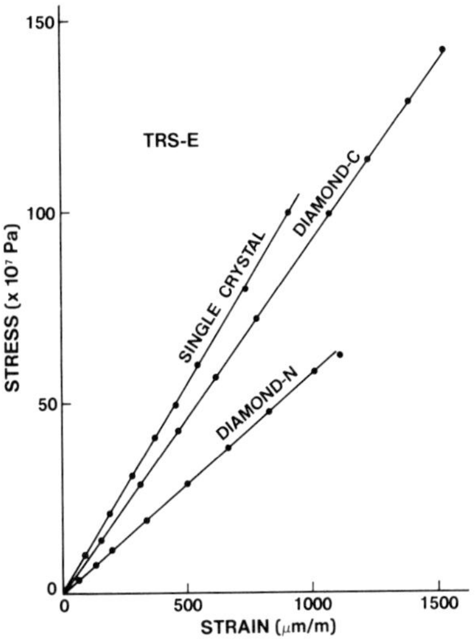

Fig. 2. Typical stress-strain curves for tension samples. Samples are as labeled in Table I.

phases, the remnant catalyst was dissolved in acid leaving the diamond behind. The point of constant weight was taken as the end point. No spalling or crystal pullout was noted. This method was very effective in that less than 1% by volume of the catalyst was estimated to remain in the diamond layer. From analysis of the fracture surfaces it was determined that the remnant catalyst was removed from the entire diamond layer. The distribution of the pores which were <1 μm in diameter was uniform from the top through the center to the bottom.

The catalyst-free diamond layers were tested for TRS and E as in the case above. The E is lower by about 12% (see Fig. 3 and Table I). The lower E is to be expected because porosity changes the effective area of the specimen in proportion to the porosity volume. As an approximation of this effect, the E was corrected for the porosity by assuming spherical voids and random distribution throughout the sample. Young's modulus (E) is calculated by the following equation:

$$EE = \frac{MC}{I} \tag{1}$$

where E is the strain, M is the moment, C is the distance to the outer fiber and I is the moment of inertia of the area. The product MC is not changed in the catalyst-free samples, but I has been reduced in proportion to the porosity. The equation becomes

$$EE = \frac{MC}{I(1-x)} \tag{2}$$

where X is the volume fraction of porosity. The E of the catalyst-free layers would become 84 x 10^{10} Pa or about 5% lower than the diamond layers plus catalyst. This is in very good agreement considering the assumption of spherical pores and shows that the presence of the remnant catalyst is a minor contributor to the strength of the layer.

The interpretation of the TRS values is different because the failure of the sample can be dependent on surface roughness or flaws. The TRS of the "as is" diamond layer had a value of 135 x 10^7 Pa, but the catalyst-free layer had a value of only 83 x 10^7 Pa. This is attributed to the porosity which in effect is adding defects to the surface. The pores act as stress concentrating sites and crack initiators which would be more detrimental to the TRS than E.

Table I. Physical Properties

Sample Type	Density, g/cm^3	TRSx10^7, Pa	Ex10^{10}, Pa	UCSx10^7, Pa	Resistivity, μΩ•cm
Diamond-C					
(tension)	4.0	135	89	-	-
(compression)	3.8	-	92	690	35
Diamond-C-R					
(tension)	3.3	83	78	-	-
Single crystal					
(tension)	3.5	105	110	-	-
Literature [2,4,5,6]	3.5	138-414*	70-110**	869	10^8 - 10^{22}
Carbonado					
(tension)	3.2-3.4	23-45	47-92	-	-
Diamond-N					
(tension)	3.3	53	60	-	-
WC-6% Co					
(tension)		204	57	-	-
(compression)	15.0	-	76	561	-
Literature [1]	15.0	221	64-65	545	17

Legend: Diamond-C = catalytically-sintered diamond compact; Diamond-C-R = catalytically-sintered diamond compact with catalyst removed; Single Crystal = single-crystal diamond; Carbonado = natural carbonado; Diamond-N = diamond compact with noncatalytic sintering aid; and WC-6% Co = cobalt-cemented tungsten carbide.

* Tensile strength.
** Calculated E.

Typical stress-strain curves as obtained in compression are given in Fig. 4. The E in compression was 92 x 10^{10} Pa which agrees closely with E obtained in tension. The compression results are more difficult to obtain because of the experimental configuration. Such critical factors as friction at the sample ends, parallelism of the ends, and alignment of the apparatus can cause errors in the results. There was some nonlinearity at the start of the loading on the compact, but this was minor and considered part of the apparatus arrangement. After this period the curve became linear until catastrophic failure which was taken as UCS. The proportional limit was never reached and this also indicates the lack of plasticity as would be expected for diamond. The UCS of 690 x 10^7 Pa is close to that obtained for single-crystal diamond [2] and this also indicates good diamond-to-diamond bonding.

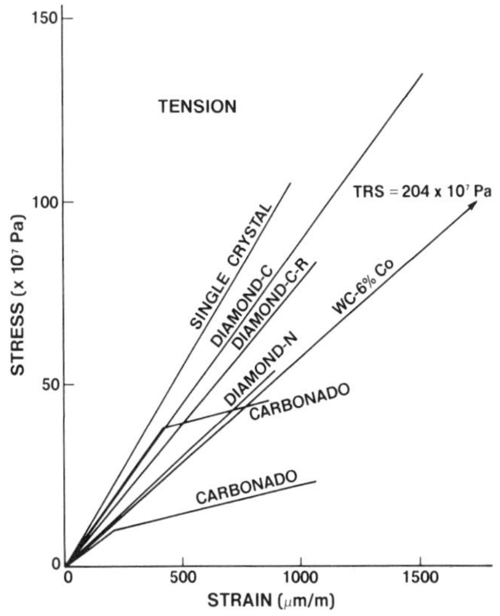

Fig. 3. Average stress-strain curves for tension samples. Samples are as labeled in Table I.

Natural Carbonado

The natural carbonado was tested for E in tension and TRS. The results were not uniform but could be correlated to the specific gravity of the specimens (see Fig. 5). For example, the closer the sample density was to the density of diamond or the less porous the material appeared, the higher the E and TRS would be. Since the inclusions and pores are normally less dense than diamond, the higher density samples will have less inclusions or voids which create nonuniform stress within the compact. These sites concentrate the stress and cause premature failure.

An unusual observation concerning the carbonados was that there was not complete failure after the first crack was heard. The samples still supported the load, but the slope of the stress-strain curve changed and the new curve appeared to be straight. The value of the new E (4-15 x 10^{10} Pa) was much lower and was in the range for normal rock types (7-14 x 10^{10} Pa) [3]. This phenomenon may indicate that the sintered-diamond network had failed and the stress was being supported by mineral inclusions. Another possibility is that the initial break was arrested by inclusions and the stress was supported by an effectively smaller cross-sectional area.

Single Crystal Diamond

A single-crystal diamond fragment was available in the approximate configuration for the TRS and E in tension tests and was shaped with difficulty into the rectangular form. The hardness and brittleness of diamond is not really appreciated until one tries to precisely remove material from its surface.

The TRS and E values are tabulated in Table I and shown in Figs. 2 and 3. The E at 110 x 10^{10} Pa was on the high end of the range calculated. The TRS value of 105 x 10^7 Pa is lower than the tensile strength obtained by other methods [4-6], but surface defects have a large effect on the TRS value and may have influenced the results. These results are probably the first direct measurements made of E and TRS on diamond.

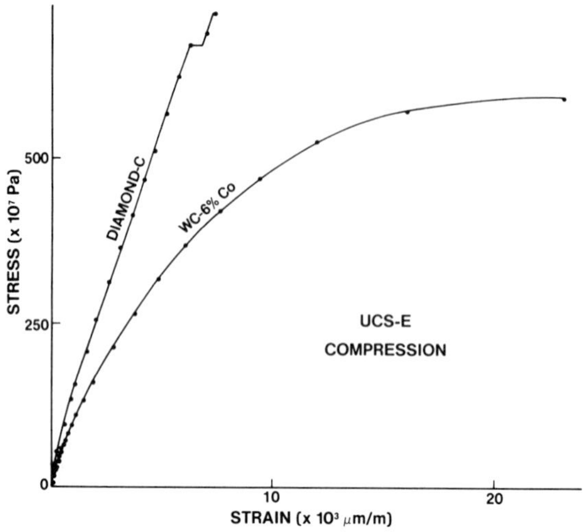

Fig. 4. Typical stress-strain curves for samples under compression. Curves as labeled in Table I.

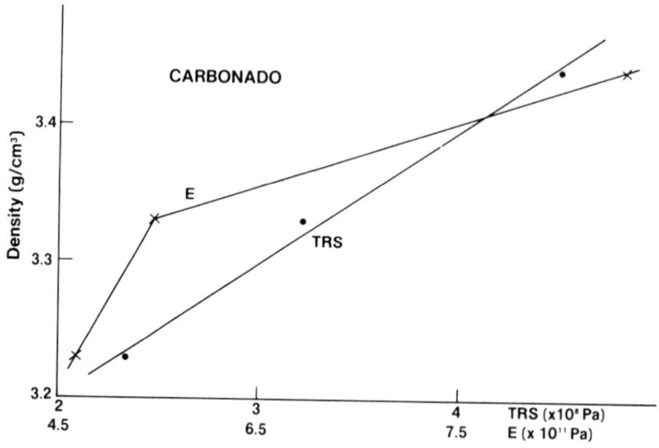

Fig. 5. The relationship between density or inclusion level and TRS or E of carbonado.

Noncatalyzed Compacts

The stress-strain curves of E in tension were linear until fracture like all the other samples (see Figs. 2 and 3). The TRS and E values are much lower than the catalytically-sintered type and single-crystal diamond (Table I). This indicates poorer bonding between the diamond particles and the difficulties of bonding diamond to diamond without the benefit of catalyzed growth.

Cemented Tungsten Carbide

The tungsten carbide tests were used as a calibration point for the test methods used here and showed good agreement with the literature values (Table I) [1]. The samples tested for E in tension and TRS exhibited brittle fracture and were within 8 and 11% of the literature values, respectively. The samples tested for E in compression and UCS exhibited plasticity before fracture. The stress-strain curve was initially linear, but the proportional limit was passed before failure and rather early when compared to that of the catalyst-sintered diamond compact. The E in compression was about 18% higher than literature which is not as good an agreement as the tension values, but as stated earlier, the test was expected to be more difficult to control. The UCS, however, was very close to the literature value. It was about 3% higher.

DISCUSSION

All the diamond samples exhibited brittle behavior and no plasticity was detected. The strength or resistance to deformation of the catalyst-sintered diamond compacts approaches that of single-crystal diamond as shown by Young's modulus in tension of 89 x 10^{10} Pa and 110 x 10^{10} Pa, respectively (see Table I and Fig. 3). The modulus of the compact was equivalent in compression and tension. Since these values are extremely high, this is a very good indication of diamond-to-diamond bonding. When the catalyst phase was removed, the diamond layer was still intact and had a very high E and almost identical to the sintered compact when porosity was taken into account. The catalyst phase, therefore, contributes very little to the strength or rigidity of the compact. Consequently, these results are a very good indication of the extensive diamond-to-diamond bonding.

The TRS measurements are a measure of the toughness or resistance to cracking of a material under tension. These results indicate that the catalytically-sintered polycrystalline-diamond compact is tougher or more resistant to cleavage than the single-crystal diamond, as may be expected. The compact is not as tough as cemented tungsten carbide, however. The UCS is a measure of this tendency under compression. The interpretation of this type of failure is difficult because of the lack of knowledge of the causes of failure in this type of test. Failure could start by shear or tension at the ends so that analogies cannot readily be made. All that can be said is that the single-crystal diamond and catalytically-sintered compacts behave about the same in this respect (see Table I). However, when compared to cemented tungsten carbide, the stress-strain curves are much different (see Fig. 4). Cemented tungsten carbide becomes plastic before failure whereas the catalytically-sintered compact does not exhibit this loss of rigidity and remains elastic until failure. This is in keeping with the expected behavior of diamond in that higher temperatures are needed to induce plastic behavior in diamond [7].

When diamond is sintered without catalysts to promote bonding as in the noncatalyzed synthetic case, the compact strength or resistance to deformation is much poorer and approaches cemented tungsten carbide (Fig. 3). However, the toughness of cemented tungsten carbide is not obtained, as shown by the difference in the TRS measurements.

The natural carbonado can have good strength, but the results are variable. Carbonado appears to be a noncatalyzed sintered type. The strength, however, can approach that of single-crystal and catalyzed diamond compacts, but again the TRS shows a much poorer toughness. The carbonado's inclusion level appears to be the controlling factor in the strength of this material.

In summary, compacts which were made under diamond-stable thermodynamic conditions, and where the presence of a catalyst metal promotes the diamond-to-diamond bonding, have properties close to single-crystal diamond. This is in sharp contrast to the other compacts tested, whether found in nature or synthesized.

ACKNOWLEDGMENTS

The author would like to thank C. R. Roberts for many informative discussions, H. P. Bovenkerk for support of this research, and R. L. Winegardner for the fabrication of the specimens.

REFERENCES

1. Grades Properties Handbook, Carboloy Systems Business Department, General Electric Company, Detroit, Michigan (1974).
2. E. H. Hull and G. T. Malloy, J. Eng. Ind. 88, 373 (1966).
3. J. C. Jaeger and N. G. W. Cook, Fundamentals of Rock Mechanics, Methuen Ltd, London (1969).
4. R. Berman, Physical Properties of Diamond, Oxford University Press, Oxford (1965).
5. Properties of Diamond, Diamond Abrasives Corporation, New York (1973).
6. R. M. Chrenko and H. M. Strong, General Electric Technical Information Series 75CRD089, Schenectady, New York (1975).
7. R. C. DeVries, Mat. Res. Bull. 10, 1193 (1975).

T-3

SOME PROPERTIES OF SINTERED DIAMOND

M. D. Horton, B. J. Pope, L. B. Horton

Megadiamond Corporation
Provo, Utah USA

and

R. P. Radtke

Rucker Hycalog
Houston, Texas USA

INTRODUCTION

Only recently was it discovered that diamond powder could be sintered into a tough, coherent mass. This can be done in spite of the fact that diamond under room pressure is thermodynamically unstable and exhibits plastic flow only at rather high temperatures [1] which increases the instability. This difficulty is overcome by the use of high pressure which not only tends to stabilize the diamond phase but also increases the sintering rate of the diamond particles.

Once it was discovered that diamond could be sintered, it became desirable to characterize the product. What were the properties of the sintered material and how did they vary with the method of preparation and the degree of sintering? The balance of this paper presents the results of a program formulated to answer these questions. The various diamond types used, along with the tests employed, and the experimental results are described.

The program was designed to measure the density, compressive strength, transverse rupture strength, and thermal conductivity of each sample variety. Also, the relation, if any, between these parameters was to be determined. Wherever possible, accepted measuring techniques were to be used although for several parameters some modification in technique was required.

DENSITY

The density of a specimen was determined by first measuring its weight in air. Then, its weight was determined while it was submerged in tetra-bromo-ethane. From these two measurements, and the known density of the heavy liquid, the density was calculated from

$$\rho_d = \left(\frac{W_a}{W_a - W_m}\right) \rho_m \qquad (1)$$

With the more porous specimens it was necessary to quickly determine the weight of the submerged sample. Otherwise, an erroneous weight would be measured due to the penetration of fluid into the pores. With care, the density could be determined with a precision of about $\pm$ 0.1%.

THERMAL CONDUCTIVITY

To determine the thermal conductivity, a cylindrical specimen was prepared and used. The specimen was painted black and then placed between two cylindrical thermodes (also painted black) of a reference material (copper) whose thermal conductivity was known. Then, with a heat source at one end of the assembly and a heat sink at the other, the temperature gradient was determined in the thermodes and the sintered diamond. Then, from the comparative slopes and cross sectional areas, the thermal conductivity was calculated as

$$k_d = \frac{k_r A_r (dT/dx)_r}{A_d (dT/dx)_d} \qquad (2)$$

The precision and accuracy of this test was not as good as the density determination; being only about $\pm$ 10%.

COMPRESSIVE STRENGTH

The test for determining compressive strength was based upon the procedure recommended in the ASTM code C773-74. There it is suggested that the specimen be 0.25 in. in diameter by 0.50 in. long. Also, it was suggested that cylinders of the test material 1 in. long by 1 in. OD be placed at the ends of a specimen during the crushing test. However, it was not possible to make sintered diamond specimens large enough to meet the specifications for either the test sample or the backup cylinder. Therefore, test specimens about 0.20 in. in diameter by 0.23 in. long were prepared. For back-up cylinders, large disks of steel tempered to a hardness of RC-55 were used. Except for these differences, the specified test procedure was followed.

The precision of the results was about $\pm$ 8%, but the accuracy was questionable. The length-to-diameter ratio was small compared to that of the standard specimen and this would cause an erroneously high measurement. On the other hand, the use of a comparatively soft backup cylinder would cause premature sample failure and an erroneously low measurement. To an unknown degree, the two errors are compensating and the measured values for compressive strength may be quite accurate.

TRANSVERSE RUPTURE STRENGTH

There is also a well-established technique for determining transverse rupture strength (ASTM B406-73). However, the suggested specimen size again exceeds manufacturing capability. A length of 0.22 in., a thickness of 0.020 in., and a width of 0.075 in. were used rather than the recommended dimensions of 0.75 in., 0.20 in. and 0.25 in. To match the reduced dimensions of the test pieces, proportionately smaller sizes were used for the carbide cylinders that supported the test specimen and the ball used to apply the breaking force. Otherwise, test procedures were standard.

This test is notoriously sensitive to the surface state of the test specimen. Therefore, to avoid uncertainties caused by surface finish, small tungsten carbide specimens of the same size were prepared by the same lapping technique. The known strength of the carbide was used to calibrate the procedure. A correction multiplier was determined with the carbide tests and then applied to the sintered diamond results.

The precision of the test was about $\pm$ 10% with the tungsten carbide and test results could be readily duplicated. However, the accuracy was not good and the uncorrected values were about 50% low. With the sintered diamond, the precision was rather poor, being only about $\pm$ 25%.

SAMPLE VARIABLES

It is possible to produce sintered diamond through the use of a wide variety of high pressure-high-temperature cycles [2-6]. Also, a considerable range of diamond powder sizes and types may be used as the raw ingredients. Furthermore, several sintering aids have been shown to produce better sintered diamonds. In this program various cycles, diamond powders, and sintering aids were used to produce the samples that were characterized according to the procedures described in the preceding sections.

First, a variety of press cycles were used to prepare several samples from a 3 micron natural diamond powder, with and without the inclusion of various sintering aids. The properties of these specimens then established a reference for other specimens made from different raw materials. These different raw materials are summarized in Table I and include natural diamond powder having a larger particle size, a finer particle size, and a bimodal blend of particle sizes. Then, two synthetic diamond powders were used as the source powder.

Table I. Source Powders for Sintered Diamond

Average Size, μ	Type	Sintering Aid*
3	Natural	SiC, B_4C, Co
3	Synthetic (explosive)	SiC
3	Synthetic	SiC
3 (Narrow dispersion)	Natural	SiC
15	Natural	SiC
3 & 30 (Bimodal blend)	Natural	SiC
1	Natural	SiC

*As well as being sintered in the pure state, these powders also contained various amounts of these sintering aids.

EXPERIMENTAL RESULTS

Since a density determination was easy, nondestructive, and did not demand a particular geometry, densities were determined for most of the sintered diamonds. Bulk densities ranged from 2.9 g/cc for slightly compacted and sintered diamond powder to 3.39 g/cc which was the highest value observed for specimens containing only a small amount of sintering aid. Obviously, the density increased with the degree of sintering, but it also increased if a larger diamond powder was used. Explosively-formed diamond produced a product of low density while other synthetic diamond yielded the same result as natural diamond. As would be expected, large amounts of sintering aid increased the density if the sintering aid had a density greater than that of diamond.

Thermal conductivities ranged from 1.3 to 5 cal/K cm. For a specified diamond powder, Fig. 1 illustrates that the conductivity increased with degree of sintering. At a given final density of diamond in the compact, all of the experimental variables except one produced material having the same conductivity. The exception was that the shock-formed diamond produced sintered diamond having a particularly low thermal conductivity. It was true, however, that some experimental approaches produced high-density pieces that therefore also had high conductivities. Either the use of a sintering aid or the inclusion of larger diamond particles in the green mix produced such a result.

The measured compressive strengths of the sintered pieces ranged from 370,000 to 1,150,000 psi. An increase in the degree of sintering for a specified starting powder produced a stronger product. Figure 2 illustrates this effect. The other important parameter was the size of the raw diamond powder. As is usually

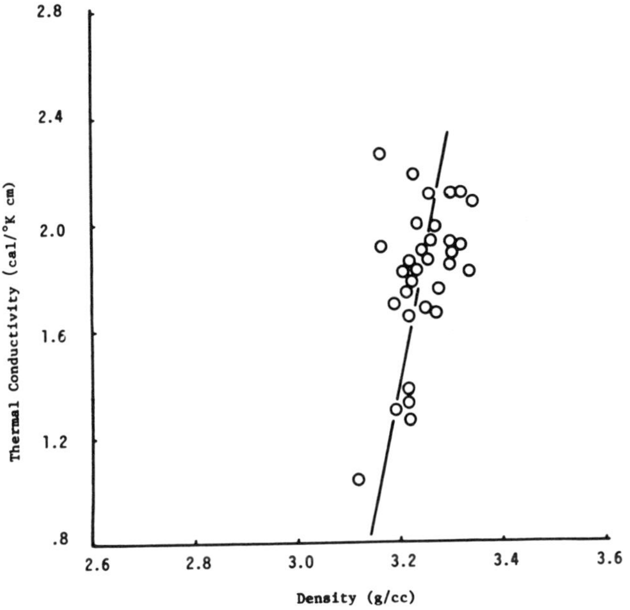

Fig. 1. Room temperature thermal conductivity as a function of density for sintered diamonds produced from 3 μ natural powder.

the case, a finer starting material produced a stronger specimen even when product densities were equal. The other variables were comparatively unimportant except for their influence on density.

The transverse rupture strength of the sintered material was about 130,000 psi and increased slightly with the degree of sintering. Figure 3 illustrates not only this result but also the large scatter in the experimental data. The values shown in the plot were calculated by comparing the observed breaking strengths with those of sintered WC/Co having a known transverse rupture strength. Due to the large scatter in the data, it was not clear that the transverse rupture strength was influenced by any of the experimental variables except degree of sintering.

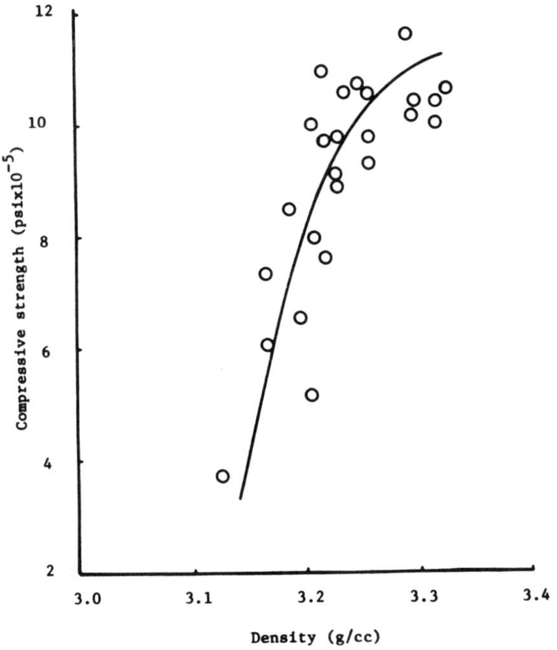

Fig. 2. Compressive strength vs. density for sintered diamonds produced from 3 μ natural powder.

DISCUSSION

The properties of sintered diamond are best understood by considering the microscopic details of the sintering process. The powder to be sintered is placed in a sample container which is in turn placed in a high-pressure cell and subjected to a high pressure. No matter how well the powder is tamped or how high the pressure, the material is not a continuum. Instead, it is an assembly of irregularly shaped particles, each with various points on the exterior in contact with the neighboring particles. Something on the order of 1/3 of the volume is not occupied by particles and most of the surface is in contact with the pores. The means that the contact points are under a very high load which usually puts them in the diamond stable regime. However, the surface adjacent to the pores is subject to a very low pressure where graphite is the stable species.

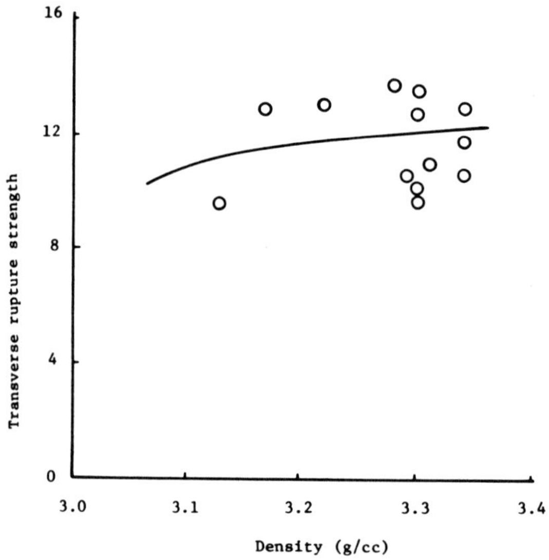

Fig. 3. Transverse rupture strength in psi x 10^4 vs. density for sintered diamonds produced from 3 μ natural powder.

As the temperature is raised so that plastic flow begins, the contact areas enlarge and the porosity decreases. However, simultaneously the diamond at the free surfaces begins to revert to graphite which weakens the structure and hinders further sintering of the powders. The art of producing sintered diamond is thus the art of maximizing plastic flow while minimizing graphitization. The use of sintering aids contributes to the ease of sintering by a mechanism that is not well understood.

The best properties for sintered diamond result when only small quantities of sintering aid are present. Thus, the vast bulk of the sintered material is diamond. Therefore, diamond represents the continuous phase with the sintering aid being present in small isolated pockets. This accounts for results that are otherwise quite inexplicable. When a large stress is applied to a test specimen, a fracture line must propagate primarily through the diamond phase and the small amount of non-diamond has a minimal influence. This explains why the compressive and transverse rupture strengths of the specimens are characterized primarily by the degree of sintering. At equal degrees of sintering, strengths will be about the same for specimens with and without the presence of a small amount of sintering aid. Or, strengths will be the same even when radically different sintering aids are used to produce a given degree of sintering.

For example, the transverse rupture strength is about 130,000 psi even if Co is used as a sintering aid. This seems illogical because the transverse rupture strength increases with tensile strength. One would expect the Co with its high tensile strength to greatly strengthen the matrix in a manner analogous to that caused by "rebar" in concrete. However, the Co is not continuous. A crack merely propagates past the Co pockets so that no significant strengthening results.

Table II lists the property values observed for the best sintered diamonds along with comparative values for other hard materials. There it can be seen that sintered diamond has characteristics which make it an exceptional material particularly suited to many applications.

Table II. Properties of Hard Materials

Material	Density, g/cc	Thermal Conductivity, cal/cm K	Compressive Strength, pxi x 10^{-6}	Transverse Rupture Strength, psi x 10^{-5}
Sintered diamond*	3.4	5	1.1	1.2
Natural diamond†	3.51	6	1.3	0.43
WC/Co 94/6	14.0	1.3	.78	2.2
B_4C	2.52	-	.26	.43
$A\ell_2O_3$ (sintered)	-	0.2	.14	.14

*Best reproducible values, not necessarily obtained simultaneously.

†Values depend on crystal axis, diamond type, and individual crystal.

NOTATION

A = cross-sectional area
k = thermal conductivity
ρ = density
T = temperature
W = weight
X = axial distance
μ = microns

Subscripts

a = in air
d = of sintered diamond
m = in or of heavy liquid
r = reference value

REFERENCES

1. T. Evans and J. Sykes, Phil. Mag. 29, 135 (1974).
2. H. T. Hall, Science 169, 868 (1970).
3. B. J. Pope, H. T. Hall, and M. D. Horton, paper presented at Industrial Diamond Association of America Conference, Scottsdale, Arizona (1972).
4. H. P. Bovenkerk, R. H. Wentorf, Jr. and R. H. Savage, U. S. Patent 3,136,615 (1964).
5. B. W. Dunnington, U. S. Patent 3,399,254 (1968).
6. H. D. Stromberg and D. R. Stephens, U. S. Patent 3,574,580 (1969).

T-4

STRESS CAPABILITY OF G.E. SINTERED DIAMOND AS OBSERVED IN OPPOSED PISTON UHP APPARATUS

F. P. Bundy and K. J. Dunn

General Electric Research and Development Center
Schenectady, New York USA

INTRODUCTION

Within recent years at General Electric, a high-pressure, high-temperature, process for sintering diamond powder into strong hard compacts was developed. This material has proved to have great industrial utility when used in tips for cutting tools and for the cores of wire drawing dies. The commercial products take the forms shown in Fig. 1, in which the sintered diamond and cemented tungsten carbide are intimately bonded together in the manufacturing process. Figure 1a shows a full disc, as manufactured, consisting of a cemented tungsten carbide base with an "icing" of sintered diamond. Such discs may be cut into segments of two, four, or six as shown in Fig. 1b. These are generally used as cutting tips for lathe tools and the like. Figure 1c shows a typical wire drawing die blank which consists of a cemented tungsten carbide ring with a core of sintered diamond. After fabrication of the blank, the core is pierced to form an axial hole which is then shaped and sized to form wire of the desired gauge. Such dies have very long service life because of their strength and resistance to wear. The diamond particles are randomly oriented

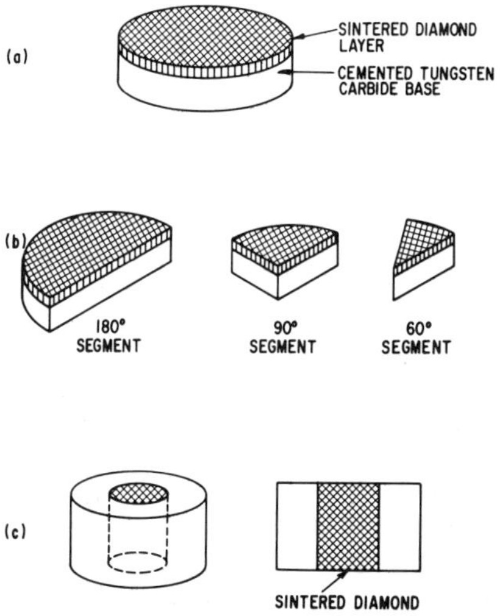

Fig. 1. Commercial forms of GE sintered diamond COMPAX.[R] (a) full disc; (b) segments of discs; (c) wire drawing die core.

so that on the average the wear surfaces contain a large percentage of diamond grains which present their "hard directions" to the wire sliding past them under great pressure.

It is quite well known that in ultra-high pressure apparatus the limits of the pressures which may be attained are set mainly by the strength and hardness of the tip region of the tapered pistons where the forces and stresses are concentrated [1]. When the sintered diamond process was attained, it became feasible to attempt making an improved opposed piston apparatus in which the main bodies of the pistons would be of cemented tungsten carbide and the highly stressed tip regions of sintered diamond, as illustrated in Fig. 2. Such an apparatus was designed and fabricated, and has proved to be capable of developing face pressures well over 500 kbars without plastic deformation or degradation [2].

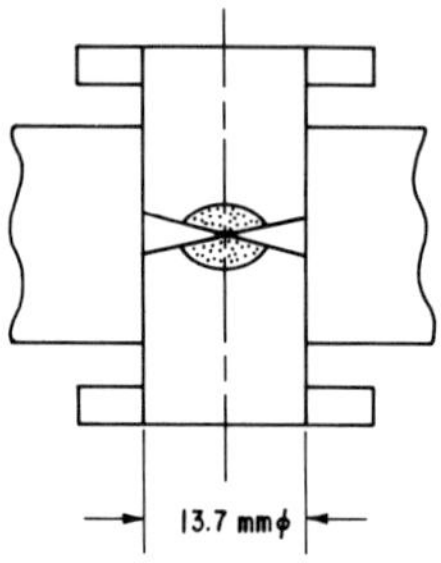

13.7mm X 1.27mm DIAMOND-TIPPED APPARATUS

Fig. 2. Section of opposed piston ultra-high pressure apparatus with cemented tungsten carbide pistons containing tips of sintered diamond.

Observation of the limit of elastic performance of this type of apparatus, in combination with a sophisticated finite element stress and strain analysis computation program applied to its particular geometry and materials, has provided an opportunity to evaluate the stress capabilities of the sintered diamond material under this type of working condition. This paper reports some of the tests and results of such an investigation.

STRESSES APPLIED TO PISTON

The detailed geometry, techniques of loading and the pressure capability of the apparatus are described elsewhere [2]. The pistons were axially loaded with forces as high as 45 metric tons, which corresponds to an average pressure on the piston base (13.7 mm diameter) of about 30 kbar. At the maximum loadings and the smallest thicknesses of cells ($\sim$0.10 mm) between the 1.27 mm diameter pressure faces, the face pressures were observed to be as high as 600 kbar. The cell pressures were calibrated by resistance-jump phase transitions in Bi, Pb, FeCo alloys, and FeV alloys [3]. The pressures of the transitions above 200 kbar were cross-checked with the NaCl compression scale of Decker [4] (up to $\sim$ 300 kbar), and with the shock compression values from Loree, et al. [5], for alpha-epsilon transitions in the FeCo and FeV alloy series. It is believed that the face pressures as calibrated in this manner are correct in absolute terms closer than 20%.

The pressures exerted by the gasket on the flank part of the piston face were calculated from the compressibility of pyrophyllite stone according to the "α = 4 formula" described by Bundy [1]. When the cell pressure and gasket flank pressure, as described above, were integrated to obtain the total force on the pressure face and flank of the piston the values obtained were quite close to the observed axial loading applied to the pistons by the hydraulic press. Therefore the face loadings used as input for the stress analysis computer program are considered to be reasonably accurate.

STRESS AND DISPLACEMENT ANALYSIS

A rather standard finite-element analysis computer program was used to calculate the stresses and displacements at various points within the body of the piston. The input included the coordinates of the elements into which the piston was divided for analysis, the modulus and Poisson ratio of the material in each element, and the forces and constraints applied to the external surfaces. Because the pistons are symmetrical about the axis the "elements" could be laid out in two dimensions on a radial plane as shown figuratively in Fig. 3. In the input to the calculation program, (1) each element was assigned the modulus and Poisson ratio of its material, (2) the axis was constrained to zero radial displacement (because of cylindrical symmetry), (3) the base was constrained to zero axial displacement, (4) on the face and flank surfaces, forces corresponding to the pressures were applied, and (5) the peripheral cylindrical surface was considered to be free of forces or constraints. The computer output yielded the displacements (radial and axial) of each nodal point, and the stresses (radial, axial, hoop, and octahedral) on each side of each element. The octahedral stress is the shearing stress on the octahedral plane associated with the three principal stresses at the site. At the yielding condition the equivalent "pure shear stress" is $1/\sqrt{3}$ or 0.577 the octahedral stress (or very roughly half).

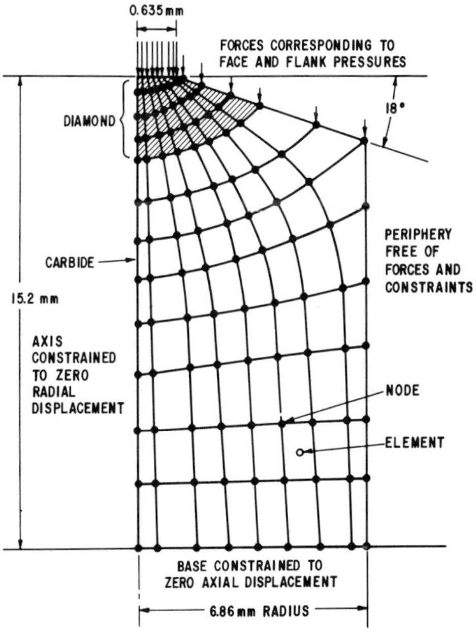

Fig. 3. Lay-out of the finite elements on the radial section of a piston.

RESULTS

A typical piston geometry for which rather complete calculations were run is shown in the half vertical section in Fig. 3. The radius of the base was 6.86 mm, and that of the pressure face, 0.635 mm. The slant angle of the flank was 18°. The sintered diamond part had a depth of a little over two pressure face diameters and an outside diameter of a little over five pressure face diameters. The axial length of the piston was 15.2 mm. The grid element lines were laid out in a roughly orthogonal system with one of the lines coinciding with the interface between the diamond and carbide parts, and with the vertical lines condensing into the pressure face zone to yield more detailed data in that region of the highest stresses.

The detailed results for the relatively "mild" case of a face pressure of 400 kbar will now be presented in a series of graphs: Figure 4a shows in an exaggerated manner (dashed lines) the displacement of the face and of the outside wall from the unstressed position. Note that the face is depressed, particularly at the center where the pressure is highest, and the wall is dilated. Figure 4b shows quantitatively the depression of the top face as a function of the radial distance out from the axis, while Fig. 4c shows the radial displacement (or dilation) of the outside wall. This dilation tends to put the carbide in this region in hoop tension, as will be shown in the stress diagrams later, and for very heavy loadings the tension can cause cracking of the carbide, a practical limitation.

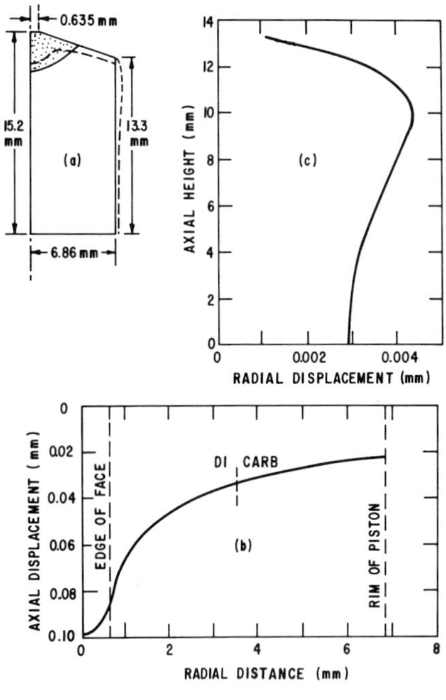

Fig. 4. (a) Exaggerated distortion of the piston by loading; (b) vertical displacement of the top surface; and (c) horizontal displacement of the side surface at 400 kbar face pressure.

Figure 5a portrays the radial displacement of points on the top surface while under load. Note that the inner parts are forced to contract, while only the outer edge suffers dilation. Figure 5b shows the axial displacement of points at various axial heights along the outer wall. In each graph a half section of the piston is shown with the path of the stress plot indicated.

Figure 6 presents the displacements of points located along the diamond/carbide interface at various distances out from the axis. Figure 6a gives the radial displacement, and Fig. 6b the

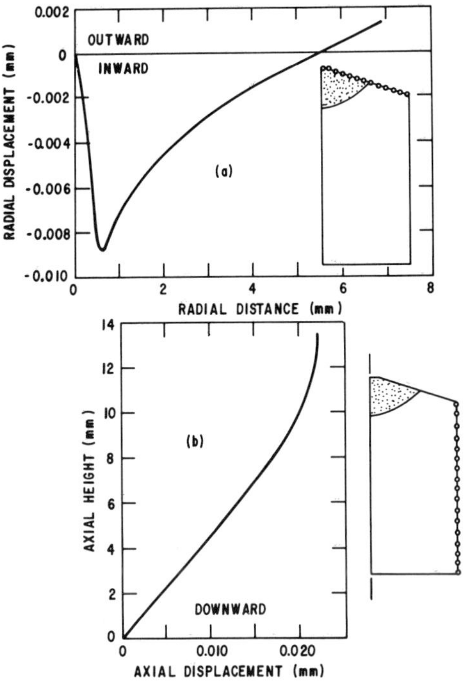

Fig. 5. Radial displacement of the top surface and vertical displacement of the side surface (400 kbar face pressure).

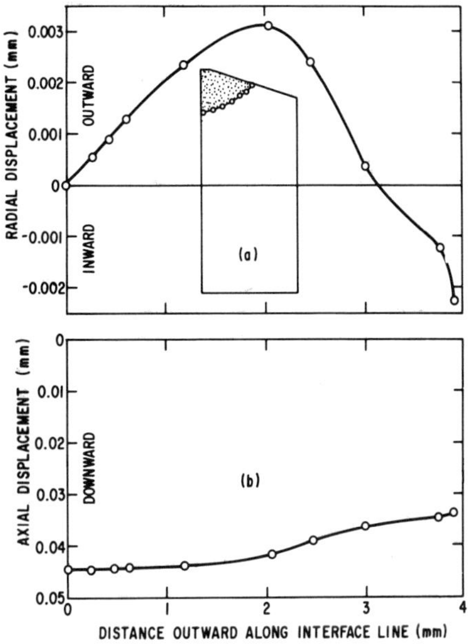

Fig. 6. Displacement of the nodes along the interface between the diamond and cemented tungsten carbide parts.

downward axial displacement. At this region it is seen that at the inner part the radial displacement is outward, while the outer part (which is near the top surface) displaces inward. Comparing Fig. 6a with Fig. 5a it is significant that at the top surface the diamond part is distorted inward while at the interface region it is distorted outward. This strain field is accompanied by large shear stresses, as will be shown. The downward displacement is fairly uniform (Fig. 6b), and causes no problems.

Figure 7 presents the distribution of axial compressive stress along the vertical axis. This is plotted with semilog coordinates because of the wide range of the stress values (11 to 400 kbar). Note that in the diamond part the stress increases from about 50 kbar at the carbide/diamond interface up to 400 kbar at the pressure face.

The axial compressive stress in the diamond part just above the diamond/carbide interface is plotted vs. the distance outward from the axis in Fig. 8. At this depth the stress has dropped to the modest value of 65 kbar near the axis, and to the low value of about 8 kbar at the flank surface by the gasket.

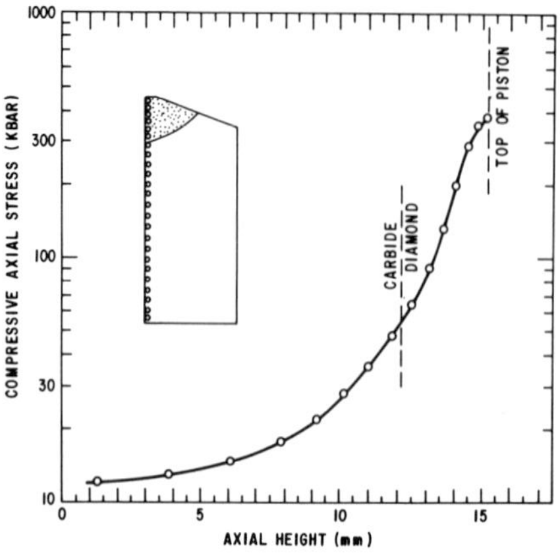

Fig. 7. Axial compressive stress along the axis of the piston.

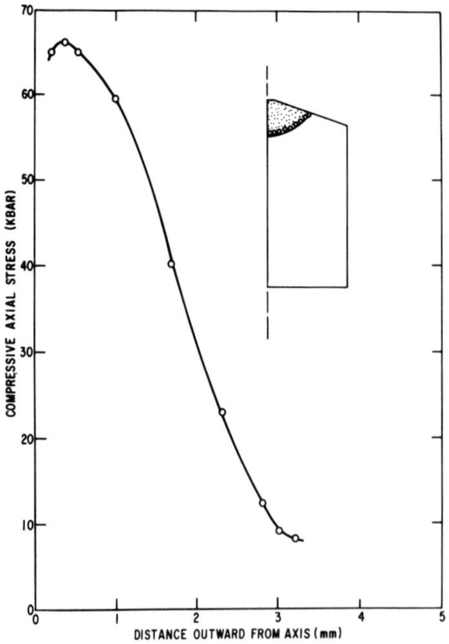

Fig. 8. Axial compressive stress vs. distance out from axis in the diamond part just above the diamond/carbide interface.

Figure 9 portrays the hoop stress at various positions along the axis. Note that this is in tension in all the carbide part, and nearly half way up into the diamond part. Above this level it becomes compressive and rapidly increases to nearly 300 kbar at the pressure face surface.

Figure 10 plots the hoop tension vs. the height along the outside wall of the carbide piston. This tension reaches a maximum value of about 1.9 kbar at a height of about 10 mm, a short distance below the shoulder. This zone is quite vulnerable to cracking at heavy loadings, and presents one of the natural limits to the loading of this type apparatus.

The next two figures present the octahedral shearing stress values along various lines within the stressed piston. The octahedra shearing stress as used here is

$$\tau_{oct} = \left[\frac{(\sigma_1-\sigma_2)^2+(\sigma_2-\sigma_3)^2+(\sigma_3-\sigma_1)^2}{2}\right]^{1/2} \tag{1}$$

where σ_1, σ_2, and σ_3 are the principal stresses at the position. This octahedral stress is equivalent to a pure shear stress of $1/\sqrt{3}$ of that value. That is

$$\tau_{simple} = 0.577\ \tau_{oct} \tag{2}$$

The equivalent simple uniaxial stress is numerically the same as this octahedral stress. Thus from the octahedral stress values, which exist at the critical positions in the piston when permanent residual distortion can barely be detected, may be translated into the elastic limit (or yield) values of shear stress, or uniaxial stress which the material can withstand.

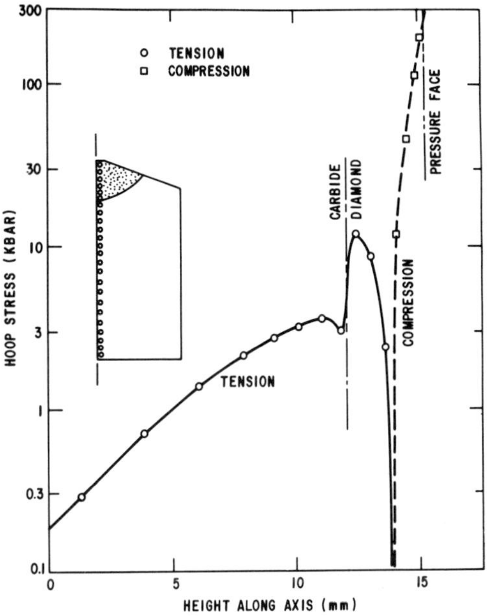

Fig. 9. Hoop stress along the axis.

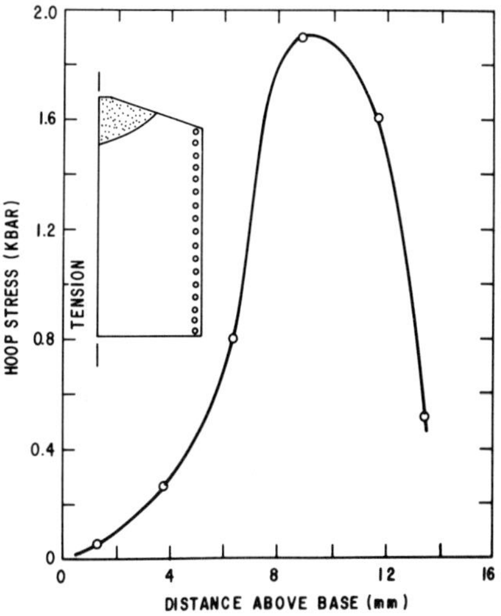

Fig. 10. Hoop stress along side of piston vs. distance above the base.

For the numerical data which will now be presented for shearing stresses the unit of kbar will be used for convenience, although strictly the kbar is a pressure unit. In our special use of it here it is meant to be the shearing force on a unit area.

Figure 11 shows the average octahedral stress in the column of elements along the axis of the piston. The stress is plotted on a logarithmic basis because of the wide range of values. Note that it reaches a maximum value of about 260 kbar just under the center of the pressure face. This corresponds approximately to a simple shear stress of about 150 kbar.

The octahedral stress in the diamond part just above the diamond/carbide interface is plotted in Fig. 12. For the first mm out from the axis it has a value of about 75 kbar and then drops off nearly linearly to less than 10 kbar at the flank surface.

DISCUSSION

The numerical stress values which have been presented in Figs. 7 through 12 above correspond to a face pressure of 400 kbar. In actual practice the apparatus has been cycled repeatedly to face pressures of 500 to 600 kbar without any visible deformation or degradation. Additional tests for deformation consisted of making traverses across the face area, along different diameters, with a recording proliferometer having a sensitivity of about 0.05 μm. For high quality pistons this

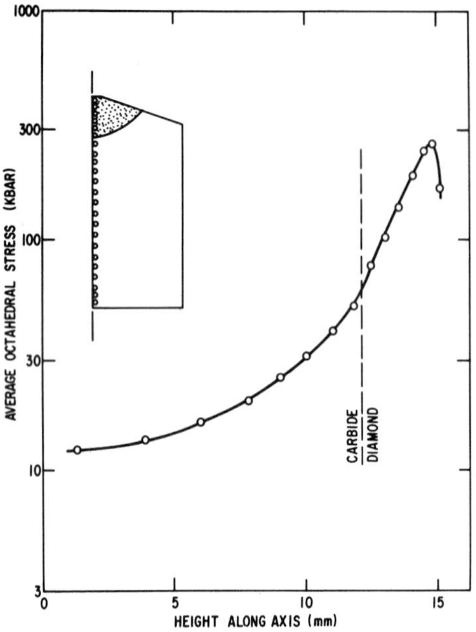

Fig. 11. Average octahedral stress along the axis.

test also shows no deformation of the diamond pressure face area whe loaded successively to over 500 kb face pressures. It may be conclud ed therefore that the piston under study is capable of supporting ela tically stresses over 25% larger than those shown in the graphs.

The maximum octahedral shearing stress occurs on the axis just under the pressure face (Fig. 11), at a value of about 260 kbar when the face pressure is 400 kbar. For a face pressure of 500 kbar the maximum octahedral shearing stress would be 1.2 x 260 or 312 kbar. Translating this octahedral shearing stress into conventional simple shearing stress gives a value of about 180 kbar for the sintered diamond as used in this application. The corresponding simple uniaxial compressive strength would be 1.2 x 260 = 312 kbar.

Actually it is known from the behavior of the apparatus that at these base loadings of 40 to 45 metric tons the carbide base of the piston is on the verge of plastically distorting. Again from Fig. 11 is is seen that the octahedral shearing stress in the carbide is about 54 kbar just under the diamond/carbide interface. For 500 kbar face pressure this would be 65 kbar, and would correspond to a simple shearing strength of 37 kbar for the carbide and simple compressive strength of 65 kbar.

It is also known from experience with this type of piston that if the sintering of the diamond part is not up to full quality the pressure face does crack and distort at face pressures of about 500 kbar, indicating that the yield strength is less than derived above, and that for a full quality piston the limits

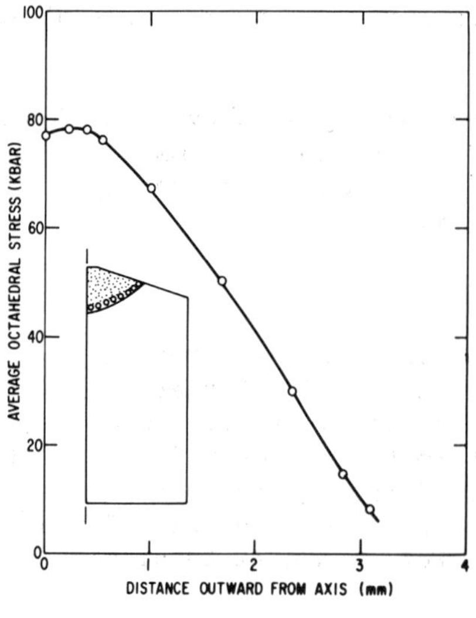

Fig. 12. Average octahedral stress in the diamond part just above the carbide/diamond interface.

of strength of both the diamond and the carbide parts are being approached, for this geometry, when the face pressure approaches 600 kbar.

Overall, from these tests and analysis it may be concluded that the simple shear strength of the sintered diamond is around 200 kbar, and its simple compressive strength around 350 kbar. The corresponding numbers for the cemented tungsten carbide are about 40 and 70 kbar, respectively, but these are not well determined because the site is deep within the piston and there was no sensitive means of detecting plastic deformation.

ACKNOWLEDGMENT

The authors gratefully acknowledge the help of R. Amory of Northeastern University in setting up the element grid and preparing the computer input format; also R. P. Messmer, GE CRD, for assistance in debugging the input format to be acceptable to the computer, and E. Plesko for tabulating, converting and graphing the output data in presentable form.

REFERENCES

1. F. P. Bundy, Rev. Sci. Instr. 48, 591 (1977).
2. F. P. Bundy, Rev. Sci. Instr. 46, 1318 (1975).
3. K. J. Dunn and F. P. Bundy, paper A-5-D presented at 6th AIRAPT Intern. High Pressure Conference, University of Colorado, Boulder, Colorado, July 25-29, 1977.
4. D. L. Decker, J. Appl. Phys. 36, 157 (1965); also J. Appl. Phys. 37, 5012 (1966).
5. T. R. Loree, C. M. Fowler, E. G. Zukas, and F. S. Minshall, J. Appl. Phys. 37, 1918 (1966).

T-5

NUCLEATION AND GROWTH OF CUBIC DIAMOND IN SHOCK WAVE EXPERIMENTS

P. S. De Carli

SRI-International
Menlo Park, California USA

INTRODUCTION

In 1959 the announcement that the presence of diamond had been detected in a sample of shock-compressed graphite was met with skepticism; it was pointed out that the calculated shock temperature was much too low to permit nucleation and growth of diamond during the microsecond available at pressure. Jamieson subsequently confirmed the presence of diamond and suggested a diffusionless mechanism, c-axis compression of rhombohedral graphite [1]. However, subsequent experiments yielded diamond far in excess of the amount of rhombohedral graphite present in the initial material. Furthermore, when shock temperature and shock pressure were varied independently, yields of diamond were found to depend more strongly on calculated shock temperature than on shock pressure, over the shock pressure range from 15 to 150 GPa. The calculated shock temperatures associated with good diamond yields were only about 1500 C, too low for nucleation and growth to be plausible.

In a subsequent paper it was suggested that a two step process might explain shock synthesis of diamond [2]. The first step would be a diffusionless collapse of graphite to a glass-like tetrahedrally bonded structure. The second step would be recrystallization, which could conceivably continue even after pressure decayed into the stability field of graphite.

If this mechanism were correct, one might expect the yield of diamond to be sensitive to pressure pulse duration. The yield of diamond was found to be independent of pressure pulse duration, over the range of 0.5 μs to about 10 μs. Furthermore, only trace yields of diamond were obtained in experiments with non-porous graphites, even though these materials had reached calculated shock temperatures believed equal to the shock temperatures calculated for high diamond yield experiments with initially porous graphites. The "equal" shock temperatures were reached by experiments with preheated samples.

Reexamination of Hugoniot data [3,4] confirmed that initially porous carbons transform more readily to diamond than do non-porous carbons, but the maximum calculable shock temperatures were too low for nucleation and growth. Indeed, the inference from Hugoniot data was that diamond formation occurred on a time scale comparable to the rise time of the shock wave, i.e., within about 10^{-8} s. However, the failure to find a substantial pressure pulse length effect would be explained if diamond growth had already ceased within the time spanned by our shortest duration experiments.

It is well known that the response of a porous material to shock compression is extremely complex when viewed on the nanosecond time scale (and μm distance scale) of the rise of the shock front. The initial distribution of shock temperature (and shock pressure) will be inhomogeneous; i.e., the material will contain hot spots. Pressure equilibrium then takes place on a nanosecond time scale; substantial temperature equilibrium can take place on a nanosecond time scale if the hot spots are very small. The standard shock temperature calculational procedure yields only the average temperature of the material, i.e., after these hot spots have equilibrated with their surroundings. No limitations on peak shock temperature are implied by this calculation. If one postulates that diamond forms only in the hot spots, the possibility of a nucleation and growth mechanism must be reconsidered.

The present paper is an attempt to demonstrate, via simple calculations, that a nucleation and growth mechanism for shock synthesis of diamond is consistent both with the experimental data and with the fundamental conservation laws. The first draft of this paper was written seven years ago. Publication has been delayed in the expectation that more rigorous arguments would be developed. Due to the press of other work, no progress has been made. However, it is hoped that the present paper will be a useful starting point for other workers.

DIAMOND GROWTH RATES

The usual product of our shock wave syntheses of diamond has been polycrystalline particles. These particles have diameters (thicknesses, in the case of flake-like particles) ranging from about 10 um downwards and have crystallite sizes in the range of 10 to 20 nm. The small crystallite size of these polycrystalline particles implies that nucleation sites are abundant. Since the effective duration of most of our recovery experiments was about a μs, one may calculate that a radial growth rate of about 1 mm/s (for spherical diamond crystals) is required.

Bundy synthesized diamond without the use of a catalyst in static high pressure flash heating experiments [5]. The effective duration of the 3000 K temperature pulse in his experiments was about one ms. The product of his synthesis was polycrystalline diamond having an average crystallite size of about 100 nm. Assuming the nucleation and growth mechanism, a radial diamond growth rate of about 50 μm/sec is required. Wentorf has inferred that the activation energy for direct growth of diamond is about 200 kcal/mol [6]. Using this activation energy, one calculates that a shock temperature of about 3600 K would yield the higher diamond growth rates of the shock wave experiments.

If the diamond remained hot as the pressure decayed into the stability field of graphite, the reverse transition to graphite would occur at a comparable rate. However, the polycrystalline diamond particles are sufficiently small to thermally equilibrate with their surroundings (cool graphite). In fact, we have found a weak experimental correlation, essentially in agreement with heat flow calculations, between pressure pulse length and the maximum diameter of diamond particle produced in any experiment.

HOT SPOT ENERGY

It should be noted that the phenomenon of hot spot formation during shock compression of porous solids is well known [7,8]. Mechanisms for hot spot formation include jetting and adiabatic shear. Both mechanisms should be operative in the shock compression of porous carbon; either mechanism can yield sufficiently hot spots to account for diamond growth. The crucial question is whether the energy content of the hot spots is large enough to account for observed yields of diamond.

In order to make the energy balance calculation, one must partition the total energy increase of shock compression into a homogeneously distributed component and a heterogeneously distributed component. One may note that both recovery measurements and Hugoniot measurements indicate that no diamond is formed from pyrolytic graphite over the shock pressure range to about 400 kbar. One may assume that the internal energy increase of shock compression is homogeneously distributed in the pyrolytic graphite, and one may further assume that the homogeneously distributed component will be the same for an initially porous carbon shocked to the same peak pressure. The difference between the total energy increase and the homogeneously distributed component should be an indication of the energy available for the hot spots.

McQueen's Hugoniot measurements [4] include two sets of data that match up, in essential parameters, with shock wave recovery experiments we have performed. We recover about 10% diamond from ATJ graphite that has been shocked to 15 GPa. From McQueen's 14.4 GPa ATJ data one can calculate a net internal energy increase of 1060 J/g. The homogeneously distributed portion is estimated to be about 100 J/g, equivalent to a temperature rise of 100°C. The remaining 960 J/g is sufficient to heat 15% of the graphite to 3500 K, the agreement with the observed 10% yield is excellent. McQueen's 23.1 GPa data for ATJ indicate a homogeneous temperature rise of about 300 K; the remainder of the internal energy increase is sufficient to heat 27% of the material to 3500 K. In recovery experiments at somewhat higher pressures of 25 GPa, we have recovered up to 25% diamond.

It must be noted that the peak temperature of the hot spots would be much higher at earlier times, i.e., within the initial ns of hot spot formation, and diamond growth rates as high as one m/s can be estimated. In fact, it would appear that diamond growth essentially ceases when the hot spot temperature has decayed to about 3500 K.

REFERENCES

1. P. S. De Carli and J. C. Jamieson, Science 133, 1821 (1961).
2. P. S. De Carli, in Science and Technology of Industrial Diamonds, Industrial Diamond Information Bureau, London (1967) p. 49.
3. B. J. Alder and R. H. Christian, Phys. Rev. Letters 7, 367 (1961).
4. R. G. Mc Queen and S. P. Marsh, in Behavior of Dense Media Under High Dynamic Pressures, Gordon and Breach, New York (1968), p. 207.
5. F. P. Bundy, J. Chem. Phys. 38, 618 (1963).
6. R. H. Wentorf, Jr., J. Phys. Chem. 69, 3063 (1965).
7. J. H. Blackburn and L. B. Seely, Nature 194, 370 (1962).
8. J. H. Blackburn and L. B. Seely, Nature 202, 276 (1964).

SHOCK COMPRESSION OF A PYROLYTIC, CEYLON NATURAL, AND HOT-PRESSED SYNTHETIC GRAPHITE TO 120 GPa*

W. H. Gust and D. A. Young

Lawrence Livermore Laboratory, University of California
Livermore, California USA

INTRODUCTION

In 1961, De Carli and Jamieson [1] recovered diamonds from shock experiments. Since then, shock-induced solid-solid phase transitions in carbon have been of continued interest because a complete description of the process is not yet available. Bundy and Kasper [2] enumerated the parameter requirements for static-pressure synthesis of diamond. However, the reaction times involved in static-pressure synthesis are much longer than those of the shocked case. Also, because temperature control is better, it is unclear whether Bundy's observations apply to shock experiments.

Pavlovskii and Drakin [3] reported shock compression data for a synthetic graphite (ρ_o = 1.85 Mg/m^3) to 300 GPa. They found no evidence for an additional transition from diamond to a close-packed liquid or metallic state. This transition would be expected if carbon exhibits the same behavior under compression as the group IV-A homologues Si and Ge [4-7].

Alder and Christian [8] reported that, in addition to the transformation to diamond, the shocked, compressed Ceylon graphite exhibited anomalous behavior at about 60 GPa. They interpreted this as a transformation to a new close-packed liquid, but this has not been confirmed by other investigators. Trunin et al. [9] reported that the data from Alder et al. [8] for P > 65 GPa are inaccurate as a result of interference by an elastic release wave caused by the very thin projectile plate that was used to obtain the higher pressures.

McQueen and Marsh [10] reported data to 80 GPa for pyrolytic and for several pressed graphites with densities varying from 2.2

*Work supported by U.S. Energy Research and Administration Agency under contract No. W-7405-Eng-48.

to 1.54 Mg/m^3. In the (U_s, U_p) plots, all exhibited discontinuities that were interpreted as being related to the transformation to the diamond phase. For the most part, the slopes for the various plots were parallel and were separated according to the initial density of each type. However, it is noteworthy that their data for a cold-pressed graphite with ρ_o = 2.13 Mg/m^3 indicate that the slope for the diamond-phase portion of the (U_s,U_p) plot (U_p > 3.0 km/s) is considerably steeper than the slopes for the diamond-phase portions for the other plots. As a matter of fact, it crosses the plot for pyrolytic graphite.

Using x-ray diffractometry on recovered specimens, Trueb [11,12] found that shock compression (100 GPa) of full-density graphite inclusions in an iron matrix produced recoverable cubic and hexagonal diamonds; similar experiments with mechanically-compacted mixtures of copper and graphite produced only cubic diamonds.

Recently Vereshchagin et al. [13] reported anomalies in measurements of the electrical resistivity of diamond at about 100 GPa which they interpret as evidence for a transformation from diamond to a metallic state.

Recent improvements have made the two-stage, light-gas gun at Lawrence Livermore Laboratory into a dependable source of very energetic planar shocks. Also, the inclined-prism technique has proved useful in observing anomalous behavior of materials, especially at low pressure. The availability of these improved techniques presented an opportunity to re-examine the low-pressure, shock-compression characteristics for graphite and, in a search for an additional transformation, to extend some of the data to higher pressures.

METHOD

We used the inclined-prism technique for experiments to 30 GPa; the response of flash gaps from 50 to 70 GPa; and coaxial, shorting-pin sensors on the two-stage, light-gas gun for pressures up to 130 GPa. These experimental systems and the Hugoniot relations necessary for their use have been described in detail elsewhere [6,10,14]. Consequently, we have restricted our remarks to those aspects of substantial concern in these experiments, namely, a detailed description of the experimental geometry and diagnostics used for cases where anomalous behavior was observed.

With the inclined-prism technique the loss of total reflection within back-lighted prisms is used to measure shock transit times and to resolve multiple wave structures, if present. Shock transit times are measured with prisms placed flat on the base plate and sample surfaces. The portion of the prism inclined at a small angle allows resolution, by continuous streak-camera scans, of the precise time that total internal reflection is lost by impact of the free surface with the inclined surface. Multiple wave structures are resolved by observing a discontinuity that occurs when $\vec{U}_{s_2} + \vec{U}_{p_1}$ overtakes the boundary moving at U_{fs_1}. The system is shown schematically in Fig. 1, and a record from a low pressure experiment is shown in Fig. 2.

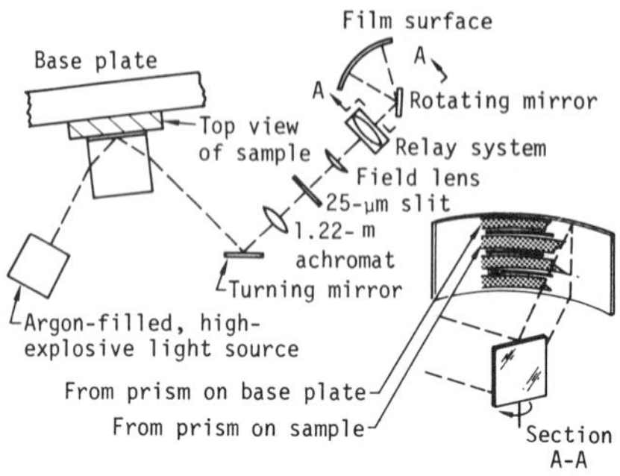

Fig. 1. Schematic of inclined-prism optics.

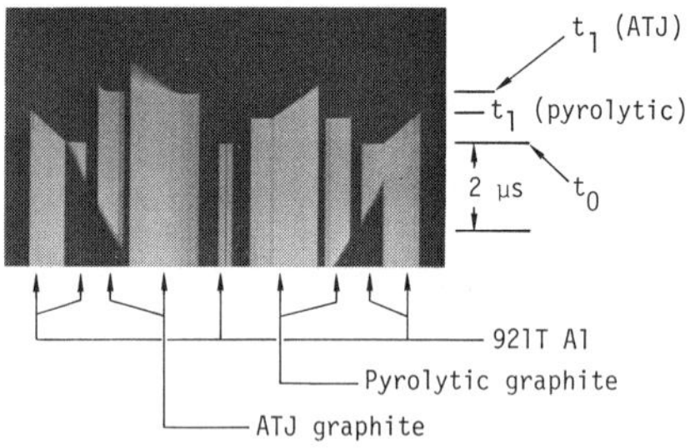

Fig. 2. Inclined-prism record for pyrolytic and ATJ graphite from a low-pressure experiment. Shock-front break out is at t_o; t_1 marks the shock front arrival at the sample surface. Free-surface velocities are obtained from the slopes of the traces.

In the gun experiments, 1.5-mm-thick plates of Ta were impacted on 25.4-mm-diam by 3.2-mm-thick graphite samples. The geometry is shown schematically in Fig. 3. Six foil switches were glued radially onto the sample at 60° intervals; they were made of 5-μm-thick manganin foils enclosed in 12-μm-thick mylar. The foil switches were glued radially onto the sample with a 1-μm-thick layer of epoxy; total thickness of a foil switch was about 30 μm. An 0.8-mm-thick Ta shim protected the foil switches from possible pre-impact damage by hot gases.

On the back side of each sample six coaxial, self-shorting pins were placed in contact with the graphite in a 12.7-mm-diameter array so that each was directly in line with a foil switch, thereby reducing the effect of impactor tilt. A seventh pin was located at the center.

Upon impact, the resulting pulses were recorded on ten oscilloscopes arranged in five master-slave systems. Oscilloscope sweep nonlinearity was accounted for by placing an accurately known sine wave (10 ns/cycle) on each shot record. Corrections were made for differences in signal-transmission line lengths and pin-closure times by conventional methods. Compression parameters were determined through impedance matches that were centered on the Hugoniot [15] for Ta, with the Ta particle velocity equal to one-half the projectile velocity. Total uncertainty in the measurements is estimated to be less than 2%.

When the material is in the diamond phase, the "catch-up" and lateral unloading waves propagate in the diamond at $\vec{C}_* + \vec{U}_p > \vec{U}_s$.

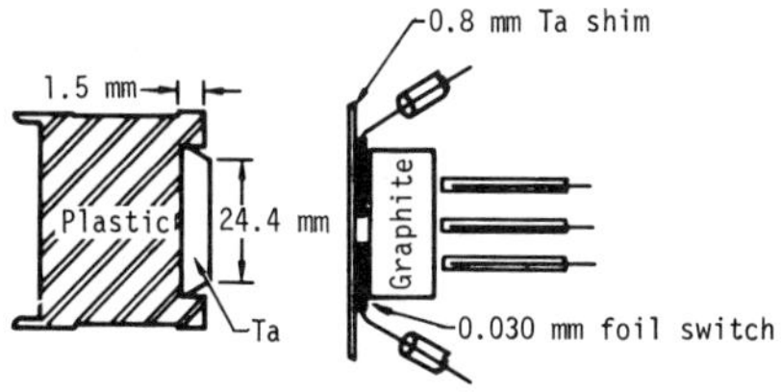

Fig. 3. Schematic of projectile and target. The Ta shim, graphite sample, and outer conductor were at ground potential. The manganin foils and pin center conductors were biased to 120 V.

The speed of sound in compressed diamond may be very high (>15 km/s). Consequently, the geometry must be designed to eliminate interaction of those waves with the shock front.

Symmetrical impact is obtained when the Ta projectile strikes the Ta shim. The impact simultaneously generates a forward-facing wave that propagates at $U_{s(Ta)}$ in the shim and a backward-facing wave that also propagates at $U_{s(Ta)}$ in the impactor. When the forward-facing wave reaches the Ta-graphite interface, part is reflected back into the compressed Ta shim and part is transmitted into the graphite where it propagates at $U_{s(gr)}$. This forward-facing wave is the one of primary interest.

Meanwhile, the backward-facing wave propagates in uncompressed Ta to the Ta-plastic sabot interface where part is transmitted and part is reflected. The wave of interest is the reflected one which now is a simple forward-facing rarefaction wave propagating in the compressed Ta at $\vec{C}_* + \vec{U}_p$. After it overtakes the Ta shim-carbon interface it propagates at $\vec{C}_* + \vec{U}_p$ (dia). In this geometry the use of 3.2-mm-thick samples effectively eliminated the interaction of catch-up waves.

The edge rarefaction wave propagates laterally through the sample at a velocity $\vec{C}_* + \vec{U}_p > \vec{U}_s$. The lateral penetration from each edge of the sample is given by

$$H \tan \alpha = H \left(\frac{C_*^2 - (U_s - U_p)^2}{U_s^2} \right)^{1/2} \qquad (1)$$

Placing the shorting pins within a diameter of 12.7 mm and on a 3.2-mm-thick by 24.4-mm-diameter sample eliminated edge rarefaction interactions from these measurements.

MATERIALS

The densities of the pyrolytic samples were about 2.2 Mg/m^3. Debye-Scherer patterns disclosed a strong orientation of the c-axis with respect to the normal of the major surfaces and a lack of order for other major axes. X-ray fluorescence analyses did not detect significant amounts of impurities.

The Ceylon graphite samples were cold-pressed from powder (Asbury Graphite Mills). Sample densities were about 2.14 Mg/m^3. X-ray fluorescence analysis of pressed samples indicated that impurities present in significant amounts were S, Fe, and Ca at about 0.3, 0.4, and 0.1 at.%, respectively. Impurity content varied considerably from sample to sample. Spectrochemical analysis of the powder showed about 0.3, 0.5, and 0.25 at.% of Si, Zr, and Fe, respectively, with small amounts (0.05 to 0.005 at.%) of Al, Mg, Cu, Ti, Mu, Ni, Mo, Cr, Sr, B, Ba, Ag, and Co.

X-ray diffraction analysis showed a strong orientation of the c-axis with the pressing axis which was also the shock propagation direction. Two discrete crystallites of FeS_2 were detected.

The hot-pressed, synthetic graphite samples were machined from UCAR-ATJ molded graphite (Union Carbide Co.). There were no significant amounts of impurities present, and to a first approximation, the crystallographic orientation was random.

RESULTS

The data are displayed in Figs. 4 and 5. For all three materials the discontinuities exhibited in the (U_s,U_p) plots are

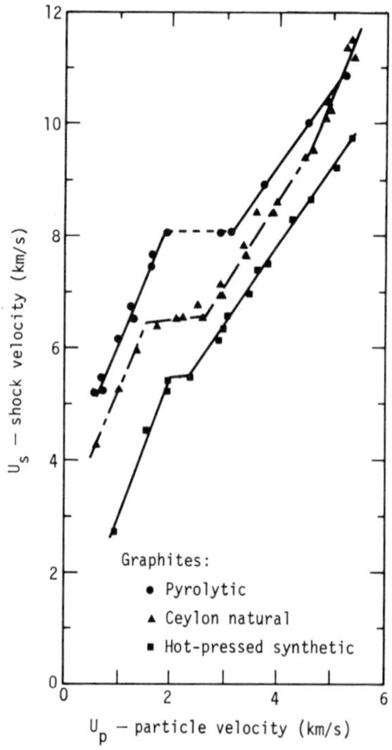

Fig. 4. Shock velocity as a function of particle velocity for shock-compressed pyrolytic, Ceylon natural, and ATJ graphites. Discontinuities at 1.5 to 2.0 km/s mark the onset of transitions from the graphite to the diamond structure; those at 2.25 to 3.0 km/s, the completions. The discontinuity in the plot for Ceylon natural graphite at $U_p = 4.5$ km/s appears to be related to an additional transformation.

taken to be related to the initiation and completion of a solid-solid phase transition (graphite to diamond forms). No related two-wave shock structures were observed for the pyrolytic or ATJ samples (inclined prism experiments were not performed on Ceylon graphite samples).

The (U_s,U_p) plot for Ceylon graphite displayed an additional discontinuity at $U_p = 4.5$ km/s (92 GPa). Above 92 GPa the material was less compressible than diamond. The samples with random crystallographic orientation and lowest initial density (ATJ) transformed to diamond at lower pressure than did the more dense, highly oriented ones.

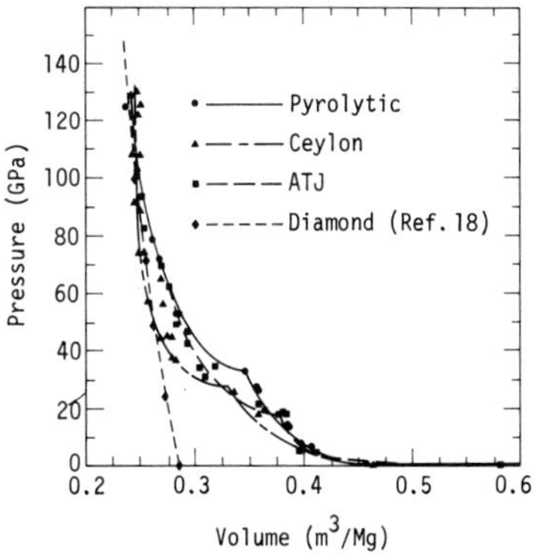

Fig. 5. Comparison of pressure vs. volume data for pyrolytic, Ceylon natural, and ATJ graphite.

DISCUSSION

Results obtained for pyrolytic graphite are in excellent agreement with those from McQueen et al. [10], with one exception. The latter reported a discontinuity, which we did not observe, at $U_p = 0.6$ km/s. Our experiments were not performed at quite as low a pressure as theirs, but utilized inclined prisms that provide very sensitive measurements of anomalous behavior at low pressure. Also, we used 921T Al for a baseplate rather than 2024 Al. At low pressure, a weak elastic wave ($C_L = 6.33$ km/s) exhibited by 2024 Al might provide a precursor that eventually interacts in the graphite to degrade flash gap response.

The discontinuity at $U_p = 1.9$ km/s is taken to be a manifestation of the initiation of a solid-solid graphite to diamond transition. The horizontal segment from $U_p = 1.9$ to 3.1 km/s apparently is a mixed-phase region; the discontinuity at 3.1 km/s marks the completion of the transition. No evidence for an additional phase change is discernible up to $U_p = 5.2$ km/s.

The response in the mixed-phase region differs from the response noted for the shock compression of pyrolytic (graphite-like) boron nitride (BN) [16]. Shock-compressed BN exhibits a somewhat similar response as it transforms to diamond-like structures except the (U_s, U_p) plot has a slope of 1.02 in the mixed-phase region. This slope indicates the possibility of a shock-induced transformation to a metastable wurtzite [17] rather than zinc-blende structure. The absence of a slope in the mixed-phase region makes it appear unlikely that there is any direct transformation from pyrolytic graphite to hexagonal diamond under the conditions of these experiments.

Our plot for ATJ graphite also exhibits characteristics indicative of a transition to diamond and agrees well with data from McQueen et al. [10]. Also, an extrapolation of the ATJ plot agrees reasonably well with the data from Pavlovskii et al. [3] for a synthetic graphite at 300 GPa. No evidence of additional anomalous behavior was seen.

Our data for cold-pressed Ceylon graphite is in excellent agreement with that of McQueen and March [10] for a pressed, pure graphite of similar density ($\rho_o = 2.13$ Mg/m^3) up to $U_p = 3.0$ km/s. A comparison indicates that as far as the graphite-diamond transition is concerned, the presence of small amounts of catalytic impurities (Fe) in our samples had no discernible effects. For

$U_p > 3.0$ km/s, the slope for McQueen and Marsh's plot is considerably steeper than ours.

In the second-phase regime, $U_p = 2.6$ to 4.5 km/s. Our data obtained through use of the two-stage, light-gas gun plotted approximately parallel to that for the pyrolytic and ATJ graphites, indicating that in that regime all three materials behave alike and probably have the same structure.

A third discontinuity was found in the (U_s, U_p) plot for Ceylon graphite at $U_p = 4.5$ km/s (see Fig. 4). This response appears to be unique and happens only to graphite with an initial density of about 2.14 Mg/m^3. Some possible causes include melting, a transition from diamond to a metallic state, and shock-front degradation by an overtaking rarefaction wave from the thin Ta impactor. Obviously, the latter must be excluded before the others are considered.

For normal materials, a simple rarefaction wave propagates in the compressed materials at $C_* + U_p = (V_o/V)\, C_o + U_p$. We have used this relation for describing rarefaction wave propagation in the compressed Ta.

However, for materials that undergo shock-induced phase changes, the method is inaccurate unless the volume ratio is related only to the high-pressure phase. Here, we have used shock-compression data by Pavlovskii [18] on single-crystal diamond to determine that at 200 GPa, $C_{*dia} = 1.3\, C_{o\,dia}$. Bulk sound speed at 1 atm is 11.2 km/s [19]. We have assumed, then, that the rarefaction wave propagates in the transformed graphite at $C_* = 14.6 + U_p$ km/s.

A time vs. distance plot that uses these assumptions is shown in Fig. 6. Similar plots made for each set of data for $U_p > 4.5$ km/s gave no indication that the discontinuity in the slope of the (U_s, U_p) plot at $U_p = 4.5$ km/s was caused by an overtaking rarefaction wave.

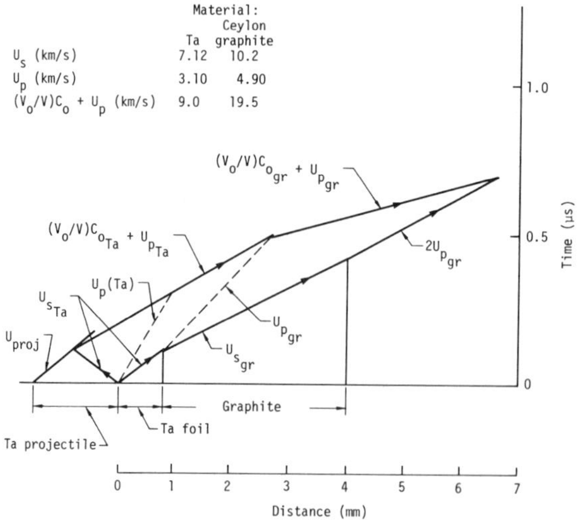

Fig. 6. Distance vs. time diagram comparing the arrival times of the main shock at the sample surface and of the rarefaction wave from the back side of the Ta flyer plate (data set for U_p of 4.5 km/s in Ceylon graphite). Note that the rarefaction wave does not overtake the main shock during the time of the measurement.

To better understand the state of carbon at the highest shock pressures, it is worthwhile to estimate the temperature along the Hugoniot. This was done for pyrolytic and Ceylon graphites using a simple Grüneisen equation of state model. Here we assume that $C_v = 3R$ and that the Grüneisen γ is a piecewise continuous function composed of a graphite segment, a diamond segment, and a two-phase-mixture segment. Each of the segments is linear in volume, and the segments are normalized to the accepted γ-values for graphite and diamond at their normal densities. Given C_v and γ, together with the Hugoniot, it is possible to compute the zero-degree isotherm and the Hugoniot temperature.

In Fig. 7, the Hugoniots for the pyrolytic and Ceylon graphites are shown in the pressure-temperature plane that is superimposed upon theoretical [20] and experimental [21] phase diagrams. Compared with the experimental phase diagram, the theoretical melting temperature appears too high. However, the experimental melt-temperatures may in fact be much too low [22].

Van Vechten [20] predicts a diamond-to-metallic transition with a diamond-metal-liquid triple point at 118 GPa and 3080 K. The computed Hugoniot for pyrolytic carbon passes close to this point, while that for the Ceylon graphite crosses the melt curve at a lower pressure.

It is perhaps significant that the kink in the (U_s, U_p) curve for Ceylon graphite occurs close to the theoretical diamond-liquid boundary. The kink could represent the shift from the diamond lattice to the more closely packed and less compressible metallic liquid. The lower-density ATJ graphite has a maximum shock pressure of 91 GPa, while the pyrolytic graphite enters the liquid (if at all) at a pressure above 110 GPa. Thus, in neither of these cases would we expect to see the effect of melting as clearly as in the Hugoniot for Ceylon graphite.

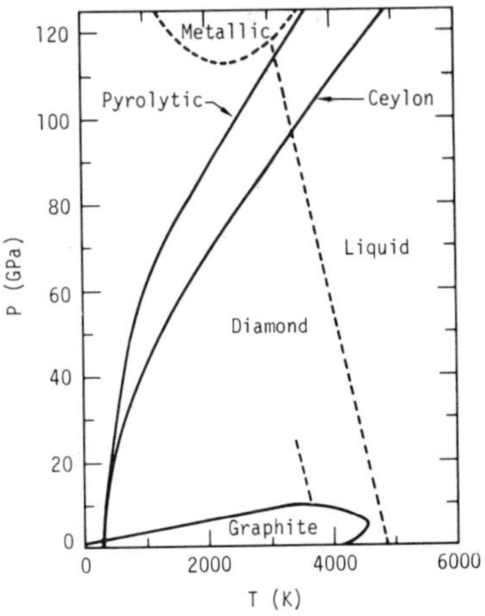

Fig. 7. Approximate Hugoniot calculations for pyrolytic and Ceylon graphite superimposed on experimental and theoretical phase diagrams. The experimental low-pressure diagram (solid lines) is from Fateeva and Vereshchagin [21] and the theoretical phase diagram (dashed line) is from Van Vechten [20].

NOTATION

C_L = longitudinal sound velocity
C_o = sound velocity in a material at initial pressure
C_* = sound velocity in a material at high pressure
H = sample thickness
P = pressure
U_{fs} = free surface velocity
U_p = particle or mass velocity behind the shock front
U_s = shock velocity
V = specific volume
α = angle between $\vec{C}_* + \vec{U}_p$ and $\vec{U}_s$
ρ = density

REFERENCES

1. P. S. DeCarli and J. C. Jamieson, Science 133, 1821 (1961).
2. F. P. Bundy and J. S. Kasper, J. Chem. Phys. 46, 3437 (1967).
3. M. N. Pavlovskii and V. P. Drakin, Zh. Eksp. Teor. Fiz. Pis'ma Red. 4, 169 (1966); English Translation JETP Lett. 4, 116 (1966).
4. S. Minomura and H. G. Drickamer, J. Phys. Chem. Solids 23, 451 (1962).
5. W. Paul and H. Brooks, Phys. Rev. 94, 1128 (1954).
6. W. H. Gust and E. B. Royce, J. Appl. Phys. 42, 1897 (1971).
7. W. H. Gust and E. B. Royce, J. Appl. Phys. 43, 4437 (1972).
8. B. J. Alder and R. H. Christian, Phys. Rev. Lett. 7, 367 (1961).
9. R. F. Trunin, G. V. Simikov, B. V. Moiseev, L. F. Popov, and M. A. Podurets, Zh. Eksp. Teor. Fiz. 56, 1161 (1969); English translation Sov. Phys.-JETP 29, 628 (1969).
10. R. G. McQueen and S. P. Marsh, in Proc. Symposium on High Dynamic Pressures, Gordon and Breach, New York (1968), p. 207.
11. L. F. Trueb, J. Appl. Phys. 39, 4707 (1968).
12. L. F. Trueb, J. Appl. Phys. 42, 503 (1971).
13. L. F. Vereshchagin, E. N. Yakovlev, B. V. Vinogradov, V. P. Sakum, and G. N. Stepanov, High Temp.-High Press. 6, 505 (1974).
14. M. Van Thiel, B. L. Hord, and K. Boutwell, in Proc. 4th Intern. Conference on High Pressure, Kawakita Printing Co., Kyoto, Japan (1975), p. 546.
15. J.J. Folkins and W. H. Gust, Bull. Am. Phys. Soc. 20, 1514 (1975).
16. W. H. Gust and D. A. Young, Phys. Rev. B 15, 5012 (1977).
17. F. R. Corrigan and F. P. Bundy, J. Chem. Phys. 63, 3812 (1975).
18. M. N. Pavlovskii, Fiz. Tverd, Tela. 13, 893 (1971); English translation Sov. Phys.-Solid State 13, 893 (1971).
19. H. J. McSkimmin and P. Andreatch, J. Appl. Phys. 43, 2944 (1972).
20. J. A. Van Vechten, Phys. Rev. B 7, 1479 (1973).
21. N. S. Fateeva and L. F. Vereschagin, Zh. Eksp. Teor. Fiz. Pis'ma Red 13, (1971).
22. G. R. Gathers, J. W. Shaner, and D. A. Young, Lawrence Livermore Laboratory, Rep. No. UCRL-51644 (1974).

T-7

CARBONADO-TYPE DIAMONDS AND POLYCRYSTALLS OF CUBIC BORON NITRIDE (β BN)*

E. N. Yakovlev

Institute of High Pressure Physics, Academy of Sciences of U.S.S.R.
Moscow, U.S.S.R.

INTRODUCTION

For a long time it was thought that there were only two ways of manufacturing diamonds: first, by growing the monocrystalls and second, by sintering diamond powder. The possibility of obtaining polycrystalline diamond and cubic boron nitride (β BN) by any other means had not been considered. However, a new scientific direction in the synthesis of superhard materials was postulated as a result of the long-standing efforts of a group headed by Academician L. F. Vereschagin. These efforts resulted in obtaining polycrystalline diamonds and polycrystalline cubic boron nitride. By the use of corresponding thermodynamic parameters, we were able to obtain strong fine-grained diamonds and polycrystalline cubic boron nitride stones with the structure of a carbonado-type diamond (see Fig. 1).

It is evident that to obtain polycrystalline diamond structures the synthesis pressure should be as far as possible above the pressures of the graphite-diamond equilibrium curve on the carbon phase diagram. Thus, to obtain polycrystalline diamonds having a microstructure of natural carbonado, it is desirable to apply pressures exceeding 80 kbar at temperatures above 1200°C, i.e., the minimum temperature associated with the synthesis of a metal catalyst. However, in order to maintain the predetermined synthesis temperature over the entire period of polycrystalline diamond formation it is necessary to remove the heat of crystallization from the reaction chamber. Removal of this heat of crystallization from the reaction chamber permits synthesis of polycrystalline structures having a fine-grained structure [1-3]. In this paper we will examine the properties of these synthesized materials used for tools and high-pressure equipment [5-10].

*Invited paper.

Fig. 1. Examples of carbonado type diamond in cylindrical form, 5 and 10 mm in diameter.

PROPERTIES OF CARBONADO-TYPE DIAMONDS AND CUBIC BORON NITRIDE

One of the main characteristics of diamond is its abrasive ability Q. The value of Q is determined by the ratio of the mass an abrasive wheel loses during the process of dressing it with diamond tool to the mass of diamond spent. To examine the abrasive ability of carbonado-type diamond, corundum wheels with average hardness (CTI) and grain (420-500 mkm) were used. The dressing regime for the wheels involved a speed up to 20 m/sec, depth to 0.02 mm and a feed of 0.1 mm/rev. Water cooling was also used in this process. The abrasive ability of carbonado-type diamond was determined to be 2-10 x 10^5. For cutting tools, carbonado-type diamonds with an abrasive ability 3 x 10^5 or greater are required.

The cutters made from carbonado-type diamonds successfully treated cemented carbides on the basis of tungsten carbide (WC + 1%Co). For example, when the working regime was a cutter speed of 12 - 14 m/min, a depth of cutting of 0.10 mm and a longitudinal feed of 0.04 mm/rev, the life of machining tools was 30 min (see Fig. 2).

Fig. 2. Different types of cutters with carbonado and polycrystalline βBN.

Another example of using the cutters concerns the machining of abrasive material "silumin" (Al + 20% Si). This material is used in manufacturing automobile motor pistons and other items. The working life of the cutters containing carbonado-type diamonds exceeds 4,000 min, while the working life of cutters containing tungsten carbide is 5 min.

It is well known that diamonds are used for measuring hardness. Tests show that carbonado-type diamond is successfully used in Rockwell devices instead of natural diamond tips.

Diamonds are also usually used in such important operations as forming smooth surfaces. Plastic deformation of surfaces during this operation permits strengthening of these materials and increasing their stability against cyclic loads (see Fig. 3). To obtain comparative data for developing smooth-surfacing tools from natural diamond and carbonado-type diamond an item was made from III-X-15 steel and treated at speeds of 20-300 m/min; the pressure used for making the smooth-surface cutting tools was 3,000 kg/cm^2. Tests of 100 tools of natural diamond gave coefficients of friction of 0.08 to 0.09. For 400 samples of carbonado tools the coefficients ranged from 0.11 to 0.12.

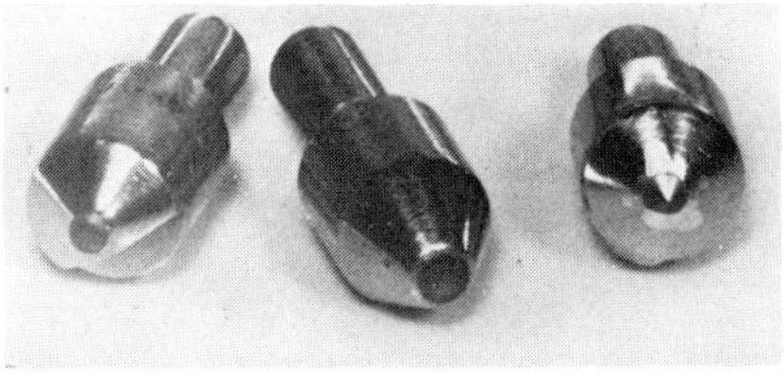

Fig. 3. Tools constructed from carbonado for making smooth surfaces.

Polycrystalline cubic boron nitride stones [4] are used for the treating of tempered to HPC 60-65 steels. Originally these steels were treated by burnishing only. Their usage of turning tools made from cubic boron nitride and smooth-surfacing tools made from carbonado-type diamond simplifies the production of these steel items.

USE OF THESE MATERIALS IN HIGH-PRESSURE WORK

Carbonado-type diamonds have opened a new stage in high-pressure techniques. Cemented carbides now permit obtaining pressures up to 300,000 kg$_f$/cm^2, while carbonado-type diamonds permit obtaining pressures up to 3 Mbar. Obtaining such high pressures makes it possible to undertake new investigations, in particular, experiments on the metallization of insulators under pressures of about 3 Mbar. These experiments are very important to fundamental physics and for its applications.

In the experiments for converting insulators into conductors under pressures of up to 1 Mbar the discovery of the transition into the conductive state for sodium chloride (NaCl), corundum (Al_2O_3), water (H_2O), silica (SiO_2), sulfur and other substances should be noted. Polycrystalline βBN was used in x-ray investigations (see Fig. 4) for pressures up to 200 kbar [9]. The experience obtained during the investigation of this wide range of insulators made it possible to perform certain experiments on the metallization of hydrogen. At the Institute of High-Pressure Physics of the USSR Academy of Sciences, L. F. Vereschagin Yu. A. Timofeev and the author have built a high pressure apparatus for studying the dielectric-metal transition of hydrogen. In these studies the conductivity of hydrogen was determined at 4.2 K and ~1 Mbar. Metallic hydrogen returns to the dielectric state as the pressure is reduced.

These examples show that the newly created materials allow us to carry out technological processes more economically than previously, and to obtain scientific information about the behavior of matter under superhigh pressures equal to 1 Mbar.

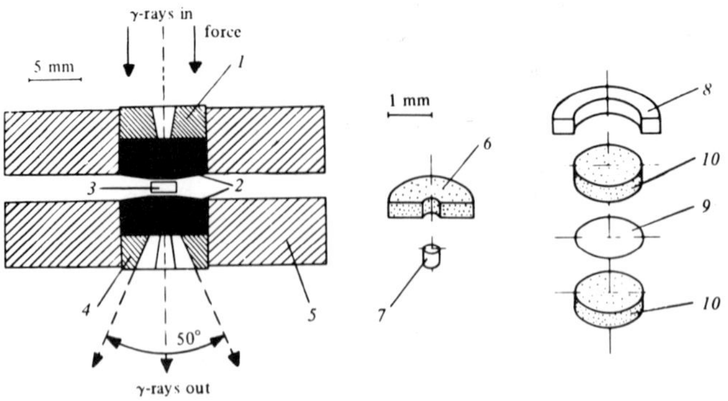

Fig. 4. (a) Schematic diagram of cubic boron nitride cell with x-ray or Mössbauer radiation path coincident with pressure axis; (b) and (c) two specimen assemblies. (1) and (4) are hard-alloy anvil seats, (2) indicates the cubic boron nitride anvils, (3) is the specimen, (5) is the steel plate, (6) is the pellet pressed from 85% B - 15% LiH mixture, (7) is the pressed NaCl plug, 0.3 mm in diameter, (8) is the mild steel compression ring, (9) is the Mössbauer absorber, and (10) indicates the disks pressed from 85% B - 15% LiH.

REFERENCES

1. L. F. Vereschagin, E. N. Yakovlev, T. D. Varfolomeeva, and L. E. Sterenberg, Dokl. Acad. Nauk SSSR 185, 555 (1969).
2. L. F. Vereschagin, E. N. Yakovlev, V. N. Slesarev, T. D. Varfolomeeva, I. S. Gladka ja, G. A. Dubitsky, and L. E. Sterenberg, Dokl. Acad. Nauk SSSR 191, 345 (1970).
3. L. F. Vereschagin, E. N. Yakovlev, T. D. Varfolomeeva, A. Ya. Preobrazhensky, V. N. Slesarev, V. A. Stepanov, and L. E. Sterenberg, Patent USSR N32971, Oat. N329760; Patent United Kingdom N1300650; Patent United Kingdom N1334846.
4. L. F. Vereschagin, F. I. Dubovitsky, and A. N. Dremin, Patent USSR N411721; Patent France N2146345.
5. L. F. Vereschagin, E. N. Yakovlev, and Yu. M. Kovaltchuk, Sb. Almazi, NiiMasch, Moscow, U.S.S.R. 1, 2 (1971).
6. L. F. Vereschagin, E. N. Yakovlev, and V. N. Slesarev, Sb. Almazi, NiiMasch, Moscow, U.S.S.R. 1, 6 (1971).
7. E. N. Yakovlev, V. N. Slesarev, and V. M. Gorelik, Sb. Almazi, NiiMasch, Moscow, U.S.S.R. 1, 17 (1971).
8. L. F. Vereschagin, E. N. Yakovlev, B. V. Vinogradov, and V. P. Sakun, in Proceedings of 4th Intern. High Pressure Conference, Kyoto, Japan (1975), p. 860.
9. E. V. Kapitanov, V. N. Paniushkin, L. F. Vereschagin, and E. N. Yakovlev, High Press.-High Temp. 7, 555 (1975).
10. L. F. Vereschagin, E. N. Yakovlev, and Yu. A. Timofeev, Pisma IETPH 21, 190 (1975).

T-8

USE OF NEW SYNTHETIC POLYCRYSTALLINE MATERIALS -- CARBONADO-TYPE DIAMOND AND CUBIC BORON NITRIDE

Yu. S. Konyaev

Institute of High Pressure Physics, USSR Academy of Sciences
Moscow, U.S.S.R.

INTRODUCTION

The design of tools using synthetic superhard materials that allow operation with minimum cutting forces and maximum dimensional stability is one of the main trends in improving the precision and quality of machining. Usually only abrasive tools have been made of synthetic superhard materials based on diamond and cubic boron nitride. The abrasive tool is used in machining hard alloys, tempered steels, and various hard-to-work alloys. However, the entire problem of the precision working of machine parts cannot be solved by using abrasive tools alone — a bladed tool of superhard materials is needed.

The synthesis of polycrystalline carbonado-type diamonds has opened up the possibility of creating fundamentally new types of metalworking, drilling, and stoneworking tools. The outstanding properties of the diamonds that have been developed also have made it possible to use them with greater effectiveness on materials that did not previously submit to cutting.

CARBONADO CUTTING TOOLS

Carbonado cutting tools have been used to work fiberglass plastics, plastics, hard alloys, nonferrous metals, titanium alloys, and some types of ceramics. (See Table I.)

Close tolerances in part dimensions have been obtained. Depending on the material being machined and the cutting conditions, machining with carbonado-fitted cutting tools produces a surface roughness of 2.5 to 0.1 μm.

An example of the effective industrial use of cutting tools is the machining of silumin (12% silicon) pistons under the following

Table I. Recommended Cutting Conditions for Carbonado Cutting Tools

Material being machined	Cutting speed, m/min	Longitudinal feed, mm/rev	Depth of cut, mm	Roughness of machined surface, microns
Fiberglass and plastics	400–600	0.04–0.07	0.5–2.0	2.5–1.6
Nonferrous materials	300–700	0.02–0.07	0.1–0.5	0.63–0.1
Ceramics	200–300	0.04–0.07	0.3–0.5	1.25–0.4
Hard alloys	10– 30	0.02–0.07	0.1–0.75	1.25–0.4
Titanium alloys	80–100	0.02–0.07	0.1–0.2	1.25–0.4

conditions:

Depth of cut, 0.15 to 0.2 mm; longitudinal feed, 0.06 mm/rev; and cutting speed, 400 m/min.

The roughness of the machined surface varies from 0.63 to 0.4 µm, and the tool life (operating time to bluntness) is 50 to 100 hr of machine time. In working fiberglass plastics (fiberglass filled) at a cutting speed of 500 m/min, a feed of 0.06 mm/rev, and a depth of cut of 1 mm, the life of the cutting tool is 300 min (100 times as great as for cutting with hard alloy cutting tools) and the surface roughness is 2.5 µm.

CARBONADO SMOOTHING TOOLS

Carbonado-fitted smoothing tools are used to smooth out the outer and inner surfaces of parts made of heat-treated steels that submit to cold plastic deformation. Under the pressure of the smoothing tool protruding microirregularities of the machined surface are plastically deformed. As this occurs, the surface roughness is reduced by a factor of 2 to 3 and the surface layers of the parts are strengthened.

Life of smoothing tools made of the new synthetic material is similar to the life of smoothing tools made of natural diamond. Practical experience with smoothing shows that this process is a highly effective type of machining. For example, when internal gages 20 mm in diameter, 15 mm long, made of steel 20 (carburized) with a hardness HRC of 62–64 are smoothed, the roughness of the machined surface varies from 0.04 to 0.02 µm.

WIRE-DRAWING DIES

Carbonado wire-drawing dies (Fig. 1) are effective when used to draw tungsten, molybdenum, and steel brass-plated wire and have a life similar to wire-drawing dies made of natural diamond. Typical operating conditions for wire-drawing in carbonado dies are presented in Table II.

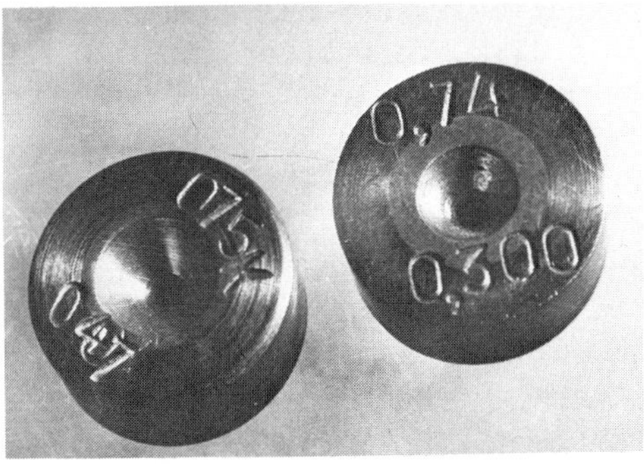

Fig. 1. Wire-drawing dies.

CARBONADO DRILLING TOOL

The high hardness, wear resistance, and impact strength of synthetic polycrystalline diamonds, as well as the possibility of developing such diamonds in required shapes, have made it possible to consider using them in a drilling tool. However, the somewhat lower heat resistance of polycrystalline diamonds compared with natural diamonds has prevented the use of traditional industrial methods of manufacturing the drilling tool (hot pressing and infiltration with low-melting metals and alloys such as copper and brass).

Table II. Wire-Drawing Conditions

Rate of wire-drawing from tungsten and molybdenum, m/min	25-100
Rate of drawing for brass-plated steel wire, m/min	400-800

A method of making a drilling tool by simultaneous application of pressure and temperature, which ensures secure locking of the diamond grains in the metal-ceramic matrix and their timely exposure as blunting occurs (i.e., self-sharpening), has been devised and proven at the Institute of High-Pressure Physics of the USSR Academy of Sciences. This method makes it possible to obtain a matrix of different hardnesses which offers the possibility of using it in making a tool for drilling extremely hard rocks (up to drillability categories of IX to XI). (See Fig. 2.) Tests of drill bits fitted with both whole and crushed carbonado-type polycrystalline diamonds (Fig. 3) fabricated by this method show that their life exceeds that of tungsten carbide bits by a factor of 10 to 20 and that of natural diamond bits by a factor of 2.

Fig. 2. Drill bits.

TOOL FOR THE CONSTRUCTION INDUSTRY

A carbonado grinding tool for evening out cement-concrete road surfaces and reinforced concrete products (wall panels, landings, and flights of stairs in residential and industrial buildings) has been designed at the Institute.

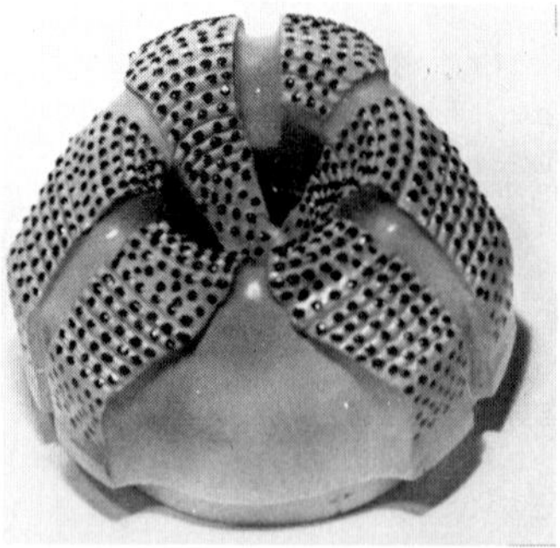

Fig. 3. Oil drill bit.

It is a cylindrical sectional cuphead in the form of a disk, reinforced with carbonado diamonds (Fig. 4). A powdered metal composite material with a hardness HRC of 25 and 35 is used as the binder.

Tests show that the design has high strength and reliability. The method of securing it to the grinder chucks ensures uniform wear of the diamond-carrying layer over the entire surface of the cupheads. During operation, the binder material is ground away to the extent that the diamonds are raised above the surface by 0.4 to 0.5 mm. This creates favorable conditions for intensive cutting. Moreover, the high strength of the binder and its dense coating of polycrystals promote their secure mounting in the tool, which accounts for its increased life. The tool has a life five times longer than that of natural diamond tools and over 100 times longer than that of an abrasive corundum tool.

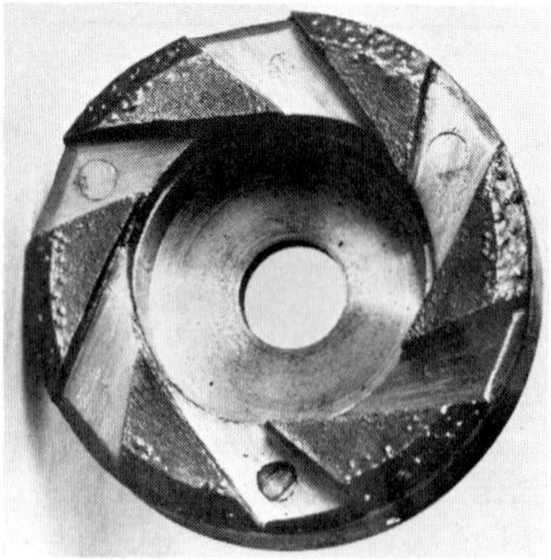

Fig. 4. Grinder for surface finishing of building materials.

DIAMOND SEGMENTS FOR THE STONEWORKING INDUSTRY

Highly effective stoneworking segments made of carbonado synthetic diamond have been created for the first time (Fig. 5). On the basis of a range of scientific studies and bench and production tests, polycrystalline diamond segments are recommended for fitting skelps and circular saws for sawing rocks, synthetic construction materials, refractory materials, reinforced concrete, etc.

The advantages of the segments are: (1) The use of a high-strength powdered metal matrix of unique composition that ensures reliable locking and holding of the polycrystals until they are completely consumed; (2) the design of the diamond segment, which is divided into a diamond-carrying and a diamond-free layer, makes it possible to provide secure attachment (sealing) of the segment to the steel holder, to make repeated use of the tool's reinforcement steel, and also permits utilization of the segments until they are completely worn, a convenient coolant supply, and the removal of cuttings; (3) the possibility of cutting a wide range of materials having a wide range of physicomechanical properties; (4) the high qualities of the kerf surface (the average roughness is 10 to 20 μm for granite, and the deviation from prescribed thickness does not exceed 2 to 3 mm); (5) low consumption of diamonds (for granite

Fig. 5. Circular saw.

cutting, the diamond consumption is about 1 carat per square meter of cut); and (6) the timely self-exposure of the diamond polycrystals ensures a high cutting speed over the entire period of operation.

TOOLS MADE OF POLYCRYSTALLINE CUBIC BORON NITRIDE

The synthesis of polycrystalline cubic boron nitride (PCBN) opens up new possibilities in the machining of carboniferous materials (steel, cast iron, alloys, etc.). In hardness, carbonado polycrystals somewhat surpass PCBN, but the PCBN are somewhat superior in heat resistance. Further, materials based on boron nitride are chemically inert with respect to ferrous metals, while carbon-base materials are chemically reactive in their presence. This difference, above all others, determines their fields of application.

Polycrystalline cubic boron nitride is used to machine steels, cast iron, and various hard-to-work alloys, while polycrystalline diamonds are used to machine nonferrous metals, fiberglass plastics, hard and titanium alloys, and a number of other materials. Cutting tools equipped with PCBN cutting elements have found extensive application in the fine finishing, finishing, and semifinishing of steel and cast iron parts of different hardness and of some hard-to-work steels and alloys. The greater the hardness of the steel and cast iron and the higher the cutting speed, the more pronounced is the advantage of PCBN cutting tools over cutting tools fitted with plates of a hard alloy or powdered ceramic material.

Thus, for sharpening parts made of tempered high-speed steels with a hardness HRC of 62-64, the life of PCBN cutting tools at cutting speeds of 50-100 m/min exceeds the life of hard alloy cutting tools by factors of ten, and the life of cutting tools made of powdered ceramic material alloyed with titanium carbide by a factor of 3 to 5. In sharpening tempered steel parts with a hardness NRC of 45-50, the life of PCBN cutting tools is 3 to 5 times greater than that of hard alloy cutting tools.

The optimal cutting speeds for PCBN in the machining of cast iron lie in the range of 300 to 600 m/min, i.e., 4 to 5 times higher than for a hard alloy, and are approximately the same as for current grades of powdered ceramic materials. The wear resistance of PCBN in this range of speeds is 2 to 3 times greater than that of powdered ceramic materials and 15 to 20 times greater than that of a hard alloy. The greater hardness and heat resistance of PCBN enables a PCBN tool to retain high cutting properties for a longer period of time, even when cutting steels with a hardness HRC of 62-66.

The low cutting forces and high wear resistance of PCBN make it possible in cutting tempered steels to obtain a surface roughness of 1.25 to 0.08 μm and class 1-2 precision of machining. Here PCBN tools ensure the absence of structural changes in the surface layer of the machined surfaces.

The possibility of combining tools made of different materials suggests the possibility of creating new industrial machining processes. For example, machining tempered steels with polycrystalline cubic boron nitride cutting tools and subsequent smoothing with a carbonado tool completely eliminates grinding, which is labor-intensive, costly, and complex. There is no doubt that other applications will be developed in the future.

U-1

RECENT PROGRESS IN HYDROTHERMAL QUARTZ CRYSTALLIZATION*

R. A. Laudise

Bell Laboratories
Murray Hill, New Jersey USA

INTRODUCTION

Because of its unique piezoelectric properties α-quartz is a singularly important material in electronics and after Si comprises the largest volume single crystal prepared commercially. The high-pressure crystallization of quartz under hydrothermal conditions has a long and distinguished history which will not be reviewed here. The reader should consult one of several recent reviews [1,2]. High-pressure conditions are required so that temperature and solvent density may be kept in a range where the solubility of α-quartz is sufficient to allow controlled crystal growth. The apparatus [3] and physical chemistry of quartz solubility [4], growth [5] and impurity partition [6] have been well studied and the reader should consult recent papers for this information. In this paper we review three new developments in quartz growth: (1) Alternative nutrient to that available from Brazil as feed stock for recrystallization; (2) the elimination of strain and cracking in quartz; and (3) the preparation of dislocation-free quartz.

ALTERNATIVE QUARTZ NUTRIENTS

Quartz is crystallized by dissolving α-SiO_2 nutrient in the hot region (typically 350 to 400°C) of a hydrothermal solution at about 25,000 psi which is typically 1.0 molar NaOH (used to increase the solubility of SiO_2 over that of pure water so as to increase crystallization rates) and recrystallizing on suitably oriented seeds in a cooler region. Supersaturation is provided by a temperature difference (ΔT, typically 20 to 50°C) between dissolving and growth regions. Supersaturated solution is continually supplied by the rapid convection of solution. In the density-temperature range

*Invited paper.

of growth (0.70 to 0.85 g/cc) the partial of density with temperature of water ($\partial\rho/\partial T$) is large enough to provide very rapid convection so that neither dissolving nor transport of solute are ordinarily rate limiting. Until recently the nutrient almost universally used was Brazilian α-quartz "lascas" (high purity vein quartz) which is also used for vitreous silica manufacture and was comparatively abundant and inexpensive. However, in view of the volatile nature of single sources of supply it became necessary to establish alternative sources [7].

Silicon dioxide is the most abundant constituent of the earth's crust and quartz is one of the most abundant minerals [8]. Yet high purity quartz particularly in crystal form is not really common. High purity single-crystal quartz occurs in veins, pegmatites, and veins associated with pegmatites. In some areas, such as Brazil [9], all three occurrences are found, and the quartz from these sources is commingled. On the North American continent there appears to be only one significant vein quartz area in Arkansas [10], but there are many pegmatitic areas.

Figures 1 and 2 illustrate schematically typical configurations of vein quartz and pegmatitic quartz deposits. Vein quartz forms by hydrothermal crystallization in veins and cavities (vugs) in the earth in a process in many ways similar to the hydrothermal crystallization of quartz in the laboratory. Deposits of vein quartz are of purity and size to make them suitable for nutrient and for the fabrication of bulk piezoelectric devices, for which they were used in large quantities before the advent of quartz synthesis.

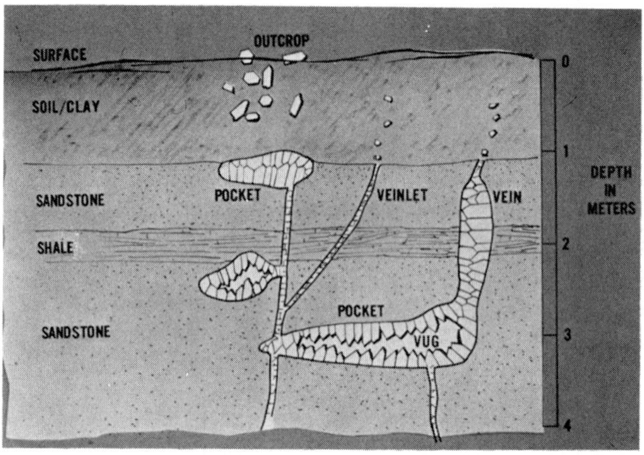

Fig. 1. Schematic of vein quartz deposit [7].

Pegmatites represent the late stages of crystallization of magmas [11-13]. Starting with a typical magma at 5 kbar containing 0.2% water at 1000°C, the residue after about 97% solidification will contain some 8% water at perhaps 700°C at 5 kbar. Continued crystallization during further cooling produces the many commercially-mined minerals found in pegmatites (mica, feldspars, Be, Li, Nb, Ta, R. E. minerals, etc.) as well as large quantities of quartz. Particularly in pegmatites of the zoned type (see Fig. 2), segregation can produce large cores of exceptionally pure, massive, coarse-grained quartz, from tens to hundreds of feet across; often

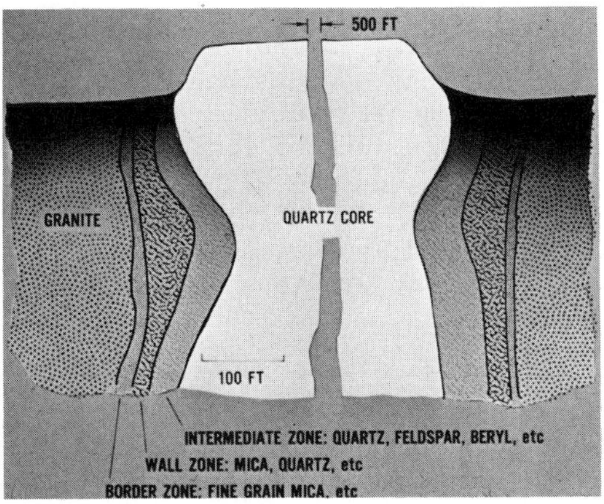

Fig. 2. Schematic of zoned granite-type pegmatitic quartz deposit [7].

constituting a whole mountain. The areas of significant concentrations of pegmatites on the North American continent are shown in Fig. 3. Because it was crystallized rapidly and is usually of poor optical quality (milky), pegmatitic core quartz has been only slightly exploited as an α-quartz source. Sometimes it is of low purity, but this is not always the case. If a given specimen is of high purity, it would generally be expected that a large part of the entire core body (the whole mountain) which crystallized with it could be of similar purity. From the above, it becomes apparent that appropriate purity deposits of pegmatitic quartz could provide an excellent nutrient and that if a new vein quartz source could be located it would also be very attractive. Elsewhere [7] we have detailed the search conducted for viable sources. Here we give only a brief summary.

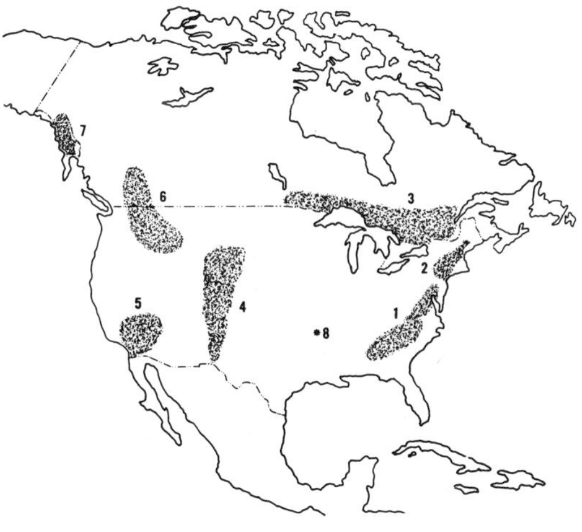

Fig. 3. Areas of significant α-quartz deposits in North America [7].

We reasoned that the crystal growth process could be practiced with no changes whatsoever provided a nutrient source chemically and physically matching those previously used could be found. Previous nutrient was lump α-quartz of relatively high chemical purity (<50 ppm of any single impurity). Common deleterious impurities are substitutional Fe^{+3} and Al^{+3} which are charge compensated by H^+ leading to OH inclusions.

Concentrations of Fe^{+3} and Al^{+3} much greater than 10 to 50 ppm can result in substantial H^+ inclusions; H^+ has been shown to severely degrade the acoustic Q of quartz, limiting its application at high frequencies [14]. Thus, impurities above about 50 ppm are to be avoided for ordinary piezoelectric applications.

A series of growth runs at standard hydrothermal conditions: 1.0M NaOH, 0.025M Li_2CO_3 (Li^+ has been shown to repress the uptake of OH and improve acoustic Q) [14], 82% fill, 25,000 psi, 350°C crystallization temperature, and 50° ΔT were made using $(01\bar{1}1)$ seeds and both vein quartz and pegmatite quartz nutrients. In all cases at least one source in each of the regions of Fig. 3 gave acceptable Q. Several of the regions gave sources with estimated deposits suitable for many years of mining. These included one vein and many pegmatite sources. Thus readily available pegmatite and a vein quartz are viable alternatives to previously-used Brazil quartz as a nutrient.

It is interesting to point out that samples of nutrient with impurity levels considerably above the 50 ppm requirement produced quite satisfactory quartz. In other experiments [7] we established that vitreous SiO_2, high purity sand, and silica gel nutrients could also be used with some modifications in growth procedure.

THE ROLE OF STRAIN IN QUARTZ GROWTH AND USEFULNESS

Quartz is usually grown on (0001) seeds. Several years ago a need based on the efficiency of cutting quartz for certain devices arose for quartz grown on $(01\bar{1}1)$ seeds. Quartz grown on this orientation often cracked during growth and device fabrication so that a systematic study of cracking was undertaken. We report here a summary of the highlights of the results of this study [15].

Cracking both during growth and during subsequent processing would be expected to be dependent upon strain in the crystals. To investigate this possibility a series of runs using seeds sorted by visual inspection in a special polariscope [15] were made. Seeds were placed in two categories: (A) good quality (comparatively moderate strain), and (B) bad quality (comparatively high strain). See Fig. 4 for examples of polariscopic views of A and B. The percent of crystals grown on seeds in categories A and B which were uncracked at the conclusion of the growth runs and the percent of resonator wafers uncracked during processing are given in Table I. As can be seen, as much as a 20% yield increase can be obtained by selecting seeds on the basis of polariscopic examination.

Since these preliminary experiments indicated that strain as observed in the polariscope seemed to correlate with cracking, a systematic study of strain was undertaken using the special polariscope.

It might be expected that strain would increase at higher growth rates. To test this hypothesis seeds of predetermined strain were used in a series of runs in laboratory vessels at various growth rates. Growth rate was altered by changing ΔT. In the first experiments (see Fig. 5) the initial seeds were visually preinspected and only those seeds qualifying as "good" (see Fig. 4 for an example) were used. Later quantitative measurements of similar seeds showed

Table I. Percent of Crystals and Wafers Uncracked

	Category A (moderate strain)	Category B (high strain)
Percent crystals uncracked after growth	88%	85%
Percent resonator wafers uncracked in processing	87–78%	78–56%

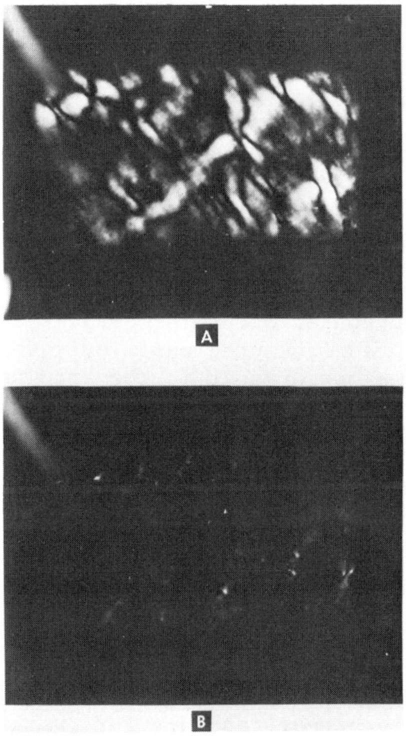

Fig. 4. Polariscopic views of unstrained A and strained B α-quartz [15].

that their "r" values* were about 40. As can be seen from Fig. 5, the strain in grown quartz increases as the growth rate increases. The region of "expected r values" based on the scatter obtained is rather wide at a given rate, but it clearly trends upward as the growth rate is increased. The scatter in r values obtained is due to several factors: (1) The initial r values for the seeds were not all identical; and (2) Random events, the inclusion of particulate matter, e.g., corrosion products flaking from the vessel wall, can introduce strain unrelated to either the original seed or to the growth process. Visual polariscope examination turned up occasional "spontaneously-generated" strain regions not associated with strain in the seed.

From the above studies and additional work [15] we were able to establish that severe strain in $(01\bar{1}1)$ growth and hence cracking could be eliminated by using specially selected low-strain seeds. Low-strain quartz suitable for seeds could be prepared by growth at reduced growth rate. Strain propagated from preexisting strain in seeds and from occasional inclusions.

*1 r = 10^{-3} waves of retardation as measured in the special polariscope [15] where the wave length is 5461 Å.

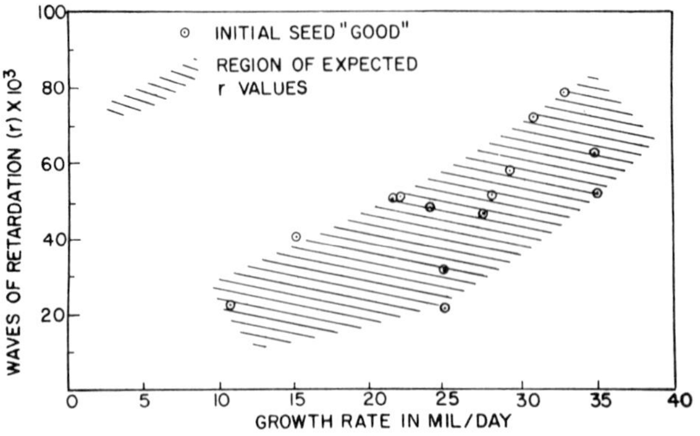

Fig. 5. Effect of growth rate on strain [15].

DISLOCATION FREE GROWTH

It was logical to associate the strain described above with the presence of dislocations so that a study of methods for the preparation of low-dislocation density and dislocation-free quartz was initiated [16]. x-ray topography was used to map natural and synthetic quartz specimens. Again we report here only highlights. Figure 6 shows a topograph of highly-dislocated and dislocation-free quartz. We were able to show that dislocation-free material could be grown on both (0001) and (01$\bar{1}$1) seeds if dislocation-free seeds were used and if the seeds were carefully etched to remove any damage introduced in their preparation. Dislocations propagated from previously existing dislocations present in the seed and from particulate inclusions grown into the quartz. These inclusions were often sodium iron silicates, so that higher yields of dislocation-free quartz were obtained when growth was in Pt- or Ag-lined vessels. Inclusions were often overgrown on (0001) seeds without dislocations or strain, while on (01$\bar{1}$1) seeds inclusions usually caused dislocations and strain. Thus the higher tendency for cracking in (01$\bar{1}$1) growth is correlated with the greater tendency for inclusion-caused dislocations and strain.

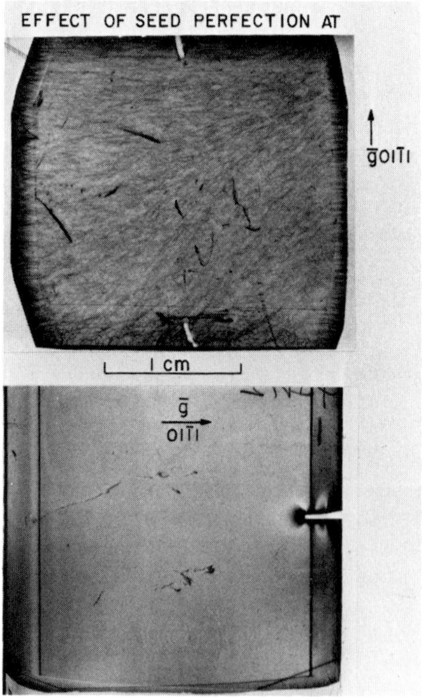

Fig. 6. X-ray topographs of highly dislocated quartz (top) and dislocation-free α-quartz (bottom).

CONCLUSIONS

The importance of commercial quartz growth has provided a driving force for a successful search for North American sources of α-quartz nutrient feed stock. Cracking in quartz has been shown to be related to strain and dislocations, and techniques for eliminating cracking and preparing dislocation-free quartz have been developed.

ACKNOWLEDGMENT

E. D. Kolb, R. L. Barns, P. E. Freeland, K. Nassau, and J. R. Patel are to be thanked for many collaborative efforts, some of which are reported here.

REFERENCES

1. A. A. Ballman and R. A. Laudise, in The Art and Science of Growing Crystals, J. J. Gilman, ed., John Wiley and Sons, New York (1963), p. 231.
2. R. A. Laudise, The Growth of Single Crystals, Prentice Hall, Englewood Cliffs, New Jersey (1970).
3. R. A. Laudise and J. W. Nielsen, in Solid State Physics, Vol. VII, F. Seitz and D. Turnbull, eds., Academic Press, New York (1961).
4. R. A. Laudise and A. A. Ballman, J. Phys. Chem. 65, 1396 (1961).
5. R. A. Laudise, J. Am. Chem. Soc. 81, 562 (1959).
6. N. C. Lias, E. E. Grudenski, E. D. Kolb, and R. A. Laudise, J. Crystal Growth 18, 1 (1973).
7. E. D. Kolb, K. Nassau, R. A. Laudise, E. E. Simpson, and K. M. Kroupa, J. Crystal Growth 36, 93 (1976).
8. W. A. Deer, R. A. Howie, and J. Zussman, Rock-Forming Minerals Vol. 4, John Wiley and Sons, New York (1963), p. 179.
9. W. D. Johnson, Jr. and R. D. Butler, Bull. Geol. Soc. Am. 57, 601 (1946).
10. R. B. Stond, R. H. Arndt, F. B. Fulkerson, and W. G. Diamond, "Mineral Resources and Industry of Arkansas," Bulletin 645, Bureau of Mines, U.S. Government Printing Office, Washington, D. C. (1969).
11. R. H. Jahns, Economic Geol. 50, 1025 (1955).
12. R. H. Jahns and C. W. Barnham, Economic Geol. 64, 843 (1969).
13. E. N. Cameron, R. H. Jahns, A. H. McNair, and L. R.Paye, "Internal Structure of Granitic Pegmatites," Economic Geol. Monograph No. 2, (1949).
14. A. A. Ballman, R. A. Laudise, and D. W. Rudd, Appl. Phys. Letters 8, 53 (1966).
15. R. L. Barns, E. D. Kolb, R. A. Laudise, E. E. Simpson, and K. M. Krupa, J. Crystal Growth 34, 189 (1976).
16. R. L. Barns, P. E. Freeland, E. D. Kolb, R. A. Laudise, and J. R. Patel, J. Crystal Growth, to be published.

U-2

HYDROTHERMAL GROWTH OF FAYALITE CRYSTALS

S. Hirano, Y. Iwai, S. Somiya, and S. Saito

Tokyo Institute of Technology
Tokyo, Japan

INTRODUCTION

Fayalite, Fe_2SiO_4, is the iron-rich end member of the olivine series, which constitutes most of the upper mantle of the earth. It is thus a most important mineral with respect to geophysical science.

Silicates of ferrous oxides have been the subject of numerous investigations of importance to geologists. Bowen and Schairer [1] synthesized fayalite in their study of the $FeO-SiO_2$ system. Flaschen and Osborn [2] completed a hydrothermal study on the system $FeO-SiO_2-H_2O$ at low oxygen partial pressures and also synthesized fayalite. However, no high-perfection single crystals of fayalite free of ferric iron have been grown, because of the difficulty of controlling the valence state of iron. Shankland [3] grew large single crystals of colorless forsterite, Mg_2SiO_4, by the flame fusion method. He then extended his research to growing olivine crystals doped with more than about 0.3 mol % iron, but obtained only partly satisfactory results due to the partial alteration of Fe^{2+} to Fe^{3+}. By the Czochralski method, Takei [4] grew large single crystals of forsterite and also intended to grow olivine crystals with ferrous iron under low oxygen partial pressure. He reported the growth of crystals with a tan color due to the inclusion of considerable amounts of ferric ion and precipitates of γ-iron due to the failure to obtain a homogeneous reduced atmosphere.

In the present study, transparent fayalite crystals free of ferric ion were successfully grown hydrothermally under the reduced conditions by adjusting the iron valance, using the same method used in the hydrothermal growth of magnetite [5]. This led to the development of a method for growing high quality fayalite crystals of sufficient size for physical measurements.

EXPERIMENTAL

A gold capsule, 5 mm in outer diameter, 4.6 mm in inner diameter and 50 mm in length, containing starting materials and mineralizer solution, was set in a test-tube type cold seal pressure vessel and subjected to the desired temperature and pressure. The pressure was measured with a Heise bourdon tube pressure gauge calibrated by the dead-weight method, and the temperature was measured at the outside wall of the vessel using a platinel thermocouple calibrated against the melting point of gold. The accuracies of pressure and temperature were within $\pm$ 20 kg/cm^2 (1.8 x 10^6 Pa) and $\pm$ 2°C.

Metallic iron powder (analytical grade, purity > 99%, Merck, Germany) and silicon powder (purity > 99.999%, Shinetsu Chem. Co., Japan) or amorphous silica (purity > 99%, Kanto Chemical Co., Japan) were used as the starting materials. These metallic powders react with supercritical water to form oxides and hydrogen sufficient to produce a reducing atmosphere. That is, the following reaction was employed to form fayalite and also control the oxygen fugacity,

$$2Fe + SiO_2 + 2H_2O \rightarrow Fe_2SiO_4 + 2H_2 \qquad (1)$$

$$2Fe + Si + 4H_2O \rightarrow Fe_2SiO_4 + 4H_2 \qquad (2)$$

Stoichiometric amounts of the starting materials and pure water necessary to form fayalite and solution were sealed hermetically without any seed crystal into the gold capsule. The amount of solution was adjusted so that the pressures inside and outside the capsule would be balanced.

The hydrothermal conditions employed in the present work were as follows; the growth temperature, up to 800°C (1073 K); the temperature difference between the growth zone and the dissolution zone, 25 or 50°C; the pressure 1500 kg/cm^2 (1.47 x 10^8 Pa); mineralizer solution was sodium hydroxide with concentrations up to 10 molar, sodium chloride up to 5.0 molar or certain mixtures of sodium hydroxide and sodium chloride solutions.

After a desired growth period, the vessel was quenched in cold water. The crystals grown in the capsule were observed with an optical microscope and identified by x-ray diffraction. The oxidation states of iron in the grown crystal were measured by the Mössbauer spectroscopic method with Co^{57} in Pd and sodium nitroprusside, $Na_2Fe_2(CN)_5NO \cdot 2H_2O$ as a reference.

RESULTS AND DISCUSSION

Fayalite single crystals could be grown under hydrothermal conditions at temperatures from 400 to 800°C. Single crystals, about 3 mm in length, were grown in the gold capsule at 650°C and 1500 kg/cm^2 for 3 days in NaCl solution, NaOH solutions, or mixture of these solutions.

The crystal morphology was found to be influenced phenomenologically by the growth temperature, the type of solution, and the oxygen fugacity. Figure 1 shows the change in morphology with the growth temperature. At temperatures below 600°C, most crystals were plate-shaped bounded mainly by {100}, {111}, and {010} as shown in Figs. 1a and 1b. As the growth temperature was raised, the {100} and {111} faces tended to become smaller, whereas the development of {021} and {110} faces was remarkable (see Figs. 1c and 1d). Similar results were also obtained in 1.0 or 2.5 molar NaOH solution. In a NaOH solution with a concentration above 2.5 molar, serpentine or a member of the mica group was formed at temperatures below 500°C by the substitutions of Fe^{2+} and Fe^{3+} for Mg^{2+} and Al^{3+}, respectively. Under the hydrothermal condition of especially low oxygen fugacity (0.005 mole hydrogen in 0.35 ml of 2.5 molar NaCl or 2.5 molar NaOH solution), serpentine crystals were found to be more stable at lower temperatures by the substitution of stable ferrous ion for Mg^{2+}.

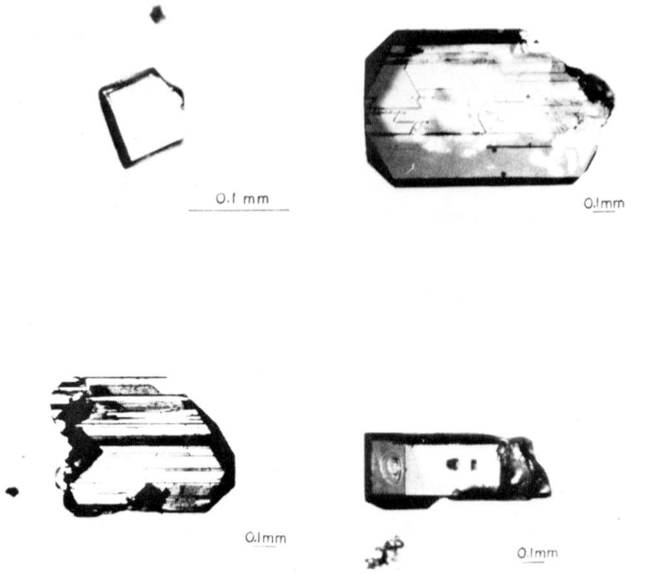

Fig. 1. Changes in the morphology of fayalite crystals with growth temperature (in 5 molar NaCl solution at 1500 kg/cm^2 for 3 days). Legend: (a) at 500°C; (b) at 650°C; (c) at 700°C; and (d) at 800°C.

The effect of the concentration of NaOH solution on the morphology of fayalite was also found to be remarkable, as shown in Fig. 2. The addition of NaOH to the NaCl solution led to the development of the {021} and {110} faces and to also increase the morphological importance of the {001} and {111} faces. Thus, the addition of NaOH changes the morphology of fayalite from platelets to needles.

The effect of the oxygen fugacity on the morphology of fayalite was examined by keeping the other variables constant except for the amount of hydrogen formed in the capsule. Since the actual value of the oxygen fugacity in hydrothermal solution is not available due

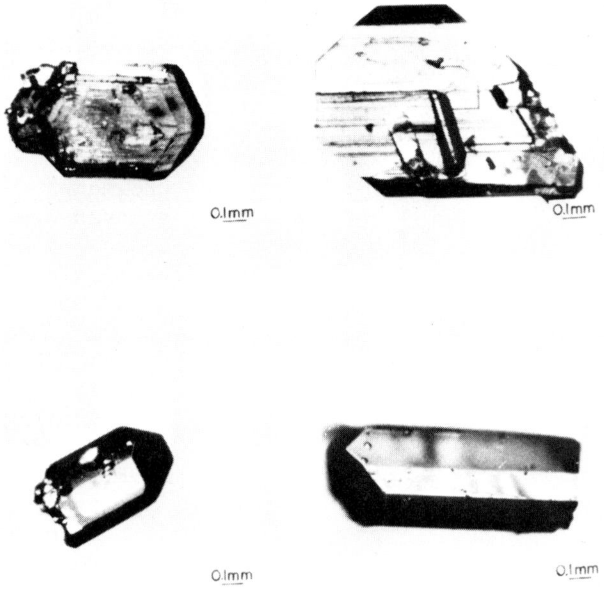

Fig. 2. Changes in the morphology of fayalite crystals prepared at 650°C and 1500 kg/cm^2 for 3 days. Legend: (a) in 5 molar NaCl solution; (2) in mixed (4.9 molar NaCl/0.1 molar NaOH) solution; (c) in mixed (4.5 molar NaCl/0.5 molar NaOH) solution; and (d) in mixed (4.0 molar NaCl/1.0 molar NaOH) solution.

to the lack of basic data, the amount of hydrogen formed by the reaction of the metallic starting substances with supercritical water was changed for comparison with the reducing environment in the present work. Figure 3 shows the effect of the oxygen fugacity (the amount of hydrogen present in the growth atmosphere) on the morphology of fayalite crystals grown at 650°C and 1500 kg/cm^2 for 3 days. At the lower oxygen fugacity, colorless plate-shaped fayalite crystals grew with the shape bounded mainly by {100}. With an increase in the oxygen fugacity, {021} and {110} faces developed to the extent comparable to {100} and {111} faces, and the crystals tended to have a bulky or needle shape.

In Fig. 4, the effects of the growth temperature, the concentration of NaOH solution, and the oxygen fugacity on the morphologies of the grown fayalite crystals are schematically summarized. Most notable are the development of {021} and {110} faces and the decrease in importance of the {111} and {100} faces as the magnitude of the growth temperature, the concentration of NaOH solution and the oxygen fugacity increased. The results may be applied to determine the growth conditions of a crystal.

The attack on the wall of the gold capsule by the concentrated hydrothermal solution was found to cause grain growth of the gold, which leads to partial leakage of hydrogen gas especially at higher temperatures. Thus, it should be understood that the phenomenon of the influence of the growth temperature on the morphology was exaggerated by the increase in the oxygen fugacity due to the partial leakage of hydrogen.

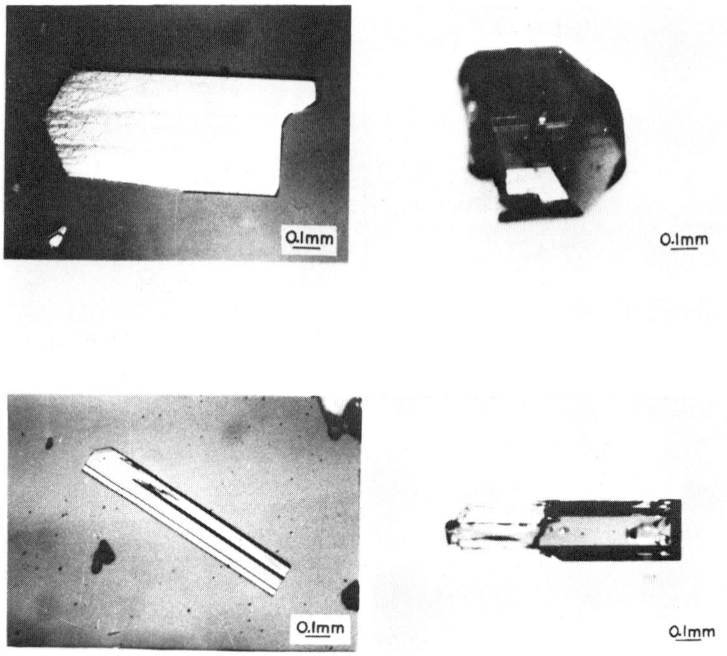

Fig. 3. Changes in the morphology of fayalite crystals prepared at 650°C and 1500 kg/cm^2 for 3 days with an amount of hydrogen in growth capsule of (a) 0.0025 mole H_2 in 0.35 ml of 2.5 m NaCl solution; (b) 0.00125 mole H_2 in 0.35 ml of 2.5 m NaCl solution; (c) 0.00125 mole H_2 in 0.35 ml of 2.5 m NaOH solution; and (d) 0.00125 mole H_2 in 0.35 ml of 2.5 m NaOH solution.

The typical growth appearances in the capsule are shown in Fig. 5. Crystals nucleated spontaneously on the inner wall of the gold capsule at the growth zone and then grew in a group of crystals as illustrated in Fig. 5a. Figures 5b and 5c represent individual crystals separated from the group of crystals grown at 650°C and 1500 kg/cm^2 in 5 molar NaCl solution for 3 days and at 450°C and 1500 kg/cm^2 in 2.5 molar NaOH solution for 3 days, respectively. The size of the crystals grown in the present work is hardly sufficient for physical measurements under very high pressure, but the basic data obtained in this study will be useful for growing larger fayalite crystals in a larger pressure vessel of glassy carbon now under construction.

Figure 6 shows a typical surface microtopograph on a (100) face of a fayalite crystal. This kind of pattern was characteristic of the (100) face of the fayalite crystals grown at a relatively low oxygen fugacity and a lower concentration of NaOH solution.

The Mössbauer spectroscopic method was applied to the fayalite crystals in order to observe the oxidation state of the iron. Figure 7 shows the spectrum taken at 25°C on a fayalite crystal grown at 650°C under 1500 kg/cm^2 in 2.5 molar NaCl solution for 3 days from iron and silicon powders as the starting materials. The

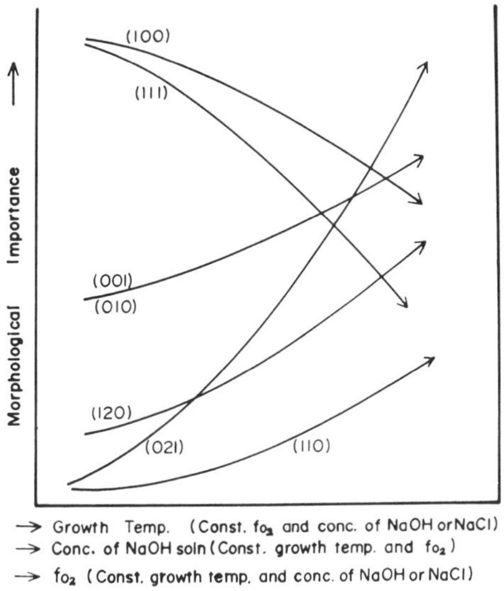

Fig. 4. Schematic diagram of the morphological changes with growth temperature, concentration of NaOH solution and oxygen fugacity.

presence of ferric ion, which substitutes for the ferrous ion in one of the two kinds of octahedral sites or which could be interstitial, could not be detected on the Mössbauer spectra of the grown fayalite crystals. The results indicate that fayalite single crystals without any detectable amount of ferric ion could be grown by the hydrothermal method under the reduced conditions established in the present study.

SUMMARY

Fayalite single crystals without any detectable inclusion of ferric ion were grown hydrothermally under controlled, reducing conditions.

The method of adjusting the reducing character of the atmos-

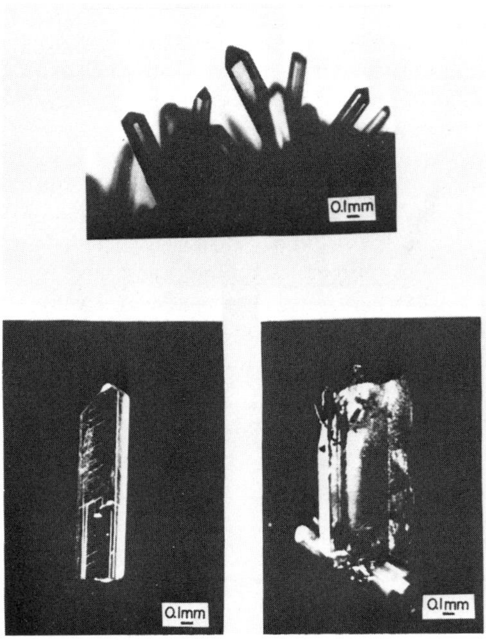

Fig. 5. Representative synthetic fayalite crystals. (a) Growth appearance in capsule (at 400°C under 1500 kg/cm ^{2} in 1 molar NaOH solution for 3 days); (b) Crystal grown at 650°C under 1500 kg/cm ^{2} in 5 molar NaCl solution for 3 days; and (c) Crystal grown at 450°C under 1500 kg/cm ^{2} in 2.5 molar NaOH solution for 3 days.

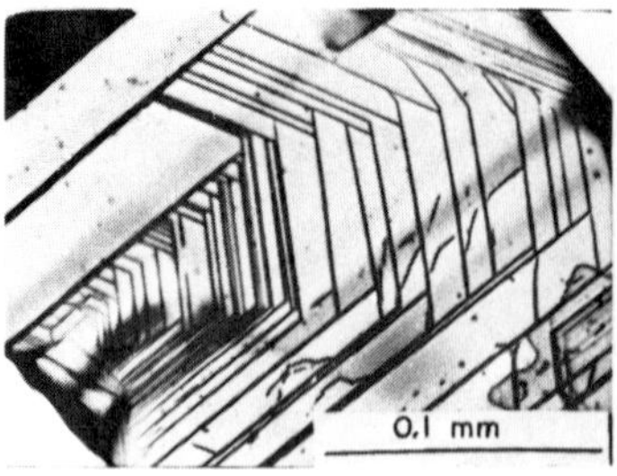

Fig. 6. Typical surface microtopograph on {100} face of fayalite crystal grown at lower oxygen fugacity.

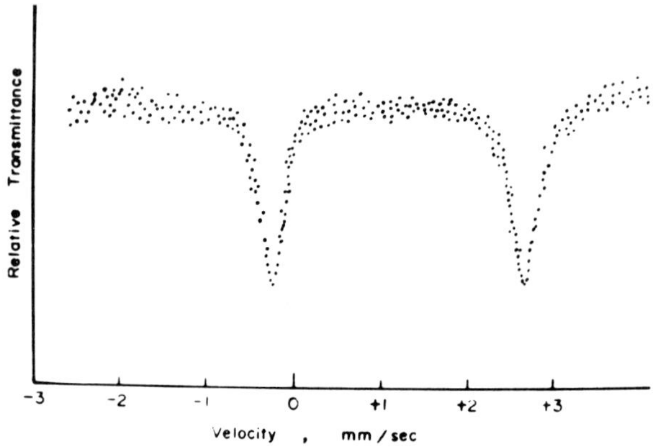

Fig. 7. Mössbauer spectrum of fayalite crystal grown at 650°C under 1500 kg/cm^2 for 3 days in 2.5 molar NaCl solution.

phere established in the present work, was found to be successful for keeping the hydrothermal growth conditions within the stability field of the divalent iron. The morphology of the fayalite crystals was markedly affected by the growth temperature, the concentration of NaOH solution, and the oxygen fugacity. The lower the growth temperature, the concentration of NaOH solution, and the oxygen fugacity, the more significant the development of {021} and {110} faces and, conversely, the decreasing importance of the {100} and {111} faces. Mössbauer spectra indicated that no detectable amount of ferric ion was included in the grown fayalite crystals. The data obtained in the present study should be useful with respect to future experiments on the growth of large fayalite crystals in a larger pressure vessel constructed of glassy carbon.

ACKNOWLEDGMENT

The authors are grateful to T. Fujii of the Fundamental Research Laboratory of Nippon Steel Company, Japan, for his kind support and use of the Mössbauer spectrometer.

REFERENCES

1. N. L. Bowen and J. F. Schairer, Am. J. Sci. 24, 176 (1932).
2. S. S. Flaschen and E. F. Osborn, Econ. Geol. 52, 923 (1957).
3. T. J. Shankland and K. N. Hemmenway, Am. Min. 48, 208 (1963). T. J. Shankland, Am. Ceram. Soc. Bull. 46, 1160 (1967).
4. F. Takei, J. Gemmol. Soc. Japan 1, 59 (1974).
5. S. Hirano and S. Somiya, J. Crystal Growth 35, 273 (1976).

U-3

CRYSTALLIZATION KINETICS OF SiO_2 AT HIGH PRESSURES

V. J. Fratello*, J. F. Hays, and D. Turnbull

Harvard University
Cambridge, Massachusetts USA

INTRODUCTION

The unusual enhancement of certain kinetic phenomena at high pressure in some silicates has long been a source of interest. Shaw [1] noted a decrease in the viscosity of obsidian glass with pressure, though he attributed it to experimental uncertainties. Kushiro et al. [2,3] found that such a decrease clearly occurs in several silicate melts. Uhlmann, Hays, and Turnbull [4] observed a large increase in the rate of crystallization of silica glass at high pressure. They showed that the increase scaled with the amount of water present and this might even account for the entire effect.

The purpose of the present work is to extend this research by examining the crystallization kinetics of ultradry silica in an attempt to separate the water effect from the intrinsic crystallization behavior. This paper will present the technique used and some preliminary results.

EXPERIMENTAL

The starting material was GE 214 fused quartz rod, chosen for its high purity and low water content. It was found, however, that the surface had large amounts of adsorbed water as well as other impurities. These impurities were removed through flame polishing. The samples and encapsulating materials were then dried at 1200°C for six hours with ultradry helium gas passing over them. The gas was predried by passing it through a desiccant and a molecular sieve and was then passed over calcium chips heated

*Allied Chemical Corporation Research Fellow, 1975-1978.

to 200°C. Samples treated in this way had water contents below the limit of detection, estimated as 1 ppm, by infrared spectroscopy. The samples were then sealed in platinum capsules in the same ultra-dry atmosphere and stored in a vacuum desiccator until just before measurement.

The pressure apparatus was a standard solid-medium piston-cylinder device with an internal graphite heater as described by Boyd and England [5] and Johannes et al. [6]. The cylinder was 1/2 in. in diameter by 1-1/4 in. in length. An 8% friction correction was used and the pressure in the ram was kept constant to within less than 1%. The error limits on the measured pressures are $\pm$ 10% [6]. Temperature variation and uncertainty in such a unit is given as $\pm$ 15° [5].

The samples were brought to the desired pressure at a temperature well below the threshold of observable crystallization. They were then raised rapidly to the run temperature. At the end of the run they were quenched to room temperature before lowering the pressure. Total run-up and run-down time never exceeded 30 sec, a negligible fraction of the total run times.

It was decided to take full advantage of the geometry of the apparatus by using cylindrical samples. Fused quartz rod of 2-mm diameter was cut to 3-mm length, the ends ground to flatness, and the entire sample flame polished for a clean surface. Capsules were made from platinum tubing including platinum discs at the ends to preserve the right circular cylinder geometry of the sample. Deviation from this geometry creates local perturbations in the rate of crystallization.

Axial sections of the charges were examined under a microscope with crossed polarizers. Nucleation appeared to be completely heterogeneous, occurring exclusively at the platinum-silica interface. Charges that conformed to the right circular geometry showed a remarkably uniform layer of crystallization adjacent to this surface for crystalline layer thicknesses up to 1/4 of the sample radius.

A series of samples was measured at 25 kbar and 1400°C for various times to determine the nucleation time. Within experimental accuracy, these runs extrapolate to zero time, indicating a virtually instantaneous saturation of all surface nucleation sites and rapid impingement of the nucleated crystals. Even at the shortest times the calculated rates are consistent. Thus the crystallization rates can be calculated by simply dividing the thickness of the crystalline layer by the run time. The indications are that a clean, smooth platinum-silica interface is important to nucleation. The rate was constant with time and displayed a maximum deviation of $\pm$ 30%. At least two samples run for different times were used for each data point to reduce uncertainty.

The quenched crystallization product was found to be all α quartz by x-ray diffraction. This is to be expected since all experiments were done within the β quartz stability field and the β→α inversion is expected at any pressure.

It is possible that even the minute amounts of residual water left by the drying process could influence the crystallization kinetics. At 1000°C, a temperature too low to observe crystallization at room pressure, substantial rates of crystallization were noted in the undried samples at 25 kbar. Drying reduced the rate of crystallization at this pressure and temperature below the limits of detection, an estimated three orders of magnitude. This result and the low concentration of water are taken as indications that the water effect has for the most part been eliminated.

To determine whether the enhancement of kinetics at high pressure is still present in the dried samples, an isothermal series of experiments was performed at 1500°C and pressures of 5, 15, and 25 kbar.

RESULTS AND DISCUSSION

The data are presented in Table I where u is the rate of crystallization.

Table I. Rates of Crystallization

T, °C	P, kbar	u, nm/sec
1400	25	20.7
1500	25	126.
1500	15	30.8
1500	5	6.18

The isothermal data are remarkably consistent. A semilogarithmic plot of these data (see Fig. 1) is nearly linear and can be fit to within 7% by the formula $u = 3.01 \exp(0.151\ P)$ where P is the pressure in kilobars, and u is in nanometers per second. The zero intercept is similar to the rate found by Ainslie, Morelock, and Turnbull [7] for devitrification of fused silica to cristobalite at room pressure. Their driest sample was one coated with pyrolitic graphite and data for this sample interpolate to 4.9 nm/sec at 1500°C.

The quantity V' given by the formula

$$V' = -RT \frac{\partial \ln u}{\partial P} \qquad (1)$$

has the dimensions of volume. Absolute rate theory evaluates this as the sum of an activation volume ΔV^* and a term involving the pressure dependence of the pre-exponential term [4]. The latter

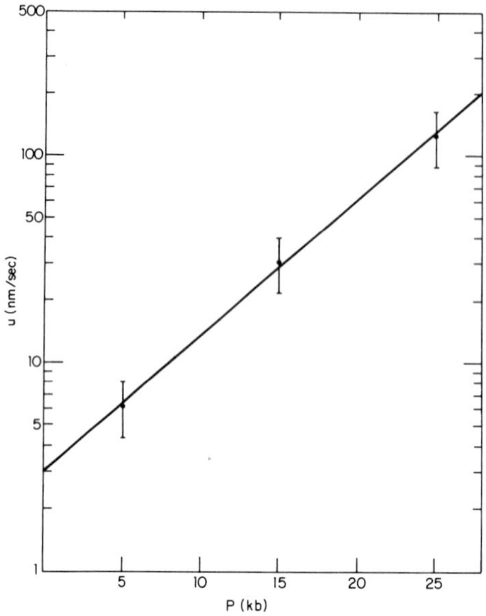

Fig. 1. Variation of crystallization rate with pressure at 1500°C.

would seem small from the nearly linear character of Fig. 1, but until more information is obtained it would not be proper to directly identify ΔV^* with V'. From the slope of the graph we find V' = -22.2 cm^3/mole. This is equal in magnitude to the molar volume of β quartz at this pressure and temperature to within two significant figures. This evidence of a large negative activation volume, reflecting the large increase of the rate of crystallization with pressure, is somewhat unexpected and as yet unexplained.

Further study of the pressure and temperature dependence of the rate of crystallization is planned to characterize this phenomenon more completely.

ACKNOWLEDGMENTS

This research was supported in part by the National Science Foundation under Materials Research Laboratory contract NSF-DMR76-01111 and by the Committee on Experimental Geology and Geophysics of Harvard University. The fused quartz used in this study was kindly supplied by General Electric Company and Quartz General Corporation. The authors wish to thank D. Walker for his assistance and many helpful discussions. One of us (V.J.F.) thanks Allied Chemical Corporation for its continuing support.

REFERENCES

1. H. R. Shaw, J. Geophys. Res. 68, 6337 (1963).
2. I. Kushiro, J. Geophys. Res. 81, 6347 (1976).
3. I. Kushiro, H. S. Yoder, Jr., and B. O. Mysen, J. Geophys. Res. 81, 6351 (1976).
4. D. R. Uhlmann, J. F. Hays, and D. Turnbull, Physics Chem. Glasses 7, 159 (1966).
5. F. R. Boyd and J. L. England, J. Geophys. Res. 65, 741 (1960).
6. W. Johannes, D. W. Chipman, J. F. Hays, P. M. Bell, H. K. Mao, R. C. Newton, A. L. Boettcher, and F. Seifert, Contrib. Mineral. Petrol. 32, 24 (1971).
7. N. G. Ainslie, C. R. Morelock, and D. Turnbull, Symposium on Nucleation and Crystallization in Glasses and Melts, American Ceramic Society, Columbus, Ohio (1962), p. 97.

V-1

INDUSTRIAL APPLICATIONS OF HIGH-PRESSURE TECHNOLOGY*

H. P. Bovenkerk

General Electric Company
Worthington, Ohio USA

INTRODUCTION

The discoveries by scientists and the technical achievements of engineers in the past 25 years have led to the creation of an industry for the routine processing of materials at pressures greater than 50 kbars combined with temperatures exceeding 1500 K. An industry such as this is also built on the foundations of the earlier findings of Bridgman and other eminent researchers who gave us the cemented carbides, high strength steels, and refractory materials without which all this would not be possible. This processing results in products of key industrial importance such as diamond and cubic boron nitride (CBN), both in single-crystal and polycrystalline-compacted forms.

SYNTHETIC DIAMOND

Synthesized diamond was first offered for commercial sale in 1957. Production of synthesized diamond on a commercial scale is now practiced in several different countries. The consumption of industrial diamond has expanded more than fourfold in this two-decade period, and the U. S. Government estimates world consumption to have been about 15 million grams for the year 1974. It is estimated that today about three quarters of the consumption of diamond for industrial abrasives is of the synthesized variety. Since the production of mined diamond for industrial use has not changed substantially in the last two decades, it is apparent that the expansion of use has resulted from the availability of manufactured diamond as well as the important performance and productivity advantage it offers. At the present time there is no technical or

*Invited paper.

economic reason why the synthesized form of diamond cannot replace naturally-occurring diamond for all abrasive uses; mined diamond is still used because it is available and there are always some traditionalists who are prejudiced in favor of mined diamond despite its generally inferior performance.

The industrial applications of diamond abrasive are primarily for the surfacing and sizing of nonmetallic materials and cemented carbides by grinding, polishing, and sawing techniques. These uses have not changed in the last two decades, but replacement of competing abrasives, which range in hardness from silica sand to silicon carbide, with diamond has caused the application expansion. In many areas this replacement is now virtually complete, in others the performance and cost effectiveness still has to be improved or major changes in machines and tooling, requiring considerable investment of capital, will be needed. When replacing other abrasive tools with diamond tools there is an initial cost factor which tends to inhibit use initially. A diamond tool may be on the order of 50 times more costly than the one it replaces made with other abrasives. A significant factor is the acceptance and expanded use of synthesized diamond to replace other abrasives as the performance improves. Diamond grinding wheels today are up to 10 times more efficient than those of two decades ago. Likewise, major performance improvements have been made in other diamond tools. These improvements have come about not only because of the significant performance improvement of the diamond abrasive itself, but also because of improvements in the bonding and matrix technology of the tools. Of course, these performance changes while they increase the capability of diamond to replace other abrasives, also decrease the consumption of diamond itself; that is, shrink the usage for a given job. Thus, there are some applications where this change in efficiency has resulted in no increase in consumption at all. However, as labor and overhead costs increase, the most cost-effective methods properly shift the burden to the tool, that is, the tool is used to remove material faster, thus wearing faster.

The progress in diamond abrasives in the past twenty years has changed significantly. One major supplier now has over 40 standard diamond product types which in combination with size give more than 400 total combinations. Additional proliferation of specialized diamond abrasives is limited mainly by the ability of the tool fabricator and end user to handle the complexities involved.

CUBIC BORON NITRIDE

Although its consumption is at a much lower rate than diamond at the present time, cubic boron nitride as an industrial abrasive for grinding, honing, and lapping hard metallic alloys and especially-hardened high-speed steels is becoming an increasingly important and accepted abrasive. This abrasive does not compete with diamond, but is complementary to it. For example, in the manufacturing and shaping of high-speed steel-cutting tools such as drills, end mills, hobs, milling cutters, broaches, and the like, aluminum oxide is the traditional abrasive used. Cubic boron nitride abrasive because of its much higher hardness and sharpness not only offers significantly higher grinding efficiency, but also greatly reduces the danger of those metallurgical changes in the workpiece that reduces the edge strength and edge wear properties of the tool. This often permits grinding with cubic boron nitride at higher removal rates than with aluminum oxide. As in the case of diamond, as labor and other costs continually increase, the productivity of man and machine is enhanced by use of the superior although more expensive abrasive.

INDUSTRIAL APPLICATIONS

In the form of polycrystalline masses shaped into cutting tools, cubic boron nitride is finding its place in industry. It is turning hard iron, nickel, and cobalt-based alloys that can be machined only very slowly with conventional carbide or oxide-based tools. The use at present has focused primarily on the alloys which are used in the parts exposed to high temperature in aircraft gas turbines. In some of these cases cutting rates with CBN tools are more than five times higher than with conventional tools. In the future it is expected that expanded use will occur in machining other hard steels with which it is often possible to cut to final size and finish rather than grind, thereby attaining greatly increased productivity of man and machine.

Likewise, polycrystalline diamond compacts have had significant industrial acceptance as cutting tools for machining nonmetallic materials, aluminum and copper based alloys, and hard cemented carbides. Other rapidly growing applications are for wire drawing dies, geophysical drilling bits, and turning and dressing stones for conventional grinding wheels.

For example, wire drawing dies are made of diamond for drawing wire in the smaller sizes and of cemented carbide for the larger sizes. Because of its hardness and wear resistance, diamond is an indispensable material for wire dies and has been so used for more than 150 years. Polycrystalline diamond, because of its isotropic structure, wears much more uniformly than single-crystal diamond and has much higher impact strength. Its wear rate is generally less than single-crystal diamond which is limited by the higher rate on the softest crystal plane on the wear surface. Because of their availability in large sizes, polycrystalline-diamond wire dies can replace a significant portion of the carbide dies with orders of magnitude increases in life without the need for resizing. Polycrystalline-diamond dies are now used for drawing wire from the smallest sizes up to about 5 mm in diameter.

A significant advance in high-pressure technology resulted from the use of diamond itself as a material for high-pressure anvil devices. This was first tried by Bridgman. The single-crystal diamond anvil is now perhaps the most popular high-pressure device used in scientific laboratories and its versatility is such that no lab should be without one. Since diamond has the highest compressive strength of any known material, this application is a natural one. The use of polycrystalline diamond for high-pressure anvils for extending the pressure range is a bootstrapping of the manufacturing technology of polycrystalline diamond itself. Pressures exceeding 500 kbars have been achieved with polycrystalline-diamond components in high-pressure apparatus with some claims extending into the megabar region. A given anvil design may have five or more times the load-carrying capability of the strongest cemented carbides since not only does diamond have a compressive strength considerably above cemented carbide, but it has a linear stress-strain curve. In other words, the proportional limit is equal to the ultimate compressive strength. Cemented carbides start to deform at a relatively low percentage of their ultimate compressive strength, hence plastic deformation severely limits the use of cemented carbide as high-pressure anvils.

Industrial manufacturing of the above material products requires more than scientific discoveries, a suitable high-pressure device, and some knowledge of the thermodynamics and kinetics of phase transformations. Much additional technology is required to find methods of controlling the consistency of the multithousands of individual high-pressure runs or microbatches which are needed to achieve viable production quantities of material. Making diamond and CBN abrasives to meet the customer's needs requires not just two processes, but a multitude of detailed and different processes, each engineered to grow crystals of controlled structure, external morphology, size, strength, and other characteristics.

The extreme conditions of pressing at high temperature and high pressure give rise to changes in the materials used in constructing the apparatus itself and the various materials such as reaction cell components. These various reactions influence both the time and functional stability of the pressure and temperature and the chemistry and kinetics of the crystal-growing processes. For example, many man-years of research and development were needed to achieve the reproducibility and time-functional stability needed to extend diamond growth periods from the initial few minutes to the much greater periods of hundreds of hours needed to grow the largest crystals made so far.

It has often been said that the resources required to advance a discovery to commercial production as related to the resources required to make the discovery may be three orders of magnitude or more. This certainly is the case for diamond making. It took a group of scientists from this company about three years to reproducibly make diamond from the time the project started. Reducing this discovery to production took another three years with a much larger effort in man-hours. Further extension of the capability in terms of products and processes has resulted from two decades of continued research and engineering effort with ever-increasing investment of time and capital.

Are diamond and cubic boron nitride products going to be the only new materials to demonstrate the commercial success of high-pressure research and technology? This question is often asked. The answer appears to be yes as far as known reactions or high-pressure products are concerned. Although many reactions and products have been discovered, they lack the high unit value and do not fulfill needs. High-pressure processing is intrinsically expensive, many thousands of dollars per kilogram, and there is little hope of changing this factor to a significant degree.

We look to scientists to help find the new materials which industry can now process as a result of the experience developed in the past two decades as outlined in this paper.

V-2

NEW TYPE OF BORON NITRIDE SINTERED UNDER VERY HIGH PRESSURE

A. Sawaoka and S. Saito

Tokyo Institute of Technology
Tokyo, Japan

and

M. Araki

Nippon Oil and Fats Company Ltd.
Taketoyo, Aichi, Japan

INTRODUCTION

Very hard materials with high ductility are very useful as high pressure anvils or cutting tools. Natural polycrystalline diamond, carbonado, has been used as a special drawing die. However, the amount of this product is limited. It is difficult to produce polycrystalline diamond without binder materials. Silicon or cobalt with other elements have a practical use as a binder.

A composite of the zinc blende type boron nitride (z-BN) and tungsten carbide has been developed for cutting tools and being used industrially [1]. Since the hardness of z-BN is comparable to diamond and since it tends not to react with transition metals at high temperatures, polycrystalline z-BN without using a binder would be a most useful material for cutting steel.

Powder crystals of boron nitride z-BN were first synthesized by Wentrof [2] using a molten catalyst. This powder has a hard sinterable property. It is assumed that the sintering is based on the phase transition from graphite-like boron nitride (g-BN) to z-BN under very high pressure. The direct conversion was first performed above 90 kbar and 1400°C, with a conversion ratio less than 30% [3]. Wakatsuki et al. [4] have lowered the pressure and temperature to around 60 kbar and 1200°C, by using poorly crystallized g-BN. They have reportedly produced a very hard sintered compact of z-BN having a Vickers hardness of 6500 kg/mm^2.

It is known that the crystal structures of z-BN and w-BN correspond to that of cubic and hexagonal diamond, respectively. The exact phase relations of the three types of BN and the kinetics of their transformation have not been clarified.

The difference of densities of z-BN and w-BN is less than 1% under normal conditions [5]. Both crystal structures similarly consist of a tetrahedron with SP^3 hybridization. Therefore, the mechanical properties such as the hardness of w-BN are expected to correspond to those of z-BN. Type w-BN can be synthesized from g-BN under a static pressure above 115 kbar [3] at relatively low temperatures or by the shock-compression method [5,6]. The maximum conversion ratio by the latter method is about 80% [6]. Since w-BN synthesized by shock compression consists of fine particles with high density defects, the powder might have active properties for sintering.

The stability of shock-synthesized w-BN at high pressure has been studied by Tani et al. [7], Hiraoka et al. [8], Corrigan and Bundy [9], and Akashi et al. [10]. Their results agree qualitatively. Type w-BN transforms to z-BN at pressures above 55 kbar and temperatures above 1300°C. Since it was reported [3] that z-BN was stable in the ranges of 1100-1500°C and above 55 kbar, w-BN would be metastable in the temperature ranges of 1100-1300°C.

This paper reports the phase relations between w-BN and z-BN, and the mechanical properties and microstructure of sintered compacts having wurtzite, zinc blende, and the mixture phases.

EXPERIMENTAL PROCEDURE

Powder Preparation

Powders of g-BN and w-BN were prepared for this study. Two kinds of g-BN powder were provided by Showa Denko Co., Ltd. One was a very fine powder with a particle size of 0.05 μm in average diameter. This was treated in a nitrogen atmosphere at 800°C for 24 hrs and used for producing z-BN compact by direct transformation. The other g-BN powder with particle sizes of 10 to 100 μm was used as raw material for synthesizing w-BN by shock compression.

The latter g-BN powder was mixed with iron powder in a weight ratio of 1:9. The mixture was then pressed into a copper tube which contained a steel rod buried in the center. The copper tube was then enclosed in a high explosive. The detonation wave generated by the explosive was transferred to the mixture via the copper wall. The shock pressure was gradually increased by the reverberation effect in the mixture. Maximum pressure was estimated to be about 300 kbar. The steel rod was useful for recovering the specimen without breakage. After the shock treatment the iron powder was dissolved in HCl. About 60% of the g-BN was transformed into w-BN. The w-BN was refined by the alkaline fusion technique. A transmission-type electron micrograph of the w-BN powder is shown in Fig. 1.

Fig. 1. Electron transmission micrograph of wurtzite type boron nitride powder synthesized by shock compression.

Camphor of 6 wt.% was added to the fine g-BN or w-BN powder and the mixture was formed into a tablet with 8.2-mm diameter and 6-mm thickness at 1.5 ton/cm^2. This tablet was treated at 600°C in a vacuum of 10^{-3} torr for 2 hrs and then placed in a high-pressure assembly in a nitrogen atmosphere.

High Pressure Sintering

High pressures and temperatures were applied by the use of a slide-type cubic anvil apparatus. The edge dimension of the anvil was 16 mm. The cross section of the high-pressure assembly used is shown in Fig. 2. Pressure was calibrated by observing the electrical resistance changes of Bi (I-II) at 25.4 kbar, Tl (II-III) at 36.7 kbar, Ba (I-II) at 55 kbar, and Bi (III-V) at 77 kbar, at room temperature. Temperature in the specimen was estimated from the electric power supplied. The relation between the power and temperature was obtained by using a chromel-alumel thermocouple under 1200°C at a corresponding pressure. The relationship between 2000 and 2300°C was estimated by identifying the phases of the specimen based on the equilibrium boundary line between g-BN and z-BN as obtained by Bundy and Wentorf [3] in which their original data were transformed to the revised NBS scale. Both curves obtained by using the chromel-alumel thermocouple and phase identifi-

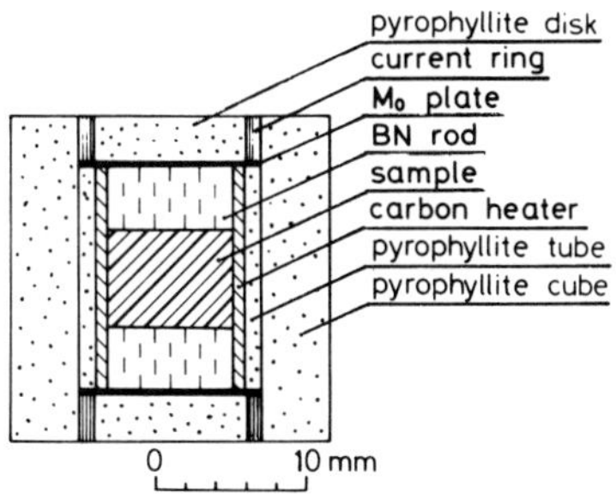

Fig. 2. Cross sectional view of the assembly used for very high pressure sintering.

cation were related by a smooth curve where the pressure effect of electromotive force was not considered.

Pressure was applied to the sample at room temperature and the specimen was heated electrically to the desired temperature. Then it was cooled either slowly to obtain a sintered compact or rapidly to examine the relation of the phase at room temperature and pressure. The weight ratio of phases present in the specimen was obtained from the relative intensities of their diffraction peaks, g-BN(002), w-BN(100), and z-BN (111). Some samples were cut by an ultrasonic cutter and polished with diamond powder for measurements of micro hardness and compressive strength.

RESULTS AND DISCUSSIONS

Transformation of g-BN to z-BN

The formation region of z-BN by the pressure treatment of g-BN for 15 min is shown in Fig. 3. The shaded area in the circles shows the proportion of the phase. The solid straight line is the phase equilibrium boundary in both forms of BN obtained by Bundy and Wentorf [3]. The minimum pressure for the transformation from g-BN to z-BN is 60 kbar between 1400-2000°C and this increases with decreasing temperature below 1400°C. The results obtained are consistent with those reported by Wakatsuki et al. [4].

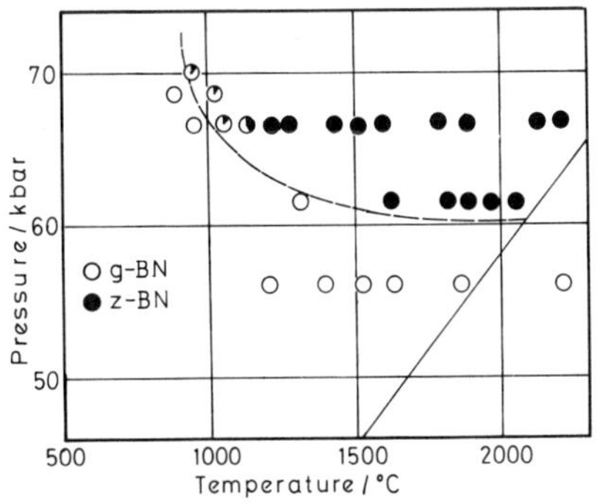

Fig. 3. Products obtained by a 15-min high pressure and temperature treatment in which fine-grained g-BN was used as the starting material. The straight line indicates the phase boundary between g-BN and z-BN used by Bundy and Wentorf [3].

Transformation of w-BN to g-BN and z-BN

The formation region of z-BN is shown in Fig. 4. In the pressure region higher than 55 kbar, w-BN transforms to z-BN at temperatures higher than 1200°C. It is an interesting phenomenon that w-BN transforms to g-BN in the stable region of z-BN. Since the transformation from w-BN to g-BN accompanies a volume increase, the reaction observed in this region is unexpected. Therefore, it appears that there are locally lower pressures at the interface of the grain and/or pore in the powder compact.

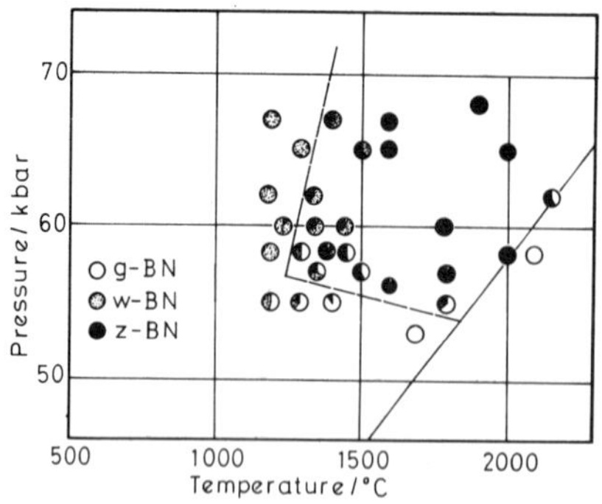

Fig. 4. Products obtained by a 15-min high pressure and temperature treatment in which a shock-synthesized w-BN was used as the starting material.

Microstructure of Sintered Compacts

Fractures of sintered compacts were observed by using transmission-type and scanning-type electron microscopes. Results of observation are summarized as follows:

1. Fine g-BN particles of 0.05-μm diameter grow to 0.1 to 0.15 μm with transformation to z-BN under pressure.
2. Grain size of z-BN in the sintered compact is about 0.3 μm after treatment of at kbar and 1600°C for 15 min as shown in Fig. 5, and 0.6 μm when treated for 30 min.
3. Particle sizes of w-BN used are less than 0.5 μm as seen in Fig. 1. But the crystallite size required for an x-ray diffraction pattern is 200-300 Å. Thus the particle will be a secondary one.
4. In spite of many efforts, the grain boundary on the fracture of w-BN sintered compact was not observed. The crystallite size could remain unchanged from that of the initial w-BN particles.
5. Rapid grain growth begins during the phase transformation from w-BN to z-BN. The grain size in the z-BN compact is about 1 μm after the 15-min treatment of w-BN at 1600°C and 66 kbar, as shown in Fig. 6. Octahedron-like grains are observed.

Mechanical Properties of Sintered Compacts

The micro-Vickers hardness of sintered compacts is measured by using a hardness tester with an indentor load of 500 gram. Since the maximum diameter of the compact obtained is 7 mm and its formation is very difficult, the exact compressive strength can not be obtained. Approximate values of the strength are summarized in Table I along with the hardness. The w-BN compact with the theoretical density has not been obtained by a pressure treatment of one hour at high temperature. Density of the compact reaches the theoretical value within an acceptable error by treatment at 1450°C and 66 kbar for 15 min in which 50% of the w-BN transforms to z-BN. The hardness and compressive strength are 7600 kg/mm^2 and about 600 kg/mm^2, respectively. Compact z-BN produced from g-BN is very hard but brittle. Compressive strengths above 200 kg/mm^2 could not be obtained for the z-BN sintered compact. The brittleness of the compact might originate from the grain shape and size.

Sintered compact with mixture phases of wurtzite and zinc blende type BN has a superior mechanical property as described above. The mixture phases of the compact were named "Wurzin".

Table I. Some Physical Properties of the Sintered Compact of BN

Starting Materials	Sintering Condition		Density g/cm^2	Product	Hardness, kg/mm^2	Compressive strength, kg/mm^2
	p,kbar	t,°C				
Fine-grained g-BN	66.5	1430	3.49	z-BN	>7500	<200
Shock-synthesized w-BN	67.0	1200	3.31	w-BN	2500	---
	67.0	1450	3.49	(w+z)-BN	7600	600
	67.0	1600	3.49	z-BN	6500	<200

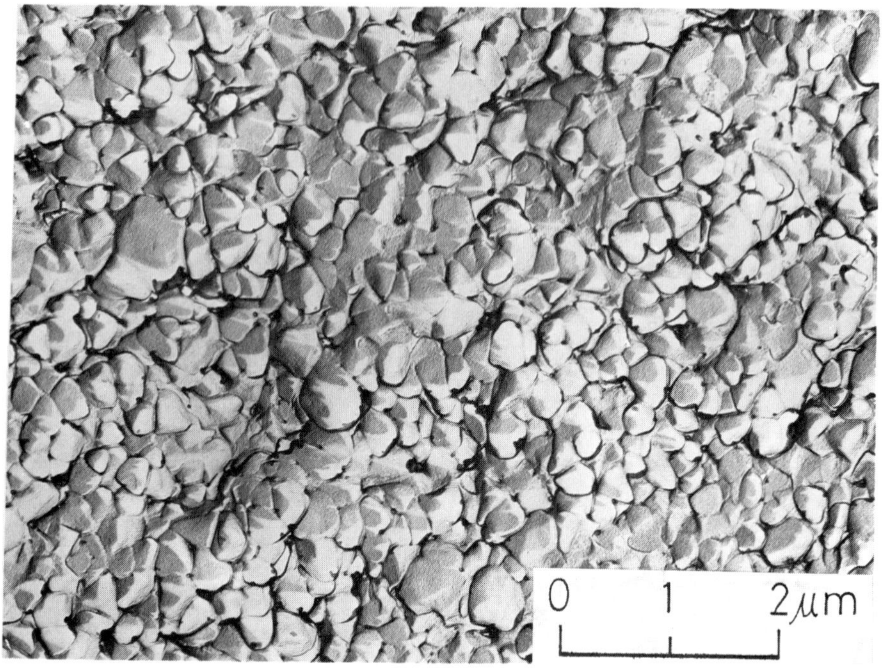

Fig. 5. Typical fracture micrography of sintered compact z-BN after g-BN has been treated at 60 kbar and 1600°C for 15 min.

Cutting Test

Preliminary cutting tests were performed on some hard steels. Wurzin, 7 mm in diameter and 5 mm in thickness, was formed into a wedge and clamped in a tool holder of a lathe machine. High speed steel (SKH-9) with a Rockwell C hardness of 55 was cut at a velocity of 300 m/min to a 0.1 mm depth. The wurzin tip did not show damage after 10 min of dry cutting in air. A view of the cutting test is shown in Fig. 7. Further cutting tests are in progress at the National Mechanical Engineering Laboratory of Japan.

ACKNOWLEDGMENTS

The authors would like to thank T. Murai of the Mechanical Engineering Laboratory and S. Okada of Mitsui Grinding Wheel Co., Ltd. for the preliminary cutting and abrasion tests. They are also grateful to K. Kondo, T. Sōma, T. Sunakawa, E. Tani, and T. Akashi for their help on the high-pressure experiments at the Tokyo Institute of Technology.

REFERENCES

1. L. E. Hibbs, Jr. and R. H. Wentorf, Jr., High Temp.-High Press. 6, 409 (1974).

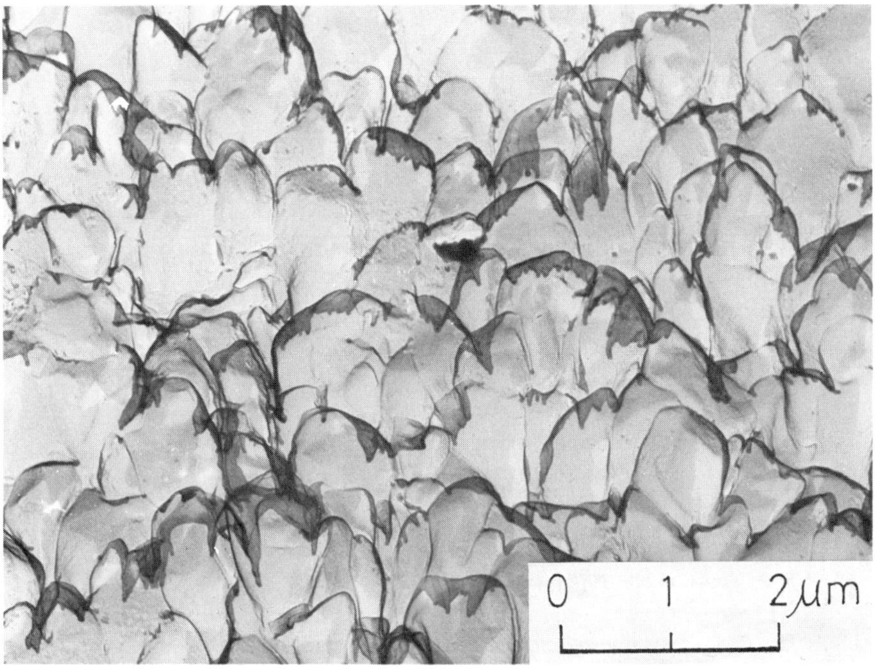

Fig. 6. Typical fracture micrography of sintered compact z-BN after w-BN has been treated at 60 kbar and 1600°C for 15 min.

Fig. 7. View of machining test with Wurzin.

2. R. H. Wentorf, Jr., J. Chem. Phys. 26, 956 (1957).
3. F. P. Bundy and R. H. Wentorf, Jr. J. Chem. Phys. 38, 1144 (1963).
4. M. Wakatsuki, K. Ichinose, and T. Aoki, Mat. Res. Bull. 7, 999 (1972).
5. T. Sōma, A. Sawaoka, and S. Saito, Mat. Res. Bull. 9, 755 (1974).
6. A. Sawaoka, T. Sōma, and S. Saito, Japan J. Appl. Phys. 13, 891 (1974).
7. E. Tani, T. Sōma, A. Sawaoka, and S. Saito, Japan. J. Appl. Phys. 14, 1605 (1975).
8. H. Hiraoka, O. Fukunaga, and M. Iwata, Yogyo-kyokai-shi 84, 163 (1976).
9. F. R. Corrigan and F. P. Bundy, J. Chem. Phys. 63, 3812 (1975).
10. T. Akashi, A. Sawaoka, and S. Saito, to be submitted.

V-3

THERMAL CONDUCTIVITY OF POLYCRYSTALLINE CUBIC BORON NITRIDE COMPACTS

F. R. Corrigan

General Electric Company
Worthington, Ohio USA

INTRODUCTION

Slack [1] predicts a room temperature thermal conductivity of about 13 W/cm°C for single-crystal cubic boron nitride with only diamond having a higher conductivity among the high thermal conductivity adamantine compounds. However, no crystals of cubic BN were available that were large enough to measure the thermal conductivity, and the highest value reported for polycrystalline compacts prepared by high-pressure/high-temperature sintering of natural isotope abundance cubic BN powder was only 1.8 W/cm°C. In this paper, natural isotope abundance polycrystalline cubic BN compacts prepared by direct high-pressure/high-temperature conversion of graphitic BN and having room temperature thermal conductivity values up to a factor of four higher than previously reported are discussed.

METHOD

Thermal diffusivity was measured on a series of polycrystalline cubic BN compacts using a flash-heating method. Briefly, the flash-heating method involves subjecting the front face of the disc-shaped sample to a short energy pulse and measuring the resultant temperature rise of the rear face [2]. Usually a solid-state laser is used as the source of the energy pulse and the thermal diffusivity is calculated from the rear surface temperature history. The thermal diffusivity measurements on the samples were made by Taylor and Groot, of the Properties Research Laboratories of Purdue University, using a Korad K2 laser as the energy pulse source with 'finite pulse time effect' corrections made by the method described by Taylor and Clark [3]. The diffusivity values were converted to thermal conductivity values using the defining relation

$$k = \alpha C_p \rho \quad (1)$$

Sample densities (tabulated in Table I) were measured by the immersion technique using water, and literature values [4,5] were used for the specific heat.

The structure of the compacts was investigated by x-ray diffraction, scanning electron microscope (SEM), and optical microscope techniques.

SAMPLE PREPARATION

The compacts were prepared by direct high-pressure/high-temperature (HP/HT) conversion of graphitic boron nitride to the cubic form at approximately 70 x 10^5 kilopascals (70 kbars) and at varying temperature conditions as listed in Table I. The maximum temperature and time at this temperature during the conversion were determined from temperature calibration runs using platinum-platinum, 10% rhodium thermocouples up to the melting point of platinum and a linear extrapolation of the temperature calibration curve at higher temperatures. The thermocouple calibration was corrected for pressure using the thermocouple pressure corrections reported by Strong, et al. [6]. The time at maximum temperature tabulated in Table I is adjusted to account for the time required to reach the vicinity of the maximum temperature after heating is started (about 3 to 4 min) as determined from the temperature calibration runs. With sample #9 the heating was varied too rapidly to allow determination of the temperature conditions during the run. The converted compacts were disc shaped, approximately 12 mm in diameter by 1.57 to 3.65 mm thick.

RESULTS

X-ray Diffraction

X-ray diffraction scans showed varying amounts of residual unconverted graphitic BN in the compacts prepared at the lower pressing temperatures, but no residual graphitic BN was detected in the higher temperature compacts. The relative intensities of the graphitic BN (002) peaks observed are tabulated in Table I.

It is well established that crystal imperfections, such as very small crystallite size or nonuniform stress distortion of the crystal lattice causes broadening of x-ray diffraction peaks (see, for example, Cullity [7]). For line width broadening due to crystallite size reduction, the crystallite size can be estimated from the measured line broadening by [7]

$$t = \frac{0.9\ \lambda}{(B_m^2 - B_s^2)^{1/2} \cos \Theta} \quad (2)$$

Table I. Thermal Conductivity Samples

Sample	Thickness, mm	Density, g/cm^3	Max. Temp., °C	Time, min	X-ray Crystallite size, μm	Peak k W/cm °C	Compressed HBN (002) Intensity
1	2.84	3.49	2440	4-6	-	9.0	0
2	3.33	3.50	2440	2-3	-	7.5	0
3	1.60	3.50	2275	2-3	-	7.0	0
4	1.57	3.48	2360	2-3	-	6.7	0
5	3.66	3.48	2440	0-1	-	5.9	0
6	1.60	3.49	2190	2-3	-	4.8	0
7	1.96	3.49	2190	2-3	0.074	4.35	30
8	3.35	3.48	2360	0-1	0.088	3.82	13
9	3.66	3.47	-	-	0.069	3.45	33
10	2.77	3.42	2025	7	0.049	3.17	73
11	3.05	3.47	1860	17	0.050	3.08	29
12	1.60	3.47	1860	17	0.048	2.95	25
13	1.93	3.36	1860	17	0.039	2.60	496
14	1.60	3.40	1780	17	0.034	2.50	242

Line width broadening measurements were made of the CBN (111) and CBN (220) reflection lines and the effective crystallite size calculated by (2). Comparable crystallite thicknesses were found in both directions and the results for the (111) planes are tabulated in Table I. For a number of the higher temperature compacts the crystallite size was too large to be determined by this technique.

SEM and Optical Microscope

Figure 1 is a scanning electron micrograph of a polished and etched surface (7-min fused sodium hydroxide etch) of a high-temperature compact in which crystallites of 10 microns and larger are observed. The line type features seen in this figure are interpreted as growth twin boundaries.

The combined x-ray diffraction and SEM crystallite size results from numerous compacts indicate the type of crystallite size/processing temperature re-

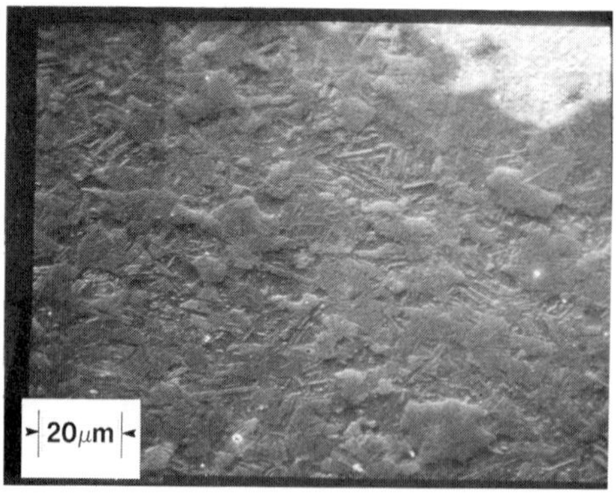

Fig. 1. Micrograph of polished and etched surface of high-temperature compact.

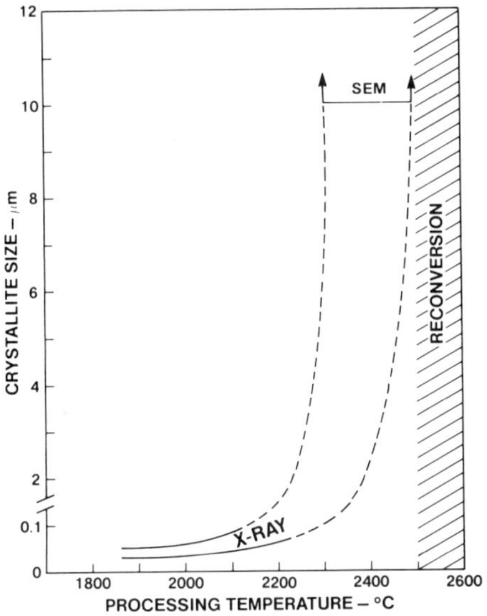

lationship shown in Fig. 2. At the lower processing temperatures a gradual increase in crystallite size is observed until at processing temperatures of 2100 to 2200°C and above a much more sharp increase in the crystallite size occurs.

Fig. 2. Dependence of crystallite size on processing temperature in the direct conversion compacts.

The large and small crystallite compacts are quite easily distinguished by the appearance of their fracture surfaces under low power magnification. Figures 3 and 4 are micrographs of the fracture surfaces of a large and small crystallite size compact, respectively, showing the contrast between the granular, sparkly appearance of the fracture surface of the large crystal compact and the smooth fracture surface typical of small crystallite size compacts.

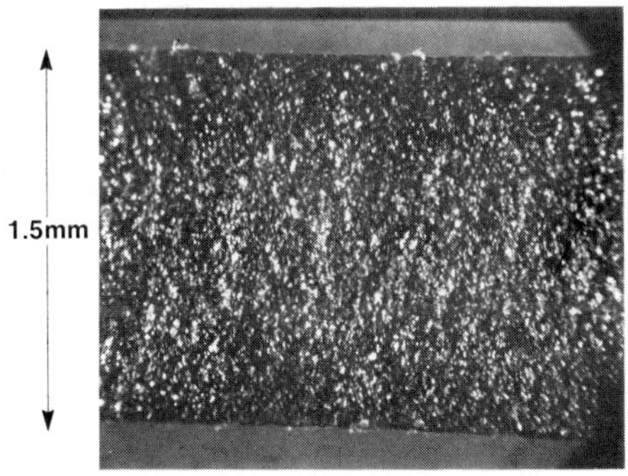

Fig. 3. Micrograph of fracture surface of high-temperature (large crystallite size) compact.

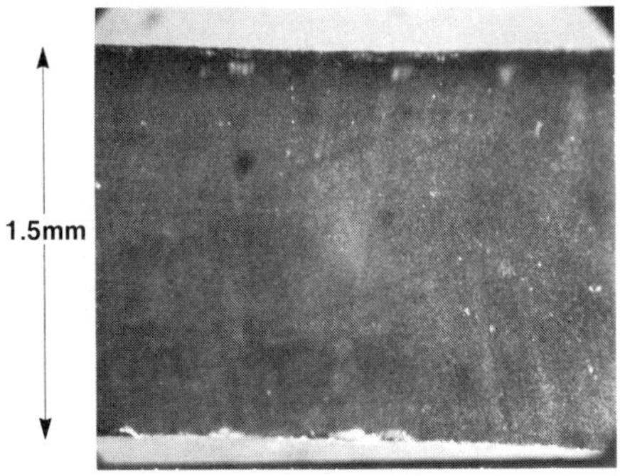

Fig. 4. Micrograph of fracture surface of low-temperature (small crystallite size) compact.

Thermal Conductivity

Thermal diffusivity measurements were made over a temperature range from -100 to 650°C. The maximum thermal conductivity values calculated for each of the samples over this temperature range are tabulated in Table I. Figure 5 shows the temperature dependence of the thermal conductivity for a number of the samples. Also shown in this figure for comparison is the temperature dependence of the thermal conductivity for high purity copper [8] and the room temperature thermal conductivities of Type 1a single crystal diamonds with nitrogen concentrations of $1.7 \times 10^{20}/cm^3$ and $3.5 \times 10^{20}/cm^3$ [9].

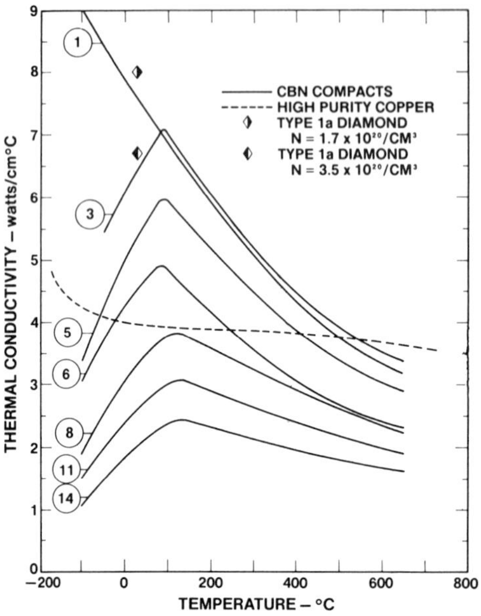

Fig. 5. Thermal conductivity for various cubic BN compacts, type 1a single-crystal diamonds and high purity copper. Sample numbers correspond to Table I.

Except for the two best conducting samples (1 and 2) all the samples show a conductivity maximum in the 50 to 150°C temperature range with the maximum shifting towards lower temperatures with increasing conductivity. A general increase in thermal conductivity with increasing compact processing temperature is observed with a factor of three to four difference in the room temperature conductivity between the high and low temperature compacts.

The wide variation in the thermal conductivity of the various compacts is attributed to differences in the microstructure and amount of residual unconverted graphitic BN in the compacts. In simple insulating crystals, such as cubic boron nitride, thermal energy is transported by lattice vibrations which can be described in terms of phonons [10]. The thermal conductivity of an infinite perfect crystal is limited only by direct interactions among the phonons themselves. Any disturbance in the perfect periodicity of the crystal lattice, including crystallite boundaries, which produces scattering of the phonons introduces additional thermal resistance. An increase in the thermal conductivity with increasing processing temperature would be expected due to the increased crystallite size and perfection of the higher temperature compacts. In addition, the thermal conductivity of the low temperature compacts is reduced further by the thermal resistance introduced by the residual graphitic BN phase present in these compacts.

ACKNOWLEDGMENTS

The author would like to thank H. P. Bovenkerk, General Electric Company for supporting this work, T. R. Lavens, General Electric Company for preparing the samples, and R. E. Taylor of Purdue University for making the thermal diffusivity measurements.

NOTATION

B_m = x-ray diffraction peak width, sample
B_s = x-ray diffraction peak width, reference standard
C_p = specific heat
k = thermal conductivity
t = crystallite thickness
α = thermal diffusivity
θ = Bragg angle
λ = wavelength
ρ = density

REFERENCES

1. G. A. Slack, J. Phys. Chem. Solids 34, 321 (1973).
2. W. J. Parker, R. J. Jenkins, C. P. Butler, and G. L. Abbott, J. Appl. Phys. 32, 1679 (1961).
3. R. E. Taylor and L. M. Clark III, High Temp.-High Press. 6, 65 (1974).
4. Y. S. Touloukian and E. H. Buyco, eds., Specific Heat - Nonmetallic Solids, Vol. 5, Thermophysical Properties of Matter, TPRC Data Series, Purdue University, West Lafayette, Indiana (1970), p. 1078.
5. N. N. Sirota and N. A. Kofman, Sov. Phys. Dokl. 21, 516 (1976).
6. H. M. Strong, R. E. Tuft, and R. E. Hanneman, Metallurgical Trans. 4, 2657 (1973).
7. B. D. Cullity, Elements of X-ray Diffraction, Addison-Wesley, Reading, Massachusetts (1959), p. 261.
8. Y. S. Touloukian, ed., Thermophysical Properties of Matter, Vol. 3, TPRC Data Series, Purdue University, West Lafayette, Indiana (1966), Table 1007R.
9. R. M. Chrenko and H. M. Strong, Rept. No. 75CRD089, General Electric Company, Research & Development Center, Schenectady, New York (1975).
10. R. Berman, Thermal Conduction in Solids, Clarendon Press, Oxford, England (1976), p. 13.

V-4

HIGH-PRESSURE SYNTHESIS OF SmB_2 AND GdB_{12}

J. F. Cannon, D. M. Cannon, and H. T. Hall

Brigham Young University
Provo, Utah USA

INTRODUCTION

Borides with a composition LnB_2 and LnB_{12} have been reported [1,2] for the smaller members of the lanthanide series. Diborides have been prepared [3-10] for Ln = Gd, Tb, Dy, Ho, Er, Tm, Yb, Lu, Y, Sc and dodecaborides are known [5,6,11-16] for Ln = Tb, Dy, Ho, Er, Tm, Yb, Lu, Y, Sc. The diborides are isomorphous, with an AlB_2-type (C32) structure. This structure is hexagonal (P6/mmm) and consists of alternating layers of metal atoms and boron atoms. The metal atoms form close-packed (3^6) layers and the boron atoms, hexagonal (6^3) layers. The layers are arranged such that boron atoms are positioned over the faces of the triangles formed by the metal atoms. The dodecaborides (except ScB_{12}) each assume a UB_{12}-type ($D2_f$) structure. This structure is cubic (Fm3m) with metal atoms at fcc positions and each metal atom surrounded by a B_{24} cubo-octahedron. The tetragonal structure of ScB_{12} is described by Matkovich, et al. [16].

Reported attempts to prepare diborides and dodecaborides of the larger lanthanide elements have not been successful [2,17,18], apparently because of the increased sizes of the metal atoms. Since the compressibilities of the lanthanides are greater than that of boron, an increase in pressure would tend to make the larger lanthanides behave more like the smaller ones. If the effect of metal-atom size is important in the formation of diborides and dodecaborides, then the application of high pressure and temperature to appropriate reaction mixtures should make it possible to extend these boride series to some of the larger lanthanides.

EXPERIMENTAL

The tetrahedral-anvil high pressure device designed by Hall [21,22] was used for the high pressure synthesis experiments. Stoichiometric mixtures of the elements were exposed to high pressure, high temperature conditions in a BN crucible surrounded by a graphite heater. Details of pressure cell construction and pressure/temperature calibration procedures may be found in earlier publications [23,24].

Boron was obtained as a -325 mesh powder of reported 99.5% purity from Research Organic/Inorganic Chemical Corporation. The lanthanides used in this work were obtained as ingots of reported 99.9% purity (metals only) from Research Organic/Inorganic Chemical Corporation and from Research Chemicals, Inc. These ingots were filed in the open atmosphere and those filings that passed a 100-mesh sieve were used. There may have been some oxide formation during filing of these ingots, but no oxide appeared visually and lines characteristic of lanthanide oxides did not appear in any of the x-ray spectra.

The product of each high pressure experiment was crushed, loaded into a 0.5 mm capillary and exposed to Ni-filtered Cu X-radiation ($\lambda(K\alpha)$ = 1.54178 Å and $\lambda(K\alpha_1)$ = 1.54051 Å) on a GE XRD-5 powder diffraction unit. A Debye-Scherrer camera with 143.2 mm diameter was used, and the sample was rotated during exposure. The Nelson-Riley [25] extrapolation procedure was used to correct for absorption. A cubic internal standard (SmN) was used with the SmB_2 spectra. Lattice parameters of the hexagonal materials were determined by a least squares refinement [26] after the absorption correction had been applied. The computer program POWDER [27] was used to calculate x-ray line intensities expected for the structure types encountered in this study. Observed x-ray line intensities were determined visually without reference to a calibration strip.

RESULTS

Lattice parameters for new compounds and for compounds prepared for comparison with the literature are found in Table I. X-ray data for the new compounds are found in Tables II and III. Results for each of the systems examined are as follows:

Nd + 12 B. Experiments made at 65 kbar and 2100°C resulted only in the preparation of NdB_6.

Sm + 12B. The only identifiable phase obtained in experiments at 65 and 70 kbar was SmB_6. At 70 kbar and temperatures of 2100 and 2700°C a minor phase that could not be identified was also obtained.

Gd + 12B. At 2100 C and pressures above 60 kbar, pure* GdB_{12} was obtained. Between 33 and 60 kbar, mixtures of GdB_{12} and GdB_6

*"Pure" in this context means that all lines in the x-ray spectrum were attributable to GdB_{12}.

Table I. Crystallographic Data

Compound	Crystal System	a, Å	c, Å	Reference
GdB_{12}	Cubic	7.524(1)*		This study
TbB_{12}	Cubic	7.509(1)		This study
		7.505		4
		7.504(1)		15
SmB_2	Hex.	3.310(1)	4.019(1)	This study
GdB_2	Hex.	3.315(3)	3.936(3)	This study
		3.318	3.933	2,10,19
		3.31	3.94	4
HoB_2	Hex.	3.279(2)	3.811(2)	This study
		3.17**	3.81	4
		3.281	3.811	8
		3.273	3.814	10
TmB_2	Hex.	3.258(3)	3.745(3)	This study
		3.250	3.739	8
		3.261	3.755	10

* Numbers in parentheses are standard deviations in last significant figure.
**This is most likely a misprint. The value meant is probably 3.27Å.

were obtained. Below 33 kbar, GdB_{12} was not found. A comparison of line intensities observed in the x-ray spectrum of GdB_{12} with those calculated assuming a UB_{12}-type structure (see Table II) establishes that the UB_{12}-type structure is correct for GdB_{12}. The calculated intensities were corrected for temperature (B = 1.0 and 1.5 for Gd and B respectively) and absorption ($\mu R = 54$). The variable B position parameter was assumed to be 0.166.

Tb + 12B. The dodecaboride was prepared at 65 kbar and 1650°C.

Ln + 2B for Ln = Gd, Ho, Tm. The diborides of these lanthanides were prepared at 60 to 70 kbar and 1240 to 1780°C. The lattice parameters obtained in this work compare favorably with those reported in the literature (see Table I). Absorption corrections were not made on the x-ray data of these compounds.

Sm + 2B. It was found that SmB_2 is best prepared at 65 kbar and 1140 to 1240°C. The best material prepared gave only a fair x-ray spectrum--high angle lines were fuzzy and had poorly resolved $K\alpha_1$-$K\alpha_2$ doublets. Extraneous lines in the spectrum were identified with SmN, apparently obtained by reaction with the BN crucible. The

Table II. X-ray Data for GdB_{12}

hkl	d_{calc}	d_{obs}	I_{obs}	I_{calc}*
1 1 1	4.344	4.313	40	31
2 0 0	3.762	3.740	30	33
2 2 0	2.660	2.642	30	33
3 1 1	2.269	2.260	100	100
2 2 2	2.172	2.161	25	31
4 0 0	1.881	1.871	10	14
3 3 1	1.726	1.721	50	48
4 2 0	1.683	1.679	30	34
4 2 2	1.536	1.533	50	46
3 3 3, 5 1 1	1.448	1.444	25	28
4 4 0	1.330	1.329	5	10
5 3 1	1.272	1.269	60	57
6 0 0, 4 4 2	1.254	1.252	40	40
6 2 0	1.1897	1.1877	15	20
5 3 3	1.1475	1.1461	20	18
6 2 2	1.1343	1.1331	10	14
4 4 4	1.0861	1.0871	5	7
5 5 1, 7 1 1	1.0536	1.0537	20	24
6 4 0	1.0434	1.0419	10	15
6 4 2	1.0055	1.0046	25	22
7 3 1, 5 5 3	.9796	.9789	80	52
8 0 0	.9405	--	N.O.	4
7 3 3	.9193	.9190	10	14
8 2 0, 6 4 4	.9125	.9126	20	22
8 2 2, 6 6 0	.8868	.8865	70	33
7 5 1, 5 5 5	.8688	.8685	30	28
6 6 2	.8631	.8630	8	15
8 4 0	.8413	.8415	5	13
7 5 3, 9 1 1	.8259	.8259	90	68
8 4 2	.8210	.8210	80	46
6 6 4	.8021	.8023	10	23
9 3 1	.7888	.7888	70	60

*Corrected for temperature (B = 1.0 for Gd and 1.5 for boron) and for absorption (μR = 54).

x-ray data shown in Table III were taken from the spectrum of material prepared at 65 kbar and 1240°C for 75 min.

Initially there was some question about whether SmB_2 had been obtained. Positions of the lines in the x-ray spectrum were consistent with an AlB_2-type material and the hexagonal lattice parameters were about what one would expect for SmB_2, but the observed line intensities did not agree properly with those calculated assuming that SmB_2 has an AlB_2-type structure. In drawing this conclusion, only adjacent lines were compared relative to one another. This is

Table III. X-ray Data for SmB_2

hkl	d_{calc}	d_{obs}	I_{obs}	I_{calc}*	I_{calc}**
0 0 1	4.020	3.991	50	19	20
1 0 0	2.867	2.872	50	52	47
1 0 1	2.334	2.332	100	100	100
0 0 2	2.010	2.007	40	14	18
1 1 0	1.655	1.652	20#	28	26
1 0 2	1.646	1.645	50#	43	55
1 1 1	1.531	1.530	20	32	31
2 0 0	1.434	1.435	5	15	13
2 0 1	1.350	1.350	15	30	30
0 0 3	1.340	1.339	5	4	8
1 1 2	1.278	1.277	40	28	36
1 0 3	1.214	1.213	80	22	44
2 0 2	1.1671	1.1667	10	16	21
2 1 0	1.0837	1.0836	8	13	11
2 1 1	1.0463	1.0458	25	29	29
1 1 3	1.0415	1.0421	25	11	22
0 0 4	1.0050	1.0042	6	2	9
2 0 3	0.9789	0.9777	25	13	26
3 0 0	0.9557	--	--	7	7
2 1 2	0.9539	0.9538	25	20	26
1 0 4	0.9484	0.9481	30	10	41
3 0 1	0.9298	0.9294	5	9	8
3 0 2	0.8631	0.8628	10	13	19
1 1 4	0.8591	0.8589	40	13	53
2 1 3	0.8426	0.8428	40	25	56
2 2 0	0.8277	0.8278	5	7	8
2 0 4	0.8229	0.8228	35	11	48
2 2 1	0.8107	0.8112	5	11	10
0 0 5	0.8040	0.8043	5	2	21
3 1 0	0.7952	0.7946	5	15	13
3 1 1	0.7801	0.7803	50	54	65
3 0 3	0.7781	0.7783	50	23	48
1 0 5	0.7742	0.7741	60	42	412 (sic)

* Corrected for absorption (μR = 130) and for temperature using isotropic temperature factors with B = 1.5.

**Corrected for absorption (μR = 130) and for temperature using anisotropic temperature factors for Sm with B_{11} = 0.10 and B_{33} = 0.005 and isotropic temperature factor for boron with B = 1.5.

Uncertain estimates because of mutual interference.

made necessary by the crude means used to determine observed line intensities. But even with this limited method of comparison it was apparent that the 103, 113, 104, 114 and 204 lines were observed to be stronger than calculations indicated they should be.

Comparisons with our data on other diborides showed the same intensity deviations for GdB_2 but not for HoB_2 and TmB_2. Among sets of published x-ray data we found similar deviations for ErB_2 and TmB_2 [7] and for GdB_2 [19] but none for LuB_2 [5]. The literature data for ErB_2 and TmB_2 were taken from Gandolfi-type spectra of single crystal platelets; thus, the intensity deviations in this case may be due to lack of completely random crystal orientation. Literature data on GdB_2 were obtained from a diffractometer and precautions were taken to insure that crystallite orientations were random [26]. To insure that our intensity deviations were not due to preferred orientation, crushed samples of SmB_2 and GdB_2 were disbursed in cornstarch and x-ray spectra redetermined. No significant changes in relative intensities were observed.

Attempts at reconciliation were made by calculating new intensity sets for structures slightly different from that of the AlB_2 type, but which retained the same hexagonal cell size. This was done by shifting the atomic positions of the B atoms. The B atoms provide so little x-ray scattering power compared to the metal atom, that relative line intensities were changed very little through this approach.

Finally, new intensity calculations were made for which the Sm atom was distorted from spherical symmetry in the manner of a prolate spheroid, with the long axis of the spheroid parallel to the crystallographic c-axis. This follows a recent suggestion in the literature [10] that the larger lanthanide atoms have such a distortion in LnB_2 compounds because of the rigidity of the B hexagonal framework. Such an hypothesis appears reasonable in view of the fact that the c/a ratio for these compounds increases as the size of the lanthanide atom increases. This distortion was introduced into the calculations via the anisotropic temperature factors. It resulted in changes that were exactly the opposite of those desired. Consequently new calculations were made for which the Sm atom was flattened at the poles like an oblate spheroid. These calculations gave intensities that were in good agreement with those observed (see Table III).

It is difficult to understand why the hexagonal-cell c-axis increases when the metal atom is extended along the a- and b-axes and shortened in the direction of the c-axis. This would seem to indicate considerable weakening of the metal-boron bond and perhaps a strengthening of the metal-metal bonds. This may account for the unexpectedly short a-axis found for SmB_2 (a for SmB_2 is actually shorter than that of GdB_2).

A few attempts were made to grow single crystals of SmB_2 by slow cooling the reaction mixture at high pressure. Unfortunately the hoped-for single crystals did not materialize. If single-crystal x-ray intensity data could be collected, a much more detailed determination of the metal-atom's departure from spherical symmetry could be made. It would be interesting to see what such data on SmB_2, GdB_2 and TbB_2 would reveal.

REFERENCES

1. G. V. Samsonov, High Temperature Compounds of Rare Earth Metals with Non Metals, Consultants Bureau, New York (1965), p. 1; English Translation of Tugoplavkie Soedineniya Redkozemel'nykh Metallov s Nemetallami, Metallurgiya, Moscow, U.S.S.R. (1964).
2. K. E. Spear in Phase Diagrams: Materials Science and Technology, Vol. 6-IV, A. M. Alper, ed., Academic Press, New York (1976), p. 91.
3. N. N. Zhuravlev and A. A. Stepanova, Sov. Phys. Crystallogr. 3, 76 (1958); English Translation of Kristallografia 3, 83 (1958).
4. B. Post in Proceedings Conference Rare Earth Research, Vol. 3, K. S. Vorres, ed., Gordon and Breach, New York (1964), p. 107.
5. M. Przybylska, A. H. Reddoch, and G. J. Ritter, J. Am. Chem. Soc. 85, 407 (1963).
6. P. Peshev, J. Etourneau and R. Naslain, Mater. Res. Bull. 5, 319 (1970).
7. R. N. Castellano, Mater. Res. Bull. 7, 261 (1972).
8. J. Bauer and J. Bebuigne, C. R. Acad. Sci. Ser. C. 277, 851 (1973).
9. J. Bauer, C. R. Acad. Sci. Ser. C. 279, 501 (1974).
10. K. E. Spear, J. Less-Common Metals 47, 195 (1976).
11. A. U. Seybolt, Trans. Am. Soc. Metals 52, 971 (1960).
12. V. V. Odintsoc and Yu. B. Paderno, Inorg. Mater. (USSR) 7, 294 (1971).
13. K. Schwetz, P. Ettmayer, R. Kieffer and A. Lipp, Radex Rundsch., 257 (1972).
14. S. LaPlaca, I. Binder and B. Post, J. Inorg. Nucl. Chem. 18, 113 (1961).
15. S. LaPlaca, D. Noonan and B. Post, Acta Crystallogr. 16, 1182 (1963).
16. V. I. Matkovich, J. Economy, R. F. Giese and R. Barrett, Acta Crystallogr. 19, 1056 (1965).
17. B. Post, D. Moskowitz and F. W. Glaser, J. Am. Chem. Soc. 78, 1800 (1956).
18. G. I. Solovyev and K. E. Spear, J. Am. Ceramic Soc. 55, 475 (1972).
19. K. E. Spear and Petsinger, JPDCS Powder File #24-1083.
20. K. E. Spear, private communication.
21. H. T. Hall, Rev. Sci. Instr. 29, 267 (1958).
22. H. T. Hall, Rev. Sci. Instr. 33, 1278 (1962).
23. J. F. Cannon and H. T. Hall, Inorg. Chem. 9, 1639 (1970).
24. J. F. Cannon and H. T. Hall, J. Less-Common Metals 40, 313 (1975).
25. J. B. Nelson and D. P. Riley, Proc. Phys. Soc. (London) 57, 160 (1945).
26. M. H. Mueller, L. Heaton and K. T. Miller, Acta Crystallogr. 13, 828 (1960).
27. D. K. Smith, "A Fortran Program for Calculating X-ray Powder Diffraction Patterns," UCRL-7196, Lawrence Radiation Laboratory, Livermore, California (1963).

V-5

DENSIFICATION AND ELECTRICAL RESISTANCE OF Y_2O_3 AT VARIOUS PRESSURES

F. W. Vahldiek

Air Force Materials Laboratory
Wright-Patterson Air Force Base, Ohio USA

INTRODUCTION

The densification of Y_2O_3 has been previously reported by various investigators [1-4]. Schieltz and Wilder [1] studied the sintering of Y_2O_3 from 1210 to 1643°C in a vacuum furnace, with difficulties encountered in its complete densification and interpretation of the sintering data in terms of sintering models based on volume and grain boundary diffusion. Jorgensen and Anderson [2] showed that the addition of up to 10 m% ThO_2 to Y_2O_3 inhibits grain growth during sintering, and allows the sintering process to proceed to essentially theoretical density of the specimens. These investigators [2] were unable to obtain theoretical density of the pure Y_2O_3 specimens heated to 2000°C in a resistance furnace. Transparent Y_2O_3 was reported by Lefever and Matsko [3] using vacuum hot pressing at 950°C and pressures up to 830 bars. Full transparency was achieved according to these investigators [3] by adding LiF to Y_2O_3. Dutta and Gazza [4] used vacuum hot pressing at temperatures ranging from 1300 to 1500°C and pressures up to 480 bars to produce transparent Y_2O_3 specimens. The present work was undertaken in order to determine the densification of Y_2O_3 with or without ThO_2 additions using vacuum sintering, regular hot pressing, and high pressure hot pressing techniques.

EXPERIMENTAL

Submicron particles of Y_2O_3 with a particle size ranging from 0.05 to 0.1 µm received from the Michigan Chemical Corporation (spectrographic grade with a purity of 99.99%) were used for the preparation of the specimens. Table I shows the emission spectrographic analyses of as-received Y_2O_3 powder, and also shows the impurities picked up during typical vacuum sintering, hot pressing, and high pressure electrical resistance experiments. Reagent grade ThO_2 powder with a particle size ranging from 0.02 to 0.3 μm and a purity of 99.92% received from Vitro Corporation was used as a sintering aid to Y_2O_3. The latter powder, with and without additions of 0.1 or 0.5 m% ThO_2, was prepressed at room temperature at 200 bars in steel dies prior to vacuum sintering, hot pressing, or high pressure experiments. The vacuum sintering was undertaken at pressures ranging from 1.2 to 2.0 x 10^{-8} bars in a Brew Ta resistance furnace at temperatures ranging from 1800 to 2000°C using Y_2O_3 powder to avoid direct contact between the Ta holder and Y_2O_3 specimens. Specimens fired in vacuum varied from 0.6 to 2 cm in diameter by 0.3 to 3 cm in length to allow for accurate bulk density and electrical conductivity determination at 1 bar.

The hot pressing of Y_2O_3 was achieved with graphite dies (coated on the inside with Y_2O_3) inductively heated to temperatures ranging from 1250 to 1600°C and pressures ranging from 210 to 500 bars. High pressure hot pressing at temperatures ranging from 1350 to 1400°C and pressures ranging from 5000 to 12,000 bars were undertaken with the previously described modified belt apparatus [5]. Y_2O_3 specimens prefired in air at 1000°C approximately 0.8 cm in diameter by 1.4 cm in length were enclosed in Pt cylinders, which in turn were placed in the high pressure cells. These cells were introduced into the modified belt device. The pressure source was a hydraulic press with a 1000 ton capacity. The high pressure system was obtained from ManLabs Inc. The specimens were heated by internal graphite resistance heaters after the desired pressures were applied. The temperature measurements were undertaken with Pt/Pt-10% Rh thermocouples with the beads placed in the outer BN sleeve surrounding the graphite heater [5]. The belt device was calibrated for actual pressure at room temperature using the well known Bi I-II, Bi II-III, Tl II-III, and Ba II-III pressure transitions prior to pressure-temperature experiments [6]. Electrical resistance measurements on previously vacuum sintered or hot pressed specimens were done in air at 1 bar and at pressures ranging from 5 to 25 kbars at temperatures up to 1200°C. Heating and cooling rates were 300°C/hr at applied pressures for the resistance measurements. A combination of 2-probe and 4-probe methods was used to determine resistance as a function of temperature at 1 bar and up to 25 kbars. Y_2O_3 specimens with platinum coated end faces and thin Pt wafer contacts welded to Pt wires were the current leads. Platinum wires wrapped around the specimens 1 cm apart were used as potential probes. These techniques using 4-probe and 2-probe methods (the latter by ac resistance measurements at 60 and 1000 Hertz) have been previously described for 1 bar and high pressure-temperature resistance determination [7-9] and were used in this work with some modifications based on specimen size. Density measurements were determined by a combination of pycnometric and x-ray measurements. Specimens were analyzed by x-ray diffraction, wet chemical, emission spectrographic, and electron microprobe techniques. X-ray diffraction studies were

Table I. Spectrochemical Analyses of Y_2O_3 Specimens

Impurity, wt.%	Powder (as-rec'd)	Sintered ($2000°C, 10^{-8}$ bars) 24 hrs	Hot-Pressed (1400°C, 300bars) 1 hr	Hot-Pressed* ($1400°C, 1.2x10^4$ bars) 1 hr	Hot-Pressed** Resistance Expt. (1400°C(max), $1.2x10^4$ bars)
Si	0.001	0.002	0.003	0.010	0.013
Mg	0.0003	0.0005	0.0005	0.001	0.002
Mn	0.0001	0.0001	0.001	0.001	0.001
B	0.0001	0.0006	0.001	0.030	0.250
Sn	0.0001	0.0001	0.001	0.0001	0.0001
Pb	0.0001	0.0001	0.001	0.0001	0.0001
Al	0.0003	0.0004	0.001	0.010	0.010
Fe	0.0003	0.0015	0.001	0.010	0.010
Cu	0.0001	0.0001	0.0005	0.001	0.001
Na	0.0004	0.0004	0.001	0.001	0.002
Zn	0.001	0.0004	0.002	0.001	0.001
Ti	0.001	0.001	0.001	0.001	0.001
Cr	0.0001	0.0002	0.0004	0.0005	0.0006
Ca	0.001	0.005	0.002	0.003	0.003
Pt	-	-	-	0.010	-

*Specimen enclosed in Pt foil

**Specimen enclosed in BN cylinder

undertaken to determine phase(s) present in Y_2O_3. Optical and electron optical studies were undertaken with a Zeiss Ultraphot II light microscope and a JEM 6A electron microscope. The photometric measurements were made with a Photovolt Corporation device using a tungsten filament at 24 W. Knoop micro hardness measurements were taken on the grains of etched specimens using a Tukon Tester Type FB at 25g loads. Y_2O_3 specimens were polished with diamond paste followed by etching with diluted HCl at 110°C for approximately 1 min.

DISCUSSION OF RESULTS

Vacuum Sintering, Hot Pressing and High Pressure Hot Pressing

Preliminary vacuum sintering data indicated that temperatures above 1800°C are needed to reach close to theoretical densities for Y_2O_3. Above 2000°C the vaporization of Y_2O_3 becomes an important factor especially for time periods up to 24 hrs. Y_2O_3 specimens vacuum sintered above 1800°C usually were gray-black in color due to oxygen loss as determined by chemical analysis. Compositional analysis showed these samples to be $Y_2O_{2.94}$. Subsequent oxidation in air at 1200°C resulted in white-yellow colored specimens. Additions of 0.1 and 0.5 m% of ThO_2 changed the color by vacuum sintering to a light grey, which by subsequent oxidation resulted in glassy-white specimens. The additions of 0.1 and 0.5 m% ThO_2 to Y_2O_3 increased the bulk density to essentially theoretical with a concomitant decrease of the average grain size of the specimens (see Table II). A detailed grain growth study was undertaken on Y_2O_3, Y_2O_3 + 0.1 m% ThO_2, and Y_2O_3 + 0.5 m% ThO_2 with the specimens rapidly cooled from 2000°C at various time intervals. Figure 1 represents the results of the grain growth of the three compositions with the smooth lines drawn through the points determined by a least squares fit of the data. The slopes of the compositions ranged from 0.46 to 0.52, which are close to the 0.50 slope predicted for grain growth [2]. Figure 2a shows a typical Y_2O_3 microstructure vacuum sintered at 2000°C for 24 hrs. Note the presence of many closed pores. Etching with HCl at 110°C will often result in pyramidal pits in Y_2O_3 specimens as can be seen in Fig. 2b. Figure 2c shows such a pyramidal etch pit at higher magnification. Figure 2d shows an essentially 100% dense microstructure of Y_2O_3 + 0.5 m% ThO_2 vacuum sintered at 2000°C for 24 hrs. Essentially pore-free Y_2O_3 specimens with or without ThO_2 additive were received by regular hot pressing up to 500 bars. Figure 3a shows a typical microstructure of hot pressed Y_2O_3 at 500 bars at 1550°C. Here again some closed pores are distinctively visible. High pressure hot pressing up to 12,000 bars and temperatures ranging from 1350 to 1400°C produced microstructures of much finer grain size ranging from 0.5 to 3 μm in diameter as shown in Fig. 3b.

X-ray diffraction and electron microprobe analyses showed all processed specimens to be single phase cubic Y_2O_3 with uniform

Table II. Preparation, Density, and Grain Size of Y_2O_3 and $Y_2O_3 + ThO_2$ Specimens

Composition	Pressure, bars	Temperature, °C	Time at Temperature and Pressure, hrs	Density, g/cm^3	Grain Size, microns
Y_2O_3*	$1.2x10^{-8}$	1800	22	4.88	15-25
Y_2O_3	$1.4x10^{-8}$	1800	22	4.83	15-30
Y_2O_3	$2.0x10^{-8}$	2000	24	4.95	10-40
Y_2O_3	$1.7x10^{-8}$	2000	24	4.97	10-30
Y_2O_3	210	1250	1	4.85	5-8
Y_2O_3	210	1400	1	4.88	5-10
Y_2O_3	300	1250	1	4.82	5-7
Y_2O_3	300	1600	1	4.96	3-10
Y_2O_3	500	1500-1550	0.5	4.99	5-20
Y_2O_3	5000	1400	1	4.92	1-4
Y_2O_3	8000	1400	1	4.95	1-4
Y_2O_3	12,000	1350	1	4.97	0.7-3
Y_2O_3+0.1 m% ThO_2**	$1.3x10^{-8}$	2000	24	5.00	5-25
Y_2O_3+0.1 m% ThO_2	210	1350	1	4.92	5-8
Y_2O_3+0.1 m% ThO_2	5000	1350	1	4.95	1-4
Y_2O_3+0.5 m% ThO_2**	$1.3x10^{-8}$	2000	24	5.02	6-25
Y_2O_3+0.5 m% ThO_2	210	1350	1	4.84	5-8
Y_2O_3+0.5 m% ThO_2	5000	1350	1	4.90	1-3
Y_2O_3+0.5 m% ThO_2	8000	1400	1	4.90	1-3
Y_2O_3+0.5 m% ThO_2	12,000	1400	1	4.96	0.5-2.5

*Theoretical density of bcc Y_2O_3 was determined to be 5.00 g/cm^3.

**For the cubic composition, Y_2O_3 + 0.1 m% ThO_2, the theoretical density = 5.01 g/cm^3, and for cubic Y_2O_3 + 0.5 m% ThO_2 the theoretical density = 5.04 g/cm^3. Theoretical density calculated from x-ray and pycnometer measurements.

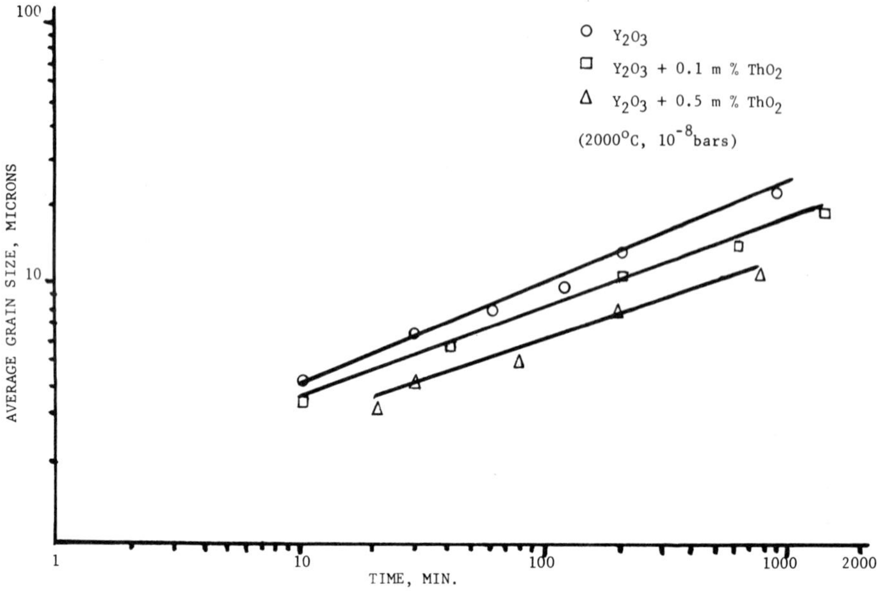

Fig. 1. Grain growth data of Y_2O_3 and Y_2O_3 + ThO_2 compositions (10^{-8} bars and 2000°C).

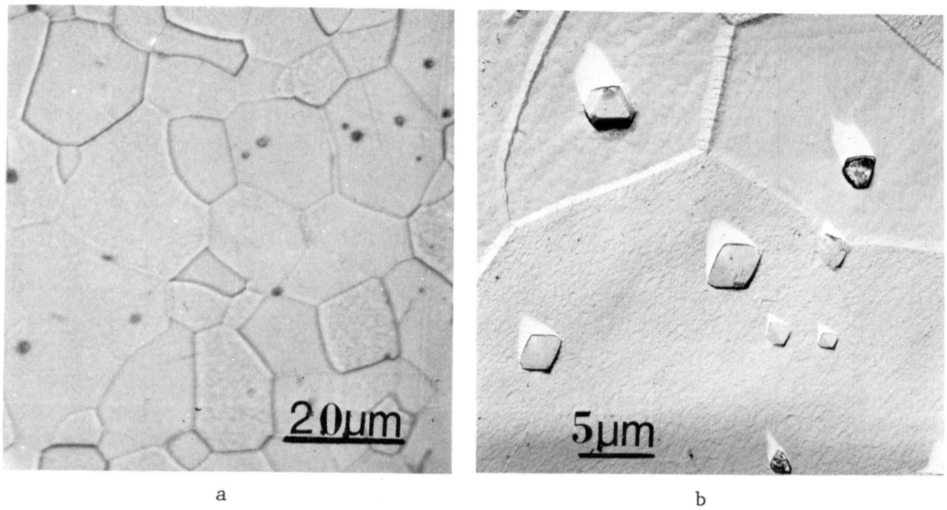

Fig. 2(a). Optical micrograph of Y_2O_3 vacuum sintered at 1.7 x 10^{-8} bar, 2000°C, 24 hrs.

Fig. 2(b). Electron micrograph (replica) of Y_2O_3 + 0.1 m% ThO_2 vacuum sintered at 2000°C showing pyramidal etch pits.

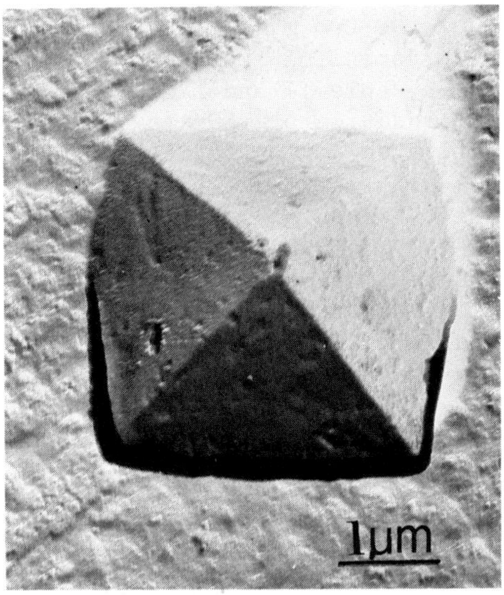

c

Fig. 2(c). Electron micrograph (same as Fig. 2b) showing pyramidal etch pit at higher magnification.

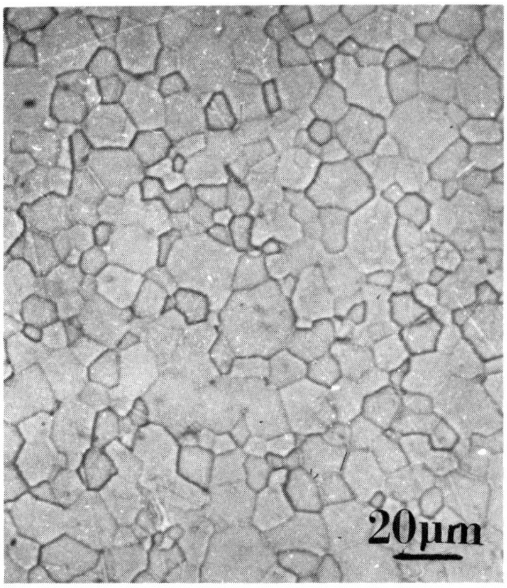

d

Fig. 2(d). Optical micrograph of Y_2O_3 + 0.5 m% ThO_2 vacuum sintered at 2000°C for 24 hrs.

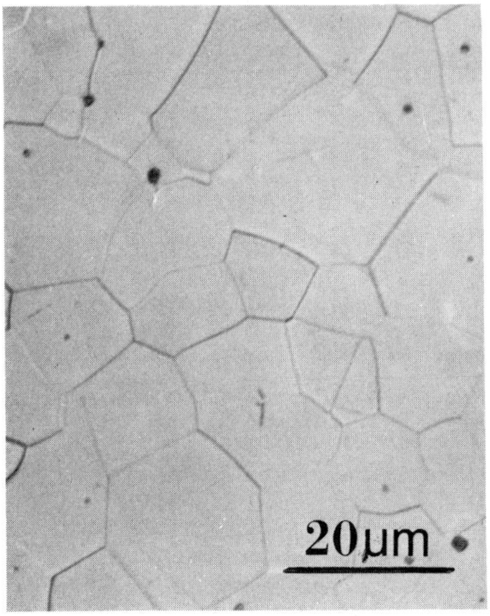

a

Fig. 3(a). Optical micrograph of Y_2O_3 hot pressed at 500 bars and 1550°C.

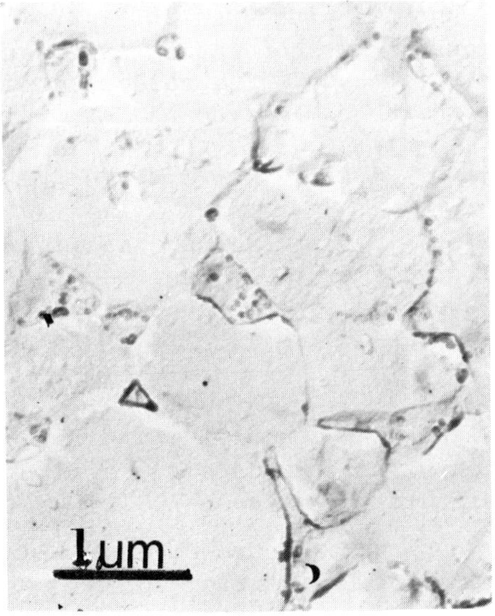

b

Fig. 3(b). Electron micrograph of Y_2O_3 pressed to 12,000 bars at 1300°C.

distribution of Y and Th in the case of the two mixtures studied. Impurities picked up during sintering or pressing are shown in Table I. From the densification data of Table II and Fig. 1, one can state that in vacuum sintering the addition of ThO_2 to Y_2O_3 seems to somewhat decrease the grain size while increasing the density. An increase of pressure will drastically decrease the grain size, however with limiting density improvements (see Table II). Melting points of Y_2O_3 and Y_2O_3-ThO_2 compositions were determined with a W resistance vacuum furnace using a W-Re thermocouple and a calibrated micro-optical pyrometer. A conical hole 0.3 cm outer diameter and 1 cm in depth was drilled into the specimens prior to melting point measurements. The following congruent melting point temperatures were found: Y_2O_3 = 2420° $\pm$ 10°C; Y_2O_3 + 0.1 mote % ThO_2 = 2435° $\pm$ 10°C; and Y_2O_3 + 0.5 mote % ThO_2 = 2470° $\pm$ 10°C. The melting point for Y_2O_3 is in good agreement with the literature [10].

Electrical Resistance

Vacuum sintered or hot pressed specimens of essentially theoretical density were used in this study. The oxygen-deficient specimens were oxidized prior to electrical resistance measurements to eliminate any error due to composition changes. First the electrical resistance of Y_2O_3 and Y_2O_3 + 0.5 m% ThO_2 were determined in air at 1 bar using the 2- and 4-probe methods. The conductivity vs. reciprocal temperature data were calculated from the resistance data, and are presented in Figs. 4 and 5, respectively. The activation energy for conduction calculated from the slope resulted in 1.96 ev for Y_2O_3. The rate equation for the conduction in Y_2O_3 at 1 bar was determined to be

$$\sigma = 110\ e^{-23{,}000/T} \tag{1}$$

For Y_2O_3 + 0.5 m% ThO_2 the activation energy for conduction at 1 bar calculated from the slope (see Fig. 5) was found to be 1.89 ev. The rate equation for the conduction in this composition resulted in

$$\sigma = 110\ e^{-21{,}900/T} \tag{2}$$

The electrical conductivity data for the above are based on four runs each with the data averaged for increasing and decreasing temperature (from about 700 to 1400°C).

The high pressure-high temperature electrical resistance experiments were carried out on Y_2O_3 at pressures ranging from 5 to 25 kbars and temperatures up to 1200°C. Sintered Y_2O_3 specimens were placed inside the high pressure resistance cells with the Y_2O_3 specimens enclosed in direct contact with BN cylinders, which resulted in B impurity pickup during the high pressure-temperature resistance runs (see Table I). The results of the high pressure-temperature resistance experiments are presented in Fig. 6. The

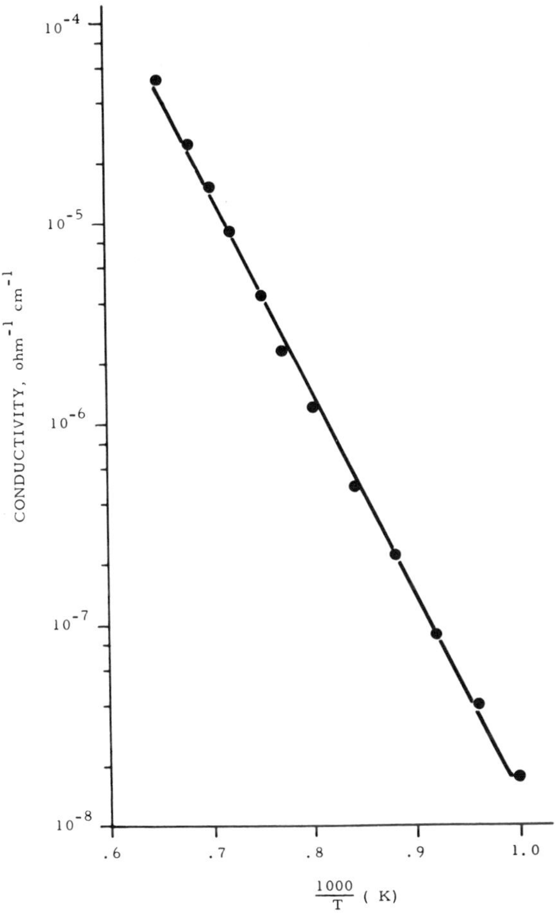

Fig. 4. Electrical conductivity of Y_2O_3.

resistance data of Y_2O_3 at 1 bar of similar sample size determined earlier are shown for comparison. The high pressure resistance data are corrected for initial temperature variations occurring during temperature increase after a desired pressure had been established. An overall decrease in resistance with increasing pressure and temperature from 5 to 25 kbars was observed for Y_2O_3. A rapid decrease in resistance was found initially going from 1 bar to 5 kbars at temperatures less than 100°C, followed by the more measurable slopes received from 5 to 25 kbars at high temperatures. For Y_2O_3 pressed at 5 kbars and temperatures ranging from 20 to 1200°C a slope of

$$dR/dT\ (P) = -0.16\Omega/°C/bar^{-1} \tag{3}$$

was found. Y_2O_3 pressed at 25 kbars and temperatures ranging from 20 to 1200°C resulted in a slope of

$$dR/dT\ (P) = -0.001\Omega/°C/bar^{-1} \tag{4}$$

The high pressure-temperature resistance data are based on three runs each with the data averaged for increasing and decreasing temperature at applied pressure.

Mechanical, Optical, and Thermal Properties

Knoop microhardness measurements were taken on grains and across grains of vacuum sintered samples only at 25g loads because of difficulties encountered with indentation size vs. grain size of Y_2O_3. Figure 7 shows a microhardness traverse on two adjacent grains (15 μm in size), and across the connecting grain boundary of Y_2O_3 and Y_2O_3 + 0.5 m% ThO_2 specimens. A scatter of less than

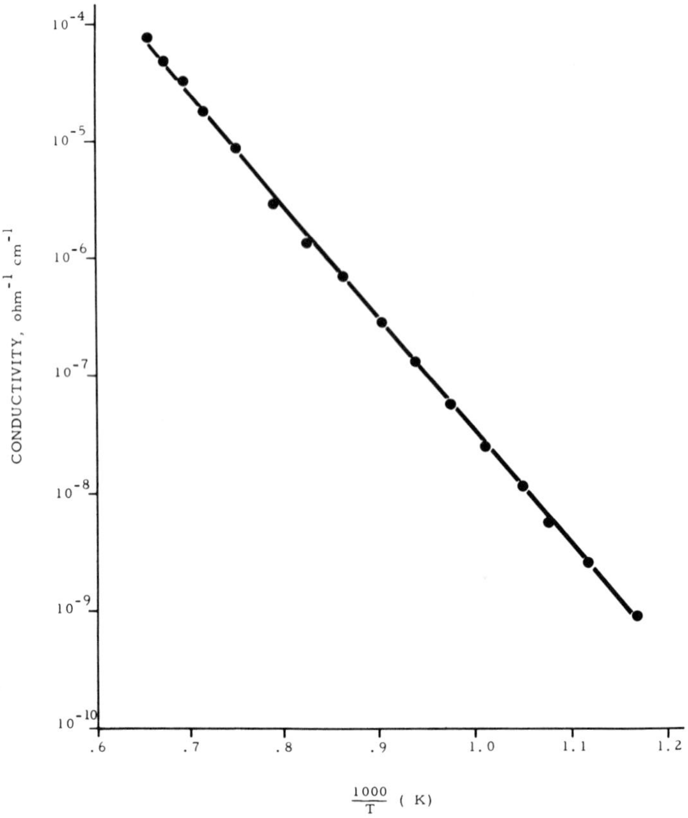

Fig. 5. Electrical conductivity of Y_2O_3 + 0.5 m% ThO_2.

10% in hardness data was encountered on the grains and across the grain boundary indicative of uniform composition with little or no substantial increase in impurities or second phase present within the grain or along the grain boundary and in good agreement with electron microprobe data. Table III shows average microhardness vs. grain size data based on 25 measurements each. Single crystal data for Y_2O_3 are included for comparison. Elastic moduli of Y_2O_3 and Y_2O_3 + 0.5 m% ThO_2 were determined with the dynamic elastic sphere technique of Anderson and Soga [11]. Spheres 0.8 cm in diameter were prepared from dense Y_2O_3 specimens using a combination cutting and grinding technique similar to the method used by the above authors [11]. The following moduli were determined for Y_2O_3: Young's modulus = 17530 kg/mm^2; Bulk modulus = 13740 kg/mm^2; Shear modulus = 6720 kg/mm^2; and Poisson's ratio = 0.296. These values compare very well with elastic modulus data reported by Manning et al. [12] on Y_2O_3. For Y_2O_3 + 0.5 m% ThO_2 the following data were determined: Young's modulus = 17750 kg/mm^2; Bulk modulus = 13910 kg/mm^2; Shear modulus = 6870 kg/mm^2; and Poisson's ratio = 0.298.

The translucency of various Y_2O_3 specimens was determined by the earlier described method. Table III summarizes the photometric data of in-line transmission in percent using filtered or unfiltered light. The data were taken on wafer-type specimens about 0.3 cm in thickness by 2 cm in diameter with the parallel surfaces polished with diamond paste. The in-line light transmission data of an almost

Table III. Knoop Microhardness and Translucence of Various Y_2O_3 Specimens

Composition	Density, g/cm^3	Avg. Grain Size, microns	KHN*-25g, kg/mm^2	Filter, None	Translucence**,% Green	Yellow	Red
Y_2O_3	4.83	22	495	10-15	2-5	5-7	2
Y_2O_3	4.95	34	500	15-20	5-6	6-7	3
Y_2O_3	4.99	20	490	15-20	4-6	5-6	3
Y_2O_3***	5.00	-	660	80-90	25	36	8
Y_2O_3+0.1 m% ThO_2	4.94	14	505	10-16	2-5	5-6	2
Y_2O_3+0.1 m% ThO_2	5.00	18	510	15-18	5-7	4-7	3
Y_2O_3+0.5 m% ThO_2	4.90	16	515	10-15	4-5	4-6	2-4
Y_2O_3+0.5 m% ThO_2	5.02	15	530	12-16	4-6	6-8	3

*The microhardness indentations were taken on grains avoiding grain boundaries.

**Photometric measurements of in-line transmission, in percent. Optical quality quartz was used as standard.

***Single crystal Y_2O_3 received from Korad Corp.

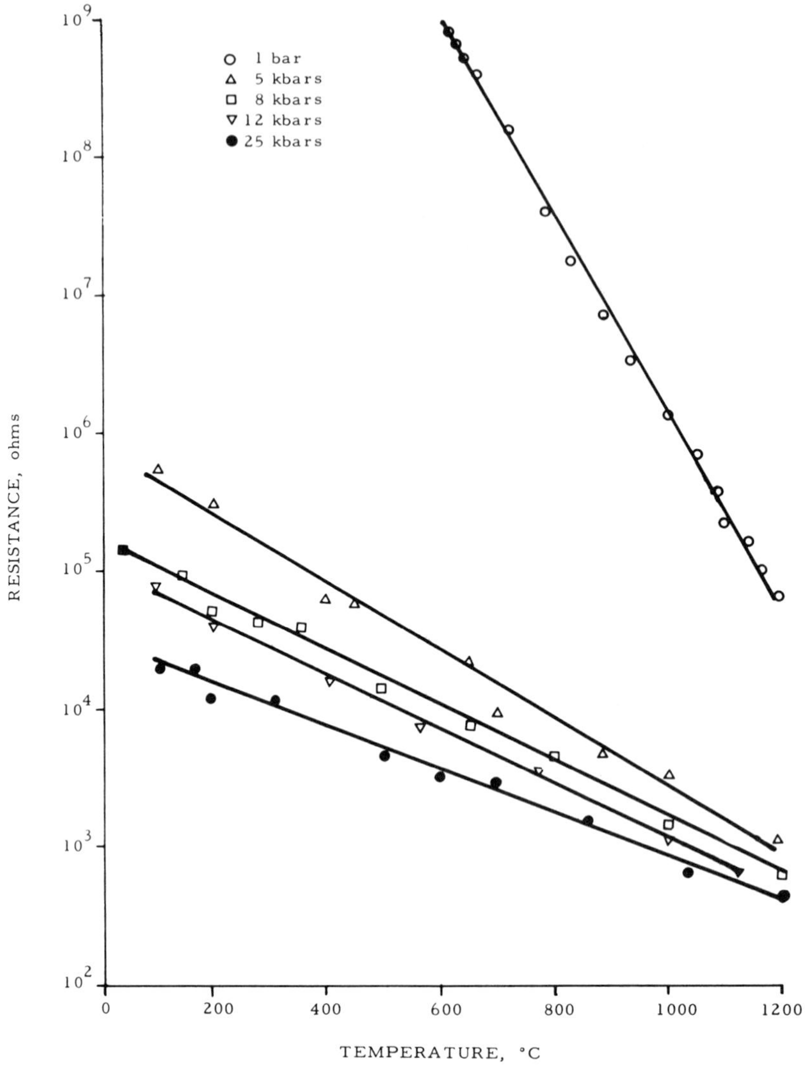

Fig. 6. Electrical resistance of Y_2O_3 at various pressures.

optical clear single crystal were added for comparison.

The thermal expansion of Y_2O_3 specimens was determined with an automated recording quartz dilatometer. The specimens were measured in argon at temperatures up to 1000°C. Figure 8 represents the average data received from four runs taking in account increasing and decreasing temperatures. The heating and cooling rates were about 150°C/hr. The thermal expansion for Y_2O_3 at 1 bar in argon was determined to be 7.1×10^{-6} cm/cm/°C.

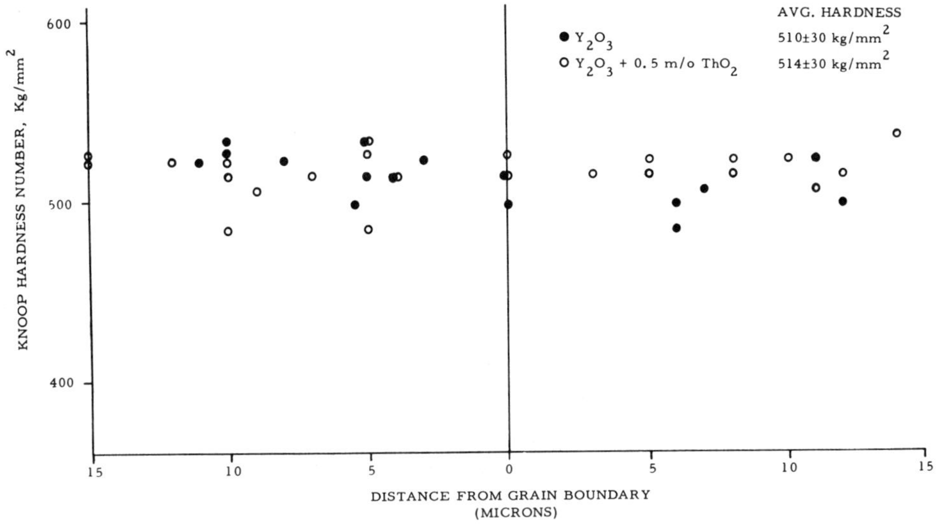

Fig. 7. Microhardness traverse of Y_2O_3 and Y_2O_3 + 0.5 m% ThO_2.

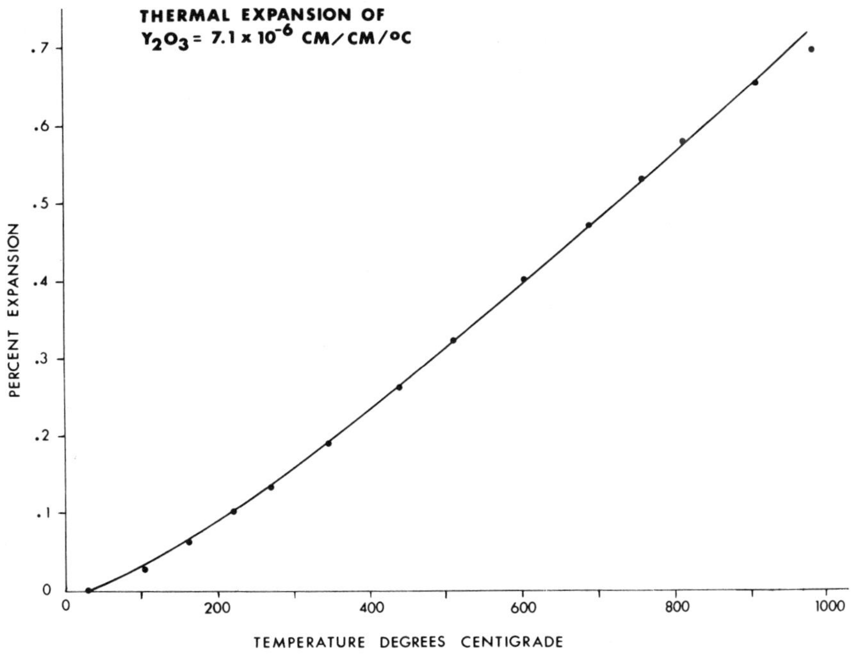

Fig. 8. Thermal expansion of Y_2O_3.

CONCLUSIONS

Translucent specimens of Y_2O_3 were prepared by vacuum sintering, hot pressing, and high pressure-hot pressing techniques. The addition of small amounts of ThO_2 to Y_2O_3 improved the density of Y_2O_3 with a small increase in mechanical properties. An overall decrease in electrical resistance with increasing temperature and pressure was determined for Y_2O_3.

REFERENCES

1. J. D. Schieltz and D. R. Wilder, J. Am. Ceram. Soc. 50, 439 (1967).
2. P. J. Jorgensen and R. C. Anderson, J. Am. Ceram. Soc. 50, 553 (1967).
3. R. A. Lefever and J. Matsko, Mat. Res. Bull. 2, 865 (1967).
4. S. K. Dutta and G. E. Gazza, Mat. Res. Bull. 4, 791 (1969).
5. F. W. Vahldiek, J. Less-Common Met. 13, 530 (1967).
6. G. C. Kennedy and P. N. Lamori, in Progress in Very High Pressure Research, F. P. Bundy, W. R. Hibbard and H. M. Strong, eds., John Wiley and Sons, New York (1961).
7. F. W. Vahldiek, J. Less-Common Met. 11, 99 (1966).
8. F. W. Vahldiek, J. Less-Common Met. 14, 133 (1968).
9. R. N. Blumenthal and J. E. Laubach, in Anisotropy in Single-Crystal Refractory Compounds, Vol. 2, F. W. Vahldiek and S. A. Mersol, eds., Plenum Press, New York (1968).
10. C. E. Curtis, J. Am. Ceram. Soc. 40, 274 (1957).
11. O. L. Anderson and N. Soga, AF Tech. Rpt. AFML TR 65-202 (1965).
12. W. R. Manning, O. Hunter, and B. R. Powell, J. Am. Ceram. Soc. 52, 436 (1969).

V-6

SYNTHESIS OF YTTERBIUM MONOXIDE UNDER HIGH PRESSURE

J. M. Léger and C. Loriers

Laboratoire d'Etude des Matériaux par Techniques Avancées
C.N.R.S., Meudon, France

and

L. Albert and J. C. Achard

Chimie Metallurgique et Spectroscopie des Terres Rares
C.N.R.S., Meudon, France

INTRODUCTION

The rare earth elements except La, Ce, Pr, Gd, Tb, and Er are known to form stable divalent halides, but only Sm, Eu, and Yb appear to have a reasonable chance of forming divalent oxides stable in the solid state. The existence of EuO is well established and it has been made by a variety of methods. The mixed valence oxide of europium, $Eu_3 O_4$, is also known.

Many attempts have been devoted to the synthesis of ytterbium monoxide. Several times success has been claimed but was never confirmed. Ytterbium with C, O, H, and N easily forms compounds and their solid solutions which have the same structure as expected for pure Yb O. In addition, the cell parameters can be quite close, so confusion is easy. In these compounds ytterbium is generally trivalent as in $Yb_2 O_3$ instead of being divalent as it should be in the monoxide, so susceptibility measurements are important to determine the actual formation of divalent ionic configuration. The existence of a mixed valence oxide such as $Eu_3 O_4$ in the system Eu O - $Eu_2 O_3$ has never been reported in the case of ytterbium.

Three types of reactions and many different techniques have been used to make ytterbium monoxide: (1) reduction of the sesquioxide $Yb_2 O_3$, (2) oxidation of Yb metal, and (3) crystal chemical stabilization.

Reduction [1] of the sesquioxide by the metal was investigated in open systems in which volatile by-products were allowed to vaporize, and it was also tried in closed systems where the vapor pressure of the metal could suppress the disproportionation of the monoxide. In both cases the synthesis of YbO could not be achieved although very different conditions of temperature, from 400 to 600°C up to the melting point of the sesquioxide, were used.

Oxidation [2] of the metal has also been carefully investigated. At low temperatures (T<400°C) in a closed system, a fcc compound (a=4.875 Å) was obtained but no analysis of C, H, and N was given. At higher temperatures (750°C), oxidation of the vapor yielded a cubic compound with a parameter equal to 4.865 Å. In both cases, metal and sesquioxide were mixed together with the fcc phase.

Crystal chemical stabilization [3] was also examined: solid solutions of $Yb_x Ca_{1-x} O$ were obtained containing up to 65 mole % YbO. In order to understand why ytterbium monoxide could not be obtained contrarily to europium monoxide, thermodynamical calculations were carried out and the Gibbs free energy variation for the reaction $Yb_{(s)} + Yb_2O_{3(s)} \rightarrow 3\ YbO_{(s)}$ has been evaluated.

All the thermodynamic data for Yb and Yb_2O_3 are known, those for YbO have been evaluated and the Gibbs free energy variation was found to be + 13 kcal/mole [4], or more recently + 8 kcal/mole [5]. The agreement is quite good because of the large possible errors due to differences of large numbers. These positive values indicate that under the given conditions the reaction is not possible: but the small values obtained do not really allow us to draw any definite conclusions. However, small but negative values were obtained in the case of EuO which agree well with the experimental results.

Use of very high pressure can make the synthesis of ytterbium monoxide possible because it is then necessary to take into account the $P\Delta V$ term in the Gibbs energy variation. This term is unusually large in the case of YbO. The volume reduction along with the above reaction is equal to 15.4 cm^3/mole at ordinary pressure. Assuming it is constant under pressure, a contribution of -15 kcal/mole at 40 kbar is to be added which is larger than necessary to make the reaction possible. Once the monoxide is made under pressure, it would perhaps be possible to bring it back to atmospheric pressure in a metastable state. No account has been taken for the variations of the bulk modulus, the thermal expansion, or the phase changes of Yb and Yb_2O_3, but that does not drastically modify the results.

EXPERIMENTAL

The high pressure experiments were conducted in a belt-type apparatus from 10 to 70 kbar and from 600 to 1300°C.

Ytterbium metal and oxide, both 99.9% pure, are treated just before the high-pressure runs in order to reduce the amounts of non-metallic impurities (UHV sublimation for Yb; high temperature treatment for Yb_2O_3). The samples are made from powders. The ytterbium powder is obtained by filing the distilled ingot. All the handling is performed under purified argon. The powders are compacted in a tungsten carbide die before introduction into the high pressure cell.

The crucible is made from boron nitride previously heated to 400°C. It is enclosed in the furnace which is a tantalum tube 0.1 mm thick. In order to prevent any pollution from the compositional water of the pyrophyllite, the inner part of the sleeve is previously heated to 1200°C under vacuum. Under these conditions no variation of the electrical resistance of the furnace could be noticed during the runs (2 hrs at 1200°C; 6 hrs at 600°C). When the pyrophyllite is not desiccated the resistance of the furnace steadily increases indicating corrosion of tantalum. The temperature is measured with a Pt-PtRh thermocouple placed in a hole in the bottom of the crucible. The pressure effect on the electromotive force has been ignored. The pressure is determined from a calibration curve made at room temperature by monitoring the discontinuities of the resistivities of Bi and Ba at 25.4 and 55 kbar as a function of the load. The samples are quenched under pressure by cutting off the electrical power supplied to the furnace. After unloading, a Debye-Scherrer pattern is recorded under a controlled atmosphere.

RESULTS

The reaction in the solid state is slow: after 6 hrs at 600°C or 2 hrs at 1200°C, diffraction lines of the sesquioxide can be seen in the x-ray pattern when using a stoechiometric starting mixture, with a large excess of ytterbium metal no such lines could be detected. The lines of ytterbium metal are always faint; partial removal of metal occurs during the crushing of the sample after the high-pressure runs; ytterbium metal being soft, small pellets are formed which are easily removed. The sesquioxide which remains in the sample can be seen in two different crystallographic structures, cubic or monoclinic, depending on the pressure and temperature of the experiments. The cubic-monoclinic boundary we found is slightly displaced toward higher pressures (about 5 kbar) than previously reported [6] during the study of this transformation in pure sesquioxide.

The reaction product diagram $Yb + Yb_2O_3$ is given in Fig. 1. We investigated this reaction in the solid state only, that is to say below the melting temperature of pure ytterbium. It is well-known that this increases rapidly with pressure: it is about 1250°C at 40 kbar instead of 824°C at normal pressure. Below a boundary defined by the two points, 10 kbar-800°C and 60 kbar-1200°C, a compound with a fcc structure is formed; above this boundary a compound with an orthorhombic structure is made.

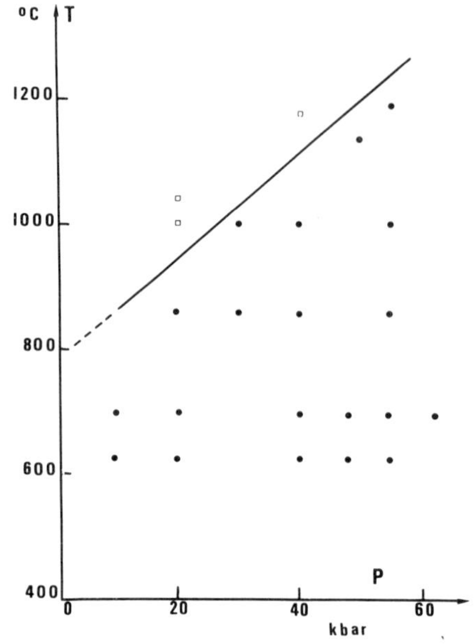

Fig. 1. Yb + Yb_2O_3 reaction product diagram. Legend: • = YbO; □ = Yb_3O_4.

The fcc structure has been attributed to the ytterbium monoxide for the following reasons: The cell parameter (4.877 ± 0.005 Å) obtained by the Nelson-Riley extrapolation is what can be expected from a linear extrapolation of the cell parameter of the monochalcogenides as a function of the anion radius (see Table I). In these compounds ytterbium is purely divalent.

In addition, it is also equal to the value obtained from the extrapolation of the values of the parameters of solid solutions [3] $Yb_{1-x}Ca_xO$ which obey the Vegard's law in an extended range $0.4 \leq x \leq 1$. The amounts of nonmetallic impurities have been determined by chemical analysis for different samples. They are less than 10 at.% for H, 2 at.% for C, and 0.5% for N. These results combined with the constancy of the cell parameters for samples made under the very different conditions of pressure and temperature we used seem to reject the presence of any ternary compounds or solid solutions with hydride, nitride, or carbide; on the other hand, stabilization by impurities cannot be completely ruled out.

Susceptibility measurements have also been carried out. The monoxide YbO is expected to be diamagnetic if all the ytterbium atoms are in the 4 f^{14} state forming purely divalent ions. The samples made under high pressure are weakly paramagnetic at room temperature ($\chi \simeq 2.10^{-6}$ emu/g) and the change of the susceptibility between 77 and 300 K shows a Curie-Weiss behavior with a Curie paramagnetic temperature of about 20 K. If we assume that the paramagnetic contribution comes only from the existence of Yb^{3+} ions we can evaluate the ratio $Yb^{3+}/Yb^{2+} + Yb^{3+}$ to be about 0.1.

The orthorhombic structure found at high temperatures is attributed to the new compound Yb_3O_4. The cell parameters determined from diffractometer data (a = 9.758 ± 0.006 Å, b = 3.333 ± 0.003 Å, c = 11.620 ± 0.008 Å) are quite close to those of $CaYb_2O_4$ [7] where the Yb^{2+} is replaced by Ca^{2+}, the ionic radius of which is very close. (These compounds are mixed valence oxides and are better described by writing CaO, Yb_2O_3 or YbO, Yb_2O_3). In addition, all the diffraction lines (46) are compatible with a space symmetry group Pnma to which both $CaYb_2O_4$ and Eu_3O_4

Table I. Parameters of Monochalcogenides and Anionic Radii

	YbTe	YbSe	YbS	YbO
a, Å	6.36	5.93	5.68	4.87 (extrapolated)
anion, Å	2.21	1.98	1.84	1.40

(calcium ferrite structure) belong. The formation of Yb_3O_4 corresponds to an incomplete reduction of the sesquioxide.

The extrapolation of the boundary YbO - Yb_3O_4 to normal pressure is doubtful. It would show that ytterbium monoxide could be made by the reduction of the sesquioxide by the metal at temperatures up to 800°C. Such a reaction could never be demonstrated. At higher temperatures no intermediate mixed valence oxide could be detected either. In addition, this monoxide formation is not correlated by thermodynamical estimations, but the possible errors are of the same order as the result.

The obtainment of the monoxide at pressures as low as 10 kbar (the smallest pressure which actually can be measured in our apparatus) is also surprising. It would show that the extra energy needed to stabilize this compound is much lower than previously estimated. Impurity stabilization cannot be completely ruled out, but obtaining compounds which can be labeled as YbO and Yb_3O_4 with good experimental evidence increases the probability that the compounds we formed under pressure are really these oxides. Under pressure, the system Yb-Yb_2O_3 appears to be similar to the system Eu - Eu_2O_3 at normal pressure. This can open the way to a high-pressure chemistry of divalent ytterbium.

REFERENCES

1. G. J. McCarthy and W. B. White, J. Less Common Metals 22, 409 (1970).
2. O. de Pous and J. C. Achard, Bull. Soc. Chim. France 3417 (1970).
3. J. C. Achard and O. de Pous, in Proceedings 8th Rare Earth Research Conference, Reno, Nevada (1970), p. 375.
4. G. Brauer, H. Barnighausen and N. Schultz, Z. Anorg. Allgem. Chem. 46, 536 (1967).
5. J. M. Haschke and H. A. Eick, Inorg. Chem. 9, 851 (1970).
6. H. R. Hoekstra, Inorg. Chem. 5, 754 (1966).
7. A. F. Reid, J. Am. Cer. Soc. 50, 491 (1967).

SYNTHESES OF $EuLn_2S_4$ AND $SrLn_2S_4$ (Ln=Lu, Yb, Er, Y) WITH Th_3P_4 TYPE STRUCTURE

Y. Ishida*, N. Kinomura, Y. Miyamoto, S. Kume, and M. Koizumi

Osaka University
Osaka, Japan

INTRODUCTION

Hirota et al. [1] have indicated that some rare earth sulfides, ALn_2S_4 (A=Mg, Mn; Ln=Tm, Yb), transform from the spinel-type to the Th_3P_4-type phase by compression. Another kind of compound exists whose composition is also denoted as ALn_2S_4 but whose structure is classified as the $CaFe_2O_4$-type [2]. The packing of constituent ions in the $CaFe_2O_4$-type is denser than that in spinel but looser than in the Th_3P_4-type. It is expected, therefore, that the $CaFe_2O_4$-type compounds transfrom to the Th_3P_4-type phase under pressure more easily than in the case of the spinel-type. This report provides some results of this kind of pressure-induced phase transformation observed in $EuLn_2S_4$ (Ln=Y, Er, Yb, Lu) and $SrLn_2S_4$ (Ln=Y, Er, Yb). A thermodynamical consideration of the phase transformation of EuY_2S_4 between the $CaFe_2O_4$- and the Th_3P_4-types is also given.

EXPERIMENTAL

The starting materials of $EuLn_2S_4$ (Ln=Y, Er, Yb, Lu) and $SrLn_2S_4$ (Ln=Y, Er, Yb) were prepared by mixing powders of either $SrCO_3$ or Eu_2O_3 and Ln_2O_3 in a molar ratio of 1 : 1 for the former and 1 : 2 for the latter. These powders were guaranteed reagents with purities of 99.9% or higher. Each mixture was heated to 1200°C for 2 hr in flowing H_2S. The material was crushed into powder and heated repeatedly until its x-ray powder pattern consisted of diffractions due to the $CaFe_2O_4$-type phase alone. The lattice parameters calculated from the patterns were in agreement with those reported by Patrie et al. [2] and by Hulliger et al. [3].

*Present address: R & D Laboratory, NGK Insulators Ltd., Nagoya, Japan.

The starting material was subjected to the high P-T condition of 1200°C and 4GPa for 1 hr. The syntheses were carried out by means of the cubic anvil-type device [4]. The products were examined with a microscope and x-ray powder diffractometer. When the Ni-filtered Cu-Kα radiation was used and the scanning of the diffractometer was 0.5°/min, the error in d-spacings was within ± 0.003Å. Densities were measured by a conventional pycnometer using a sample of about 100mg. Magnetic susceptibilities were measured using a device described by Shimada et al. [5] and electric conductivities were obtained by means of a four-probe apparatus in the temperature range from 77 to 300 K.

In the experiments concerned with the equilibrium of two phases of EuY_2S_4, the synthesized samples were treated once more under the various combinations of temperature and pressure up to 1500°C and 5GPa, and the products were examined by x-ray diffractions. Temperature dependencies of specific heat were obtained at atmospheric pressure by means of a Rigaku Denki scanning calorimeter. The measurements were undertaken by running an electric current through a furnace containing a sample container. First, the temperature of the container without the sample was raised at a constant rate of 2°C/min and the electric power input was recorded. Secondly, the same procedure was repeated after having placed the sample in the container. In the latter case more power was necessary to maintain the same rate of heating. Since the latent heat corresponding to the difference of power in the two procedures was used to elevate the temperature of the sample, the heat capacity was evaluated from the record. The error of measurement was 10% when pure corundum was used as a standard.

RESULTS

Microscopic observation showed that every product synthesized under high temperature and pressure consisted of a single phase. The x-ray powder diffraction patterns which were assigned as the Th_3P_4-type structure were used to calculate the lattice parameters. The results are shown in the Table I in which the calculated and the observed d-values are also listed. The densities were tabulated in Table II.

The magnetic susceptibilities of $EuLn_2S_4$ (Ln=Y, Er, Yb, Lu) and $SrLn_2S_4$ (Ln=Er, Yb) of both the $CaFe_2O_4$- and the Th_3P_4-type phases increase linearly as the reciprocal of temperature decreases in the range from 77 to 300 K. The conductivity measurements indicate that the samples, regardless of composition and structure, possess high electric resistivities on the order of 10^6Ω-cm at room temperature which steadily increases to 10^8Ω-cm as the temperatures were lowered to 77 K.

The experimental determinations of stable phases under various combinations of temperature and pressure made it possible to draw an equilibrium boundary between the $CaFe_2O_4$- and the Th_3P_4-type

Table I. X-ray Powder Data for Th_3P_4-type $EuLn_2S_4$ and $SrLn_2S_4$

hkl	I/I_0	$EuLu_2S_4$ d_{obs}	$EuLu_2S_4$ d_{calc}	$EuYb_2S_4$ d_{obs}	$EuYb_2S_4$ d_{calc}	$EuEr_2S_4$ d_{obs}	$EuEr_2S_4$ d_{calc}	EuY_2S_4 d_{obs}	EuY_2S_4 d_{calc}	$SrYb_2S_4$ d_{obs}	$SrYb_2S_4$ d_{calc}
211	VS	3.429	3.430	3.432	3.433	3.437	3.437	3.450	3.452	3.437	3.439
310	VS	2.656	2.657	2.660	2.660	2.663	2.663	2.673	2.674	2.664	2.664
321	S	2.245	2.246	2.246	2.248	2.248	2.250	2.260	2.260	2.250	2.251
420	S	1.879	1.879	1.880	1.881	1.883	1.883	1.891	1.891	1.883	1.884
332	W	1.791	1.791	1.792	1.793	1.795	1.795	1.802	1.803	1.795	1.796
510	W	1.649	1.648	1.650	1.649	1.652	1.651	1.659	1.658	1.652	1.652
611	W	2.364	3.363	1.364	1.364	1.366	1.366	1.372	1.372	1.367	1.367
		a=8.402Å		a=8.410Å		a=8.420Å		a=8.456Å		a=8.425Å	

hkl	I/I_0	$SrEr_2S_4$ d_{obs}	$SrEr_2S_4$ d_{calc}	SrY_2S_4 d_{obs}	SrY_2S_4 d_{calc}
211	VS	3.451	3.453	3.471	3.472
310	VS	2.672	2.674	2.689	2.689
321	S	2.259	2.260	2.272	2.273
420	S	1.891	1.891	1.902	1.902
332	W	1.803	1.803	1.813	1.813
510	W	1.658	1.659	1.668	1.668
611	W	1.372	1.372	1.380	1.380
		a=8.457Å		a=8.504Å	

Table II. Observed and Calculated Densities of Th_3P_4-type $EuLn_2S_4$ and $SrLn_2S_4$.

	ρ_{obs}, g/cm^3	ρ_{calc}, g/cm^3	ΔV, %*
$EuLu_2S_4$	7.0	7.06	-8.5
$EuYb_2S_4$	6.9	6.99	-8.8
$EuEr_2S_4$	6.9	6.84	-10.0
EuY_2S_4	5.0	5.12	-10.2
$SrYb_2S_4$	6.2	6.24	-7.7
$SrEr_2S_4$	5.9	6.04	-9.5
SrY_2S_4	4.2	4.25	-10.2

$*\Delta V = (V_{Th_3P_4} - V_{CaFe_2O_4})/V_{CaFe_2O_4}$ accompanied by the transformation from $CaFe_2O_4$-type to the Th_3P_4-type phases.

phases of EuY_2S_4 as seen in Fig. 1. The experimental temperature dependencies of the specific heat are summarized in Fig. 2.

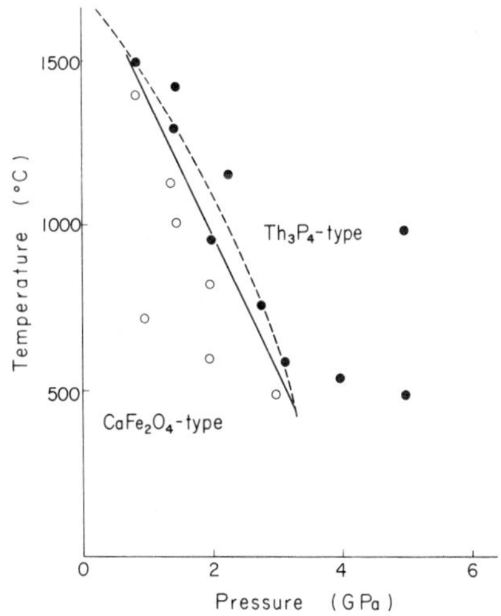

Fig. 1. Phase diagram for EuY_2S_4: ● synthesized Th_3P_4-type phase; o synthesized $CaFe_2O_4$- type phase; — experimental phase boundary; — thermodynamic phase boundary.

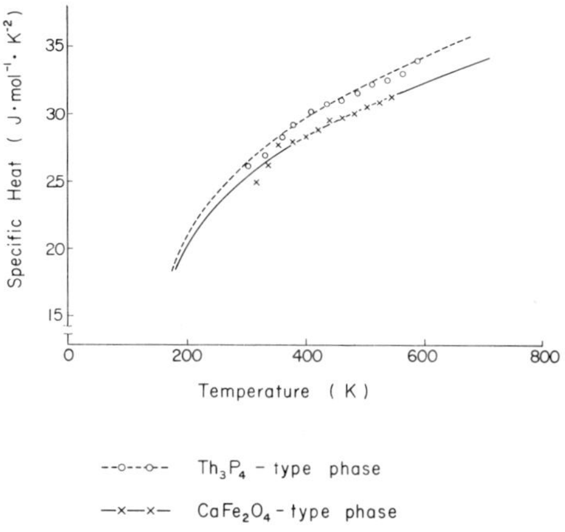

Fig. 2. Temperature dependencies of specific heat C_p of EuY_2S_4.

DISCUSSION

The experimental results show that $CaFe_2O_4$-type $EuLn_2S_4$ (Ln=Lu, Yb, Er, Y) and $SrLn_2S_4$(Ln=Yb, Er, Y) transform to the

Th_3P_4-type phase by compression. As shown in Table II, the decreases of volume accompanied by this transformation are about 10%. It reaches more than 15% in the transformation of sulfospinel to the Th_3P_4-type [1] and this difference is explained by the denser packing of constituent ions in the $CaFe_2O_4$-type structure than in spinel. The coordination numbers of cations are (8) in Th_3P_4, (8, 6) in $CaFe_2O_4$ and (6, 4) in spinel-type, respectively. Consequently, the transformation of $CaFe_2O_4$ to the Th_3P_4-type phase is expected to occur more easily than that from spinel to the Th_3P_4-type. The threshold pressure required to form Th_3P_4-type ALn_2S_4 is about 5.5GPa at 1000°C for the spinel-type [1] and 2GPa for EuY_2S_4 as seen in Fig. 1.

Paramagnetism was observed in all the samples investigated. As noted in Table III the effective number of Bohr magnetons per

Table III. Magnetic Data for the $CaFe_2O_4$- and Th_3P_4-type Compounds

		$CaFe_2O_4$		Th_3P_4	
	P_{calc}	P_{obs}	θ, K	P_{obs}	θ,K
EuY_2S_4	7.94	7.4	-1	6.9	14
$EuEr_2S_4$	15.70	14.6	-12	13.3	2
$EuYb_2S_4$	10.21	9.6	-23	8.8	-8
$EuLu_2S_4$	7.94	7.4	-4	6.8	5
$SrEr_2S_4$	14.37	12.3	-4	12.7	-5
$SrYb_2S_4$	9.07	8.5	-38	8.6	-47

chemical formula obtained from the measurements shows fairly good agreement with the sum of the individual magnetic ions in each compound, indicating that no magnetic ordering of long range exists in these compounds. In the above calculation Eu is assumed to be divalent. The consistency between the observed and the calculated values indicates that this assumption is correct and that Eu is in the divalent state and Ln (Ln=Lu, Yb, Er) is in the trivalent state.

Since the volume change accompanied by the phase transformation is the largest in EuY_2S_4, a thermodynamical consideration is made using this substance. When two phases are connected with an allotropic relation the difference in G is expressed approximately by

$$\Delta G(T, P) = \Delta G(T, 0) + \Delta V(T, 0)\ P \qquad (1)$$

while $\Delta G = 0$ when equilibrium is achieved. When the following relations are substituted in (1),

$$\Delta G(T, 0) = \Delta H(T, 0) - T\Delta S(T, 0)$$
$$\Delta H(T, 0) = \Delta H(T_0, 0) + \int_{T_0}^{T} \Delta C_p dT \qquad (2)$$
$$\Delta S(T, 0) = \Delta S(T_0, 0) + \int_{T_0}^{T} (\Delta C_p/T) dT$$

we obtain

$$P=-[\Delta G(T_0, 0) + \int_{T_0}^{T} \Delta C_p dT - T \int_{T_0}^{T} (\Delta C_p/T)dT]/\Delta V(T, 0) \qquad (3)$$

No sign of the phase transformation is observed in the $CaFe_2O_4$-type EuY_2S_4 but it melts at 1600°C at atmospheric pressure. However, if the phase boundary in Fig. 1 is interpolated leftwards it intersects the temperature axis around 1700°C (about 1970 K). Therefore, it can be assumed that the phase transformation would occur at 1970 K and atmospheric pressure if the melting point of the compound were higher than this temperature.

Specific heat is denoted by the Einstein formula

$$C_p = 3Rx^2 e^x/(e^x-1)^2 + AT \qquad (4)$$

where $x = \theta_E/T$. The Einstein formula approximation is not accurate at low temperature but it is useful in the present case because only the state at high temperatures is under consideration. When appropriate values for θ_E and A are chosen the mean deviation of the calculated C_p from the observed is minimized and the best fitted values are determined. These are given in Table IV.

Table IV. Thermodynamic Data for EuY_2S_4

	θ_E, K	A, J/mol-K^2
$CaFe_2O_4$-type	400	13.4×10^{-3}
Th_3P_4-type	450	16.7×10^{-3}

Entropy of a solid is normally the sum of four terms: vibrational, atom-configurational, magnetic, and electric. In EuY_2S_4 the contributions due to magnetism and electricity are small because of paramagnetism and high resistivity in both the $CaFe_2O_4$- and the Th_3P_4-type phases. The atomic configuration depends on the number of sites in a lattice and on the number and kinds of constituent ions. In the $CaFe_2O_4$-type structure divalent cations locate in 8-fold coordination and trivalent locate in 6-fold coordination [6]. In the Th_3P_4-type all cations occupy equivalent sites of the 8-fold coordination [7]. Subsequently, the difference in the configurational entropy accompanied by the transformation from the $CaFe_2O_4$- to the Th_3P_4-type is given by

$$\Delta S^* = -k\ln \frac{(3N_0)!}{(2N_0)!N_0!} \simeq R\ln \frac{3^3}{2^2} \simeq 15.9 \text{ J/mol-K} \qquad (5)$$

Since ΔS in (2) represents only the contribution due to the lattice vibration, ΔS* should be taken into account in a more accurate treatment, and (3) is rewritten in the following way:

$$P=\left[3RT\ln\left(\frac{1-e^{x_1}}{1-e^{x_2}}\right) - \frac{1}{2}(A_2-A_1)T^2 - T\Delta S^*\right]_T^{T_0}/\Delta V \qquad (6)$$

where suffixes 1 and 2 denote $CaFe_2O_4$- and Th_3P_4-types, respectively. This correction should bring a closer approach of the calculated result to the real thermodynamical equilibrium. If all the practical values for A, S, and θ_E are used, to set equal to 1970 K, and $\Delta V=V_1-V_2$ at room temperature and pressure are substituted in (6), the dashed line of Fig. 1 results. A good coincidence is seen between the experimental and the theoretical phase boundaries.

NOTATION

A	= constant α^2V/β	S	= entropy
C_p	= specific heat at constant pressure	T	= temperature
G	= Gibbs free energy	V	= molar volume
H	= enthalpy	k	= Boltzmann's constant
N_0	= Avogadro's number	α	= coefficient of thermal expansion
P	= pressure	β	= compressibility
R	= gas constant	Δ	= difference
		θ_E	= Einstein temperature

REFERENCES

1. K. Hirota, N. Kinomura, S. Kume, and M. Koizumi, Mat. Res. Bull. 11, 227 (1976).
2. M. Patrie, S. M. Golabi, J. Flahaut, and L. Domange, C. R. Acad. Sci. 259, 4039 (1964).
3. F. Hulliger and O. Vogt, Phys. Lett. 21, 138 (1966).
4. Y. Yanagisawa and S. Kume, Mat. Res. Bull. 8, 1241 (1973).
5. M. Shimada, S. Kume, and M. Koizumi, J. Am. Ceram. Soc. 51, 713 (1968).
6. B. F. Decker and J. S. Kasper, Acta Cryst. 10, 332 (1957).
7. P. I. Kripyakevich, Soviet Phys.-Crystal. 7, 556 (1963).

CONTENTS OF VOLUME 2

B. Properties of Mantle Materials

C. Properties of Crystal Silicates and Melts

G. Rock Mechanics

II. HIGH PRESSURE IN ENERGY RESOURCE RECOVERY

H. Energy Resource Recovery

O. Design, Analysis and Monitoring of High Pressure Vessels

IV. INDUSTRIAL APPLICATIONS OF HIGH PRESSURE

P. Hot Isostatic Processing: Process and Equipment

Q. Hydrostatic Extrusion

R. High Pressure Metal Working

S. Explosive Forming and Welding

V. FUTURE DIRECTIONS IN HIGH PRESSURE APPLICATIONS

T. Investigations of Extremes of Pressure and Temperature

U. Shock Wave Methods and Investigations

AUTHOR INDEX
Volume 1 and Volume 2

SUBJECT INDEX, VOLUME 1